Foundations of Constructive Probability Theory

Using Bishop's work on constructive analysis as a framework, this monograph gives a systematic, detailed, and general constructive theory of probability theory and stochastic processes. It is the first extended account of this theory: Almost all of the constructive existence and continuity theorems that permeate the book are original. It also contains results and methods hitherto unknown in the constructive and nonconstructive settings. The text features logic only in the common sense and, beyond a certain mathematical maturity, requires no prior training in either constructive mathematics or probability theory. It will thus be accessible and of interest to both probabilists interested in the foundations of their specialty and constructive mathematicians who wish to see Bishop's theory applied to a particular field.

YUEN-KWOK CHAN completed a PhD in constructive mathematics with Errett Bishop before leaving academia for a career in private industry. He is now an independent researcher in probability and its applications.

ENCYCLOPEDIA OF MATHEMATICS AND ITS APPLICATIONS

This series is devoted to significant topics or themes that have wide application in mathematics or mathematical science and for which a detailed development of the abstract theory is less important than a thorough and concrete exploration of the implications and applications.

Books in the **Encyclopedia of Mathematics and Its Applications** cover their subjects comprehensively. Less important results may be summarized as exercises at the ends of chapters. For technicalities, readers can be referred to the bibliography, which is expected to be comprehensive. As a result, volumes are encyclopedic references or manageable guides to major subjects

ENCYCLOPEDIA OF MATHEMATICS AND ITS APPLICATIONS
All the titles listed below can be obtained from good booksellers or from Cambridge University Press. For a complete series listing, visit www.cambridge.org/mathematics

129 J. Berstel, D. Perrin, and C. Reutenauer *Codes and Automata*
130 T. G. Faticoni *Modules over Endomorphism Rings*
131 H. Morimoto *Stochastic Control and Mathematical Modeling*
132 G. Schmidt *Relational Mathematics*
133 P. Kornerup and D. W. Matula *Finite Precision Number Systems and Arithmetic*
134 Y. Crama and P. L. Hammer (eds.) *Boolean Models and Methods in Mathematics, Computer Science, and Engineering*
135 V. Berthé and M. Rigo (eds.) *Combinatorics, Automata, and Number Theory*
136 A. Kristály, V. D. Rădulescu, and C. Varga *Variational Principles in Mathematical Physics, Geometry, and Economics*
137 J. Berstel and C. Reutenauer *Noncommutative Rational Series with Applications*
138 B. Courcelle and J. Engelfriet *Graph Structure and Monadic Second-Order Logic*
139 M. Fiedler *Matrices and Graphs in Geometry*
140 N. Vakil *Real Analysis through Modern Infinitesimals*
141 R. B. Paris *Hadamard Expansions and Hyperasymptotic Evaluation*
142 Y. Crama and P. L. Hammer *Boolean Functions*
143 A. Arapostathis, V. S. Borkar, and M. K. Ghosh *Ergodic Control of Diffusion Processes*
144 N. Caspard, B. Leclerc, and B. Monjardet *Finite Ordered Sets*
145 D. Z. Arov and H. Dym *Bitangential Direct and Inverse Problems for Systems of Integral and Differential Equations*
146 G. Dassios *Ellipsoidal Harmonics*
147 L. W. Beineke and R. J. Wilson (eds.) with O. R. Oellermann *Topics in Structural Graph Theory*
148 L. Berlyand, A. G. Kolpakov, and A. Novikov *Introduction to the Network Approximation Method for Materials Modeling*
149 M. Baake and U. Grimm *Aperiodic Order I: A Mathematical Invitation*
150 J. Borwein et al. *Lattice Sums Then and Now*
151 R. Schneider *Convex Bodies: The Brunn-Minkowski Theory (Second Edition)*
152 G. Da Prato and J. Zabczyk *Stochastic Equations in Infinite Dimensions (Second Edition)*
153 D. Hofmann, G. J. Seal, and W. Tholen (eds.) *Monoidal Topology*
154 M. Cabrera García and Á. Rodríguez Palacios *Non-Associative Normed Algebras I: The Vidav-Palmer and Gelfand-Naimark Theorems*
155 C. F. Dunkl and Y. Xu *Orthogonal Polynomials of Several Variables (Second Edition)*
156 L. W. Beineke and R. J. Wilson (eds.) with B. Toft *Topics in Chromatic Graph Theory*
157 T. Mora *Solving Polynomial Equation Systems III: Algebraic Solving*
158 T. Mora *Solving Polynomial Equation Systems IV: Buchberger Theory and Beyond*
159 V. Berthé and M. Rigo (eds.) *Combinatorics, Words, and Symbolic Dynamics*
160 B. Rubin *Introduction to Radon Transforms: With Elements of Fractional Calculus and Harmonic Analysis*
161 M. Ghergu and S. D. Taliaferro *Isolated Singularities in Partial Differential Inequalities*
162 G. Molica Bisci, V. D. Radulescu, and R. Servadei *Variational Methods for Nonlocal Fractional Problems*
163 S. Wagon *The Banach-Tarski Paradox (Second Edition)*
164 K. Broughan *Equivalents of the Riemann Hypothesis I: Arithmetic Equivalents*
165 K. Broughan *Equivalents of the Riemann Hypothesis II: Analytic Equivalents*
166 M. Baake and U. Grimm (eds.) Aperiodic Order II: Crystallography and Almost Periodicity
167 M. Cabrera García and Á. Rodríguez Palacios *Non-Associative Normed Algebras II: Representation Theory and the Zel'manov Approach*
168 A. Yu. Khrennikov, S. V. Kozyrev, and W. A. Zúñiga-Galindo *Ultrametric Pseudodifferential Equations and Applications*
169 S. R. Finch *Mathematical Constants II*
170 J. Krajíček *Proof Complexity*
171 D. Bulacu, S. Caenepeel, F. Panaite and F. Van Oystaeyen *Quasi-Hopf Algebras*
172 P. McMullen *Geometric Regular Polytopes*
173 M. Aguiar and S. Mahajan *Bimonoids for Hyperplane Arrangements*
174 M. Barski and J. Zabczyk *Mathematics of the Bond Market: A Lévy Processes Approach*
175 T. R. Bielecki, J. Jakubowski, and M. Nieweglowski *Fundamentals of the Theory of Structured Dependence between Stochastic Processes*
176 A. A. Borovkov *Asymptotic Analysis of Random Walks: Light-Tailed Distributions*
177 Y.-K. Chan *Foundations of Constructive Probability Theory*

Foundations of Constructive Probability Theory

YUEN-KWOK CHAN
Citigroup

CAMBRIDGE
UNIVERSITY PRESS

University Printing House, Cambridge CB2 8BS, United Kingdom

One Liberty Plaza, 20th Floor, New York, NY 10006, USA

477 Williamstown Road, Port Melbourne, VIC 3207, Australia

314–321, 3rd Floor, Plot 3, Splendor Forum, Jasola District Centre, New Delhi – 110025, India

79 Anson Road, #06–04/06, Singapore 079906

Cambridge University Press is part of the University of Cambridge.

It furthers the University's mission by disseminating knowledge in the pursuit of education, learning, and research at the highest international levels of excellence.

www.cambridge.org
Information on this title: www.cambridge.org/9781108835435
DOI: 10.1017/9781108884013

First published 2021

A catalogue record for this publication is available from the British Library.

Library of Congress Cataloging-in-Publication Data
Names: Chan, Yuen-Kwok, author.
Title: Foundations of constructive probability theory / Yuen-Kwok Chan.
Description: Cambridge, UK ; New York, NY : Cambridge University Press, 2021. |
Series: Encyclopedia of mathematics and its applications |
Includes bibliographical references and index.
Identifiers: LCCN 2020046705 (print) | LCCN 2020046706 (ebook) |
ISBN 9781108835435 (hardback) | ISBN 9781108884013 (epub)
Subjects: LCSH: Probabilities. | Stochastic processes. | Constructive mathematics.
Classification: LCC QA273 .C483 2021 (print) | LCC QA273 (ebook) |
DDC 519.2–dc23
LC record available at https://lccn.loc.gov/2020046705
LC ebook record available at https://lccn.loc.gov/2020046706

ISBN 978-1-108-83543-5 Hardback

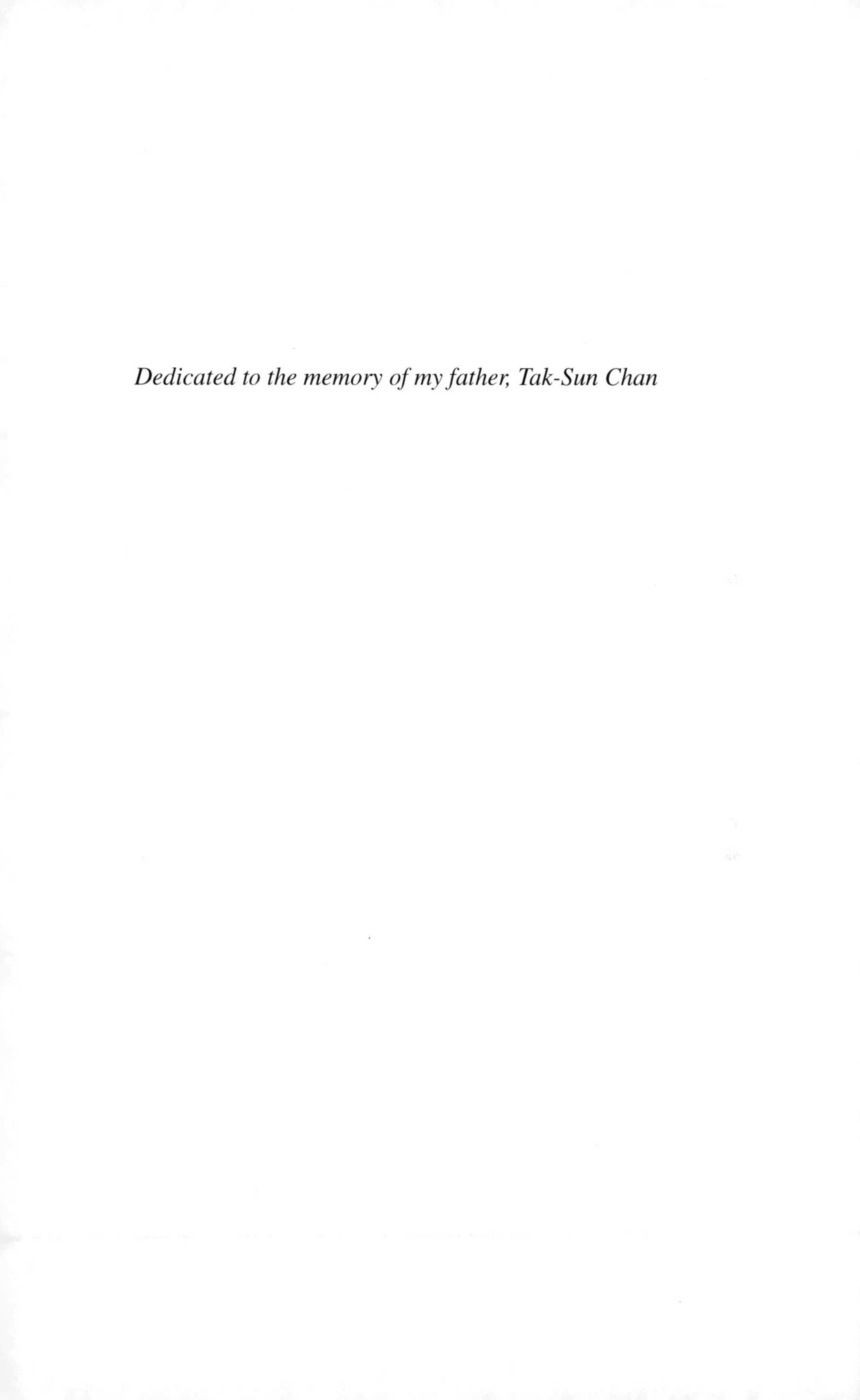

Dedicated to the memory of my father, Tak-Sun Chan

Contents

Acknowledgments

Yuen-Kowk Chan is retired from Citigroup's Mortgage Analytics unit. All opinions expressed by the author are his own. The author is grateful to the late Professor E. Bishop for teaching him constructive mathematics, to the late Professors R. Getoor and R. Blumenthal for teaching him probability and for mentoring him, and to the late Professors R. Pyke and W. Birnbaum and the other statisticians in the Mathematics Department of the University of Washington, circa 1970s, for their moral support. The author is also thankful to the constructivists in the Mathematics Department of New Mexico State University, circa 1975, for hosting a sabbatical visit and for valuable discussions, especially to Professors F. Richman, D. Bridges, M. Mandelkern, W. Julian, and the late Professor R. Mines. Professors Melody Chan and Fritz Scholz provided incisive and valuable critiques of the introduction chapter of an early draft of this book. Professor Douglas Bridges gave many thoughtful comments of the draft. The author also wishes to thank Ms Jill Hobbs for her meticulous copyediting, and Ms Niranjana Harikrishnan for her aesthetically pleasing typography.

Nomenclature

$\equiv$ ... by definition equal to, 8

R ... set of real numbers, 8

d_{ecld} ... Euclidean metric, 8

$a \vee b$ ... $\max(a,b)$, 8

$a \wedge b$ ... $\min(a,b)$, 8

a_+ ... $\max(a,0)$, 8

a_- ... $\min(a,0)$, 8

$A \cup B$ ... union of sets A and B, 8

$A \cap B, AB$ ... intersection of sets A and B, 8

$[a]_1$ ... an integer $[a]_1 \in (a, a+2)$ for given $a \in R$, 9

$X|A$ ... restriction of function X on a set to a subset A, 9

$X' \circ X, X'(X)$ ... composite of functions X' and X, 10

$(X \leq a)$ ... $\{\omega \in domain(X) : X(\omega) \leq a\}$, 11

$X(\cdot, \omega'')$ ... function of first variable, given value of second variable for a function X of two variables, 12

$T^*(Y) \equiv T(\cdot, Y)$ ... dual function of Y relative to a certain mapping T, 12

(S,d) ... metric space, with metric d on set S, 12

$x \neq y$ ... $d(x,y) > 0$, where x, y are in some metric space, 13

J_c ... metric complement of subset J in a metric space, 12

$\bigotimes$ ... direct product of functions or sets, 14

$C_u(S,d), C_u(S)$ ... space of uniformly continuous real-valued functions on metric space (S,d), 14

$C_{ub}(S,d), C_{ub}(S)$ ... subspace of $C_u(S,d)$ whose members are bounded, 14

$C_0(S,d)$, $C_0(S)$ ... subspace of $C_u(S,d)$ whose members vanish at infinity, 15

$C(S,d), C(S)$ ... subspace of $C_u(S,d)$ whose members have bounded supports, 15

$\widehat{d}$ ... $1 \wedge d$, 15

O, o ... bounds for real-valued function, 16

$\square$ ... mark for end of proof or end of definition, 16

ξ ... binary approximation of a metric space, 20

$\|\xi\|$ ... modulus of local compactness corresponding to ξ, 20

$(\overline{S}, \overline{d})$ ... one-point compactification of (S,d), 34

Δ ... point at infinity, 34

$F|B$ ... $\{f|B : f \in F\}$, 35

$\int_{-\infty}^{+\infty} X(x) dF(x)$ ... Riemann–Stieljes integral, 46

1_A ... indicator of measurable set A, 67

A^c ... measure-theoretic complement of measurable set A, 67

$\diamond$ ... ordering between certain real numbers and functions, 73

(G,λ) .... profile system, 73
$(a,b) \ll \alpha$ .... the interval (a,b) is bounded in profile by α, 74
(Ω, L, E) .... probability space, 138
$\int E(d\omega)X(\omega)$ .... $E(X)$, 139
ρ_{Prob} .... the probability metric on r.v.'s, 145
$L(G)$ .... probability subspace generated by the family G of r.v.'s, 150
$\widehat{J}(S,d)$ .... set of distributions on complete metric space (S,d), 151
$\Rightarrow$ .... weak convergence of distributions, or convergence in distributions of r.r.v.'s, 155
$\rho_{Dist,\xi}$ .... metric on distributions on a locally compact metric space relative to binary approximation ξ, 156
F_X .... P.D.F. induced on R by an r.r.v. X, 165
$\Phi_{Sk,\xi}$ .... Skorokhod representation of distributions on (S,d), determined by ξ, 170
$L_{|L'}$ .... subspace of conditionally integrable r.r.v.'s given the subspace L', 184
$L_{|G}$ .... subspace of conditionally integrable r.r.v.'s, given $L(G)$, 184
E_A .... conditional expectation given an event A with positive probability, 184
$\varphi_{\overline{\mu},\overline{\sigma}}$ .... multivariate normal p.d.f., 192
$\Phi_{\overline{\mu},\overline{\sigma}}$ .... multivariate normal distribution, 193
$\varphi_{0,I}$ .... multivariate standard normal p.d.f., 193
$\Phi_{0,I}$ .... multivariate standard normal distribution, 193
Ψ .... tail of univariate standard normal distribution, 193
ψ_X .... characteristic function of r.v. X with values in R^n, 204
ψ_J .... characteristic function of distribution J on R^n, 204
$\widehat{g}$ .... Fourier transform of complex-valued function g on R^n, 204
$f \star g$ .... Convolution of complex-valued functions f and g on R^n, 204
ρ_{char} .... metric on characteristic functions on R^n, 210
$\widehat{R}(Q \times \Omega, S)$ .... set of r.f.'s with parameter set Q, state space (S,d), and sample space (Ω, L, E), 227
$X|K$ .... restriction of $X \in \widehat{R}(Q \times \Omega, S)$ to parameter subset $K \subset Q$, 227
$\delta_{Cp,K}$ .... modulus of continuity in probability of $X|K$, 228
$\delta_{cau,K}$ .... modulus of continuity a.u. of $X|K$, 228
$\delta_{auc,K}$ .... modulus of a.u. continuity of $X|K$, 228
$\widehat{F}(Q,S)$ .... set of consistent families of f.j.d.'s with parameter set Q and state space S, 232
$\widehat{\rho}_{Marg,\xi,Q}$ .... marginal metric for the set $\widehat{F}(Q,S)$ relative to the binary approximation ξ, 237
$\widehat{F}_{Cp}(Q,S)$ .... subset of $\widehat{F}(Q,S)$ whose members are continuous in probability, 238
$\widehat{\rho}_{Cp,\xi,Q,Q(\infty)}$ .... metric on $\widehat{F}_{Cp}(Q,S)$ relative to dense subset $Q_\infty$of parameter metric space Q, 240
$\widehat{\rho}_{Prob,Q}$ .... probability metric on $\widehat{R}(Q \times \Omega, S)$, 265
$\rho_{Sup,Prob}$ .... metric on $\widehat{F}_{Cp}(Q,S)$, 277
$\widehat{R}_{Cp}(Q \times \Omega, S)$ .... subset of $\widehat{R}(Q \times \Omega, S)$ whose members are continuous in probability, 228
$\widehat{R}_{Meas}(Q \times \Omega, S)$ .... subset of $\widehat{R}(Q \times \Omega, S)$ whose members are measurable, 276
$\widehat{R}_{Meas,Cp}(Q \times \Omega, S)$ .... $\widehat{R}_{Meas}(Q \times \Omega, S) \cap \widehat{R}_{Cp}(Q \times \Omega, S)$, 276
$\mathcal{L}$ .... filtration in probability space (Ω, L, E), 300
$\mathcal{L}_{\mathcal{X}}$ .... natural filtration of a process X, 300
$\mathcal{L}^+$ .... right-limit extension of filtration $\mathcal{L}$, 301
$L^{(\tau)}$ .... probability subspace of observables at stopping time τ relative to filtration $\mathcal{L}$, 302
$\overline{\lambda}$ .... the special convex function on R, 316
$Q_m, \overline{Q}_m, \widetilde{Q}_m, Q_\infty, \overline{Q}_\infty$ .... certain subsets of dyadic rationals in $[0,\infty)$, 332
$(\widehat{C}[0,1], \rho_{\widehat{C}[0,1]})$ .... metric space of a.u. continuous processes on $[0,1]$, 334

Φ_{Lim} .... extension by limit of a process with parameter set Q_∞ to parameter set $[0,1]$, 335
$\widehat{D}[0,1]$ .... set of all a.u. càdlàg processes on $[0,1]$, 406
δ_{aucl} .... modulus of a.u. càdlàg, 406
$\widehat{D}_{\delta(aucl),\delta(cp)}[0,1]$ .... subset of $\widehat{D}[0,1]$ whose members have moduli δ_{Cp}, and δ_{aucl}, 406
$\rho_{\widehat{D}[0,1]}$ .... metric on $\widehat{D}[0,1]$, 409
$(\widehat{R}_{Dreg}(Q_\infty \times \Omega, S), \widehat{\rho}_{Prob,Q(\infty)})$ .... metric space of D-regular processes, 410
Φ_{rLim} .... extension by right-limit of a process with parameter set Q_∞ to parameter set $[0,1]$, 418
β_{auB} .... modulus of a.u. boundedness, 445
δ_{SRCp} .... modulus of strong right continuity in probability, 445
$\overline{\tau}_{f,a,N}(X)$ .... certain first exit times by the process X, 488
$\mathbf{T}$ .... a Markov semigroup, 500
$\delta_{\mathbf{T}}$ .... a modulus of strong continuity of $\mathbf{T}$, 500
$\alpha_{\mathbf{T}}$ .... a modulus of smoothness of $\mathbf{T}$, 500
$F^{*,\mathbf{T}}_{r(1),\cdots,r(m)}$ .... a finite joint transition distribution generated by $\mathbf{T}$, 502
$(\mathscr{T}, \rho_{\mathscr{T}})$ .... metric space of Markov semigroups. 525
$\mathbf{V}$ .... a Feller semigroup, 538
$\delta_{\mathbf{V}}$ .... a modulus of strong continuity of $\mathbf{V}$, 538
$\alpha_{\mathbf{V}}$ .... a modulus of smoothness of $\mathbf{V}$, 538
$\kappa_{\mathbf{V}}$ .... a modulus of nonexplosion of $\mathbf{V}$, 538
$F^{*,\mathbf{V}}_{r(1),\cdots,r(m)}$ .... a finite joint transition distribution generated by $\mathbf{V}$, 539
$((S,d),(\Omega,L,E),\{U^{x,\mathbf{V}} : x \in S\})$ .... Feller process, 543
$((R^m,d^m),(\Omega,L,E),\{B^x : x \in R^m\})$ .... Brownian motion as a Feller process, 568

Part I

Introduction and Preliminaries

1

Introduction

The present work on probability theory is an outgrowth of the constructive analysis in [Bishop 1967] and [Bishop and Bridges 1985].

Perhaps the simplest explanation of constructive mathematics is by way of focusing on the following two commonly used theorems. The first, the *principle of finite search*, states that, given a finite sequence of 0-or-1 integers, either all members of the sequence are equal to 0, or there exists a member that is equal to 1. We use this theorem without hesitation because, given the finite sequence, a finite search would determine the result.

The second theorem, which we may call the *principle of infinite search*, states that, given an infinite sequence of 0-or-1 integers, either all members of the sequence are equal to 0, or there exists a member that is equal to 1. The name "infinite search" is perhaps unfair, but it brings into sharp focus the point that the computational meaning of this theorem is not clear. The theorem is tantamount to an infinite loop in computer programming without any assurance of termination.

Most mathematicians acknowledge the important distinction between the two theorems but regard the principle of infinite search as an expedient tool to prove theorems, with the belief that theorems so proved can then be specialized to constructive theorems, when necessary. Contrary to this belief, many classical theorems proved directly or indirectly via the principle of infinite search are actually equivalent to the latter: as such, they do not have a constructive proof. Oftentimes, not even the numerical meaning of the theorems in question is clear.

We believe that, for the constructive formulations and proofs of even the most abstract theorems, the easiest way is to employ a disciplined and systematic approach, by using only finite searches and by quantifying mathematical objects and theorems at each and every step, with natural numbers as a starting point. The references cited earlier show that this approach is not only possible but also fruitful.

It should be emphasized that we do not claim that theorems whose proofs require the principle of infinite search are untrue or incorrect. They are certainly correct and consistent derivations from commonly accepted axioms. There is indeed no reason why we cannot discuss such classical theorems alongside their constructive counterparts. The term "nonconstructive mathematics" is not meant to be pejorative. We will use, in its place, the more positive term "classical mathematics."

Moreover, it is a myth that constructivists use a different system of logic. The only logic we use is commonsense logic; no formal language is needed. The present author considers himself a mathematician who is neither interested in nor equipped to comment on the formalization of mathematics, whether classical or constructive.

Since a constructively valid argument is also correct from the classical viewpoint, a reader of the classical persuasion should have no difficulties understanding our proofs. Proofs using only finite searches are surely agreeable to any reader who is accustomed to infinite searches.

Indeed, the author would consider the present book a success if the reader, but for this introduction and occasional remarks in the text, finishes reading without realizing that this is a constructive treatment. At the same time, we hope that a reader of the classical persuasion might consider the more disciplined approach of constructive mathematics for his or her own research an invitation to a challenge.

Cheerfully, we hasten to add that we do not think that finite computations in constructive mathematics are the end. We would prefer a finite computation with n steps to one with $n!$ steps. We would be happy to see a systematic and general development of mathematics that is not only constructive but also computationally efficient. That admirable task will, however, be left to abler hands.

Probability theory, which is rooted in applications, can naturally be expected to be constructive. Indeed, the crowning achievements of probability theory – the laws of large numbers, the central limit theorems, the analysis of Brownian motion processes and their stochastic integrals, and the analysis of Levy processes, to name just a few – are exemplars of constructive mathematics. Kolmogorov, the grandfather of modern probability theory, actually took an interest in the formalization of general constructive mathematics.

Nevertheless, many a theorem in modern probability actually implies the principle of infinite search. The present work attempts a systematic constructive development. Each existence theorem will be a construction. The input data, the construction procedure, and the output objects are the essence and integral parts of the theorem. Incidentally, by inspecting each step in the procedure, we can routinely observe how the output varies with the input. Thus a continuity theorem in epsilon–delta terms routinely follows an existence theorem. For example, we will construct a Markov process from a given semigroup and prove that the resulting Markov process varies continuously with the semigroup.

The reader familiar with the probability literature will notice that our constructions resemble Kolmogorov's construction of the Brownian motion process, which is replete with Borel–Cantelli estimates and rates of convergence. This is in contrast to popular proofs of existence via Prokhorov's theorem. The reader can regard Part III of this book, Chapters 6–11, the part on stochastic processes, as an extension of Kolmogorov's constructive methods to stochastic processes: Daniell–Kolmogorov–Skorokhod construction of random fields, measurable random fields, a.u. continuous processes, a.u. càdlàg processes, martingales, strong Markov processes, and Feller processes, all with locally compact state spaces.

Such a systematic, constructive, and general treatment of stochastic processes, we believe, has not previously been attempted.

The purpose of this book is twofold. First, a student with a general mathematics background at the first-year graduate-school level can use it as an introduction to probability or to constructive mathematics. Second, an expert in probability can use it as a reference for further constructive development in his or her own research specialties.

Part II of this book, Chapters 3–5, is a repackaging and expansion of the measure theory in [Bishop and Bridges 1985]. It enables us to have a self-contained probability theory in terms familiar to probabilists.

For expositions of constructive mathematics, see the first chapters of the last cited reference. See also [Richman 1982] and [Stolzenberg 1970]. We give a synopsis in Chapter 2, along with basic notations and terminologies for later reference.

2

Preliminaries

2.1 Natural Numbers

We start with the natural numbers as known in elementary schools. All mathematical objects are constructed from natural numbers, and every theorem is ultimately a calculation on the natural numbers. From natural numbers are constructed the integers and the rational numbers, along with the arithmetical operations, in the manner taught in elementary schools.

We claim to have a natural number only when we have provided a finite method to calculate it, i.e., to find its decimal representation. This is the fundamental difference from classical mathematics, which requires no such finite method; an infinite procedure in a proof is considered just as good in classical mathematics.

The notion of a finite natural number is so simple and so immediate that no attempt is needed to define it in even simpler terms. A few examples would suffice as clarification: $1, 2$, and 3 are natural numbers. So are 9^9 and 9^{9^9}; the multiplication method will give, at least in principle, their decimal expansion in a finite number of steps. In contrast, the "truth value" of a particular mathematical statement is a natural number only if a finite method has been supplied that, when carried out, would prove or disprove the statement.

2.2 Calculation and Theorem

An algorithm or a calculation means any finite, step-by-step procedure. A mathematical object is defined when we specify the calculations that need to be done to produce this object. We say that we have proved a theorem if we have provided a step-by-step method that translates the calculations doable in the hypothesis to a calculation in the conclusion of the theorem. The statement of the theorem is merely a summary of the algorithm contained in the proof.

Although we do not, for good reasons, write mathematical proofs in a computer language, the reader would do well to compare constructive mathematics to the development of a large computer software library, with successive objects and library functions being built from previous ones, each with a guarantee to finish in a finite number of steps.

2.3 Proof by Contradiction

There is a trivial form of proof by contradiction that is valid and useful in constructive mathematics. Suppose we have already proved that one of two given alternatives, A and B, must hold, meaning that we have given a finite method, that, when unfolded, gives either a proof for A or a proof for B. Suppose subsequently we also prove that A is impossible. Then we can conclude that we have a proof of B; we need only exercise said finite method, and see that the resulting proof is for B.

2.4 Recognizing Nonconstructive Theorems

Consider the simple theorem "if a is a real number, then $a \leq 0$ or $0 < a$," which may be called the principle of excluded middle for real numbers. We can see that this theorem implies the principle of infinite search by the following argument. Let $(x)_{i=1,2,\ldots}$ be any given sequence of 0-or-1 integers. Define the real number $a = \sum_{i=1}^{\infty} x_i 2^{-i}$. If $a \leq 0$, then all members of the given sequence are equal to 0; if $0 < a$, then some member is equal to 1. Thus the theorem implies the principle of infinite search, and therefore cannot have a constructive proof.

Consequently, any theorem that implies this limited principle of excluded middle cannot have a constructive proof. This observation provides a quick test to recognize certain theorems as nonconstructive. Then it raises the interesting task of examining the theorem for constructivization of a part or the whole, or the task of finding a constructive substitute of the theorem that will serve all future purposes in its stead.

For the aforementioned principle of excluded middle of real numbers, an adequate constructive substitute is the theorem "if a is a real number, then, for arbitrarily small $\varepsilon > 0$, we have $a < \varepsilon$ or $0 < a$." Heuristically, this is a recognition that a general real number a can be computed with arbitrarily small, but nonzero, error.

2.5 Prior Knowledge

We assume that the reader of this book has familiarity with calculus, real analysis, and metric spaces, as well as some rudimentary knowledge of complex analysis. These materials are presented in the first chapters of [Bishop and Bridges 1985]. We will also quote results from typical undergraduate courses in calculus or linear algebra, with the minimal constructivization wherever needed.

We assume also that the reader has had an introductory course in probability theory at the level of [Feller I 1971] or [Ross 2003]. The reader should have no difficulty in switching back and forth between constructive mathematics and classical mathematics, or at least no more than in switching back and forth between classical mathematics and computer programming. Indeed, the reader is urged to read, concurrently with this book if not before delving into it, the many classical texts in probability.

2.6 Notations and Conventions

If x, y are mathematical objects, we write $x \equiv y$ to mean "x is defined as y," "x, which is defined as y," "x, which has been defined earlier as y," or any other grammatical variation depending on the context.

2.6.1 Numbers

Unless otherwise indicated, N, Q, and R will denote the set of integers, the set of rational numbers in the decimal or binary system, and the set of real numbers, respectively. We will also write $\{1,2,\ldots\}$ for the set of positive integers. The set R is equipped with the Euclidean metric $d \equiv d_{ecld}$. Suppose $a, b, a_i \in R$ for $i = m, m+1, \ldots$ for some $m \in N$. We will write $\lim_{i\to\infty} a_i$ for the limit of the sequence $a_m, a_{m+1}, \ldots$ if it exists, without explicitly referring to m. We will write $a \vee b, a \wedge b, a_+$, and a_- for $\max(a,b)$, $\min(a,b), a \vee 0$, and $a \wedge 0$, respectively. The sum $\sum_{i=m}^{n} a_i \equiv a_m + \cdots + a_n$ is understood to be 0 if $n < m$. The product $\prod_{i=m}^{n} a_i \equiv a_m \cdots a_n$ is understood to be 1 if $n < m$. Suppose $a_i \geq 0$ for $i = m$, $m+1, \ldots$ We write $\sum_{i=m}^{\infty} a_i < \infty$ if and only if $\sum_{i=m}^{\infty} |a_i| < \infty$, in which case $\sum_{i=m}^{\infty} a_i$ is taken to be $\lim_{n\to\infty} \sum_{i=m}^{n} a_i$. In other words, unless otherwise specified, convergence of a series of real numbers means absolute convergence.

Regarding real numbers, we quote Lemma 2.18 from [Bishop and Bridges 1985], which will be used, extensively and without further comments, in the present book.

Limited proof by contradiction of an inequality of real numbers. Let x, y be real numbers such that the assumption $x > y$ implies a contradiction. Then $x \leq y$. This lemma remains valid if the relations $>$ and $\leq$ are replaced by $<$ and $\geq$, respectively.

We note, however, that if the relations $>$ and $\leq$ are replaced by $\geq$ and $<$, respectively, then the lemma would not have a constructive proof. Roughly speaking, the reason is that a constructive proof of $x < y$ implies the calculation of a positive $\varepsilon > 0$ such that $y - x > \varepsilon$, which is more than a proof of $x \leq y$; the latter requires only a proof that $x > y$ is impossible and does not require the calculation of anything. The reader should ponder on the subtle but important difference.

2.6.2 Set, Operation, and Function

Set. In general, a *set* is a collection of objects equipped with an equality relation. To define a set is to specify how to construct an element of the set, and how to prove that two elements are equal. A set is also called a *family*.

A member ω in the collection Ω is called an element of the latter, or, in symbols, $\omega \in \Omega$.

The usual set-theoretic notations are used. Let two subsets A and B of a set Ω be given. We will write $A \cup B$ for the union, and $A \cap B$ or AB for the intersection. We write $A \subset B$ if each member ω of A is a member of B. We write $A \supset B$ for

$B \subset A$. The *set-theoretic complement* of a subset A of the set Ω is defined as the set $\{\omega \in \Omega : \omega \in A$ implies a contradiction$\}$. We write $\omega \notin A$ if $\omega \in A$ implies a contradiction.

Nonempty set. A set Ω is said to be *nonempty* if we can construct some element $\omega \in \Omega$.

Empty set. A set Ω is said to be *empty* if it is impossible to construct an element $\omega \in \Omega$. We will let ϕ denote an empty set.

Operation. Suppose A, B are sets. A finite, step-by-step, method X that produces an element $X(x) \in B$ given any $x \in A$ is called an *operation* from A to B. The element $X(x)$ need not be unique. Two different applications of the operation X with the same input element x can produce different outputs. An example of an operation is $[\cdot]_1$, which assigns to each $a \in R$ an integer $[a]_1 \in (a, a+2)$. This operation is a substitute of the classical operation $[\cdot]$ and will be used frequently in the present work.

Function. Suppose Ω, Ω' are sets. Suppose X is an operation that, for each ω in some nonempty subset A of Ω, constructs a unique member $X(\omega)$ in Ω'. Then the operation X is called a *function* from Ω to Ω', or simply a function on Ω. The subset A is called the *domain* of X. We then write $X : \Omega \to \Omega'$, and write $domain(X)$ for the set A. Thus a function X is an operation that has the additional property that if $\omega_1 = \omega_2$ in $domain(X)$, then $X(\omega_1) = X(\omega_2)$ in Ω'. To specify a function X, we need to specify its domain as well as the operation that produces the image $X(\omega)$ from each given member ω of $domain(X)$.

Two functions X, Y are considered equal, $X = Y$ in symbols, if

$$domain(X) = domain(Y),$$

and if $X(\omega) = Y(\omega)$ for each $\omega \in domain(X)$. When emphasis is needed, this equality will be referred to as the *set-theoretic equality*, in contradistinction to almost everywhere equality, to be defined later.

A function is also called a *mapping*.

Domain of function on a set need not be the entire set. The nonempty $domain(X)$ is not required to be the whole set Ω. This will be convenient when we work with functions defined only almost everywhere, in a sense to be made precise later in the setting of a measure/integration space. Henceforth, we write $X(\omega)$ only with the implicit or explicit condition that $\omega \in domain(X)$.

Miscellaneous set notations and function notations. Separately, we sometimes use the expression $\omega \to X(\omega)$ for a function X whose domain is understood. For example, the expression $\omega \to \omega^2$ stands for the function $X : R \to R$ defined by $X(\omega) \equiv \omega^2$ for each $\omega \in R$.

Let $X : \Omega \to \Omega'$ be a function, and let A be a subset of Ω such that $A \cap domain(X)$ is nonempty. Then the *restriction* $X|A$ of X to A is defined as the function from A to Ω' with $domain(X|A) \equiv A \cap domain(X)$ and $(X|A)(\omega)$ for each $\omega \in domain(X|A)$. The set

$$B \equiv \{\omega' \in \Omega' : \omega' = X(\omega) \text{ for some } \omega \in domain(X)\}$$

is called the *range* of the function X, and is denoted by $range(X)$.

A function $X: A \to B$ is called a *surjection* if $range(X) = B$; in that case, there exists an operation $Y: B \to A$, not necessarily a function, such that $X(Y(b) = b$ for each $b \in B$. The function X is called an *injection* if for each $a, a' \in domain(X)$ with $X(a) = X(a')$, we have $a = a'$. It is called a *bijection* if $domain(X) = A$ and if X is both a surjection and an injection.

Let $X : B \to A$ be a surjection with $domain(X) = B$. Then the triple (A, B, X) is called an *indexed set*. In that case, we write $X_b \equiv X(b)$ for each $b \in B$. We will, by abuse of notations, call A or $\{X_b : b \in B\}$ an *indexed set*, and write $A \equiv \{X_b : b \in B\}$. We will call B the index set, and say that A is indexed by the members b of B.

Finite set, enumerated set, countable set. A set A is said to be *finite* if there exists a bijection $v : \{1, \ldots, n\} \to A$, for some $n \geq 1$, in which case we write $|A| \equiv n$ and call it the *size* of A. We will then call v an *enumeration* of the set A, and call the pair (A, v) an *enumerated set*. When the enumeration v is understood from the context, we will abuse notations and simply call the set $A \equiv \{v_1, \ldots, v_n\}$ an enumerated set.

A set A is said to be *countable* if there exists a surjection $v : \{1, 2, \ldots\} \to A$. A set A is said to be *countably infinite* if there exists a bijection $v : \{1, 2, \ldots\} \to A$. We will then call v an *enumeration* of the set A, and call the pair (A, v) an enumerated set. When the enumeration v is understood from the context, we will abuse notations and simply call the set $A \equiv \{v_1, v_2, \ldots\}$ an enumerated set.

Composite function. Suppose $X : \Omega \to \Omega'$ and $X' : \Omega' \to \Omega''$ are such that the set A defined by

$$A \equiv \{\omega \in domain(X) : X(\omega) \in domain(X')\}$$

is nonempty. Then the *composite function* $X' \circ X : \Omega \to \Omega''$ is defined to have $domain(X' \circ X) = A$ and $(X' \circ X)(\omega) = X'(X(\omega))$ for $\omega \in A$. The alternative notation $X'(X)$ will also be used for $X' \circ X$.

Sequence. Let Ω be a set and let $n \geq 1$ be an arbitrary integer. A function $\omega : \{1, \ldots, n\} \to \Omega$ that assigns to each $i \in \{1, \ldots, n\}$ an element $\omega(i) \equiv \omega_i \in \Omega$ is called *a finite sequence* of elements in Ω. A function $\omega : \{1, 2, \ldots,\} \to \Omega$ that assigns to each $i \in \{1, 2, \ldots\}$ an element $\omega(i) \equiv \omega_i \in \Omega$ is called *an infinite sequence* of elements in Ω. We will then write $\omega \equiv (\omega_1, \ldots, \omega_n) \equiv$ or $(\omega_i)_{i=1,\ldots,n}$ in the first case, and write $(\omega_1, \omega_2, \ldots)$ or $(\omega_i)_{i=1,2,\ldots}$ in the second case, for the sequence ω. If, in addition, j is a sequence of integers in $domain(\omega)$, such that $j_k < j_h$ for each $k < h$ in $domain(j)$, then the sequence $\omega \circ j : domain(j) \to \Omega$ is called a *subsequence* of ω. Throughout this book, we will write a subscripted symbol a_b interchangeably with $a(b)$ to lessen the burden on subscripts. Thus, $a_{b(c)}$ stands for a_{b_c}. Similarly, $\omega_{j_k} \equiv \omega_{j(k)} \equiv \omega(j(k))$ for each $k \in domain(j)$, and we write $(\omega_{j(1)}, \omega_{j(2)}, \ldots)$ or $(\omega_{j(k)})_{k=1,2,\ldots}$, or simply $(\omega_{j(k)})$, for the subsequence when the domain of j is clear. If $(\omega_1, \ldots, \omega_n)$ is a sequence, we will write $\{\omega_1, \ldots, \omega_n\}$ for the range of ω. Thus an element $\omega_0 \in \Omega$ is in $\{\omega_1, \ldots, \omega_n\}$ if and only if there exists $i = 1, \ldots, n$ such that $\omega_0 = \omega_i$.

Suppose $(\omega_i)_{i=1,2,\ldots}$ and $(\omega'_i)_{i=1,2,\ldots}$ are two infinite sequences. We will write $(\omega_i, \omega'_i)_{i=1,2,\ldots}$ for the merged sequence $(\omega_1, \omega'_1, \omega_2, \omega'_2, \ldots)$. Similar notations are used for several sequences.

Cartesian product of sequence of sets. Let $(\Omega_n)_{n=0,1,\ldots}$ be a sequence of nonempty sets. Consider any $0 \leq n \leq \infty$, i.e., n is a nonnegative integer or the symbol ∞. We will let $\Omega^{(n)}$ denote the Cartesian product $\prod_{j=0}^{n} \Omega_j$. Consider $0 \leq k < \infty$ with $k \leq n$. The *coordinate function* π_k is the function with $domain(\pi_k) = \Omega^{(n)}$ and $\pi_k(\omega_0, \omega_1, \ldots) = \omega_k$. If $\Omega_n = \Omega$ for each $n \geq 0$, then we will write Ω^n for $\Omega^{(n)}$ for each $n \geq 0$. Let X be a function on Ω_k and let Y be a function on $\Omega^{(k)}$. When confusion is unlikely, we will use the same symbol X also for the function $X \circ \pi_k$ on $\Omega^{(n)}$, which depends only on the kth coordinate. Likewise, we will use Y also for the function $Y \circ (\pi_0, \ldots, \pi_k)$ on $\Omega^{(n)}$, which depends only on the first $k+1$ coordinates. Thus every function on Ω_k or $\Omega^{(k)}$ is identified with a function on $\Omega^{(\infty)}$. Accordingly, sets of functions on $\Omega_k, \Omega^{(k)}$ are regarded also as sets of functions on $\Omega^{(n)}$.

Function of several functions. Let M be the family of all real-valued functions on Ω, equipped with the set-theoretic equality for functions. Suppose $X, Y \in M$ and suppose f is a function on $R \times R$ such that the set

$$D \equiv \{\omega \in domain(X) \cap domain(Y) : (X(\omega), Y(\omega)) \in domain(f)\}$$

is nonempty. Then $f(X,Y)$ is defined as the function with $domain(f(X,Y)) \equiv D$ and $f(X,Y)(\omega) \equiv f(X(\omega), Y(\omega))$ for each $\omega \in D$. The definition extends to a sequence of functions in the obvious manner. Examples are where $f(x,y) \equiv x+y$ for each $(x,y) \in R \times R$, or where $f(x,y) \equiv xy$ for each $(x,y) \in R \times R$.

Convergent series of real-valued functions. Suppose $(X_i)_{i=m,m+1,\ldots}$ is a sequence of real-valued functions on a set Ω. Suppose the set

$$D \equiv \left\{\omega \in \bigcap_{i=m}^{\infty} domain(X_i) : \sum_{i=m}^{\infty} |X_i(\omega)| < \infty\right\}$$

is nonempty. Then $\sum_{i=m}^{\infty} X_i$ is defined as the function with $domain\left(\sum_{i=m}^{\infty} X_i\right) \equiv D$ and with value $\sum_{i=m}^{\infty} X_i(\omega)$ for each $\omega \in D$. This function $\sum_{i=m}^{\infty} X_i$ is then called a convergent series. Thus convergence for series means absolute convergence.

Ordering of functions. Suppose $X, Y \in M$ and A is a subset of Ω, and suppose $a \in R$. We say $X \leq Y$ on A if (i) $A \cap domain(X) = A \cap domain(Y)$ and (ii) $X(\omega) \leq Y(\omega)$ for each $\omega \in A \cap domain(X)$. If $X \leq Y$ on Ω, we will simply write $X \leq Y$. Thus $X \leq Y$ implies $domain(X) = domain(Y)$. We write $X \leq a$ if $X(\omega) \leq a$ for each $\omega \in domain(X)$. We will write

$$(X \leq a) \equiv \{\omega \in domain(X) : X(\omega) \leq a\}.$$

We make similar definitions when the relation $\leq$ is replaced by $<$, $\geq$, $>$, or $=$. We say X is nonnegative if $X \geq 0$.

Suppose $a \in R$. We will abuse notations and write a also for the constant function X with $domain(X) = \Omega$ and with $X(\omega) = a$ for each $\omega \in domain(X)$.

Regarding one of several variables as a parameter. Let X be a function on the product set $\Omega' \times \Omega''$. Let $\omega' \in \Omega'$ be such that $(\omega', \omega'') \in domain(X)$ for some $\omega'' \in \Omega''$. Define the function $X(\omega', \cdot)$ on Ω'' by

$$domain(X(\omega', \cdot)) \equiv \{\omega'' \in \Omega'' : (\omega', \omega'') \in domain(X)\},$$

and by $X(\omega', \cdot)(\omega'') \equiv X(\omega', \omega'')$ for each $\omega'' \in domain(X(\omega', \cdot))$. Similarly, let $\omega'' \in \Omega''$ be such that $(\omega', \omega'') \in domain(X)$ for some $\omega' \in \Omega'$. Define the function $X(\cdot, \omega'')$ on Ω' by

$$domain(X(\cdot, \omega'')) \equiv \{\omega' \in \Omega' : (\omega', \omega'') \in domain(X)\},$$

and by $X(\cdot, \omega'')(\omega') \equiv X(\omega', \omega'')$ for each $\omega' \in domain(X(\cdot, \omega''))$.

More generally, given a function X on the Cartesian product $\Omega' \times \Omega'' \times \cdots \times \Omega^{(n)}$, for each $(\omega', \omega'', \ldots, \omega^{(n)}) \in domain(X)$, we define similarly the functions $X(\cdot, \omega'', \omega''', \ldots, \omega^{(n)})$, $X(\omega', \cdot, \omega''', \ldots, \omega^{(n)})$, $\ldots$,$X(\omega', \omega'', \ldots, \omega^{(n-1)}, \cdot)$ on the sets $\Omega', \Omega'', \ldots, \Omega^{(n)}$, respectively.

Let M', M'' denote the families of all real-valued functions on two sets Ω', Ω'', respectively, and let L'' be a subset of M''. Suppose

$$T : \Omega' \times L'' \to R \tag{2.6.1}$$

is a real-valued function. We can define the function

$$T^* : L'' \to M'$$

with

$$domain(T^*) \equiv \{Y \in L'' : domain(T(\cdot, Y')) \text{ is nonempty}\}$$

and by $T^*(Y) \equiv T(\cdot, Y)$. When there is no risk of confusion, we write T also for the function T^*, we write TY for $T(\cdot, Y)$, and we write

$$T : L'' \to M'$$

interchangeably with the expression 2.6.1. Thus the duality

$$T(\cdot, Y)(\omega') \equiv T(\omega', Y) \equiv T(\omega', \cdot)(Y). \tag{2.6.2}$$

2.6.3 Metric Space

The definitions and notations related to metric spaces in [Bishop and Bridges 1985], with few exceptions, are familiar to readers of classical texts. A summary of these definitions and notations follows.

Metric complement. Let (S, d) be a *metric space*. If J is a subset of S, its *metric complement* is the set $\{x \in S : d(x, y) > 0 \text{ for all } y \in J\}$. Unless otherwise specified, J_c will denote the metric complement of J.

Condition valid for all but countably many points in metric space. A condition is said to hold for *all but countably many* members of S if it holds for each member in the metric complement J_c of some countable subset J of S.

Inequality in a metric space. We will say that two elements $x, y \in S$ are *unequal*, and write $x \neq y$, if $d(x, y) > 0$.

Metrically discrete subset of a metric space. We will call a subset A of S *metrically discrete* if, for each $x, y \in A$ we have $x = y$ or $d(x, y) > 0$. Classically, each subset A of S is metrically discrete.

Limit of a sequence of functions with values in a metric space. Let $(f_n)_{n=1,2,...}$ be a sequence of functions from a set Ω to S such that the set

$$D \equiv \left\{\omega \in \bigcap_{i=1}^{\infty} domain(f_i) : \lim_{i\to\infty} f_i(\omega) \text{ exists in } S\right\}$$

is nonempty. Then $\lim_{i\to\infty} f_i$ is defined as the function with *domain* $(\lim_{i\to\infty} f_i) \equiv D$ and with value

$$\left(\lim_{i\to\infty} f_i\right)(\omega) \equiv \lim_{i\to\infty} f_i(\omega)$$

for each $\omega \in D$. We emphasize that $\lim_{i\to\infty} f_i$ is well defined only if it can be shown that D is nonempty.

Continuous function. A function $f : S \to S'$ is said to be *uniformly continuous* on a subset $A \subset domain(f)$, relative to the metrics d, d' on S, S' respectively, if there exists an operation $\delta : (0,\infty) \to (0,\infty)$ such that $d'(f(x), f(y)) < \varepsilon$ for each $x, y \in A$ with $d(x, y) < \delta(\varepsilon)$, for each $\varepsilon > 0$. When there is a need to be precise as to the metrics d, d', we will say that $f : (S,d) \to (S',d')$ is uniformly continuous on A. The operation δ is called a *modulus of continuity* of f on A.

Lipschitz continuous function. If there exists a coefficient $c \geq 0$ such that $d'(f(x), f(y)) \leq cd(x, y)$ for all $x, y \in A$, then the function f is said to be *Lipschitz continuous* on A, and the constant c is then called a Lipschitz constant of f on A. In that case, we will say simply that f has Lipschitz constant c.

Totally bounded metric space, compact metric space. A metric space (S,d) is said to be *totally bounded* if, for each $\varepsilon > 0$, there exists a finite subset $A \subset S$ such that for each $x \in S$ there exists $y \in A$ with $d(x, y) < \varepsilon$. The subset A is then called an *ε-approximation* of S. A *compact metric space K* is defined as a complete and totally bounded metric space.

Locally compact metric space. A subset $A \subset S$ is said to be *bounded* if there exists $x \in S$ and $a > 0$ such that $A \subset (d(\cdot, x) \leq a)$. A subset $S' \subset S$ is said to be *locally compact* if every bounded subset of S' is contained in some compact subset. The metric space (S,d) is said to be locally compact if the subset S is locally compact.

Continuous function on metric space. A function $f : (S,d) \to (S',d')$ is said to be continuous if $domain(f) = S$ and if it is uniformly continuous on each bounded subset K of S.

Product of a finite sequence of metric spaces. Suppose $(S_n,d_n)_{n=1,2,\ldots}$ is a sequence of metric spaces. For each integer $n \geq 1$, define

$$d^{(n)}(x,y) \equiv \left(\bigotimes_{i=1}^{n} d_i\right)(x,y) \equiv (d_1 \otimes \cdots \otimes d_n)(x,y) \equiv \bigvee_{i=1}^{n} d_i(x_i,y_i)$$

for each $x,y \in \prod_{i=1}^{n} S_i$. Then

$$(S^{(n)},d^{(n)}) \equiv \bigotimes_{i=1}^{n}(S_i,d_i) \equiv \left(\prod_{i=1}^{n} S_i, \bigotimes_{i=1}^{n} d_i\right)$$

is a metric space called the *product metric space* of $S_1,\ldots,S_n$.

Product of an infinite sequence of metric spaces. Define the infinite product metric $\bigotimes_{i=1}^{\infty} d_i$ on $\prod_{i=1}^{\infty} S_i$ by

$$d^{(\infty)}(x,y) \equiv \left(\bigotimes_{i=1}^{\infty} d_i\right)(x,y) \equiv \sum_{i=1}^{\infty} 2^{-i}(1 \wedge d_i(x_i,y_i))$$

for each $x,y \in \prod_{i=1}^{\infty} S_i$. Define the infinite product metric space

$$(S^{(\infty)},d^{(\infty)}) \equiv \bigotimes_{i=1}^{\infty}(S_i,d_i) \equiv \left(\prod_{i=1}^{\infty} S_i, \bigotimes_{i=1}^{\infty} d_i\right).$$

Powers of a sequence of metric spaces. Suppose, in addition, (S_n,d_n) is a copy of the same metric space (S,d) for each $n \geq 1$. Then we simply write $(S^n,d^n) \equiv (S^{(n)},d^{(n)})$ and $(S^\infty,d^\infty) \equiv (S^{(\infty)},d^{(\infty)})$. Thus, in this case,

$$d^n(x,y) \equiv \bigvee_{i=1}^{n} d(x_i,y_i)$$

for each $x = (x_1,\ldots,x_n), y = (y_1,\ldots,y_n) \in S^n$, and

$$d^\infty(x,y) \equiv \sum_{i=1}^{\infty} 2^{-i}(1 \wedge d(x_i,y_i))$$

for each $x = (x_1,x_2,\ldots), y = (y_1,y_2,\ldots) \in S^\infty$.

If, in addition, (S_n,d_n) is locally compact for each $n \geq 1$, then the finite product space $(S^{(n)},d^{(n)})$ is locally compact for each $n \geq 1$, while the infinite product space $(S^{(\infty)},d^{(\infty)})$ is complete but not necessarily locally compact. If (S_n,d_n) is compact for each $n \geq 1$, then both the finite and infinite product spaces are compact.

Spaces of real-valued continuous functions. Suppose (S,d) is a metric space. We will write $C_u(S,d)$, or simply $C_u(S)$, for the space of real-valued functions functions on (S,d) with $domain(f) = S$ that are *uniformly continuous on* S. We will write $C_{ub}(S,d)$, or simply $C_{ub}(S)$, for the subspace of $C_u(S)$ whose members are bounded. Let $x_\circ$ be an arbitrary, but fixed, reference point in (S,d). A uniformly continuous function f on (S,d) is then said to vanish at infinity if, for each $\varepsilon > 0$, there exists $a > 0$ such that $|f| \leq \varepsilon$ for each $x \in S$ with $d(x,x_\circ) > a$. Write $C_0(S,d)$, or simply $C_0(S)$, for the space of continuous functions on (S,d) that vanish at infinity.

Space of real-valued continuous functions with bounded support. A real-valued function f on S is said to have a subset $A \subset S$ as *support* if $x \in domain(f)$ and $|f(x)| > 0$ together imply $x \in A$. Then we also say that f is supported by A, or that A supports f. We will write $C(S,d)$, or simply $C(S)$, for the subspace of $C_u(S,d)$ whose members have bounded supports. In the case where (S,d) is locally compact, $C(S)$ consists of continuous functions on (S,d) with compact supports. Summing up,

$$C(S) \subset C_0(S) \subset C_{ub}(S) \subset C_u(S).$$

Infimum and supremum. Suppose a subset A of R is nonempty. A number $b \in R$ is called a lower bound of A, and A said to bounded from below, if $b \leq a$ for each $a \in A$. A lower bound b of A is called the *greatest lower bound*, or *infimum*, of A if $b \geq b'$ for each lower bound b' of A. In that case, we write $\inf A \equiv b$.

Similarly, a number $b \in R$ is called an upper bound of A, and A said to be bounded from above, if $b \geq a$ for each $a \in A$. An upper bound b of A is called the *least upper bound*, or *supremum*, of A if $b \leq b'$ for each upper bound b' of A. In that case, we write $\sup A \equiv b$.

In contrast to classical mathematics, there is no constructive general proof for the existence of a infimum for an subset of R that is bounded from below. Existence needs to be proved before each usage for each special case, much as in the case of limits.

In that regard, [Bishop and Bridges 1985] prove that if a nonempty subset A of R is totally bounded, then both $\inf A$ and $\sup A$ exist. Moreover, suppose f is a continuous real-valued function on a compact metric space (K,d). Then the last cited text proves that $\inf_K f \equiv \inf\{f(x) : x \in K\}$ and $\sup_K f \equiv \sup\{f(x) : x \in K\}$ exist.

2.6.4 Miscellaneous

Notations for if, only if, etc. The symbols $\Rightarrow$, $\Leftarrow$, and $\Leftrightarrow$ will in general stand for "only if," "if," and "if and only if," respectively. An exception will be made where the symbol $\Rightarrow$ is used for weak convergence, as defined later. The intended meaning will be clear from the context.

Capping a metric at 1. If (S,d) is a metric space, we will write $\widehat{d} \equiv 1 \wedge d$.

Abbreviation for subsets. We will often write "$x,y,\ldots,z \in A$" as an abbreviation for "$\{x,y,\ldots,z\} \subset A$."

Default notations for numbers. Unless otherwise indicated by the context, the symbols i,j,k,m,n,p will denote integers, the symbols a,b will denote real numbers, and the symbols ε, δ will denote positive real numbers. For example, the statement "for each $i \geq 1$" will mean "for each integer $i \geq 1$."

Notations for convergence. Suppose $(a_n)_{n=1,2,...}$ is a sequence of real numbers. Then $a_n \to a$ stands for $\lim_{n\to\infty} a_n = a$. We write $a_n \uparrow a$ if (a_n) is a nondecreasing sequence and $a_n \to a$. Similarly, we write $a_n \downarrow a$ if (a_n) is a nonincreasing sequence and $a_n \to a$. More generally, suppose f is a function on some subset $A \subset R$. Then $f(x) \to a$ stands for $\lim_{x\to x_0} f(x) = a$, where x_0 can stand for a real number or for one of the symbols ∞ or $-\infty$.

Big O and small o. Suppose f and g are functions with domains equal to some subset $A \subset R$. Let $x_0 \in A$ be arbitrary. If for some $c > 0$, we have $|f(x)| \leq c|g(x)|$ for all $x \in A$ in some neighborhood B of x_0, then we write $f(x) = O(g(x))$. If for each $\varepsilon > 0$, we have $|f(x)| \leq \varepsilon|g(x)|$ for each $x \in A$ in some neighborhood B of x_0, then we write $f(x) = o(g(x))$. Here, a subset $B \subset R$ is a neighborhood of x_0 if there exists an open interval (a,b) such that $x_0 \in (a,b)$.

End-of-proof or end-of-definition marker. Finally, we sometimes use the symbol $\square$ to mark the end of a proof or a definition.

3

Partition of Unity

In the previous chapter, we summarized the basic concepts and theorems about metric spaces from [Bishop and Bridges 1985]. Locally compact metric space was introduced. It is a simple and wide-ranging generalization of the real line. With few exceptions, the metric spaces used in the present book are locally compact.

In the present chapter, we will define and construct binary approximations of a locally compact metric space (S,d), then define and construct a partition of unity relative to each binary approximation. Roughly speaking, a binary approximation is a digitization of (S,d), a generalization of the dyadic rationals that digitize the space R of real numbers. A partition of unity is then a sequence in $C(S,d)$ that serves as a basis for the linear space $C(S,d)$ of continuous functions on (S,d) with compact supports, in the sense that each $f \in C(S,d)$ can be approximated by linear combinations of members in the partition of unity.

A partition of unity provides a countable set of basis functions for the metrization of probability distributions on the space (S,d). Because of that important role, we will study binary approximations and partitions of unity in detail in this chapter.

3.1 Abundance of Compact Subsets

First we cite a theorem from [Bishop and Bridges 1985] that guarantees an abundance of compact subsets.

Theorem 3.1.1. Abundance of compact sets. *Let $f : K \to R$ be a continuous function on a compact metric space (K,d) with $domain(f) = K$. Then, for all but countably many real numbers $\alpha > \inf_K f$, the set*

$$(f \leq \alpha) \equiv \{x \in K : f(x) \leq \alpha\}$$

is compact.

Proof. See theorem (4.9) in chapter 4 of [Bishop and Bridges 1985]. □

Classically, the set $(f \leq \alpha)$ is compact for each $\alpha \geq \inf_K f$, without exception. Such a general theorem would, however, imply the principle of infinite search and is therefore nonconstructive. Theorem 3.1.1 is sufficient for all our purposes.

Definition 3.1.2. Convention for compact sets $(f \leq \alpha)$**.** We hereby adopt the convention that if the compactness of the set $(f \leq \alpha)$ is required in a discussion, compactness has been explicitly or implicitly verified, usually by proper prior selection of the constant α, enabled by an application of Theorem 3.1.1. □

The following simple corollary of Theorem 3.1.1 guarantees an abundance of compact neighborhoods of a compact set.

Corollary 3.1.3. Abundance of compact neighborhoods. *Let (S,d) be a locally compact metric space, and let K be a compact subset of S. Then the subset*

$$K_r \equiv (d(\cdot, K) \leq r) \equiv \{x \in S : d(x,K) \leq r\}$$

is compact for all but countably many $r > 0$.

Proof. 1. Let $n \geq 1$ be arbitrary. Then $A_n \equiv (d(\cdot, K) \leq n)$ is a bounded set. Since (S,d) is locally compact, there exists a compact set K_n such that $A_n \subset K_n \subset S$. The continuous function f on the compact metric space (K_n, d) defined by $f \equiv d(\cdot, K)$ has infimum 0. Hence, by Theorem 3.1.1, the set $\{x \in K_n : d(x,K) \leq r\}$ is compact for all but countably many $r \in (0,\infty)$. In other words, there exists a countable subset J of $(0,\infty)$ such that for each r in the metric complement J_c of J in $(0,\infty)$, the set

$$\{x \in K_n : d(x,K) \leq r\}$$

is compact.

2. Now let $r \in J_c$ be arbitrary. Take $n \geq 1$ so large that $r \in (0,n)$. Then $K_r \subset A_n$. Hence the set

$$K_r = K_r A_n = K_r K_n = \{x \in K_n : d(x,K) \leq r\}$$

is compact according to Step 1. Since J is countable and $r \in J_c$ is arbitrary, we see that K_r is compact for all but countably many $r \in (0,\infty)$. □

Separately, the next elementary metric-space lemma will be convenient.

Lemma 3.1.4. If $(S,\, d)$ is compact, then the subspace of $C(S^\infty,\, d^\infty)$ consisting of members that depend on finitely many coordinates is dense in $C(S^\infty,\, d^\infty)$. *Suppose (S,d) is a compact metric space. Let $x_\circ$ be an arbitrary but fixed reference point in (S,d).*

Let $n \geq 1$ be arbitrary. Define the projection mapping $j_n^* : S^\infty \to S^\infty$ by

$$j_n^*(x_1, x_2, \ldots) \equiv (x_1, x_2, \ldots, x_n, x_\circ, x_\circ, \ldots)$$

for each $(x_1, x_2, \ldots) \in S^\infty$. Then $j_n^* \circ j_m^* = j_n^*$ for each $m \geq n$. Let

$$L_{0,n} \equiv \{f \in C(S^\infty, d^\infty) : f = f \circ j_n^*\}. \tag{3.1.1}$$

Then $L_{0,n} \subset L_{0,n+1}$.

Let $L_{0,\infty} \equiv \bigcup_{n=1}^{\infty} L_{0,n}$. *Then the following conditions hold:*

1. $L_{0,n}$ *and* $L_{0,\infty}$ *are linear subspaces of* $C(S^\infty, d^\infty)$*, and consist of functions that depend, respectively, on the first* n *coordinates and on finitely many coordinates.*

2. The subspace $L_{0,\infty}$ *is dense in* $C(S^\infty, d^\infty)$ *relative to the supremum norm* $\|\cdot\|$. *Specifically, let* $f \in C(S^\infty, d^\infty)$ *be arbitrary, with a modulus of continuity* δ_f. *Then* $f \circ j_n^* \in L_{0,n}$. *Moreover, for each* $\varepsilon > 0$ *we have* $\|f - f \circ j_n^*\| \leq \varepsilon$ *if* $n > -\log_2(\delta_f(\varepsilon))$. *In particular, if* f *has Lipschitz constant* $c > 0$*, then* $\|f - f \circ j_n^*\| \leq \varepsilon$ *if* $n > \log_2(c\varepsilon^{-1})$.

Proof. Let $m \geq n \geq 1$ and $w \in S^\infty$ be arbitrary. Then, for each $(x_1, x_2, \ldots) \in S^\infty$, we have

$$\begin{aligned} j_n^*(j_m^*(x_1, x_2, \ldots)) &= j_n^*(x_1, x_2, \ldots, x_m, x_\circ, x_\circ, \ldots) \\ &= (x_1, x_2, \ldots, x_n, x_\circ, x_\circ, \ldots) = j_n^*(x_1, x_2, \ldots). \end{aligned}$$

Thus $j_n^* \circ j_m^* = j_n^*$.

1. It is clear from the defining equality 3.1.1 that $L_{0,n}$ is a linear subspace of $C(S^\infty, d^\infty)$. Let $f \in L_{0,n}$ be arbitrary. Then

$$f = f \circ j_n^* = f \circ j_n^* \circ j_m^* = f \circ j_m^*.$$

Hence $f \in L_{0,m}$. Thus $L_{0,n} \subset L_{0,m}$. Consequently, $L_{0,\infty} \equiv \bigcup_{p=1}^{\infty} L_{0,p}$ is a union of a nondecreasing sequence of linear subspaces of $C(S^\infty, d^\infty)$ and, therefore, is also a linear subspace of $C(S^\infty, d^\infty)$.

2. Let $f \in C(S^\infty, d^\infty)$ be arbitrary, with a modulus of continuity δ_f. Let $\varepsilon > 0$ be arbitrary. Suppose $n > -\log_2(\delta_f(\varepsilon))$. Then $2^{-n} < \delta_f(\varepsilon)$. Let $(x_1, x_2, \ldots) \in S^\infty$ be arbitrary. Then

$$\begin{aligned} &d^\infty((x_1, x_2, \ldots), j_n^*(x_1, x_2, \ldots)) \\ &\quad = d^\infty((x_1, x_2, \ldots), (x_1, x_2, \ldots, x_n, x_\circ, x_\circ, \ldots)) \\ &\quad \equiv \sum_{k=1}^{n} 2^{-k}\widehat{d}(x_k, x_k) + \sum_{k=n+1}^{\infty} 2^{-k}\widehat{d}(x_k, x_\circ) \leq 0 + 2^{-n} < \delta_f(\varepsilon), \end{aligned}$$

where $\widehat{d} \equiv 1 \wedge d$. Hence

$$|f(x_1, x_2, \ldots) - f \circ j_n^*(x_1, x_2, \ldots)| < \varepsilon,$$

where $(x_1, x_2, \ldots) \in S^\infty$ is arbitrary. We conclude that $\|f - f \circ j_n^*\| \leq \varepsilon$, as alleged. □

3.2 Binary Approximation

Let (S, d) be an arbitrary locally compact metric space. Then S contains a countable dense subset. A binary approximation, defined presently, is a structured and well-quantified countable dense subset.

Recall that (i) $|A|$ denotes the number of elements in an arbitrary finite set A; (ii) a subset A of S is said to be metrically discrete if for each $y,z \in A$, either $y = z$ or $d(y,z) > 0$; and (iii) a finite subset A of a subset $K \subset S$ is called an ε-approximation of K if for each $x \in K$, there exists $y \in A$ with that $d(x,y) \leq \varepsilon$. Classically, each subset of (S,d) is metrically discrete. Condition (iii) can be written more succinctly as

$$K \subset \bigcup_{x \in A} (d(\cdot,x) \leq \varepsilon).$$

Definition 3.2.1. Binary approximation and modulus of local compactness. Let (S,d) be a locally compact metric space, with an arbitrary but fixed reference point $x_\circ$. Let $A_0 \equiv \{x_\circ\} \subset A_1 \subset A_2 \subset \ldots$ be a sequence of metrically discrete and finite subsets of S. For each $n \geq 1$, write $\kappa_n \equiv |A_n|$. Suppose

$$(d(\cdot,x_\circ) \leq 2^n) \subset \bigcup_{x \in A(n)} (d(\cdot,x) \leq 2^{-n}) \tag{3.2.1}$$

and

$$\bigcup_{x \in A(n)} (d(\cdot,x) \leq 2^{-n+1}) \subset (d(\cdot,x_\circ) \leq 2^{n+1}) \tag{3.2.2}$$

for each $n \geq 1$. Then the sequence $\xi \equiv (A_n)_{n=1,2,\ldots}$ of subsets is called a *binary approximation* for (S,d) relative to $x_\circ$, and the sequence of integers

$$\|\xi\| \equiv (\kappa_n)_{n=1,2,\ldots} \equiv (|A_n|)_{n=1,2,\ldots}$$

is called the *modulus of local compactness* of (S,d) corresponding to ξ.

Thus a binary approximation is an expanding sequence of 2^{-n}-approximation for $(d(\cdot,x_\circ) \leq 2^n)$ as $n \to \infty$. The next proposition shows that the definition is not vacuous. □

First note that $\bigcup_{n=1}^{\infty} A_n$ is dense in (S,d) in view of relation 3.2.1. In the case where (S,d) is compact, for $n \geq 1$ so large that $S = (d(\cdot,x_\circ) \leq 2^n)$, relation 3.2.1 says that we need at most κ_n points to make a 2^{-n}-approximation of S.[1]

Lemma 3.2.2. Existence of metrically discrete ε-approximations. *Let K be a compact subset of the locally compact metric space (S,d). Let $A_0 \equiv \{x_1,\ldots,x_n\}$ be an arbitrary metrically discrete finite subset of K. Let $\varepsilon > 0$ be arbitrary. Then the following conditions hold:*

1. There exists a metrically discrete finite subset A of K such that (i) $A_0 \subset A$ and (ii) A is an ε-approximation of K.

2. In particular, there exists a metrically discrete finite set A that is an ε-approximation of K.

[1] Incidentally, the number $\log \kappa_n$ is a bound for Kolmogorov's 2^{-n}-*entropy* of the compact metric space (S,d), which represents the informational content in a 2^{-n}-approximation of S. See [Lorentz 1966] for a definition of ε-entropy.

Proof. 1. By hypothesis, the set $A_0 \equiv \{x_1, \ldots, x_n\}$ is a metrically discrete finite subset of the compact set K. Let $\varepsilon > 0$ be arbitrary. Take an arbitrary $\varepsilon_0 \in (0, \varepsilon)$. Take any ε_0-approximation $\{y_1, \ldots, y_m\}$ of K. Write $\alpha \equiv m^{-1}(\varepsilon - \varepsilon_0)$.

2. Trivially, the set

$$A_0 \cup \{y_1, \ldots, y_m\}$$

is an ε_0-approximation of K. Moreover, A_0 is metrically discrete.

3. Consider y_1. Then either (i) there exists $x \in A_0$ such that $d(x, y_1) < \alpha$ or (ii) for each $x \in A_0$ we have $d(x, y_1) > 0$. In case (i), define $A_1 \equiv A_0$. In case (ii), define $A_1 \equiv A_0 \cup \{y_1\}$. Then, in either case, A_1 is a metrically discrete finite subset of K. Moreover, $A_0 \subset A_1$. By the assumption in Step 2, the set $A_0 \cup \{y_1, \ldots, y_m\}$ is an ε_0-approximation of K. Hence there exists $z \in A_0 \cup \{y_1, \ldots, y_m\}$ such that

$$d(z, y) < \varepsilon_0.$$

There are three possibilities: (i′) $z \in A_0$, (ii′) $z \in \{y_2, \ldots, y_m\}$, or (iii′) $z = y_1$.

4. In cases (i′) and (ii′), we have, trivially, $z \in A_1 \cup \{y_2, \ldots, y_m\}$ and $d(z, y) < \varepsilon_0 + \alpha$. Let $w \equiv z \in A_0 \cup \{y_1, \ldots, y_m\}$.

5. Consider case (iii′), where $z = y_1$. Then, according to Step 3, either (i) or (ii) holds. In case (ii), we have, again trivially,

$$z = y_1 \in A_0 \cup \{y_1\} \equiv A_1,$$

and $d(z, y) < \varepsilon_0$. Consider case (i). Then $A_1 \equiv A_0$, and there exists $x \in A_0$ such that $d(x, y_1) < \alpha$. Consequently, $x \in A_1$ and

$$d(x, y) \leq d(x, y_1) + d(y_1, y) = d(x, y_1) + d(z, y) < \alpha + \varepsilon_0.$$

Let $w \equiv x \in A_0 \cup \{y_1, \ldots, y_m\}$.

6. Combining Steps 4 and 5, we see that, in any case, there exists $w \in A_1 \cup \{y_2, \ldots, y_m\}$ such that $d(w, y) < \varepsilon_0 + \alpha$. Since $y \in K$ is arbitrary, we conclude that the set

$$A_1 \cup \{y_2, \ldots, y_m\}$$

is an $(\varepsilon_0 + \alpha)$-approximation of K, and that the set A_1 is metrically discrete. Moreover, $A_0 \subset A_1$.

7. Repeat Steps 3–6 with $A_1 \cup \{y_2, \ldots, y_m\}$ and $\varepsilon_0 + \alpha$ in the roles of $A_0 \cup \{y_1, \ldots, y_m\}$ and ε_0, respectively. We obtain a metrically discrete subset A_2 such that

$$A_2 \cup \{y_3, \ldots, y_m\}$$

is an $(\varepsilon_0 + 2\alpha)$-approximation of K. Moreover $A_0 \subset A_2$.

8. Recursively on $y_3, \ldots, y_m$, we obtain a metrically discrete subset A_m that is an $(\varepsilon_0 + m\alpha)$-approximation of K. Moreover, $A_0 \subset A_m$. Since

$$\varepsilon_0 + m\alpha \equiv \varepsilon_0 + (\varepsilon - \varepsilon_0) = \varepsilon,$$

it follows that A_m is an ε-approximation of K. In conclusion, the set $A \equiv A_m$ has the desired properties in Assertion 1.

9. To prove Assertion 2, take an arbitrary $x_1 \in K$ and define $A_0 \equiv \{x_1\}$. Then, by Assertion 1, there exists a metrically discrete finite subset A of K such that (i) $A_0 \subset A$ and (ii) A is an ε-approximation of K. Thus the set A has the desired properties in Assertion 2. □

Proposition 3.2.3. Existence of binary approximations. *Each locally compact metric space (S,d) has a binary approximation.*

Proof. Let $x_\circ \in S$ be an arbitrary but fixed reference point. Proceed inductively on $n \geq 0$ to construct a metrically discrete and finite subset A_n of S to satisfy Conditions 3.2.1 and 3.2.2 in Definition 3.2.1.

To start, let $A_0 \equiv \{x_\circ\}$. Suppose the set A_n has been constructed for some $n \geq 0$ such that if $n \geq 1$, then (i) A_n is metrically discrete and finite and (ii) Conditions 3.2.1 and 3.2.2 in Definition 3.2.1 are satisfied. Proceed to construct A_{n+1}.

To that end, write $\varepsilon \equiv 2^{-n-2}$ and take any $r \in [2^{n+1}, 2^{n+1} + \varepsilon)$ such that

$$K \equiv (d(\cdot, x_\circ) \leq r)$$

is compact. This is possible in view of Corollary 3.1.3. If $n = 0$, then $A_n \equiv \{x_\circ\} \subset K$ trivially. If $n \geq 1$, then, according to the induction hypothesis, the set A_n is metrically discrete, and by Condition 3.2.2 in Definition 3.2.1, we have

$$A_n \subset \bigcup_{x \in A(n)} (d(\cdot, x) \leq 2^{-n+1}) \subset (d(\cdot, x_\circ) \leq 2^{n+1}) \subset K.$$

Hence we can apply Lemma 3.2.2 to construct a 2^{-n-1}-approximation A_{n+1} of K, which is metrically discrete and finite, such that $A_n \subset A_{n+1}$. It follows that

$$(d(\cdot, x_\circ) \leq 2^{n+1}) \subset K \subset \bigcup_{x \in A(n+1)} (d(\cdot, x) \leq 2^{-n-1}),$$

proving Condition 3.2.1 in Definition 3.2.1 for $n + 1$.

Now let

$$y \in \bigcup_{x \in A(n+1)} (d(\cdot, x) \leq 2^{-n})$$

be arbitrary. Then $d(y, x) \leq 2^{-n}$ for some $x \in A_{n+1} \subset K$. Hence

$$d(x, x_\circ) \leq r < 2^{n+1} + \varepsilon.$$

Consequently,

$$\begin{aligned} d(y, x_\circ) &\leq d(y, x) + d(x, x_\circ) \\ &\leq 2^{-n} + 2^{n+1} + \varepsilon \equiv 2^{-n} + 2^{n+1} + 2^{-n-2} \leq 2^{n+2}. \end{aligned}$$

Thus we see that

$$\bigcup_{x\in A(n+1)} (d(\cdot,x) \le 2^{-n}) \subset (d(\cdot,x_\circ) \le 2^{n+2}),$$

proving Condition 3.2.2 in Definition 3.2.1 for $n+1$. Induction is completed. The sequence $\xi \equiv (A_n)_{n=1,2,...}$ satisfies all the conditions in Definition 3.2.1 to be a binary approximation of (S,d). □

Definition 3.2.4. Finite product and power of binary approximations. Let $n \ge 1$ be arbitrary. For each $i = 1,\dots,n$, let (S_i,d_i) be a locally compact metric space, with a reference point $x_{i,\circ} \in S_i$ and with a binary approximation $\xi_i \equiv (A_{i,p})_{p=1,2,...}$ relative to $x_{i,\circ}$. Let $(S^{(n)},d^{(n)}) \equiv \left(\prod_{i=1}^n S_i, \bigotimes_{i=1}^n d_i\right)$ be the product metric space, with $x_\circ^{(n)} \equiv (x_{1,\circ},\dots,x_{n,\circ})$ designated as the reference point in $(S^{(n)},d^{(n)})$.

For each $p \ge 1$, let $A_p^{(n)} \equiv A_{1,p} \times \cdots \times A_{n,p}$. The next lemma proves that $(A_p^{(n)})_{p=1,2,...}$ is a binary approximation of $(S^{(n)},d^{(n)})$ relative to $x_\circ^{(n)}$. We will call $\xi^{(n)} \equiv (A_p^{(n)})_{p=1,2,...}$ the *product binary approximation* of $\xi_1,\dots,\xi_n$, and write

$$\xi^{(n)} \equiv \xi_1 \otimes \cdots \otimes \xi_n.$$

If $(S_i,d_i) = (S,d)$ for some locally compact metric space, with $x_{i,\circ} = x_\circ$ and $\xi_i = \xi$ for each $i = 1,\dots,n$, then we will call $\xi^{(n)}$ the nth *power* of the binary approximation ξ, and write $\xi^n \equiv \xi^{(n)}$. □

Lemma 3.2.5. Finite product binary approximation is indeed a binary approximation. *Use the assumptions and notations in Definition 3.2.4. Then the finite product binary approximation $\xi^{(n)}$ is indeed a binary approximation of $(S^{(n)},d^{(n)})$ relative to $x_\circ^{(n)}$.*

Let $\|\xi_i\| \equiv (\kappa_{i,p})_{p=1,2,...} \equiv (|A_{i,p}|)_{p=1,2,...}$ be the modulus of local compactness of (S_i,d_i) corresponding to ξ_i, for each $i = 1,\dots,n$. Let $\left\|\xi^{(n)}\right\|$ be the modulus of local compactness of $(S^{(n)},d^{(n)})$ corresponding to $\xi^{(n)}$. Then

$$\left\|\xi^{(n)}\right\| = \left(\prod_{i=1}^n \kappa_{i,p}\right)_{p=1,2,...}.$$

In particular, if $\xi_i \equiv \xi$ for each $i = 1,\dots,n$ for some binary approximation ξ of some locally compact metric space (S,d), then the finite power binary approximation is indeed a binary approximation of (S^n,d^n) relative to $x_\circ^n$. Moreover, the modulus of local compactness of (S^n,d^n) corresponding to ξ^n is given by

$$\|\xi^n\| = (\kappa_p^n)_{p=1,2,...}.$$

Proof. Recall that $A_p^{(n)} \equiv A_{1,p} \times \cdots \times A_{n,p}$ for each $p \ge 1$. Hence $A_1^{(n)} \subset A_2^{(n)} \subset \cdots$.

1. Let $p \ge 1$ be arbitrary. Let

$$x \equiv (x_1,\dots,x_n), y \equiv (y_1,\dots,y_n) \in A_p^{(n)} \equiv A_{1,p} \times \cdots \times A_{n,p}$$

be arbitrary. For each $i = 1, \ldots, n$, because $(A_{i,q})_{q=1,2,\ldots}$ is a binary approximation of (S_i, d_i), the set $A_{i,p}$ is metrically discrete. Hence either (i) $x_i = y_i$ for each $i = 1, \ldots, n$ or (ii) $d_i(x_i, y_i) > 0$ for some $i = 1, \ldots, n$. In case (i), we have $x = y$. In case (ii), we have

$$d^{(n)}(x, y) \equiv \bigvee_{j=1}^{n} d_j(x_j, y_j) \geq d_i(x_i, y_i) > 0.$$

Thus the set $A_p^{(n)}$ is metrically discrete.

2. Next note that

$$\begin{aligned}
(d^{(n)}(\cdot, x_\circ^{(n)}) \leq 2^p) &\equiv \left\{(y_1, \ldots, y_n) \in S^{(n)} : \bigvee_{i=1}^{n} d_i(y_i, x_{i,\circ}) \leq 2^p\right\} \\
&= \bigcap_{i=1}^{n} \{(y_1, \ldots, y_n) \in S^{(n)} : d_i(y_i, x_{i,\circ}) \leq 2^p\} \\
&\subset C \equiv \bigcap_{i=1}^{n} \bigcup_{x(i) \in A(i,p)} \{(y_1, \ldots, y_n) \in S^{(n)} : d_i(y_i, x_i) \leq 2^{-p}\},
\end{aligned} \tag{3.2.3}$$

where the last inclusion is due to Condition 3.2.1 applied to the binary approximation $(A_{i,q})_{q=1,2,\ldots}$. Basic Boolean operations then yield

$$\begin{aligned}
C &= \bigcup_{(x(1),\ldots,x(n)) \in A(1,p) \times \cdots \times A(n,p)} \bigcap_{i=1}^{n} \{(y_1, \ldots, y_n) \in S^{(n)} : d_i(y_i, x_i) \leq 2^{-p}\} \\
&= \bigcup_{(x(1),\ldots,x(n)) \in A(1,p) \times \cdots \times A(n,p)} \left\{(y_1, \ldots, y_n) \in S^{(n)} : \bigvee_{i=1}^{n} d_i(y_i, x_i) \leq 2^{-p}\right\} \\
&\equiv \bigcup_{(x(1),\ldots,x(n)) \in A(1,p) \times \cdots \times A(n,p)} \\
&\quad \times \{(y_1, \ldots, y_n) \in S^{(n)} : d^{(n)}((y_1, \ldots, y_n), (x_1, \ldots, x_n)) \leq 2^{-p}\} \\
&\equiv \bigcup_{(x(1),\ldots,x(n)) \in A(1,p) \times \cdots \times A(n,p)} (d^{(n)}(\cdot, (x_1, \ldots, x_n)) \leq 2^{-p}) \\
&= \bigcup_{x \in A_p^{(n)}} (d^{(n)}(\cdot, x) \leq 2^{-p}).
\end{aligned} \tag{3.2.4}$$

Combining with relation 3.2.3, this yields

$$(d^{(n)}(\cdot, x_\circ^{(n)}) \leq 2^p) \subset \bigcup_{x \in A_p^{(n)}} (d^{(n)}(\cdot, x) \leq 2^{-p}),$$

where $p \geq 1$ is arbitrary. Thus Condition 3.2.1 has been verified for the sequence $\xi^{(n)} \equiv (A_p^{(n)})_{p=1,2,\ldots}$.

3. In the other direction, we have, similarly,

$$\bigcup_{x \in A_p^{(n)}} (d^{(n)}(\cdot,x) \leq 2^{-p+1})$$

$$\equiv \bigcup_{x \in A_p^{(n)}} \bigcap_{i=1}^{n} (d_i(\cdot,x) \leq 2^{-p+1})$$

$$= \bigcap_{i=1}^{n} \bigcup_{x(i) \in A(i,p)} \{(y_1,\ldots,y_n) \in S^{(n)} : d_i(y_i,x_i) \leq 2^{-p+1}\}$$

$$\subset \bigcap_{i=1}^{n} \{(y_1,\ldots,y_n) \in S^{(n)} : d_i(y_i,x_{i,\circ}) \leq 2^{p+1}\}$$

$$= \left\{ (y_1,\ldots,y_n) \in S^{(n)} : \bigvee_{i=1}^{n} d_i(y_i,x_{i,\circ}) \leq 2^{p+1} \right\}$$

$$= (d^{(n)}(\cdot,x_\circ^{(n)}) \leq 2^{p+1}),$$

where $p \geq 1$ is arbitrary. This verifies Condition 3.2.2 for the sequence $\xi^{(n)} \equiv (A_p^{(n)})_{p=1,2,\ldots}$. Thus all the conditions in Definition 3.2.1 have been proved for the sequence $\xi^{(n)}$ to be a binary approximation of $(S^{(n)},d^{(n)})$ relative to $x_\circ^{(n)}$. Moreover,

$$\|\xi^{(n)}\| \equiv (|A_q^{(n)}|)_{q=1,2,\ldots} = \left(\prod_{i=1}^{n} |A_{i,q}|\right)_{q=1,2,\ldots} \equiv \left(\prod_{i=1}^{n} \kappa_{i,q}\right)_{q=1,2,\ldots}. \quad \square$$

We next extend the construction of the powers of binary approximations to the infinite power (S^∞,d^∞) of a compact metric space (S,d). Recall that $\widehat{d} \equiv 1 \wedge d$.

Definition 3.2.6. Countable power of binary approximation for a compact metric space. Suppose (S,d) is a compact metric space, with a reference point $x_\circ \in S$ and a binary approximation $\xi \equiv (A_n)_{n=1,2,\ldots}$ relative to $x_\circ$. Let (S^∞,d^∞) be the countable power of the metric space (S,d), with $x_\circ^\infty \equiv (x_\circ,x_\circ,\ldots)$ designated as the reference point in (S^∞,d^∞).

For each $n \geq 1$, define the subset

$$B_n \equiv A_{n+1}^{n+1} \times \{x_\circ^\infty\}$$
$$= \{(x_1,\ldots,x_{n+1},x_\circ,x_\circ \cdots) : x_i \in A_{n+1} \text{ for each } i = 1,\ldots,n+1\}.$$

The next lemma proves that $\xi^\infty \equiv (B_n)_{n=1,2,\ldots}$ is a binary approximation of (S^∞,d^∞) relative to $x_\circ^\infty$. We will call ξ^∞ the *countable power of the binary approximation* ξ. $\square$

Lemma 3.2.7. Countable power of binary approximation of a compact metric space is indeed a binary approximation. *Suppose* (S,d) *is a compact* *metric space, with a reference point* $x_\circ \in S$ *and a binary approximation* $\xi \equiv (A_n)_{n=1,2,...}$ *relative to* $x_\circ$. *Without loss of generality, assume that* $d \leq 1$. *Then the sequence* $\xi^\infty \equiv (B_n)_{n=1,2,...}$ *in Definition 3.2.6 is indeed a binary approximation of* (S^∞, d^∞) *relative to* $x_\circ^\infty$.

Let $\|\xi\| \equiv (\kappa_n)_{n=1,2,...} \equiv (|A_n|)_{n=1,2,...}$ *denote the modulus of local compactness of* (S,d) *corresponding to* ξ. *Then the modulus of local compactness of* (S^∞, d^∞) *corresponding to* ξ^∞ *is given by*

$$\|\xi^\infty\| = (\kappa_{n+1}^{n+1})_{n=1,2,...}.$$

Proof. Let $n \geq 1$ be arbitrary.

1. Let

$$x \equiv (x_1, \ldots, x_{n+1}, x_\circ, x_\circ, \ldots), y \equiv (y_1, \ldots, y_{n+1}, x_\circ, x_\circ, \ldots) \in B_n$$

be arbitrary. Since A_{n+1} is metrically discrete, we have either (i) $x_i = y_i$ for each $i = 1, \ldots, n+1$ or (ii) $\widehat{d}(x_i, y_i) > 0$ for some $i = 1, \ldots, n+1$. In case (i), we have $x = y$. In case (ii), we have

$$d^\infty(x,y) \equiv \sum_{j=1}^{\infty} 2^{-j}\widehat{d}(x_j, y_j) \geq 2^{-i}\widehat{d}(x_i, y_i) > 0.$$

Thus we see that B_n is metrically discrete.

2. Next, let $y \equiv (y_1, y_2, \ldots) \in S^\infty$ be arbitrary. Let $j = 1, \ldots, n+1$ be arbitrary. Then

$$y_j \in (d(\cdot, x_\circ) \leq 2^{n+1}) \subset \bigcup_{z \in A(n+1)} (d(\cdot, z) \leq 2^{-n-1}),$$

where the first containment relation is a trivial consequence of the hypothesis that $d \leq 1$, and where the second is an application of Condition 3.2.1 of Definition 3.2.1 to the binary approximation $\xi \equiv (A_n)_{n=1,2,...}$. Hence there exists some $u_j \in A_{n+1}$ with $d(y_j, u_j) \leq 2^{-n-1}$, where $j = 1, \ldots, n+1$ is arbitrary. It follows that

$$u \equiv (u_1, \ldots, u_{n+1}, x_\circ, x_\circ, \ldots) \in A_{n+1}^{n+1} \times \{x_\circ^\infty\} \equiv B_n,$$

and that

$$\begin{aligned} d^\infty(y,u) &\leq \sum_{j=1}^{n+1} 2^{-j}\widehat{d}(y_j, u_j) + \sum_{j=n+2}^{\infty} 2^{-j} \\ &\leq \sum_{j=1}^{n+1} 2^{-j}2^{-n-1} + 2^{-n-1} < 2^{-n-1} + 2^{-n-1} = 2^{-n}, \end{aligned}$$

where $y \in S^\infty$ is arbitrary. We conclude that

$$(d^\infty(\cdot, x_\circ^\infty) \leq 2^n) = S^\infty \subset \bigcup_{u \in B(n)} (d^\infty(\cdot, u) \leq 2^{-n}).$$

where the equality is trivial because $d^\infty \leq 1$. Thus Condition 3.2.1 of Definition 3.2.1 is verified for the sequence $(B_n)_{n=1,2,...}$. At the same time, again because because $d^\infty \leq 1$, we have trivially

$$\bigcup_{u \in B(n)} (d^\infty(\cdot,u) \leq 2^{-n+1}) \subset S^\infty = (d^\infty(\cdot,x_\circ^\infty) \leq 2^{n+1}).$$

Thus Condition 3.2.2 of Definition 3.2.1 is also verified for the sequence $(B_n)_{n=1,2,...}$. All the conditions in Definition 3.2.1 have been verified for the sequence $\xi^\infty \equiv (B_n)_{n=1,2,...}$ to be a binary approximation of (S^∞,d^∞) relative to $x_\circ^\infty$.

Moreover,

$$\|\xi^\infty\| \equiv (|B_n|)_{n=1,2,...} = (|A_{n+1}^{n+1}|)_{n=1,2,...} \equiv (\kappa_{n+1}^{n+1})_{n=1,2,...}.$$

The lemma is proved. □

3.3 Partition of Unity

In this section, we define and construct a partition of unity relative to a binary approximation of a locally compact metric space (S,d).

There are many different versions of partitions of unity in the mathematics literature, providing approximate linear bases in the analysis of various linear spaces of functions. The present version, roughly speaking, furnishes an approximate linear basis for $C(S,d)$, the space of continuous functions with compact supports on a locally compact metric space. In this version, the basis functions will be endowed with specific properties that make later applications simpler. For example, each basis function will be Lipschitz continuous.

First we prove an elementary lemma for Lipschitz continuous functions.

Lemma 3.3.1. Definition and basics for Lipschitz continuous functions. *Let (S,d) be an arbitrary metric space. A real-valued function f on S is said to be* Lipschitz continuous, *with Lipschitz constant $c \geq 0$, if $|f(x) - f(y)| \leq cd(x,y)$ for each $x,y \in S$. We will then also say that the function has* Lipschitz constant c.

Let $x_\circ \in S$ be an arbitrary but fixed reference point. Let f,g be real-valued functions with Lipschitz constants a,b, respectively, on S. Then the following conditions hold:

1. $d(\cdot,x_\circ)$ has Lipschitz constant 1.

2. $\alpha f + \beta g$ has Lipschitz constant $|\alpha|a + |\beta|b$ for each $\alpha,\beta \in R$. If, in addition, $|f| \leq 1$ and $|g| \leq 1$, then fg has Lipschitz constant $a+b$.

3. $f \vee g$ and $f \wedge g$ have Lipschitz constant $a \vee b$.

4. $1 \wedge (1 - cd(\cdot,x_\circ))_+$ has Lipschitz constant c for each $c > 0$.

5. If $\|f\| \vee \|g\| \leq 1$, then fg has Lipschitz constant $a+b$.

6. Suppose (S',d') is a locally compact metric space. Suppose f' is a real-valued function on S', with Lipschitz constant $a' > 0$. Suppose $\|f\| \vee \|f'\| \leq 1$.

Then $f \otimes f' : S \times S' \to R$ has Lipschitz constant $a + a'$, where $S \times S'$ is equipped with the product metric $\widetilde{d} \equiv d \otimes d'$, and where $f \otimes f'(x,x') \equiv f(x)f'(x')$ for each $(x,x') \in S \times S'$.

7. Assertion 6 can be generalized to a p-fold product $f \otimes f' \otimes \cdots \otimes f^{(p)}$.

Proof. Let $x, y \in S$ be arbitrary.

1. By the triangle inequality, we have $|d(x,x_\circ) - d(y,x_\circ)| \leq d(x,y)$. Assertion 1 follows.

2. Note that

$$
\begin{aligned}
&|\alpha f(x) + \beta g(x) - (\alpha f(y) + \beta g(y))| \\
&\quad \leq |\alpha(f(x) - f(y))| + |\beta(g(x) - g(y))| \leq (|\alpha| a + |\beta| b) d(x,y)
\end{aligned}
$$

and that if $|f| \leq 1$ and $|g| \leq 1$, then

$$
\begin{aligned}
&|f(x)g(x) - f(y)g(y)| \\
&\quad \leq |(f(x) - f(y))g(x)| + |(g(x) - g(y))f(y)|. \\
&\quad \leq |f(x) - f(y)| + |g(x) - g(y)| \leq (a + b)d(x,y).
\end{aligned}
$$

Assertion 2 is proved.

3. To prove Assertion 3, first consider arbitrary $r, s, t, u \in R$. We will show that

$$
\alpha \equiv |r \vee s - t \vee u| \leq \beta \equiv |r - t| \vee |s - u|. \tag{3.3.1}
$$

To see this, we may assume, without loss of generality, that $r \geq s$. If $t \geq u$, then $\alpha = |r - t| \leq \beta$. On the other hand, if $t \leq u$, then

$$
-\beta \leq -|s - u| \leq s - u \leq r - u \leq r - t \leq |r - t| \leq \beta,
$$

whence $\alpha = |r - u| \leq \beta$. By continuity, we see that inequality 3.3.1 holds for each $r, s, t, u \in R$.

Consequently,

$$
|f(x) \vee g(x) - f(y) \vee g(y)| \leq |f(x) - f(y)| \vee |g(x) - g(y)| \leq (a \vee b) d(x,y)
$$

for each $x, y \in S$. Thus $f \vee g$ has Lipschitz constant $a \vee b$. Assertion 2 then implies that the function $f \wedge g = -((-f) \vee (-g))$ also has Lipschitz constant $a \vee b$. Assertion 3 is proved.

4. Assertion 4 follows immediately from Assertions 1, 2, and 3.

5. Assertion 5 follows from

$$
\begin{aligned}
&|f(x)g(x) - f(y)g(y)| \\
&\quad \leq |f(x)(g(x) - g(y))| + |(f(x) - f(y))g(y)| \leq (b + a)d(x,y).
\end{aligned}
$$

6. Suppose $\|f\| \vee \|f'\| \leq 1$. Then for each $(x,x'), (y,y') \in S \times S'$.

$$
\begin{aligned}
&|f(x)f'(x') - f(y)f'(y')| \\
&\quad \leq |f(x)(f'(x') - f'(y'))| + |(f(x) - f(y))f'(y')| \\
&\quad \leq (a' + a)(d(x',y') \vee d(x,y)) \equiv (a' + a) d \otimes d'((x,x')), d(y,y'))
\end{aligned}
$$

Thus the function $f \otimes f' : S \times S' \to R$ has Lipschitz constant $a + a'$. Assertion 6 follows.

7. The proof of Assertion 7 is omitted. □ □

The next definitions and propositions embellish proposition 6.15 on page 119 of [Bishop and Bridges 1985].

Definition 3.3.2. ε-Partition of unity. Let A be an arbitrary metrically discrete and finite subset of a locally compact metric space (S,d). Because the set A is finite, we can write $A = \{x_1, \ldots, x_\kappa\}$ for some sequence $x \equiv (x_1, \ldots, x_\kappa)$, where $x : \{1, \ldots, \kappa\} \to A$ is an enumeration of the finite set A. Thus $|A| \equiv \kappa$. Let $\varepsilon > 0$ be arbitrary. Define, for each $k = 1, \ldots, \kappa$, the function

$$\eta_k \equiv 1 \wedge (2 - \varepsilon^{-1} d(\cdot, x_k))_+ \in C(S,d) \tag{3.3.2}$$

and

$$g_k^+ \equiv \eta_1 \vee \cdots \vee \eta_k \in C(S,d). \tag{3.3.3}$$

In addition, define $g_0^+ \equiv 0$. Also, for each $k = 1, \ldots, \kappa$, define

$$g_{x(k)} \equiv g_k^+ - g_{k-1}^+. \tag{3.3.4}$$

Then the subset $\{g_y : y \in A\}$ of $C(S,d)$ is called the *ε-partition of unity* of (S,d), determined by the enumerated set A. The members of $\{g_y : y \in A\}$ are called the *basis functions* of the ε-partition of unity. □

Proposition 3.3.3. Properties of ε-partition of unity. *Let $A = \{x_1, \ldots, x_\kappa\}$ be an arbitrary metrically discrete and enumerated finite subset of a locally compact metric space (S,d). Let $\varepsilon > 0$ be arbitrary. Let $\{g_x : x \in A\}$ be the ε-partition of unity determined by the enumerated set A. Then the following conditions hold:*

1. g_x has values in $[0,1]$ and has the set $(d(\cdot,x) < 2\varepsilon)$ as support, for each $x \in A$.

2. $\sum_{x \in A} g_x \leq 1$ on S.

3. $\sum_{x \in A} g_x = 1$ on $\bigcup_{x \in A} (d(\cdot,x) \leq \varepsilon)$.

4. For each $x \in A$, the functions g_x, $\sum_{y \in A; y < x} g_y$, and $\sum_{y \in A} g_y$ have Lipschitz constant $2\varepsilon^{-1}$. Here $y < x$ means $y = x_i$ and $x = x_j$ for some $i, j \in \{1, \ldots, \kappa\}$ with $i < j$.

Proof. 1. Use the notations in Definition 3.3.2. Let $k = 1, \ldots, \kappa$ be arbitrary. Suppose $y \in S$ is such that $g_{x(k)}(y) > 0$. By the defining equality 3.3.4, it follows that $g_k^+(y) > g_{k-1}^+(y)$. Hence $\eta_k(y) > 0$ by equality 3.3.3. Equality 3.3.2 then implies that $d(y, x_k) < 2\varepsilon$. Thus we see that the function $g_{x(k)}$ has $(d(\cdot, x_k) < 2\varepsilon)$ as support. In general, $g_{x(k)}$ has values in $[0,1]$, thanks to equalities 3.3.2, 3.3.3, and 3.3.4. Assertion 1 is proved.

2. Note that $\sum_{x \in A} g_x = g_\kappa^+ \equiv \eta_1 \vee \cdots \vee \eta_\kappa \leq 1$. Assertion 2 is verified.

3. Suppose $y \in S$ is such that $d(y, x_k) \leq \varepsilon$ for some $k = 1, \ldots, \kappa$. Then $\eta_k(y) = 1$ according to equality 3.3.2. Hence $\sum_{x \in A} g_x(y) = g_\kappa^+(y) = 1$ by equality 3.3.3. Assertion 3 is proved.

4. Now let $k = 1, \ldots, \kappa$ be arbitrary. Refer to Lemma 3.3.1 for the basic properties of Lipschitz constants. Then, in view of the defining equality 3.3.2, the function η_k has Lipschitz constant ε^{-1}. Hence $g_k^+ \equiv \eta_0 \vee \cdots \vee \eta_k$ has Lipschitz constant ε^{-1}. In particular, $\sum_{y \in A} g_y \equiv g_\kappa^+$ has Lipschitz constant ε^{-1}. Moreover, for each $k = 1, \ldots, \kappa$, the function

$$\sum_{y \in A;\, y < x(k)} g_y \equiv \sum_{i=1}^{k-1} g_{x(i)} = g_k^+$$

has Lipschitz constant ε^{-1}, whence $g_{x(k)} \equiv g_k^+ - g_{k-1}^+$ has Lipschitz constant $2\varepsilon^{-1}$. Summing up, for each $x \in A$, the functions $\sum_{y \in A} g_y$, $\sum_{y \in A;\, y<x} g_y$, and g_x have Lipschitz constant $c \equiv 2\varepsilon^{-1}$. Assertion 4 and the proposition are proved. □

Recall that if $f \in C(S,d)$, then $\sup_{x \in S} |f(x)|$ exists and is denoted by $\|f\|$.

Definition 3.3.4. Partition of unity of a locally compact metric space. Let (S,d) be a locally compact metric space, with a reference point $x_\circ \in S$. Let the nondecreasing sequence $\xi \equiv (A_n)_{n=1,2,\ldots}$ of enumerated finite subsets of (S,d) be a binary approximation of (S,d) relative to $x_\circ$.

For each $n \geq 1$, let $\{g_{n,x} : x \in A_n\}$ be the 2^{-n}-partition of unity of (S,d) determined by the metrically discrete and finite subset A_n, as in Definition 3.3.2. Then the sequence

$$\pi \equiv (\{g_{n,x} : x \in A_n\})_{n=1,2,\ldots}$$

is called a *partition of unity* of (S,d) determined by the binary approximation ξ. □

Proposition 3.3.5. Properties of partition of unity. *Let $\xi \equiv (A_n)_{n=1,2,\ldots}$ be a binary approximation of the locally compact metric space (S,d) relative to a reference point $x_\circ$. Let $\pi \equiv (\{g_{n,x} : x \in A_n\})_{n=1,2,\ldots}$ be the partition of unity determined by ξ. Let $n \geq 1$ be arbitrary. Then the following conditions hold:*

1. $g_{n,x} \in C(S,d)$ has values in $[0,1]$ and has support $(d(\cdot,x) \leq 2^{-n+1})$, for each $x \in A_n$.

2. $\sum_{x \in A(n)} g_{n,x} \leq 1$ on S.

3. $\sum_{x \in A(n)} g_{n,x} = 1$ on $\bigcup_{x \in A(n)} (d(\cdot,x) \leq 2^{-n})$.

4. For each $x \in A_n$, the functions $g_{n,x}$, $\sum_{y \in A(n);\, y<x} g_{n,y}$, and $\sum_{y \in A(n)} g_{n,y}$ have Lipschitz constant 2^{n+1}. Here $y < x$ means $y = x_i$ and $x = x_j$ for some $i, j \in \{1, \ldots, |A_n|\}$ with $i < j$.

5. For each $x \in A_n$,

$$g_{n,x} = \sum_{y \in A(n+1)} g_{n,x} g_{n+1,y} \tag{3.3.5}$$

on S.

Proof. Assertions 1–4 are restatements of their counterparts in Proposition 3.3.3 for the case $\varepsilon \equiv 2^{-n}$.

5. Now let $x \in A_n$ be arbitrary. By Assertion 1,

$$(g_{n,x} > 0) \subset (d(\cdot,x) \le 2^{-n+1}).$$

At the same time,

$$(d(\cdot,x) \le 2^{-n+1}) \subset (d(\cdot,x_\circ) \le 2^{n+1})$$

$$\subset \bigcup_{y \in A(n+1)} (d(\cdot,y) \le 2^{-n-1}) \subset \left(\sum_{y \in A(n+1)} g_{n+1,y} = 1 \right),$$

where the first inclusion is by Condition 3.2.2 of Definition 3.2.1, the second by Condition 3.2.1 of Definition 3.2.1 applied to $n+1$, and the third by Assertion 3 applied to $n+1$. Combining,

$$(g_{n,x} > 0) \subset \left(\sum_{y \in A(n+1)} g_{n+1,y} = 1 \right).$$

The desired equality 3.3.5 in Assertion 5 follows. □

The following proposition is a first application of partitions of unity.

Proposition 3.3.6. Interpolation with linear combination of Lipschitz continuous functions. *Let (S,d) be an arbitrary locally compact metric space, with an arbitrary but fixed reference point $x_\circ$. Let $\xi \equiv (A_k)_{k=1,2,...}$ be a binary approximation of (S,d) relative $x_\circ$. Let $\pi \equiv (\{g_{k,x} : x \in A_k\})_{k=1,2,...}$ be the partition of unity determined by ξ. Let $n \ge 1$ be arbitrary but fixed. Let $f \in C(S^n,d^n)$ be arbitrary, with a modulus of continuity δ_f.*

Let $\varepsilon > 0$ be arbitrary. Then the following conditions hold:

1. There exists $k \ge 1$ so large that (i) the function *f has the set*

$$B \equiv (d(\cdot,x_\circ) \le 2^k)^n \subset S^n$$

as support, and that (ii) $2^{-k} < 2^{-1}\delta_f(3^{-1}\varepsilon)$.

2. Take any $k \ge 1$ that satisfies Conditions (i) *and* (ii). *Then*

$$\sup_{(y,y',...,y^{(n)}) \in S^n} |f(y,y',\dots,y^{(n)}) - \sum_{(x,x',...,x^{(n)}) \in A(k)^n} f(x,x',\dots,x^{(n)}) g_x(y) g_{x'}(y') \cdots g_{x^{(n)}}(y^{(n)})| \le \varepsilon. \tag{3.3.6}$$

In other words,

$$\left\| f - \sum_{(x,x',...,x^{(n)}) \in A(k)^n} f(x,x',\dots,x^{(n)}) g_x \otimes g_{x'} \otimes \cdots \otimes g_{x^{(n)}} \right\| \le \varepsilon, \tag{3.3.7}$$

where $\|\cdot\|$ signifies the supremum norm in $C(S^n,d^n)$.

3. The function represented by the sum in inequality 3.3.7 is Lipschitz continuous.

4. In particular, suppose $n = 1$ and $|f| \leq 1$. Take any $k \geq 1$ that satisfies Conditions (i) *and* (ii). *Then*

$$\|f - g\| \leq \varepsilon$$

for some Lipschitz continuous function with Lipschitz constant $2^{k+1}|A_k|$.

Proof. We will give the proof only for the case where $n = 2$; the other cases are similar.

1. Let K be a compact subset of $C(S^2, d^2)$ such that the function $f \in C(S^2, d^2)$ has K as support. Since the compact set K is bounded, there exists $k \geq 1$ so large that

$$K \subset B \equiv (d(\cdot, x_\circ) \leq 2^k) \times (d(\cdot, x_\circ) \leq 2^k)$$

and that (ii) $2^{-k} < 2^{-1}\delta_f(3^{-1}\varepsilon)$. Conditions (i) and (ii) follow. Assertion 1 is proved.

2. Now fix any $k \geq 1$ that satisfies Conditions (i) and (ii). Note that $(d(\cdot, x_\circ) \leq 2^k) \subset \bigcup_{x \in A(k)}(d(\cdot, x) \leq 2^{-k})$ by relation 3.2.1 of Definition 3.2.1. Hence Condition (i) implies that (i′) the function f has the set

$$\bigcup_{x \in A(k)} (d(\cdot, x) \leq 2^{-k}) \times \bigcup_{x' \in A(k)} (d(\cdot, x') \leq 2^{-k})$$

as support. For abbreviation, write $\alpha \equiv 2^{-k}$, $A \equiv A_k$, and $g_x \equiv g_{k,x}$ for each $x \in A$.

3. Let $(y, y') \in S^2$ be arbitrary. Suppose $x, x' \in A$ are such that $g_x(y)g_{x'}(y') > 0$. Then, since $g_x, g_{x'}$ have $(d(\cdot, x) < 2^{-k+1}), (d(\cdot, x') < 2^{-k+1})$, respectively, as support, according to Assertion 1 of Proposition 3.3.5, it follows that $d^2((y, y'), (x, x')) < 2\alpha < \delta_f(3^{-1}\varepsilon)$. Consequently, the inequality

$$|f(y, y') - f(x, x')|g_x(y)g_{x'}(y') \leq 3^{-1}\varepsilon g_x(y)g_{x'}(y') \tag{3.3.8}$$

holds for arbitrary $(x, x') \in A^2$ such that $g_x(y)g_{x'}(y') > 0$ and, consequently, holds for arbitrary $(x, x') \in A^2$.

4. There are two possibilities: (i″) $|f(y, y')| > 0$ or (ii″) $|f(y, y')| < 3^{-1}\varepsilon$.

5. First consider case (i″). Then, by Condition (i′), we have

$$\begin{aligned}(y, y') &\in (d(\cdot, x_\circ) \leq 2^k) \times (d(\cdot, x_\circ) \leq 2^k) \subset B \\ &\equiv \bigcup_{(x, x') \in A \times A} (d(\cdot, x) < 2^{-k}) \times (d(\cdot, x') < 2^{-k}).\end{aligned}$$

Hence, by Assertion 3 of Proposition 3.3.3, we have $\sum_{x \in A} g_x(y) = 1$ and $\sum_{x' \in A} g_{x'}(y') = 1$. Therefore

$$\left| f(y,y') - \sum_{x\in A}\sum_{x'\in A} f(x,x')g_x(y)g_{x'}(y') \right|$$

$$= \left| \sum_{x\in A}\sum_{x'\in A} f(y,y')g_x(y)g_{x'}(y') - \sum_{x\in A}\sum_{x'\in A} f(x,x')g_x(y)g_{x'}(y') \right|$$

$$\leq \sum_{x\in A}\sum_{x'\in A} |f(y,y') - f(x,x')| g_x(y)g_{x'}(y')$$

$$\leq 3^{-1}\varepsilon \sum_{x\in A}\sum_{x'\in A} g_x(y)g_{x'}(y') \leq 3^{-1}\varepsilon < \varepsilon,$$

where the second inequality follows from inequality 3.3.8, and where the third inequality is thanks to Assertion 2 of Proposition 3.3.3.

6. Now consider case (ii), where

$$|f(y,y')| < 3^{-1}\varepsilon. \tag{3.3.9}$$

Then

$$\left| f(y,y') - \sum_{x\in A}\sum_{x'\in A} f(x,x')g_x(y)g_{x'}(y') \right| < 3^{-1}\varepsilon + \sum_{x\in A}\sum_{x'\in A} |f(x,x')| g_x(y)g_{x'}(y')$$

$$\leq 3^{-1}\varepsilon + \sum_{x\in A}\sum_{x'\in A} (|f(y,y')| + 3^{-1}\varepsilon) g_x(y)g_{x'}(y')$$

$$\leq 3^{-1}\varepsilon + \sum_{x\in A}\sum_{x'\in A} (3^{-1}\varepsilon + 3^{-1}\varepsilon) g_x(y)g_{x'}(y')$$

$$\leq 3^{-1}\varepsilon + (3^{-1}\varepsilon + 3^{-1}\varepsilon) = \varepsilon, \tag{3.3.10}$$

where the first and third inequalities are by inequality 3.3.9, where the second inequality is by inequality 3.3.8, and the last inequality is thanks to Assertion 2 of Proposition 3.3.3.

7. Summing up, Steps 5 and 6 show that

$$\left| f(y,y') - \sum_{x\in A}\sum_{x'\in A} f(x,x')g_x(y)g_{x'}(y') \right| \leq \varepsilon,$$

where $(y,y') \in S^2$ is arbitrary. The desired inequality 3.3.6 follows for the case where $n = 2$. The proof for the general case $n \geq 1$ is similar. Assertion 2 of the present proposition is verified.

8. Let $x, x' \in A$ be arbitrary. Then the functions $g_x, g_{x'}$ are Lipschitz, according to Assertion 4 of Proposition 3.3.3. Hence the function $g_x \otimes g_{x'}$ is Lipschitz, by Assertion 6 of Lemma 3.3.1. Assertion 3 of the present proposition follows.

9. Now suppose, in addition, that $n = 1$ and $|f| \leq 1$. According to Proposition 3.3.3, each of the functions g_x in the last sum has Lipschitz constant 2^{k+1}, while $f(x)$ is bounded by 1 by hypothesis. By Assertion 2, we have

$$\left\| f - \sum_{x \in A(k)} f(x) g_x \right\| \leq \varepsilon, \tag{3.3.11}$$

where the function $g \equiv \sum_{x \in A(k)} f(x) g_x$ has Lipschitz constant

$$\sum_{x \in A(k)} |f(x)| 2^{k+1} \leq 2^{k+1} |A_k|,$$

according to Assertion 2 of Lemma 3.3.1. Assertion 4 of the present proposition is proved. □

3.4 One-Point Compactification

The countable power of a locally compact metric space (S,d) is not necessarily locally compact, while the countable power of a compact metric space remains compact. For that reason, we will often find it convenient to embed a locally compact metric space into a compact metric space such that while the metric is not preserved, the continuous functions are. This embedding is made precise in the present section, by an application of partitions of unity.

The next definition is an embellishment of definition 6.6, proposition 6.7, and theorem 6.8 of [Bishop and Bridges 1985].

Definition 3.4.1. One-point compactification. A *one-point compactification* of a locally compact metric space (S,d) is a compact metric space $(\overline{S},\overline{d})$ with an element Δ, called the *point at infinity*, such that the following five conditions hold:

1. $\overline{d} \leq 1$ and $S \cup \{\Delta\}$ is a dense subset of $(\overline{S},\overline{d})$.

2. Let K be an arbitrary compact subset of (S,d). Then there exists $c > 0$ such that $\overline{d}(x,\Delta) \geq c$ for each $x \in K$.

3. Let K be an arbitrary compact subset of (S,d). Let $\varepsilon > 0$ be arbitrary. Then there exists $\delta_K(\varepsilon) > 0$ such that for each $y \in K$ and $z \in S$ with $\overline{d}(y,z) < \delta_K(\varepsilon)$, we have $d(y,z) < \varepsilon$. In particular, the identity mapping $\bar{\iota} : (S,\overline{d}) \to (S,d)$, defined by $\bar{\iota}(x) \equiv x$ for each $x \in S$, is uniformly continuous on each compact subset K of (S,d).

4. The identity mapping $\iota : (S,d) \to (S,\overline{d})$, defined by $\iota(x) \equiv x$ for each $x \in S$, is uniformly continuous on (S,d). In other words, for each $\varepsilon > 0$, there exists $\delta_{\overline{d}}(\varepsilon) > 0$ such that $\overline{d}(x,y) < \varepsilon$ for each $x,y \in S$ with $d(x,y) < \delta_{\overline{d}}(\varepsilon)$.

5. For each $n \geq 1$, we have

$$(d(\cdot,x_\circ) > 2^{n+1}) \subset (\overline{d}(\cdot,\Delta) \leq 2^{-n}).$$

Thus, as a point $x \in S$ moves away from $x_\circ$ relative to d, it converges to the point Δ at infinity relative to $\overline{d}$. □

First we provide some convenient notations.

Definition 3.4.2. Restriction of a family of functions. Let A, A' be arbitrary sets and let B be an arbitrary subset of A. Recall that the restriction of a function

$f : A \to A'$ to a subset $B \subset A$ is denoted by $f|B$. Suppose F is a family of functions from A to A' and suppose $B \subset A$. Then we call the family

$$F|B \equiv \{f|B : f \in F\}$$

the *restriction of F to B*. □

The next theorem constructs a one-point compactification. The proof loosely follows the lines of theorem 6.8 in chapter 4 of [Bishop and Bridges 1985].

Theorem 3.4.3. Construction of a one-point compactification from a binary approximation. *Let (S,d) be a locally compact metric space. Let the sequence $\xi \equiv (A_n)_{n=1,2,...}$ of subsets be a binary approximation of (S,d) relative to $x_\circ$. Then there exists a one-point compactification $(\overline{S},\overline{d})$ of (S,d) such that the following conditions hold:*

(i) For each $p \geq 1$ and for each $y,z \in S$ with

$$d(y,z) < p^{-1}2^{-p-1},$$

we have

$$\overline{d}(y,z) < 2^{-p+1}.$$

(ii) For each $n \geq 1$, for each $y \in (d(\cdot,x_\circ) \leq 2^n)$, and for each $z \in S$ with

$$\overline{d}(y,z) < 2^{-n-1}|A_n|^{-2},$$

we have

$$d(y,z) < 2^{-n+2}.$$

The one-point compactification $(\overline{S},\overline{d})$ constructed in the proof is said to be relative to the binary approximation ξ.

Proof. 1. Let

$$\pi \equiv (\{g_{n,x} : x \in A_n\})_{n=1,2,...}$$

be the partition of unity of (S,d) determined by ξ.

Let $n \geq 1$ be arbitrary. Then $\{g_{n,x} : x \in A_n\}$ is a 2^{-n}-partition of unity corresponding to the metrically discrete and enumerated finite set A_n. Moreover, by Assertion 4 of Proposition 3.3.5, the function $g_{n,x}$ has Lipschitz constant 2^{n+1} for each $x \in A_n$.

2. Define

$$\widetilde{S} \equiv \{(x,i) \in S \times \{0,1\} : i = 0 \text{ or } (x,i) = (x_\circ,1)\}$$

and define $\Delta \equiv (x_\circ,1)$. Identify each $x \in S$ with $\bar{x} \equiv (x,0) \in \widetilde{S}$. Thus $\widetilde{S} = S \cup \{\Delta\}$. Extend each function $f \in C(S,d)$ to a function on $\widetilde{S}$ by defining $f(\Delta) \equiv 0$. In particular, the function $g_{n,x}$ is extended to a function on $\widetilde{S}$, with $g_{n,x}(\Delta) \equiv 0$, for each $x \in A_n$. Define a function $\overline{d}$ on $\widetilde{S} \times \widetilde{S}$ by

$$\overline{d}(y,z) \equiv \sum_{n=1}^{\infty} 2^{-n}|A_n|^{-1} \sum_{x\in A(n)} |g_{n,x}(y) - g_{n,x}(z)| \tag{3.4.1}$$

for each $y,z \in \widetilde{S}$. Then $\overline{d}(y,y) = 0$ for each $y \in \widetilde{S}$. Symmetry and triangle inequality of the function $\overline{d}$ are immediate consequences of equality 3.4.1. Moreover, $\overline{d} \le 1$ since the functions $g_{n,x}$ have values in $[0,1]$.

3. Let K be an arbitrary compact subset of (S,d) and let $y \in K$ be arbitrary. Let $n \ge 1$ be so large that

$$y \in K \subset (d(\cdot,x_\circ) \le 2^n).$$

Then

$$y \in \bigcup_{x\in A(n)} (d(\cdot,x) \le 2^{-n}) \subset \left(\sum_{x\in A(n)} g_{n,x} = 1\right),$$

where the membership relation is by Condition 3.2.1 in Definition 3.2.1, and where the set inclusion is according to Assertion 3 of Proposition 3.3.3. Hence the defining equality 3.4.1, where z is replaced by Δ, yields

$$\overline{d}(y,\Delta) \ge 2^{-n}|A_n|^{-1} \sum_{x\in A(n)} g_{n,x}(y) = c \equiv 2^{-n}|A_n|^{-1} > 0, \tag{3.4.2}$$

establishing Condition 2 in Definition 3.4.1. Note that the constant c is independent of $y \in K$.

4. Let $n \ge 1$ be arbitrary. Let $y \in (d(\cdot,x_\circ) \le 2^n)$ and $z \in S$ be arbitrary such that

$$\overline{d}(y,z) < \delta_{\xi,n} \equiv 2^{-n-1}|A_n|^{-2}.$$

As seen in Step 3,

$$\sum_{x\in A(n)} g_{n,x}(y) = 1.$$

Hence there exists $x \in A_n$ such that

$$g_{n,x}(y) > 2^{-1}A_n|^{-1} > 0. \tag{3.4.3}$$

At the same time,

$$|g_{n,x}(y) - g_{n,x}(z)| \le \sum_{u\in A(n)} |g_{n,u}(y) - g_{n,u}(z)|$$
$$\le 2^n|A_n|\overline{d}(y,z) < 2^n|A_n|\delta_{\xi,n} \equiv 2^{-1}|A_n|^{-1}.$$

Hence inequality 3.4.3 implies that $g_{n,x}(z) > 0$. Recall that $g_{n,x} \in C(S,d)$ has support $(d(\cdot,x) \le 2^{-n+1})$. Consequently, $y,z \in (d(\cdot,x) < 2^{-n+1})$. Thus $d(y,z) < 2^{-n+2}$. This establishes Assertion (ii) of the present theorem.

Now let K be an arbitrary compact subset of (S,d) and let $\varepsilon > 0$ be arbitrary. Let $n \ge 1$ be so large that $K \subset (d(\cdot,x_\circ) \le 2^n)$ and that $2^{-n+2} < \varepsilon$. Let $\delta_K(\varepsilon) \equiv \delta_{\xi,n}$.

Then, by the preceding paragraph, for each $y \in K$ and $z \in S$ with $\overline{d}(y,z) < \delta_K(\varepsilon) \equiv \delta_{\xi,n}$, we have $d(y,z) < \varepsilon$. Condition 3 in Definition 3.4.1 has been verified.

In particular, suppose $y,z \in \widetilde{S}$ are such that $\overline{d}(y,z) = 0$. Then either $y = z = \Delta$ or $y,z \in S$, in view of inequality 3.4.2. Suppose $y,z \in S$. Then the preceding paragraph applied to the compact set $K \equiv \{y,z\}$, implies that $d(y,z) = 0$. Since (S,d) is a metric space, we conclude that $y = z$. In view of the last two statements of Step 2, $(\widetilde{S},\overline{d})$ is a metric space.

Let $(\overline{S},\overline{d})$ be the completion of the metric space $(\widetilde{S},\overline{d})$. Then $(\overline{S},\overline{d})$ is a complete metric space, with $\overline{d} \leq 1$ on $\overline{S} \times \overline{S}$, and $\widetilde{S}$ is a dense subset of $(\overline{S},\overline{d})$.

5. Recall that $g_{n,x}$ has values in $[0,1]$ and, as remarked earlier, has Lipschitz constant 2^{n+1}, for each $x \in A_n$, for each $n \geq 1$. Now let $p \geq 1$ be arbitrary. Let $y,z \in S$ be arbitrary such that $d(y,z) < p^{-1}2^{-p-1}$. Then

$$\begin{aligned}
\overline{d}(y,z) &\equiv \sum_{n=1}^{\infty} 2^{-n}|A_n|^{-1} \sum_{x \in A(n)} |g_{n,x}(y) - g_{n,x}(z)| \\
&\leq \sum_{n=1}^{p} 2^{-n}2^{n+1}d(y,z) + \sum_{n=p+1}^{\infty} 2^{-n} \\
&< p \cdot 2 \cdot p^{-1}2^{-p-1} + 2^{-p} = 2^{-p} + 2^{-p} = 2^{-p+1}.
\end{aligned}$$

In short, for arbitrary $y,z \in S$ such that $d(y,z) < p^{-1}2^{-p-1}$, we have

$$\overline{d}(y,z) < 2^{-p+1}. \tag{3.4.4}$$

This establishes Assertion (i) of the present theorem.

Since 2^{-p+1} is arbitrarily small for sufficiently large $p \geq 1$, we see that the identity mapping $\iota : (S,d) \to (S,\overline{d})$ is uniformly continuous. This establishes Condition 4 in Definition 3.4.1.

6. Let $n \geq 1$ be arbitrary. Consider each $y \in (d(\cdot,x_\circ) > 2^{n+1})$. Let $m = 1,\ldots,n$ be arbitrary. Then

$$y \in (d(\cdot,x_\circ) > 2^{m+1}) \subset \bigcap_{x \in A(m)} (d(\cdot,x) \geq 2^{-m+1}).$$

Let $x \in A_m$ be arbitrary. Suppose $d(y,x) < 2^{-m+1}$. Then

$$y \in \bigcup_{x \in A(m)} (d(\cdot,x) \leq 2^{-m+1}) \subset (d(\cdot,x_\circ) \leq 2^{m+1}) \tag{3.4.5}$$

by Condition 3.2.2 in Definition 3.2.1 of a binary approximation. This is a contradiction. Hence $d(y,x) \geq 2^{-m+1}$. Since $g_{m,x}$ has support $(d(\cdot,x) \leq 2^{-m+1})$ according to Assertion 1 of Proposition 3.3.5, we infer that $g_{m,x}(y) = 0$, where $x \in A_m$ and $m = 1,\ldots,n$ are arbitrary. Hence the defining equality 3.4.1, where z is replaced by Δ, reduces to

$$\overline{d}(y,\Delta) \equiv \sum_{m=1}^{\infty} 2^{-m}|A_m|^{-1} \sum_{x\in A(m)} g_{m,x}(y)$$
$$= \sum_{m=n+1}^{\infty} 2^{-m}|A_m|^{-1} \sum_{x\in A(m)} g_{m,x}(y)$$
$$\leq \sum_{m=n+1}^{\infty} 2^{-m} = 2^{-n}. \quad (3.4.6)$$

Since $y \in (d(\cdot,x_\circ) > 2^{n+1})$ is arbitrary, we conclude that

$$(d(\cdot,x_\circ) > 2^{n+1}) \subset (\overline{d}(\cdot,\Delta) \leq 2^{-n}). \quad (3.4.7)$$

This proves Condition 5 in Definition 3.4.1.

7. We will prove next that $(\widetilde{S},\overline{d})$ is totally bounded. To that end, let $p \geq 1$ be arbitrary. Let

$$m \equiv m_p \equiv [(p+2) + \log_2(p+2)]_1,$$

where we recall that $[\cdot]_1$ is the operation that assigns to each $a \in [0,\infty)$ an integer $[a]_1$ in $(a,a+2)$. Then

$$2^{-m} < \overline{\delta}_p \equiv (p+2)^{-1}2^{-p-2}.$$

Note that

$$\widetilde{S} \equiv S \cup \{\Delta\} \subset (d(\cdot,x_\circ) < 2^m) \cup (d(\cdot,x_\circ) > 2^{m-1}) \cup \{\Delta\}$$
$$\subset \bigcup_{x\in A(m)} (d(\cdot,x) \leq 2^{-m}) \cup (\overline{d}(\cdot,\Delta) \leq 2^{-m+2}) \cup \{\Delta\},$$

where the second set inclusion is due to Condition 3.2.1 in Definition 3.2.1, and to relation 3.4.7 where n is replaced by $m-2$. Continuing,

$$\widetilde{S} \subset \bigcup_{x\in A(m)} (d(\cdot,x) \leq (p+1)^{-1}2^{-p-2}) \cup (\overline{d}(\cdot,\Delta) < (p+2)^{-1}2^{-p}) \cup \{\Delta\}$$
$$\subset \bigcup_{x\in A(m)} (\overline{d}(\cdot,x) < 2^{-p}) \cup (\overline{d}(\cdot,\Delta) < 2^{-p}) \cup \{\Delta\},$$

where the last set inclusion is by applying inequality 3.4.4 with p replaced by $p+1$. Consequently, the set

$$\overline{A}_p \equiv A_{m(p)} \cup \{\Delta\}$$

is a metrically discrete 2^{-p}-approximation of $(\widetilde{S},\overline{d})$. Since 2^{-p} is arbitrarily small, the metric space $(\widetilde{S},\overline{d})$ is totally bounded. Hence its completion $(\overline{S},\overline{d})$ is compact, and $\widetilde{S}$ is dense in $(\overline{S},\overline{d})$, proving Condition 1 in Definition 3.4.1.

8. Incidentally, since $\widetilde{S} \equiv S \cup \{\Delta\}$ is a dense subset of $(\overline{S},\overline{d})$, the metrically discrete finite subset $\overline{A}_p$ is a 2^{-p}-approximation of $(\overline{S},\overline{d})$.

Summing up, $(\overline{S},\overline{d})$ satisfies all the conditions in Definition 3.4.1 to be a one-point compactification of (S,d). □

Corollary 3.4.4. Compactification of a binary approximation. *Use the same notations and assumptions as in Theorem 3.4.3. Let (S,d) be a locally compact metric space. Let the sequence $\xi \equiv (A_n)_{n=1,2,...}$ of subsets be a binary approximation of (S,d) relative to the reference point $x_\circ$. For each $n \geq 1$, let $A_n \equiv \{x_{n,1},\ldots,x_{n,\kappa(n)}\}$. Thus*

$$\|\xi\| \equiv (|A_n|)_{n=1,2,...} = (|\kappa_n|)_{n=1,2,...}.$$

For each $p \geq 1$, write $m_p \equiv [(p+2)+\log_2(p+2)]_1$. Define

$$\overline{A}_p \equiv A_{m(p)} \cup \{\Delta\} \equiv \{x_{m(p),1},\ldots,x_{m(p),\kappa(m(p))},\Delta\}.$$

Then $\overline{\xi} \equiv (\overline{A}_p)_{p=1,2,...}$ is a binary approximation of $(\overline{S},\overline{d})$ relative to $x_\circ$, called the compactification of the binary approximation ξ.

The corresponding modulus of local compactness of $(\overline{S},\overline{d})$ is given by

$$\|\overline{\xi}\| \equiv (|\overline{A}_p|)_{p=1,2,...} = (|A_{m(p)}|+1)_{p=1,2,...}$$

and is determined by $\|\xi\|$.

Proof. Let $p \geq 1$ be arbitrary. According to Step 8 of the proof of Theorem 3.4.3, the finite set $\overline{A}_p$ is a metrically discrete 2^{-p}-approximation of $(\overline{S},\overline{d})$. Hence

$$(\overline{d}(\cdot,x_\circ) \leq 2^p) \subset \overline{S} \subset \sum_{x\in\overline{A}(p)} (\overline{d}(\cdot,x) \leq 2^{-p}).$$

At the same time, Condition 1 of Definition 3.4.1 says that $\overline{d} \leq 1$. Hence, trivially,

$$\bigcup_{x\in\overline{A}(p)} (\overline{d}(\cdot,x) \leq 2^{-p+1}) \subset \overline{S} \subset (\overline{d}(\cdot,x_\circ) \leq 1) \subset (\overline{d}(\cdot,x_\circ) \leq 2^{p+1}).$$

Thus all the conditions in Definition 3.2.1 have been verified for $\overline{\xi} \equiv (\overline{A}_p)_{p=1,2,...}$ to be a binary approximation of $(\overline{S},\overline{d})$ relative to $x_\circ$. □

Let (S,d) be an arbitrary locally compact metric space, and let $(\overline{S},\overline{d})$ be a one-point compactification of (S,d). Let $n \geq 1$ be arbitrary. Note that if $n \geq 2$, then the power metric space $(\overline{S}^n,\overline{d}^n)$ is compact, but need not be a one-point compactification of (S^n,d^n).

Recall that $C_{ub}(S,d)$ denotes the space of bounded and uniformly continuous functions on a locally compact metric space (S,d). The next proposition proves the promised preservation of continuous functions for a one-point compactification.

Proposition 3.4.5. Continuous functions on (S,d) and continuous functions on $(\overline{S},\overline{d})$. *Let (S,d) be a locally compact metric space, with a fixed reference point $x_\circ \in S$. Let $(\overline{S},\overline{d})$ be a one-point compactification of (S,d), with the point at infinity Δ. Then the following conditions hold:*

1. Each compact subset K of (S,d) is also a compact subset of $(\overline{S},\overline{d})$.

2. Let $n \geq 1$ be arbitrary. Then

$$C(S^n, d^n) \subset C(\overline{S}^n, \overline{d}^n)|S^n \subset C_{ub}(S^n, d^n). \tag{3.4.8}$$

Moreover, for each $f \in C(S^n, d^n)$, there exists $\widetilde{f} \in C(\overline{S}^n, \overline{d}^n)$ such that $\widetilde{f}(x_1, \ldots, x_n) \equiv f(x_1, \ldots, x_n)$ or $\widetilde{f}(x_1, \ldots, x_n) \equiv 0$ according as $(x_1, \ldots, x_n) \in S^n$ or $x_i = \Delta$ for some $i = 1, \ldots, n$, for each $(x_1, \ldots, x_n) \in \widetilde{S}^n$.

Proof. 1. Suppose K is a compact subset of (S,d). By Conditions 3 and 4 of Definition 3.4.1, the identity mapping $\iota : (K,d) \to (K,\overline{d})$ and its inverse are uniformly continuous. Since, by assumption, (K,d) is compact, so is $(K,\overline{d})$. Assertion 1 is proved.

2. Let $n \geq 1$ be arbitrary. Consider each $f \in C(S^n, d^n)$, with a modulus of continuity δ_f. Let the compact subset K of (S,d) be such that K^n is a compact support of f. Write $\widetilde{S} \equiv S \cup \{\Delta\}$. Extend f to a function $\widetilde{f} : \widetilde{S}^n \to R$ by defining $\widetilde{f}(x_1, \ldots, x_n) \equiv f(x_1, \ldots, x_n)$ or $\widetilde{f}(x_1, \ldots, x_n) \equiv 0$ according as $(x_1, \ldots, x_n) \in S^n$ or $x_i = \Delta$ for some $i = 1, \ldots, n$, for each $(x_1, \ldots, x_n) \in \widetilde{S}^n$. We will show that $\widetilde{f}$ is uniformly continuous on $(\overline{S}^n, \overline{d}^n)$.

3. To that end, let $\varepsilon > 0$ be arbitrary. By Condition 3 in Definition 3.4.1, there exists $\delta_K(\delta_f(\varepsilon)) > 0$ such that for each $y \in K$ and $z \in S$ with $\overline{d}(y,z) < \delta \equiv \delta_K(\delta_f(\varepsilon))$, we have $d(y,z) < \delta_f(\varepsilon)$.

4. By Condition 2 in Definition 3.4.1, we have $\overline{d}(K, \Delta) > 0$. Hence $\overline{\delta} \equiv \delta \wedge \overline{d}(K, \Delta) > 0$.

5. Now consider each $(x_1, \ldots, x_n), (y_1, \ldots, y_n) \in \widetilde{S}^n$ with

$$\overline{d}^n((x_1, \ldots, x_n), (y_1, \ldots, y_n)) < \overline{\delta} \leq \delta. \tag{3.4.9}$$

There are four possibilities:

(i) $|\widetilde{f}(x_1, \ldots, x_n) - \widetilde{f}(y_1, \ldots, y_n)| > 0$ and $x_i = \Delta$ for some $i = 1, \ldots, n$.

(ii) $|\widetilde{f}(x_1, \ldots, x_n) - \widetilde{f}(y_1, \ldots, y_n)| > 0$ and $y_i = \Delta$ for some $i = 1, \ldots, n$.

(iii) $|\widetilde{f}(x_1, \ldots, x_n) - \widetilde{f}(y_1, \ldots, y_n)| > 0$ and $(x_1, \ldots, x_n), (y_1, \ldots, y_n) \in S^n$.

(iv) $|\widetilde{f}(x_1, \ldots, x_n) - \widetilde{f}(y_1, \ldots, y_n)| < \varepsilon$.

6. Consider case (i). Then $\widetilde{f}(x_1, \ldots, x_n) = 0$ by the definition of the function $\widetilde{f}$. Hence $|\widetilde{f}(y_1, \ldots, y_n)| > 0$. Therefore $(y_1, \ldots, y_n) \in S^n$ by the definition of the function $\widetilde{f}$, and $|f(y_1, \ldots, y_n)| \equiv |\widetilde{f}(y_1, \ldots, y_n)| > 0$. Since, by assumption, the compact set K^n is a support of the function f, we see that $(y_1, \ldots, y_n) \in K^n$. Hence $y_i \in K$. Therefore

$$\overline{d}(y_i, x_i) = \overline{d}(y_i, \Delta) \geq \overline{d}(K, \Delta) \geq \delta \wedge \overline{d}(K, \Delta) \equiv \overline{\delta}.$$

Consequently,

$$\overline{d}^n((x_1, \ldots, x_n), (y_1, \ldots, y_n)) \equiv \bigvee_{j=1}^{n} \overline{d}(y_j, x_j) \geq \overline{d}(y_i, x_i) \geq \overline{\delta},$$

a contradiction to inequality 3.4.9.

7. Similarly, case (ii) leads to a contradiction. Only cases (iii) and (iv) remain possible.

8. Consider case (iii). Then

$$|f(x_1,\ldots,x_n) - f(y_1,\ldots,y_n)| \equiv |\widetilde{f}(x_1,\ldots,x_n) - \widetilde{f}(y_1,\ldots,y_n)| > 0. \tag{3.4.10}$$

Hence $|f(x_1,\ldots,x_n)| > 0$ or $|f(y_1,\ldots,y_n)| > 0$. Assume, without loss of generality, that $|f(x_1,\ldots,x_n)| > 0$. Since the compact set K^n is a support of f, we have $(x_1,\ldots,x_n) \in K^n$, while $(y_1,\ldots,y_n) \in S^n$ by assumption. Let $i = 1,\ldots,n$ be arbitrary. Then $x_i \in K$ and $y_i \in S$. At the same time, inequality 3.4.9 implies that $\overline{d}(y_i,x_i) < \delta$. Therefore, according to Step 3, we have $d(y_i,x_i) < \delta_f(\varepsilon)$, where $i = 1,\ldots,n$ is arbitrary. Combining, we obtain

$$d^n((x_1,\ldots,x_n),(y_1,\ldots,y_n)) \equiv \bigvee_{i=1}^{n} d(y_i,x_i) < \delta_f(\varepsilon). \tag{3.4.11}$$

It follows that

$$|\widetilde{f}(x_1,\ldots,x_n) - \widetilde{f}(y_1,\ldots,y_n)| \equiv |f(x_1,\ldots,x_n) - f(y_1,\ldots,y_n)| < \varepsilon. \tag{3.4.12}$$

9. Summing up, we see that each of the only possible cases (iii) and (iv) implies that

$$|\widetilde{f}(x_1,\ldots,x_n) - \widetilde{f}(y_1,\ldots,y_n)| < \varepsilon,$$

where $\varepsilon > 0$ and $(x_1,\ldots,x_n),(y_1,\ldots,y_n) \in \widetilde{S}^n$ are arbitrary with

$$\overline{d}^n((x_1,\ldots,x_n),(y_1,\ldots,y_n)) < \overline{\delta}. \tag{3.4.13}$$

Thus the function $\widetilde{f} : (\widetilde{S}^n,\overline{d}^n) \to R$ is uniformly continuous. Since $\widetilde{S}^n$ is a dense subset of the compact metric space $(\overline{S}^n,\overline{d}^n)$, the function $\widetilde{f}$ can be extended to a continuous function $\overline{f} : (\overline{S}^n,\overline{d}^n) \to R$. Thus $f = \widetilde{f}|S^n = (\overline{f}|\widetilde{S}^n)|S^n = \overline{f}|S^n$, where $\overline{f} \in C(\overline{S}^n,\overline{d}^n)$. In short, $f \in C(\overline{S}^n,\overline{d}^n)|S^n$, where $f \in C(S^n,d^n)$ is arbitrary. We conclude that $C(S^n,d^n) \subset C(\overline{S}^n,\overline{d}^n)|S^n$. This proves the first half of the desired Condition 3.4.8 in Assertion 2.

10. Now consider each $\overline{f} \in C(\overline{S}^n,\overline{d}^n)$, with a modulus of continuity $\delta_{\overline{f}}$. Then $\overline{f}|S^n : (S^n,\overline{d}^n) \to R$ is bounded and uniformly continuous with the modulus of continuity $\delta_{\overline{f}}$. At the same time, by Condition 4 in Definition 3.4.1, the identity mapping $\iota: (S,d) \to (S,\overline{d})$, defined by $\iota(x) \equiv x$ for each $x \in S$, is uniformly continuous on (S,d). Consequently, the mapping $\iota^n : (S^n,d^n) \to (S^n,\overline{d}^n)$, defined by $\iota^n(x_1,\ldots,x_n) \equiv (x_1,\ldots,x_n)$ for each $(x_1,\ldots,x_n) \in S^n$, is uniformly continuous. Hence the composite mapping $(\overline{f}|S^n) \circ \iota^n : (S^n,d^n) \to R$ is bounded and uniformly continuous. Since $(\overline{f}|S^n) \circ \iota^n = \overline{f}|S^n$ on S^n, we see that the function $\overline{f}|S^n : (S^n,d^n) \to R$ is bounded and uniformly continuous.

In short, $\overline{f}|S^n \in C_{ub}(S^n,d^n)$, where $\overline{f} \in C(\overline{S}^n,\overline{d}^n)$ is arbitrary. We conclude that $C(\overline{S}^n,\overline{d}^n)|S^n \subset C_{ub}(S^n,d^n)$. This proves the second half of the desired Condition 3.4.8 in Assertion 2.

The proposition is proved. □

Proposition 3.4.6. Continuous functions on (S^∞, d^∞) and continuous functions on $(\overline{S}^\infty, \overline{d}^\infty)$. *Let $\overline{g} \in C(\overline{S}^\infty,\overline{d}^\infty)$ be arbitrary, with modulus of continuity $\delta_{\overline{g}}$. Define $g \equiv \overline{g}|S^\infty$. Then $g \in C_{ub}(S^\infty,d^\infty)$. In short, $C(\overline{S}^\infty,\overline{d}^\infty)|S^\infty \subset C_{ub}(S^\infty,d^\infty)$.*

Proof. Let $\varepsilon > 0$ be arbitrary. Take $n \geq 1$ so large that $2^{-n} \leq \varepsilon' \equiv 2^{-1}\delta_{\overline{g}}(\varepsilon)$. By Condition 4 of Definition 3.4.1, there exists $\delta_{\overline{d}}(\varepsilon') \in (0,1)$ such that $\overline{d}(x',y') < \varepsilon'$ for each $x',y' \in S$ with $d(x',y') < \delta_{\overline{d}}(\varepsilon')$. Define $\delta_g(\varepsilon) \equiv 2^{-n}\delta_{\overline{d}}(\varepsilon')$. Let $x,y \in S^\infty$ be arbitrary such that

$$d^\infty(x,y) \equiv \sum_{i=1}^{\infty} 2^{-i}(1 \wedge d(x_i,y_i)) < \delta_g(\varepsilon).$$

Then, for each $i = 1,\ldots,n$, we have $1 \wedge d(x_i,y_i) < 2^n\delta_g(\varepsilon) \equiv \delta_{\overline{d}}(\varepsilon')$, whence $d(x_i,y_i) < \delta_{\overline{d}}(\varepsilon')$ and so $\overline{d}(x_i,y_i) < \varepsilon'$. Hence

$$\begin{aligned}\overline{d}^\infty(x,y) &\equiv \sum_{i=1}^{\infty} 2^{-i}(1 \wedge \overline{d}(x_i,y_i)) \\ &\leq \sum_{i=1}^{n} 2^{-i}(1 \wedge \overline{d}(x_i,y_i)) + 2^{-n} \leq \sum_{i=1}^{n} 2^{-i}\varepsilon' + \varepsilon' = 2\varepsilon' \equiv \delta_{\overline{g}}(\varepsilon).\end{aligned}$$

Hence

$$|g(x) - g(y)| = |\overline{g}(x) - \overline{g}(y)| < \varepsilon.$$

Thus $g \in C_{ub}(S^\infty,d^\infty)$, as alleged. □

Part II

Probability Theory

4

Integration and Measure

We introduce next the Riemann–Stieljes integral on R. Then we present a general treatment of integration and measure theory in terms of Daniell integration, adapted from [Bishop and Bridges 1985].

The standard course in measure theory usually starts with measurable sets, before defining a measure function and an integration function. In contrast, the Daniell integration theory starts with the integration function and the integrable functions. Thus we discuss the computation of the integration early on.

The end products of the measure- and integration functions are essentially the same in both approaches; the dissimilarity is only superficial. However, Daniell integrals are more natural, and cleaner, in the constructive approach.

4.1 Riemann–Stieljes Integral

Definition 4.1.1. Distribution function. A *distribution function* is a nondecreasing real-valued function F on R, with $domain(F)$ dense in R. □

Definition 4.1.2. Riemann–Stieljes sum relative to a distribution function and a partition of the real line R. Let F be a distribution function, and let $X \in C(R)$ be arbitrary. An arbitrary finite and increasing sequence $(x_0, \ldots, x_n)$ in $domain(F)$ is called a *partition* of R relative to the distribution function F. One partition is said to be a *refinement* of another if the former contains the latter as a subsequence.

For any partition $(x_1, \ldots, x_n)$ relative to the distribution function F, define its *mesh* as

$$\text{mesh}(x_1, \ldots, x_n) \equiv \left(\bigvee_{i=1}^{n} (x_i - x_{i-1}) \right),$$

and define the *Riemann–Stieljes sum* as

$$S(x_0, \ldots, x_n) \equiv \sum_{i=1}^{n} X(x_i)(F(x_i) - F(x_{i-1})). \qquad \square$$

Theorem 4.1.3. Existence of Riemann–Stieljes integral. *Let F be a distribution function, and let $X \in C(R)$ be arbitrary. Then the Riemann–Stieljes sum converges as* $\text{mesh}(x_1,\ldots,x_n) \to 0$ *with $x_0 \to -\infty$ and $x_n \to +\infty$. The limit will be called the* Riemann–Stieljes integral *of X with respect to the distribution function F, and will be denoted by $\int_{-\infty}^{+\infty} X(x)dF(x)$, or simply by $\int X(x)dF(x)$.*

Proof. 1. Suppose X has modulus of continuity δ_X and vanishes outside the compact interval $[a,b]$, where $a,b \in domain(F)$. Let $\varepsilon > 0$ be arbitrary. Consider an arbitrary partition $(x_0,\ldots,x_n)$ with (i) $x_0 < a-2 < b+2 < x_n$ and (ii) $\text{mesh}(x_1,\ldots,x_n) < 1 \wedge \delta_X(\varepsilon)$, where δ_X is a modulus of continuity for X.

2. Let i be any index with $0 < i \le n$. Suppose we insert m points between (x_{i-1},x_i) and make a refinement $(\ldots,x_{i-1},y_1,\ldots,y_{m-1},x_i,\ldots)$. Let y_0 and y_m denote x_{i-1} and x_i, respectively. Then the difference in Riemann–Stieljes sums for the new and old partitions is bounded by

$$\left|X(x_i)(F(x_i)-F(x_{i-1})) - \sum_{j=1}^{m} X(y_j)(F(y_j)-F(y_{j-1})\right|$$

$$= \left|\sum_{j=1}^{m}(X(x_i)-X(y_j))(F(y_j)-F(y_{j-1})\right|$$

$$\le \left|\sum_{j=1}^{m}\varepsilon(F(y_j)-F(y_{j-1})\right| = \varepsilon(F(x_i)-F(x_{i-1})$$

Moreover, the difference is 0 if $x_i < a-2$ or $x_{i-1} > b+2$. Since $x_i - x_{i-1} < 1$, the difference is 0 if $x_{i-1} < a-1$ or $x_i > b+1$.

3. Since any refinement of $(x_0,\ldots,x_n)$ can be obtained by inserting points between the pairs (x_{i-1},x_i), we see that the Riemann–Stieljes sum of any refinement differs from that for $(x_0,\ldots,x_n)$ by at most $\sum \varepsilon(F(x_i)-F(x_{i-1}))$, where the sum is over all i for which $a < x_{i-1}$ and $x_i < b$. The difference is therefore at most $\varepsilon(F(b)-F(a))$.

4. Consider a second partition $(u_0,\ldots,u_p)$ satisfying Conditions (i) and (ii). Because the domain of F is dense, we can find a third partition $(v_0,\ldots,v_q)$ satisfying the same conditions and the additional condition that $|v_k - x_i| > 0$ and $|v_k - u_j| > 0$ for all i,j,k. Then $(v_0,\ldots,v_q)$ and $(x_0,\ldots,x_n)$ have a common refinement – namely, the merged sequence rearranged in increasing order. Thus their Riemann–Stieljes sums differ from each other by at most $2\varepsilon(F(b)-F(a))$ according to Step 3. Similarly, the Riemann–Stieljes sum for $(u_0,\ldots,u_p)$ differs from that of $(v_0,\ldots,v_q)$ by at most $2\varepsilon(F(b)-F(a))$. Hence the Riemann–Stieljes sums for $(u_0,\ldots,u_p)$ and $(x_0,\ldots,x_n)$ differ by at most $4\varepsilon(F(b)-F(a))$.

5. Since ε is arbitrary, the asserted convergence is proved. □

Theorem 4.1.4. Basic properties of the Riemann–Stieljes integral. *The Riemann-Stieljes integral is linear on $C(R)$. It is also positive in the sense that if $\int X(x)dF(x) > 0$, then there exists $x \in R$ such that $X(x) > 0$.*

Proof. Linearity follows trivially from the defining formulas. Suppose $a,b \in domain(F)$ are such that X vanishes outside $[a,b]$. If the integral is greater than some positive real number c, then the Riemann–Stieljes sum $S(x_0,\ldots,x_n)$ for some partition with $x_1 = a$ and $x_n = b$ is greater than c. It follows that $X(x_i)(F(x_i) - F(x_{i-1}))$ is greater than or equal to $n^{-1}c$ for some index i. Hence

$$X(x_i)(F(b) - F(a)) \geq X(x_i)(F(x_i) - F(x_{i-1})) \geq n^{-1}c.$$

This implies $X(x_i) > n^{-1}c(F(b) - F(a)) > 0$. □

Definition 4.1.5. Riemann sums and Riemann integral. In the special case where $domain(F) \equiv R$ and $F(x) \equiv x$ for each $x \in R$, the Riemann–Stieljes sums and Riemann–Stieljes integral of a function $X \in C(R)$ relative to the distribution F are called the Riemann sums and the Riemann integral of X, respectively. □

4.2 Integration on a Locally Compact Metric Space

In this section, the Riemann–Stieljes integration is generalized to functions $X \in C(S,d)$, where (S,d) is a locally compact metric space (S,d).

Traditionally, integration is usually defined in terms of a measure, a function on a family of subsets that is closed relative to the operations of countable unions, countable intersections, and relative to complements. In the case of a metric space, one such family can be generated via these three operations from the family of all open subsets. Members of the family thus generated are called Borel sets. In the special case of R, the open sets can in turn be generated from a countable subfamily of intervals in successive partitions of R, wherein ever smaller intervals cover any compact interval in R. The intervals in the countable family can thus serve as building blocks in the analysis of measures on R.

The Daniell integration theory is a more natural choice for the constructive development. Integrals of functions, rather than measures of sets, are the starting point. In the special case of a locally compact metric space (S,d), the family $C(S,d)$ of continuous functions with compact supports supplies the basic integrable functions.

Definition 4.2.1. Integration on a locally compact metric space. An *integration* on a locally compact metric space (S,d) is a real-valued linear function I on the linear space $C(S,d)$ such that (i) $I(X) > 0$ for some $X \in C(S,d)$ and (ii) for each $X \in C(S,d)$ with $I(X) > 0$, there exists a point x in S for which $X(x) > 0$. Condition (ii) will be called the *positivity condition* of the integration I.

It immediately follows that if $X \in C(S,d)$ is such that $X \leq 0$, then $I(X) \leq 0$. By the linearity and the positivity of the function I, we see that if $X \in C(S,d)$ is such that $X \geq 0$, then $I(X) \geq 0$. □

The Riemann–Stieljes integration defined for a distribution function F on R is an integration on (R,d), where d is the Euclidean metric and is denoted by

$\int \cdot dF$, with $I(X)$ written as $\int X(x)dF(x)$ for each $X \in C(S)$. Riemann–Stieljes integrals provide an abundance of examples for integration on locally compact metric spaces.

It follows from the positivity Condition (ii) and from the linearity of I that if $X,Y \in C(S,d)$ are such that $I(X) > I(Y)$, then there exists a point x in S for which $X(x) > Y(x)$.

The positivity Condition (ii), extended in the next proposition, is a powerful tool in proving existence theorems. It translates a condition on integrals into the existence of a point in S with certain properties. To prove the next proposition, we need the following lemma, which will be used again in a later chapter. This lemma, obtained from [Chan 1975], is a pleasant surprise because, in general, the convergence of a series of nonnegative real numbers does not follow constructively from the boundedness of partial sums.

Lemma 4.2.2. Positivity of a linear function on a linear space of functions. *Suppose I is a linear function on a linear space L of functions on a set S such that $I(X) \geq 0$ for each nonnegative function $X \in L$.*

Suppose I satisfies the following condition: for each $X_0 \in L$, there exists a non-negative function $Z \in L$ such that for each sequence $(X_i)_{i=1,2,\ldots}$ of nonnegative functions in L with $\sum_{i=1}^{\infty} I(X_i) < I(X_0)$, there exists $x \in S$ with (i) $Z(x) = 1$ *and* (ii) $\sum_{i=1}^{p} X_i(x) \leq X_0(x)$ *for each $p > 0$.*

Then, for each $X_0 \in L$ and for each sequence $(X_i)_{i=1,2,\ldots}$ of nonnegative functions in L with $\sum_{i=1}^{\infty} I(X_i) < I(X_0)$, there exists $x \in S$ such that $\sum_{i=1}^{\infty} X_i(x)$ converges and is less than $X_0(x)$.

Proof. Classically, the convergence of $\sum_{i=1}^{\infty} X_i(x)$ follows trivially from the boundedness of the partial sums. Note that if the constant function 1 is a member of L, then the lemma can be simplified with $Z \equiv 1$ or, equivalently, with Z omitted altogether.

1. Suppose $X_0 \in L$ and suppose $(X_i)_{i=1,2,\ldots}$ is a sequence of nonnegative functions in L with $\sum_{i=1}^{\infty} I(X_i) < I(X_0)$. Let $Z \in L$ be as given in the hypothesis. Take a positive real number α so small that

$$\alpha I(Z) + \sum_{i=1}^{\infty} I(X_i) + \alpha < I(X_0). \tag{4.2.1}$$

Take an increasing sequence $(n_k)_{k=1,2,\ldots}$ of integers such that

$$\sum_{i=n(k)}^{\infty} I(X_i) < 2^{-2k}\alpha$$

for each $k \geq 1$.

2. Consider the sequence of functions

$$\left(\alpha Z, X_1, 2\sum_{i=n(1)}^{n(2)} X_i, X_2, 2^2\sum_{i=n(2)}^{n(3)} X_i, X_3, \ldots\right)$$

It can easily be verified that the series of the corresponding values for the function I then converges to a sum less than $\alpha I(Z) + \sum_{i=1}^{\infty} I(X_i) + \alpha$, which is in turn less than $I(X_0)$ according to inequality 4.2.1.

3. Hence, by Conditions (i) and (ii) in the hypothesis, there exists a point $x \in S$ with $Z(x) = 1$ such that

$$\alpha Z(x) + X_1(x) + \cdots + X_k(x) + 2^k \sum_{i=n(k)}^{n(k+1)} X_i(x) \leq X_0(x)$$

for each $k \geq 1$. In particular, $\sum_{i=n(k)}^{n(k+1)} X_i(x) \leq 2^{-k} X_0(x)$ so $\sum_{i=1}^{\infty} X_i(x) < \infty$. The last displayed inequality implies also that

$$\alpha Z(x) + \sum_{i=1}^{\infty} X_i(x) \leq X_0(x).$$

Because $Z(x) = 1$, we obtain

$$\alpha + \sum_{i=1}^{\infty} X_i(x) \leq X_0(x)$$

as desired. □

Proposition 4.2.3. Positivity of an integration on a locally compact metric space. *Let I be an integration on a locally compact metric space (S,d). Let $(X_i)_{i=0,1,2,\ldots}$ be a sequence in $C(S,d)$ such that X_i is nonnegative for $i \geq 1$, and such that $\sum_{i=1}^{\infty} I(X_i) < I(X_0)$. Then there exists $x \in S$ such that $\sum_{i=1}^{\infty} X_i(x) < X_0(x)$.*

Proof. 1. Let K be a compact support of X_0. The set $B \equiv \{x \in S : d(x,K) \leq 2\}$ is bounded. Hence, since S is locally compact, there exists a compact subset $\overline{K}$ such that $B \subset \overline{K}$. Define $Z \equiv (1 - d(\cdot, \overline{K}))_+$.

2. Let $\varepsilon \in (0,1)$ be arbitrary. By Lemma 3.2.2, there exists a metrically discrete and enumerated finite set $A \equiv \{y_1, \ldots, y_n\}$ that is an ε-approximation of K. Let $\{Y_{y(1)}, \ldots, Y_{y(n)}\}$ be the ε-partition of unity determined by A, as in Definition 3.3.2. For short, we abuse notations and write $Y_k \equiv Y_{y(k)}$ for each $k = 1, \ldots, n$. By Proposition 3.3.3, we have $Y_k \geq 0$ for each $k = 1, \ldots, n$, and $\sum_{k=1}^{n} Y_k \leq 1$, with equality prevailing on $K \subset \bigcup_{x \in A}(d(\cdot, x) \leq \varepsilon)$. It follows that $\sum_{k=1}^{n} X_i Y_k \leq X_i$, and $\sum_{k=1}^{n} X_0 Y_k = X_0$ since K is a support of the

function X_0. Consequently, $\sum_{k=1}^n I(X_iY_k) \leq I(X_i)$ for each $i \geq 0$, with equality in the case $i = 0$. Therefore

$$\sum_{k=1}^{n}\sum_{i=1}^{\infty} I(X_iY_k) \leq \sum_{i=1}^{\infty} I(X_i) < I(X_0) = \sum_{k=1}^{n} I(X_0Y_k)$$

3. Hence there exists some $k = 1, \ldots, n$ for which

$$\sum_{i=1}^{\infty} I(X_iY_k) < I(X_0Y_k)$$

Again by Proposition 3.3.3, for each $x \in S$ with $Y_k(x) > 0$, we have $d(x, y_k) < 2\varepsilon$, whence $d(x, K) \leq 2$ and $x \in B$. Therefore

$$(Y_k(x) > 0 \text{ and } Y_k(x') > 0) \Rightarrow (d(x,x') \leq 4\varepsilon \text{ and } x,x' \in B \subset \overline{K})$$

for each $x, x' \in S$. Define $Z_1 \equiv Y_k$.

4. Let $m \geq 1$ be arbitrary, By repeating the previous argument with $\varepsilon_m = (4m)^{-1}$, we can construct inductively a sequence of nonnegative continuous functions $(Z_m)_{m=1,2,...}$ such that for each $m \geq 1$ and for each $x, x' \in S$, we have

$$(Z_m(x) > 0 \text{ and } Z_m(x') > 0) \Rightarrow (d(x,x') \leq m^{-1} \text{ and } x,x' \in \overline{K}) \tag{4.2.2}$$

and such that

$$\sum_{i=1}^{\infty} I(X_iZ_1 \ldots Z_m) < I(X_0Z_1 \ldots Z_m). \tag{4.2.3}$$

5. Since all terms in relation 4.2.3 are nonnegative, the same inequality holds if the infinite sum is replaced by the partial sum of the first m terms. By the positivity of I, this implies, for each $m \geq 1$, that there exists $x_m \in (S,d)$ such that

$$\sum_{i=1}^{m} X_iZ_1 \ldots Z_m(x_m) < X_0Z_1 \ldots Z_m(x_m) \tag{4.2.4}$$

6. Inequality 4.2.4 immediately implies that $Z_p(x_m) > 0$ for each $p \leq m$. Therefore the inference 4.2.2 yields $x_p \in \overline{K}$ and $d(x_p, x_m) \leq p^{-1}$ for each $p \leq m$. Hence $(x_m)_{m=1,2,...}$ is a Cauchy sequence in $\overline{K}$ and converges to some point $x \in \overline{K}$. By the definition of the function Z at the beginning of this proof, we have $Z(x) = 1$.

Canceling positive common factors on both sides of inequality 4.2.4, we obtain $\sum_{i=1}^{p} X_i(x_m) < X_0(x_m)$ for each $p \leq m$. Letting $m \to \infty$ yields $\sum_{i=1}^{p} X_i(x) \leq X_0(x)$ for each $p \geq 1$.

7. The conditions in Lemma 4.2.2 have thus been established. Accordingly, there exists $x \in S$ such that $\sum_{i=1}^{\infty} X_i(x)$ converges and is less than $X_0(x)$. The proposition is proved. □

4.3 Integration Space: The Daniell Integral

Integration on a locally compact space is a special case of Daniell integration, introduced next.

Definition 4.3.1. Integration Space. An *integration space* is a triple (Ω, L, I) where Ω is a nonempty set, L is a set of real-valued functions on Ω, and I is a real-valued function with $domain(I) = L$, satisfying the following conditions:

1. If $X, Y \in L$ and $a, b \in R$, then $aX + bY, |X|$, and $X \wedge 1$ belong to L, and $I(aX + bY) = aI(X) + bI(Y)$. In particular, if $X, Y \in L$, then there exists $\omega \in domain(X) \cap domain(Y)$.

2. If a sequence $(X_i)_{i=0,1,2,...}$ of functions in L is such that X_i is nonnegative for each $i \geq 1$ and such that $\sum_{i=1}^{\infty} I(X_i) < I(X_0)$, then there exists a point $\omega \in \bigcap_{i=0}^{\infty} domain(X_i)$ such that $\sum_{i=1}^{\infty} X_i(\omega) < X_0(\omega)$. This condition will be referred to as the *positivity condition* for I.

3. There exists $X_0 \in L$ such that $IX_0 = 1$.

4. For each $X \in L$, we have $I(X \wedge n) \to I(X)$ and $I(|X| \wedge n^{-1}) \to 0$ as $n \to \infty$.

Then the function I is called an *integration* on (Ω, L), and $I(X)$ is called the *integral* of X, for each $X \in L$. □

Given $X \in L$, the function $X \wedge n = n((n^{-1}X) \wedge 1)$ belongs to L by Condition 1 of Definition 4.3.1. Similarly, the function $|X| \wedge n^{-1}$ belongs to L. Hence the integrals $I(X \wedge n)$ and $I(|X| \wedge n^{-1})$ in Condition 4 make sense.

In the following discussion, to minimize clutter, we will write IX for $I(X)$ and IXY etc. for $I(XY)$ etc. when there is no risk of confusion.

The positivity condition, the innocent-looking Condition 2, is a powerful tool that is useful in many constructive existence proofs.

Lemma 4.3.2. An existence theorem in an integration space. *Let (Ω, L, I) be an arbitrary integration space.*

1. Let $(Y_i)_{i=1,2,...}$ be an arbitrary sequence in L such that $\sum_{i=1}^{\infty} I|Y_i| < \infty$. Then there exists an element ω in the set

$$D \equiv \left\{ \omega \in \bigcap_{i=1}^{\infty} domain(Y_i) : \sum_{i=1}^{\infty} |Y_i(\omega)| < \infty \right\}.$$

In other words, the set D is nonempty.

2. Let $(Y_i)_{i=1,2,...}$ be an arbitrary sequence in L. Then the set $\bigcap_{i=1}^{\infty} domain(Y_i)$ is nonempty.

Proof. 1. Let $(Y_i)_{i=1,2,...}$ be an arbitrary sequence in L such that $\sum_{i=1}^{\infty} I|Y_i| < \infty$. By Condition 3 in Definition 4.3.1, there exists $X_0 \in L$ such that $IX_0 = 1$. Hence $\sum_{i=1}^{\infty} I|Y_i| < IY_0$, where $Y_0 = aX_0$ for some sufficiently large $a > 0$. It follows from Condition 2 in Definition 4.3.1 that there exists a point

$$\omega \in \bigcap_{i=0}^{\infty} domain(|Y_i|) = \bigcap_{i=0}^{\infty} domain(Y_i)$$

such that $\sum_{i=1}^{\infty} |Y_i(\omega)| < Y_0(\omega)$. Assertion 1 is proved.

2. Let $(Y_i)_{i=1,2,...}$ be an arbitrary sequence in L. Then the sequence $(0Y_1, 0Y_2, \ldots)$ satisfies the hypothesis in Assertion 1. 4.3.1. Accordingly, there exists a an element ω in the set

$$\left\{\omega \in \bigcap_{i=1}^{\infty} domain(Y_i) : \sum_{i=1}^{\infty} |0Y_i(\omega)| < \infty\right\} = \bigcap_{i=1}^{\infty} domain(Y_i).$$

Assertion 2 is proved. □

One trivial but useful example of an integration space is the triple $(\Omega, L, \delta_\omega)$, where ω is a given point in a given set Ω, L is the set of all fu ıctions X on Ω whose domains contain ω, and δ_ω is defined on L by $\delta_\omega(X) = X(\omega)$. The integration δ_ω is called the *point mass* at ω. The next proposition gives an abundance of nontrivial examples.

Proposition 4.3.3. An integration on a locally compact space yields an integration space. *Let I be an integration on the locally compact metric space (S,d) as defined in Definition 4.2.1. Then $(S, C(S,d), I)$ is an integration space.*

Proof. The positivity condition in Definition 4.3.1 has been proved for $(S, C(S,d), I)$ in Proposition 4.2.3. The other conditions in Definition 4.3.1 are easily verified. □

The next proposition collects some more useful properties of integration spaces.

Proposition 4.3.4. Basic properties of an integration space. *Let (Ω, L, I) be an integration space. Then the following hold:*

1. If $X, Y \in L$, then $X \vee Y, X \wedge Y \in L$. If, in addition, $a > 0$, then $X \wedge a \in L$ and $I(X \wedge a)$ is continuous in a.

2. If $X \in L$, then $X_+, X_- \in L$ and $IX = IX_+ + IX_-$.

3. For each $X \in L$ with $IX > 0$, there exists ω such that $X(\omega) > 0$.

4. Suppose $X \in L$ is such that $X \geq 0$. Then $IX \geq 0$. Let $X, Y, Z \in L$ be arbitrary. Then

$$I|X - Z| \leq I|X - Y| + I|Z - Y| \tag{4.3.1}$$

and $I|Y - Z| = I|Z - Y|$. In particular, we obtain $I|X - Z| \leq I|X| + I|Z|$ when we set $Y \equiv 0X$.

5. There exists a nonnegative $X \in L$ such that $IX = 1$.

Proof. 1. The first part of Assertion 1 follows from $X \vee Y = (X + Y + |X - Y|)/2$ and $X \wedge Y = (X + Y - |X - Y|)/2$. The second part follows from $X \wedge a = a(a^{-1}X \wedge 1)$ and, in view of Condition 4 in Definition 4.3.1, from $I|X \wedge a - X \wedge b| \leq I(|b - a| \wedge |X|)$ for $a, b > 0$.

2. Assertion 2 follows from $X_+ = X \vee 0X$, $X_- = X \wedge 0X$, and $X = X_+ + X_-$.

3. Suppose $X \in L$ has integral $IX > 0$. The positivity condition in Definition 4.3.1, when applied to the sequence $(X, 0X, 0X, \ldots)$, guarantees an $\omega \in domain(X)$ such that $X(\omega) > 0$. Assertion 3 is proved.

4. Suppose $X(\omega) \geq 0$ for each $\omega \in domain(X)$. Suppose $IX < 0$. Then $I(-X) > 0$, and Assertion 3 would yield $\omega \in domain(X)$ with $-X(\omega) > 0$ or, equivalently, $X(\omega) < 0$, a contradiction. We conclude that $IX \geq 0$. The first part of Assertion 4 is verified. Now let $X, Y, Z \in L$ be arbitrary. Then $|X - Y| + |Z - Y| - |X - Z| \geq 0$. Hence by the first part of Assertion 4, and by linearity, we obtain

$$I|X - Y| + I|Z - Y| - I|X - Z| \geq 0,$$

which proves inequality 4.3.1 in Assertion 4. In the particular case where $X \equiv Y$, this reduces to $I|Z - Y| \geq I|Y - Z|$. Similarly, $I|Y - Z| \geq I|Z - Y|$. Hence $I|Y - Z| = I|Z - Y|$. Assertion 4 is proved.

5. By Condition 3 of Definition 4.3.1, there exists $X \in L$ such that $I(X) > 0$. By Assertion 4, and by the linearity of I, we see that $IX \leq I|X|$ and so $I|X| > 0$. Let X_0 denote the function $|X|/I|X|$. Then $X_0 \in L$ is nonnegative and $IX_0 = 1$. Assertion 5 and the proposition are proved. □

Definition 4.3.5. Integration subspace. Let (Ω, L, I) be an integration space. Let L' be a linear subspace of L such that (Ω, L', I) is an integration space. We will then call (Ω, L', I) an integration subspace of (Ω, L, I). When confusion is unlikely, we will abuse terminology and simply call L' an integration subspace of L, with Ω and I understood. □

Proposition 4.3.6. A linear subspace closed to absolute values and minimum with constants is an integration subspace. *Let (Ω, L, I) be an integration space. Let L' be a linear subspace of L such that if $X, Y \in L'$, then $|X|, X \wedge 1 \in L'$. Then (Ω, L', I) is an integration subspace of (Ω, L, I).*

Proof. By hypothesis, L' is closed to linear operations, absolute values, and the operation of taking minimum with the constant 1. Condition 1 in Definition 4.3.1 for an integration space is thus satisfied by L'. Conditions 2 and 3 are trivially inherited by (Ω, L', I) from (Ω, L, I). □

Proposition 4.3.7. Integration induced by a surjection. *Let (Ω, L, I) be an integration space. Let $\pi : \overline{\Omega} \to \Omega$ be a surjection. For each $f \in L$, write $f(\pi) \equiv f \circ \pi$. Define $\overline{L} \equiv \{f(\pi) : f \in L\}$ and define $\overline{I} : \overline{L} \to R$ by $\overline{I}X \equiv I(f)$ for each $f \in L$ and for each $X \equiv f(\pi) \in \overline{L}$. Then $(\overline{\Omega}, \overline{L}, \overline{I})$ is an integration space.*

Proof. Suppose $X = f(\pi) = g(\pi)$ for some $f, g \in L$. Let $\omega \in domain(f)$ be arbitrary. Since π is a surjection, there exists $\varpi \in \overline{\Omega}$ such that $\pi(\varpi) = \omega \in domain(f)$. It follows that $\varpi \in domain(f(\pi)) = domain(g(\pi))$ and so

$\omega = \pi(\varpi) \in domain(g)$. Since $\omega \in domain(f)$ is arbitrary, we see that $domain(f) \subset domain(g)$ and, by symmetry, $domain(f) = domain(g)$. Moreover, $f(\omega) = f(\pi(\varpi)) = g(\pi(\varpi)) = g(\omega)$. We conclude that $f = g$.

Next let and $a, b \in R$ be arbitrary. Let $X \equiv f(\pi)$ and $Y \equiv h(\pi)$, where $f, h \in L$ are arbitrary. Then $af + bh \in L$ and so $aX + bY = (af + bh)(\pi) \in \overline{L}$. Furthermore

$$\overline{I}(aX + bY) = I(af + bh) = aI(f) + bI(h) \equiv a\overline{I}X + b\overline{I}Y.$$

Thus $\overline{L}$ is a linear space and $\overline{I}$ is a linear function. Similarly, $|X| = |f|(\pi) \in \overline{L}$ and $a \wedge X = (a \wedge f)(\pi) \in \overline{L}$. Furthermore, $\overline{I}(a \wedge X) \equiv I(a \wedge f) \to I(f) \equiv \overline{I}X$ as $a \to \infty$, while $\overline{I}(a \wedge |X|) \equiv I(a \wedge |f|) \to 0$ as $a \to 0$. Thus Conditions 1, 3, and 4 in Definition 4.3.1 for an integration space are verified for the triple $(\overline{\Omega}, \overline{L}, \overline{I})$.

It remains to prove the positivity condition (Condition 2) in Definition 4.3.1. To that end, let $(X_i)_{i=0,1,2,...}$ be a sequence in $\overline{L}$ such that X_i is nonnegative for each $i \geq 1$ and such that $\sum_{i=1}^{\infty} \overline{I}X_i < \overline{I}X_0$. For each $i \geq 0$, let $f_i \in L$ be such that $X_i = f_i(\pi)$. Then, since π is a surjection, $f_i \geq 0$ for each $i \geq 1$. Moreover, $\sum_{i=1}^{\infty} I(f_i) \equiv \sum_{i=1}^{\infty} \overline{I}X_i < \overline{I}X_0 \equiv I(f_0)$. Since I is an integration, there exists $\omega \in \bigcap_{i=0}^{\infty} domain(f_i)$ such that $\sum_{i=1}^{\infty} f_i(\omega) < f_0(\omega)$. Let $\varpi \in \overline{\Omega}$ be such that $\pi(\varpi) = \omega$. Then

$$\varpi \in \bigcap_{i=0}^{\infty} domain(f_i(\pi)) = \bigcap_{i=0}^{\infty} domain(X_i).$$

By hypothesis, $\sum_{i=1}^{\infty} X_i(\varpi) = \sum_{i=1}^{\infty} f_i(\omega) < f_0(\omega) = X_0(\varpi)$. All the conditions in Definition 4.3.1 have been established. Accordingly, $(\overline{\Omega}, \overline{L}, \overline{I})$ is an integration space. □

4.4 Complete Extension of Integration

A common and useful integration space is $(S, C(S,d), I)$, where (S,d) is a locally compact metric space and where I is an integration in the sense of Definition 4.2.1. However, the family $C(S,d)$ of continuous functions with compact support is too narrow to hold all the interesting integrable functions in applications. For example, in the case where (S,d) is the unit interval $[0,1]$ with the Euclidean metric and where I is the Lebesgue integral, it will be important to be able to integrate simple step functions like the indicator $1_{[2^{-1},1]}$. For that reason, we will expand the family $C(S,d)$ to a family of integrable functions that includes functions which need not be continuous, and that includes an abundance of indicators.

More generally, given an integration space (Ω, L, I), we will expand the family L to a larger family L_1 and extend the integration I to L_1. We will do so by summing certain series of small pieces in L, in a sense to be made precise presently. This is analogous to of the expansion of the set of rational numbers to the set of real numbers by representing a real number as the sum of a convergent series of rational numbers.

Definition 4.4.1. Integrable functions and complete extension of an integration space. Let (Ω, L, I) be an arbitrary integration space. Recall that each function X on Ω is required to have a nonempty $domain(X)$.

A function X on Ω is called an *integrable function* if there exists a sequence $(X_n)_{n=1,2,\ldots}$ in L satisfying the following two conditions:

(i) $\sum_{i=1}^{\infty} I|X_i| < \infty$.

(ii) For each $\omega \in \bigcap_{i=1}^{\infty} domain(X_i)$ such that $\sum_{i=1}^{\infty} |X_i(\omega)| < \infty$, we have $\omega \in domain(X)$ and

$$X(\omega) = \sum_{i=1}^{\infty} X_i(\omega).$$

The sequence $(X_n)_{n=1,2,\ldots}$ is then called a *representation* of the integrable function X by elements of L, relative to I. The set of integrable functions will be denoted by L_1. The sum

$$I_1(X) \equiv \sum_{i=1}^{\infty} IX_i \tag{4.4.1}$$

is then called the *integral* of X. The function $I_1 : L_1 \to R$ is called the *complete extension*, or simply *completion,* of I. Likewise, L_1 and (Ω, L_1, I_1) are called the *complete extensions*, or simply *completions,* of L and (Ω, L, I), respectively.

The next proposition and theorem prove that (i′) the function I_1 is well defined on L_1 and (ii′) the completion (Ω, L_1, I_1) is an integration space, with $L \subset L_1$ and $I = I_1|L$. □

Proposition 4.4.2. Complete extension of an integration is well defined. *Let (Ω, L, I) be an arbitrary integration space. Let X be an arbitrary integrable function. If $(X_n)_{n=1,2,\ldots}$ and $(Y_n)_{n=1,2,\ldots}$ are two representations of the integrable function X, then $\sum_{i=1}^{\infty} IX_i = \sum_{i=1}^{\infty} IY_i$.*

Proof. By Condition (i) of Definition 4.4.1 for a representation, we have $\sum_{i=1}^{\infty} I|X_i| < \infty$ and $\sum_{i=1}^{\infty} I|Y_i| < \infty$. Therefore, by Assertion 4 of Proposition 4.3.4, we have

$$\sum_{i=1}^{\infty} I|Y_i - X_i| \le \sum_{i=1}^{\infty} I|X_i| + \sum_{i=1}^{\infty} I|Y_i| < \infty.$$

Suppose, for the sake of a contradiction, that $\sum_{i=1}^{\infty} IX_i < \sum_{i=1}^{\infty} IY_i$. Then, for some sufficiently large number $m \ge 1$, we have

$$\sum_{i=m+1}^{\infty} I|X_i| + \sum_{i=m+1}^{\infty} I|Y_i| < \sum_{i=1}^{m} IY_i - \sum_{i=1}^{m} IX_i = I\sum_{i=1}^{m}(Y_i - X_i).$$

The positivity condition (Condition 2) of Definition 4.3.1 for an integration space therefore implies the existence of a point $\omega \in \bigcap_{i=1}^{\infty}(domain(X_i) \cap domain(Y_i))$ such that

$$\sum_{i=m+1}^{\infty} |X_i(\omega)| + \sum_{i=m+1}^{\infty} |Y_i(\omega)| < \sum_{i=1}^{m} Y_i(\omega) - \sum_{i=1}^{m} X_i(\omega).$$

It follows that

$$\begin{aligned}
\sum_{i=m+1}^{\infty} |X_i(\omega)| + \sum_{i=m+1}^{\infty} |Y_i(\omega)| &< \left| \sum_{i=1}^{m} Y_i(\omega) - \sum_{i=1}^{m} X_i(\omega) \right| \\
&= \left| \sum_{i=1}^{\infty} Y_i(\omega) - \sum_{i=1}^{\infty} X_i(\omega) - \left(\sum_{i=m+1}^{\infty} Y_i(\omega) - \sum_{i=m+1}^{\infty} X_i(\omega) \right) \right| \\
&\leq \left| \sum_{i=1}^{\infty} Y_i(\omega) - \sum_{i=1}^{\infty} X_i(\omega) \right| + \left| \sum_{i=m+1}^{\infty} (Y_i(\omega) - X_i(\omega)) \right| \\
&= |X(\omega) - X(\omega)| + \left| \sum_{i=m+1}^{\infty} (Y_i(\omega) - X_i(\omega)) \right| = \left| \sum_{i=m+1}^{\infty} (Y_i(\omega) - X_i(\omega)) \right| \\
&\leq \sum_{i=m+1}^{\infty} |X_i(\omega)| + \sum_{i=m+1}^{\infty} |Y_i(\omega)|,
\end{aligned} \tag{4.4.2}$$

where the next-to-last equality is because the sequences $(Y_n)_{n=1,2,\ldots}$ and $(X_n)_{n=1,2,\ldots}$ are, by hypothesis, both representations of X. Note that the expressions at both ends of the strict inequality 4.4.2 are the same, which is a contradiction. Thus the assumption $\sum_{i=1}^{\infty} IX_i < \sum_{i=1}^{\infty} IY_i$ leads to a contradiction. Therefore $\sum_{i=1}^{\infty} IX_i \geq \sum_{i=1}^{\infty} IY_i$. Similarly, $\sum_{i=1}^{\infty} IX_i \leq \sum_{i=1}^{\infty} IY_i$. Hence $\sum_{i=1}^{\infty} IX_i = \sum_{i=1}^{\infty} IY_i$. □

Theorem 4.4.3. A complete extension of an integration space is an integration space. *Let (Ω, L, I) be an arbitrary integration space. Then its complete extension (Ω, L_1, I_1) is an integration space. Moreover, $L \subset L_1$, and $I_1 X = IX$ for each $X \in L$.*

Proof. 1. Let $X \in L$ be arbitrary. Then $(X, 0X, 0X, \ldots)$ is a representation of X. Hence $X \in L_1$ and $I_1 X = IX$. Thus $L \subset L_1$.

2. It remains to verify, for the triple (Ω, L_1, I_1), Conditions 1–4 in Definition 4.3.1 of an integration space.

3. To that end, let $X, Y \in L_1$ be arbitrary, with representations of $(X_n)_{n=1,2,\ldots}$ and $(Y_n)_{n=1,2,\ldots}$, respectively, by elements of L. Let $a, b \in R$.

First consider the sequence

$$(V_n)_{n=1,2,\ldots} \equiv (aX_n + bY_n, X_n, -X_n, Y_n, -Y_n)_{n=1,2,\ldots}. \tag{4.4.3}$$

Then

$$\sum_{n=1}^{\infty} I|aX_n + bY_n| + 2I|X_n| + 2I|Y_n| < \infty.$$

Suppose $\omega \in \bigcap_{n=1}^{\infty} domain(V_n) = \bigcap_{n=1}^{\infty} domain(X_n) \cap domain(Y_n)$ is such that

$$\sum_{n=1}^{\infty}(|aX_n(\omega) + bY_n(\omega)| + |X_n(\omega)| + |-X_n(\omega)| + |Y_n(\omega)| + |-Y_n(\omega)|) < \infty.$$

Then $\sum_{n=1}^{\infty} |X_n(\omega)| < \infty$ and $\sum_{n=1}^{\infty} |Y_n(\omega)| < \infty$. Hence $\omega \in domain(X) \cap domain(Y)$, with $X(\omega) = \sum_{n=1}^{\infty} X_n(\omega)$ and $Y(\omega) = \sum_{n=1}^{\infty} Y_n(\omega)$. Consequently, $\omega \in domain(aX + bY)$, with

$$\begin{aligned}(aX + bY)(\omega) &\equiv aX(\omega) + bY(\omega)\\ &= \sum_{n=1}^{\infty}(aX_n(\omega) + bY_n(\omega) + X_n(\omega) - X_n(\omega) + Y_n(\omega) - Y_n(\omega))\\ &= \sum_{n=1}^{\infty} V_n(\omega).\end{aligned}$$

Thus Conditions (i) and (ii) of Definition 4.4.1 have been verified for the sequence $(V_n)_{n=1,2,...}$ to be a representation of the function $aX + bY$. Accordingly, $aX + bY \in L_1$ and

$$I_1(aX + bY) = \sum_{n=1}^{\infty} IV_n = \sum_{n=1}^{\infty} I(aX_n + bY_n) = aI_1X + bI_1Y.$$

The seemingly redundant terms $X_n, -X_n, Y_n, -Y_n, \ldots$ are included to ensure that the absolute convergence of the series of the resulting sequence $(V_n)_{n=1,2,...}$ at some ω implies the absolute convergence of the series of the sequences $(X_n)_{n=1,2,...}$ and $(Y_n)_{n=1,2,...}$. Similar tricks will be used repeatedly in the following discussion without further comments.

4. Because

$$I\left(\left|\sum_{i=1}^{n} X_i\right| - \left|\sum_{i=1}^{n-1} X_i\right|\right) \leq I|X_n|$$

for each $n \geq 1$, the sequence

$$\left(\left|\sum_{i=1}^{n} X_i\right| - \left|\sum_{i=1}^{n-1} X_i\right|, X_n, -X_n\right)_{n=1,2,...}$$

is a representation of $|X|$. Hence $|X|$ belongs to to L_1, with

$$I_1|X| = \sum_{n=1}^{\infty}\left(I\left|\sum_{i=1}^{n} X_i\right| - I\left|\sum_{i=1}^{n-1} X_i\right|\right) = \lim_{n\to\infty} I\left|\sum_{i=1}^{n} X_i\right|. \tag{4.4.4}$$

Here an empty sum $\sum_{i=1}^{0}$ is, by convention, equal to 0.

5. Similarly, because

$$I\left|1 \wedge \sum_{i=1}^{n} X_i - 1 \wedge \sum_{i=1}^{n-1} X_i\right| \le I\left|\sum_{i=1}^{n} X_i - \sum_{i=1}^{n-1} X_i\right| = I|X_n|$$

for each $n \ge 1$, the sequence

$$\left(1 \wedge \sum_{i=1}^{n} X_i - 1 \wedge \sum_{i=1}^{n-1} X_i, X_n, -X_n\right)_{n=1,2,\ldots}$$

is a representation of $1 \wedge X$. Hence $1 \wedge X$ belongs to L_1, with

$$\begin{aligned} I_1(1 \wedge X) &= \lim_{n\to\infty} I\left(1 \wedge \sum_{i=1}^{n} X_i\right) \le \lim_{n\to\infty} I\left(\sum_{i=1}^{n} X_i\right) \\ &= \lim_{n\to\infty} \sum_{i=1}^{n} IX_i = \sum_{i=1}^{\infty} IX_i \equiv I_1 X. \end{aligned} \tag{4.4.5}$$

Combining with Steps 3 and 4, we see that Condition 1 of Definition 4.3.1 has been verified for the triple (Ω, L_1, I_1).

6. It remains to verify Conditions 2–4 in Definition 4.3.1 for the triple (Ω, L_1, I_1). To that end, let $(X_i)_{i=0,1,2,\ldots}$ be a sequence of functions in L_1 such that $X_i \ge 0$ for each $i \ge 1$, and such that

$$\sum_{i=1}^{\infty} I_1 X_i < I_1 X_0. \tag{4.4.6}$$

We need to find some $\omega \in \bigcap_{i=0}^{\infty} domain(X_i)$ such that $\sum_{i=1}^{\infty} X_i(\omega) < X_0(\omega)$. For that purpose, take any $\alpha \in (0, 2^{-1})$ so small that $\sum_{i=1}^{\infty} I_1 X_i + 2\alpha I_1 X_0 < I_1 X_0$. This is possible because of inequality 4.4.6. Then $(1-2\alpha)X_i \in L_1$ according to the linearity of L_1 established in Step 3.

Let $i \ge 0$ be arbitrary. Let $(X_{i,k})_{k=1,2,\ldots}$ be a sequence in L that is a representation of $(1-2\alpha)X_i \in L_1$. Then

$$\sum_{k=1}^{\infty} I|X_{i,k}| < \infty \tag{4.4.7}$$

and $(1-2\alpha)I_1 X_i \equiv \sum_{k=1}^{\infty} IX_{i,k}$. Suppose $i \ge 1$. Then $X_i \ge 0$ by assumption. Equality 4.4.4, where X and $(X_i)_{i=1,\ldots}$ are replaced by $(1-2\alpha)X_i$ and $(X_{i,k})_{k=1,2,\ldots}$, respectively, yields

$$\lim_{n\to\infty} I\left|\sum_{k=1}^{n} X_{i,k}\right| = (1-2\alpha)I_1|X_i| = (1-2\alpha)I_1 X_i. \tag{4.4.8}$$

Note that, by inequality 4.4.6, we have $I_1 X_0 > 0$. Hence equalities 4.4.8 and 4.4.7 together imply that there exists $m_i \geq 1$ so large that

$$I\left|\sum_{k=1}^{m(i)} X_{i,k}\right| < (1-2\alpha)I_1 X_i + 2^{-i}\alpha I_1 X_0$$

and that

$$\sum_{k=m(i)+1}^{\infty} I|X_{i,k}| < 2^{-i}\alpha I_1 X_0.$$

Summing over $i \geq 1$, we obtain

$$\begin{aligned}
&\sum_{i=1}^{\infty} I\left|\sum_{k=1}^{m(i)} X_{i,k}\right| + \sum_{i=1}^{\infty}\sum_{k=m(i)+1}^{\infty} I|X_{i,k}| \\
&\quad < (1-2\alpha)\sum_{i=1}^{\infty} I_1 X_i + \alpha I_1 X_0 + \alpha I_1 X_0 \\
&\quad < (1-2\alpha)I_1 X_0 + 2\alpha I_1 X_0 = I_1 X_0 \equiv \sum_{k=1}^{\infty} I X_{0,k},
\end{aligned}$$

where the last inequality is thanks to inequality 4.4.6. Hence there exists $m_0 \geq 1$ so large that

$$\sum_{i=1}^{\infty} I\left|\sum_{k=1}^{m(i)} X_{i,k}\right| + \sum_{i=1}^{\infty}\sum_{k=m(i)+1}^{\infty} I|X_{i,k}| + \sum_{k=m(0)+1}^{\infty} I|X_{0,k}| < \sum_{k=1}^{m(0)} I X_{0,k}.$$

In short,

$$\sum_{i=1}^{\infty} I\left|\sum_{k=1}^{m(i)} X_{i,k}\right| + \sum_{i=0}^{\infty}\sum_{k=m(i)+1}^{\infty} I|X_{i,k}| < \sum_{k=1}^{m(0)} I X_{0,k}.$$

The positivity condition of Definition 4.3.1 for (Ω, L, I) therefore guarantees the existence of some

$$\omega \in \bigcap_{i=1}^{\infty}\bigcap_{k=1}^{m(i)} domain(X_{i,k}) \cap \bigcap_{i=0}^{\infty}\bigcap_{k=m(i)+1}^{\infty} domain(X_{i,k}) \cap \bigcap_{k=1}^{m(0)} domain(X_{0,k}),$$

such that

$$\sum_{i=1}^{\infty}\left|\sum_{k=1}^{m(i)} X_{i,k}(\omega)\right| + \sum_{i=0}^{\infty}\sum_{k=m(i)+1}^{\infty} |X_{i,k}(\omega)| < \sum_{k=1}^{m(0)} X_{0,k}(\omega). \tag{4.4.9}$$

Consider each $i \geq 0$. Then

$$\sum_{k=m(i)+1}^{\infty} |X_{i,k}(\omega)| < \infty.$$

by inequality 4.4.9. Hence

$$\sum_{k=1}^{\infty} |X_{i,k}(\omega)| < \infty.$$

At the same time, by assumption, the sequence $(X_{i,k})_{k=1,2,\ldots}$ is a representation of X_i. It follows from Definition 4.4.1 that $\omega \in domain(X_i)$ and that

$$\sum_{k=1}^{\infty} X_{i,k}(\omega) = X_i(\omega),$$

where $i \geq 0$ is arbitrary. Hence

$$\begin{aligned}
\sum_{i=1}^{\infty} X_i(\omega) &= \sum_{i=1}^{\infty}\sum_{k=1}^{\infty} X_{i,k}(\omega) = \sum_{i=1}^{\infty}\sum_{k=1}^{m(i)} X_{i,k}(\omega) + \sum_{i=1}^{\infty}\sum_{k=m(i)+1}^{\infty} X_{i,k}(\omega) \\
&\leq \sum_{i=1}^{\infty}\left|\sum_{k=1}^{m(i)} X_{i,k}(\omega)\right| + \sum_{i=1}^{\infty}\sum_{k=m(i)+1}^{\infty} |X_{i,k}(\omega)| \\
&\leq \sum_{i=1}^{\infty}\left|\sum_{k=1}^{m(i)} X_{i,k}(\omega)\right| + \sum_{i=0}^{\infty}\sum_{k=m(i)+1}^{\infty} |X_{i,k}(\omega)| - \sum_{k=m(0)+1}^{\infty} |X_{0,k}(\omega)| \\
&< \sum_{k=1}^{m(0)} X_{0,k}(\omega) + \sum_{k=m(0)+1}^{\infty} X_{0,k}(\omega) = \sum_{k=1}^{\infty} X_{0,k}(\omega) = X_0(\omega),
\end{aligned}$$

where the last inequality follows from inequality 4.4.9.

Summing up, for each sequence $(X_i)_{i=0,1,2,\ldots} L_1$ such that $X_i \geq 0$ for each $i \geq 1$, and such that $\sum_{i=1}^{\infty} I_1 X_i < I_1 X_0$, there exists $\omega \in \bigcap_{i=0}^{\infty} domain(X_i)$ such that $\sum_{i=1}^{\infty} X_i(\omega) < X_0(\omega)$. This proves the positivity condition (Condition 2) of Definition 4.3.1 for the triple (Ω, L_1, I_1).

7. Since (Ω, L, I) is, by hypothesis, an integration space, there exists $X_0 \in L$ such that $IX_0 = 1$, according to Definition 4.3.1. Since $L \subset L_1$by Step 1, we have $X_0 \in L_1$ with $I_1 X_0 = IX_0 = 1$. Thus Condition 3 of Definition 4.3.1 is verified for the triple (Ω, L_1, I_1).

8. Condition 4 of Definition 4.3.1 remains. Let $X \in L_1$ be arbitrary, with a representation $(X_i)_{i=1,2,\ldots}$ in L. Then, for every $m > 0$, the sequence

$$(U_i)_{i=1,2,\ldots} \equiv (X_1, -X_1, X_2, -X_2, \ldots, X_m, -X_m, X_{m+1}, X_{m+2}, X_{m+3}, \ldots)$$

is a representation of the function $X - \sum_{i=1}^{m} X_i \in L_1$. Therefore, equation 4.4.4 implies that

$$I_1\left|X - \sum_{i=1}^{m} X_i\right| = \lim_{n\to\infty} I\left|\sum_{i=1}^{n} U_i\right| = \lim_{n\to\infty} I\left|\sum_{i=m+1}^{n} X_i\right| \leq \sum_{i=m+1}^{\infty} I|X_i| \to 0 \tag{4.4.10}$$

as $m \to \infty$.

9. Let $\varepsilon > 0$ be arbitrary. Then it follows from the convergence condition 4.4.10 that there exists $m \geq 1$ so large that $I_1|X - \sum_{i=1}^{m} X_i| < \varepsilon$. Write $Y \equiv \sum_{i=1}^{m} X_i \in L$. Then

$$I_1|X - Y| < \varepsilon.$$

Moreover, since $Y \in L$, there exists, according to Condition 4 of Definition 4.3.1 for (Ω, L, I), some $p \geq 1$ so large that

$$|I_1(Y \wedge n) - I_1 Y| = |I(Y \wedge n) - IY| < \varepsilon$$

for each $n \geq p$. Consider each $n \geq p$. Then, since $|X \wedge n - Y \wedge n| \leq |X - Y|$, we have

$$I_1|X \wedge n - Y \wedge n| \leq I_1|X - Y| < \varepsilon.$$

Hence

$$I_1 X \geq I_1(X \wedge n) > I_1(Y \wedge n) - \varepsilon > I_1 Y - 2\varepsilon > I_1 X - 3\varepsilon.$$

Since $\varepsilon > 0$ is arbitrary, the last displayed inequality implies that $I_1(X \wedge n) \to I_1 X$ as $n \to \infty$.

10. Separately, again by Condition 4 of Definition 4.3.1 for (Ω, L, I), there exists $p \geq 1$ so large that $I(|Y| \wedge n^{-1}) < \varepsilon$ for each $n \geq p$. Therefore, using the elementary inequality $a \wedge c - b \wedge c \leq |a - b|$ for each $a, b, c \in (0, \infty)$, we obtain

$$\begin{aligned} I_1(|X| \wedge n^{-1}) &\leq I_1(|Y| \wedge n^{-1}) + I_1|(|X| - |Y|)| \\ &\leq I_1(|Y| \wedge n^{-1}) + I_1|X - Y| < 2\varepsilon, \end{aligned}$$

for each $n \geq p$. Since $\varepsilon > 0$ is arbitrary, the last displayed inequality implies that $I_1(|X| \wedge n^{-1}) \to 0$ as $n \to \infty$.

11. In Steps 9 and 10, we have verified Condition 4 in Definition 4.3.1 for (Ω, L_1, I_1). Summing up, all the conditions in Definition 4.3.1 have been verified for (Ω, L_1, I_1) to be an integration space. The theorem is proved. □

Corollary 4.4.4. L is dense in L_1. *Let (Ω, L, I) be an arbitrary integration space. Let (Ω, L_1, I_1) be its completion. Let $X \in L_1$ be arbitrary, with a representation $(X_i)_{i=1,2,\ldots}$ in L. Then $I_1|X - \sum_{i=1}^{m} X_i| \to 0$ as $m \to \infty$.*

Proof. See inequality 4.4.10 in Step 8 of the proof of Theorem 4.4.3. □

Definition 4.4.5. Notational convention for a complete integration space. From this point forward, we will use the same symbol I to denote both the given integration and its complete extension. Thus we write I also for its I_1.

An integration space (Ω, L, I) is said to be *complete* if $(\Omega, L, I) = (\Omega, L_1, I)$. If, in addition, $1 \in L$ and $I1 = 1$, then (Ω, L, I) is called a *probability integration space*. □

Lemma 4.4.6. Integrable function from the sum of a series in L. *Let (Ω, L, I) be an arbitrary integration space. Let $(X_n)_{n=1,2,\ldots}$ be an arbitrary sequence in L such that $\sum_{i=1}^{\infty} I|X_i| < \infty$. Define the function X on Ω by*

$$domain(X) \equiv \left\{ \omega \in \bigcap_{i=1}^{\infty} domain(X_i) : \sum_{i=1}^{\infty} |X_i(\omega)| < \infty \right\}$$

and by $X(\omega) \equiv \sum_{i=1}^{\infty} X_i(\omega)$ for each $\omega \in domain(X)$.

Then $domain(X)$ is nonempty, and the function X is integrable, with $(X_n)_{n=1,2,...}$ as a representation in L, and with $IX = \sum_{i=1}^{\infty} IX_i$. We will then call $X \in L_1$ the sum of the sequence $(X_n)_{n=1,2,...}$ *and define $\sum_{i=1}^{\infty} X_i \equiv X \in L_1$.*

Proof. By hypothesis, we have $\sum_{i=1}^{\infty} I|X_i| < \infty$. Thus Condition (i) of Definition 4.4.1 is satisfied by the sequence $(X_n)_{n=1,2,...}$. Moreover, by Assertion 1 of Lemma 4.3.2, $domain(X)$ is nonempty. Now consider each $\omega \in \bigcap_{i=1}^{\infty} domain(X_i)$ such that $\sum_{i=1}^{\infty} |X_i(\omega)| < \infty$. Then, by the definition of the function X in the hypothesis, we have $\omega \in domain(X)$ and

$$X(\omega) = \sum_{i=1}^{\infty} X_i(\omega).$$

Thus Condition (ii) of Definition 4.4.1 is also satisfied for the sequence $(X_n)_{n=1,2,...}$ to be a representation of the function X in L. Accordingly, $X \in L_1$ and $IX = \sum_{i=1}^{\infty} IX_i$. □

Theorem 4.4.7. Nothing is gained from further complete extension. *Let (Ω, L, I) be an arbitrary integration space. Let (Ω, L_1, I) denote its complete extension. In turn, let $(\Omega, (L_1)_1, I)$ denote the complete extension of (Ω, L_1, I).*

Then $(\Omega, (L_1)_1, I) = (\Omega, L_1, I)$. In other words, the complete extension of an arbitrary integration space is complete, and further completion yields nothing new.

Proof. 1. Because $L_1 \subset (L_1)_1$ according to Theorem 4.4.3, it remains only to prove that $(L_1)_1 \subset L_1$.

2. To that end, let $Z \in (L_1)_1$ be arbitrary. Then there exists a sequence $(Z_k)_{k=1,2,...}$ in L_1 that is a representation of the function Z. Hence, by Definition 4.4.1, the following two conditions hold:

(i) $\sum_{k=1}^{\infty} I|Z_k| < \infty$.

(ii) For each $\omega \in \bigcap_{k=1}^{\infty} domain(Z_k)$ such that $\sum_{k=1}^{\infty} |Z_k(\omega)| < \infty$, we have $\omega \in domain(Z)$ and

$$Z(\omega) = \sum_{k=1}^{\infty} Z_k(\omega).$$

3. Let $k \geq 1$ be arbitrary. Let the sequence $(X_{k,m})_{m=1,2,...}$ in L be a representation of the function $Z_k \in L_1$. Then

(i′) $\sum_{m=1}^{\infty} I|X_{k,m}| < \infty$.

(ii′) For each $\omega \in \bigcap_{m=1}^{\infty} domain(X_{k,m})$ such that $\sum_{m=1}^{\infty} |X_{k,m}(\omega)| < \infty$, we have $\omega \in domain(Z_k)$ and

$$Z_k(\omega) = \sum_{m=1}^{\infty} X_{k,m}(\omega).$$

4. By Corollary 4.4.4, we have $I_1|Z_k - \sum_{m=1}^{p} X_{k,m}| \to 0$ as $p \to \infty$. Hence there exists $m_k \geq 1$ so large that $I|Z_k - \sum_{m=1}^{m(k)} X_{k,m}| < 2^{-k}$ and $\sum_{m=m(k)+1}^{\infty} I|X_{k,m}| < I|Z_k| + 2^{-k}$. Consider the sequence

$$(V_{k,1}, V_{k,2}, \ldots) \equiv \left(\sum_{m=1}^{m(k)} X_{k,m}, X_{k,m(k)+1}, X_{k,m(k)+2}, \ldots \right)$$

in L. Let the sequence $(U_n)_{n=1,2,\ldots}$ be an enumeration of the double sequence $(V_{k,m})_{k,m=1,2,\ldots}$ in L. We will verify that the sequence $(U_n)_{n=1,2,\ldots}$ in L is a representation of the function Z.

5. Note that

(i″)

$$\sum_{n=1}^{\infty} I|U_n| = \sum_{k=1}^{\infty}\sum_{m=1}^{\infty} I|V_{k,m}| = \sum_{k=1}^{\infty} I \left| \sum_{m=1}^{m(k)} X_{k,m} \right| + \sum_{k=1}^{\infty} \sum_{m=m(k)+1}^{\infty} I|X_{k,m}|$$

$$\leq \sum_{k=1}^{\infty}(I|Z_k| + 2^{-k}) + \sum_{k=1}^{\infty}(I|Z_k| + 2^{-k}) = 2\sum_{k=1}^{\infty} I|Z_k| + 2 < \infty.$$

6. Consider each $\omega \in \bigcap_{n=1}^{\infty} domain(U_n)$ such that $\sum_{n=1}^{\infty} |U_n(\omega)| < \infty$. Then

$$\omega \in \bigcap_{n=1}^{\infty} domain(U_n) = \bigcap_{k=1}^{\infty}\bigcap_{m=1}^{\infty} domain(V_{k,m}) = \bigcap_{k=1}^{\infty}\bigcap_{m=1}^{\infty} domain(X_{k,m}). \tag{4.4.11}$$

Moreover,

$$\sum_{k=1}^{\infty} \left| \sum_{m=1}^{m(k)} X_{k,m}(\omega) \right| + \sum_{k=1}^{\infty} \sum_{m=m(k)+1}^{\infty} |X_{k,m}(\omega)| = \sum_{k=1}^{\infty}\sum_{m=1}^{\infty} |V_{k,m}(\omega)| = \sum_{n=1}^{\infty} |U_n(\omega)| < \infty. \tag{4.4.12}$$

Let $k \geq 1$ be arbitrary. Then inequality 4.4.12 implies that

$$\sum_{m=m(k)+1}^{\infty} |X_{k,m}(\omega)| < \infty,$$

whence $\sum_{m=1}^{\infty} |X_{k,m}(\omega)| < \infty$. Consequently, by Condition (ii′), we have $\omega \in domain(Z_k)$ with

$$Z_k(\omega) = \sum_{m=1}^{\infty} X_{k,m}(\omega). \tag{4.4.13}$$

Therefore

$$
\begin{aligned}
\sum_{n=1}^{\infty} U_n(\omega) &= \sum_{k=1}^{\infty}\sum_{m=1}^{\infty} V_{k,m}(\omega) \\
&= \sum_{k=1}^{\infty}\left(\sum_{m=1}^{m(k)} X_{k,m}(\omega) + \sum_{m=m(k)+1}^{\infty} X_{k,m}(\omega)\right) \\
&= \sum_{k=1}^{\infty}\sum_{m=1}^{\infty} X_{k,m}(\omega) = \sum_{k=1}^{\infty} Z_k(\omega) = Z(\omega), \qquad (4.4.14)
\end{aligned}
$$

where the last two equalities are due to equality 4.4.13 and Condition (ii), respectively. Summing up, we have proved that:

(ii$''$) for each $\omega \in \bigcap_{n=1}^{\infty} domain(U_n)$ such that $\sum_{n=1}^{\infty} |U_n(\omega)| < \infty$, we have $\omega \in domain(Z)$ and

$$Z(\omega) = \sum_{n=1}^{\infty} U_n(\omega).$$

7. Conditions (i$''$) and (ii$''$) together show that the sequence $(U_n)_{n=1,2,\ldots}$ in L is a representation of the function Z. Thus we have proved that $Z \in L_1$, where $Z \in (L_1)_1$ is arbitrary. We conclude that $(L_1)_1 \subset L_1$, as alleged. □

Following is a powerful tool for the construction of integrable functions on a complete integration space.

Theorem 4.4.8. Monotone Convergence Theorem. *Let (Ω, L, I) be a complete integration space. Then the following holds:*

1. Suppose $(X_i)_{i=1,2,\ldots}$ is a sequence in L such that $X_{i-1} \leq X_i$ for each $i \geq 2$, and such that $\lim_{i\to\infty} I(X_i)$ exists. Then there exists $X \in L$ such that (i) *$domain(X) \subset \bigcap_{i=1}^{\infty} domain(X_i)$ and* (ii) *$X = \lim_{i\to\infty} X_i$ on $domain(X)$. Moreover, $\lim_{i\to\infty} I|X - X_i| = 0$ and $IX_i \uparrow IX$ as $i \to \infty$.*

2. Suppose $(Y_i)_{i=1,2,\ldots}$ is a sequence in L such that $Y_{i-1} \geq Y_i$ for each $i \geq 2$, and such that $\lim_{i\to\infty} I(Y_i)$ exists. Then there exists $Y \in L$ such that (iii) *$domain(Y) \subset \bigcap_{i=1}^{\infty} domain(Y_i)$ and* (iv) *$Y = \lim_{i\to\infty} Y_i$ on $domain(Y)$. Moreover, $\lim_{i\to\infty} I|Y - Y_i| = 0$ and $IY_i \downarrow IY$ as $i \to \infty$.*

Proof. 1. Suppose $(X_i)_{i=1,2,\ldots}$ is a sequence in L such that $X_{i-1} \leq X_i$ for each $i \geq 2$, and such that $\lim_{i\to\infty} I(X_i)$ exists. Consider the sequence $(V_n)_{n=1,2,\ldots} \equiv (X_1, X_2 - X_1, X_3 - X_2, \ldots)$ in L. Then $\sum_{n=1}^{\infty} I|V_n| = \lim_{i\to\infty} I(X_i) < \infty$. Hence, according to Lemma 4.4.6, the function X defined by

$$domain(X) \equiv \left\{\omega \in \bigcap_{i=1}^{\infty} domain(V_i) : \sum_{i=1}^{\infty} |V_i(\omega)| < \infty\right\},$$

and by $X(\omega) \equiv \sum_{i=1}^{\infty} V_i(\omega)$ for each $\omega \in domain(X)$, is integrable. Moreover,

$$domain(X) \subset \bigcap_{i=1}^{\infty} domain(V_i) = \bigcap_{i=1}^{\infty} domain(X_i),$$

which proves the desired Condition (i). Furthermore, consider each $\omega \in domain(X)$. Then $X(\omega) \equiv \sum_{i=1}^{\infty} V_i(\omega) = \lim_{i\to\infty} X_i(\omega)$, which proves the desired Condition (ii). In addition, according to Corollary 4.4.4, we have

$$I|X - X_m| = I\left|X - \sum_{n=1}^{m} V_n\right| \to 0$$

as $m \to \infty$. Finally, $|IX - IX_m| \le I|X - X_m| \to 0$. Hence $IX_m \uparrow IX$. Assertion 1 of the theorem is proved.

2. Suppose $(Y_i)_{i=1,2,\ldots}$ is a sequence in L such that $Y_{i-1} \ge Y_i$ for each $i \ge 2$, and such that $\lim_{i\to\infty} I(Y_i)$ exists. For each $i \ge 1$, define $X_i \equiv -Y_i$. Then $(X_i)_{i=1,2,\ldots}$ is a sequence in L such that $X_{i-1} \le X_i$ for each $i \ge 2$, and such that $\lim_{i\to\infty} I(X_i)$ exists. Hence, by Assertion 1, there exists $X \in L$ such that (i) $domain(X) \subset \bigcap_{i=1}^{\infty} domain(X_i)$ and (ii) $X = \lim_{i\to\infty} X_i$ on $domain(X)$. Moreover, $\lim_{i\to\infty} I|X - X_i| = 0$ and $IX_i \uparrow IX$ as $i \to \infty$. Now define $Y \equiv -X \in L$. Then (iii)

$$domain(Y) = domain(X) \subset \bigcap_{i=1}^{\infty} domain(X_i) = \bigcap_{i=1}^{\infty} domain(Y_i)$$

and (iv) $Y = -X = -\lim_{i\to\infty} X_i = \lim_{i\to\infty} Y_i$ on $domain(Y)$. Moreover, $\lim_{i\to\infty} I|Y - Y_i| = \lim_{i\to\infty} I|X - X_i| = 0$ and $IY_i = -IX_i \downarrow -IX = IY$ as $i \to \infty$. Assertion 2 is proved. □

4.5 Integrable Set

To model an event in an experiment of chance that may or may not occur depending on the outcome, we can use a function of the outcome with only two possible values, 1 or 0. Equivalently, we can specify the subset of those outcomes that realize the event. We make these notions precise in the present section.

Definition 4.5.1. Indicators and mutually exclusive subsets. A function X on a set Ω with only two possible values, 1 or 0, is called an *indicator*.

Subsets $A_1, \ldots, A_n$ of a set Ω are said to be *mutually exclusive* if $A_i A_j = \phi$ for each $i, j = 1, \ldots, n$ with $i \ne j$. Indicators $X_1, \ldots, X_n$ are said to be *mutually exclusive* if the sets $\{\omega \in domain(X_i) : X_i(\omega) = 1\}$ $(i = 1, \ldots, n)$ are mutually exclusive. □

In the remainder of this section, let (Ω, L, I) be a complete integration space. Recall that an integrable function need not be defined everywhere. However, such functions are defined almost everywhere in the sense of the next definition.

Definition 4.5.2. Full set, and conditions holding almost everywhere. A subset D of Ω is called a *full set* if $D \supset domain(X)$ for some integrable function

$X \in L = L_1$. By Definition 4.4.1, $domain(X)$ is nonempty for each integrable function $X \in L_1$. Hence each full set D is nonempty.

Two integrable functions $Y, Z \in L_1$ are said to be equal *almost everywhere* (or $Y = Z$ a.e. in symbols) if there exists a full set $D \subset domain(Y) \cap domain(Z)$ such that $Y = Z$ on D.

More generally, a condition about a general element ω of Ω is said to hold *almost everywhere* (a.e. for short) if it holds for each ω in a full set D. □

For example, according to the terminology established in the Introduction of this book, the statement $Y \leq Z$ means that for each $\omega \in \Omega$ we have (i′) $\omega \in domain(Y)$ iff $\omega \in domain(Z)$, and (ii′) $Y(\omega) \leq Z(\omega)$ if $\omega \in domain(Y)$. Hence the statement $Y \leq Z$ a.e. means that there exists some full set D such that, for each $\omega \in D$, Conditions (i′) and (ii′) hold.

A similar argument holds when $\leq$ is replaced by $\geq$ or by $=$.

If A, B are subsets of Ω, then $A \subset B$ a.e. iff $AD \subset BD$ for some full set D.

Proposition 4.5.3. Properties of full sets. *Let $X, Y, Z \in L$ denote integrable functions.*

1. *A subset that contains a full set is a full set. The intersection of a sequence of full sets is again a full set.*
2. *Suppose W is a function on Ω and $W = X$ a.e. Then W is an integrable function with $IW = IX$.*
3. *If D is a full set, then $D = domain(X)$ for some $X \in L$*
4. *$X = Y$ a.e. if and only if $I|X - Y| = 0$.*
5. *If $X \leq Y$ a.e., then $IX \leq IY$.*
6. *If $X \leq Y$ a.e. and $Y \leq Z$ a.e., then $X \leq Z$ a.e., Moreover, if $X \leq Y$ a.e. and $X \geq Y$ a.e., then $X = Y$ a.e.*
7. *Almost everywhere equality is an equality relation in L. In other words, for all $X, Y, Z \in L$, we have* (i) *$X = X$ a.e.;* (ii) *if $X = Y$ a.e., then $Y = X$ a.e.; and* (iii) *if $X = Y$ a.e. and $Y = Z$ a.e., then $X = Z$ a.e.*

Proof. 1. Suppose a subset A contains a full set D. By definition, $D \supset domain(X)$ for some $X \in L$. Hence $A \supset domain(X)$. Thus A is a full set.

Now suppose $D_n \supset domain(X_n)$, where $X_n \in L$ for each $n \geq 1$. Define the function X by $domain(X) = \bigcap_{n=1}^{\infty} domain(X_n)$, and by $X(\omega) = X_1(\omega)$ for each $\omega \in domain(X)$. Then the function X has the sequence $(X_1, 0X_2, 0X_3, \ldots)$ as a representation. Hence X is integrable, and so $domain(X)$ is a full set. Since

$$\bigcap_{n=1}^{\infty} D_n \supset \bigcap_{n=1}^{\infty} domain(X_n) = domain(X),$$

we see that $\bigcap_{n=1}^{\infty} D_n$ is a full set.

2. By the definition of a.e. equality, there exists a full set D such that $D \cap domain(W) = D \cap domain(X)$ and $W(\omega) = X(\omega)$ for each $\omega \in D \cap domain(X)$. By the definition of a full set, $D \supset domain(Z)$ for some $Z \in L$. It follows that the sequence $(X, 0Z, 0Z, \ldots)$ is a representation of the function W. Therefore $W \in L_1 = L$ with $IW = IX$.

3. Suppose D is a full set. By definition, $D \supset domain(X)$ for some $X \in L$. Define a function W by $domain(W) \equiv D$ and $W(\omega) \equiv 0$ for each $\omega \in D$. Then $W = 0X$ on the full set $domain(X)$. Hence by Assertion 2, W is an integrable function, with $D = domain(W)$.

4. Suppose $X = Y$ a.e. Then $|X - Y| = 0(X - Y)$ a.e. Hence $I|X - Y| = 0$ according to Assertion 2. Suppose, conversely, that $I|X - Y| = 0$. Then the function defined by $Z \equiv \sum_{n=1}^{\infty} |X - Y|$ is integrable. By definition,

$$domain(Z) \equiv \left\{ \omega \in domain(X - Y) : \sum_{n=1}^{\infty} |X(\omega) - Y(\omega)| < \infty \right\}$$

$$= \{\omega \in domain(X) \cap domain(Y) : X(\omega) = Y(\omega)\}.$$

Thus we see that $X = Y$ on the full set $domain(Z)$.

5. Suppose $X \leq Y$ a.e. Then $Y - X = |Y - X|$ a.e. Hence, by Assertion 4, $I(Y - X) = I|Y - X| \geq 0$.

6. Suppose $X \leq Y$ a.e. and $Y \leq Z$ a.e. Then there exists a full set D such that $D \cap domain(X) = D \cap domain(Y)$ and $X(\omega) \leq Y(\omega)$ for each $\omega \in D \cap domain(X)$. Similarly, there exists a full set D' such that $D' \cap domain(Y) = D' \cap domain(Z)$ and $Y(\omega) \leq Z(\omega)$ for each $\omega \in D' \cap domain(Y)$. By Assertion 1, the set DD' is a full set. Furthermore,

$$DD' \cap domain(X) = DD' \cap domain(Y) = DD' \cap domain(Z)$$

and $X(\omega) \leq Y(\omega) \leq Z(\omega)$ for each $\omega \in DD' \cap domain(X)$. It follows that $X \leq Z$ a.e. The remainder of Assertion 6 is similarly proved.

7. This is a trivial consequence of Assertion 4. □

Definition 4.5.4. Integrable set, measure of integrable set, measure-theoretic complement, and null set. A subset A of Ω is called an *integrable set* if there exists an indicator X that is an integrable function such that $A = (X = 1)$. We then call X an indicator of A and define $1_A \equiv X$; we also define the meas*ure* of A to be $\mu(A) \equiv IX$. We will call the set $A^c \equiv (X = 0)$ a *measure-theoretic complement* of the set A.

An arbitrary subset of an integrable set A with measure $\mu(A) = 0$ is called a *null set*.

Note that two distinct integrable indicators X and Y can be indicators of the same integrable set A. Hence 1_A and A^c are not uniquely defined relative to the set-theoretic equality for functions. However, the next proposition shows that they are uniquely defined relative to a.e. equality. □

Proposition 4.5.5. Uniqueness of integrable indicators and measure-theoretic complement relative to a.e. equality. *Let A and B be integrable sets. Let X, Y be integrable indicators of A, B, respectively. Then the following conditions hold:*

1. $A = B$ a.e. iff $X = Y$ a.e. Hence 1_A is a well-defined integrable function relative to a.e. equality, and the measure $\mu(A)$ is well defined. Note that, in general, the set A^c need not be an integrable set.

2. If $A = B$ a.e., then $(X = 0) = (Y = 0)$ a.e. Hence A^c is a well-defined subset relative to equality a.e.

3. Ω is a full set. The empty set ϕ is a null set.

4. Suppose $\mu(A) = 0$. Then A^c is a full set.

5. If C is a subset of Ω such that $C = A$ a.e., then C is an integrable set, with $1_C = 1_A$ a.e. and $\mu(C) = \mu(A)$.

Proof. By the definition of an indicator for an integrable set, we have $A = (X = 1)$ and $B = (Y = 1)$. Let D be an arbitrary full set. Then the intersection $D' \equiv D \cap domain(X) \cap domain(Y)$ is a full set. Since $D'D = D'$, we have $D'DA = D'D(X = 1)$ and $D'DB = D'D(Y = 1)$.

1. Suppose $A = B$ a.e. Then $A = B$ on some full set D. Then $DA = DB$. It follows from the previous paragraph that $D'(X = 1) = D'D(X = 1) = D'D(Y = 1) = D'(Y = 1)$. By the remark following Definition 4.5.1, we see that for each $\omega \in D'$, $X(\omega)$ and $Y(\omega)$ are defined and equal. Hence $X = Y$ a.e. Moreover, it follows from Proposition 4.5.3 that $\mu(A) \equiv IX = IY \equiv \mu(B)$. Conversely, suppose $X = Y$ a.e. with $D \cap domain(X) = D \cap domain(Y)$ and $X(\omega) = Y(\omega)$ for each $\omega \in D \cap domain(X)$. Then

$$D'A = D'DA = D'D(X = 1) = D'D(Y = 1) = D'(Y = 1) = D'B.$$

Hence $A = B$ a.e.

2. Suppose $A = B$ a.e. In the proof for Assertion 1, we saw that for each ω in the full set D', we have $X(\omega) = 0$ iff $Y(\omega) = 0$. In short, $(X = 0) = (Y = 0)$ a.e. Thus Assertion 2 is proved.

3. Let X be an arbitrary integrable function. Trivially $\Omega \supset domain(X)$. Hence Ω is a full set. Now define a function $Y \equiv 0X$. Then Y is an integrable indicator, with $IY = 0$. Moreover,

$$\begin{aligned}(Y = 1) &= \{\omega \in domain(Y) : Y(\omega) = 1\} \\ &= \{\omega \in domain(Y) : 0X(\omega) \equiv Y(\omega) = 1\} = \phi.\end{aligned}$$

Thus Y is an indicator of the empty subset ϕ. Consequently, $\mu(\phi) \equiv IY = 0$, whence ϕ is a null set.

4. Suppose $\mu(A) = 0$. Then $IX = \mu(A) = 0$. Hence the function $Z \equiv \sum_{i=1}^{\infty} X$ is integrable. Moreover,

$$domain(Z) = \left\{ \omega \in domain(X) : \sum_{i=1}^{\infty} |X(\omega)| = 0 \right\} = (X = 0) = A^c.$$

Thus A^c contains the domain of the integrable function Z, and is therefore a full set by Definition 4.5.2.

5. Suppose $C = A$ on some full set D. Define a function W by $domain(W) \equiv C \cup (X = 0)$, and by $W(\omega) \equiv 1$ or 0 according to $\omega \in C$ or $\omega \in (X = 0)$, respectively. Then $W = 1 = X$ on $D(X = 1)$, and $W = 0 = X$ on $D(X = 10)$. Hence $W = X$ on the full set

$$D \cap domain(X) = D(X = 1) \cup D(X = 0).$$

In short, $W = X$ a.e. Therefore, by Assertion 2 of Proposition 4.5.3, the function W is integrable. Hence the set $C = (W = 1)$ has the integrable indicator, and is therefore an integrable set, with $1_C \equiv W = X \equiv 1_A$ a.e. Moreover, $\mu(C) = IW = IX = \mu(A)$. □

Definition 4.5.6. Convention: unless otherwise specified, equality of integrable functions will mean equality a.e.; equality of integrable sets will mean equality a.e. Let (Ω, L, I) be an arbitrary complete integration space. Let $X, Y \in L$ and $A, B \subset \Omega$ be arbitrary integrable sets. Henceforth, unless otherwise specified, the statement $X = Y$ will mean $X = Y$ a.e. Similarly, the statements $X \leq Y$, $X < Y$, $X \geq Y$, and $X > Y$ will mean $X \leq Y$ a.e., $X < Y$ a.e., $X \geq Y$ a.e., and $X > Y$ a.e., respectively. Likewise, the statements $A = B$, $A \subset B$, and $A \supset B$ will mean $A = B$ a.e., $A \subset B$ a.e., and $A \supset B$ a.e., respectively. □

Suppose each of a sequence of statements is valid a.e. Then, in view of Assertion 1 of Proposition 4.5.3, there exists a full set on which all of these statements are valid; in other words, a.e., we have the validity of all the statements. For example if $(A_n)_{n=1,2,...}$ is a sequence of integrable sets with $A_n \subset A_{n+1}$ a.e. for each $n \geq 1$, then $A_1 \subset A_2 \subset \cdots$ a.e.

Proposition 4.5.7. Basics of measures of integrable sets. *Let (Ω, L, I) be an arbitrary complete integration space. Let A, B be arbitrary integrable sets, with measure-theoretic complements A^c, B^c, respectively. Then the following conditions hold:*

1. $A \cup A^c$ is a full set, and $AA^c = \phi$.

2. $A \cup B$ is an integrable set, with integrable indicator $1_{A \cup B} = 1_A \vee 1_B$.

3. AB is an integrable set, with indicator $1_{AB} = 1_A \wedge 1_B$.

4. AB^c is an integrable set, with $1_{AB^c} = 1_A - 1_A \wedge 1_B$. Furthermore, $A(AB^c)^c = AB$.

5. $\mu(A \cup B) + \mu(AB) = \mu(A) + \mu(B)$.

6. If $A \supset B$, then $\mu(AB^c) = \mu(A) - \mu(B)$.

Proof. Let A, B be arbitrary integrable sets, with integrable indicators X, Y, respectively.

1. We have $A = (X = 1)$ and $A^c = (X = 0)$. Hence $AA^c = \phi$. Moreover, $A \cup A^c = domain(X)$ is a full set.

2. By Assertion 1 of Proposition 4.3.4, the function $X \vee Y$ is integrable. At the same time, $A \cup B = (X \vee Y = 1)$. Hence the set $A \cup B$ is integrable, with integrable indicator $1_{A \cup B} = X \vee Y = 1_A \vee 1_B$.

3. By Assertion 1 of Proposition 4.3.4, the function $X \wedge Y$ is integrable. At the same time, $AB = (X \wedge Y = 1)$. Hence the set AB is integrable, with integrable indicator $1_{AB} = X \wedge Y = 1_A \wedge 1_B$

4. By Assertion 1 of Proposition 4.3.4, the function $X - X \wedge Y$ is integrable. Hence the set

$$AB^c = (X = 1)(Y = 0) = (X - X \wedge Y = 1)$$

is integrable, with integrable indicator $1_{AB^c} = X - X \wedge Y = 1_A - 1_A \vee 1_B$.

Furthermore, on the full set $domain(X) \cap domain(Y)$, we have

$$A(AB^c)^c = (X = 1)(X - X \wedge Y = 0) = (X = 1)(Y = 1) = AB.$$

Thus $A(AB^c)^c = AB$.

5. Since $1_A \vee 1_B + 1_A \wedge 1_B = 1_A + 1_B$, Assertion 6 follows from the linearity of I.

6. Suppose $AD \supset BD$ for some full set D. Then $X \geq Y$ on D. Hence $X \wedge Y = Y$. Consequently, $IX \wedge Y = IY$. By Assertion 5, we have $\mu(AB^c) = I(X - X \wedge Y) = IX - IY = \mu(A) - \mu(B)$. Assertion 7 is proved. □

Proposition 4.5.8. Sequence of integrable sets. *For each $n \geq 1$, let A_n be an integrable set with a measure-theoretic complement A_n^c. Then the following conditions hold:*

1. If $A_n \subset A_{n+1}$ for each $n \geq 1$, and if $\mu(A_n)$ converges, then $\bigcup_{n=1}^{\infty} A_n$ is an integrable set with $\mu\left(\bigcup_{n=1}^{\infty} A_n\right) = \lim_{n\to\infty} \mu(A_n)$ and $\left(\bigcup_{n=1}^{\infty} A_n\right)^c = \bigcap_{n=1}^{\infty} A_n^c$.

2. If $A_n \supset A_{n+1}$ for each $n \geq 1$, and if $\mu(A_n)$ converges, then $\bigcap_{n=1}^{\infty} A_n$ is an integrable set with $\mu\left(\bigcap_{n=1}^{\infty} A_n\right) = \lim_{n\to\infty} \mu(A_n)$ and $\left(\bigcap_{n=1}^{\infty} A_n\right)^c = \bigcup_{n=1}^{\infty} A_n^c$.

3. If $A_n A_m = \phi$ for each $n > m \geq 1$, and if $\sum_{n=1}^{\infty} \mu(A_n)$ converges, then $\bigcup_{n=1}^{\infty} A_n$ is an integrable set with $\mu\left(\bigcup_{n=1}^{\infty} A_n\right) = \sum_{n=1}^{\infty} \mu(A_n)$.

4. If $\sum_{n=1}^{\infty} \mu(A_n)$ converges, then $\bigcup_{n=1}^{\infty} A_n$ is an integrable set with $\mu\left(\bigcup_{n=1}^{\infty} A_n\right) \leq \sum_{n=1}^{\infty} \mu(A_n)$.

Proof. For each $n \geq 1$, let $1_{A(n)}$ be the integrable indicator of A_n. Then $A_n^c = (1_{A(n)} = 0)$.

1. Define a function Y by

$$domain(Y) \equiv \left(\bigcup_{n=1}^{\infty} A_n\right) \cup \left(\bigcap_{n=1}^{\infty} A_n^c\right)$$

with $Y(\omega) \equiv 1$ or 0 according to $\omega \in \bigcup_{n=1}^{\infty} A_n$ or $\omega \in \bigcap_{n=1}^{\infty} A_n^c$, respectively. Then $(Y = 1) = \bigcup_{n=1}^{\infty} A_n$ and $(Y = 0) = \bigcap_{n=1}^{\infty} A_n^c$. For each $n \geq 1$, we have $A_n \subset A_{n+1}$ and so $1_{A(n+1)} \geq 1_{A(n)}$. By assumption, we have the convergence of

$$\begin{aligned} &I(1_{A(1)}) + I(1_{A(2)} - 1_{A(1)}) + \cdots + I(1_{A(n)} - 1_{A(n-1)}) \\ &= I1_{A(n)} = \mu(A_n) \to \lim_{k\to\infty} \mu(A_k) \end{aligned}$$

as $n \to \infty$. Hence $X \equiv 1_{A(1)} + (1_{A(2)} - 1_{A(1)}) + (1_{A_3} - 1_{A_2}) + \cdots$ is an integrable function. Consider an arbitrary $\omega \in domain(X)$. The limit

$$\begin{aligned} &\lim_{n\to\infty} 1_{A(n)}(\omega) \\ &= \lim_{n\to\infty} (1_{A(1)}(\omega) + (1_{A(2)} - 1_{A(1)})(\omega) + \cdots + (1_{A(n)} - 1_{A(n-1)})(\omega)) = X(\omega) \end{aligned}$$

exists, and is either 0 or 1 since it is the limit of a sequence in $\{0, 1\}$. Suppose $X(\omega) = 1$. Then $1_{A(n)}(\omega) = 1$ for some $n \geq 1$. Hence $\omega \in \bigcup_{n=1}^{\infty} A_n$ and so $Y(\omega) \equiv 1 = X(\omega)$. Suppose $X(\omega) = 0$. Then $1_{A(n)}(\omega) = 0$ for each $n \geq 1$. Hence $\omega \in \bigcap_{n=1}^{\infty} A_n^c$ and so $Y(\omega) \equiv 0 = X(\omega)$. Combining, we see that $Y = X$ on the full set $domain(X)$. According to Proposition 4.5.3, we therefore have $Y \in L$. Thus $\bigcup_{n=1}^{\infty} A_n = (Y = 1)$ is an integrable set with Y as its indicator, and has measure equal to

$$IY = IX = \lim_{n\to\infty} I1_{A(n)} = \lim_{n\to\infty} \mu(A_n).$$

Moreover $\left(\bigcup_{n=1}^{\infty} A_n\right)^c = (Y = 0) = \bigcap_{n=1}^{\infty} A_n^c$.

2. Similar.

3. Write $B_n = \bigcup_{i=1}^{n} A_i$. Repeated application of Proposition 4.5.7 leads to $\mu(B_n) = \sum_{i=1}^{n} \mu(A_i)$. From Assertion 1 we see that $\bigcup_{n=1}^{\infty} A_n = \bigcup_{n=1}^{\infty} B_n$ is an integrable set with $\mu\left(\bigcup_{n=1}^{\infty} A_n\right) = \mu\left(\bigcup_{n=1}^{\infty} B_n\right) = \lim_{n\to\infty} \mu(B_n) = \sum_{i=1}^{\infty} \mu(A_i)$.

4. Define $B_1 = A_1$ and $B_n = \left(\bigcup_{k=1}^{n} A_k\right)\left(\bigcup_{k=1}^{n-1} A_k\right)^c$ for $n > 1$. Let D denote the full set $\bigcap_{k=1}^{\infty}(A_k \cup A_k^c)(B_k \cup B_k^c)$. Clearly, $B_n B_k = \phi$ on D for each positive integer $k < n$. This implies $\mu(B_n B_k) = 0$ for each positive integer $k < n$. Furthermore, for every $\omega \in D$, we have $\omega \in \bigcup_{k=1}^{\infty} A_k$ iff there is a smallest $n > 0$ such that $\omega \in \bigcup_{k=1}^{n} A_k$. Since for every $\omega \in D$ either $\omega \in A_k$ or $\omega \in A_k^c$, we have $\omega \in \bigcup_{k=1}^{\infty} A_k$ iff there is an $n > 0$ such that $\omega \in B_n$. In other words, $\bigcup_{k=1}^{\infty} A_k = \bigcup_{k=1}^{\infty} B_k$. Moreover, $\mu(B_n) = \mu\left(\bigcup_{k=1}^{n} A_k\right) - \mu\left(\bigcup_{k=1}^{n-1} A_k\right)$. Hence the sequence (B_n) of integrable sets satisfies the hypothesis in Assertion 3. Therefore $\bigcup_{k=1}^{\infty} B_k$ is an integrable set, with

$$\begin{aligned} \mu\left(\bigcup_{k=1}^{\infty} A_k\right) = \mu\left(\bigcup_{k=1}^{\infty} B_k\right) &= \lim_{n\to\infty} \sum_{k=1}^{n} \mu(B_k) \leq \lim_{n\to\infty} \sum_{k=1}^{n} \mu(A_k) \\ &= \sum_{n=1}^{\infty} \mu(A_n). \end{aligned}$$

□

Proposition 4.5.9. Convergence in L implies an a.e. convergent subsequence. *Let $X \in L$ and let $(X_n)_{n=1,2,...}$ be a sequence in L. If $I|X_n - X| \to 0$, then there exists a subsequence $(Y_n)_{n=1,2,...}$ such that $Y_n \to X$ a.e.*

Proof. Let $(Y_n)_{n=1,2,...}$ be a subsequence such that $I|Y_n - X| < 2^{-n}$. Define the sequence

$$(Z_n)_{n=1,2,...} \equiv (X, -X + Y_1, X - Y_1, -X + Y_2, X - Y_2, \ldots).$$

Then $\sum_{n=1}^{\infty} I|Z_n| < I|X| + 2 < \infty$. Hence the set

$$\left\{\omega \in \bigcap domain(Z_n) : \sum_{n=1}^{\infty} |Z_n(\omega)| < \infty\right\}$$

is a full set. Moreover, for each $\omega \in D$, we have

$$X(\omega) = \sum_{n=1}^{\infty} Z_n(\omega) = \lim_{n\to\infty} (Z_1(\omega) + \cdots + Z_{2n}(\omega)) = \lim_{n\to\infty} Y_n(\omega). \qquad \square$$

We will use the next theorem many times to construct integrable functions.

Theorem 4.5.10. A sufficient condition for a function to be integrable. *Suppose X is a function defined a.e. on Ω. Suppose there exist two sequences $(Y_n)_{n=1,2,...}$ and $(Z_n)_{n=1,2,...}$ in L such that $|X - Y_n| \le Z_n$ a.e. for each $n \ge 1$ and such that $IZ_n \to 0$ as $n \to \infty$. Then $X \in L$. Moreover, $I|X - Y_n| \to 0$ as $n \to \infty$.*

Proof. According to Proposition 4.5.9, there exists a subsequence $(Z_{n(k)})_{k=1,2,...}$ such that $Z_{n(k)} \to 0$ a.e. and such that $IZ_{n(k)} \le 2^{-k}$ for each $k \ge 1$. Since, by assumption, $|X - Y_{n(k)}| \le Z_{n(k)}$ a.e. for each $k \ge 1$, it follows that $Y_{n(k)} \to X$ a.e. In other words, $Y_{n(k)} \to X$ on $domain(V)$ for some $V \in L$.

Moreover, $|Y_{n(k+1)} - Y_{n(k)}| \le |X - Y_{n(k)}| + |X - Y_{n(k+1)}| \le Z_{n(k)} + Z_{n(k+1)}$ a.e. Hence

$$2I|V| + \sum_{k=1}^{\infty} I|Y_{n(k+1)} - Y_{n(k)}| \le 2I|V| + \sum_{k=1}^{\infty}(IZ_{n(k)} + IZ_{n(k+1)})$$

$$\le 2I|V| + \sum_{k=1}^{\infty}(2^{-k} + 2^{-k+1}) < 2I|V| + 2 < \infty.$$

Therefore the sequence

$$(V, -V, Y_{n(1)}, Y_{n(2)} - Y_{n(1)}, Y_{n(3)} - Y_{n(2)}, \ldots)$$

in L is a representation of the function X. Consequently, $X \in L_1 = L$.

Moreover, since $|X - Y_n| \le Z_n$ a.e. for each $n \ge 1$, we have $I|X - Y_n| \le IZ_n \to 0$ as $n \to \infty$. $\square$

4.6 Abundance of Integrable Sets

In this section, let (Ω, L, I) be a complete integration space.

Let X be any function defined on a subset of Ω and let t be a real number. Recall from the notations and conventions described in the Introduction that we use the abbreviation $(t \leq X)$ for the subset $\{\omega \in domain(X) : t \leq X(\omega)\}$. Similar notations are used for $(X < t)$, $(X \leq t)$, and $(X < t)$. We will also write $(t < X \leq u)$ and similar for the intersection $(t < X)(X \leq u)$ and similar. Recall also the definition of the metric complement J_c of a subset J of a metric space.

In the remainder of this section, let X be an arbitrary but fixed integrable function. We will show that the sets $(t \leq X)$ and $(t < X)$ are integrable sets for each positive t in the metric complement of some countable subset of R. In other words, $(t \leq X)$ and $(t < X)$ are integrable sets for all but countably many $t \in (0, \infty)$.

First define some continuous functions that will serve as surrogates for step functions on R. Specifically, for real numbers s,t with $0 < s < t$, define $g_{s,t}(x) \equiv (t-s)^{-1}(x \wedge t - x \wedge s)$. Then, by Assertion 1 of Proposition 4.3.4 and by linearity, the function $g_{s,t}(X) \equiv (t-s)^{-1}(X \wedge t - X \wedge s)$ is integrable for each $s,t \in R$ with $0 < s < t$. Moreover, $1 \geq g_{t',t} \geq g_{s,s'} \geq 0$ for each $t',t,s,s' \in R$ with $t' < t \leq s < s'$. If we can prove that $\lim_{s \uparrow t} I g_{s,t}(X)$ exists, then we can use the Monotone Convergence Theorem to show that the limit function $\lim_{s \uparrow t} g_{s,t}(X)$ is integrable and is an indicator of $(t \leq X)$, proving that the latter set is integrable.

Classically the existence of $\lim_{s \uparrow t} I g_{s,t}(X)$ is trivial since, for fixed t, the integral $I g_{s,t}(X)$ is nonincreasing in s and bounded from below by 0. A nontrivial constructive proof that the limit exists for all but countably many t's is given by [Bishop and Bridges 1985], who devised a theory of profiles for that purpose. In the following, we give a succinct presentation of this theory.

Definition 4.6.1. Profile system. Let K be a nonempty open interval in R. Let G be a family of continuous functions on R with values in $[0,1]$. Let $t \in K$ and $g \in G$ be arbitrary. If $g = 0$ on $(-\infty, t] \cap K$, then we say that t *precedes* g and write $t \diamond g$. If $g = 1$ on $[t, \infty) \cap K$, then we say that g *precedes* t and write $g \diamond t$.

Let $t,s \in K$ and $g \in G$ be arbitrary with $t < s$. If both $t \diamond g$ and $g \diamond s$, then we say that the function g *separates* the points t and s, and write $t \diamond g \diamond s$. If for each $t,s \in K$ with $t < s$, there exists $g \in G$ such that $t \diamond g \diamond s$, then we say that the family G *separates points* in K.

Suppose the family G separates points in K, and suppose λ is a real-valued function on G that is nondecreasing in the sense that, for each g, g' with $g \leq g'$ on K, we have $\lambda(g) \leq \lambda(g')$. Then we say that (G, λ) is a *profile system* on the interval K. □

Definition 4.6.2. Profile bound. Let (G, λ) be an arbitrary profile system on a nonempty open interval K. Then we say that a closed interval $[t,s] \subset K$ has a positive real number α as a *profile bound*, and write $[t,s] \ll \alpha$, if there exist $t',s' \in K$ and $f,g \in G$ such that (i) $f \diamond t'$, $t' < t \leq s < s'$, $s' \diamond g$ and (ii) $\lambda(f) - \lambda(g) < \alpha$.

Suppose $a,b \in R$ and $a \leq b$ are such that $(a,b) \subset K$. Then we say an open interval $(a,b) \subset K$ has a positive real number α as a *profile bound*, and write $(a,b) \ll \alpha$, if $[t,s] \ll \alpha$ for each closed subinterval $[t,s]$ of (a,b). Note that the open interval (a,b), defined as the set $\{x \in R : a < x < b\}$, can be empty. □

Note that $t \diamond g$ is merely an abbreviation for $1_{[t,\infty)} \geq g$ and $g \diamond t$ is an abbreviation for $g \geq 1_{[t,\infty)}$.

The motivating example of a profile system is when $K \equiv (0,\infty), G \equiv \{g_{s,t} : s,t \in K$ and $0 < s < t\}$, and the function λ is defined on G by $\lambda(g) \equiv Ig(X)$ for each $g \in G$. It can easily be verified that (G,λ) is then a profile system on K.

The next lemma lists some basic properties of a profile system.

Lemma 4.6.3. Basics of a profile system. *Let (G,λ) be an arbitrary profile system on an open interval K in R. Then the following conditions hold:*

1. If $f \diamond t$, $t \leq s$, and $s \diamond g$, then $f \geq g$ and $\lambda(f) \geq \lambda(g)$.

2. If $t \leq s$ and $s \diamond g$, then $t \diamond g$.

3. If $g \diamond t$ and $t \leq s$, then $g \diamond s$.

4. In view of the transitivity in Assertions 2 *and* 3, *we can rewrite, without ambiguity, Condition* (i) *in Definition 4.6.1 as $f \diamond t' < t \leq s < s' \diamond g$.*

5. Suppose $[t,s] \ll \alpha$ and $t_0 < t \leq s < s_0$. Let $\varepsilon > 0$ be arbitrary. Then there exist $t_1,s_1 \in K$ and $f_1,g_1 \in G$ such that (i) $t_0 \diamond f_1 \diamond t_1 < t \leq s < s_1 \diamond g_1 \diamond s_0$, (ii) $\lambda(f_1) - \lambda(g_1) < \alpha$, *and* (iii) $t - \varepsilon < t_1 < t$ *and* $s < s_1 < s + \varepsilon$.

6. Every closed subinterval of K has a finite profile bound.

Proof. We will prove only Assertions 5 and 6, as the rest of the proofs are trivial.

1. Suppose $[t,s] \ll \alpha$ and $t_0 < t \leq s < s_0$. Then, according to Definition 4.6.2, there exist $t',s' \in K$ and $f,g \in G$ such that (i′) $f \diamond t' < t \leq s < s' \diamond g$ and (ii′) $\lambda(f) - \lambda(g) < \alpha$.

Let $\varepsilon > 0$ be arbitrary. Then $t_0 \vee t' \vee (t - \varepsilon) < t$. Hence there exist real numbers t'',t_1 such that

$$t_0 \vee t' \vee (t - \varepsilon) < t'' < t_1 < t. \tag{4.6.1}$$

Since G separates points in K, there exists $f_1 \in G$ such that

$$t_0 < t'' \diamond f_1 \diamond t_1 < t. \tag{4.6.2}$$

Then $f \diamond t' < t'' \diamond f_1$. Hence, in view of Assertion 1, we have $\lambda(f) \geq \lambda(f_1)$. Similarly, we obtain $s'',s_1 \in K$ and $g_1 \in G$ such that

$$s < s_1 < s'' < s_0 \wedge s' \wedge (s + \varepsilon) \tag{4.6.3}$$

and

$$s < s_1 \diamond g_1 \diamond s'' < s_0 \tag{4.6.4}$$

with $\lambda(g_1) \geq \lambda(g)$. Hence $\lambda(f_1) - \lambda(g_1) \leq \lambda(f) - \lambda(g) < \alpha$. Condition (ii) in Assertion 5 is established.

Condition (i) in Assertion 5 follows from relations 4.6.2 and 4.6.4. Condition (iii) in Assertion 5 follows from relations 4.6.1 and 4.6.3. Assertion 5 is proved.

2. Given any interval $[t,s] \subset K$, let t'',t',s',s'' be members of K such that $t'' < t' < t \leq s < s' < s''$. Since G separates points in K, there exist $f,g \in G$ such that $t'' \diamond f \diamond t' < t \leq s < s' \diamond g \diamond s''$. Take any real number α such that $\lambda(f) - \lambda(g) < \alpha$. Then $[t,s] \ll \alpha$. Assertion 6 is proved. □

Lemma 4.6.4. Subintervals with small profile bounds of a given interval with a finite profile bound. *Let (G,λ) be a profile system on a proper open interval K in R. Let $[a,b]$ be a closed subinterval of K with $[a,b] \ll \alpha$. Let $\varepsilon > 0$ be arbitrary. Let q be an arbitrary integer with $q \geq \alpha\varepsilon^{-1}$.*

Then there exists a sequence $s_0 = a \leq s_1 \leq \ldots \leq s_q = b$ of points in the interval $[a,b]$ such that $(s_{k-1},s_k) \ll \varepsilon$ for each $k = 1,\ldots,q$.

Proof. 1. As an abbreviation, write $d_n \equiv 2^{-n}(b-a)$ for each $n \geq 1$. By hypothesis, $[a,b] \ll \alpha$. Hence there exist $a',b' \in K$ and $f',f'' \in G$ such that (i) $f' \diamond a' < a \leq b < b' \diamond f''$ and (ii) $\lambda(f') - \lambda(f'') < \alpha \leq q\varepsilon$.

2. Let $n \geq 1$ be arbitrary. For each $i = 0,\ldots,2^n$, define

$$t_{n,i} \equiv a + id_n.$$

Thus $t_{n,0} = a$ and $t_{n,2^n} = b$. Define

$$D_n \equiv \{t_{n,i} : 0 \leq i \leq 2^n\}.$$

Then, for each $i = 0,\ldots,2^n$, we have

$$t_{n,i} \equiv a + id_n = a + 2id_{n+1} \equiv t_{n+1,2i} \in D_{n+1}.$$

Hence $D_n \subset D_{n+1}$.

3. Let $n \geq 1$ be arbitrary. Let $i = 1,\ldots,2^n$ be arbitrary. Then we have $t_{n,i-1}, t_{n,i} \in [a,b] \subset K$, with $t_{n,i-1} < t_{n,i}$. By hypothesis, the pair (G,λ) is a profile system on the interval K. Hence the family G of functions separates points in K. Therefore there exists a function $f_{n,i} \in G$ with $t_{n,i-1} \diamond f_{n,i} \diamond t_{n,i}$.

In addition, define $f_{n,0} \equiv f'$ and $f_{n,2^n+1} \equiv f''$. Then, according to Condition (i) in Step 1, we have $f_{n,0} \diamond t_{n,0} = a$ and $b = t_{n,2^n} \diamond f_{n,2^n+1}$. Combining with Step 3, we obtain

$$f_{n,0} \diamond t_{n,0} \diamond f_{n,1} \diamond t_{n,1} \cdots \diamond f_{n,2^n} \diamond t_{n,2^n} \diamond f_{n,2^n+1}. \tag{4.6.5}$$

By Assertion 1 of Lemma 4.6.3, relation 4.6.5 implies that

$$\lambda(f_{n,0}) \geq \lambda(f_{n,1}) \geq \cdots \geq \lambda(f_{n,2^n}) \geq \lambda(f_{n,2^n+1}). \tag{4.6.6}$$

4. Next, let $t = t_{n,i} \in D_n$ and $s \equiv t_{n+1,j} \in D_{n+1}$ be arbitrary with $s \leq t - d_n$. Then $t > a$. Hence $i \geq 1$ and $s \leq t_{n,i-1}$. Moreover,

$$f_{n+1,j} \diamond t_{n+1,j} = s \leq t_{n,i-1} \diamond f_{n,i}.$$

Therefore, by Assertion 1 of Lemma 4.6.3, we obtain

$$\lambda(f_{n+1,j}) \geq \lambda(f_{n,i}), \tag{4.6.7}$$

where $t = t_{n,i} \in D_n$ and $s \equiv t_{n+1,j} \in D_{n+1}$ are arbitrary with $s \leq t - d_n$.

5. Similarly, let $t = t_{n,i} \in D_n$ and $s \equiv t_{n+1,j} \in D_{n+1}$ be arbitrary with $s \geq t + d_{n+1}$. Then $s > a$. Hence $j \geq 1$ and $t_{n+1,j-1} \geq t$. Hence

$$f_{n,i} \diamond t_{n,i} = t \leq t_{n+1,j-1} \diamond f_{n+1,j}.$$

Therefore, by Assertion 1 of Lemma 4.6.3, we obtain

$$\lambda(f_{n,i}) \geq \lambda(f_{n+1,j}), \tag{4.6.8}$$

for each $t = t_{n,i} \in D_n$ and $s \equiv t_{n+1,j} \in D_{n+1}$ with $s \geq t + d_{n+1}$.

6. For each $n \geq 1$ and for each $i = 0, \ldots, 2^n$, write, as an abbreviation,

$$S_{n,i} \equiv \lambda(f') - \lambda(f_{n,i}).$$

In view of Condition (ii) in Step 1, there exists $\varepsilon' > 0$ such that

$$q\varepsilon > q\varepsilon' > \lambda(f') - \lambda(f'') \tag{4.6.9}$$

and such that

$$|\varepsilon' - S_{n,i}k^{-1}| > 0 \tag{4.6.10}$$

for each $k = 1, \ldots, q$, for each $i = 0, \ldots, 2^n$, for each $n \geq 1$.

7. Let $n \geq 1$ be arbitrary. Note that

$$S_{n,0} \equiv \lambda(f') - \lambda(f_{n,0}) \equiv \lambda(f') - \lambda(f') = 0.$$

and

$$S_{n,2^n} \equiv \lambda(f') - \lambda(f_{n,2^n}) \leq \lambda(f') - \lambda(f'') < q\varepsilon'.$$

Hence inequality 4.6.6 implies that

$$0 = S_{n,0} \leq S_{n,1} \leq \cdots \leq S_{n,2^n} < q\varepsilon'. \tag{4.6.11}$$

8. Now let $k = 1, \ldots, q$ be arbitrary. Let $n \geq 1$ be arbitrary. Note from inequality 4.6.10 that $|k\varepsilon' - S_{n,i}| > 0$ for each $i = 0, \ldots, 2^n$. Therefore, from inequality 4.6.11, we see that there exists a largest integer $i_{n,k} = 0, \ldots, 2^n$ such that

$$S_{n,i(n,k)} < k\varepsilon'. \tag{4.6.12}$$

In other words, $i_{n,k}$ is the largest integer among $0, \ldots, 2^n$ such that

$$\lambda(f') - \lambda(f_{n,i(n,k)}) < k\varepsilon'.$$

Define

$$s_{n,k} \equiv t_{n,i(n,k)} \in [a,b].$$

Note that $i_{n,q} = 2^n$ by inequality 4.6.11. Hence

$$s_{n,q} \equiv t_{n,2^n} = b.$$

For convenience, define $i_{n,0} \equiv 0$.

9. Suppose $k \leq q - 1$. Then $i_{n,k+1}$ is, by definition, the largest integer among $0, \ldots, 2^n$ such that $S_{n,i(n,k+1)} < (k+1)\varepsilon'$. At the same time, by inequality 4.6.12, we have

$$S_{n,i(n,k)} < k\varepsilon' < (k+1)\varepsilon'.$$

Hence $i_{n,k} \leq i_{n,k+1}$. Consequently,

$$s_{n,k} \equiv t_{n,i(n,k)} \leq t_{n,i(n,k+1)} \equiv s_{n,k+1}, \tag{4.6.13}$$

provided that $k < q$.

10. With $k = 1, \ldots, q$ arbitrary but fixed until further notice, we will show that the sequence $(s_{n,k})_{n=1,2,\ldots}$ converges as $n \to \infty$. To that end, let $n \geq 1$ be arbitrary, and proceed to estimate $|s_{n,k} - s_{n+1,k}|$. As an abbreviation, write $i \equiv i_{n,k}$ and $j \equiv i_{n+1,k}$.

11. First suppose, for the sake of a contradiction, that $s_{n+1,k} - s_{n,k} > d_n$. Then

$$t_{n,i} + d_n \equiv s_{n,k} + d_n < s_{n+1,k} \equiv t_{n+1,j} \leq b.$$

Hence $i \leq 2^n - 1$ and

$$t_{n,i+1} = t_{n,i} + d_n < t_{n+1,j}.$$

Therefore, since $t_{n,i+1}, t_{n+1,j} \in D_{n+1}$, we obtain

$$t_{n,i+1} + d_{n+1} \leq t_{n+1,j}.$$

Inequality 4.6.8, where i is replaced by $i + 1$, therefore implies that

$$\lambda(f_{n,i+1}) \geq \lambda(f_{n+1,j}). \tag{4.6.14}$$

At the same time, by the definitions of $i_{n,k}$ and $i_{n+1,k}$, we have

$$S_{n+1,j} < k\varepsilon' \leq S_{n,i+1},$$

whence

$$\lambda(f_{n+1,j}) > \lambda(f_{n,i+1}),$$

contradicting inequality 4.6.14. We conclude that

$$s_{n+1,k} - s_{n,k} \leq d_n. \tag{4.6.15}$$

12. Now suppose, again for the sake of a contradiction, that $s_{n,k} - s_{n+1,k} > d_n + d_{n+1}$. Then

$$t_{n+1,j} + d_{n+1} \equiv s_{n+1,k} + d_{n+1} < s_{n,k} \equiv t_{n,i} \leq b.$$

Hence $j \leq 2^{n+1} - 1$ and

$$t_{n+1,j+1} + d_n = t_{n+1,j} + d_{n+1} + d_n = s_{n+1,k} + d_{n+1} + d_n < s_{n,k} = t_{n,i}.$$

Inequality 4.6.7, where j is replaced by $j+1$, therefore implies that

$$\lambda(f_{n+1,j+1}) \geq \lambda(f_{n,i}). \tag{4.6.16}$$

At the same time, by the definition of $i_{n,k}$ and $i_{n+1,k}$, we have

$$S_{n,i} < k\varepsilon' \leq S_{n+1,j+1},$$

whence

$$\lambda(f_{n,i}) > \lambda(f_{n+1,j+1}),$$

contradicting inequality 4.6.16. We conclude that

$$s_{n,k} - s_{n+1,k} \leq d_n + d_{n+1}. \tag{4.6.17}$$

13. Combining inequalities 4.6.15 and 4.6.17, we obtain

$$|s_{n,k} - s_{n+1,k}| \leq d_n + d_{n+1} \equiv (2^{-n} + 2^{-n-1})(b-a) < 2^{-n+1}(b-a).$$

Hence, for each $p \geq n$, we can inductively obtain

$$\begin{aligned} |s_{n,k} - s_{p,k}| &= |(s_{n,k} - s_{n+1,k}) + (s_{n+1,k} - s_{n+2,k}) + \cdots + (s_{p-1,k} - s_{p,k})| \\ &\leq (2^{-n+1} + 2^{-n} + \ldots)(b-a) = 2^{-n+2}(b-a) = 4d_n. \end{aligned} \tag{4.6.18}$$

Thus we see that the sequence $(s_{n,k})_{n=1,2,\ldots}$ is Cauchy, and converges to some $s_k \in [a,b]$ as $n \to \infty$. Letting $p \to \infty$ in inequality 4.6.18, we obtain

$$|s_{n,k} - s_k| \leq 4d_n, \tag{4.6.19}$$

where $n \geq 1$ is arbitrary.

14. For ease of notations, we will also define $s_{n,0} \equiv s_0 \equiv a$. Then inequality 4.6.13 implies that

$$a \equiv s_{n,0} \leq s_{n,1} \leq \cdots \leq s_{n,q} = b. \tag{4.6.20}$$

Letting $n \to \infty$, we obtain

$$a = s_0 \leq s_1 \leq \cdots \leq s_q = b. \tag{4.6.21}$$

15. Now let $k = 1, \ldots, q$ be arbitrary. Suppose $[u,v] \subset (s_{k-1}, s_k)$ for some real numbers $u \leq v$. We will show that $[u,v] \ll \varepsilon$. To that end, let $n \geq 1$ be so large that

$$s_{k-1} + 5d_n < u \leq v < s_k - 5d_n.$$

This implies, in view of inequality 4.6.19, that

$$s_{n,k-1} + d_n \leq s_{k-1} + 5d_n < u \leq v < s_k - 5d_n < s_{n,k} - d_n. \tag{4.6.22}$$

As abbreviations, write $i \equiv i_{n,k-1}$ and $j \equiv i_{n,k}$. Then $t_{n,i} \equiv s_{n,k-1} < s_{n,k} \leq b$. Hence $i_{n,k-1} \equiv i < 2^n$.

By the definitions of the integers $i_{n,k-1}$ and $i_{n,k}$, we then have

$$(k-1)\varepsilon' \leq S_{n,i(n,k-1)+1} \tag{4.6.23}$$

and

$$S_{n,i(n,k)} < k\varepsilon'. \tag{4.6.24}$$

Consequently,

$$t_{n,i+1} = t_{n,i} + d_n = s_{n,k-1} + d_n < u, \tag{4.6.25}$$

where the inequality is by inequality 4.6.22. At the same time,

$$v < s_{n,k} - d_n = t_{n,j} - d_n = t_{n,j-1}, \tag{4.6.26}$$

where the inequality is by inequality 4.6.22.

Combining inequalities 4.6.25 and 4.6.26, we obtain

$$f_{n,i+1} \diamond t_{n,i+1} < u \leq v < t_{n,j-1} \diamond f_{n,j}. \tag{4.6.27}$$

Furthermore, inequalities 4.6.23 and 4.6.24 together imply that

$$S_{n,j} - S_{n,i+1} < k\varepsilon' - (k-1)\varepsilon' = \varepsilon'.$$

Equivalently,

$$\lambda(f_{n,i+1}) - \lambda(f_{n,j}) < \varepsilon'.$$

Thus Conditions (i) and (ii) in Definition 4.6.2, where $[t,s]$, t',s',f,g, and α are replaced by $[u,v]$, $t_{n,i+1}, t_{n,j-1}, f_{n,i+1}, f_{n,j}$, and ε', respectively, are satisfied. Accordingly, $[u,v] \ll \varepsilon' < \varepsilon$.

Since $[u,v]$ is an arbitrary closed subinterval of (s_{k-1}, s_k), we have established that $(s_{k-1}, s_k) \ll \varepsilon$ according to Definition 4.6.1, where $k = 1, \ldots, q$ is arbitrary. This, together with inequality 4.6.21, prove the lemma. □

Theorem 4.6.5. All but countably many points have an arbitrarily low profile bound. *Let (G,λ) be an arbitrary profile system on a proper open interval K. Then there exists a countable subset J of K such that for each $t \in K \cap J_c$, we have $[t,t] \ll \varepsilon$ for arbitrarily small $\varepsilon > 0$.*

Proof. Let $[a,b] \subset [a_2,b_2] \subset \cdots$ be a sequence of subintervals of K such that $K = \bigcup_{p=1}^{\infty}[a_p,b_p]$. Let $p \geq 1$ be arbitrary. Then $[a_p,b_p] \ll \alpha_p$ for some real number α_p, according to Assertion 6 of Lemma 4.6.3. Now let q_p be an arbitrary integer with $q_p \geq \alpha_p p$. Then, according to Lemma 4.6.4, there exists a finite sequence

$$s_0^{(p)} = a_p \leq s_1^{(p)} \leq \cdots \leq s_{q(p)}^{(p)} = b_p \tag{4.6.28}$$

such that $(s_{k-1}^{(p)}, s_k^{(p)}) \ll \frac{1}{p}$ for each $k = 1, \ldots, q_p$.

Define $J \equiv \{s_k^{(p)} : 1 \leq k \leq q_p; p \geq 1\}$. Suppose $t \in K \cap J_c$. Let $\varepsilon > 0$ be arbitrary. Let $p \geq 1$ be so large that $t \in [a_p, b_p]$ and that $\frac{1}{p} < \varepsilon$. By the definition of the metric complement J_c, we have $|t - s_k^{(p)}| > 0$ for each $k = 1, \ldots, q_p$. Hence, by inequality 4.6.28, we have $t \in (s_{k-1}^{(p)}, s_k^{(p)})$ for some $k = 1, \ldots, q_p$. Since $(s_{k-1}^{(p)}, s_k^{(p)}) \ll \frac{1}{p}$, we have $[t,t] \ll \frac{1}{p} < \varepsilon$ according to Definition 4.6.2. □

An immediate application of Theorem 4.6.4 is to establish the abundance of integrable sets, in the following main theorem of this section.

Theorem 4.6.6. Abundance of integrable sets. *Given an integrable function X on the complete integration space (Ω, L, I), there exists a countable subset J of $(0,\infty)$ such that for each positive real number t in the metric complement J_c of J, the following conditions hold:*

1. The sets $(t \leq X)$ and $(t < X)$ are integrable sets, with $(t \leq X)^c = (X < t)$ and $(t < X)^c = (X \leq t)$.

2. The measures $\mu(t \leq X)$ and $\mu(t < X)$ are equal and are continuous at each $t > 0$ with $t \in J_c$.

Proof. 1. Let $K \equiv (0,\infty)$. Define the family of functions

$$G \equiv \{g_{s,t} : s,t \in K \text{ and } 0 < s < t\},$$

where $g_{s,t}$ denotes the function defined on R by $g_{s,t}(x) \equiv (t-s)^{-1}(x \wedge t - x \wedge s)$ for each $x \in R$. Define the real-valued function λ on G by

$$\lambda(g) \equiv Ig(X)$$

for each $g \in G$.

For each $t,s \in K$ with $t < s$, we have $t \diamond g_{s,t} \diamond s$ in the sense of Definition 4.6.1. In other words, the family G separates points in K. As observed in the beginning of this section, the function $g_{s,t}(X) \equiv (t-s)^{-1}(X \wedge t - X \wedge s)$ is integrable, for each $s,t \in R$ with $0 < s < t$, by Assertion 1 of Proposition 4.3.4 and by linearity. Hence the function λ is welldefined. Moreover, by Assertion 4 of Proposition 4.3.4, for each g, g' with $g \leq g'$ on K, we have $\lambda(g) \leq \lambda(g')$. Thus the function λ is is nondecreasing. Summing up, the couple (G,λ) is a profile system according to Definition 4.6.1.

2. Let the countable subset J of K be constructed for the profile system (G,λ) as in Theorem 4.6.5.

3. Let $t \in K \cap J_c$ be arbitrary. Then, by Theorem 4.6.5, we have $[t,t] \ll \frac{1}{p}$ for each $p \geq 1$. Recursively applying Assertion 5 of Lemma 4.6.3, we can construct two sequences $(u_p)_{p=0,1,\ldots}$ and $(v_p)_{p=0,1,\ldots}$ in K, and two sequences $(f_p)_{p=1,2,\ldots}$ and $(g_p)_{p=1,2,\ldots}$ in G, such that for each $p \geq 1$ we have (i) $u_{p-1} \diamond f_p \diamond u_p < t < v_p \diamond g_p \diamond v_{p-1}$, (ii) $\lambda(f_p) - \lambda(g_p) < \frac{1}{p}$, and (iii) $t - \frac{1}{p} < u_p < v_p < t + \frac{1}{p}$. Then Condition (i) implies that $f_p \diamond u_p \diamond f_{p+1}$ for each $p > 1$. Hence, by Assertion 1 of Lemma 4.6.3, we have $f_p \geq f_{p+1}$ for each $p \geq 1$.

4. Consider $p,q \geq 1$. We have $f_q \diamond u_q < t < v_p \diamond g_p$. Hence

$$\lambda(f_q) \geq \lambda(g_p) > \lambda(f_p) - \frac{1}{p}, \tag{4.6.29}$$

where the second inequality is by Condition (ii). By symmetry, we also have $\lambda(f_p) > \lambda(f_q) - \frac{1}{q}$. Combining, we obtain $|\lambda(f_q) - \lambda(f_p)| < \frac{1}{p} + \frac{1}{q}$. Hence $(\lambda(f_p))_{p=1,2,...}$ is a Cauchy sequence and converges. Similarly, $(\lambda(g_p))_{p=1,2,...}$ converges. In view of Condition (ii), the two limits are equal.

5. By the definition of λ, we see that $\lim_{p\to\infty} If_p(X) \equiv \lim_{p\to\infty} \lambda(f_p)$ exists. Since $f_p \geq f_{p+1}$ for each $p > 1$, as proved in Step 3, the Monotone Convergence Theorem (Theorem 4.4.8) implies that $Y \equiv \lim_{p\to\infty} f_p(X)$ is an integrable function, with $If_p(X) \downarrow IY$. Likewise, $Z \equiv \lim_{p\to\infty} g_p(X)$ is an integrable function, with $Ig_p(X) \uparrow IZ$. Furthermore,

$$I|Y - Z| = IY - IZ = \lim_{p\to\infty}(\lambda(f_p) - \lambda(g_p)) = 0, \tag{4.6.30}$$

where the last equality is thanks to equality 4.6.29. Hence, according to Assertion 4 of Proposition 4.5.3, we have $Y = Z$ a.e.

6. We next show that Y is an indicator with $(Y = 1) = (t \leq X)$. To that end, consider each $\omega \in domain(Y)$. Suppose $Y(\omega) > 0$. Let $p \geq 2$ be arbitrary. Then $\omega \in domain(X)$ and $f_p(X(\omega)) \geq Y(\omega) > 0$. Condition (i) in Step 3 implies that $u_{p-1} \diamond f_p$. Therefore we have, according to Definition 4.6.1, $f_p(u) = 0$ for each $u \in (-\infty, u_{p-1}] \cap K$. Since $f_p(X(\omega)) > 0$, we infer that $X(\omega) \in [u_{p-1}, \infty)$. At the same time, Condition (i) in Step 3 implies that $f_{p-1} \diamond u_{p-1}$. Combining, $f_{p-1}(X(\omega)) = 1$, where $p \geq 2$ is arbitrary. Letting $p \to \infty$, we conclude that $X(\omega) \in [t, \infty)$ and $Y(\omega) = 1$, assuming $Y(\omega) > 0$. In particular, Y can have only two possible values, 0 or 1. Thus Y is an indicator. We have also seen that

$$(Y = 1) \subset (Y > 0) \subset (t \leq X). \tag{4.6.31}$$

7. Conversely, consider each $\omega \in domain(X)$ such that $t \leq X(\omega)$. Let $p \geq 1$ be arbitrary. Condition (i) in Step 3 implies that $f_p \diamond t$. Therefore we have, according to Definition 4.6.1, $f_p(u) = 1$ for each $u \in [t, \infty) \cap K$. Combining, we conclude that $f_p(X(\omega)) = 1$, where $p \geq 1$ is arbitrary. It follows that $\lim_{p\to\infty} f_p(X(\omega)) = 1$. Thus $\omega \in domain(Y)$ and $Y(\omega) = 1$. Thus $(t \leq X) \subset (Y = 1)$.

8. Combined with relation 4.6.31, this implies that $(t \leq X) = (Y = 1)$. In other words, the set $(X \geq t)$ has the integrable indicator Y as an indicator and is therefore an integrable set,with $(X \geq t)^c \equiv (Y = 0)$. Now consider each $\omega \in (Y = 0)$. Then $\lim_{p\to\infty} f_p(X(\omega)) = Y(\omega) = 0$. Hence $X(\omega) < t$, and so $\omega \in (X < t)$. Conversely, consider each $\omega \in (X < t)$. Then $Y(\omega) = \lim_{p\to\infty} f_p(X(\omega)) = 0$. Hence $\omega \in (Y = 0)$. Combining, we obtain

$$(X \geq t)^c \equiv (Y = 0) = (X < t).$$

Summing up, the set $(t \leq X)$ is an integrable set, with Y as an integrable indicator and with $(t \leq X)^c = (X < t)$.

9. Similarly, we can prove that the set $(t < X)$ is an integrable set, with Z as an integrable indicator and with

$$(X > t)^c \equiv (Z = 0) = (X \leq t).$$

$(t < X)^c = (X \leq t)$. Thus Assertion 1 of the theorem is proved.

10. To prove Assertion 2, first note that

$$\mu(t \leq X) = IY = IZ = \mu(t < X),$$

where the second equality follows from equality 4.6.30.

11. It remains for us to show that $\mu(t \leq X)$ is continuous at t. To that end, let $p > 1$ be arbitrary. Recall Conditions (i–iii) in Step 3 – to wit, (i) $u_{p-1} \diamondsuit f_p \diamondsuit u_p < t < v_p \diamondsuit g_p \diamondsuit v_{p-1}$, (ii) $\lambda(f_p) - \lambda(g_p) < \frac{1}{p}$, and (iii) $t - \frac{1}{p} < u_p < v_p < t + \frac{1}{p}$. By the monotone convergence in Step 5, we have

$$\lambda(f_p) \equiv If_p(X) \geq \mu(t \leq X) \geq Ig_p(X) \equiv \lambda(g_p).$$

Hence

$$\lambda(f_p) - \mu(t \leq X) \leq \lambda(f_p) - \lambda(g_p) \leq \frac{1}{p}.$$

12. With $p > 1$ arbitrary but fixed, consider any $t' \in K \cap J_c$ such that $t' \in (u_p, v_p)$. Similarly to Step 3, by recursively applying Assertion 5 of Lemma 4.6.3, we can construct two sequences $(u'_q)_{q=0,1,\ldots}$ and $(v'_q)_{q=0,1,\ldots}$ in K, and two sequences $(f'_q)_{q=1,2,\ldots}$ and $(g'_q)_{q=1,2,\ldots}$ in G, such that for each $q \geq 1$ we have (i) $u'_{q-1} \diamondsuit f'_q \diamondsuit u'_q < t' < v'_q \diamondsuit g'_q \diamondsuit v'_{q-1}$, (ii) $\lambda(f'_q) - \lambda(g'_q) < \frac{1}{q}$, and (iii) $t' - \frac{1}{q} < u'_q < v'_q < t' + \frac{1}{q}$. Then Condition (i) implies that $f'_q \diamondsuit u'_q \diamondsuit f'_{q+1}$ for each $q > 1$.

13. Let $\overline{q} \geq 1$ be so large that $u'_{q-1}, v'_{q-1} \in (u_p, v_p)$ for each $q \geq \overline{q}$. Consider each $q \geq \overline{q}$. Because $u'_{q-1}, v'_{q-1} \in (u_p, v_p)$, Condition (i') can be extended to

$$f_p \diamondsuit u_p < u'_{q-1} \diamondsuit f'_q \diamondsuit u'_q < t' < v'_q \diamondsuit g'_q \diamondsuit v'_{q-1} < v_p \diamondsuit g_p.$$

It follows that

$$0 \leq \lambda(f_p) - \lambda(f'_q) \leq \lambda(f_p) - \lambda(g_p) \leq \frac{1}{p}.$$

By Step 5, $(t \leq X)$ is an integrable set, with

$$0 \leq \lambda(f_p) - \mu(t \leq X) \leq \frac{1}{p}.$$

Similarly, $(t' \leq X)$ is an integrable set, with

$$0 \leq \lambda(f'_q) - \mu(t' \leq X) \leq \frac{1}{q}.$$

Combining the three last displayed inequalities, we obtain

$$|\mu(t \le X) - \mu(t' \le X)| \le \frac{2}{p} + \frac{1}{q}.$$

Since $q \ge \overline{q}$ is arbitrarily large, the last displayed inequality implies

$$|\mu(t \le X) - \mu(t' \le X)| \le \frac{2}{p},$$

where $t' \in (u_p, v_p) \cap J_c$ is arbitrary. Since $p \ge 1$ is arbitrary, continuity of $\mu(t \le X)$ at t has thus been established. □

Corollary 4.6.7. Abundance of integrable sets. *Let X be an arbitrary integrable function. There exists a countable subset J of R such that for each t in the metric complement J_c of J, the following conditions hold:*

1. If $t > 0$, then the sets $(t < X)$ and $(t \le X)$ are integrable, with equal measures that are continuous at t.

2. If $t < 0$, then the sets $(X < t)$ and $(X \le t)$ are integrable, with equal measures that are continuous at t.

Proof. Assertion 1 is a restatement of Theorem 4.6.6. To prove Assertion 2, apply Assertion 1 to X and $-X$. □

Definition 4.6.8. Regular and continuity points of an integrable function relative to an integrable set. Let X be an integrable function, let A be an integrable set, and let $t \in R$.

1. We say that t is a *regular point* of X relative to A if (i) there exists a sequence $(s_n)_{n=1,2,...}$ of real numbers decreasing to t such that $(s_n < X)A$ is integrable for each $n \ge 1$ and such that $\lim_{n\to\infty} \mu(s_n < X)A$ exists, and (ii) there exists a sequence $(r_n)_{n=1,2,...}$ of real numbers increasing to t such that $(r_n < X)A$ is integrable for each $n \ge 1$ and such that $\lim_{n\to\infty} \mu(r_n < X)A$ exists.

2. If, in addition, the two limits in (i) and (ii) are equal, then we call t a continuity point of X relative to A.

3. We say that a positive real number $t > 0$ is a regular point of X if Conditions (i) and (ii), with A replaced by Ω, are satisfied. We say that a negative real number $t < 0$ is a regular point of X if $-t$ is a regular point of $-X$. □

Corollary 4.6.9. Simple properties of regular and continuity points. *Let X be an integrable function, let A be an integrable set, and let t be a regular point of X relative to A. Then the following conditions hold:*

1. If u is a regular point of X, then u is a regular point for X relative to any integrable set B. If u is a continuity point of X, then u is a continuity point for X relative to any integrable set B.

2. All but countably many positive real numbers are continuity points of X.

3. All but countably many real numbers are continuity points of X relative to A. Hence all but countably many real numbers are regular points of X relative to A.

4. The sets $A(t < X), A(t \leq X), A(X < t)$, $A(X \leq t)$, *and* $A(X = t)$ *are integrable sets.*

5. $(X \leq t)A = A((t < X)A)^c$ *and* $(t < X)A = A((X \leq t)A)^c$.

6. $(X < t)A = A((t \leq X)A)^c$ *and* $(t \leq X)A = A((X < t)A)^c$.

7. For a.e. $\omega \in A$, *we have* $t < X(\omega)$, $t = X(\omega)$, *or* $t > X(\omega)$. *Thus we have a limited, but very useful, version of the principle of excluded middle.*

8. Let $\varepsilon > 0$ *be arbitrary. There exists* $\delta > 0$ *such that if* $r \in (t - \delta, t]$ *and* $A(X < r)$ *is integrable, then* $\mu(A(X < t)) - \mu(A(X < r)) < \varepsilon$. *There exists* $\delta > 0$ *such that if* $s \in [t, t + \delta)$ *and* $A(X \leq s)$ *is integrable, then* $\mu(A(X \leq s)) - \mu(A(X \leq t)) < \varepsilon$.

9. If t *is a continuity point of* X *relative to* A, *then* $\mu((t < X)A) = \mu((t \leq X)A)$.

Proof. 1. Suppose $u > 0$ and u is a regular point of X. Then by Definition 4.6.8, (i′) there exists a sequence $(s_n)_{n=1,2,\ldots}$ of real numbers decreasing to u such that $(s_n < X)$ is integrable and $\lim_{n\to\infty} \mu(s_n < X)$ exists and (ii′) there exists a sequence $(r_n)_{n=1,2,\ldots}$ of positive real numbers increasing to u such that $(r_n < X)$ is integrable and $\lim_{n\to\infty} \mu(r_n < X)$ exists. Now let B be any integrable set. Then for all $m > n \geq 1$, we have

$$\begin{aligned} 0 \leq \mu((s_m < X)B) - \mu((s_n < X)B) &= \mu(B(s_m < X)(s_n < X)^c) \\ &\leq \mu((s_m < X)(s_n < X)^c) = \mu(s_m < X) - \mu(s_n < X) \downarrow 0 \end{aligned}$$

as $n \to \infty$. Therefore $(\mu((s_n < X)B))_{n=1,2,\ldots}$ is a Cauchy sequence and converges, verifying Condition (i) of Definition 4.6.8. Condition (ii) of Definition 4.6.8 is similarly verified. Hence u is a regular point of X relative to B. Suppose, in addition, that u is a continuity point of X. Then for each $n \geq 1$, we have

$$\begin{aligned} 0 \leq \mu((r_n < X)B) - \mu((s_n < X)B) &= \mu(B(r_n < X)(s_n < X)^c) \\ &\leq \mu((r_n < X)(s_n < X)^c) = \mu(r_n < X) - \mu(s_n < X) \downarrow 0. \end{aligned}$$

Therefore the two sequences $(\mu((r_n < X)B))_{n=1,2,\ldots}$ and $(\mu((s_n < X)B))_{n=1,2,\ldots}$ have the same limit. Thus t is a continuity point of X relative to B, according to Definition 4.6.8. Thus Assertion 1 is proved for the case $t > 0$. The case $t < 0$ is similar. Assertion 1 is proved.

2. By Corollary 4.6.7, there exists a countable subset J of R such that for each $t > 0$ in the metric complement J_c of J, the sets $(t < X)$ and $(t \leq X)$ are integrable, with equal measures that are continuous at t. Consider each $t \in (0, \infty)J_c$. Take any sequence $(s_n)_{n=1,2,\ldots}$ in $(0, \infty)J_c$ such that $s_n \downarrow t$. Then $(s_n < X)$ is integrable for each $n \geq 1$, with $\mu(s_n < X) \downarrow \mu(t < X)$ by continuity. Similarly, we can take a sequence $(r_n)_{n=1,2,\ldots}$ in $(0, \infty)J_c$ such that $r_n \uparrow t$. Then $(r_n < X)$ is integrable for each $n \geq 1$, with $\mu(r_n < X) \uparrow \mu(t < X)$ by continuity. Combining, we see that the conditions in Definition 4.6.8 are satisfied for t to be a continuity points of X. Assertion 2 is proved.

3. By Assertion 2, there exists a countable subset J of R such that each $t \in (0,\infty)J_c$ is a continuity point of both X and $-X$. There is no loss of generality in assuming that $0 \in J$. Consider each $r \in (0,\infty)J_c$. Then r is a continuity point of X. Hence, by Assertion 1, r is a continuity point of X relative to A. Now consider each $r \in (-\infty,0)J_c$. Then $t \equiv -r \in (0,\infty)J_c$ is a continuity point of $-X$. Hence by Assertion 1, t is a continuity point of $-X$ relative to A. It follows that r is a continuity point of X relative to A. Thus Assertion 3 is verified.

4. By hypothesis, t is a regular point of X relative to the integrable A. Hence there exist two sequences of real numbers $(s_n)_{n=1,2,...}$ and $(r_n)_{n=1,2,...}$ satisfying Conditions (i) and (ii) in Definition 4.6.8. Then $(s_n < X)A$ is an integrable set for $n \geq 1$ and $a = \lim_{n\to\infty} \mu(s_n < X)A$ exists. Since (s_n) decreases to t, we have

$$A(t < X) = \bigcup_{n=1}^{\infty} A(s_n < X).$$

Hence the Monotone Convergence Theorem implies that the union $A(t < X)$ is integrable, with

$$\mu(A(t < X)) = a. \tag{4.6.32}$$

Consequently, the set

$$A(X \leq t) = A((t < X)A)^c$$

is integrable by Proposition 4.5.7.

Similarly, the set $(r_n < X)A$ is an integrable set for $n \geq 1$, and $b = \lim_{n\to\infty} \mu(r_n < X)A$ exists. Since (r_n) increases to t, we have

$$A(t \leq X) = \bigcap_{n=1}^{\infty} A(r_n < X).$$

Hence the Monotone Convergence Theorem implies that the intersection $A(t \leq X)$ is integrable, with

$$\mu A(t \leq X) = b. \tag{4.6.33}$$

Hence, by Assertion 3 of Proposition 4.5.7, the set

$$A(X < t) = A((t \leq X)A)^c \quad \text{a.e} \tag{4.6.34}$$

is integrable, with $\mu A(X < t) = \mu A - b$.

Likewise, the Monotone Convergence Theorem implies that the intersection

$$A(X = t) = \bigcap_{n=1}^{\infty} A(r_n < X \leq s_n)$$

is integrable, with

$$\mu(A(X = t)) = \lim_{n\to\infty} (\mu(r_n < X) - \mu(s_n < X)) = b - a. \tag{4.6.35}$$

Assertion 4 is proved.

5. Using Proposition 4.5.7, we obtain

$$(X \leq t)A = A((t < X)A)^c$$

and

$$(t < X)A = A((X \leq t)A)^c.$$

This proves Assertion 5.

6. Similar.

7. Combining equalities 4.6.33, 4.6.32, and 4.6.35, we see that

$$\mu A(t \leq X) - \mu A(t < X) = b - a = \mu(A(X = t)).$$

Hence

$$\begin{aligned}
\mu A &= \mu A(t \leq X) + \mu A(X < t) \\
&= \mu A(t < X) + b - a + \mu A(X < t) \\
&= \mu(A(t < X)) + \mu(A(X = t)) + \mu A(X < t) \\
&= \mu(A((t < X) \cup (X = t) \cup (X < t)).
\end{aligned}$$

Since $A \supset A((t < X) \cup (X = t) \cup (X < t))$, the last displayed equality implies that

$$A = A((t < X) \cup (X = t) \cup (X < t)) \quad \text{a.e.}$$

This proves Assertion 7.

8. Let $\varepsilon > 0$ be arbitrary. Let $n \geq 1$ be so large that $\mu((t < X)A) - \mu(s_n < X)A < \varepsilon$. Define $\delta = s_n - t$. Suppose $s \in [t, t + \delta)$ and $A(X \leq s)$ are integrable. Then $s < s_n$ and so

$$\begin{aligned}
&\mu(A(X \leq s)) - \mu(A(X \leq t)) \\
&\quad = (\mu(A) - \mu(s < X)A) - (\mu(A) - \mu((t < X)A)) \\
&\quad = \mu((t < X)A) - \mu(s < X)A \leq \mu((t < X)A) - \mu(s_n < X)A < \varepsilon.
\end{aligned}$$

This proves the second half of Assertion 8, the first half being similar.

9. Suppose t is a continuity point of X relative to A. Then the limits $a = \lim_{n\to\infty} \mu(s_n < X)A$ and $b = \lim_{n\to\infty} \mu(r_n < X)A$ are equal. Hence by equalities 4.6.32 and 4.6.33

$$\mu((t \leq X)A) = b = a = \mu((t < X)A).$$

Assertion 9 and the corollary are proved. □

Theorem 4.6.10. Chebychev's inequality. *Let $X \in L$ be arbitrary. Then the following conditions hold:*

1. (First and common version of Chebychev's inequality.) *If $t > 0$ is a regular point of the integrable function $|X|$, then we have $\mu(|X| > t) \leq t^{-1} I|X|$.*

2. (Second version.) *If* $I|X| < b$ *for some* $b > 0$, *then for each* $s > 0$, *we have* $(|X| > s) \subset B$ *for some integrable set* B *with* $\mu(B) < s^{-1}b$. *This second version of* Chebychev's inequality *is useful when a real number* $s > 0$ *is given without any assurance that the set* $(|X| > s)$ *is integrable.*

Proof. 1. $1_{(|X|>t)} \leq t^{-1}|X|$. Assertion 1 follows.

2. Take an arbitrary regular point t of the integrable function X in the open interval $(b^{-1}I|X|s, s)$. Let $B \equiv (|X| > t)$. By Assertion 1, we then have $\mu(B) \leq t^{-1}I|X| < s^{-1}b$. Moreover, $(|X| > s) \subset (|X| > t) \equiv B$. Assertion 2 is proved. □

Definition 4.6.11. Convention: implicit assumption of regular points of integrable functions. Let X be an integrable function, and let A be an integrable set. Henceforth, if the integrability of the set $(X < t)A$ or $(X \leq t)A$, for some $t \in R$, is required in a discussion, then it is understood that the real number t can be chosen, and has been chosen, from the regular points of the integrable function X relative to the integrable set A.

Likewise, if the integrability of the set $(t < X)$ or $(t \leq X)$, for some $t > 0$, is required in a discussion, then it is understood that the number $t > 0$ has been chosen from the regular points of the integrable function X.

For brevity, we will sometimes write $(X < t; Y \leq s; \ldots)$ for $(X < t)$ $(Y \leq s)\ldots$ □

Recall that $C_{ub}(R)$ is the space of bounded and uniformly continuous functions on R.

Proposition 4.6.12. The product of a bounded continuous function of an integrable function and an integrable indicator is integrable. *Suppose* $X \in L$, A *is an integrable set, and* $f \in C_{ub}(R)$. *Then* $f(X)1_A \in L$. *In particular, if* $X \in L$ *is bounded, then* $X1_A$ *is integrable.*

Proof. 1 By Assertion 3 of Corollary 4.6.9, all but countably many real numbers are regular points of X relative to A. In other words, there exists a countable subset J of R such that each $t \in J_c$ is a regular point of the integrable function X relative to the integrable set A. Here J_c denotes the metric complement of the set J in R.

2. Let $c > 0$ be so large that $|f| \leq c$ on R. Let δ_f be a modulus of continuity of the function f. Let $\varepsilon > 0$ be arbitrary. Since X is integrable, there exists $a > 0$ so large that

$$I|X| - I|X| \wedge (a-1) < \varepsilon.$$

Since f is uniformly continuous, there exists a sequence $-a = t_0 < t_1 < \cdots < t_n = a$ such that $\bigvee_{i=1}^{n}(t_i - t_{i-1}) < \delta_f(\varepsilon)$. Then

$$Y \equiv \sum_{i=1}^{n} f(t_i)1_{(t(i-1)<X\leq t(i))A}$$

is an integrable function, where we recall the convention established in Definition 4.6.11 regarding the choice of regular points. Moreover, since $1_{(|X|>a)A} \leq |X| - |X| \wedge (a-1)$, we have

$$
\begin{aligned}
|f(X)1_A - Y| &\equiv \left| f(X)1_A - \sum_{i=1}^{n} f(t_i)1_{(t(i-1)<X\leq t(i))A} \right| \\
&= \left| \sum_{i=1}^{n} (f(X) - f(t_i))1_{(t(i-1)<X\leq t(i))A} \right| + |f(X)|1_{(|X|>a)A} \\
&\leq \left| \sum_{i=1}^{n} (f(X) - f(t_i))1_{(t(i-1)<X\leq t(i))A} \right| + c1_{(|X|>a)A} \\
&\leq \varepsilon 1_A + c(|X| - |X| \wedge (a-1)),
\end{aligned}
$$

where

$$
\begin{aligned}
&I(\varepsilon 1_A + c(|X| - |X| \wedge (a-1))) \\
&\quad = \varepsilon\mu(A) + c(I|X| - I|X| \wedge (a-1)) < \varepsilon\mu(A) + c\varepsilon \to 0
\end{aligned}
$$

as $\varepsilon \to 0$. Hence, by Theorem 4.5.10, $f(X)1_A \in L$. This proves the first part of the proposition.

3. Now suppose $X \in L$ is bounded. Let $b > 0$ be such that $|X| \leq b$. Define $f \in C_{ub}(R)$ by $f(x) \equiv -b \vee x \wedge b$ for each $x \in R$. Then $X = f(X)$ and so, according to the first part of this proposition, $X1_A \in L$. The second part of the proposition is also proved. □

4.7 Uniform Integrability

In this section, let (Ω, L, I) be a complete integration space.

Proposition 4.7.1. Moduli of integrability for integrable functions. *Let $X \in L$ be arbitrary. Recall Definition 4.6.11. Then the following conditions hold:*

1. Let A be an arbitrary integrable set. Then $X1_A \in L$.

2. $I(|X|1_A) \to 0$ as $\mu(A) \to 0$, where A is an arbitrary integrable set. More precisely, there exists an operation $\delta : (0,\infty) \to (0,\infty)$ such that $I(|X|1_A) \leq \varepsilon$ for each integrable set A with $\mu(A) < \delta(\varepsilon)$, for each $\varepsilon > 0$.

3. $I(|X|1_{(|X|>a)}) \to 0$ as $a \to \infty$. More precisely, suppose $I|X| \leq b$ for some $b > 0$, and let operation δ be as described in Assertion 2. For each $\varepsilon > 0$, if we define $\eta(\varepsilon) \equiv b/\delta(\varepsilon)$, then $I(|X|1_{(|X|>a)}) \leq \varepsilon$ for each $a > \eta(\varepsilon)$.

4. Suppose an operation $\eta > 0$ is such that $I(|X|1_{(|X|>a)}) \leq \varepsilon$ for each $a > \eta(\varepsilon)$, for each $\varepsilon > 0$. Then the operation δ defined by $\delta(\varepsilon) \equiv \frac{\varepsilon}{2}/\eta(\frac{\varepsilon}{2})$ satisfies the conditions in Assertion 2.

Proof. 1. Let $n \geq 1$ be arbitrary. Then $n1_A$ is integrable. Hence by Assertion 1 of Proposition 4.3.4, the function $|X| \wedge (n1_A)$ is integrable. Now consider each $\omega \in domain(X) \cap domain(1_A)$. Suppose

$$|X(\omega)1_A(\omega)| \wedge n \neq |X(\omega)| \wedge (n1_A(\omega)).$$

Then $1_A(\omega) \neq 0$. Hence $1_A(\omega) = 1$. It follows that $|X(\omega)| \wedge n \neq |X(\omega)| \wedge n)$, which is a contradiction. Thus

$$|X(\omega)1_A(\omega)| \wedge n = |X(\omega)| \wedge (n1_A(\omega))$$

for each $\omega \in domain(X) \cap domain(1_A)$. In other words, $|X1_A| \wedge n = |X| \wedge (n1_A)$. We saw earlier that $|X| \wedge (n1_A)$ is integrable. Hence $|X1_A| \wedge n$ is integrable.

Moreover, let $n, p \geq 1$ be arbitrary with $n > p$. Consider each $\omega \in domain(X) \cap domain(1_A)$. Suppose

$$|X(\omega)1_A(\omega)| \wedge n - |X(\omega)1_A(\omega)| \wedge p > |X(\omega)| \wedge n - |X(\omega)| \wedge p.$$

Then $1_A(\omega) \neq 0$. Hence $1_A(\omega) = 1$. It follows that

$$|X(\omega)| \wedge n - |X(\omega)| \wedge p > |X(\omega)| \wedge n - |X(\omega)| \wedge p,$$

which is a contradiction. Thus

$$|X(\omega)1_A(\omega)| \wedge n - |X(\omega)1_A(\omega)| \wedge p \leq |X(\omega)| \wedge n - |X(\omega)| \wedge p$$

for each $\omega \in domain(X) \cap domain(1_A)$. In other words,

$$|X1_A| \wedge n - |X1_A| \wedge p \leq |X| \wedge n - |X| \wedge p.$$

Consequently, since $|X| \in L$, we have

$$I(|X1_A| \wedge n - |X1_A| \wedge p) \leq I(|X| \wedge n - |X| \wedge p) \to 0$$

as $p \to \infty$. Thus $\lim_{n\to\infty} I(|X1_A| \wedge n)$ exists. By the Monotone Convergence Theorem, the limit function $|X1_A| = \lim_{n\to\infty} |X1_A| \wedge n$ is integrable. Similarly, $|X_+1_A|$ is integrable, and so is $X1_A = 2|X_+1_A| - |X1_A|$. Assertion 1 is proved.

2. Let $\varepsilon > 0$ be arbitrary. Since X is integrable, there exists $a > 0$ so large that $I|X| - I(|X| \wedge a) < 2^{-1}\varepsilon$. Define $\delta(\varepsilon) \equiv 2^{-1}\varepsilon/a$. Now let A be an arbitrary integrable set with $\mu(A) < \delta(\varepsilon) \equiv 2^{-1}\varepsilon/a$. Then, since $|X|1_A \leq (|X| - |X| \wedge a)1_A + a1_A$, we have

$$I(|X|1_A) \leq I|X| - I(|X| \wedge a) + a\mu(A) < 2^{-1}\varepsilon + a2^{-1}\varepsilon/a = \varepsilon.$$

Assertion 2 is proved.

3. Take any $b \geq I|X|$. Define $\eta(\varepsilon) \equiv b/\delta(\varepsilon)$, where δ is an operation as in Assertion 2. Let $a > \eta(\varepsilon)$ be arbitrary. Define the integrable set $A \equiv (|X| > a)$. Chebychev's inequality then gives $\mu(A) = \mu(|X| > a) \leq I|X|/a \leq b/a < \delta(\varepsilon)$. Hence $I(1_A|X|) < \varepsilon$ by Assertion 2. Thus Assertion 3 is verified.

4. Suppose an operation $\eta > 0$ is such that $I(|X|1_{(|X|>a)}) \leq \varepsilon$ for each $a > \eta(\varepsilon)$, for each $\varepsilon > 0$. Define the operation δ by $\delta(\varepsilon) \equiv \frac{\varepsilon}{2}/\eta\left(\frac{\varepsilon}{2}\right)$. Suppose an integrable set A is such that $\mu(A) < \delta(\varepsilon) \equiv \frac{\varepsilon}{2}/\eta\left(\frac{\varepsilon}{2}\right)$. Take any $a > \eta\left(\frac{\varepsilon}{2}\right)$. Then

$$
\begin{aligned}
I(|X|1_A) &\leq I(a1_{A(X\leq a)} + 1_{(|X|>a)}|X|) \\
&\leq I(a1_A + 1_{(|X|>a)}|X|) \leq a\mu(A) + \frac{\varepsilon}{2} \leq a\frac{\varepsilon}{2}/\eta\left(\frac{\varepsilon}{2}\right) + \frac{\varepsilon}{2}.
\end{aligned}
$$

By taking a arbitrarily close to $\eta\left(\frac{\varepsilon}{2}\right)$, we obtain

$$
I(|X|1_A) \leq \eta\left(\frac{\varepsilon}{2}\right)\frac{\varepsilon}{2}/\eta\left(\frac{\varepsilon}{2}\right) + \frac{\varepsilon}{2} = \varepsilon.
$$

Assertion 4 and the proposition are proved. □

Note that in the proof for Assertion 4 of Proposition 4.7.1, we use a real number $a > \eta(\varepsilon)$ arbitrarily close to $\eta(\varepsilon)$ rather than simply $a = \eta(\varepsilon)$. This ensures that a can be a regular point of $|X|$, as required in Definition 4.6.11.

Definition 4.7.2. Uniform integrability and simple modulus of integrability. A family G of integrable functions is said to be *uniformly integrable* if for each $\varepsilon > 0$, there exists $\eta(\varepsilon) > 0$ so large that $I(|X|1_{(|X|>a)}) \leq \varepsilon$ for each $a > \eta(\varepsilon)$, for each $X \in G$. The operation η is then called a *simple modulus of integrability* of G.

Proposition 4.7.1 ensures that each family G consisting of finitely many integrable functions is uniformly integrable. □

Proposition 4.7.3. Alternative definition of uniform integrability, and modulus of integrability, in the special case of a probability space. *Suppose the integration space (Ω, L, I) is such that $1 \in L$ and $I1 = 1$. Then a family G of integrable r.r.v.'s is uniformly integrable in the sense of Definition 4.7.2 iff* (i) *there exists $b \geq 0$ such that $I|X| \leq b$ for each $X \in G$, and* (ii) *for each $\varepsilon > 0$, there exists $\delta(\varepsilon) > 0$ such that $I|X|1_A \leq \varepsilon$ for each integrable set A with $\mu(A) < \delta(\varepsilon)$, and for each $X \in G$. The operation δ is then called a* modulus of integrability *of G.*

Proof. 1. First suppose the family G is uniformly integrable. In other words, for each $\varepsilon > 0$, there exists $\eta(\varepsilon)$ such that

$$
I(|X|1_{(|X|>a)}) \leq \varepsilon
$$

for each $a > \eta(\varepsilon)$, and for each $X \in G$. Define $b \equiv \eta(1) + 2$. Let $X \in G$ be arbitrary. Take any $a \in (\eta(1), \eta(1) + 1)$. Then

$$
I|X| = I(1_{(|X|>a)}|X|) + I(1_{(|X|\leq a)}|X|) \leq 1 + aI1 = 1 + a < 1 + \eta(1) + 1 \equiv b,
$$

where the second equality follows from the hypothesis that $I1 = 1$. This verifies Condition (i) of the present proposition. Now let $\varepsilon > 0$ be arbitrary. Define $\delta(\varepsilon) \equiv \frac{\varepsilon}{2}/\eta(\frac{\varepsilon}{2})$. Then Assertion 4 of Proposition 4.7.1 implies that $I(|X|1_A) \leq \varepsilon$ for each integrable set A with $\mu(A) < \delta(\varepsilon)$, for each $X \in G$. This verifies Condition (ii) of the present proposition.

2. Conversely, suppose Conditions (i) and (ii) of the present proposition hold. For each $\varepsilon > 0$, define $\eta(\varepsilon) \equiv b/\delta(\varepsilon)$. Then, according to Assertion 3 of Proposition 4.7.1, we have $I(|X|1_{(|X|>a)}) \leq \varepsilon$ for each $a > \eta(\varepsilon)$. Thus the family G is uniformly integrable in the sense of Definition 4.7.2. □

Proposition 4.7.4. Dominateduniform integrability. *If there exists an integrable function Y such that $|X| \leq Y$ for each X in a family G of integrable functions, then G is uniformly integrable.*

Proof. Note that $b \equiv I|Y|$ satisfies Condition (i) in Definition 4.7.2. Then Assertion 3 of Proposition 4.7.1 guarantees an operation η such that for each $\varepsilon > 0$, we have $I(1_{(|Y|>a)}|Y|) \leq \varepsilon$ for each $a > \eta(\varepsilon)$. Hence, for each $\varepsilon > 0$, for each $X \in G$, and for each $a > \eta(\varepsilon)$, we have

$$I(1_{(|X|>a)}|X|) \leq I(1_{(Y>a)}Y) \leq \varepsilon.$$

Thus η is a common simple modulus of integrability for members X of G. The conditions in Definition 4.7.2 have been verified for the family G to be uniformly integrable. □

Proposition 4.7.5. Each integrable function is the L_1 limit of some sequence of linear combinations of integrable indicators. *The following conditions hold:*

1. Suppose X is an integrable function with $X \geq 0$. Then there exists a sequence $(Y_k)_{k=1,2,...}$ such that for each $k \geq 1$ we have (i) $Y_k \equiv \sum_{i=1}^{n(k)-1} t_{k,i}1_{(t(k,i)<X\leq t(k,i+1))} \in L$ *for some sequence* $0 < t_{k,1} < \cdots < t_{k,n(k)}$, (ii) $I|Y_k - X| \to 0$ *as* $k \to \infty$, *and* (iii) $Y_k \uparrow X$ *as* $k \to \infty$, *on* $D \equiv \bigcap_{k=1}^{\infty} domain(Y_k)$. *Moreover, we can take $n_k \equiv 2^{2k}$ and $t_{k,i} \equiv i2^{-k}a$ for each $i = 1, \ldots, n_k$, where a is some positive real number.*

2. Suppose X is an integrable function. Then there exists a sequence $(Z_k)_{k=1,2,...}$ of linear combinations of mutually exclusive integrable indicators such that $I|X - Z_k| \leq 2^{-k}$ and such that $Z_k \to X$ on $\bigcap_{k=1}^{\infty} domain(Z_k)$ as $k \to \infty$. Furthermore, there exists a sequence $(U_k)_{k=1,2,...}$ of linear combinations of integrable indicators that is a representation of X in L.

3. Suppose X and X' are bounded integrable functions. Then XX' is integrable.

Proof. 1. By Assertion 2 of Corollary 4.6.9, there exists a countable subset J of $(0,\infty)$ such that each member of J_c is a regular point of the integrable functions in the countable set $\{i^{-1}2^kX : k \geq 1, i \geq 0\}$. Take an arbitrary $a \in (0,\infty)J_c$. Then the set $(a < i^{-1}2^kX)$ is integrable for each $k \geq 1, i \geq 0$.

2. Let $k \geq 1$ be arbitrary. Define $n_k \equiv 2^{2k}$, and define $t_{k,i} \equiv i2^{-k}a$ for each $i = 0, \ldots, n_k$. Then the set

$$(t_{k,i} < X) \equiv (i2^{-k}a < X) = (a < i^{-1}2^kX)$$

is integrable for each $i = 0, \ldots, n_k$. Hence the set

$$(t_{k,i} < X \leq t_{k,i+1}) \equiv (t_{k,i} < X)(t_{k,i+1} < X)^c$$

is integrable for each $i = 0, \ldots, n_k - 1$. Consequently, the function

$$Y_k \equiv \sum_{i=0}^{n(k)-1} t_{k,i} 1_{(t(k,i) < X \leq t(k,i+1))}$$

is integrable.

3. We will verify that $Y_k \uparrow X$ on the full set D. To that end, note that, by definition, $t_{k,n(k)} \equiv 2^k a \to \infty$ and $t_{k,i+1} - t_{k,i} = 2^{-k} a \to 0$ for each $i = 0, \ldots, n_k - 1$, as $k \to \infty$. Let $h > k \geq 1$ be arbitrary. Consider any

$$\omega \in D \equiv \bigcap_{k=1}^{\infty} domain(Y_k).$$

Suppose $Y_k(\omega) > 0$. Then

$$0 < Y_k(\omega) = t_{k,i} 1_{(t(k,i) < X(\omega) \leq t(k,i+1))} \tag{4.7.1}$$

for some $i = 0, \ldots, n_k - 1$. Write $p \equiv i 2^{h-k}$ and write

$$q \equiv 2^{h-k}(i+1) \leq 2^{h-k} n_k \equiv 2^{h+k} < 2^{2h} \equiv n_h.$$

Then

$$t_{h,q} \equiv q 2^{-h} a = 2^{h-k}(i+1) 2^{-h} a = (i+1) 2^{-k} a \equiv t_{k,i+1}. \tag{4.7.2}$$

Moreover,

$$t_{h,p} \equiv 2^{-h} pa \equiv 2^{-h}(i 2^{h-k}) a = t_{k,i} < X(\omega) \leq t_{k,i+1} = t_{h,q},$$

where the two inequalities are thanks to inequality 4.7.1, and where the last equality is from equality 4.7.2. In short, we have

$$t_{h,p} < X(\omega) \leq t_{h,q}.$$

It follows that there exists j with $p \leq j < q$ such that $t_{h,j} < X(\omega) \leq t_{h,j+1}$. Consequently, $Y_h(\omega) = t_{h,j}$ and

$$Y_k(\omega) = t_{k,i} \equiv t_{h,p} \leq t_{h,j} = Y_h(\omega) < X(\omega) \leq t_{h,j+1} \leq t_{h,q} = t_{k,i+1}. \tag{4.7.3}$$

Thus we see that $0 \leq Y_k \leq Y_h < X$ on D for each $h > k \geq 1$. Next, let $\varepsilon > 0$ be arbitrary. Then either (i′) $X(\omega) > 0$ or (ii′) $X(\omega) < \varepsilon$. In case (i′), we have $t_{m,i} < X(\omega)$ for some $m \geq 1$ and $i \geq 0$ with $i \leq n_m - 1$. Hence $Y_m(\omega) > 0$. Therefore $Y_k(\omega) \geq Y_m(\omega) > 0$ for each $k \geq m$. Therefore, for each

$$k \geq k_0 \equiv m \vee \log_2(a\varepsilon^{-1}),$$

inequality 4.7.3 implies that

$$X(\omega) - Y_k(\omega) \leq t_{k,i+1} - t_{k,i} = 2^{-k} a \leq 2^{-k(0)} a < \varepsilon. \tag{4.7.4}$$

In case (ii′), we have, trivially, $X(\omega) - Y_k(\omega) < \varepsilon$ for each $k \geq k_0$. Combining, we see that

$$X(\omega) - Y_k(\omega) < \varepsilon$$

in either case, for each $k \geq k_0$. Since $\varepsilon > 0$ is arbitrarily small, we see that $Y_k \uparrow X$ on D. Moreover, we see that

$$X - Y_k < \varepsilon$$

for each $k \geq k_0$.

4. Consequently,

$$I(X - Y_k) \leq \varepsilon I1 = \varepsilon$$

for each $k \geq k_0$. Since $\varepsilon > 0$ is arbitrary, we conclude that $I|Y_k - X| \to 0$. Assertion 1 is proved.

5. By Assertion 1, we see that there exists a sequence $(Y_k^+)_{k=1,2,...}$ of linear combinations of mutually exclusive indicators such that $I|X_+ - Y_k^+| < 2^{-k-1}$ for each $k \geq 1$ and such that $Y_k^+ \uparrow X_+ \; on \; D^+ \equiv \bigcap_{k=1}^{\infty} domain(Y_k^+)$. By the same token, there exists a sequence $(Y_k^-)_{k=1,2,...}$ of linear combinations of mutually exclusive indicators such that $I|X_- - Y_k^-| < 2^{-k-1}$ for each $k \geq 1$ and such that $Y_k^- \uparrow X_- \; on \; D^- \equiv \bigcap_{k=1}^{\infty} domain(Y_k^-)$.

Let $k \geq 1$ be arbitrary. Define $Z_k \equiv Y_k^+ - Y_k^-$. Then

$$I|X - Z_k| \leq I|X_+ - Y_k^+| + I|X_- - Y_k^-| < 2^{-k}.$$

Moreover, we see from the proof of Assertion 1 that Y_k^+ can be taken to be a linear combination of mutually exclusive indicators of subsets of $(X_+ > 0)$. By the same token, Y_k^- can be taken to be a linear combination of mutually exclusive indicators of subsets of $(X_- > 0)$. Since $(X_+ > 0)$ and $(X_- > 0)$ are disjoint, $Z_k \equiv Y_k^+ - Y_k^-$ is a linear combination of mutually exclusive indicators. Since $Y_k^+ \uparrow X_+$ on D^+ and $Y_k^- \uparrow X_- \; on \; D^-$, we have $Z_k \to X = X_+ - X_-$ on $\bigcap_{k=1}^{\infty} domain(Z_k) = D^+ \cap D^-$.

Next, define $Z_0 \equiv 0$ and define $U_k \equiv Z_k - Z_{k-1}$ for each $k \geq 1$. Then $\sum_{k=1}^{\infty} I(U_k) < \infty$ and $\sum_{k=1}^{\infty} U_k = X$ on $\bigcap_{k=1}^{\infty} domain(U_k)$. Hence $(U_k)_{k=1,2,...}$ is a representation of X in L. Assertion 2 is proved.

6. Assertion 3 is trivial if X and X' are integrable indicators. Hence it is also valid when X and X' are linear combinations of integrable indicators. Now suppose X and X' are arbitrary integrable functions bounded in absolute value by some $a > 0$. By Assertion 2, there exist sequences $(Z_n)_{n=1,2,...}$ and $(Z'_n)_{n=1,2,...}$ of linear combinations of integrable indicators such that $I|X - Z_n| \to 0$ and $I|X' - Z'_n| \to 0$. Then, for each $n \geq 1$, the product function $Z_n Z'_n$ is integrable by the second statement in this paragraph. Moreover,

$$|XX' - Z_n Z'_n| \leq a|X - Z_n| + a|X' - Z'_n|$$

for each $n \geq 1$. Therefore, by Theorem 4.5.10, the function XX' is integrable. Assertion 3 and the proposition are proved. □

4.8 Measurable Function and Measurable Set

In this section, let (Ω, L, I) be a complete integration space, and let (S,d) be a complete metric space with a fixed reference point $x_\circ \in S$. In the case where $S = R$, it is understood that d is the Euclidean metric and that $x_\circ = 0$.

Recall that, as an abbreviation, we write $AB \equiv A \cap B$ for subsets A and B of Ω. Recall from the notations and conventions described in the Introduction that if X is a real-valued function on Ω and if $t \in R$, then we use the abbreviation $(t \leq X)$ for the subset $\{\omega \in domain(X) : t \leq X(\omega)\}$. Similarly, "$\leq$" may be replaced by "$\geq$," "$<$," or "$=$." As usual, we write a_b interchangeably with $a(b)$ to lessen the burden on subscripts.

Recall that $\mu A \equiv \mu(A)$ stands for the measure of an integrable subset A of Ω.

We next generalize the theory of real-valued measurable functions in [Bishop and Bridges 1985] to a theory of measurable functions with values in the complete metric space (S,d). Recall that $C_{ub}(S)$ stands for the space of bounded and uniformly continuous real-valued functions on S.

Definition 4.8.1. Measurable function. A function X from (Ω, L, I) to the complete metric space (S,d) is called a *measurable function* if

(i) for each integrable set A and each $f \in C_{ub}(S)$, we have $f(X)1_A \in L$, and

(ii) for each integrable set A, there exists a countable subset J of R such that $\mu(d(x_\circ, X) > a)A) \to 0$ as $a \to \infty$ while a remains in the metric complement J_c of J.

In the special case where the constant function 1 is integrable, then Conditions (i) and (ii) reduce to (i′) $f(X) \in L$ and (ii′) $\mu(d(x_\circ, X) > a) \to 0$ as $a \to \infty$ while $a \in J_c$. □

It is obvious that if Condition (ii) holds for one point $x_\circ \in S$, then it holds for any point $x'_\circ \in S$. The next lemma shows that, given Condition (i) and given an arbitrary integrable set A, the measure in Condition (ii) is welldefined for all but countably many $a \in R$. Hence Condition (ii) makes sense.

Let X be a function defined a.e. on a complete integration space (Ω, L, I). Then it can be proved that X is measurable iff it satisfies Condition (*): for each integrable set A and $\varepsilon > 0$, there exist an integrable set B and an integrable function Y such that $B \subset A$, $\mu(AB^c) < \varepsilon$, and $|X - Y| < \varepsilon$ on B. Condition (*) is used as the definition of a measurable function in [Bishop and Bridges 1985]. Thus Definition 4.8.1 of a measurable function, which we find more convenient, is equivalent to the one in [Bishop and Bridges 1985].

Definition 4.8.2. Measurable set and measure-theoretic complement. Suppose a subset B of Ω is such that $B = (X = 1)$ for some real-valued measurable indicator function X. Then we say that B is a *measurable set*, and call the function $1_B \equiv X$ the measurable indicator of B. The set $B^c \equiv (X = 0)$ is then called the *measure-theoretic complement* of B. Lemma 4.8.4 proves that 1_B and B^c are uniquely defined relative to a.e. equality. □

Lemma 4.8.3. Integrability of some basic sets. *Let X be a function from Ω to S that satisfies Condition* (i) *in Definition 4.8.1. Let A be an arbitrary integrable set. Then the set $(d(x_\circ, X) > a)A$ is integrable for all but countably many $a \in R$. Thus $\mu(d(x_\circ, X) > a)A$ is well defined for all but countably many $a \in R$.*

Proof. Let $n \geq 0$ be arbitrary. Then $h_n \equiv 1 \wedge (n+1-d(x_\circ,\cdot))_+ \in C_{ub}(S)$ and so $h_n(X)1_A \in L$ by hypothesis. Hence all but countably many $b \in (0,1)$ are regular points of $h_n(X)1_A$. Therefore the set

$$(d(x_\circ, X) > n+1-b)A = (h_n(X)1_A < b)A$$

is integrable for all but countably many $b \in (0,1)$. Equivalently, $(d(x_\circ, X) > a)A$ is integrable for all but countably many $a \in (n, n+1)$. Since $n \geq 0$ is arbitrary, we see that $(d(x_\circ, X) > a)A$ is integrable for all but countably many points $a > 0$. For each $a \leq 0$, the set $(d(x_\circ, X) > a)A = A$ is integrable by hypothesis. □

Lemma 4.8.4. The indicator and measure-theoretic complement of a measurable set are welldefined relative to a.e. equality. *Let B, C be arbitrary measurable subsets of Ω such that $B = C$ a.e. Let X, Y be measurable indicators of B, C, respectively.*

1. $B = C$ a.e. iff $X = Y$ a.e. Hence 1_B is a well-defined measurable relative to a.e. equality.

2. If $B = C$ a.e., then $(X = 0) = (Y = 0)$ a.e. Hence B^c is a well-defined subset relative to equality a.e.

Proof. 1. Suppose $B = C$ a.e. Then $BD = CD$ for some full set D. Let A be an arbitrary integrable set. Then $X1_A$ and $Y1_A$ are integrable indicators. Hence $D' \equiv D \cap domain(X1_A) \cap domain(Y1_A)$ is a full set. Note that, on the full set D',) the functions X and Y can have only two possible values, 1 or 0. Moreover, $D'(X = 1) = D'B = D'C = D'(Y = 1)$. Hence $X = Y$ on the full set D'. Thus $X = Y$ a.e.

Conversely, suppose $X = Y$ a.e. Then $X = Y$ on some full set D. Hence $(X = 0)D = (Y = 0)D$. Since X, Ycan have only two possible values, 1 or 0, it follows that $X = Y$ on D. In short, $X = Y$ a.e. Assertion 1 is proved.

2. Suppose $B = C$ a.e. Then Assertion 1 implies that $X = Y$ a.e. Then the previous paragraph shows that $(X = 0) = (Y = 0)$ on the full set D. Thus $(X = 0) = (Y = 0)$ a.e. The lemma is proved. □

Definition 4.8.5. Convention: unless otherwise specified, equality of measurable functions will mean equality a.e.; equality of measurable sets will mean equality a.e. Let (Ω, L, I) be an arbitrary complete integration space. Let X, Y be measurable functions with values in some metric space (S, d). Let $A, B \subset \Omega$ be arbitrary measurable sets. Henceforth, unless otherwise specified, the statement $X = Y$ will mean $X = Y$ a.e., Similarly, the statements $X \leq Y$, $X < Y$, $X \geq Y$, and $X > Y$ will mean $X \leq Y$ a.e., $X < Y$ a.e., $X \geq Y$ a.e., and

$X > Y$ a.e., respectively. Likewise, the statements $A = B$, $A \subset B$, and $A \supset B$ will mean $A = B$ a.e., $A \subset B$ a.e., and $A \supset B$ a.e., respectively.

The next proposition gives an equivalent condition to Condition (ii) in Definition 4.8.1.

Proposition 4.8.6. Alternative definition of measurable functions. *For each $n \geq 0$, define the function $h_n \equiv 1 \wedge (n+1-d(x_\circ,\cdot))_+ \in C_{ub}(S)$. Then a function X from (Ω, L, I) to the complete metric space (S,d) is a measurable function iff, for each integrable set A and for each $f \in C_{ub}(S)$, we have (i′) $f(X)1_A \in L$, and (ii′) $Ih_n(X)1_A \to \mu(A)$ as $n \to \infty$. In the anticipated case where (Ω, L, I) is a probability space, where $\mu(\Omega) = 1$, conditions (i′) and (ii′) can be replaced by (i′) $f(X) \in L$ and (ii′) $Ih_n(X) \to 1$ as $n \to \infty$.*

Proof. 1. Suppose Conditions (i′) and (ii′) hold. Let $n \geq 0$ be arbitrary. We need to verify that the function X is measurable. Note that since $h_n \in C_{ub}(S)$, we have $h_n(X)1_A \in L$ by Condition (i′). Now let A be an arbitrary integrable set. Then, for each $n \geq 1$ and $a > n+1$, we have

$$Ih_n(X)1_A \leq \mu(d(x_\circ, X) \leq a)A \leq \mu(A).$$

Letting $n \to \infty$, Condition (ii′) and the last displayed inequality imply that $\mu(d(x_\circ, X) \leq a)A \to \mu(A)$. Equivalently, $\mu(d(x_\circ, X) > a)A \to 0$ as $a \to \infty$. The conditions in Definition 4.8.1 are satisfied for X to be measurable.

2. Conversely, suppose X is measurable. Then Definition 4.8.1 of measurable functions implies Condition (i′) of the present lemma. It implies also that $\mu(d(x_\circ, X) \leq a)A \to \mu(A)$ as $a \to \infty$. At the same time, for each $a > 0$ and $n > a$, we have

$$\mu(d(x_\circ, X) \leq a)A \leq Ih_n(X)1_A \leq \mu(A).$$

Letting $a \to \infty$, we see that $Ih_n(X)1_A \to \mu(A)$. Thus Condition (ii′) of the present lemma is also proved. □

Proposition 4.8.7. Basics of a measurable function. *Let $X : (\Omega, L, E) \to S$ be an arbitrary measurable function with values in the complete metric space (S,d).*

1. Then $domain(X)$ is a full set. In other words, X is defined a.e. In particular, if A is a measurable set, then $A \cup A^c$ is a full set.

2. Each function $Y : (\Omega, L, E) \to S$ that is equal a.e. to the measurable function X is itself measurable.

3. Each integrable function $Z : (\Omega, L, E) \to R$ is a real-valued measurable function. Each integrable set is measurable.

Proof. 1. Let X be an arbitrary measurable function. Let $f \equiv 0$ denote the constant 0 function. Then $f \in C_{ub}(S)$. Hence, by Condition (i) in Definition 4.8.1, we have $f(X)1_A \in L$. Consequently, $D \equiv domain(f(X)1_A)$ is a full set. Since $domain(X) = domain(f(X)) \supset D$, we see that $domain(X)$ is a full set.

Next, let A be an arbitrary measurable set. In other words, its indicator 1_A is measurable. Hence the set $A \cup A^c = domain(1_A)$ is a full set according to the previous paragraph. Assertion 1 is proved.

2. Now suppose Y is a function on Ω, with values in S such that $Y = X$ a.e., where X is some measurable function. Let A be any integrable set. Let $f \in C_{ub}(S)$ be arbitrary. Then, by Condition (i) in Definition 4.8.1, we have $f(X)1_A \in L$. Moreover, because $Y = X$ a.e., we have $f(Y)1_A = f(X)1_A$ a.e. Consequently, $f(Y)1_A \in L$. Again because $Y = X$ a.e.,

$$\mu(d(x_\circ, Y) > a)A = \mu(d(x_\circ, X) > a)A \to 0$$

as $a \to \infty$. Thus the conditions in Definition 4.8.1 are verified for Y to be measurable.

3. Next, let Z be any integrable function. Let $f \in C_{ub}(R)$ be arbitrary and let A be an arbitrary integrable set. By Proposition 4.6.12, we have $f(Z)1_A \in L$, which establishes Condition (i) of Definition 4.8.1 for the function Z. By Chebychev's inequality,

$$\mu(|Z| > a)A \le \mu(|Z| > a) \le a^{-1}I|Z| \to 0$$

as $a \to \infty$. Condition (ii) of Definition 4.8.1 follows. Hence Z is measurable. In particular, 1_A and A are measurable. □

The next proposition will be used repeatedly to construct measurable functions from given ones.

Proposition 4.8.8. Construction of a measurable function from pieces of given measurable functions on measurable sets in a disjoint union. *Let (S,d) be a complete metric space. Let $(X_i, A_i)_{i=1,2,\ldots}$ be a sequence where for each $i, j \ge 1$, X_i is a measurable function on (Ω, L, I) with values in S,* and where (i) *A_i is a measurable subset of Ω,* (ii) *if $i \neq j$, then $A_i A_j = \phi$,* (iii) *$\bigcup_{k=1}^{\infty} A_k$ is a full set, and* (iv) *$\sum_{k=1}^{\infty} \mu(A_k A) = \mu(A)$ for each integrable set A.*

Define a function X by

$$domain(X) \equiv \bigcup_{i=1}^{\infty} domain(X_i)A_i$$

and by $X \equiv X_i$ on $domain(X_i)A_i$, for each $i \ge 1$. Then X is a measurable function on Ω with values in S.

The same conclusion holds for a finite sequence $(X_i, A_i)_{i=1,\ldots,n}$.

Proof. We will give the proof for the infinite sequence only. For each $n \ge 1$, define $h_n \equiv 1 \wedge (n + 1 - d(x_\circ, \cdot))_+ \in C_{ub}(S)$.

Let $f \in C_{ub}(S)$ be arbitrary, with $|f| \le c$ on S for some $c > 0$. Let A be an arbitrary integrable set. Since

$$\sum_{i=1}^{\infty} If(X_i)1_{A(i)A} \le c \sum_{i=1}^{\infty} \mu(A_i A) < \infty,$$

the function $Y \equiv \sum_{i=1}^{\infty} f(X_i)1_{A(i)A}$ is integrable. At the same time, $f(X)1_A = Y$ on the full set

$$\left(\bigcup_{i=1}^{\infty} A_i\right) \cap \left(\bigcap_{i=1}^{\infty} domain(X_i)\right).$$

Hence $f(X)1_A$ is integrable. In particular, $h_n(X)1_A$ is integrable for each $n \geq 1$. Moreover, $Ih_n(X_i)1_{A(i)A} \uparrow \mu(A_iA)$ as $n \to \infty$, for each $i \geq 1$, according to Condition (ii$'$) of Lemma 4.8.6. Consequently,

$$Ih_n(X)1_A = \sum_{i=1}^{\infty} Ih_n(X_i)1_{A(i)A} \uparrow \sum_{i=1}^{\infty} \mu(A_iA) = \mu(A).$$

Hence, by Lemma 4.8.6, X is a measurable function. □

We next provide a metric space lemma.

Lemma 4.8.9. Sufficient condition for uniform continuity on a metric space. *Let (S,d) be an arbitrary metric space. Let A, B be arbitrary subsets of S, and let $a > 0$ be arbitrary such that for each $x \in S$, we have either* (i) $(d(\cdot,x) < a) \subset A$ *or* (ii) $(d(\cdot,x) < a) \subset B$. *Suppose $\lambda : S \to R$ is a function with $domain(\lambda) = S$ such that λ is uniformly continuous on each of A and B. Then λ is uniformly continuous on S.*

Proof. Let $\varepsilon > 0$ be arbitrary. Since λ is uniformly continuous on each of A and B, there exists $\delta_0 > 0$ so small that $|\lambda(x) - \lambda(y)| < \varepsilon$ for each x, y with $d(x,y) < \delta_0$, provided that either $x, y \in A$ or $x, y \in B$.

Let $\delta \equiv a \wedge \delta_0$. Consider each $x, y \in S$ with $d(x,y) < \delta$. By hypothesis, either Condition (i) or Condition (ii) holds. Assume that Condition (i) holds. Then since $d(x,x) = 0 < a$ and $d(y,x) < \delta \leq a$, we have $x, y \in A$. Hence, since $d(y,x) < \delta \leq \delta_0$, we have $|\lambda(x) - \lambda(y)| < \varepsilon$. Similarly, if Condition (ii) holds, then $|\lambda(x) - \lambda(y)| < \varepsilon$, where $x, y \in S$ are arbitrary with $d(x,y) < \delta$. Thus λ is uniformly continuous on S. □

Proposition 4.8.10. A continuous function of a measurable function is measurable. *Let (S,d) and $(S'd')$ be complete metric spaces. Let $X : (\Omega, L, I) \to (S,d)$ be an arbitrary measurable function. Suppose a function $f : (S,d) \to (S',d')$ with $domain(f) = S$ is uniformly continuous and bounded on each bounded subset of S.*

Then the composite function $f(X) \equiv f \circ X : (\Omega, L, I) \to (S',d')$ is measurable. In particular, $d(x,X)$ is a real-valued measurable function for each $x \in S$.

Proof. We need to prove that $Y \equiv f(X)$ is measurable.

1. To that end, let $g \in C_{ub}(S')$ be arbitrary, with $|g| \leq b$ for some $b > 0$. Consider an arbitrary integrable set A and an arbitrary $\varepsilon > 0$. Since X is measurable by hypothesis, there exists $a > 0$ so large that $\mu(B) < \varepsilon$, where $B \equiv (d(x_\circ, X) > a)A$. Define $h \equiv 1 \wedge (a - d(x_\circ, \cdot))_+ \in C_{ub}(S)$.

The function f is, by hypothesis, uniformly continuous on the bounded set $G \equiv (d(\cdot,x_\circ) < 2+a)$. By assumption, g is uniformly continuous. Therefore $(g \circ f)$ and $(g \circ f)h$ are uniformly continuous on G.

At the same time, $h = 0$ on $H \equiv (d(\cdot,x_\circ) > a)$. Hence $(g \circ f)h = 0$ on H. Thus $(g \circ f)h$ is uniformly continuous on H.

Now consider each $x \in S$. Either (i) $d(x,x_\circ) < a + \frac{3}{2}$ or (ii) $d(x,x_\circ) > a + \frac{1}{2}$. In case (i), we have $\left(d(\cdot,x) < \frac{1}{2}\right) \subset (d(\cdot,x_\circ) < 2+a) \equiv G$. In case (ii), we have $\left(d(\cdot,x) < \frac{1}{2}\right) \subset (d(\cdot,x_\circ) > a) \equiv H$. Combining, Lemma 4.8.9 implies that $(g \circ f)h$ is uniformly continuous on S. Moreover, since $(g \circ f)h$ is bounded on G by hypothesis, and is equal to 0 on H, it is bounded on S. In short, $(g \circ f)h \in C_{ub}(S)$. Since X is measurable, the function $g(Y)h(X)1_A = (g \circ f)(X)h(X)1_A$ is integrable. At the same time,

$$|g(Y)1_A - g(Y)h(X)1_A| \le b(1-h(X))1_A, \tag{4.8.1}$$

where

$$I(1-h(X))1_A \le \mu(d(x_\circ,X) > a)A = \mu(B) < \varepsilon.$$

Since $\varepsilon > 0$ is arbitrary, by Theorem 4.5.10, inequality 4.8.1 implies that the function $g(Y)1_A$ is integrable. We have verified Condition (i) of Definition 4.8.1 for the function Y.

2. Now let $c > a$ be arbitrary. By hypothesis, there exists $c' > 0$ so large that $d'(x'_\circ, f) \le c'$ on $(d(x_\circ,\cdot) < c)$. Then, for each $x \in S$ with $d'(x'_\circ, f(x)) > c'$, we have $d(x_\circ,x) \ge c$. Hence

$$d(x'_\circ, f(X)) > c') \subset (d(x_\circ,X) \ge c).$$

Therefore

$$\begin{aligned}\mu(d(x_\circ,Y) > c')A &\equiv \mu(d(x'_\circ, f(X)) > c')A \\ &\le \mu(d(x_\circ,X) \ge c)A \le \mu(d(x_\circ,X) > a)A < \varepsilon.\end{aligned}$$

By Lemma 4.8.3, the left-hand side of the last displayed inequality is well defined for all but countably many $c' > 0$. Since $\varepsilon > 0$ is arbitrary, we conclude that $\mu(d(x_\circ,Y) > c')A \to 0$ as $c' \to \infty$. Thus we have verified all the conditions of Definition 4.8.1 for the function Y to be measurable. In other words, $f(X)$ is measurable. □

Corollary 4.8.11. Condition for measurability of the identity function, and of a continuous function of a measurable function. *Let (S,d) be a complete metric space. Suppose $(S, C_{ub}(S), I)$ is an integration space, with completion (S,L,I). Define the identity function $X : S \to S$ by $X(x) \equiv x$ for each $x \in S$. Define $h_k \equiv 1 \wedge (k+1-d(x_\circ,\cdot))_+ \in C_{ub}(S)$ for each $k \ge 1$. Suppose $Ih_k1_A \uparrow \mu(A)$ for each integrable set A, Then the following conditions hold:*

1. The identity function $X : (S,L,I) \to (S,d)$ is a measurable function.

2. Let (S',d') be a second complete metric space. Suppose a function $f: (S,d) \to (S',d')$ with $domain(f) = S$ is uniformly continuous and bounded on each bounded subset of S. Then the function $f: (S,L,I) \to (S',d')$ is measurable. In particular, $d(x,\cdot)$ is a real-valued measurable function for each $x \in S$.

Proof. 1. Let $f \in C_{ub}(S)$ be arbitrary, and let A be an arbitrary integrable set. Then $f(X) \equiv f \in C_{ub}(S) \subset L$. Hence $f(X)1_A \in L$. Moreover, $Ih_k(X)1_A = Ih_k 1_A \uparrow \mu(A)$ by hypothesis. Hence X is measurable according to Lemma 4.8.6.

2. The conditions in the hypothesis of Proposition 4.8.10 are satisfied by the functions $X : (S,L,I) \to (S,d)$ and $f : (S,d) \to (S',d')$. Accordingly, the function $f \equiv f(X)$ is measurable. □

The next proposition says that in the case where (S,d) is locally compact, the conditions for measurability in Definition 4.8.1 can be relaxed by replacing $C_{ub}(S)$ with the subfamily $C(S)$.

Proposition 4.8.12. Sufficient condition for measurability in case S is locally compact. *Let (S,d) be a locally compact metric space. Define $h_n \equiv 1 \wedge (n+1-d(x_\circ,\cdot))_+ \in C(S)$ for each $n \geq 1$. Let X be a function from (Ω,L,I) to (S,d) such that $f(X)1_A \in L$ for each integrable set A and each $f \in C(S)$. Then the following conditions hold:*

1. If $Ih_n(X)1_A \uparrow \mu(A)$ for each integrable set A, then X is measurable.

2. If $\mu(d(x_\circ,X) > a)A \to 0$ as $a \to \infty$ for each integrable set A, then X is measurable.

Proof. Let $n \geq 1$ be arbitrary. Note that the function h_n has the bounded set $(d(x_\circ,\cdot) < n+1)$ as support, which is contained in some compact subset because the metric space (S,d) is locally compact. Hence $h_n \in C(S)$.

1. Now let A be an arbitrary integrable set. Let $g \in C_{ub}(S)$ be arbitrary, with $|g| \leq c$ on S. Since $h_n \in C(S)$, we have $h_n, h_n g \in C(S)$, Hence, by the hypothesis of the proposition, we have $h_n(X)g(X)1_A \in L$. Moreover, $Ih_n(X)1_A \uparrow \mu(A)$ by the hypothesis of Assertion 1. Hence

$$I|g(X)1_A - h_n(X)g(X)1_A| \leq cI(1_A - h_n(X)1_A) = c(\mu(A) - Ih_n(X)1_A) \to 0 \tag{4.8.2}$$

as $n \to \infty$. Therefore Theorem 4.5.10 is applicable; it implies that $g(X)1_A \in L$, where $g \in C_{ub}(S)$ is arbitrary. Thus the conditions in Proposition 4.8.6 are satisfied for X to be measurable. Assertion 1 is proved.

2. For each $a > 0$ and for each $n > a$,

$$0 \leq \mu(A) - Ih_n(X)1_A = I(1 - h_n(X))1_A \leq \mu(d(x_\circ,X) > a)A,$$

which, by the hypothesis of Assertion 2, converges to 0 as $a \to \infty$. Hence, by Assertion 1, the function X is measurable. Assertion 2 and the proposition are proved. □

Definition 4.8.13. Regular and continuity points of a measurable function relative to each integrable set. Suppose X is a real-valued measurable function on (Ω, L, I). We say that $t \in R$ is a *regular point of X relative to an integrable set A* if

(i) there exists a sequence $(s_n)_{n=1,2,...}$ of real numbers decreasing to t such that $(s_n < X)A$ is an integrable set for each $n \geq 1$ and such that $\lim_{n\to\infty} \mu(s_n < X)A$ exists, and

(ii) there exists a sequence $(r_n)_{n=1,2,...}$ of real numbers increasing to t such that $(r_n < X)A$ is integrable for each $n \geq 1$ and such that $\lim_{n\to\infty} \mu(r_n < X)A$ exists.

If, in addition, the two limits in (i) and (ii) are equal, then we call t a *continuity point* of X *relative to* A.

If a real number t is a regular point of X relative to each integrable set A, then we call t a *regular point* of X. If a real number t is a continuity point of X relative to each integrable set A, then we say t is a *continuity point* of X. □

The next proposition shows that regular points and continuity points of a real-valued measurable function are abundant, and that they inherit the properties of regular points and continuity points of integrable functions.

Proposition 4.8.14. All but countably many real numbers are continuous points of a real-valued measurable function, relative to each given integrable set A. *Let X be a real-valued measurable function on (Ω, L, I). Let A be an integrable set and let t be a regular point of X relative to A. Then the following conditions hold:*

1. All but countably many $u \in R$ are continuity points of X relative to A. Hence all but countably many $u \in R$ are regular points of X relative to A.

2. A^c is a measurable set.

3. The sets $A(t < X), A(t \leq X), A(X < t), A(X \leq t)$, and $A(X = t)$ are integrable sets.

4. $(X \leq t)A = A((t < X)A)^c$ and $(t < X)A = A((X \leq t)A)^c$.

5. $(X < t)A = A((t \leq X)A)^c$ and $(t \leq X)A = A((X < t)A)^c$.

6. For a.e. $\omega \in A$, we have $t < X(\omega)$, $t = X(\omega)$, or $t > X(\omega)$.

7. Let $\varepsilon > 0$ be arbitrary. There exists $\delta > 0$ such that if $r \in (t - \delta, t]$ and if $A(X < r)$ is integrable, then $\mu(A(X < t)) - \mu(A(X < r)) < \varepsilon$. Similarly, there exists $\delta > 0$ such that if $s \in [t, t + \delta)$ and if $A(X \leq s)$ is integrable, then $\mu(A(X \leq s)) - \mu(A(X \leq t)) < \varepsilon$.

8. If t is a continuity point of X relative to A, then $\mu((t < X)A) = \mu((t \leq X)A)$.

Proof. In the special case where X is an integrable function, Assertions 1, 3, 4, 5, 6, 7, and 8 of the present proposition are restatements of Assertions 3, 4, 5, 6, 7, 8, and 9, respectively, of Corollary 4.6.9.

1. In general, suppose X is a real-valued measurable function. Let $n \geq 1$ be arbitrary. Then, by Definition 4.8.1, $((-n)\vee X \wedge n)1_A \in L$. Hence all but countably many $u \in (-n, n)$ are continuity points of the integrable function $((-n)\vee X\wedge n)1_A$

relative to A. At the same time, for each $t \in (-n,n)$, we have $(X < t)A = (((-n) \vee X \wedge n)1_A < t)A$. Hence, a point $u \in (-n,n)$ is a continuity point of $((-n) \vee X \wedge n)1_A$ relative to A iff it is a continuity point of X relative to A. Combining, we see that all but countably many points in the interval $(-n,n)$ are continuity points of X relative to A. Therefore all but countably many points in R are continuity points of X relative to A. This proves Assertion 1 of the proposition.

2. To prove that A^c is a measurable set, consider the function $1_{A^c} : (\Omega, L, I) \rightarrow R$. Let $x_\circ \equiv 0$ be the reference point in (R,d), where d is the Euclidean metric. Let B be an arbitrary integrable set and let $f \in C_{ub}(R)$ be arbitrary. Then the sets B and BA are integrable. Hence, by Assertion 3 of 4.5.7, the set $B(BA)^c$ is integrable. Since $BA^c = B(BA)^c$, the set BA^c is integrable. Consequently, the function

$$f(1_{A^c})1_B = f(1)1_{BA^c} + f(0)1_{BA}$$

is integrable. At the same time, we have $(|0 - 1_{A^c}|) > a) = (|1_{A^c}|) > a) = \phi$ for each $a > 1$. Hence, trivially, $\mu(|0 - 1_{A^c}|) > a) \rightarrow 0$. Thus the conditions in Definition 4.8.1 are satisfied for the function 1_{A^c} to be measurable and the set A^c is measurable. Assertion 2 is proved.

3. The remaining assertions are proved similarly to Step 1 – that is, by reducing the measurable function X to the integrable functions $((-n) \vee X \wedge n)1_A$ and then applying Corollary 4.6.9. □

Let X be an arbitrary real-valued measurable function. We have proved that, given each integrable set A, there exists a countable subset J_A of R such that each point t in the metric complement $J_{A,c}$ is a regular point of X relative to A.

In the special case where the integration space Ω is the union of a sequence $(A_k)_{k=1,2,...}$ of integrable sets, it follows that each point t in the metric complement $\left(\bigcup_{k=1}^{\infty} J_{A(k)}\right)_c$ is a regular point of X relative to A_k, for each $k \geq 1$. We will show that as a consequence, all but countably many points $t \in R$ are regular points of the measurable function X. The following definition and proposition will make this precise.

Definition 4.8.15. Finite or σ-finite integration space. The complete integration space (Ω, L, I) is said to be *finite* if the constant function 1 is integrable. It is said to be sigma finite, or *σ-finite*, if there exists a sequence $(A_k)_{k=1,2,...}$ of integrable sets with positive measures such that (i) $A_k \subset A_{k+1}$ for $k = 1, 2, \ldots$, (ii) $\bigcup_{k=1}^{\infty} A_k$ is a full set, and (iii) for any integrable set A we have $\mu(A_k A) \rightarrow \mu(A)$. The sequence $(A_k)_{k=1,2,...}$ is then called an *I-basis* for (Ω, L, I). □

If (Ω, L, I) is finite, then it is σ-finite, with an I-basis given by $(A_k)_{k=1,2,...}$, where $A_k \equiv \Omega$ for each $k \geq 1$. An example is where $(S, C_{ub}(S), I)$ is an integration space with completion (S, L, I). Then the constant function 1 is integrable, and so (S, L, I) is finite. The next lemma provides other examples that include the Lebesgue integration space $(R, L, \int \cdot dx)$.

Lemma 4.8.16. Completion of an integration on a locally compact metric space results in a σ-finite integration space. *Suppose (S,d) is a locally compact metric space. Let (S,L,I) be the completion of some integration space $(S,C(S),I)$. Then $(\Omega,L,I) \equiv (S,L,I)$ is σ-finite. Specifically, there exists an increasing sequence $(a_k)_{k=1,2,...}$ of positive real numbers with $a_k \uparrow \infty$ such that $(A_k)_{k=1,2,...}$ is an I-*basis *for (S,L,I), where $A_k \equiv (d(x_\circ,\cdot) \leq a_k)$ for each $k \geq 1$.*

Proof. Consider each $k \geq 1$. Define $X_k \equiv 1 \wedge (k+1-d(x_\circ,\cdot))_+ \in C(S) \subset L$. Let $c \in (0,1)$ be arbitrary and let $a_k \equiv k+1-c$. Then the set $A_k \equiv (d(x_\circ,\cdot) \leq a_k) = (X_k \geq c)$ is integrable. Conditions (i) and (ii) of Definition 4.8.15 are easily verified. For Condition (iii) of Definition 4.8.15, consider any integrable set A. According to Assertion 2 of Corollary 4.8.11, the real-valued function $d(x_\circ,\cdot)$ is measurable on $(\Omega,L,I) \equiv (S,L,I)$. Hence

$$\mu A - \mu A_k A = \mu(d(x_\circ,\cdot) > a_k)A \to 0.$$

Thus Condition (iii) in Definition 4.8.15 is also verified for $(A_k)_{k=1,2,...}$ to be an I-basis. □

Proposition 4.8.17. In the case of a σ-finite integration space, all but countably many points are continuous points of an arbitrary given real-valued measurable function. *Suppose X is a real-valued measurable function on a σ-finite integration space (Ω,L,I). Then all but countably many real numbers t are continuity points, and hence regular points, of X.*

Proof. Let $(A_k)_{k=1,2,...}$ be an I-basis for (Ω,L,I). According to Proposition 4.8.14, for each k there exists a countable subset J_k of R such that if $t \in (J_k)_c$, where $(J_k)_c$ stands for the metric complement of J_k in R, then t is a continuity point of X relative to A_k. Define $J \equiv \bigcup_{k=1}^{\infty} J_k$.

Consider each $t \in J_c$. Let the integrable set A be arbitrary. According to Condition (iii) in Definition 4.8.15, we can select a subsequence $(A_{k(n)})_{n=1,2,...}$ of (A_k) such that $\mu(A) - \mu(AA_{k(n)}) < \frac{1}{n}$ for each $n \geq 1$. Let $n \geq 1$ be arbitrary. Write $B_n \equiv A_{k(n)}$. Then $t \in (J_{k(n)})_c$, and so t is a continuity point of X relative to B_n. Consequently, according to Proposition 4.8.14, the sets $(X < t)B_n$ and $(X \leq t)B_n$ are integrable, with

$$\mu(X < t)B_n = \mu(X \leq t)B_n.$$

Furthermore, according to Assertion 7 of Proposition 4.8.14, there exists $\delta_n > 0$ such that (i) if $r \in (t-\delta_n,t]$ and if $(X < r)B_n$ is integrable, then $\mu(X < t)B_n - \mu(X < r)B_n < \frac{1}{n}$, and (ii) if $s \in [t,t+\delta_n)$ and if $(X \leq s)B_n$ is integrable, then $\mu(X \leq s)B_n - \mu(X \leq t)B_n < \frac{1}{n}$.

Let $r_0 \equiv t-1$ and $s_0 \equiv t+1$. Inductively, we can select $r_n \in (t-\delta_n,t) \cap (r_{n-1},t) \cap \left(t-\frac{1}{n},t\right)$ such that r_n is a regular point of X relative to both B_n and A. Similarly, we can select $s_n \in (t,t+\delta_n) \cap (t,s_{n-1}) \cap \left(t,t+\frac{1}{n}\right)$ such that s_n is a regular point of X relative to both B_n and A. Then, for each $n \geq 1$, we have

$$\begin{aligned}
&\mu(r_n < X)A - \mu(s_n < X)A \\
&\quad = \mu A(r_n < X)(X \le s_n)A \le \mu(r_n < X)(X \le s_n)B_n + \mu(AB_n^c) \\
&\quad = \mu(X \le s_n)B_n - \mu(X \le r_n)B_n + \mu(A) - \mu(AB_n) \\
&\quad \le \left(\mu(X \le t)B_n + \frac{1}{n}\right) - \left(\mu(X < t)B_n - \frac{1}{n}\right) + (\mu(A) - \mu(AA_{k(n)})) \\
&\quad \le \frac{1}{n} + \frac{1}{n} + \frac{1}{n}.
\end{aligned} \tag{4.8.3}$$

Since the sequence $(\mu(r_n < X)A)$ is nonincreasing and the sequence $(\mu(s_n < X)A)$ is nondecreasing, inequality 4.8.3 implies that both sequences converge, and to the same limit. By Definition 4.8.13, t is a continuity point of X relative to A, where $t \in J_c$ is arbitrary. The proposition is proved. □

We now expand the convention in Definition 4.6.11 to cover measurable functions.

Definition 4.8.18. Convention regarding regular points of measurable functions. Let X be a real-valued measurable function, and let A be an integrable set. Henceforth, when the integrability of the set $(X < t)A$ or $(X \le t)A$ is required in a discussion, for some $t \in R$, then it is understood that the real number t has been chosen from the regular points of the measurable function X relative to the given integrable set A.

Furthermore, if (Ω, L, I) is a σ-finite integration space, when the measurability of the set $(X < t)$ or $(X \le t)$ is required in a discussion, for some $t \in R$, then it is understood that the real number t has been chosen from the regular points of the measurable functions X.□

Corollary 4.8.19. Properties of regular points. *Let X be a real-valued measurable function on a σ-finite integration space (Ω, L, I), and let t be a regular point of X. Then*

$$(X \le t) = (t < X)^c \tag{4.8.4}$$

and $(t < X) = (X \le t)^c$ are measurable sets. Similarly, $(X < t) = (t \le X)^c$ and $(t \le X) = (X < t)^c$ are measurable sets. According to the convention in Definition 4.8.5, the equalities here are understood to be a.e. equalities.

Proof. We will prove only equality 4.8.4, with the proofs of the rest being similar. Define an indicator function Y, with $domain(Y) = (X \le t) \cup (t < X)$, by $Y = 1$ on $(X \le t)$ and $Y = 0$ on $(t < X)$. It suffices to show that Y satisfies Conditions (i) and (ii) of Definition 4.8.1 for a measurable function. To that end, consider an arbitrary $f \in C_{ub}(R)$ and an arbitrary integrable subset A. By hypothesis, and by Definition 4.8.18, t is a regular point of X relative to A. Moreover,

$$f(Y)1_A = f(1)1_{(X \le t)A} + f(0)1_{(t<X)A},$$

which is integrable according to Assertion 3 of Proposition 4.8.14. Thus Condition (i) of Definition 4.8.1 is verified. Moreover, since Y has only the possible values of 0 and 1, the set $(|Y| > a)$ is empty for each $a > 1$. Hence, trivially, $\mu(|Y| > a)A \to 0$ as $a \to \infty$, thereby establishing Condition (ii) of Definition 4.8.1. Consequently, Y is a measurable indicator for $(X \leq t)$, and $(t < X) = (Y = 0) = (X \leq t)^c$. This proves equality 4.8.4. □

Proposition 4.8.20. A vector of measurable functions with values in locally compact metric spaces constitutes a measurable function with values in the product metric space. *Let (Ω, L, I) be a complete integration space.*

1. Let $n \geq 2$ be arbitrary. For each $i = 1, \ldots, n$, let $(S^{(i)}, d^{(i)})$ be an arbitrary locally compact metric space, with some arbitrary but fixed reference point $x_\circ^{(i)} \in S^{(i)}$, and let $X^{(i)} : (\Omega, L, I) \to (S^{(i)}, d^{(i)})$ be an arbitrary measurable function.

Let $(\widetilde{S}, \widetilde{d}) \equiv \bigotimes_{i=1}^n (S^{(i)}, d^{(i)})$ be the product locally compact metric space. Define the function

$$X : (\Omega, L, I) \to (\widetilde{S}, \widetilde{d})$$

by $domain(X) \equiv \bigcap_{i=1}^n domain(X^{(i)})$ and by $X(\omega) \equiv (X^{(1)}(\omega), \ldots, X^{(n)}(\omega)) \in \widetilde{S}$ for each $\omega \in domain(X)$. Then X is a measurable function.

2. In addition, let (S,d) be an arbitrary complete metric space, and let $g : (\widetilde{S}, \widetilde{d}) \to (S,d)$ be an arbitrary function that is uniformly continuous on compact subsets of $(\widetilde{S}, \widetilde{d})$. Then

$$g(X) \equiv g \circ X : (\Omega, L, I) \to (S,d)$$

is a measurable function.

3. As a special case of Assertion 2, let (S,d) be an arbitrary locally compact metric space. Let $X^{(1)}, X^{(2)} : (\Omega, L, I) \to (S,d)$ be arbitrary measurable functions. Then the function

$$d(X^{(1)}, X^{(2)}) : (\Omega, L, I) \to R$$

is measurable.

Proof. 1. Let $i = 1, \ldots, n$ be arbitrary. Let $\xi^{(i)} \equiv (A_k^{(i)})_{k=1,2,\ldots}$ be an arbitrary but fixed binary approximation of the locally compact metric space $(S^{(i)}, d^{(i)})$ relative to the reference point $x_\circ^{(i)} \in S^{(i)}$, in the sense of Definition 3.2.1. Let

$$\pi^{(i)} \equiv (\{g_{k,x(i)}^{(i)} : x_i \in A_k^{(i)}\})_{k=1,2,\ldots}$$

be the partition of unity of $(S^{(i)}, d^{(i)})$ relative to the binary approximation $\xi^{(i)}$, in the sense of Definition 3.3.4. To lessen the burden on subscripts, write $A(i,k) \equiv A_k^{(i)}$ for each $k \geq 1$. Then, by Definition 3.2.1, we have

$$(d^{(i)}(\cdot, x_\circ^{(i)}) \leq 2^k) \subset \bigcup_{x(i) \in A(i,k)} (d^{(i)}(\cdot, x_i) \leq 2^{-k}) \tag{4.8.5}$$

for each $k \geq 1$. Moreover, by Proposition 3.3.5, $g_{k,x(i)}^{(i)} \in C(S,d)$ has values in $[0,1]$ and has support $(d^{(i)}(\cdot,x_i) \leq 2^{-k+1})$, for each $x_i \in A_k^{(i)}$. Furthermore, by the same Proposition 3.3.5,

$$\sum_{x(i)\in A(i,k)} g_{k,x(i)}^{(i)} \leq 1 \tag{4.8.6}$$

on $S^{(i)}$.

2. For ease of notations, assume that $n = 2$ for the remainder of this proof. The proof for the general case is similar.

3. We need to show that the function X with values in the product space $(\widetilde{S},\widetilde{d})$ is measurable. To that end, first let $f \in C(\widetilde{S},\widetilde{d})$ be arbitrary, with a modulus of continuity δ_f. Let A be an arbitrary integrable subset of Ω. Let $\alpha > 0$ be arbitrary. Let $k \geq 1$ be so large that (i) the function f has the set $(\widetilde{d}(\cdot,(x_\circ^{(1)},x_\circ^{(2)})) \leq 2^{k-1})$ as support and (ii) $2^{-k+1} < \delta_f(\alpha)$. Define the function g on $(\widetilde{S},\widetilde{d})$ by

$$g(y_1,y_2) \equiv \sum_{x(1)\in A(1,k)} \sum_{x(i2)\in A(2,g)} f(x_1,x_2) g_{k,x(1)}^{(1)}(y_1) g_{k,x(2)}^{(2)}(y_2) \tag{4.8.7}$$

for each $(y_1,y_2) \in (\widetilde{S},\widetilde{d})$.

4. By hypothesis, the function $X^{(1)}$ is measurable. Hence, for each $x_1 \in A_k^{(1)}$, the function $g_{k,x(1)}^{(1)}(X^{(1)})1_A$ is bounded and integrable. Similarly, for each $x_2 \in A_k^{(2)}$, the function $g_{k,x(2)}^{(2)}(X^{(2)})1_A$ is bounded and integrable. Thus equality 4.8.7 implies that the function $g(X^{(1)},X^{(2)})1_A$ is integrable.

5. Let $(y_1,y_2) \in (\widetilde{S},\widetilde{d})$ be arbitrary. We will show that

$$f(y_1,y_2) = h(y_1,y_2) \equiv \sum_{x(1)\in A(1,k)} \sum_{x(i2)\in A(2,g)} f(y_1,y_2) g_{k,x(1)}^{(1)}(y_1) g_{k,x(2)}^{(2)}(y_2).$$

Suppose, for the sake of a contradiction, that $|f(y_1,y_2) - h(y_1,y_2)| > 0$. Then $|f(y_1,y_2)| > 0$. Hence, according to Condition (i) in Step 3, we have

$$d^{(1)}(y_1,x_\circ^{(1)}) \vee d^{(2)}(y_2,x_\circ^{(2)}) \equiv \widetilde{d}((y_1,y_2),(x_\circ^{(1)},x_\circ^{(2)})) \leq 2^{k-1} < 2^k.$$

Therefore, by relation 4.8.5, there exists $(x_1,x_2) \in A_k^{(1)} \times A_k^{(2)}$ such that

$$y_i \in \bigcup_{x(i)\in A(i,k)} (d(\cdot,x_i) \leq 2^{-k})$$

for each $i = 1,2$. Therefore, according to Assertion 3 of Proposition 3.3.5, we have

$$\sum_{x(i)\in A(i,k)} g_{k,x(i)}^{(i)}(y_i) = 1$$

for each $i = 1,2$. It follows that

$$h(y_1,y_2) = f(y_1,y_2)\left(\sum_{x(1)\in A(1,k)} g^{(1)}_{k,x(1)}(y_1)\right)\left(\sum_{x(2)\in A(2,k)} g^{(2)}_{k,x(2)}(y_2)\right)$$
$$= f(y_1,y_2),$$

which is a contradiction. We conclude that $f(y_1,y_2) = h(y_1,y_2)$ for each $(y_1,y_2) \in (\widetilde{S},\widetilde{d})$.

6. Hence

$$\begin{aligned}
&|f(y_1,y_2) - g(y_1,y_2)| \\
&= |h(y_1,y_2) - g(y_1,y_2)| \\
&= \left|\sum_{x(1)\in A(1,k)}\ \sum_{x(i2)\in A(2,g)} (f(y_1,y_2) - f(x_1,x_2))g^{(1)}_{k,x(1)}(y_1)g^{(2)}_{k,x(2)}(y_2)\right| \\
&\le \sum_{x(1)\in A(1,k)}\ \sum_{x(i2)\in A(2,g)} |f(y_1,y_2) - f(x_1,x_2)|g^{(1)}_{k,x(1)}(y_1)g^{(2)}_{k,x(2)}(y_2).
\end{aligned} \tag{4.8.8}$$

Suppose, in the last displayed sum, the summand corresponding to x_1,x_2 is positive. Then $g^{(1)}_{k,x(1)}(y_1) > 0$. Hence, according to the last statement in Step 1, we have

$$d^{(i)}(y_i,x_i) \le 2^{-k+1} < \delta_f(\alpha)$$

for each $i = 1,2$, where the last inequality is by Condition (ii) in Step 3. It follows that $\widetilde{d}((y_1,y_2),(x_1,x_2)) < \delta_f(\alpha)$ and therefore that $|f(y_1,y_2) - f(x_1,x_2)| < \alpha$. Thus

$$|f(y_1,y_2) - f(x_1,x_2)|g^{(1)}_{k,x(1)}(y_1)g^{(2)}_{k,x(2)}(y_2) \le \alpha g^{(1)}_{k,x(1)}(y_1)g^{(2)}_{k,x(2)}(y_2).$$

Consequently, inequality 4.8.8 implies that

$$|f(y_1,y_2) - g(y_1,y_2)| \le \alpha \sum_{x(1)\in A(1,k)} g^{(1)}_{k,x(1)}(y_1) \sum_{x(i2)\in A(2,g)} g^{(2)}_{k,x(2)}(y_2) \le \alpha,$$

where the last inequality is thanks to inequality 4.8.6, and where $(y_1,y_2) \in (\widetilde{S},\widetilde{d})$ is arbitrary.

7. It follows that

$$|f(X^{(1)},X^{(2)})1_A - g(X^{(1)},X^{(2)})1_A| \le \alpha 1_A.$$

Since $\alpha > 0$ is arbitrarily small, and since $g(X^{(1)},X^{(2)})1_A$ is integrable, as established in Step 4, Theorem 4.5.10 implies that the function $f(X^{(1)},X^{(2)})1_A$ is integrable, where the integrable set A is arbitrary.

8. At the same time,

$$\begin{aligned}
&\mu(d((X_1,X_2),(x^{(1)}_\circ,x^{(2)}_\circ)) > a)A \\
&\quad\le \mu((d^{(1)}(X_1,x^{(1)}_\circ) > a)A) + \mu((d^{(2)}(X_1,x^{(2)}_\circ) > a)A) \to 0,
\end{aligned}$$

because each measure on the right-hand side converges to 0 as $a \to \infty$, thanks to the measurability of the functions X_1, X_2.

9. In view of Steps 7 and 8, Proposition 4.8.12 implies that the function

$$X \equiv (X^{(1)}, X^{(2)}) : (\Omega, L, I) \to (\widetilde{S}, \widetilde{d})$$

is measurable. Assertion 1 is proved.

10. Next, suppose (S,d) is a complete metric space and suppose a function $g : (\widetilde{S}, \widetilde{d}) \to (S,d)$ is uniformly continuous on compact subsets of $(\widetilde{S}, \widetilde{d})$. Since the product space $(\widetilde{S}, \widetilde{d})$ is locally compact, each bounded subset is contained in a compact subset. Hence the function g is uniformly continuous and bounded on bounded subsets. Therefore, according to Proposition 4.8.10, the composite function $g(X) : (\Omega, L, I) \to (S,d)$ is measurable. Assertion 2 is proved.

11. Now let (S,d) be an arbitrary locally compact metric space. Let $X^{(1)}, X^{(2)} \colon (\Omega, L, I) \to (S,d)$ be arbitrary measurable functions. Then the function $(X^{(1)}, X^{(2)}) \colon (\Omega, L, I) \to (\widetilde{S}, \widetilde{d})$ is measurable, by Assertion 1. At the same time, the distance function $d : (\widetilde{S}, \widetilde{d}) \to R$ is uniformly continuous on compact subsets of $(\widetilde{S}, \widetilde{d})$. Hence the composite function $d(X^{(1)}, X^{(2)})$ is measurable by Assertion 2. Thus Assertion 3 and the proposition are proved. □

Corollary 4.8.21. Some operations that preserve measurability for real-valued measurable functions. *Let $X, Y : (\Omega, L, I) \to R$ be real-valued measurable functions. Then $aX + bY$, 1, $X \vee Y$, $X \wedge Y$, $|X|$, and X^α are measurable functions for any real numbers a, b, α with $\alpha \geq 0$. Let A, B be measurable sets. Then $A \cup B$ and AB are measurable. Moreover, $(A \cup B)^c = A^c B^c$ and $(AB)^c = A^c \cup B^c$.*

Proof. Each of the real-valued functions $aX+bY$, 1, $X \vee Y$, $X \wedge Y$, $|X|$, and X^α can be written as $g(X,Y)$ for some function $g : R^2 \to R$ that is uniformly continuous on compact subsets of R^2. Thus these functions are measurable by Assertion 2 of Proposition 4.8.20.

Let A, B be measurable sets with indicators U, V, respectively. Then $U \vee V$ is a measurable indicator, with

$$(U \vee V = 1) = (U = 1) \cup (V = 1) = A \cup B.$$

Hence $A \cup B$ is a measurable set, with $U \vee V$ as a measurable indicator. Moreover, $(A \cup B)^c = (U \vee V = 0) = (U = 0)(V = 0) = A^c B^c$. Similarly, AB is measurable, with $(AB)^c = A^c \cup B^c$. □

Proposition 4.8.22. A real-valued measurable function dominated by an integrable function is integrable. *If X is a real-valued measurable function such that $|X| \leq Y$ for some nonnegative integrable function Y, then X is integrable. In particular, if A is a measurable set and Z is an integrable function, then $Z1_A$ is integrable.*

Proof. Let $(a_n)_{n=1,2,...}$ be an increasing sequence of positive real numbers with $a_n \to \infty$. Let $(b_n)_{n=1,2,...}$ be a decreasing sequence of positive real numbers with $b_n \to 0$. Let $n \geq 1$ be arbitrary. Then the function $f : R \to R$ defined by $f(x) \equiv (-a_n) \wedge x \vee a_n$ for each $x \in R$ that is a member of $C_{ub}(R)$. Hence the function

$$X_n \equiv ((-a_n) \wedge X \vee a_n)1_{(b(n)<Y)} = f(X)1_{(b(n)<Y)} \tag{4.8.9}$$

is integrable, thanks to the measurability of the functions X and Y.

Define the full set

$$D \equiv (|X| \leq Y) \cap ((b_n < Y) \cup (b_n \geq Y)) \cap ((|X| \leq a_n) \cup (|X| > a_n)). \tag{4.8.10}$$

Consider each $\omega \in D$.

Suppose $|X(\omega) - X_n(\omega)| > 0$. Suppose also that $b_n < Y(\omega)$ and $|X(\omega)| \leq a_n$. Then, by equality 4.8.9, we obtain $X_n(\omega) = X(\omega)$, which is a contradiction. Hence (i) $b_n \geq Y(\omega)$, or (ii) $b_n < Y(\omega)$ and $X(\omega) > a_n$, or (iii) $b_n < Y(\omega)$ and $X(\omega) < -a_n$. Consider case (i). Then $1_{(Y\leq b(n))}(\omega) = 1$, and $X_n(\omega) = 0$ by equality 4.8.9. Hence

$$|X(\omega) - X_n(\omega)| = |X(\omega)| \leq Y(\omega) \leq Y(\omega)1_{(Y\leq b(n))}(\omega),$$

where the first inequality is by the defining equality 4.8.10. Next consider case (ii). Then $1_{(b(n)<Y)}(\omega) = 1$ and $X_n(\omega) = a_n 1_{(b(n)<Y)}(\omega) = a_n$. Hence

$$|X(\omega) - X_n(\omega)| = |X(\omega) - a_n| \leq |X(\omega)| \leq Y(\omega)1_{(|X|>a(n))}(\omega).$$

Now consider case (iii). Then $1_{(b(n)<Y)}(\omega) = 1$ and $X_n(\omega) = -a_n$. Hence

$$|X(\omega) - X_n(\omega)| = |X(\omega) + a_n| \leq |X(\omega)| \leq Y(\omega)1_{(|X|>a(n))}(\omega).$$

Combining, we see that, in each of the cases (i–iii), we have

$$|X - X_n| \leq Y(1_{(Y\leq b(n))} + 1_{(|X|>a(n))}). \tag{4.8.11}$$

Now note that $\mu(|X| > a_n) \to 0$ as $n \to \infty$. Hence $IY1_{(|X|>a(n))} \to 0$ as $n \to \infty$, by Assertion 2 of Proposition 4.7.1. At the same time, by Assertion 3 of Proposition 4.7.1, we have $I(Y1_{(Y>a(n))}) \to 0$. Combining, we obtain $IY(1_{(Y\leq b(n))} + 1_{(|X|>a(n))}) \to 0$. Hence, by Theorem 4.5.10, inequality 4.8.11 implies that the real-valued measurable function X is integrable, as alleged. □

4.9 Convergence of Measurable Functions

In this section, let (Ω, L, I) be a complete integration space, and let (S,d) be a complete metric space, with a fixed reference point $x_\circ \in S$. In the case where $S = R$, it is understood that d is the Euclidean metric and that $x_\circ = 0$. We will introduce several notions of convergence of measurable functions on (Ω, L, I) with values in (S,d). We will sometimes write (a_i) for short for a given sequence $(a_i)_{i=1,2,\ldots}$.

Recall the following definition.

Definition 4.9.1. Limit of a sequence of functions. If $(Y_i)_{i=1,2,\ldots}$ is a sequence of functions from a set Ω' to the metric space (S,d), and if the set

$$D \equiv \left\{ \omega \in \bigcap_{i=1}^{\infty} domain(Y_i) : \lim_{i\to\infty} Y_i(\omega) \text{ exists in } (S,d) \right\}$$

is nonempty, then the function $\lim_{i\to\infty} Y_i$ is defined by $domain(\lim_{i\to\infty} Y_i) \equiv D$ and by $(\lim_{i\to\infty} Y_i)(\omega) \equiv \lim_{i\to\infty} Y_i(\omega)$ for each $\omega \in D$. □

Definition 4.9.2. Convergence in measure, a.u., a.e., and in L_1. For each $n \geq 1$, let X, X_n be functions on the complete integration space (Ω, L, I), with values in the complete metric space (S,d).

1. The sequence (X_n) is said to *converge to X uniformly on a subset A* of Ω if for each $\varepsilon > 0$, there exists $p \geq 1$ so large that $A \subset \bigcap_{n=p}^{\infty}(d(X_n, X) \leq \varepsilon)$.

2. The sequence (X_n) is said to *converge to X almost uniformly* (a.u.) if for each integrable set A and for each $\varepsilon > 0$, there exists an integrable set B with $\mu(B) < \varepsilon$ such that X_n converges to X uniformly on AB^c.

3. The sequence (X_n) is said to *converge to X in measure* if for each integrable set A and for each $\varepsilon > 0$, there exists $p \geq 1$ so large that for each $n \geq p$, there exists an integrable set B_n with $\mu(B_n) < \varepsilon$ and $AB_n^c \subset (d(X_n, X) \leq \varepsilon)$.

4. The sequence (X_n) is said to be *Cauchy in measure* if for each integrable set A and for each $\varepsilon > 0$, there exists $p \geq 1$ so large that for each $m, n \geq p$ there exists an integrable set $B_{m,n}$ with $\mu(B_{m,n}) < \varepsilon$ and $AB_{m,n}^c \subset (d(X_n, X_m) \leq \varepsilon)$.

5. Suppose $S = R$ and $X, X_n \in L$ for $n \geq 1$. The sequence (X_n) is said to *converge to X in L_1* if $I|X_n - X| \to 0$.

6. The sequence (X_n) is said to converge to X on an integrable subset A if $A \subset domain(\lim_{n\to\infty} X_n)$ and if $X = \lim_{n\to\infty} X_n$ on A. The sequence (X_n) is said to converge to X *almost everywhere* (a.e.) on an integrable subset A if (X_n) converges to X on DA for some full set D. □

We will use the abbreviation $X_n \to X$ to stand for "(X_n) converges to X" in whichever sense specified.

Proposition 4.9.3. a.u. Convergence implies convergence in measure, etc. *For each $n \geq 1$, let X, X_n be functions on the complete integration space (Ω, L, I), with values in the complete metric space (S,d). Then the following conditions hold:*

1. If $X_n \to X$ a.u. then (i) *X is defined a.e. on each integrable set A,* (ii) *$X_n \to X$ in measure, and* (iii) *$X_n \to X$ a.e. on each integrable set A.*

2. If (i′) *X_n is a measurable function for each $n \geq 1$* and (ii′) *$X_n \to X$ in measure, then X is a measurable function.*

3. If (i$''$) X_n *is a measurable function for each* $n \geq 1$, *and* (ii$''$) $X_n \to X$ a.u., *then* X *is a measurable function.*

Proof. 1. Suppose $X_n \to X$ a.u. Let the integrable set A and $n \geq 1$ be arbitrary. Then, by Definition 4.9.2 for a.u. convergence, there exists an integrable set B_n with $\mu(B_n) < 2^{-n}$ such that $X_h \to X$ uniformly on AB_n^c. Hence, by Definition 4.9.2 for uniform convergence, there exists $p \equiv p_n \geq 1$ so large that

$$AB_n^c \subset \bigcap_{h=p(n)}^{\infty} (d(X_h, X) \leq 2^{-n}). \tag{4.9.1}$$

Consequently, $AB_n^c \subset domain(X)$. Define the integrable set $B \equiv \bigcap_{k=1}^{\infty} \bigcup_{n=k}^{\infty} B_n$. Then

$$\mu(B) \leq \sum_{n=k}^{\infty} \mu(B_n) < \sum_{n=k}^{\infty} 2^{-n} = 2^{-k+1}$$

for each $k \geq 1$. Hence B is a null set, and $D \equiv B^c$ is a full set. Moreover,

$$AD \equiv AB^c = \bigcup_{k=1}^{\infty} \bigcap_{n=k}^{\infty} AB_n^c \subset \bigcup_{k=1}^{\infty} \bigcap_{n=k}^{\infty} domain(X) \subset domain(X).$$

Thus the function X is defined a.e. on the integrable set A. Part (i) of Assertion 1 is proved.

2. Let $\varepsilon > 0$ be arbitrary. Then, by Definition 4.9.2 for a.u. convergence, there exists an integrable set C with $\mu(C) < \varepsilon$ such that $X_n \to X$ uniformly on AC^c. Hence there exists $p \geq 1$ so large that $AC^c \subset \bigcap_{n=p}^{\infty} (d(X_n, X) \leq \varepsilon)$.

Next let $n \geq p$ be arbitrary. Define $C_n \equiv C$. Then $\mu(C_n) < \varepsilon$ and $AC_n^c = AC^c \subset (d(X_n, X) < \varepsilon)$. Thus the condition in Definition 4.9.2 is verified for the convergence $X_n \to X$ in measure. Part (ii) of Assertion 1 is proved.

3. By Step 1, we have $X_h \to X$ uniformly on AB_n^c, for each $n \geq 1$. It follows that $X_h \to X$ at each point in $AD = \bigcup_{k=1}^{\infty} \bigcap_{n=k} AB_n^c$, where $D \equiv B^c$ is a full set. Thus $X_n \to X$ a.e. on A, where A is an arbitrary integrable set. Part (iii) of Assertion 1 is also verified. Assertion 1 is proved.

4. Next, suppose (i$'$)X_n is measurable for each $n \geq 1$, and suppose (ii$'$) $X_n \to X$ in measure. We need to prove that X is a measurable function. To that end, let $f \in C_{ub}(S)$ be arbitrary. Then $|f| \leq c$ on S for some $c > 0$, and f has a modulus of continuity δ_f. Let A be an arbitrary integrable set. Let $k \geq 1$ be arbitrary. Because $X_n \to X$ in measure, there exists $n \equiv n_k \geq 1$ and an integrable set $B_n \equiv B_{n(k)}$, with

$$\mu(B_n) < k^{-1} \wedge \delta_f(k^{-1}) \tag{4.9.2}$$

and with

$$AB_n^c \subset (d(X_n, X) < k^{-1} \wedge \delta_f(k^{-1})) \subset (|f(X) - f(X_n)| < k^{-1}).$$

This implies that

$$(|f(X) - f(X_n)| \geq k^{-1}) \cap (AB_n \cup AB_n^c) \subset AB_n$$

and that

$$AB_n^c \subset (d(X_n, X) < 1). \tag{4.9.3}$$

Because X_n is measurable, we have $f(X_{n(k)})1_A \in L$. Moreover, on the full set $B_n^c \cup B_n$, we have

$$\begin{aligned} |f(X)1_A - f(X_{n(k)})1_A| &= |f(X) - f(X_n)|1_{AB(n)^c} + |f(X) - f(X_n)|1_{AB(n)} \\ &\leq k^{-1}1_{AB(n)^c} + 2c1_{AB(n)} \leq k^{-1}1_A + 2c1_{B(n(k))}, \end{aligned} \tag{4.9.4}$$

where

$$I(k^{-1}1_A + 2c1_{B(n(k))}) = k^{-1}\mu(A) + 2c\mu(B_n) < k^{-1}(\mu(A) + 2c) \rightarrow 0$$

as $k \rightarrow \infty$. Hence, by Theorem 4.5.10, inequality 4.9.4 implies that $f(X)1_A \in L$. Condition (i) in Definition 4.8.1 has been proved for X. Hence, by Lemma 4.8.3, there exists a countable subset J of R such that the set $(d(X, x_\circ) > a)A$ is integrable for each $a \in J_c$.

5. It remains to verify Condition (ii) in Definition 4.8.1 for X to be measurable. To that end, consider each $\varepsilon > 0$. Suppose $k \geq 1$ is so large that $k^{-1} < \varepsilon$. Let $n \equiv n_k$, as defined in Step 4. Since X_n is a measurable function, there exists $\alpha_k > 0$ so large that

$$\mu(d(X_n, x_\circ) > \alpha_k)A < k^{-1}. \tag{4.9.5}$$

Consider each $b \in J_c$ such that $b > \alpha_k + 1$. Then

$$\begin{aligned} (d(X, x_\circ) > b)A &\subset (d(X_n, x_\circ) > \alpha_k)A \cup (d(X_n, X) > 1)A \\ &\subset (d(X_n, x_\circ) > \alpha_k)A \cup (d(X_n, X) > 1)AB_n^c \cup AB_n \\ &\subset (d(X_n, x_\circ) > \alpha_k)A \cup (d(X_n, X) > 1)(d(X_n, X) < 1) \cup AB_n \\ &\subset (d(X_n, x_\circ) > \alpha_k)A \cup \phi \cup B_n, \end{aligned}$$

where the third set-inclusion relation follows from relation 4.9.3. Hence, in view of inequalities 4.9.5 and 4.9.2,

$$\mu(d(X, x_\circ) > b)A \leq \mu(d(X_n, x_\circ) > \alpha_k)A + \mu(B_n) < k^{-1} + k^{-1} = 2k^{-1} < 2\varepsilon,$$

where $b \in J_c$ is arbitrary with $b > \alpha_k + 1$ and where $\varepsilon > 0$ is arbitrary. We conclude that $\mu(d(X, x_\circ) > b)A \rightarrow 0$ as $b \rightarrow \infty$. Thus Condition (ii) in Definition 4.8.1 is verified. Accordingly, the function X is measurable. Assertion 2 is proved.

6. Suppose (i) X_n is measurable for each $n \geq 1$ and (ii) $X_n \rightarrow X$ a.u. Then, by Assertion 1, we have $X_n \rightarrow X$ in measure. Hence, by Assertion 2, the function X is measurable. Assertion 3 and the proposition are proved. □

Proposition 4.9.4. In the case of σ-finite (Ω, L, I), each sequence Cauchy in measure converges in measure and contains an a.u. convergent subsequence. *Suppose (Ω, L, I) is σ-finite. For each $n \geq 1$, let X_n be a function on (Ω, L, I), with values in the complete metric space (S,d). Suppose the sequence $(X_n)_{n=1,2,\ldots}$ is Cauchy in measure. Then there exists a subsequence $(X_{n(k)})_{k=1,2,\ldots}$ such that $X \equiv \lim_{k\to\infty} X_{n(k)}$ is a measurable function, with $X_{n(k)} \to X$ a.u. and $X_{n(k)} \to X$ a.e. Moreover, $X_n \to X$ in measure.*

Proof. 1. Let $(A_k)_{k=1,2,\ldots}$ be a sequence of integrable sets that is an I-basis of (Ω, L, I). Thus (i) $A_k \subset A_{k+1}$ for each $k \geq 1$, (ii) $\bigcup_{k=1}^{\infty} A_k$ is a full set, and (iii) for each integrable set A we have $\mu(A_k A) \to \mu(A)$.

By hypothesis, (X_n) is Cauchy in measure. By Definition 4.9.2, for each $k \geq 1$ there exists $n_k \geq 1$ such that for each $m, n \geq n_k$, there exists an integrable set $B_{m,n,k}$ with

$$\mu(B_{m,n,k}) < 2^{-k} \tag{4.9.6}$$

and with

$$A_k B_{m,n,k}^c \subset (d(X_n, X_m) \leq 2^{-k}). \tag{4.9.7}$$

By inductively replacing n_k with $n_1 \vee \ldots \vee n_k$, we may assume that $n_{k+1} \geq n_k$ for each $k \geq 1$. Define

$$B_k \equiv B_{n(k+1),n(k),k}$$

for each $k \geq 1$. Then $\mu(B_k) < 2^{-k}$, and

$$A_k B_k^c \subset (d(X_{n(k+1)}, X_{n(k)}) \leq 2^{-k}) \tag{4.9.8}$$

for each $k \geq 1$. Let $i \geq 1$ be arbitrary. Define

$$C_i \equiv \bigcup_{k=i}^{\infty} B_k. \tag{4.9.9}$$

Then

$$\mu(A_i^c C_i) \leq \mu(C_i) \leq \sum_{k=i}^{\infty} 2^{-k} = 2^{-i+1}. \tag{4.9.10}$$

Hence $\bigcap_{i=1}^{\infty} A_i^c C_i$ is a null set and $D \equiv \bigcup_{i=1}^{\infty} A_i C_i^c$ is a full set. Moreover,

$$A_i C_i^c \subset \bigcap_{k=i}^{\infty} A_i B_k^c \subset \bigcap_{k=i}^{\infty} A_k B_k^c \subset \bigcap_{k=i}^{\infty} (d(X_{n(k+1)}, X_{n(k)}) \leq 2^{-k}), \tag{4.9.11}$$

in view of relation 4.9.8. Here the second inclusion is because $A_i \subset A_k$ for each $k \geq i$. Therefore, since (S,d) is complete, the sequence $(X_{n(k)})_{k=1,2,\ldots}$ converges uniformly on $A_i C_i^c$. In other words, $X_{n(k)} \to X$ uniformly on $A_i C_i^c$, where $X \equiv \lim_{k\to\infty} X_{n(k)}$.

Next let A be an arbitrary integrable set. Let $\varepsilon > 0$ be arbitrary. In view of Condition (iii), there exists $i \geq 1$ so large that $2^{-i+1} < \varepsilon$ and

$$\mu(AA_i^c) = \mu(A) - \mu(A_iA) < \varepsilon. \tag{4.9.12}$$

Define $B \equiv AA_i^c \cup C_i$. Then $\mu(B) < 2\varepsilon$. Moreover, $AB^c \subset A_iC_i^c$. Hence relation 4.9.11 implies that $X_{n(k)} \to X$ uniformly on AB^c. Since $\varepsilon > 0$ is arbitrary, we conclude that $X_{n(k)} \to X$ a.u. It then follows from Assertion 1 of Proposition 4.9.3 that $X_{n(k)} \to X$ in measure, $X_{n(k)} \to X$ a.e. on each integrable set, and X is measurable.

2. It remains only to prove that $X_n \to X$ in measure. To that end, define, for each $m \geq n_i$,

$$\overline{B}_m \equiv AA_i^c \cup B_{m,n(i),i} \cup C_i.$$

Then, in view of inequalities 4.9.12, 4.9.6, and 4.9.10, we have, for each $m \geq n_i$,

$$\mu(\overline{B}_m) = \mu(AA_i^c \cup B_{m,n(i),i} \cup C_i) < \varepsilon + 2^{-i} + 2^{-i+1} < 3\varepsilon. \tag{4.9.13}$$

Moreover,

$$\begin{aligned}
A\overline{B}_m^c &= A(A^c \cup A_i)B_{m,n(i),i}^c C_i^c \\
&= AA_iB_{m,n(i),i}^c C_i^c = (A_iB_{m,n(i),i}^c)(AA_iC_i^c) \\
&\subset (d(X_m, X_{n(i)}) \leq 2^{-i})(d(X, X_{n(i)}) \leq 2^{-i+1}) \\
&\subset (d(X, X_m) \leq 2^{-i} + 2^{-i+1}) \\
&\subset (d(X, X_m) < 2\varepsilon)
\end{aligned} \tag{4.9.14}$$

for each $m \geq n(i)$, where the first inclusion is because of relations 4.9.7 and 4.9.11. Since the integrable set A and the positive real number $\varepsilon > 0$ are arbitrary, we have verified the condition in Definition 4.9.2 for $X_m \to X$ in measure. The proposition is proved. □

Proposition 4.9.5. Convergence in measure in terms of convergence of integrals. *For each $n \geq 1$, let X, X_n be functions on (Ω, L, I), with values in the complete metric space (S,d), such that $d(X_n, X_m)$ and $d(X_n, X)$ are measurable for each $n,m \geq 1$. Then the following conditions hold:*

1. If $I(1 \wedge d(X_n, X))1_A \to 0$ for each integrable set A, then $X_n \to X$ in measure.

2. Conversely, if $X_n \to X$ in measure, then $I(1 \wedge d(X_n, X))1_A \to 0$ for each integrable set A.

3. The sequence (X_n) is Cauchy in measure iff $I(1 \wedge d(X_n, X_m))1_A \to 0$ as $n,m \to \infty$ for each integrable set A.

Proof. Let the integrable set A and the positive real number $\varepsilon \in (0,1)$ be arbitrary.

1. Suppose $I(1 \wedge d(X_n, X))1_A \to 0$. Then, by Chebychev's inequality,

$$\mu(d(X_n, X) > \varepsilon)A) \leq \mu(1 \wedge d(X_n, X))1_A > \varepsilon) \leq \varepsilon^{-1}I(1 \wedge d(X_n, X))1_A \to 0$$

as $n \to \infty$. In particular, there exists $p \geq 1$ so large that $\mu(1 \wedge d(X_n, X) > \varepsilon)A < \varepsilon$ for each $n \geq p$. Now consider each $n \geq p$. Define the integrable set $B_n \equiv (1 \wedge d(X_n, X) > \varepsilon)A$. Then $\mu(B_n) < \varepsilon$ and $AB_n^c \subset (d(X_n, X) \leq \varepsilon)$. Thus $X_n \to X$ in measure.

2. Conversely, suppose $X_n \to X$ in measure. Let $\varepsilon > 0$ be arbitrary. Then there exists $p \geq 1$ so large that for each $n \geq p$, there exists an integrable set B_n with $\mu(B_n) < \varepsilon$ and $AB_n^c \subset (d(X_n, X) \leq \varepsilon)$. Hence

$$\begin{aligned} I(1 \wedge d(X_n, X))1_A &= I(1 \wedge d(X_n, X))1_{AB(n)} + I(1 \wedge d(X_n, X))1_{AB(n)^c} \\ &\leq I1_{B(n)} + I\varepsilon 1_A < \varepsilon + \varepsilon\mu(A), \end{aligned}$$

where $\varepsilon > 0$ is arbitrary. Thus $I(1 \wedge d(X_n, X))1_A \to 0$.

3. The proof of Assertion 3 is similar to that of Assertions 1 and 2, and is omitted.

□

The next proposition will be convenient for establishing a.u. convergence.

Proposition 4.9.6. Sufficient condition for a.u. convergence. Suppose (Ω, L, I) is σ-finite, with an I-basis $(A_i)_{i=1,2,\ldots}$. *For each $n, m \geq 1$, let X_n be a function on Ω, with values in the complete metric space (S, d), such that $d(X_n, X_m)$ is measurable.* Suppose that for each $i \geq 1$, there exists a sequence $(\varepsilon_n)_{n=1,2,\ldots}$ of positive real numbers such that $\sum_{n=1}^{\infty} \varepsilon_n < \infty$ and such that $I(1 \wedge d(X_n, X_{n+1}))1_{A(i)} < \varepsilon_n^2$ for each $n \geq 1$. Then $X \equiv \lim_{n\to\infty} X_n$ exists on a full set, and $X_n \to X$ *a.u. If, in addition, X_n is measurable for each $n \geq 1$, then the limit X is measurable.*

Proof. 1. As an abbreviation, write $Z_n \equiv 1 \wedge d(X_{n+1}, X_n)$ for each $n \geq 1$. Let A be an arbitrary integrable set and let $\varepsilon > 0$ be arbitrary. Take $i \geq 1$ so large that $\mu(AA_i^c) < \varepsilon$. By hypothesis, there exists a sequence $(\varepsilon_n)_{n=1,2,\ldots}$ of positive real numbers such that $\sum_{n=1}^{\infty} \varepsilon_n < \infty$ and such that $IZ_n 1_{A(i)} < \varepsilon_n^2$ for each $n \geq 1$. Chebychev's inequality then implies that

$$\mu(Z_n > \varepsilon_n)A_i \leq \mu(Z_n 1_{A(i)} > \varepsilon_n) \leq I(\varepsilon_n^{-1} Z_n 1_{A(i)}) < \varepsilon_n$$

for each $n \geq 1$. Let $p \geq 1$ be so large that $\sum_{n=p}^{\infty} \varepsilon_n < 1 \wedge \varepsilon$. Let $C \equiv \bigcup_{n=p}^{\infty} (Z_n > \varepsilon_n)A_i$ and let $B \equiv AA_i^c \cup C$. Then $\mu(B) < \varepsilon + \sum_{n=p}^{\infty} \varepsilon_n < 2\varepsilon$. Moreover,

$$\begin{aligned} AB^c = A((A^c \cup A_i)C^c) = AA_iC^c &= AA_i \bigcap_{n=p}^{\infty} (Z_n \leq \varepsilon_n) \\ &\subset \bigcap_{n=p}^{\infty} (Z_n \leq \varepsilon_n) \subset \bigcap_{n=p}^{\infty} (d(X_{n+1}, X_n) \leq \varepsilon_n). \end{aligned}$$

Since $\sum_{n=p}^{\infty} \varepsilon_n < \infty$, it follows that $X_n \to X$ uniformly on AB^c, where $X \equiv \lim_{n\to\infty} X_n$. Since A and $\varepsilon > 0$ are arbitrary, we see that $X_n \to X$ a.u.

2. If, in addition, X_n is measurable for each $n \geq 1$, then X is a measurable function by Assertion 3 of Proposition 4.9.3.

□

Proposition 4.9.7. A continuous function preserves convergence in measure. *Let (Ω, L, I) be a complete integration space. Let $(S',d'),(S'',d'')$ be locally compact metric spaces and let $(\widetilde{S},\widetilde{d}) \equiv (S',d') \otimes (S'',d'')$ denote the product metrics space. Let $X', X'_1 X'_2, \ldots$ be a sequence of measurable functions with values in S' such that $X'_n \rightarrow X'$ in measure. Similarly, let $X'', X''_1 X''_2, \ldots$ be a sequence of measurable functions with values in S'' such that $X''_n \rightarrow X''$ in measure.*

Let $f : (\widetilde{S},\widetilde{d}) \rightarrow (S,d)$ be a continuous function with values in a complete metric space (S,d) that is uniformly continuous on compact subsets of $(\widetilde{S},\widetilde{d})$. Then $f(X'_n, X''_n) \rightarrow f(X', X'')$ in measure as $n \rightarrow \infty$.

Generalization to $m \geq 2$ sequences of measurable functions is similar.

Proof. 1. Let $x'_\circ$ and $x''_\circ$ be fixed reference points in $S' and S''$, respectively. Write $x_\circ \equiv (x'_\circ, x''_\circ) \in \widetilde{S}$. For each $x, y \in \widetilde{S}$, write $x \equiv (x', x'')$ and $y \equiv (y', y'')$. Likewise, write $X \equiv (X', X'')$ and $X_n \equiv (X'_n, X''_n)$ for each $n \geq 1$. For each $n \geq 1$, the functions $f(X), f(X_n)$ are measurable functions with values in S, thanks to Assertion 2 of Proposition 4.8.20.

2. Let A be an arbitrary integrable set. Let $\varepsilon > 0$ be arbitrary. By Condition (ii) in Definition 4.8.1, there exists $a > 0$ so large that $\mu(B') < \varepsilon$ and $\mu(B'') < \varepsilon$, where $B' \equiv (d'(x'_\circ, X') > a)A$ and $B'' \equiv (d''(x''_\circ, X'') > a)A$. Since $(\widetilde{S},\widetilde{d})$ is locally compact, the bounded subset $(\widetilde{d}(x_\circ, \cdot) \leq a + 1)$ is contained in some compact subset. On the other hand, by hypothesis, the function $f : (\widetilde{S},\widetilde{d}) \rightarrow (S,d)$ is uniformly continuous on each compact subset of $\widetilde{S}$. Hence there exists $\delta_1 > 0$ so small that we have

$$d(f(x), f(y)) < \varepsilon \tag{4.9.15}$$

for each

$$x, y \in (\widetilde{d}(x_\circ, \cdot) \leq a + 1)$$

with $\widetilde{d}(x,y) < \delta_1$. Define $\delta \in (0, 1 \wedge \delta_1)$. For each $n \geq 1$, define $C'_n \equiv (d'(X'_n, X') \geq \delta)A$ and $C''_n \equiv (d''(X''_n, X'') \geq \delta)A$. By hypothesis, $X'_n \rightarrow X'$, and $X''_n \rightarrow X''$ in measure as $n \rightarrow \infty$. Hence there exists $p \geq 1$ so large that $\mu(C'_n) < \varepsilon$ and $\mu(C''_n) < \varepsilon$ for each $n \geq p$. Consider any $n \geq p$. Then

$$\mu(B' \cup B'' \cup C'_n \cup C''_n) < 4\varepsilon.$$

Moreover,

$$\begin{aligned}
&A(B' \cup B'' \cup C'_n \cup C''_n)^c \\
&\quad = AB'^c B''^c C'^c_n C''^c_n \\
&\quad = A(d'(x'_\circ, X') \leq a; d''(x''_\circ, X'') \leq a; d'(X'_n, X') < \delta; d''(X''_n, X'') < \delta) \\
&\qquad \subset A(d'(x'_\circ, X') \leq a; d'(x'_\circ, X'_n) \leq a + 1; d''(x''_\circ, X'') \leq a; d''(x''_\circ, X''_n) \\
&\quad \leq a + 1; d'(X'_n, X') < \delta; d''(X''_n, X'') < \delta)
\end{aligned}$$

$$= A(\widetilde{d}(x_\circ, X) \le a)(\widetilde{d}(x_\circ, X_n) \le a+1)(\widetilde{d}(X_n, X) < \delta_1)$$
$$\subset (\widetilde{d}(x_\circ, X) \le a+1)(\widetilde{d}(x_\circ, X_n) \le a+1)(\widetilde{d}(X_n, X) < \delta_1)$$
$$\subset (d(f(X), f(X_n)) < \varepsilon),$$

where the last set-inclusion relation is thanks to inequality 4.9.15. Since $\varepsilon > 0$ and A are arbitrary, the condition in Definition 4.9.2 is verified for $f(X_n) \to f(X)$ in measure. □

Theorem 4.9.8. Dominated Convergence Theorem. *Let $(X_n)_{n=1,2,...}$ be a sequence of real-valued measurable functions on the complete integration space (Ω, L, I), and let X be a real-valued function defined a.e. on Ω, with $X_n \to X$ in measure. Suppose there exists an integrable function Y such that $|X| \le Y$ a.e. and $|X_n| \le Y$ a.e. for each $n \ge 1$. Then X, X_n are integrable for each $n \ge 1$, and $I|X_n - X| \to 0$.*

Proof. 1. By Assertion 2 of Proposition 4.9.3, the function X is a measurable function. Since $|X| \le Y$ a.e. and $|X_n| \le Y$ a.e. for each $n \ge 1$, Proposition 4.8.22 implies that X, X_n are integrable for each $n \ge 1$.

2. Let $\varepsilon > 0$ be arbitrary. Since Y is integrable and is nonnegative a.e., there exists $a > 0$ so small that

$$I(Y \wedge a) < \varepsilon. \tag{4.9.16}$$

Define $A \equiv (Y > a)$. Then $|X_n - X| \le 2Y = 2(Y \wedge a)$ a.e. on A^c, for each $n \ge 1$. By Proposition 4.7.1, there exists $\delta \in (0, \varepsilon/(1+\mu A))$ so small that $IY1_B < \varepsilon$ for each integrable set B with $\mu(B) < \delta$. On the other hand, by hypothesis, $X_n \to X$ in measure. Hence there exists $m > 0$ so large that for each $n \ge m$, we have $AB_n^c \subset (|X_n - X| \le \delta)$ for some integrable set B_n with $\mu(B_n) < \delta$. Then

$$IY1_{B(n)} < \varepsilon. \tag{4.9.17}$$

Combining, for each $n \ge m$, we have, on the full set $AB_n \cup AB_n^c \cup A^c$,

$$|X - X_n| \le |X - X_n|1_{AB(n)} + |X - X_n|1_{AB(n)^c} + |X - X_n|1_{A^c}$$
$$\le 2Y1_{B(n)} + \delta 1_A + 2(Y \wedge a). \tag{4.9.18}$$

In view of inequalities 4.9.17 and 4.9.16, it follows that

$$I|X - X_n| \le I(2Y1_{B(n)} + \delta 1_A + 2(Y \wedge a))$$
$$\le 2\varepsilon + \delta\mu(A) + 2\varepsilon \le 2\varepsilon + \varepsilon + 2\varepsilon. \tag{4.9.19}$$

Since $\varepsilon > 0$ is arbitrary, we see that $I|X_n - X| \to 0$. □

The next definition introduces Newton's notation for the Riemann–Stieljes integration relative to a distribution function.

Definition 4.9.9. Lebesgue integration. Suppose F is a distribution function on R. Let I be the Riemann–Stieljes integration with respect to F, and let (R, L, I)

be the completion of $(R, C(R), I)$. We will use the notation $\int \cdot dF$ for I. For each $X \in L$, we write $\int X dF$ or $\int X(x) dF(x)$ for IX. An integrable function in L is then said to be integrable relative to the distribution function F, and a measurable function on (R, L, I) is said to be measurable relative to F.

Suppose X is a measurable function relative to F, and suppose $s, t \in R$ such that the functions $1_{(s\wedge t, s]}X$ and $1_{(s\wedge t, t]}X$ are integrable relative to F. Then we write

$$\int_s^t X dF \equiv \int_s^t X(x) dF(x) \equiv \int X 1_{(s\wedge t, t]} dF - \int X 1_{(s\wedge t, s]} dF.$$

Thus

$$\int_s^t X dF = -\int_t^s X dF.$$

If A is a measurable set relative to F such that $X 1_A$ is integrable, then we write

$$\int_A X dF \equiv \int_{x\in A} X(x) dF(x) \equiv \int X 1_A dF.$$

In the special case where $F(x) \equiv x$ for $x \in R$, we write $\int \cdot dx$ for $\int \cdot dF$. Let $s < t$ in R be arbitrary. The integration spaces $(R, L, \int \cdot dx)$ and $([s,t], L_{[s,t]}, \int_s^t \cdot dx)$ are called the *Lebesgue integration spaces* on R and $[s,t]$, respectively, and $\int \cdot dx$ and $\int_s^t \cdot dx$ are called the *Lebesgue integrations*. Then an integrable function in L or $L_{[s,t]}$ is said to be *Lebesgue integrable* and a measurable function is said to be *Lebesgue measurable*. □

Since the identity function Z, defined by $Z(x) \equiv x$ for each $x \in R$, is continuous and therefore a measurable function on $(R, L, \int \cdot dF)$, all but countably many $t \in R$ are regular points of Z. Hence $(s,t] = (s < Z \le t)$ is a measurable set in $(R, L, \int \cdot dF)$ for all but countably many $s, t \in R$. In other words, $1_{(s,t]}$ is measurable relative to F for all but countably many $s, t \in R$. Therefore the definition of $\int_s^t X(x) dF(x)$ is not vacuous.

Proposition 4.9.10. Intervals are Lebesgue integrable. *Let $s, t \in R$ be arbitrary with $s \le t$. Then each of the intervals $[s,t]$, (s,t), $(s,t]$, and $[s,t)$ is Lebesgue integrable, with the same Lebesgue measure equal to $t - s$, and with measure-theoretic complements $(-\infty, s) \cup (t, \infty)$, $(-\infty, s] \cup [t, \infty)$, $(-\infty, s] \cup (t, \infty)$, and $(-\infty, s) \cup [t, \infty)$, respectively. Each of the intervals $(-\infty, s)$, $(-\infty, s]$, (s, ∞), and $[s, \infty)$ is Lebesgue measurable.*

Proof. Consider the Lebesgue integration $\int \cdot dx$ and the Lebesgue measure μ.

Let $a, b \in R$ be such that $a < s \le t < b$. Define $f \equiv f_{a,s,t,b} \in C(R)$ such that $f \equiv 1$ on $[s,t]$, $f \equiv 0$ on $(-\infty, a] \cup [b, \infty)$, and f is linear on $[a,s]$ and on $[t,b]$. Let $t_0 < \cdots < t_n$ be any partition in the definition of a Riemann–Stieljes sum $S \equiv \sum_{i=1}^n f(t_i)(t_i - t_{i-1})$ such that $a = t_j$ and $b = t_k$ for some $j, k = 1, \ldots, n$

with $j \le k$. Then $S = \sum_{i=j+1}^{k} f(t_i)(t_i - t_{i-1})$ since f has $[a,b]$ as support. Hence $0 \le S \le \sum_{i=j+1}^{k}(t_i - t_{i-1}) = t_k - t_j = b - a$. Now let $n \to \infty, t_0 \to -\infty$, $t_n \to \infty$, and let the mesh of the partition approach 0. It follows from the last inequality that $\int f(x)dx \le b - a$. Similarly, $t - s \le \int f(x)dx$. Now, with s,t fixed, let $(a_k)_{k=1,2,...}$ and $(b_k)_{k=1,2,...}$ be sequences in R such that $a_k \uparrow s$ and $b_k \downarrow t$, and let $g_k \equiv f_{a_k,s,t,b_k}$ Then, by the previous argument, we have $t - s \le \int g_k(x)dx \le b_k - a_k \downarrow t - s$. Hence, by the Monotone Convergence Theorem, the limit $g \equiv \lim_{k\to\infty} g_k$ is integrable, with integral $t - s$. It is obvious that $g = 1$ or 0 on $domain(g)$. In other words, g is an indicator function. Moreover, $[s,t] = (g = 1)$. Hence $[s,t]$ is an integrable set, with $1_{[s,t]} = g$, with measure $\mu([s,t]) = \int g(x)dx = t - s$, and with measure-theoretic complement $[s,t]^c = (-\infty,s) \cup (t,\infty)$.

Next consider the half-open interval $(s,t]$. Since $(s,t] = \bigcup_{k=1}^{\infty}\left[s+\frac{1}{k},t\right]$, where $\left[s + \frac{1}{k},t\right]$ is integrable for each $k \ge 1$ and $\mu\left(\left[s + \frac{1}{k},t\right]\right) = t - s - \frac{1}{k} \uparrow t - s$ as $k \to \infty$, we have the integrability of $(s,t]$, and $\mu([s,t]) = \lim_{k\to\infty} \mu\left(\left[s+\frac{1}{k},t\right]\right) = t - s$. Moreover,

$$
\begin{aligned}
(s,t]^c &= \bigcap_{k=1}^{\infty}\left[s + \frac{1}{k},t\right]^c \\
&= \bigcap_{k=1}^{\infty}\left(\left(-\infty,s + \frac{1}{k}\right) \cup (t,\infty)\right) = (-\infty,s] \cup (t,\infty).
\end{aligned}
$$

The proofs for the intervals (s,t) and $[s,t)$ are similar.

Now consider the interval $(-\infty,s)$. Define the function X on the full set D by $X(x) = 1$ or 0 according as $x \in (-\infty,s)$ or $x \in [s,\infty)$. Let A be any integrable subset of R. Then, for each $n \ge 1$ with $n > -s$, we have $|X1_A - 1_{[-n,s)}1_A| \le 1_{(-\infty,-n)}1_A$ on the full set $D(A \cup A^c)([-n,s) \cup [-n,s)^c)$. At the same time, $\int 1_{(-\infty,-n)}(x)1_A(x)dx \to 0$. Therefore, by Theorem 4.5.10, the function $X1_A$ is integrable. It follows that for any $f \in C(R)$, the function $f(X)1_A = f(1)X1_A + f(0)(1_A - X1_A)$ is integrable. We have thus verified Condition (i) in Definition 4.8.1 for X to be measurable. At the same time, since $|X| \le 1$, we have trivially $\mu(|X| > a) = 0$ for each $a > 1$. Thus Condition (ii) in Definition 4.8.1 is also verified. We conclude that $(-\infty,s)$ is measurable. Similarly, we can prove that $(-\infty,s]$, (s,∞), and $[s,\infty)$ are all measurable. □

4.10 Product Integration and Fubini's Theorem

In this section, let (Ω',L',I') and (Ω'',L'',I'') be two arbitrary but fixed complete integration spaces. Let $\Omega \equiv \Omega' \times \Omega''$ denote the product set. We will construct the product integration space and embed the given integration spaces in it. The definitions and results can easily be generalized to more than two given integration spaces.

Definition 4.10.1. Direct product of functions. Let X', X'' be arbitrary members of L', L'', respectively. Define the function $X' \otimes X'' : \Omega \to R$ by $domain(X' \otimes X'') \equiv domain(X') \times domain(X'')$ and by $(X' \otimes X'')(\omega', \omega'') \equiv X'(\omega')X''(\omega'')$ for each $\omega \in \Omega$. The function $X' \otimes X''$ is then called the *direct product of the functions* X' and X''. When the risk of confusion is low, we will write $X' \otimes X''$ and $X'X''$ interchangeably. □

Definition 4.10.2. Simple functions. Let $n, m \geq 1$ be arbitrary. Let $X'_1, \ldots, X'_n \in L'$ be mutually exclusive indicators, and let $X''_1, \ldots, X''_m \in L''$ be mutually exclusive indicators. For each $i = 1, \ldots, n$ and $j = 1, \ldots, m$, let $c_{i,j} \in R$ be arbitrary. Then the real-valued function

$$X = \sum_{i=1}^{n} \sum_{j=1}^{m} c_{i,j} X'_i X''_j$$

is called a *simple function relative to* L', L''. Let L_0 denote the set of simple functions on $\Omega' \times \Omega''$. Two simple functions are said to be equal if they have the same domain and the same values on the common domain. In other words, equality in L_0 is the set-theoretic equality:

$$I(X) = \sum_{i=1}^{n} \sum_{j=1}^{m} c_{i,j} I'(X'_i) I''(X''_j). \tag{4.10.1}$$

□

Lemma 4.10.3. Simple functions constitute a linear space. *As in Definition 4.10.2, let L_0 be the set of simple functions on $\Omega' \times \Omega''$ relative to L', L''. Then the following conditions hold:*

1. *If $X \in L_0$, then $|X|, a \wedge X \in L_0$ for each $a > 0$.*
2. *L_0 is a linear space.*
3. *The function I on L_0 is linear.*

Proof. 1. Let $X, Y \in L_0$ be arbitrary. We may assume that X is the linear combination in Definition 4.10.2, and that Y is a similar linear combination $Y \equiv \sum_{k=1}^{p} \sum_{h=1}^{q} b_{k,h} Y'_k Y''_h$ in the notations of Definition 4.10.2. Define $X'_0 \equiv 1 - \sum_{i=1}^{n} X'_i$ and $X''_0 \equiv 1 - \sum_{j=1}^{m} X''_j$. Similarly, define $Y'_0 \equiv 1 - \sum_{k=1}^{p} Y'_k$ and $Y''_0 \equiv 1 - \sum_{h=1}^{q} Y''_h$. For convenience, define $c_{i,0} \equiv c_{0,j} \equiv 0$ for each $i = 0, \ldots, n$ and $j = 0, \ldots, m$. Define $b_{k,0} \equiv b_{0,h} \equiv 0$ for each $k = 0, \ldots, p$ and $h = 0, \ldots, q$.

2. Let $a > 0$ be arbitrary. Consider each $(\omega', \omega'') \in domain(X)$. We have either (i) $X'_i(\omega')X''_j(\omega'') = 0$ for each $i = 1, \ldots, n$ and $j = 1, \ldots, m$, or (ii) $X'_k(\omega')X''_h(\omega'') = 1$ for exactly one pair of k, h with $k = 1, \ldots, n$ and $h = 1, \ldots, m$. In case (i), we have

$$|X(\omega', \omega'')| = \left| \sum_{i=1}^{n} \sum_{j=1}^{m} c_{i,j} X'_i(\omega') X''_j(\omega'') \right| = 0 = \sum_{i=1}^{n} \sum_{j=1}^{m} |c_{i,j}| X'_i(\omega') X''_j(\omega'')$$

and

$$a \wedge X(\omega',\omega'') = a \wedge \sum_{i=1}^{n}\sum_{j=1}^{m} c_{i,j} X_i'(\omega')X_j''(\omega'')|$$

$$= a \wedge 0 = \sum_{i=1}^{n}\sum_{j=1}^{m} (a \wedge c_{i,j}) X_i'(\omega')X_j''(\omega'').$$

Consider case (ii). Then

$$|X(\omega',\omega'')| = \left|\sum_{i=1}^{n}\sum_{j=1}^{m} c_{i,j} X_i'(\omega')X_j''(\omega'')\right| = |c_{k,h} X_k'(\omega')X_h''(\omega'')|$$

$$= |c_{k,h}| X_k'(\omega')X_h''(\omega'') = \sum_{i=1}^{n}\sum_{j=1}^{m} |c_{i,j}| X_i'(\omega')X_j''(\omega''),$$

and

$$a \wedge X(\omega',\omega'') = a \wedge \sum_{i=1}^{n}\sum_{j=1}^{m} c_{i,j} X_i'(\omega')X_j''(\omega'') = a \wedge (c_{k,h} X_k'(\omega')X_h''(\omega''))$$

$$= (a \wedge c_{k,h}) X_k'(\omega')X_h''(\omega'') = \sum_{i=1}^{n}\sum_{j=1}^{m} (a \wedge c_{i,j}) X_i'(\omega')X_j''(\omega'').$$

Thus, in either case, the conditions in Definition 4.10.2 are satisfied for the functions $|X|, a \wedge X$ to be simple functions. Assertion 1 is proved.

3. Similarly, we can prove that L_0 is closed under scalar multiplication. It remains to show that L_0 is also closed under addition,

4. To that end, first note that if $Y' \in L'$ is an arbitrary integrable indicator, then we have $Y'X_0' \equiv Y' - \sum_{i=1}^{n} Y'X_i' \in L'$, whence $Y'X_i' \in L'$ is an integrable indicator for each $i = 0, \ldots, n$. Similarly for L''.

5. Then, in view of the observation in the previous paragraph,

$$(X_i'Y_k')_{i=0,\ldots,n;\, k=0,\ldots,p;\, (i,k)\neq(0,0)}$$

is a double sequence of N mutually exclusive integrable indicators in L', where $N \equiv (n+1)(p+1) - 1$. Let

$$(Z_\nu')_{\nu=0,\ldots,N} \equiv (X_{i(\nu)}'Y_{k(\nu)}')_{\nu=0,\ldots,N}$$

be an arbitrary but fixed rearrangement of the double sequence $(X_i'Y_k')_{i=0,\ldots,n;\; k=0,\ldots,p}$ into a single sequence, such that $i_0 = k_0 = 0$. Similarly,

$$(X_j''Y_h'')_{j=0,\ldots,m;\; h=0,\ldots,q;\, (j,h)\neq(0,0)}$$

is a sequence of M mutually exclusive integrable integrable indicators in L'', where $M \equiv (m+1)(q+1) - 1$. Let

$$(Z_\mu'')_{\mu=0,\ldots,M} \equiv (X_{j(\mu)}''Y_{h(\mu)}'')_{\mu=0,\ldots,M}$$

be an arbitrary but fixed rearrangement of the double sequence $(X''_j Y''_h)_{j=0,\dots,m;h=0,\dots,q}$ into a single sequence, such that $j_0 = h_0 = 0$.

6. For each $\nu = 0, \dots, N$ and for each $\mu = 0, \dots, M$, define the real number

$$a_{\nu,\mu} \equiv c_{i(\nu),j(\mu)} + b_{k(\nu),h(\mu)}.$$

Then, for each $\nu = 0, \dots, N$, we have

$$a_{\nu,0} = c_{i(\nu),j(0)} + b_{k(\nu),h(0)} = c_{i(\nu),0} + b_{k(\nu),0} = 0 + 0 = 0.$$

Similarly, for each $\mu = 0, \dots, M$, we have $a_{0,\mu} = 0$.

7. Then

$$\begin{aligned}
X + Y &\equiv \sum_{i=0}^{n}\sum_{j=0}^{m} c_{i,j} X'_i X''_j + \sum_{k=0}^{p}\sum_{h=0}^{q} b_{k,h} Y'_k Y''_h \\
&= \sum_{i=0}^{n}\sum_{j=0}^{m} c_{i,j} X'_i X''_j \left(\sum_{k=0}^{p}\sum_{h=0}^{q} Y'_k Y''_h\right) \\
&\quad + \sum_{k=0}^{p}\sum_{h=0}^{q} b_{k,h} Y'_k Y''_h \left(\sum_{i=0}^{n}\sum_{j=0}^{m} X'_i X''_j\right) \\
&= \sum_{i=0}^{n}\sum_{k=0}^{p}\sum_{j=0}^{m}\sum_{h=0}^{q} (c_{i,j} + b_{k,h})(X'_i Y'_k)(X''_j Y''_h) \\
&= \sum_{\nu=0}^{N}\sum_{\mu=0}^{M} (c_{i(\nu),j(\mu)} + b_{k(\nu),h(\mu)})(X'_{i(\nu)} Y'_{k(\nu)})(X''_{j(\mu)} Y''_{h(\mu)}) \\
&= \sum_{\nu=0}^{N}\sum_{\mu=0}^{M} a_{\nu,\mu} Z'_\nu Z''_\mu = \sum_{\nu=1}^{N}\sum_{\mu=1}^{M} a_{\nu,\mu} Z'_\nu Z''_\mu.
\end{aligned} \tag{4.10.2}$$

8. Summing up, we see that the two sequences $(Z'_1, \dots, Z'_N)$ and $(Z''_1, \dots, Z''_M)$ of mutually exclusive integrable indicators in L' and L'', respectively, together with the sequence $(a_{\nu,\kappa})_{\nu=1,\dots,N;\mu=1,\dots,M}$ of real numbers, satisfy the conditions in Definition 4.10.2 for the function

$$X + Y = \sum_{\nu=1}^{N}\sum_{\mu=1}^{M} a_{\nu,\mu} Z'_\nu Z''_\mu$$

to be a member of L_0, where $X, Y \in L_0$ are arbitrary. Thus L_0 is also closed relative to addition. We conclude that L_0 is a linear space. Assertion 2 is proved.

9. To see that the operation I is additive, we work backward from equality 4.10.2. More precisely, by the definition of the function I, we have, in view of equality 4.10.2,

$$
\begin{aligned}
I(X+Y) &\equiv \sum_{\nu=1}^{N}\sum_{\mu=1}^{M} a_{\nu,\mu} I'(Z'_{\nu}) I''(Z''_{\mu}) \\
&= \sum_{\nu=0}^{N}\sum_{\mu=0}^{M} a_{\nu,\mu} I'(Z'_{\nu}) I''(Z''_{\mu}) \\
&= \sum_{\nu=0}^{N}\sum_{\mu=0}^{M} (c_{i(\nu),j(\mu)} + b_{k(\nu),h(\mu)}) I'(X'_{i(\nu)} Y'_{k(\nu)}) I''(X''_{j(\mu)} Y''_{h(\mu)}) \\
&= \sum_{i=0}^{n}\sum_{k=0}^{p}\sum_{j=0}^{m}\sum_{h=0}^{q} (c_{i,j} + b_{k,h}) I'(X'_i Y'_k) I''(X''_j Y''_h) \\
&= \sum_{i=0}^{n}\sum_{j=0}^{m} c_{i,j} I'\left(X'_i \sum_{k=0}^{p} Y'_k\right) I''\left(X''_j \sum_{h=0}^{q} Y''_h\right) \\
&\quad + \sum_{k=0}^{p}\sum_{h=0}^{q} b_{k,h} I'\left(Y'_k \sum_{i=0}^{n} X'_i\right) I''\left(Y''_h \sum_{j=0}^{m} X''_j\right) \\
&= \sum_{i=0}^{n}\sum_{j=0}^{m} c_{i,j} I'(X'_i) I''(X''_j) + \sum_{k=0}^{p}\sum_{h=0}^{q} b_{k,h} I'(Y'_k) I''(Y''_h) \\
&= \sum_{i=1}^{n}\sum_{j=1}^{m} c_{i,j} I'(X'_i) I''(X''_j) + \sum_{k=1}^{p}\sum_{h=1}^{q} b_{k,h} I'(Y'_k) I''(Y''_h) \\
&\equiv I(X) + I(Y).
\end{aligned}
$$

Thus the operation I is additive. Finally, let $a \in R$ be arbitrary. Then $aX \equiv \sum_{i=1}^{n}\sum_{j=1}^{m}(ac_{i,j})X'_i X''_j$. Hence

$$
\begin{aligned}
I(aX) &\equiv \sum_{i=1}^{n}\sum_{j=1}^{m} (ac_{i,j}) I'(X'_i) I''(X''_j) \\
&= a\left(\sum_{i=1}^{n}\sum_{j=1}^{m} c_{i,j} I'(X'_i) I''(X''_j)\right) = aI(X).
\end{aligned}
$$

Combining, we see that the operation I is linear.

10. Now suppose two simple functions U, V are equal. Suppose $|I(U) - I(V)| > 0$. Then, by linearity, we know that $X \equiv U - V$ is a simple function, with $|I(X)| > 0$. Let $X \equiv \sum_{i=1}^{n}\sum_{j=1}^{m} c_{i,j} X'_i X''_j$ in the notations of Definition 4.10.2. Then

$$
\left|\sum_{i=1}^{n}\sum_{j=1}^{m} c_{i,j} I'(X'_i) I''(X''_j)\right| \equiv |I(X)| > 0.
$$

Hence $|c_{i,j}I'(X'_i)I''(X''_j)| > 0$ for some $i = 1,\dots,n$ and $j = 1,\dots,m$. Consequently, $|c_{i,j}| > 0$ and $X'_i(\omega') = 1$ and $X''_j(\omega'') = 1$ for some $\omega \equiv (\omega',\omega'') \in domain(X'_iX''_j)$, thanks to the positivity of the integrations I', I''. Therefore

$$|U(\omega) - V(\omega)| \equiv |X(\omega)| = |c_{i,j}X'_i(\omega')X''_j(\omega'')| = |c_{i,j}| > 0,$$

which is a contradiction. Thus we see that $I(U) = I(V)$ for arbitrary simple functions with $U = V$. Thus I is a well-defined function, and we saw earlier that it is linear. Assertion 3 and the lemma are proved. □

Theorem 4.10.4. Integration on space of simple functions. *Let the set L_0 of simple functions and the function I on L_0 be defined as in Definition 4.10.2. Then the triple (Ω, L_0, I) is an integration space.*

Proof. We need to verify the three conditions in Definition 4.3.1.

1. The linearity of the space L_0 and the linearity of I were proved in Lemma 4.10.3.

2. Next consider any $X \in L_0$, with $X = \sum_{i=0}^{n}\sum_{j=0}^{m} c_{i,j}X'_iX''_j$ in the notations of Definition 4.10.2. By Step 2 in the proof of Lemma 4.10.3, we see that

$$|X| = \sum_{i=0}^{n}\sum_{j=0}^{m}|c_{i,j}|X'_iX''_j \in L_0$$

and

$$a \wedge X = \sum_{i=0}^{n}\sum_{j=0}^{m}(a \wedge c_{i,j})X'_iX''_j \in L_0$$

for each $a > 0$. Hence

$$I(X \wedge a) = \sum_{i=0}^{n}\sum_{j=0}^{m}(a \wedge c_{i,j})I'(X'_i)I''(X''_j)$$

$$\to \sum_{i=0}^{n}\sum_{j=0}^{m}c_{i,j}I'(X'_i)I''(X''_j) \equiv I(X)$$

as $a \to \infty$. Likewise,

$$I(|X| \wedge a) = \sum_{i=0}^{n}\sum_{j=0}^{m}(a \wedge |c_{i,j}|)I'(X'_i)I''(X''_j) \to 0$$

as $a \to 0$. Conditions 1 and 4 in Definition 4.3.1 are thus satisfied by the triple (Ω, L_0, I).

3. Next, since I' is an integration, there exists, according to Proposition 4.3.4, some nonnegative $X' \in L'$ such that $I'X' = 1$. By Condition 4 in Definition 4.3.1, there exists $n \geq 2$ so large that $I'X' \wedge n - I'X' \wedge n^{-1} > 0$. By Assertion 1 of Corollary 4.6.7, there exists $t \in (0, n^{-1})$ such that $(t < X')$ is an integrable set.

Let X_1' be the integrable indicator of the integrable set $(t < X')$ in (Ω', L', I'). Suppose $X'(\omega') \wedge n - X''(\omega') \wedge n^{-1} > 0$. Then $X'(\omega') \geq n^{-1} > t$, whence $X_1'(\omega') = 1$ by the definition of an indicator. Thus $X' \wedge n - X' \wedge n^{-1} \leq X_1'$. Hence $I'X_1' \geq I'X' \wedge n - I'X' \wedge n^{-1} > 0$. Similarly, there exists an integrable indicator X_1'' in (Ω'', L'', I'') such that $I''X_1' > 0$. Now define the simple function $X \equiv X_1'X_1''$. Then $IX \equiv I'(X_1')I''(X_1'') > 0$. This proves Condition 3 in Definition 4.3.1 for the triple (Ω, L_0, I).

4. It remains to prove Condition 2 in Definition 4.3.1, the positivity condition. To that end, suppose $(X_k)_{k=0,1,2,\ldots}$ is a sequence of simple functions in L_0, such that $X_k \geq 0$ for $k \geq 1$ and such that $\sum_{k=1}^{\infty} I(X_k) < I(X_0)$. For each $k \geq 0$, we have

$$X_k = \sum_{i=1}^{n(k)} \sum_{j=1}^{m(k)} c_{k,i,j} X'_{k,i} X''_{k,j}$$

as in Definition 4.10.2. It follows that

$$\sum_{k=1}^{\infty} \sum_{i=1}^{n(k)} \sum_{j=1}^{m(k)} c_{k,i,j} I'(X'_{k,i}) I''(X''_{k,j}) \equiv \sum_{k=1}^{\infty} I(X_k) < I(X_0)$$
$$= \sum_{i=1}^{n(0)} \sum_{j=1}^{m(0)} c_{0,i,j} I'(X'_{0,i}) I''(X''_{0,j}).$$

In view of the positivity condition on the integration I', there exists $\omega' \in \Omega'$ such that

$$\sum_{k=1}^{\infty} \sum_{i=1}^{n(k)} \sum_{j=1}^{m(k)} c_{k,i,j} X'_{k,i}(\omega') I''(X''_{k,j}) < \sum_{i=1}^{n(0)} \sum_{j=1}^{m(0)} c_{0,i,j} X'_{0,i}(\omega') I''(X''_{0,j}).$$

In view of the positivity condition on the integration I'', the last inequality in turn yields some $\omega'' \in \Omega''$ such that

$$\sum_{k=1}^{\infty} \sum_{i=1}^{n(k)} \sum_{j=1}^{m(k)} c_{k,i,j} X'_{k,i}(\omega') X''_{k,j}(\omega'') < \sum_{i=1}^{n(0)} \sum_{j=1}^{m(0)} c_{0,i,j} X'_{0,i}(\omega') X''_{0,j}(\omega'').$$

Equivalently, $\sum_{k=1}^{\infty} X_k(\omega', \omega'') < X_0(\omega', \omega'')$. The positivity condition for (Ω, L_0, I) has thus also been verified. We conclude that (Ω, L_0, I) is an integration space. □

Definition 4.10.5. Product integration space. The completion (Ω, L, I) of the integration space (Ω, L_0, I) will be called the *product integration space* of (Ω', L', I') *and* (Ω'', L'', I''), and we define

$$(\Omega, L' \otimes L'', I' \otimes I'') \equiv (\Omega, L, I).$$

The integration $I' \otimes I'' \equiv I$ is called the *product integration* . By abuse of notations, the space $L' \otimes L'' \equiv L$ is also called the *product integration space.* □

Proposition 4.10.6. The direct product of integrable functions is integrable. *Let (Ω, L, I) denote the product integration space of (Ω', L', I') and (Ω'', L'', I'').*

1. Let $X' \in L'$ and $X'' \in L''$ be arbitrary. Then $X' \otimes X'' \in L$ and $I(X' \otimes X'') = (I'X')(I''X'')$.

2. Moreover, if D' and D'' are full subsets of Ω' and Ω'', respectively, then $D' \times D''$ is a full subset of Ω.

Proof. 1. First suppose $X' = \sum_{i=1}^{n} a_i' X_i'$ and $X'' = \sum_{i=1}^{m} a_i'' X_i''$, where (i) $X_1', \ldots, X_n'$ are mutually exclusive integrable indicators in (Ω', L', I'), (ii) $X_1'', \ldots, X_m''$ are mutually exclusive integrable integrators in (Ω'', L'', I''), and (iii) $a_1', \ldots, a_n', a_1'', \ldots, a_m'' \in R$. Then $X'X'' \equiv \sum_{i=1}^{n} \sum_{j=1}^{m} a_i' a_j'' X_i' X_j'' \in L_0 \subset L$. Moreover

$$I(X'X'') = \sum_{i=1}^{n} \sum_{j=1}^{m} a_i' a_j'' I'(X_i') I''(X_j'')$$

$$= \left(\sum_{i=1}^{n} a_i' I' X_i' \right) \left(\sum_{j=1}^{m} a_j'' I'' X_j'' \right) = (I'X')(I''X''). \tag{4.10.3}$$

2. Next let D', D'' be full subsets of Ω', Ω', respectively. Define the measurable function Y' on Ω' by $domain(Y') \equiv D'$ and $Y' \equiv 1$ on D'. Thus Y' is a measurable indicator of D', and $X' \equiv 1 - Y' = 0Y'$ is an integrable indicator. Similarly, define the measurable indicator Y'' of D'' and $X'' \equiv 0Y''$ on Ω''. Then X'' is an integrable indicator on Ω''. Hence $X \equiv 0Y'X'' + 0X'Y'' + X'X''$ is a simple function, with $IX = (I'X')(I''X'') = 0$. On the other hand, $domain(X) = domain(Y') \times domain(Y'') \equiv D' \times D''$. Therefore $D' \times D''$ is a full subset of Ω, proving Assertion 2 of the proposition.

3. Now consider arbitrary $X' \in L'$ and $X'' \in L''$. We need to show that $X'X'' \in L$. By linearity, there is no loss of generality in assuming that $X' \geq 0$ and $X'' \geq 0$. By Assertion 1 of Proposition 4.7.5, there exist sequences $(X_k')_{k=1,2,\ldots}$ and $(X_k'')_{k=1,2,\ldots}$, where (i) for each $n \geq 1$, the functions X_k' and X_k'' are linear combinations of mutually exclusive integrable indicators in (Ω', L', I') and (Ω'', L'', I''), respectively; (ii) $0 \leq X_k' \uparrow X'$ and $0 \leq X_k'' \uparrow X''$ on $D' \equiv \bigcap_{k=1}^{\infty} domain(X_k')$ and $D'' \equiv \bigcap_{k=1}^{\infty} domain(X_k'')$, respectively; and (iii) $I'X_k' \uparrow IX'$ and $I'X_k'' \uparrow IX''$.

4. Let $k \geq 1$ be arbitrary. By equality 4.10.3, we have $I(X_k'X_k'') = I'(X_k')I''(X_k'') \uparrow I'(X')I''(X'')$. Therefore, by the Monotone Convergence Theorem, $X_k'X_k'' \uparrow X$ a.e. relative to I for some $X \in L$, and $IX = (I'X')(I''X'')$. On the other hand, $X_k'X_k'' \uparrow X'X''$ on the full set $D' \times D''$. Thus $X'X'' = X$ a.e. Consequently, $X'X'' \in L$ and $I(X'X'') = IX = (I'X')(I''X'')$. □

Next is Fubini's Theorem, which enables the calculation of the product integral as iterated integrals.

Theorem 4.10.7. Fubini's Theorem. *Let $(\Omega, L, I) \equiv (\Omega, L' \otimes L'', I' \otimes I'')$ be the product integration space of (Ω', L', I') and (Ω'', L'', I''). Let $X \in L' \otimes L''$ be arbitrary. Then the following conditions hold:*

1. There exists a full subset D' of Ω' such that (i′) *for each $\omega' \in D'$, the function $X(\omega', \cdot)$ is a member of L'',* (ii′) *the function $I''X$ defined by $domain(I''X) \equiv D'$ and $(I''X)(\omega') \equiv I''(X(\omega', \cdot))$ for each $\omega' \in D'$ is a member of L', and* (iii′) *$IX = I'(I''X)$.*

2. Similarly, there exists a full subset D'' of Ω'' such that (i″) *for each $\omega'' \in D''$, the function $X(\cdot, \omega'')$ is a member of L',* (ii″) *the function $I'X$ defined by $domain(I'X) \equiv D''$ and $(I'X)(\omega'') \equiv I'(X(\cdot, \omega''))$ for each $\omega'' \in D''$ is a member of L'', and* (iii′) *$IX = I''(I'X)$.*

Proof. 1. First consider a simple function $X = \sum_{i=1}^{n} \sum_{j=1}^{m} c_{i,j} X'_i X''_j$, using the notations of Definition 4.10.2. Define $D' \equiv \bigcap_{i=0}^{n} domain(X'_i)$. Then D' is a full subset of Ω'. Let $\omega' \in D'$ be arbitrary. Then $X(\omega', \cdot) = \sum_{i=0}^{n} \sum_{j=0}^{m} c_{i,j} X'_i(\omega') X''_j \in L''$, verifying Condition (i′). Define the function $I''X$ as in Condition (ii′). Then

$$(I''X)(\omega') \equiv I''(X(\omega', \cdot)) = \sum_{i=1}^{n} \sum_{j=1}^{m} c_{i,j} X'_i(\omega') I'' X''_j$$

for each $\omega' \in D'$. Thus $I''X = \sum_{i=1}^{n} \sum_{j=1}^{m} (c_{i,j} I'' X''_j) X'_i \in L'$, which verifies Condition (ii′). Moreover, it follows from the last equality that

$$I'(I''X) = \sum_{i=1}^{n} \sum_{j=1}^{m} (c_{i,j} I'' X''_j)(I' X'_i) = IX,$$

proving also Condition (iii′). Thus Conditions (i′–iii′) are proved in the case of a simple function X.

2. Next let $X \in L' \otimes L''$ be arbitrary. Then there exists a sequence $(X_k)_{k=1,2,...}$ of simple functions that is a representation of X relative to the integration I. Let $k \geq 1$ be arbitrary. Then

$$X_k = \sum_{i=1}^{n(k)} \sum_{j=1}^{m(k)} c_{k,i,j} X'_{k,i} X''_{k,j},$$

in the notations of Definition 4.10.2. Then

$$|X_k| = \sum_{i=1}^{n(k)} \sum_{j=1}^{m(k)} |c_{k,i,j}| X'_{k,i} X''_{k,j}.$$

3. Consequently, $IX_k = I'(I''X_k)$ and $I|X_k| = I'(I''|X_k|)$ by the first part of this proof. Therefore

$$\sum_{k=1}^{\infty} I'|I''X_k| \leq \sum_{k=1}^{\infty} I'(I''|X_k|) = \sum_{k=1}^{\infty} I|X_k| < \infty.$$

Hence the functions $Y \equiv \sum_{k=1}^{\infty} I''X_k$ and $Z \equiv \sum_{k=1}^{\infty} I''|X_k|$ are in L', with

$$I'Y = \sum_{k=1}^{\infty} I'(I''X_k) = \sum_{k=1}^{\infty} IX_k = IX. \tag{4.10.4}$$

Consider any $\omega' \in D' \equiv domain(Z)$. Then $\sum_{k=1}^{\infty} I''|X_k(\omega',\cdot)| < \infty$. Moreover, if $\omega'' \in \Omega''$ is such that $\sum_{k=1}^{\infty} |X_k(\omega',\cdot)|(\omega'') < \infty$, then $\sum_{k=1}^{\infty} |X_k(\omega',\omega'')| < \infty$ and

$$X(\omega',\cdot)(\omega'') \equiv X(\omega',\omega'') = \sum_{k=1}^{\infty} X_k(\omega',\omega'') \equiv \sum_{k=1}^{\infty} X_k(\omega',\cdot)(\omega'').$$

In other words, for each $\omega' \in D'$, the sequence $(X_k(\omega',\cdot))_{k=1,2,\ldots}$ is a representation of $X(\omega',\cdot)$ in L'', in the sense of Definition 4.4.1, whence $X(\omega',\cdot) \in L''$ with

$$(I''X)(\omega') \equiv I''(X(\omega',\cdot)) = \sum_{k=1}^{\infty} I''X_k(\omega',\cdot) = Y(\omega').$$

Thus we see that $I''X = Y$ on the full set D'. Since $Y \in L'$, so also $I''X \in L'$. Moreover, $I'(I''X) = I'(Y) = IX$, where the last equality is by equality 4.10.4. Conditions (i′–iii′) have thus been verified for an arbitrary $X \in L$. Assertion 1 is proved.

4. Assertion 2, in which the roles of I' and I'' are reversed, is proved similarly. □

Following is the straightforward generalization of Fubini's Theorem to product integration with more than two factors.

Definition 4.10.8. Product of several integration spaces. Let $n \geq 1$ be arbitrary. Let $(\Omega^{(1)}, L^{(1)}, I^{(1)}), \ldots, (\Omega^{(n)}, L^{(n)}, I^{(n)})$ be complete integration spaces. If $n = 1$, let

$$\left(\prod_{i=1}^{n} \Omega^{(i)}, \bigotimes_{i=1}^{n} L^{(i)}, \bigotimes_{i=1}^{n} I^{(i)}\right) \equiv (\Omega^{(1)}, L^{(1)}, I^{(1)}).$$

Inductively for $n \geq 2$, define

$$\left(\prod_{i=1}^{n} \Omega^{(i)}, \bigotimes_{i=1}^{n} L^{(i)}, \bigotimes_{i=1}^{n} I^{(i)}\right) \equiv \left(\prod_{i=1}^{n-1} \Omega^{(i)}, \bigotimes_{i=1}^{n-1} L^{(i)}, \bigotimes_{i=1}^{n-1} I^{(i)}\right) \bigotimes (\Omega^{(n)}, L^{(n)}, I^{(n)}),$$

where the product of the two integration spaces $\left(\prod_{i=1}^{n-1} \Omega^{(i)}, \bigotimes_{i=1}^{n-1} L^{(i)}, \bigotimes_{i=1}^{n-1} I^{(i)}\right)$ and $(\Omega^{(n)}, L^{(n)}, I^{(n)})$ on the right-hand side is as described in Definition 4.10.5. Then, the complete integration space

$$\left(\prod_{i=1}^{n} \Omega^{(i)}, \bigotimes_{i=1}^{n} L^{(i)}, \bigotimes_{i=1}^{n} I^{(i)}\right)$$

is called the *product integration space* of the given integration spaces.

In the special case where $(\Omega^{(1)}, L^{(1)}, I^{(1)}) = \cdots = (\Omega^{(n)}, L^{(n)}, I^{(n)})$ are all equal to the same integration space $(\Omega_\circ, L_\circ, I_\circ)$, we write

$$(\Omega_\circ^n, L_\circ^{\otimes n}, I_\circ^{\otimes n}) \equiv \left(\prod_{i=1}^{n} \Omega^{(i)}, \bigotimes_{i=1}^{n} L^{(i)}, \bigotimes_{i=1}^{n} I^{(i)}\right)$$

and call it the nth *power of the integration space* $(\Omega_\circ, L_\circ, I_\circ)$. □

Theorem 4.10.9. Fubini's Theorem for the product of several integration spaces. *Let $n \geq 1$ be arbitrary. Let $(\Omega^{(1)}, L^{(1)}, I^{(1)}), \ldots, (\Omega^{(n)}, L^{(n)}, I^{(n)})$ be complete integration spaces. Let $(\Omega, L, I) \equiv \left(\prod_{i=1}^{n} \Omega^{(i)}, \bigotimes_{i=1}^{n} L^{(i)}, \bigotimes_{i=1}^{n} I^{(i)}\right)$ be their product space. Let $X \in \bigotimes_{i=1}^{n} L^{(i)}$ be arbitrary. Let $k \in \{1, \ldots, n\}$ be arbitrary. Then there exists a full subset $D^{(k)}$ of $\prod_{i=1; i \neq k}^{n} \Omega^{(i)}$ such that* (i) *for each $\widehat{\omega}_k \equiv (\omega_1, \ldots, \omega_{k-1}, \omega_{k+1}, \ldots, \omega_n) \in D^{(k)}$, the function $X_{\widehat{\omega}(k)} : \Omega^{(k)} \to R$, defined by*

$$X_{\widehat{\omega}(k)}(\omega_k) \equiv X(\omega_1, \ldots, \omega_{k-1}, \omega_k, \omega_{k+1}, \ldots, \omega_n),$$

is a member of $L^{(k)}$; (ii) *the function $\widehat{X}_k : \prod_{i=1; i \neq k}^{n} \Omega^{(i)} \to R$, defined by $domain(\widehat{X}_k) \equiv D^{(k)}$ and by $\widehat{X}_k(\widehat{\omega}_k) \equiv I^{(k)}(X_{\widehat{\omega}(k)})$ for each $\widehat{\omega}_k \in D^{(k)}$, is a member of $\bigotimes_{i=1; i \neq k}^{n} L^{(i)}$; and* (iii) $\left(\bigotimes_{i=1}^{n} I^{(i)}\right) X = \left(\bigotimes_{i=1; i \neq k}^{n} I^{(i)}\right) \widehat{X}_k$.

A special example is where $Y_k \in L^{(k)}$ is given for each $k = 1, \ldots, n$, and where we define the function $X : \Omega \to R$ by

$$X(\omega_1, \ldots, \omega_n) \equiv Y_1(\omega_1) \ldots Y_n(\omega_n)$$

for each $(\omega_1, \ldots, \omega_n) \in \Omega$ such that $\omega_k \in domain(Y_k)$ for each $k = 1, \ldots, n$. Then $X \in \bigotimes_{i=1}^{n} L^{(i)}$ and

$$\left(\left(\bigotimes_{i=1}^{n} I^{(i)}\right)\left(\bigotimes_{i=1}^{n} Y_i\right)\right) \equiv IX = \prod_{i=1}^{n} I_i(Y_i).$$

Proof. The case where $n = 1$ is trivial. The case where $n = 2$ is proved in Theorem 4.10.7 and Proposition 4.10.10. The proof of the general case is by induction on n; it is straightforward and therefore omitted. □

Theorem 4.10.10. A measurable function on one factor of a product integration space is equivalent to a measurable function on the product.

1. Let (Ω, L, I) be the product integration space of (Ω', L', I') and (Ω'', L'', I''). Let X' be an arbitrary measurable function on (Ω', L', I') with values in some complete metric space (S,d). Define $X : \Omega \to S$ by $X(\omega) \equiv X'(\omega')$ for each $(\omega', \omega'') \in \Omega$ such that $\omega' \in domain(X')$. Then X is measurable on the product space (Ω, L, I) with values in (S,d). Moreover, $If(X)1_{A' \times \Omega''} = I'f(X')1_{A'}$ for each $f \in C_{ub}(S)$ and for each integrable subset A' of Ω'.

2. Similarly, let X'' be an arbitrary measurable function on (Ω'', L'', I'') with values in some complete metric space (S,d). Define $X : \Omega \to S$ by $X(\omega) \equiv X''(\omega'')$ for each $(\omega',\omega'') \in \Omega$ such that $\omega'' \in domain(X'')$. Then X is measurable on (Ω, L, I) with values in (S,d). Moreover, $If(X)1_{\Omega'\times A''} = I''f(X'')1_{A''}$ for each $f \in C_{ub}(S)$ and for each integrable subset A'' of Ω''.

3. More generally, let $n \geq 1$ be arbitrary. Let $(\Omega^{(1)}, L^{(1)}, I^{(1)}), \ldots, (\Omega^{(n)}, L^{(n)}, I^{(n)})$ be complete integration spaces. Let $(\Omega, L, I) \equiv \left(\prod_{i=1}^{n} \Omega^{(i)}, \bigotimes_{i=1}^{n} L^{(i)}, \bigotimes_{i=1}^{n} I^{(i)}\right)$ be their product space. Let $i = 1,\ldots,n$ be arbitrary, and let $X^{(i)}$ be an arbitrary measurable function on $(\Omega^{(i)}, L^{(i)}, I^{(i)})$ with values in some complete metric space (S,d). Define the function $X : \Omega \to S$ by $X(\omega) \equiv X^{(i)}(\omega_i)$ for each $\omega \equiv (\omega_1,\ldots,\omega_n) \in \Omega$ such that $\omega_i \in domain(X^{(i)})$. Then X is a measurable function on (Ω, L, I) with values in (S,d). Moreover, $If(X)1_A = I^{(i)}f(X^{(i)})1_{A(i)}$ for each $f \in C_{ub}(S)$ and each integrable subset A_i of $\Omega^{(i)}$, where $A \equiv \prod_{k=1}^{n} A_k$, where $A_k \equiv \Omega^{(k)}$ for each $k = 1,\ldots,n$ with $k \neq i$.

4. Suppose, in addition, that $\Omega^{(k)}$ is an integrable set with $I^{(k)}\Omega^{(k)} = 1$, for each $k = 1,\ldots,n$. Then X is a measurable function with values in (S,d), such that $If(X) = I^{(i)}f(X^{(i)})$ for each $f \in C_{ub}(S)$. Anticipating a definition later, we say that the measurable function X has the same distribution as $X^{(i)}$.

Proof. Let $x_\circ \in S$ be an arbitrary but fixed reference point. For each $n \geq 0$, define the function $h_n \equiv 1 \wedge (n+1-d(x_\circ,\cdot))_+ \in C_{ub}(S)$.

1. As in the hypothesis of Assertion 1, let X' be an arbitrary measurable function on (Ω', L', I') with values in some complete metric space (S,d). Define $X : \Omega \to S$ by $X(\omega) \equiv X'(\omega')$ for each $(\omega',\omega'') \in \Omega$ such that $\omega' \in domain(X')$. We need to prove that X is a measurable function. To that end, let $f \in C_{ub}(S)$ and $Y \in L$ be arbitrary. Then $|f| \leq b$ for some $b > 0$. First assume that $Y \equiv 1_{A'\times A''}$ where A', A'' are integrable subsets of Ω', Ω'', respectively. Then $f(X)Y = (f(X')1_{A'})1_{A''}$ is integrable on the product space (Ω, L, I), according to Proposition 4.10.6. Moreover,

$$Ih_n(X)1_{A'\times A''} = (I'h_n(X')1_{A'})(I''1_{A''}) \uparrow (I'1_{A'})(I''1_{A''})$$

as $n \to \infty$. Hence, by linearity, if Y is a simple function in L, then we have (i) $f(X)Y \in L$ and (ii) $Ih_n(X)Y \to IY$. Now let $(Y_k)_{k=1,2,\ldots}$ be a sequence of simple functions in L, which is a representation of $Y \in L$. Then

$$\sum_{k=1}^{\infty} I|f(X)Y_k| \leq b\sum_{k=1}^{\infty} I|Y_k| < \infty, \tag{4.10.5}$$

where $b > 0$ is any bound for $f \in C_{ub}(S)$. Hence

$$f(X)Y = \sum_{k=1}^{\infty} f(X)Y_k \in L.$$

Similarly, $h_n(X)Y \in L$ and

$$Ih_n(X)Y = \sum_{k=1}^{\infty} Ih_n(X)Y_k$$

for each $n \geq 0$. Now $I|h_n(X)Y_k| \leq I|Y_k|$ and, by Condition (ii), $Ih_n(X)Y_k \to Y$ as $n \to \infty$, for each $k \geq 1$. Hence $Ih_n(X)Y \to \sum_{k=1}^{\infty} IY_k = IY$ as $n \to \infty$, where $Y \in L$ is arbitrary.

2. In particular, if A is an arbitrary integrable subset of Ω, then $Ih_n(X)1_A \uparrow I1_A \equiv \mu(A)$, where μ is the measure function relative to the integration I. We have thus verified the conditions in Proposition 4.8.6 for X to be measurable. Assertion 1 is proved. Assertion 2 is proved similarly by symmetry. Assertion 3 follows from Assertions 1 and 2 by induction. Assertion 4 is a special case of Assertion 3, where $A_k \equiv \Omega^{(k)}$ for each $k = 1, \ldots, n$. □

For products of integrations based on locally compact spaces, the following proposition will be convenient.

Proposition 4.10.11. Product of integration spaces based on locally compact metric spaces. *Let (S_1, d_1) be an arbitrary locally compact metric space. Let $(S_1, C(S_1, d_1), I_1)$ be an integration space, with completion (S_1, L_1, I_1). Let $n \geq 2$ be arbitrary. Let $(S, d) \equiv (S_1^n, d_1^n)$ be the nth power metric space of (S_1, d_1). Let*

$$(S, L, I) \equiv (S_1^n, L_1^{\otimes n}, I_1^{\otimes n})$$

be the nth power integration space of (S_1, L_1, I_1). Then $C(S, d) \subset L$, and $(S, C(S, d), I)$ is an integration space with (S, L, I) as its completion.

Proof. Consider only the case $n = 2$, with the general case being similar. By Definition 4.10.5, the product integration space (S, L, I) is the completion of the subspace (S, L_0, I), where L_0 is the space of simple functions relative to L_1, L_1. Let X', X'' be arbitrary members of L_1. When the risk of confusion is low, we will write $X' \otimes X''$ and $X'X''$ interchangeably.

1. Let $X \in C(S, d)$ be arbitrary. We need to show that $X \in L$. Since X has compact support, there exists $V_1, V_2 \in C(S_1, d_1)$ such that (i) $0 \leq V_1, V_2 \leq 1$, (ii) if $x \equiv (x_1, x_2) \in S$ is such that $|X(x)| > 0$, then $V_1(x_1) = 1 = V_2(x_2)$, and (iii) $IV_1 > 0, IV_2 > 0$.

2. Let $\varepsilon > 0$ be arbitrary. Then, according to Assertion 2 of Proposition 3.3.6, there exists a family $\{g_x : x \in A\}$ of Lipschitz continuous functions indexed by some discrete finite subset A of S_1 such that

$$\left\| X - \sum_{(x,x') \in A^2} X(x, x') g_x \otimes g_{x'} \right\| \leq \alpha, \tag{4.10.6}$$

where $\|\cdot\|$ signifies the supremum norm in $C(S^n,d^n)$. Multiplication by $V_1 \otimes V_2$ yields, in view of Condition (ii),

$$\left\| X - \sum_{(x,x')\in A^2} X(x,x')(g_x V_1 \otimes g_{x'} V_2) \right\| < \varepsilon V_1 \otimes V_2. \qquad (4.10.7)$$

Since $C(S_1,d_1) \subset L_1$, we have $V_1 \otimes V_2 \in L$ and $g_x V_1 \otimes g_{x'} V_2 \in L$ for each $(x,x') \in A^2$, according to Proposition 4.10.6. Since $I(\varepsilon V_1 V_2) > 0$ is arbitrarily small, inequality 4.10.7 implies that $X \in L$, thanks to Theorem 4.5.10. Since $X \in C(S,d)$ is arbitrary, we conclude that $C(S,d) \subset L$.

3. Since I is a linear function on L, and since $C(S,d)$ is a linear subspace of L, the function I is a linear function on $C(S,d)$. Since $I(V_1 V_2) = (I_1 V_1)(I_1 V_2) > 0$, the triple $(S,C(S,d),I)$ satisfies Condition (i) of Definition 4.2.1. Condition (ii) of Definition 4.2.1, the positivity condition, follows trivially from the positivity condition of (S,L,I). Hence $(S,C(S,d),I)$ is an integration space. Since $C(S,d) \subset L$ and since (S,L,I) is complete, the completion $\overline{L}$ of $C(S,d)$ relative to I is such that $\overline{L} \subset L$.

4. We will now show that, conversely, $L \subset \overline{L}$. To that end, consider any $Y_1 Y_2 \in L_1$. Let $\varepsilon > 0$ be arbitrary. Then there exists $U_i \in C(S_1,d_1)$ such that $I_1|U_i - Y_i| < \varepsilon$ for each $i = 1,2$ because L_1 is the completion of $C(S_1,d_1)$ relative to I_1, by hypothesis. Consequently,

$$\begin{aligned}
I|Y_1Y_2 - U_1U_2| &\le I|Y_1(Y_2 - U_2)| + I|(Y_1 - U_1)U_2| \\
&= I_1|Y_1| \cdot I_1|Y_2 - U_2| + I_1|Y_1 - U_1| \cdot I_1|U_2| \\
&< I_1|Y_1|\varepsilon + \varepsilon(I_1|Y_2| + \varepsilon),
\end{aligned}$$

where the equality is provided by Fubini's Theorem. Since $\varepsilon > 0$ is arbitrarily small, while $U_1U_2 \in C(S,d) \subset \overline{L}$, we see that $Y_1Y_2 \in \overline{L}$. Each simple function on S, as in Definition 4.10.2, is a linear combination of functions of the form Y_1Y_2, where $Y_1, Y_2 \in L_1$. Thus we conclude that $L_0 \subset \overline{L}$. At the same time, $\overline{L}$ is complete relative to I. Hence the completion L of L_0 is a subspace of $\overline{L}$. Thus $L \subset \overline{L}$, as alleged.

5. Summing up the results of Steps 3 and 4, we obtain $\overline{L} = L$. In other words, the completion of $(S,C(S,d),I)$ is (S,L,I). The proposition is proved. □

Proposition 4.10.12. The product of σ-finite integration spaces is σ-finite. *Let (Ω',L',I') and (Ω'',L'',I'') be arbitrary integration spaces that are σ-finite, with I-bases $(A'_k)_{k=1,2,\ldots}$ and $(A''_k)_{k=1,2,\ldots}$, respectively. Then the product integration space*

$$(\Omega,L,I) \equiv (\Omega' \times \Omega'', L' \otimes L'', I' \otimes I'')$$

is σ-finite, with an I-basis $(A_k)_{k=1,2,\ldots} \equiv (A'_k \times A''_k)_{k=1,2,\ldots}$.

Proof. By the definition of an I-basis, we have $A'_k \subset A'_{k+1}$ and $A''_k \subset A''_{k+1}$ for each $k \geq 1$. Hence, $A_k \equiv A'_k \times A''_k \subset A'_{k+1} \times A''_{k+1}$ for each $k \geq 1$. Moreover,

$$\bigcup_{k=1}^{\infty}(A'_k \times A''_k) = \left(\bigcup_{k=1}^{\infty} A'_k\right) \times \left(\bigcup_{k=1}^{\infty} A''_k\right).$$

Again by the definition of an I-basis, the two unions on the right-hand side are full subsets in Ω', Ω'', respectively. Hence the union on the left-hand side is, according to Assertion 2 of Proposition 4.10.6, a full set in Ω.

Now let $f \equiv 1_{B'}1_{B''}$, where B', B'' are arbitrary integrable subsets in Ω', Ω'', respectively. Then

$$I(1_{A(k)}f) = I'(1_{A'(k)}1_{B'})I''(1_{A''(k)}1_{B''}) \to I'(1_{B'})I''(1_{B''}) = If$$

as $k \to \infty$. By linearity, it follows that

$$I(1_{A(k)}g) \to Ig$$

for each simple function g on $\Omega' \times \Omega''$ relative to L', L''. Consider each $h \in L$. Let $\varepsilon > 0$ be arbitrary. Since (Ω, L, I) is the completion of (Ω, L_0, I), where L_0 is the space of simple functions on $\Omega' \times \Omega''$ relative to L', L'', it follows that $I|h - g| < \varepsilon$ for some $g \in L_0$. Hence

$$\begin{aligned}|I1_{A(k)}f - If| &\leq |I1_{A(k)}f - I1_{A(k)}g| + |I1_{A(k)}g - Ig| + |Ig - If| < \varepsilon \\ &\leq I|f - g| + |I1_{A(k)}g - Ig| + |Ig - If| < 3\varepsilon\end{aligned}$$

for sufficiently large $k \geq 1$. Since $\varepsilon > 0$ is arbitrary, we conclude that $I(1_{A(k)}h) \to Ih$. In particular, if A is an arbitrary integrable subset of Ω, we have $I(1_{A(k)}1_A) \to I1_A$. In other words, $\mu(A_kA) \to \mu(A)$ for each integrable set $A \subset \Omega$. We have verified the conditions in Definition 4.8.15 for (Ω, L, I) to be σ-finite, with $(A_k)_{k=1,2,\ldots} \equiv (A'_k \times A''_k)_{k=1,2,\ldots}$ as an I-basis. □

The next definition establishes some familiar notations for the special cases of the Lebesgue integration space on R^n.

Definition 4.10.13. Lebesgue integration on R^n. Let $(R, L, \int \cdot dx)$ denote the Lebesgue integration space on R, as in Definition 4.9.9. The product integration space

$$\left(R^n, \overline{L}, \int \cdots \int \cdot dx_1 \ldots dx_n\right) \equiv \left(\prod_{i=1}^{n} R, \bigotimes_{i=1}^{n} L, \bigotimes_{i=1}^{n} \int \cdot dx\right)$$

is called the *Lebesgue integration space of dimension* n. Similar terminology applies when R^n is replaced by an interval $\prod_{i=1}^{n}[s_i, t_i] \subset R^n$. When confusion is unlikely, we will also abbreviate $\int \cdots \int \cdot dx_1 \ldots dx_n$ to $\int \cdot dx$, with the understanding that the dummy variable x is now a member of R^n. An integrable function relative to $\int \cdots \int \cdot dx_1 \ldots dx_n$ will be called *Lebesgue integrable* on R^n. □

Corollary 4.10.14. Restriction of the Lebesgue integration on R^n to $C(R^n)$ is an integration whose completion is the Lebesgue integration on R^n. *Let $n \geq 1$ be arbitrary. Then, in the notations of Definition 4.10.13, we have $C(R^n) \subset \overline{L}$. Moreover, $(R^n, C(R^n), \int \cdots \int \cdot dx_1 \ldots dx_n)$ is an integration space, and its completion is equal to the Lebesgue integration space $(R^n, \overline{L}, \int \cdots \int \cdot dx_1 \ldots dx_n)$.*

Proof. Let $S_i \equiv R$ for each $i = 1, \ldots, n$. Let $S \equiv R^n$. Proposition 4.10.11 then applies and yields the desired conclusions. □

Definition 4.10.15. Product of countably many probability integration spaces. For each $n \geq 1$, let $(\Omega^{(n)}, L^{(n)}, I^{(n)})$ be a complete integration space such that $1 \in L^{(n)}$ with $I^{(n)}1 = 1$. Consider the Cartesian product $\overline{\Omega} \equiv \prod_{i=1}^{\infty} \Omega^{(i)}$. Let $n \geq 1$ be arbitrary. Let $\left(\prod_{i=1}^{n} \Omega^{(i)}, \bigotimes_{i=1}^{n} L^{(i)}, \bigotimes_{i=1}^{n} I^{(i)}\right)$ be the product of the first n complete integration spaces. For each $g \in \bigotimes_{i=1}^{n} L^{(i)}$, define a function $\overline{g}$ on $\overline{\Omega}$ by $domain(\overline{g}) \equiv domain(g) \times \prod_{i=n+1}^{\infty} \Omega^{(i)}$, and by $\overline{g}(\omega_1, \omega_2, \ldots) \equiv g(\omega_1, \ldots, \omega_n)$ for each $(\omega_1, \omega_2, \ldots) \in domain(\overline{g})$. Let

$$G_n \equiv \left\{ \overline{g} : g \in \bigotimes_{i=1}^{n} L^{(i)} \right\}.$$

Then $G_n \subset G_{n+1}$. Let $\overline{L} \equiv \bigcup_{n=1}^{\infty} G_n$ and define a function $\overline{I} : \overline{L} \to R$ by $\overline{I}(\overline{g}) \equiv \left(\bigotimes_{i=1}^{n} I^{(i)}\right)(g)$ if $\overline{g} \in G_n$, for each $\overline{g} \in \overline{L}$. Theorem 4.10.16 says that $\overline{L}$ is a linear space, that $\overline{I}$ is a well-defined linear function, and that $(\overline{\Omega}, \overline{L}, \overline{I})$ is an integration space.

Let $\left(\prod_{i=1}^{\infty} \Omega^{(i)}, \bigotimes_{i=1}^{\infty} L^{(i)}, \bigotimes_{i=1}^{\infty} I^{(i)}\right)$ denote the completion of $(\overline{\Omega}, \overline{L}, \overline{I})$, and call it the *product of the sequence* $(\Omega^{(n)}, L^{(n)}, I^{(n)})_{n=1,2,\ldots}$ *of complete integration spaces*.

In the special case where $(\Omega^{(i)}, L^{(i)}, I^{(i)}) = (\Omega_0, L_0, I_0)$ for each $i \geq 1$, for some complete integration space (Ω_0, L_0, I_0), then we write

$$(\Omega_0^{\infty}, L_0^{\otimes\infty}, I_0^{\otimes\infty}) \equiv \left(\prod_{i=1}^{\infty} \Omega^{(i)}, \bigotimes_{i=1}^{\infty} L^{(i)}, \bigotimes_{i=1}^{\infty} I^{(i)}\right)$$

and call it the *countable power of the integration space* (Ω_0, L_0, I_0). □

Theorem 4.10.16. The countable product of probability integration spaces is a well-defined integration space. *For each $n \geq 1$, let $(\Omega^{(n)}, L^{(n)}, I^{(n)})$ be a complete integration space such that $1 \in L^{(n)}$ with $I^{(n)}1 = 1$. Then the following conditions hold:*

1. The set $\overline{L}$ of functions is a linear space. Moreover, $\overline{I}$ is a well-defined linear function and $(\overline{\Omega}, \overline{L}, \overline{I})$ is an integration space.

2. Let $N \geq 1$ be arbitrary. Let $Z^{(N)}$ be a measurable function on $(\Omega^{(N)}, L^{(N)}, I^{(N)})$ with values in some complete metric space (S, d). Define the function $\overline{Z}^{(N)} : \Omega \to S$ by $\overline{Z}^{(N)}(\omega) \equiv Z^{(N)}(\omega_i)$ for each $\omega \equiv (\omega_1, \omega_2, \ldots) \in \overline{\Omega}$ such *that* $\omega_N \equiv (\omega_1, \omega_2, \ldots, \omega_N) \in domain(Z^{(N)})$. *Let* $M \geq 1$ *be arbitrary. Let* $f_j \in C_{ub}(S,d)$ *be arbitrary for each* $j \leq M$. *Then*

$$\overline{I}\left(\prod_{j=1}^{M} f_j(\overline{Z}^{(j)})\right) = \prod_{j=1}^{M} I^{(j)} f_j(\overline{Z}^{(j)}).$$

3. For each $N \geq 1$, *the function* $\overline{Z}^{(N)}$ *is measurable on the countable product space* $\left(\prod_{i=1}^{\infty} \Omega^{(i)}, \bigotimes_{i=1}^{\infty} L^{(i)}, \overline{I}\right)$.

Proof. 1. Obviously, G_n and $\overline{L}$ are linear spaces. Suppose $\overline{g} = \overline{h}$ for some $\overline{g} \in G_n$ and $\overline{h} \in G_m$ with $n \leq m$. Then

$$\begin{aligned} h(\omega_1, \ldots, \omega_m) &\equiv \overline{h}(\omega_1, \omega_2, \ldots) = \overline{g}(\omega_1, \omega_2, \ldots) \\ &\equiv g(\omega_1, \ldots, \omega_n) = g(\omega_1, \ldots, \omega_n) 1_A(\omega_{n+1}, \ldots, \omega_m), \end{aligned}$$

where $A \equiv \prod_{i=n+1}^{m} \Omega^{(i)}$. Then $1_A = \bigotimes_{i=n+1}^{m} 1_{\Omega^{(i)}}$. Since, by hypothesis, $1_{\Omega^{(i)}} = 1 \in L^{(i)}$ with $I^{(i)} 1 = 1$, for each $i \geq 1$, Fubini's Theorem implies that $1_A \in \bigotimes_{i=n+1}^{m} L^{(i)}$, with $\left(\bigotimes_{i=n+1}^{m} I^{(i)}\right)(1_A) = 1$, and that

$$\begin{aligned} \overline{I}(\overline{h}) &\equiv \left(\bigotimes_{i=1}^{m} I^{(i)}\right)(h) = \left(\bigotimes_{i=1}^{m} I^{(i)}\right)(g \otimes 1_A) \\ &= \left(\bigotimes_{i=1}^{n} I^{(i)} \otimes \bigotimes_{i=n+1}^{m} I^{(i)}\right)(g \otimes 1_A) \\ &= \left(\left(\bigotimes_{i=1}^{n} I^{(i)}\right)(g)\right) \cdot \left(\bigotimes_{i=n+1}^{m} I^{(i)}\right)(1_A) = \left(\left(\bigotimes_{i=1}^{n} I^{(i)}\right)(g)\right) \cdot 1 = \overline{I}(\overline{g}). \end{aligned}$$

Thus the function $\overline{I}$ is well defined. The linearity of $\overline{I}$ is obvious. The verification of the other conditions in Definition 4.3.1 is straightforward. Accordingly, $(\overline{\Omega}, \overline{L}, \overline{I})$ is an integration space.

2. In view of Fubini's Theorem (Theorem 4.10.7), the proofs of Assertions 2 and 3 are straightforward and omitted. □

Following are two results that will be convenient.

Proposition 4.10.17. The region below the graph of a nonnegative integrable function is an integrable set in the product space. *Let* (Q, L, I) *be a complete integration space that is* σ*-finite. Let* $(\Theta, \Lambda, I_0) \equiv (\Theta, \Lambda, \int \cdot dr)$ *be the Lebesgue integration space based on* $\Theta \equiv R$ *or* $\Theta \equiv [0,1]$. *Let* $\lambda : Q \to R$ *be an arbitrary measurable function on* (Q, L, I). *Then the following conditions hold:*

1. The sets

$$A_\lambda \equiv \{(t,r) \in Q \times \Theta : r \leq \lambda(t)\}$$

and

$$A'_\lambda \equiv \{(t,r) \in Q \times \Theta : r < \lambda(t)\}$$

are measurable on $(Q,L,I) \otimes (\Theta,\Lambda,I_0)$.

2. *Suppose, in addition, that* λ *is a nonnegative integrable function. Then the sets*

$$B_\lambda \equiv \{(t,r) \in Q \times \Theta : 0 \leq r \leq \lambda(t)\}$$

and

$$B'_\lambda \equiv \{(t,r) \in Q \times \Theta : 0 \leq r < \lambda(t)\}$$

are integrable, with

$$(I \otimes I_0)B_\lambda = I\lambda = (I \otimes I_0)B'_\lambda. \tag{4.10.8}$$

Proof. 1. Let g be the identity function on Θ, with $g(r) \equiv r$ for each $r \in \Theta$. By Proposition 4.10.10, g and λ can be regarded as measurable functions on $Q \times \Theta$, with values in Θ. Define the function $f : Q \times \Theta \to R$ by $f(t,r) \equiv g(r) - \lambda(t) \equiv r - \lambda(t)$ for each $(t,r) \in Q \times \Theta$. Then f is the difference of two real-valued measurable functions on $Q \times \Theta$. Hence f is measurable. Therefore there exists a sequence $(a_n)_{n=1,2,\ldots}$ in $(0,\infty)$ with $a_n \downarrow 0$ such that $(f \leq a_n)$ is measurable for each $n \geq 1$. We will write a_n and $a(n)$ interchangeably.

2. Let $A \subset Q$ and $B \subset \Theta$ be arbitrary integrable subsets of Q and Θ, respectively. Let $h : \Theta \to \Theta$ be the identity function, with $h(r) \equiv r$ for each $r \in \Theta$. Let $m \geq n$ be arbitrary. Then

$$\begin{aligned} &I \otimes I_0\big(1_{(f \leq a(n))(A\times B)} - 1_{(f\leq a(m))(A\times B)}\big) \\ &= I \otimes I_0(1_{(a(m)<f\leq a(n))(A\times B)}) \\ &= I \otimes I_0(1_{(\lambda - a(n)\leq h<\lambda-a(m))(A\times B)}) \\ &= I\left(1_A\left(\int_{[\lambda-a(n),\lambda-a(m))} 1_B(r)dr\right)\right). \end{aligned} \tag{4.10.9}$$

Since for each $t \in Q$, the Lebesgue measure

$$I_0[\lambda(t) - a_n, \lambda(t) - a_m) = (a_n - a_m) \downarrow 0$$

as $n \to \infty$, and since 1_B is integrable, Proposition 4.7.1 implies that

$$\int_{[\lambda(t)-a(n),\lambda(t)-a(m))} 1_B(r)dr \downarrow 0$$

uniformly in $t \in Q$. Since 1_A is integrable, the Dominated Convergence Theorem implies that the right-hand side of equality 4.10.9 converges to 0 as $n \to \infty$. Consequently, $I \otimes I_0(1_{(f\leq a(n))(A\times B)})$ converges as $n \to \infty$. Therefore, by the Monotone Convergence Theorem, the limit $1_{(f\leq 0)(A\times B)}$ is integrable on $Q \times \Theta$.

3. Now let C be an arbitrary integrable subset of $Q \times \Theta$. Let $(A_i)_{i=1,2,...}$ and $(B_i)_{i=1,2,...}$ be I-bases of the σ-finite integration spaces (Q, L, I) and (Θ, Λ, I_0), respectively. Then, by Proposition 4.10.12, $Q \times \Theta$ is σ-finite with an I-basis $(A_i \times B_i)_{i=1,2,...}$. By Step 2, we see that the function $1_{(f\leq 0)(A(i)\times B(i))C} = 1_{(f\leq 0)(A(i)\times B(i))}1_C$ is integrable on $Q\times\Theta$, for each $i \geq 1$. Moreover, as $i, j \to \infty$ with $j \geq i$, we have

$$\begin{aligned} 0 &\leq I1_{(f\leq 0)C(A(j)\times B(j))} - I1_{(f\leq 0)C(A(i)\times B(i))} \\ &\leq I1_{C(A(j)\times B(j))} - I1_{C(A(i)\times B(i))} \to 0. \end{aligned}$$

Hence, by the Monotone Convergence Theorem, $1_{(f\leq 0)CD}$ is integrable, where $D \equiv \bigcup_{i=1}^{\infty}(A_i \times B_i)$ is a full set. Consequently, $1_{(f\leq 0)C}$ is integrable. In other words, $1_{(f\leq 0)}1_C$ is integrable. Since the integrable subset C of $Q\times\Theta$ is arbitrary, we conclude that $1_{(f\leq 0)}$ is measurable. Equivalently, $(f \leq 0)$ is measurable.

4. Recalling the definition of f at the beginning of this proof, we obtain

$$\begin{aligned} A_\lambda &\equiv \{(t,r) \in Q \times \Theta : r - \lambda(t) \leq 0\} \\ &= \{(t,r) \in Q \times \Theta : 0 \leq f(t,r)\} \equiv (f \leq 0), \end{aligned}$$

whence A_λ is measurable. Similarly, A'_λ is measurable. Assertion 1 is proved.

5. Suppose, in addition, that λ is a nonnegative integrable function. Then for each $t \in Q$, we have

$$\int 1_{B(\lambda)}(t,r)dr = \int (1_{A(\lambda)}(t,r) - 1_{A(0)}(t,r)dr = \int_{(0,\lambda(t)]} dr = \lambda(t).$$

Fubini's Theorem therefore yields $(I \otimes I_0)B_\lambda = I\lambda$, the first half of equality 4.10.8. The second half is similarly verified. Assertion 2 is proved. □

Proposition 4.10.18. Regions between graphs of nonnegative integrable functions. *Let (Q, L, I) be a complete integration space that is σ-finite. Suppose $\lambda_0 \equiv 0 \leq \lambda_1 \leq \lambda_2 \leq \cdots \leq \lambda_n$ are integrable functions with $\lambda_n \leq 1$. Define $\lambda_{n+1} \equiv 1$. For each $k = 1, \ldots, n+1$, define*

$$\Delta_k \equiv \{(t,r) \in Q \times R : r \in (\lambda_{k-1}(t), \lambda_k(t))\}.$$

Then $\Delta_1, \ldots, \Delta_{n+1}$ are mutually exclusive measurable subsets in $(Q, L, I) \otimes (R, M, J)$, whose union is a full set. Moreover, $\Delta_1, \ldots, \Delta_n$ are integrable in $(Q, L, I) \otimes (R, M, J)$, with integrals equal to $I\lambda_1$, $I\lambda_2 - I\lambda_1, \ldots, I\lambda_n - I\lambda_{n-1}$, respectively.

Proof. Use Proposition 4.10.17. □

5

Probability Space

In this chapter, we specialize the study of complete integration spaces to probability integration spaces, introduced in Definition 4.3.1, where the constant function 1 is integrable with integral equal to 1. An integrable function can then be interpreted as an observable in a probabilistic experiment that, on repeated observations, has an expected value given by its integral. Likewise, an integrable set can be interpreted as an event, and its measure as the probability of said event occurring. We will transition from terms used in measure theory to commonly used terms in probability theory. Then we will introduce and study more concepts and tools common in probability theory.

In this chapter, unless otherwise specified, (S,d) will denote a complete metric space, mostly but not always locally compact. Let $x_\circ \in S$ be an arbitrary but fixed reference point. Recall that $C_{ub}(S) \equiv C_{ub}(S,d)$ stands for the space of bounded and uniformly continuous functions on (S,d), and that $C(S) \equiv C(S,d)$ stands for the space of continuous functions on (S,d) with compact support.

Let $n \geq 1$ be arbitrary. Define the auxiliary function $h_n \equiv 1 \wedge (1 + n - d(\cdot, x_\circ))_+ \in C_{ub}(S)$. Note that the function h_n has bounded support. Hence $h_n \in C(S)$ if (S,d) is locally compact.

Separately, for each integration space (Ω, L, J), we will let $(\Omega, \overline{L}, J)$ denote its complete extension.

5.1 Random Variable

Definition 5.1.1. Probability space and r.v.'s. Henceforth, unless otherwise specified, (Ω, L, E) will denote a probability integration space, i.e., a complete integration space in which the constant function 1 is integrable with $E1 = 1$. Then (Ω, L, E) will simply be called a *probability space*. The integration E will be called an *expectation*, and the integral EX of each $X \in L$ will be called the *expected value* of X.

A measurable function X on (Ω, L, E) with values in a complete metric space (S,d) is called a *random variable*, or r.v. for abbreviation. Two r.v.'s are considered equal if they have equal values on a full subset of Ω. A real-valued measurable function X on (Ω, L, E) is then called a *real random variable*, or r.r.v. for abbreviation. An integrable real-valued function X is called an *integrable real random variable*, its integral EX called its *expected value*.

A measurable set is sometimes called an *event*. It is then integrable because $1_A \leq 1$, and its measure $\mu(A)$ is called its *probability* and denoted by $P(A)$ or PA. The function P on the set of measurable sets is called the *probability function* corresponding to the expectation E. Sometimes we will write $E(A)$ for $P(A)$. The set Ω is called the *sample space*, and a point $\omega \in \Omega$ is called a *sample* or an *outcome*. If an outcome ω belongs to an event A, the event A is said to *occur* for ω, and ω is said to *realize* A.

The terms "*almost surely*," "*almost sure*," and the abbreviation "a.s." will stand for "almost everywhere" or its abbreviation "a.e." Henceforth, unless otherwise specified, equality of r.v.'s and equality of events will mean a.s. equality, and the term "complement" for events will stand for "measure-theoretic complement." If X is an integrable r.r.v. and $A, B, \ldots$ are events, we will sometimes write $E(X; A, B, \ldots)$ for $EX1_{AB\ldots}$.

Let $X \in L$ be arbitrary. We will sometimes use the more suggestive notation

$$\int E(d\omega)X(\omega) \equiv EX,$$

where ω is a dummy variable. For example, if $Y \in L \otimes L \otimes L$, we can define a function $Z \in L \otimes L$ by the formula

$$Z(\omega_1, \omega_3) \equiv \int E(d\omega_2)Y(\omega_1, \omega_2, \omega_3) \equiv EY(\omega_1, \cdot, \omega_3)$$

for each $(\omega_1, \omega_3) \in \Omega^2$ for which the right-hand side is defined. □

In [Billingsley 1968], an r.v. is called a random element, and an r.r.v. is called a random variable. Our usage of the two terms follows [Neveu 1965], for the benefit of both acronyms.

Being a measurable function, an r.v. inherits all the definitions and properties for measurable functions developed in the preceding chapters. In particular, since the constant function 1 is integrable, Ω is an integrable set, with 1 as its probability. A probability space is trivially σ-finite. Therefore r.v.'s inherit the theorems on measurable functions that require a σ-finite integration space.

First we restate Definition 4.8.13 of regular points in a simpler form, now in the context of a probability space. The reader can verify that, in the present context, the restated Definition 5.1.2 is indeed equivalent to Definition 4.8.13.

Definition 5.1.2. Regular and continuity points of an r.r.v. Let (Ω, L, E) be an arbitrary probability space. Let X be an r.r.v. on (Ω, L, E). Then a point $t \in R$ is a regular point of X if (i) there exists a sequence $(s_n)_{n=1,2,\ldots}$ of real numbers decreasing to t such that $(X \leq s_n)$ is a measurable set for each $n \geq 1$ and such that $\lim_{n\to\infty} P(X \leq s_n)$ exists, and (ii) there exists a sequence $(r_n)_{n=1,2,...}$ of real numbers increasing to t such that $(X \leq r_n)$ is measurable for each $n \geq 1$ and such that $\lim_{n\to\infty} P(X \leq r_n)$ exists. If, in addition, the two limits in (i) and (ii) are equal, then we call t a continuity point of the r.r.v. X.

Note that Condition (i) implies that, if $t \in R$ is a regular point of an r.r.v. X, then $(X \leq t)$ is measurable, with

$$P(X \leq s_n) \downarrow P(X \leq t)$$

for each sequence $(s_n)_{n=1,2,...}$ satisfying Condition (i), thanks to the Monotone Convergence Theorem. Similarly, in that case $(X < t)$ is measurable, with

$$P(X \leq r_n) \uparrow P(X < t)$$

for each sequence $(r_n)_{n=1,2,...}$ satisfying Condition (ii). Consequently, $(X = t)$ is measurable, with measure $P(X = t) = 0$ if t is a continuity point. □

We reiterate the convention in Definition 4.8.18 regarding regular points, in the context of a probability space and r.r.v.'s.

Definition 5.1.3. Convention regarding regular points of r.r.v.'s. Let X be an arbitrary r.r.v. When the measurability of the set $(X < t)$ or $(X \leq t)$ is required in a discussion for some $t \in R$, it is understood that the real number t has been chosen from the regular points of the r.r.v. X.

For example, a sequence of statements like "Let $t \in R$ be arbitrary. ... Then $P(X_i > t) < a$ for each $i \geq 1$" means "Let $t \in R$ be arbitrary, such that t is a regular point for X_i for each $i \geq 1$ ... Then $P(X_i > t) < a$ for each $i = 1,2,\ldots$." The purpose of this convention is to obviate unnecessary distraction from the main arguments.

If, for another example, the measurability of the set $(X \leq 0)$ is required in a discussion, we would need to first supply a proof that 0 is a regular point of X or, instead of $(X \leq 0)$, use $(X \leq a)$ as a substitute, where a is some regular point near 0. Unless the exact value 0 is essential to the discussion, the latter, usually effortless alternative will be used. The implicit assumption of regularity of the point a is clearly possible, for example, when we have the freedom to pick the number a from some open interval. This is thanks to Proposition 4.8.14, which says that all but countably many real numbers are regular points of X.

Classically, all $t \in R$ are regular points for each r.r.v. X, so this convention would be redundant classically. □

In the case of a measurable indicator X, the only possible values, 0 and 1, are regular points. Separately, recall that the indicator 1_A and the complement A^c of an event are uniquely defined relative to a.s. equality.

Proposition 5.1.4. Basic properties of r.v.'s. *Let (Ω, L, E) be a probability space.*

1. Suppose A is an event. Then A^c is an event. Moreover, $(A^c)^c = A$ and $P(A^c) = 1 - P(A)$.

2. A subset A of Ω is a full set iff it is an event with probability 1.

3. Let (S,d) be a complete metric space. A function $X : \Omega \to S$ is an r.v. with values in (S,d) iff (i) *$f(X) \in L$ for each $f \in C_{ub}(S,d)$ and* (ii) *$P(d(X,x_\circ) \geq a) \to 0$ as $a \to \infty$. Note that if the metric d is bounded, then Condition* (ii) *is automatically satisfied.*

4. Let (S,d) be a complete metric space, with a reference point $x_\circ$. For each $n \geq 1$, define $h_n \equiv 1 \wedge (1+n-d(\cdot,x_\circ))_+ \in C_{ub}(S)$. Then a function $X : \Omega \to S$ is an r.v. iff (i) *$f(X) \in L$ for each $f \in C_{ub}(S)$ and* (iii) *$Eh_n(X) \uparrow 1$ as $n \to \infty$. In that case, we have $E|f(X) - f(X)h_n(X)| \to 0$, where $fh_n \in C(S)$*

5. Let (S,d) be a locally compact metric space, with a reference point $x_\circ$. For each $n \geq 1$, define the function h_n as in Assertion 4. Then $h_n \in C(S)$. A function $X : \Omega \to S$ is an r.v. iff (iv) *$f(X) \in L$ for each $f \in C(S)$ and* (iii) *$Eh_n(X) \uparrow 1$ as $n \to \infty$. In that case, for each $f \in C_{ub}(S)$, there exists a sequence $(g_n)_{n=1,2,\ldots}$ in $C(S)$ such that $E|f(X) - g_n(X)| \to 0$.*

6. If X is an integrable r.r.v. and A is an event, then $EX = E(X;A)+E(X;A^c)$.

7. A point $t \in R$ is a regular point of an r.r.v. X iff it is a regular point of X relative to Ω.

8. If X is an r.r.v. such that $(t-\varepsilon < X < t) \cup (t < X < t+\varepsilon)$ is a null set for some $t \in R$ and $\varepsilon > 0$, then the point $t \in R$ is a regular point of X.

Proof. 1. Suppose A is an event with indicator 1_A and complement $A^c = (1_A = 0)$. Because 1 is integrable, so is $1 - 1_A$. At the same time, $A^c = (1_A = 0) = (1 - 1_A = 1)$. Hence A^c is an event with indicator $1 - 1_A$. Moreover, $P(A^c) = E(1 - 1_A) = 1 - P(A)$. Repeating the argument with the event A^c, we see that

$$(A^c)^c = (1 - (1 - 1_A) = 1) = (1_A = 1) = A.$$

2. Suppose A is a full set. Since any two full sets are equal a.s., we have $A = \Omega$ a.s. Hence $P(A) = P(\Omega) = 1$. Conversely, if A is an event with $P(A) = 1$, then according to Assertion 1, A^c is a null set with $A = (A^c)^c$. Hence by Assertion 4 of Proposition 4.5.5, A is a full set.

3. Suppose X is an r.v. Since Ω is an integrable set, Conditions (i) and (ii) of the present proposition hold as special cases of Conditions (i) and (ii) in Definition 4.8.1 when we take $A = \Omega$.

Conversely, suppose Conditions (i) and (ii) of the present proposition hold. Let $f \in C_{ub}(S)$ be arbitrary and let A be an arbitrary integrable set. Then $f(X) \in L$ by Condition (i), and so $f(X)1_A \in L$. Moreover $P(d(x_\circ,X) \geq a)A \leq P(d(x_\circ,X) \geq a) \to 0$ as $a \to \infty$. Thus Conditions (i) and (ii) in Definition 4.8.1 are established for X to be a measurable function. In other words, X is an r.v. Assertion 3 is proved.

4. Given Condition (i), then Conditions (ii) and (iii) are equivalent to each other, according to Proposition 4.8.6, whence Condition (ii) holds iff the function X is an r.v. Thus Assertion 4 follows from Assertion 3.

5. Suppose (S,d) is locally compact. Suppose Condition (iii) holds. We will first verify that Conditions (i) and (iv) are then equivalent. Trivially, Condition (i) implies Condition (iv). Conversely, suppose Condition (iv) holds. Let $f \in C_{ub}(S)$ be arbitrary. We need to prove that $f(X) \in L$. There is no loss of generality in assuming that $0 \leq f \leq b$ for some $b > 0$. Then

$$E(f(X)h_m(X) - f(X)h_n(X)) \leq bE(h_m(X) - h_n(X)) \to 0$$

as $m \geq n \to \infty$, thanks to Condition (iii). Thus $Ef(X)h_n(X)$ converges as $n \to \infty$. Hence the Monotone Convergence Theorem implies that $\lim_{n\to\infty} f(X)h_n(X)$ is integrable. Since $\lim_{n\to\infty} fh_n = f$ on S, it follows that $f(X) = \lim_{n\to\infty} f(X)h_n(X) \in L$. Thus Condition (i) holds. Summing up, given Condition (iii), the Conditions (i) and (iv) are equivalent to each other, as alleged. Hence Conditions (iii) and (iv) together are equivalent to Conditions (iii) and (i) together, which is equivalent to the function X being an r.v. Moreover, in that case, the Monotone Convergence Theorem implies that $E|f(X)h_n(X) - f(X)| \to 0$, where $fh_n \in C(S)$ for each $n \geq 1$. Assertion 5 is proved.

6. $EX = EX(1_A + 1_{A^c}) = EX1_A + EX1_{A^c} \equiv E(X;A) + E(X;A^c)$. Assertion 6 is verified.

7. Assertion 7 follows easily from Definition 4.6.8.

8. Suppose X is an r.r.v. such that $B \equiv (t-\varepsilon < X < t) \cup (t < X < t+\varepsilon)$ is a null set for some $t \in R$ and $\varepsilon > 0$. Let $(s_n)_{n=1,2,...}$ be a sequence of regular points of X in $(t, t+\varepsilon)$ that decreases to t. Then $(s_n < X) = (s_{n+1} < X)$ a.s. for each $n \geq 1$, because $(s_{n+1} < X \leq s_n) \subset B$ is a null set. Hence $P(s_n < X) = P(s_{n+1} < X)$ for each $n \geq 1$. Therefore $\lim_{n\to\infty} P(s_n < X)$ trivially exists. Similarly, there exists a sequence $(r_n)_{n=1,2,...}$ of regular points of X in $(t-\varepsilon, t)$ that increases to t such that $\lim_{n\to\infty} P(r_n < X)$ exists. The conditions in Definition 4.8.13 have been proved for t to be a regular point of X. The proposition is proved. □

We will make heavy use of the following Borel–Cantelli Lemma.

Proposition 5.1.5. First Borel–Cantelli Lemma. *Suppose $(A_n)_{n=1,2,...}$ is a sequence of events such that $\sum_{n=1}^{\infty} P(A_n)$ converges. Then a.s. only a finite number of the events A_n occur. More precisely, we have $P\left(\bigcup_{k=1}^{\infty}\bigcap_{n=k}^{\infty} A_n^c\right) = 1$.*

Proof. By Assertion 4 of Proposition 4.5.8, for each $k \geq 1$, the union $B_k \equiv \bigcup_{n=k}^{\infty} A_n$ is an event, with $P(B_k) \leq \sum_{n=k}^{\infty} P(A_n) \to 0$. Then $\bigcap_{n=k}^{\infty} A_n^c \subset \bigcap_{n=k+1}^{\infty} A_n^c$ for each $k \geq 1$, and

$$P\left(\bigcap_{n=k}^{\infty} A_n^c\right) = P(B_k^c) \to 1.$$

Therefore, by Assertion 1 of Proposition 4.5.8, the union $\bigcup_{k=1}^{\infty}\bigcap_{n=k}^{\infty} A_n^c$ is an event, with $P\left(\bigcup_{k=1}^{\infty}\bigcap_{n=k}^{\infty} A_n^c\right) = 1$. □

Definition 5.1.6. **L_p space.** Let X, Y be arbitrary r.r.v.'s. Let $p \in [1, \infty)$ be arbitrary. If X^p is integrable, define $\|X\|_p \equiv (E|X|^p)^{1/p}$. Define L_p to be the family of all r.r.v.'s X such that X^p is integrable. We will refer to $\|X\|_p$ as the L_p*-norm* of X. Let $n \geq 1$ be an integer. If $X \in L_n$, then $E|X|^n$ is called the n-th *absolute moment*, and EX^n the n-th *moment*, of X. If $X \in L_1$, then EX is also called the *mean* of X.

If $X, Y \in L_2$, then according to Proposition 5.1.7 (proved next), X, Y, and $(X - EX)(Y - EY)$ are integrable. Then $E(X - EX)^2$ and $E(X - EX)(Y - EY)$ are called the *variance* of X and the *covariance* of X and Y, respectively. The square root of the variance of X is called the *standard deviation* of X. □

Next are several basic inequalities for L_p.

Proposition 5.1.7. **Basic inequalities in L_p.** *Let $p, q \in [1, \infty)$ be arbitrary.*

1. **Hoelder's inequality.** *Suppose $p, q > 1$ and $\frac{1}{p} + \frac{1}{q} = 1$. If $X \in L_p$ and $Y \in L_q$, then $XY \in L_1$ and $E|XY| \leq \|X\|_p \|Y\|_q$. The special case where $p = q = 2$ is referred to as the Cauchy–Schwarz inequality.*
2. **Minkowski's inequality.** *If $X, Y \in L_p$, then $X + Y \in L_p$ and $\|X + Y\|_p \leq \|X\|_p + \|Y\|_p$.*
3. **Lyapunov's inequality.** *If $p \leq q$ and $X \in L_q$, then $X \in L_p$ and $\|X\|_p \leq \|X\|_q$.*

Proof. 1. Write α, β for $\frac{1}{p}, \frac{1}{q}$, respectively. Then $x^\alpha y^\beta \leq \alpha x + \beta y$ for nonnegative x, y. This can be seen by noting that with y fixed, the function f defined by $f(x) \equiv \alpha x + \beta y - x^\alpha y^\beta$ is equal to 0 at $x = y$, is decreasing for $x < y$, and is increasing for $x > y$. Let $a, b \in R$ be arbitrary with $a > \|X\|_p$ and $b > \|Y\|_q$. Replacing x, y by $|X/a|^p, |Y/b|^q$, respectively, we see that

$$|XY| \leq (\alpha |X/a|^p + \beta |Y/b|^q)ab.$$

It follows that $|XY|$ is integrable, with the integral bounded by

$$E|XY| \leq (\alpha \|X\|_p^p / a^p + \beta \|Y\|_q^q / b^q)ab.$$

As $a \to \|X\|_p$ and $b \to \|Y\|_q$, the last bound approaches $\|X\|_p \|Y\|_q$.

2. Suppose first that $p > 1$. Let $q \equiv \frac{p}{p-1}$. Then $\frac{1}{p} + \frac{1}{q} = 1$. Because $|X+Y|^p \leq (2(|X| \vee |Y|))^p \leq 2^p(|X|^p + |Y|^p)$, we have $X + Y \in L_p$. It follows trivially that $|X + Y|^{p-1} \in L_q$. Applying Hoelder's inequality, we estimate

$$\begin{aligned} E|X + Y|^p &\leq E|X + Y|^{p-1}|X| + E|X + Y|^{p-1}|Y| \\ &\leq (E|X + Y|^{(p-1)q})^{1/q}(\|X\|_p + \|Y\|_p) \\ &= (E|X + Y|^p)^{1/q}(\|X\|_p + \|Y\|_p). \end{aligned} \tag{5.1.1}$$

Suppose $\|X + Y\|_p > \|X\|_p + \|Y\|_p$. Then inequality 5.1.1, when divided by $(E|X + Y|^p)^{1/q}$, would imply $\|X + Y\|_p = (E|X + Y|^p)^{1-1/q} \leq \|X\|_p + \|Y\|_p$, which is a contradiction. This proves Minkowski's inequality for $p > 1$. Suppose now $p \geq 1$. Then $|X|^r, |Y|^r \in L_{p/r}$ for any $r < 1$. The preceding proof of the special case of Minkowski's inequality for the exponent $\frac{p}{r} > 1$ therefore implies

$$(E(|X|^r + |Y|^r)^{p/r})^{r/p} \leq (E(|X|^p)^{r/p} + (E(|Y|^p)^{r/p}. \tag{5.1.2}$$

Since

$$\begin{aligned}(|X|^r + |Y|^r)^{p/r} &\leq 2^{p/r}(|X|^r \vee |Y|^r)^{p/r} \\ &= 2^{p/r}(|X|^p \vee |Y|^p) \leq 2^{p/r}(|X|^p + |Y|^p) \in L,\end{aligned}$$

we can let $r \to 1$ and apply the Dominated Convergence Theorem to the left-hand side of inequality 5.1.2. Thus we conclude that $(|X| + |Y|)^p \in L$ and that $(E(|X| + |Y|)^p)^{1/p} \leq (E(|X|^p)^{1/p} + (E(|X|^p)^{1/p}$. Minkowski's inequality is proved.

3. Since $|X|^p \leq 1 \vee |X|^q \in L$, we have $X \in L_p$. Suppose $E|X|^p > (E|X|^q)^{p/q}$. Let $r \in (0, p)$ be arbitrary. Clearly, $|X|^r \in L_{q/r}$. Applying Hoelder's inequality to $|X|^r$ and 1, we obtain

$$E|X|^r \leq (E|X|^q)^{r/q}.$$

At the same time, $|X|^r \leq 1 \vee |X|^q \in L$. As $r \to p$, the Dominated Convergence Theorem yields $E|X|^p \leq (E|X|^q)^{p/q}$, establishing Lyapunov's inequality. □

Next we restate and simplify some definitions and theorems of convergence of measurable functions, in the context of r.v.'s.

Definition 5.1.8. Convergence in probability, a.u., a.s., and in L_1. For each $n \geq 1$, let X_n, X be functions on the probability space (Ω, L, E), with values in the complete metric space (S, d).

1. The sequence (X_n) is said to *converge to X almost uniformly* (*a.u.*) on the probability space (Ω, L, E) if $X_n \to X$ a.u. on the integration space (Ω, L, E). In that case, we write $X = \text{a.u.}\lim_{n\to\infty} X_n$. Since (Ω, L, E) is a probability space, Ω is a full set. It can therefore be easily verified that $X_n \to X$ a.u. iff for each $\varepsilon > 0$, there exists a measurable set B with $P(B) < \varepsilon$ such that X_n converges to X uniformly on B^c.
2. The sequence (X_n) is said to *converge to X in probability* on the probability space (Ω, L, E) if $X_n \to X$ in measure. Then we write $X_n \to X$ in probability. It can easily be verified that $X_n \to X$ in probability iff for each $\varepsilon > 0$, there exists $p \geq 1$ so large that for each $n \geq p$, there exists a measurable set B_n with $P(B_n) < \varepsilon$ such that $B_n^c \subset (d(X_n, X) \leq \varepsilon)$.
3. The sequence (X_n) is said to be *Cauchy in probability* if it is Cauchy in measure. It can easily be verified that (X_n) is Cauchy in probability iff for each $\varepsilon > 0$ there exists $p \geq 1$ so large that for each $m, n \geq p$, there exists a measurable set $B_{m,n}$ with $P(B_{m,n}) < \varepsilon$ such that $B_{m,n}^c \subset (d(X_n, X_m) \leq \varepsilon)$.
4. The sequence (X_n) is said to converge to X *almost surely* (a.s.) if $X_n \to X$ a.e. □

Proposition 5.1.9. a.u. Convergence implies convergence in probability, etc. *For each $n \geq 1$, let X, X_n be functions on the probability space (Ω, L, E), with values in the complete metric space (S,d). Then the following conditions hold:*

1. If $X_n \to X$ a.u., then (i) *X is defined a.e.,* (ii) *$X_n \to X$ in probability, and (iii) $X_n \to X$ a.s.*

2. If (i) *X_n is an r.v. for each $n \geq 1$ and* (ii) *$X_n \to X$ in probability, then X is an r.v.*

3. If (i) *X_n is an r.v. for each $n \geq 1$ and* (ii) *$X_n \to X$ a.u., then X is an r.v.*

4. If (i) *X_n is an r.v. for each $n \geq 1$ and* (ii) *$(X_n)_{n=1,2,\ldots}$ is Cauchy in probability, then there exists a subsequence $(X_{n(k)})_{k=1,2,\ldots}$ such that $X \equiv \lim_{k\to\infty} X_{n(k)}$ is an r.v., with $X_{n(k)} \to X$ a.u. and $X_{n(k)} \to X$ a.s. Moreover, $X_n \to X$ in probability.*

5. Suppose (i) *X_n, X are r.r.v.'s for each $n \geq 1$,* (ii) *$X_n \uparrow X$ in probability, and* (iii) *$a \in R$ is a regular point of X_n, X for each $n \geq 0$. Then $P((X_n > a)B) \uparrow P((X > a)B)$ as $n \to \infty$, for each measurable set B.*

Proof. Assertions 1–3 are trivial consequences of the corresponding assertions in Proposition 4.9.3. Assertion 4 is a trivial consequence of Proposition 4.9.4. It remains to prove Assertion 5. To that end, let $\varepsilon > 0$ be arbitrary. Then, by Condition (iii), $a \in R$ is a regular point of the r.r.v. X. Hence there exists $a' > a$ such that $P(a' \geq X > a) < \varepsilon$. Since by Condition (ii), $X_n \uparrow X$ in probability, there exists $m \geq 1$ so large that $P(X - X_n > a' - a) < \varepsilon$ for each $n \geq m$. Now let $n \geq m$ be arbitrary. Let $A \equiv (a' \geq X > a) \cup (X - X_n > a' - a)$. Then $P(A) < 2\varepsilon$. Moreover,

$$
\begin{aligned}
P((X > a)B) - P((X_n > a)B) &\leq P(X > a; X_n \leq a) \\
&= P((X > a; X_n \leq a)A^c) + P(A) \\
&< P((X > a) \cap (X_n \leq a) \cap ((a' < X) \cup (X \leq a)) \\
&\quad \cap (X - X_n \leq a' - a)) + 2\varepsilon \\
&= P((X_n \leq a) \cap (a' < X) \cap (X - X_n \leq a' - a)) + 2\varepsilon \\
&= 0 + 2\varepsilon = 2\varepsilon.
\end{aligned}
$$

Since $P(A) < 2\varepsilon$ is arbitrarily small, we see that $P((X_n > a)B) \uparrow P((X > a)B)$, as alleged in Assertion 5. □

The next definition and proposition say that convergence in probability can be metrized.

Definition 5.1.10. Probability metric on the space of r.v.'s. Let (Ω, L, E) be a probability space. Let (S,d) be a complete metric space. We will let $M(\Omega, S)$ denote the space of r.v.'s on (Ω, L, E) with values in (S,d), where two r.v.'s are considered equal if they are equal a.s. Define the metric

$$
\rho_{Prob}(X,Y) \equiv E(1 \wedge d(X,Y)) \tag{5.1.3}
$$

for each $X, Y \in M(\Omega, S)$. The next proposition proves that ρ_{Prob} is indeed a metric. We will call ρ_{Prob} *the probability metric on the space $M(\Omega, S)$* of r.v.'s. □

Proposition 5.1.11. Basics of the probability metric ρ_{Prob} on the space $M(\Omega,S)$ of r.v.'s. *Let (Ω,L,E) be a probability space. Let $X,X_1,X_2,\ldots$ be arbitrary r.v.'s with values in the complete metric space (S,d). Then the following conditions hold:*

1. The pair $(M(\Omega,S),\rho_{Prob})$ is a metric space. Note that $\rho_{Prob} \leq 1$.

2. $X_n \to X$ in probability iff for each $\varepsilon > 0$, there exists $p \geq 1$ so large that $P(d(X_n,X) > \varepsilon) < \varepsilon$ for each $n \geq p$.

3. Sequential convergence relative to ρ_{Prob} is equivalent to convergence in probability.

4. The metric space $(M(\Omega,S),\rho_{Prob})$ is complete.

5. Suppose there exists a sequence $(\varepsilon_n)_{n=1,2,\ldots}$ of positive real numbers such that $\sum_{n=1}^{\infty}\varepsilon_n < \infty$ and such that $\rho_{Prob}(X_n,X_{n+1}) \equiv E(1\wedge d(X_n,X_{n+1})) < \varepsilon_n^2$ for each $n \geq 1$. Then $Y \equiv \lim_{n\to\infty} X_n$ is an r.v., and $X_n \to Y$ a.u.

Proof. 1. Let $X,Y \in M(\Omega,S)$ be arbitrary. Then $d(X,Y)$ is an r.r.v according to Proposition 4.8.20. Hence $1\wedge d(X,Y)$ is an integrable function, and ρ_{Prob} is well defined in equality 5.1.3. Symmetry and triangle inequality for the function ρ_{Prob} are obvious from its definition. Suppose $\rho_{Prob}(X,Y) \equiv E(1\wedge d(X,Y)) = 0$. Let $(\varepsilon_n)_{n=1,2,\ldots}$ be a sequence in $(0,1)$ with $\varepsilon_n \downarrow 0$. Then Chebychev's inequality implies

$$P(d(X,Y) > \varepsilon_n) = P(1\wedge d(X,Y) > \varepsilon_n) \leq \varepsilon_n^{-1}E(1\wedge d(X,Y)) = 0$$

for each $n \geq 1$. Hence $A \equiv \bigcup_{n=1}^{\infty}(d(X,Y) > \varepsilon_n)$ is a null set. On the full set A^c, we have $d(X,Y) \leq \varepsilon_n$ for each $n \geq 1$. Therefore $d(X,Y) = 0$ on the full set A^c. Thus $X = Y$ in $M(\Omega,S)$. Summing up, ρ_{Prob} is a metric.

2. Suppose $X_n \to X$ in probability. Let $\varepsilon > 0$ be arbitrary. Then, according to Definition 5.1.8, there exists $p \geq 1$ so large that for each $n \geq p$, there exists an integrable set B_n with $P(B_n) < \varepsilon$ and $B_n^c \subset (d(X_n,X) \leq \varepsilon)$. Now consider each $n \geq p$. Then $P(d(X_n,X) > \varepsilon) \leq P(B_n) < \varepsilon$ for each $n \geq p$. Conversely, suppose, for each $\varepsilon > 0$, there exists $p \geq 1$ so large that $P(d(X_n,X) > \varepsilon) < \varepsilon$ for each $n \geq p$. Let $\varepsilon > 0$ be arbitrary. Consider each $n \geq 1$. Define the integrable set $B_n \equiv (d(X_n,X) > \varepsilon)$. Then $P(B_n) < \varepsilon$ and $B_n^c \subset (d(X_n,X) \leq \varepsilon)$. Hence $X_n \to X$ in probability according to Definition 5.1.8. Assertion 2 is proved.

3. Suppose $\rho_{Prob}(X_n,X) \equiv E(1\wedge d(X_n,X)) \to 0$. Let $\varepsilon > 0$ be arbitrary. Take $p \geq 1$ so large that $E(1\wedge d(X_n,X)) < \varepsilon(1\wedge\varepsilon)$ for each $n \geq p$. Then Chebychev's inequality implies that

$$P(d(X_n,X) > \varepsilon) \leq P(1\wedge d(X_n,X) \geq 1\wedge\varepsilon) \leq (1\wedge\varepsilon)^{-1}E(1\wedge d(X_n,X)) < \varepsilon$$

for each $n \geq p$. Thus $X_n \to X$ in probability, by Assertion 2. Conversely, suppose $X_n \to X$ in probability. Then, by Assertion 2, there exists $p \geq 1$ so large that $P(d(X_n,X) > \varepsilon) < \varepsilon$ for each $n \geq p$. Hence

$$
\begin{aligned}
E(1 \wedge d(X_n, X)) &= E(1 \wedge d(X_n, X))1_{(d(X(n),X)>\varepsilon)} \\
&\quad + E(1 \wedge d(X_n, X))1_{(d(X(n),X)\leq\varepsilon)} \\
&\leq E1_{(d(X(n),X)>\varepsilon)} + \varepsilon = P(d(X_n, X) > \varepsilon) + \varepsilon < 2\varepsilon
\end{aligned}
$$

for each $n \geq p$. Thus $\rho_{Prob}(X_n, X) \equiv E(1 \wedge d(X_n, X)) \to 0$. Assertion 3 is proved.

4. Suppose $\rho_{Prob}(X_n, X_m) \equiv E(1 \wedge d(X_n, X_m)) \to 0$ *as* $n, m \to \infty$. Let $\varepsilon > 0$ be arbitrary. Take $p \geq 1$ so large that $E(1 \wedge d(X_n, X_m)) < \varepsilon(1 \wedge \varepsilon)$ for each $n, m \geq p$. Then Chebychev's inequality implies that

$$
\begin{aligned}
P(d(X_n, X_m) > \varepsilon) &\leq P(1 \wedge d(X_n, X_m) \geq 1 \wedge \varepsilon) \\
&\leq (1 \wedge \varepsilon)^{-1} E(1 \wedge d(X_n, X_m)) < \varepsilon
\end{aligned}
$$

for each $n, m \geq p$. Thus the sequence $(X_n)_{n=1,2,\ldots}$ of functions is Cauchy in probability. Hence Proposition 4.9.4 implies that $X \equiv \lim_{k\to\infty} X_{n(k)}$ is an r.v. for some subsequence $(X_{n(k)})_{k=1,2,\ldots}$ of $(X_n)_{n=1,2,\ldots}$, and that $X_n \to X$ in probability. By Assertion 3, it then follows that $\rho_{Prob}(X_n, X) \to 0$. Thus the metric space $(M(\Omega, S), \rho_{Prob})$ is complete, and Assertion 4 is proved.

5. Suppose there exists a sequence $(\varepsilon_n)_{n=1,2,\ldots}$ of positive real numbers such that $\sum_{n=1}^{\infty} \varepsilon_n < \infty$, and such that $\rho_{Prob}(X_n, X_{n+1}) \equiv E(1 \wedge d(X_n, X_{n+1})) < \varepsilon_n^2$ for each $n \geq 1$. Trivially, the probability space (Ω, L, E) is σ-finite with the sequence $(A_i)_{i=1,2,\ldots}$ as an I-basis, where $A_i \equiv \Omega$ for each $i \geq 1$. Then, for each $i \geq 1$, we have $I(1 \wedge d(X_n, X_{n+1}))1_{A(i)} < \varepsilon_n^2$ for each $n \geq 1$. Hence Proposition 4.9.6 is applicable; it implies that $X \equiv \lim_{n\to\infty} X_n$ is an r.v., and $X_n \to X$ a.u. Assertion 5 and the proposition are proved. □

Corollary 5.1.12. Reciprocal of an a.s. positive r.r.v. *Let X be a nonnegative r.r.v. such that $P(X < a) \to 0$ as $a \to 0$. Define the function X^{-1} by $domain(X^{-1}) \equiv D \equiv (X > 0)$ and $X^{-1}(\omega) \equiv (X(\omega))^{-1}$ for each $\omega \in D$.* Then X^{-1} is an r.r.v.

Proof. Let $a_1 > a_2 > \cdots > 0$ be a sequence such that $PD_k^c \to 0$, where $D_k \equiv P(X \geq a_k)$ for each $k \geq 1$. Then $D = \bigcup_{k=1}^{\infty} D_k$, whence D is a full set. Let $j \geq k \geq 1$ be arbitrary. Define the r.r.v. $Y_k \equiv (X \vee a_k)^{-1} 1_{D(k)}$. Then $X^{-1} 1_{D(k)} = Y_k$. Moreover, $Y_j \geq Y_k$ and

$$
(Y_j - Y_k > 0) \subset (1_{D(j)} - 1_{D(k)} > 0) = D_j D_k^c \subset D_k^c.
$$

Consequently, since $PD_k^c \to 0$ as $k \to \infty$, the sequence $(Y_k)_{k=1,2,\ldots}$ converges a.u. Hence, according to Proposition 5.1.9, $Y \equiv \lim_{k\to\infty} Y_k$ is an r.r.v. Since $X^{-1} = Y$ on the full set D, the function X^{-1} is an r.r.v. □

We saw in Proposition 5.1.11 that convergence in L_1 of r.r.v.'s implies convergence in probability. The next proposition gives the converse in the case of uniform integrability.

Proposition 5.1.13. Uniform integrability of a sequence of r.r.v.'s and convergence in probability together imply convergence in L_1. *Suppose $(X_n)_{n=1,2,...}$ is a uniformly integrable sequence of r.r.v.'s. Suppose $X_n \to X$ in probability for some r.r.v. X. Then the following conditions hold:*

1. $\lim_{n\to\infty} E|X_n|$ exists.

2. X is integrable and $X_n \to X$ in L_1.

Proof. 1. Let $\varepsilon > 0$ be arbitrary. By hypothesis, the sequence $(X_n)_{n=1,2,...}$ is uniformly integrable. Hence, by Proposition 4.7.3, there exist (i) $b \geq 0$ so large that $E|X_n| \leq b$ for each $n \geq 1$ and (ii) $\delta(\varepsilon) \in (0,\varepsilon)$ so small that $E(|X_n|; A) < \varepsilon$ for each $n \geq 1$ and for each event A with $PA < \delta(\varepsilon)$.

2. Since, again by hypothesis, $X_n \to X$ in probability, there exists $p(\varepsilon) \geq 1$ so large that $P(|X - X_n| > \delta(\varepsilon)) < \delta \equiv \delta(\varepsilon)$ for each $n \geq p(\varepsilon)$. Let $m,n \geq p(\varepsilon)$ be arbitrary. Then

$$\begin{aligned}
|E|X_m| - E|X_n|| &\leq E|X_m - X_n| \\
&\leq (E|X_m - X_n|; |X_m - X_n| \leq 2\delta) + (E|X_m - X_n|; |X_m - X_n| > 2\delta) \\
&\leq 2\delta + E(|X_m| + |X_n|; |X - X_n| \vee |X - X_m| > \delta) \\
&\leq 2\varepsilon + E(|X_m| + |X_n|; (|X - X_n| > \delta) \cup (|X - X_m| > \delta)) \\
&= 2\varepsilon + E(|X_m|; |X - X_n| > \delta) + E(|X_m|; |X - X_m| > \delta) \\
&\quad + E(|X_n|; |X - X_n| > \delta) + E(|X_n|; |X - X_m| > \delta) \\
&< 2\varepsilon + \varepsilon + \varepsilon + \varepsilon + \varepsilon = 6\varepsilon.
\end{aligned} \tag{5.1.4}$$

Thus $(E|X_n|)_{n=1,2,...}$ is a Cauchy sequence. Consequently, $c \equiv \lim_{n\to\infty} E|X_n|$ exists. Assertion 1 is proved.

3. Letting $m \to \infty$ while keeping $n = p(\varepsilon)$ in inequality 5.1.4, we obtain

$$|c - E|X_{p(\varepsilon)}|| \leq 6\varepsilon. \tag{5.1.5}$$

Moreover, inequality 5.1.4 implies that

$$|E(|X_m| \wedge a) - E(|X_n| \wedge a)| \leq |E|X_m| - E|X_n|| < 6\varepsilon \tag{5.1.6}$$

for each $m,n \geq p(\varepsilon)$, for each $a > 0$. Again, let $m \to \infty$, while keeping $n = p(\varepsilon)$. Then $E(|X_m| \wedge a) \to E(|X| \wedge a)$ as $m \to \infty$, by the Dominated Convergence Theorem. Hence inequality 5.1.6 yields

$$|E(|X| \wedge a) - E(|X_{p(\varepsilon)}| \wedge a)| \leq 6\varepsilon. \tag{5.1.7}$$

4. Next note that $X_{p(\varepsilon)}$ is, by hypothesis, an integrable r.r.v. Hence there exists $a_\varepsilon > 0$ so large that

$$|E|X_{p(\varepsilon)}| - E(|X_{p(\varepsilon)}| \wedge a)| < \varepsilon \tag{5.1.8}$$

for each $a > a_\varepsilon$. Then, for each $a > a_\varepsilon$, we have

$$|c - E(|X| \wedge a)| \leq |c - E|X_{p(\varepsilon)}| + |E|X_{p(\varepsilon)}| - E(|X_{p(\varepsilon)}| \wedge a)|$$
$$+ |E(|X_{p(\varepsilon)}| \wedge a) - E(|X| \wedge a)|$$
$$\leq 6\varepsilon + \varepsilon + 6\varepsilon = 13\varepsilon,$$

where the second inequality is thanks to inequalities 5.1.5, 5.1.7, and 5.1.8. Since $\varepsilon > 0$ is arbitrary, we see that $E(|X| \wedge a) \uparrow c$ as $a \to \infty$.

5. Therefore, by the Monotone Convergence Theorem, the r.r.v. $|X|$ is integrable, with expectation equal to $c \equiv \lim_{n\to\infty} E|X_n|$. Consequently, the r.r.v. X is integrable.

6. At the same time, inequality 5.1.4 yields $E|X_m - X_n| \leq 6\varepsilon$ for each $m, n \geq p(\varepsilon)$. Now consider each fixed $n \geq p(\varepsilon)$. Then the nonnegative sequence $((X_m - X_n) \vee 0)_{m=1,2,\ldots}$ of r.r.v.'s converges in probability to the r.r.v. $(X - X_n) \vee 0$. Hence, by Step 5, $E((X - X_n) \vee 0) = \lim_{m\to\infty} E((X_m - X_n) \vee 0)$. Similarly, $E(-((X - X_n) \wedge 0)) = \lim_{m\to\infty} E(-((X_m - X_n) \wedge 0))$. Combining,

$$\begin{aligned} E|X - X_n| &= E((X - X_n) \vee 0) - E((X - X_n) \wedge 0) \\ &= \lim_{m\to\infty} E((X_m - X_n) \vee 0) - \lim_{m\to\infty} E((X_m - X_n) \wedge 0) \\ &= \lim_{m\to\infty} E((X_m - X_n) \vee 0) - E((X_m - X_n) \wedge 0) \\ &= \lim_{m\to\infty} E|X_m - X_n| \leq 6\varepsilon, \end{aligned}$$

where $n \geq p(\varepsilon)$ is arbitrary. Thus $E|X - X_n| \to 0$ as $n \to \infty$. In other words, $X_n \to X$ in L_1. The proposition is proved. □

Proposition 5.1.14. Necessary and sufficient condition for a.u. convergence. *For $n \geq 1$, let X, X_n be r.v.'s with values in the locally compact metric space (S,d). Then the following two conditions are equivalent:* (i) *for each $\varepsilon > 0$, there exist an integrable set B with $P(B) < \varepsilon$ and an integer $m \geq 1$ such that for each $n \geq m$, we have $d(X, X_n) \leq \varepsilon$ on B^c, and* (ii) *$X_n \to X$ a.u.*

Proof. Suppose Condition (i) holds. Let $(\varepsilon_k)_{k=1,2,\ldots}$ be a sequence of positive real numbers with $\sum_{k=1}^{\infty} \varepsilon_k < \infty$. By hypothesis, for each $k \geq 1$ there exist an integrable set B_k with $P(B_k) < \varepsilon_k$ and an integer $m_k \geq 1$ such that for each $n \geq m_k$, we have $d(X, X_n) \leq \varepsilon_k$ on B_k^c. Let $\varepsilon > 0$ be arbitrary. Let $p \geq 1$ be so large that $\sum_{k=p}^{\infty} \varepsilon_k < \varepsilon$ and define $A \equiv \bigcup_{k=p}^{\infty} B_k$. Then $P(A) \leq \sum_{k=p}^{\infty} \varepsilon_k < \varepsilon$. Moreover, on $A^c = \bigcap_{k=p}^{\infty} B_k^c$, we have $d(X, X_n) \leq \varepsilon_k$ for each $n \geq m_k$ and each $k \geq p$. Therefore $X_n \to X$ uniformly on A^c. Since $P(A)$ is arbitrarily small, we have $X_n \to X$ a.u. Thus Condition (ii) is verified.

Conversely, suppose Condition (ii) holds. Let $\varepsilon > 0$ be arbitrary. Then, by Definition 4.9.2, there exists a measurable set B with $P(B) < \varepsilon$ such that X_n converges to X uniformly on B^c. Hence there exists $m \geq 1$ so large that $\bigcup_{n=m}^{\infty} (d(X_n, X) > \varepsilon) \subset B$. In particular, for each $n \geq m$, we have $(d(X_n, X) > \varepsilon) \subset B$, whence $d(X, X_n) \leq \varepsilon$ on B^c. Condition (i) is established. □

Definition 5.1.15. Probability subspace generated by a family of r.v.'s with values in a complete metric space. Let (Ω, L, E) be a probability space and let L' be a subset of L. If (Ω, L', E) is a probability space, then we call (Ω, L', E) a *probability subspace* of (Ω, L, E). When confusion is unlikely, we will abuse terminology and simply call L' a probability subspace of L, with Ω and E understood.

Let G be a nonempty family of r.v.'s with values in a complete metric space (S,d). Define

$$L_{Cub}(G) \equiv \{f(X_1,\ldots,X_n) : n \geq 1; f \in C_{ub}(S^n,d^n); X_1,\ldots,X_n \in G\}.$$

Then $(\Omega, L_{Cub}(G), E)$ is an integration subspace *of* (Ω, L, E). Its completion

$$\begin{aligned} L(G) &\equiv L(X : X \in G) \\ &\equiv \{f(X_1,\ldots,X_n) : n \geq 1; f \in C_{ub}(S^n,d^n); X_1,\ldots,X_n \in G\}^- \end{aligned}$$

will be called the *probability subspace of* L *generated by the family* G.

If G is a finite or countably infinite set $\{X_1, X_2, \ldots\}$, we write $L(X_1, X_2, \ldots)$ for $L(G)$. □

Note that $L_{Cub}(G)$ is a linear subspace of L containing constants and is closed to the operation of maximum and absolute values. Hence $(\Omega, L_{Cub}(G), E)$ is indeed an integration space, according to Proposition 4.3.6. Since $1 \in L_{Cub}(G)$ with $E1 = 1$, the completion $(\Omega, L(G), E)$ is a probability space. Any r.r.v. in $L(G)$ has its value determined once all the values of the r.v.'s in the generating family G have been observed. Intuitively, $L(G)$ contains all the information obtainable by observing the values of all $X \in G$.

Proposition 5.1.16. Probability subspace generated by a family of r.v.'s with values in a locally compact metric space. *Let* (Ω, L, E) *be a probability space. Let* G *be a nonempty family of r.v.'s with values in a locally compact metric space* (S,d). *Let*

$$L_C(G) \equiv \{f(X_1,\ldots,X_n) : n \geq 1; f \in C(S^n); X_1,\ldots,X_n \in G\}.$$

Then $(\Omega, L_C(G), E)$ *is an integration subspace* of (Ω, L, E). *Moreover, its completion* $\overline{L}_C$ *is equal to* $L(G) \equiv \overline{L}_{Cub}$.

Proof. Note first that $L_C(G) \subset L_{Cub}(G)$, and that $L_C(G)$ is a linear subspace of $\overline{L}_{Cub}$ such that if $U, V \in L_C(G)$, then $|U|, U \wedge 1 \in L_C(G)$. Hence $L_C(G)$ is an integration subspace of the complete integration space $(\Omega, \overline{L}_{Cub}, E)$ according to Proposition 4.3.6. Consequently, $\overline{L}_C \subset \overline{L}_{Cub}$.

Conversely, let $U \in L_{Cub}(G)$ be arbitrary. Then $U = f(X_1, \ldots, X_n)$ for some $f \in C_{ub}(S^n,d^n)$ and some $X_1, \ldots, X_n \in G$. Then, by Assertion 4 of Proposition 5.1.4, there exists a sequence $(g_k)_{k=1,2,\ldots}$ in $C(S^n,d^n)$ such that $E|f(X) - g_k(X)| \to 0$, where we write $X \equiv (X_1, \ldots, X_n)$. Since $g_k(X) \in L_{C(G)} \subset \overline{L}_C$

for each $k \geq 1$, and since $\overline{L}_C$ is complete, we see that $U = f(X) \in \overline{L}_C$. Since $U \in L_{Cub}(G)$ is arbitrary, we have $L_{Cub}(G) \subset \overline{L}_C$. Consequently, $\overline{L}_{Cub} \subset \overline{L}_C$.

Summing up, $\overline{L}_C = \overline{L}_{Cub} \equiv L(G)$, as alleged. □

The next lemma sometimes comes in handy.

Lemma 5.1.17. The intersection of probability subspaces is a probability subspace. *Let (Ω, L, E) be a probability space. Let $\widehat{L}$ be a nonempty family of probability subspaces L' of L. Then $L'' \equiv \bigcap_{L' \in \widehat{L}} L'$ is a probability subspace of L.*

Proof. Clearly, the intersection L'' is a linear subspace of L, contains the constant function 1 with $E1 = 1$, and is such that if $X, Y \in L''$, then $|X|, X \wedge 1 \in L''$. Hence it is an integration subspace of L, according to Proposition 4.3.6. At the same time, since the sets L' in the family $\widehat{L}$ are closed in the space L relative to the norm $E|\cdot|$, so is their intersection L''. Since L is complete relative to E, so is the closed subspace L''. Summing up, (Ω, L'', I) is a probability subspace of (Ω, L, I). □

5.2 Probability Distribution on Metric Space

Definition 5.2.1. Distribution on a complete metric space. Suppose (S,d) is a complete metric space. Let $n \geq 1$ be arbitrary. Recall the function $h_n \equiv 1 \wedge (1 + n - d(\cdot, x_\circ))_+ \in C_{ub}(S,d)$, where $x_\circ \in S$ is an arbitrary but fixed reference point. Note that the function h_n has bounded support. Hence $h_n \in C(S,d)$ if (S,d) is locally compact.

Let J be an integration on $(S, C_{ub}(S,d))$, in the sense of Definition 4.3.1. Suppose $Jh_n \uparrow 1$ as $n \to \infty$. Then the integration J is called a *probability distribution,* or simply a *distribution, on* (S,d). We will let $\widehat{J}(S,d)$ denote the set of distributions on the complete metric space (S,d). □

Lemma 5.2.2. Distribution basics. *Suppose (S,d) is a complete metric space. Then the following conditions hold:*

1. Let J be an arbitrary distribution on (S,d). Let $(S,L,J) \equiv (S, \overline{C_{ub}(S)}, J)$ be the complete extension of the integration space $(S, C_{ub}(S), J)$. Then (S,L,J) is a probability space.

2. Suppose the metric space (S,d) is bounded. Let J be an integration on $(S, C_{ub}(S))$ such that $J1 = 1$. Then the integration J is a distribution on (S,d).

3. Suppose (S,d) is locally compact. Let J be an integration on $(S, C(S,d))$ in the sense of Definition 4.2.1. Suppose $Jh_n \uparrow 1$ as $n \to \infty$. Then J is a distribution on (S,d).

Proof. 1. By Definition 5.2.1, $Jh_n \uparrow 1$ as $n \to \infty$. At the same time, $h_n \uparrow 1$ on S. The Monotone Convergence Theorem therefore implies that $1 \in L$ and $J1 = 1$. Thus (S,L,J) is a probability space.

2. Suppose (S,d) is bounded. Then $h_n = 1$ on S, for sufficiently large $n \geq 1$. Hence, trivially $Jh_n \uparrow J1 = 1$, where the equality is by assumption. Therefore the integration J ion $(S, C_{ub}(S))$ satisfies the conditions in Definition 5.2.1 to be a distribution.

3. Since (S,d) is locally compact, we have $h_n \in C(S)$ for each $n \geq 1$. Moreover, $Jh_n \uparrow 1$ by hypothesis. Let (S,L,J) denote the completion of $(S, C(S), J)$. Let $f \in C_{ub}(S)$ be arbitrary, with some bound $b \geq 0$ for $|f|$. Then

$$J|h_m f - h_n f| \leq bJ|h_m - h_n| = bJ(h_m - h_n) \to 0$$

as $m \geq n \to \infty$. Hence the sequence $(h_n f)_{n=1,2,\ldots}$ is Cauchy in the complete integration space L relative to J. Therefore $g \equiv \lim_{n\to\infty}(h_n f) \in L$, with $Jg = \lim_{n\to\infty} Jh_n f$. At the same time, $\lim_{n\to\infty}(h_n f) = f$ on S. Hence $f = g \in L$, with $Jf = Jg = \lim_{n\to\infty} Jh_n f$. Since $f \in C_{ub}(S)$ is arbitrary, we conclude that $C_{ub}(S) \subset L$. Consequently, $(S, C_{ub}(S), J)$ is an integration subspace of (S,L,J). Moreover, in the special case $f \equiv 1$, we obtain $1 \in L$ with $J1 = \lim_{n\to\infty} Jh_n = 1$. Thus the integration J on $C_{ub}(S)$ satisfies the conditions in Definition 5.2.1 to be a distribution. □

Definition 5.2.3. Distribution induced by an r.v. Let X be an r.v. on a probability space (Ω, L, E) with values in the complete metric space (S,d). For each $f \in C_{ub}(S)$, define $E_X f \equiv Ef(X)$. Lemma 5.2.4 (next) proves that E_X is a distribution on (S,d). We will call E_X the *distribution induced on the complete metric space* (S,d) *by the r.v.* X. The completion $(S, L_X, E_X) \equiv (S, \overline{C_{ub}(S)}, E_X)$ of $(S, C_{ub}(S), E_X)$ is a probability space, called the *probability space induced on the complete metric space* (S,d) *by the r.v.* X. □

Lemma 5.2.4. Distribution induced by an r.v. is indeed a distribution. *Let X be an arbitrary r.v. on a probability space (Ω, L, E) with values in the complete metric space (S,d). Then the function E_X introduced in Definition 5.2.3 is indeed a distribution.*

Proof. Let $f \in C_{ub}(S)$ be arbitrary. By Assertion 4 of Proposition 5.1.4, we have $f(X) \in L$. Hence $E_X f \equiv Ef(X)$ is well defined. The space $C_{ub}(S)$ is linear, contains constants, and is closed to absolute values and taking minimums. The remaining conditions in Definition 4.3.1 for E_X to be an integration on $(S, C_{ub}(S))$ follow from the corresponding conditions for E. Moreover, $E_X h_n \equiv Eh_n(X) \uparrow 1$ as $n \to \infty$, where the convergence is again by Assertion 4 of Proposition 5.1.4. All the conditions in Definition 5.2.1 have been verified for E_X to be a distribution. □

Proposition 5.2.5. Each distribution is induced by some r.v. *Suppose J is a distribution on a complete metric space (S,d). Let (S,L,J) denote the completion of the integration space $(S, C_{ub}(S), J)$. Then the following conditions hold:*

1. The identity function $X : (S,L,J) \to (S,d)$, defined by $X(x) = x$ for each $x \in S$, is an r.v.

2. The function $d(\cdot, x_\circ)$ is an r.r.v. on (S, L, J).

3. $J = E_X$. Thus each distribution on a complete metric space is induced by some r.v.

Proof. By Lemma 5.2.2, (S, L, J) is a probability space and $Jh_n \uparrow 1$ as $n \to \infty$. Hence the hypothesis in Corollary 4.8.11 is satisfied. Accordingly, the identity function X is an r.v. on $(\Omega, L, E) \equiv (S, L, J)$, and $d(\cdot, x_\circ)$ is an r.r.v. Moreover, for each $f \in C_{ub}(S)$, we have $Jf \equiv Ef \equiv Ef(X) \equiv E_X f$. Hence the integration spaces $(S, C_{ub}(S, J), J)$ and $(S, C_{ub}(S, J), E_X)$ are equal. Therefore their completions are the same. In other words, $(S, L, J) = (S, L_X, E_X)$. □

Proposition 5.2.6. Relation between probability spaces generated and induced by an r.v. *Suppose X is an r.v. on the probability space (Ω, L, E) with values in a complete metric space (S,d). Let $(\Omega, L(X), E)$ be the probability subspace of (Ω, L, E) generated by $\{X\}$. Let (S, L_X, E_X) be the probability space induced on (S,d) by X. Let $f : S \to R$ be an arbitrary function. Then the following conditions hold:*

1. $f \in L_X$ iff $f(X) \in L(X)$, in which case $E_X f = Ef(X)$. In words, a function f is integrable relative to E_X iff $f(X)$ is integrable relative to E, in which case $E_X f = Ef(X)$.

2. f is an r.r.v. on (S, L_X, E_X) iff $f(X)$ is an r.r.v. on $(\Omega, L(X), E)$.

Proof. 1. Suppose $f \in L_X$. Then, by Corollary 4.4.4, there exists a sequence $(f_n)_{n=1,2,\ldots}$ in $C_{ub}(S)$ such that $E_X|f - f_n| \to 0$ and $f = \lim_{n\to\infty} f_n$. Consequently,

$$E|f_n(X) - f_m(X)| \equiv E_X|f_n - f_m| \to 0.$$

Thus $(f_n(X))_{n=1,2,\ldots}$ is a Cauchy sequence in $L(X)$ relative to the expectation E. Since $L(X)$ is complete, we have $Y \equiv \lim_{n\to\infty} f_n(X) \in L(X)$ with

$$E|f_n(X) - Y| \to 0,$$

whence

$$EY = \lim_{n\to\infty} Ef_n(X) \equiv \lim_{n\to\infty} E_X f_n = E_X f.$$

Since $f(X) = \lim_{n\to\infty} f_n(X) = Y$ on the full set $domain(Y)$, it follows that $f(X) \in L(X)$, with $Ef(X) = EY = E_X f$.

Conversely, suppose $Z \in L(X)$. We will show that $Z = f(X)$ for some integrable function f relative to E_X. Since $L(X)$ is, by definition, the completion of $L_{Cub}(X) \equiv \{f(X) : f \in C_{ub}(S)\}$, the latter is dense in the former, relative to the norm $E|\cdot|$. Hence there exists a sequence $(f_n)_{n=1,2,\ldots}$ in $C_{ub}(S)$ such that

$$E|Z - f_n(X)| \to 0. \tag{5.2.1}$$

Consequently,

$$E_X|f_n - f_m| \equiv E|f_n(X) - f_m(X)| \to 0.$$

Hence $E_X|f_n - f| \to 0$ where $f \equiv \lim_{n\to\infty} f_n \in \overline{C_{ub}(S)} \equiv L_X$. By the first part of this proof in the previous paragraph, we have

$$E|f_n(X) - Y| \to 0, \tag{5.2.2}$$

where

$$Y = f(X) \qquad \text{a.s.} \tag{5.2.3}$$

Convergence expressions 5.2.1 and 5.2.2 together imply that $Z = Y$ a.s., which, together with equality 5.2.3, yields $Z = f(X)$, where $f \in L_X$. Assertion 1 is proved.

2. For each $n \geq 1$, define $g_n \equiv 1 \wedge (1 + n - |\cdot|)_+ \in C_{ub}(R)$. Suppose the function f is an r.r.v. on (S, L_X, E_X). Then, by Proposition 5.1.4, we have (i) $g \circ f \in L_X$ for each $g \in C_{ub}(R)$ and (ii) $E_X g_n \circ f \uparrow 1$ as $n \to \infty$.

In view of Condition (i), we have $g(f(X)) \equiv g \circ f(X) \in L(X)$ for each $g \in C_{ub}(R)$ by Assertion 1. Moreover, $Eg_n(f(X)) = E_X g_n \circ f \uparrow 1$ as $n \to \infty$. Combining, we can apply Assertion 4 of Proposition 5.1.4 to the function $f(X) : \Omega \to R$ in the place of $X : \Omega \to S$, and we conclude that $f(X)$ is an r.r.v. on $(\Omega, L(X), E)$.

Conversely, suppose $f(X)$ is an r.r.v. on $(\Omega, L(X), E)$. Then, again by Assertion 4 of Proposition 5.1.4, we have (i′) $g(f(X)) \in L(X)$ for each $g \in C_{ub}(R)$ and (ii′) $E(g_n(f(X)) \uparrow 1$ as $n \to \infty$. In view of Condition (i′), we have $g \circ f \in L_X$ for each $g \in C_{ub}(R)$ by Assertion 1 of the present proposition. Moreover, $E_X g_n \circ f = Eg_n(f(X)) =\uparrow 1$ as $n \to \infty$. Combining, we see that f is an r.r.v. on (S, L_X, E_X), again by Assertion 4 of Proposition 5.1.4. □

Proposition 5.2.7. Regular points of an r.r.v. f relative to induced distribution by an r.v. X are the same as regular points of $f(X)$. *Suppose X is an r.v. on the probability space (Ω, L, E) with values in a complete metric space (S, d). Suppose f is an r.r.v. on (S, L_X, E_X). Then $t \in R$ is a regular point of the r.r.v. f on (S, L_X, E_X) iff it is a regular point of the r.r.v. $f(X)$ on (Ω, L, E). Similarly, $t \in R$ is a continuity point of f iff it is a continuity point of $f(X)$.*

Proof. Suppose f is an r.r.v. on (S, L_X, E_X). By Definition 4.8.13, t is a regular point of f iff (i) there exists a sequence $(s_n)_{n=1,2,\ldots}$ of real numbers decreasing to t such that $(s_n < f)$ is integrable relative to E_X for each $n \geq 1$, and $\lim_{n\to\infty} P_X(s_n < f)$ exists, and (ii) there exists a sequence $(r_n)_{n=1,2,\ldots}$ of real numbers increasing to t such that $(r_n < f)$ is integrable relative to E_X for each $n \geq 1$, and $\lim_{n\to\infty} P_X(r_n < f)$ exists. In view of Proposition 5.2.6, Conditions (i) and (ii) are equivalent to (i′) there exists a sequence $(s_n)_{n=1,2,\ldots}$ of real numbers decreasing to t such that $(s_n < f(X))$ is integrable relative to E for each $n \geq 1$, and $\lim_{n\to\infty} P(s_n < f(X))$ exists, and (ii′) there exists a sequence $(r_n)_{n=1,2,\ldots}$ of real numbers increasing to t such that $(r_n < f(X))$ is integrable relative to E for each $n \geq 1$, and $\lim_{n\to\infty} P(r_n < f(X))$ exists. In other words, t is a regular point of f iff t is a regular point of $f(X)$.

Moreover, a regular point t of f is a continuity point of f iff the two limits in Conditions (i) and (ii) exist and are equal. Equivalently, t is a continuity point of f iff the two limits in Conditions (i′) and (ii′) exist and are equal. Combining, we conclude that t is a continuity point of f iff it is a continuity point of $f(X)$. □

5.3 Weak Convergence of Distributions

Recall that if X is an r.v. on a probability space (Ω, L, E) with values in S, then E_X denotes the distribution induced on S by X.

Definition 5.3.1. Weak convergence of distributions on a complete metric space. Recall that $\widehat{J}(S,d)$ denotes the set of distributions on a complete metric space (S,d). A sequence $(J_n)_{n=1,2,\ldots}$ in $\widehat{J}(S,d)$ is said to *converge weakly* to $J \in \widehat{J}(S,d)$ if $J_n f \to Jf$ for each $f \in C_{ub}(S)$. We then write $J_n \Rightarrow J$. Suppose $X, X_1, X_2, \ldots$ are r.v.'s with values in S, not necessarily on the same probability space. The sequence $(X_n)_{n=1,2,\ldots}$ of r.v.'s is said to *converge weakly* to r.v. X, or to *converge in distribution*, if $E_{X(n)} \Rightarrow E_X$. We then write $X_n \Rightarrow X$. □

Proposition 5.3.2. Convergence in probability implies weak convergence. *Let $(X_n)_{n=0,1,\ldots}$ be a sequence of r.v.'s on the same probability space (Ω, L, E), with values in a complete metric space (S,d). If $X_n \to X_0$ in probability, then $X_n \Rightarrow X_0$.*

Proof. Suppose $X_n \to X_0$ in probability. Let $f \in C_{ub}(S)$ be arbitrary, with $|f| \leq c$ for some $c > 0$, and with a modulus of continuity δ_f. Let $\varepsilon > 0$ be arbitrary. By Definition 5.1.8 of convergence in probability, there exists $p \geq 1$ so large that for each $n \geq p$, there exists an integrable set B_n with $P(B_n) < \varepsilon$ and

$$B_n^c \subset (d(X_n, X_0) < \delta_f(\varepsilon)) \subset (|f(X_n) - f(X_0)| < \varepsilon).$$

Consider each $n \geq p$. Then

$$\begin{aligned}|Ef(X_n) - Ef(X_0)| &= E|f(X_n) - f(X_0)|1_{B(n)} + E|f(X_n) - f(X_0)|1_{B(n)^c} \\ &\leq 2cP(B_n) + \varepsilon < 2c\varepsilon + \varepsilon.\end{aligned}$$

Since $\varepsilon > 0$ is arbitrarily small, we conclude that $Ef(X_n) \to Ef(X_0)$. Equivalently, $J_{X(n)} f \to J_{X(0)} f$. Since $f \in C_{ub}(S)$ is arbitrary, we have $J_{X(n)} \Rightarrow J_{X(0)}$. In other words, $X_n \Rightarrow X_0$. □

Lemma 5.3.3. Weak convergence of distributions on a locally compact metric space. *Suppose (S,d) is locally compact. Suppose $J, J', J_p \in \widehat{J}(S,d)$ for each $p \geq 1$. Then $J_p \Rightarrow J$ iff $J_p f \to Jf$ for each $f \in C(S)$. Moreover, $J = J'$ if $Jf = J'f$ for each $f \in C(S)$. Consequently, a distribution on a locally compact metric space is uniquely determined by the expectation of continuous functions with compact supports.*

Proof. Since $C(S) \subset C_{ub}(S)$, it suffices to prove the "if" part. To that end, suppose that $J_p f \to Jf$ for each $f \in C(S)$. Let $g \in C_{ub}(S)$ be arbitrary. We need to prove that $J_p g \to Jg$. We assume, without loss of generality, that $0 \leq g \leq 1$. Let $\varepsilon > 0$ be arbitrary. Since J is a distribution, there exists $n \geq 1$ so large that $J(1 - h_n) < \varepsilon$, where $h_n \in C_{ub}(S)$ as defined at the beginning of this chapter. Since $h_n, gh_n \in C(S)$, we have, by hypothesis, $J_m h_n \to Jh_n$ and $J_m gh_n \to Jgh_n$ as $m \to \infty$. Hence

$$|J_m g - Jg| \leq |J_m g - J_m gh_n| + |J_m gh_n - Jgh_n| + |Jgh_n - Jg|$$
$$\leq |1 - J_m h_n| + |J_m gh_n - Jgh_n| + |Jh_n - 1| < \varepsilon + \varepsilon + \varepsilon$$

for sufficiently large $m \geq 1$. Since $\varepsilon > 0$ is arbitrary, we conclude that $J_m g \to Jg$, where $g \in C_{ub}(S)$ is arbitrary. Thus $J_p \Rightarrow J$.

Now suppose $Jf = J'f$ for each $f \in C(S)$. Define $J_p \equiv J'$ for each $p \geq 1$. Then $J_p f \equiv J'f = Jf$ for each $f \in C(S)$. Hence by the previous paragraph, $J'g \equiv J_p g \to Jg$ for each $g \in C_{ub}(S)$. Thus $J'g = Jg$ for each $g \in C_{ub}(S)$. In other words, $J = J'$ on $C_{ub}(S)$. We conclude that $J = J'$ as distributions. □

In the important special case of a locally compact metric space (S,d), the weak convergence of distributions on (S,d) can be metrized, as in the next definition and proposition.

Definition 5.3.4. Distribution metric for distributions on a locally compact metric space. Suppose the metric space (S,d) is locally compact, with an arbitrary but fixed reference point $x_\circ \in S$. Let $\xi \equiv (A_n)_{n=1,2,...}$ be an arbitrary but fixed binary approximation of (S,d) relative to $x_\circ$. Let

$$\pi \equiv (\{g_{n,x} : x \in A_n\})_{n=1,2,...}$$

be the partition of unity of (S,d) determined by ξ, as in Definition 3.3.4.

Let $\widehat{J}(S,d)$ denote the set of distributions on the locally compact metric space (S,d). Let $J, J' \in \widehat{J}(S,d)$ be arbitrary. Define

$$\rho_{Dist,\xi}(J,J') \equiv \sum_{n=1}^{\infty} 2^{-n}|A_n|^{-1} \sum_{x \in A(n)} |Jg_{n,x} - J'g_{n,x}| \tag{5.3.1}$$

and call $\rho_{Dist,\xi}$ the *distribution metric on* $\widehat{J}(S,d)$ *relative to the binary approximation* ξ. The next proposition shows that the function $\rho_{Dist,\xi}$ is indeed a metric, and that sequential convergence relative to $\rho_{Dist,\xi}$ is equivalent to weak convergence. Note that $\rho_{Dist,\xi} \leq 1$. □

In the following, recall that $[\cdot]_1$ is the operation that assigns to each $a \in [0,\infty)$ an integer $[a]_1$ in $(a, a+2)$.

Proposition 5.3.5. Sequential metrical convergence is equivalent to weak convergence on a locally compact metric space. *Suppose the metric space (S,d) is locally compact, with the reference point $x_\circ \in S$. Let $\xi \equiv (A_n)_{n=1,2,...}$ be a* *binary approximation of (S,d) relative to $x_\circ$, with a corresponding modulus of local compactness $\|\xi\| \equiv (|A_n|)_{n=1,2,...}$ of (S,d). Let $\rho_{Dist,\xi}$ be the operation introduced in Definition 5.3.4.*

Let $J_p \in \widehat{J}(S,d)$ for $p \geq 1$. Let $f \in C(S)$ be arbitrary, with a modulus of continuity δ_f, with $|f| \leq 1$, and with $(d(\cdot,x_\circ) \leq b)$ as support for some $b > 0$. Then the following conditions hold:

1. Let $\alpha > 0$ be arbitrary. Then there exists $\delta_{\widehat{J}}(\alpha) \equiv \delta_{\widehat{J}}(\alpha,\delta_f,b,\|\xi\|) > 0$ such that for each $J, J' \in \widehat{J}(S,d)$ with $\rho_{Dist,\xi}(J,J') < \delta_{\widehat{J}}(\alpha)$, we have $|Jf - J'f| < \alpha$.

2. Let

$$\pi \equiv (\{g_{n,x} : x \in A_n\})_{n=1,2,...}$$

be the partition of unity of (S,d) determined by ξ. Suppose $J_p g_{n,x} \to J g_{n,x}$ as $p \to \infty$, for each $x \in A_n$, for each $n \geq 1$. Then $\rho_{Dist,\xi}(J_p, J) \to 0$.

3. $J_p f \to Jf$ for each $f \in C(S)$ iff $\rho_{Dist,\xi}(J_p, J) \to 0$. Thus

$$J_p \Rightarrow J \qquad \text{iff} \qquad \rho_{Dist,\xi}(J_p, J) \to 0.$$

4. The function $\rho_{Dist,\xi}$ is a metric.

Proof. 1. Let $\alpha > 0$ be arbitrary. Let $\varepsilon \equiv 3^{-1}\alpha$. Let $n \equiv \left[0 \vee \left(1 - \log_2 \delta_f\left(\frac{\varepsilon}{3}\right)\right) \vee \log_2 b\right]_1$. We will show that

$$\delta_{\widehat{J}}(\alpha) \equiv \delta_{\widehat{J}}(\alpha,\delta_f,b,\|\xi\|) \equiv 2^{-n}|A_n|^{-1}\varepsilon$$

has the desired property in Assertion 1. To that end, suppose $J, J' \in \widehat{J}(S,d)$ are such that $\rho_{Dist,\xi}(J,J') < \delta_{\widehat{J}}(\alpha)$. By Definition 3.3.4 of π, the sequence $\{g_{n,x} : x \in A_n\}$ is a 2^{-n}-partition of unity of (S,d) determined by A_n.

2. Separately, by hypothesis, the function f has support

$$(d(\cdot,x_\circ) \leq b) \subset (d(\cdot,x_\circ) \leq 2^n) \subset \bigcup_{x \in A(n)} (d(\cdot,x) \leq 2^{-n}),$$

where the first inclusion is because $b < 2^n$ and the second inclusion is by Definition 3.2.1. Since $2^{-n} < \frac{1}{2}\delta_f\left(\frac{1}{3}\varepsilon\right)$, Proposition 3.3.6 then implies that

$$\|f - g\| \leq \varepsilon, \tag{5.3.2}$$

where

$$g \equiv \sum_{x \in A(n)} f(x) g_{n,x}.$$

3. By the definition of $\rho_{Dist,\xi}$, we have

$$2^{-n}|A_n|^{-1} \sum_{x \in A(n)} |J g_{n,x} - J' g_{n,x}| \leq \rho_{Dist,\xi}(J,J') < \delta_{\widehat{J}}(\alpha).$$

Therefore

$$|Jg - J'g| \equiv \left| \sum_{x \in A(n)} f(x)(Jg_{n,x} - J'g_{n,x}) \right|$$
$$\leq \sum_{x \in A(n)} |Jg_{n,x} - J'g_{n,x}| < 2^n |A_n| \delta_{\widehat{J}}(\alpha) \equiv \varepsilon.$$

Combining with inequality 5.3.2, we obtain

$$|Jf - J'f| \leq |Jg - J'g| + 2\varepsilon < \varepsilon + 2\varepsilon = 3\varepsilon \equiv \alpha,$$

where $J, J' \in \widehat{J}(S,d)$ are arbitrary such that $\rho_{Dist,\xi}(J, J') < \delta_{\widehat{J}}(\alpha)$. Assertion 1 is proved.

4. Suppose $J_p g_{n,x} \to J g_{n,x}$ as $p \to \infty$, for each $x \in A_n$, for each $n \geq 1$. Let $\varepsilon > 0$ be arbitrary. Note that

$$\rho_{Dist,\xi}(J, J_p) \equiv \sum_{n=1}^{\infty} 2^{-n} |A_n|^{-1} \sum_{x \in A(n)} |Jg_{n,x} - J_p g_{n,x}|$$
$$\leq \sum_{n=1}^{m} 2^{-n} |A_n|^{-1} \sum_{x \in A(n)} |Jg_{n,x} - J_p g_{n,x}| + 2^{-m}.$$

We can first fix $m \geq 1$ so large that $2^{-m} < 2^{-1}\varepsilon$. Then, for sufficiently large $p \geq 1$, the last double summation yields a sum also less than $2^{-1}\varepsilon$, whence $\rho_{Dist,\xi}(J, J_p) < \varepsilon$. Since $\varepsilon > 0$ is arbitrary, we see that $\rho_{Dist,\xi}(J, J_p) \to 0$. Assertion 2 is proved.

5. Suppose $\rho_{Dist,\xi}(J_p, J) \to 0$. Then Assertion 1 implies that $J_p f \to Jf$ for each $f \in C(S)$. Hence $J_p \Rightarrow J$, thanks to Lemma 5.3.3. Conversely, suppose $J_p f \to Jf$ for each $f \in C(S)$. Then, in particular, $J_p g_{n,x} \to J g_{n,x}$ as $p \to \infty$, for each $x \in A_n$, for each $n \geq 1$. Hence $\rho_{Dist,\xi}(J_p, J) \to 0$ by Assertion 2. Assertion 3 is verified. Applying it to the special case where $J_p = J'$ for each $p \geq 1$, we obtain $\rho_{Dist,\xi}(J', J) = 0$ iff $J = J'$.

6. Symmetry and the triangle inequality required for a metric follow trivially from the defining equality 5.3.1. Hence $\rho_{Dist,\xi}$ is a metric. □

From the defining equality 5.3.1, we have $\rho_{Dist,\xi}(J, J') \leq 1$ for each $J, J' \in \widehat{J}(S,d)$. Hence the metric space $(\widehat{J}(S,d), \rho_{Dist,\xi})$ is bounded. It is not necessarily complete. An easy counterexample is found by taking $S \equiv R$ with the Euclidean metric, and taking J_p to be the point mass at p for each $p \geq 0$. In other words $J_p f \equiv f(p)$ for each $f \in C(R)$. Then $\rho_{Dist,\xi}(J_p, J_q) \to 0$ as $p,q \to \infty$. On the other hand $J_p f \to 0$ for each $f \in C(R)$. Hence if $\rho_{Dist,\xi}(J_p, J) \to 0$ for some $J \in \widehat{J}(S,d)$, then $Jf = 0$ for each $f \in C(R)$, and so $J = 0$, contradicting the condition for J to be a distribution and an integration. The obvious problem here is that the mass of the distributions J_p escapes to infinity as $p \to \infty$. The condition of tightness, defined next for a subfamily of $\widehat{J}(S,d)$, prevents this from happening.

Definition 5.3.6. Tightness. Suppose the metric space (S,d) is locally compact. Let $\beta : (0,\infty) \to [0,\infty)$ be an operation. Let $\overline{J}$ be a subfamily of $\widehat{J}(S,d)$ such that for each $\varepsilon > 0$ and for each $J \in \overline{J}$, we have $P_J(d(\cdot,x_\circ) > a) < \varepsilon$ for each $a > \beta(\varepsilon)$, where P_J is the probability function of the distribution J. Then we say the subfamily $\overline{J}$ is *tight*, with β as a *modulus of tightness* relative to the reference point $x_\circ$. We say that a distribution J has modulus of tightness β if the singleton family $\{J\}$ has modulus of tightness β.

A family M of r.v.'s with values in the locally compact metric space (S,d), not necessarily on the same probability space, is said to be *tight*, with modulus of tightness β, if the family $\{E_X : X \in M\}$ is tight with modulus of tightness β. We will say that an r.v. X has modulus of tightness β if the singleton$\{X\}$ family has modulus of tightness β. □

We emphasize that we have defined tightness of a subfamily $\overline{J}$ of $\widehat{J}(S,d)$ only when the metric space (S,d) is locally compact, even as weak convergence in $\widehat{J}(S,d)$ is defined for the more general case of any complete metric space (S,d).

Note that, according to Proposition 5.2.5, $d(\cdot,x_\circ)$ is an r.r.v. relative to each distribution J. Hence, given each $J \in \overline{J}$, the set $(d(\cdot,x_\circ) > a)$ is integrable relative to J for all but countably many $a > 0$. Therefore the probability $P_J(d(\cdot,x_\circ) > a)$ makes sense for all but countably many $a > 0$. However, the countable exceptional set of values of a depends on J.

A modulus of tightness for a family M of r.v.'s gives the uniform rate of convergence $P(d(x_\circ,X) > a)) \to 0$ as $a \to \infty$, independent of $X \in M$, where the probability function P and the corresponding expectation E are specific to X. This is analogous to a modulus of uniform integrability for a family G of integrable r.r.v.'s, which gives the rate of convergence $E(|X|;|X| > a) \to 0$ as $a \to \infty$, independent of $X \in G$.

Lemma 5.3.7. A family of r.r.v.'s bounded in L_p is tight. *Let $p > 0$ be arbitrary. Let M be a family of r.r.v.'s such that $E|X|^p \le b$ for each $X \in M$, for some $b \ge 0$. Then the family M is tight, with a modulus of tightness β relative to $0 \in R$ defined by $\beta(\varepsilon) \equiv b^{\frac{1}{p}}\varepsilon^{-\frac{1}{p}}$ for each $\varepsilon > 0$.*

Proof. Let $X \in M$ be arbitrary. Let $\varepsilon > 0$ be arbitrary. Then, for each $a > \beta(\varepsilon) \equiv b^{\frac{1}{p}}\varepsilon^{-\frac{1}{p}}$, we have

$$P(|X| > a) = P(|X|^p > a^p) \le a^{-p}E|X|^p \le a^{-p}b < \varepsilon,$$

where the first inequality is Chebychev's inequality, and the second is by the definition of the constant b in the hypothesis. Thus X has the operation β as a modulus of tightness relative to $0 \in R$. □

If a family $\overline{J}$ of distributions is tight relative to a reference point $x_\circ$, then it is tight relative to any other reference point x_0', thanks to the triangle inequality. Intuitively, tightness limits the escape of mass to infinity as we go through distributions in $\overline{J}$. Therefore a tight family of distributions remains so after a finite-distance shift of the reference point.

Proposition 5.3.8. Tightness, in combination with convergence of a sequence of distributions at each member of $C(S)$, implies weak convergence to some distribution. *Suppose the metric space (S,d) is locally compact. Let $\{J_n : n \geq 1\}$ be a tight family of distributions, with a modulus of tightness β relative to the reference point $x_\circ$.*

Suppose $J(f) \equiv \lim_{n\to\infty} J_n(f)$ exists for each $f \in C(S)$. Then J is a distribution, and $J_n \Rightarrow J$. Moreover, J has the modulus of tightness $\beta + 2$.

Proof. 1. Clearly, J is a linear function on $C(S)$. Suppose $f \in C(S)$ is such that $Jf > 0$. Then, in view of the convergence in the hypothesis, there exists $n \geq 1$ such that $J_n f > 0$. Since J_n is an integration, there exists $x \in S$ such that $f(x) > 0$. We have thus verified Condition (ii) in Definition 4.2.1 for J.

2. Next let $\varepsilon \in (0,1)$ be arbitrary, and take any $a > \beta(\varepsilon)$. Then $P_n(d(\cdot,x_\circ) > a) < \varepsilon$ for each $n \geq 1$, where $P_n \equiv P_{J(n)}$ is the probability function for J_n. Define $h_k \equiv 1 \wedge (1 + k - d(\cdot,x_\circ))_+ \in C(S)$ for each $k \geq 1$. Take any $m \equiv m(\varepsilon,\beta) \in (a, a+2)$. Then $h_m \geq 1_{(d(\cdot,x(\circ))\leq a)}$, whence

$$J_n h_m \geq P_n(d(\cdot,x_\circ) \leq a) > 1 - \varepsilon \tag{5.3.3}$$

for each $n \geq 1$. By hypothesis, $J_n h_m \to J h_m$ as $n \to \infty$. Inequality 5.3.3 therefore yields

$$J h_m \geq 1 - \varepsilon > 0. \tag{5.3.4}$$

We have thus verified Condition (i) in Definition 4.2.1 for J to be an integration on (S,d). Therefore, by Proposition 4.3.3, $(S, C(S), J)$ is an integration space. At the same time, inequality 5.3.4 implies that $J h_m \uparrow 1$. We conclude that J is a distribution. Since $J_n f \to Jf$ for each $f \in C(S)$ by hypothesis, Lemma 5.3.3 implies that $J_n \Rightarrow J$.

3. Now note that inequality 5.3.4 implies that

$$P_J(d(\cdot,x_\circ) \leq a+2) = J1_{(d(\cdot,x_\circ)\leq a+2)} \geq J h_m \geq 1 - \varepsilon > 0, \tag{5.3.5}$$

where $a > \beta(\varepsilon)$ is arbitrary. Thus J is tight with the modulus of tightness $\beta + 2$. The proposition is proved. □

Corollary 5.3.9. A tight $\rho_{Dist,\xi}$-Cauchy sequence of distributions converges. *Let ξ be a binary approximation of a locally compact metric space (S,d) relative to a reference point $x_\circ \in S$. Let $\rho_{Dist,\xi}$ be the distribution metric on the space $\widehat{J}(S,d)$ of distributions on (S,d), determined by ξ. Suppose the subfamily $\{J_n : n \geq 1\} \subset \widehat{J}(S,d)$ of distributions is tight, with a modulus of tightness β relative to $x_\circ$.*

Suppose $\rho_{Dist,\xi}(J_n, J_m) \to 0$ as $n,m \to \infty$. Then $J_n \Rightarrow J$ and $\rho_{Dist,\xi}(J_n, J) \to 0$, for some $J \in \widehat{J}(S,d)$ with the modulus of tightness $\beta + 2$.

Proof. Suppose $\rho_{Dist,\xi}(J_n, J_m) \to 0$ as $n,m \to \infty$. Let $f \in C(S)$ be arbitrary. We will prove that $J(f) \equiv \lim_{n\to\infty} J_n(f)$ exists. Let $\varepsilon > 0$ be arbitrary. Then there exists $a > \beta(\varepsilon)$ such that $P_n(d(\cdot,x_\circ) > a) < \varepsilon$ for each $n \geq 1$, where $P_n \equiv P_{J(n)}$ is the probability function for J_n. Let $k \geq 1$ be so large that $k \geq a$, and recall that

$$h_k \equiv 1 \wedge (1 + k - d(\cdot,x_\circ))_+.$$

Then

$$J_n h_k \geq P_n(d(\cdot,x_\circ) \leq a) > 1 - \varepsilon$$

for each $n \geq 1$. At the same time, $fh_k \in C(S)$. Since $\rho_{Dist,\xi}(J_n, J_m) \to 0$, it follows that $(J_n fh_k)_{n=1,2,...}$ is a Cauchy sequence of real numbers, according to Assertion 1 of Proposition 5.3.5. Hence $Jfh_k \equiv \lim_{n\to\infty} J_n(fh_k)$ exists. Consequently,

$$\begin{aligned}|J_n f - J_m f| &\leq |J_n f - J_n fh_k| + |J_n fh_k - J_m fh_k| + |J_m fh_k - J_m f| \\ &\leq |1 - J_n h_k| + |J_n fh_k - J_m fh_k| + |J_m h_k - 1| \\ &\leq \varepsilon + |J_n fh_k - J_m fh_k| + \varepsilon < \varepsilon + \varepsilon + \varepsilon\end{aligned}$$

for sufficiently large $n,m \geq 1$. Since $\varepsilon > 0$ is arbitrary, we conclude that $J(f) \equiv \lim_{n\to\infty} J_n f$ exists for each $f \in C(S)$. By Proposition 5.3.8, J is a distribution with the modulus of tightness $\beta + 2$ and $J_n \Rightarrow J$. Proposition 5.3.5 then implies that $\rho_{Dist,\xi}(J_n, J) \to 0$. □

Proposition 5.3.10. A weakly convergent sequence of distributions on a locally compact metric space is tight. *Suppose the metric space (S,d) is locally compact. Let J, J_n be distributions for each $n \geq 1$. Suppose $J_n \Rightarrow J$. Then the family $\{J, J_1, J_2, \ldots\}$ is tight. In particular, any finite family of distributions on S is tight, and any finite family of r.v.'s with values in S is tight.*

Proof. For each $n \geq 1$, write P and P_n for P_J and $P_{J(n)}$, respectively. Since J is a distribution, we have $P(d(\cdot,x_\circ) > a) \to 0$ as $a \to \infty$. Thus any family consisting of a single distribution J is tight. Let β_0 be a modulus of tightness of $\{J\}$ with reference to $x_\circ$, and for each $k \geq 1$, let β_k be a modulus of tightness of $\{J_k\}$ with reference to $x_\circ$. Let $\varepsilon > 0$ be arbitrary. Let $a > \beta_0(\frac{\varepsilon}{2})$ and define $f \equiv 1 \wedge (a + 1 - d(\cdot,x_\circ))_+$. Then $f \in C(S)$ with

$$1_{(d(\cdot,x(\circ))>a+1)} \leq 1 - f \leq 1_{(d(\cdot,x(\circ))>a)}.$$

Hence $1 - Jf \leq P(d(\cdot,x_\circ) > a) < \frac{\varepsilon}{2}$. By hypothesis, we have $J_n \Rightarrow J$. Hence there exists $m \geq 1$ so large that $|J_n f - Jf| < \frac{\varepsilon}{2}$ for each $n > m$. Consequently,

$$P_n(d(\cdot,x_\circ) > a + 1) \leq 1 - J_n f < 1 - Jf + \frac{\varepsilon}{2} < \varepsilon$$

for each $n > m$. Define $\beta(\varepsilon) \equiv (a + 1) \vee \beta_1(\varepsilon) \vee \cdots \vee \beta_m(\varepsilon)$. Then, for each $a' > \beta(\varepsilon)$ we have

(i) $P(d(\cdot,x_\circ) > a') \leq P(d(\cdot,x_\circ) > a) < \frac{\varepsilon}{2}$.
(ii) $P_n(d(\cdot,x_\circ) > a') \leq P_n(d(\cdot,x_\circ) > a+1) < \varepsilon$ for each $n > m$.
(iii) $a' > \beta_n(\varepsilon)$ and so $P_n(d(\cdot,x_\circ) > a') \leq \varepsilon$ for each $n = 1,\ldots,m$.
Since $\varepsilon > 0$ is arbitrary, the family $\{J, J_1, J_2 \ldots\}$ is tight. □

The next proposition provides alternative characterizations of weak convergence in the case of locally compact (S,d).

Proposition 5.3.11. Modulus of continuity of the function $J \to Jf$, for a fixed Lipschitz continuous function f. *Suppose (S,d) is locally compact, with a reference point $x_\circ$. Let $\xi \equiv (A_n)_{n=1,2,\ldots}$ be a binary approximation of (S,d) relative to $x_\circ$, with a corresponding modulus of local compactness $\|\xi\| \equiv (|A_n|)_{n=1,2,\ldots}$ of (S,d). Let $\rho_{Dist,\xi}$ be the distribution metric on the space $\widehat{J}(S,d)$ of distributions on (S,d), determined by ξ, as introduced in Definition 5.3.4.*

Let J, J', J_p be distributions on (S,d), for each $p \geq 1$. Let β be a modulus of tightness of $\{J, J'\}$ relative to $x_\circ$. Then the following conditions hold:

1. Let $f \in C(S,d)$ be arbitrary with $|f| \leq 1$ and with modulus of continuity δ_f. Then, for each $\varepsilon > 0$, there exists $\widetilde{\Delta}(\varepsilon,\delta_f,\beta,\|\xi\|) > 0$ such that if $\rho_{Dist,\xi}(J,J') < \widetilde{\Delta}(\varepsilon,\delta_f,\beta,\|\xi\|)$, then $|Jf - J'f| < \varepsilon$.

2. The following three conditions are equivalent: (i) *$J_p f \to Jf$ for each Lipschitz continuous $f \in C(S)$,* (ii) *$J_p \Rightarrow J$, and* (iii) *$J_p f \to Jf$ for each Lipschitz continuous f that is bounded.*

Proof. 1. By Definition 3.2.1, we have

$$(d(\cdot,x_\circ) \leq 2^n) \subset \bigcup_{x \in A(n)} (d(\cdot,x) \leq 2^{-n}) \tag{5.3.6}$$

and

$$\bigcup_{x \in A(n)} (d(\cdot,x) \leq 2^{-n+1}) \subset (d(\cdot,x_\circ) \leq 2^{n+1}) \tag{5.3.7}$$

for each $n \geq 1$.

2. Let $\varepsilon > 0$ be arbitrary. As an abbreviation, write $\alpha \equiv \frac{\varepsilon}{2}$. Let

$$b \equiv \left[1 + \beta\left(\frac{\varepsilon}{4}\right)\right]_1.$$

Define $h \equiv 1 \wedge (b - d(\cdot,x_\circ))_+ \in C(S)$. Then the functions h and fh have support $(d(\cdot,x_\circ) \leq b)$, and we have $h = 1$ on $(d(\cdot,x_\circ) \leq b-1)$.

3. Since the function h has a Lipschitz constant of 1, the function fh has a modulus of continuity δ_{fh} defined by $\delta_{fh}(\gamma) \equiv \frac{\gamma}{2} \wedge \delta_f\left(\frac{\gamma}{2}\right)$ for each $\gamma > 0$. As in Step 1 of the proof of Proposition 5.3.5, let $n \equiv [0 \vee (1 - \log_2 \delta_{fh}(3^{-2}\alpha)) \vee \log_2 b]_1$, and define

$$\delta_{\widehat{J}}(\alpha) \equiv \delta_{\widehat{J}}(\alpha,\delta_{fh},b,\|\xi\|) \equiv 3^{-1}2^{-n}|A_n|^{-1}\alpha.$$

Then, by Proposition 5.3.5,

$$|Jfh - J'fh| < \alpha, \tag{5.3.8}$$

provided that $J, J' \in \widehat{J}$ are such that $\rho_{Dist,\xi}(J, J') < \delta_{\widehat{J}}(\alpha)$.

4. Now define

$$\widetilde{\Delta}(\varepsilon) \equiv \widetilde{\Delta}(\varepsilon, \delta_f, \beta, \|\xi\|) \equiv \delta_{\widehat{J}}(\alpha).$$

Suppose $\rho_{Dist,\xi}(J, J') < \widetilde{\Delta}(\varepsilon)$. We need to prove that $|Jf - J'f| < \varepsilon$. To that end, note that since J, J' have tightness modulus β, and since $1 - h = 0$ on $(d(\cdot, x_\circ) \leq b - 1)$ where $b - 1 > \beta(\frac{\varepsilon}{4})$, we have

$$J(1 - h) \leq P_J(d(\cdot, x_\circ) \geq b - 1) \leq \frac{\varepsilon}{4}.$$

Consequently,

$$|Jf - Jfh| = |Jf(1 - h)| \leq J(1 - h) \leq \frac{\varepsilon}{4}. \tag{5.3.9}$$

Similarly,

$$|J'f - J'fh| \leq \frac{\varepsilon}{4}. \tag{5.3.10}$$

Combining inequalities 5.3.8, 5.3.9, and 5.3.10, we obtain

$$\begin{aligned}|Jf - J'f| &\leq |Jf - Jfh| + |Jfh - J'fh| + |J'f - J'fh| \\ &< \frac{\varepsilon}{4} + \alpha + \frac{\varepsilon}{4} \equiv \frac{\varepsilon}{4} + \frac{\varepsilon}{2} + \frac{\varepsilon}{4} = \varepsilon,\end{aligned}$$

as desired. Assertion 1 is thus proved.

5. We need to prove that Conditions (i–iii) are equivalent. To that end, first suppose (i) $J_p f \to Jf$ for each Lipschitz continuous $f \in C(S)$. Let

$$\pi \equiv (\{g_{n,x} : x \in A_n\})_{n=1,2,\ldots}$$

be the partition of unity of (S, d) determined by ξ. Then, for each $n \geq 1$ and each $x \in A_n$, we have $J_p g_{n,x} \to J g_{n,x}$ as $p \to \infty$, because $g_{n,x} \in C(S)$ is Lipschitz continuous by Proposition 3.3.5. Hence $\rho_{Dist,\xi}(J_p, J) \to 0$ by Assertion 2 of Proposition 5.3.5. Consequently, by Assertion 3 of Proposition 5.3.5, we have $J_p \Rightarrow J$. Thus we have proved that Condition (i) implies Condition (ii).

6. Suppose next that $J_p \Rightarrow J$. Then since (S, d) is locally compact, we have $\rho_{Dist,\xi}(J_p, J) \to 0$ by Assertion 3 of Proposition 5.3.5. Separately, in view of Proposition 5.3.10, the family $\{J, J_1, J_2, \ldots\}$ is tight, with some modulus of tightness β. Let $f \in C(S)$ be arbitrary and Lipschitz continuous. We need to prove that $J_p f \to Jf$. By linearity, we may assume that $|f| \leq 1$, whence $J_p f \to Jf$ by Assertion 1. Thus Condition (ii) implies Condition (iii).

7. Finally, Condition (iii) trivially implies Condition (i). Assertion 2 and the proposition are proved. □

5.4 Probability Density Function and Distribution Function

In this section, we discuss two simple and useful methods to construct distributions on a locally compact metric space (S,d). The first starts with one integration I on (S,d) in the sense of Definition 4.2.1, where the full set S need not be integrable. Then, for each nonnegative integrable function g with integral 1, we can construct a distribution on (S,d) using g as a density function. A second method is for the special case where $(S,d) = (R,d)$ is the real line, equipped with the Euclidean metric. Let F be an arbitrary distribution function on R, in the sense of Definition 4.1.1, such that $F(t) \to 0$ as $t \to -\infty$, and $F(t) \to 1$ as $t \to \infty$. Then the Riemann–Stieljes integral corresponding to F constitutes a distribution on (R,d).

Definition 5.4.1. Probability density function. Let I be an integration on a locally compact metric space (S,d) in the sense of Definition 4.2.1. Let (S,Λ,I) denote the completion of the integration space $(S,C(S),I)$. Let $g \in \Lambda$ be an arbitrary nonnegative integrable function with $Ig = 1$. Define

$$I_g f \equiv Igf \tag{5.4.1}$$

for each $f \in C(S,d)$. According to the following lemmas, the function I_g is a probability distribution on (S,d), in the sense of Definition 5.2.1.

In such a case, g will be called a *probability density function*, or *p.d.f.* for short, relative to the integration I, and the completion (S,Λ_g,I_g) of $(S,C(S,d),I_g)$ will be called the probability space generated by the p.d.f. g.

Suppose, in addition, that X is an arbitrary r.v. on some probability space (Ω,L,E) with values in S such that $E_X = I_g$, where E_X is the distribution induced on the metric space (S,d) by the r.v. X, in the sense of Definition 5.2.3. Then the r.v. *X is said to have the p.d.f. g relative to I.* □

Frequently used p.d.f.'s are defined on $(S,\Lambda,I) \equiv (R^n,\Lambda,\int \cdot dx)$, the n-dimensional Euclidean space equipped with the Lebesgue integral, and on $(S,\Lambda,I) \equiv (\{1,2,\ldots\},\Lambda,I)$ with the counting measure I defined by $If \equiv \sum_{n=1}^{\infty} f(n)$ for each $f \in C(S)$.

In the following discussion, for each $n \geq 1$, define $h_n \equiv 1 \wedge (1+n-d(\cdot,x_\circ))_+ \in C(S,d)$.

Lemma 5.4.2. I_g is indeed a probability distribution. *Let I be an arbitrary integration on a locally compact metric space (S,d) in the sense of Definition 4.2.1. Let g be a p.d.f. relative to the integration I. Then I_g is indeed a probability distribution on (S,d).*

Proof. 1. By Definition 5.4.1, we have $g \in \Lambda$ with $g \geq 0$ and $Ig = 1$. Let $\varepsilon > 0$ be arbitrary. By Corollary 4.4.4, there exists $f \in C(S,d)$ such that $I|f-g| < \varepsilon$.

Let $n \geq 1$ be so large that the compact support of f is contained in the set $(d(\cdot, x_\circ) \leq n)$. Then (i) $fh_n = f$, (ii) $|Ifh_n - Igh_n| \leq I|f - g| < \varepsilon$, and (iii) $|If - 1| = |If - Ig| < \varepsilon$. Combining,

$$|Igh_n - 1| \leq |Igh_n - Ifh_n| + |Ifh_n - If| + |If - 1| < \varepsilon + 0 + \varepsilon = 2\varepsilon.$$

Since $\varepsilon > 0$ is arbitrarily small, we see that $Igh_n \uparrow 1$ as $n \to \infty$.

2. By the defining equality 5.4.1, the function I_g is linear on the space $C(S,d)$. Moreover, $I_g h_n \equiv Igh_n > 0$ for some sufficiently large $n \geq 1$. This verifies Condition (i) of Definition 4.2.1 for the linear function I_g on $C(S,d)$. Next suppose $f \in C(S,d)$ with $I_g f > 0$. Then $Igf \equiv I_g f > 0$. Hence, by the positivity condition, Condition (ii) of Definition 4.2.1 for the integration I, there exists $x \in S$ such that $g(x)f(x) > 0$. Consequently, $f(x) > 0$. Thus Condition (ii) of Definition 4.2.1 is also verified for the function I_g on $C(S,d)$.

3. Accordingly, $(S, C(S,d), I_g)$ is an integration space. Since $I_g h_n \equiv Ih_n g \uparrow 1$, Assertion 3 of Lemma 5.2.2 implies that I_g is a probability distribution on (S,d). The lemma is proved. □

Distributions on R can be studied in terms of their corresponding distribution functions, as introduced in Definition 4.1.1 and specialized to probability distribution functions.

Recall the convention that if F is a function, then we write $F(t)$ only with the explicit or implicit assumption that $t \in domain(F)$.

Definition 5.4.3. Probability distribution functions. Suppose F is a distribution function on R satisfying the following conditions: (i) $F(t) \to 0$ as $t \to -\infty$, and $F(t) \to 1$ as $t \to \infty$; (ii) for each $t \in domain(F)$, the left limit $\lim_{r<t;r\to t} F(s)$ exist; (iii) for each $t \in domain(F)$, the right limit $\lim_{s>t;s\to t} F(s)$ exists and is equal to $F(t)$; (iv) $domain(F)$ contains the metric complement A_c of some countable subset A of R; and (v) if $t \in R$ is such that both the left and right limits exist as just defined then $t \in domain(F)$. Then F is called a *probability distribution function,* or a P.D.F. for short. A point $t \in domain(F)$ is called a regular point of F. A point $t \in domain(F)$ at which the just-defined left and right limits exist and are equal is called a *continuity point* of F.

Suppose X is an r.r.v. on some probability space (Ω, L, E). Let F_X be the function defined by (i′) $domain(F_X) \equiv \{t \in R : t \text{ is a regular point of } X\}$ and (ii′) $F_X(t) \equiv P(X \leq t)$ for each $t \in domain(F_X)$. Then F_X is called the P.D.F. of X. □

Recall in the following that $\int \cdot dF$ denotes the Riemann–Stieljes integration relative to a distribution function F on R.

Proposition 5.4.4. F_X is indeed a P.D.F. *Let X be an r.r.v. on a probability space (Ω, L, E) with $F_X : R \to [0,1]$ as in Definition 5.4.3. Let $E_X : L_X \to R$ denote* *the distribution induced on R by X, in the sense of Definition 5.2.3. Then the following conditions hold:*

1. *F_X is a P.D.F.*
2. *$\int \cdot dF_X = E_X$.*

Proof. As an abbreviation, write $J \equiv E_X$ and $F \equiv F_X$, and write P for the probability function associated to the probability expectation E.

1. We need to verify Conditions (i) through (v) in Definition 5.4.3 for F. Condition (i) holds because $P(X \leq t) \to 0$ as $t \to -\infty$ and $P(X \leq t) = 1 - P(X > t) \to 1$ as $t \to \infty$, by the definition of a measurable function. Next consider any $t \in domain(F)$. Then t is a regular point of X, by the definition of F_X. Hence there exists a sequence $(s_n)_{n=1,2,...}$ of real numbers decreasing to t such that $(X \leq s_n)$ is integrable for each $n \geq 1$ and such that $\lim_{n\to\infty} P(X \leq s_n)$ exists. Since F is a nondecreasing function, it follows that $\lim_{s>t;s\to t} F(s)$ exists. Similarly, $\lim_{r<t;r\to t} F(s)$ exists. Moreover, by Definition 5.1.2, we have

$$\lim_{s>t;s\to t} F(s) = \lim_{n\to\infty} P(X \leq s_n) = P(X \leq t) \equiv F(t).$$

Conditions (ii) and (iii) in Definition 5.4.3 have thus been verified. Condition (iv) in Definition 5.4.3 follows from Assertion 1 of Proposition 4.8.14. Condition (v) remains. Suppose $t \in R$ is such that both $\lim_{r<t;r\to t} F(r)$ and $\lim_{s>t;s\to t} F(s)$ exist. Then there exists a sequence $(s_n)_{n=1,2,...}$ in $domain(F)$ decreasing to t such that $F(s_n)$ converges. This implies that $(X \leq s_n)$ is an integrable set, for each $n \geq 1$, and that $P(X \leq s_n)$ converges. Hence $(X > s_n)$ is an integrable set, and $P(X > s_n)$ converges. Similarly, there exists a sequence $(r_n)_{n=1,2,...}$ increasing to t such that $(X > r_n)$ is an integrable set and $P(X > r_n)$ converges. We have thus verified the conditions in Definition 4.8.13 for t to be a regular point of X. In other words, $t \in domain(F)$. Condition (v) in Definition 5.4.3 has also been verified. Summing up, $F \equiv F_X$ is a P.D.F.

2. Note that both $\int \cdot dF_X$ and E_X are complete extensions of integrations defined on $(R, C(R))$. Hence, to show that they are equal, it suffices to prove that they are equal on $C(R)$. To that end, let $f \in C(R)$ be arbitrary, with a modulus of continuity for δ_f. Let $\varepsilon > 0$ be arbitrary. The Riemann–Stieljes integral $\int f(t)dF_X(t)$ is, by definition, the limit of Riemann–Stieljes sums $S(t_1, \ldots, t_n) = \sum_{i=1}^{n} f(t_i)(F_X(t_i) - F_X(t_{i-1}))$ as $t_1 \to -\infty$ and $t_n \to \infty$, with the mesh of the partition $t_1 < \cdots < t_n$ approaching 0. Consider such a Riemann–Stieljes sum where the mesh is smaller than $\delta_f(\varepsilon)$, and where $[t_1, t_n]$ contains a support of f. Then

$$|S(t_1, \ldots, t_n) - Ef(X)| = \left| E \sum_{i=1}^{n} (f(t_i) - f(X)) 1_{(t_{i-1}<X\leq t_i)} \right| \leq \varepsilon.$$

Passing to the limit, we have $\int f(t)dF_X(t) = Ef(X) \equiv E_X f$. □

Proposition 5.4.4 says that F_X is a P.D.F. for each r.r.v. X. The next proposition gives the converse.

Proposition 5.4.5. Basics of P.D.F. The following conditions hold:

1. Let J be any distribution on R, and let (R, L, J) denote the completion of $(R, C_{ub}(R), J)$. Then $J = \int \cdot dF_X$, where F_X is the P.D.F. of the r.r.v. X on (R, L, J) defined by $X(x) \equiv x$ for each $x \in R$.

2. Let F be a P.D.F. For each $t \in domain(F)$, the interval $(-\infty, t]$ is integrable relative to $\int \cdot dF$, and $\int 1_{(-\infty,t]}dF = F(t)$.

3. If two P.D.F.'s F and F' are equal on some dense subset D of $domain(F) \cap domain(F')$, then $F = F'$.

4. If two P.D.F.'s F and F' are such that $\int \cdot dF = \int \cdot dF'$, then $F = F'$.

5. Let F be an arbitrary P.D.F. Then $F = F_X$ for some r.r.v. X.

6. Let F be an arbitrary P.D.F. Then all but countably many $t \in R$ are continuity points of F in the sense of Definition 5.4.3.

7. Let J be an arbitrary distribution on R. Then there exists a unique P.D.F. F such that $J = \int \cdot dF$. Thus there is a bijection between distributions on R and P.D.F.'s. For that reason, we will often abuse terminology and refer to F as a distribution, and write F for J.

Proof. 1. According to Proposition 5.2.5, the identity function $X : (R, L, J) \to R$, defined by $X(x) = x$ for each $x \in R$, is an r.r.v. Moreover, for each $f \in C_{ub}(R)$, we have $f(X) = f \in C_{ub}(R)$. Hence, in view of Proposition 5.4.4, we have $Jf = Jf(X) \equiv E_X f = \int f(x)dF_X(x)$. Assertion 1 is verified.

2. Next, let F be a P.D.F. Define $J \equiv \int \cdot dF$. Consider any $t, s \in domain(F)$ with $t < s$, and any $f \in C(R)$ with $0 \leq f \leq 1$ such that $[t, s]$ is a support of f. Then $Jf \equiv \int f(x)dF(x)$ is the limit of Riemann–Stieljes sums $\sum_{i=1}^{n} f(t_i)(F(t_i) - F(t_{i-1}))$, where the sequence $t_0 < \cdots < t_n$ includes the points t, s.

Consider any such Riemann–Stieljes sum. If $i = 0, \ldots, n$ is such that $t_i < t$ or $t_i > s$, then $f(t_i) = 0$. We may assume, without loss of generality, that $t_0 = t$ and $t_n = s$. It follows that the Riemann–Stieljes sums in question are bounded by $\sum_{i=1}^{n}(F(t_i) - F(t_{i-1})) = F(s) - F(t)$. Passing to the limit, we see that $Jf \leq F(s) - F(t)$ for each $f \in C(R)$ with $0 \leq f \leq 1$ such that $[t, s]$ is a support of f. A similar argument shows that $Jf \geq F(s) - F(t)$ for each $f \in C(R)$ with $0 \leq f \leq 1$ such that $f = 1$ on $[t, s]$.

By Condition (i) in Definition 5.4.3, there exists a decreasing sequence $(r_k)_{k=1,2,\ldots}$ in $domain(F)$ such that $r_1 < t$, $r_k \to -\infty$ and $F(r_k) \to 0$. By Condition (iii) in Definition 5.4.3, there exists a decreasing sequence $(s_n)_{n=1,2,\ldots}$ in $domain(F)$ such that $F(s_n) \to F(t)$ and $s_n \downarrow t$.

For each $k, n \geq 1$, let $f_{k,n} \in C(R)$ be defined by $f_{k,n} = 1$ on $[r_k, s_{n+1}]$, $f_{k,n} = 0$ on $(-\infty, r_{k+1}] \cup [s_n, \infty)$, and $f_{k,n}$ be linear on $[r_{k+1}, r_k]$ and on $[s_{n+1}, s_n]$. Consider any $n \geq 1$ and $j > k \geq 1$. Then $0 \leq f_{k,n} \leq f_{j,n} \leq 1$, and $f_{j,n} - f_{k,n}$ has $[r_{j+1}, r_k]$ as support. Therefore, as seen in the previous paragraph, we have $Jf_{j,n} - Jf_{k,n} \leq F(r_k) - F(r_{j+1}) \to 0$ as $j \geq k \to \infty$. Hence the Monotone

Convergence Theorem implies that $f_n \equiv \lim_{k\to\infty} f_{k,n}$ is integrable, with $Jf_n = \lim_{k\to\infty} Jf_{k,n}$. Moreover, $f_n = 1$ on $(-\infty, s_{n+1}]$, $f_n = 0$ on $[s_n, \infty)$, and f_n is linear on $[s_{n+1}, s_n]$.

Now consider any $m \geq n \geq 1$. Then $0 \leq f_m \leq f_n \leq 1$, and $f_n - f_m$ has $[t, s_n]$ as support. Therefore, as seen earlier, $Jf_n - Jf_m \leq F(s_n) - F(t) \to 0$ as $m \geq n \to \infty$. Hence, the Monotone Convergence Theorem implies that $g \equiv \lim_{n\to\infty} f_n$ is integrable, with $Jg = \lim_{n\to\infty} Jf_n$. It is evident that on $(-\infty, t]$ we have $f_n = 1$ for each $n \geq 1$. Hence g is defined and equal to 1 on $(-\infty, t]$. Similarly, g is defined and equal to 0 on (t, ∞).

Consider any $x \in domain(g)$. Then either $g(x) > 0$ or $g(x) < 1$. Suppose $g(x) > 0$. Then the assumption $x > t$ would imply $g(x) = 0$, which is a contradiction. Hence $x \in (-\infty, t]$ and so $g(x) = 1$. On the other hand, suppose $g(x) < 1$. Then $f_n(x) < 1$ for some $n \geq 1$, whence $x \geq s_{n+1}$ for some $n \geq 1$. Hence $x \in (t, \infty)$ and so $g(x) = 0$. Combining, we see that 1 and 0 are the only possible values of g. In other words, g is an integrable indicator. Moreover, $(g = 1) = (-\infty, t]$ and $(g = 0) = (t, \infty)$. Thus the interval $(-\infty, t]$ is an integrable set with $1_{(-\infty,t]} = g$.

Finally, for any $k, n \geq 1$, we have $F(s_{n+1}) - F(r_k) \leq Jf_{k,n} \leq F(s_n) - F(r_{k+1})$. Letting $k \to \infty$, we obtain $F(s_{n+1}) \leq Jf_n \leq F(s_n)$ for $n \geq 1$. Letting $n \to \infty$, we obtain, in turn, $Jg = F(t)$. In other words $J1_{(-\infty,t]} = F(t)$. Assertion 2 is proved.

3. Suppose two P.D.F.'s F and F' are equal on some dense subset D of $domain(F) \cap domain(F')$. We will show that then $F = F'$. To that end, consider any $t \in domain(F)$. Let $(s_n)_{n=1,2,\ldots}$ be a decreasing sequence in D converging to t. By hypothesis, $F'(s_n) = F(s_n)$ for each $n \geq 1$. At the same time, $F(s_n) \to F(t)$ since $t \in domain(F)$. Therefore $F'(s_n) \to F(t)$. By the monotonicity of F', it follows that $\lim_{s>t;s\to t} F'(s) = F(t)$. Similarly, $\lim_{r<t;r\to t} F'(r)$ exists. Therefore, according to Definition 5.4.3, we have $t \in domain(F')$, and $F'(t) = \lim_{s>t;s\to t} F'(s) = F(t)$. We have thus proved that $domain(F) \subset domain(F')$ and $F' = F$ on $domain(F)$. By symmetry, $domain(F) = domain(F')$. We conclude that $F = F'$. Assertion 3 is verified.

4. Suppose two P.D.F.'s F and F' are such that $\int \cdot dF = \int \cdot dF'$. Write $J \equiv \int \cdot dF = \int \cdot dF'$. Consider any $t \in D \equiv domain(F) \cap domain(F')$. By Assertion 2, the interval $(-\infty, t]$ is integrable relative to J, with $F(t) = J1_{(-\infty,t]} = F'(t)$. Since D is a dense subset of R, we have $F = F'$ by Assertion 3. This proves Assertion 4.

5. Let F be any P.D.F. By Assertion 1, we have $\int \cdot dF = \int \cdot dF_X$ for some r.r.v. X. Therefore $F = F_X$ according to Assertion 4. Assertion 5 is proved.

6. Let F be any P.D.F. By Assertion 5, $F = F_X$ for some r.r.v. X. Hence $F(t) = F_X(t) \equiv P(X \leq t)$ for each regular point t of X. Consider any continuity point t of X. Then, by Definition 4.8.13, we have $\lim_{n\to\infty} P(X \leq s_n) = \lim_{n\to\infty} P(X \leq r_n)$ for some decreasing sequence (s_n) with $s_n \to t$ and for

some increasing sequence (r_n) with $r_n \to t$. Since the function is nondecreasing, it follows that $\lim_{s>t;s\to t} F(s) = \lim_{r<t;r\to t} F(r)$. Summing up, every continuity point of X is a continuity point of F. By Proposition 4.8.14, all but countably many $t \in R$ are continuity points of the r.r.v. X. Hence all but countably many $t \in R$ are continuity points of F. This validates Assertion 6.

7. Let J be arbitrary. By Assertion 1, there exists a P.D.F. F such that $J = \int \cdot dF$. The uniqueness of F follows from Assertion 4. The proposition is proved. □

5.5 Skorokhod Representation

In this section, let (S,d) be a locally compact metric space with an arbitrary but fixed reference point $x_\circ \in S$. Let

$$(\Theta_0, L_0, I) \equiv \left([0,1], L_0, \int \cdot dx\right)$$

denote the Lebesgue integration space based on the unit interval $[0,1]$, and let μ be the corresponding Lebesgue measure. Then we will call I the *uniform distribution on* $[0,1]$.

Given two distributions E and E' on the locally compact metric space (S,d), we saw in Proposition 5.2.5 that they are equal to the distributions induced by some S-valued r.v.'s X and X', respectively. The underlying probability spaces on which X and X' are defined can, in general, be different. Therefore functions of both X and X', such as $d(X,X')$, and their associated probabilities are, up to this point, undefined. Additional conditions on joint probabilities are needed to construct one common probability space on which both X and X' are defined.

One such condition is independence, to be made precise in a later section, where the observed value of X has no effect whatsoever on the probabilities related to X'.

In some other situations, it is desirable, instead, to have models where $X = X'$ if $E = E'$, and more generally where $d(X,X')$ is small when E is close to E'. In this section, we construct the Skorokhod representation, which, to each distribution E on S, assigns a unique r.v. $X: \Theta_0 \to S$ that induces E. In the context of applications to random fields, theorem 3.1.1 of [Skorokhod 1956] introduced this representation and proves that it is continuous relative to weak convergence of E and a.u. convergence of X. We will prove this result for applications in Chapter 6.

In addition, we will prove that, when restricted to a tight subset of distributions, the Skorokhod representation is uniformly continuous relative to the distribution metric $\rho_{Dist,\xi}$ on the space of distributions E, and the metric ρ_{Prob} on the space of r.v.'s $X: \Theta_0 \to S$. The metrics $\rho_{Dist,\xi}$ and ρ_{Prob} were introduced in Definition 5.3.4 and in Proposition 5.1.11, respectively.

The Skorokhod representation is a generalization of the *quantile mapping*, which to each P.D.F. F assigns the r.r.v. $Y \equiv F^{-1}: \Theta_0 \to R$ on the probability space Θ_0, where Y can be shown to induce the P.D.F. F.

Skorokhod's proof, in terms of Borel sets, is recast here in terms of continuous basis functions in a partition of unity π, in the sense of Definition 3.3.4. The use of a partition of unity facilitates the proof of the aforementioned metrical continuity.

Recall that $[\cdot]_1$ is an operation that assigns to each $r \in (0,\infty)$ an integer $[r]_1 \in (r,r+2)$.

Theorem 5.5.1. Construction of the Skorokhod representation. *Let $\xi \equiv (A_n)_{n=1,2,\ldots}$ be a binary approximation of the locally compact metric space (S,d), relative to the reference point $x_\circ \in S$. Recall that $\widehat{J}(S,d)$ stands for the space of distributions on (S,d), endowed with the distribution $\rho_{Dist,\xi}$, as in Definition 5.3.4. Recall that $M(\Theta_0,S)$ stands for the space of r.v.'s on the probability space (Θ_0,L_0,I) with values in (S,d), endowed with the probability metric ρ_{Prob}, as in Definition 5.1.10.*

Then there exists a mapping

$$\Phi_{Sk,\xi} : \widehat{J}(S,d) \to M(\Theta_0,S)$$

such that for each $E \in \widehat{J}(S,d)$, the r.v. $X \equiv \Phi_{Sk,\xi}(E) : \Theta_0 \to S$ induces the distribution E, or $I_X = E$ in symbols.

The mapping $\Phi_{Sk,\xi}$ will be called the Skorokhod representation *of distributions on (S,d), determined by ξ.*

Proof. Let $E \in \widehat{J}(S,d)$ be an arbitrary distribution on (S,d), with a modulus of tightness β relative to $x_\circ$. Let

$$\pi \equiv (\{g_{n,x} : x \in A_n\})_{n=1,2,\ldots}$$

be the partition of unity of (S,d) determined by ξ, as in Definition 3.3.4.

1. Let $n \geq 1$ be arbitrary. By Definition 3.2.1, the enumerated finite set $A_n \equiv \{x_{n,1},\ldots,x_{n,\kappa(n)}\}$ is a 2^{-n}–approximation of $(d(\cdot,x_\circ) \leq 2^n)$. In other words,

$$A_n \subset (d(\cdot,x_\circ) \leq 2^n) \tag{5.5.1}$$

and

$$(d(\cdot,x_\circ) \leq 2^n) \subset \bigcup_{x\in A(n)} (d(\cdot,x) \leq 2^{-n}). \tag{5.5.2}$$

Recall from Proposition 3.3.5 that $0 \leq g_{n,x} \leq \sum_{x\in A(n)} g_{n,x} \leq 1$,

$$(g_{n,x} > 0) \subset (d(\cdot,x) \leq 2^{-n+1}) \tag{5.5.3}$$

for each $x \in A_n$, and

$$\bigcup_{x\in A(n)} (d(\cdot,x) \leq 2^{-n}) \subset \left(\sum_{x\in A(n)} g_{n,x} = 1 \right). \tag{5.5.4}$$

Define $K_n \equiv \kappa_n + 1$, and define the sequence

$$(f_{n,1}, \ldots, f_{n,K(n)}) \equiv \left(g_{n,x(n,1)}, \ldots, g_{n,x(n,\kappa(n))}, \left(1 - \sum_{x \in A(n)} g_{n,x} \right) \right) \tag{5.5.5}$$

of nonnegative continuous functions on S. Then

$$\sum_{k=1}^{K(n)} f_{n,k} = 1 \tag{5.5.6}$$

on S, where $n \geq 1$ is arbitrary.

2. For the purpose of this proof, an open interval (a,b) in $[0,1]$ is defined by an arbitrary pair of endpoints a,b, where $0 \leq a \leq b \leq 1$. Two open intervals $(a,b),(a',b')$ are considered equal if $a = a'$ and $b = b'$. For arbitrary open intervals $(a,b),(a',b') \subset [0,1]$ we will write, as an abbreviation, $(a,b) < (a',b')$ if $b \leq a'$.

3. Let $n \geq 1$ be arbitrary. Define the product set

$$B_n \equiv \{1, \ldots, K_1\} \times \cdots \times \{1, \ldots, K_n\}.$$

Let μ denote the Lebesgue measure on $[0,1]$. Define the open interval $\Theta \equiv (0,1)$. Then, since

$$\sum_{k=1}^{K(1)} Ef_{n,k} = E \sum_{k=1}^{K(1)} f_{n,k} = E1 = 1,$$

we can subdivide the open interval Θ into mutually exclusive open subintervals $\Theta_1, \ldots, \Theta_{K(1)}$, such that

$$\mu\Theta_k = Ef_{n,k}$$

for each $k = 1, \ldots, K_1$, and such that $\Theta_k < \Theta_j$ for each $k = 1, \ldots, \kappa_1$ with $k < j$.

4. We will construct, for each $n \geq 1$, a family of mutually exclusive open subintervals

$$\{\Theta_{k(1),\ldots,k(n)} : (k_1, \ldots, k_n) \in B_n\}$$

of $(0,1)$ such that for each $(k_1, \ldots, k_n) \in B_n$, we have

(i) $\mu\Theta_{k(1),\ldots,k(n)} = Ef_{1,k(1)} \cdots f_{n,k(n)}$.

(ii) $\Theta_{k(1),\ldots,k(n)} \subset \Theta_{k(1),\ldots,k(n-1)}$ if $n \geq 2$.

(iii) $\Theta_{k(1),\ldots,k(n-1),k} < \Theta_{k(1),\ldots,k(n-1),j}$ for each $k,j = 1, \ldots, \kappa_n$ with $k < j$.

5. Proceed inductively. Step 3 gave the construction for $n = 1$. Now suppose the construction has been carried out for some $n \geq 1$ such that Conditions (i–iii) are satisfied. Consider each $(k_1, \ldots, k_n) \in B_n$. Then

$$\sum_{k=1}^{K(n+1)} Ef_{1,k(1)} \cdots f_{n,k(n)} f_{n+1,k} = Ef_{1,k(1)} \cdots f_{n,k(n)} \sum_{k=1}^{K(n+1)} f_{n+1,k}$$

$$= Ef_{1,k(1)} \cdots f_{n,k(n)} = \mu\Theta_{k(1),\ldots,k(n)},$$

where the last equality is because of Condition (i) in the induction hypothesis. Hence we can subdivide $\Theta_{k(1),\ldots,k(n)}$ into K_{n+1} mutually exclusive open subintervals

$$\Theta_{k(1),\ldots,k(n),1}, \ldots, \Theta_{k(1),\ldots,k(n),K(n+1)}$$

such that

$$\mu\Theta_{k(1),\ldots,k(n),k(n+1)} = Ef_{1,k(1)} \cdots f_{n,k(n)} f_{n+1,k(n+1)} \tag{5.5.7}$$

for each $k_{n+1} = 1, \ldots, K_{n+1}$. Thus Condition (i) holds for $n+1$. In addition, we can arrange these open subintervals such that

$$\Theta_{k(1),\ldots,k(n),k} < \Theta_{x(1),\ldots,x(n),j}$$

for each $k, j = 1, \ldots, K_{n+1}$ with $k < j$. This establishes Condition (iii) for $n+1$. Condition (ii) also holds for $n+1$ since, by construction, $\Theta_{k(1),\ldots,k(n),k(n+1)}$ is a subinterval of $\Theta_{k(1),\ldots,k(n)}$ for each $(k_1, \ldots, k_{n+1}) \in B_{n+1}$. Induction is completed.

6. Note that mutual exclusiveness and Condition (i) together imply that

$$\mu \bigcup_{(k(1),\ldots,k(n))\in B(n)} \Theta_{k(1),\ldots,k(n)} = \sum_{(k(1),\ldots,k(n))\in B(n)} \mu\Theta_{k(1),\ldots,k(n)}$$

$$= E\left(\sum_{k(1)=1}^{K(1)} f_{1,K(1)}\right) \cdots \left(\sum_{k(n)=1}^{K(n)} f_{n,K(n)}\right)$$

$$= E1 = 1 \tag{5.5.8}$$

for each $n \geq 1$. Hence the set

$$D \equiv \bigcap_{n=1}^{\infty} \bigcup_{(k(1),\ldots,k(n))\in B(n)} \Theta_{k(1),\ldots,k(n)} \tag{5.5.9}$$

is a full subset of $[0, 1]$.

7. Let $\theta \in D$ be arbitrary. Consider each $n \geq 1$. Then $\theta \in \Theta_{k(1),\ldots,k(n)}$ for some unique sequence $(k_1, \ldots, k_n) \in B_n$ since the intervals in each union in equality 5.5.9 are mutually exclusive. By the same token, $\theta \in \Theta_{j(1),\ldots,j(n+1)}$ for some unique $(j_1, \ldots, j_{n+1}) \in B_{n+1}$. Then $\theta \in \Theta_{j(1),\ldots,j(n)}$ in view of Condition (ii) in Step 4. Hence, by uniqueness of the sequence $(k_1, \ldots, k_n)$, we have $(j_1, \ldots, j_n) = (k_1, \ldots, k_n)$. Now define $k_{n+1} \equiv j_{n+1}$. It follows that $\theta \in \Theta_{k(1),\ldots,k(n+1)}$. Thus we obtain inductively a unique sequence $(k_p)_{p=1,2,\ldots}$ such that $k_p \in \{1, \ldots, K_p\}$ and $\theta \in \Theta_{k(1),\ldots,k(p)}$, for each $p \geq 1$.

Since the open interval $\Theta_{k(1),\ldots,k(n)}$ contains the given point θ, it has positive Lebesgue measure. Hence, in view of Condition (i) in Step 4, it follows that

$$Ef_{1,k(1)} \cdots f_{n,k(n)} > 0, \tag{5.5.10}$$

where $n \geq 1$ is arbitrary.

8. Define the function $X_n : [0,1] \to (S,d)$ by

$$domain(X_n) \equiv \bigcup_{(k(1),\ldots,k(n)) \in B(n)} \Theta_{k(1),\ldots,k(n)}$$

and by

$$X_n \equiv x_{n,k(n)} \text{ or } x_\circ \quad \text{on } \Theta_{k(1),\ldots,k(n)}, \text{ based on either } \mathrm{k_n} \leq \kappa_\mathrm{n} \text{ or } k_n = K_n, \tag{5.5.11}$$

for each $(k_1,\ldots,k_n) \in B_n$. Then, according to Proposition 4.8.8, X_n is a r.v. with values in the metric space (S,d). In other words, $X_n \in M(\Theta_0, S)$.

Now define the function $X \equiv \lim_{n\to\infty} X_n$. We proceed to prove that the function X is a well-defined r.v. by showing that $X_n \to X$ a.u.

9. To that end, let $n \geq 1$ be arbitrary, but fixed until further notice. Define

$$m \equiv m_n \equiv n \vee [\log_2(1 \vee \beta(2^{-n}))]_1, \tag{5.5.12}$$

where β is the assumed modulus of tightness of the distribution E relative to the reference point $x_\circ \in S$. Then

$$2^m > \beta(2^{-n}).$$

Take an arbitrary $\alpha_n \in (\beta(2^{-n}), 2^m)$. Then

$$E(d(\cdot,x_\circ) > \alpha_n) \leq 2^{-n}, \tag{5.5.13}$$

because β is a modulus of tightness of E. At the same time,

$$(d(\cdot,x_\circ) \leq \alpha_n) \subset (d(\cdot,x_\circ) \leq 2^m) \subset \bigcup_{x\in A(m)} (d(\cdot,x) \leq 2^{-m}) \subset \left(\sum_{x\in A(m)} g_{m,x} = 1\right), \tag{5.5.14}$$

where the second and third inclusions are by relations 5.5.2 and 5.5.4, respectively. Define the Lebesgue measurable set

$$D_n \equiv \bigcup_{(k(1),\ldots,k(m))\in B(m);\, k(m)\leq\kappa(m)} \Theta_{k(1),\ldots,k(m)} \subset [0,1]. \tag{5.5.15}$$

Then, in view of equality 5.5.8, we have

$$\begin{aligned} \mu(D_n^c) &= \sum_{(k(1),\ldots,k(m))\in B(m);\, k(m)=K(m)} \mu\Theta_{k(1),\ldots,k(m)} \\ &= \sum_{(k(1),\ldots,k(m))\in B(m);\, k(m)=K(m)} Ef_{1,k(1)} \cdots f_{m,k(m)} \end{aligned}$$

$$= \sum_{k(1)=1}^{K(1)} \cdots \sum_{k(m-1)=1}^{K(m-1)} Ef_{1,k(1)} \cdots f_{m-1,k(m)-1} f_{m,K(m)}$$

$$= Ef_{m,K(m)} = E\left(1 - \sum_{x \in A(m)} g_{m,x}\right)$$

$$\leq E(d(\cdot, x_\circ) > \alpha_n) \leq 2^{-n}, \tag{5.5.16}$$

where the first inequality is thanks to relation 5.5.14, and the second is by inequality 5.5.13.

10. Consider each $\theta \in D$. By Step 7, there exists a unique sequence $(k_p)_{p=1,2,\ldots}$ such that $k_p \in \{1, \ldots, K_p\}$ and $\theta \in \Theta_{k(1),\ldots,k(p)}$ for each $p \geq 1$. In particular, $\theta \in \Theta_{k(1),\ldots,k(m)}$. Suppose $k_m \leq \kappa_m$. Then

$$f_{m,k(m)} \equiv g_{m,x(m,k(m))},$$

according to the defining equality 5.5.5. Moreover, by Condition (ii) in Step 4, we have $\theta \in \Theta_{k(1),\ldots,k(p)} \subset \Theta_{k(1),\ldots,k(m)}$ for each $p \geq m$. Suppose, for the sake of a contradiction, that $k_{m+1} = K_{m+1}$. Then, by the defining equality 5.5.5, we have

$$f_{m+1,k(m+1)} \equiv 1 - \sum_{x \in A(m+1)} g_{m+1,x}.$$

Separately, inequality 5.5.10 applied to $m + 1$ yields

$$0 < Ef_{1,k(1)} \cdots f_{m,k(m)} f_{m+1,k(m+1)} \tag{5.5.17}$$

On the other hand, using successively the relations 5.5.3, 5.5.1, 5.5.2, and 5.5.4, we obtain

$$(f_{m,k(m)} > 0) = (g_{m,x(m,k(m))} > 0)$$
$$\subset (d(\cdot, x_{m,k(m)}) \leq 2^{-m+1}) \subset (d(\cdot, x_\circ) \leq 2^m + 2^{-m+1})$$
$$\subset (d(\cdot, x_\circ) \leq 2^{m+1}) \subset \bigcup_{x \in A(m+1)} (d(\cdot, x) \leq 2^{-m-1})$$
$$\subset \left(\sum_{x \in A(m+1)} g_{m+1,x} = 1\right) = (f_{m+1,k(m+1)} = 0).$$

Consequently, $f_{m,k(m)} f_{m+1,k(m+1)} = 0$. Hence the right-hand side of inequality 5.5.17 vanishes, while the left-hand side is 0, which is a contradiction. We conclude that $k_{m+1} \leq \kappa_{m+1}$. Summing up, from the assumption that $k_m \leq \kappa_m$, we can infer that $k_{m+1} \leq \kappa_{m+1}$.

Repeating these steps, we obtain, for each $q \geq m$, the inequality

$$k_q \leq \kappa_q,$$

whence

$$f_{q,k(q)} \equiv g_{q,x(q,k(q))}, \tag{5.5.18}$$

where $q \geq m$ is arbitrary. It follows from the defining equality 5.5.11 that

$$X_q(\theta) = x_{q,k(q)} \tag{5.5.19}$$

for each $q \geq m \equiv m_n$, provided that $k_m \leq \kappa_m$.

11. Next, suppose $\theta \in DD_n$. Then $k_m \leq \kappa_m$ according to the defining equality 5.5.15 for the set D_n. Hence equality 5.5.19 holds for each $q \geq m$. Thus we see that

$$DD_n \subset \bigcap_{q=m(n)}^{\infty} (X_q = x_{q,k(q)}). \tag{5.5.20}$$

Now let $q \geq p \geq m$ be arbitrary. As an abbreviation, write $y_p \equiv x_{p,k(p)}$. Then inequality 5.5.10 and equality 5.5.18 together imply that

$$Ef_{1,k(1)} \cdots f_{m-1,k(m-1)} g_{m,y(m)} \cdots g_{p,y(p)} \cdots g_{q,y(q)} = Ef_{1,k(1)} \cdots f_{q,k(q)} > 0.$$

Hence there exists $z \in S$ such that

$$(f_{1,k(1)} \cdots f_{m-1,k(m-1)} g_{m,y(m)} \cdots g_{p,y(p)} \cdots g_{q,y(q)})(z) > 0,$$

whence $g_{p,y(p)}(z) > 0$ and $g_{q,y(q)}(z) > 0$. Consequently, by relation 5.5.3, we obtain

$$d(y_p, y_q) \leq d(y_p, z) + d(z, y_q) \leq 2^{-p+1} + 2^{-q+1} \to 0 \tag{5.5.21}$$

as $p, q \to \infty$. Since (S, d) is complete, we have

$$X_p(\theta) \equiv x_{p,k(p)} \equiv y_p \to y$$

as $p \to \infty$, for some $y \in S$. Hence $\theta \in domain(X)$, with $X(\theta) \equiv y$, where we recall that $X \equiv \lim_{n\to\infty} X_n$.. Moreover, with $q \to \infty$ in inequality 5.5.21, we obtain

$$d(X_p(\theta), X(\theta)) \leq 2^{-p+1}, \tag{5.5.22}$$

where $p \geq m \equiv m_n$ and $\theta \in DD_n$ are arbitrary. Since $\mu(DD_n)^c = \mu D_n^c \leq 2^{-n}$ is arbitrarily small when $n \geq 1$ is sufficiently large, we conclude that $X_n \to X$ a.u. relative to the Lebesgue measure I, as $n \to \infty$. It follows that the function $X : [0,1] \to S$ is measurable. In other words, X is a r.v.

12. It remains to verify that $I_X = E$, where I_X is the distribution induced by X on S. For that purpose, let $h \in C(S)$ be arbitrary. We need to prove that $Ih(X) = Eh$. Without loss of generality, we assume that $|h| \leq 1$ on S. Let δ_h be a modulus of continuity of the function h. Let $\varepsilon > 0$ be arbitrary. Let $n \geq 1$ be so large that (i′)

$$2^{-n} < \varepsilon \wedge \frac{1}{2}\delta_h\left(\frac{\varepsilon}{3}\right) \tag{5.5.23}$$

and (ii′) h is supported by $(d(\cdot,x_\circ) \leq 2^n)$. Then relation 5.5.2 implies that h is supported by $\bigcup_{x\in A(n)}(d(\cdot,x) \leq 2^{-n})$. At the same time, by the defining equality 5.5.11 of the simple r.v. $X_n : [0,1] \to S$, we have

$$\begin{aligned}
Ih(X_n) &= \sum_{(k(1),\dots,k(n))\in B(n);\, k(n)\leq\kappa(n)} h(x_{n,k(n)})\mu\Theta_{k(1),\dots,k(n)} + h(x_\circ)\mu(D_n^c) \\
&= \sum_{k(1)=1}^{K(1)} \cdots \sum_{k(n-1)=1}^{K(n-1)} \sum_{k(n)=1}^{\kappa(n)} h(x_{n,k(n)})Ef_{1,k(1)}\cdots f_{n-1,k(n-1)}f_{n,k(n)} \\
&\quad + h(x_\circ)\mu(D_n^c) \\
&= \sum_{k=1}^{\kappa(n)} h(x_{n,k})Ef_{n,k} + h(x_\circ)\mu(D_n^c) \\
&= \sum_{k=1}^{\kappa(n)} h(x_{n,k(n)})Eg_{n,x(n,k)} + h(x_\circ)\mu(D_n^c) \\
&= \sum_{x\in A(n)} h(x)Eg_{n,x} + h(x_\circ)\mu(D_n^c),
\end{aligned}$$

where the third equality is thanks to equality 5.5.6. Hence

$$\left| Ih(X_n) - E\sum_{x\in A(n)} h(x)g_{n,x} \right| \leq |h(x_\circ)\mu(D_n^c)| \leq \mu(D_n^c) \leq 2^{-n} < \varepsilon. \quad (5.5.24)$$

At the same time, since A_n is a 2^{-n}-partition of unity of (S,d), with $2^{-n} < \frac{1}{2}\delta_h\left(\frac{\varepsilon}{3}\right)$, Assertion 2 of Proposition 3.3.6 implies that $\left\| \sum_{x\in A(n)} h(x)g_{n,x} - h \right\| \leq \varepsilon$. Hence

$$\left| E\sum_{x\in A(n)} h(x)g_{n,x} - Eh \right| \leq \varepsilon.$$

Inequality 5.5.24 therefore yields

$$|Ih(X_n) - Eh| < 2\varepsilon.$$

Since $\varepsilon > 0$ is arbitrarily small, we have $Ih(X_n) \to Eh$ as $n \to \infty$. At the same time, $h(X_n) \to h(X)$ a.u. relative to I. Hence the Dominated Convergence Theorem implies that $Ih(X_n) \to Ih(X)$. It follows that $Eh = Ih(X)$, where $h \in C(S)$ is arbitrary. We conclude that $E = I_X$.

Define $\Phi_{Sk,\xi}(E) \equiv X$, and the theorem is proved. □

Theorem 5.5.2. Metrical continuity of Skorokhod representation. *Let $\xi \equiv (A_n)_{n=1,2,\dots}$ be a binary approximation of the locally compact metric space (S,d) relative to the reference point $x_\circ \in S$. Let $\|\xi\| \equiv (\kappa_n)_{n=1,2,\dots}$ be the modulus of local compactness of (S,d) corresponding to ξ. In other words, $\kappa_n \equiv |A_n|$ is the number of elements in the enumerated finite set A_n, for each $n \geq 1$.*

Let $\widehat{J}(S,d)$ be the set of distributions on (S,d). Let $\widehat{J}_\beta(S,d)$ be an arbitrary tight subset of $\widehat{J}(S,d)$, with a modulus of tightness β relative to $x_\circ$. Recall the probability metric ρ_{Prob} on $M(\Theta_0, S)$ defined in Definition 5.1.10. Then the Skorokhod representation

$$\Phi_{Sk,\xi} : (\widehat{J}(S,d), \rho_{Dist,\xi}) \to (M(\Theta_0, S), \rho_{Prob})$$

constructed in Theorem 5.5.1 is uniformly continuous on the subset $\widehat{J}_\beta(S,d)$, with a modulus of continuity $\delta_{Sk}(\cdot, \|\xi\|, \beta)$ depending only on $\|\xi\|$ and β.

Proof. Refer to the proof of Theorem 5.5.1 for notations. In particular, let

$$\pi \equiv (\{g_{n,x} : x \in A_n\})_{n=1,2,\ldots}$$

denote the partition of unity of (S,d) determined by ξ.

1. Let $n \geq 1$ be arbitrary. Recall from Proposition 3.3.5 that for each $x \in A_n$, the functions $g_{n,x}$ and $\sum_{y \in A(n)} g_{n,y}$ in $C(S)$ have Lipschitz constant 2^{n+1} and have values in $[0,1]$. Consequently, each of the functions $f_{n,1}, \ldots, f_{n,K(n)}$ defined in formula 5.5.5 has a Lipschitz constant 2^{n+1} and has values in $[0,1]$.

2. Let

$$(k_1, \ldots, k_n) \in B_n \equiv \{1, \ldots, K_1\} \times \cdots \times \{1, \ldots, K_n\}$$

be arbitrary. Then, by Lemma 3.3.1 on the basic properties of Lipschitz functions, the function

$$h_{k(1),\ldots,k(n)} \equiv n^{-1} \sum_{i=1}^{n} \sum_{k=1}^{k(i)-1} f_{1,k(1)} \cdots f_{i-1,k(i-1)} f_{i,k} \in C_{ub}(S)$$

has a Lipschitz constant given by

$$\begin{aligned} n^{-1} &\sum_{i=1}^{n} \sum_{k=1}^{k(i)-1} (2^{1+1} + 2^{2+1} + \cdots + 2^{i+1}) \\ &\leq 2^2 n^{-1} \sum_{i=1}^{n} \kappa_i (1 + 2 + \cdots + 2^{i-1}) \\ &< 2^2 n^{-1} \sum_{i=1}^{n} \kappa_n 2^i < 2^2 n^{-1} \kappa_n 2^{n+1} = n^{-1} 2^{n+3} \kappa_n, \end{aligned}$$

where $n \geq 1$ is arbitrary. Moreover,

$$0 \leq h_{k(1),\ldots,k(n)} \leq n^{-1} \sum_{i=1}^{n} \sum_{k=1}^{k(i)-1} f_{i,k} \leq n^{-1} \sum_{i=1}^{n} 1 = 1.$$

3. Now let $E, E' \in \widehat{J}_\beta(S,d)$ be arbitrary. Let the objects $\{\Theta_{k(1),\ldots,k(n)} : (k_1, \ldots, k_n) \in B_n; n \geq 1\}$, D, $(X_n)_{n=1,2,\ldots}$, and let X be constructed as in Theorem 5.5.1 relative to E. Let the objects $\{\Theta'_{k(1),\ldots,k(n)} : (k_1, \ldots, k_n) \in B_n; n \geq 1\}$, D', $(X'_n)_{n=1,2,\ldots}$, and let X' be similarly constructed relative to E'.

4. Let $\varepsilon > 0$ be arbitrary. Fix

$$n \equiv [3 - \log_2 \varepsilon]_1.$$

Thus $2^{-n+3} < \varepsilon$. As in the proof of Theorem 5.5.1, let

$$m \equiv m_n \equiv n \vee [\log_2(1 \vee \beta(2^{-n}))]_1, \tag{5.5.25}$$

and let

$$c \equiv m^{-1} 2^{m+3} \kappa_m, \tag{5.5.26}$$

where β is the given modulus of tightness of the distributions E in $\widehat{J}_\beta(S,d)$ relative to the reference point $x_\circ \in S$. Let

$$\alpha \equiv 2^{-n} \prod_{i=1}^{m} K_i^{-1} = 2^{-n} |B_m|^{-1}. \tag{5.5.27}$$

By Assertion 1 of Proposition 5.3.11, there exists $\widetilde{\Delta}(m^{-1}\alpha, c, \beta, \|\xi\|) > 0$ such that if

$$\rho_{Dist,\xi}(E, E') < \delta_{Sk}(\varepsilon, \|\xi\|, \beta) \equiv \widetilde{\Delta}(m^{-1}\alpha, c, \beta, \|\xi\|),$$

then

$$|Ef - E'f| < m^{-1}\alpha \tag{5.5.28}$$

for each $f \in C_{ub}(S)$ with Lipschitz constant $c > 0$ and with $|f| \leq 1$.

5. Now suppose

$$\rho_{Dist,\xi}(E, E') < \delta_{Sk}(\varepsilon, \|\xi\|, \beta). \tag{5.5.29}$$

We will prove that

$$\rho_{Prob}(X, X') < \varepsilon.$$

To that end, consider each $(k_1, \ldots, k_m) \in B_m$. We will calculate the endpoints of the corresponding open interval

$$(a_{k(1),\ldots,k(m)}, b_{k(1),\ldots,k(m)}) \equiv \Theta_{k(1),\ldots,k(m)}.$$

Recall that, by construction, $\{\Theta_{k(1),\ldots,k(m-1),k} : 1 \leq k \leq K_m\}$ is the set of subintervals in a partition of the open interval $\Theta_{k(1),\ldots,k(m-1)}$ into mutually exclusive open subintervals, with

$$\Theta_{k(1),\ldots,k(m-1),k} < \Theta_{k(1),\ldots,k(m-1),j}$$

if $1 \leq k < j \leq K_m$. Hence the left endpoint of $\Theta_{k(1),\ldots,k(m)}$ is

$$\begin{aligned} a_{k(1),\ldots,k(m)} &= a_{k(1),\ldots,k(m-1)} + \sum_{k=1}^{k(m)-1} \mu\Theta_{k(1),\ldots,k(m-1),k} \\ &= a_{k(1),\ldots,k(m-1)} + \sum_{k=1}^{k(m)-1} Ef_{1,k(1)} \cdots f_{m-1,k(m-1)} f_{m,k}, \end{aligned} \tag{5.5.30}$$

where the second equality is due to Condition (i) in Step 4 of the proof of Theorem 5.5.1. Recursively, we then obtain

$$a_{k(1),\dots,k(m)} = a_{k(1),\dots,k(m-2)} + \sum_{k=1}^{k(m-1)-1} Ef_{1,k(1)} \cdots f_{m-2,k(m-2)} f_{m-1,k}$$

$$+ \sum_{k=1}^{k(m)-1} Ef_{1,k(1)} \cdots f_{m-1,k(m-1)} f_{m,k}$$

$$= \cdots = \sum_{i=1}^{m} \sum_{k=1}^{k(i)-1} Ef_{1,k(1)} \cdots f_{i-1,k(i-1)} f_{i,k} \equiv mEh_{k(1),\dots,k(m)}.$$

6. Similarly, write

$$(a'_{k(1),\dots,k(m)}, b'_{k(1),\dots,k(m)}) \equiv \Theta'_{k(1),\dots,k(m)}.$$

Then

$$a'_{k(1),\dots,k(m)} = mE'h_{k(1),\dots,k(m)}.$$

Combining,

$$|a_{k(1),\dots,k(m)} - a'_{k(1),\dots,k(m)}| = m|Eh_{k(1),\dots,k(m)} - E'h_{k(1),\dots,k(m)}| < mm^{-1}\alpha = \alpha, \tag{5.5.31}$$

where the inequality is obtained by applying inequality 5.5.28 to the function $f \equiv h_{k(1),\dots,k(m)}$; in Step 2, the latter was observed to have values in $[0,1]$ and to have Lipschitz constant $c \equiv m^{-1}2^{m+3}\kappa_m$. By symmetry, we can similarly prove that

$$|b_{k(1),\dots,k(m)} - b'_{k(1),\dots,k(m)}| < \alpha. \tag{5.5.32}$$

7. Inequality 5.5.22 in Step 11 of the proof of Theorem 5.5.1 gives

$$d(X_m, X) \le 2^{-m+1} \tag{5.5.33}$$

on DD_n. Now partition the set $B_m \equiv B_{m,0} \cup B_{m,1} \cup B_{m,2}$ into three disjoint subsets such that

$$B_{m,0} = \{(k_1,\dots,k_m) \in B_m : k_m = K_m\},$$
$$B_{m,1} = \{(k_1,\dots,k_m) \in B_m : k_m \le \kappa_m; \mu\Theta_{k(1),\dots,k(m)} > 2\alpha\},$$
$$B_{m,2} = \{(k_1,\dots,k_m) \in B_m : k_m \le \kappa_m; \mu\Theta_{k(1),\dots,k(m)} < 3\alpha\}.$$

8. For each $(k_1,\dots,k_m) \in B_{m,1}$, define the α-interior

$$\widetilde{\Theta}_{k(1),\dots,k(m)} \equiv (a_{k(1),\dots,k(m)} + \alpha, b_{k(1),\dots,k(m)} - \alpha)$$

of the open interval $\Theta_{k(1),\dots,k(m)}$. Define the set

$$H \equiv \bigcup_{(k(1),\dots,k(m)) \in B(m,1)} \widetilde{\Theta}_{k(1),\dots,k(m)} \subset \Theta.$$

Then

$$H^c = \bigcup_{(k(1),\dots,k(m))\in B(m,1)} \Theta_{k(1),\dots,k(m)}\widetilde{\Theta}^c_{k(1),\dots,k(m)} \cup \bigcup_{(k(1),\dots,k(m))\in B(m,2)} \Theta_{k(1),\dots,k(m)} \cup \bigcup_{(k(1),\dots,k(m))\in B(m,0)} \Theta_{k(1),\dots,k(m)}.$$

Hence

$$\begin{aligned}\mu H^c &= \sum_{(k(1),\dots,k(m))\in B(m,1)} 2\alpha + \sum_{(k(1),\dots,k(m))\in B(m,2)} \mu\Theta_{k(1),\dots,k(m)} \\ &\quad + \mu \bigcup_{(k(1),\dots,k(m))\in B(m);\, k(m)=K(m)} \Theta_{k(1),\dots,k(m)} \\ &< \sum_{(k(1),\dots,k(m))\in B(m,1)} 2\alpha + \sum_{(k(1),\dots,k(m))\in B(m,2)} 3\alpha + \mu D_n^c \\ &< |B_{m,1}|2\alpha + |B_{m,2}|3\alpha + 2^{-n} \\ &\leq |B_m|3\alpha + 2^{-n} \equiv 3\cdot 2^{-n} + 2^{-n} = 2^{-n+2}, \end{aligned} \tag{5.5.34}$$

where the set D_n was defined in equality 5.5.15 in the proof of Theorem 5.5.1, where the second inequality is from inequality 5.5.16, and where next-to-last equality is by the defining equality 5.5.27. Note for later reference that the set H depends only on E, and not on E'.

9. Now let $\theta \in HDD'$ be arbitrary. Then, according to the definitions of H_n, D, and D', we have

$$\theta \in \widetilde{\Theta}_{k(1),\dots,k(m)}\Theta'_{j(1),\dots,j(m)}$$

for some $(k_1,\dots,k_m) \in B_{m,1}$ and some $(j_1,\dots,j_m) \in B_m$. Hence, in view of inequalities 5.5.31 and 5.5.32, we have

$$\begin{aligned}\theta &\in (a_{k(1),\dots,k(m)} + \alpha, b_{k(1),\dots,k(m)} - \alpha) \\ &\subset (a'_{k(1),\dots,k(m)}, b'_{k(1),\dots,k(m)}) \equiv \Theta'_{k(1),\dots,k(m)}.\end{aligned}$$

Consequently,

$$\theta \in \Theta'_{k(1),\dots,k(m)}\Theta'_{j(1),\dots,j(m)}. \tag{5.5.35}$$

At the same time, the intervals in the set $\{\Theta'_{i(1),\dots,ik(m)} : (i_1,\dots,i_m) \in B_m\}$ are mutually exclusive. Hence the intersection of the two open intervals on the right-hand side of relation 5.5.35 will be empty unless their subscripts are identical. Consequently, $(k_1,\dots,k_m) = (j_1,\dots,j_m)$. In particular, $j_m = k_m \leq \kappa_m$, where the inequality is because $(k_1,\dots,k_m) \in B_{m,1}$. Hence $\theta \in D_n D'_n$ by the defining equality 5.5.15 for the sets D_n and D'_n. At the same time, by the defining equality 5.5.11 for the r.v.'s X_m, and by a similar defining equality for the r.v.'s X'_m, we have

$$X_m(\theta) = x_{m,k(m)} = x_{m,j(m)} = X'_m(\theta).$$

Since $\theta \in HDD'$ is arbitrary, we have proved that (i) $HDD' \subset D_n D'_n DD'$ and (ii) $X_m = X'_m$ on HDD'.

10. By inequality 5.5.33 and a similar inequality for the r.v.'s X'_m, X', we have

$$d(X_m, X) \vee d(X'_m, X') \leq 2^{-m+1} \tag{5.5.36}$$

on $D_n D'_n DD'$. Combining with Conditions (i) and (ii) in Step 9, we obtain

$$\begin{aligned} HDD' &\subset (X_m = X'_m) \cap (d(X_m, X) \leq 2^{-m+1}) \cap (d(X'_m, X') \leq 2^{-m+1}) \\ &\subset (d(X, X') \leq 2^{-m+2}), \end{aligned} \tag{5.5.37}$$

where D, D' are full sets. For later reference, note that this implies

$$HDD' \subset (d(X, X') \leq 2^{-m+2}) \subset (d(X, X') \leq 2^{-n+2}) \subset (d(X, X') < \varepsilon). \tag{5.5.38}$$

11. From relation 5.5.37 and inequality 5.5.34, we infer that

$$\begin{aligned} \rho_{Prob}(X, X') &= I(1 \wedge d(X, X'); H) + I(1 \wedge d(X, X'); H^c) \leq 2^{-m+2} + \mu H^c \\ &< 2^{-m+2} + 2^{-n+2} \leq 2^{-n+2} + 2^{-n+2} = 2^{-n+3} < \varepsilon, \end{aligned}$$

where $E, E' \in \widehat{J}_\beta(S, d)$ are arbitrary such that $\rho_{Dist,\xi}(E, E') < \delta_{Sk}(\varepsilon, \|\xi\|, \beta)$, where $X \equiv \Phi_{Sk,\xi}(E)$ and $X' \equiv \Phi_{Sk,\xi}(E')$, and where $\varepsilon > 0$ is arbitrary. Thus the mapping $\Phi_{Sk,\xi} : (\widehat{J}(S, d), \rho_{Dist,\xi}) \to (M(\Theta_0, S), \rho_{Prob})$ is uniformly continuous on the subspace $\widehat{J}_\beta(S, d)$, with $\delta_{Sk}(\cdot, \|\xi\|, \beta)$ as a modulus of continuity. The theorem is proved. □

Skorokhod's continuity theorem in [Skorokhod 1956], in terms of a.u. convergence, is a consequence of the preceding proof.

Theorem 5.5.3. Continuity of Skorokhod representation in terms of weak convergence and a.u. convergence. *Let ξ be a binary approximation of the locally compact metric space (S, d), relative to the reference point $x_\circ \in S$.*

Let $E, E^{(1)}, E^{(2)}, \ldots$ be a sequence of distributions on (S, d) such that $E^{(n)} \Rightarrow E$. Let $X \equiv \Phi_{Sk,\xi}(E)$ and $X^{(n)} \equiv \Phi_{Sk,\xi}(E^{(n)})$ for each $n \geq 1$. Then $X^{(n)} \to X$ a.u. on (Θ_0, L_0, I).

Proof. Let $\|\xi\|$ be the modulus of local compactness of (S, d) corresponding to ξ. By hypothesis, $E^{(n)} \Rightarrow E$. Hence $\rho_{Dist,\xi}(E, E^{(n)}) \to 0$ by Proposition 5.3.5. By Proposition 5.3.10, the family $\widehat{J}_\beta(S, d) \equiv \{E, E^{(1)}, E^{(2)}, \ldots\}$ is tight, with some modulus of tightness β. Let $\varepsilon > 0$ be arbitrary. Let $\delta_{Sk}(\varepsilon, \|\xi\|, \beta) > 0$ be defined as in Theorem 5.5.2. By relation 5.5.38 in Step 10 of the proof of Theorem 5.5.2, we see that there exists a Lebesgue measurable subset H of $[0, 1]$ which depends only on E, with $\mu H^c < \varepsilon$, such that for each $E' \in \widehat{J}_\beta(S, d)$ we have

$$H \subset (d(X, X') < \varepsilon) \quad \text{a.s.}, \tag{5.5.39}$$

where $X' \equiv \Phi_{Sk,\xi}(E')$, provided that $\rho_{Dist,\xi}(E, E') < \delta_{Sk}(\varepsilon, \|\xi\|, \beta)$. Take any $p \geq 1$ so large that $\rho_{Dist,\xi}(E, E^{(n)}) < \delta_{Sk}(\varepsilon, \|\xi\|, \beta)$ for each $n \geq p$. Then

$$H \subset \bigcap_{n=p}^{\infty}(d(X, X^{(n)}) < \varepsilon) \quad \text{a.s.}$$

Consequently, $X_n \to X$ a.u. according to Proposition 5.1.14. □

5.6 Independence and Conditional Expectation

The product space introduced in Definition 4.10.5 gives a model for compounding two independent experiments into one. This section introduces the notion of conditional expectations, which is a more general method of compounding probability spaces.

Definition 5.6.1. Independent set of r.v.'s. Let (Ω, L, E) be a probability space. A finite set $\{X_1, \ldots, X_n\}$ of r.v.'s where X_i has values in a complete metric space (S_i, d_i), for each $i = 1, \ldots, n$, is said to be *independent* if

$$Ef_1(X_1)\ldots f_n(X_n) = Ef_1(X_1)\ldots Ef_n(X_n) \tag{5.6.1}$$

for each $f_1 \in C_{ub}(S_1), \ldots, f_n \in C_{ub}(S_n)$. In that case, we will also simply say that $X_1, \ldots, X_n$ are independent r.v.'s. A sequence of events $A_1, \ldots, A_n$ is said to be independent if their indicators $1_{A(1)}, \ldots, 1_{A(n)}$ are independent r.r.v.'s.

An arbitrary set of r.v.'s is said to be independent if every finite subset is independent. □

Proposition 5.6.2. Independent r.v.'s from product space. *Let $F_1, \ldots, F_n$ be distributions on the locally compact metric spaces $(S_1, d_1), \ldots, (S_n, d_n)$, respectively. Let $(S, d) \equiv (S_1 \times \ldots, S_n, d_1 \otimes \ldots \otimes d_n)$ be the product metric space. Consider the product integration space*

$$(\Omega, L, E) \equiv (S, L, F_1 \otimes \cdots \otimes F_n) \equiv \bigotimes_{j=1}^{n}(S_j, L_j, F_j),$$

where (S_i, L_i, F_i) is the probability space that is the completion of $(S_i, C_{ub}(S_i), F_i)$, for each $i = 1, \ldots, n$. Then the following conditions hold:

1. Let $i = 1, \ldots, n$ be arbitrary. Define the coordinate r.v. $X_i : \Omega \to S_i$ by *$X_i(\omega) \equiv \omega_i$ for each $\omega \equiv (\omega_1, \ldots, \omega_n) \in \Omega$. Then the r.v.'s $X_1, \ldots, X_n$* are *independent. Moreover, X_i induces the distribution F_i on (S_i, d_i) for each $i = 1, \ldots, n$.*

2. $F_1 \otimes \cdots \otimes F_n$ is a distribution on (S, d). Specifically, it is the distribution F induced on (S, d) by the r.v. $X \equiv (X_1, \ldots, X_n)$.

Proof. 1. By Proposition 4.8.10, the continuous functions $X_1, \ldots, X_n$ on (S, L, E) are measurable. Let $f_i \in C_{ub}(S_i)$ be arbitrary, for each $i = 1, \ldots, n$. Then

$$Ef_1(X_1)\ldots f_n(X_n) = (F_1 \otimes \cdots \otimes F_n)(f_1 \otimes \cdots \otimes f_n) = F_1 f_1 \ldots F_n f_n \tag{5.6.2}$$

by Fubini's Theorem (Theorem 4.10.9). Let $i = 1, \ldots, n$ be arbitrary. In the special case where $f_j \equiv 1$ for each $j = 1, \ldots, n$ with $j \neq i$, we obtain, from equality 5.6.2,

$$Ef_i(X_i) = F_i f_i. \tag{5.6.3}$$

Hence equality 5.6.2 yields

$$Ef_1(X_1) \ldots f_n(X_n) = Ef_1(X_1) \ldots Ef_n(X_n)$$

where $f_i \in C_{ub}(S_i)$ is arbitrary for each $i = 1, \ldots, n$. Thus the r.v.'s $X_1, \ldots, X_n$ are independent. Moreover, equality 5.6.3 shows that the r.v. X_i induces the distribution F_i on (S_i, d_i) for each $i = 1, \ldots, n$. Assertion 1 is proved.

2. Since X is an r.v. with values in S, it induces a distribution E_X on (S, d). Hence

$$E_X f \equiv Ef(X) = (F_1 \otimes \cdots \otimes F_n) f$$

for each $f \in C_{ub}(S)$. Thus $F_1 \otimes \cdots \otimes F_n = E_X$ is a distribution F on (S, d). Assertion 2 is proved. □

Proposition 5.6.3. Basics of independence. *Let (Ω, L, E) be a probability space. For $i = 1, \ldots, n$, let X_i be an r.v. with values in a complete metric space (S_i, d_i), and let $(S_i, L_{X(i)}, E_{X(i)})$ be the probability space it induces on (S_i, d_i). Suppose the r.v.'s $X_1, \ldots, X_n$ are independent. Then, for arbitrary $f_1 \in L_{X(1)}, \ldots, f_n \in L_{X(n)}$, we have*

$$E \prod_{i=1}^{n} f_i(X_i) = \prod_{i=1}^{n} Ef_i(X_i). \tag{5.6.4}$$

Proof. 1. Consider each $i = 1, \ldots, n$. Let $f_i \in L_{X(i)}$ be arbitrary. By Definition 5.2.3, $L_{X(i)}$ is the completion of $(\Omega, C_{ub}(S_i), E_{X(i)})$. The r.r.v. $f_i \in L_{X(i)}$ is therefore the L_1-limit relative to $E_{X(i)}$ of a sequence $(f_{i,h})_{h=1,2,\ldots}$ in $C_{ub}(S_i)$ as $h \to \infty$. Moreover, according to Assertion 1 of Proposition 5.2.6, we have $f_i(X_i) \in L(X_i)$ and

$$E|f_{i,h}(X_i) - f_i(X_i)| = E_{X(i)}|f_{i,h} - f_i| \to 0$$

as $h \to \infty$. By passing to subsequences if necessary, we may assume that

$$f_{i,h}(X_i) \to f_i(X_i) \quad \text{a.u.} \tag{5.6.5}$$

as $h \to \infty$, for each $i = 1, \ldots, n$.

2. Consider the special case where $f_i \geq 0$ for each $i = 1, \ldots, n$. Then we may assume that $f_{i,h} \geq 0$ for each $h \geq 1$, for each $i = 1, \ldots, n$. Let $a > 0$ be arbitrary. By the independence of the r.v.'s $X_1, \ldots, X_n$, we then have

$$E \prod_{i=1}^{n} (f_{i,h}(X_i) \wedge a) = \prod_{i=1}^{n} E(f_{i,h}(X_i) \wedge a) \equiv \prod_{i=1}^{n} E_{X(i)}(f_{i,h} \wedge a).$$

In view of the a.u. convergence 5.6.5, we can let $h \to \infty$ and apply the Dominated Convergence Theorem to obtain

$$E\prod_{i=1}^{n}(f_i(X_i) \wedge a) = \prod_{i=1}^{n} E_{X(i)}(f_i \wedge a).$$

Now let $a \to \infty$ and apply the Monotone Convergence Theorem to obtain

$$E\prod_{i=1}^{n} f_i(X_i) = \prod_{i=1}^{n} E_{X(i)}(f_i) = \prod_{i=1}^{n} Ef_i(X_i).$$

3. The same equality for arbitrary $f_1 \in L_{X(1)}, \ldots, f_n \in L_{X(n)}$ follows by linearity. □

We next define the conditional expectation of an r.r.v. as the revised expectation given the observed values of all the r.v.'s in a family G.

Definition 5.6.4. Conditional expectation. Let (Ω, L, E) be a probability space, and let L' be a probability subspace of L. Let $Y \in L$ be arbitrary. If there exists $X \in L'$ such that $EZY = EZX$ for each indicator $Z \in L'$, then we say that X is the *conditional expectation* of Y given L', and define $E(Y|L') \equiv X$. We will call $L_{|L'} \equiv \{Y \in L : E(Y|L') \quad \text{exists}\}$ the subspace of *conditionally integrable r.r.v.'s* given the subspace L'.

In the special case where $L' \equiv L(G)$ is the probability subspace generated by a given family of r.v.'s with values in some complete metric space (S,d), we will simply write $E(Y|G) \equiv E(Y|L')$ and say that $L_{|G} \equiv L_{|L'}$ is the subspace of *conditionally integrable r.r.v.'s* given the family G. In the special case where $G \equiv \{V_1, \ldots, V_m\}$ for some $m \geq 1$, we write also $E(Y|V_1, \ldots, V_m) \equiv E(Y|G) \equiv E(Y|L')$.

In the further special case where $m = 1$, and where $V_1 = 1_A$ for some measurable set A with $P(A) > 0$, it can easily be verified that, for arbitrary $Y \in L$, the conditional $E(Y|1_A)$ exists and is given by $E(Y|1_A) = P(A)^{-1}E(Y1_A)1_A$. In that case, we will write

$$E_A(Y) \equiv P(A)^{-1}E(Y1_A)$$

for each $Y \in L$, and write $P_A(B) \equiv E_A(1_B)$ for each measurable set B. The next lemma proves that (Ω, L, E_A) is a probability space, called the *conditional probability space given the event A*.

More generally, if $Y_1, \ldots, Y_n \in L_{|L'}$ then we define the vector

$$E((Y_1, \ldots, Y_n)|L') \equiv (E(Y_1|L'), \ldots, E(Y_n|L'))$$

of integrable r.r.v.'s in L'.

Let A be an arbitrary measurable subset of (Ω, L, E). If $1_A \in L_{|L'}$ we will write $P(A|L') \equiv E(1_A|L')$ and call $P(A|L')$ the *conditional probability* of the event A given the probability subspace L'. If $1_A \in L_{|G}$ for some give family of r.v.'s with values in some complete metric space (S,d), we will simply write $P(A|G) \equiv E(1_A|G)$. In the case where $G \equiv \{V_1,\ldots,V_m\}$, we write also $P(A|V_1,\ldots,V_m) \equiv E(1_A|V_1,\ldots,V_m)$. □

Before proceeding, note that the statement $E(Y|L') = X$ asserts two things: $E(Y|L')$ exists, and it is equal to X. We have defined the conditional expectation without the sweeping classical assertion of its existence. Before we use a particular conditional expectation, we will first supply a proof of its existence.

Lemma 5.6.5. Conditional probability space given an event is indeed a probability space. *Let the measurable set A be arbitrary, with $P(A) > 0$. Then* the triple (Ω, L, E_A) is indeed a a probability space.

Proof. We need to verify the conditions in Definition 4.3.1 for an integration space.

1. Clearly, E_A is a linear function on the linear L.

2. Let $(Y_i)_{i=0,1,2,\ldots}$ be an arbitrary sequence of functions in L such that Y_i is nonnegative for each $i \geq 1$ and such that $\sum_{i=1}^{\infty} E_A(Y_i) < E_A(Y_0)$. Then $\sum_{i=1}^{\infty} E(Y_i 1_A) < E(Y_0 1_A)$ by the definition of the function E_A. Hence, since E is an integration, there exists $\omega \in \bigcap_{i=0}^{\infty} domain(Y_i 1_A)$ such that $\sum_{i=1}^{\infty} Y_i(\omega)1_A(\omega) < Y_0(\omega)1_A(\omega)$. It follows that $1_A(\omega) > 0$. Dividing by $1_A(\omega)$, we obtain $\sum_{i=1}^{\infty} Y_i(\omega) < Y_0(\omega)$.

3. Trivially, $E_A(1) \equiv P(A)^{-1}E(1_A) = 1$.

4. Let $Y \in L$ be arbitrary. Then

$$E_A(Y \wedge n) \equiv E(Y \wedge n)1_A \rightarrow E(Y)1_A \equiv E_A(Y)$$

as $n \rightarrow \infty$. Similarly, $E_A(|Y| \wedge n^{-1}) \equiv E(|Y| \wedge n^{-1})1_A \rightarrow 0$ as $n \rightarrow \infty$.

Summing up, all four conditions in Definition 4.3.1 are satisfied by the triple (Ω, L, E_A). Because L is complete relative to the integration E in the sense of Definition 4.4.1, it is also complete relative to the integration E_A. Because $1 \in L$ with $E_A(1) = 1$, the complete integration space (Ω, L, E_A) is a probability space. □

We will show that the conditional expectation, if it exists, is unique in the sense of equality a.s. The next two propositions prove basic properties of conditional expectations.

Proposition 5.6.6. Basics of conditional expectation. *Let (Ω, L', E) be a probability subspace of a probability space (Ω, L, E). Then the following conditions hold:*

1. *Suppose $Y_1 = Y_2$ a.s. in L, and suppose $X_1, X_2 \in L'$ are such that $EZY_j = EZX_j$ for each $j = 1,2$, for each indicator $Z \in L'$. Then $X_1 = X_2$ a.s. Consequently, the conditional expectation, if it exists, is uniquely defined.*
2. *Suppose $X, Y \in L_{|L'}$. Then $aX + bY \in L_{|L'}$, and*

$$E(aX + bY|L') = aE(X|L') + bE(Y|L')$$

for each $a, b \in R$. *If, in addition,* $X \leq Y$ *a.s., then* $E(X|L') \leq E(Y|L')$ *a.s. In particular, if, in addition,* $|X| \in L_{|L'}$, then $|E(X|L')| \leq E(|X||L')$ *a.s.*

3. $E(E(Y|L')) = E(Y)$ *for each* $Y \in L_{|L'}$. *Moreover,* $L' \subset L_{|L'}$, *and* $E(X|L') = X$ *for each* $X \in L'$.
4. *Let* $U \in L'$ *and* $Y \in L_{|L'}$ *be arbitrary. In other words, suppose the conditional expectation* $E(Y|L')$ *exists. Then* $UY \in L_{|L'}$ *and* $E(UY|L') = UE(Y|L')$.
5. *Let* $Y \in L$ *be arbitrary. Let* G *be an arbitrary set of r.v.'s with values in some complete metric space* (S, d). *Suppose there exists* $X \in L(G)$ *such that*

$$EYh(V_1, \ldots, V_k) = EXh(V_1, \ldots, V_k) \tag{5.6.6}$$

for each $h \in C_{ub}(S^k)$, *for each finite subset* $\{V_1, \ldots, V_k\} \subset G$, *and for each* $k \geq 1$; *then* $E(Y|G) = X$.

6. Let L' be a probability subspace with $L' \subset L''$. *Suppose* $Y \in L_{|L''}$ *with* $X \equiv E(Y|L'')$. *Then* $Y \in L_{|L'}$ *iff* $X \in L'$, *in which case* $E(Y|L') = E(Y|L'')$. *Moreover,*

$$E(E(Y|L'')|L') = E(Y|L').$$

7. *If* $Y \in L$, *and if* Y, Z *are independent for each indicator* $Z \in L'$, *then* $E(Y|L') = EY$.
8. Let Y be an r.r.v.. with $Y^2 \in L$. Suppose $X \equiv E(Y|L')$ exists. Then $X^2, (Y - X)^2 \in L$ and $EY^2 = EX^2 + E(Y - X)^2$. Consequently, $EX^2 \leq EY^2$ and $E(Y - X)^2 \leq EY^2$.

Proof. 1. Suppose $Y_1 = Y_2$ a.s. in L, and suppose $X_1, X_2 \in L'$ are such that $EZY_j = EZX_j$ for each $j = 1, 2$, for each indicator $Z \in L'$. Let $t > 0$ be arbitrary and let $Z \equiv 1_{(t < X_1 - X_2)} \in L'$. Then

$$tP(t < X_1 - X_2) \leq EZ(X_1 - X_2) = EZ(Y_1 - Y_2) = 0,$$

whence $P(t < X_1 - X_2) = 0$. It follows that that $(0 < X_1 - X_2)$ is a null set, and that $X_1 \geq X_2$ a.s. By symmetry, $X_1 = X_2$ a.s.

2. Suppose $X, Y \in L_{|L'}$. Let $Z \in L'$ be an arbitrary indicator. Then

$$\begin{aligned} E(Z(aX + bY)) &= aE(ZX) + bE(ZY) \\ &= aE(ZE(X|L')) + bE(ZE(Y|L')) \\ &= E(Z(aE(X|L') + bE(Y|L'))). \end{aligned}$$

Hence $aX + bY \in L_{|L'}$ and

$$E(aX + bY|L') = aE(X|L') + bE(Y|L').$$

Suppose, in addition, that $X \leq Y$ a.s. Let $U \equiv E(X|L')$ and $V \equiv E(Y|L')$. Then $U - V \in L'$. Consider each $t > 0$, with $Z \equiv 1_{(U-V>t)} \in L'$. Hence

$$\begin{aligned} tP(U - V > t) &= Et1_{(U-V>t)} \leq E1_{(U-V>t)}(U - V) \\ &= EZ(U - V) = EZ(X - Y) \leq 0. \end{aligned}$$

Consequently, $(U - V > t)$ is a null set and $(U - V \leq t)$ is a full set, for each $t > 0$. It follows that $D \equiv \bigcap_{n=1}^{\infty}(U - V \leq t_n)$ is a full set, where we let $(t_n)_{n=1,2,\ldots}$ be an arbitrary decreasing sequence in $(0, \infty)$ with $t_n \downarrow 0$. Moreover, $E(X|L') - E(Y|L') \equiv U - V \leq 0$ on the full set D. Assertion 2 is proved.

3. If $Y \in L_{|L'}$, then

$$E(E(Y|L')) = E(1E(Y|L')) = E(1Y) = E(Y).$$

Moreover, if $Y \in L'$, then for each indicator $Z \in L'$, we have trivially $E(ZY) = E(ZY)$, whence $E(Y|L') = Y$. Assertion 3 is proved.

4. Suppose $Y \in L_{|L'}$, with $X \equiv E(Y|L')$. Then, by definition,

$$EZY = EZX \tag{5.6.7}$$

for each indicator $Z \in L'$. The equality extends to all linear combinations of integrable indicators in L'. The reader can verify that such linear combinations are dense in L' relative to the L_1-norm, relative to the integration E. Hence equality 5.6.7 extends, by the Dominated Convergence Theorem, to all such bounded integrable r.r.v.'s $Z \in L'$. Moreover, if $U, Z \in L'$ are bounded and integrable r.r.v.'s, so is UZ, and equality 5.6.7 implies that $E(UZY) = E(UZX)$. It follows that $E(UY|L') = UX = UE(Y|L')$. This verifies Assertion 4.

5. Let $Y \in L$ be arbitrary, such that equality 5.6.6 holds for some $X \in L(G)$. Let Z be an arbitrary indicator in $L' \equiv L(G)$. Then Z is the L_1-limit, relative to E, of some sequence $(h_n(V_{n,1}, \ldots, V_{n,k(n)}))_{n=1,2,\ldots}$, where $h_n \in C_{ub}(S^{k(n)})$ with $0 \leq h_n \leq 1$, for each $n \geq 1$. It follows that

$$EYZ = \lim_{n \to \infty} EYh_n(V_{n,1}, \ldots, V_{n,k(n)}) = \lim_{n \to \infty} EXh_n(V_{n,1}, \ldots, V_{n,k(n)}) = EXZ,$$

where the second equality is due to equality 5.6.6. Thus $E(Y|G) \equiv E(Y|L') = X$, proving Assertion 5.

6. Let L' be a probability subspace with $L' \subset L''$. Suppose $Y \in L_{|L''}$. Let $X \equiv E(Y|L'')$.

Suppose, in addition, $Y \in L_{|L'}$. Let $W \equiv E(Y|L') \in L'$. Then $E(UY) = E(UW)$ for each $U \in L'$. Hence $E(UW) = E(UY) = E(UX)$ for each $U \in L' \subset L''$. Thus $X \in L_{|L'}$ with $E(X|L') = W \equiv E(Y|L')$. We have proved the "only if" part of Assertion 6. Conversely, suppose $X \in L_{|L'}$. Consider each $U \in L' \subset L''$. Then, because $X \equiv E(Y|L'')$, we have $EUY \equiv EUX$. Thus $Y \in L_{|L'}$. We have proved also the "if" part of Assertion 6.

7. Suppose $Y \in L$, and suppose Y, Z are independent for each indicator $Z \in L'$. Then, for each indicator $Z \in L'$, we have

$$E(ZY) = (EZ)(EY) = E(ZEY).$$

Since trivially $EY \in L'$, it follows that $E(Y|L') = EY$.

8. Let Y be an r.r.v. with $Y^2 \in L$. Suppose $X = E(Y|L')$ exists. Since $Y \subset Y^2 \subset L$, there exists a deceasing sequence $\varepsilon_1 > \varepsilon_2 > \cdots$ of positive real numbers such that $EY^2 1_A < 2^{-k}$ for each measurable set A with $P(A) < \varepsilon_k$, for each $k \geq 1$.

Since X is an r.r.v., there exists a sequence $0 \equiv a_0 < a_1 < a_2 < \cdots$ of positive real numbers with $a_k \to \infty$ such that $P(|X| \geq a_k) < \varepsilon_k$. Let $k \geq 1$ be arbitrary. Then

$$EY^2 1_{(|X| \geq a(k))} < 2^{-k}.$$

Write $Z_k \equiv 1_{(a(k+1)>|X|\geq a(k))}$. Then $Z_k, XZ_k, X^2Z_k \in L'$ are bounded in absolute value by $1, a_{k+1}, a_{k+1}^2$, respectively. Hence

$$\begin{aligned}
EY^2 1_{(a(k+1)>|X|\geq a(k))} &= E((Y-X)+X)^2 Z_k \\
&= E(Y-X)^2 Z_k + 2E(Y-X)XZ_k + EX^2 Z_k \\
&= E(Y-X)^2 Z_k + 2E(YXZ_k) - 2E(X^2 Z_k) + EX^2 Z_k \\
&= E(Y-X)^2 Z_k + 2E(E(Y|L')XZ_k) - EX^2 Z_k \\
&\equiv E(Y-X)^2 Z_k + 2E(XXZ_k) - EX^2 Z_k \\
&= E(Y-X)^2 Z_k + EX^2 Z_k, \qquad (5.6.8)
\end{aligned}$$

where the fourth equality is because $XZ_k \in L'$. Since $Y^2 \in L$ by assumption, equality 5.6.8 implies that

$$\sum_{k=0}^{\infty} EX^2 Z_k \leq \sum_{k=0}^{\infty} EY^2 1_{(a(k+1)>|X|\geq a(k))} = EY^2.$$

Consequently,

$$X^2 = \sum_{k=0}^{\infty} X^2 1_{(a(k+1)>|X|\geq a(k))} \equiv \sum_{k=0}^{\infty} X^2 Z_k \in L.$$

Similarly, $(Y - X) \in L^2$. Moreover, summing equality 5.6.8 over $k = 0, 1, \ldots,$ we obtain

$$EY^2 = E(Y-X)^2 + EX^2.$$

Assertion 8 and the proposition are proved. □

Proposition 5.6.7. The space of conditionally integrable functions given a probability subspace is closed relative to the L_1-norm. *Let (Ω, L, E) be a probability space. Let (Ω, L', E) be a probability subspace of (Ω, L, E). Let $L_{|L'}$ be the space of r.r.v.'s conditionally integrable given L'. Then the following conditions hold:*

1. Let $X, Y \in L$ be arbitrary. Suppose $EUX \leq EUY$ for each indicator $U \in L'$. Then $EZX \leq EZY$ for each bounded nonnegative r.r.v. $Z \in L'$.

2. Suppose $Y \in L_{|L'}$. Then $E|E(Y|L')| \leq E|Y|$.

3. The linear subspace $L_{|L'}$ of L is closed relative to the L_1-norm.

Proof. 1. Suppose $EUX \leq EUY$ for each indicator $U \in L'$. Then, by linearity, $EVX \leq EVY$ for each nonnegative linear combination Y of indicators in L'. Now consider each bounded nonnegative r.r.v. $Z \in L'$. We may assume, without loss of generality, that Z has values in $[0,1]$. Then $E|Z - V_n| \to 0$ for some sequence $(V_n)_{n=1,2,\ldots}$ of nonnegative linear combinations of indicators in L', with values in $[0,1]$. By passing to a subsequence, we may assume that $V_n \to Z$ a.s. Hence, by the Dominated Convergence Theorem, we have

$$EZX = \lim_{n\to\infty} EV_nX \leq \lim_{n\to\infty} EV_nY = EZY.$$

2. Suppose $Y \in L_{|L'}$, with $E(Y|L') = X \in L'$. Let $\varepsilon > 0$ be arbitrary. Then, since X is integrable, there exists $a > 0$ such that $EX1_{(|X|\leq a)} < \varepsilon$. Hence

$$\begin{aligned} E|X| &= EX1_{(X>a)} - EX1_{(X<-a)} + EX1_{(|X|\leq a)} \\ &< EX1_{(X>a)} - EX1_{(X<-a)} + \varepsilon \\ &= EY1_{(Y>a)} - EY1_{(Y<-a)} + \varepsilon \\ &\leq E|Y|1_{(X>a)} + E|Y|1_{(X<-a)} + \varepsilon \leq E|Y| + \varepsilon, \end{aligned}$$

where the second equality is because $Y = E(X|L')$. Since $\varepsilon > 0$ is arbitrarily small, we conclude that $E|X| \leq E|Y|$, as alleged.

3. Let $(Y_n)_{n=1,2,\ldots}$ be a sequence in $L_{|L'}$ such that $E|Y_n - Y| \to 0$ for some $Y \in L$. For each $n \geq 1$, let $X_n \equiv E(Y_n|L')$. Then, by Assertion 2, we have

$$E|X_n - X_m| = E|E(Y_n - Y_m|L')| \leq E|Y_n - Y_m| \to 0$$

as $n, m \to \infty$. Thus $(X_n)_{n=1,2,\ldots}$ is a Cauchy sequence in the complete metric space L' relative to the L_1-norm. It follows that $E|X_n - X| \to 0$ for some $X \in L'$, as $n \to \infty$. Hence, for each indicator $Z \in L'$, we have

$$EXZ = \lim_{n\to\infty} EX_nZ = \lim_{n\to\infty} EY_nZ = EYZ.$$

It follows that $E(Y|L') = X$ and $Y \in L_{|L'}$. □

5.7 Normal Distribution

The classical development of the topics in the remainder of this chapter is an exemplar of constructive mathematics. However, some tools in this development have been given many proofs – some constructive and others not. An example is the spectral theorem for symmetric matrices discussed in this section. For ease of reference, we therefore present some of these topics here, using only constructive proofs.

Recall some notations and basic theorems from matrix algebra.

Definition 5.7.1. Matrix notations. For an arbitrary $m \times n$ matrix

$$\theta \equiv [\theta_{i,j}]_{i=1,\ldots,m;\, j=1,\ldots,n} \equiv \begin{bmatrix} \theta_{1,1}, & \ldots, & \theta_{1,n} \\ \cdot & & \cdot \\ \cdot & \ldots & \cdot \\ \cdot & \ldots & \cdot \\ \theta_{m,1}, & \ldots, & \theta_{m,n} \end{bmatrix},$$

of real or complex elements $\theta_{i,j}$, we will let

$$\theta^T \equiv [\theta_{j,i}]_{j=1,\ldots,n;\,1=1,\ldots m} = \begin{bmatrix} \theta_{1,1}, & \ldots, & \theta_{m,1} \\ \cdot & \cdots & \cdot \\ \cdot & \cdots & \cdot \\ \cdot & \cdots & \cdot \\ \theta_{1,n} & ,\ldots, & \theta_{m,n} \end{bmatrix}$$

denote the transpose, which is an $n \times m$ matrix. If $n = m$ and $\theta = \theta^T$, then θ is said to be symmetric. If $\theta_{i,j} = 0$ for each $i, j = 1, \ldots, n$ with $i \neq j$, then θ is called a diagonal matrix. For each sequence of complex numbers $(\lambda_1, \ldots, \lambda_n)$, write $diag(\lambda_1, \ldots, \lambda_n)$ for the diagonal matrix θ with $\theta_{i,i} = \lambda_i$ for each $i = 1, \ldots, n$. A matrix θ is said to be real if all its elements $\theta_{i,j}$ are real numbers. Unless otherwise specified, all matrices in the following discussion are assumed to be real.

For an arbitrary sequence $\overline{\mu} \equiv (\mu_1, \ldots, \mu_n) \in R^n$, we will abuse notations and let $\overline{\mu}$ denote also the column vector

$$\overline{\mu} \equiv (\mu_1, \ldots, \mu_n) \equiv \begin{bmatrix} \mu_1 \\ \cdot \\ \cdot \\ \cdot \\ \mu_n \end{bmatrix}.$$

Thus $\overline{\mu}^T = [\mu_1, \ldots, \mu_n]$. A 1×1 matrix is identified with its only entry. Hence, if $\overline{\mu} \in R^n$, then

$$|\mu| \equiv \|\overline{\mu}\| \equiv \sqrt{\overline{\mu}^T \overline{\mu}} = \sqrt{\sum_{i=1}^{n} \mu_i^2}.$$

We will let I_n denote the $n \times n$ diagonal matrix $diag(1, \ldots, 1)$. When the dimension n is understood, we write simply $I \equiv I_n$. Likewise, we will write 0 for any matrix whose entries are all equal to the real number 0, with dimensions understood from the context.

The determinant of an $n \times n$ matrix θ is denoted by $\det \theta$. The n complex roots $\lambda_1, \ldots, \lambda_n$ of the polynomial $\det(\theta - \lambda I)$ of degree n are called the *eigenvalues* of θ. Then $\det \theta = \lambda_1 \ldots \lambda_n$. Let $j = 1, \ldots, n$ be arbitrary. Then there exists a nonzero column vector x_j, whose elements are in general complex, such that $\theta x_j = \lambda_j x_j$. The vector x_j is called an *eigenvector* for the eigenvalue λ_j. If θ is real and symmetric, then the n eigenvalues $\lambda_1, \ldots, \lambda_n$ are real.

Let $\overline{\sigma}$ be a symmetric $n \times n$ matrix whose elements are real. Then $\overline{\sigma}$ is said to be *nonnegative definite* if $x^T \overline{\sigma} x \geq 0$ for each $x \in R^n$. In that case, all its eigenvalues are nonnegative and, for each eigenvalue, there exists a real eigenvector whose elements are real. It is said to be *positive definite* if $x^T \overline{\sigma} x > 0$ for each nonzero $x \in R^n$. In that case, all its eigenvalues are positive, whence $\overline{\sigma}$ is nonsingular, with an inverse $\overline{\sigma}^{-1}$. An $n \times n$ real matrix U is said to be orthogonal if $U^T U = I$. This is equivalent to saying that the column vectors of U form an orthonormal basis of R^n. □

Theorem 5.7.2. Spectral theorem for symmetric matrices. *Let θ be an arbitrary $n \times n$ symmetric matrix. Then the following conditions hold:*

1. There exists an orthogonal matrix U such that $U^T \theta U = \Lambda$, where

$$\Lambda \equiv diag(\lambda_1, \ldots, \lambda_n)$$

and $\lambda_1, \ldots, \lambda_n$ are eigenvalues of θ.

2. Suppose, in addition, that $\lambda_1, \ldots, \lambda_n$ are nonnegative. Define the symmetric matrix $A \equiv U \Lambda^{\frac{1}{2}} U^T$, where $\Lambda^{\frac{1}{2}} = diag(\lambda_1^{\frac{1}{2}}, \ldots, \lambda_n^{\frac{1}{2}})$. Then $\theta = AA^T$.

Proof. 1. Proceed by induction on n. The assertion is trivial if $n = 1$. Suppose the assertion has been proved for $n - 1$. Recall that for an arbitrary unit vector v_n, there exist $v_1, \ldots, v_{n-1} \in R^n$ such that $v_1, \ldots, v_{n-1}, v_n$ form an orthonormal basis of R^n. Now let v_n be an eigenvector of θ corresponding to λ_n. Let V be the $n \times n$ matrix whose ith column is v_i for each $i = 1, \ldots, n$. Then V is an orthogonal matrix. Define an $(n-1) \times (n-1)$ symmetric matrix η by $\eta_{i,j} \equiv v_i^T \theta v_j$ for each $i, j = 1, \ldots, n-1$. By the induction hypothesis, there exists an $(n-1) \times (n-1)$ orthogonal matrix

$$W \equiv \begin{bmatrix} w_{1,1}, & \ldots, & w_{1,n-1} \\ \cdot & \ldots & \cdot \\ \cdot & \ldots & \cdot \\ \cdot & \ldots & \cdot \\ w_{n-1,1}, & \ldots, & w_{n-1,n-1} \end{bmatrix}$$

such that

$$W^T \eta W = \Lambda_{n-1} = diag(\lambda_1, \ldots, \lambda_{n-1}) \tag{5.7.1}$$

for some $\lambda_1, \ldots, \lambda_{n-1} \in R$. Define the $n \times n$ matrices

$$W' \equiv \begin{bmatrix} w_{1,1}, & \ldots, & w_{1,n-1}, & 0 \\ \cdot & \ldots & \cdot & \cdot \\ \cdot & \ldots & \cdot & \cdot \\ \cdot & \ldots & \cdot & \cdot \\ w_{n-1,1}, & \ldots, & w_{n-1,n-1}, & 0 \\ 0, & \ldots, & 0, & 1 \end{bmatrix}$$

and $U \equiv VW'$. Then it is easily verified that U is orthogonal. Moreover,

$$U^T\theta U = W'^T V^T \theta V W'$$

$$= W'^T \begin{bmatrix} v_1^T\theta v_1, & \dots, & v_1^T\theta v_{n-1}, & v_1^T\theta v_n \\ \cdot & \dots & \cdot & \cdot \\ \cdot & \dots & \cdot & \cdot \\ \cdot & \dots & \cdot & \cdot \\ v_{n-1}^T\theta v_1, & \dots, & v_{n-1}^T\theta v_{n-1}, & v_{n-1}^T\theta v_n \\ v_n^T\theta v_1, & \dots, & v_n^T\theta v_{n-1}, & v_n^T\theta v_n \end{bmatrix} W'$$

$$= \begin{bmatrix} w_{1,1}, & \dots, & w_{1,n-1}, & 0 \\ \cdot & \dots & \cdot & \cdot \\ \cdot & \dots & \cdot & \cdot \\ \cdot & \dots & \cdot & \cdot \\ w_{n-1,1}, & \dots, & w_{n-1,n-1}, & 0 \\ 0, & \dots, & 0, & 1 \end{bmatrix}^T \begin{bmatrix} \eta_{1,1}, & \dots, & \eta_{1,n-1}, & 0 \\ \cdot & \dots & \cdot & \cdot \\ \cdot & \dots & \cdot & \cdot \\ \cdot & \dots & \cdot & \cdot \\ \eta_{n-1,1}, & \dots, & \eta_{n-1,n-1}, & 0 \\ 0, & \dots, & 0, & \lambda_n \end{bmatrix}$$

$$\begin{bmatrix} w_{1,1}, & \dots, & w_{1,n-1}, & 0 \\ \cdot & \dots & \cdot & \cdot \\ \cdot & \dots & \cdot & \cdot \\ \cdot & \dots & \cdot & \cdot \\ w_{n-1,1}, & \dots, & w_{n-1,n-1}, & 0 \\ 0, & \dots, & 0, & 1 \end{bmatrix}$$

$$= \begin{bmatrix} \lambda_1, & \dots, & 0, & 0 \\ \cdot & \dots & \cdot & \cdot \\ \cdot & \dots & \cdot & \cdot \\ \cdot & \dots & \cdot & \cdot \\ 0 & \dots, & \lambda_{n-1}, & 0 \\ 0 & \dots, & 0, & \lambda_n \end{bmatrix} \equiv \Lambda \equiv diag(\lambda_1, \dots, \lambda_n),$$

where the fourth equality is thanks to equality 5.7.1. Induction is completed. The equality $U^T\theta U = \Lambda$ implies that $\theta U = U\Lambda$ and that λ_i is an eigenvalue of θ with an eigenvector given by the ith column of U. Assertion 1 is thus proved.

Since

$$\theta = U\Lambda U^T = U\Lambda^{\frac{1}{2}}\Lambda^{\frac{1}{2}}U^T = U\Lambda^{\frac{1}{2}}U^T U\Lambda^{\frac{1}{2}}U^T = AA^T,$$

Assertion 2 is proved. □

Definition 5.7.3. Normal distribution with positive definite covariance. Let $n \geq 1$ and $\overline{\mu} \in R^n$ be arbitrary. Let $\overline{\sigma}$ be an arbitrary positive definite $n \times n$ matrix. Then the function defined on R^n by

$$\varphi_{\overline{\mu},\overline{\sigma}}(y) \equiv (2\pi)^{-\frac{n}{2}}(\det\overline{\sigma})^{-\frac{1}{2}} \exp\left(-\frac{1}{2}(y-\overline{\mu})^T\overline{\sigma}^{-1}(y-\overline{\mu})\right) \tag{5.7.2}$$

for each $y \in R^n$ is a p.d.f. Let $\Phi_{\overline{\mu},\overline{\sigma}}$ be the corresponding distribution on R^n, and let $Y \equiv (Y_1, \ldots, Y_n)$ be any r.v. with values in R^n and with $\Phi_{\overline{\mu},\overline{\sigma}}$ as its distribution. Then $\varphi_{\overline{\mu},\overline{\sigma}}$, $\Phi_{\overline{\mu},\overline{\sigma}}$, Y, and the sequence $Y_1, \ldots, Y_n$ are said to be the *normal p.d.f., the normal distribution, normally distributed, and jointly normal*, respectively, with mean $\overline{\mu}$ and covariance matrix $\overline{\sigma}$. Proposition 5.7.6 later in this section justifies the terminology. The p.d.f. $\varphi_{0,I}$ and the distribution $\Phi_{0,I}$ are said to be *standard normal*, where I is the identity matrix.

In the case where $n = 1$, define $\sigma \equiv \sqrt{\overline{\sigma}}$ and write Φ_{μ,σ^2} also for the P.D.F. associated with the distribution Φ_{μ,σ^2}, and call it a *normal P.D.F.* Thus $\Phi_{0,1}(x) = \int_{-\infty}^{x} \varphi_{0,1}(u)du$ for each $x \in R$.

In Definition 5.7.7, we will generalize the definition of normal distributions to an arbitrary nonnegative definite matrix $\overline{\sigma}$. □

Proposition 5.7.4. Basics of standard normal distribution. *Consider the case $n = 1$. Then the following conditions hold:*

1. The function $\varphi_{0,1}$ on R defined by

$$\varphi_{0,1}(x) \equiv \frac{1}{\sqrt{2\pi}} \exp\left(-\frac{1}{2}x^2\right)$$

is a p.d.f. on R relative to the Lebesgue measure. Thus $\Phi_{0,1}$ is a P.D.F. on R.

2. Write $\Phi \equiv \Phi_{0,1}$. We will call the function $\Psi \equiv 1 - \Phi : [0,\infty) \to \left(0, \frac{1}{2}\right]$ the tail of Φ. Then $\Phi(-x) = 1 - \Phi(x)$ for each $x \in R$. Moreover,

$$\Psi(x) \leq e^{-x^2/2}$$

for each $x \geq 0$.

3. The inverse function $\overline{\Psi} : \left(0, \frac{1}{2}\right] \to [0,\infty)$ of Ψ is a decreasing function from $(0,1)$ to R such that $\overline{\Psi}(\varepsilon) \to \infty$ as $\varepsilon \to 0$. Moreover, $\overline{\Psi}(\varepsilon) \leq \sqrt{-2\log\varepsilon}$ for each $\varepsilon \in \left(0, \frac{1}{2}\right]$.

Proof. 1. We calculate

$$\left(\frac{1}{\sqrt{2\pi}} \int_{-\infty}^{+\infty} e^{-x^2/2}dx\right)^2 = \frac{1}{2\pi} \int_{-\infty}^{+\infty} \int_{-\infty}^{+\infty} e^{-(x^2+y^2)/2}dxdy = 1, \quad (5.7.3)$$

where the last equality is by Corollary A.0.13, which also says that the function $(x,y) \to e^{-(x^2+y^2)/2}$ on R^2 is Lebesgue integrable. Thus $\varphi_{0,1}$ is Lebesgue integrable, with an integral equal to 1, and hence is a p.d.f. on R.

2. First,

$$\begin{aligned}\Phi(-x) &\equiv \int_{-\infty}^{-x} \varphi(u)du \\ &= \int_{x}^{\infty} \varphi(-v)dv = \int_{x}^{\infty} \varphi(v)dv = 1 - \Phi(x),\end{aligned}$$

where we made a change of integration variables $v = -u$ and noted that $\varphi(-v) = \varphi(v)$.

Next, if $x \in \left[\frac{1}{\sqrt{2\pi}}, \infty\right)$, then

$$\Psi(x) \equiv \frac{1}{\sqrt{2\pi}} \int_x^\infty e^{-u^2/2} du$$
$$\leq \frac{1}{\sqrt{2\pi}} \int_x^\infty \frac{u}{x} e^{-u^2/2} du = \frac{1}{\sqrt{2\pi}} \frac{1}{x} e^{-x^2/2} \leq e^{-x^2/2}.$$

On the other hand, if $x \in \left[0, \frac{1}{\sqrt{2\pi}}\right)$, then

$$\Psi(x) \leq \Psi(0) = \frac{1}{2} < \exp\left(-\left(\frac{1}{\sqrt{2\pi}}\right)^2 / 2\right) \leq e^{-x^2/2}.$$

Therefore, by continuity, $\Psi(x) \leq e^{-x^2/2}$ for each $x \in [0, \infty)$.

3. Consider any $\varepsilon \in (0,1)$. Define $x \equiv \sqrt{-2 \log \varepsilon}$. Then $\Psi(x) \leq e^{-x^2/2} = \varepsilon$ by Assertion 2. Since $\overline{\Psi}$ is a decreasing function, it follows that

$$\sqrt{-2 \log \varepsilon} \equiv x = \overline{\Psi}(\Psi(x)) \geq \overline{\Psi}(\varepsilon).$$ □

Proposition 5.7.5. Moments of standard normal r.r.v. *Suppose an r.r.v. X has the standard normal distribution* $\Phi_{0,1}$, *with p.d.f.* $\varphi_{0,1}(x) \equiv \frac{1}{\sqrt{2\pi}} e^{-x^2/2}$. *Then* X^m *is integrable for each* $m \geq 0$. *Moreover, for each even integer* $m \equiv 2k \geq 0$, *we have*

$$EX^m = EX^{2k} = (2k-1)(2k-3)\ldots 3 \cdot 1 = (2k)!\,2^{-k}/k!,$$

while $EX^m = 0$ *for each odd integer* $m > 0$.

Proof. Let $m \geq 0$ be any even integer. Let $a > 0$ b arbitrary. Then, integrating by parts, we have

$$\frac{1}{\sqrt{2\pi}} \int_{-a}^{a} x^{m+2} e^{-x^2/2} dx = \frac{1}{\sqrt{2\pi}} (-x^{m+1} e^{-x^2/2})|_{-a}^{a} + (m+1) \frac{1}{\sqrt{2\pi}} \int_{-a}^{a} x^m e^{-x^2/2} dx. \quad (5.7.4)$$

Since the function g_m defined by $g_m(x) \equiv 1_{[-a,a]}(x) x^m$ for each $x \in R$ is Lebesgue measurable and is bounded by the integrable function x^m relative to the P.D.F. $\Phi_{0,1}$, the function g_m is integrable relative to the P.D.F. $\Phi_{0,1}$, which has $\varphi_{0,1}$ as its p.d.f. Moreover, equality 5.7.4 can be rewritten as

$$\int g_{m+2}(x) d\Phi_{0,1}(x) = \frac{1}{\sqrt{2\pi}} (-x^{m+1} e^{-x^2/2})|_{-a}^{a} + (m+1) \int g_m(x) d\Phi_{0,1}(x)$$

or, in view of Proposition 5.4.4, as

$$Eg_{m+2}(X) = \frac{1}{\sqrt{2\pi}} (-x^{m+1} e^{-x^2/2})|_{-a}^{a} + (m+1) Eg_m(X). \quad (5.7.5)$$

The lemma is trivial for $m = 0$. Suppose the lemma has been proved for integers up to and including the even integer $m \equiv 2k - 2$. By the induction hypothesis,

X^m is integrable. At the same time, $g_m(X) \to X^m$ in probability as $a \to \infty$. Hence, by the Dominated Convergence Theorem, we have $Eg_m(X) \uparrow EX^m$ as $a \to \infty$. Since $|a|^{m+1}e^{-a^2/2} \to 0$ as $a \to \infty$, equality 5.7.5 yields $Eg_{m+2}(X) \uparrow (m+1)EX^m$ as $a \to \infty$. The Monotone Convergence Theorem therefore implies that X^{m+2} is integrable, with $EX^{m+2} = (m+1)EX^m$, or

$$EX^{2k} = (2k-1)EX^{2k-2} = \cdots = (2k-1)(2k-3)\ldots 1 = (2k)!\,2^{-k}/k!$$

Since X^{m+2} is integrable, so is X^{m+1}, according to Lyapunov's inequality. Moreover,

$$EX^{m+1} = \int x^{m+1} d\Phi_{0,1}(x) = \int x^{m+1}\varphi_{0,1}(x)dx = 0$$

since $x^{m+1}\varphi_{0,1}(x)$ is an odd function of $x \in R$. Induction is completed. □

The next proposition shows that $\varphi_{\overline{\mu},\overline{\sigma}}$ and $\Phi_{\overline{\mu},\overline{\sigma}}$ in Definition 5.7.3 are well defined.

Proposition 5.7.6. Basics of normal distributions with positive definite covariance. *Let $n \geq 1$ and $\overline{\mu} \in R^n$ be arbitrary. Let $\overline{\sigma}$ be an arbitrary positive definite $n \times n$ matrix. Use the notations in Definition 5.7.3. Then the following conditions hold:*

1. $\varphi_{\overline{\mu},\overline{\sigma}}$ is indeed a p.d.f. on R^n, i.e., $\int \varphi_{\overline{\mu},\overline{\sigma}}(x)dx = 1$, where $\int \cdot dx$ stands for the Lebesgue integration on R^n. Thus the corresponding distribution $\Phi_{\overline{\mu},\overline{\sigma}}$ on R^n is well defined. Moreover, $\Phi_{\overline{\mu},\overline{\sigma}}$ is equal to the distribution of the r.v. $Y \equiv \overline{\mu}+AX$, where A is an arbitrary $n \times n$ matrix with $\overline{\sigma} = AA^T$, and where X is an arbitrary r.v. with values in R^n and with the standard normal distribution $\Phi_{0,I}$. In short, linear combinations of a finite set of standard normal r.r.v.'s and the constant 1 are jointly normal. More generally, linear combinations of a finite set of jointly normal r.r.v.'s are jointly normal.

2. Let $Z \equiv (Z_1, \ldots, Z_n)$ be an r.v. with values in R^n with distribution $\Phi_{\overline{\mu},\overline{\sigma}}$. Then $EZ = \overline{\mu}$ and $E(Z-\overline{\mu})(Z-\overline{\mu})^T = \overline{\sigma}$.

3. Let $Z_1, \ldots, Z_n$ be jointly normal r.r.v.'s. Then $Z_1, \ldots, Z_n$ are independent iff they are pairwise uncorrelated. In particular, if $Z_1, \ldots, Z_n$ are jointly standard normal, then they are independent.

Proof. For each $x \equiv (x_1, \ldots, x_n) \in R^n$, we have, by Definition 5.7.3,

$$\varphi_{0,I}(x_1, \ldots, x_n) \equiv \varphi_{0,I}(x) \equiv (2\pi)^{-\frac{n}{2}} \exp\left(-\frac{1}{2}x^T x\right)$$

$$= \prod_{i=1}^{n}\left(\frac{1}{\sqrt{2\pi}}\exp\left(-\frac{1}{2}x_i^2\right)\right) = \varphi_{0,1}(x_1)\ldots\varphi_{0,1}(x_n).$$

Since $\varphi_{0,1}$ is a p.d.f. on R according to Proposition 5.7.4, Proposition 4.10.6 implies that the Cartesian product $\varphi_{0,I}$ is a p.d.f. on R^n. Let $X \equiv (X_1, \ldots, X_n)$ be an arbitrary r.v. with values in R^n and with p.d.f. $\varphi_{0,I}$. Then

$$Ef_1(X_1)\dots f_n(X_n) = \int\cdots\int f_1(x_1)\dots f_n(x_n)\varphi_{0,1}(x_1)\dots\varphi_{0,1}(x_n)dx_1\dots dx_n$$
$$= \prod_{i=1}^{n}\int f_i(x_i)\varphi_{0,1}(x_i)dx_i = Ef_1(X_1)\dots Ef_n(X_n) \tag{5.7.6}$$

for each $f_1,\dots,f_n \in C(R)$. Separately, for each $i = 1,\dots,n$, the r.r.v. X_i has distribution $\varphi_{0,1}$, whence X_i has an mth moment for each $m \geq 0$, with $EX_i^m = 0$ if m is odd, according to Proposition 5.7.5.

1. Next let $\overline{\sigma},\overline{\mu}$ be as given. Let A be an arbitrary $n\times n$ matrix such that $\overline{\sigma} = AA^T$. By Theorem 5.7.2, such a matrix A exists. Then $\det(\overline{\sigma}) = \det(A)^2$. Since $\overline{\sigma}$ is positive definite, it is nonsingular and so is A. Let X be an arbitrary r.v. with values in R^n and with the standard normal distribution $\Phi_{0,I}$. Define the r.v. $Y \equiv \overline{\mu} + AX$. Then, for arbitrary $f \in C(R^n)$, we have

$$Ef(Y) = Ef(\overline{\mu}+AX) = \int f(\overline{\mu}+Ax)\varphi_{0,I}(x)dx$$
$$\equiv (2\pi)^{-\frac{n}{2}}\int f(\overline{\mu}+Ax)\exp\left(-\frac{1}{2}x^Tx\right)dx$$
$$= (2\pi)^{-\frac{n}{2}}\det(A)^{-1}\int f(y)\exp\left(-\frac{1}{2}(y-\overline{\mu})^T(A^{-1})^TA^{-1}(y-\overline{\mu})\right)dy$$
$$= (2\pi)^{-\frac{n}{2}}\det(\overline{\sigma})^{-\frac{1}{2}}\int f(y)\exp\left(-\frac{1}{2}(y-\overline{\mu})^T\overline{\sigma}^{-1}(y-\overline{\mu})\right)dy$$
$$\equiv \int f(y)\varphi_{\overline{\mu},\overline{\sigma}}(y)dy,$$

where the fourth equality is by the change of integration variables $y = \overline{\mu} + Ax$. Thus $\varphi_{\overline{\mu},\overline{\sigma}}$ is the p.d.f. on R^n of the r.v. Y, and $\Phi_{\overline{\mu},\overline{\sigma}}$ is the distribution of Y.

2. Next, let $Z_1,\dots,Z_n$ be jointly normal r.r.v.'s with distribution $\Phi_{\overline{\mu},\overline{\sigma}}$. By Assertion 1, there exist a standard normal r.v. $X \equiv (X_1,\dots,X_n)$ on some probability space (Ω',L',E'), and an $n\times n$ matrix $AA^T = \overline{\sigma}$, such that $E'f(\overline{\mu}+AX) = \Phi_{\overline{\mu},\overline{\sigma}}(f) = Ef(Z)$ for each $f \in C(R^n)$. Thus Z and $Y \equiv \overline{\mu}+AX$ induce the same distribution on R^n. Let $i,j = 1,\dots,n$ be arbitrary. Since X_i,X_j,X_iX_j and, therefore, Y_i,Y_j,Y_iY_j are integrable, so are Z_i,Z_j,Z_iZ_j, with

$$EZ = E'Y = \overline{\mu} + AE'X = \overline{\mu}$$

and

$$E(Z-\overline{\mu})(Z-\overline{\mu})^T = E'(Y-\overline{\mu})(Y-\overline{\mu})^T = AE'XX^TA^T = AA^T = \overline{\sigma}.$$

3. Suppose $Z_1,\dots,Z_n$ are pairwise uncorrelated. Then $\overline{\sigma}_{i,j} = E(Z_i-\overline{\mu}_i)(Z_j-\overline{\mu}_j) = 0$ for each $i,j = 1,\dots,n$ with $i\neq j$. Thus $\overline{\sigma}$ and $\overline{\sigma}^{-1}$ are diagonal matrices, with $(\overline{\sigma}^{-1})_{i,j} = \overline{\sigma}_{i,i}$ or 0 depending on whether $i = j$ or not. Hence, for each $f_1,\dots,f_n \in C(R)$, we have

$$Ef_1(Z_1)\ldots f_n(Z_n)$$
$$= (2\pi)^{-\frac{n}{2}}(\det\overline{\sigma})^{-\frac{1}{2}}$$
$$\times \int \cdots \int f(z_1)\ldots f(z_n)\exp\left(-\frac{1}{2}(z-\overline{\mu})^T\overline{\sigma}^{-1}(z-\overline{\mu})\right)dz_1\ldots dz_n$$
$$= (2\pi)^{-\frac{n}{2}}(\overline{\sigma}_{1,1}\ldots\overline{\sigma}_{n,n})^{-\frac{1}{2}}$$
$$\times \int \cdots\cdots \int f(z_1)\ldots f(z_n)\exp\sum_{i=1}^{n}\left(-\frac{1}{2}(z_i-\overline{\mu}_i)\overline{\sigma}_{i,i}^{-1}(z_i-\overline{\mu}_i)\right)dz_1\ldots dz_n$$
$$= (2\pi\overline{\sigma}_{i,i})^{-\frac{1}{2}}\int f(z_i)\exp\left(-\frac{1}{2}(z_i-\overline{\mu}_i)\overline{\sigma}_{i,i}^{-1}(z_i-\overline{\mu}_i)\right)dz_i$$
$$= Ef_1(Z_1)\ldots Ef_n(Z_n).$$

We conclude that $Z_1,\ldots,Z_n$ are independent if they are pairwise uncorrelated. The converse is trivial. □

Next we generalize the definition of a normal distribution to include the case where the covariance matrix is nonnegative definite.

Definition 5.7.7. Normal distribution with nonnegative definite covariance. Let $n \geq 1$ and $\overline{\mu} \in R^n$ be arbitrary. Let $\overline{\sigma}$ be an arbitrary nonnegative definite $n \times n$. Define the *normal distribution* $\Phi_{\overline{\mu},\overline{\sigma}}$on R^n by

$$\Phi_{\overline{\mu},\overline{\sigma}}(f) \equiv \lim_{\varepsilon\to 0}\Phi_{\overline{\mu},\overline{\sigma}+\varepsilon I}(f) \tag{5.7.7}$$

for each $f \in C(R^n)$, where for each $\varepsilon > 0$, the function $\Phi_{\overline{\mu},\overline{\sigma}+\varepsilon I}$ is the normal distribution on R^n introduced in Definition 5.7.3 for the positive definite matrix $\overline{\sigma} + \varepsilon I$. Lemma 5.7.8 (next) proves that $\Phi_{\overline{\mu},\overline{\sigma}}$ is well defined and is indeed a distribution.

A sequence $Z_1,\ldots,Z_n$ of r.r.v.'s is said to be *jointly normal,* with $\Phi_{\overline{\mu},\overline{\sigma}}$ as its distribution, if $Z \equiv (Z_1,\ldots,Z_n)$ has the distribution $\Phi_{\overline{\mu},\overline{\sigma}}$ on R^n.

Lemma 5.7.8. Normal distribution with nonnegative definite covariance is well defined. *Use the notations and assumptions in Definition 5.7.7. Then the following conditions hold:*

1. The limit $\lim_{\varepsilon\to 0}\Phi_{\overline{\mu},\overline{\sigma}+\varepsilon I}(f)$ *in equality 5.7.7 exists for each* $f \in C(R^n)$*. Moreover,* $\Phi_{\overline{\mu},\overline{\sigma}}$ *is the distribution of* $Y \equiv \overline{\mu} + AX$ *for some standard normal* $X \equiv (X_1,\ldots,X_n)$ *and some* $n \times n$ *matrix* A *with* $AA^T = \overline{\sigma}$.

2. If $\overline{\sigma}$ *is positive definite, then* $\Phi_{\overline{\mu},\overline{\sigma}}(f) = \int f(y)\varphi_{\overline{\mu},\overline{\sigma}}(y)dy$*, where* $\varphi_{\overline{\mu},\overline{\sigma}}$ *was defined in Definition 5.7.3. Thus Definition 5.7.7 of* $\Phi_{\overline{\mu},\overline{\sigma}}$ *for a nonnegative definite* $\overline{\sigma}$ *is consistent with Definition 5.7.3 for a positive definite* $\overline{\sigma}$.

3. Let $Z \equiv (Z_1,\ldots,Z_n)$ *be an arbitrary r.v. with values in* R^n *and with distribution* $\Phi_{\overline{\mu},\overline{\sigma}}$*. Then* $Z_1^{k(1)}\ldots Z_n^{k(n)}$ *is integrable for each* $k_1,\ldots,k_n \geq 0$*. In particular,* Z *has mean* $\overline{\mu}$ *and covariance matrix* $\overline{\sigma}$.

Proof. 1. Let $\varepsilon > 0$ be arbitrary. Then $\overline{\sigma} + \varepsilon I$ is positive definite. Hence, the normal distribution $\Phi_{\overline{\mu},\overline{\sigma}+\varepsilon I}$ has been defined. Separately, Theorem 5.7.2 implies that there exists an orthogonal matrix U such that $U^T\overline{\sigma}U = \Lambda$, where $\Lambda \equiv diag(\lambda_1, \ldots, \lambda_n)$ is a diagonal matrix whose diagonal elements consist of the eigenvalues $\lambda_1, \ldots, \lambda_n$ of $\overline{\sigma}$. These eigenvalues are nonnegative since $\overline{\sigma}$ is nonnegative definite. Hence, again by Theorem 5.7.2, we have

$$\overline{\sigma} + \varepsilon I = A_\varepsilon A_\varepsilon^T, \tag{5.7.8}$$

where

$$A_\varepsilon \equiv U\Lambda_\varepsilon^{\frac{1}{2}}U^T, \tag{5.7.9}$$

where $\Lambda_\varepsilon^{\frac{1}{2}} = diag(\sqrt{\lambda_1 + \varepsilon}, \ldots, \sqrt{\lambda_n + \varepsilon})$.

Now let X be an arbitrary r.v. on R^n with the standard normal distribution $\Phi_{0,I}$. In view of equality 5.7.8, Proposition 5.7.6 implies that $\Phi_{\overline{\mu},\overline{\sigma}+\varepsilon I}$ is equal to the distribution of the r.v.

$$Y^{(\varepsilon)} \equiv \overline{\mu} + A_\varepsilon X.$$

Define $A \equiv U\Lambda^{\frac{1}{2}}U^T$, where $\Lambda^{\frac{1}{2}} = diag(\sqrt{\lambda_1}, \ldots, \sqrt{\lambda_n})$ and define $Y \equiv \overline{\mu} + AX$. Then

$$\begin{aligned}
E|A_\varepsilon X - AX|^2 &= EX^T(A_\varepsilon - A)^T(A_\varepsilon - A)X \\
&= \sum_{i=1}^{n}\sum_{j=1}^{n}\sum_{k=1}^{n} EX_iU_{i,j}(\sqrt{\lambda_j + \varepsilon} - \sqrt{\lambda_j})^2U_{j,k}X_k \\
&= \sum_{i=1}^{n}\sum_{j=1}^{n} U_{i,j}(\sqrt{\lambda_j + \varepsilon} - \sqrt{\lambda_j})^2U_{j,i} \\
&= \sum_{j=1}^{n}(\sqrt{\lambda_j + \varepsilon} - \sqrt{\lambda_j})^2\sum_{i=1}^{n} U_{i,j}U_{j,i} \\
&= \sum_{j=1}^{n}(\sqrt{\lambda_j + \varepsilon} - \sqrt{\lambda_j})^2 \to 0
\end{aligned}$$

as $\varepsilon \to 0$. Lyapunov's inequality then implies that

$$E|Y^{(\varepsilon)} - Y| = E|A_\varepsilon X - AX| \le (E|A_\varepsilon X - AX|^2)^{\frac{1}{2}} \to 0$$

as $\varepsilon \to 0$. In other words, $Y^{(\varepsilon)} \to Y$ in probability. Consequently, the distribution $\Phi_{\overline{\mu},\overline{\sigma}+\varepsilon I}$ converges to the distribution F_Y of Y. We conclude that the limit $\Phi_{\overline{\mu},\overline{\sigma}}(f)$ in equality 5.7.7 exists and is equal to $EF(Y)$. In other words, $\Phi_{\overline{\mu},\overline{\sigma}}$ is the distribution of the r.v. $Y \equiv \overline{\mu} + AX$. Moreover,

$$AA^T = U\Lambda^{\frac{1}{2}}U^TU\Lambda^{\frac{1}{2}}U^T = U\Lambda U^T = \overline{\sigma}.$$

Assertion 1 is proved.

2. Next suppose $\overline{\sigma}$ is positive definite. Then $\varphi_{\overline{\mu},\overline{\sigma}+\varepsilon I} \to \varphi_{\overline{\mu},\overline{\sigma}}$ uniformly on compact subsets of R^n. Hence

$$\lim_{\varepsilon\to 0} \Phi_{\overline{\mu},\overline{\sigma}+\varepsilon I}(f) = \lim_{\varepsilon\to 0} \int f(y)\varphi_{\overline{\mu},\overline{\sigma}+\varepsilon I}(y)dy = \int f(y)\varphi_{\overline{\mu},\overline{\sigma}}(y)dy$$

for each $f \in C(R^n)$. Therefore Definition 5.7.7 is consistent with Definition 5.7.3, proving Assertion 2.

3. Now let $Z \equiv (Z_1, \ldots, Z_n)$ be any r.v. with values in R^n and with distribution $\Phi_{\overline{\mu},\overline{\sigma}}$. By Assertion 1, $\Phi_{\overline{\mu},\overline{\sigma}}$ is the distribution of $Y \equiv \overline{\mu} + AX$ for some standard normal $X \equiv (X_1, \ldots, X_n)$ and some $n \times n$ matrix A with $AA^T = \overline{\sigma}$. Thus Z and Y has the same distribution. Let $k_1, \ldots, k_n \geq 0$ be arbitrary. Then the r.r.v. $Y_1^{k(1)} \ldots Y_n^{k(n)}$ is a linear combination of products $X_1^{j(1)} \ldots X_n^{j(n)}$ integrable where $j_1, \ldots, j_n \geq 0$, each of which is integrable in view of Proposition 5.7.5 and Proposition 4.10.6. Hence $Y_1^{k(1)} \ldots Y_n^{k(n)}$ is integrable. It follows that $Z_1^{k(1)} \ldots Z_n^{k(n)}$ is integrable. $EZ = EY = \overline{\mu}$ and

$$E(Z - \overline{\mu})(Z - \overline{\mu})^T = E(Y - \overline{\mu})(Y - \overline{\mu})^T = EAXX^TA^T = AA^T = \overline{\sigma}.$$

In other words, Z has mean $\overline{\mu}$ and covariance matrix $\overline{\sigma}$, proving Assertion 3. $\square$

We will need some bounds related to the normal p.d.f. in later sections.

Recall from Proposition 5.7.4 the standard normal P.D.F. Φ on R, its tail Ψ, and the inverse function $\overline{\Psi}$ of the latter.

Lemma 5.7.9. Some bounds for normal probabilities.

1. *Suppose h is a measurable function on R relative to the Lebesgue integration. If $|h| \leq a$ on $[-\alpha,\alpha]$ and $|h| \leq b$ on $[-\alpha,\alpha]^c$ for some $a,b,\alpha > 0$, then*

$$\int |h(x)|\varphi_{0,\sigma}(x)dx \leq a + 2b\Psi\left(\frac{\alpha}{\sigma}\right)$$

for each $\sigma > 0$.

2. *In general, let $n \geq 1$ be arbitrary. Let I denote the $n \times n$ identity matrix. Suppose f is a Lebesgue integrable function on R^n, with $|f| \leq 1$. Let $\sigma > 0$ be arbitrary. Define a function f_σ on R^n by*

$$f_\sigma(x) \equiv \int_{y\in R^n} f(x-y)\varphi_{0,\sigma I}(y)dy$$

for each $x \in R^n$. Suppose f is continuous at some $t \in R^n$. In other words, suppose, for arbitrary $\varepsilon > 0$, there exists $\delta_f(\varepsilon,t) > 0$ such that $|f(t)-f(r)| < \varepsilon$ for each $r \in R^n$ with $|r-t| < \delta_f(\varepsilon,t)$. Let $\varepsilon > 0$ be arbitrary. Let $\alpha \equiv \delta_f\left(\frac{\varepsilon}{2},t\right) > 0$ and let

$$\sigma < \alpha/\overline{\Psi}\left(\frac{1}{2}\left(1-\left(1-\frac{\varepsilon}{4}\right)^{\frac{1}{n}}\right)\right). \tag{5.7.10}$$

Then

$$|f_\sigma(t) - f(t)| \le \varepsilon.$$

3. *Again consider the case* $n = 1$. *Let* $\varepsilon > 0$ *be arbitrary. Suppose* $\sigma > 0$ *is so small that* $\sigma < \varepsilon/\overline{\Psi}(\frac{\varepsilon}{8})$. *Let* $r, s \in R$ *be arbitrary with* $r + 2\varepsilon < s$. *Let* $f \equiv 1_{(r,s]}$. *Then* $1_{(r+\varepsilon,s-\varepsilon]} - \varepsilon \le f_\sigma \le 1_{(r-\varepsilon,s+\varepsilon]} + \varepsilon$.

Proof. 1. We estimate

$$\frac{1}{\sqrt{2\pi}\sigma}\int |h(x)|e^{-x^2/(2\sigma^2)}dx$$
$$\le \frac{1}{\sqrt{2\pi}\sigma}\int_{-\alpha}^{\alpha} ae^{-x^2/(2\sigma^2)}dx + \frac{1}{\sqrt{2\pi}\sigma}\int_{|x|>\alpha} be^{-x^2/(2\sigma^2)}dx$$
$$\le a + \frac{1}{\sqrt{2\pi}}\int_{|u|>\frac{\alpha}{\sigma}} be^{-u^2/2}du = a + 2b\Psi\left(\frac{\alpha}{\sigma}\right).$$

2. Let $f, t, \varepsilon, \delta_f, \alpha$, and σ be as given. Then inequality 5.7.10 implies that

$$\left(1 - \left(1 - \frac{\varepsilon}{4}\right)^{\frac{1}{n}}\right) > 2\Psi\left(\frac{\alpha}{\sigma}\right),$$

whence

$$2\left(1 - \left(1 - 2\Psi\left(\frac{\alpha}{\sigma}\right)\right)^n\right) < \frac{\varepsilon}{2}. \tag{5.7.11}$$

Then, $|f(t-u) - f(t)| < \frac{\varepsilon}{2}$ for $u \in R^n$ with $\|u\| \equiv |u_1| \vee \cdots \vee |u_n| < \alpha$. By hypothesis, $\sigma \le \alpha/\overline{\Psi}(\frac{\varepsilon}{8})$. Hence $\frac{\alpha}{\sigma} \ge \overline{\Psi}(\frac{\varepsilon}{8})$ and so $\Psi(\frac{\alpha}{\sigma}) \le \frac{\varepsilon}{8}$. Hence, by Assertion 1, we have

$$|f_\sigma(t) - f(t)| = \left|\int (f(t-u) - f(t))\varphi_{0,\sigma I}(u)du\right|$$
$$\le \int_{u:\|u\|<\alpha} |f(t-u) - f(t)|\varphi_{0,\sigma I}(u)du$$
$$+ \int_{u:\|u\|\ge\alpha} |f(t-u) - f(t)|\varphi_{0,\sigma I}(u)du$$
$$\le \frac{\varepsilon}{2} + 2\left(1 - \int_{u:\|u\|<\alpha} \varphi_{0,\sigma I}(u)du\right)$$
$$= \frac{\varepsilon}{2} + 2\left(1 - \left(\Phi\left(\frac{\alpha}{\sigma}\right) - \Phi\left(\frac{\alpha}{\sigma}\right)\right)^n\right)$$
$$= \frac{\varepsilon}{2} + 2\left(1 - \left(1 - 2\Psi\left(\frac{\alpha}{\sigma}\right)\right)^n\right) < \frac{\varepsilon}{2} + \frac{\varepsilon}{2} = \varepsilon,$$

as desired, where the last inequality is from inequality 5.7.11.

3. Define $f_\sigma(x) \equiv \int f(x-y)\varphi_{0,\sigma}(y)dy$ for each $x \in R$. Consider each $t \in (r+\varepsilon, s-\varepsilon]$. Then f is constant in a neighborhood of t, and hence continuous at t. More precisely, let $\delta_f(\theta, t) \equiv \varepsilon$ for each $\theta > 0$. Then

$$(t - \delta_f(\theta), t + \delta_f(\theta)) = (t - \varepsilon, t + \varepsilon) \subset (r,s] \subset (f = 1)$$

for each $\theta > 0$. Let $\alpha \equiv \delta_f(\frac{\varepsilon}{2}, t) \equiv \varepsilon$. Then, by hypothesis,

$$\sigma < \varepsilon\overline{\Psi}\left(\frac{\varepsilon}{8}\right)^{-1} = \alpha/\overline{\Psi}\left(\frac{1}{2}\left(1 - \left(1 - \frac{\varepsilon}{4}\right)\right)\right). \tag{5.7.12}$$

Hence, by Assertion 2, we have

$$|f(t) - f_\sigma(t)| \leq \varepsilon,$$

where $t \in (r + \varepsilon, s - \varepsilon]$ is arbitrary. Since $1_{(r+\varepsilon,s-\varepsilon]}(t) \leq f(t)$, it follows that

$$1_{(r+\varepsilon,s-\varepsilon]}(t) - \varepsilon \leq f_\sigma(t) \tag{5.7.13}$$

for each $t \in (r + \varepsilon, s - \varepsilon]$. Since $f_\sigma \geq 0$, inequality 5.7.13 is trivially satisfied for $t \in (-\infty, r + \varepsilon] \cup (s - \varepsilon, \infty)$. We have thus proved that inequality 5.7.13 holds on $domain(1_{(r+\varepsilon,s-\varepsilon]})$. Next consider any $t \in (-\infty, r - \varepsilon] \cup (s + \varepsilon, \infty)$. Again, for arbitrary $\theta > 0$, we have $|f(t) - f(u)| = 0 < \theta$ for each $u \in (t - \delta_f(\theta), t + \delta_f(\theta))$. Hence, by Assertion 2, we have $f_\sigma(t) = f_\sigma(t) - f(t) < \varepsilon$. It follows that

$$f_\sigma(t) \leq 1_{(r-\varepsilon,s+\varepsilon]}(t) + \varepsilon \tag{5.7.14}$$

for each $t \in (-\infty, r - \varepsilon] \cup (s + \varepsilon, \infty)$. Since $f_\sigma \leq 1$, inequality 5.7.14 is trivially satisfied for $t \in (r - \varepsilon, s + \varepsilon]$. We have thus proved that inequality 5.7.14 holds on $domain(1_{(r-\varepsilon,s+\varepsilon]})$. Assertion 3 is proved. □

5.8 Characteristic Function

In previous sections we analyzed distributions J on a locally compact metric space (S,d) in terms of their values Jg at basis functions g in a partition of unity. In the special case where (S,d) is the Euclidean space R, the basis functions can be replaced by the exponential functions h_λ, where $\lambda \in R$, where $h_\lambda(x) \equiv e^{i\lambda x}$ for each $x \in R$, and where $i \equiv \sqrt{-1}$. The result is characteristic functions, which are most useful in the study of distributions of r.r.v.'s.

The classical development of this tool, such as in [Chung 1968] or [Loeve 1960], is constructive, except for infrequent and nonessential appeals to the principle of infinite search. The bare essentials of this material are presented here for completeness and for ease of reference. The reader who is familiar with the topic and is comfortable with the notion that the classical treatment is constructive, or easily made so, can skip over this and the next section and come back only for reference.

We will be working with complex-valued measurable functions. Let $\mathbb{C}$ denote the complex plane equipped with the usual metric.

Definition 5.8.1. Complex-valued integrable function. Let I be an integration on a locally compact metric space (S,d), and let (S,Λ,I) denote the completion of the integration space $(S,C(S),I)$. A function $X \equiv IU + iIV : S \to \mathbb{C}$ whose real part U and imaginary part V are measurable on (S, Λ, I) is said to be measurable on (S, Λ, I). If both U, V are integrable, then X is said to be *integrable*, with integral $IX \equiv IU + iIV$. □

By separation into real and imaginary parts, the complex-valued functions immediately inherit the bulk of the theory of integration developed hitherto in this book for real-valued functions. One exception is the very basic inequality $|IX| \leq I|X|$ when $|X|$ is integrable. Its trivial proof in the case of real-valued integrable functions relies on the linear ordering of R, which is absent in $\mathbb{C}$. The next lemma gives a proof for complex-valued integrable functions.

Lemma 5.8.2. $|IX| \leq I|X|$ **for complex-valued integrable function** X. *Use the notations in Definition 5.8.1. Let* $X : S \to \mathbb{C}$ *be an arbitrary complex-valued function. Then the function* X *is measurable in the sense of Definition 5.8.1 iff it is measurable in the sense of Definition 5.8.1. In other words, the former is consistent with the latter. Moreover, if* X *is measurable and if* $|X| \in L$*, then* X *is integrable with* $|IX| \leq I|X|$.

Proof. Write $X \equiv IU + iIV$, where U, V are the real and imaginary parts of X, respectively.

1. Suppose X is measurable in the sense of Definition 5.8.1. Then $U, V : (S, \Lambda, I) \to R$ are measurable functions. Therefore the function $(U, V) : (S, \Lambda, I) \to R^2$ is measurable. At the same time, we have $X = f(U, V)$, where the continuous function $f : R^2 \to \mathbb{C}$ is defined by $f(u, v) \equiv u + iv$. Hence X is measurable in the sense of Definition 4.8.1, according to Proposition 4.8.10.

Conversely, suppose $X(S, \Lambda, I) \to \mathbb{C}$ is measurable in the sense of Definition 4.8.1. Note that U, V are continuous functions of X. Hence, again by Proposition 4.8.10, both U, V are measurable. Thus X is measurable in the sense of Definition 5.8.1.

2. Suppose X is measurable and $|X| \in L$. Then, by Definition 5.8.1, both U and V are measurable, with $|U| \vee |V| \leq |X| \in L$. It follows that $U, V \in L$. Thus X is integrable according to Definition 5.8.1.

Let $\varepsilon > 0$ be arbitrary. Then either (i) $I|X| < 3\varepsilon$ or (ii) $I|X| > 2\varepsilon$.

First consider case (i). Then

$$|IX| = |IU + iIV| \leq |IU| + |iIV| \leq I|U| + I|V| \leq 2I|X| < I|X| + 3\varepsilon.$$

Now consider case (ii). By the Dominated Convergence Theorem, there exists $a > 0$ so small that $I(|X| \wedge a) < \varepsilon$. Then

$$I|X|1_{(|X| \leq a)} \leq I(|X| \wedge a) < \varepsilon. \tag{5.8.1}$$

Write $A \equiv (a < |X|)$. Then

$$|IU1_A - IU| = |IU1_{(|X| \leq a)}| \leq I|U|1_{(|X| \leq a)} \leq I|X|1_{(|X| \leq a)} < \varepsilon.$$

Similarly, $|IV1_A - IV| < \varepsilon$. Hence

$$|I(X1_A) - IX| = |I(U1_A - IU) + i(IV1_A - IV)| < 2\varepsilon. \tag{5.8.2}$$

Write $c \equiv I|X|1_A$. Then it follows that

$$c \equiv I|X|1_A = I|X| - I|X|1_{(|X|\leq a)} > 2\varepsilon - 2\varepsilon = 0,$$

where the inequality is on account of Condition (ii) and inequality 5.8.1. Now define a probability integration space (S, L, E) using $g \equiv c^{-1}|X|1_A$ as a probability density function on the integration space (S, Λ, I). Thus

$$E(Y) \equiv c^{-1}I(Y|X|1_A)$$

for each $Y \in L$. Then

$$|c^{-1}I(X1_A)| \equiv \left|E\left(\frac{X}{|X|\vee a}1_A\right)\right| = \left|E\left(\frac{U}{|X|\vee a}1_A\right) + iE\left(\frac{V}{|X|\vee a}1_A\right)\right|$$

$$= \left(\left(E\left(\frac{U}{|X|\vee a}1_A\right)\right)^2 + \left(E\left(\frac{V}{|X|\vee a}1_A\right)\right)^2\right)^{\frac{1}{2}}$$

$$\leq \left(E\left(\frac{U^2}{(|X|\vee a)^2}1_A\right) + E\left(\frac{V^2}{(|X|\vee a)^2}1_A\right)\right)^{\frac{1}{2}}$$

$$= \left(E\left(\frac{|X|^2}{(|X|\vee a)^2}1_A\right)\right)^{\frac{1}{2}} \leq 1,$$

where the first inequality is thanks to Lyapunov's inequality. Hence $|I(X1_A)| \leq c \equiv I|X|1_A$. Inequality 5.8.2 therefore yields

$$|IX| < |I(X1_A)| + 2\varepsilon \leq I|X|1_A + 2\varepsilon < I|X| + 3\varepsilon.$$

Summing up, we have $|IX| < I|X| + 3\varepsilon$ regardless of case (i) or case (ii), where $\varepsilon > 0$ is arbitrary. We conclude that $I|X| \leq I|X|$. □

Lemma 5.8.3. Basic inequalities for exponentials. *Let $x, y, x', y' \in R$ be arbitrary, with $y \leq 0$ and $y' \leq 0$. Then*

$$|e^{ix} - 1| \leq 2 \wedge |x| \tag{5.8.3}$$

and

$$|e^{ix+y} - e^{ix'+y'}| \leq 2 \wedge |x - x'| + 1 \wedge |y - y'|.$$

Proof. If $x \geq 0$, then

$$|e^{ix} - 1|^2 = |\cos x - 1 + i\sin x|^2 = 2(1 - \cos x)$$

$$= 2\int_0^x \sin u\,du \leq 2\int_0^x u\,du \leq x^2. \tag{5.8.4}$$

Hence, by symmetry and continuity, $|e^{ix} - 1| \leq |x|$ for arbitrary $x \in R$. At the same time, $|e^{ix} - 1| \leq 2$. Equality 5.8.3 follows.

Now assume $y \geq y'$.

$$|e^{ix+y} - e^{ix'+y'}| \leq |e^{ix+y} - e^{ix'+y}| + |e^{ix'+y} - e^{ix'+y'}|$$
$$\leq |e^{ix} - e^{ix'}|e^{y} + |e^{y} - e^{y'}|$$
$$\leq |e^{i(x-x')} - 1|e^{y} + e^{y}(1 - e^{-(y-y')})$$
$$\leq (2 \wedge |x - x'|)e^{y} + (1 - e^{-(y-y')})$$
$$\leq 2 \wedge |x - x'| + 1 \wedge |y - y'|,$$

where the last inequality is because $y' \leq y \leq 0$ by assumption. Hence, by symmetry and continuity, the same inequality holds for arbitrary $y, y' \leq 0$. □

Recall the matrix notations and basics from Definition 5.7.1. Moreover, we will write $|x| \equiv (x_1^2 + \cdots + x_n^2)^{\frac{1}{2}}$ and write $\|x\| \equiv |x_1| \vee \cdots \vee |x_n|$ for each $x \equiv (x_1, \ldots, x_n) \in R^n$.

Definition 5.8.4. Characteristic function, Fourier transform, and convolution. Let $n \geq 1$ be arbitrary.

1. Let $X \equiv (X_1, \ldots, X_n)$ be an r.v. with values in R^n. The *characteristic function* of X is the complex-valued function ψ_X on R^n defined by

$$\psi_X(\lambda) \equiv E \exp i\lambda^T X \equiv E \cos(\lambda^T X) + iE \sin(\lambda^T X).$$

for each $\lambda \in R^n$.

2. Let J be an arbitrary distribution on R^n. The *characteristic function* of J is defined to be $\psi_J \equiv \psi_X$, where X is any r.v. with values in R^n such that $E_X = J$. Thus $\psi_J(\lambda) \equiv Jh_\lambda$, where $h_\lambda(x) \equiv \exp i\lambda^T X$ for each $\lambda, x \in R^n$.

3. If g is a complex-valued integrable function on R^n relative to the Lebesgue integration, the *Fourier transform* of g is defined to be the complex-valued function $\widehat{g}$ on R^n with

$$\widehat{g}(\lambda) \equiv \int_{x \in R^n} (\exp i\lambda^T x) g(x) dx$$

for $\lambda \in R^n$, where $\int \cdot dx$ signifies the Lebesgue integration on R^n, and where $x \in R^n$ is the integration variable. The *convolution* of two complex-valued Lebesgue integrable functions f, g on R^n is the complex-valued function $f \star g$ defined by

$$domain(f \star g) \equiv \{x \in R^n : f(x - \cdot)g \in L\}$$

and by $(f \star g)(x) \equiv \int_{u \in R^n} f(x - y)g(y)dy$ for each $x \in domain(f \star g)$.

4. Suppose $n = 1$. Let F be an P.D.F. on R. The *characteristic function* of F is defined as $\psi_F \equiv \psi_J$, where $J \equiv \int \cdot dF$. If, in addition, F has a p.d.f. f on R, then the characteristic function of f is defined as $\psi_f \equiv \psi_F$. In that case, $\psi_F(\lambda) = \int e^{i\lambda t} f(t)dt \equiv \widehat{f}(\lambda)$ for each $\lambda \in R$. □

We can choose to express the characteristic function in terms of the r.v. X, or in terms of the distribution J, or, in the case $n = 1$, in terms of the P.D.F., as a matter of convenience. A theorem proved in one set of notations will be used in another set without further comment.

Lemma 5.8.5. Basics of convolution. *Let f, g, h be complex-valued Lebesgue integrable functions on R^n. Then the following conditions hold:*

1. *$f \star g$ is Lebesgue integrable.*
2. *$f \star g = g \star f$*
3. *$(f \star g) \star h = f \star (g \star h)$*
4. *$(af + bg) \star h = a(f \star h) + b(g \star h)$ for all complex numbers a, b.*
5. *Suppose $n = 1$, and suppose g is a p.d.f. If $|f| \leq a$ for some $a \in R$, then $|f \star g| \leq a$. If f is real-valued with $a \leq f \leq b$ for some $a, b \in R$, then $a \leq f \star g \leq b$.*
6. *$\widehat{f \star g} = \widehat{f}\widehat{g}$*
7. *$|\widehat{f}| \leq \|f\|_1 \equiv \int_{x \in R^n} |f(x)|dx$*

Proof. If f and g are real-valued, then the integrability of $f \star g$ follows from Corollary A.0.9 in Appendix A. Assertion 1 then follows by linearity. We will prove Assertions 6 and 7; the remaining assertions are left as an exercise. For Assertion 6, for each $\lambda \in R^n$, we have

$$\begin{aligned}
\widehat{f \star g}(\lambda) &\equiv \int (\exp i\lambda^T x)\left(\int f(x-y)g(y)dy\right)dx \\
&= \int_{-\infty}^{\infty}\left(\int_{-\infty}^{\infty}(\exp i\lambda^T(x-y))f(x-y)dx\right)(\exp i\lambda^T y)g(y)dy \\
&= \int_{-\infty}^{\infty}\left(\int_{-\infty}^{\infty}(\exp i\lambda^T u)f(u))du\right)(\exp i\lambda^T y)g(y)dy \\
&= \int \widehat{f}(\lambda)(\exp i\lambda^T y)g(y)dy = \widehat{f}(\lambda)\widehat{g}(\lambda),
\end{aligned}$$

as asserted. At the same time, for each $\lambda \in R^n$, we have

$$\begin{aligned}
|\widehat{f}(\lambda)| &\equiv \left|\int_{x \in R^n}(\exp i\lambda^T x)f(x)dx\right| \leq \int_{x \in R^n}|(\exp i\lambda^T x)f(x)|dx| \\
&= \int_{x \in R^n}|f(x)|dx,
\end{aligned}$$

where the inequality is by Lemma 5.8.2. Assertion 7 is verified. □

Proposition 5.8.6. Uniform continuity of characteristic functions. *Let X be an r.v. with values in R^m. Let β_X be a modulus of tightness of X. Then the following conditions hold:*

1. *$|\psi_X(\lambda)| \leq 1$ and $\psi_{a+BX}(\lambda) = \exp(i\lambda^T a)\psi_X(\lambda^T B)$ for each $a, \lambda \in R^n$ and for each $n \times m$ matrix B*
2. *ψ_X is uniformly continuous. More precisely, ψ_X has a modulus of continuity given by $\delta(\varepsilon) \equiv \frac{\varepsilon}{3}/\beta\left(\frac{\varepsilon}{3}\right)$ for $\varepsilon > 0$.*

3. *If g is a Lebesgue integrable function on R^n, then $\widehat{g}$ is uniformly continuous. More precisely, for each $\varepsilon > 0$, there exists $\gamma \equiv \gamma_g(\varepsilon) > 0$ such that $\int 1_{(|x|>\gamma)}|g(x)|dx < \varepsilon$. Then a modulus of continuity of $\widehat{g}$ is given by $\delta(\varepsilon) \equiv \frac{\varepsilon}{\|g\|+2}/\gamma_g\left(\frac{\varepsilon}{\|g\|+2}\right)$ for $\varepsilon > 0$, where $\|g\| \equiv \int |g(t)|dt$.*

Proof. 1. For each $\lambda \in R^n$, we have $|\psi_X(\lambda)| \equiv |E\exp(i\lambda^T X)| \leq E|\exp(i\lambda^T X)| = E1 = 1$. Moreover,

$$\psi_{a+BX}(\lambda) = \exp(i\lambda^T a)E(i\lambda^T BX) = \exp(i\lambda^T a)\psi_X(\lambda^T B)$$

for each $a, \lambda \in R^n$ and for each $n \times m$ matrix B.

2. Let $\varepsilon > 0$. Let $\delta(\varepsilon) \equiv \frac{\varepsilon}{3}/\beta\left(\frac{\varepsilon}{3}\right)$. Suppose $h \in R^n$ is such that $|h| < \delta(\varepsilon)$. Then $\beta\left(\frac{\varepsilon}{3}\right) < \frac{\varepsilon}{3|h|}$. Pick $a \in \left(\beta\left(\frac{\varepsilon}{3}\right), \frac{\varepsilon}{3|h|}\right)$. Then $P(|X| > a) < \frac{\varepsilon}{3}$ by the definition of β. On the other hand, for each $x \in R^n$ with $|x| \leq a$, we have $|\exp(ih^T x) - 1| \leq |h^T x| \leq |h|a < \frac{\varepsilon}{3}$. Hence, for each $\lambda \in R^n$,

$$\begin{aligned}|\psi_X(\lambda + h) - \psi_X(\lambda)| &\leq E|\exp(i\lambda^T X)(\exp(ih^T X) - 1)| \\ &\leq E(|\exp(ih^T X) - 1|; |X| \leq a) + 2P(|X| > a) \\ &< \frac{\varepsilon}{3} + 2\frac{\varepsilon}{3} = \varepsilon.\end{aligned}$$

3. Proceed in the same manner as in Step 2. Let $\varepsilon > 0$. Write $\varepsilon' \equiv \frac{\varepsilon}{\|g\|+2}$. Let $\delta(\varepsilon) \equiv \frac{\varepsilon'}{\gamma_g(\varepsilon')}$. Suppose $h \in R^n$ is such that $|h| < \delta(\varepsilon)$. Then $\gamma_g(\varepsilon') < \frac{\varepsilon'}{|h|}$. Pick $a \in \left(\gamma_g(\varepsilon'), \frac{\varepsilon'}{|h|}\right)$. Then $\int_{|x|>a} |g(x)|dx < \varepsilon'$ by the definition of γ_g. Moreover, for each $x \in R^n$ with $|x| \leq a$, we have $|\exp(ih^T x) - 1| \leq |h^T x| \leq |h|a < \varepsilon'$. Hence, for each $\lambda \in R^n$,

$$\begin{aligned}|\widehat{g}(\lambda + h) - \widehat{g}(\lambda)| &\leq \int |\exp(i\lambda^T x)(\exp(ih^T x) - 1)g(x)|dx \\ &\leq \int_{|x|\leq a} |(\exp(ih^T x) - 1)g(x)|dx + \int_{|x|>a} |2g(x)|dx \\ &\leq \varepsilon' \int |g(x)|dx + 2\varepsilon' = \varepsilon'(\|g\| + 2) = \varepsilon.\end{aligned}$$

□

Lemma 5.8.7. Characteristic function of normal distribution. *Let $\Phi_{\overline{\mu},\overline{\sigma}}$ be an arbitrary normal distribution on R^n, with mean $\overline{\mu}$ and covariance matrix $\overline{\sigma}$. Then the characteristic function of $\Phi_{\overline{\mu},\overline{\sigma}}$ is given by*

$$\psi_{\overline{\mu},\overline{\sigma}}(\lambda) \equiv \exp\left(i\overline{\mu}^T\lambda - \frac{1}{2}\lambda^T\overline{\sigma}\lambda\right)$$

for each $\lambda \in R^n$.

Proof. 1. Consider the special case where $n = 1$, $\overline{\mu} = 0$, and $\overline{\sigma} = 1$. Let X be an r.r.v. with the standard normal distribution $\Phi_{0,1}$. By Proposition 5.7.5, X^p is integrable for each $p \geq 0$, with $m_p \equiv EX^p = (2k)!2^{-k}/k!$ if p is equal to some even integer $2k$, and with $m_p \equiv EX^p = 0$ otherwise. Using these moment formulas, we compute the characteristic function

$$\psi_{0,1}(\lambda) = \frac{1}{\sqrt{2\pi}} \int e^{i\lambda x} e^{-x^2/2} dx = \frac{1}{\sqrt{2\pi}} \int \sum_{p=0}^{\infty} \frac{(i\lambda x)^p}{p!} e^{-x^2/2} dx$$

$$= \sum_{p=0}^{\infty} \frac{(i\lambda)^p}{p!} m_p = \sum_{k=0}^{\infty} \frac{(-1)^k \lambda^{2k}}{(2k)!} m_{2k}$$

$$= \sum_{k=0}^{\infty} \frac{(-1)^k \lambda^{2k}}{(2k)!} (2k)! \, 2^{-k}/k! = \sum_{k=0}^{\infty} \frac{(-\lambda^2/2)^k}{k!}$$

$$= e^{-\lambda^2/2},$$

where Fubini's Theorem justifies any change in the order of integration and summation.

2. Now consider the general case. By Lemma 5.7.8, $\Phi_{\overline{\mu},\overline{\sigma}}$ is the distribution of an r.v. $Y = \overline{\mu} + AX$ for some matrix A with $\overline{\sigma} \equiv AA^T$ and for some r.v. X with the standard normal p.d.f. $\varphi_{0,I}$ on R^n, where I is the $n \times n$ identity matrix. Let $\lambda \in R^n$ be arbitrary. Write $\theta \equiv A^T \lambda$. Then

$$\psi_{\overline{\mu},\overline{\sigma}}(\lambda) \equiv E \exp(i\lambda^T Y) \equiv E \exp(i\lambda^T \overline{\mu} + i\lambda^T AX)$$

$$= \int_{x \in R^n} \exp(i\lambda^T \overline{\mu} + i\theta^T x) \varphi_{0,I}(x) dx$$

$$= \exp(i\lambda^T \overline{\mu}) \int \cdots \int \exp\left(i \sum_{j=1}^{n} \theta_j x_j\right) \varphi_{0,1}(x_1) \ldots \varphi_{0,1}(x_n) dx_1 \ldots dx_n.$$

By Fubini's Theorem and by the first part of this proof, this reduces to

$$\psi_{\overline{\mu},\overline{\sigma}}(\lambda) = \exp(i\lambda^T \overline{\mu}) \prod_{j=1}^{n} \left(\int \exp(i\theta_j x_j) \varphi_{0,1}(x_j) dx_j \right)$$

$$= \exp(i\lambda^T \overline{\mu}) \prod_{j=1}^{n} \exp\left(-\frac{1}{2}\theta_j^2\right) = \exp(i\lambda^T \overline{\mu}) \exp\left(-\frac{1}{2}\theta^T \theta\right)$$

$$= \exp\left(i\lambda^T \overline{\mu} - \frac{1}{2}\lambda^T AA^T \lambda\right) \equiv \exp\left(i\lambda^T \overline{\mu} - \frac{1}{2}\lambda^T \overline{\sigma} \lambda\right).$$

□

Corollary 5.8.8. Convolution with normal density. *Suppose f is a Lebesgue integrable function on R^n. Let $\sigma > 0$ be arbitrary. Write $\overline{\sigma} \equiv \sigma^2 I$, where I is the $n \times n$ identity matrix. Define $f_\sigma \equiv f \star \varphi_{0,\overline{\sigma}}$. Then*

$$f_\sigma(t) = (2\pi)^{-n} \int \exp\left(i\lambda^T t - \frac{1}{2}\sigma^2 \lambda^T \lambda\right) \widehat{f}(-\lambda) d\lambda$$

for each $t \in R^n$.

Proof. In view of Lemma 5.8.7, we have, for each $t \in R^n$,

$$
\begin{aligned}
f_\sigma(t) &\equiv \int \varphi_{0,\overline{\sigma}}(t-x)f(x)dx \\
&= (2\pi\sigma^2)^{-\frac{n}{2}} \int \exp\left(-\frac{1}{2\sigma^2}(t-x)^T(t-x)\right) f(x)dx \\
&= (2\pi\sigma^2)^{-\frac{n}{2}} \int \psi_{0,I}(\sigma^{-1}(t-x))f(x)dx \\
&= (2\pi\sigma^2)^{-\frac{n}{2}} \int\int (2\pi)^{-\frac{n}{2}} \exp\left(i\sigma^{-1}(t-x)^T y - \frac{1}{2}y^T y\right) f(x)dxdy \\
&= (2\pi\sigma)^{-n} \int \exp\left(-i\sigma^{-1}y^T t - \frac{1}{2}y^T y\right) \widehat{f}(-\sigma^{-1}y)dy \\
&= (2\pi)^{-n} \int \exp\left(-i\lambda^T t - \frac{1}{2}\sigma^2\lambda^T\lambda\right) \widehat{f}(-\lambda)d\lambda.
\end{aligned}
$$

Note that in the double integral, the integrand is a continuous function in (x, y) and is bounded in absolute value by a constant multiple of $\exp\left(-\frac{1}{2}y^T y\right)f(x)$ that is, by Proposition 4.10.6, Lebesgue integrable on R^{2n}. This justifies the changes in order of integration, thanks to Fubini. □

The next theorem recovers a distribution on R^n from its characteristic function.

Theorem 5.8.9. Inversion formula for characteristic functions. *Let J, J' be a distribution on R^n, with characteristic functions $\psi_J, \psi_{J'}$ respectively. Let f be an arbitrary Lebesgue integrable function on R^n. Let $\widehat{f}$ denote the Fourier transform of f. Let $\sigma > 0$ be arbitrary. Write $\overline{\sigma} \equiv \sigma^2 I$, where I is the $n \times n$ identity matrix. Define $f_\sigma \equiv f \star \varphi_{0,\overline{\sigma}}$. Then the following conditions hold:*

1. *We have*

$$J f_\sigma = (2\pi)^{-n} \int \exp\left(-\frac{1}{2}\sigma^2\lambda^T\lambda\right) \widehat{f}(-\lambda)\psi_J(\lambda)d\lambda. \tag{5.8.5}$$

2. *Suppose $f \in C_{ub}(R^n)$ and $|f| \le 1$. Let $\varepsilon > 0$ be arbitrary. Suppose $\sigma > 0$ is so small that*

$$\sigma \le \delta_f\left(\frac{\varepsilon}{2}\right) \Big/ \sqrt{-2\log\left(\frac{1}{2}\left(1-\left(1-\frac{\varepsilon}{4}\right)^{\frac{1}{n}}\right)\right)}.$$

Then

$$|Jf - Jf_\sigma| \le \varepsilon.$$

Consequently, $Jf = \lim_{\sigma\to 0} Jf_\sigma$.

3. *Suppose $f \in C_{ub}(R)$ is arbitrary such that $\widehat{f}$ is Lebesgue integrable on R^n. Then*

$$Jf = (2\pi)^{-n} \int \widehat{f}(-\lambda)\psi_J(\lambda)d\lambda.$$

4. *If ψ_J is Lebesgue integrable on R^n, then J has a p.d.f. Specifically, then*

$$Jf = (2\pi)^{-n} \int f(x)\hat{\psi}_J(-x)dx$$

for each $f \in C(R^n)$.

5. $J = J'$ *iff* $\psi_J = \psi_{J'}$.

Proof. Write $\|f\| \equiv \int |f(t)|dt < \infty$. Then $|\widehat{f}| \le \|f\|$.

1. Consider the function $Z(\lambda, x) \equiv \exp(i\lambda^T x - \frac{1}{2}\sigma^2\lambda^T\lambda)\widehat{f}(-\lambda)$ on the product space $(R^n, L_0, I_0) \otimes (R^n, L, J)$, where $(R^n, L_0, I_0) \equiv (R^n, L_0, \int \cdot dx)$ is the Lebesgue integration space and where (R^n, L, J) is the probability space that is the completion of $(R^n, C_{ub}(R^n), J)$. The function Z is a continuous function of (λ, x). Hence Z is measurable. Moreover, $|Z| \le U$ where $U(\lambda, x) \equiv \|f\| e^{-\sigma^2\lambda^2/2}$ is integrable. Hence Z is integrable by the Dominated Convergence Theorem.

Define $h_\lambda(t) \equiv \exp(it^T\lambda)$ for each $t, \lambda \in R^n$. Then $J(h_\lambda) = \psi_J(\lambda)$ for each $\lambda \in R^n$. Corollary 5.8.8 then implies that

$$\begin{aligned} Jf_\sigma &= (2\pi)^{-n} \int J(h_\lambda) \exp\left(-\frac{1}{2}\sigma^2\lambda^T\lambda\right) \widehat{f}(-\lambda)d\lambda \\ &\equiv (2\pi)^{-n} \int \psi_J(\lambda) \exp\left(-\frac{1}{2}\sigma^2\lambda^T\lambda\right) \widehat{f}(-\lambda)d\lambda, \end{aligned} \tag{5.8.6}$$

proving Assertion 1.

2. Now suppose $f \in C_{ub}(R^n)$ with modulus of continuity δ_f with $|f| \le 1$. Recall that $\overline{\Psi} : \left(0, \frac{1}{2}\right] \to [0, \infty)$ denotes the inverse of the tail function $\Psi \equiv 1 - \Phi$ of the standard normal P.D.F. Φ. Proposition 5.7.4 says that $\overline{\Psi}(\alpha) < \sqrt{-2\log\alpha}$ for each $\alpha \in \left(0, \frac{1}{2}\right]$. Hence

$$\begin{aligned} \sigma &\le \delta_f\left(\frac{\varepsilon}{2}\right) \Big/ \sqrt{-2\log\left(\frac{1}{2}\left(1 - \left(1 - \frac{\varepsilon}{4}\right)^{\frac{1}{n}}\right)\right)} \\ &\le \delta_f\left(\frac{\varepsilon}{2}\right) \Big/ \overline{\Psi}\left(\frac{1}{2}\left(1 - \left(1 - \frac{\varepsilon}{4}\right)^{\frac{1}{n}}\right)\right), \end{aligned}$$

where the first inequality is by hypothesis. Therefore Lemma 5.7.9 implies that $|f_\sigma - f| \le \varepsilon$. Consequently, $|Jf - Jf_\sigma| \le \varepsilon$. Hence $Jf = \lim_{\sigma\to 0} Jf_\sigma$. This proves Assertion 2.

3. Now let $f \in C_{ub}(R^n)$ be arbitrary. Then, by linearity, Assertion 2 implies that

$$\lim_{\sigma\to 0} Jf_\sigma = Jf \tag{5.8.7}$$

$$Jf_\sigma = (2\pi)^{-n} \int \exp\left(-\frac{1}{2}\sigma^2\lambda^T\lambda\right) \widehat{f}(-\lambda)\psi_J(\lambda)d\lambda. \tag{5.8.8}$$

Suppose $\widehat{f}$ is Lebesgue integrable on R^n. Then the integrand in equality 5.8.5 is dominated in absolute value by the integrable function $|\widehat{f}|$, and converges a.u. on

R^n to the function $\widehat{f}(-\lambda)\psi_J(\lambda)$ as $\sigma \to 0$. Hence the Dominated Convergence Theorem implies that

$$\lim_{\sigma\to 0} Jf_\sigma = (2\pi)^{-n}\int \widehat{f}(-\lambda)\psi_J(\lambda)d\lambda.$$

Combining with equality 5.8.7, Assertion 3 is proved.

4. Next consider the case where ψ_J is Lebesgue integrable. Suppose $f \in C(R^n)$ with $|f| \le 1$. Then the function $U_\sigma(x,\lambda) \equiv f(x)\psi_J(\lambda)e^{-i\lambda x-\sigma^2\lambda^2/2}$ is an integrable function relative to the product Lebesgue integration on R^{2n}, and is dominated in absolute value by the integrable function $f(x)\psi_J(\lambda)$. Moreover, $U_\sigma \to U_0$ uniformly on compact subsets of R^{2n} where $U_0(x,\lambda) \equiv f(x)\psi_J(\lambda)e^{-i\lambda x}$. Hence $U_\sigma \to U_0$ in measure relative to $I_0 \otimes I_0$. The Dominated Convergence Theorem therefore yields, as $\sigma \to 0$,

$$\begin{aligned}
Jf_\sigma &= (2\pi)^{-n}\int \exp\left(-\frac{1}{2}\sigma^2\lambda^T\lambda\right)\widehat{f}(-\lambda)\psi_J(\lambda)d\lambda\\
&= (2\pi)^{-n}\int \exp\left(-\frac{1}{2}\sigma^2\lambda^T\lambda\right)\psi_J(\lambda)\int \exp(-i\lambda x)f(x)dxd\lambda\\
&\to (2\pi)^{-n}\int \psi_J(\lambda)\int \exp(-i\lambda x)f(x)dxd\lambda\\
&= (2\pi)^{-n}\int \hat{\psi}(-x)f(x)dx.
\end{aligned}$$

On the other hand, by Assertion 2, we have $If_\sigma \to If$ as $\sigma \to 0$. Assertion 4 is proved.

5. Assertion 5 follows from Assertion 4. □

Definition 5.8.10. Metric of characteristic functions. Let $n \ge 1$ be arbitrary. Let ψ, ψ' be arbitrary characteristic functions on R^n. Define

$$\rho_{char}(\psi,\psi') \equiv \rho_{char,n}(\psi,\psi') \equiv \sum_{j=1}^{\infty} 2^{-j}\sup_{|\lambda|\le j}|\psi(\lambda)-\psi'(\lambda)|. \tag{5.8.9}$$

Then ρ_{char} is a metric. □

We saw earlier that characteristic functions are continuous and bounded in absolute values by 1. Hence the supremum inside the parentheses in equality 5.8.9 exists and is bounded by 2. Thus ρ_{char} is well defined. In view of Theorem 5.8.9, it is easily seen that ρ_{char} is a metric. Convergence relative to ρ_{char} is equivalent to uniform convergence on each compact subset of R^n.

The next theorem shows that the correspondence between distributions on R^n and their characteristic functions is uniformly continuous when restricted to a tight subset.

Theorem 5.8.11. Continuity theorem for characteristic functions. *Let ξ be an arbitrary binary approximation of R. Let $n \ge 1$ be arbitrary but fixed. Let*

$\xi^n \equiv (A_p)_{p=1,2,...}$ be the binary approximation of R^n that is the nth power of ξ. Let $\|\xi^n\|$ be the modulus of local compactness of R^n associated with ξ^n. Let ρ_{Dist,ξ^n} be the corresponding distribution metric on the space of distributions on R^n, as in Definition 5.3.4. Let $\widehat{J_0}$ be a family of distributions on R^n.

Let $J, J' \in \widehat{J_0}$ be arbitrary, with corresponding characteristic functions ψ, ψ'. Then the following conditions hold:

1. For each $\varepsilon > 0$, there exists $\delta_{ch,dstr}(\varepsilon, n) > 0$ such that if $\rho_{char}(\psi, \psi') < \delta_{ch,dstr}(\varepsilon, n)$, then $\rho_{Dist,\xi^n}(J, J') < \varepsilon$.

2. Suppose $\widehat{J_0}$ is tight, with some modulus of tightness β. Then, for each $\varepsilon > 0$, there exists $\delta_{dstr,ch}(\varepsilon, \beta, \|\xi^n\|) > 0$ such that if $\rho_{Dist,\xi^n}(J, J') < \delta_{dstr,ch}(\varepsilon, \beta)$, then $\rho_{char}(\psi, \psi') < \varepsilon$.

3. If $(J_m)_{m=0,1,...}$ is a sequence of distributions on R^n with a corresponding sequence $(\psi_m)_{m=0,1,...}$ of characteristic functions such that $\rho_{char}(\psi_m, \psi_0) \to 0$, then $J_m \Rightarrow J_0$.

Proof. Let

$$\pi_{R^n} \equiv (\{g_{p,x} : x \in A_p\})_{p=1,2,...} \tag{5.8.10}$$

be the partition of unity of R^n determined by ξ^n, as introduced in Definition 3.3.4. Thus $\|\xi^n\| \equiv (|A_p|)_{p=1,2,...}$. Let

$$v_n \equiv \int_{|y|\le 1} dy,$$

the volume of the unit n-ball $\{y \in R^n : |y| \le 1\}$ in R^n.

1. Let $\varepsilon > 0$ be arbitrary. As an abbreviation, write

$$\alpha \equiv \frac{1}{8}\varepsilon.$$

Let $p \equiv [0 \vee (1 - \log_2 \varepsilon)]_1$. Thus

$$2^{-p} < \frac{\varepsilon}{2}.$$

For each $\theta > 0$, define

$$\delta_p(\theta) \equiv 2^{-p-1}\theta > 0. \tag{5.8.11}$$

Recall from Proposition 5.7.4 the standard normal P.D.F. Φ on R, its decreasing tail function $\Psi : [0,\infty) \to \left(0, \frac{1}{2}\right]$, and the inverse function $\overline{\Psi} : \left(0, \frac{1}{2}\right] \to [0,\infty)$ of the latter. Define

$$\sigma \equiv \delta_p\left(\frac{\alpha}{2}\right) \Big/ \sqrt{-2\log\left(\frac{1}{2}\left(1 - \left(1 - \frac{\alpha}{4}\right)^{\frac{1}{n}}\right)\right)} > 0. \tag{5.8.12}$$

Define

$$m \equiv \left[\sigma^{-1} n^{\frac{1}{2}} \overline{\Psi}\left(\frac{1}{2} \wedge v_n^{-1} \varepsilon 2^{-5} (2\pi)^{\frac{n}{2}} \sigma^n n^{-1}\right)\right]_1.$$

Thus $m \geq 1$ is so large that

$$v_n 2^2 (2\pi)^{-\frac{n}{2}} \sigma^{-n} n \Psi(\sigma n^{-\frac{1}{2}} m) < \frac{1}{8}\varepsilon. \tag{5.8.13}$$

Finally, define

$$\delta_{ch,dstr}(\varepsilon) \equiv \delta_{ch,dstr}(\varepsilon, n) \equiv v_n^{-1} \varepsilon 2^{-m-3} (2\pi)^{\frac{n}{2}} \sigma^n > 0. \tag{5.8.14}$$

Now suppose the characteristic functions ψ, ψ' on R^n are such that

$$\rho_{char}(\psi, \psi') \equiv \sum_{j=1}^{\infty} 2^{-j} \sup_{|\lambda| \leq j} |\psi(\lambda) - \psi'(\lambda)| < \delta_{ch,dstr}(\varepsilon). \tag{5.8.15}$$

We will prove that $\rho_{Dist,\xi^n}(J, J') < \varepsilon$. To that end, first note that with $m \geq 1$ as defined earlier, the last displayed inequality implies

$$\sup_{|\lambda| \leq m} |\psi(\lambda) - \psi'(\lambda)| < 2^m \delta_{ch,dstr}(\varepsilon). \tag{5.8.16}$$

Next, let $k = 1, \ldots, p$ and $x \in A_k$ be arbitrary. Write $f \equiv g_{k,x}$ as an abbreviation. Then, by Proposition 3.3.5, f has values in $[0, 1]$ and has a Lipschitz constant $2^{k+1} \leq 2^{p+1}$. Consequently, f has the modulus of continuity δ_p defined in equality 5.8.11. Hence, in view of equality 5.8.12, Theorem 5.8.9 implies that

$$|Jf - Jf_\sigma| \leq \alpha \equiv \frac{1}{8}\varepsilon, \tag{5.8.17}$$

and that

$$Jf_\sigma = (2\pi)^{-n} \int \exp\left(-\frac{1}{2}\sigma^2 \lambda^T \lambda\right) \widehat{f}(-\lambda) \psi(\lambda) d\lambda, \tag{5.8.18}$$

where $f_\sigma \equiv f \star \varphi_{0,\sigma^2 I}$, where I is the $n \times n$ identity matrix, and where $\widehat{f}$ stands for the Fourier transform of f. Moreover, by Proposition 3.3.5, the function $f \equiv g_{k,x}$ has the sphere $\{y \in R^n : |y - x| \leq 2^{-k+1}\}$ as support. Therefore

$$|\widehat{f}| \leq \int f(y) dy \leq v_n (2^{-k+1})^n < v_n, \tag{5.8.19}$$

where v_n is the volume of the unit n-sphere in R^n, as defined previously.

By equality 5.8.18 for J and a similar equality for J', we have

$$\begin{aligned} |Jf_\sigma - J'f_\sigma| &= (2\pi)^{-n} \left| \int \exp\left(-\frac{1}{2}\sigma^2 \lambda^T \lambda\right) \widehat{f}(-\lambda)(\psi(\lambda) - \psi'(\lambda)) d\lambda \right| \\ &\leq (2\pi)^{-n} \int_{|\lambda| \leq m} \exp\left(-\frac{1}{2}\sigma^2 \lambda^T \lambda\right) |\widehat{f}(-\lambda)(\psi(\lambda) - \psi'(\lambda)| d\lambda \\ &\quad + (2\pi)^{-n} \int_{|\lambda| > m} \exp\left(-\frac{1}{2}\sigma^2 \lambda^T \lambda\right) |\widehat{f}(-\lambda)(\psi(\lambda) - \psi'(\lambda)| d\lambda). \end{aligned} \tag{5.8.20}$$

In view of inequalities 5.8.16 and 5.8.19, the first summand in the last sum is bounded by

$$(2\pi)^{-n} v_n 2^m \delta_{ch,dstr}(\varepsilon) \int \exp\left(-\frac{1}{2}\sigma^2\lambda^T\lambda\right) d\lambda$$
$$\leq (2\pi)^{-n} v_n 2^m \delta_{ch,dstr}(\varepsilon)(2\pi)^{\frac{n}{2}}\sigma^{-n} = \frac{1}{8}\varepsilon,$$

where the last equality is from the defining equality 5.8.14. The second summand is bounded by

$$(2\pi)^{-n} v_n 2 \int_{|\lambda|>m} \exp\left(-\frac{1}{2}\sigma^2\lambda^T\lambda\right) d\lambda$$
$$\leq (2\pi)^{-n} v_n 2 \int \cdots \int_{|\lambda_1|\vee\cdots\vee|\lambda_n|>m/\sqrt{n}} \exp\left(-\frac{1}{2}\sigma^2(\lambda_1^2+\cdots+\lambda_n^2)\right)$$
$$\times\, d\lambda_1 \ldots d\lambda_n$$
$$\leq (2\pi)^{-n} v_n 2 (2\pi)^{\frac{n}{2}}\sigma^{-n} \int \cdots \int_{|\lambda_1|\vee\cdots\vee|\lambda_n|>\sigma m/\sqrt{n}} \varphi_{0,1}(\lambda_1)\ldots\varphi_{0,1}(\lambda_n)$$
$$\times\, d\lambda_1 \ldots d\lambda_n$$
$$\leq v_n 2 (2\pi)^{-\frac{n}{2}}\sigma^{-n} \sum_{j=1}^{n} \int \cdots \int_{|\lambda_j|>\sigma m/\sqrt{n}} \varphi_{0,1}(\lambda_1)\ldots\varphi_{0,1}(\lambda_n) d\lambda_1 \ldots d\lambda_n$$
$$= v_n 2 (2\pi)^{-\frac{n}{2}}\sigma^{-n} \sum_{j=1}^{n} \int_{|\lambda_j|>\sigma m/\sqrt{n}} \varphi_{0,1}(\lambda_j) d\lambda_j$$
$$= v_n 2^2 (2\pi)^{-\frac{n}{2}}\sigma^{-n} n \Psi(\sigma n^{-\frac{1}{2}} m) < \frac{1}{8}\varepsilon,$$

where the last inequality follows from inequality 5.8.13. Hence inequality 5.8.20 yields

$$|Jf_\sigma - J'f_\sigma| \leq \frac{1}{8}\varepsilon + \frac{1}{8}\varepsilon = \frac{1}{4}\varepsilon.$$

Combining with inequality 5.8.17 for J and a similar inequality for J', we obtain

$$|Jf - J'f| \leq |Jf - Jf_\sigma| + |Jf_\sigma - J'f_\sigma| + |J'f - J'f_\sigma|$$
$$\leq \frac{1}{8}\varepsilon + \frac{1}{4}\varepsilon + \frac{1}{8}\varepsilon = \frac{\varepsilon}{2},$$

where $f \equiv g_{k,x}$, where $k = 1, \ldots, p$ and $x \in A_p$ are arbitrary. Hence

$$\rho_{Dist,\xi^n}(J, J') \equiv \sum_{k=1}^{\infty} 2^{-k}|A_k|^{-1} \sum_{x\in A(k)} |Jg_{k,x} - J'g_{k,x}|$$

$$\leq \sum_{k=1}^{p} 2^{-k}\frac{\varepsilon}{2} + \sum_{k=p+1}^{\infty} 2^{-k}$$

$$\leq \frac{\varepsilon}{2} + 2^{-p} < \frac{\varepsilon}{2} + \frac{\varepsilon}{2} = \varepsilon.$$

Assertion 1 has been proved.

2. Conversely, let $\varepsilon > 0$ be arbitrary. Write $p \equiv [0 \vee (2 - \log_2 \varepsilon)]_1$. For each $\theta > 0$, define $\delta_p(\theta) \equiv p^{-1}\theta$. By Proposition 5.3.11, there exists $\widetilde{\Delta}(\frac{\varepsilon}{4}, \delta_p, \beta, \|\xi_{R^n}\|) > 0$ such that if

$$\rho_{Dist,\xi^n}(J, J') < \widetilde{\Delta}\left(\frac{\varepsilon}{4}, \delta_p, \beta, \|\xi^n\|\right)$$

then for each $f \in C_{ub}(R^n)$ with modulus of continuity δ_p and with $|f| \leq 1$, we have

$$|Jf - J'f| < \frac{\varepsilon}{4}. \tag{5.8.21}$$

Define

$$\delta_{dstr,ch}(\varepsilon, \beta, \|\xi^n\|) \equiv \widetilde{\Delta}\left(\frac{\varepsilon}{4}, \delta_p, \beta, \|\xi^n\|\right). \tag{5.8.22}$$

We will prove that $\delta_{dstr,ch}(\varepsilon, \beta, \|\xi^n\|)$ has the desired properties. To that end, suppose

$$\rho_{Dist,\xi^n}(J, J') < \delta_{dstr,ch}(\varepsilon, \beta, \|\xi^n\|) \equiv \widetilde{\Delta}\left(\frac{\varepsilon}{4}, \delta_p, \beta, \|\xi^n\|\right).$$

Let $\lambda \in R^n$ be arbitrary with $|\lambda| \leq p$. Define the function

$$h_\lambda(x) \equiv \exp(i\lambda^T x) \equiv \cos\lambda^T x + i \sin\lambda^T x$$

for each $x \in R^n$. Then, using inequality 5.8.3, we obtain

$$|\cos\lambda^T x - \cos\lambda^T y| \leq |\exp(i\lambda^T x) - \exp(i\lambda^T y)|$$
$$= |\exp(i\lambda^T (x - y)) - 1| \leq |\lambda^T (x - y)| \leq p|x - y|,$$

for each $x, y \in R^n$. Hence the function $\cos(\lambda^T \cdot)$ on R^n has modulus of continuity δ_p. Moreover, $|\cos(\lambda^T \cdot)| \leq 1$. Hence, inequality 5.8.21 is applicable and yields

$$|J\cos(\lambda^T \cdot) - J'\cos(\lambda^T \cdot)| < \frac{\varepsilon}{4}.$$

Similarly,

$$|J\sin(\lambda^T \cdot) - J'\sin(\lambda^T \cdot)| < \frac{\varepsilon}{4}.$$

Combining,

$$|\psi(\lambda) - \psi'(\lambda)| = |Jh_\lambda - J'h_\lambda|$$
$$\leq |J\cos(\lambda^T \cdot) - J'\cos(\lambda^T \cdot)| + |J\sin(\lambda^T \cdot) - J'\sin(\lambda^T \cdot)| < \frac{\varepsilon}{2}, \tag{5.8.23}$$

where $\lambda \in R^n$ is arbitrary with $|\lambda| \leq p$. We conclude that

$$\begin{aligned}\rho_{char}(\psi,\psi') &\equiv \sum_{j=1}^{\infty} 2^{-j} \sup_{|\lambda|\leq j} |\psi(\lambda) - \psi'(\lambda)| \\ &\leq \sum_{j=1}^{p} 2^{-j} \sup_{|\lambda|\leq p} |\psi(\lambda) - \psi'(\lambda)| + \sum_{j=p+1}^{\infty} 2^{-j}2 \\ &\leq \frac{\varepsilon}{2} + 2^{-p+1} \leq \frac{\varepsilon}{2} + \frac{\varepsilon}{2} = \varepsilon.\end{aligned}$$

3. Finally, suppose $\rho_{char}(\psi_m,\psi_0) \to 0$. Then Assertion 1 yields $\rho_{Dist,\xi^n}(J_m, J_0) \to 0$ as $m \to \infty$. Hence Proposition 5.3.5 implies that $J_m \Rightarrow J_0$.

The theorem is proved. □

The following propositions relate the moments of an r.r.v. X to the derivatives of its characteristic function.

Proposition 5.8.12. Taylor expansion of characteristic functions. *Let $n \geq 1$ be arbitrary, and let X be an arbitrary r.r.v. Suppose X^n is integrable, with a simple modulus of integrability η in the sense of Definition 4.7.2, and with $E|X|^n \leq b$ for some $b > 0$. Let ψ denote the characteristic function of X. Define the remainder $r_n(\lambda)$ by*

$$\psi(\lambda) \equiv \sum_{k=0}^{n} (i\lambda)^k EX^k/k! + r_n(\lambda)$$

for each $\lambda \in R$. Then the following conditions hold:

1. *The characteristic function ψ has a continuous derivative of order n on R, with*

$$\psi^{(k)}(\lambda) = i^k EX^k e^{i\lambda X} \tag{5.8.24}$$

for each $\lambda \in R$, for each $k = 0,\ldots,n$. In particular, the kth moment of X is given by $EX^k = (-i)^k\psi^{(k)}(0)$, for each $k = 0,\ldots,n$. Moreover, $\psi^{(n)}$ is uniformly continuous on R, with a modulus of continuity $\delta_{\psi,n}$ on R defined by

$$\delta_{\psi,n}(\varepsilon) \equiv \frac{\varepsilon}{2b}\left(\eta\left(\frac{\varepsilon}{2}\right)\right)^{-\frac{1}{n}}$$

for each $\varepsilon > 0$.

2. *For each $\lambda \in R$ with*

$$|\lambda| < \frac{n!\,\varepsilon}{2b}\left(\eta\left(\frac{n!\,\varepsilon}{2}\right)\right)^{-\frac{1}{n}},$$

we have $|r_n(\lambda)| < \varepsilon|\lambda|^n$.

3. *Suppose X^{n+1} is integrable. Then, for each $t \in R$, we have*

$$\psi(t) \equiv \sum_{k=0}^{n} \psi^{(k)}(t_0)(t-t_0)^k/k! + \bar{r}_n(t),$$

where

$$|\bar{r}_n(t)| \leq |t - t_0|^{n+1} E|X|^{n+1}/(n+1)!.$$

Proof. We first observe that $(e^{iax} - 1)/a \to ix$ uniformly for x in any compact interval $[-t,t]$ as $a \to 0$. This can be shown by first noting that, for arbitrary $\varepsilon > 0$, Taylor's Theorem (Theorem B.0.1 in Appendix B), implies that $|e^{iax} - 1 - iax| \leq a^2x^2/2$ and so $|a^{-1}(e^{iax} - 1) - ix| \leq ax^2/2 < \varepsilon$ for each $x \in [-t,t]$, provided that $a < 2\varepsilon/t^2$.

1. Let $\lambda \in R$ be arbitrary. Proceed inductively. The assertion is trivial if $n=0$. Suppose the assertion has been proved for $k=0,\ldots n-1$. Let $\varepsilon>0$ be arbitrary, and let t be so large that $P(|X| > t) < \varepsilon$. For $a > 0$, define $D_a \equiv i^k X^k e^{i\lambda X}(e^{iaX} - 1)/a$. By the observation at the beginning of this proof, D_a converges uniformly to $i^{k+1}X^{k+1}e^{i\lambda X}$ on $(|X| \leq t)$ as $a \to 0$. Thus we see that D_a converges a.u. to $i^{k+1}X^{k+1}e^{i\lambda X}$. At the same time, $|D_a| \leq |X|^k|(e^{iaX} - 1)/a| \leq |X|^{k+1}$ where $|X|^{k+1}$ is integrable. The Dominated Convergence Theorem applies, yielding $\lim_{a\to 0} ED_a = i^{k+1}EX^{k+1}e^{i\lambda X}$. On the other hand, by the induction hypothesis, $ED_a \equiv a^{-1}(i^kEX^ke^{i(\lambda+a)X} - i^kEX^ke^{i\lambda X}) = a^{-1}(\psi^{(k)}(\lambda + a) - \psi^{(k)}(\lambda))$. Combining, we see that $\frac{d}{d\lambda}\psi^{(k)}(\lambda)$ exists and is equal to $i^{k+1}EX^{k+1}e^{i\lambda X}$. Induction is completed.

We next prove the continuity of $\psi^{(n)}$. To that end, let $\varepsilon > 0$ be arbitrary. Let $\lambda, a \in R$ be arbitrary with

$$|a| < \delta_{\psi,n}(\varepsilon) \equiv \frac{\varepsilon}{2b}\left(\eta\left(\frac{\varepsilon}{2}\right)\right)^{-\frac{1}{n}}.$$

Then

$$\begin{aligned}
|\psi^{(n)}(\lambda + a) - \psi^{(n)}(\lambda)| &= |EX^n e^{i(\lambda+a)X} - EX^n e^{i\lambda X}| \\
&\leq E|X|^n|e^{iaX} - 1| \leq E|X|^n(2 \wedge |aX|) \\
&\leq 2E|X|^n 1_{(|X|^n > \eta(\frac{\varepsilon}{2}))} + E|X|^n(|aX|)1_{(|X|^n \leq \eta(\frac{\varepsilon}{2}))} \\
&\leq \frac{\varepsilon}{2} + |a|\left(\eta\left(\frac{\varepsilon}{2}\right)\right)^{\frac{1}{n}} E|X|^n < \frac{\varepsilon}{2} + \frac{\varepsilon}{2b}E|X|^n \\
&\leq \frac{\varepsilon}{2} + \frac{\varepsilon}{2} = \varepsilon.
\end{aligned}$$

Thus $\delta_{\psi,n}$ is the modulus of continuity of $\psi^{(n)}$ on R. Assertion 1 is verified.

2. Assertion 2 is an immediate consequence of Assertion 1.

3. Suppose X^{n+1} is integrable. Then $\psi^{(n+1)}$ exists on R, with $|\psi^{(n+1)}| \leq E|X|^{n+1}$ by equality 5.8.24. Hence $\bar{r}_n(t) \leq E|X|^{n+1}|t-t_0|^{n+1}/(n+1)!$ according to Taylor's Theorem. □

For the proof of a partial converse, we need some basic equalities for binomial coefficients.

Lemma 5.8.13. Binomial coefficients. *For each $n \geq 1$, we have*

$$\sum_{k=0}^{n} \binom{n}{k} (-1)^k k^j = 0$$

for $j = 0, \ldots, n-1$, and

$$\sum_{k=0}^{n} \binom{n}{k} (-1)^k k^n = (-1)^n n!.$$

Proof. Differentiate j times the binomial expansion

$$(1 - e^t)^n = \sum_{k=0}^{n} \binom{n}{k} (-1)^k e^{kt}$$

to get

$$n(n-1)\ldots(n-j+1)(-1)^j (1-e^t)^{n-j} = \sum_{k=0}^{n} \binom{n}{k} (-1)^k k^j e^{kt},$$

and then set t to 0. □

Classical proofs for the next theorem in familiar texts rely on Fatou's Lemma, which is not constructive because it trivially implies the principle of infinite search. The following proof contains an easy fix.

Proposition 5.8.14. Moments of r.r.v. and derivatives of its characteristic function. *Let ψ denote the characteristic function of X. Let $n \geq 1$ be arbitrary. If ψ has a continuous derivative of order $2n$ in some neighborhood of $\lambda = 0$, then X^{2n} is integrable.*

Proof. Write $\lambda_k \equiv 2^{-k}$ for each $k \geq 1$. Then

$$\frac{\sin^2(\lambda_k X)}{\lambda_k^2} = \frac{\sin^2(2\lambda_{k+1} X)}{(2\lambda_{k+1})^2} = \frac{(2\sin(\lambda_{k+1}X)\cos(\lambda_{k+1}X))^2}{(2\lambda_{k+1})^2}$$

$$= \frac{\sin^2(\lambda_{k+1}X)}{\lambda_{k+1}^2} \cos^2(\lambda_{k+1}X) \leq \frac{\sin^2(\lambda_{k+1}X)}{\lambda_{k+1}^2}$$

for each $k \geq 1$. Thus we see that the sequence $\left(\left(\frac{\sin^2(\lambda_k X)}{\lambda_k^2}\right)^n\right)_{k=1,2,\ldots}$ of integrable r.r.v.'s is nondecreasing. Since $\psi^{(2n)}$ exists, we have, by Taylor's Theorem

$$\psi(\lambda) = \sum_{j=0}^{2n} \frac{\psi^{(j)}(0)}{j!} \lambda^j + o(\lambda^{2n})$$

as $\lambda \to 0$. Hence for any $\lambda \in R$ we have

$$
\begin{aligned}
E\left(\frac{\sin\lambda X}{\lambda}\right)^{2n} &= E\left(\frac{e^{i\lambda X} - e^{-i\lambda X}}{2i\lambda}\right)^{2n} \\
&= (2i\lambda)^{-2n} E \sum_{k=0}^{2n} \binom{2n}{k} (-1)^k e^{i(2n-2k)\lambda X} \\
&= (2i\lambda)^{-2n} \sum_{k=0}^{2n} \binom{2n}{k} (-1)^k \psi((2n-2k)\lambda) \\
&= (2i\lambda)^{-2n} \sum_{k=0}^{2n} \binom{2n}{k} (-1)^k \left\{ \sum_{j=0}^{2n} \frac{\psi^{(j)}(0)}{j!} (2n-2k)^j \lambda^j + o(\lambda^{2n}) \right\} \\
&= (2i\lambda)^{-2n} \left\{ o(\lambda^{2n}) + \sum_{j=0}^{2n} \frac{\psi^{(j)}(0)\lambda^j}{j!} \sum_{k=0}^{2n} \binom{2n}{k} (-1)^k (2n-2k)^j \right\} \\
&= o(1) + (2i\lambda)^{-2n} \{\psi^{(2n)}(0)\lambda^{2n} 2^{2n}\} = (-1)^n \psi^{(2n)}(0)
\end{aligned}
$$

in view of Lemma 5.8.13. Consequently, $E(\frac{\sin\lambda_k X}{\lambda_k})^{2n} \to (-1)^n \psi^{(2n)}(0)$. At the same time, $\frac{\sin\lambda_k t}{\lambda_k} \to t$ uniformly for t in any compact interval. Hence $(\frac{\sin\lambda_k X}{\lambda_k})^{2n} \uparrow X^{2n}$ a.u. as $k \to \infty$. Therefore, by the Monotone Convergence Theorem, the limit r.r.v. $X^{2n} = \lim_{k\to\infty} (\frac{\sin\lambda_k X}{\lambda_k})^{2n}$ is integrable. □

Proposition 5.8.15. Product distribution and direct product of characteristic function. *Let F_1, F_2 be distributions on R^n and R^m, respectively, with the characteristic functions ψ_1, ψ_2, respectively. Let the function $\psi_1 \otimes \psi_2$ be defined by*

$$
(\psi_1 \otimes \psi_2)(\lambda) \equiv \psi_1(\lambda_1)\psi_2(\lambda_2)
$$

for each $\lambda \equiv (\lambda_1, \lambda_2) \in R^{n+m}$, where$\lambda_1 \in R^n$ and $\lambda_2 \in R^m$. Let F be a distribution on R^{n+m} with characteristic function ψ. Then $F = F_1 \otimes F_2$ iff $\psi = \psi_1 \otimes \psi_2$.

Proof. Suppose $F = F_1 \otimes F_2$. Let $\lambda \equiv (\lambda_1, \lambda_2) \in R^{n+m}$ be arbitrary, where $\lambda_1 \in R^n$ and $\lambda_2 \in R^m$. Let $\exp(i\lambda^T \cdot)$ be the function on R^{n+m} whose value at arbitrary $x \equiv (x_1, x_2) \in R^{n+m}$, where $x_1 \in R^n$ and $x_2 \in R^m$, is $\exp(i\lambda^T x)$. Similarly, let $\exp(i\lambda_1^T \cdot)$, $\exp(i\lambda_2^T \cdot)$ be the functions whose values at $(x_1, x_2) \in R^{n+m}$ are $\exp(i\lambda_1^T x_1)$, $\exp(i\lambda_2^T x_2)$, respectively. Then

$$
\begin{aligned}
\psi(\lambda) \equiv F\exp(i\lambda^T \cdot) &= F\exp(i\lambda_1^T \cdot)\exp(i\lambda_2^T \cdot) \\
&= (F_1 \otimes F_2)\exp(i\lambda_1^T \cdot)\exp(i\lambda_2^T \cdot) \\
&= (F_1 \exp(i\lambda_1^T \cdot))(F_2 \exp(i\lambda_2^T \cdot)) \\
&= \psi_1(\lambda_1)\psi_2(\lambda_2) = (\psi_1 \otimes \psi_2)(\lambda).
\end{aligned}
$$

Thus $\psi = \psi_1 \otimes \psi_2$.

Conversely, suppose $\psi = \psi_1 \otimes \psi_2$. Let $G \equiv F_1 \otimes F_2$. Then G has characteristic function $\psi_1 \otimes \psi_2$ by the previous paragraph. Thus the distributions F and G have the same characteristic function ψ. By Theorem 5.8.9, it follows that $F = G \equiv F_1 \otimes F_2$. □

Corollary 5.8.16. Independence in terms of characteristic functions. *Let $X_1 : \Omega \to R^n$ and $X_2 : \Omega \to R^m$ be r.v.'s on a probability space (Ω, L, E), with characteristic functions ψ_1, ψ_2, respectively. Let ψ be the characteristic function of the r.v. $X \equiv (X_1, X_2) : \Omega \to R^{n+m}$. Then X_1, X_2 are independent iff $\psi = \psi_1 \otimes \psi_2$.*

Proof. Let F, F_1, F_2 be the distributions induced by X, X_1, X_2 on R^{n+m}, R^n, R^m, respectively. Then X_1, X_2 are independent iff $F = F_1 \otimes F_2$, by Definition 5.6.1. Since $F = F_1 \otimes F_2$ iff $\psi = \psi_1 \otimes \psi_2$, according to Proposition 5.8.15, the corollary is proved. □

Proposition 5.8.17. Conditional expectation of jointly normal r.r.v.'s. Let $Z_1, \ldots, Z_n, Y_1, \ldots, Y_m$ *be arbitrary jointly normal r.r.v.'s with mean* 0. *Suppose the covariance matrix $\overline{\sigma}_Z \equiv EZZ^T$ of $Z \equiv (Z_1, \ldots, Z_n)$ is positive definite. Let $\overline{\sigma}_Y \equiv EYY^T$ be the covariance matrix of $Y \equiv (Y_1, \ldots, Y_m)$. Define the $n \times m$ cross-covariance matrix $c_{Z,Y} \equiv EZY^T$, and define the $n \times m$ matrix $b_Y \equiv \overline{\sigma}_Z^{-1} c_{Z,Y}$. Then the following conditions hold:*

1. *The $m \times m$ matrix $\sigma_{Y|Z} \equiv \overline{\sigma}_Y - c_{Z,Y}^T \overline{\sigma}_Z^{-1} c_{Z,Y}$ is nonnegative definite.*
2. *For each $f \in L_Y$, we have*

$$E(f(Y)|Z) = \Phi_{b_Y^T Z, \sigma_{Y|Z}} f.$$

   Heuristically, given Z, the conditional distribution of Y is normal with mean $b_Y^T Z$ and covariance matrix $\sigma_{Y|Z}$. In particular, $E(Y|Z) = b_Y^T Z = c_{Z,Y}^T \overline{\sigma}_Z^{-1} Z$.
3. *The r.v.'s $V \equiv E(Y|Z)$ and $X \equiv Y - E(Y|Z)$ are independent normal r.v.'s with values in R^m.*
4. *$EY^T Y = EV^T V + EX^T X$.*

Proof. 1. Let $X \equiv (X_1, \ldots, X_m) \equiv Y - b_Y^T Z$. Thus $Y = b_Y^T Z + X$. Then $Z_1, \ldots, Z_n, X_1, \ldots, X_m$ are jointly normal according to Proposition 5.7.6. Furthermore,

$$EZX^T = EZY^T - EZZ^T b_Y \equiv c_{Z,Y} - \overline{\sigma}_Z b_Y = 0,$$

while the covariance matrix of X is given by

$$\begin{aligned}
\overline{\sigma}_X \equiv EXX^T &= EYY^T - EYZ^T b_Y - Eb_Y^T ZY^T + Eb_Y^T ZZ^T b_Y \\
&= \overline{\sigma}_Y - c_{Z,Y}^T b_Y - b_Y^T c_{Z,Y} + b_Y^T \overline{\sigma}_Z b_Y \\
&= \overline{\sigma}_Y - c_{Z,Y}^T \overline{\sigma}_Z^{-1} c_{Z,Y} - b_Y^T c_{Z,Y} + b_Y^T c_{Z,Y} \\
&= \overline{\sigma}_Y - c_{Z,Y}^T \overline{\sigma}_Z^{-1} c_{Z,Y} \equiv \sigma_{Y|Z},
\end{aligned}$$

whence $\sigma_{Y|Z}$ is nonnegative definite.

2. Hence the r.v. $U \equiv (Z,X)$ in R^{n+m} has mean 0 and covariance matrix

$$\overline{\sigma}_U \equiv \begin{bmatrix} \overline{\sigma}_Z & 0 \\ 0 & \overline{\sigma}_X \end{bmatrix} \equiv \begin{bmatrix} \overline{\sigma}_Z & 0 \\ 0 & \sigma_{Y|Z} \end{bmatrix}.$$

Accordingly, U has the characteristic function

$$\begin{aligned} E\exp(i\lambda^T U) &= \psi_{0,\overline{\sigma}_U}(\lambda) \equiv \exp\left(-\frac{1}{2}\lambda^T \overline{\sigma}_U \lambda\right) \\ &= \exp\left(-\frac{1}{2}\theta^T \overline{\sigma}_Z \theta\right)\exp\left(-\frac{1}{2}\gamma^T \overline{\sigma}_X \gamma\right) \\ &= E\exp(i\theta^T Z)E\exp(i\gamma^T X), \end{aligned}$$

for each $\lambda \equiv (\theta_1,\ldots,\theta_n,\gamma_1,\ldots,\gamma_m) \in R^{n+m}$. It follows from Corollary 5.8.16 that Z,X are independent. In other words, the distribution $E_{(Z,X)}$ induced by (Z,X) on R^{n+m} is given by the product distribution

$$E_{(Z,X)} = E_Z \otimes E_X$$

of E_Z, E_X induced on R^n, R^m, respectively, by Z,X, respectively.

Now let $f \in L_Y$ be arbitrary. Thus $f(Y) \in L$. Let $z \equiv (z_1,\ldots,z_n) \in R^n$ and $z \equiv (x_1,\ldots,x_m) \in R^m$ be arbitrary. Define

$$\widetilde{f}(z,x) \equiv f(b_Y^T z + x)$$

and

$$\overline{f}(z) \equiv E_X \widetilde{f}(z,\cdot) \equiv E\widetilde{f}(z,X) \equiv Ef(b_Y^T z + X) = \Phi_{b_Y^T z,\sigma_{Y|Z}} f.$$

We will prove that the r.r.v. $\overline{f}(Z)$ is the conditional expectation of $f(Y)$ given Z. To that end, let $g \in C(R^n)$ be arbitrary. Then, by Fubini's Theorem,

$$\begin{aligned} Ef(Y)g(Z) &= E\widetilde{f}(Z,X)g(Z) = E_{(Z,X)}(\widetilde{f}g) = E_Z \otimes E_X(\widetilde{f}g) \\ &= E_Z(E_X(\widetilde{f}g)) = E_Z\overline{f}g = E\overline{f}(Z)g(Z). \end{aligned}$$

It follows that $E(f(Y)|Z) = \overline{f}(Z) \equiv \Phi_{b_Y^T Z,\sigma_{Y|Z}} f$. In particular, $E(Y|Z) = b_Y^T Z = c_{Z,Y}^T \overline{\sigma}_Z^{-1} Z$.

3. By Step 2, the r.v.'s Z,X are independent normal. Hence the r.v.'s $V \equiv E(Y|Z) = b_Y^T Z$ and $X \equiv Y - E(Y|Z)$ are independent normal.

4. Hence $EV^T X = (EV^T)(EX) = 0$. It follows that

$$EY^T Y = E(V+X)^T(V+X) = EV^T V + EX^T X. \qquad \square$$

5.9 Central Limit Theorem

Let $X_1,\ldots,X_n$ be independent r.r.v.'s with mean 0 and standard deviations $\sigma_1,\ldots,\sigma_n$, respectively. Define σ by $\sigma^2 = \sigma_1^2 + \cdots + \sigma_n^2$ and consider the distribution F of the scaled sum $X = (X_1 + \cdots + X_n)/\sigma$. By replacing X_i with X_i/σ we may assume that $\sigma = 1$. The Central Limit Theorem says that if each individual summand X_i is small relative to the sum X, then F is close to the standard normal distribution $\Phi_{0,1}$.

One criterion, due to Lindberg and Feller, for the summands X_k $(k = 1, \ldots, n)$ to be individually small relative to the sum, is for

$$\theta(r) \equiv \sum_{k=1}^{n} (E1_{|X_k|>r} X_k^2 + E1_{|X_k|\le r}|X_k|^3)$$

to be small for some $r \ge 0$.

Lemma 5.9.1. Lindberg–Feller bound. *Suppose $r \ge 0$ is such that $\theta(r) < \frac{1}{8}$. Then*

$$\sum_{k=1}^{n} \sigma_k^3 \le \theta(r) \tag{5.9.1}$$

Proof. Consider each $k = 1, \ldots, n$. Then, since $\theta(r) < \frac{1}{8}$ by hypothesis, we have $z \equiv E1_{|X_k|>r} X_k^2 < \frac{1}{8}$ and $a \equiv E1_{|X_k|\le r}|X_k|^3 < \frac{1}{8}$. A consequence is that $(z + a^{2/3})^{3/2} \le z + a$, which can be seen by noting that the two sides are equal at $z = 0$ and by comparing first derivatives relative to z on $[0, \frac{1}{8}]$. Lyapunov's inequality then implies that

$$\begin{aligned}
\sigma_k^3 &= (EX_k^2 1_{(|X_k|>r)} + EX_k^2 1_{(|X_k|\le r)})^{3/2} \\
&\le (EX_k^2 1_{(|X_k|>r)} + (E|X_k|^3 1_{(|X_k|\le r)})^{2/3})^{3/2} \\
&\equiv (z + a^{2/3})^{3/2} \le z + a \equiv EX_k^2 1_{(|X_k|>r)} + E|X_k|^3 1_{(|X_k|\le r)}.
\end{aligned}$$

Summing over k, we obtain inequality 5.9.1. □

Theorem 5.9.2. Central Limit Theorem. *Let $f \in C(R)$ and $\varepsilon > 0$ be arbitrary. Then there exists $\delta > 0$ such that if $\theta(r) < \delta$ for some $r \ge 0$, then*

$$\left| \int f(x)dF(x) - \int f(x)d\Phi_{0,1}(x) \right| < \varepsilon. \tag{5.9.2}$$

Proof. Let ξ_R be an arbitrary but fixed binary approximation of R relative to the reference point 0. We assume, without loss of generality, that $|f(x)| \le 1$. Let δ_f be a modulus of continuity of f, and let $b > 0$ be so large that f has $[-b, b]$ as support. Let $\varepsilon > 0$ be arbitrary. By Proposition 5.3.5, there exists $\delta_{\widehat{J}}(\varepsilon, \delta_f, b, \|\xi_R\|) > 0$ such that if the distributions $F, \Phi_{0,1}$ satisfy

$$\rho_{\xi(R)}(F, \Phi_{0,1}) < \varepsilon' \equiv \delta_{\widehat{J}}(\varepsilon, \delta_f, b, \|\xi_R\|), \tag{5.9.3}$$

then inequality 5.9.2 holds. Separately, according to Corollary 5.8.11, there exists $\delta_{ch,dstr}(\varepsilon') > 0$ such that if the characteristic functions $\psi_F, \psi_{0,1}$ of $F, \Phi_{0,1}$, respectively, satisfy

$$\rho_{char}(\psi_F, \psi_{0,1}) \equiv \sum_{j=1}^{\infty} 2^{-j} \sup_{|\lambda| \le j} |\psi_F(\lambda) - \psi_{0,1}(\lambda)| < \varepsilon'' \equiv \delta_{ch,dstr}(\varepsilon'),$$

(5.9.4)

then inequality 5.9.3 holds.

Now take $m \ge 1$ be so large that $2^{-m+2} < \varepsilon''$, and define

$$\delta \equiv \frac{1}{8} \wedge \frac{1}{6} m^{-3} \varepsilon''.$$

Suppose

$$\theta(r) < \delta$$

for some $r \ge 0$. Then $\theta(r) < \frac{1}{8}$. We will show that inequality 5.9.2 holds.

To that end, let $\lambda \in [-m, m]$ and $k = 1, \ldots, n$ be arbitrary. Let φ_k denote the characteristic function of X_k, and let Y_k be a normal r.r.v. with mean 0, variance σ_k^2, and characteristic function $e^{-\sigma_k^2 \lambda^2/2}$. Then

$$E|Y_k|^3 = \frac{2}{\sqrt{2\pi}\sigma_k} \int_0^{\infty} y^3 \exp\left(-\frac{1}{2\sigma_k^2} y^2\right) dy.$$

$$= \frac{4\sigma_k^3}{\sqrt{2\pi}} \int_0^{\infty} u \exp(-u) du = \frac{4\sigma_k^3}{\sqrt{2\pi}} = 2\sqrt{\frac{2}{\pi}} \sigma_k^3,$$

where we made a change of integration variables $u \equiv -\frac{1}{2\sigma_k^2} y^2$. Moreover, since $\sum_{k=1}^{n} \sigma_k^2 = 1$ by assumption, and since all characteristic functions have absolute value bounded by 1, we have

$$|\varphi_F(\lambda) - e^{-\lambda^2/2}| = \left| \prod_{k=1}^{n} \varphi_k(\lambda) - \prod_{k=1}^{n} e^{-\sigma_k^2 \lambda^2/2} \right|$$

$$\le \sum_{k=1}^{n} |\varphi_k(\lambda) - e^{-\sigma_k^2 \lambda^2/2}|. \tag{5.9.5}$$

By Proposition 5.8.12, the Taylor expansions up to degree 2 for the characteristic functions $\varphi_k(\lambda)$ and $e^{-\sigma_k^2 \lambda^2/2}$ are equal because the two corresponding distributions have equal first and second moments. Hence the difference of the two functions is equal to the difference of the two remainders in their respective Taylor expansions. Again by Proposition 5.8.12, the remainder for $\varphi_k(\lambda)$ is bounded by

$$\lambda^2 E X_k^2 1_{(|X_k| > r)} + \frac{|\lambda|^3}{3!} E|X_k|^3 1_{(|X_k| \le r)} \le m^3 (E X_k^2 1_{(|X_k| > r)} + E|X_k|^3 1_{(|X_k| \le r)}).$$

By the same token, the remainder for $e^{-\sigma_k^2 \lambda^2/2}$ is bounded by a similar expression, where X_k is replaced by Y_k and where $r \ge 0$ is replaced by $s \ge 0$, which becomes, as $s \to \infty$,

$$m^3 E|Y_k|^3 = 2\sqrt{\frac{2}{\pi}} m^3 \sigma_k^3 < 2m^3 \sigma_k^3.$$

Combining, inequality 5.9.5 yields, for each $\lambda \in [-m, m]$,

$$|\varphi_F(\lambda) - e^{-\lambda^2/2}| \le m^3 \sum_{k=1}^{n} (EX_k^2 1_{(|X_k|>r)} + E|X_k|^3 1_{(|X_k|\le r)}) + 2m^3 \sum_{k=1}^{n} \sigma_k^3$$

$$\le 3m^3\theta(r) \le 3m^3\delta \le \frac{\varepsilon''}{2},$$

where the second inequality follows from the definition of $\theta(r)$ and from Lemma 5.9.1. Hence, since $|\psi_F - \psi_{0,1}| \le 2$, we obtain

$$\rho_{char}(\psi_F, \psi_{0,1}) \le \sum_{j=1}^{m} 2^{-j} \sup_{|\lambda|\le j} |\psi_F(\lambda) - \psi_{0,1}(\lambda)| + 2^{-m+1}$$

$$\le \frac{\varepsilon''}{2} + \frac{\varepsilon''}{2} = \varepsilon'' \equiv \delta_{ch,dstr}(\varepsilon'),$$

establishing inequality 5.9.4. Consequently, inequality 5.9.3 and, in turn, inequality 5.9.2 follow. The theorem is proved. □

Corollary 5.9.3. Lindberg's Central Limit Theorem. *For each* $p = 1, 2, \ldots$, *let* $n_p \ge 1$ *be arbitrary, and let* $(X_{p,1}, \ldots, X_{p,n(p)})$ *be an independent sequence of r.r.v.'s with mean* 0 *and variance* $\sigma_{p,k}^2$ *such that* $\sum_{k=1}^{n(p)} \sigma_{p,k}^2 = 1$. *Suppose for each* $r > 0$ *we have*

$$\lim_{p\to\infty} \sum_{k=1}^{n(p)} EX_{p,k}^2 1_{(|X(p,k)|>r)} = 0. \tag{5.9.6}$$

Then $\sum_{k=1}^{n(p)} X_{p,k}$ *converges in distribution to the standard normal distribution* $\Phi_{0,1}$ *as* $p \to \infty$.

Proof. Let $\delta > 0$ be arbitrary. According to Theorem 5.9.2, it suffices to show that there exists $r > 0$ such that, for sufficiently large p, we have

$$\sum_{k=1}^{n(p)} EX_{p,k}^2 1_{(|X(p,k)|>r)} < \frac{\delta}{2}$$

and

$$\sum_{k=1}^{n(p)} E|X_{p,k}|^3 1_{(|X(p,k)|<r)} < \frac{\delta}{2}.$$

For that purpose, take any $r \in \left(0, \frac{\delta}{2}\right)$. Then the first of the last two inequalities holds for sufficiently large p, in view of inequality 5.9.6 in the hypothesis. The second follows from

$$\sum_{k=1}^{n(p)} E|X_{p,k}|^3 1_{(|X(p,k)|<r)} \leq r \sum_{k=1}^{n(p)} EX_{p,k}^2 = r < \frac{\delta}{2}. \qquad \square$$

Because of the importance of the Central Limit Theorem, much work since the early development of probability theory has been dedicated to an optimal rate of convergence, culminating in Feller's bound: $\sup_{t \in R} |F(t) - \Phi(t)| \leq 6\theta(r)$ for each $r \geq 0$. The proof on pages 544–546 of [Feller II 1971], which is a careful analysis of the difference $\varphi_k(\lambda) - e^{-\sigma_k^2 \lambda^2/2}$, contains a few typos and omitted steps that serve to keep the reader on his or her toes. That proof contains also a superfluous assumption that the maximum distance between two P.D.F.'s is always attained at some point in R. There is no constructive proof for the general validity of that assumption. There is, however, an easy constructive substitute that says if one of the two P.D.F.s is continuously differentiable, then the supremum distance exists: there is a sequence (t_n) in R such that $\lim_{n \to \infty} |F(t_n) - \Phi_{0,1}(t_n)|$ exists and bounds any $|F(t) - \Phi_{0,1}(t)|$. This is sufficient for Feller's proof.

Part III

Stochastic Process

6

Random Field and Stochastic Process

In this chapter, unless otherwise specified, (S,d) will denote a locally compact metric space, with an arbitrary but fixed reference point $x_\circ$. Let Q be an arbitrary set. Let (Ω, L, E) be an arbitrary probability space.

6.1 Random Field and Finite Joint Distributions

In this section, we introduce random fields, their marginal distributions, and some notions of their continuity.

Definition 6.1.1. Random field. Suppose a function

$$X : Q \times \Omega \to S$$

is such that, for each $t \in Q$, the function $X_t \equiv X(t,\cdot)$ is an r.v. on Ω with values in S. Then X is called a *random field*, or r.f. for short, with *sample space* (Ω, L, E), with *parameter set* Q, and with *state space* (S,d). To be precise, we will sometimes write

$$X : Q \times (\Omega, L, E) \to (S,d).$$

We will let $\widehat{R}(Q \times \Omega, S)$ denote the set of all such r.f.'s. Two r.f.'s $X, Y \in \widehat{R}(Q \times \Omega, S)$ are considered equal if $X_t = Y_t$ a.s. on Ω, for each $t \in Q$.

Let $X \in \widehat{R}(Q \times \Omega, S)$ be arbitrary. For each $\omega \in \Omega$ such that $domain(X(\cdot,\omega))$ is nonempty, the function $X(\cdot,\omega)$ is called a *sample function*. If K is a subset of Q, then we write

$$X|K \equiv X|(K \times \Omega) : K \times \Omega \to S,$$

and call the r.f. $X|K$ the *restriction* of X to K.

In the special case where the parameter set Q is a subset of R, then the r.f. X is called a *stochastic process*, or simply a *process*. In that case, the variable $t \in Q$ is often called the *time parameter.* □

When the parameter set Q is countably infinite, we can view an r.f. with state space (S,d) as an r.v. with values in (S^∞, d^∞).

When the parameter set Q is a metric space, we introduce three notions of continuity of an r.f. that correspond to the terminology in [Neveu 1965]. For ease of presentation, we restrict our attention to the special case where Q is bounded. The generalization to a locally compact metric space Q is straightforward.

Definition 6.1.2. Continuity of r.f. on a bounded metric parameter space. Let $X : (Q,d_Q) \times (\Omega, L, E) \to (S,d)$ be an r.f., where (S,d) is a locally compact metric space, and where (Q,d_Q) is a bounded metric space. Thus $d_Q \leq b$ for some $b \geq 0$.

1. Suppose, for each $\varepsilon > 0$, there exists $\delta_{Cp}(\varepsilon) > 0$ such that

$$E(1 \wedge d(X_t, X_s)) \leq \varepsilon$$

for each $s,t \in Q$ with $d_Q(t,s) < \delta_{Cp}(\varepsilon)$. Then the r.f. X is said to be *continuous in probability*, with the operation δ_{Cp} as a *modulus of continuity in probability*. We will let $\widehat{R}_{Cp}(Q \times \Omega, S)$ denote the set of r.f.'s that are continuous in probability, with the given bounded metric space as the parameter space.

2. Suppose $domain(X(\cdot,\omega))$ is dense in Q for a.e. $\omega \in \Omega$. Suppose, in addition, that for each $\varepsilon > 0$, there exists $\delta_{cau}(\varepsilon) > 0$ such that for each $s \in Q$, there exists a measurable set $D_s \subset domain(X_s)$ with $P(D_s^c) < \varepsilon$, such that for each $\omega \in D_s$ and for each $t \in domain(X(\cdot,\omega))$ with $d_Q(t,s) < \delta_{cau}(\varepsilon)$, we have

$$d(X(t,\omega), X(s,\omega)) \leq \varepsilon.$$

Then the r.f. X is said to be *continuous a.u.*, with the operation δ_{cau} as a *modulus of continuity a.u.* on Q.

3. Suppose, for each $\varepsilon > 0$, there exist $\delta_{auc}(\varepsilon) > 0$ and a measurable set D with $P(D^c) < \varepsilon$ such that for each $\omega \in D$, we have (i) $domain(X(\cdot,\omega)) = Q$ and (ii)

$$d(X(t,\omega), X(s,\omega)) \leq \varepsilon,$$

for each $t,s \in domain(X(\cdot,\omega))$ with $d_Q(t,s) < \delta_{auc}(\varepsilon)$. Then the r.f. X is said to be *a.u. continuous*, with the operation δ_{auc} as a *modulus of a.u. continuity*. □

The reader can give simple examples of stochastic processes that are continuous in probability but not continuous a.u., and of processes that are continuous a.u. but not a.u. continuous.

Definition 6.1.3. Continuity of r.f. on an arbitrary metric parameter space. Let $X : Q \times \Omega \to S$ be an r.f., where (S,d) is a locally compact metric space and where (Q,d_Q) is an arbitrary metric space.

The r.f. X is said to be *continuous in probability* if, for each bounded subset K of Q, the restricted r.f. $X|K : K \times \Omega \to S$ is continuous in probability with modulus of continuity in probability $\delta_{Cp,K}$.

The r.f. X is said to be *continuous a.u.* if, for each bounded subset K of Q, the restricted r.f. $X|K : K \times \Omega \to S$ is continuous a.u., with some modulus of continuity a.u. $\delta_{cau,K}$.

The r.f. X is said to be *a.u. continuous* if, for each bounded subset K of Q, the restricted r.f. $X|K : K \times \Omega \to S$ is a.u. continuous, with some modulus of a.u. continuity $\delta_{auc,K}$. If, in addition, $Q = [0,\infty)$, and if $\delta_{auc,[M,M+1]} = \delta_{auc,[0,1]}$ for each $M \geq 0$, then X is said to be *time-uniformly a.u. continuous.* □

Proposition 6.1.4. Alternative definitions of r.f. continuity. *Let $X : Q \times \Omega \to S$ be an r.f., where (S,d) is a locally compact metric space and where (Q,d_Q) is a bounded metric space. Then the following conditions hold:*

1. Suppose the r.f. X is continuous in probability, with a modulus of continuity in probability δ_{Cp}. Let $\varepsilon > 0$ be arbitrary. Define $\delta_{cp}(\varepsilon) \equiv \delta_{Cp}(2^{-2}(1 \wedge \varepsilon)^2) > 0$. Then for each $s,t \in Q$ with $d_Q(t,s) < \delta_{cp}(\varepsilon)$, there exists measurable set $D_{t,s}$ with $PD_{t,s}^c \leq \varepsilon$ such that

$$d(X(t,\omega),X(s,\omega)) \leq \varepsilon$$

for each $\omega \in D_{t,s}$. Conversely, if there exists an operation δ_{cp} with the just described properties, then the r.f. X is continuous in probability, with a modulus of continuity in probability δ_{Cp} defined by $\delta_{Cp}(\varepsilon) \equiv \delta_{cp}(2^{-1}\varepsilon)$ for each $\varepsilon > 0$.

2. The r.f. X is continuous a.u. iff for each $\varepsilon > 0$ and $s \in Q$, there exists a measurable set $D_s \subset domain(X_s)$ with $P(D_s^c) < \varepsilon$, such that for each $\alpha > 0$, there exists $\delta'_{cau}(\alpha,\varepsilon) > 0$ such that

$$d(X(t,\omega),X(s,\omega)) \leq \alpha$$

for each $t \in Q$ with $d_Q(t,s) < \delta'_{cau}(\alpha,\varepsilon)$, for each $\omega \in D_s$.

3. Suppose the r.f. X is a.u. continuous. Then for each $\varepsilon > 0$, there exists a measurable set D with $P(D^c) < \varepsilon$ such that for each $\alpha > 0$, there exists $\delta'_{auc}(\alpha,\varepsilon) > 0$ such that

$$d(X(t,\omega),X(s,\omega)) \leq \alpha$$

for each $s,t \in Q$ with $d_Q(t,s) < \delta'_{auc}(\alpha,\varepsilon)$, for each $\omega \in D$. Conversely, if such an operation δ'_{auc} exists, then X has a modulus of a.u. continuity given by $\delta_{auc}(\varepsilon) \equiv \delta'_{auc}(\varepsilon,\varepsilon)$ for each $\varepsilon > 0$.

Proof. As usual, write $\widehat{d} \equiv 1 \wedge d$.

1. Suppose X is continuous in probability, with a modulus of continuity in probability δ_{Cp}. Let $\varepsilon > 0$ be arbitrary. Write $\varepsilon' \equiv 1 \wedge \varepsilon$. Suppose $s,t \in Q$ are arbitrary

$$d_Q(t,s) < \delta_{cp}(\varepsilon) \equiv \delta_{Cp}(2^{-2}(1 \wedge \varepsilon)^2) \equiv \delta_{Cp}(2^{-2}\varepsilon'^2).$$

Then, by Definition 6.1.2 of δ_{Cp} as a modulus of continuity in probability, we have $E\widehat{d}(X_t,X_s) \leq 2^{-2}\varepsilon'^2 < \varepsilon'^2$. Take any $\alpha \in (2^{-1}\varepsilon',\varepsilon')$ such that the set $D_{t,s} \equiv (\widehat{d}(X_t,X_s) \leq \alpha)$ is measurable. Then Chebychev's inequality implies that $P(D_{t,s}^c) < \alpha^{-1}2^{-2}\varepsilon'^2 < \varepsilon' \leq \varepsilon$. Moreover, for each $\omega \in D_{t,s}$, we have $\widehat{d}(X(t,\omega),X(s,\omega)) \leq \alpha < \varepsilon' \leq \varepsilon$. Thus the operation δ_{cp} has the properties described in Assertion 1.

Conversely, suppose δ_{cp} is an operation with the properties described in Assertion 1. Let $\varepsilon > 0$ be arbitrary. Let $s,t \in Q$ be arbitrary with $d_Q(t,s) < \delta_{cp}(\frac{1}{2}\varepsilon)$. Then by hypothesis, there exists a measurable subset $D_{t,s}$ with $PD_{t,s}^c \leq \frac{1}{2}\varepsilon$ such that for each $\omega \in D_{t,s}$, we have $d(X_t(\omega), X_s(\omega)) \leq \frac{1}{2}\varepsilon$. It follows that $E(1 \wedge d(X_t, X_s)) \leq \frac{1}{2}\varepsilon + PD_{t,s}^c \leq \varepsilon$. Thus X is continuous in probability. Assertion 1 is proved.

2. Suppose X is continuous a.u., with δ_{cau} as a modulus of continuity a.u. Let $\varepsilon > 0$ and $s \in Q$ be arbitrary. Then there exists, for each $k \geq 1$, a measurable set $D_{s,k}$ with $P(D_{s,k}^c) < 2^{-k}\varepsilon$ such that for each $\omega \in D_{s,k}$, we have

$$d(X(t,\omega), X(s,\omega)) \leq 2^{-k}\varepsilon$$

for each $t \in Q$ with $d_Q(t,s) < \delta_{cau}(2^{-k}\varepsilon)$. Let $D_s \equiv \bigcap_{k=1}^{\infty} D_{s,k}$. Then $P(D_s^c) < \sum_{k=1}^{\infty} 2^{-k}\varepsilon = \varepsilon$. Now let $\alpha > 0$ be arbitrary. Let $k \geq 1$ be so large that $2^{-k} < \alpha$, and let

$$\delta'_{cau}(\alpha, \varepsilon) \equiv \delta_{cau}(2^{-k}\varepsilon).$$

Consider each $\omega \in D_s$ and $t \in Q$ with $d_Q(t,s) < \delta'_{cau}(\alpha, \varepsilon)$. Then $\omega \in D_{s,k}$, and $d_Q(t,s) < \delta_{cau}(2^{-k}\varepsilon)$. Hence

$$d(X(t,\omega), X(s,\omega)) \leq 2^{-k}\varepsilon < \alpha.$$

Thus the operation δ'_{cau} has the described properties in Assertion 2.

Conversely, let δ'_{cau} be an operation with the properties described in Assertion 2. Let $\varepsilon > 0$ be arbitrary. Let $s \in Q$ be arbitrary. Then there exists a measurable set D_s with $P(D_s^c) < \varepsilon$ such that for each $\omega \in D_s$, and $t \in Q$ with $d_Q(t,s) < \delta_{cau}(\varepsilon) \equiv \delta'_{cau}(\varepsilon, \varepsilon)$, we have $d(X(t,\omega), X(s,\omega)) \leq \varepsilon$. Thus the r.f. X is continuous a.u., with the operation δ_{cau} as a modulus of continuity a.u. Assertion 2 is proved.

3. For Assertion 3, proceed almost verbatim as in the proof of Assertion 2. Suppose the r.f. X is a.u. continuous, with δ_{auc} as a modulus of a.u. continuity. Let $\varepsilon > 0$ be arbitrary. Then there exists, for each $k \geq 1$, a measurable set D_k with $P(D_k^c) < 2^{-k}\varepsilon$ such that for each $\omega \in D_k$, we have

$$d(X(t,\omega), X(s,\omega)) \leq 2^{-k}\varepsilon$$

for each $s,t \in Q$ with $d_Q(t,s) < \delta_{auc}(2^{-k}\varepsilon)$. Let $D \equiv \bigcap_{k=1}^{\infty} D_k$. Then $P(D^c) < \sum_{k=1}^{\infty} 2^{-k}\varepsilon = \varepsilon$. Now let $\alpha > 0$ be arbitrary. Let $k \geq 1$ be so large that $2^{-k} < \alpha$, and let

$$\delta'_{auc}(\alpha, \varepsilon) \equiv \delta_{auc}(2^{-k}\varepsilon).$$

Consider each $\omega \in D$ and $s,t \in Q$ with $d_Q(t,s) < \delta'_{auc}(\alpha, \varepsilon)$. Then $\omega \in D_k$ and $d_Q(t,s) < \delta_{auc}(2^{-k}\varepsilon)$. Hence

$$d(X(t,\omega), X(s,\omega)) \leq 2^{-k}\varepsilon < \alpha.$$

Thus the operation δ'_{auc} has the properties described in Assertion 3.

Conversely, suppose there exists an operation δ'_{auc} with the properties described in Assertion 3. Let $\varepsilon > 0$ be arbitrary. Define $\delta_{auc}(\varepsilon) \equiv \delta'_{auc}(\varepsilon,\varepsilon)$. Then there exists a measurable set D with $P(D^c) < \varepsilon$ such that for each $\omega \in D$ and $s,t \in Q$ with $d_Q(t,s) < \delta_{auc}(\varepsilon) \equiv \delta'_{auc}(\varepsilon,\varepsilon)$, we have $d(X(t,\omega),X(s,\omega)) \leq \varepsilon$. Thus the r.f. X is a.u. continuous, with the operation δ_{auc} as a modulus of a.u. continuity. Assertion 3 is proved. □

Proposition 6.1.5. a.u. Continuity implies continuity a.u., etc. *Let* $X : (Q,d_Q)\times \Omega \to (S,d)$ *be an r.f., where* (S,d) *is a locally compact metric space and where* (Q,d_Q) *is a bounded metric space. Then a.u. continuity of* X *implies continuity a.u., which in turn implies continuity in probability.*

Proof. Let $\varepsilon > 0$ be arbitrary. Suppose X is a.u. continuous, with modulus of a.u. continuity given by δ_{auc}. Define $\delta_{cau} \equiv \delta_{auc}$. Let D be a measurable set satisfying Condition 3 in Definition 6.1.2. Then $D_s \equiv D$ satisfies Condition 2 in Definition 6.1.2. Accordingly, X is continuous a.u.

Now suppose X is continuous a.u., with modulus of continuity a.u. given by δ_{cau}. Let D_s be a measurable set satisfying Condition 2 in Definition 6.1.2. Then $D_{t,s} \equiv D_s$ satisfies the conditions in Assertion 1 of Proposition 6.1.4, provided that we define $\delta_{cp} \equiv \delta_{cau}$. Accordingly, X is continuous in probability. □

Definition 6.1.6. Marginal distributions of an r.f. Let (S,d) be an arbitrary complete metric space. Let $X : Q \times (\Omega,L,E) \to (S,d)$ be an r.f. Let $n \geq 1$ be arbitrary, and let $t \equiv (t_1,\ldots,t_n)$ be an arbitrary sequence in the parameter set Q. Let $F_{t(1),\ldots,t(n)}$ denote the distribution induced on (S^n,d^n) by the r.v. $(X_{t(1)},\ldots,X_{t(n)})$. Thus

$$F_{t(1),\ldots,t(n)}f \equiv Ef(X_{t(1)},\ldots,X_{t(n)}) \tag{6.1.1}$$

for each $f \in C_{ub}(S^n)$. Then we call the indexed family

$$F \equiv \{F_{t(1),\ldots,t(n)} : n \geq 1 \text{ and } t_1,\ldots,t_n \in Q\}$$

the family of *marginal distributions of* X. We will say that the r.f. X extends the family F of finite joint distributions and that X is an *extension* of F.

Let $X' : Q \times (\Omega',L',E') \to (S,d)$ be an r.f. with sample space (Ω',L',E'). Then X and X' are said to be *equivalent* if their marginal distributions at each finite sequence in Q are the same. In other words, X and X' are said to be equivalent if

$$Ef(X_{t(1)},\ldots,X_{t(n)}) = E'f(X'_{t(1)},\ldots,X'_{t(n)})$$

for each $f \in C_{ub}(S^n)$, for each sequence $(t_1,\ldots,t_n)$ in Q, for each $n \geq 1$. In short, two r.f.'s are equivalent if they extend the same family of finite joint distributions. □

6.2 Consistent Family of f.j.d.'s

In the last section, we saw that each r.f. gives rise to a family of marginal distributions. Conversely, in this section, we seek conditions for a family F of finite joint distributions to be the family of marginal distributions of some r.f. We will show that a necessary condition is consistency, to be defined next. In the following chapters we will present various sufficient conditions on F for the construction of r.f.'s with F as the family of marginal distributions of r.f.'s and processes with various desired properties of sample functions.

Recall that, unless otherwise specified, (S,d) will denote a locally compact metric space, with an arbitrary but fixed reference point $x_\circ$, and (Ω, L, E) will denote an arbitrary probability space.

Definition 6.2.1. Consistent family of f.j.d.'s. Let Q be a set. Suppose, for each $n \geq 1$ and for each finite sequence $t_1, \ldots, t_n$ in Q, a distribution $F_{t(1),\ldots,t(n)}$ is given on the locally compact metric space $(S^n, d^{(n)})$, which will be called a *finite joint distribution*, or *f.j.d.* for short. Then the indexed family

$$F \equiv \{F_{t(1),\ldots,t(n)} : n \geq 1 \text{ and } t_1, \ldots, t_n \in Q\}$$

is said to be a *consistent family* of *f.j.d.'s* with parameter set Q and state space S, if the following *Kolmogorov consistency condition* is satisfied.

Let $n, m \geq 1$ be arbitrary. Let $t \equiv (t_1, \ldots, t_m)$ be an arbitrary sequence in Q, and let $i \equiv (i_1, \ldots, i_n)$ be an arbitrary sequence in $\{1, \ldots, m\}$. Define the continuous function $i^* : S^m \to S^n$ by

$$i^*(x_1, \ldots, x_m) \equiv (x_{i(1)}, \ldots, x_{i(n)}) \tag{6.2.1}$$

for each $(x_1, \ldots, x_m) \in S^m$, and call i^* the *dual function of the sequence* i. Then, for each $f \in C_{ub}(S^n)$, we have

$$F_{t(1),\ldots,t(m)}(f \circ i^*) = F_{t(i(1)),\ldots,t(i(n))} f \tag{6.2.2}$$

or, in short,

$$F_t(f \circ i^*) = F_{t \circ i}(f).$$

We will let $\widehat{F}(Q,S)$ denote the set of consistent families of *f.j.d.'s* with parameter set Q and state space S. Two consistent families $F \equiv \{F_{t(1),\ldots,t(n)} : n \geq 1$ and $t_1, \ldots, t_n \in Q\}$ and $F' \equiv \{F'_{t(1),\ldots,t(n)} : n \geq 1$ and $t_1, \ldots, t_n \in Q\}$ are considered equal if $F_{t(1),\ldots,t(n)} = F'_{t(1),\ldots,t(n)}$ as distributions for each $n \geq 1$ and $t_1, \ldots, t_n \in Q$. In that case, we write simply $F = F'$. When there is little risk of confusion, we will call a consistent family of f.j.d.'s simply a *consistent family*. □

Note that for an arbitrary $f \in C_{ub}(S^n)$, we have $f \circ i^* \in C_{ub}(S^m)$, so $f \circ i^*$ is integrable relative to $F_{t(1),\ldots,t(m)}$. Hence the left-hand side of equality 6.2.2 makes sense.

Definition 6.2.2. Notations for sequences. Given any sequence $(a_1, \ldots, a_m)$ of objects, we will use the shorter notation a to denote the sequence. When there is little risk of confusion, we will write $\kappa\sigma \equiv \kappa \circ \sigma$ for the composite of two functions $\sigma : A \to B$ and $\kappa : B \to C$. Separately, for each $m \geq n \geq 1$, define the sequence

$$\kappa \equiv \kappa_{n,m} : \{1,\ldots,m-1\} \to \{1,\ldots,m\}$$

by

$$(\kappa_1,\ldots,\kappa_{m-1}) \equiv (1,\ldots,\widehat{n},\ldots,m) \equiv (1,\ldots,n-1,n+1,\ldots,m),$$

where the caret on the top of an element in a sequence signifies the omission of that element. Let $\kappa^* \equiv \kappa^*_{n,m}$ denote the dual function of sequence κ. Thus

$$\kappa^* x = \kappa^*(x_1,\ldots,x_m) \equiv x\kappa = (x_{\kappa(1)},\ldots,x_{\kappa(m-1)}) = (x_1,\ldots,\widehat{x_n},\ldots,x_m)$$

for each $x \equiv (x_1,\ldots,x_m) \in S^m$. In words, the function $\kappa^*_{n,m}$ deletes the nth entry of the sequence $(x_1,\ldots,x_m)$.

Lemma 6.2.3. Consistency when parameter set is a metrically discrete subset of R. *Let (S,d) be a locally compact metric space. Let Q be an arbitrary metrically discrete subset of R. Suppose, for each $m \geq 1$ and nonincreasing sequence $r \equiv (r_1,\ldots,r_m)$ in Q, there exists a distribution $F_{r(1),\ldots,r(m)}$ on (S^m,d^m) such that*

$$F_{r(1),\ldots,\widehat{r(n)}\ldots,r(m)} f = F_{r(1),\ldots,r(m)}(f \circ \kappa^*_{n,m}) \tag{6.2.3}$$

or, equivalently,

$$F_{r\circ\kappa(n,m)} f = F_r(f \circ \kappa^*_{n,m}), \tag{6.2.4}$$

for each $f \in C_{ub}(S^{m-1})$, for each $n = 1,\ldots,m$. Then the family

$$F \equiv \{F_{r(1),\ldots,r(n)} : n \geq 1; r_1 \leq \ldots \leq r_n \ in \ Q\}$$

of f.j.d.'s can be uniquely extended to a consistent family of f.j.d.'s

$$F = \{F_{s(1),\ldots,s(m)} : m \geq 1; s_1,\ldots,s_m \in Q\} \tag{6.2.5}$$

with parameter Q.

Proof. 1. Let the integers m,n, with $m \geq n \geq 1$, and the increasing sequence $r \equiv (r_1,\ldots,r_m)$ in Q be arbitrary. Let $r' \equiv (r'_1,\ldots,r'_n)$ be an arbitrary subsequence of r. Then $m > h \equiv m-n \geq 0$. Moreover, r' can be obtained by deleting h elements in the sequence r. Specifically, $r' = \kappa^* r = r\kappa$, where

$$\kappa \equiv \kappa_{n(h),m}\kappa_{n(h-1),m-1}\ldots\kappa_{n(1),m-h} : \{1,\ldots,m-h\} \to \{1,\ldots,m\} \tag{6.2.6}$$

if $h > 0$, and where κ is the identity function if $h = 0$. Call such an operation κ a *deletion*. Then, by repeated application of equality 6.2.3 in the hypothesis, we obtain

$$F_{r'} f = F_{r\kappa} f = F_r(f \circ \kappa^*) \tag{6.2.7}$$

for each $f \in C(\overline{S}^{m-h})$, for each deletion κ, for each increasing sequence $r \equiv (r_1,\ldots,r_m)$ in Q.

2. Let the sequence $s \equiv (s_1,\ldots,s_p)$ in Q be arbitrary. Let $r \equiv (r_1,\ldots,r_m)$ be an arbitrary increasing sequence in Q such that s is a sequence in $\{r_1,\ldots,r_m\}$.

Then because the sequence r is increasing, there exists a unique function $\sigma :$ $\{1, \ldots, p\} \to \{1, \ldots, m\}$ such that $s = r\sigma$. Let $f \in C_{ub}(S^p, d^p)$ be arbitrary. Define

$$\overline{F}_{s(1),\ldots,s(p)} f \equiv \overline{F}_s f \equiv F_r(f \circ \sigma^*). \tag{6.2.8}$$

We will verify that $\overline{F}_s f$ is well defined. To that end, let $r' \equiv (r'_1, \ldots, r'_{m'})$ be a second increasing sequence in Q such that s is a sequence in $\{r'_1, \ldots, r'_{m'}\}$, and let $\sigma' : \{1, \ldots, p\} \to \{1, \ldots, m'\}$ be the corresponding function such that $s = r'\sigma'$. We need to verify that $F_r(f \circ \sigma^*) = F_{r'}(f \circ \sigma'^*)$. To that end, let $\bar{r} \equiv (\bar{r}_1, \ldots, \bar{r}_{\overline{m}})$ be an arbitrary supersequence of r and r'. Then $r = \bar{r}\kappa$ and $r' = \bar{r}\kappa'$ for some deletions κ and κ', respectively. Hence $s = r\sigma = \bar{r}\kappa\sigma$ and $s = r'\sigma' = \bar{r}\kappa'\sigma'$. Then by the uniqueness alluded to in the previous paragraph, we have $\kappa\sigma = \kappa'\sigma'$ and so $\sigma^* \circ \kappa^* = \sigma'^* \circ \kappa'^*$. Consequently,

$$\begin{aligned} F_r(f \circ \sigma^*) &= F_{\bar{r}\kappa}(f \circ \sigma^*) = F_{\bar{r}}(f \circ \sigma^* \circ \kappa^*) = F_{\bar{r}}(f \circ \sigma'^* \circ \kappa'^*) \\ &= F_{\bar{r}\kappa'}(f \circ \sigma'^*) = F_{r'}(f \circ \sigma'^*), \end{aligned}$$

where the second and fourth equalities are thanks to equality 6.2.7. This shows that $\overline{F}_s f$ is well defined in equality 6.2.8. The same equality says that $\overline{F}_s$ is the distribution induced by the r.v. $\sigma^* : (S^m, \overline{C_{ub}}(S^m, d^m), F_r) \to (S^p, d^p)$, where $\overline{C_{ub}}(S^m, d^m)^-$ stands for the completion of $C(S^m, d^m)$ relative to the distribution F_r. In particular, $\overline{F}_{s(1),\ldots,s(p)} \equiv \overline{F}_s$ is a distribution.

3. Next, let $s \equiv (s_1, \ldots, s_q)$ be an arbitrary sequence in Q, and let $(s_{i(1)}, \ldots, s_{i(p)})$ be an arbitrary subsequence of s. Write $i \equiv (i_1, \ldots, i_p)$. Let the increasing sequence $r \equiv (r_1, \ldots, r_m)$ be arbitrary such that s is a sequence in $\{r_1, \ldots, r_m\}$, and let $\sigma : \{1, \ldots, q\} \to \{1, \ldots, m\}$ such that $s = r\sigma$. Then $si = r\sigma i$. Hence, for each $f \in C_{ub}(S^p, d^p)$, we have

$$\overline{F}_{si} f = \overline{F}_{r\sigma i} f \equiv F_r(f \circ i^* \circ \sigma^*) \equiv \overline{F}_s(f \circ i^*).$$

Thus the family

$$\overline{F} \equiv \{\overline{F}_{s(1),\ldots,s(p)} : p \geq 1; s_1, \ldots, s_p \in Q\}$$

of f.j.d.'s is consistent.

4. Lastly, let $s \equiv (s_1, \ldots, s_q)$ be an arbitrary increasing sequence in Q. Write $r \equiv s$. Then $s = r\sigma$, where $\sigma : \{1, \ldots, q\} \to \{1, \ldots, q\}$ is the identity function. Hence

$$\overline{F}_{s(1),\ldots,s(q)} f \equiv F_{r(1),\ldots,r(q)} f \circ \sigma^* = F_{s(1),\ldots,s(q)} f$$

for each $f \in \overline{C_{ub}}(S^q, d^q)$. In other words, $\overline{F}_{s(1),\ldots,s(q)} \equiv F_{s(1),\ldots,s(q)}$. Thus the family $\overline{F}$ is an extension of the family F, and we can simply write F for $\overline{F}$. The lemma is proved. □

Lemma 6.2.4. Two consistent families of f.j.d.'s are equal if their corresponding f.j.d.'s on initial sections are equal. *Let (S,d) be a locally compact metric space. Let $Q \equiv \{t_1,t_2,\ldots\}$ be a countably infinite set. Let $F \equiv \{F_{s(1),\ldots,s(m)} : m \geq 1; s_1,\ldots,s_m \in Q\}$ and $F' \equiv \{F'_{s(1),\ldots,s(m)} : m \geq 1; s_1,\ldots,s_m \in Q\}$ be two members of $\widehat{F}(Q,S)$ such that $F_{t(1),\ldots,t(n)} = F'_{t(1),\ldots,t(n)}$ for each $n \geq 1$. Then $F = F'$.*

Proof. Let $m \geq 1$ be arbitrary, and let $(s_1,\ldots,s_m)$ be an arbitrary sequence in Q. Then $(s_1,\ldots,s_m)$ is a subsequence of $(t_1,\ldots,t_n)$ for some sufficiently large $n \geq 1$. Hence $(s_1,\ldots,s_m) = (t_{i(1)},\ldots,t_{i(m)})$ for some sequence $i \equiv (i_1,\ldots,i_m)$ in $\{1,\ldots,n\}$. Let $i^* : S^n \to S^m$ be the dual function of the sequence i. Let $f \in C_{ub}(S^m,d^m)$ be arbitrary. Then

$$F_{s(1),\ldots,s(m))}f = F_{t(i(1)),\ldots,t(i(m))}f = F_{t(1),\ldots,t(n)}(f \circ i^*),$$

where the last equality is by the consistency of the family F. Similarly,

$$F'_{s(1),\ldots,s(m))}f = F'_{t(i(1)),\ldots,t(i(m))}f = F'_{t(1),\ldots,t(k)}(f \circ i^*).$$

Since the right-hand sides of the two last displayed equalities are equal by hypothesis, the left-hand sides are equal – namely,

$$F_{s(1),\ldots,s(m))}f = F'_{s(1),\ldots,s(m))}f,$$

where $f \in C_{ub}(S^m,d^m)$, $m \geq 1$, and the sequence $(s_1,\ldots,s_m)$ in Q is arbitrary. We conclude that $F = F'$. □

The next lemma extends the consistency condition 6.2.2 to integrable functions.

Proposition 6.2.5. The consistency condition extends to cover integrable functions. *Suppose the consistency condition 6.2.2 holds for each $f \in C_{ub}(S^n)$, for the family F of f.j.d.'s. Then a real-valued function f on S^n is integrable relative to $F_{t(i(1)),\ldots,t(i(n))}$ iff $f \circ i^*$ is integrable relative to $F_{t(1),\ldots,t(m)}$, in which case equality 6.2.2 also holds for f.*

Proof. 1. Write $E \equiv F_{t(1),\ldots,t(m)}$. Since $i^* : (S^m,d^{(m)}) \to (S^n,d^{(n)})$ is uniformly continuous, i^* is an r.v. on the completion (S^m,L,E) of $(S^m,C(S^m),E)$ with values in S^n, whence it induces a distribution E_{i^*} on S^n. Equality 6.2.2 then implies that

$$E_{i^*}f \equiv E(f \circ i^*) \equiv F_{t(1),\ldots,t(m)}(f \circ i^*) = F_{t(i(1)),\ldots,t(i(n))}f \qquad (6.2.9)$$

for each $f \in C_{ub}(S^n,d^{(n)})$. Hence the set of integrable functions relative to E_{i^*} is equal to the set of integrable functions relative to $F_{t(i(1)),\ldots,t(i(n))}$. Moreover, $E_{i^*} = F_{t(i(1)),\ldots,t(i(n))}$ on this set.

2. According to Proposition 5.2.6, a function $f : S^n \to R$ is integrable relative to E_{i^*} iff the function $f \circ i^*$ is integrable relative to E, in which case

$$E_{i^*} f = E(f \circ i^*).$$

In other words, a function $f : S^n \to R$ is integrable relative to $F_{t(i(1)),\ldots,t(i(n))}$ iff the function $f \circ i^*$ is integrable relative to $F_{t(1),\ldots,t(m)}$, in which case

$$F_{t(i(1)),\ldots,t(i(n))} f = F_{t(1),\ldots,t(m)} f \circ i^*. \qquad \square$$

Proposition 6.2.6. Marginal distributions are consistent. *Let $X : Q \times \Omega \to S$ be an r.f. Then the family F of marginal distributions of X is consistent.*

Proof. Let $n, m \geq 1$ and $f \in C_{ub}(S^n)$ be arbitrary. Let $t \equiv (t_1, \ldots, t_m)$ be an arbitrary sequence in Q, and let $i \equiv (i_1, \ldots, i_n)$ be an arbitrary sequence in $\{1, \ldots, m\}$. Using the defining equalities 6.1.1 and 6.2.1, we obtain

$$F_{t(1),\ldots,t(m)}(f \circ i^*) \equiv E((f \circ i^*)(X_{t(1)}, \ldots, X_{t(m)}))$$
$$\equiv Ef(X_{t(i(1))}, \ldots, X_{t(i(n))}) \equiv F_{t(i(1)),\ldots,t(i(n))} f.$$

Thus the consistency condition 6.2.2 holds. $\square$

Definition 6.2.7. Restriction to a subset of the parameter set. Let (S, d) be a locally compact metric space. Recall that $\widehat{F}(Q, S)$ is the set of consistent families of f.j.d.'s with parameter set Q and state space S. Let Q' be any subset of Q. For each $F \in \widehat{F}(Q, S)$ define

$$F|Q' \equiv \Phi_{Q,Q'}(F) \equiv \{F_{s(1),\ldots,s(n)} : n \geq 1; s_1, \ldots, s_n \in Q'\} \tag{6.2.10}$$

and call $F|Q'$ the *restriction* of the consistent family F to Q'. The function

$$\Phi_{Q,Q'} : \widehat{F}(Q, S) \to \widehat{F}(Q', S)$$

will be called the restriction mapping of consistent families with parameter set Q to consistent families with parameter set Q'.

Let $\widehat{F}_0 \subset \widehat{F}(Q, S)$ be arbitrary. Denote its image under the mapping $\Phi_{Q,Q'}$ by

$$\widehat{F}_0|Q' \equiv \Phi_{Q,Q'}(\widehat{F}_0) = \{F|Q' : F \in \widehat{F}_0\}, \tag{6.2.11}$$

and call $\widehat{F}_0|Q'$ the restriction of the set $\widehat{F}_0$ of consistent families to Q'. $\square$

We next introduce a metric on the set $\widehat{F}(Q, S)$ when Q is countably infinite.

Definition 6.2.8. Marginal metric on the space of consistent families of f.j.d.'s with a countably infinite parameter set. Let (S, d) be a locally compact metric space, with a binary approximation ξ relative to some fixed reference point $x_\circ$. Let $n \geq 1$ be arbitrary. Recall that ξ^n is the nth power of ξ and a binary approximation of (S^n, d^n) relative to $x_\circ^{(n)} \equiv (x_\circ, \ldots, x_\circ) \in S^n$, as in Definition 3.2.4. Recall, from Definition 5.3.4, the distribution metric ρ_{Dist,ξ^n} on the set of distributions on (S^n, d^n). Recall, from Assertion 3 of Proposition 5.3.5, that sequential convergence relative to ρ_{Dist,ξ^n} is equivalent to weak convergence.

Let $Q \equiv \{t_1, t_2, \ldots\}$ be a countably infinite parameter set. Recall that $\widehat{F}(Q,S)$ is the set of consistent families of f.j.d.'s with parameter set Q and state space S. Define a metric $\widehat{\rho}_{Marg,\xi,Q}$ on $\widehat{F}(Q,S)$ by

$$\widehat{\rho}_{Marg,\xi,Q}(F,F') \equiv \sum_{n=1}^{\infty} 2^{-n} \rho_{Dist,\xi^n}(F_{t(1),\ldots,t(n)}, F'_{t(1),\ldots,t(n)}) \qquad (6.2.12)$$

for each $F, F' \in \widehat{F}(Q,S)$. The next lemma proves that $\widehat{\rho}_{Marg,\xi,Q}$ is indeed a metric. We will call $\widehat{\rho}_{Marg,\xi,Q}$ the *marginal metric* for the space $\widehat{F}(Q,S)$ of consistent families of f.j.d.'s, relative to the binary approximation ξ of the locally compact state space (S,d). Note that $\widehat{\rho}_{Marg,\xi,Q} \leq 1$ because $\rho_{Dist,\xi^n} \leq 1$ for each $n \geq 1$. We emphasize that the metric $\widehat{\rho}_{Marg,\xi,Q}$ depends on the enumeration t of the set Q. Two different enumerations lead to two different metrics, though those metrics are equivalent.

As observed earlier, sequential convergence relative to ρ_{Dist,ξ^n} is equivalent to weak convergence of distributions on (S^n, d^n), for each $n \geq 1$. Hence, for each sequence $(F^{(m)})_{m=0,1,2,\ldots}$ in $\widehat{F}(Q,S)$, we have $\widehat{\rho}_{Marg,\xi,Q}(F^{(m)}, F^{(0)}) \to 0$ iff $F^{(m)}_{t(1),\ldots,t(n)} \Rightarrow F^{(0)}_{t(1),\ldots,t(n)}$ as $m \to \infty$, for each $n \geq 1$. □

Lemma 6.2.9. A marginal metric is indeed a metric. *Let (S,d) be a locally compact metric space, with a binary approximation ξ relative to some fixed reference point $x_\circ$. Let $Q \equiv \{t_1, t_2, \ldots\}$ be a countably infinite parameter set.. Then the marginal metric $\widehat{\rho}_{Marg,\xi,Q}$ defined on $\widehat{F}(Q,S)$ in Definition 6.4.2 is indeed a metric.*

Proof. 1. Symmetry and triangle inequality for $\widehat{\rho}_{Marg,\xi,Q}$ follow from their respective counterparts for ρ_{Dist,ξ^n} for each $n \geq 1$ in the defining equality 6.2.12.

2. Let $F, F' \in \widehat{F}(Q,S)$ be arbitrary such that $\widehat{\rho}_{Marg,\xi,Q}(F,F') = 0$. Then equality 6.2.12 implies that $\rho_{Dist,\xi^n}(F_{t(1),\ldots,t(n)}, F'_{t(1),\ldots,t(n)}) = 0$ for each $n \geq 1$. Hence, since ρ_{Dist,ξ^n} is a metric, we have $F_{t(1),\ldots,t(n)} = F'_{t(1),\ldots,t(n)}$, for each $n \geq 1$. Hence $F = F'$ by Lemma 6.2.4.

3. Summing up, $\widehat{\rho}_{Marg,\xi,Q}$ is a metric. □

Definition 6.2.10. Continuity in probability of consistent families. Let (S,d) be a locally compact metric space. Write $\widehat{d} \equiv 1 \wedge d$. Let (Q, d_Q) be a metric space. Recall that $\widehat{F}(Q,S)$ is the set of consistent families of f.j.d.'s with parameter space Q and state space (S,d). Let $F \in \widehat{F}(Q,S)$ be arbitrary.

1. Suppose (Q, d_Q) is bounded. Suppose, for each $\varepsilon > 0$, there exists $\delta_{Cp}(\varepsilon) > 0$ such that

$$F_{s,t}\widehat{d} \leq \varepsilon$$

for each $s,t \in Q$ with $d_Q(s,t) < \delta_{Cp}(\varepsilon)$. Then the consistent family F of f.j.d.'s is said to be *continuous in probability*, with δ_{Cp} as a *modulus of continuity in probability*.

2. More generally, let the metric space (Q, d_Q) be arbitrary, not necessarily bounded. Then the consistent family F of f.j.d.'s is said to be *continuous in*

probability if, for each bounded subset K of Q, the restricted consistent family $F|K$ is continuous in probability. We will let $\widehat{F}_{Cp}(Q,S)$ denote the subset of $\widehat{F}(Q,S)$ whose members are continuous in probability. □

Lemma 6.2.11. Continuity in probability extends to f.j.d.'s of higher dimensions. *Let (S,d) be a locally compact metric space. Let (Q,d_Q) be a bounded metric space. Suppose F is a consistent family of f.j.d.'s with state space S and parameter space Q that is continuous in probability, with a modulus of continuity in probability δ_{Cp}. Then the following conditions hold:*

1. Let $m \geq 1$ be arbitrary. Let $f \in C_{ub}(S^m,d^m)$ be arbitrary with a modulus of continuity δ_f and with $|f| \leq 1$. Let and $\varepsilon > 0$ be arbitrary. Then there exists $\delta_{fjd}(\varepsilon,m,\delta_f,\delta_{Cp}) > 0$ such that for each $s_1,\ldots,s_m,t_1,\ldots,t_m \in Q$ with

$$\bigvee_{k=1}^{m} d_Q(s_k,t_k) < \delta_{fjd}(\varepsilon,m,\delta_f,\delta_{Cp}), \tag{6.2.13}$$

we have

$$|F_{s(1),\ldots,s(m)}f - F_{t(1),\ldots,t(m)}f| \leq \varepsilon. \tag{6.2.14}$$

2. Suppose, in addition, F' is a consistent family of f.j.d.'s with state space S and parameter space Q that is continuous in probability. Suppose there exists a dense subset $\widetilde{Q}$ of (Q,d_Q) such that $F_{s(1),\ldots,s(m)} = F'_{s(1),\ldots,s(m)}$ for each $s_1,\ldots,s_m \in \widetilde{Q}$. Then $F = F$.

Proof. 1. Let $m \geq 1$ and $f \in C_{ub}(S^m,d^m)$ be as given. Write

$$\alpha \equiv 2^{-3}m^{-1}\varepsilon(1 \wedge \delta_f(2^{-1}\varepsilon))$$

and define

$$\delta_{fjd}(\varepsilon,m,\delta_f,\delta_{Cp}) \equiv \delta_{Cp}(\alpha).$$

Suppose $s_1,\ldots,s_m,t_1,\ldots,t_m \in Q$ satisfy inequality 6.2.13. Then

$$\bigvee_{k=1}^{m} d_Q(s_k,t_k) < \delta_{Cp}(\alpha). \tag{6.2.15}$$

Let $i \equiv (1,\ldots,m)$ and $j \equiv (m+1,\ldots,2m)$. Thus i and j are sequences in $\{1,\ldots,2m\}$. Let $x \in S^{2m}$ be arbitrary. Then

$$(f \circ i^*)(x_1,\ldots,x_{2m}) \equiv f(x_{i(1)},\ldots,x_{i(m)}) = f(x_1,\ldots,x_m)$$

and

$$(f \circ j^*)(x_1,\ldots,x_{2m}) \equiv f(x_{j(1)},\ldots,x_{j(m)}) = f(x_{m+1},\ldots,x_{2m}),$$

where i^*, j^* are the dual functions of the functions i, j, respectively, in the sense of Definition 6.2.1. Consider each $k = 1,\ldots,m$. Let $h \equiv (h_1,h_2) \equiv (k,m+k)$.

Thus $h : \{1,2\} \to \{1,\ldots,2m\}$ is a sequence in $\{1,\ldots,2m\}$. Let h^* denote the dual function. Define

$$(r_1,\ldots,r_{2m}) \equiv (s_1,\ldots,s_m,t_1,\ldots,t_m).$$

Then

$$\begin{aligned} F_{r(1),\ldots,r(2m)}(\widehat{d} \circ h^*) &= F_{r(h(1)),r(h(2))}\widehat{d} \\ &= F_{r(k),r(m+k)}\widehat{d} = F_{s(k),t(k)}\widehat{d} < \alpha, \end{aligned} \tag{6.2.16}$$

where the inequality follows from inequality 6.2.15 in view of the definition of δ_{Cp} as a modulus of continuity in probability of the family F. Now take any

$$\delta_0 \in (2^{-1}(1 \wedge \delta_f(2^{-1}\varepsilon)), 1 \wedge \delta_f(2^{-1}\varepsilon)).$$

Let

$$A_k \equiv \{x \in S^{2m} : \widehat{d}(x_k,x_{m+k}) > \delta_0\} = (\widehat{d} \circ h^* > \delta_0) \subset S^{2m}.$$

In view of inequality 6.2.16, Chebychev's inequality yields

$$\begin{aligned} F_{r(1),\ldots,r(2m)}(A_k) &= F_{r(1),\ldots,r(2m)}(\widehat{d} \circ h^* > \delta_0) < \delta_0^{-1}\alpha \\ &< 2(1 \wedge \delta_f(2^{-1}\varepsilon))^{-1}\alpha \equiv 2(1 \wedge \delta_f(2^{-1}\varepsilon))^{-1}\alpha \\ &\equiv 2(1 \wedge \delta_f(2^{-1}\varepsilon))^{-1}2^{-3}m^{-1}\varepsilon(1 \wedge \delta_f(2^{-1}\varepsilon)) = 2^{-2}m^{-1}\varepsilon. \end{aligned}$$

Let $A \equiv \bigcup_{k=1}^{m} A_k \subset S^{2m}$. Then

$$F_{r(1),\ldots,r(2m)}(A) \le \sum_{k=1}^{m} F_{r(1),\ldots,r(2m)}(A_k) \le 2^{-2}\varepsilon.$$

Now consider each $x \in A^c$. We have $x \in A_k^c$, whence

$$1 \wedge d(x_k,x_{m+k}) \equiv \widehat{d}(x_k,x_{m+k}) \le \delta_0 < 1$$

for each $k = 1,\ldots,m$. Therefore

$$d^m((x_1,\ldots,x_m),(x_{m+1},\ldots,x_{2m})) \equiv \bigvee_{k=1}^{m} d(x_k,x_{m+k}) \le \delta_0 < \delta_f(2^{-1}\varepsilon).$$

Consequently,

$$|(f \circ i^*)(x) - (f \circ j^*)(x)| = |f(x_1,\ldots,x_m) - f(x_{m+1},\ldots,x_{2m})| < 2^{-1}\varepsilon$$

where $x \in A^c$ is arbitrary. By hypothesis, $|f| \le 1$. Hence

$$\begin{aligned} &|F_{s(1),\ldots,s(m)}f - F_{t(1),\ldots,t(m)}f| \\ &\quad = |F_{r(i(1)),\ldots,r(i(n))}f - F_{r(j(1)),\ldots,r(j(n))}f| \\ &\quad = |F_{r(1),\ldots,r(2m)}(f \circ i^*) - F_{r(1),\ldots,r(2m)}(f \circ j^*)| \end{aligned}$$

$$= |F_{r(1),\ldots,r(2m)}(f \circ i^* - f \circ j^*)|$$
$$= |F_{r(1),\ldots,r(2m)}(f \circ i^* - f \circ j^*)|1_{A^c} + |F_{r(1),\ldots,r(2m)}(f \circ i^* - f \circ j^*)|1_A$$
$$\leq 2^{-1}\varepsilon + 2F_{r(1),\ldots,r(2m)}(A)$$
$$\leq 2^{-1}\varepsilon + 2 \cdot 2^{-2}\varepsilon = \varepsilon, \tag{6.2.17}$$

as desired. Assertion 1 is proved.

2. From relations 6.2.13 and 6.2.14 in Assertion 1, we see that for each arbitrary but fixed $m \geq 1$ and $f \in C_{ub}(S^m, d^m)$, the expectation $F_{s(1),\ldots,s(m)} f$ is a continuous function of $(s_1, \ldots, s_m) \in (Q^m, d^m)$. Similarly, $F'_{s(1),\ldots,s(m)} f$ is a continuous function of $(s_1, \ldots, s_m) \in (Q^m, d^m)$. At the same time, by the assumption in Assertion 2, we have $F_{s(1),\ldots,s(m)} f = F'_{s(1),\ldots,s(m)} f$ for each $(s_1, \ldots, s_m)$ in the dense subset $\widetilde{Q}^m$ of (Q^m, d^m). Consequently, $F_{s(1),\ldots,s(m)} f = F'_{s(1),\ldots,s(m)} f$ for each $(s_1, \ldots, s_m) \in (Q^m, d^m)$, where $m \geq 1$ and $f \in C_{ub}(S^m, d^m)$ are arbitrary. Thus $F = F'$. Assertion 2 and the lemma are proved. □

Definition 6.2.12. Metric space of consistent families that are continuous in probability. Let (S, d) be a locally compact metric space, with a reference point $x_\circ \in S$ and a binary approximation $\xi \equiv (A_n)_{n=1,2,\ldots}$ relative to $x_\circ$. Let (Q, d_Q) be a locally compact metric space. Let $Q_\infty \equiv \{q_1, q_2, \ldots\}$ be an arbitrary countably infinite and dense subset of Q.

Recall that $\widehat{F}(Q, S)$ is the set of consistent families of f.j.d.'s with parameter set Q and state space S. Let $\widehat{F}_{Cp}(Q, S)$ denote the subset of $\widehat{F}(Q, S)$ whose members are continuous in probability.

Relative to the countably infinite parameter subset Q_∞ and the binary approximation ξ, define a metric $\widehat{\rho}_{Cp,\xi,Q,Q(\infty)}$ on $\widehat{F}_{Cp}(Q, S)$ by

$$\widehat{\rho}_{Cp,\xi,Q,Q(\infty)}(F, F') \equiv \widehat{\rho}_{Marg,\xi,Q(\infty)}(F|Q_\infty, F'|Q_\infty)$$
$$\equiv \sum_{n=1}^{\infty} 2^{-n} \rho_{Dist,\xi^n}(F_{q(1),\ldots,q(n)}, F'_{q(1),\ldots,q(n)}) \tag{6.2.18}$$

for each $F, F' \in \widehat{F}_{Cp}(Q, S)$, where $\widehat{\rho}_{Marg,\xi,Q(\infty)}$ is the marginal metric on $\widehat{F}(Q_\infty, S)$ introduced in Definition 6.2.8. In other words,

$$\widehat{\rho}_{Cp,\xi,Q,Q(\infty)}(F, F') \equiv \widehat{\rho}_{Marg,\xi,Q(\infty)}(\Phi_{Q,Q(\infty)}(F), \Phi_{Q,Q(\infty)}(F'))$$

for each $F, F' \in \widehat{F}_{Cp}(Q, S)$. Lemma 6.2.13 (next) shows that $\widehat{\rho}_{Cp,\xi,Q,Q(\infty)}$ is indeed a metric. Then, trivially, the mapping

$$\Phi_{Q,Q(\infty)} : (\widehat{F}_{Cp}(Q, S), \widehat{\rho}_{Cp,\xi,Q,Q(\infty)}) \to (\widehat{F}_{Cp}(Q_\infty, S), \widehat{\rho}_{Marg,\xi,Q(\infty)})$$

is an isometry. Note that $0 \leq \widehat{\rho}_{Cp,\xi,Q,Q(\infty)} \leq 1$. □

Lemma 6.2.13. $\widehat{\rho}_{Cp,\xi,Q,Q(\infty)}$ **is indeed a metric.** *Let* $Q_\infty \equiv \{q_1, q_2, \ldots\}$ *be an arbitrary countably infinite and dense subset of* Q. *Then the function* $\widehat{\rho}_{Cp,\xi,Q,Q(\infty)}$ *defined in Definition 6.2.12 is a metric on* $\widehat{F}_{Cp}(Q, S)$.

Proof. Suppose $F, F' \in \widehat{F}_{Cp}(Q,S)$ are such that $\widehat{\rho}_{Cp,\xi,Q,Q(\infty)}(F,F') = 0$. By the defining equality 6.2.18, we then have $\widehat{\rho}_{Marg,\xi,Q(\infty)}(F|Q_\infty, F'|Q_\infty) = 0$. Hence, since $\widehat{\rho}_{Marg,\xi,Q(\infty)}$ is a metric on $\widehat{F}(Q_\infty,S)$, we have $F|Q_\infty = F'|Q_\infty$. In other words,

$$F_{q(1),\ldots,q(n)} = F'_{q(1),\ldots,q(n)}$$

for each $n \geq 1$. Consider each $m \geq 1$ and each $s_1,\ldots,s_m \in Q_\infty$. Let $n \geq 1$ be so large that $\{s_1,\ldots,s_m\} \subset \{q_1,\ldots,q_n\}$ and obtain, by the consistency of F and F',

$$F_{s(1),\ldots,s(m)} = F'_{s(1),\ldots,s(m)}. \tag{6.2.19}$$

Now let $m \geq 1$ and $t_1,\ldots,t_m \in Q$ be arbitrary. Let $f \in C(S^m)$ be arbitrary. For each $i = 1,\ldots,m$, let $(s_i^{(p)})_{p=1,2,\ldots}$ be a sequence in Q_∞ with $d(s_i^{(p)},t_i) \to 0$ as $p \to \infty$. Then there exists a bounded subset $K \subset Q$ such that the sequences $(t_1,\ldots,t_m)$ and $(s_i^{(p)})_{p=1,2,\ldots;i=1,\ldots,m}$ are in K. Since $F|K$ is continuous in probability, we have, by Lemma 6.2.11,

$$F_{s(p,1),\ldots,s(p,m)}f \to F_{t(1),\ldots,t(m)}f,$$

where we write $s(p,i) \equiv s_i^{(p)}$ to lessen the burden on subscripts. Similarly,

$$F'_{s(p,1),\ldots,s(p,m)}f \to F'_{t(1),\ldots,t(m)}f.$$

On the other hand,

$$F_{s(p,1),\ldots,s(p,m)}f = F'_{s(p,1),\ldots,s(p,m)}f$$

thanks to equality 6.2.19. Combining,

$$F'_{t(1),\ldots,t(m)}f = F_{t(1),\ldots,t(m)}f.$$

We conclude that $F = F'$.

Conversely, suppose $F = F'$. Then, trivially, $\widehat{\rho}_{Cp,\xi,Q,Q(\infty)}(F,F') = 0$ from equality 6.2.18. Separately, the triangle inequality and symmetry of $\widehat{\rho}_{Cp,\xi,Q,Q(\infty)}$ follow from equality 6.2.18 and from the fact that ρ_{Dist,ξ^n} is a metric for each $n \geq 1$. Summing up, $\widehat{\rho}_{Cp,\xi,Q,Q(\infty)}$ is a metric. □

6.3 Daniell–Kolmogorov Extension

In this and the next section, unless otherwise specified, (S,d) will be a locally compact metric space, and $Q \equiv \{t_1,t_2,\ldots\}$ will denote an enumerated countably infinite parameter set, enumerated by the bijection $t : \{1,2,\ldots\} \to Q$. Let $x_\circ$ be an arbitrary but fixed reference point in (S,d).

Recall that $\widehat{F}(Q,S)$ is the set of consistent families of f.j.d.'s with parameter set Q and the locally compact state space (S,d). We will prove the Daniell–Kolmogorov Extension Theorem, which constructs, for each member $F \in \widehat{F}(Q,S)$, a probability space (S^Q,L,E) and an r.f. $U : Q \times (S^Q,L,E) \to (S,d)$ with marginal distributions given by F.

Furthermore, we will prove the uniform metrical continuity of the Daniell–Kolmogorov Extension on each pointwise tight subset of $\widehat{F}(Q,S)$, in a sense to be made precise in the following discussion. This metrical continuity implies sequential continuity relative to weak convergence.

Recall that $[\cdot]_1$ is an operation that assigns to each $c \geq 0$ an integer $[c]_1$ in the interval $(c, c+2)$. As usual, for arbitrary symbols a and b, we will write a_b and $a(b)$ interchangeably.

Definition 6.3.1. Path space, coordinate function, and distributions on path space. Let $S^Q \equiv \prod_{t \in Q} S$ denote the space of functions from Q to S, called the *path space*. Relative to the enumerated set Q, define a complete metric d^Q on S^Q by

$$d^Q(x,y) \equiv d^\infty(x \circ t, y \circ t) \equiv \sum_{i=0}^{\infty} 2^{-i}(1 \wedge d(x_{t(i)}, y_{t(i)}))$$

for arbitrary $x,y \in S^Q$. Define the function $U : Q \times S^Q \rightarrow S$ by $U(r,x) \equiv x_r$ for each $(r,x) \in Q \times S^Q$. The function U is called the *coordinate function of* $Q \times S^Q$. Note that the function $t^* : (S^\infty, d^\infty) \rightarrow (S^Q, d^Q)$, defined by $t^*(x_1, x_2, \ldots) \equiv (x_{t(1)}, x_{t(2)}, \ldots)$ for each $(x_1, x_2, \ldots) \in S^\infty$, is an isometry.

Note that $d^Q \leq 1$ and that the path space (S^Q, d^Q) is complete. (S^Q, d^Q) need not be locally compact. However, (S^Q, d^Q) is compact if (S,d) is compact.

In keeping with the terminology used in Definition 5.2.1, we will let $\widehat{J}(S^Q, d^Q)$ denote the set of distributions on the complete path space (S^Q, d^Q). □

Theorem 6.3.2. Compact Daniell–Kolmogorov Extension. *Suppose the metric space (S,d) is compact. Then there exists a function*

$$\overline{\Phi}_{DK} : \widehat{F}(Q,S) \rightarrow \widehat{J}(S^Q, d^Q)$$

such that for each consistent family of f.j.d.'s $F \in \widehat{F}(Q,S)$, the distribution $E \equiv \overline{\Phi}_{DK}(F)$ on the path space satisfies the following two conditions:

(i) The coordinate function

$$U : Q \times (S^Q, L, E) \rightarrow (S,d)$$

is an r.f., where (S^Q, L, E) is the probability space that is the completion of $(S^Q, C(S^Q, d^Q), E)$, and

(ii) the r.f. U has marginal distributions given by the family F.

The function $\overline{\Phi}_{DK}$ will be called the Compact Daniell–Kolmogorov Extension.

Proof. Without loss of generality, assume that $Q \equiv \{t_1, t_2, \ldots\} \equiv \{1, 2, \ldots\}$. Then $(S^Q, d^Q) = (S^\infty, d^\infty)$. Since (S,d) is compact by hypothesis, its countable power $(S^Q, d^Q) = (S^\infty, d^\infty)$ is also compact. Hence $C_{ub}(S^Q, d^Q) = C(S^Q, d^Q)$.

1. Consider each $F \equiv \{F_{s(1),\ldots,s(m)} : m \geq 1; s_1, \ldots, s_m \in Q\} \in \widehat{F}(Q,S)$. Let $f \in C(S^\infty, d^\infty)$ be arbitrary, with a modulus of continuity δ_f. For each $n \geq 1$, define the function $f_n \in C(S^n, d^n)$ by

$$f_n(x_1,x_2,\ldots,x_n) \equiv f(x_1,x_2,\ldots,x_n,x_\circ,x_\circ,\ldots) \tag{6.3.1}$$

for each $(x_1,x_2,\ldots,x_n) \in S^n$. Let $\varepsilon > 0$. Consider each $m \geq n \geq 1$ so large that $2^{-n} < \delta_f(\varepsilon)$. Define the function $f_{n,m} \in C(S^m,d^m)$ by

$$f_{n,m}(x_1,x_2,\ldots,x_m) \equiv f_n(x_1,x_2,\ldots,x_n) \equiv f(x_1,x_2,\ldots,x_n,x_\circ,x_\circ,\ldots) \tag{6.3.2}$$

for each $(x_1,x_2,\ldots,x_m) \in S^m$. Consider the initial-section subsequence $i \equiv (i_1,\ldots,i_n) \equiv (1,\ldots,n)$ of the sequence $(1,\ldots,m)$. Let $i^* : S^m \to S^n$ be the dual function of the sequence i, as in Definition 6.2.1. Then for each $(x_1,\ldots,x_m) \in S^m$, we have

$$\begin{aligned} f_{n,m}(x_1,x_2,\ldots,x_m) &\equiv f_n(x_1,x_2,\ldots,x_n) \\ &= f_n(x_{i(1)},x_{i(2)},\ldots,x_{i(n)}) \equiv f_n \circ i^*(x_1,x_2,\ldots,x_m). \end{aligned}$$

In short,

$$f_{n,m} = f_n \circ i^*,$$

whence, by the consistency of the family F of f.j.d.'s, we obtain

$$F_{1,\ldots,m} f_{n,m} = F_{1,\ldots,m}(f_n \circ i^*) = F_{i(1),\ldots,i(n)} f_n = F_{1,\ldots,n} f_n. \tag{6.3.3}$$

At the same time,

$$\begin{aligned} &d^\infty((x_1,x_2,\ldots,x_m,x_\circ,x_\circ,\ldots),(x_1,x_2,\ldots,x_n,x_\circ,x_\circ,\ldots)) \\ &\equiv \sum_{i=1}^{n} 2^{-i}(1 \wedge d((x_i,x_i)) + \sum_{i=n+1}^{m} 2^{-i}(1 \wedge d((x_i,x_\circ)) \\ &\quad + \sum_{i=m+1}^{\infty} 2^{-i}(1 \wedge d((x_\circ,x_\circ)) \\ &= 0 + \sum_{i=n+1}^{m} 2^{-i}(1 \wedge d((x_i,x_\circ)) + 0 \leq 2^{-n} < \delta_f(\varepsilon) \end{aligned}$$

for each $(x_1,x_2,\ldots,x_m) \in S^m$. Hence

$$\begin{aligned} &|f_m(x_1,x_2,\ldots,x_m) - f_{n,m}(x_1,x_2,\ldots,x_m)| \\ &\quad \equiv |f(x_1,x_2,\ldots,x_m,x_\circ,x_\circ,\ldots) - f(x_1,x_2,\ldots,x_n,x_\circ,x_\circ,\ldots)| < \varepsilon \end{aligned}$$

for each $(x_1,x_2,\ldots,x_m) \in S^m$. Consequently, $|F_{1,\ldots,m} f_m - F_{1,\ldots,m} f_{n,m}| \leq \varepsilon$. Combined with equality 6.3.3, this yields

$$|F_{1,\ldots,m} f_m - F_{1,\ldots,n} f_{n,}| \leq \varepsilon, \tag{6.3.4}$$

where $m \geq n \geq 1$ are arbitrary with $2^{-n} < \delta_f(\varepsilon)$. Thus we see that the sequence $(F_{1,\ldots,n} f_{n,})_{n=1,2,\ldots}$ of real numbers is Cauchy and has a limit. Define

$$Ef \equiv \lim_{n\to\infty} F_{1,\ldots,n} f_{n,}. \tag{6.3.5}$$

2. Letting $m \to \infty$ in inequality 6.3.4, we obtain

$$|Ef - F_{1,\ldots,n} f_{n,}| \leq \varepsilon, \tag{6.3.6}$$

where $n \geq 1$ is arbitrary with $2^{-n} < \delta_f(\varepsilon)$.

3. Proceed to prove that E is an integration on the compact metric space (S^Q, d^Q), in the sense of Definition 4.2.1. We will first verify that the function E is linear. To that end, let $f, g \in C(S^\infty, d^\infty)$ and $a, b \in R$ be arbitrary. For each $n \geq 1$, define the function f_n relative to f as in equality 6.3.1. Similarly, define the functions $g_n, (af + bg)_n$ relative to the functions $g, af + bg$, respectively, for each $n \geq 1$. Then the defining equality 6.3.1 implies that $(af + bg)_n = af_n + bg_n$ for each $n \geq 1$. Hence

$$\begin{aligned} E(af + bg) &\equiv \lim_{n\to\infty} F_{1,\ldots,n}(af + bg)_n = \lim_{n\to\infty} F_{1,\ldots,n}(af_n + bg_n) \\ &= a \lim_{n\to\infty} F_{1,\ldots,n} f_n + b \lim_{n\to\infty} F_{1,\ldots,n} g_n \\ &\equiv aEf + bEg. \end{aligned}$$

Thus E is a linear function. Moreover, in the special case where $f \equiv 1$, we have

$$E1 \equiv Ef \equiv \lim_{n\to\infty} F_{1,\ldots,n} f_n = \lim_{n\to\infty} F_{1,\ldots,n} 1 = 1 > 0. \tag{6.3.7}$$

Inequality 6.3.7 shows that the triple $(S^Q, C(S^Q, d^Q), E)$ satisfies Condition (i) of Definition 4.2.1. It remains to verify Condition (ii), the positivity condition, of Definition 4.2.1. To that end, let $f \in C(S^\infty, d^\infty)$ be arbitrary with $Ef > 0$. Then by the defining equality 6.3.5, we have $F_{1,\ldots,n} f_n > 0$ for some $n \geq 1$. Hence, since $F_{1,\ldots,n}$ is a distribution, there exists $(x_1, x_2, \ldots, x_n) \in S^n$ such that $f_{n,}(x_1, x_2, \ldots, x_n) > 0$. Therefore

$$f(x_1, x_2, \ldots, x_n, x_\circ, x_\circ, \ldots) \equiv f_{n,}(x_1, x_2, \ldots, x_n) > 0.$$

Thus the positivity condition of Definition 4.2.1 is also verified. Accordingly, E is an integration on the compact metric space (S^Q, d^Q).

4. Since the compact metric space (S^Q, d^Q) is bounded, and since $E1 = 1$, Assertion 2 of Lemma 5.2.2 implies that the integration E is a distribution on (S^Q, d^Q), and that the completion (S^Q, L, E) of the integration space $(S^Q, (S^Q, d^Q), E)$ is a probability space. In symbols, $E \in \widehat{J}(S^Q, d^Q)$. Define $\overline{\Phi}_{DK}(F) \equiv E$. Thus we have constructed a function $\overline{\Phi}_{DK} : \widehat{F}(Q, S) \to \widehat{J}(S^Q, d^Q)$.

5. It remains to show that the coordinate function $U : Q \times (S^Q, L, E) \to S$ is an r.f. with marginal distributions given by the family F. To that end, let $n \geq 1$ be arbitrary. Let $\varepsilon \in (0, 1)$ be arbitrary. By Definition 6.3.1, we have $U_n(x) \equiv U(n, x) \equiv x_n$ for each $x \equiv (x_1, x_2, \ldots) \in S^\infty$. Hence, for each $x \equiv (x_1, x_2, \ldots), y \equiv (y_1, y_2, \ldots) \in S^\infty$ such that $d^\infty(x, y) < 2^{-n}\varepsilon$, we have

$$1 \wedge d(U_n(x), U_n(y)) = 1 \wedge d(x_n, y_n) \leq 2^n \sum_{i=1}^{\infty} 2^{-i}(1 \wedge d_i(x_i, y_i))$$

$$\equiv 2^n d^{\infty}(x, y) < 2^n 2^{-n}\varepsilon = \varepsilon.$$

Since $\varepsilon < 1$, it follows that $d(U_n(x), U_n(y)) < \varepsilon$. We conclude that $U_n \in C(S^Q, d^Q) \subset L$, where $n \in Q$ is arbitrary. Thus $U : Q \times (S^Q, L, E) \to S$ is an r.f.

6. To prove that the r.f. U has marginal distributions given by the family F, let $n \geq 1$ and let $g \in C(S^n, d^n)$ be arbitrary. Define the function $f \in C(S^{\infty}, d^{\infty})$ by

$$f(x_1, x_2, \ldots) \equiv g(x_1, \ldots, x_n) \tag{6.3.8}$$

for each $x \equiv (x_1, x_2, \ldots) \in S^{\infty}$. Let $m \geq n$ be arbitrary. As in Step 1, define the function $f_m \in C(S^m, d^m)$ by

$$f_m(x_1, x_2, \ldots, x_m) \equiv f(x_1, x_2, \ldots, x_m, x_\circ, x_\circ, \ldots) \equiv g(x_1, \ldots, x_n), \tag{6.3.9}$$

for each $(x_1, x_2, \ldots, x_m) \in S^m$. Then

$$f_m(x_1, x_2, \ldots, x_m) = g(x_1, \ldots, x_n) = f_n(x_1, x_2, \ldots, x_n) = f_n \circ i^*(x_1, x_2, \ldots, x_m)$$

for each $(x_1, x_2, \ldots, x_m) \in S^m$, where $i \equiv (i_1, \ldots, i_n) \equiv (1, \ldots, n)$ is the initial-section subsequence of the sequence $(1, \ldots, m)$, and where $i^* : S^m \to S^n$ is the dual function of the sequence i. In short, $f_m = f_n \circ i^*$. At the same time,

$$g(U_1, \ldots, U_n)(x) \equiv g(U_1(x), \ldots, U_n(x)) = g(x_1, \ldots, x_n) = f(x)$$

for each $x \equiv (x_1, x_2, \ldots) \in S^{\infty}$. Hence

$$g(U_1, \ldots, U_n) = f \in C(S^{\infty}, d^{\infty}). \tag{6.3.10}$$

Combining,

$$Eg(U_1, \ldots, U_n) = Ef \equiv \lim_{m \to \infty} F_{1,\ldots,m} f_m$$

$$= \lim_{m \to \infty} F_{1,\ldots,m} f_n \circ i^* = F_{1,\ldots,n} f_n = F_{1,\ldots,n} g, \tag{6.3.11}$$

where $n \geq 1$ and $g \in C(S^n, d^n)$ are arbitrary, and where the fourth equality is by the consistency of the family F of f.j.d.'s. Thus the r.f. U has marginal distributions given by the family F. The theorem is proved. □

Proceed to prove the metrical continuity of the Compact Daniell–Kolmogorov Extension. First, we will specify the metrics.

Definition 6.3.3. Specification of the binary approximation of the compact path space, and the distribution metric on the set of distributions on said path space. Suppose the state space (S, d) is compact. Recall that $Q \equiv \{t_1, t_2, \ldots\}$ is an enumerated countable parameter set. Let $\xi \equiv (A_k)_{k=1,2,\ldots}$ be an arbitrary binary approximation of (S,d) relative to the reference point $x_\circ$. Since the metric space (S,d) is compact, the countable power $\xi^\infty \equiv (B_k)_{k=1,2,}$ of ξ is defined and is a binary approximation of (S^∞,d^∞), according to Definition 3.2.6. For each $k \geq 1$, define

$$\widetilde{B}_k \equiv \{x \in S^Q : x \circ t \in B_k\} \equiv (t^*)^{-1} B_k.$$

Because the function $t^* : (S^\infty,d^\infty) \to (S^Q,d^Q)$ is an isometry, the sequence $\xi^Q \equiv (\widetilde{B}_k)_{k=1,2,}$ is a binary approximation of (S^Q,d^Q).

Moreover, since (S^Q,d^Q) is compact, and therefore locally compact and complete, the set $\widehat{J}(S^Q,d^Q)$ of distributions on (S^Q,d^Q) is defined and is equipped with the distribution metric ρ_{Dist,ξ^Q} relative to ξ^Q in Definition 5.3.4. Recall that sequential convergence relative to the metric ρ_{Dist,ξ^Q} is equivalent to weak convergence. □

Theorem 6.3.4. Continuity of the Compact Daniell–Kolmogorov Extension. *Suppose the state space (S,d) is compact. Recall that $Q \equiv \{t_1,t_2,\ldots\}$is an enumerated countable parameter set. Let $\xi \equiv (A_k)_{k=1,2,}$ be an arbitrary binary approximation of (S,d) relative to the reference point $x_\circ$. Let the corresponding objects $(S^Q,d^Q), \xi^Q \equiv (\widetilde{B}_k)_{k=1,2,}$, $\widehat{J}(S^Q,d^Q)$, and ρ_{Dist,ξ^Q} be as in Definition 6.3.3.*

Recall that the set $\widehat{F}(Q,S)$ of consistent families of f.j.d.'s is equipped with the marginal metric $\widehat{\rho}_{Marg,\xi,Q}$ relative to ξ, as in Definition 6.2.8.

Then the Compact Daniell–Kolmogorov Extension

$$\overline{\Phi}_{DK} : (\widehat{F}(Q,S), \widehat{\rho}_{Marg,\xi,Q}) \to (\widehat{J}(S^Q,d^Q), \rho_{Dist,\xi^Q})$$

constructed in Theorem 6.3.2 is uniformly continuous, with a modulus of continuity $\overline{\delta}_{DK}(\cdot, \|\xi\|)$ dependent only on the modulus of local compactness $\|\xi\| \equiv (|A_k|)_{k=1,2,\ldots}$ of the compact metric space (S,d).

Proof. Without loss of generality, assume that $d \leq 1$, and that $Q \equiv \{t_1,t_2,\ldots\} = \{1,2,\ldots\}$. Then the objects (S^Q,d^Q) and $\xi^Q \equiv (\widetilde{B}_k)_{k=1,2,\ldots}$ are equal to (S^∞,d^∞) and $\xi^\infty \equiv (B_k)_{k=1,2,\ldots}$.

1. Let $\varepsilon \in (0,1)$ be arbitrary. As an abbreviation, write $c \equiv 2^4\varepsilon^{-1}$ and $\alpha \equiv 2^{-1}\varepsilon$. Let $m \equiv [\log_2 2\varepsilon^{-1}]_1$. Define the operation δ_c by

$$\delta_c(\varepsilon') \equiv c^{-1}\varepsilon' \tag{6.3.12}$$

for each $\varepsilon' > 0$. Fix $n \geq 1$ so large that

$$2^{-n} < 3^{-1}c^{-1}\alpha \equiv \delta_c(3^{-1}\alpha).$$

By the definition of the operation $[\cdot]_1$, we have $2\varepsilon^{-1} < 2^m < 2\varepsilon^{-1} \cdot 2^2 \equiv 2^{-1}c$. Hence

$$2^{m+1} < c$$

and

$$2^{-m} < 2^{-1}\varepsilon.$$

2. Let $F, F' \in \widehat{F}(Q,S)$ be arbitrary, with $F \equiv \{F_{s(1),\dots,s(m)} : m \geq 1; s_1, \dots, s_m \in Q\}$ and $F' \equiv \{F'_{s(1),\dots,s(m)} : m \geq 1; s_1, \dots, s_m \in Q\}$. Consider the distributions $F_{1,\dots,n} \in F$ and $F'_{1,\dots,n} \in F$'. Since, by assumption, $d \leq 1$, we have $d^n \leq 1$ as well. Hence, the distributions $F_{1,\dots,n}, F'_{1,\dots,n}$ on (S^n, d^n) have modulus of tightness equal to $\beta \equiv 1$. Let ξ^n be the nth power of ξ, as in Definition 3.2.4. Thus ξ^n is a binary approximation for (S^n, d^n). Recall the distribution metric ρ_{Dist,ξ^n} relative to ξ^n on the set of distributions on (S^n, d^n), as introduced in Definition 5.3.4. Then Assertion 1 of Proposition 5.3.11 applies to the compact metric space (S^n, d^n) and the distribution metric ρ_{Dist,ξ^n}, to yield

$$\widetilde{\Delta} \equiv \widetilde{\Delta}(3^{-1}\alpha, \delta_c, 1, \|\xi^n\|) > 0 \tag{6.3.13}$$

such that, if

$$\rho_{Dist,\xi^n}(F_{1,\dots,n}, F'_{1,\dots,n}) < \widetilde{\Delta},$$

then

$$|F_{1,\dots,n}g - F'_{1,\dots,n}g| < 3^{-1}\alpha \tag{6.3.14}$$

for each $g \in C(S^n, d^n)$ with $|g| \leq 1$ and with modulus of continuity δ_c. Recall from Lemma 3.2.5 that the modulus of local compactness $\|\xi^n\|$ of (S^n, d^n) is determined by the modulus of local compactness $\|\xi\|$ of (S,d). Hence all four variables in the operation $\widetilde{\Delta}$ in equality 6.3.13 are dependent only on ε and $\|\xi\|$. In addition, the integer n is dependent only on ε. Therefore we can define

$$\overline{\delta}_{DK}(\varepsilon) \equiv \overline{\delta}_{DK}(\varepsilon, \|\xi\|) \equiv 2^{-n}\widetilde{\Delta} > 0. \tag{6.3.15}$$

We will prove that the operation $\overline{\delta}_{DK}$ is a modulus of continuity of the Compact Daniell–Kolmogorov Extension $\overline{\Phi}_{DK}$.

3. Suppose, for that purpose, that

$$\widehat{\rho}_{Marg,\xi,Q}(F,F') \equiv \sum_{k=1}^{\infty} 2^{-k}\rho_{Dist,\xi^k}(F_{1,\dots,k}, F'_{1,\dots,k}) < \overline{\delta}_{DK}(\varepsilon). \tag{6.3.16}$$

We need to show that

$$\rho_{Dist,\xi^\infty}(E, E') < \varepsilon,$$

where $E \equiv \overline{\Phi}_{DK}(F)$ and $E' \equiv \overline{\Phi}_{DK}(F')$.

4. To that end, let

$$\pi \equiv (\{g_{k,x} : x \in B_k\})_{k=1,2,\dots}$$

be the partition of unity of the compact metric space (S^∞, d^∞) determined by its binary approximation $\xi^\infty \equiv (B_k)_{k=1,2,}$, in the sense of Definition 3.3.4. Thus the family $\{g_{k,x} : x \in B_k\}$ of basis functions is the 2^{-k}-partition of unity of (S^∞, d^∞)

determined by the enumerated finite subset B_k, for each $k \geq 1$. According to Definition 5.3.4, we have

$$\rho_{Dist,\xi^\infty}(E,E') \equiv \sum_{k=1}^{\infty} 2^{-k}|B_k|^{-1} \sum_{x\in B(k)} |Eg_{k,x} - E'g_{k,x}|. \tag{6.3.17}$$

5. The assumed inequality 6.3.16 yields

$$\rho_{Dist,\xi^n}(F_{1,\ldots,n}, F'_{1,\ldots,n}) < 2^n\overline{\delta}_{DK}(\varepsilon) \equiv \widetilde{\Delta}. \tag{6.3.18}$$

Consider each $k = 1,\ldots,m$. Let $x \in B_k$ be arbitrary. Proposition 3.3.3 says that the basis function $g_{k,x}$ has values in $[0,1]$ and has Lipschitz constant 2^{k+1} on (S^∞,d^∞), where $2^{k+1} \leq 2^{m+1} < c$. Hence the function $g_{k,x}$ has Lipschitz constant c and, equivalently, has modulus of continuity δ_c. Now define the function $g_{k,x,n} \in C(S^n,d^n)$ by

$$g_{k,x,n}(y_1,\ldots,y_n) \equiv g_{k,x}(y_1,\ldots,y_n,x_\circ,x_\circ,\ldots)$$

for each $(y_1,\ldots,y_n) \in S^n$. Then, for each $(z_1,z_2,\ldots,z_n),(y_1,y_2,\ldots,y_n) \in (S^n,d^n)$, we have

$$\begin{aligned}
&|g_{k,x,n}(z_1,z_2,\ldots,z_n) - g_{k,x,n}(y_1,y_2,\ldots,y_n)| \\
&\quad\equiv |g_{k,x}(z_1,z_2,\ldots,z_n,x_\circ,x_\circ,\ldots) - g_{k,x}(y_1,y_2,\ldots,y_n,x_\circ,x_\circ,\ldots)| \\
&\quad\leq cd^\infty((z_1,z_2,\ldots,z_n,x_\circ,x_\circ,\ldots),(y_1,y_2,\ldots,y_n,x_\circ,x_\circ,\ldots)) \\
&\quad= c\sum_{k=1}^{n} 2^{-k}d(z_k,y_k) \leq c\bigvee_{k=1}^{n} d(z_k,y_k) \\
&\quad\equiv cd^n((z_1,z_2,\ldots,z_n),(y_1,y_2,\ldots,y_n)).
\end{aligned}$$

Thus the function $g_{k,x,n}$ also has Lipschitz constant c and, equivalently, has modulus of continuity δ_c. In addition, $|g_{k,x}| \leq 1$, whence $|g_{k,x,n}| \leq 1$. In view of inequality 6.3.18, inequality 6.3.14 therefore holds for $g_{k,x,n}$, to yield

$$|F_{1,\ldots,n}g_{k,x,n} - F'_{1,\ldots,n}g_{k,x,n}| < 3^{-1}\alpha.$$

At the same time, since $2^{-n} < \delta_c(3^{-1}\alpha)$, where δ_c is a modulus of continuity of the function $g_{k,x,n}$, inequality 6.3.6 in the proof of Theorem 6.3.2 applies to the functions $g_{k,x}, g_{k,x,n}$ in the place of f, f_n, and to the constant $3^{-1}\alpha$ in the place of ε. This yields

$$|Eg_{k,x} - F_{1,\ldots,n}g_{k,x,n}| \leq 3^{-1}\alpha, \tag{6.3.19}$$

with a similar inequality when E,F are replaced by E',F', respectively. The triangle inequality therefore leads to

$$\begin{aligned}
|Eg_{k,x} - E'g_{k,x}| &\leq |F_{1,\ldots,n}g_{k,x,n} - F'_{1,\ldots,n}g_{k,x,n}| + 3^{-1}\alpha + 3^{-1}\alpha \\
&< 3^{-1}\alpha + 2\cdot 3^{-1}\alpha = \alpha,
\end{aligned} \tag{6.3.20}$$

where $k = 1, \ldots, m$ and $x \in B_k$ are arbitrary. It follows that

$$\begin{aligned}
&\rho_{Dist,\xi^\infty}(\overline{\Phi}_{DK}(F), \overline{\Phi}_{DK}(F')) \\
&\quad \equiv \rho_{Dist,\xi^\infty}(E, E') \\
&\quad \equiv \sum_{k=1}^{\infty} 2^{-k} |B_k|^{-1} \sum_{x \in B(k)} |Eg_{k,x} - E'g_{k,x}| \\
&\quad \leq \sum_{k=1}^{m} 2^{-k} \alpha + \sum_{k=m+1}^{\infty} 2^{-k} < \alpha + 2^{-m} < \alpha + 2^{-1}\varepsilon \equiv \varepsilon,
\end{aligned}$$

where $F, F' \in \widehat{F}(Q,S)$ are arbitrary, provided that $\widehat{\rho}_{Marg,\xi,Q}(F,F') < \overline{\delta}_{DK}(\varepsilon, \|\xi\|)$. Since $\varepsilon > 0$ is arbitrary, we conclude that the Compact Daniell–Kolmogorov Extension

$$\overline{\Phi}_{DK} : (\widehat{F}(Q,S), \widehat{\rho}_{Marg,\xi,Q}) \to (\widehat{J}(S^Q, d^Q), \rho_{Dist,\xi^Q})$$

is uniformly continuous on $\widehat{F}(Q,S)$, with modulus of continuity $\overline{\delta}_{DK}(\cdot, \|\xi\|)$. The theorem is proved. □

To generalize Theorems 6.3.2 and 6.3.4 to a locally compact, but not necessarily compact, state space (S,d), we (i) map each consistent family of f.j.d.'s on the latter to a corresponding family on the one-point compactification $(\overline{S}, \overline{d})$, whose f.j.d.'s assign probability 1 to powers of S; (ii) apply Theorems 6.3.2 and 6.3.4 to the compact state space $(\overline{S}, \overline{d})$, resulting in a distribution on the path space $(\overline{S}^Q \overline{d}^Q)$; and (iii) prove that this latter distribution assigns probability 1 to the path subspace S^Q and is a distribution on the complete metric space (S^Q, d^Q). The remainder of this section makes this precise.

First we define the modulus of pointwise tightness for an arbitrary consistent family of f.j.d.'s.

Definition 6.3.5. Modulus of pointwise tightness. Let (S,d) be a locally compact metric space with the reference point $x_\circ$. For each $n \geq 1$, define the function $h_n \equiv 1 \wedge (1 + n - d(\cdot, x_\circ))_+ \in C(S,d)$. Let Q be an arbitrary set. Recall that $\widehat{F}(Q,S)$ is the set of consistent families of f.j.d.'s with parameter set Q and state space S. Let $F \in \widehat{F}(Q,S)$ be arbitrary. Let $t \in Q$ be arbitrary, and consider the distribution F_t on (S,d). According to Definition 5.2.1, we have $F_t h_n \uparrow 1$ as $n \to \infty$. Hence, for each $\varepsilon > 0$, there exists some $\beta(\varepsilon, t) > 0$ so large that $F_t(1 - h_n) < \varepsilon$ for each $n \geq \beta(\varepsilon, t)$. Note that $\beta(\cdot, t)$ is equivalent to a modulus of tightness of the distribution F_t, in the sense of Definition 5.3.6.

We will call such an operation $\beta : (0,\infty) \times Q \to (0,\infty)$ a *modulus of pointwise tightness* of the family F. Define $\widehat{F}_\beta(Q,S)$ as the subset of $\widehat{F}(Q,S)$ whose members share a common modulus of pointwise tightness β. □

Lemma 6.3.6. Mapping each consistent family of f.j.d.'s with a locally compact state space S to a consistent family of f.j.d.'s with the one-point compactification $\overline{S}$ as state space. *Suppose (S,d) is locally compact, not necessarily* *compact, with a one-point compactification* $(\overline{S},\overline{d})$. *Recall that* $Q \equiv \{t_1,t_2,\ldots\}$ *is an enumerated countably infinite parameter set. Define the function*

$$\psi : \widehat{F}(Q,S) \to \widehat{F}(Q,\overline{S}) \tag{6.3.21}$$

as follows. Let $F \equiv \{F_{s(1),\ldots,s(m)} : m \geq 1; s_1,\ldots,s_m \in Q\} \in \widehat{F}(Q,S)$ *be arbitrary. For each* $m \geq 1$ *and* $\overline{f} \in C(\overline{S}^m,\overline{d}^m)$, *define*

$$\overline{F}_{s(1),\ldots,s(m)}\overline{f} \equiv F_{s(1),\ldots,s(m)}(\overline{f}|S^m). \tag{6.3.22}$$

Define $\psi(F) \equiv \overline{F}$. *Then* $\overline{F} \equiv \{\overline{F}_{s(1),\ldots,s(m)} : m \geq 1; s_1,\ldots,s_m \in Q\} \in \widehat{F}(Q,\overline{S})$. *Hence the function* ψ *is well defined.*

Proof. Assume, without loss of generality, that $Q \equiv \{1,2,\ldots\}$.

1. Consider each $F \equiv \{F_{s(1),\ldots,s(m)} : m \geq 1; s_1,\ldots,s_m \in Q\} \in \widehat{F}(Q,S)$. Let $m \geq 1$ be arbitrary. Let $\overline{f} \in C(\overline{S}^m,\overline{d}^m)$ be arbitrary. Then $\overline{f}|S^m \in C_{ub}(S^m,d^m)$ by Assertion 2 of Proposition 3.4.5. Hence $\overline{f}|S^m$ is integrable relative to the distribution $F_{s(1),\ldots,s(m)}$ on (S^m,d^m). Therefore equality 6.3.22 makes sense. Since $F_{s(1),\ldots,s(m)}$ is a distribution, the right-hand side of equality 6.3.22 is a linear function of $\overline{f}$. Hence $\overline{F}_{s(1),\ldots,s(m)}$ is a linear function on $C(\overline{S}^m,\overline{d}^m)$. Suppose $\overline{F}_{s(1),\ldots,s(m)}\overline{f} > 0$. Then $F_{s(1),\ldots,s(m)}(\overline{f}|S^m) > 0$. Again, since $F_{s(1),\ldots,s(m)}$ is a distribution, it follows that there exists $x \in S^m$ such $\overline{f}(x) = (\overline{f}|S^m)(x) > 0$. Thus $\overline{F}_{s(1),\ldots,s(m)}$ is an integration on the compact metric space $(\overline{S}^m,\overline{d}^m)$. Moreover, $\overline{F}_{s(1),\ldots,s(m)}1 \equiv F_{s(1),\ldots,s(m)}(1) = 1$. Therefore $\overline{F}_{s(1),\ldots,s(m)}$ is a distribution.

2. Next, we need to verify that the family $\overline{F} \equiv \{\overline{F}_{s(1),\ldots,s(m)} : m \geq 1; s_1,\ldots,s_m \in Q\}$ is consistent. To that end, let $n,m \geq 1$ be arbitrary. Let $s \equiv (s_1,\ldots,s_m)$ be an arbitrary sequence in Q, and let $i \equiv (i_1,\ldots,i_n)$ be an arbitrary sequence in $\{1,\ldots,m\}$. Define the dual function $i^* : \overline{S}^m \to \overline{S}^n$ by

$$i^*(x_1,\ldots,x_m) \equiv (x_{s(i(1))},\ldots,x_{s(i(n))}) \tag{6.3.23}$$

for each $(x_1,\ldots,x_m) \in \overline{S}^m$. Define $j \equiv (j_1,\ldots,j_n) \equiv (i_1,\ldots,i_n)$ and define the dual function $j^* : S^m \to S^n$ by

$$j^*(x_1,\ldots,x_m) \equiv (x_{s(j(1))},\ldots,x_{s(j(n))}) \equiv (x_{s(i(1))},\ldots,x_{s(i(n))}) \tag{6.3.24}$$

for each $(x_1,\ldots,x_m) \in S^m$. Consider each $\overline{f} \in C(\overline{S}^n,\overline{d}^n)$. Then, for each $(x_1,\ldots,x_m) \in S^m$, we have

$$\begin{aligned}
((\overline{f} \circ i^*)|S^m)(x_1,\ldots,x_m) &= (\overline{f} \circ i^*)(x_1,\ldots,x_m) = \overline{f}(x_{s(i(1))},\ldots,x_{s(i(n))}) \\
&= (\overline{f}|S^n)(x_{s(i(1))},\ldots,x_{s(i(n))}) \\
&= (\overline{f}|S^n)(x_{s(j(1))},\ldots,x_{s(j(n))}) \\
&= (\overline{f}|S^m) \circ j^*(x_1,\ldots,x_m).
\end{aligned}$$

In short, $(\overline{f} \circ i^*)|S^m = (\overline{f}|S^m) \circ j^*$. It follows that

$$\begin{aligned}\overline{F}_{s(1),\ldots,s(m)}(\overline{f} \circ i^*) &\equiv F_{s(1),\ldots,s(m)}((\overline{f} \circ i^*)|S^m) = F_{s(1),\ldots,s(m)}((\overline{f}|S^m) \circ j^*)\\ &= F_{s(j(1)),\ldots,s(j(n))}(\overline{f}|S^m) = F_{s(i(1)),\ldots,s(i(n))}(\overline{f}|S^m)\\ &\equiv \overline{F}_{s(i(1)),\ldots,s(i(n))}\overline{f}, \qquad (6.3.25)\end{aligned}$$

where the first and last equalities are by the defining equality 6.3.22, and where the third equality is thanks to the consistency of the family $F \in \widehat{F}(Q,S)$. Equality 6.3.25 shows that the family $\overline{F} \equiv \psi(F)$ is consistent. In short, $\overline{F} \in \widehat{F}(Q,\overline{S})$. Thus the function $\psi : \widehat{F}(Q,S) \to \widehat{F}(Q,\overline{S})$ is well defined. The lemma is proved. □

We are now ready to construct, and prove the metrical continuity of, the Daniell–Kolmogorov Extension, where the state space is required only to be locally compact.

Theorem 6.3.7. Construction, and metrical continuity, of the Daniell–Kolmogorov Extension, with locally compact state space. *$Q \equiv \{t_1,t_2,\ldots\}$ will denote an enumerated countably infinite parameter set, enumerated by the bijection $t : \{1,2,\ldots\} \to Q$. Suppose (S,d) is a locally compact metric space, not necessarily compact, with a one-point compactification $(\overline{S},\overline{d})$, relative to some arbitrary but fixed binary approximation ξ, and with the point Δ at infinity. Recall that $\widehat{J}(S^Q,d^Q)$ denotes the set of distributions on the complete metric space (S^Q,d^Q). Then the following conditions hold:*

1. There exists a function

$$\Phi_{DK} : \widehat{F}(Q,S) \to \widehat{J}(S^Q,d^Q)$$

such that for each $F \in \widehat{F}(Q,S)$, the coordinate function $U : Q \times (S^Q,L,E) \to (S,d)$ is an r.f. with marginal distributions given by the family F, where $E \equiv \Phi_{DK}(F)$, and where (S^Q,L,E) is the probability space that is the completion of the integration space $(S^Q,C_{ub}(S^Q,d^Q),E)$. The function Φ_{DK} is called the Daniell–Kolmogorov Extension.

2. Let $\widehat{F}_\beta(Q,S)$ be an arbitrary subset $\widehat{F}(Q,S)$ whose members share a common modulus of pointwise tightness β. Define

$$\widehat{J}_\beta(S^Q,d^Q) \equiv \Phi_{DK}(\widehat{F}_\beta(Q,S)) \subset \widehat{J}(S^Q,d^Q).$$

Then the Daniell–Kolmogorov Extension

$$\Phi_{DK} : (\widehat{F}_\beta(Q,S),\widehat{\rho}_{Marg,\xi,Q}) \to \left(\widehat{J}_\beta(S^Q,d^Q),\rho_{Dist,\overline{\xi}^Q,*}\right) \qquad (6.3.26)$$

is uniformly continuous relative to the marginal metric $\widehat{\rho}_{Marg,\xi,Q}$ on $\widehat{F}(Q,S)$, as defined in Definition 6.2.8, and relative to the metric $\rho_{Dist,\overline{\xi}^Q,*}$ on $\widehat{J}_\beta(S^Q,d^Q)$ to be defined in the following proof.

Proof. Assume, without loss of generality, that $Q \equiv \{1,2,\ldots\}$.

1. Consider the function

$$\psi : \widehat{F}(Q,S) \to \widehat{F}(Q,\overline{S}) \tag{6.3.27}$$

defined in Lemma 6.3.6. Separately, by applying Theorem 6.3.4, to the compact metric space $(\overline{S},\overline{d})$, we have the uniformly continuous Compact Daniell–Kolmogorov Extension

$$\overline{\Phi}_{DK} : (\widehat{F}(Q,\overline{S}),\widehat{\rho}_{Marg,\overline{\xi},Q}) \to (\widehat{J}(\overline{S}^Q,\overline{d}^Q),\rho_{Dist,\overline{\xi}^Q}). \tag{6.3.28}$$

Define

$$\widehat{J}_{DK}(\overline{S}^Q,\overline{d}^Q) \equiv \overline{\Phi}_{DK}(\psi(\widehat{F}(Q,S))).$$

2. Let $\overline{E} \in \widehat{J}_{DK}(\overline{S}^Q,\overline{d}^Q)$ be arbitrary. Then there exists $F \in \widehat{F}(Q,S)$ such that $\overline{E} = \overline{\Phi}_{DK}(\overline{F})$, where $\overline{F} \equiv \psi(F) \in \widehat{F}(Q,\overline{S})$. Theorem 6.3.2, when applied to the compact metric space $(\overline{S},\overline{d})$ and the consistent family $\overline{F}$ of f.j.d.'s with state space $(\overline{S},\overline{d})$, says that the coordinate function

$$\overline{U} : Q \times (\overline{S}^Q,\overline{L},\overline{E}) \to (\overline{S},\overline{d})$$

is an r.f. with marginal distributions given by the family $\overline{F}$, where $(\overline{S}^Q,\overline{L},\overline{E})$ is the completion of $(\overline{S}^Q,C(\overline{S}^Q,\overline{d}^Q),\overline{E})$. In the next several steps, we will prove, successively, that (i) S^Q is a full set in $(\overline{S}^Q,\overline{L},\overline{E})$, (ii) $C_{ub}(S^Q,d^Q) \subset \overline{L}$, and (iii) the restricted function $E \equiv \overline{E}|C_{ub}(S^Q,d^Q)$ is a distribution on (S^Q,d^Q). Note that each function $f \in C_{ub}(S^Q,d^Q)$ can be regarded as a function on $\overline{S}^Q$ with $domain(f) \equiv S^Q \subset \overline{S}^Q$.

3. To proceed, let $m \geq 1$ be arbitrary. Define

$$h_m \equiv 1 \wedge (1 + m - d(\cdot,x_\circ))_+ \in C(S,d).$$

Note that $|h_m(z) - h_m(z)| \leq d(x,y)$ for each $z,y \in S$. Separately, by Proposition 3.4.5, we have $C(S,d) \subset C(\overline{S},\overline{d})|S$. Hence there exists $\overline{h}_m \in C(\overline{S},\overline{d})$ such that $h_m = \overline{h}_m|S$.

4. Now let $n \geq 1$ be arbitrary. Consider the r.r.v. $\overline{h}_m(\overline{U}_n)$, with values in $[0,1]$. Then

$$\overline{E}\overline{h}_m(\overline{U}_n) = \overline{F}_n\overline{h}_m \equiv F_n(\overline{h}_m|S) = F_nh_m \uparrow 1 \tag{6.3.29}$$

as $m \to \infty$, where the first equality is because the r.f. $\overline{U}$ has marginal distributions given by the family $\overline{F}$, where the second equality is by the defining equality 6.3.22 in Lemma 6.3.6, and where the convergence is because F_n is a distribution on the locally compact metric space (S,d). The Monotone Convergence Theorem therefore implies that the limit $Y_n \equiv \lim_{m\to\infty} \overline{h}_m(\overline{U}_n)$ is integrable on $(\overline{S}^\infty,\overline{L},\overline{E})$, with values in $[0,1]$ and with integral $\overline{E}Y_n = 1$. It follows that $\overline{P}(Y_n = 1) = 1$. Here $\overline{P}$ denotes the probability function associated with the probability integration $\overline{E}$.

5. With $n \geq 1$ be arbitrary, but fixed until further notice, consider each $x \equiv (x_1, x_2, \ldots) \in (Y_n = 1)$. Then

$$\lim_{m\to\infty} \overline{h}_m(x_n) = \lim_{m\to\infty} \overline{h}_m(\overline{U}_n(x)) \equiv Y_n(x) = 1. \tag{6.3.30}$$

Let $\alpha \in (0,1)$ be arbitrary. Then there exists $m \geq 1$ such that

$$\overline{h}_m(x_n) > \alpha.$$

Note that the bounded subset $(d(\cdot, x_\circ) \leq 1 + m)$ of (S, d) is contained in some compact subset $K_m \subset S$. By Assertion 1 of Proposition 3.4.5, the set K_m is also a compact subset of $(\overline{S}, \overline{d})$. Now, because $\widetilde{S} \equiv S \cup \{\Delta\}$ is dense in $(\overline{S}, \overline{d})$, and because $\overline{h}_m \in C(\overline{S}, \overline{d})$, there exists a sequence $(y_k)_{k=1,2,\ldots}$ in $\widetilde{S} \equiv S \cup \{\Delta\}$ such that $\overline{d}(y_k, x_n) \to 0$ as $k \to \infty$, and such that $|\overline{h}_m(y_k) - \overline{h}_m(x_n)| < \alpha$ for each $k \geq 1$. Then for each $k \geq 1$, we have $\overline{h}_m(y_k) > \overline{h}_m(x_n) - \alpha > 0$, whence $y_k \neq \Delta$, $y_k \in S$ and $h_m(y_k) = \overline{h}_m(y_k) > 0$. It follows that $y_k \in (d(, x_\circ) \leq 1 + m) \subset K_m$. Since $\overline{d}(y_k, x_n) \to 0$ as $k \to \infty$, and since K_m is a compact subset of $(\overline{S}, \overline{d})$, we infer that $x_n \in K_m \subset S$. Summing up,

$$(Y_n = 1) \subset \{x \equiv (x_1, x_2, \ldots) \in \overline{S}^\infty; x_n \in S\},$$

where $n \geq 1$ is arbitrary.

6. It follows that

$$\bigcap_{n=1}^{\infty}(Y_n = 1) \subset \bigcap_{n=1}^{\infty}\{(x_1, x_2, \ldots) \in \overline{S}^\infty : x_n \in S\} = S^\infty.$$

Hence S^∞ contains the full set that is the intersection of the sequence of full subsets of $(\overline{S}^\infty, \overline{L}, \overline{E})$ on the left-hand side. Thus S^∞ is itself a full subset of $(\overline{S}^\infty, \overline{L}, \overline{E})$. This proves the desired Condition (i) in Step 2.

7. Proceed to verify Condition (ii). To that end, consider the coordinate function

$$U : Q \times S^\infty \to S$$

as in Definition 6.3.1.

First, let $f \in C(S^n, d^n)$ be arbitrary. By Proposition 3.4.5, we have $C(S^n, d^n) \subset C(\overline{S}^n, \overline{d}^n)|S^n$. Hence there exists $\overline{f} \in C(\overline{S}^n, \overline{d}^n)$ such that $f = \overline{f}|S^n$. Because $\overline{U}$ is an r.f. with sample space $(\overline{S}^\infty, \overline{L}, \overline{E})$ and with values in $(\overline{S}, \overline{d})$, the function $\overline{f}(\overline{U}_1, \ldots, \overline{U}_n)$ on $(\overline{S}^\infty, \overline{L}, \overline{E})$ is an r.r.v. Consequently, $\overline{f}(\overline{U}_1, \ldots, \overline{U}_n) \in \overline{L}$ because $\overline{f}$ is bounded. At the same time, we have $f(U_1, \ldots, U_n) = \overline{f}(\overline{U}_1, \ldots, \overline{U}_n)1_{S^\infty}$ on the full subset S^∞ of $(\overline{S}^\infty, \overline{L}, \overline{E})$. It follows that

$$f(U_1, \ldots, U_n) = \overline{f}(\overline{U}_1, \ldots, \overline{U}_n)1_{S^\infty} \in \overline{L} \tag{6.3.31}$$

and that

$$\begin{aligned} \overline{E} f(U_1, \ldots, U_n) &= \overline{E}\overline{f}(\overline{U}_1, \ldots, \overline{U}_n) \\ &= \overline{F}_{1,\ldots,n}\overline{f} \equiv F_{1,\ldots,n}(\overline{f}|S^n) = F_{1,\ldots,n}(f), \end{aligned} \tag{6.3.32}$$

where $f \in C(S^n, d^n)$ is arbitrary, where the first equality is from equality 6.3.31, where the second equality is because the r.f. $\overline{U} : Q \times (\overline{S}^Q, \overline{L}, \overline{E}) \to (\overline{S}, \overline{d})$ has marginal distributions given by the family $\overline{F}$, and where the third equality is by the defining equality 6.3.22 in Lemma 6.3.6.

8. Next, let $f \in C_{ub}(S^n, d^n)$ be arbitrary, with $0 \leq f \leq b$ for some $b > 0$. Let $m \geq 1$ be arbitrary. From Step 3, we have the function

$$h_m \equiv 1 \wedge (1 + m - d(\cdot, x_\circ))_+ \in C(S, d).$$

Define the function by g_m on S^n by $g_m(x_1, \ldots, x_n) \equiv h_m(x_1) \ldots h_m(x_n)$ for each $(x_1, \ldots, x_n) \in S^n$. Then $g_m \equiv h_m \otimes \cdots \otimes h_m \in C(S^n, d^n)$. Hence $fg_m \in C(S^n, d^n)$. Moreover, $g_m(U_1, \ldots, U_n) \in \overline{L}$ and

$$\overline{E} g_m(U_1, \ldots, U_n) = F_{1,\ldots,n}(g_m), \tag{6.3.33}$$

according to equalities 6.3.31 and 6.3.32. Let $\varepsilon > 0$ be arbitrary. Because $F_{1,\ldots,n}$ is a distribution on (S^n, d^n), there exists $m \geq 1$ so large that

$$F_{1,\ldots,n} \widetilde{h}_m > 1 - \varepsilon, \tag{6.3.34}$$

where we define

$$\widetilde{h}_m \equiv 1 \wedge (m - d^n(\cdot, (x_\circ, \ldots, x_\circ)))_+ \in C(S^n, d^n).$$

Let $(x_1, \ldots, x_n) \in S^n$ be arbitrary such that $\widetilde{h}_m(x_1, \ldots, x_n) > 0$. Then

$$\bigvee_{i=1}^{n} d(x_i, x_\circ) \equiv d^n((x_1, \ldots, x_n), (x_\circ, \ldots, x_\circ)) \leq m,$$

whence $g_m(x_1, \ldots, x_n) \equiv h_m(x_1) \ldots h_m(x_n) = 1$. We conclude that $\widetilde{h}_m \leq g_m$ and therefore that

$$1 \geq F_{1,\ldots,n} g_m \geq F_{1,\ldots,n} \widetilde{h}_m > 1 - \varepsilon, \tag{6.3.35}$$

where the last inequality is from inequality 6.3.34. Hence, using the monotonicity of the sequence $(g_m)_{m=1,2,\ldots}$ and the nonnegativity of the function f, we obtain

$$F_{1,\ldots,n} |fg_{m'} - fg_m| \leq bF_{1,\ldots,n}(g_{m'} - g_m) < b\varepsilon, \tag{6.3.36}$$

for each $m' \geq m$. Thus $F_{1,\ldots,n} fg_m$ converges as $m \to \infty$. It follows from the Monotone Convergence Theorem that the function $f = \lim_{m \to \infty} F_{1,\ldots,n} fg_m$ is integrable relative to $F_{1,\ldots,n}$, and that $F_{1,\ldots,n} f = \lim_{m \to \infty} F_{1,\ldots,n} fg_m$.

Similarly, note that $fg_m \in C(S^n, d^n)$. Hence relation 6.3.31 yields

$$(fg_m)(U_1, \ldots, U_n) \in \overline{L}.$$

At the same time,

$$\overline{E}(1 - g_m(U_1, \ldots, U_n)) = 1 - F_{1,\ldots,n}(g_m) < \varepsilon,$$

according to equality 6.3.33. Consequently,

$$\overline{E}|(fg_{m'})(U_1,\ldots,U_n) - (fg_m)(U_1,\ldots,U_n)|$$
$$\leq b\overline{E}|g_{m'}(U_1,\ldots,U_n) - g_m(U_1,\ldots,U_n)| < b\varepsilon, \quad (6.3.37)$$

for each $m' \geq m$. Hence $\overline{E}(fg_m)(U_1,\ldots,U_n)$ converges as $m \to \infty$. It follows from the Monotone Convergence Theorem that

$$f(U_1,\ldots,U_n) = \lim_{m\to\infty}(fg_m)(U_1,\ldots,U_n) \in \overline{L} \quad (6.3.38)$$

and that

$$\overline{E}f(U_1,\ldots,U_n) = \lim_{m\to\infty}\overline{E}(fg_m)(U_1,\ldots,U_n)$$
$$= \lim_{m\to\infty}F_{1,\ldots,n}(fg_m) = F_{1,\ldots,n}(f), \quad (6.3.39)$$

where the second equality is thanks to equality 6.3.32, and where $f \in C_{ub}(S^n,d^n)$ is arbitrary with $f \geq 0$. By linearity, relation 6.3.38 and equality 6.3.39 hold for each $f \in C_{ub}(S^n,d^n)$.

9. Now let $j \in C_{ub}(S^\infty,d^\infty)$ be arbitrary, with a modulus of continuity δ_j and with $|j| \leq b$ for some $b > 0$. We will prove that $j \in \overline{L}$. To that end, let $\varepsilon > 0$ be arbitrary. Let $n \geq 1$ be arbitrary with $2^{-n} < \delta_j(\varepsilon)$. Define $j_n \in C_{ub}(S^n,d^n)$ by

$$j_n(x_1,x_2,\ldots,x_n) \equiv j(x_1,x_2,\ldots,x_n,x_\circ,x_\circ,\ldots) \quad (6.3.40)$$

for each $(x_1,x_2,\ldots,x_n) \in S^n$. Then

$$d^\infty((x_1,x_2,\ldots,x_n,x_\circ,x_\circ,\ldots),(x_1,x_2,\ldots)) \leq \sum_{i=n+1}^{\infty} 2^{-i} = 2^{-n} < \delta_j(\varepsilon).$$

for each $x = (x_1,x_2,\ldots) \in S^\infty$. Hence

$$|j_n(U_1(x),\ldots,U_n(x)) - j(x_1,x_2,\ldots)|$$
$$= |j_n(x_1,x_2,\ldots,x_n) - j(x_1,x_2,\ldots)|$$
$$= |j(x_1,x_2,\ldots,x_n,x_\circ,x_\circ,\ldots) - j(x_1,x_2,\ldots)| < \varepsilon.$$

for each $x = (x_1,x_2,\ldots) \in S^\infty$. In other words,

$$|j_n(U_1,\ldots,U_n) - j| < \varepsilon \quad (6.3.41)$$

on the full set S^∞, whence

$$\overline{E}|j_n(U_1,\ldots,U_n) - j| \leq \varepsilon, \quad (6.3.42)$$

where $n \geq 1$ is arbitrary with $2^{-n} < \delta_j(\varepsilon)$. Inequality 6.3.41 trivially implies that $j_n(U_1,\ldots,U_n) \to j$ in probability on $(\overline{S}^Q,\overline{L},\overline{E})$, as $n \to \infty$. At the same time, $j_n \in C_{ub}(S^n,d^n)$, whence $j_n(U_1,\ldots,U_n) \in \overline{L}$ according to relation 6.3.38.

The Dominated Convergence Theorem therefore implies that $j \in \overline{L}$ and that

$$\overline{E}j = \lim_{n\to\infty}\overline{E}j_n(U_1,\ldots,U_n), \quad (6.3.43)$$

where $j \in C_{ub}(S^\infty, d^\infty)$ is arbitrary. We conclude that $C_{ub}(S^\infty, d^\infty) \subset \overline{L}$. This proves the desired Condition (ii) in Step 2.

10. In view of Condition (ii), we can define the function

$$E \equiv \overline{E}|C_{ub}(S^\infty, d^\infty). \tag{6.3.44}$$

In other words,

$$Ej \equiv \overline{E}j = \lim_{n\to\infty} \overline{E}j_n(U_1,\ldots,U_n) \tag{6.3.45}$$

for each $j \in C_{ub}(S^\infty, d^\infty)$. In particular, for each $f \in C_{ub}(S^n, d^n)$, we have

$$Ef(U_1,\ldots,U_n) = \overline{E}f(U_1,\ldots,U_n) = F_{1,\ldots,n}(f), \tag{6.3.46}$$

where the second equality is from equality 6.3.39.

11. Next observe that $C_{ub}(S^\infty, d^\infty)$ is a linear subspace of $\overline{L}$ such that if $f, g \in C_{ub}(S^\infty, d^\infty)$, then $|f|, f \wedge 1 \in C_{ub}(S^\infty, d^\infty)$. Proposition 4.3.6 therefore implies that $(S^\infty, C_{ub}(S^\infty, d^\infty), E)$ is an integration space. Since $d^\infty \leq 1$ and $E1 = 1$, Assertion 2 of Lemma 5.2.2 implies that the integration E is a distribution on (S^∞, d^∞). This proves the desired Condition (iii) in Step 2.

12. Thus we can define a function

$$\varphi : \widehat{J}_{DK}(\overline{S}^Q, \overline{d}^Q) \to \widehat{J}(S^Q, d^Q)$$

by

$$\varphi(\overline{E}) \equiv E \equiv \overline{E}|C_{ub}(S^\infty, d^\infty) \in \widehat{J}(S^Q, d^Q)$$

for each $\overline{E} \in \widehat{J}_{DK}(\overline{S}^Q, \overline{d}^Q)$.

13. Now define the composite function

$$\Phi_{DK} \equiv \varphi \circ \overline{\Phi}_{DK} \circ \psi : \widehat{F}(Q,S) \to \widehat{J}(S^Q, d^Q).$$

Let $F \in \widehat{F}(Q,S)$ be arbitrary. Let $E \equiv \Phi_{DK}(F) \in \widehat{J}(S^Q, d^Q)$. Then $E = \varphi(\overline{E})$, where $\overline{E} \equiv \overline{\Phi}_{DK}(\overline{F})$ and $\overline{F} \equiv \psi(F)$.

Let (S^Q, L, E) be the probability space that is the completion of the integration space $(S^Q, C_{ub}(S^Q, d^Q), E)$. Equality 6.3.46 shows that the coordinate function $U : Q \times (S^Q, L, E) \to (S, d)$ is an r.f. with marginal distributions given by the family F. We conclude that the function Φ_{DK} satisfies all the conditions of Assertion 1 of the present theorem. Thus Assertion 1 is proved.

14. To prove Assertion 2, let $\widehat{F}_\beta(Q,S)$ be an arbitrary subset $\widehat{F}(Q,S)$ whose members share a common modulus of pointwise tightness β. We will first show that the restricted mapping

$$\psi : (\widehat{F}_\beta(Q,S), \widehat{\rho}_{Marg,\xi,Q}) \to (\widehat{F}(Q,\overline{S}), \widehat{\rho}_{Marg,\overline{\xi},Q}) \tag{6.3.47}$$

is uniformly continuous, where the marginal metrics $\widehat{\rho}_{Marg,\xi,Q}$ and $\widehat{\rho}_{Marg,\overline{\xi},Q}$ are as in Definition 6.2.8.

15. Recall from Definition 6.3.3 that $\xi \equiv (A_n)_{n=1,2,\ldots}$ is a binary approximation of (S,d), relative to some fixed reference point $x_\circ$. For each $n \geq 1$, let

$A_n \equiv \{x_{n,1}, \ldots, x_{n,\kappa(n)}\} \subset S$. Let $\overline{\xi} \equiv (\overline{A}_p)_{p=1,2,\ldots}$ be the compactification of the binary approximation ξ. Thus there exists an increasing sequence $(m_n)_{n=1,2,\ldots}$ of integers such that

$$\overline{A}_n \equiv A_{m(n)} \cup \{\Delta\} \equiv \{x_{m(n),1}, \ldots, x_{m(n),\kappa(m(n))}, \Delta\},$$

for each $n \geq 1$, where Δ is the point at infinity of the one-point compactification $(\overline{S}, \overline{d})$ of (S,d) relative to ξ. Then $\overline{\xi}$ is a binary approximation of $(\overline{S}, \overline{d})$ relative to $x_\circ$, according to Corollary 3.4.4.

16. Now let $\nu \geq 1$ be arbitrary but fixed until further notice. For each $n \geq 1$, let $A_n^\nu \equiv A_n \times \cdots \times A_n \subset S^\nu$. Then $\xi^\nu \equiv (A_n^\nu)_{n=1,2,\ldots}$ is the binary approximation of (S^ν, d^ν) relative to the reference point $(x_\circ, \ldots, x_\circ) \in S^\nu$. Similarly, $\overline{\xi}^\nu \equiv (\overline{A}_n^\nu)_{n=1,2,\ldots}$ is the binary approximation of $(\overline{S}^\nu, \overline{d}^\nu)$ relative to the fixed reference point $(x_\circ, \ldots, x_\circ) \in \overline{S}^\nu$. To lessen the burden on subscripts, we will write A_n^ν and $A(\nu,n)$ interchangeably, and write $\overline{A}_n^\nu$ and $\overline{A}(\nu,n)$ interchangeably.

Let the sequence

$$\pi^\nu \equiv (\{g_{\nu,n,x} : x \in A_n^\nu\})_{n=1,2,\ldots}$$

in $C(S^\nu, d^\nu)$ be the partition of unity of (S^ν, d^ν) determined by the binary approximation ξ^ν, in the sense of Definition 3.3.4. Likewise, let the sequence

$$\overline{\pi}^\nu \equiv (\{\overline{g}_{\nu,n,x} : x \in \overline{A}_n^\nu\})_{n=1,2,\ldots}$$

in $C(\overline{S}^\nu, \overline{d}^\nu)$ be the partition of unity of $(\overline{S}^\nu, \overline{d}^\nu)$ determined by the binary approximation $\overline{\xi}^\nu$.

17. Let $F \equiv \{F_{s(1),\ldots,s(m)} : m \geq 1; s_1, \ldots, s_m \in Q\}$, $F' \equiv \{F'_{s(1),\ldots,s(m)} : m \geq 1; s_1, \ldots, s_m \in Q\} \in \widehat{F}_\beta(Q,S)$ be arbitrary. Let $\overline{F} \equiv \{\overline{F}_{s(1),\ldots,s(m)} : m \geq 1; s_1, \ldots, s_m \in Q\} \equiv \psi(F)$, $\overline{F}' \equiv \{\overline{F'}_{s(1),\ldots,s(m)} : m \geq 1; s_1, \ldots, s_m \in Q\} \equiv \psi(F')$. Then, according to Definition 5.3.4, we have

$$\rho_{Dist,\xi^\nu}(F_{1,\ldots,\nu}, F'_{1,\ldots,\nu}) \equiv \sum_{n=1}^{\infty} 2^{-n} |A_n^\nu|^{-1} \sum_{x \in A(\nu,n)} |F_{1,\ldots,\nu} g_{\nu,n,x} - F'_{1,\ldots,\nu} g_{\nu,n,x}|. \tag{6.3.48}$$

Similarly,

$$\begin{aligned} &\rho_{Dist,\overline{\xi}^\nu}(\overline{F}_{1,\ldots,\nu}, \overline{F'}_{1,\ldots,\nu}) \\ &\equiv \sum_{n=1}^{\infty} 2^{-n} |\overline{A}_n^\nu|^{-1} \sum_{x \in \overline{A}(\nu,n)} |\overline{F}_{1,\ldots,\nu} \overline{g}_{\nu,n,x} - \overline{F'}_{1,\ldots,\nu} \overline{g}_{\nu,n,x}| \\ &= \sum_{n=1}^{\infty} 2^{-n} |\overline{A}_n^\nu|^{-1} \sum_{x \in \overline{A}(\nu,n)} |F_{1,\ldots,\nu}(\overline{g}_{\nu,n,x}|S^\nu) - F'_{1,\ldots,\nu}(\overline{g}_{\nu,n,x}|S^\nu)|, \end{aligned} \tag{6.3.49}$$

where the last equality is due to the defining equality 6.3.22.

18. Let $\varepsilon > 0$ be arbitrary. Take $N \geq \nu$ so large that $2^{-N} < \varepsilon$. Suppose

$$\rho_{Dist,\xi^\nu}(F_{1,\ldots,\nu}, F'_{1,\ldots,\nu}) < 2^{-N}|A_N^\nu|^{-1}\varepsilon.$$

Consider each $n = 1,\ldots,N$. Then it follows from equality 6.3.48 that

$$\begin{aligned}|F_{1,\ldots,\nu}g_{\nu,n,x} - F'_{1,\ldots,\nu}g_{\nu,n,x}| &\leq 2^n|A_n^\nu|\rho_{Dist,\xi^\nu}(F_{1,\ldots,\nu}, F'_{1,\ldots,\nu}) \\ &< 2^n|A_n^\nu|2^{-N}|A_N^\nu|^{-1}\varepsilon \leq \varepsilon, \qquad (6.3.50)\end{aligned}$$

for each $x \in A_n^\nu$.

19. Take any $m \geq \bigvee_{i=1}^\nu \beta(N^{-1}\varepsilon, i)$, where, by hypothesis, β is a modulus of pointwise tightness of the consistent family $F \in \widehat{F}_\beta(Q,S)$. From Step 8, the function $g_m \equiv h_m \otimes \cdots \otimes h_m \in C(S^n, d^n)$. Thus $g_m(x_1,\ldots,x_\nu) \equiv h_m(x_1)\ldots h_m(x_\nu)$ for each $(x_1,\ldots,x_\nu) \in (S^\nu, d^\nu)$.

Let $i = 1,\ldots,\nu$ be arbitrary. Then $m \geq \beta(N^{-1}\varepsilon, i)$. Hence, by Definition 6.3.5, we have

$$1 - F_i h_m = F_i(1 - h_m) < N^{-1}\varepsilon.$$

Therefore

$$\begin{aligned}0 &\leq 1 - F_{1,\ldots,\nu}g_m \\ &= 1 - Eg_m(U_1,\ldots,U_\nu) \equiv 1 - Eh_m(U_1)\ldots h_m(U_\nu) \\ &= (1 - Eh_m(U_1)) + E(h_m(U_1) - h_m(U_1)h_m(U_2)) \\ &\quad + \cdots + E(h_m(U_1)\ldots h_m(U_{\nu-1}) - h_m(U_1)\ldots h_m(U_\nu)) \\ &\leq (1 - Eh_m(U_1)) + E(1 - h_m(U_2)) + \cdots + E(1 - h_m(U_\nu)) \\ &= (1 - F_1h_m) + (1 - F_2h_m) + \cdots + (1 - F_\nu h_m) < \nu N^{-1}\varepsilon < \varepsilon, \qquad (6.3.51)\end{aligned}$$

where the first equality is because the coordinate function $U : Q \times (S^Q, L, E) \to (S,d)$ is an r.f. with marginal distributions given by the family F.

Consider each $x \in \overline{A}_n^\nu$. It follows that

$$\begin{aligned}0 &\leq F_{1,\ldots,\nu}(\overline{g}_{\nu,n,x}|S^\nu) - F_{1,\ldots,\nu}(\overline{g}_{\nu,n,x}|S^\nu)g_m \\ &= F_{1,\ldots,\nu}(\overline{g}_{\nu,n,x}|S^\nu)(1 - g_m) \leq 1 - F_{1,\ldots,\nu}g_m < \varepsilon. \qquad (6.3.52)\end{aligned}$$

Similarly,

$$0 \leq F'_{1,\ldots,\nu}(\overline{g}_{\nu,n,x}|S^\nu) - F'_{1,\ldots,\nu}(\overline{g}_{\nu,n,x}|S^\nu)g_m < \varepsilon. \qquad (6.3.53)$$

Moreover, the function $f \equiv (\overline{g}_{\nu,n,x}|S^\nu) \in C_{ub}(S^\nu, d^\nu)$ has some modulus of continuity $\delta_{\nu,n,x}$. Suppose $z \equiv (x_1,\ldots,x_\nu), y \equiv (y_1,\ldots,y_\nu) \in S^\nu$ are such that

$$d^\nu(z,y) < \widetilde{\delta}_{\nu,n}(\varepsilon) \equiv (2^{-1}\nu^{-1}\varepsilon) \wedge \bigwedge_{x \in \overline{A}(\nu,n)} \delta_{\nu,n,x}(2^{-1}\varepsilon).$$

Then

$$|g_m(z) - g_m(y)| \equiv |h_m(z_1)\ldots h_m(z_\nu) - h_m(y_1)\ldots h_m(y_\nu)|$$
$$\leq |h_m(z_1) - h_m(y_1)| + \cdots + |h_m(z_\nu) - h_m(y_\nu)|$$
$$\leq \nu \bigvee_{i=1}^{\nu} d(z_i, y_i) \equiv \nu d^\nu(z,y) < \nu(2^{-1}\nu^{-1}\varepsilon) < 2^{-1}\varepsilon$$

and

$$|f(z) - f(y)| < 2^{-1}\varepsilon.$$

Hence

$$|fg_m(z) - fg_m(y)| \leq |f(z)g_m(z) - f(y)g_m(z)| + |f(y)g_m(z) - f(y)g_m(y)|$$
$$\leq |f(z) - f(y)| + |g_m(z) - g_m(y)| < 2^{-1}\varepsilon + 2^{-1}\varepsilon = \varepsilon.$$

We conclude that the function $fg_m \equiv (\overline{g}_{\nu,n,x}|S^\nu)g_m$ has the modulus of continuity $\widetilde{\delta}_{\nu,n}$. We emphasize that the modulus $\widetilde{\delta}_{\nu,n}$ is regardless of m.

20. Therefore we can take

$$m \equiv m_\nu \equiv \left[\bigvee_{i=1}^{\nu} \beta(N^{-1}\varepsilon, i) \vee \bigvee_{k=1}^{N}(-\log_2(2^{-1}\widetilde{\delta}_{\nu,k}(3^{-1}\varepsilon))\right]_1.$$

Then $m > \bigvee_{i=1}^{\nu} \beta(N^{-1}\varepsilon, i)$ and $2^{-m} < 2^{-1}\widetilde{\delta}_{\nu,n}(3^{-1}\varepsilon)$.

Suppose $F, F' \in \widehat{F}_\beta(Q,S)$ are such that

$$\rho_{Dist,\xi^\nu}(F_{1,\ldots,\nu}, F'_{1,\ldots,\nu}) \leq 2^{-m}|A_m^\nu|^{-1}\varepsilon \equiv 2^{-m(\nu)}|A_{m(\nu)}^\nu|^{-1}\varepsilon. \tag{6.3.54}$$

Note that the function $h_m \in C(S,d)$ has the set $(d(\cdot,x_\circ) \leq m+1)$ as support. Hence the function $g_m \equiv h_m \otimes \cdots \otimes h_m \in C(S^\nu, d^\nu)$ has the set

$$(d^\nu(\cdot,(x_\circ,\ldots,x_\circ)) \leq m+1) \subset B \equiv (d^\nu(\cdot,(x_\circ,\ldots,x_\circ)) \leq 2^m) \subset S^\nu$$

as support. Let $x \in \overline{A}_k^\nu$ be arbitrary, and write $f \equiv \overline{g}_{\nu,n,x}|S^\nu$. Then (i) the function $fg_m \in C(S^\nu, d^\nu)$ has the set B as support and (ii) $2^{-m} < 2^{-1}\widetilde{\delta}_{\nu,n}(3^{-1}\varepsilon)$, where $\widetilde{\delta}_{\nu,n}$ is a modulus of continuity of fg_m.

Hence Conditions (i) and (ii) in Assertion 2 of Proposition 3.3.6 are satisfied, where we replace the objects n, (S,d), ξ, f, δ_f, and π by 1, (S^ν, d^ν), $\xi^\nu \equiv (A_k^\nu)_{k=1,2,\ldots}$, fg_m, $\widetilde{\delta}_{\nu,n}$, and $\pi^\nu \equiv (\{g_{\nu,k,y} : y \in A_k^\nu\})_{k=1,2,\ldots}$, respectively. Accordingly,

$$\left\| fg_m - \sum_{x \in A(\nu,m)} (fg_m)(y) g_{\nu,m,y} \right\| \leq \varepsilon. \tag{6.3.55}$$

Consequently,

$$0 \leq \left| F_{1,\ldots,\nu} fg_m - \sum_{x \in A(\nu,m)} (fg_m)(y) F_{1,\ldots,\nu} g_{\nu,m,y} \right| \leq \varepsilon$$

and

$$0 \leq \left| F'_{1,\dots,\nu} f g_m - \sum_{x \in A(\nu,m)} (f g_m)(y) F'_{1,\dots,\nu} g_{\nu,m,y} \right| \leq \varepsilon.$$

Combining the two last displayed inequalities, we obtain

$$\begin{aligned}
&|F_{1,\dots,\nu}(\overline{g}_{\nu,n,x}|S^\nu)g_m - F'_{1,\dots,\nu}(\overline{g}_{\nu,n,x}|S^\nu)g_m| \\
&\quad \equiv |F_{1,\dots,\nu} f g_m - F'_{1,\dots,\nu} f g_m| \\
&\quad \leq \left| \sum_{y \in A(\nu,m)} (f g_m)(y)(F_{1,\dots,\nu} g_{\nu,m,y} - F'_{1,\dots,\nu} g_{\nu,m,y}) \right| + 2\varepsilon \\
&\quad \leq \sum_{y \in A(\nu,m)} |F_{1,\dots,\nu} g_{\nu,m,y} - F'_{1,\dots,\nu} g_{\nu,m,y}| + 2\varepsilon \\
&\quad \leq 2^m |A_m^\nu| \rho_{Dist,\xi^\nu}(F_{1,\dots,\nu}, F'_{1,\dots,\nu}) + 2\varepsilon \\
&\quad \leq \varepsilon + 2\varepsilon = 3\varepsilon,
\end{aligned} \tag{6.3.56}$$

where the third inequality is from the defining equality 6.3.48, where the last inequality is by inequality 6.3.54, and where $n = 1, \dots, N$ is arbitrary. From equality 6.3.49, we therefore obtain

$$\begin{aligned}
&\rho_{Dist,\overline{\xi}^\nu}(\overline{F}_{1,\dots,\nu}, \overline{F'}_{1,\dots,\nu}) \\
&\quad \leq \sum_{n=1}^{N} 2^{-n} |\overline{A}_n^\nu|^{-1} \sum_{x \in \overline{A}(\nu,n)} |F_{1,\dots,\nu}(\overline{g}_{\nu,n,x}|S^\nu) - F'_{1,\dots,\nu}(\overline{g}_{\nu,n,x}|S^\nu)| + 2^{-N} \\
&\quad \leq \sum_{n=1}^{N} 2^{-n} |\overline{A}_n^\nu|^{-1} \sum_{x \in \overline{A}(\nu,n)} (|F_{1,\dots,\nu}(\overline{g}_{\nu,n,x}|S^\nu)g_m \\
&\qquad - F'_{1,\dots,\nu}(\overline{g}_{\nu,n,x}|S^\nu)g_m| + 2\varepsilon) + 2^{-N} \\
&\quad < \sum_{n=1}^{N} 2^{-n} |\overline{A}_n^\nu|^{-1} \sum_{x \in \overline{A}(\nu,n)} (3\varepsilon) + 2\varepsilon + 2^{-N} = \sum_{n=1}^{N} 2^{-n}(3\varepsilon) + 2\varepsilon + 2^{-N} \\
&\quad < 3\varepsilon + 2\varepsilon + 2^{-N} < 3\varepsilon + 2\varepsilon + \varepsilon = 6\varepsilon,
\end{aligned}$$

where the second inequality is thanks to inequalities 6.3.52 and 6.3.53, and where the third inequality is by inequality 6.3.56.

21. Summing up, for arbitrary $\varepsilon > 0$ and $\nu \geq 1$, we have $\rho_{Dist,\overline{\xi}^\nu}(\overline{F}_{1,\dots,\nu}, \overline{F'}_{1,\dots,\nu}) < 6\varepsilon$, provided that

$$\rho_{Dist,\xi^\nu}(F_{1,\dots,\nu}, F'_{1,\dots,\nu}) \leq 2^{-m(\nu)} |A_{m(\nu)}^\nu|^{-1} \varepsilon. \tag{6.3.57}$$

22. Now let $\alpha > 0$ be arbitrary. Write $\varepsilon \equiv 2^{-3}\alpha$. Take $\kappa \geq 1$ so large that $2^\kappa < \varepsilon$. Let $F, F' \in \widehat{F}_\beta(Q, S)$ be arbitrary such that

$$\widehat{\rho}_{Marg,\xi,Q}(F,F') < \delta_\psi(\alpha) \equiv 2^{-\kappa} \bigwedge_{\nu=1}^{\kappa} 2^{-m(\nu)}|A^\nu_{m(\nu)}|^{-1}\varepsilon. \tag{6.3.58}$$

Let $\overline{F} \equiv \psi(F)$, and $\overline{F'} \equiv \psi(F')$. Then, for each $\nu = 1,\ldots,\kappa$, we have

$$\rho_{Dist,\xi^\nu}(F_{1,\ldots,\nu},F'_{1,\ldots,\nu}) \leq 2^\kappa \widehat{\rho}_{Marg,\xi,Q}(F,F') \leq 2^{-m(\nu)}|A^\nu_{m(\nu)}|^{-1}\varepsilon,$$

whence, according to Step 21, we have $\rho_{Dist,\overline{\xi}^\nu}(\overline{F}_{1,\ldots,\nu},\overline{F'}_{1,\ldots,\nu}) < 6\varepsilon$. Consequently,

$$\begin{aligned}
\widehat{\rho}_{Marg,\overline{\xi},Q}(\psi(F),\psi(F')) &\equiv \widehat{\rho}_{Marg,\overline{\xi},Q}(\overline{F},\overline{F'}) \\
&\equiv \sum_{\nu=1}^{\infty} 2^{-\nu}\rho_{Dist,\overline{\xi}^\nu}(\overline{F}_{1,\ldots,\nu},\overline{F'}_{1,\ldots,\nu}) \\
&\leq \sum_{\nu=1}^{\kappa} 2^{-\nu}\rho_{Dist,\overline{\xi}^\nu}(\overline{F}_{1,\ldots,\nu},\overline{F'}_{1,\ldots,\nu}) + 2^{-\kappa} \\
&\leq \sum_{\nu=1}^{\kappa} 2^{-\nu}(6\varepsilon) + \varepsilon \leq 6\varepsilon + \varepsilon < \alpha.
\end{aligned}$$

Since $\alpha > 0$ is arbitrary, we see that the mapping

$$\psi : (\widehat{F}_\beta(Q,S),\widehat{\rho}_{Marg,\xi,Q}) \to (\widehat{F}(Q,\overline{S}),\widehat{\rho}_{Marg,\overline{\xi},Q}) \tag{6.3.59}$$

in relation 6.3.47 is uniformly continuous, as alleged, with a modulus of continuity δ_ψ defined in equality 6.3.58.

23. Next consider the function

$$\varphi : \widehat{J}_{DK}(\overline{S}^Q,\overline{d}^Q) \to \widehat{J}(S^Q,d^Q).$$

We will define a metric on its range $\varphi(\widehat{J}_{DK}(\overline{S}^Q,\overline{d}^Q))$. Since the metric space $(\overline{S},\overline{d})$ is compact, the countable power $\overline{\xi}^Q = \overline{\xi}^\infty$ of $\overline{\xi}$ is a binary approximation of $(\overline{S}^Q,\overline{d}^Q)$, according to Definition 3.2.6. Hence the distribution metric $\rho_{Dist,\overline{\xi}^Q}$ on the set $\widehat{J}_{DK}(\overline{S}^Q,\overline{d}^Q)$ of distributions is defined according to Definition 5.3.4. Now consider each $E,E' \in \varphi(\widehat{J}_{DK}(\overline{S}^Q,\overline{d}^Q))$. Let $\overline{E},\overline{E'} \in \widehat{J}_{DK}(\overline{S}^Q,\overline{d}^Q)$ be such that $E = \varphi(\overline{E})$ and $E = \varphi(\overline{E}')$. Suppose $\overline{E} = \overline{E'}$. Let $\overline{f} \in C(\overline{S}^\infty,\overline{d}^\infty)$ be arbitrary. Then $f \equiv \overline{f}|S^\infty \in C_{ub}(S^\infty,d^\infty)$. Hence

$$\overline{E}\overline{f} = \overline{E}f = Ef = E'f = \overline{E}'f = \overline{E}'\overline{f},$$

where the first and last equalities are because $\overline{f} = f$ on the full subset S^∞relative to $\overline{E}$ and to $\overline{E}'$, and where the second and fourth equalities are thanks to the defining equality 6.3.45. Thus $\overline{E} = \overline{E}'$ as distributions on the compact metric space $(\overline{S}^\infty,\overline{d}^\infty)$, provided that $\varphi(\overline{E}) = \varphi(\overline{E}')$. We conclude that the function φ is an injection. Now define

$$\rho_{Dist,\overline{\xi}^Q,*}(E,E') \equiv \rho_{Dist,\overline{\xi}^Q}(\overline{E},\overline{E'}). \tag{6.3.60}$$

Suppose $\rho_{Dist,\overline{\xi}^Q,*}(E,E') = 0$. Then $\rho_{Dist,\overline{\xi}^Q}(\overline{E},\overline{E'}) = 0$. Hence $\overline{E} = \overline{E'}$ because $\rho_{Dist,\overline{\xi}^Q}$ is a metric. Moreover, the defining equality 6.3.45 immediately implies the symmetry of, and triangular inequality for, the function $\rho_{Dist,\overline{\xi}^Q,*}$, which is therefore a metric on $\varphi(\widehat{J}_{DK}(\overline{S}^Q,\overline{d}^Q))$.

24. It is easy to see that the function

$$\varphi : (\widehat{J}_{DK}(\overline{S}^Q,\overline{d}^Q),\rho_{Dist,\overline{\xi}^Q}) \to (\varphi(\widehat{J}_{DK}(\overline{S}^Q,\overline{d}^Q)),\rho_{Dist,\overline{\xi}^Q,*}) \tag{6.3.61}$$

is continuous. To be precise, let $\overline{E},\overline{E'} \in \widehat{J}_{DK}(\overline{S}^Q,\overline{d}^Q)$ arbitrary. Let $E = \varphi(\overline{E})$ and $E = \varphi(\overline{E}')$. Then

$$\rho_{Dist,\overline{\xi}^Q,*}(\varphi(\overline{E}),\varphi(\overline{E}')) = \rho_{Dist,\overline{\xi}^Q,*}(E,E') \equiv \rho_{Dist,\overline{\xi}^\infty}(\overline{E},\overline{E'}).$$

Thus φ is an isometry, and hence continuous.

25. Now define the composite

$$\Phi_{DK} \equiv \varphi \circ \overline{\Phi}_{DK} \circ \psi : (\widehat{F}_\beta(Q,S),\widehat{\rho}_{Marg,\xi,Q}) \to (\varphi(\widehat{J}_{DK}(\overline{S}^Q,\overline{d}^Q)),\rho_{Dist,\overline{\xi}^Q,*}) \tag{6.3.62}$$

of the three uniformly continuous functions in formulas 6.3.47, 6.3.28, and 6.3.61. Call Φ_{DK} the Daniell–Kolmogorov Extension. Then Φ_{DK} is itself uniformly continuous. Moreover, its range

$$\widehat{J}_\beta(S^Q,d^Q) \equiv \Phi_{DK}(\widehat{F}_\beta(Q,S)) \subset \varphi(\overline{\Phi}_{DK}(\psi(\widehat{F}(Q,S)))) \equiv \varphi(\widehat{J}_{DK}(\overline{S}^Q,\overline{d}^Q))$$

is a subset of $\varphi(\widehat{J}_{DK}(\overline{S}^Q,\overline{d}^Q))$. Hence formula 6.3.62 can be rewritten as the uniformly continuous function

$$\Phi_{DK} : (\widehat{F}_\beta(Q,S),\widehat{\rho}_{Marg,\xi,Q}) \to (\widehat{J}_\beta(S^Q,d^Q),\rho_{Dist,\overline{\xi}^Q,*}). \tag{6.3.63}$$

26. Let $F \in \widehat{F}(Q,S)$ be arbitrary. Let $\overline{F} \equiv \psi(F)$, $\overline{E} \equiv \Phi_{DK}(\overline{F})$, and $E \equiv \varphi(E)$. Then $\Phi_{DK}(F) = E$. Let (S^Q,L,E) be the completion of $(S^Q,C_{ub}(S^Q,d^Q),E)$. Then, according to equality 6.3.39, we have

$$Ef(U_1,\ldots,U_n) = F_{1,\ldots,n}(f)$$

where $f \in C_{ub}(S^n,d^n)$ is arbitrary. Thus the coordinate function $U : Q \times (S^Q, L,E) \to (S,d)$ is an r.f. with marginal distributions given by the family F. Assertion 1 of the present theorem is proved.

27. Assertion 2 has been proved in Step 25. The theorem is proved. □

6.4 Daniell–Kolmogorov–Skorokhod Extension

Use the notations as in the previous section. Unless otherwise specified, (S,d) will be a locally compact metric space, and $Q \equiv \{t_1,t_2,\ldots\}$ will denote an enumerated countably infinite parameter set, enumerated by the bijection $t : \{1,2,\ldots\} \to Q$. Let $x_\circ$ be an arbitrary but fixed reference point in (S,d).

For two consistent families F and F' of f.j.d.'s with the parameter set Q and the locally compact state space (S,d), the Daniell–Kolmogorov Extension in Section 6.3 constructs two corresponding distributions E and E' on the path space (S^Q,d^Q), such that the families F and F' of f.j.d.'s are the marginal distributions of the r.f.'s $U : Q \times (S^Q,L,E) \to (S,d)$ and $U : Q \times (S^Q,L',E') \to S$, respectively, even as the underlying coordinate function U remains the same.

In contrast, theorem 3.1.1 in [Skorokhod 1956] combines the Daniell–Kolmogorov Extension with Skorokhod's Representation Theorem, presented as Theorem 5.5.1 in this book. It produces (i) as the sample space, the fixed probability space

$$(\Theta_0, L_0, I_0) \equiv \left([0,1], L_0, \int \cdot dx\right)$$

based on the uniform distribution $\int \cdot dx$ on the unit interval $[0,1]$, and (ii) for each $F \in \widehat{F}(Q,S)$, an r.f. $Z : Q \times (\Theta_0, L_0, I_0) \to (S,d)$ with marginal distributions given by F. The sample space is fixed, but two different families F and F' of f.j.d.'s result in two r.f.'s Z and Z' that are distinct functions on the same sample space. Theorem 3.1.1 in [Skorokhod 1956] shows that the Daniell–Kolmogorov–Skorokhod Extension thus obtained is continuous relative to weak convergence in $\widehat{F}(Q,S)$. Because the r.f.'s produced can be regarded as r.v.'s on one and the same probability space (Θ_0, L_0, I_0) with values in the path space (S^Q,d^Q), we will have at our disposal the familiar tools of making new r.v.'s, including taking a continuous function of r.v.'s Z and Z', and taking limits of r.v.'s in various senses.

In our Theorem 5.5.1, we recast Skorokhod's Representation Theorem in terms of partitions of unity in the sense of Definition 3.3.4. Namely, where Borel sets are used in [Skorokhod 1956], we use continuous basis functions with compact support. This will facilitate the subsequent proof of metrical continuity of the Daniell–Kolmogorov–Skorokhod Extension, as well as the derivation of an accompanying modulus of continuity. This way, we gain the advantage of metrical continuity, which is stronger than sequential weak convergence.

Recall from Definition 6.1.1 that $\widehat{R}(Q \times \Omega, S)$ denotes the set of r.f.'s with parameter set Q, sample space (Ω, L, E), and state space (S,d). We will next identify each member of $\widehat{R}(Q \times \Omega, S)$ with an r.v. on Ω with values in the path space (S^Q,d^Q), with the latter having been introduced in Definition 6.3.1. As observed in Definition 6.3.1, the path space (S^Q,d^Q) is complete, but not necessarily locally compact. However, it is compact if (S,d) is compact.

Lemma 6.4.1. A random field with a countable parameter set can be regarded as an r.v. with values in the path space. Let (S,d) *be a locally compact metric space. Let* $Q = \{t_1, t_2, \ldots\}$ *be an enumerated countably infinite parameter set. Let* (Ω, L, E) *be a probability space. Then the following holds.*

1. Let $X : Q \times (\Omega, L, E) \to (S,d)$ *be an arbitrary r.f. Define the function*

$$\widetilde{X} : (\Omega, L, E) \to (S^Q, d^Q)$$

by $domain(\widetilde{X}) \equiv \bigcap_{i=1}^{\infty} domain(X_{t(i)})$, *and by* $\widetilde{X}(\omega)(t_i) \equiv X_{t(i)}(\omega)$ *for each* $i \geq 1$, *for each* $\omega \in domain(\widetilde{X})$. *Then the function* $\widetilde{X}$ *is an r.v. on* (Ω, L, E) *with values in* (S^Q, d^Q).

2. *Conversely, let* $\widetilde{X} : (\Omega, L, E) \to (S^Q, d^Q)$ *be an arbitrary r.v. For each* $s \in Q$, *define the function* $X_s : (\Omega, L, E) \to (S, d)$ *by* $domain(X_s) \equiv domain(\widetilde{X})$ *and by* $X_s(\omega) \equiv X(s,\omega) \equiv (\widetilde{X}(\omega))(s)$. *Then* $X : Q \times (\Omega, L, E) \to (S, d)$ *is an r.f.*

Convention: Henceforth, for brevity of notations, we will write X *also for* $\widetilde{X}$, *when it is clear from context that we intend to mean the latter.*

Proof. 1. Let $x_\circ$ be an arbitrary but fixed reference point in S.

2. Let $m \geq 1$ be arbitrary. Let $x, y \in S^Q$ be arbitrary. Define $j_m(x) \equiv x^{(m)} \in S^Q$ by $x^{(m)}(t_i) \equiv x(t_i)$ or $x^{(m)}(t_i) \equiv x_\circ$ according as $i \leq m$ or $i > m$, for each $i \geq 1$. Then, by Definition 6.3.1, we have

$$\begin{aligned} d^Q(j_m(x), j_m(y)) \equiv d^Q(x^{(m)}, y^{(m)}) &\equiv \sum_{i=1}^{\infty} 2^{-i}(1 \wedge d(x^{(m)}(t_i), y^{(m)}(t_i))) \\ &= \sum_{i=1}^{m} 2^{-i}(1 \wedge d(x(t_i), y(t_i))) + \sum_{i=m+1}^{\infty} 2^{-i}(1 \wedge d(x_\circ, x_\circ)) \\ &\leq \sum_{i=1}^{\infty} 2^{-i}(1 \wedge d(x(t_i), y(t_i))) \equiv d^Q(x, y). \end{aligned}$$

We see that the function $j_m : (S^Q, d^Q) \to (S^Q, d^Q)$ is a contraction, and hence continuous.

3. Similarly, let $u \equiv (u_1, \ldots, u_m), v \equiv (v_1, \ldots, v_m) \in S^m$ be arbitrary. Define $k_m(u) \equiv u^{(m)} \in S^Q$ by $u^{(m)}(t_i) \equiv u_i$ or $u^{(m)}(t_i) \equiv x_\circ$ according as $i \leq m$ or $i > m$, for each $i \geq 1$. Then

$$\begin{aligned} d^Q(k_m(u), k_m(v)) \equiv d^Q(u^{(m)}, v^{(m)}) &\equiv \sum_{i=1}^{\infty} 2^{-i}(1 \wedge d(u^{(m)}(t_i), v^{(m)}(t_i))) \\ &= \sum_{i=1}^{m} 2^{-i} 1 \wedge d(u_i, v_i) + \sum_{i=m+1}^{\infty} 2^{-i}(1 \wedge d(x_\circ, x_\circ)) \\ &\leq \sum_{i=1}^{m} 2^{-i} 1 \wedge d(u_i, v_i) \leq \bigvee_{i=1}^{m} d(u_i, v_i) \equiv d^m(u, v). \end{aligned}$$

We see that the function $k_m : (S^m, d^m) \to (S^Q, d^Q)$ is a contraction, and hence continuous.

4. Consider each $f \in C_{ub}(S^Q, d^Q)$, with a modulus of continuity δ_f. Then $f \circ k_m \in C_{ub}(S^m, d^m)$. Since $X : Q \times (\Omega, L, E) \to (S, d)$ is an r.f., it follows that $(f \circ k_m)(X_{t(1)}, \ldots, X_{t(m)}) \in L$. Let $\varepsilon > 0$ be arbitrary. Let $m \geq 1$ be so large that $2^{-m} < \delta_f(\varepsilon)$.

5. Let $x \in S^Q$ be arbitrary. Define $u \equiv (x_{t(1)}, \ldots, x_{t(m)}) \in S^m$. Then we have $x^{(m)}(t_i) \equiv x(t_i) = u_i \equiv u^{(m)}(t_i)$ or $x^{(m)}(t_i) \equiv x_\circ \equiv u^{(m)}(t_i)$ depending on whether $i \leq m$ or $i > m$, for each $i \geq 1$. It follows that $x^{(m)} = u^{(m)} \equiv k_m(u) = k_m(x_{t(1)}, \ldots, x_{t(m)})$, for each $x \in S^Q$. Consequently,

$$f(X^{(m)}) = f \circ k_m(X_{t(1)}, \ldots, X_{t(m)}) \in L.$$

At the same time,

$$d^Q(\widetilde{X}, X^{(m)}) \equiv d^\infty(\widetilde{X} \circ t, X^{(m)} \circ t)$$
$$= \sum_{i=m+1}^{\infty} 2^{-i}(1 \wedge d(X_{t(i)}, x_\circ)) \leq \sum_{i=m+1}^{\infty} 2^{-i} = 2^{-m} < \delta_f(\varepsilon).$$

Consequently,

$$|f(\widetilde{X}) - f(X^{(m)})| \leq \varepsilon.$$

Since $\varepsilon > 0$ is arbitrary, the function $f(\widetilde{X})$ is the uniform limit of a sequence in L. Hence $f(\widetilde{X}) \in L$, where $f \in C_{ub}(S^Q, d^Q)$ is arbitrary. Since the complete metric space (S^Q, d^Q) is bounded, Assertion 4 of Proposition 5.1.4 implies that the function $\widetilde{X}$ is an r.v. on (Ω, L, E). Assertion 1 is proved.

6. Conversely, let $\widetilde{X} : (\Omega, L, E) \to (S^Q, d^Q)$ be an arbitrary r.v. Consider each $s \in Q$. Then the function $j_s : (S^Q, d^Q) \to (S, d)$, defined by $j_s(x) \equiv x(s)$ for each $x \in S^Q$, is uniformly continuous. Hence $j_s(\widetilde{X}) : (\Omega, L, E) \to (S, d)$ is an r.v. At the same time, for each $\omega \in domain(\widetilde{X})$, we have $X_s(\omega) \equiv X(s, \omega) \equiv (\widetilde{X}(\omega))(s) = j_s(\widetilde{X}(\omega))$. Hence $X_s = j_s(\widetilde{X})$ and X_s is an r.v. In other words, X is an r.f. Assertion 2 and the lemma are proved. □

Definition 6.4.2. Metric space of r.f.'s with a countable parameter set. Let (S, d) be a locally compact metric space, not necessarily compact. Let $Q = \{t_1, t_2, \ldots\}$ be an enumerated countably infinite parameter set. Let (Ω, L, E) be a probability space.

Let $Z, Z' : Q \times (\Omega, L, E) \to (S, d)$ be arbitrary r.f.'s. In symbols, $Z, Z' \in \widehat{R}(Q \times \Omega, S)$. Lemma 6.4.1, with the notations therein, says that the functions

$$\widetilde{Z}, \widetilde{Z}' : (\Omega, L, E) \to (S^Q, d^Q)$$

are r.v.'s on (Ω, L, E) with values in the complete metric space (S^Q, d^Q). In symbols, $\widetilde{Z}, \widetilde{Z}' \in M(\Omega, S^Q)$.

Define

$$\widehat{\rho}_{Prob,Q}(Z, Z') \equiv \rho_{Prob}(\widetilde{Z}, \widetilde{Z}') \equiv E(1 \wedge d^Q(\widetilde{Z}, \widetilde{Z}'))$$
$$= Ed^Q(\widetilde{Z}, \widetilde{Z}') = \sum_{n=1}^{\infty} 2^{-n} E(1 \wedge d(Z_{t(n)}, Z'_{t(n)})). \qquad (6.4.1)$$

Note that $\widehat{\rho}_{Prob,Q} \leq 1$. Then $\widehat{\rho}_{Prob,Q}$ is a metric, called the *probability metric* on the space $\widehat{R}(Q \times \Omega, S)$ of r.f.'s.

In view of the right-hand side of the defining equality 6.4.1, the metric $\widehat{\rho}_{Prob,Q}$ is determined by the enumeration $\{t_1,t_2,\ldots\}$ of the countably infinite set Q. A adifferent enumeration would produce a different, albeit equivalent, metric.

Equality 6.4.1 implies that convergence of a sequence $(Z^{(k)})$ of r.f.'s in $\widehat{R}(Q \times \Omega, S)$ relative to the metric $\widehat{\rho}_{Prob,Q}$ is equivalent to convergence in probability and, therefore, to the weak convergence of the sequence $Z_s^{(k)}$, for each $s \in Q$. However, the metrical continuity is stronger. □

Theorem 6.4.3. Compact Daniell–Kolmogorov–Skorokhod Extension. Let *(S,d) be a compact metric space. Let $Q = \{t_1,t_2,\ldots\}$ be an enumerated countably infinite parameter set. Let $\xi \equiv (A_k)_{k=1,2,\ldots}$ be an arbitrary binary approximation of (S,d) relative to the reference point $x_\circ$.*

Then there exists a function

$$\overline{\Phi}_{DKS,\xi} : \widehat{F}(Q,S) \to \widehat{R}(Q \times \Theta_0, S)$$

such that for each $F \in \widehat{F}(Q,S)$, the r.f. $Z \equiv \overline{\Phi}_{DKS,\xi}(F) : Q \times \Theta_0 \to (S,d)$ has marginal distributions given by the family F. The function $\Phi_{DKS,\xi}$ constructed in the proof that follows will be called the Daniell–Kolmogorov–Skorokhod Extension *relative to the binary approximation ξ of (S,d).*

Proof. Let the corresponding objects (S^Q,d^Q), $\xi^Q \equiv (\widetilde{B}_k)_{k=1,2,}$, $\widehat{J}(S^Q,d^Q)$, and ρ_{Dist,ξ^Q} be as in Definition 6.3.3. Without loss of generality, assume that $Q \equiv \{t_1,t_2,\ldots\} \equiv \{1,2,\ldots\}$.

1. Consider the Compact Daniell–Kolmogorov Extension

$$\overline{\Phi}_{DK} : \widehat{F}(Q,S) \to \widehat{J}(S^Q,d^Q),$$

in Theorem 6.3.2, which maps each consistent family $F \in \widehat{F}(Q,S)$ of f.j.d.'s to a distribution $E \equiv \overline{\Phi}_{DK}(F)$ on (S^Q,d^Q), such that (i) the coordinate function $U : Q \times (S^Q,L,E) \to S$ is an r.f., where L is the completion of $C(S^Q,d^Q)$ relative to the distribution E, and (ii) U has marginal distributions given by the family F.

2. Since the state space (S,d) is compact by hypothesis, the countable power $\xi^Q \equiv \xi^\infty \equiv (B_k)_{k=1,2,}$ of ξ is defined and is a binary approximation of $(S^Q,d^Q) = (S^\infty,d^\infty)$, according to Definition 3.2.6. Recall that the space $\widehat{J}(S^Q,d^Q)$ is then equipped with the distribution metric ρ_{Dist,ξ^Q} defined relative to ξ^∞, according to Definition 5.3.4, and that convergence of a sequence of distributions on (S^Q,d^Q) relative to the metric ρ_{Dist,ξ^Q} is equivalent to weak convergence.

3. Recall from Definition 5.1.10 that $M(\Theta_0,S^Q)$ denotes the set of r.v.'s $Z \equiv (Z_{t(1)},Z_{t(2)},\ldots) \equiv (Z_1,Z_2,\ldots)$ on (Θ_0,L_0,I_0), with values in the compact path space (S^Q,d^Q). Theorem 5.5.1 constructed the Skorokhod representation

$$\Phi_{Sk,\xi^\infty} : \widehat{J}(S^Q,d^Q) \to M(\Theta_0,S^Q)$$

such that for each distribution $E \in \widehat{J}(S^Q, d^Q)$, with $Z \equiv \Phi_{Sk,\xi^\infty}(E) : \Theta_0 \to S^Q$, we have

$$E = I_{0,Z}, \tag{6.4.2}$$

where $I_{0,Z}$ is the distribution induced on the compact metric space (S^Q, d^Q) by the r.v. Z, in the sense of Definition 5.2.3.

4. We will now verify that the composite function

$$\overline{\Phi}_{DKS,\xi} \equiv \Phi_{Sk,\xi^\infty} \circ \overline{\Phi}_{DK} : \widehat{F}(Q,S) \to M(\Theta_0, S^Q) \tag{6.4.3}$$

has the desired properties. To that end, let the consistent family $F \in \widehat{F}(Q,S)$ of f.j.d.'s be arbitrary. Let $Z \equiv \overline{\Phi}_{DKS,\xi}(F)$. Then $Z \equiv \Phi_{Sk,\xi^\infty}(E)$, where $E \equiv \overline{\Phi}_{DK}(F)$. We need only verify that the r.v. $Z \in M(\Theta_0, S^Q)$, when viewed as an r.f. $Z : Q \times (\Theta_0, L_0, I_0) \to (S,d)$, has marginal distributions given by the family F.

5. For that purpose, let $n \geq 1$ and $g \in C(S^n, d^n)$ be arbitrary. Define the function $f \in C(S^\infty, d^\infty)$ by

$$f(x_1, x_2, \ldots) \equiv g(x_1, \ldots, x_n),$$

for each $(x_1, x_2, \ldots) \in S^\infty$. Then, for each $x \equiv (x_1, x_2, \ldots) \in S^\infty$, we have, by the definition of the coordinate function U,

$$f(x) = f(x_1, x_2, \ldots) = f(U_1(x), U_2(x), \ldots) = g(U_1, \ldots, U_n)(x). \tag{6.4.4}$$

Therefore

$$\begin{aligned} I_0 g(Z_1, \ldots, Z_n) &= I_0 f(Z_1, Z_2 \ldots) = I_0 f(Z) = I_{0,Z} f \\ &= Ef = Eg(U_1, \ldots, U_n) = F_{1,\ldots,n} g, \end{aligned} \tag{6.4.5}$$

where the third equality is by the definition of the induced distribution $I_{0,Z}$, where the fourth equality follows from equality 6.4.2, where the fifth equality is by equality 6.4.4, and where the last equality is by Condition (ii) in Step 1. Since $n \geq 1$ and $g \in C(S^n, d^n)$ are arbitrary, we conclude that the r.f. $\overline{\Phi}_{DKS,\xi}(F) = Z : Q \times (\Theta_0, L_0, I_0) \to (S,d)$ has marginal distributions given by the family F. The theorem is proved. □

Theorem 6.4.4. Continuity of Compact Daniell–Kolmogorov–Skorokhod Extension. *Use the same assumptions and notations as in Theorem* 6.4.3. *In particular, suppose the state space* (S,d) *is compact. Recall that the modulus of local compactness of* (S,d) *corresponding to the binary approximation* $\xi \equiv (A_p)_{p=1,2,\ldots}$ *is defined as the sequence*

$$\|\xi\| \equiv (|A_p|)_{p=1,2,\ldots}$$

of integers. Then the Compact Daniell–Kolmogorov–Skorokhod Extension

$$\overline{\Phi}_{DKS,\xi} : (\widehat{F}(Q,S), \widehat{\rho}_{Marg,\xi,Q}) \to (\widehat{R}(Q \times \Theta_0, S), \widehat{\rho}_{Prob,Q}) \tag{6.4.6}$$

is uniformly continuous with a modulus of continuity $\overline{\delta}_{DKS}(\cdot, \|\xi\|)$ *dependent only on* $\|\xi\|$. *The marginal metric* $\widehat{\rho}_{Marg,\xi,Q}$ *and the probability metric* $\widehat{\rho}_{Prob,Q}$ *were introduced in Definitions 6.2.8 and 6.4.2, respectively.*

Proof. 1. By the defining equality 6.4.3 in Theorem 6.4.3, we have

$$\overline{\Phi}_{DKS,\xi} \equiv \Phi_{Sk,\xi^\infty} \circ \overline{\Phi}_{DK}, \tag{6.4.7}$$

where the Compact Daniell–Kolmogorov Extension

$$\overline{\Phi}_{DK} : (\widehat{F}(Q,S), \widehat{\rho}_{Marg,\xi,Q}) \to (\widehat{J}(S^Q, d^Q), \rho_{Dist,\xi^Q})$$

is uniformly continuous according to Theorem 6.3.4, with modulus of continuity

$$\overline{\delta}_{DK}(\cdot, \|\xi\|)$$

dependent only on the modulus of local compactness $\|\xi\| \equiv (|A_k|)_{k=1,2,...}$ of the compact metric space (S,d).

2. Separately, the metric space (S,d) is compact by hypothesis, and hence its countable power (S^Q, d^Q) is compact. Moreover, the countable power ξ^Q is defined and is a binary approximation of (S^Q, d^Q). Furthermore, since $d^Q \leq 1$, the set $\widehat{J}(S^Q, d^Q)$ of distributions on (S^Q, d^Q) is trivially tight, with a modulus of tightness $\beta \equiv 1$. Hence Theorem 5.5.2 is applicable to the metric space (S^Q, d^Q), along with its binary approximation ξ^Q, and implies that the Skorokhod representation

$$\Phi_{Sk,\xi^\infty} : (\widehat{J}(S^Q, d^Q), \rho_{Dist,\xi^Q}) \to (M(\Theta_0, S^Q), \rho_{Prob})$$

is uniformly continuous, with a modulus of continuity $\delta_{Sk}(\cdot, \|\xi^Q\|, 1)$ depending only on $\|\xi^Q\|$.

With $M(\Theta_0, S^Q)$ identified with $\widehat{R}(Q \times \Theta_0, S)$, this is equivalent to saying that

$$\Phi_{Sk,\xi^\infty} : (\widehat{J}(S^Q, d^Q), \rho_{Dist,\xi^Q}) \to (\widehat{R}(Q \times \Theta_0, S), \widehat{\rho}_{Prob,Q})$$

is uniformly continuous, with a modulus of continuity $\delta_{Sk}(\cdot, \|\xi^Q\|, 1)$.

3. Combining, we see that the composite function $\overline{\Phi}_{DKS,\xi}$ in equality 6.4.6 is uniformly continuous, with a modulus of continuity given by the composite operation

$$\overline{\delta}_{DKS}(\cdot, \|\xi\|) \equiv \overline{\delta}_{DK}(\delta_{Sk}(\cdot, \left\|\xi^Q\right\|, 1), \|\xi\|),$$

where we observe that the modulus of local compactness $\|\xi^Q\|$ of the countable power (S^Q, d^Q) is determined by the modulus of local compactness $\|\xi\|$ of the compact metric space (S,d), according to Lemma 3.2.7. The theorem is proved. □

We will next prove the Daniell–Kolmogorov–Skorokhod Extension Theorem for the general case of a state space is required only to be locally compact, not necessarily compact.

In the remainder of this section, let $Q \equiv \{t_1, t_2, \ldots\}$ denote a countable parameter set. For simplicity of presentation, we will assume, in the proofs, that $t_n = n$ for each $n \geq 1$.

Theorem 6.4.5. Daniell–Kolmogorov–Skorokhod Extension and its continuity. *Suppose (S,d) is locally compact, not necessarily compact, with a binary approximation ξ. Let $Q = \{t_1, t_2, \ldots\}$ be an enumerated countably infinite parameter set. Then the following conditions hold:*

1. **Existence:** *There exists a function*

$$\Phi_{DKS,\xi} : \widehat{F}(Q,S) \rightarrow \widehat{R}(Q \times \Theta_0, S)$$

such that for each $F \in \widehat{F}(Q,S)$, the r.f. $Z \equiv \Phi_{DKS,\xi}(F) : Q \times \Theta_0 \rightarrow S$ has marginal distributions given by the family F. The function $\Phi_{DKS,\xi}$ will be called the Daniell–Kolmogorov–Skorokhod Extension relative to the binary approximation ξ of (S,d).

2. **Continuity:** *Let $\widehat{F}_\beta(Q,S)$ be a pointwise tight subset of $\widehat{F}(Q,S)$, with a modulus of pointwise tightness $\beta : (0,\infty) \times Q \rightarrow (0,\infty)$. Then the Daniell–Kolmogorov–Skorokhod Extension*

$$\Phi_{DKS,\xi} : (\widehat{F}_\beta(Q,S), \widehat{\rho}_{Marg,\xi,Q}) \rightarrow (\widehat{R}(Q \times \Theta_0, S), \widehat{\rho}_{Prob,Q}), \tag{6.4.8}$$

is uniformly continuous, where $\widehat{\rho}_{Marg,\xi,Q}$ is the marginal metric introduced in Definition 6.2.8. Moreover, the uniformly continuous function $\Phi_{DKS,\xi}$ has a modulus of continuity $\delta_{DKS}(\cdot, \|\xi\|, \beta)$ dependent only on the modulus of local compactness $\|\xi\| \equiv (|A_k|)_{k=1,2,\ldots}$ of the locally compact state space (S,d), and on the modulus of pointwise tightness β.

Proof. Without loss of generality, assume that $Q \equiv \{t_1, t_2, \ldots\} \equiv \{1, 2, \ldots\}$.

1. Let $\overline{\xi}$ be the compactification of the given binary approximation ξ, as constructed in Corollary 3.4.4. Thus $\overline{\xi}$ is a binary approximation of $(\overline{S}, \overline{d})$ relative to the fixed reference point $x_\circ \in S$. Since the metric space $(\overline{S}, \overline{d})$ is compact, the countable power $\overline{\xi}^Q$ of $\overline{\xi}$ is defined and is a binary approximation of $(\overline{S}^Q, \overline{d}^Q)$, according to Definition 6.3.3.

2. Recall from Lemma 6.3.6 the mapping

$$\psi : \widehat{F}(Q,S) \rightarrow \widehat{F}(Q,\overline{S}). \tag{6.4.9}$$

3. Apply Theorems 6.4.3 and 6.4.4 to the compact metric space $(\overline{S}, \overline{d})$ to obtain the Compact Daniell–Kolmogorov–Skorokhod Extension

$$\overline{\Phi}_{DKS,\overline{\xi}} \equiv \Phi_{Sk,\overline{\xi}^\infty} \circ \overline{\Phi}_{DK} : (\widehat{F}(Q,\overline{S}), \widehat{\rho}_{Marg,\overline{\xi},Q}) \rightarrow (\widehat{R}(Q \times \Theta_0, \overline{S}), \widehat{\rho}_{Prob,Q}), \tag{6.4.10}$$

which is uniformly continuous with a modulus of continuity $\overline{\delta}_{DKS}(\cdot, \|\overline{\xi}\|)$ dependent only on $\|\overline{\xi}\|$.

4. Separately, let

$$\overline{Z} \equiv \overline{\Phi}_{DKS,\overline{\xi}}(\widehat{F}(Q,S)) \subset \widehat{R}(Q \times \Theta_0, \overline{S}).$$

be arbitrary. Then there exists $F \in \widehat{F}(Q,S)$ such that $\overline{F} \equiv \psi(F) \in \widehat{F}(Q,\overline{S})$, $\overline{E} \equiv \overline{\Phi}_{DK}(\overline{F})$, and $\overline{Z} = \Phi_{Sk,\overline{\xi}^\infty}(\overline{E})$.

5. According to Step 2 of of the proof of Theorem 6.3.7, we have (i) S^Q is a full set in $(\overline{S}^Q, \overline{L}, \overline{E})$, (ii) $C_{ub}(S^Q, d^Q) \subset \overline{L}$, and (iii) the restricted function $E \equiv \varphi(\overline{E}) \equiv \overline{E}|C_{ub}(S^Q, d^Q)$ is a distribution on (S^Q, d^Q). Moreover, according to Theorem 6.3.7, the coordinate function $U : Q \times (S^Q, L, E) \to (S,d)$ is an r.f. with marginal distributions given by the family F.

6. By the definition of the Skorokhod representation $\Phi_{Sk,\overline{\xi}^\infty}$, we have $\overline{E} = I_{0,\overline{Z}}$. At the same time, S^Q is a full set in $(\overline{S}^Q, \overline{L}, \overline{E})$. Hence S^Q is a full subset of $(\overline{S}^Q, \overline{L}, I_{0,\overline{Z}})$. Equivalently, $I_0 1_{(\overline{Z} \in S^Q)} = I_{0,\overline{Z}} 1_{S^Q} = 1$, where we view the r.f. $\overline{Z}$ as a r.v. on (Θ_0, L_0, I_0). In other words, $D \equiv \{\theta \in \Theta_0 : \overline{Z}(\theta) \in S^Q\}$ is a full subset of (Θ_0, L_0, I_0). For each $t \in Q$, define $Z_t \equiv \overline{Z}_t|D$. Then

$$Z_t = \overline{Z}_t \quad \text{a.s.}$$

on (Θ_0, L_0, I_0).

Let $f \in C_{ub}(S^n, d^n)$ be arbitrary. Define $g \in C_{u,b}(S^Q, d^Q)$ by

$$g(x) \equiv f(x(t_1), \ldots, x(t_n)) = f(U_{t(1)}, \ldots, U_{t(n)})(x)$$

for each $x \in S^Q$. Then $g = f(U_{t(1)}, \ldots, U_{t(n)})$. Hence

$$\begin{aligned} I_0 f(Z_{t(1)}, \ldots, Z_{t(n)}) &= I_0 g(Z) = I_0 g(\overline{Z}) = I_{0,\overline{Z}} g \\ &= \overline{E} g \equiv E g = E f(U_{t(1)}, \ldots, U_{t(n)}) = F_{t(1),\ldots,t(n)} f. \end{aligned}$$

Thus $Z \in \widehat{R}(Q \times \Theta_0, S)$ and has marginal distribution given by the family F. Define $\gamma(\overline{Z}) \equiv Z$. Thus we have a function

$$\gamma : \overline{\Phi}_{DKS,\overline{\xi}}(\widehat{F}(Q,\overline{S})) \to \widehat{R}(Q \times \Theta_0, S). \tag{6.4.11}$$

Define the composite function

$$\Phi_{DKS,\xi} \equiv \gamma \circ \overline{\Phi}_{DKS,\overline{\xi}} \circ \psi.$$

Let $F \in \widehat{F}(Q,S)$ be arbitrary. Let $Z \equiv \Phi_{DKS,\xi}(F)$. We have just shown that, then, $Z \in \widehat{R}(Q \times \Theta_0, S)$ and has marginal distribution given by the family F. Thus Assertion 1 is proved.

7. It remains to prove that the composite function $\Phi_{DKS,\xi}$ is uniformly continuous on $\widehat{F}_\beta(Q,S)$. From Step 3, we saw that the function

$$\overline{\Phi}_{DKS,\overline{\xi}} : (\widehat{F}(Q,\overline{S}), \widehat{\rho}_{Marg,\overline{\xi},Q}) \to (\widehat{R}(Q \times \Theta_0, \overline{S}), \widehat{\rho}_{Prob,Q}) \tag{6.4.12}$$

is uniformly continuous. From Step 22 in the proof of Theorem 6.3.7, we saw that the function

$$\psi : (\widehat{F}_\beta(Q,S), \widehat{\rho}_{Marg,\xi,Q}) \to (\widehat{F}(Q,\overline{S}), \widehat{\rho}_{Marg,\overline{\xi},Q}) \tag{6.4.13}$$

is uniformly continuous. Hence we need only show that the function γ is uniformly continuous on $\overline{\Phi}_{DKS,\overline{\xi}} \circ \psi(\widehat{F}_\beta(Q,S))$.

To that end, let $\overline{Z}, \overline{Z'} \in \overline{\Phi}_{DKS,\overline{\xi}} \circ \psi(\widehat{F}_\beta(Q,S))$ be arbitrary. Let $Z \equiv \gamma(\overline{Z})$ and $Z' \equiv \gamma(\overline{Z'})$. Then there exist $F \in \widehat{F}_\beta(Q,S)$ such that $\overline{F} \equiv \psi(F)$, $\overline{E} \equiv \overline{\Phi}_{DK}(\overline{F}), \overline{Z} = \Phi_{Sk,\overline{\xi}^\infty}(\overline{E})$, and $Z \equiv \gamma(\overline{Z}) = \Phi_{DKS,\xi}(F)$. Similarly, there exist $F' \in \widehat{F}_\beta(Q,S)$ such that $\overline{F'} \equiv \psi(F')$, $\overline{E'} \equiv \overline{\Phi}_{DK}(\overline{F'}), \overline{Z'} = \Phi_{Sk,\overline{\xi}^\infty}(\overline{E'})$, and $Z' \equiv \gamma(\overline{Z'}) = \Phi_{DKS,\xi}(F')$.

Let $\varepsilon > 0$ be arbitrary. Let $\nu \geq 1$ be so large that $2^{-\nu} < \varepsilon$. Take any $b > \beta(\varepsilon,\nu)$. Then $(d(\cdot,x_\circ) \leq b) \subset K$ for some compact subset K of (S,d). Then, by Condition 3 of Definition 3.4.1, there exists $\delta_K(\varepsilon) > 0$ such that for each $y,z \in K$ with $\overline{d}(y,z) < \delta_K(\varepsilon)$, we have $d(y,z) < \varepsilon$.

8. Take any $\alpha \in (\varepsilon, 2\varepsilon)$. Suppose $\widehat{\rho}_{Prob,Q}(\overline{Z},\overline{Z'}) < 2^{-\nu}\alpha\delta_K(\varepsilon)$. Consider each $n = 1,\ldots,\nu$. Then

$$
\begin{aligned}
2^{-n} I_0 \overline{d}(Z_n, Z'_n) &= 2^{-n} I_0 \overline{d}(\overline{Z}_n, \overline{Z}'_n) \\
&\leq \sum_{m=1}^{\infty} 2^{-m} I_0 \overline{d}(\overline{Z}_m, \overline{Z}'_m) \equiv \widehat{\rho}_{Prob,Q}(\overline{Z},\overline{Z'}) < 2^{-\nu}\alpha\delta_K(\varepsilon),
\end{aligned}
$$

whence $I_0\overline{d}(Z_n,Z'_n) < \alpha\delta_K(\varepsilon)$. Chebychev's inequality therefore implies that

$$
\mu_0(\overline{d}(Z_n,Z'_n) \geq \delta_K(\varepsilon)) < \alpha < 2\varepsilon, \tag{6.4.14}
$$

where μ_0 is the Lebesgue measure function on Θ_0. We estimate

$$
\begin{aligned}
&I_0(1 \wedge d(Z_n, Z'_n)) \\
&\quad \leq I_0(1 \wedge d(Z_n, Z'_n)) 1_{(d(Z(n),x(\circ)) \vee d(Z(n),x(\circ)) \leq b)} \\
&\qquad + I_0(d(Z_n,x_\circ) > b) + I_0(d(Z_n,x_\circ) > b) \\
&\quad \leq I_0(1 \wedge d(Z_n, Z'_n)) 1_{(d(Z(n),x(\circ)) \vee d(Z(n),x(\circ)) \leq b)} + \varepsilon + \varepsilon \\
&\quad \leq I_0(1 \wedge d(Z_n, Z'_n)) 1_{(\overline{d}(Z(n),Z'(n)) < \delta_K(\varepsilon))} 1_{(d(Z(n),x(\circ)) \vee d(Z(n),x(\circ)) \leq b)} \\
&\qquad + \mu_0(\overline{d}(Z_n,Z'_n) \geq \delta_K(\varepsilon)) + \varepsilon + \varepsilon \\
&\quad \leq I_0\{(1 \wedge d(Z_n, Z'_n)) 1_{(\overline{d}(Z(n),Z'(n)) < \delta_K(\varepsilon))} 1_{(d(Z(n),x(\circ)) \vee d(Z'(n),x(\circ)) \leq b)}\} \\
&\qquad + 2\varepsilon + \varepsilon + \varepsilon,
\end{aligned} \tag{6.4.15}
$$

where the second inequality is because $b > \beta(\varepsilon;\nu)$ and where the last inequality is thanks to inequality 6.4.14. Now suppose the integrable function in braces on the right-hand side of inequality 6.4.15 is greater than 0 at some $\theta \in \Theta_0$. Then $d(Z_n(\theta),x_\circ) \vee d(Z'_n(\theta),x_\circ) \leq b$, whence $Z_n(\theta), Z'_n(\theta) \in K$. At the same time $\overline{d}(Z_n(\theta),Z'_n(\theta)) < \delta_K(\varepsilon)$. Hence $d(Z_n(\theta),Z'_n(\theta)) < \varepsilon$ by the last remark of Step 7. Thus we see that this integrable function is bounded by ε. Consequently, inequality 6.4.15 yields

$$
I_0(1 \wedge d(Z_n,Z'_n)) \leq \{\varepsilon\} + 2\varepsilon + \varepsilon + \varepsilon = 5\varepsilon,
$$

where $n = 1, \ldots, \nu$ is arbitrary. Hence

$$\begin{aligned}\widehat{\rho}_{Prob,Q}(\gamma(\overline{Z}), \gamma(\overline{Z'})) = \widehat{\rho}_{Prob,Q}(Z, Z') &\equiv \sum_{n=1}^{\infty} 2^{-n} I_0(1 \wedge d(Z_n, Z_n')) \\ &\leq \sum_{n=1}^{\nu} 2^{-n} I_0(1 \wedge d(Z_n, Z_n')) \\ &\quad + 2^{-\nu} \leq \sum_{n=1}^{\nu} 2^{-n} 5\varepsilon + \varepsilon < 6\varepsilon,\end{aligned}$$

where $\overline{Z}, \overline{Z'} \in \overline{\Phi}_{DKS,\overline{\xi}} \circ \psi(\widehat{F}_\beta(Q,S))$ are arbitrary with $\widehat{\rho}_{Prob,Q}(\overline{Z}, \overline{Z'}) < 2^{-\nu}\alpha\delta_K(\varepsilon)$. Thus the mapping

$$\gamma : (\overline{\Phi}_{DKS,\overline{\xi}} \circ \psi(\widehat{F}_\beta(Q,S)), \widehat{\rho}_{Prob,Q}) \to (\widehat{R}(Q \times \Theta_0, S), \widehat{\rho}_{Prob,Q})$$

is uniformly continuous. Combining this with the uniformly continuous functions $\overline{\Phi}_{DKS,\overline{\xi}}$ and ψ in formulas 6.4.10 and 6.4.13, respectively, we see that the composite function

$$\Phi_{DKS,\xi} \equiv \gamma \circ \overline{\Phi}_{DKS,\overline{\xi}} \circ \psi : (\widehat{F}_\beta(Q,S), \widehat{\rho}_{Marg,\xi,Q}) \to (\widehat{R}(Q \times \Theta_0, S), \widehat{\rho}_{Prob,Q}) \tag{6.4.16}$$

is uniformly continuous. Assertion 2 and the theorem are proved. □

As a corollary, we prove Skorokhod's sequential continuity theorem, essentially theorem 3.1.1 in [Skorokhod 1956].

Theorem 6.4.6. Sequential continuity of the Daniell–Kolmogorov–Skorokhod Extension. *Let $F^{(0)}, F^{(1)}, F^{(2)}, \ldots$ be an arbitrary sequence in $\widehat{F}(Q,S)$ such that*

$$\widehat{\rho}_{Marg,\xi,Q}(F^{(p)}, F^{(0)}) \to 0. \tag{6.4.17}$$

For each $p \geq 0$, write

$$Z^{(p)} \equiv \Phi_{DKS,\xi}(F^{(p)}) \in (\widehat{R}(Q \times \Theta_0, S), \widehat{\rho}_{Prob,Q})$$

Then

$$Z^{(p)} \to Z^{(0)} \qquad \text{a.u.}$$

as r.v.'s on (Θ_0, L_0, I_0) with values in the path space (S^Q, d^Q).

Proof. 1. Let $p \geq 0$ be arbitrary. Write $\overline{F}^{(p)} \equiv \psi(F^{(p)})$, $\overline{E}^{(p)} \equiv \overline{\Phi}_{DK}(\overline{F}^{(p)}) \in \widehat{J}(\overline{S}^Q, \overline{d}^Q)$, and $\overline{Z}^{(p)} \equiv \Phi_{Sk,\overline{\xi}^\infty}(\overline{E}^{(p)})$. Then $Z^{(p)} \equiv \gamma(\overline{Z}^{(p)})$. Since the function

$$\overline{\Phi}_{DK} : (\widehat{F}(Q,S), \widehat{\rho}_{Marg,\xi,Q}) \to (\widehat{J}(\overline{S}^Q, \overline{d}^Q), \rho_{Dist,\overline{\xi}^Q})$$

is uniformly continuous, the convergence relation 6.4.17 implies $\rho_{Dist,\overline{\xi}^Q}(\overline{E}^{(p)}, \overline{E}^{(0)}) \to 0$. At the same time, the metric space $(\overline{S}^Q, \overline{d}^Q)$ is compact. Hence Theorem 5.5.3 is applicable, and implies that

$$\overline{Z}^{(p)} \equiv \Phi_{Sk,\overline{\xi}^Q}(\overline{E}^{(p)}) \to \Phi_{Sk,\overline{\xi}^Q}(\overline{E}^{(0)}) \equiv \overline{Z}^{(0)} \qquad \text{a.u.}$$

as r.v.'s on (Θ_0, L_0, I_0) with values in $(\overline{S}^Q, \overline{d}^Q)$. In other words,

$$\overline{d}^Q(Z^{(p)}, Z^{(0)}) \to 0 \quad \text{a.u.} \tag{6.4.18}$$

2. We proceed to show that

$$d^Q(Z^{(p)}, Z^{(0)}) \to 0 \quad \text{a.u.} \tag{6.4.19}$$

To that end, let $n \geq 1$ be arbitrary. Then there exists $b > 0$ so large that

$$I_0 B^c < 2^{-n},$$

where

$$B \equiv \left(\bigvee_{k=1}^{n} d(Z^{(0)}_{t(k)}, x_\circ) \leq b \right). \tag{6.4.20}$$

Let $h \equiv n \vee [\log_2 b]_1$. In view of the a.u. convergence 6.4.18, there exist $m \geq h$ and a measurable subset A of (Θ_0, L_0, I_0) with

$$I_0 A^c < 2^{-n}$$

such that

$$1_A \sum_{k=1}^{\infty} 2^{-k} \overline{d}(Z^{(p)}_{t(k)}, Z^{(0)}_{t(k)}) \equiv 1_A \overline{d}^Q(Z^{(p)}, Z^{(0)}) \leq 2^{-2h-1} |A_h|^{-2} \tag{6.4.21}$$

for each $p \geq m$.

3. Now consider each $\theta \in AB$ and each $k = 1, \ldots, n$. Then, $k \leq h$ and, by inequality 6.4.21, we have

$$\overline{d}(Z^{(p)}_{t(k)}(\theta), Z^{(0)}_{t(k)}(\theta)) \leq 2^k 2^{-2h-1} |A_h|^{-2} \leq 2^{-h-1} |A_h|^{-2}, \tag{6.4.22}$$

while equality 6.4.20 yields

$$d(Z^{(0)}_{t(k)}(\theta), x_\circ) \leq b < 2^h. \tag{6.4.23}$$

4. According to Assertion (ii) of Theorem 3.4.3, regarding the one-point compactification $(\overline{S}, \overline{d})$ of (S, d) relative to the binary approximations $\xi \equiv (A_p)_{p=1,2,\ldots}$, if $y \in (d(\cdot, x_\circ) \leq 2^h)$ and $z \in S$ are such that

$$\overline{d}(y, z) < 2^{-h-1} |A_h|^{-2},$$

then

$$d(y, z) < 2^{-h+2}.$$

Hence inequalities 6.4.22 and 6.4.23 together yield

$$d\big(Z_{t(k)}^{(p)}(\theta), Z_{t(k)}^{(0)}(\theta)\big) < 2^{-h+2} \leq 2^{-n+2},$$

where $\theta \in AB$, and $k = 1, \ldots, n$ are arbitrary. Consequently,

$$\begin{aligned} 1_{AB}\sum_{k=1}^{\infty} 2^{-k}\big(1 \wedge d\big(Z_{t(k)}^{(p)}, Z_{t(k)}^{(0)}\big)\big) &\leq 1_{AB}\sum_{k=1}^{n} 2^{-k}\big(1 \wedge di\big(Z_{t(k)}^{(p)}, Z_{t(k)}^{(0)}\big)\big) + \sum_{k=n+1}^{\infty} 2^{-k} \\ &\leq 1_{AB}\sum_{k=1}^{n} 2^{-k}2^{-n+2} + \sum_{k=n+1}^{\infty} 2^{-k} \\ &< 2^{-n+2} + 2^{-n} < 2^{-n+3}, \end{aligned}$$

where 2^{-n+3} and $I_0(AB)^c < 2^{-n+1}$ are arbitrarily small if $n \geq 1$ is sufficiently large. Hence, by Proposition 5.1.14, we have

$$\sum_{k=1}^{\infty} 2^{-k}\big(1 \wedge d\big(Z_{t(k)}^{(p)}, Z_{t(k)}^{(0)}\big)\big) \to 0 \quad \text{a.u.}$$

Equivalently,

$$d^{Q}(Z^{(p)}, Z^{(0)}) \to 0 \quad \text{a.u.} \tag{6.4.24}$$

In other words, $Z^{(p)} \to Z^{(0)}$ a.u. as r.v.'s on (Θ_0, L_0, I_0), as alleged. □

7

Measurable Random Field

7.1 Measurable r.f. That Is Continuous in Probability

In this chapter, let (S,d) be a locally compact metric space, but not necessarily a linear space or ordered. Let (Q,d_Q) be a compact metric space endowed with an arbitrary but fixed integration. (Q,d_Q) need not have an ordering or a linear structure.

Consider each consistent family F of f.j.d.'s with parameter space Q and state space S that is continuous in probability. We will construct a measurable r.f. $X : Q \times \Omega \rightarrow S$ whose marginal distributions are given by the family F, and which is measurable as a function on $Q \times \Omega$. We will prove that the construction is metrically continuous.

In the special case where $S \equiv [-\infty, \infty]$ and where Q is a subinterval of $[-\infty, \infty]$, where the symbol $[-\infty, \infty]$ stands for a properly defined and metrized two-point compactification of the real line, the main theorem in section III.4 of [Neveu 1965] gives a classical construction. The construction in [Neveu 1965] uses a sequence of step processes on half-open intervals, and then uses the limit supremum of this sequence as the desired measurable process X. Existence of a limit supremum is, however, by invoking principle of infinite search, and is not constructive.

[Potthoff 2009] gives a constructive proof of existence in the case where $S \equiv [-\infty, \infty]$ and where (Q,d_Q) is a metric space, by using linear combinations, with stochastic coefficients, of certain deterministic basis functions that serve as successive L_1–approximations to the desired measurable random field X, obviating the use of any limit supremum. These deterministic basis functions are continuous on (Q,d_Q) with values in the state space $[0,1]$, and are from a partition of unity of (Q,d_Q).

Construction of measurable processes in the general case has not been done in the past in classical texts. In the general case, where neither an ordering nor a linear structure is available on the state space (S,d) or the parameter space (Q,d_Q), the aforementioned limit supremum or linear combinations of basis functions would not be available.

We will get around these difficulties by replacing the linear combinations with *stochastic approximations* by basis functions, essentially by continuously varying the probability weighting for picking a basis function near each parameter point $t \in Q$. This method of construction of measurable processes, and the subsequent theorem of metrical continuity of the construction, in epsilon–delta terms, also seem hitherto unknown.

In the rest of this section, we will make the statements in the last paragraph precise.

Definition 7.1.1. Specification of locally compact state space and compact parameter space, and their binary approximations. In this section, let (S,d) be a locally compact metric space, with a binary approximation $\xi \equiv (A_n)_{n=1,2,...}$ relative to some arbitrary but fixed reference point $x_\circ \in S$.

Let (Q,d_Q) be a compact metric space, with $d_Q \leq 1$ and with a binary approximation $\xi_Q \equiv (B_n)_{n=1,2,...}$ relative to some arbitrary but fixed reference point $q_\circ \in Q$. Let I be an arbitrary but fixed distribution on (Q,d_Q), and let (Q,Λ,I) denote the probability space that is the completion of $(Q,C(Q,d_Q),I)$.

The assumption of compactness of (Q,d_Q) simplifies presentation. The generalization of the results to a locally compact parameter space (Q,d_Q) is possible by considering each member in a sequence $(Q_i)_{i=1,2,...}$ of compact and integrable subsets that forms an I-basis of Q. □

Definition 7.1.2. Probability metric for measurable r.f.'s. Let (Ω,L,E) be an arbitrary probability space. Recall from Definition 6.1.1 the space $\widehat{R}(Q \times \Omega, S)$ of r.f.'s $X : Q \times (\Omega,L,E) \to (S,d)$, with sample space (Ω,L,E), with parameter space Q, and with the locally compact state space (S,d). Recall from Definition 5.1.10 the metric space $(M(Q \times \Omega,S),\rho_{Prob})$ of r.v.'s on the product probability space $(Q \times \Omega, \Lambda \otimes L, I \otimes E)$, and with values in the state space (S,d). We will say that an r.f. $X \in \widehat{R}(Q \times \Omega, S)$ is *measurable* if $X \in M(Q \times \Omega, S)$. We will write

$$\widehat{R}_{Meas}(Q \times \Omega, S) \equiv \widehat{R}(Q \times \Omega, S) \cap M(Q \times \Omega, S)$$

for the space of such measurable r.f.'s. Note that $\widehat{R}_{Meas}(Q \times \Omega, S)$ then inherits the probability metric ρ_{Prob} on $M(Q \times \Omega, S)$, which is defined, according to Definition 5.1.10, by

$$\rho_{Prob}(X,Y) \equiv (I \otimes E)(1 \wedge d(X,Y)) \tag{7.1.1}$$

for each $X,Y \in M(Q \times \Omega, S)$. □

Definition 7.1.3. Supremum-probability metric for measurable r.f.'s that are continuous in probability. Recall from Definition 6.1.2 the set $\widehat{R}_{Cp}(Q \times \Omega, S)$ of r.f.'s that are continuous in probability. Let

$$\widehat{R}_{Meas,Cp}(Q \times \Omega, S) \equiv \widehat{R}_{Meas}(Q \times \Omega, S) \cap \widehat{R}_{Cp}(Q \times \Omega, S)$$

denote the subset of the metric space $(\widehat{R}_{Meas}(Q \times \Omega, S), \rho_{Prob})$ whose members are continuous in probability. As a subset, it inherits the probability metric ρ_{Prob} from the latter. Define a second metric $\rho_{Sup,Prob}$ on this set $\widehat{R}_{Meas,Cp}(Q\times\Omega, S)$ by

$$\rho_{Sup,Prob}(X,Y) \equiv \sup_{t\in Q} E(1 \wedge d(X_t,Y_t)) \tag{7.1.2}$$

for each $X,Y \in \widehat{R}_{Meas,Cp}(Q \times \Omega, S)$. In this definition, $E(1 \wedge (X_t,Y_t))$ is a continuous function on the compact metric space (Q,d_Q), on account of continuity in probability, whence the supremum exists. Moreover, the defining formulas 7.1.1 and 7.1.2 imply that $\rho_{Prob} \leq \rho_{Sup,Prob}$. In words, $\rho_{Sup,Prob}$ is a stronger metric than ρ_{Prob} on the space of measurable r.f.'s that are continuous in probability. □

Definition 7.1.4. Specification of a countable dense subset of the parameter space and a partition of unity of Q. By Definition 7.1.1, $\xi_Q \equiv (B_n)_{n=1,2,...}$ is an arbitrary but fixed binary approximation of the compact metric space (Q,d_Q) relative to the reference point $q_\circ \in Q$. Thus $B_1 \subset B_2 \subset \cdots$ is a sequence of metrically discrete and enumerated finite subsets of Q, with $B_n \equiv \{q_{n,1},\ldots,q_{n,\gamma(n)}\}$ for each $n \geq 1$. Then $\|\xi_Q\| \equiv (|B_n|)_{n=1,2,...} = (|\gamma_n|)_{n=1,2,...}$.

1. Define the set

$$Q_\infty \equiv \{t_1,t_2,\ldots\} \equiv \bigcup_{n=1}^{\infty} B_n. \tag{7.1.3}$$

By assumption, $d_Q \leq 1$. Hence, for each $n \geq 1$, we have, by Definition 3.2.1 of a binary approximation,

$$Q = (d_Q(\cdot,q_\circ) \leq 2^n) \subset \bigcup_{q\in B(n)} (d_Q(\cdot,q) \leq 2^{-n}). \tag{7.1.4}$$

Hence $Q_\infty \equiv \bigcup_{n=1}^{\infty} B_n$ is a metrically discrete, countably infinite, and dense subset of (Q,d_Q). Moreover, we can fix an enumeration of Q_∞ in such a manner that

$$\{t_1,t_2,\ldots,t_{\gamma(n)}\} = B_n \equiv \{q_{n,1},\ldots,q_{n,\gamma(n)}\}, \tag{7.1.5}$$

where $\gamma_n \equiv |B_n|$, for each $n \geq 1$.

2. Let

$$\pi_Q \equiv (\{\lambda_{n,q} : q \in B_n\})_{n=1,2,...}$$

be the partition of unity of (Q,d_Q) determined by ξ_Q. Let $n \geq 1$ be arbitrary. Then, for each $q \in B_n$, the basis function $\lambda_{n,q} \in C(Q,d_Q)$ has values in $[0,1]$ and has support $(d_Q(\cdot,q) \leq 2^{-n+1})$. Moreover,

$$Q \subset \bigcup_{q\in B(n)} (d_Q(\cdot,q) \leq 2^{-n}) \subset \left(\sum_{q\in B(n)} \lambda_{n,q} = 1\right), \tag{7.1.6}$$

where the second inclusion is according to Condition 3 of Proposition 3.3.5. Define the auxiliary continuous functions $\lambda_{n,0}^{+} \equiv 0$ and

$$\lambda_{n,k}^{+} \equiv \sum_{i=1}^{k} \lambda_{n,q(n,i)},$$

for each $k = 1, \ldots, \gamma_n$. Then

$$0 \equiv \lambda_{n,0}^{+} \leq \lambda_{n,1}^{+} \leq \cdots \leq \lambda_{n,\gamma(n)}^{+} = 1$$

on Q. □

Recall some short-hand notations. For an arbitrary integrable set A in a complete integration space (Ω, L, E), we write EA, $E(A)$, and $E1_A$ interchangeably. Thus, if (Ω, L, E) is a probability space, then $EA \equiv P(A)$ is the probability of A. Recall also that $[\cdot]_1$ is the operation that assigns to each $a \in R$ an integer $[a]_1 \in (a, a+2)$. We will write $\widehat{d} \equiv 1 \wedge d$, and write a subscripted expression x_y interchangeably with $x(y)$. Caution: $\xi_Q \equiv \xi(Q)$ here is a fixed binary approximation of the parameter space, not to be confused with the previously defined ξ^Q.

Lemma 7.1.5. Extension of a measurable r.f. with parameter set Q_∞ to the full parameter set Q, given continuity in probability. *Let*

$$(\Theta_1, L_1, I_1) \equiv \left([0,1], L_1, \int \cdot d\theta\right)$$

denote the Lebesgue integration space based on the interval $[0,1]$. *Let the compact metric parameter space* (Q, d_Q), *its dense subset* Q_∞, *and related auxiliary objects be as specified as in Definition 7.1.4.*

Consider the locally compact metric space (S, d), *without necessarily any linear structure or ordering. Let* (Ω_0, L_0, E_0) *be an arbitrary probability space. Recall the space* $\widehat{R}_{Cp}(Q_\infty \times \Omega_0, S)$ *of r.f.'s that are continuous in probability over the parameter subspace* (Q_∞, d_Q).

Define the product sample space $(\Omega, L, E) \equiv (\Theta_1, L_1, I_1) \otimes (\Omega_0, L_0, E_0)$. *Recall the space* $\widehat{R}_{Meas,Cp}(Q \times \Omega, S)$ *of r.f.'s with parameter space* Q *that are measurable and continuous in probability. Then there exists a function*

$$\Psi_{meas,\xi(Q)} : \widehat{R}_{Cp}(Q_\infty \times \Omega_0, S) \to \widehat{R}_{Meas,Cp}(Q \times \Omega, S),$$

such that for each $Z \in \widehat{R}_{Cp}(Q_\infty \times \Omega_0, S)$ *with a modulus of continuity in probability* δ_{Cp}, *the r.f.*

$$X \equiv \Psi_{meas,\xi(Q)}(Z) : Q \times (\Omega, L, E) \to S$$

satisfies the following conditions.

1. For a.e. $\theta \in \Theta_1$, *we have* $X_s(\theta, \cdot) = Z_s$ *a.s. on* Ω_0 *for each* $s \in Q_\infty$.

2. The r.f. $X|Q_\infty$ *is equivalent to* Z, *in the sense that they share the same marginal distributions.*

3. The r.f. X is measurable and continuous in probability, with the same modulus of continuity in probability δ_{Cp} as Z.

4. There exists a full subset D of (Ω, L, E) such that for each $\omega \in D$, and for each $t \in domain(X(\cdot,\omega))$, there exists a sequence (s_j) in Q_∞ with $d_Q(t,s_j) \to 0$ and with $d(X(t,\omega), X(s_j,\omega)) \to 0$ as $j \to \infty$.

Proof. 1. Consider each r.f. $Z \in \widehat{R}_{Cp}(Q_\infty \times \Omega_0, S)$, with a modulus of continuity in probability δ_{Cp}. Define the full subset

$$D_0 \equiv \bigcap_{q \in Q(\infty)} domain(Z_q)$$

of (Ω_0, L_0, E_0).

2. Augment each sample point $\omega_0 \in D_0$ with a secondary sample point $\theta \in \Theta_1$. More precisely, define a function

$$\widetilde{Z} : Q_\infty \times \Omega \to S$$

by

$$domain(\widetilde{Z}) \equiv Q_\infty \times \Theta_1 \times D_0$$

and

$$\widetilde{Z}(q,\theta,\omega_0) \equiv Z(q,\omega_0)$$

for each $(q,\theta,\omega_0) \in domain(\widetilde{Z})$.

3. Let $q \in Q_\infty$ be arbitrary. Note that $\widetilde{Z}_q(\theta,\omega_0) = Z_q(\omega_0)$ for each $(\theta,\omega_0) \in \Omega$ such that $\omega_0 \in domain(Z_q)$. Therefore we can apply Assertion 2 of Theorem 4.10.10, where (Ω', L', I'), (Ω'', L'', I''), (ω', ω''), X'', and X are replaced by (Θ_1, L_1, I_1), (Ω_0, L_0, E_0), (θ, ω_0), Z_q, and $\widetilde{Z}_q$, respectively, to infer that the function $\widetilde{Z}_q$ is a measurable function on (Ω, L, I). In other words, $\widetilde{Z}_q$ is an r.v. on (Ω, L, I), where $q \in Q_\infty$ is arbitrary. Thus $\widetilde{Z} : Q_\infty \times \Omega \to S$ is an r.f.

4. Proceed to extend the r.f. $\widetilde{Z}$, by a sequence of stochastic approximations, to a measurable r.f. $X : Q \times \Omega \to S$. To be precise, let $m \geq 1$ and $k = 1, \ldots, \gamma_m$ be arbitrary, where the integer γ_m is as specified in Definition 7.1.4. Define

$$\Delta_{m,k} \equiv \{(t,\theta) \in Q \times \Theta_1 : \theta \in (\lambda^+_{m,k-1}(t), \lambda^+_{m,k}(t))\}.$$

Relation 7.1.6 says that $\lambda^+_{m,\gamma(m)} = 1$. Hence Proposition 4.10.18 implies that the sets $\Delta_{m,1}, \ldots, \Delta_{m,\gamma(m)}$ are mutually disjoint integrable subsets of $Q \times \Theta_1$, and that their union $\bigcup_{i=1}^{\gamma(m)} \Delta_{m,i}$ is a full subset. Define a function $X^{(m)} : Q \times \Omega \to S$ by

$$domain(X^{(m)}) \equiv \bigcup_{i=1}^{\gamma(m)} \Delta_{m,i} \times D_0$$

and by

$$X^{(m)}(t,\omega) \equiv \widetilde{Z}(q_{m,i},\omega) \equiv Z(q_{m,i},\omega_0) \tag{7.1.7}$$

for each $(t,\omega) \equiv (t,\theta,\omega_0) \in \Delta_{m,i} \times D_0$, for each $i = 1,\ldots,\gamma_m$. Here, the function $Y^{(m,i)}$, defined by $Y^{(m,i)}(t,\theta,\omega_0) \equiv Z_{q(m,i)}(\omega_0)$ for each $(t,\theta,\omega_0) \in Q \times \Theta_1 \times D_0$, is measurable by Assertion 3 of Theorem 4.10.10. Since the measurable sets $\Delta_{m,1} \times D_0, \ldots, \Delta_{m,\gamma(m)} \times D_0$ are mutually exclusive in $(Q \times \Omega, \Lambda \otimes L, I \otimes E)$, with union equal to a full subset by Fubini's Theorem, the function $X^{(m)} : Q \times \Omega \to S$ is measurable on $(Q \times \Omega, \Lambda \otimes L, I \otimes E)$, according to Proposition 4.8.8.

5. Now let $t \in Q$ be arbitrary. Define the open interval

$$\Delta_{m,k,t} \equiv (\lambda^+_{m,k-1}(t), \lambda^+_{m,k}(t)) \equiv \{\theta \in \Theta_1 : \theta \in (\lambda^+_{m,k-1}(t), \lambda^+_{m,k}(t))\}. \quad (7.1.8)$$

Then $\bigcup_{i=1}^{\gamma(m)} \Delta_{m,i,t}$ is a full subset of $\Theta_1 \equiv [0,1]$. Hence $\Delta_{m,1,t} \times D_0, \ldots, \Delta_{m,\gamma(m),t} \times D_0$ are mutually exclusive measurable subsets of Ω whose union is a full subset of Ω. Furthermore, by the definition of $X^{(m)}$ in Step 2, we have

$$X_t^{(m)}(\omega) \equiv \widetilde{Z}(q_{m,i},\omega) \equiv Z(q_{m,i},\omega_0), \quad (7.1.9)$$

for each $\omega \equiv (\theta,\omega_0) \in \Delta_{m,i,t} \times D_0$, for each $i = 1,\ldots,\gamma_m$. Hence $X_t^{(m)} : \Omega \to S$ is an r.v. by Proposition 4.8.8. Thus we see that $X^{(m)} : Q \times \Omega \to S$ is an r.f. By Step 4, $X^{(m)}$ is a measurable function. Therefore $X^{(m)}$ is a measurable r.f.

6. We will next construct an a.u. convergent subsequence of $(X^{(m)})_{m=1,2,\ldots}$. Let $m_0 \equiv 0$. Let $j \geq 1$ be arbitrary. Write, as an abbreviation,

$$\varepsilon_j \equiv 2^{-j}. \quad (7.1.10)$$

Let

$$n_j \equiv j \vee [(2 - \log_2 \delta_{Cp}(\varepsilon_j))]_1. \quad (7.1.11)$$

Recursively on $j \geq 1$, define

$$m_j \equiv m_{j-1} \vee n_j. \quad (7.1.12)$$

Define

$$X \equiv \lim_{j\to\infty} X^{(m(j))}. \quad (7.1.13)$$

A priori, the limit need not exist anywhere. We will show that actually $X_t^{(m(j))} \to X_t$ a.u. for each $t \in Q$, and that therefore X is a well-defined function and is an r.f.

7. To that end, let $j \geq 1$ be arbitrary. Then $m_j \geq n_j \geq j$. Consider each $t \in Q$ and $s \in Q_\infty$ with

$$d_Q(s,t) < 2^{-1}\delta_{Cp}(\varepsilon_j). \quad (7.1.14)$$

Consider each $\theta \in \bigcup_{i=1}^{\gamma(m(j))} \Delta_{m(j),i,t}$. Then, for each $\omega_0 \in D_0$, we have

$$(\theta,\omega_0) \in \bigcup_{i=1}^{\gamma(m(j))} \Delta_{m(j),i,t} \times D_0 \equiv domain(X_t^{(m(j))}),$$

and

$$
\begin{aligned}
&\widehat{d}(\widetilde{Z}_s(\theta,\omega_0), X_t^{(m(j))}(\theta,\omega_0)) \\
&= \sum_{i=1}^{\gamma(m(j))} 1_{\Delta(m(j),i,t)}(\theta)\widehat{d}(\widetilde{Z}_s(\theta,\omega_0), X_t^{(m(j))}(\theta,\omega_0)) \\
&= \sum_{i=1}^{\gamma(m(j))} 1_{\Delta(m(j),i,t)}(\theta)\widehat{d}(Z_s(\omega_0), X_t^{(m(j))}(\theta,\omega_0)) \\
&= \sum_{i=1}^{\gamma(m(j))} 1_{\Delta(m(j),i,t)}(\theta)\widehat{d}(Z_s(\omega_0), Z_{q(m(j),i)}(\omega_0)), \qquad (7.1.15)
\end{aligned}
$$

where the last inequality follows from the defining formula 7.1.9. Hence, since D_0 is a full set, we obtain

$$
E_0\widehat{d}(\widetilde{Z}_s(\theta,\cdot), X_t^{(m(j))}(\theta,\cdot)) = \sum_{i=1}^{\gamma(m(j))} 1_{\Delta(m(j),i,t)}(\theta)E_0\widehat{d}(Z_s, Z_{q(m(j),i)}). \qquad (7.1.16)
$$

Suppose the summand with index i on the right-hand side is positive. Then $\Delta_{m(j),i,t} \equiv (\lambda^+_{m(j),i-1}(t), \lambda^+_{m(j),i}(t))$ is a nonempty open interval. Hence

$$
\lambda^+_{m(j),i-1}(t) < \lambda^+_{m(j),i}(t).
$$

Equivalently, $\lambda_{m(j),q(m(j),i)}(t) > 0$. At the same time, the continuous function $\lambda_{m(j),q(m(j),i)}$ on Q has support $(d_Q(\cdot, q_{m(j),i}) \leq 2^{-m(j)+1})$, as specified in Step 2 of Definition 7.1.4. Consequently,

$$
d_Q(t, q_{m(j),i}) \leq 2^{-m(j)+1}. \qquad (7.1.17)
$$

Inequalities 7.1.14 and 7.1.17 together imply that

$$
\begin{aligned}
d_Q(s, q_{m(j),i}) &\leq d_Q(s,t) + d_Q(t, q_{m(j),i}) < \frac{1}{2}\delta_{Cp}(\varepsilon_j) + 2^{-m(j)+1} \\
&< \frac{1}{2}\delta_{Cp}(\varepsilon_j) + \frac{1}{2}\delta_{Cp}(\varepsilon_j) = \delta_{Cp}(\varepsilon_j), \qquad (7.1.18)
\end{aligned}
$$

where the third inequality follows from defining formulas 7.1.11 and 7.1.12. Hence, by the definition of δ_{Cp} as a modulus of continuity in probability of the r.f. Z, inequality 7.1.18 yields

$$
E_0\widehat{d}(Z_s, Z_{q(m(j),i)}) \leq \varepsilon_j.
$$

Summing up, this inequality holds for the ith summand in the right-hand side of equality 7.1.16, if that ith summand is positive at all. Equality 7.1.16 therefore reduces to

$$E_0\widehat{d}(\widetilde{Z}_s(\theta,\cdot),X_t^{(m(j))}(\theta,\cdot)) \leq \varepsilon_j \sum_{i=1}^{\gamma(m(j))} 1_{\Delta(m(j),i,t)}(\theta) = \varepsilon_j, \tag{7.1.19}$$

where θ is an arbitrary member of the full set $\bigcup_{i=1}^{\gamma(m(j))} \Delta_{m(j),i,t}$. Therefore, by Fubini's Theorem,

$$\begin{aligned} E\widehat{d}(\widetilde{Z}_s, X_t^{(m(j))}) &\equiv (I_1 \otimes E_0)\widehat{d}(\widetilde{Z}_s, X_t^{(m(j))}) \\ &= \int_0^1 d\theta\, E_0\widehat{d}(\widetilde{Z}_s(\theta,\cdot), X_t^{(m(j))}(\theta,\cdot)) \leq \int_0^1 d\theta \varepsilon_j = \varepsilon_j, \end{aligned} \tag{7.1.20}$$

where $t \in Q$ and $s \in Q_\infty$ are arbitrary with

$$d_Q(s,t) < \frac{1}{2}\delta_{Cp}(\varepsilon_j).$$

8. Now let $j \geq 1$ and $t \in Q$ be arbitrary. Take any $s \in Q_\infty$ such that

$$d_Q(s,t) < \frac{1}{2}\delta_{Cp}(\varepsilon_j) \wedge \frac{1}{2}\delta_{Cp}(\varepsilon_{j+1}).$$

Then it follows from inequality 7.1.20 that

$$\begin{aligned} E\widehat{d}(X_t^{(m(j))}, X_t^{(m(j+1))}) &\leq E\widehat{d}(\widetilde{Z}_s, X_t^{(m(j))}) + E\widehat{d}(\widetilde{Z}_s, X_t^{(m(j+1))}) \\ &\leq \varepsilon_j + \varepsilon_{j+1} \equiv 2^{-j} + 2^{-j-1} < 2^{-j+1}, \end{aligned} \tag{7.1.21}$$

where $t \in Q$ and $j \geq 1$ are arbitrary. Hence, by Assertion 5 of Proposition 5.1.11, the function $X_t \equiv \lim_{j\to\infty} X_t^{(m(j))}$ is an r.v. and

$$X_t^{(m(j))} \to X_t \quad \text{a.u.}$$

We conclude that

$$X : Q \times \Omega \to S$$

is an r.f.

9. We will now show that X is a measurable r.f. By Fubini's Theorem, inequality 7.1.21 implies

$$(I \otimes E)\widehat{d}(X^{(m(j))}, X^{(m(j+1))}) \leq 2^{-j+1}, \tag{7.1.22}$$

for each $j \geq 1$. Since $X^{(m(j))}$ is a measurable function on $(Q\times\Omega, \Lambda\otimes L, I\otimes E)$, for each $j \geq 1$, Assertion 5 of Proposition 5.1.11 implies that $X_t^{(m(j))} \to X$ a.u. on $(Q \times \Omega, \Lambda \otimes L, I \otimes E)$, and that $X \equiv \lim_{j\to\infty} X_t^{(m(j))}$ is a measurable function on $(Q \times \Omega, \Lambda \otimes L, I \otimes E)$. Thus X is a measurable r.f.

10. Define the full subset

$$D_1 \equiv \bigcap_{s\in Q(\infty)} \bigcap_{n=1}^{\infty} \bigcup_{i=1}^{\gamma(n)} \Delta_{n,i,s}$$

of $\Theta_1 \equiv [0,1]$. Consider each $\theta \in D_1$. Let $s \in Q_\infty$be arbitrary. Inequality 7.1.19 says

$$E_0\widehat{d}(\widetilde{Z}_s(\theta,\cdot), X_s^{(m(j))}(\theta,\cdot)) \le \varepsilon_j. \tag{7.1.23}$$

Hence $X_s^{(m(j))}(\theta,\cdot) \to \widetilde{Z}_s(\theta,\cdot)$ a.s. on (Ω_0, L_0, E_0). In other words, $\widetilde{Z}_s(\theta,\cdot) = \lim_{j\to\infty} X_s^{(m(j))}(\theta,\cdot) \equiv X_s(\theta,\cdot)$ a.s. on (Ω_0, L_0, E_0). At the same time, since $s \in Q_\infty$, we have $\widetilde{Z}_s(\theta,\cdot) = Z_s$ on D_0. Combining, we see that $X_s(\theta,\cdot) = Z_s$ a.s. on (Ω_0, L_0, E_0). The desired Condition 1 in the conclusion of the present lemma is proved.

11. Now let $K \ge 1$, $f \in C_{ub}(S^K)$, and $s_1, \ldots, s_K \in Q_\infty$ be arbitrary. Then, in view of Condition 1, Fubini's Theorem implies

$$\begin{aligned}
Ef(X_{s(1)}, \ldots, X_{s(K)}) &\equiv (I_1 \otimes E_0) f(X_{s(1)}, \ldots, X_{s(K)}) \\
&= \int_0^1 d\theta\, E_0 f(X_{s(1)}(\theta,\cdot), \ldots, X_{s(K)}(\theta,\cdot)) \\
&= \int_0^1 d\theta\, E_0 f(Z_{s(1)}, \ldots, Z_{s(K)}) = E_0 f(Z_{s(1)}, \ldots, Z_{s(K)}).
\end{aligned}$$

Thus the r.f.'s $X|Q_\infty$ and Z are equivalent, establishing the desired Condition 2 in the conclusion of the lemma.

12. We will next prove that X is continuous in probability. For that purpose, let $\varepsilon > 0$ be arbitrary. Let $j \ge 1$ and $t, t' \in Q$ be arbitrary with

$$d_Q(t,t') < \delta_{Cp}(\varepsilon).$$

Since Q_∞ is dense in Q, there exist $s, s' \in Q_\infty$ with $d_Q(s,s') < \delta_{Cp}(\varepsilon)$ and

$$d_Q(t,s) \vee d_Q(t',s') < \frac{1}{2}\delta_{Cp}(\varepsilon_j).$$

It follows that $E_0\widehat{d}(Z_s, Z_{s'}) \le \varepsilon$. We can then apply inequality 7.1.20 to obtain

$$\begin{aligned}
E\widehat{d}(X_t^{(m(j))}, X_{t'}^{(m(j))}) &\le E\widehat{d}(X_t^{(m(j))}, \widetilde{Z}_s) + E\widehat{d}(\widetilde{Z}_s, \widetilde{Z}_{s'}) + E\widehat{d}(\widetilde{Z}_{s'}, X_{t'}^{(m(j))}) \\
&\le \varepsilon_j + (I \otimes E_0)\widehat{d}(\widetilde{Z}_s, \widetilde{Z}_{s'}) + \varepsilon_j \\
&= \varepsilon_j + \int_0^1 d\theta\, E_0\widehat{d}(\widetilde{Z}_s(\theta), \widetilde{Z}_{s'}(\theta)) + \varepsilon_j \\
&= \varepsilon_j + \int_0^1 d\theta\, E_0\widehat{d}(Z_s, Z_{s'}) + \varepsilon_j \le \varepsilon_j + \varepsilon + \varepsilon_j,
\end{aligned}$$

where the first equality is thanks to Fubini's Theorem. Letting $j \to \infty$ yields

$$E\widehat{d}(X_t, X_{t'}) \le \varepsilon,$$

where $\varepsilon > 0$ is arbitrarily small. Summing up, the r.f. X is continuous in probability on Q, with δ_{Cp} as a modulus of continuity in probability. Condition 3 has been established.

13. It remains to prove Condition 4. For each $s \in Q_\infty$, letting $j \to \infty$ in inequality 7.1.20 with $t = s$, we obtain

$$E\widehat{d}(\widetilde{Z}_s, X_s) \equiv I_1 \otimes E_0\widehat{d}(\widetilde{Z}_s, X_s) = 0. \tag{7.1.24}$$

Hence

$$D_2 \equiv \bigcap_{s \in Q(\infty)} (\widetilde{Z}_s = X_s) \tag{7.1.25}$$

is a full subset of (Ω, L, E). Define the full subset

$$D \equiv D_2 \cap (D_1 \times D_0)$$

of the sample space (Ω, L, E).

14. Consider each $\omega \equiv (\theta, \omega_0) \in D \subset D_2$. Then $\theta \in D_1$ and $\omega_0 \in D_0$. Let

$$t \in domain(X(\cdot, \omega))$$

be arbitrary. Then $(t, \omega) \in domain(X)$. Hence, by the defining equality 7.1.13, we have

$$X(t, \omega) = \lim_{j \to \infty} X^{(m(j))}(t, \omega). \tag{7.1.26}$$

Let $\varepsilon > 0$ be arbitrary. Then there exists $J \geq 1$ so large that

$$d(X(t, \omega), X^{(m(j))}(t, \omega)) < \varepsilon$$

for each $j \geq J$. Consider each $j \geq J$. Then

$$(t, \omega) \in domain(X^{(m(j))}) \equiv \bigcup_{i=1}^{\gamma(m(j))} \Delta_{m(j),i} \times D_0. \tag{7.1.27}$$

Hence there exists $i_j = 1, \ldots, \gamma_{m(j)}$ such that

$$(t, \theta, \omega_0) \equiv (t, \omega) \in \Delta_{m(j),i(j)} \times D_0. \tag{7.1.28}$$

Consequently,

$$X^{(m(j))}(t, \omega) = \widetilde{Z}(q_{m(j),i(j)}, \omega) = X(q_{m(j),i(j)}, \omega), \tag{7.1.29}$$

where the first equality follows from equality 7.1.7, and the second from equality 7.1.25 in view of the relation $\omega \in D_2$. At the same time, since $(t, \theta) \in \Delta_{m(j),i(j)}$ by equality 7.1.28, we have $\lambda_{m(j),q(m(j),i(j))}(t) > 0$. In addition, the continuous function $\lambda_{m(j),q(m(j),i(j))}$ on Q has support $(d_Q(\cdot, q_{m(j),i(j)}) \leq 2^{-m(j)+1})$, as specified in Step 2 of Definition 7.1.4. Consequently,

$$d_Q(t, q_{m(j),i(j)}) \leq 2^{-m(j)+1}. \tag{7.1.30}$$

Summing up, for each ω in the full set D, and for each $t \in domain(X(\cdot,\omega))$, the sequence $(s_j)_{j=‘1,2,...} \equiv (q_{m(j),i(j)})_{j=‘1,2,...}$ in Q_∞ is such that

$$d_Q(t,s_j) \to 0$$

and

$$\begin{aligned} d(X(t,\omega),X(s_j,\omega)) &\equiv d(X(t,\omega),X(q_{m(j),i(j)},\omega)) \\ &= d(X(t,\omega),X^{(m(j))}(t,\omega)) \to 0, \end{aligned}$$

as $j \to \infty$, where the second equality is from equality 7.1.29 and where the convergence is from formula 7.1.26. Since $\omega \in D$ and $t \in domain(X(\cdot,\omega))$ are arbitrary, Condition 4 of the conclusion of the lemma has also been proved. □

We are now ready to construct the extension of an arbitrary family of f.j.d.'s that is continuous in probability to a measurable r.f. We will prove, at the same time, that this construction is uniformly metrically continuous on any subset of consistence families whose members share some common modulus of pointwise tightness.

Definition 7.1.6. Specification of a space of consistent families whose members share a common modulus of pointwise tightness as well as a common modulus of continuity in probability. Recall the set $\widehat{F}(Q,S)$ of consistent families of f.j.d.'s with the parameter set Q and state space S. Recall also the subset $\widehat{F}_{Cp}(Q,S)$, whose members are continuous in probability, equipped with the metric $\widehat{\rho}_{Cp,\xi,Q,Q(\infty)}$ defined in Definition 6.2.12 by

$$\begin{aligned} \widehat{\rho}_{Cp,\xi,Q,Q(\infty)}(F,F') &\equiv \widehat{\rho}_{Marg,\xi,Q(\infty)}(F|Q_\infty,F'|Q_\infty) \\ &\equiv \sum_{n=1}^{\infty} 2^{-n}\rho_{Dist,\xi^n}(F_{q(1),\dots,q(n)},F'_{q(1),\dots,q(n)}) \end{aligned} \quad (7.1.31)$$

for each $F,F' \in \widehat{F}_{Cp}(Q,S)$. Recall the subset $\widehat{F}_\beta(Q,S)$ of $\widehat{F}(Q,S)$, whose members share some common modulus of pointwise tightness β. Define $\widehat{F}_{Cp,\beta}(Q,S) \equiv \widehat{F}_{Cp}(Q,S) \cap \widehat{F}_\beta(Q,S)$. Then $\widehat{F}_{Cp,\beta}(Q,S)$ inherits the metric $\widehat{\rho}_{Cp,\xi,Q,Q(\infty)}$.

Recall, from Definition 7.1.3, the metric space $(\widehat{R}_{Meas,Cp}(Q\times\Omega,S),\rho_{Sup,Prob})$ of measurable r.f.'s which are continuous in probability.

Theorem 7.1.7. Construction of a measurable r.f. from a family of consistent f.j.d.'s that is continuous in probability. *Let the compact metric parameter space (Q,d_Q), its dense subset Q_∞, and related auxiliary objects be as specified as in Definition 7.1.4. Consider the locally compact metric space (S,d), which is not required to have any linear structure or ordering. Let*

$$(\Theta_1,L_1,I_1) \equiv (\Omega_0,L_0,E_0) \equiv \left([0,1],L_0,\int \cdot d\theta\right)$$

denote the Lebesgue integration space based on the interval $[0,1]$. *Define the product sample space*

$$(\Omega, L, E) \equiv (\Theta_1, L_1, I_1) \otimes (\Omega_0, L_0, E_0).$$

Let the objects δ_{Cp}, β, $\widehat{F}_{Cp,\beta}, \widehat{\rho}_{Cp,\xi,Q,Q(\infty)}, \widehat{R}_{Meas,Cp}(Q \times \Omega, S)$, *and* $\rho_{Sup,Prob}$ *be as specified in Definition 7.1.6.*

Then there exists a uniformly continuous mapping

$$\Phi_{meas,\xi,\xi(Q)} : (\widehat{F}_{Cp,\beta}(Q,S), \widehat{\rho}_{Cp,\xi,Q,Q(\infty)}) \rightarrow (\widehat{R}_{Meas,Cp}(Q \times \Omega, S), \rho_{Sup,Prob}) \tag{7.1.32}$$

such that for each $F \in \widehat{F}_{Cp,\beta}(Q,S)$, *the measurable r.f.*

$$X \equiv \Phi_{meas,\xi,\xi(Q)}(F) : Q \times (\Omega, L, E) \rightarrow S$$

has marginal distributions given by the family F.

Moreover, the mapping has a modulus of continuity $\delta_{meas,\xi,\xi(Q}(\cdot, \delta_{Cp}, \beta, \|\xi\|, \|\xi_Q\|)$ *determined by the parameters* $\delta_{Cp}, \beta, \|\xi\|, \|\xi_Q\|$.

We will refer to the mapping $\Phi_{meas,\xi,\xi(Q)}$ *as the* measurable extension *relative to the binary approximations* ξ *and* ξ_Q *of* (S,d) *and* (Q,d_q), *respectively.*

Proof. 1. Recall from Definition 6.2.12 the isometry

$$\Phi_{Q,Q(\infty)} : (\widehat{F}_{Cp}(Q,S), \widehat{\rho}_{Cp,\xi,Q,Q(\infty)}) \rightarrow (\widehat{F}_{Cp}(Q_\infty,S), \widehat{\rho}_{Marg,\xi,Q(\infty)}),$$

where $\Phi_{Q,Q(\infty)}(F) \equiv F|Q_\infty$ for each $\widehat{F}_{Cp}(Q,S)$. If the family $F \in \widehat{F}_{Cp}(Q,S)$ is pointwise tight with a modulus of pointwise tightness β, then so is $\Phi_{Q,Q(\infty)}(F) \equiv F|Q_\infty$. In other words, we have the isometry

$$\Phi_{Q,Q(\infty)} : (\widehat{F}_{Cp,\beta}(Q,S), \widehat{\rho}_{Cp,\xi,Q,Q(\infty)}) \rightarrow (\widehat{F}_{Cp,\beta}(Q_\infty,S), \widehat{\rho}_{Marg,\xi,Q(\infty)}). \tag{7.1.33}$$

Likewise, by assumption, members of $\widehat{F}_{Cp,\beta}(Q,S)$ share a common modulus of continuity in probability δ_{Cp}. Hence, so do members of $\Phi_{Q,Q(\infty)}(\widehat{F}_{Cp,\beta}(Q,S))$.

2. Theorem 6.4.5 describes the uniformly continuous Daniell–Kolmogorov–Skorokhod Extension

$$\Phi_{DKS,\xi} : (\widehat{F}_\beta(Q_\infty,S), \widehat{\rho}_{Marg,\xi,Q(\infty)}) \rightarrow (\widehat{R}(Q_\infty \times \Theta_0, S), \widehat{\rho}_{Prob,Q(\infty)}). \tag{7.1.34}$$

The uniformly continuous mapping $\Phi_{DKS,\xi}$ has a modulus of continuity $\delta_{DKS}(\cdot, \|\xi\|, \beta)$ dependent only on the modulus of local compactness $\|\xi\| \equiv (|A_k|)_{k=1,2,\ldots}$ of the locally compact state space (S,d), and on the modulus of pointwise tightness β. Hence

$$\Phi_{DKS,\xi} : (\widehat{F}_{Cp,\beta}(Q_\infty,S), \widehat{\rho}_{Marg,\xi,Q(\infty)}) \rightarrow (\widehat{R}(Q_\infty \times \Theta_0, S), \widehat{\rho}_{Prob,Q(\infty)}) \tag{7.1.35}$$

is uniformly continuous. Consider each $F \in \widehat{F}_{Cp,\beta}(Q_\infty,S)$. Let $Z \equiv \Phi_{DKS,\xi}(F|Q_\infty)$. Then $E_0\widehat{d}(Z_s,Z_t) = F_{t,s}\widehat{d}$ for each $s,t \in Q_\infty$. It follows that the r.f.

Z is continuous in probability. In other words, $Z \in \widehat{R}_{Cp}(Q_\infty \times \Omega_0, S)$. Thus we have the uniformly continuous Daniell–Kolmogorov–Skorokhod Extension

$$\Phi_{DKS,\xi} : (\widehat{F}_{Cp,\beta}(Q_\infty, S), \widehat{\rho}_{Marg,\xi,Q(\infty)}) \to (\widehat{R}_{Cp}(Q_\infty \times \Theta_0, S), \widehat{\rho}_{Prob,Q(\infty)}) \tag{7.1.36}$$

with the modulus of continuity $\delta_{DKS}(\cdot, \|\xi\|, \beta)$.

3. Now let

$$\Psi_{meas,\xi(Q)} : \widehat{R}_{Cp}(Q_\infty \times \Omega_0, S) \to \widehat{R}_{Meas,Cp}(Q \times \Omega, S)$$

be the mapping constructed in Theorem 7.1.5.

4. We will show that the composite mapping

$$\Phi_{meas,\xi,\xi(Q)} \equiv \Psi_{meas,\xi(Q)} \circ \Phi_{DKS,\xi} \circ \Phi_{Q,Q(\infty)} \tag{7.1.37}$$

has the desired properties. To that end, let $F \in \widehat{F}_{Cp,\beta}(Q, S)$ be arbitrary. Let $X \equiv \Phi_{meas,\xi,\xi(Q)}(F)$. Then $X \equiv \Psi_{meas,\xi(Q)}(Z)$, where $Z \equiv \Phi_{DKS,\xi}(\Phi_{Q,Q(\infty)}(F)) = \Phi_{DKS,\xi}(F|Q_\infty)$. Consequently, Z has marginal distributions given by $F|Q_\infty$. According to Theorem 7.1.5, X is continuous in probability. Moreover, $X|Q_\infty$ is equivalent to Z. Hence $X|Q_\infty$ has marginal distributions given by $F|Q_\infty$.

5. We need to prove that the r.f. X has marginal distributions given by F. To that end, consider each $k \geq 1, r_1, \ldots, r_k \in Q, s_1, \ldots, s_k \in Q_\infty$, and $f \in C_{ub}(S^k, d^k)$. Then

$$Ef(X_{s(1)}, \ldots, X_{s(k)}) = F_{s(1),\ldots,s(k)}f.$$

Now let $s_k \to r_k$ in Q for each $k = 1, \ldots, k$. Then the left-hand side converges to $Ef(X_{r(1)}, \ldots, X_{r(k)})$, on account of the continuity in probability of X. The right-hand side converges to $F_{r(1),\ldots,r(k)}f$ by the continuity in probability of F, according to Lemma 6.2.11. Hence

$$Ef(X_{r(1)}, \ldots, X_{r(k)}) = F_{r(1),\ldots,r(k)}f.$$

We conclude that the measurable r.f. $X \equiv \Phi_{meas,\xi,\xi(Q)}(F)$ has marginal distributions given by the family F.

6. It remains to prove that the composite mapping $\Phi_{meas,\xi,\xi(Q)}$ is continuous. Since the two constituent mappings $\Phi_{Q,Q(\infty)}$and $\Phi_{DKS,\xi}$ in expressions 7.1.33 and 7.1.36, respectively, are uniformly continuous, it suffices to prove that the remaining constituent mapping

$$\Psi_{meas,\xi(Q)} : (\widehat{R}_{Cp}(Q_\infty \times \Omega_0, S), \widehat{\rho}_{Prob,Q(\infty)}) \\ \to (\widehat{R}_{Meas,Cp}(Q \times \Omega, S), \rho_{Sup,Prob})$$

also is uniformly continuous.

7. To that end, define, as in the proof of Lemma 7.1.5, for each $j \geq 1$,

$$\varepsilon_j \equiv 2^{-j}, \tag{7.1.38}$$

$$n_j \equiv j \vee [(2 - \log_2 \delta_{Cp}(\varepsilon_j))]_1, \tag{7.1.39}$$

and

$$m_j \equiv m_{j-1} \vee n_j, \tag{7.1.40}$$

where $m_0 \equiv 0$.

8. Now let $\varepsilon > 0$ be arbitrary. Fix

$$j \equiv [0 \vee (5 - \log_2 \varepsilon)]_1.$$

Then $2^{-j} < 2^{-5}\varepsilon$. Define

$$\delta_{meas}(\varepsilon, \delta_{Cp}, \|\xi_Q\|) \equiv 2^{-\gamma(m(j))-4}\varepsilon^2,$$

where $(|\gamma_k|)_{k=1,2,...} \equiv (|B_k|)_{k=1,2,...} \equiv \|\xi_Q\|$. Let $Z, Z' \in \widehat{R}_{Cp}(Q_\infty \times \Omega_0, S)$ be arbitrary such that

$$\widehat{\rho}_{Prob,Q(\infty)}(Z, Z') < \delta_{meas}(\varepsilon, \delta_{Cp}, \|\xi_Q\|). \tag{7.1.41}$$

We will verify that $\rho_{Sup,Prob}(\Psi_{meas,\xi(Q)}(Z), \Psi_{meas,\xi(Q)}(Z')) < \varepsilon$.

9. Write $X \equiv \Psi_{meas,\xi(Q)}(Z)$, and $X' \equiv \Psi_{meas,\xi(Q)}(Z')$. Thus $Z, Z' : Q_\infty \times \Omega_0 \to S$ are r.f.'s, and $X, X' : Q \times \Omega \to S$ are measurable r.f.'s. As in the proof of Lemma 7.1.5, define the full subset

$$D_0 \equiv \bigcap_{q \in Q(\infty)} domain(Z_q)$$

of (Ω_0, L_0, E_0). Similarly, define the full subset $D_0' \equiv \bigcap_{q \in Q(\infty)} domain(Z_q')$ of (Ω_0, L_0, E_0). Then $D_0 D_0'$ is a full subset of Ω_0. Note that inequality 7.1.41 is equivalent to

$$E_0 \sum_{i=1}^{\infty} 2^{-i} \widehat{d}(Z_{t(i)}, Z'_{t(i)}) < 2^{-\gamma(m(j))-4}\varepsilon^2. \tag{7.1.42}$$

Hence, by Chebychev's inequality, there exists a measurable set $A \subset \Omega_0$ with

$$E_0 A^c < 2^{-2}\varepsilon,$$

such that

$$\sum_{i=1}^{\infty} 2^{-i} \widehat{d}(Z(t_i, \omega_0), Z'(t_i, \omega_0)) \leq 2^{-\gamma(m(j))-2}\varepsilon, \tag{7.1.43}$$

for each $\omega_0 \in A$.

10. Consider each $t \in Q$. According to Step 5 in the proof of Lemma 7.1.5, the set

$$\widetilde{\Delta}_t \equiv \bigcup_{i=1}^{\gamma(m(j))} \Delta_{m(j),i,t} \equiv \bigcup_{i=1}^{\gamma(m(j))} (\lambda^+_{m(j),i-1}(t), \lambda^+_{m(j),i}(t))$$

is a full subset of $\Theta_1 \equiv [0,1]$. Let $(\theta, \omega_0) \in \widetilde{\Delta}_t \times A D_0 D_0'$ be arbitrary. Then $(\theta, \omega_0) \in \Delta_{m(j),i,t} \times A D_0 D_0'$ for some $i = 1, \ldots, \gamma_{m(j)}$. Therefore the defining equality 7.1.9 in the proof of Lemma 7.1.5 says that

$$X_t^{(m(j))}(\theta,\omega_0) = Z(q_{m(j),i},\omega_0) \tag{7.1.44}$$

and

$$X_t^{\prime(m(j))}(\theta,\omega_0) = Z'(q_{m(j),i},\omega_0). \tag{7.1.45}$$

Separately, inequality 7.1.43 implies that

$$\bigvee_{k=1}^{\gamma(m(j))} \widehat{d}(Z(t_k,\omega_0), Z'(t_k,\omega_0)) \le 2^{-2}\varepsilon. \tag{7.1.46}$$

Consequently,

$$\begin{aligned}\widehat{d}(X_t^{(m(j))}(\theta,\omega_0), X_t^{\prime(m(j))}(\theta,\omega_0)) &= \widehat{d}(Z(q_{m(j),i},\omega_0), Z'(q_{m(j),i},\omega_0))\\ &\le \bigvee_{k=1}^{\gamma(m(j))} \widehat{d}(Z(q_{m(j),k},\omega_0), Z'(q_{m(j),k},\omega_0))\\ &= \bigvee_{k=1}^{\gamma(m(j))} \widehat{d}(Z(t_k,\omega_0), Z'(t_k,\omega_0)) \le 2^{-2}\varepsilon,\end{aligned} \tag{7.1.47}$$

where the second equality is thanks to the enumeration equality 7.1.5 in Definition 7.1.4, where the last inequality is inequality 7.1.46, and where $(\theta,\omega_0) \in \widetilde{\Delta}_t \times AD_0D_0'$ is arbitrary. Hence, since $\widehat{d} \le 1$, it follows from inequality 7.1.47 that

$$\begin{aligned}E\widehat{d}(X_t^{(m(j))}, X_t^{\prime(m(j))}) &\le E\widehat{d}(X_t^{(m(j))}, X_t^{\prime(m(j))})1_{\widetilde{\Delta}(t)\times AD(0)D'(0)}\\ &\quad + E1_{\widetilde{\Delta}(t)\times(AD(0)D'(0))^c}\\ &\le 2^{-2}\varepsilon + (I_1 \otimes E_0)1_{\widetilde{\Delta}(t)\times A^c}\\ &= 2^{-2}\varepsilon + E_0 1_{A^c} < 2^{-2}\varepsilon + 2^{-2}\varepsilon = 2^{-1}\varepsilon.\end{aligned} \tag{7.1.48}$$

11. Separately, inequality 7.1.21 says that

$$E\widehat{d}(X_t^{(m(k))}, X_t^{(m(k+1))}) < 2^{-k+1}, \tag{7.1.49}$$

for arbitrary $k \ge 1$. Hence

$$\begin{aligned}E\widehat{d}(X_t^{(m(j))}, X_t) &\le E\widehat{d}(X_t^{(m(j))}, X_t^{(m(j+1))}) + E\widehat{d}(X_t^{(m(j+1))}, X_t^{(m(j+2))}) + \cdots\\ &\le 2^{-j+1} + 2^{-j} + \cdots = 2^{-j+2}.\end{aligned} \tag{7.1.50}$$

Similarly,

$$E\widehat{d}(X_t^{\prime(m(j))}, X_t') \le 2^{-j+2}. \tag{7.1.51}$$

Combining inequalities 7.1.48, 7.1.50, and 7.1.51, we obtain

$$E\widehat{d}(X_t, X_t') < 2^{-1}\varepsilon + 2^{-j+2} + 2^{-j+2} = 2^{-1}\varepsilon + 2^{-j+3} < 2^{-1}\varepsilon + 2^{-1}\varepsilon = \varepsilon,$$

where $t \in Q$ is arbitrary. Therefore

$$\rho_{Sup,Prob}(X,X') \equiv \sup_{t \in Q} E\widehat{d}(X_t, X'_t) < \varepsilon.$$

In other words,

$$\rho_{Sup,Prob}(\Psi_{meas,\xi(Q)}(Z), \Psi_{meas,\xi(Q)}(Z')) < \varepsilon,$$

provided that

$$\widehat{\rho}_{Prob,Q(\infty)}(Z,Z') < \delta_{meas}(\varepsilon, \delta_{Cp}, \|\xi_Q\|). \tag{7.1.52}$$

Thus $\delta_{meas}(\cdot, \delta_{Cp}, \|\xi_Q\|)$ is a modulus of continuity of $\Psi_{meas,\xi(Q)}$ on

$$\Phi_{DKS,\xi}(\widehat{F}_{Cp,\delta(Cp)}(Q,S)|Q_\infty) = \Phi_{DKS,\xi}(\Phi_{Q,Q(\infty)}(\widehat{F}_{Cp,\delta(Cp)}(Q,S))).$$

12. Combining, we conclude that the composite function $\Phi_{meas,\xi,\xi(Q)}$ in expression 7.1.32 is indeed uniformly continuous, with the composite modulus of continuity

$$\begin{aligned}\delta_{meas,\xi,\xi(Q}(\cdot, \delta_{Cp}, \beta, \|\xi\|, \|\xi_Q\|) &\equiv \delta_{Q,Q(\infty)}(\delta_{DKS}(\delta_{meas}(\cdot, \delta_{Cp}, \|\xi_Q\|), \|\xi\|, \beta)) \\ &= \delta_{DKS}(\delta_{meas}(\cdot, \delta_{Cp}, \|\xi_Q\|), \|\xi\|, \beta),\end{aligned}$$

as alleged. □

7.2 Measurable Gaussian Random Field

In this section, let $(S,d) \equiv (R, d_{ecld})$, where d_{ecld} is the Euclidean metric. Let (Q, d_Q) be a compact metric space with an arbitrary but fixed distribution I_Q.

As an application of Theorem 7.1.7, we will construct a measurable Gaussian r.f. $X : Q \times \Omega \to R$ from its continuous mean and covariance functions; we will then prove the continuity of this construction. For that purpose we need only prove that from the mean and covariance functions we can construct a consistent family of normal f.j.d.'s that is continuous in probability, and that the construction is continuous.

Let ξ, and $\xi_Q \equiv (B_n)_{n=1,2,...}$ be arbitrary but fixed binary approximations of the Euclidean state space $(S,d) \equiv (R,d)$, and of the compact parameter space (Q, d_Q), respectively, as specified in Definitions 7.1.1 and 7.1.4, respectively, along with associated objects. In particular, recall the enumerated, countably infinite, dense subset

$$Q_\infty \equiv \{t_1, t_2, \ldots\} \equiv \bigcup_{n=1}^{\infty} B_n \tag{7.2.1}$$

of Q, where $B_n \equiv \{q_{n,1}, \ldots, q_{n,\gamma(n)}\} = \{t_1, \ldots, t_{\gamma(n)}\}$ are sets, for each $n \geq 1$.

Definition 7.2.1. Gaussian r.f. Let (Ω, L, E) be an arbitrary probability space. A real-valued r.f. $X : Q \times \Omega \to R$ is said to be *Gaussian* if its marginal distributions are normal. The functions $\mu(t) \equiv EX_t$ and $\sigma(t,s) \equiv E(X_t - EX_t)(X_s - EX_s)$ for each $t,s \in Q$ are then called the mean and covariance functions, respectively, of the r.f. X. A Gaussian r.f. is said to be *centered* if $\mu(t) \equiv EX_t = 0$ for each $t \in Q$. Without loss of generality, we treat only Gaussian r.f.'s that are centered.

Let $\widehat{R}_{Gauss}(Q \times \Omega, S)$ denote the subset of $\widehat{R}_{Meas,Cp}(Q \times \Omega, S)$ whose members are Gaussian r.f.'s. □

Recall the matrix terminologies in Definition 5.7.1, especially those related to nonnegative definite matrices.

Definition 7.2.2. Nonnegative definite functions. Let $\sigma : Q^2 \to [0,\infty)$ be an arbitrary symmetric function. If, for each $m \geq 1$ and for each $r_1, \ldots, r_m \in Q$, the square matrix $[\sigma(r_k, r_h)]_{k=1,\ldots,m; h=1,\ldots,m}$ is nonnegative definite, then σ is said to be a *nonnegative definite function* on the set Q^2. If, for each $m \geq 1$ and for each $r_1, \ldots, r_m \in Q$ such that $d_Q(r_k, r_h) > 0$ for each $k, h = 1, \ldots, m$ with $k \neq h$, the matrix

$$[\sigma(r_k, r_h)]_{k=1,\ldots,m; h=1,\ldots,m}$$

is positive definite, then the function σ is said to be *positive definite* on the set Q^2.

Let $G_{\delta(cov),b}(Q)$ denote the set of continuous nonnegative definite functions on the metric space (Q^2, d_Q^2) whose members share some common modulus of continuity δ_{cov}, as well as some common bound $b > 0$, on (Q^2, d_Q^2). Equip $G_{\delta(cov),b}(Q)$ with the metric $\widehat{\rho}_{cov}$ defined by

$$\widehat{\rho}_{cov}(\sigma, \sigma') \equiv \sup_{(t,s) \in Q \times Q} |\sigma(t,s) - \sigma'(t,s)|$$

for each $\sigma, \sigma' \in G_{\delta(cov),b}(Q)$. □

Proposition 7.2.3. Consistent family of normal f.j.d.'s from covariance function. *Let $G_{\delta(cov),b}(Q)$ be as described in Definition 7.2.2. Let $\sigma \in G_{\delta(cov),b}(Q)$ be arbitrary. Thus $\sigma : Q \times Q \to [0,\infty)$ is a continuous nonnegative definite function.*

Let $m \geq 1$ and $r \equiv (r_1, \ldots, r_m) \in Q^m$ be arbitrary. Write the nonnegative definite matrix

$$\overline{\sigma}(r) \equiv [\sigma(r_k, r_h)]_{k=1,\ldots,m; h=1,\ldots,m} \tag{7.2.2}$$

and define the normal distribution

$$F^{\sigma}_{r(1),\ldots,r(m)} \equiv \Phi_{0,\overline{\sigma}(r)} \tag{7.2.3}$$

on R^n with mean 0 *and covariance matrix $\overline{\sigma}(r)$. Then the following conditions hold:*

1. The family

$$F^{\sigma} \equiv \Phi_{covar,fjd}(\sigma) \equiv \{F^{\sigma}_{r(1),\ldots,r(m)} : m \geq 1; r_1, \ldots, r_m \in Q\} \tag{7.2.4}$$

of f.j.d.'s is consistent.

2. The consistent family F^{σ} is continuous in probability, with a modulus of continuity in probability $\delta_{Cp} \equiv \delta_{Cp,\delta(\cdot,cov)}$ defined by

$$\delta_{Cp}(\varepsilon) \equiv \delta_{Cp,\delta(\cdot,cov)}(\varepsilon) \equiv \delta_{cov}(2^{-1}\varepsilon^2)$$

for each $\varepsilon > 0$.

3. The consistent family F^σ *is pointwise tight on* Q*, with a modulus of pointwise tightness* $\beta \equiv \beta_b$ *defined by*

$$\beta_b(\varepsilon, r) \equiv \sqrt{\varepsilon^{-1}b}$$

for each $\varepsilon > 0$*, for each* $r \in Q$.

4. Thus we have a function

$$\Phi_{covar,fjd} : G_{\delta(cov),b}(Q) \to \widehat{F}_{Cp,\beta}(Q,R),$$

where $\widehat{F}_{Cp,\beta}(Q,R)$ *is a subset of* $\widehat{F}(Q,R)$ *whose members share the common modulus of continuity in probability* $\delta_{Cp} \equiv \delta_{Cp,\delta(\cdot,cov)}$ *as well as the common modulus of pointwise tightness* $\beta \equiv \beta_b$.

Proof. 1. First prove the consistency of the family F^σ. To that end, let $n, m \geq 1$ be arbitrary. Let $r \equiv (r_1, \ldots, r_m)$ be an arbitrary sequence in Q, and let $j \equiv (j_1, \ldots, j_n)$ be an arbitrary sequence in $\{1, \ldots, m\}$. Let the matrix $\overline{\sigma}(r)$ be defined as in equality 7.2.2. By Lemma 5.7.8, $F^\sigma_{r(1),\ldots,r(m)} \equiv \Phi_{0,\overline{\sigma}(r)}$ is the distribution of an r.v. $Y \equiv AZ$, where A is an $m \times m$ matrix such that $\overline{\sigma}(r) = AA^T$, and where Z is a standard normal r.v. on some probability space (Ω, L, E), with values in R^m.

Let the dual function $j^* : R^m \to R^n$ be defined by

$$j^*(x_1, \ldots, x_m) \equiv (x_{j(1)}, \ldots, x_{j(n)}),$$

for each $x \equiv (x_1, \ldots, x_m) \in R^m$. Then $j^*(x) = Bx$ for each $x \in R^m$, where the $n \times m$ matrix

$$B \equiv [b_{k,h}]_{k=1,\ldots,n;\, h=1,\ldots,m}$$

is defined by $b_{k,h} \equiv 1$ or 0 depending on whether $h = j_k$ or $h \neq j_k$. Let $\widetilde{A} \equiv BA$. Define the $n \times n$ matrix

$$\begin{aligned}\widetilde{\sigma}(r) &\equiv \widetilde{A}\widetilde{A}^T = BAA^TB^T = B\overline{\sigma}(r)B^T \\ &= [\overline{\sigma}_{j(k),j(h)}]_{k=1,\ldots,n;h=1,\ldots,n} = [\sigma(r_{j(k)}, r_{j(h)})]_{k=1,\ldots,n;h=1,\ldots,n},\end{aligned}$$

where the fourth equality is by using the definition of the matrix B and carrying out the matrix multiplication. Then, by the defining formula 7.2.3,

$$F^\sigma_{r(j(1)),\ldots,r(j(n))} = \Phi_{0,\widetilde{\sigma}(r)}.$$

At the same time, the r.v.

$$\widetilde{Y} \equiv j^*(Y) = BY = BAZ \equiv \widetilde{A}Z$$

has the normal characteristic function defined by

$$E(\exp i\lambda^T \widetilde{A}Z) = \exp\left(-\frac{1}{2}\lambda^T \widetilde{A}\widetilde{A}^T\lambda\right) = \exp\left(-\frac{1}{2}\lambda^T \widetilde{\sigma}(r)\lambda\right)$$

for each $\lambda \in R^n$. Hence $\widetilde{Y}$ has the normal distribution $\Phi_{0,\widetilde{\sigma}(r)}$. Combining, we see that for each $f \in C(R^n)$,

$$F^\sigma_{r(1),\dots,r(m)}(f \circ j^*) = E(f \circ j^*(Y)) = Ef(\widetilde{Y})$$
$$= \Phi_{0,\widetilde{\sigma}(r)}(f) = F^\sigma_{r(j(1)),\dots,r(j(n))}f.$$

We conclude that the family F^σ of f.j.d.'s is consistent. Assertion 1 is verified.

2. To prove the continuity in probability of the family F^σ, consider the case where $m = 2$. Consider each $r \equiv (r_1, r_2) \in Q^2$ with

$$d_Q(r_1, r_2) < \delta_{Cp}(\varepsilon) \equiv \delta_{cov}\left(\frac{1}{2}\varepsilon^2\right). \tag{7.2.5}$$

Let d denote the Euclidean metric for R. As in Step 1, there exists an r.v. $Y \equiv (Y_1, Y_2)$ with values in R^2 with the normal distribution $\Phi_{0,\overline{\sigma}(r)}$, where $\overline{\sigma}(r) \equiv [\sigma(r_k, r_h)]_{k=1,2;h=1,2}$. Then

$$F^\sigma_{r(1),r(2)}(1 \wedge d) = E1 \wedge |Y_1 - Y_2| = \Phi_{0,\overline{\sigma}(r)}(1 \wedge d) \le \Phi_{0,\overline{\sigma}(r)}d$$
$$= \sqrt{\sigma(r_1,r_1) - 2\sigma(r_1,r_2) + \sigma(r_2,r_2)}$$
$$\le \sqrt{|\sigma(r_1,r_1) - \sigma(r_1,r_2)| + |\sigma(r_2,r_2) - \sigma(r_1,r_2)|}$$
$$\le \sqrt{\frac{1}{2}\varepsilon^2 + \frac{1}{2}\varepsilon^2} = \varepsilon,$$

where the second inequality is Lyapunov's inequality, and where the last inequality is thanks to inequality 7.2.5. Thus F^σ is continuous in probability, with $\delta_{Cp}(\cdot, \delta_{cov})$ as a modulus of continuity in probability. Assertion 2 is proved.

3. Let $r \in Q$ be arbitrary. Let $\varepsilon > 0$ be arbitrary. Let $n \ge 1$ be arbitrary and so large that $n > \beta_b(\varepsilon, r) \equiv \sqrt{\varepsilon^{-1}b}$. Define the function $h_n \equiv 1 \wedge (1 + n - |\cdot|)_+ \in C(R, d)$. Then there exists an r.r.v. Y with the normal distribution $\Phi_{0,[r]} \equiv \Phi_{0,\sigma(r,r)}$. Take any $\alpha \in (\sqrt{\varepsilon^{-1}b}, n)$. Then

$$1 - F^\sigma_r h_n \equiv 1 - \Phi_{0,[r]}h_n \equiv E(1 - h_n(Y))$$
$$\le E1_{(|Y|>\alpha)} \le \alpha^{-2}EY^2 = \alpha^{-2}\sigma(r,r) \le (\varepsilon b^{-1})b = \varepsilon,$$

where the second inequality is by Chebychev's inequality. Thus β_b is a modulus of pointwise tightness of the family F^σ in the sense of Definition 6.3.5, as alleged in Assertion 3.

4. Assertion 4 is a summary of Assertions 1–3. □

Proposition 7.2.4. Normal f.j.d.'s depend continuously on a covariance function. *Let $G_{\delta(cov),b}(Q)$ be as in Definition 7.2.2. Equip $G_{\delta(cov),b}(Q)$ with the metric $\widehat{\rho}_{cov}$ defined by*

$$\widehat{\rho}_{cov}(\sigma,\sigma') \equiv \sup_{(t,s)\in Q\times Q} |\sigma(t,s) - \sigma'(t,s)|$$

for each $\sigma,\sigma' \in G_{\delta(cov),b}(Q)$. Recall from Definition 7.1.6 the metric space

$$(\widehat{F}_{Cp,\beta}(Q,R), \widehat{\rho}_{Cp,\xi,Q,Q(\infty)})$$

of consistent families of f.j.d.'s with parameter space (Q,d_Q) and state space R, whose members share the common modulus of continuity in probability $\delta_{Cp} \equiv \delta_{Cp,\delta(\cdot,cov)}$ as well as the common modulus of pointwise tightness $\beta \equiv \beta_b$.

Then the function

$$\Phi_{covar,fjd} : (G_{\delta(cov),b}(Q), \widehat{\rho}_{cov}) \to (\widehat{F}_{Cp,\beta}(Q,R), \widehat{\rho}_{Cp,\xi,Q,Q(\infty)})$$

in Proposition 7.2.3 is uniformly continuous, with a modulus of continuity $\delta_{covar,fjd}$ defined in equality 7.2.8 in the following proof.

Proof. Recall that $Q_\infty \equiv \{t_1, t_2, \ldots\}$ is an enumeration of the dense subset of Q.

1. Let $\varepsilon > 0$ be arbitrary. Let $n \geq 1$ be arbitrary. By Theorem 5.8.11, there exists $\delta_{ch,dstr}(\varepsilon,n) > 0$ such that, for arbitrary distributions J, J' on R^n whose respective characteristic functions ψ, ψ' satisfy

$$\rho_{char,n}(\psi,\psi') \equiv \sum_{j=1}^{\infty} 2^{-j} \sup_{|\lambda|\leq j} |\psi(\lambda) - \psi'(\lambda)| < \delta_{ch,dstr}(\varepsilon,n), \tag{7.2.6}$$

we have

$$\rho_{Dist,\xi^n}(J,J') < \varepsilon, \tag{7.2.7}$$

where ρ_{Dist,ξ^n} is the metric on the space of distributions on R^n, as in Definition 5.3.4.

2. Let $\varepsilon > 0$ be arbitrary. Let $m \geq 1$ be so large that $2^{-m+1} < \varepsilon$. Let $K \geq 1$ be so large that

$$2^{-K} < \alpha \equiv 2^{-1} \bigwedge_{n=1}^{m} \delta_{ch,dstr}(2^{-1}\varepsilon, n).$$

Define

$$\delta_{covar,fjd}(\varepsilon) \equiv 2K^{-2}m^{-2}\alpha. \tag{7.2.8}$$

We will verify that $\delta_{covar,fjd}$ is a desired modulus of continuity of the function $\Phi_{covar,fjd}$.

3. To that end, let $\sigma,\sigma' \in G_{\delta(cov),b}(Q)$ be arbitrary such that

$$\widehat{\rho}_{cov}(\sigma,\sigma') < \delta_{covar,fjd}(\varepsilon). \tag{7.2.9}$$

Let $F^{\sigma} \equiv \Phi_{covar,fjd}(\sigma)$ and $F^{\sigma'} \equiv \Phi_{covar,fjd}(\sigma')$ be constructed as in Theorem 7.2.3. We will show that $\widehat{\rho}_{Cp,\xi,Q,Q(\infty)}(F^{\sigma}, F^{\sigma'}) < \varepsilon$.

4. First note that inequality 7.2.9 is equivalent to

$$\sup_{(t,s)\in Q\times Q} |\sigma(t,s) - \sigma'(t,s)| < 2K^{-2}m^{-2}\alpha. \tag{7.2.10}$$

Next, consider each $n = 1, \ldots, m$. By Theorem 7.2.3, the joint normal distribution $F^{\sigma}_{t(1),\ldots,t(n)}$ has mean 0 and covariance matrix

$$\overline{\sigma} \equiv [\sigma(t_k, t_h)]_{k=1,\ldots,n;h=1,\ldots,n}. \tag{7.2.11}$$

Hence it has a characteristic function defined by

$$\chi^{\sigma}_{t(1),\ldots,t(n)}(\lambda) \equiv \exp\left(-\frac{1}{2}\sum_{k=1}^{n}\sum_{h=1}^{n}\lambda_k\sigma(t_k,t_h)\lambda_h\right)$$

for each $\lambda \equiv (\lambda_1, \ldots, \lambda_n) \in R^n$, with a similar equality for σ'. Let $\lambda \equiv (\lambda_1, \ldots, \lambda_n) \in R^n$ be arbitrary. As an abbreviation, write

$$\gamma(\lambda) \equiv -\frac{1}{2}\sum_{k=1}^{n}\sum_{h=1}^{n}\lambda_k\sigma(t_k,t_h)\lambda_h$$

and

$$\gamma'(\lambda) \equiv -\frac{1}{2}\sum_{k=1}^{n}\sum_{h=1}^{n}\lambda_k\sigma'(t_k,t_h)\lambda_h.$$

Then, because the function σ is nonnegative definite, we have $\gamma(\lambda) \leq 0$ and $\gamma'(\lambda) \leq 0$. Consequently,

$$|e^{\gamma(\lambda)} - e^{\gamma'(\lambda)}| \leq |\gamma(\lambda) - \gamma'(\lambda)|.$$

5. Suppose $|\lambda| \leq K$. Then

$$\begin{aligned}
|e^{\gamma(\lambda)} - e^{\gamma'(\lambda)}| &\leq |\gamma(\lambda) - \gamma'(\lambda)| \\
&\leq \frac{1}{2}\sum_{k=1}^{n}\sum_{h=1}^{n}|\lambda_k\sigma(t_k,t_h)\lambda_h - \lambda_k\sigma'(t_k,t_h)\lambda_h| \\
&= \frac{1}{2}\sum_{k=1}^{n}\sum_{h=1}^{n}|\lambda_k| \cdot |\sigma(t_k,t_h) - \lambda_k\sigma'(t_k,t_h)| \cdot |\lambda_h| \\
&\leq \frac{1}{2}\sum_{k=1}^{n}\sum_{h=1}^{n}|\lambda_k| \cdot 2K^{-2}m^{-2}\alpha \cdot |\lambda_h| \\
&\leq n^2K \cdot K^{-2}m^{-2}\alpha \cdot K \leq \alpha,
\end{aligned}$$

where the third inequality is by inequality 7.2.10.

6. We estimate

$$\rho_{char,n}(\chi^{\sigma}_{t(1),\dots,t(n)}, \chi^{\sigma'}_{t(1),\dots,t(n)})$$
$$\equiv \sum_{j=1}^{\infty} 2^{-j} \sup_{|\lambda|\leq j} |\chi^{\sigma}_{t(1),\dots,t(n)}(\lambda) - \chi^{\sigma'}_{t(1),\dots,t(n)}(\lambda)|$$
$$\leq \sup_{|\lambda|\leq K} |e^{\gamma(\lambda)} - e^{\gamma'(\lambda)}| + \sum_{j=K+1}^{\infty} 2^{-j} \leq \alpha + \sum_{j=K+1}^{\infty} 2^{-j}$$
$$= \alpha + 2^{-K} < \alpha + \alpha \leq \delta_{ch,dstr}(2^{-1}\varepsilon, n).$$

Hence, according to inequality 7.2.7, we have

$$\rho_{Dist,\xi^n}(F^{\sigma}_{t(1),\dots,t(n)}, F^{\sigma'}_{t(1),\dots,t(n)}) < 2^{-1}\varepsilon,$$

where $n = 1, \dots, m$ is arbitrary. Therefore, according to Definition 6.2.12,

$$\widehat{\rho}_{Cp,\xi,Q,Q(\infty)}(F^{\sigma}, F^{\sigma'}) \equiv \sum_{n=1}^{\infty} 2^{-n} \rho_{Dist,\xi^n}(F^{\sigma}_{t(1),\dots,t(n)}, F^{\sigma}_{t(1),\dots,t(n)})$$
$$\leq \sum_{n=1}^{m} 2^{-n} \rho_{Dist,\xi^n}(F^{\sigma}_{t(1),\dots,t(n)}, F^{\sigma}_{t(1),\dots,t(n)}) + \sum_{n=m+1}^{\infty} 2^{-n}$$
$$\leq \sum_{n=1}^{m} 2^{-n} 2^{-1}\varepsilon + 2^{-m} < 2^{-1}\varepsilon + 2^{-1}\varepsilon = \varepsilon, \qquad (7.2.12)$$

where we used the bounds $0 \leq \rho_{Dist,\xi^n} \leq 1$ for each $n \geq 1$.

Since $\varepsilon > 0$ is arbitrarily small, we conclude that the function $\Phi_{covar,fjd}$ is uniformly continuous, with modulus of continuity $\delta_{covar,fjd}$. The proposition is proved. □

Now we can apply the theorems from Section 7.1. As before, let

$$(\Theta_0, L_0, I_0) \equiv (\Theta_1, L_1, I_1) \equiv \left([0,1], L_1, \int \cdot d\theta\right)$$

be the Lebesgue integration space based on the interval $[0,1]$, and let

$$(\Omega, L, E) \equiv (\Theta_1, L_1, E_1) \otimes (\Omega_0, L_0, E_0).$$

Theorem 7.2.5. Construction of a measurable Gaussian r.f. from a continuous covariance function, and continuity of said construction. *Let $G_{\delta(cov),b}(Q)$ be as in Definition 7.2.2. Equip $G_{\delta(cov),b}(Q)$ with the metric $\widehat{\rho}_{cov}$ defined by*

$$\widehat{\rho}_{cov}(\sigma, \sigma') \equiv \sup_{(t,s)\in Q\times Q} |\sigma(t,s) - \sigma'(t,s)|$$

for each $\sigma, \sigma' \in G_{\delta(cov),b}(Q)$. Recall from Definition 7.1.3 the metric space $(\widehat{R}_{Meas,Cp}(Q \times \Omega, R), \rho_{Sup,Prob})$ of measurable r.f.'s $X : Q \times \Omega \to R$ that are continuous in probability.

Then there exists a uniformly continuous mapping

$$\Phi_{cov,gauss,\xi,\xi(Q)} : (G_{\delta(cov),b}(Q), \widehat{\rho}_{cov}) \to (\widehat{R}_{Meas,Cp}(Q \times \Omega, R), \rho_{Sup,Prob}),$$

with a modulus of continuity $\delta_{cov,gauss}(\cdot, \delta_{cov}, b, \|\xi\|, \|\xi_Q\|)$*, with the following properties:*

1. Let $\sigma \in G_{\delta(cov),b}(Q)$ *be arbitrary. Then* $X \equiv \Phi_{cov,gauss,\xi,\xi(Q)}(\sigma)$ *is a Gaussian r.f..*

2. Let $r_1, r_2 \in Q$ *be arbitrary. Then* $EX_{r(1)} = 0$*, and* $EX_{r(1)}X_{r(2)} = \sigma(r_1, r_2)$.

We will call the function $\Phi_{cov,gauss,\xi,\xi(Q)}$ *the measurable Gaussian extension relative to the binary approximations* ξ *and* ξ_Q.

Proof. 1. Recall from Definition 7.1.6 the metric space $(\widehat{F}_{Cp,\beta}(Q,R), \widehat{\rho}_{Cp,\xi,Q,Q(\infty)})$ of consistent families of f.j.d.'s with parameter space (Q, d_Q) and state space R, whose members share the common modulus of continuity in probability $\delta_{Cp} \equiv \delta_{Cp,\delta(\cdot,cov)}$ as well as the common modulus of pointwise tightness $\beta \equiv \beta_b$.

2. Recall, from Theorem 7.1.7, the uniformly continuous mapping

$$\Phi_{meas,\xi,\xi(Q)} : (\widehat{F}_{Cp,\beta}(Q,S), \widehat{\rho}_{Cp,\xi,Q,Q(\infty)}) \to (\widehat{R}_{Meas,Cp}(Q \times \Omega, S), \rho_{Sup,Prob}), \tag{7.2.13}$$

with a modulus of continuity $\delta_{meas,\xi,\xi(Q}(\cdot, \delta_{Cp}, \beta, \|\xi\|, \|\xi_Q\|)$.

3. Recall, from Proposition 7.2.4, the uniformly continuous mapping

$$\Phi_{covar,fjd} : (G_{\delta(cov),b}(Q), \widehat{\rho}_{cov}) \to (\widehat{F}_{Cp,\beta}(Q,R), \widehat{\rho}_{Cp,\xi,Q,Q(\infty)}), \tag{7.2.14}$$

with a modulus of continuity $\delta_{covar,fjd}$.

4. We will verify that the composite mapping

$$\Phi_{cov,gauss,\xi,\xi(Q)} \equiv \Phi_{meas,\xi,\xi(Q)} \circ \Phi_{covar,fjd} : (G_{\delta(cov),b}(Q), \widehat{\rho}_{cov}) \to (\widehat{R}_{Meas,Cp}(Q \times \Omega, R), \rho_{Sup,Prob})$$

has the desired properties.

5. First note that the composite function $\Phi_{cov,gauss,\xi,\xi(Q)}$ is uniformly continuous, with a modulus of continuity defined by the composite operation

$$\delta_{cov,gauss}(\cdot, \delta_{cov}, b, \|\xi\|, \|\xi_Q\|) \equiv \delta_{covar,fjd}(\delta_{meas,\xi,\xi(Q}(\cdot, \delta_{Cp}, \beta, \|\xi\|, \|\xi_Q\|)).$$

6. Next, let $\sigma \in G_{\delta(cov),b}(Q)$ be arbitrary. Let

$$X \equiv \Phi_{cov,gauss,\xi,\xi(Q)}(\sigma) \equiv \Phi_{meas,\xi,\xi(Q)} \circ \Phi_{covar,fjd}(\sigma) : Q \times \Omega \to R.$$

Then

$$X \equiv \Phi_{meas,\xi,\xi(Q)}(F^{\sigma}) \equiv \Phi_{meas,\xi,\xi(Q)} \circ \Phi_{covar,fjd}(\sigma),$$

where $F^{\sigma} \equiv \Phi_{covar,fjd}(\sigma)$. By Proposition 7.2.3, $F^{\sigma} \equiv \Phi_{covar,fjd}(\sigma)$ is a consistent family of of normal f.j.d.'s and is continuous in probability. According to Theorem 7.1.7, the measurable r.f. X has marginal distributions given by F^{σ}. Hence X is a Gaussian r.f and is continuous in probability. Assertion 1 is proved.

7. Now let $r_1, r_2 \in Q$. Then the r.r.v.'s $X_{r(1)}, X_{r(2)}$ have joint normal distribution $F^{\sigma}_{r(1),r(2)} = \Phi_{0,\overline{\sigma}}$, where $\overline{\sigma} \equiv [\sigma(r_i, r_j)]_{i,j=1,2}$. Hence, by Lemma 5.7.8, we have $EX_{r(1)} = 0$, and $EX_{r(1)}X_{r(2)} = \sigma(r_1, r_2)$. Assertion 2 and the theorem are proved. $\square$

8

Martingale

In this chapter, we define a martingale $X \equiv \{X_t : t = 1,2,\ldots\}$ for modeling one's fortune in a fair game of chance. Then we will prove the basic theorems on martingales, which have wide-ranging applications. Among these is the a.u. convergence of X_t as $t \to \infty$. Our proof is constructive and quantifies rates of convergence by means of a maximal inequality. There are proofs in traditional texts that also are constructive and quantify rates similarly by means of maximal inequalities. These traditional maximal inequalities, however, require the integrability of $|X_t|^p$ for some $p > 1$, or at least the integrability of $|X_t|\log|X_t|$. For the separate case of $p = 1$, the classical proof of a.u. convergence is by a separate inference from certain upcrossing inequalities. Such inference is essentially equivalent to the principle of infinite search and is not constructive.

In contrast, the maximal inequality we present requires only the integrability of $|X_t|$. Therefore, thanks to Lyapunov's inequality, it is at once applicable to the case of integrable $|X_t|^p$ for any given $p \geq 1$, without having to first determine whether $p > 1$ or $p = 1$.

For readers who are uninitiated in the subject, the previous paragraphs are perhaps confusing, but will become clear as we proceed. For the rich body of classical results on, and applications of, martingales, see, e.g., [Doob 1953; Chung 1968; Durret 1984].

8.1 Filtration and Stopping Time

Definition 8.1.1. Assumptions and notations. In this chapter, let (S,d) be a locally compact metric space with an arbitrary but fixed reference point $x_\circ$. Let (Ω, L, E) be an arbitrary probability space. Unless otherwise specified, an r.v. refers to a measurable function with values in S. Let Q denote an arbitrary nonempty subset of R, called the time parameter set.

If (Ω, L', E) is a probability subspace of (Ω, L, E), we will simply call L' a probability subspace of L when Ω and E are understood.

As an abbreviation, we will write $A \in L$ if A is a measurable subset of (Ω, L, E). Thus $A \in L$ iff $1_A \in L$, in which case we will write $P(A)$, PA, $E1_A$, and EA

interchangeably, and write $E(X; A) \equiv EX1_A$ for each $X \in L$. As usual, we write a subscripted expression x_y interchangeably with $x(y)$. Recall the convention in Definition 5.1.3 regarding regular points of r.r.v.'s. □

Definition 8.1.2. Filtration and adapted process. Suppose that for each $t \in Q$, there exists a probability subspace $(\Omega, L^{(t)}, E)$ of (Ω, L, E), such that $L^{(t)} \subset L^{(s)}$ for each $t, s \in Q$ with $t \leq s$. Then the family $\mathcal{L} \equiv \{L^{(t)} : t \in Q\}$ is called a *filtration* in (Ω, L, E). The filtration $\mathcal{L}$ is said to be *right continuous* if, for each $t \in Q$, we have

$$L^{(t)} = \bigcap_{s \in Q; s > t} L^{(s)}.$$

Separately, suppose $\widetilde{Q}$ is a subset of Q. Then a stochastic process $X : \widetilde{Q} \times \Omega \to S$ is said to be *adapted* to the filtration $\mathcal{L}$ if X_t is a r.v. on $(\Omega, L^{(t)}, E)$ for each $t \in \widetilde{Q}$. □

The probability space $L^{(t)}$ can be regarded as the observable history up to the time t. Thus a process X adapted to $\mathcal{L}$ is such that X_t is observable at the time t, for each $t \in \widetilde{Q}$.

Definition 8.1.3. Natural filtration of a stochastic process. Let $X : Q \times \Omega \to S$ be an arbitrary stochastic process. For each $t \in Q$, define the set

$$G^{(X,t)} \equiv \{X_r : r \in Q;\ r \leq t\},$$

and let

$$L^{(X,t)} \equiv L(X_r : r \in Q;\ r \leq t) \equiv L(G^{(X,t)})$$

be the probability subspace of L generated by the set $G^{(X,t)}$ of r.v.'s. Then the family $\mathcal{L}_{\mathcal{X}} \equiv \{L^{(X,t)} : t \in Q\}$ is called the *natural filtration of the process* X. □

Lemma 8.1.4. A natural filtration is indeed a filtration. *Let $X : Q \times \Omega \to S$ be an arbitrary stochastic process. Then the natural filtration $\mathcal{L}_X$ of X is a filtration to which the process X is adapted.*

Proof. For each $t \leq s$ in Q, we have $G^{(X,t)} \subset G^{(X,s)}$, whence $L^{(X,t)} \subset L^{(X,s)}$. Thus $\mathcal{L}_X$ is a filtration. Let $t \in Q$ be arbitrary. Then $f(X_t) \in L(G^{(X,t)}) \equiv L^{(X,t)}$ for each $f \in C_{ub}(S,d)$. At the same time, because X_t is an r.v. on (Ω, L, E), we have $P(d(X_t, x_\circ) \geq a) \to 0$ *as* $a \to \infty$. Hence X_t is an r.v. on $(\Omega, L^{(X,t)}, E)$ according to Proposition 5.1.4. Thus the process X is adapted to its natural filtration $\mathcal{L}_X$. □

Definition 8.1.5. Right-limit extension and right continuity of a filtration. Suppose (i) $\overline{Q} = [0, \infty)$ or (ii) $\overline{Q} \equiv [0, a]$ for some $a > 0$. Suppose Q is a subset that is dense in $\overline{Q}$ and that, in case (ii), contains the endpoint a. Let $\mathcal{L} \equiv \{L^{(t)} : t \in Q\}$ be an arbitrary filtration of a given probability space (Ω, L, E).

In case (i), define, for each $t \in \overline{Q}$, the probability subspace

$$L^{(t+)} \equiv \bigcap \{L^{(s)} : s \in Q \cap (t,\infty)\}. \quad (8.1.1)$$

of L. In case (ii), define, for each $t \in \overline{Q}$, the probability subspace

$$L^{(t+)} \equiv \bigcap \{L^{(s)} : s \in Q \cap (t,a] \cup \{a\})\}. \quad (8.1.2)$$

Then the filtration $\mathcal{L}^+ \equiv \{L^{(t+)} : t \in \overline{Q}\}$ is called the *right-limit extension* of the filtration $\mathcal{L}$.

If $Q = \overline{Q}$ and $L^{(t)} = L^{(t+)}$for each $t \in \overline{Q}$, then $\mathcal{L}$ is said to be *a right continuous filtration.* □

Lemma 8.1.6. Right-limit extension of a filtration is right continuous. *In the notations of Definition 8.1.5, we have $(\mathcal{L}^+)^+ = \mathcal{L}^+$. In words, the right-limit extension of the filtration $\mathcal{L} \equiv \{L^{(t)} : t \in Q\}$ is right continuous.*

Proof. We will give the proof only for the case where $\overline{Q} = [0,\infty)$, with the proof for the case where $\overline{Q} \equiv [0,a]$ being similar. To that end, let $t \in \overline{Q}$ be arbitrary. Then

$$\begin{aligned}(L^{(t+)+}) &\equiv \bigcap \{L^{(s+)} : s \in \overline{Q} \cap (t,\infty)\} \\ &\equiv \bigcap \left\{\bigcap \{L^{(u)} : u \in Q \cap (s,\infty)\} : s \in \overline{Q} \cap (t,\infty)\right\} \\ &= \bigcap \{L^{(u)} : u \in Q \cap (t,\infty)\} \equiv L^{(t+)},\end{aligned}$$

where the third equality is because $u \in Q \cap (t,\infty)$ iff $u \in Q \cap (s,\infty)$ for some $s \in Q \cap (t,\infty)$, thanks to the assumption that Q is dense in $\overline{Q}$. □

Definition 8.1.7. r.r.v. with values in a subset of R. Let A denote an arbitrary nonempty subset of R. We say that an r.r.v. η has values in the subset A if $(\eta \in A)$ is a full set. □

Lemma 8.1.8. r.r.v. with values in an increasing sequence in R. *Let the subset $A \equiv \{t_0, t_1, \ldots\} \subset R$ be arbitrary such that $t_{n-1} < t_n$ for each $n \geq 1$. Then an r.r.v. η has values in A iff* (i) *$(\eta = t_n)$ is measurable for each $n \geq 0$ and* (ii) *$\sum_{n=1}^{\infty} P(\eta = t_n) = 1$.*

Proof. Recall Definition 4.8.13 of regular points of a real-valued measurable function. Suppose the r.r.v. η has values in A. For convenience, write $t_{-1} \equiv t_0 - (t_1 - t_0)$. Consider each $n \geq 0$. Write $\Delta_n \equiv 2^{-1}((t_n - t_{n-1}) \wedge (t_{n+1} - t_n)) > 0$. Then there exist regular points t, s of the r.r.v. η such that

$$t_{n-1} < t_n - \Delta_n < s < t_n < t < t + \Delta_n < t_{n+1}.$$

Then $(\eta = t_n) = (\eta \leq t)(\eta \leq s)^c(\eta \in A)$. Since $(\eta \leq t)$, $(\eta \leq s)$, and $(\eta \in A)$ are measurable subsets, it follows that the set $(\eta = t_n)$ is measurable. At the same time, $P(\eta \leq t_m) \uparrow 1$ as $m \to \infty$ since η is an r.r.v. Hence

$$\sum_{n=1}^{m} P(\eta = t_n) = P(\eta \le t_m) \uparrow 1$$

as $m \to \infty$. In other words, $\sum_{n=1}^{\infty} P(\eta = t_n) = 1$. Thus we have proved that if the r.r.v. η has values in A, then Conditions (i) and (ii) hold.

Conversely, if Conditions (i) and (ii) hold, then $\bigcup_{n=1}^{\infty}(\eta = t_n)$ is a full set that is contained in $(\eta \in A)$, whence the latter set is a full set. Thus the r.r.v. η has values in A. □

Definition 8.1.9. Stopping time, space of integrable observables at a stopping time, and simple stopping time. Let Q denote an arbitrary nonempty subset of R. Let $\mathcal{L}$ be an arbitrary filtration with time parameter set Q. Then an r.r.v. τ with values in Q is called a *stopping time* relative to the filtration $\mathcal{L}$ if

$$(\tau \le t) \in L^{(t)} \tag{8.1.3}$$

for each regular point $t \in Q$ of the r.r.v. τ. We will omit the reference to $\mathcal{L}$ when it is understood from the context, and simply say that τ is a stopping time. Each r.v. relative to the probability subspace

$$L^{(\tau)} \equiv \{Y \in L : Y1_{(\tau \le t)} \in L^{(t)} \text{ for each regular point } t \in Q \text{ of } \tau\}$$

is said to be *observable at the stopping time* τ. Each member of $L^{(\tau)}$ is called an *integrable observable at the stopping time* τ.

Let $X : Q \times \Omega \to S$ be an arbitrary stochastic process adapted to the filtration $\mathcal{L}$. Define the function X_τ by

$$domain(X_\tau) \equiv \{\omega \in domain(\tau) : (\tau(\omega), \omega) \in domain(X)\}$$

and by

$$X_\tau(\omega) \equiv X(\tau(\omega), \omega) \tag{8.1.4}$$

for each $\omega \in domain(X_\tau)$. Then the function X_τ is called the *observation of the process X at the stopping time* τ. In general, X_τ need not be a well-defined r.v. We will need to prove that X_τ is a well-defined r.v. in each application before using it as such.

A stopping time τ with values in some metrically discrete finite subset of Q is called a *simple stopping time*. □

We leave it as an exercise to verify that $L^{(\tau)}$ is indeed a probability subspace. A trivial example of a stopping time is a deterministic time $\tau \equiv s$, where $s \in Q$ is arbitrary.

Definition 8.1.10. Specialization to a metrically discrete parameter set. In the remainder of this section, unless otherwise specified, assume that the parameter set $Q \equiv \{0, \Delta, 2\Delta, \ldots\}$ is equally spaced, with some fixed $\Delta > 0$, and let $\mathcal{L} \equiv \{L^{(t)} : t \in Q\}$ be an arbitrary but fixed filtration in (Ω, L, E) with parameter Q. □

Proposition 8.1.11. Basic properties of stopping times: a metrically discrete case. *Let τ and τ' be stopping times with values in $Q \equiv \{0, \Delta, 2\Delta, \ldots\}$, relative to the filtration $\mathcal{L}$. For each $n \geq -1$, write $t_n \equiv n\Delta$ for convenience. Then the following conditions hold:*

1. Let η be an r.r.v. with values in Q . Then η is a stopping time iff $(\eta = t_n) \in L^{(t(n))}$ for each $n \geq 0$.

2. $\tau \wedge \tau'$, $\tau \vee \tau'$ are stopping times.

3. If $\tau \leq \tau'$, then $L^{(\tau)} \subset L^{(\tau')}$.

4. Let $X : Q \times \Omega \to S$ be an arbitrary stochastic process adapted to the filtration $\mathcal{L}$. Then X_τ is a well-defined r.v. on the probability space $(\Omega, L^{(\tau)}, E)$.

Proof. 1. By Lemma 8.1.8, the set $(\eta = t_n)$ is measurable for each $n \geq 0$, and $\sum_{n=1}^{\infty} P(\eta = t_n) = 1$. Suppose η is a stopping time. Let $n \geq 0$ be arbitrary. Then $(\eta \leq t_n) \in L^{(t(n))}$. Moreover, if $n \geq 1$, then $(\eta \leq t_{n-1})^c \in L^{(t(n-1))} \subset L^{(t(n))}$. If $n = 0$, then $(\eta \leq t_{n-1})^c = (\eta \geq 0)$ is a full set, whence $(\eta \geq 0) \in L^{(t(n))}$. Combining, we see that $(\eta = t_n) = (\eta \leq t_n)(\eta \leq t_{n-1})^c \in L^{(t(n))}$. We have proved the "only if" part of Assertion 1.

Conversely, suppose $(\eta = t_n) \in L^{(t(n))}$ for each $n \geq 0$. Let $t \in Q$ be arbitrary. Then $t = t_m$ for some $m \geq 0$. Hence $(\eta \leq t) = \bigcup_{n=0}^{m}(\eta = t_n)$, where, by assumption, $(\eta = t_n) \in L^{(t(n))} \subset L^{(t(m))}$ for each $n = 0, \ldots, m$. Thus we see that $(\eta \leq t)$ is observable at time t, where $t \in Q$ is arbitrary. We conclude that η is a stopping time.

2. Let $t \in Q$ be arbitrary. Then, since Q is countable and discrete, we have

$$(\tau \wedge \tau' \leq t) = (\tau \leq t) \cup (\tau' \leq t) \in L^{(t)}$$

and

$$(\tau \vee \tau' \leq t) = (\tau \leq t)(\tau' \leq t) \in L^{(t)}.$$

Thus $\tau \wedge \tau'$ and $\tau \vee \tau'$ are stopping times.

3. Let $Y \in L^{(\tau)}$ be arbitrary. Consider each $t \in Q$. Then, since $\tau \leq \tau'$,

$$Y 1_{(\tau' \leq t)} = \sum_{s \in Q} Y 1_{(\tau = s)} 1_{(\tau' \leq t)} = \sum_{s \in [0,t]Q} Y 1_{(\tau = s)} 1_{(\tau' \leq t)} \in L^{(t)}.$$

Thus $Y \in L^{(\tau')}$, where $Y \in L^{(\tau)}$ is arbitrary. We conclude that $L^{(\tau)} \subset L^{(\tau')}$.

4. Let $X : Q \times \Omega \to S$ be an arbitrary stochastic process adapted to the filtration $\mathcal{L}$. Define the full sets $A \equiv \bigcap_{n=0}^{\infty} domain(X_{t(n)})$ and $B \equiv \bigcup_{n=0}^{\infty}(\tau = t_n)$. Consider each $\omega \in AB$. Then $(\tau(\omega), \omega) = (t_n, \omega) \in domain(X)$ on $(\tau = t_n)$, for each $n \geq 0$. In short, X_τ is defined and is equal to the r.v. $X_{t(n)}$ on $(\tau = t_n)$, for each $n \geq 0$. Since $\bigcup_{n=0}^{\infty}(\tau = t_n)$ is a full set, the function X_τ is therefore an r.v. according to Proposition 4.8.8. Assertion 4 is proved. □

Simple first exit times from a time-varying neighborhood, introduced next, are examples of simple stopping times.

Definition 8.1.12. Simple first exit time. Let $Q' \equiv \{s_0, \ldots, s_n\}$ be a finite subset of $Q \equiv \{0, \Delta, 2\Delta, \ldots\}$, where $(s_0, \ldots, s_n)$ is an increasing sequence. Recall that $\mathcal{L} \equiv \{L^{(t)} : t \in Q\}$ is a filtration.

1. Let $x : Q' \to S$ be an arbitrary given function. Let $b : Q' \to (0, \infty)$ be an arbitrary function such that for each $t, r, s \in Q$', we have either $b(s) \le d(x_t, x_r)$ or $b(s) > d(x_t, x_r)$. Let $t \in Q'$ be arbitrary. Define

$$\eta_{t,b,Q'}(x) \equiv \sum_{r \in Q'; t \le r} r 1_{(d(x(t),x(r))>b(r))} \prod_{s \in Q'; t \le s < r} 1_{(d(x(t),x(s)) \le b(s))}$$

$$+ s_n \prod_{s \in Q'; t \le s} 1_{(d(x(s),x(t)) \le b(s))}. \tag{8.1.5}$$

In words, $\eta_{t,b,Q'}(x)$ is the first time $r \in [t, s_n]Q'$ such that x_r is at a distance greater than $b(r)$ from the initial position x_t, with $\eta_{t,b,Q'}(x)$ set to the final time $s_n \in Q'$ if no such r exists. Then $\eta_{t,b,Q'}(x)$ is called the *simple first exit time* for the function $x|[t, s_n]Q'$ to exit the time-varying b-neighborhood of x_t. In the special case where $b(r) = \alpha$ for each $r \in Q'$ for some constant $\alpha > 0$, we will write simply $\eta_{t,\alpha,Q'}(x)$ for $\eta_{t,b,Q'}(x)$.

2. More generally, let $X : Q' \times \Omega \to S$ be an arbitrary process adapted to the filtration $\mathcal{L}$. Let $b : Q' \to (0, \infty)$ be an arbitrary function such that for each t, r, $s \in Q'$, the real number $b(s)$ is a regular point for the r.r.v. $d(X_t, X_r)$. Let $t \in Q'$ be arbitrary. Define the r.r.v. $\eta_{t,b,Q'}(X)$ on Ω as

$$\eta_{t,b,Q'}(X) \equiv \sum_{r \in Q'; t \le r} r 1_{(d(X(t),X(r))>b(r))} \prod_{s \in Q'; t \le s < r} 1_{(d(X(t),X(s)) \le b(s))}$$

$$+ s_n \prod_{s \in Q'; t \le s} 1_{(d(X(s),X(t)) \le b(s))}. \tag{8.1.6}$$

Then $\eta_{t,b,Q'}(X)$ is called the *simple first exit time* for the process $X|[t, s_n]Q'$ to exit the time-varying b-neighborhood of X_t. When there is little risk of confusion as to the identity of the process X, we will omit the reference to X, write $\eta_{t,b,Q'}$ for $\eta_{t,b,Q'}(X)$, and abuse notations by writing $\eta_{t,b,Q'}(\omega)$ for $\eta_{t,b,Q'}(X(\omega))$, for each $\omega \in \bigcap_{u \in Q'} domain(X_u)$. $\square$

The next proposition verifies that $\eta_{t,b,Q'}(X)$ is a simple stopping time relative to the filtration $\mathcal{L}$. It also proves some simple properties that are intuitively obvious when described in words.

Proposition 8.1.13. Basic properties of simple first exit times. *Let $Q' \equiv \{s_0, \ldots, s_n\}$ be a finite subset of Q, where $(s_0, \ldots, s_n)$ is an increasing sequence. Use the assumptions and notations in Part 2 of Definition 8.1.12. Let $t, s \in Q' \equiv \{s_0, \ldots, s_n\}$ be arbitrary. Let $\omega \in domain(\eta_{t,b,Q'})$ be arbitrary. Then the following conditions hold:*

1. $t \le \eta_{t,b,Q'}(\omega) \le s_n$.

2. The r.r.v. $\eta_{t,b,Q'}$ is a simple stopping time relative to the filtration $\mathcal{L}$.

3. If $\eta_{t,b,Q'}(\omega) < s_n$, then $d(X(t,\omega), X(\eta_{t,b,Q'}(\omega),\omega)) > b(\eta_{t,b,Q'}(\omega))$. In words, if the simple first exit time occurs before the final time, then the sample path exits successfully at the simple first exit time.

4. If $t \leq s < \eta_{t,b,Q'}(\omega)$, then $d(X(t,\omega), X(s,\omega)) \leq b(s)$. In words, before the simple first exit time, the sample path on $[t, s_n]Q'$ remains in the b-neighborhood of the initial point. Moreover, if

$$d(X(\eta_{t,b,Q'}(\omega),\omega), X(t,\omega)) \leq b(\eta_{t,b,Q'}(\omega)),$$

then

$$d(X(s,\omega), X(t,\omega)) \leq b(s)$$

for each $s \in Q'$ with $t \leq s \leq \eta_{t,b,Q'}(\omega)$. In words, if the sample path is in the b-neighborhood at the simple first exit time, then it is in the b-neighborhood at any time prior to the simple first exit time.

Conversely, if $r \in [t, s_n)Q'$ is such that $d(X(t,\omega), X(s,\omega)) \leq b(s)$ for each $s \in (t,r]Q'$, then $r < \eta_{t,b,Q'}(\omega)$. In words, if the sample path stays within the b-neighborhood up to and including a certain time, then the simple first exit time can come only after that certain time.

5. Suppose $s_0 = s_{k(0)} < s_{k(1)} < \cdots < s_{k(p)} = s_n$ is a subsequence of $s_0 < s_1 < \cdots < s_n$. Define $Q'' \equiv \{s_{k(1)}, \ldots, s_{k(p)}\}$. Let $t \in Q'' \subset Q'$ be arbitrary. Then $\eta_{t,b,Q'} \leq \eta_{t,b,Q''}$. In other words, if the process X is sampled at more time points, then the more densely sampled simple first exit time can occur no later.

Proof. By hypothesis, the process X is adapted to the filtration $\mathcal{L}$.

1. Assertion 1 is obvious from the defining equality 8.1.6.

2. By equality 8.1.6, for each $r \in \{t, \ldots, s_{n-1}\}$, we have

$$(\eta_{t,b,Q'} = r) = (d(X_r, X_t) > b(r)) \bigcap_{s \in Q'; t \leq s < r} (d(X_s, X_t) \leq b(s)) \in L^{(r)} \subset L^{(s(n))}. \tag{8.1.7}$$

Consequently,

$$(\eta_{t,b,Q'} = s_n) = \bigcap_{r \in Q'; r < t(n)} (\eta_{t,b,Q'} = r)^c \in L^{(s(n))}.$$

Hence $\eta_{t,b,Q'}$ is a simple stopping time relative to $\mathcal{L}'$ with values in Q', according to Proposition 8.1.11. Assertion 2 is proved.

3. Suppose $\eta_{t,b,Q'}(\omega) < s_n$. Then the last term on the right-hand side of the defining equality 8.1.6 is 0, whence some unique summand in the sum is nonzero at ω and is equal to $\eta_{t,b,Q'}(\omega)$. In other words, there exists $r \in Q'$ with $t \leq r$ such that

$$\eta_{t,b,Q'}(X)(\omega) \equiv r 1_{(d(X(t),X(r))>b(r))}(\omega) \prod_{s \in Q'; t \leq s < r} 1_{(d(X(t),X(s)) \leq b(s))}(\omega).$$

It follows that $d(X(t,\omega),X(r,\omega)) > b(r)$ and that $\eta_{t,b,Q'}(X)(\omega) = r$. Hence

$$d(X(t,\omega),X(\eta_{t,b,Q'}(\omega),\omega)) > b(\eta_{t,b,Q'}(\omega)),$$

as alleged in Assertion 3.

4. Suppose $t < s < r \equiv \eta_{t,b,Q'}(\omega)$. Then

$$d(X(t,\omega),X(s,\omega)) \leq b(s) \tag{8.1.8}$$

by equality 8.1.7. The last inequality is trivially satisfied if $t = s$. Hence if $r \equiv \eta_{t,b,Q'}(\omega) = s_n$ with $d(X(t,\omega),X(r,\omega)) \leq b(r)$, then inequality 8.1.8 holds for each $s \in Q'$ with $t \leq s \leq r$.

Conversely, suppose $r \in Q'$ is such that $t \leq r < s_n$ and such that

$$d(X(t,\omega),X(s,\omega)) \leq b(s)$$

for each $s \in (t,r]Q'$. Suppose $s \equiv \eta_{t,b,Q'}(\omega) \leq r < s_n$. Then $d(X(t,\omega), X(s,\omega)) > b(s)$ by Assertion 3, which is a contradiction. Hence $\eta_{t,b,Q'}(\omega) > r$. Assertion 4 is verified.

5. Let $t \in Q'' \subset Q'$ be arbitrary. Suppose, for the sake of contradiction, that $s \equiv \eta_{t,b,Q''}(\omega) < \eta_{t,b,Q'}(\omega) \leq s_n$. Then $t < s$ and $s \in Q'' \subset Q'$. Hence, by Assertion 4 applied to the time s and to the simple first exit time $\eta_{t,b,Q'}$, we have

$$d(X(t,\omega),X(s,\omega)) \leq b(s).$$

On the other hand, by Assertion 3 applied to the time s and to the simple first exit time $\eta_{t,b,Q''}$, we have

$$d(X(t,\omega),X(s,\omega)) > b(s),$$

which is a contradiction. Hence $\eta_{t,b,Q''}(\omega) \geq \eta_{t,b,Q'}(\omega)$. Assertion 5 is proved. □

8.2 Martingale

Definition 8.2.1. Martingale and submartingale. Let Q be an arbitrary nonempty subset of R. Let $\mathcal{L} \equiv \{L^{(t)} : t \in Q\}$ be an arbitrary filtration in (Ω, L, E). Let $X : Q \times \Omega \to R$ be a stochastic process such that $X_t \in L^{(t)}$ for each $t \in Q$.

1. The process X is called a *martingale* relative to $\mathcal{L}$ if, for each $t,s \in Q$ with $t \leq s$, we have $EZX_t = EZX_s$ for each indicator $Z \in L^{(t)}$. According to Definition 5.6.4, the last condition is equivalent to $E(X_s|L^{(t)}) = X_t$ for each $t,s \in Q$ with $t \leq s$.

2. The process X is called a *wide-sense submartingale* relative to $\mathcal{L}$ if, for each $t,s \in Q$ with $t \leq s$, we have $EZX_t \leq EZX_s$ for each indicator $Z \in L^{(t)}$. If, in addition, $E(X_s|L^{(t)})$ exists for each $t,s \in Q$ with $t \leq s$, then X is called a *submartingale* relative to $\mathcal{L}$.

3. The process X is called a *wide-sense supermartingale* relative to $\mathcal{L}$ if, for each $t,s \in Q$ with $t \leq s$, we have $EZX_t \geq EZX_s$ for each indicator $Z \in L^{(t)}$. If, in addition, $E(X_s|L^{(t)})$ exists for each $t,s \in Q$ with $t \leq s$, then X is called a *supermartingale* relative to $\mathcal{L}$.

When there is little risk of confusion, we will omit the explicit reference to the given filtration $\mathcal{L}$. □

With a martingale, the r.r.v. X_t can represent a gambler's fortune at the current time t. Then the conditional expectation of said fortune at a later time s, given all information up to and including the current time t, is exactly the gambler's current fortune. Thus a martingale X can be used as a model for a fair game of chance. Similarly, a submartingale can be used to model a favorable game.

Clearly, a submartingale is also a wide-sense submartingale. The two notions are classically equivalent because, classically, with the benefit of the principle of infinite search, the conditional expectation always exists. Hence, any result that we prove for wide-sense submartingales holds classically for submartingales.

Proposition 8.2.2. Martingale basics. *Let $X : Q \times \Omega \to R$ be an arbitrary process adapted to the filtration $\mathcal{L} \equiv \{L^{(t)} : t \in Q\}$. Unless otherwise specified, all martingales and wide-sense submartingales are relative to the filtration $\mathcal{L}$. Then the following conditions hold:*

1. The process X is a martingale iff it is both a wide-sense submartingale and a wide-sense supermartingale.

2. The process X is a wide-sense supermartingale iff $-X$ is a wide-sense submartingale.

3. The expectation EX_t is constant for $t \in Q$ if X is a martingale. The expectation EX_t is nondecreasing in t if X is a wide-sense submartingale.

4. Suppose X is a martingale. Then $|X|$ is a wide-sense submartingale. In particular, $E|X_t|$ is nondecreasing in $t \in Q$.

5. Suppose X is a martingale. Let $a \in Q$ be arbitrary. Then the family $\{X_t : t \in (-\infty,a]Q\}$ is uniformly integrable.

6. Let $\overline{\mathcal{L}} \equiv \{\overline{L}^{(t)} : t \in Q\}$ be an arbitrary filtration such that $L^{(t)} \subset \overline{L}^{(t)}$ for each $t \in Q$. Suppose X is a wide-sense submartingale relative to the filtration $\overline{\mathcal{L}}$. Then X is a wide-sense submartingale relative to the filtration $\mathcal{L}$. The same assertion holds for martingales.

7. Suppose X is a wide-sense submartingale relative to the filtration $\mathcal{L}$. Then it is a wide-sense submartingale relative to the natural filtration $\mathcal{L}_X \equiv \{L^{(X,t)} : t \in Q\}$ of the process X.

Proof. 1. Assertions 1–3 being trivial, we will prove Assertions 4–7 only.

2. To that end, let $t,s \in Q$ with $t \leq s$ be arbitrary. Let the indicator $Z \in L^{(t)}$ and the real number $\varepsilon > 0$ be arbitrary. Then

$$E(|X_s|Z; X_t > \varepsilon) \geq E(X_sZ; X_t > \varepsilon) = E(X_tZ; X_t > \varepsilon) = E(|X_t|Z; X_t > \varepsilon),$$

where the first equality is from the definition of a martingale. Since $-X$ is also a martingale, we have similarly

$$E(|X_s|Z; X_t < -\varepsilon) \geq E(-X_s Z; X_t < -\varepsilon) = E(-X_t Z; X_t < -\varepsilon)$$
$$= E(|X_t|Z; X_t < -\varepsilon).$$

Adding the last two displayed inequalities, we obtain

$$E(|X_s|; Z) \geq E(|X_s|Z; X_t > \varepsilon) + E(|X_s|Z; X_t < -\varepsilon)$$
$$\geq E(|X_t|Z; X_t > \varepsilon) + E(|X_t|Z; X_t < -\varepsilon)$$
$$= E(|X_t|Z) - E(|X_t|Z; |X_t| \leq \varepsilon).$$

Since

$$E(|X_t|Z; |X_t| \leq \varepsilon) \leq E(|X_t|; |X_t| \leq \varepsilon) \to 0$$

as $\varepsilon \to 0$, we conclude that

$$E(|X_s|; Z) \geq E(|X_t|; Z),$$

where $t,s \in Q$ with $t \leq s$ and the indicator $Z \in L^{(t)}$ are arbitrary. Thus the process $|X|$ is a wide-sense submartingale. Assertion 4 is proved.

3. Suppose X is a martingale. Consider each $a \in Q$. Let $t \in Q$ be arbitrary with $t \leq a$, and let $\varepsilon > 0$ be arbitrary. Then, since X_a is integrable, there exists $\delta \equiv \delta_{X(a)}(\varepsilon) > 0$ so small that $E|X_a|1_A < \varepsilon$ for each measurable set A with $P(A) < \delta$. Now let $\gamma > \beta(\varepsilon) \equiv E|X_a|\delta^{-1}$ be arbitrary. Then by Chebychev's inequality,

$$P(|X_t| > \gamma) \leq E|X_t|\gamma^{-1} \leq E|X_a|\gamma^{-1} < \delta,$$

where the second inequality is because $|X|$ is a wide-sense submartingale by Assertion 4. Hence

$$E|X_t|1_{(X(t)>\gamma)} \leq E|X_a|1_{(X(t)>\gamma)} < \varepsilon,$$

where the first inequality is because $|X|$ is a wide-sense submartingale. Since $t \in Q$ is arbitrary with $t \leq a$, we conclude that the family $\{X_t : t \in (-\infty, a]Q\}$ is uniformly integrable, with a simple modulus of uniform integrability β, in the sense of Definition 4.7.2. Assertion 5 has been verified.

4. To prove Assertion 6, assume that the process X is a wide-sense submartingale relative to some filtration $\overline{\mathcal{L}}$ such that $L^{(t)} \subset \overline{L}^{(t)}$ for each $t \in Q$. Let $t,s \in Q$ with $t \leq s$ be arbitrary. Consider each indicator $Z \in L^{(t)}$. Then $Z \in \overline{L}^{(t)}$. Hence $EZX_t \leq EZX_s$, where the indicator $Z \in L^{(t)}$ is arbitrary. Thus X is a wide-sense submartingale relative to $\mathcal{L}$. The proof for martingales is similar. Assertion 6 is verified.

5. To prove Assertion 7, suppose X is a wide-sense submartingale relative to $\mathcal{L}$. Consider each $t \in Q$. We have $X_r \in L^{(t)}$ for each $r \in [0,t]Q$. Hence $L^{(X,t)} \equiv L(X_r : r \in [0,t]Q) \subset L^{(t)}$. Therefore Assertion 6 implies that X is a wide-sense

submartingale relative to $\mathcal{L}_X$, and similarly for martingales. Assertion 7 and the proposition are proved. □

Definition 8.2.3. Specialization to uniformly spaced parameters. Recall that, unless otherwise specified, we assume in the remainder of this section that $Q \equiv \{0, \Delta, 2\Delta, \ldots\}$ with some fixed $\Delta > 0$, and that $\mathcal{L} \equiv \{L^{(t)} : t \in Q\}$ denotes an arbitrary but fixed filtration in (Ω, L, E) with parameter Q. For ease of notations, we will further assume, by a change of units if necessary, that $\Delta = 1$. □

Theorem 8.2.4. Doob decomposition. *Let*

$$Y : \{0, 1, 2, \ldots\} \times (\Omega, L, E) \to R$$

be a process that is adapted to the filtration $\mathcal{L} \equiv \{L^{(n)} : n \geq 0\}$. *Suppose the conditional expectation* $E(Y_m|L^{(n)})$ *exists for each* $m, n \geq 0$ *with* $n \leq m$. *For each* $n \geq 0$, *define*

$$X_n \equiv Y_0 + \sum_{k=1}^{n}(Y_k - E(Y_k|L^{(k-1)})) \qquad (8.2.1)$$

and

$$A_n \equiv \sum_{k=1}^{n}(E(Y_k|L^{(k-1)}) - Y_{k-1}), \qquad (8.2.2)$$

where an empty sum is by convention equal to 0. *Then the process*

$$X : \{0, 1, 2, \ldots\} \times \Omega \to R$$

is a martingale relative to the filtration $\mathcal{L}$. *Moreover,* $A_n \in L^{(n-1)}$ *and* $Y_n = X_n + A_n$ *for each* $n \geq 1$.

Proof. From the defining equality 8.2.1, we see that $X_n \in L^{(n)}$ for each $n \geq 1$. Hence the process $X : \{0, 1, 2, \ldots\} \times \Omega \to R$ is adapted to the filtration $\mathcal{L}$. Let $m > n \geq 1$ be arbitrary. Then

$$\begin{aligned} E(X_m|L^{(n)}) &= E\left(\left\{X_n + \sum_{k=n+1}^{m}(Y_k - E(Y_k|L^{(k-1)}))\right\}|L^{(n)}\right) \\ &= X_n + \sum_{k=n+1}^{m}\{E(Y_k|L^{(n)}) - E(E(Y_k|L^{(k-1)})|L^{(n)})\} \\ &= X_n + \sum_{k=n+1}^{m}\{E(Y_k|L^{(n)}) - E(Y_k|L^{(n)})\} = X_n, \end{aligned}$$

where we used basic properties of conditional expectations in Proposition 5.6.6. Thus the process X is a martingale relative to the filtration $\mathcal{L}$. Moreover, $A_n \in L^{(n-1)}$ because all the summands in the defining equality 8.2.2 are members of $L^{(n-1)}$. □

Intuitively, Theorem 8.2.4 says that a multi-round game Y can be turned into a fair game X by charging a fair price determined at each round as the conditional expectation of payoff at the next round, with the cumulative cost of entry equal to A_n by the time n.

The next theorem of Doob and its corollary are key to the analysis of martingales. It proves that under reasonable conditions, a fair game can never be turned into a favorable one by sampling at a sequence of stopping times, or by stopping at some stopping time, with an honest scheme, short of peeking into the future. The reader can look up "gambler's ruin" in the literature for a counterexample where these reasonable conditions are not assumed, where a fair coin tossing game can be turned into an almost sure win by stopping when and only when the gambler is ahead by one dollar. This latter strategy sounds intriguing except for the lamentable fact that to achieve almost sure winning against a house with infinite capital, the strategy would require the gambler both to stay in the game for an unbounded number of rounds and to have infinite capital to avoid bankruptcy first.

The next theorem and its proof are essentially restatements of parts of theorems 9.3.3 and 9.3.4 in [Chung 1968], except that for the case of wide-sense submartingales, we add a condition to make the theorem constructive.

Theorem 8.2.5. Doob's optional sampling theorem. *Let*

$$X : \{0,1,2,\ldots\} \times (\Omega, L, E) \to R$$

be a wide-sense submartingale relative to a filtration $\mathcal{L} \equiv \{L^{(k)} : k \geq 0\}$. Let $\overline{\tau} \equiv (\tau_n)_{n=1,2,\ldots}$ be a nondecreasing sequence of stopping times with values in $\{0,1,2,\ldots\}$ relative to the filtration $\mathcal{L}$. Define the function $X_{\overline{\tau}} : \{0,1,2,\ldots\} \times (\Omega, L, E) \to R$ by $X_{\overline{\tau},n} \equiv X_{\tau(n)}$ for each $n \geq 0$. Suppose one of the following three conditions holds:

(i) The function $X_{\tau(n)}$ is an integrable r.r.v. for each $n \geq 0$, and the family $\{X_n : n \geq 0\}$ of r.r.v.'s is uniformly integrable.

(ii) For each $m \geq 1$, there exists some $M_m \geq 0$ such that $\tau_m \leq M_m$.

(iii) The given process X is a martingale, and the family $\{X_n : n \geq 0\}$ of r.r.v.'s is uniformly integrable.

Then $X_{\overline{\tau}}$ is a wide-sense submartingale relative to the filtration $\mathcal{L}^{\overline{\tau}} \equiv \{L^{(\tau(n))} : n \geq 0\}$. If, in addition, the given process X is a martingale, then $X_{\overline{\tau}}$ is a martingale relative to the filtration $\mathcal{L}^{\overline{\tau}}$.

Proof. Recall that $Q \equiv \{0,1,2,\ldots\}$. Let $m \geq n$ and the indicator $Z \in L^{(\tau(n))}$ be arbitrary. We need to prove that the function $X_{\tau(n)}$ is integrable and that

$$E(X_{\tau(m)}Z) \geq E(X_{\tau(n)}Z). \tag{8.2.3}$$

First we will prove that $X_{\tau(n)}$ is integrable.

1. Suppose Condition (i) holds. Then the function $X_{\tau(n)}$ is integrable by assumption.

2. Suppose Condition (ii) holds. Then the function

$$X_{\tau(n)} = \sum_{u=0}^{M(n)} X_{\tau(n)} 1_{(\tau(n)=u)} = \sum_{u=0}^{M(n)} X_u 1_{(\tau(n)=u)}$$

is a finite sum of integrable r.r.v.'s. Hence $X_{\tau(n)}$ is itself an integrable r.r.v.

3. Suppose Condition (iii) holds. Then X is a martingale. Hence $|X|$ is a wide-sense submartingale and $E|X_t|$ is nondecreasing in $t \in Q$, according to Assertion 4 of Proposition 8.2.2. Consider each $v, v' \in Q$ with $v \leq v'$. Then it follows that

$$\sum_{u=v}^{v'} E|X_{\tau(n)}| 1_{(\tau(n)=u)} = \sum_{u=v}^{v'} E|X_u| 1_{(\tau(n)=u)} \leq \sum_{u=v}^{v'} E|X_{v'}| 1_{(\tau(n)=u)}$$

$$= E|X_{v'}| 1_{(v \leq \tau(n) \leq v')} \leq \alpha_{v,v'} \equiv E|X_{v'}| 1_{(v \leq \tau(n))}. \quad (8.2.4)$$

Let $v \to \infty$. Since τ_n is a nonnegative r.r.v., we have $P(v \leq \tau_n) \to 0$. Therefore $\alpha_{v,v'} \to 0$, thanks to the uniform integrability of the family $\{X_t : t \in Q\}$ of r.r.v.'s under Condition (iii). Summing up, we conclude that $\sum_{u=v}^{v'} E|X_{\tau(n)}| 1_{(\tau(n)=u)} \to 0$ as $v \to \infty$. Thus $\sum_{u=0}^{\infty} E|X_{\tau(n)}| 1_{(\tau(n)=u)} < \infty$. Consequently, the Monotone Convergence Theorem implies that the function $X_{\tau(n)} = \sum_{u=0}^{\infty} X_{\tau(n)} 1_{(\tau(n)=u)}$ is an integrable r.r.v.

4. Summing up, we see that $X_{\tau(n)}$ is an integrable r.r.v. under each one of the three Conditions (i–iii). It remains to prove the relation 8.2.3. To that end, let $u, v \in Q$ be arbitrary with $u \leq v$. Then $Z 1_{\tau(n)=u} \in L^{(u)} \subset L^{(v)}$. Hence

$$Y_{u,v} \equiv X_v Z 1_{\tau(n)=u} \in L^{(v)}.$$

Moreover,

$$\begin{aligned} EY_{u,v} 1_{\tau(m) \geq v} &\equiv EX_v Z 1_{\tau(n)=u} 1_{\tau(m) \geq v} \\ &= EX_v Z 1_{\tau(n)=u} 1_{\tau(m)=v} + EX_v Z 1_{\tau(n)=u} 1_{\tau(m) \geq v+1} \\ &= EX_{\tau(m)} Z 1_{\tau(n)=u} 1_{\tau(m)=v} + EX_v Z 1_{\tau(n)=u} 1_{\tau(m) \geq v+1} \\ &\leq EX_{\tau(m)} Z 1_{\tau(n)=u} 1_{\tau(m)=v} + EX_{v+1} Z 1_{\tau(n)=u} 1_{\tau(m) \geq v+1}, \end{aligned}$$

where the inequality is because the indicator

$$Z 1_{\tau(n)=u} 1_{\tau(m) \geq v+1} = Z 1_{\tau(n)=u} (1 - 1_{\tau(m) \leq v})$$

is a member of $L^{(v)}$, and because X is, by hypothesis, a wide-sense submartingale. In short,

$$EY_{u,v} 1_{\tau(m) \geq v} \leq EX_{\tau(m)} Z 1_{\tau(n)=u} 1_{\tau(m)=v} + EY_{u,v+1} 1_{\tau(m) \geq v+1}, \quad (8.2.5)$$

where $v \in [u, \infty)Q$ is arbitrary. Let $\kappa \geq 0$ be arbitrary. Applying inequality 8.2.5 successively to $v = u, u+1, u+2, \ldots, u+\kappa$, we obtain

$$
\begin{aligned}
&EY_{u,u}1_{\tau(m)\geq u}\\
&\quad\leq EX_{\tau(m)}Z1_{\tau(n)=u}1_{\tau(m)=u}+EY_{u,u+1}1_{\tau(m)\geq u+1}\\
&\quad\leq EX_{\tau(m)}Z1_{\tau(n)=u}1_{\tau(m)=u}+EX_{\tau(m)}Z1_{\tau(n)=u}1_{\tau(m)=u+1}\\
&\qquad+EY_{u,u+2}1_{\tau(m)\geq u+2}\leq\cdots\\
&\quad\leq EX_{\tau(m)}Z1_{\tau(n)=u}\sum_{v\in[u,u+\kappa]Q}1_{\tau(m)=v}+EY_{u,u+(\kappa+1)}1_{\tau(m)\geq u+(\kappa+1)}\\
&\quad=EX_{\tau(m)}Z1_{\tau(n)=u}1_{u\leq\tau(m)\leq u+\kappa}+EX_{u+(\kappa+1)}Z1_{\tau(n)=u}1_{\tau(m)\geq u+(\kappa+1)}\\
&\quad=EX_{\tau(m)}Z1_{\tau(n)=u}1_{\tau(m)\leq u+\kappa}+EX_{u+(\kappa+1)}Z1_{\tau(n)=u}1_{\tau(m)\geq u+(\kappa+1)}.\\
&\quad=EZX_{\tau(m)}Z1_{\tau(n)=u}-EX_{\tau(m)}Z1_{\tau(n)=u}1_{\tau(m)\geq u+(\kappa+1)}\\
&\qquad+EX_{u+(\kappa+1)}Z1_{\tau(n)=u}1_{\tau(m)\geq u+(\kappa+1)}.\\
&\quad\equiv EX_{\tau(m)}Z1_{\tau(n)=u}-EX_{\tau(m)}1_{A(\kappa)}+EX_{u+(\kappa+1)}1_{A(\kappa)},
\end{aligned}
\tag{8.2.6}
$$

where the second equality is because $\tau_n \leq \tau_m$, and where A_κ is the measurable set whose indicator is $1_{A(\kappa)} \equiv Z1_{\tau(n)=u}1_{\tau(m)\geq u+(\kappa+1)}$ and whose probability is therefore bounded by

$$P(A_\kappa) \leq P(\tau_m \geq u + (\kappa + 1)).$$

Now let $\kappa \to \infty$. Then $P(A_\kappa) \to 0$ because $X_{\tau(m)}$ is an integrable r.r.v., as proved in Steps 1–3. Consequently, the second summand on the right-hand side of inequality 8.2.6 tends to 0. Now consider the third summand on the right-hand side of inequality 8.2.6. Suppose Condition (ii) holds. Then, as soon as κ is so large that $u + (\kappa + 1) \geq M_m$, we have $P(A_\kappa) = 0$, whence said two summands vanish as $\kappa \to \infty$. Suppose, alternatively, that Condition (i) or (iii) holds. Then the last summand tends to 0, thanks to the uniform integrability of the family $\{X_t : t \in [0,\infty)\}$ of r.r.v.'s guaranteed by Condition (i) or (iii). Summing up, the second and third summands both tend to 0 as $\kappa \to \infty$, with only the first summand on the right-hand side of inequality 8.2.6 surviving, to yield

$$EY_{u,u}1_{\tau(m)\geq u} \leq EX_{\tau(m)}Z1_{\tau(n)=u}.$$

Equivalently,

$$EX_u Z1_{\tau(n)=u}1_{\tau(m)\geq u} \leq EX_{\tau(m)}Z1_{\tau(n)=u}.$$

Since $(\tau_n = u) \subset (\tau_m \geq u)$, this last inequality simplifies to

$$EX_{\tau(n)}Z1_{\tau(n)=u} \leq EX_{\tau(m)}Z1_{\tau(n)=u},$$

where $u \in Q \equiv \{0,1,2,\ldots\}$ is arbitrary. Summation over $u \in Q$ then yields the desired equality 8.2.3. Thus $X_{\overline{\tau}}$ is a wide-sense submartingale relative to the filtration $\mathcal{L}^{\overline{\tau}} \equiv \{L^{(\tau(n))} : n = 0,1,\ldots\}$. The first part of the conclusion of the theorem, regarding wide-sense submartingales, has been proved.

5. Finally, suppose the given wide-sense submartingale X is actually a martingale. Then $-X$ is a wide-sense submartingale, and so by the preceding arguments, both processes $X_{\overline{\tau}}$ and $-X_{\overline{\tau}}$ are a wide-sense submartingale relative to the filtration $\mathcal{L}^{\overline{\tau}}$. Combining, we conclude that $X_{\overline{\tau}}$ is a martingale if X is a martingale, provided that one of the three Conditions (i–iii) holds. The theorem is proved. □

Corollary 8.2.6. Doob's optional stopping theorem for a finite game. *Let $n \geq 1$ be arbitrary. Write $Q' \equiv \{0, 1, \ldots, n\} \equiv \{t_0, t_1, \ldots, t_n\} \subset Q$. Let $X : Q' \times \Omega \to R$ be a process adapted to the filtration $\mathcal{L} \equiv \{L^{(t)} : t \in Q\}$. Let τ be an arbitrary simple stopping time relative to $\mathcal{L}$ with values in Q'. Define the r.r.v.*

$$X_\tau \equiv \sum_{t \in Q'} X_t 1_{(\tau=t)} \in L'^{(\tau)}.$$

Define the process $X' : \{0, 1, 2\} \times \Omega \to R$ by

$$(X'_0, X'_1, X'_2) \equiv (X_{t(0)} X_\tau, X_{t(n)}).$$

Define the filtration $\mathcal{L}' \equiv \{L'^{(i)} : i = 0, 1, 2\}$ by

$$(L'^{(0)}, L'^{(1)}, L'^{(2)}) \equiv (L^{(t(0))}, L^{(\tau)}, L^{(t(n))}).$$

Then the following conditions hold:

1. If the process X is a wide-sense submartingale relative to $\mathcal{L}$, then the process X' is a wide-sense submartingale relative to the filtration $\mathcal{L}'$.

2. If the process X is a martingale relative to $\mathcal{L}$, then the process X' is a martingale relative to $\mathcal{L}'$.

Proof. Extend the process X to the parameter set $Q \equiv \{0, 1, \ldots\}$ by $X_t \equiv X_{t \wedge n}$ for each $t \in \{0, 1, \ldots\}$. Likewise, extend the filtration $\mathcal{L}$ by defining $L^{(t)} \equiv L^{(t \wedge n)}$ for each $t \in \{0, 1, \ldots\}$. We can verify that the extended process $X : \{0, 1, \ldots\} \times \Omega \to R$ retains the same property of being a martingale or wide-sense submartingale, respectively, as the given process being a martingale or wide-sense submartingale, relative to the extended filtration $\mathcal{L}$. Now define a sequence $\overline{\tau} \equiv (\tau_0, \tau_1, \ldots)$ of stopping times by $\tau_0 \equiv t_0$, $\tau_1 \equiv \tau$, and $\tau_m \equiv t_n$ for each $m \geq 2$. Then it can easily be verified that the sequence $\overline{\tau}$ satisfies Condition (ii) of Theorem 8.2.5. Hence the process $X_{\overline{\tau}}$ defined in Theorem 8.2.5 is a martingale if X is a martingale, and it is a wide-sense submartingale if X is a wide-sense submartingale. Since

$$(X'_0, X'_1, X'_2) \equiv (X_{t(0)} X_\tau, X_{t(n)}) = (X_{\tau(0)} X_{\tau(1)}, X_{\tau(2)})$$

and

$$(L'^{(0)}, L'^{(1)}, L'^{(2)}) \equiv (L^{(t(0))}, L^{(\tau)}, L^{(t(n))}) = (L^{(\tau(0))}, L^{(\tau(1))}, L^{(\tau(2))}),$$

the conclusion of the corollary follows. □

8.3 Convexity and Martingale Convergence

In this section, we consider the a.u. convergence of martingales. Suppose $X : \{1,2,\ldots\} \times \Omega \to R$ is a martingale relative to some filtration $\mathcal{L} \equiv \{L^{(n)} : n = 1, 2,\ldots\}$. A classical theorem says that if $E|X_n|$ is bounded as $n \to \infty$, then X_n converges a.u. as $n \to \infty$. The theorem can be proved, classically, by the celebrated upcrossing inequality of Doob, thanks to the principle of infinite search. See, for example, [Durret 1984]. While the upcrossing inequality is constructive, the inference of a.u. convergence from it is not.

The following example shows that the martingale convergence theorem, as stated earlier, actually implies the principle of infinite search. Let $(a_n)_{n=1,2,\ldots}$ be an arbitrary nondecreasing sequence in $\{0,1\}$. Let Y be an arbitrary r.r.v. that takes the value -1 or $+1$ with equal probabilities. For each $n \geq 1$, define $X_n \equiv 1 + a_n Y$. Then the process $X : \{1,2,\ldots\} \times \Omega \to \{0,1,2\}$ is a martingale relative to its natural filtration, with $E|X_n| = EX_n = 1$ for each $n \geq 1$. Suppose $X_n \to U$ a.u. for some r.r.v. U. Then there exists $b \in (0,1)$ such that the set $(U < b)$ is measurable. Either (i) $P(U < b) < \frac{1}{2}$ or (ii) $P(U < b) > 0$. In case (i), we must have $a_n = 0$ for each $n \geq 1$. In case (ii), because of a.u. convergence, there exists $b' \in (b,1)$ such that $P(X_n < b') > 0$ for some $n \geq 1$, whence $a_n = 1$ for some $n \geq 1$. Since the nondecreasing sequence $(a_n)_{n=1,2,\ldots}$ is arbitrary, we have deduced the principle of infinite search from the classical theorem of martingale convergence.

Thus the boundedness X_n of together with the constancy of $E|X_n|$ is not sufficient for the constructive a.u. convergence. Boundedness is not the issue; convexity is. The function $|x|$ simply does not have any positive convexity away from $x = 0$.

Bishop's maximal inequality for martingales, given as theorem 3 in chapter 8 of [Bishop 1967], uses an admissible symmetric convex function $\lambda(x)$ in the place of the function $|x|$. It proves a.u. convergence of martingales under a condition in terms of the convergence of $E\lambda(X_n)$, thus obviating the use of upcrossing inequalities.

We will modify Bishop's theorem by using strictly convex functions $\lambda(x)$ that are not necessarily symmetric, but which have positive and continuous second derivatives, as a natural alternative to the function $|x|$. This allows the use of a special strictly convex function $\overline{\lambda}$ such that $|\overline{\lambda}(x)| \leq 3|x|$ for each $x \in R$. Then the boundedness and convergence of $E\overline{\lambda}(X_n)$ follow, classically, from the boundedness of $E|X_n|$. Thus we will have a criterion for constructive a.u. convergence that, from the classical view point, imposes no additional condition beyond the boundedness of $E|X_n|$. The proof, being constructive, produces rates of a.u. convergence.

Definition 8.3.1. Strictly convex function. A continuous function $\lambda : R \to R$ is said to be *strictly convex* if it has a positive continuous second derivative λ'' on R. □

This definition generalizes the admissible functions in chapter 8 of [Bishop 1967]. The conditions of symmetry and nonnegativity of λ are dropped here. Correspondingly, we need to generalize Bishop's version of Jensen's inequality, given as lemma 2 in chapter 8 of [Bishop 1967], to the following version.

Theorem 8.3.2. Bishop–Jensen inequality. *Let $\lambda : R \to R$ be a strictly convex function. Define the continuous function θ on $[0,\infty)$ by $\theta(x) \equiv \inf_{y\in[-x,x]} \lambda''(y) > 0$ for each $x > 0$. Define the continuous function $g : R^2 \to [0,\infty)$ by*

$$g(x_0,x_1) \equiv \frac{1}{2}(x_1 - x_0)^2\theta(|x_0| \vee |x_1|) \tag{8.3.1}$$

for each $(x_0,x_1) \in R^2$.

Let X_0 and X_1 be integrable r.r.v.'s on (Ω,L,E) such that $\lambda(X_0),\lambda(X_1)$ are integrable. Suppose either (i) $E(X_1|X_0) = X_0$ *or* (ii) *the strictly convex function $\lambda : R \to R$ is nondecreasing and $EUX_0 \le EUX_1$ for each indicator $U \in L(X_0)$.*

Then the r.r.v. $g(X_0,X_1)$ is integrable and

$$0 \le Eg(X_0,X_1) \le E\lambda(X_1) - E\lambda(X_0). \tag{8.3.2}$$

Proof. 1. Let $(x_0,x_1) \in R^2$ be arbitrary. Then

$$\begin{aligned} 0 \le g(x_0,x_1) &= \theta(|x_0| \vee |x_1|) \int_{v=x(0)}^{x(1)} \int_{u=x(0)}^{v} du dv \\ &\le \int_{v=x(0)}^{x(1)} \left(\int_{u=x(0)}^{v} \lambda''(u)du \right) dv = \int_{v=x(0)}^{x(1)} (\lambda'(v) - \lambda'(x_0))dv \\ &= \lambda(x_1) - \lambda(x_0) - \lambda'(x_0)(x_1 - x_0), \end{aligned} \tag{8.3.3}$$

where the last equality is by the Fundamental Theorem of Calculus.

2. Let $(\Omega,L(X_0),E)$ denote the probability subspace of (Ω,L,E) generated by the r.r.v. X_0. Let $V \in L(X_0)$ be an arbitrary indicator such that $\lambda'(X_0)V$ is bounded. Suppose Condition (i) in the hypothesis holds: $E(X_1|X_0) = X_0$. Then, by the properties of conditional expectations and the assumed boundedness of the r.r.v. $\lambda'(X_0)V \in L(X_0)$, we obtain

$$E(X_1 - X_0)\lambda'(X_0)V = E(X_0 - X_0)\lambda'(X_0)V = 0.$$

Suppose Condition (ii) holds. Then $EUX_0 \le EUX_1$ for each indicator $U \in L(X_0)$, and the function $\lambda : R \to R$ is nondecreasing. Hence $\lambda' \ge 0$ and the bounded r.r.v. $\lambda'(X_0)V \in L(X_0)$ is nonnegative. Therefore, by Assertion 1 of Proposition 5.6.7, where X,Y,Z,L' are replaced by $X_0,X_1,\lambda'(X_0)V,L(X_0)$, respectively, we have

$$E\lambda'(X_0)VX_0 \le E\lambda'(X_0)VX_1.$$

Summing up, under either Condition (i) or (ii), we have

$$E(X_1 - X_0)\lambda'(X_0)V \ge 0, \tag{8.3.4}$$

for each indicator $V \in L(X_0)$ such that $\lambda'(X_0)V$ is bounded.

3. Now the r.r.v.'s $\lambda(X_0),\lambda(X_1),X_0,X_1$ are integrable by hypothesis. Let $b > a > 0$ be arbitrary. Since the function λ' is continuous, it is bounded on $[-b,b]$.

Hence the r.r.v. $\lambda'(X_0)1_{(a\geq|X(0|)}$ is bounded. Consequently, inequality 8.3.3 implies that the r.r.v. $g(X_0,X_1)1_{(a\geq|X(0|)}$ is bounded by the integrable r.r.v.

$$\lambda(X_1)1_{(a\geq|X(0|)} - \lambda(X_0)1_{(a\geq|X(0|)} - \lambda'(X_0)(X_1 - X_0)1_{(a\geq|X(0|)}$$

and is therefore itself integrable.

4. At the same time, $\lambda'(X_0)1_{(b\geq|X(0)|>a)}$ is bounded. Therefore inequality 8.3.4 holds with $V \equiv 1_{(b\geq|X(0)|>a)}$. Combining,

$$\begin{aligned}
0 &\leq Eg(X_0,X_1)1_{(b\geq|X(0|)} - Eg(X_0,X_1)1_{(a\geq|X(0|)} \\
&= Eg(X_0,X_1)V \leq E(\lambda(X_1) - \lambda(X_0) - \lambda'(X_0)(X_1 - X_0))V \\
&\leq E(\lambda(X_1) - \lambda(X_0))V \equiv E(\lambda(X_1) - \lambda(X_0))1_{(b\geq|X(0)|>a)} \to 0
\end{aligned}$$

as $b > a \to \infty$, where the second inequality is due to inequality 8.3.3, and where the third inequality is thanks to inequality 8.3.4. Hence the integral $Eg(X_0,X_1)1_{(a\geq|X(0|)}$ converges as $a \to \infty$. It follows from the Monotone Convergence Theorem that the r.r.v.

$$g(X_0,X_1) = \lim_{a\to\infty} g(X_0,X_1)1_{(a\geq|X(0|)}$$

is integrable, with

$$\begin{aligned}
Eg(X_0,X_1) &= \lim_{a\to\infty} Eg(X_0,X_1)1_{(a\geq|X(0|)} \\
&\leq \lim_{a\to\infty} E(\lambda(X_1) - \lambda(X_0) - \lambda'(X_0)(X_1 - X_0))1_{(a\geq|X(0|)} \\
&\leq \lim_{a\to\infty} E(\lambda(X_1) - \lambda(X_0))1_{(a\geq|X(0|)} = E\lambda(X_1) - E\lambda(X_0),
\end{aligned}$$

where the first inequality follows from inequality 8.3.3, and where the second inequality follows from inequality 8.3.4 in which V is replaced by $1_{(a\geq|X(0)|)}$. The desired inequality 8.3.2 is proved. □

Now we are ready to formulate and prove the advertised maximal inequality.

Definition 8.3.3. The special convex function. Define the continuous function $\overline{\lambda} : R \to R$ by

$$\overline{\lambda}(x) \equiv 2x + (e^{-|x|} - 1 + |x|) \tag{8.3.5}$$

for each $x \in R$. We will call $\overline{\lambda}$ the *special convex function.* □

Theorem 8.3.4. A maximal inequality for martingales.

1. The special convex function $\overline{\lambda}$ is increasing and strictly convex, with

$$|x| \leq |\overline{\lambda}(x)| \leq 3|x| \tag{8.3.6}$$

for each $x \in R$.

2. Let $Q \equiv \{t_0, t_1, \ldots, t_n\}$ be an arbitrary enumerated finite subset of R, with $t_0 < t_1 < \cdots < t_n$. Let $X : Q \times \Omega \to R$ be an arbitrary martingale relative to the filtration $\mathcal{L} \equiv \{L^{(t(i))} : i = 1, \ldots, n\}$. Let $\varepsilon > 0$ be arbitrary. Suppose

$$E\overline{\lambda}(X_{t(n)}) - E\overline{\lambda}(X_{t(0)}) < \frac{1}{6}\varepsilon^3 \exp(-3(E|X_{t(0)}| \vee E|X_{t(n)}|)\varepsilon^{-1}). \tag{8.3.7}$$

Then

$$P\left(\bigvee_{k=0}^{n} |X_{t(k)} - X_{t(0)}| > \varepsilon\right) < \varepsilon. \tag{8.3.8}$$

We emphasize that the last two displayed inequalities are regardless of how large $n \geq 0$ is. We also note that in view of inequality 8.3.6, the r.r.v. $\overline{\lambda}(Y)$ is integrable for each integrable r.r.v. Y. Thus inequality 8.3.7 makes sense when $|X_{t(n)}|$ is integrable, in contrast to the classical counterpart, which requires either $X_{t(n)}^p$ is integrable for some $p > 1$ or $|X_{t(n)}| \log |X_{t(n)}|$ is integrable.

Proof. 1. First note that $\overline{\lambda}(0) = 0$. Elementary calculus yields a continuous first derivative $\overline{\lambda}'$ on R such that

$$\overline{\lambda}'(x) = 2 + (-e^{-x} + 1) \geq 2 \tag{8.3.9}$$

for each $x \geq 0$, and such that

$$\overline{\lambda}'(x) = 2 + (e^{x} - 1) \geq 1 \tag{8.3.10}$$

for each $x \leq 0$. Therefore the function $\overline{\lambda}$ is increasing. Moreover, $\overline{\lambda}$ has a positive and continuous second derivative

$$\overline{\lambda}''(x) = e^{-|x|} > 0$$

for each $x \in R$. Thus the function $\overline{\lambda}$ is strictly convex, and

$$\theta(x) \equiv \inf_{y \in [-x,x]} \overline{\lambda}''(y) = e^{-x} > 0$$

for each $x > 0$. Furthermore, since $0 \leq e^{-r} - 1 + r \leq r$ for each $r \geq 0$, the triangle inequality yields, for each $x \in R$,

$$\begin{aligned} |x| &= 2|x| - |x| \leq |2x| - |(e^{-|x|} - 1 + |x|)| \\ &\leq |2x + (e^{-|x|} - 1 + |x|)| \equiv |\overline{\lambda}(x)| \\ &\leq |2x| + |(e^{-|x|} - 1 + |x|)| \leq 2|x| + |x| = 3|x|. \end{aligned}$$

This establishes the desired inequality 8.3.6 in Assertion 1.

2. As in Theorem 8.3.2, define the continuous function $g : R^2 \to [0, \infty)$ by

$$g(x_0, x_1) \equiv \frac{1}{2}(x_1 - x_0)^2 \theta(|x_0| \vee |x_1|) = \frac{1}{2}(x_1 - x_0)^2 \exp(-(|x_0| \vee |x_1|)) \tag{8.3.11}$$

for each $(x_0, x_1) \in R^2$.

3. By relabeling if necessary, we assume, without loss of generality, that

$$Q \equiv \{t_0, t_1, \ldots, t_n\} = \{0, 1, \ldots, n\}$$

as enumerated sets. Let $\varepsilon > 0$ be arbitrary. For abbreviation, write $K \equiv E|X_0| \vee E|X_n|$, $b \equiv 3K\varepsilon^{-1}$, and

$$\gamma \equiv \frac{1}{2}\varepsilon^2\theta(b) = \frac{1}{2}\varepsilon^2 e^{-b} \equiv \frac{1}{2}\varepsilon^2 \exp(-3K\varepsilon^{-1}).$$

Then inequality 8.3.7 in the hypothesis can be rewritten as

$$E\overline{\lambda}(X_n) - E\overline{\lambda}(X_0) < \frac{1}{3}\varepsilon\gamma. \tag{8.3.12}$$

4. Let $\tau \equiv \eta_{0,\varepsilon,Q}$ be the simple first exit time of the process X to exit, after $t = 0$, from the ε-neighborhood of X_0, in the sense of Definition 8.1.12. Define the probability subspace $L^{(\tau)}$ relative to the simple stopping time τ, as in Definition 8.1.9. Define the r.r.v.

$$X_\tau \equiv \sum_{t\in Q} X_t 1_{(\tau=t)} \in L^{(\tau)}.$$

As in Corollary 8.2.6, define the process $X' : \{0,1,2\} \times \Omega \to R$ by

$$(X'_0, X'_1, X'_2) \equiv (X_0, X_\tau, X_n)$$

and define the filtration $\mathcal{L}' \equiv \{L'^{(i)} : i = 0,1,2\}$ by

$$(L'^{(0)}, L'^{(1)}, L'^{(2)}) \equiv (L^{(0)}, L^{(\tau)}, L^{(n)}).$$

Then, by Assertion 1 of Corollary 8.2.6, the process X' is a martingale relative to the filtration $\mathcal{L}'$. In other words, $EUX'_{i-1} \leq EUX'_i$ for each indicator $U \in L(X'_{i-1}) \subset L'^{(i-1)}$, for each $i = 1,2$.

5. Thus Condition (ii) in Theorem 8.3.2 is satisfied by the pair X'_{i-1}, X'_i of r.r.v.'s and by the strictly convex function $\overline{\lambda}$, for each $i = 1,2$. Accordingly, for each $i = 1,2$, the nonnegative r.r.v. $Y_i \equiv g(X'_{i-1}, X'_i)$ is integrable, with

$$0 \leq EY_i \leq E\overline{\lambda}(X'_i) - E\overline{\lambda}(X'_{i-1}). \tag{8.3.13}$$

Consequently,

$$0 \leq EY_1 \leq E\overline{\lambda}(X'_1) - E\overline{\lambda}(X'_0) \leq E\overline{\lambda}(X'_2) - E\overline{\lambda}(X'_0)$$
$$\equiv E\overline{\lambda}(X_n) - E\overline{\lambda}(X_0) < \frac{1}{3}\varepsilon\gamma,$$

where the last inequality is inequality 8.3.12. Chebychev's inequality therefore yields a measurable set A with $P(A^c) < \frac{1}{3}\varepsilon$ such that

$$A \subset (Y_1 \leq \gamma).$$

6. Next, note that

$$E|X'_0| \vee E|X'_1| \leq E|X'_0| \vee E|X'_2| \equiv E|X_0| \vee E|X_n| \equiv K \equiv \frac{1}{3}\varepsilon b,$$

where the inequality is because X' is a martingale. Chebychev's inequality therefore yields, for each $i = 0,1$, a measurable set B_i with $P(B_i^c) < \frac{1}{3}\varepsilon$ such that $B_i \subset (|X'_i| \leq b)$.

7. Now consider each $\omega \in AB_0B_1$. Then $|X'_0(\omega)| \vee |X'_1(\omega)| \leq b$. Hence

$$\begin{aligned}\frac{1}{2}(X_\tau(\omega) - X_0(\omega))^2\theta(b) &\equiv \frac{1}{2}(X'_1(\omega) - X'_0(\omega))^2\theta(b)\\ &\leq \frac{1}{2}(X'_1(\omega) - X'_0(\omega))^2\theta(|X'_0(\omega)| \vee |X'_1(\omega)|)\\ &\equiv g(X'_0(\omega), X'_1(\omega))\\ &\equiv Y_1(\omega) \leq \gamma \equiv \frac{1}{2}\varepsilon^2\theta(b),\end{aligned}$$

where the first inequality is because θ is a decreasing function, and where the last inequality is because $\omega \in A \subset (Y_1 \leq \gamma)$. Dividing by $\frac{1}{2}\theta(b)$ and taking square roots, we obtain

$$|X_{\eta(0,\varepsilon,Q)}(\omega) - X_0(\omega)| \equiv |X_\tau(\omega) - X_0(\omega)| \leq \varepsilon$$

for each $\omega \in AB_0B_1$. It follows from Assertion 4 of Proposition 8.1.13, which describes the basic properties of the simple first exit time $\eta_{0,\varepsilon,Q}$, that on AB_0B_1, we have

$$\bigvee_{t\in Q} |X_t - X_0| \leq \varepsilon.$$

Summing up,

$$\begin{aligned}P\left(\bigvee_{k=0}^{n} |X_{t(k)} - X_{t(0)}| > \varepsilon\right) &= P\left(\bigvee_{t\in Q} |X_t - X_0| > \varepsilon\right)\\ &\leq P(AB_1B_2)^c < \frac{1}{3}\varepsilon + \frac{1}{3}\varepsilon + \frac{1}{3}\varepsilon = \varepsilon,\end{aligned}$$

as alleged. □

Theorem 8.3.4 leads to the following a.u. convergence theorem for martingales. We emphasize that while constructive proofs of a.u. convergence for a martingale $X : \{1,2,\ldots\} \times \Omega \to R$ are well known if $\lim_{k\to\infty} E|X_k|^p$ exists for some $p > 1$, the following theorem requires no L_p-integrability for $p > 1$.

Theorem 8.3.5. a.u. Convergence of a martingale, and rate of said a.u. convergence. *Suppose the parameter set Q is either $Q \equiv \{1,2,\ldots\}$ or $Q \equiv \{\cdots, -2, -1\}$. Let $X : Q \times \Omega \to R$ be an arbitrary martingale relative to its natural filtration $\mathcal{L}$. Then the following conditions hold:*

1. **a.u. convergence.** *Suppose $\beta \equiv \lim_{|t|\to\infty} E|X_t|$ exists, and suppose $\alpha \equiv \lim_{|t|\to\infty} E\exp(-|X_t|)$ exists. Then $X_t \to Y$ a.u. as $|t| \to \infty$ in Q, for some r.r.v. Y.*

2. **Rate of a.u. convergence.** *Specifically, for each $h \geq 0$, define*

$$b_h \equiv \sup_{t\in Q;|t|\geq h} |(E|X_t| - \beta)|, \tag{8.3.14}$$

$$a_h \equiv \sup_{t\in Q;|t|\geq h} |(E\exp(-|X_t|) - \alpha)|,$$

and

$$\delta_h \equiv a_h + b_h.$$

Then $a_h, b_h, \delta_h \to 0$ as $h \to \infty$. Define $k_0 \equiv 0$. Inductively, for each $m \geq 1$, take any $k_m \geq k_{m-1}$ so large that

$$\delta_{k(m)} \leq \frac{1}{12} 2^{-3m} \exp(-2^m 3(\beta + b_{k(m-1)})). \tag{8.3.15}$$

Let $\varepsilon > 0$ be arbitrary. Take $m \geq 1$ so large that $2^{-m+2} < \varepsilon$. Then there exists a measurable set A with $P(A^c) < \varepsilon$ where

$$A \subset \bigcap_{p=m}^{\infty} \bigcap_{t\in Q;|t|\geq k(p)} (|X_t - Y| \leq 2^{-p+3}). \tag{8.3.16}$$

Proof. 1. Recall the special convex function $\overline{\lambda} : R \to R$, defined by

$$\overline{\lambda}(x) \equiv 2x + (e^{-|x|} - 1 + |x|) \tag{8.3.17}$$

for each $x \in R$. Define, for convenience, $\iota \equiv 1$ or $\iota \equiv -1$ according as $Q \equiv \{1, 2, \ldots\}$ or $Q \equiv \{\cdots, -2, -1\}$. Define

$$\gamma \equiv 2EX_\iota + \alpha - 1 + \beta.$$

Then, since $X : Q \times \Omega \to R$ is a martingale, the r.r.v.

$$\overline{\lambda}(X_t) \equiv 2X_t + (e^{-|X_t|} - 1 + |X_t|) \tag{8.3.18}$$

is integrable for each $t \in Q$, with $E\overline{\lambda}(X_t) \equiv 2EX_\iota + (Ee^{-|X_t|} - 1 + E|X_t|)$. Hence

$$\lim_{|t|\to\infty} E\overline{\lambda}(X_t) = 2EX_\iota + \alpha - 1 + \beta \equiv \gamma.$$

Moreover, for each $h \geq 0$, we have

$$\sup_{t\in Q;|t|\geq h} |E\overline{\lambda}(X_t) - \gamma| = \sup_{t\in Q;|t|\geq h} |(Ee^{-|X_t|} + E|X_t|) - (\alpha + \beta)|$$

$$\leq a_h + b_h \equiv \delta_h.$$

2. Let $m \geq 1$ be arbitrary. Then it follows that

$$\sup_{t\in Q;|t|\geq k(m)} |E\overline{\lambda}(X_t) - \gamma| \leq \delta_{k(m)} \leq \frac{1}{12} 2^{-3m} \exp(-2^m 3(\beta + b_{k(m-1)})), \tag{8.3.19}$$

where the second inequality is by inequality 8.3.15.

3. To finish the proof, consider first the special case where $Q \equiv \{0,1,2,\ldots\}$. Then inequality 8.3.19 yields

$$|E\overline{\lambda}(X_{k(m+1)}) - E\overline{\lambda}(X_{k(m)})| \leq 2\delta_{k(m)} \leq \frac{1}{6}2^{-3m}\exp(-2^m 3(\beta + b_{k(m-1)})). \tag{8.3.20}$$

4. Separately, from equality 8.3.14, we then have

$$E|X_{k(m)}| \vee E|X_{k(m+1)}| \leq \beta + b_{k(m-1)}. \tag{8.3.21}$$

Take any $\varepsilon_m \in (2^{-m}, 2^{-m+1})$. The last two displayed inequalities combine to yield

$$\begin{aligned} E\overline{\lambda}(X_{k(m+1)}) - E\overline{\lambda}(X_{k(m)}) &\leq \frac{1}{6}2^{-3m}\exp(-2^m 3(\beta + b_{k(m-1)})) \\ &\leq \frac{1}{6}2^{-3m}\exp(-2^m 3(E|X_{k(m)}| \vee E|X_{k(m+1)}|)) \\ &< \frac{1}{6}\varepsilon_m^3 \exp(-3\varepsilon_m^{-1}(E|X_{k(m)}| \vee E|X_{k(m+1)}|)), \end{aligned} \tag{8.3.22}$$

where the last inequality is because $2^{-m} < \varepsilon_m$.

5. In view of inequality 8.3.22, Theorem 8.3.4 is applicable and implies that $P(B_m) < \varepsilon_m$, where

$$B_m \equiv \left(\bigvee_{i=k(m)}^{k(m+1)} |X_i - X_{k(m)}| > \varepsilon_m \right), \tag{8.3.23}$$

where $m \geq 1$ is arbitrary.

6. Now define $A_m \equiv \bigcap_{h=m}^{\infty} B_h^c$. Then

$$P(A_m^c) \leq \sum_{h=m}^{\infty} P(B_h) < \sum_{h=m}^{\infty} \varepsilon_h < \sum_{h=m}^{\infty} 2^{-h+1} = 2^{-m+2}.$$

Consider each $\omega \in A_m$. Let $p \geq m$ be arbitrary. Consider each $j \geq i \geq k_p$. Then $k_h \leq i \leq k_{h+1}$ and $k_n \leq j \leq k_{n+1}$ for some $n \geq h \geq p$. Consequently,

$$\begin{aligned} &|X_i(\omega) - X_j(\omega)| \\ &\leq |X_i(\omega) - X_{k(h)}(\omega)| + |X_{k(h)}(\omega) - X_{k(h+1)}(\omega)| + \cdots + |X_{k(n)}(\omega) - X_j(\omega)| \\ &\leq \varepsilon_h + (\varepsilon_h + \cdots + \varepsilon_n) < 2^{-h+1} + \sum_{\kappa=h}^{\infty} 2^{-\kappa+1} \\ &= 2^{-h+1} + 2^{-h+2} < 2^{-h+3} \leq 2^{-p+3}, \end{aligned} \tag{8.3.24}$$

where the second inequality is because $\omega \in A_m \subset B_h^c B_{h+1}^c \cdots B_n^c$.

Since 2^{-p+3} is arbitrarily small for sufficiently large $p \geq m$, we see that the sequence $(X_i(\omega))_{i=1,2,\ldots}$ of real numbers is Cauchy, and so $Y(\omega) \equiv \lim_{i\to\infty} X_i(\omega)$ exists. Fixing i and letting $j \to \infty$ in inequality 8.3.24, we obtain

$$|X_i(\omega) - Y(\omega)| \leq 2^{-p+3}$$

for each $i \geq k_p$, where $p \geq m$ is arbitrary, for each $\omega \in A_m$. In short,

$$A_m \subset \bigcap_{p=m}^{\infty} \bigcap_{t\in Q;\,|t|\geq k(p)} (|X_t - Y| \leq 2^{-p+3}). \tag{8.3.25}$$

Thus $X_i \to Y$ uniformly on the measurable set A_m, where $P(A_m^c) < 2^{-m+2}$ is arbitrarily small for sufficiently large $m \geq 1$. In other words, $X_i \to Y$ a.u. By Proposition 5.1.9, the function Y is an r.r.v. The theorem has been proved for the case where $Q \equiv \{0,1,2,\ldots\}$.

7. Now consider the case where $Q \equiv \{\cdots, -2, -1, 0\}$. At the risk of boredom, we give a proof that is a mirror image of the preceding paragraphs. To be precise, inequality 8.3.19 yields

$$|E\overline{\lambda}(X_{-k(m+1)}) - E\overline{\lambda}(X_{-k(m)})| \leq 2\delta_{k(m)} \leq \frac{1}{6}2^{-3m}\exp(-2^m 3(\beta + b_{k(m-1)})). \tag{8.3.26}$$

8. Separately, from equality 8.3.14, we then have

$$E|X_{-k(m)}| \vee E|X_{-k(m+1)}| \leq \beta + b_{k(m-1)}. \tag{8.3.27}$$

Take any $\varepsilon_m \in (2^{-m}, 2^{-m+1})$. The last two displayed inequalities combine to yield

$$\begin{aligned} E\overline{\lambda}(X_{-k(m)}) - E\overline{\lambda}(X_{-k(m+1)}) &\leq \frac{1}{6}2^{-3m}\exp(-2^m 3(\beta + b_{k(m-1)})) \\ &\leq \frac{1}{6}2^{-3m}\exp(-2^m 3(E|X_{-k(m+1)}| \vee E|X_{-k(m)}|)) \\ &< \frac{1}{6}\varepsilon_m^3\exp(-3\varepsilon_m^{-1}(E|X_{-k(m+1)}| \vee E|X_{-k(m)}|)). \end{aligned} \tag{8.3.28}$$

9. In view of inequality 8.3.28, Theorem 8.3.4 is applicable and implies that $P(B_m) < \varepsilon_m$, where

$$B_m \equiv \left(\bigvee_{i=-k(m+1)}^{-k(m)} |X_i - X_{-k(m+1)}| > \varepsilon_m \right), \tag{8.3.29}$$

where $m \geq 1$ is arbitrary.

10. Now define $A_m \equiv \bigcap_{h=m}^{\infty} B_h^c$. Then

$$P(A_m^c) \leq \sum_{h=m}^{\infty} P(B_h) < \sum_{h=m}^{\infty} \varepsilon_h < \sum_{h=m}^{\infty} 2^{-h+1} = 2^{-m+2}.$$

Consider each $\omega \in A_m$. Let $p \geq m$ be arbitrary. Consider each $j \leq i \leq -k_p$. Then $-k_{h+1} \leq i \leq -k_h$ and $-k_{n+1} \leq j \leq -k_n$ for some $n \geq h \geq p$. Consequently,

$$\begin{aligned}|X_i(\omega) - X_j(\omega)| &\leq |X_i(\omega) - X_{-k(h+1)}(\omega)| + |X_{-k(h+1)}(\omega) - X_{-k(h)}(\omega)| \\ &\quad + \cdots + |X_{-k(n+1)}(\omega) - X_j(\omega)| \\ &\leq \varepsilon_h + (\varepsilon_h + \cdots + \varepsilon_n) < 2^{-h+1} + \sum_{\kappa=h}^{\infty} 2^{-\kappa+1} \\ &= 2^{-h+1} + 2^{-h+2} < 2^{-h+3} \leq 2^{-p+3}, \end{aligned} \tag{8.3.30}$$

where the second inequality is because $\omega \in A_m \subset B_h^c B_{h+1}^c \cdots B_n^c$. Since 2^{-p+3} is arbitrarily small for sufficiently large $p \geq m$, we see that the sequence $(X_i(\omega))_{i=-1,-2,\ldots}$ of real numbers is Cauchy, and so $Y(\omega) \equiv \lim_{i\to-\infty} X_i(\omega)$ exists. Fixing i and letting $j \to -\infty$ in inequality 8.3.30, we obtain

$$|X_i(\omega) - Y(\omega)| \leq 2^{-p+3}$$

for each $i \leq -k_p$, where $p \geq m$ is arbitrary, for each $\omega \in A_m$. In short,

$$A_m \subset \bigcap_{p=m}^{\infty} \bigcap_{t\in Q;|t|\geq k(p)} (|X_t - Y| \leq 2^{-p+3}). \tag{8.3.31}$$

Thus $X_i \to Y$ uniformly as $i \to -\infty$, on the measurable set A_m, where $P(A_m^c) < 2^{-m+2}$ is arbitrarily small for sufficiently large $m \geq 1$. In other words, $X_i \to Y$ a.u. as $i \to -\infty$. By Proposition 5.1.9, the function Y is an r.r.v. In other words, the theorem has also been proved in the case where $Q \equiv \{\cdots, -2, -1, 0\}$. □

8.4 Strong Law of Large Numbers

Applications of martingales are numerous. One application is to prove the Strong Law of Large Numbers (SLLN). This theorem says that if $Z_1, Z_2, \ldots$ is a sequence of integrable independent and identically distributed r.r.v.'s with mean 0, then $n^{-1}(Z_1+\cdots+Z_n) \to 0$ a.u. Historically, the first proof of this theorem in its generality, due to Kolmogorov, is constructive, complete with rates of convergence. See, for example, theorem 5.4.2 of [Chung 1968]. Subsequently, remarkable proofs are also given in terms of a.u. martingale convergence via Doob's upcrossing inequality. See, for example, theorem 9.4.1 of [Chung 1968]. As observed earlier, the theorem that deduces a.u. convergence from upcrossing inequalities actually implies the principle of infinite search, and cannot be made constructive. In this section, we present a constructive proof by a simple application of our Theorem 8.3.5.

First we discuss the Weak Law of Large Numbers, with a well-known proof by characteristic functions.

Theorem 8.4.1. Weak Law of Large Numbers. *Suppose $Z_1, Z_2, \ldots$ is a sequence of integrable, independent, and identically distributed r.r.v's with* *mean 0, on some probability space (Ω, L, E). Let η be a simple modulus of integrability of Z_1, in the sense of Definition 4.7.2. For each $m \geq 1$, let $S_m \equiv m^{-1}(Z_1 + \cdots + Z_m)$. Then*

$$E|S_m| \to 0$$

as $m \to \infty$. More precisely, there exists a sequence $(q_m)_{m=1,2,\ldots}$ of positive integers, which depends only on η and is such that

$$b_m \equiv \sup_{k \geq q(m)} E|S_k| \leq 2^{-m} \tag{8.4.1}$$

for each $m \geq 1$.

Proof. 1. First note that, by Proposition 4.7.1, for each $k \geq 1$, the r.r.v. Z_k has a modulus of integrability δ defined by

$$\delta(\varepsilon) \equiv \frac{\varepsilon}{2}/\eta\left(\frac{\varepsilon}{2}\right)$$

for each $\varepsilon > 0$.

2. By hypothesis, the independent r.r.v.'s $Z_1, Z_2, \ldots$ have a common distribution J on R. Hence they share a common characteristic function ψ. Therefore, for each $k \geq 1$, the characteristic function of the r.r.v. S_k is given by

$$\psi_k \equiv \psi^k\left(\frac{\cdot}{k}\right).$$

Let J_n denote the distribution of S_n. Let J_0 denote the distribution on R that assigns probability 1 to the point $0 \in R$. Then the characteristic function of J_0 is the constant function $\psi_0 \equiv 1$ on R. Define the remainder function r_1 of the first-degree Taylor expansion of the characteristic function ψ by

$$\psi(u) \equiv 1 + iuEZ_1 + r_1(u) = 1 + r_1(u)$$

for each $u \in R$, where the second equality is because the mean EZ_1 vanishes by hypothesis.

3. Separately, take an arbitrary $a > \eta(1)$. Then $E|Z_1| \leq E|Z_1|1_{(Z(1)>a)} + a \leq 1 + a$, by Definition 4.7.2 for η as a simple modulus of integrability. Letting $a \downarrow \eta(1)$ then yields

$$E|Z_k| = E|Z_1| \leq b \equiv 1 + \eta(1) \tag{8.4.2}$$

for each $k \geq 1$.

4. Let $n \geq 1$ be arbitrary, but fixed until further notice. As an abbreviation, define the positive real numbers

$$c \equiv c_n \equiv \pi^{-1}2^{2n+4} \tag{8.4.3}$$

and the positive integer

$$p_n \equiv p_{n,\eta} \equiv \left[\frac{8bc^3}{\pi}\eta\left(\frac{\pi}{8c^2}\right)\right]_1. \tag{8.4.4}$$

5. Let $m \geq 1$ be arbitrary. Let

$$n_m \equiv n_m(\eta) \equiv (m+2) \vee [1 - \log_2(\delta(2^{-m-1}))]_1. \tag{8.4.5}$$

Define

$$q_m \equiv q_{m,\eta} \equiv p_{n(m)}. \tag{8.4.6}$$

We will show that the sequence $(q_m)_{m=1,2,\ldots}$ of positive integers has the desired properties.

6. Consider each $k \geq p_n$ and each $u \in [-c,c] \equiv [-c_n, c_n]$. As an abbreviation, write $z \equiv r_1(\frac{u}{k})$ and $\alpha \equiv \frac{\pi}{4c^2}$. Then

$$\left|\frac{u}{k}\right| \leq \frac{c}{k} \leq \frac{c}{p_n} < \frac{\pi c}{8bc^3}\left(\eta\left(\frac{\pi}{8c^2}\right)\right)^{-1} \equiv \frac{\alpha}{2b}\left(\eta\left(\frac{\alpha}{2}\right)\right)^{-1}.$$

Hence, by Assertion 2 of Proposition 5.8.12, where $n, X, \lambda, \varepsilon$ are replaced by $1, Z_1, \frac{u}{k}, \alpha$, respectively, we obtain

$$\left|r_1\left(\frac{u}{k}\right)\right| < \alpha\left|\frac{u}{k}\right| \leq \alpha\frac{c}{k} \equiv \frac{\pi}{4kc}.$$

Therefore $|z| < \frac{\pi}{4kc}$. Consequently, the binomial expansion yields

$$\begin{aligned}
\left|\left(1 + r_1\left(\frac{u}{k}\right)\right)^k - 1\right| &\equiv |(1+z)^k - 1| \\
&= |(1 + C_1^k z + C_2^k z^2 + \cdots + C_k^k z^k) - 1| \\
&\leq C_1^k \frac{\pi}{4kc} + C_2^k\left(\frac{\pi}{4kc}\right)^2 + \cdots + C_k^k\left(\frac{\pi}{4kc}\right)^k \\
&\leq \left(\frac{\pi}{4c}\right)^1 + \left(\frac{\pi}{4c}\right)^2 + \cdots + \left(\frac{\pi}{4c}\right)^k < \frac{\pi}{4c}\left(1 - \frac{\pi}{4c}\right)^{-1} \\
&\equiv \frac{\pi}{4c}\left(1 - \frac{\pi}{4\pi^{-1}2^{2n+4}}\right)^{-1} \\
&= \frac{\pi}{4c}(1 - \pi^2 2^{-2n-6})^{-1} < \frac{\pi}{4c}\cdot 2 = \frac{\pi}{2c},
\end{aligned}$$

where

$$C_j^k \equiv \frac{k(k-1)\cdots(k-j+1)}{j!} \leq k^j$$

for each $j = 1, \ldots, k$. Consequently,

$$|\psi_k(u) - \psi_0(u)| = \left|\psi^k\left(\frac{u}{k}\right) - 1\right| = \left|\left(1 + r_1\left(\frac{u}{k}\right)\right)^k - 1\right| < \frac{\pi}{2c}, \tag{8.4.7}$$

where $u \in [-c,c]$, $k \geq p_n$, and $n \geq 1$ are arbitrary.

7. Now take an arbitrary $a \in (2^{-n}, 2^{-n+1})$. Define the function $f \in C(R)$ by

$$f(x) \equiv a^{-1}(1 - a^{-1}|x|)_+ \tag{8.4.8}$$

for each $x \in R$. Then the Fourier transform $\widehat{f}$ of the function f is bounded in absolute value by

$$|\widehat{f}(u)| \equiv \left| \int_{x \in R} e^{iux} f(x) dx \right| = \frac{2}{a} \int_0^a (\cos ux) \left(1 - \frac{x}{a}\right) dx$$

$$= \frac{2}{a} \int_0^a \frac{\sin ux}{u} \frac{1}{a} dx = 2 \frac{(1 - \cos au)}{a^2 u^2} \leq 1 \wedge \frac{4}{a^2 u^2} \qquad (8.4.9)$$

at each $u \in R$, where the third equality is by integration by parts. Therefore $\widehat{f}$ is Lebesgue integrable on R. Hence Assertion 3 of Theorem 5.8.9 implies that

$$J_k f = (2\pi)^{-1} \int \widehat{f}(-u) \psi_k(u) du,$$

with a similar equality when k is replaced by 0. Consequently,

$$2\pi |J_k f - J_0 f| = \left| \int \widehat{f}(-u)(\psi_k(u) - \psi_0(u)) du \right|$$

$$\leq \int_{|u| \leq c} |\widehat{f}(-u)| \frac{\pi}{2c} du + \int_{|u| > c} |\widehat{f}(-u)| \cdot |\psi_k(u) - \psi_0(u)| du$$

$$\leq \int_{|u| \leq c} |\widehat{f}(-u)| \frac{\pi}{2c} du + \int_{|u| > c} |\widehat{f}(-u)| \cdot 2du$$

$$\leq \int_{|u| \leq c} \frac{\pi}{2c} du + \int_{|u| > c} \frac{4}{a^2 u^2} \cdot 2du = \pi + \frac{2^4}{a^2 c}$$

$$< \pi + \frac{2^4}{2^{-2n} c} \equiv \pi + \frac{2^4}{2^{-2n} \pi^{-1} 2^{2n+4}} = \pi + \pi = 2\pi,$$

where the third inequality is thanks to inequality 8.4.9, and where the last inequality is because $a \in (2^{-n}, 2^{-n+1})$. Hence

$$|J_k f - J_0 f| \leq 1.$$

At the same time, from the defining formula 8.4.8, we see that $1_{[-a,a]} \geq af$. Hence

$$P(|S_k| > a) = J_k(1 - 1_{[-a,a]}) \leq 1 - aJ_k f \leq 1 - a(J_0 f - 1)$$

$$= 1 - a(f(0) - 1) = 1 - a(a^{-1} - 1) = a, \qquad (8.4.10)$$

where $a \in (2^{-n}, 2^{-n+1})$, $k \geq p_n$, and $n \geq 1$ are arbitrary.

8. Now let $m \geq 1$ be arbitrary. Let $n \equiv n_m$. Take an arbitrary $a \in (2^{-n}, 2^{-n+1})$. Consider each $k \geq q_m \equiv p_{n(m)} \equiv p_n$. Then, by inequality 8.4.10, we have

$$P(|S_k| > a) < a < 2^{-n+1} \equiv 2^{-n(m)+1} < \delta(2^{-m-1}),$$

where the last inequality is by the defining equality 8.4.5. Hence, since δ is a modulus of integrability of Z_κ for each $\kappa = 1, \ldots, k$, it follows that

$$E|S_k|1_{(|S(k)|>a)} \le k^{-1}\sum_{\kappa=1}^{k} E|Z_\kappa|1_{(|S(k)|>a)} \le k^{-1}\sum_{\kappa=1}^{k} 2^{-m-1} = 2^{-m-1}.$$

Consequently,

$$E|S_k| \le E|S_k|1_{(|S(k)|\le a)} + E|S_k|1_{(|S(k)|>a)} \le a + 2^{-m-1}$$
$$< 2^{-n(m)+1} + 2^{-m-1} \le 2^{-m-1} + 2^{-m-1} = 2^{-m}, \qquad (8.4.11)$$

where the last inequality is, again, by the defining equality 8.4.5, and where $m \ge 1$ and $k \ge q_m$ are arbitrary. We conclude that $E|S_k| \to 0$ as $k \to \infty$. Moreover, inequality 8.4.11 shows that

$$b_m \equiv \sup_{k\ge q(m)} E|S_k| \le 2^{-m} \qquad (8.4.12)$$

for each $m \ge 1$. The theorem is proved. □

Theorem 8.4.2. Strong Law of Large Numbers. *Suppose $Z_1, Z_2, \ldots$ is a sequence of integrable, independent, and identically distributed r.r.v.'s with mean 0, on some probability space (Ω, L, E). Let η be a simple modulus of integrability of Z_1 in the sense of Definition 4.7.2. Then the following conditions hold:*

1. The partial sums

$$S_k \equiv k^{-1}(Z_1 + \cdots + Z_k) \to 0 \qquad \text{a.u.}$$

as $k \to \infty$.

2. More precisely, for each $m \ge 1$ there exists an integer $k_{m,\eta}$ and a measurable set A, with $P(A^c) < 2^{-m+2}$ and with

$$A \subset \bigcap_{p=m}^{\infty} \bigcap_{k=k(p,\eta)}^{\infty} (|S_k| \le 2^{-p+3}). \qquad (8.4.13)$$

Proof. 1. Let $m \ge j \ge 1$ be arbitrary, and let I_j denote the distribution of Z_j on R. Then, in view of the hypothesis of independence and identical distribution, the r.v. $(Z_1, \ldots, Z_j, \cdots, Z_m)$ with values in R^m has the same distribution as the r.v $(Z_j, \ldots, Z_1, \cdots, Z_m)$, where, for brevity, the latter stands for the sequence obtained from $(Z_1, \ldots, Z_j, \cdots, Z_m)$ by swapping the first and the jth members. Thus

$$Eh(Z_1, \ldots, Z_j, \ldots, Z_m) = Eh(Z_j, \ldots, Z_1, \ldots, Z_m) \qquad (8.4.14)$$

for each integrable function h on R^m relative to the joint distribution $E_{Z(1),\ldots,Z(m)}$.

2. Let $Q \equiv \{\cdots, -2, -1\}$. For each $t \equiv -k \in Q$, define

$$X_t \equiv S_{|t|} \equiv S_k.$$

Let $\mathcal{L} \equiv \mathcal{L}_X \equiv \{L^{(X,t)} : t \in Q\}$ be the natural filtration of the process $X : Q \times \Omega \to R$. Let $t \in Q$ be arbitrary. Then $t = -n$ for some $n \ge 1$. Hence $(\Omega, L^{(X,t)}, E)$ is the probability subspace of (Ω, L, E) generated by the family

$$G^{(X,t)} \equiv \{X_r : r \in Q;\ r \leq t\} = \{S_m : m \geq n\}.$$

In other words, $(\Omega, L^{(X,t)}, E)$ is the completion of the integration space

$$(\Omega, L_{Cub}(G^{(X,t)}), E),$$

where

$$L_{Cub}(G^{(X,t)}) = \{f(S_n, \ldots, S_m) : m \geq n; f \in C_{ub}(R^{m-n+1})\}. \qquad (8.4.15)$$

By Lemma 8.1.4, the process X is adapted to its natural filtration $\mathcal{L} \equiv \mathcal{L}_X$.

3. We will prove that the process X is a martingale relative to the filtration $\mathcal{L}$. To that end, let $s,t \in Q$ be arbitrary with $t \leq s$. Then $t = -n$ and $s = -k$ for some $n \geq k \geq 1$. Let $Y \in L_{Cub}(G^{(X,t)})$ be arbitrary. Then, in view of equality 8.4.15, we have

$$\begin{aligned} Y &= f(S_n, \ldots, S_m) \\ &\equiv f((Z_1 + \cdots + Z_n)n^{-1}, \ldots, (Z_1 + \cdots + Z_m)m^{-1}) \end{aligned}$$

for some $f \in C_{ub}(R^{m-n+1})$, for some $m \geq n$. Let $j = 1, \ldots, n$ be arbitrary. Then, since the r.r.v. Y is bounded, the r.r.v. YZ_j is integrable. Hence,

$$\begin{aligned} EYZ_j &\equiv Ef((Z_1 + \cdots + Z_j + \cdots + Z_n)n^{-1}, \ldots, \\ &\qquad (Z_1 + \cdots + Z_j + \cdots + Z_m)m^{-1})Z_j \\ &= Ef((Z_j + \cdots + Z_1 + \cdots + Z_n)n^{-1}, \ldots, \\ &\qquad (Z_j + \cdots + Z_1 + \cdots + Z_m)m^{-1})Z_1 \\ &= Ef((Z_1 + \cdots + Z_j + \cdots + Z_n)n^{-1}, \ldots, \\ &\qquad (Z_1 + \cdots + Z_j + \cdots + Z_m)m^{-1})Z_1 \equiv EYZ_1, \end{aligned}$$

where the second equality is by equality 8.4.14, and where the third equality is because addition is commutative. In short,

$$EYZ_j = EYZ_1$$

for each $j = 1, \ldots, n$. Therefore, since $k \leq n$, we have

$$EYS_k = k^{-1}E(YZ_1 + \cdots + YZ_k) = EYZ_1.$$

In particular, $EYS_n = EYZ_1 = EYS_k$. In other words, $EYX_t = EYX_s$, where $X_t \in L^{(X,t)}$ and where $Y \in L_{Cub}(G^{(X,t)})$ is arbitrary. Hence, according to Assertion 5 of Proposition 5.6.6, we have

$$E(X_s | L^{(X,t)}) = X_t.$$

Since $s,t \in Q$ are arbitrary with $t \leq s$, the process X is a martingale relative to its natural filtration $\mathcal{L} \equiv \{L^{(X,t)} : t \in Q\}$.

4. Thanks to Theorem 8.4.1, the Weak Law of Large Numbers, the supremum

$$b_h \equiv \sup_{t\in Q;|t|\geq h} E|X_t| \tag{8.4.16}$$

is well defined for each $h \geq 1$. Moreover, by Theorem 8.4.1, there exists a sequence $(q_h)_{h=1,2,...} \equiv (q_{h,\eta})_{h=1,2,...}$ of positive integers, which depends only on η, and is such that

$$b_{q(h)} \equiv \sup_{k\geq q(h)} E|S_k| \leq 2^{-h}. \tag{8.4.17}$$

Thus

$$\lim_{|t|\to\infty} E|X_t| \equiv \lim_{k\to\infty} E|S_k| = \beta \equiv 0. \tag{8.4.18}$$

Moreover, inequality 8.4.17 can be rewritten as

$$b_{q(h)} \equiv \sup_{t\in Q;|t|\geq q(h)} |(E|X_t| - \beta)| \leq 2^{-h}, \tag{8.4.19}$$

for each $h \geq 1$.

5. Next, let $h \geq 1$ be arbitrary, and consider each $k \geq q_h$. Then

$$E|e^{-|S(k)|} - 1| \leq E|S_k| \leq 2^{-h}.$$

Thus

$$\lim_{|t|\to\infty} Ee^{-|X(t)|} \equiv \lim_{k\to\infty} Ee^{-|S(k)|} = \alpha \equiv 1, \tag{8.4.20}$$

with

$$\begin{aligned} a_{q(h)} &\equiv \sup_{t\in Q;|t|\geq q(h)} |(E\exp(-|X_t|) - \alpha)| \\ &\equiv \sup_{t\in Q;|t|\geq q(h)} |(E\exp(-|X_t|) - 1)| \leq \sup_{t\in Q;|t|\geq q(h)} E|X_t| \leq 2^{-h}. \end{aligned} \tag{8.4.21}$$

6. In view of equalities 8.4.18 and 8.4.20, Theorem 8.3.5 implies that $X_t \to Y$ a.u. as $|t| \to \infty$ in Q, for some r.r.v. Y. In other words, $S_k \to Y$ a.u. as $k \to \infty$. Hence, by the Dominated Convergence Theorem, we have

$$E(1 \wedge |Y|) = \lim_{k\to\infty} E(1 \wedge |S_k|) \leq \lim_{k\to\infty} E|S_k| = 0.$$

It follows that $Y = 0$. Thus $S_k \to 0$ a.u. as $k \to \infty$. This proves Assertion 1 of the present theorem.

7. For the rate of a.u. convergence, we follow Assertion 2 of Theorem 8.3.5 and its proof. Specifically, define

$$\delta_h \equiv a_h + b_h$$

for each $h \geq 1$. Define $k_0 \equiv 0$. Inductively, for each $m \geq 1$, take any $k_m \equiv k_{m,\eta} \geq k_{m-1}$ so large that

$$\delta_{k(m)} \leq \frac{1}{12} 2^{-3m} \exp(-2^m 3(\beta + b_{k(m-1)})). \tag{8.4.22}$$

Let $\varepsilon > 0$ be arbitrary. Take $m \geq 1$ so large that $2^{-m+2} < \varepsilon$. Then there exists a measurable set A with $P(A^c) < \varepsilon$ where

$$A \subset \bigcap_{p=m}^{\infty} \bigcap_{t \in Q; |t| \geq k(p)} (|X_t - Y| \leq 2^{-p+3}). \tag{8.4.23}$$

In other words,

$$A \subset \bigcap_{p=m}^{\infty} \bigcap_{k=k(p)}^{\infty} (|S_k| \leq 2^{-p+3}), \tag{8.4.24}$$

as desired. □

9

a.u. Continuous Process

In this chapter, let (S,d) be a locally compact metric space. Unless otherwise specified, this will serve as the state space for the processes in this chapter. Consider an arbitrary consistent family F of f.j.d.'s that is continuous in probability, with state space S and parameter set $[0,1]$. We will find conditions on the f.j.d.'s in F under which an a.u. continuous process X can be constructed with marginal distributions given by the family F.

A classical proof of the existence of such processes X, as elaborated in [Billingsley 1974], uses the following theorem.

Theorem 9.0.1. Prokhorov's Relative Compactness Theorem. *Each tight family $\overline{J}$ of distributions on a locally compact metric space (H,d_H) is relatively compact, in the sense that each sequence in $\overline{J}$ contains a subsequence that converges weakly to some distribution on (H,d_H).*

Prokhorov's theorem implies the principle of infinite search, and is therefore not constructive. This can be seen as follows. Let $(r_n)_{n=1,2,...}$ be an arbitrary nondecreasing sequence in $H \equiv \{0,1\}$. Let the doubleton H be endowed with the Euclidean metric d_H defined by $d_H(x,y) = |x-y|$ for each $x,y \in H$. For each $n \geq 1$, let J_n be the distribution on (H,d_H) that assigns unit mass to r_n; in other words, $J_n(f) \equiv f(r_n)$ for each $f \in C(H,d_H)$. Then the family $\overline{J} \equiv \{J_1,J_2,\ldots\}$ is tight, and Prokhorov's theorem implies that J_n converges weakly to some distribution J on (H,d_H). It follows that $J_n g$ converges as $n \to 0$, where $g \in C(H,d_H)$ is defined by $g(x) = x$ for each $x \in H$. Thus $r_n \equiv g(r_n) \equiv J_n g$ converges as $n \to 0$. Since $(r_n)_{n=1,2,...}$ is an arbitrary nondecreasing sequence in $\{0,1\}$, the principle of infinite search follows.

In our constructions, we will bypass any use of Prokhorov's theorem or of unjustified supremums, in favor of direct proofs using Borel–Cantelli estimates. We will give a necessary and sufficient condition on the f.j.d.'s in the family F for F to be extendable to an a.u. continuous process X. We will call this condition C-regularity. We will derive a modulus of a.u. continuity of the process X from a given modulus of continuity in probability and a given modulus of C-regularity of the consistent family F, in a sense defined presently. We will also prove that the extension is uniformly metrically continuous on an arbitrary set $\widehat{F}_0$ of such consistent families F that share a common modulus of C-regularity.

In essence, the material presented in Sections 9.1 and 9.2 is a constructive and more general version of material from section 7 of chapter 2 of [Billingsley 1974], though the latter treats only the special case where $S = R$. We remark that the generalization to the arbitrary locally compact state space (S,d) is not entirely trivial, because we forego the convenience of linear interpolation in R.

Chapter 10 of this book will introduce the condition of D-regularity, analogous to C-regularity, for the treatment of processes that are, almost uniformly, right continuous with left limits, again with a general locally compact metric space as state space.

In Section 9.3, we will prove a generalization of Kolmogorov's theorem for a.u. Hoelder continuity, to the case of the state space (S,d), in a sense to be made precise.

In Section 9.4, we will apply this result to construct a Brownian motion.

In Section 9.5, in the case of Gaussian processes, we will present a hitherto-unknown sufficient condition on the covariance function to guarantee a.u. Hoelder continuity. Then we will present the sufficient condition on the covariance function, due to [Garsia, Rodemich, and Rumsey 1970], for a.u. continuity along with a modulus of a.u. continuity. We will present their proof with a minor modification to make it strictly constructive.

For a more general parameter space that is a subset of R^m for some $m \geq 0$, with some restriction on its local ε-entropy, [Potthoff 2009-2] gives sufficient conditions on the pair distributions $F_{t,s}$ to guarantee the construction of an a.u. continuous or an a.u. Hoelder, real-valued, random field.

In this and later chapters we will use the following notations for the dyadic rationals.

Definition 9.0.2. Notations for dyadic rationals. For each $m \geq 0$, define $p_m \equiv 2^m$, $\Delta_m \equiv 2^{-m}$, and the enumerated sets of dyadic rationals

$$\begin{aligned} Q_m &\equiv \{t_0, t_1, \ldots, t_{p(m)}\} = \{q_{m,0}, \ldots, q_{m,p(m)}\} \\ &\equiv \{0, \Delta_m, 2\Delta_m, \ldots, p_m\Delta_m\} \equiv \{0, \Delta_m, 2\Delta_m, \ldots, 1\} \subset [0,1], \\ \overline{Q}_m &\equiv \{u_0, u_1, \ldots, u_{p(2m)}\} \equiv \{0, 2^{-m}, 2 \cdot 2^{-m}, \ldots, 2^m\} \subset [0, 2^m], \\ \widetilde{Q}_m &\equiv \{0, \Delta_m, 2\Delta_m, \ldots\} \equiv \{0, 2^{-m}, 2 \cdot 2^{-m}, \ldots\} \subset [0, \infty). \end{aligned}$$

Define

$$Q_\infty \equiv \bigcup_{m=0}^{\infty} Q_m = \{t_0, t_1, \ldots\}.$$

Let $m \geq 0$ be arbitrary. Then the enumerated set Q_m is a 2^{-m}-approximation of $[0,1]$, with $Q_m \subset Q_{m+1}$. Conditions in Definition 3.2.1 can easily be verified for the sequence

$$\xi_{[0,1]} \equiv (Q_m)_{m=1,2,...}$$

to be a binary approximation of $[0,1]$ relative to the reference point $q_\circ \equiv 0$.

In addition, define

$$\overline{Q}_\infty \equiv \bigcup_{m=0}^{\infty} \overline{Q}_m = \{u_0, u_1, \ldots\}. \square$$

Definition 9.0.3. Miscellaneous notations and conventions. As usual, to lighten the notational burden, we will write an arbitrary subscripted symbol a_b interchangeably with $a(b)$. We will write TU for a composite function $T \circ U$. If $f : A \to B$ is a function from a set A to a set B, and if A' is a nonempty subset A, then the restricted function $f|A' : A' \to B$ will also be denoted simply by f when there is little risk of confusion. If A is a measurable subset on a probability space (Ω, L, E), then we will write PA, $P(A)$, $E(A)$, EA, or $E1_A$ interchangeably. For arbitrary r.r.v. $Y \in L$ and measurable subsets A, B, we will write $(A;B) \equiv AB$, $E(Y;A) \equiv E(Y1_A)$, and $A \in L$. As a further abbreviation, we drop the parentheses when there is little risk of confusion. For example, we write $1_{Y \leq \beta; Z > \alpha} \equiv 1_{(Y \leq \beta)(Z > \alpha)}$. For an arbitrary integrable function $X \in L$, we will sometimes use the more suggestive notation $\int E(d\omega)X(\omega)$ for EX, where ω is a dummy variable.

Let Y be an arbitrary r.r.v. Recall from Definition 5.1.3 the convention that if measurability of the set $(Y < \beta)$ or $(Y \leq \beta)$ is required in a discussion, for some $\beta \in R$, then it is understood that the real number β has been chosen from the regular points of the r.r.v. Y.

Separately, recall that $\widehat{R}(Q \times \Omega, S)$ denotes the set of r.f.'s with an arbitrary parameter set Q , sample space (Ω, L, E), and state space (S, d).

Recall that $[\cdot]_1$ is an operation that assigns to each $a \in R$ an integer $[a]_1 \in (a, a+2)$. As usual, write $\widehat{d} \equiv 1 \wedge d$. $\square$

9.1 Extension from Dyadic Rational Parameters to Real Parameters

Our approach to extend a given family F of f.j.d.'s that is continuous in probability on the parameter set $[0,1]$ is as follows. First note that F carries no more useful information than its restriction $F|Q_\infty$, where Q_∞ is the dense subset of dyadic rationals in $[0,1]$, because the family can be recovered from the $F|Q_\infty$, thanks to continuity in probability. Hence we can first extend the family $F|Q_\infty$ to a process $Z : Q_\infty \times \Omega \to S$ by applying the Daniell–Kolmogorov Theorem or the Daniell–Kolmogorov–Skorokhod Theorem. Then any condition of the family F is equivalent to a condition to Z.

In particular, in the current context, any condition on f.j.d.'s to make F extendable to an a.u. continuous process $X : [0,1] \times \Omega \to S$ can be stated in terms of a process $Z : Q_\infty \times \Omega \to S$, with the latter to be extended by limit to the process X. It is intuitively obvious that any a.u. continuous process $Z : Q_\infty \times \Omega \to S$ is extendable to an a.u. continuous process $X : [0,1] \times \Omega \rightarrow S$, because Q_∞ is dense [0, 1]. In this section, we will make this precise, and we will define metrics such that the extension operation is itself a metrically continuous construction.

Definition 9.1.1. Metric space of a.u. continuous processes. Let $C[0,1]$ be the space of continuous functions $x : [0,1] \rightarrow (S,d)$, endowed with the *uniform metric* defined by

$$d_{C[0,1]}(x,y) \equiv \sup_{t\in[0,1]} d(x(t),y(t)) \tag{9.1.1}$$

for each $x,y \in C[0,1]$. Write $\widehat{d}_{C[0,1]} \equiv 1 \wedge d_{C[0,1]}$.

Let (Ω, L, E) be an arbitrary probability space. Let $\widehat{C}[0,1]$ denote the set of stochastic processes $X : [0,1] \times (\Omega, L, E) \rightarrow (S,d)$ that are a.u. continuous on $[0,1]$. Define a metric $\rho_{\widehat{C}[0,1]}$ on $\widehat{C}[0,1]$ by

$$\rho_{\widehat{C}[0,1]}(X,Y) \equiv E \sup_{t\in[0,1]} \widehat{d}(X_t,Y_t) \equiv E\widehat{d}_{C[0,1]}(X,Y) \tag{9.1.2}$$

for each $X,Y \in \widehat{C}[0,1]$. Lemma 9.1.2 (next) says that $(\widehat{C}[0,1], \rho_{\widehat{C}[0,1]})$ is a well-defined metric space. □

Lemma 9.1.2. $\rho_{\widehat{C}[0,1]}$ **is a metric.** *The function* $\sup_{t\in[0,1]} \widehat{d}(X_t,Y_t)$ *is an r.r.v. The function* $\rho_{\widehat{C}[0,1]}$ *is well defined and is a metric.*

Proof. Let $X,Y \in \widehat{C}[0,1]$ be arbitrary, with moduli of a.u. continuity $\delta^X_{auc}, \delta^Y_{auc}$, respectively. First note that the function $\sup_{t\in[0,1]} \widehat{d}(X_t,Y_t)$ is defined a.s., on account of continuity on $[0,1]$ of X,Y on a full subset of Ω. We need to prove that it is measurable, so that the expectation in the defining formula 9.1.2 makes sense.

To that end, let $\varepsilon > 0$ be arbitrary. Then there exist measurable sets D_X, D_Y with $P(D^c_X) \vee P(D^c_Y) < \varepsilon$, on the intersection of which we have

$$d(X_t,X_s) \vee d(Y_t,Y_s) \leq \varepsilon$$

for each $t,s \in [0,1]$ with $|t-s| < \delta \equiv \delta^X_{auc}(\varepsilon) \wedge \delta^Y_{auc}(\varepsilon)$. Now let the sequence $s_0,\ldots,s_n$ be an arbitrary δ-approximation of $[0,1]$, for some $n \geq 1$. Then, for each $t \in [0,1]$, we have $|t - s_k| < \delta$ for some $k = 0,\ldots,n$, whence

$$d(X_t,X_{s(k)}) \vee d(Y_t,Y_{s(k)}) \leq \varepsilon \tag{9.1.3}$$

on $D \equiv D_X D_Y$. This, in turn, implies

$$|d(X_t,Y_t) - d(X_{s(k)},Y_{s(k)})| \leq 2\varepsilon$$

on D. It follows that

$$\left| \sup_{t\in[0,1]} \widehat{d}(X_t,Y_t) - \bigvee_{k=1}^{n} \widehat{d}(X_{s(k)},Y_{s(k)}) \right| \leq 2\varepsilon$$

on D, where $Z \equiv \bigvee_{k=1}^{n} \widehat{d}(X_{t(k)}, Y_{t(k)})$ is an r.r.v., and where $P(D^c) < 2\varepsilon$. For each $p \geq 1$, we can repeat this argument with $\varepsilon \equiv \frac{1}{p}$. Thus we obtain a sequence Z_p of r.r.v.'s with $Z_p \to \sup_{t \in [0,1]} \widehat{d}(X_t, Y_t)$ in probability as $p \to \infty$. The function $\sup_{t \in [0,1]} \widehat{d}(X_t, Y_t)$ is accordingly an r.r.v. and, being bounded by 1, is integrable. Summing up, the expectation in equality 9.1.2 exists, and $\rho_{\widehat{C}[0,1]}$ is well defined.

Verification of the conditions for the function $\rho_{\widehat{C}[0,1]}$ to be a metric is straightforward and is omitted here. $\square$

Definition 9.1.3. Extension by limit, if possible, of a process with parameter set Q_∞. Let $Z : Q_\infty \times \Omega \to S$ be an arbitrary process. Define a function $X \equiv \Phi_{Lim}(Z)$ by

$$domain(X) \equiv \{(r, \omega) \in [0,1] \times \Omega : \lim_{s \to r; s \in Q(\infty)} Z(s, \omega) \text{ exists}\}$$

and

$$X(r, \omega) \equiv \lim_{s \to r; s \in Q(\infty)} Z(s, \omega)$$

for each $(r, \omega) \in domain(X)$. We will call X the *extension-by-limit of the process Z to the parameter set* $[0, 1]$. A similar definition is made where the interval $[0, 1]$ is replaced by the interval $[0, \infty)$, and where the set Q_∞ of dyadic rationals in $[0, 1]$ is replaced by the set $\overline{Q}_\infty$ of dyadic rationals in $[0, \infty)$.

We emphasize that, absent any additional conditions on the process Z, the function X need not be a process. Indeed, need not even be a well-defined function. $\square$

Theorem 9.1.4. Extension by limit of a.u. continuous process on Q_∞ to a.u. continuous process on $[0, 1]$; and metrical continuity of said extension. *Recall from Definition 6.4.2 the metric space $(\widehat{R}(Q_\infty \times \Omega, S), \rho_{Prob,Q(\infty)})$ of processes $Z : Q_\infty \times \Omega \to S$. Let $\widehat{R}_0$ be a subset of $\widehat{R}(Q_\infty \times \Omega, S)$ whose members are a.u. continuous with a common modulus of a.u. continuity δ_{auc}. Then the following conditions hold:*

1. Let $Z \in \widehat{R}_0$ be arbitrary. Then its extension-by-limit $X \equiv \Phi_{Lim}(Z)$ is an a.u. continuous process such that $X_t = Z_t$ on $domain(X_t)$ for each $t \in Q_\infty$. Moreover, the process X has the same modulus of a.u. continuity δ_{auc} as Z.

2. The extension-by-limit

$$\Phi_{Lim} : (\widehat{R}_0, \rho_{Prob,Q(\infty)}) \to (\widehat{C}[0,1], \rho_{\widehat{C}[0,1]})$$

is uniformly continuous, with a modulus of continuity $\delta_{Lim}(\cdot, \delta_{auc})$.

Proof. 1. Let $Z \in \widehat{R}_0$ be arbitrary. Let $\varepsilon > 0$ be arbitrary. Then, by hypothesis, there exists $\delta_{auc}(\varepsilon) > 0$ and a measurable set $D \subset \Omega$ with $P(D^c) < \varepsilon$ such that for each $\omega \in D$ and for each $s, s' \in Q_\infty$ with $|s - s'| < \delta_{auc}(\varepsilon)$, we have

$$d(Z(s, \omega), Z(s', \omega)) \leq \varepsilon. \tag{9.1.4}$$

Next let $\omega \in D$ and $r, r' \in [0,1]$ be arbitrary with $|r - r'| < \delta_{auc}(\varepsilon)$. Letting $s.s' \to r$ with $s, s' \in Q_\infty$, we have $|s - s'| \to 0$, and so $d(Z(s,\omega), Z(s',\omega)) \to 0$ as $s, s' \to r$. Since (S,d) is complete, we conclude that the limit

$$X(r,\omega) \equiv \lim_{s \to r; s \in Q(\infty)} Z(s,\omega)$$

exists. Moreover, letting $s' \to r$ with $s, s' \in Q_\infty$, inequality 9.1.4 yields

$$d(Z(s,\omega), X(r,\omega)) \leq \varepsilon. \tag{9.1.5}$$

Since $\varepsilon > 0$ is arbitrary, we see that $Z_s \to X_r$ a.u. as $s \to r$. Hence X_r is an r.v. Thus $X : [0,1] \times \Omega \to S$ is a stochastic process.

Now let $s \to r$ and $s' \to r'$ with $s, s' \in Q_\infty$ in inequality 9.1.4. Then we obtain

$$d(X(r,\omega), X(r',\omega)) \leq \varepsilon, \tag{9.1.6}$$

where $\omega \in D$ and $r, r' \in [0,1]$ are arbitrary with $|r - r'| < \delta_{auc}(\varepsilon)$. Thus X has the same modulus of a.u. continuity δ_{auc} as Z.

2. It remains to verify that the mapping Φ_{Lim} is a continuous function. To that end, let $\varepsilon > 0$ be arbitrary. Write $\alpha \equiv \frac{1}{3}\varepsilon$. Let $m \equiv m(\varepsilon, \delta_{auc}) \geq 1$ be so large that $2^{-m} < \delta_{auc}(\alpha)$. Define

$$\delta_{Lim}(\varepsilon, \delta_{auc}) \equiv 2^{-p(m)-1}\alpha^2.$$

Let $Z, Z' \in \widehat{R}_0$ be arbitrary such that

$$\rho_{Prob,Q(\infty)}(Z, Z') < \delta_{Lim}(\varepsilon, \delta_{auc}).$$

Equivalently,

$$E \sum_{n=0}^{\infty} 2^{-n-1} \widehat{d}(Z_{t(n)}, Z'_{t(n)}) < 2^{-p(m)-1}\alpha^2. \tag{9.1.7}$$

Then, by Chebychev's inequality, there exists a measurable set A with $P(A^c) < \alpha$ such that for each $\omega \in A$, we have

$$\sum_{n=0}^{\infty} 2^{-n-1} \widehat{d}(Z(t_n,\omega), Z'(t_n,\omega)) < 2^{-p(m)-1}\alpha,$$

whence

$$\widehat{d}(Z(t_n,\omega), Z'(t_n,\omega)) < \alpha \tag{9.1.8}$$

for each $n = 0, \ldots, p_m$.

Now let $X \equiv \Phi_{Lim}(Z)$ and $X' \equiv \Phi_{Lim}(Z')$. By Assertion 1, the processes X and X' have the same modulus of a.u. continuity δ_{auc} as Z and Z'. Hence, there exist measurable sets D, D' with $P(D^c) \vee P(D'^c) < \alpha$ such that for each $\omega \in DD'$, we have

$$\widehat{d}(X(r,\omega), X(s,\omega)) \vee \widehat{d}(X'(r,\omega), X'(s,\omega)) \leq \alpha \tag{9.1.9}$$

for each $r, s \in [0,1]$ with $|r - s| < \delta_{auc}(\alpha)$.

Now consider each $\omega \in ADD'$. Let $r \in [0,1]$ be arbitrary. Since $t_0, \ldots, t_{p(m)}$ is a 2^{-m}-approximation of $[0,1]$, there exists $n = 0, \ldots, p_m$ such that $|r - t_n| < 2^{-m} < \delta_{auc}(\alpha)$. Then inequality 9.1.9 holds with $s \equiv t_n$. Combining inequalities 9.1.8 and 9.1.9, we obtain

$$\begin{aligned}
&\widehat{d}(X(r,\omega), X'(r,\omega)) \\
&\leq \widehat{d}(X(r,\omega), X(s,\omega)) + \widehat{d}(X(s,\omega), X'(s,\omega)) + \widehat{d}(X'(r,\omega), X'(s,\omega)) \\
&= \widehat{d}(X(r,\omega), X(s,\omega)) + \widehat{d}(Z(s,\omega), Z'(s,\omega)) + \widehat{d}(X'(r,\omega), X'(s,\omega)) < 3\alpha,
\end{aligned}$$

where $\omega \in ADD'$ and $r \in [0,1]$ are arbitrary. It follows that

$$\begin{aligned}
\rho_{\widehat{C}[0,1]}(X, X') &\equiv E \sup_{r \in [0,1]} \widehat{d}(X_r, X'_r) \\
&\leq E \sup_{r \in [0,1]} \widehat{d}(X_r, X'_r) 1_{ADD'} + P(ADD')^c \\
&< 3\alpha + 3\alpha = 6\alpha \equiv \varepsilon.
\end{aligned}$$

We conclude that $\delta_{Lim}(\cdot, \delta_{auc})$ is a modulus of continuity of the function Φ_{Lim}. □

9.2 C-Regular Family of f.j.d.'s and C-Regular Process

Definition 9.2.1. C-regularity. Let (Ω, L, E) be an arbitrary sample space. Let $Z : Q_\infty \times \Omega \to S$ be an arbitrary process. We will say that Z is a *C-regular* process if there exists an increasing sequence $\overline{m} \equiv (m_n)_{n=0,1,\ldots}$ of positive integers, called the *modulus of C-regularity of the process Z, such that* for each $n \geq 0$ and for each $\beta_n > 2^{-n}$ such that the set

$$A_{t,s}^{(n)} \equiv (d(Z_t, Z_s) > \beta_n)$$

is measurable for each $s, t \in Q_\infty$, we have

$$P(C_n) < 2^{-n}, \tag{9.2.1}$$

where

$$C_n \equiv \bigcup_{t \in Q(m(n))} \bigcup_{s \in [t,t']Q(m(n+1))} A_{t,s}^{(n)} \cup A_{s,t'}^{(n)}. \tag{9.2.2}$$

Here, for each $t \in Q_{m(n)}$, we abuse notations and write $t' \equiv 1 \wedge (t + 2^{-m(n)}) \in Q_{m(n)}$.

Let F be an arbitrary consistent family of f.j.d.'s that is continuous in probability on $[0,1]$. Then the family F of consistent f.j.d.'s is said to be *C-regular* with the sequence $\overline{m} \equiv (m_n)_{n=0,1,\ldots}$ as a *modulus of C-regularity if $F|Q_\infty$ is the family of marginal distributions of some C-regular process Z.*

Let $X : [0,1] \times \Omega \to S$ be an arbitrary process that is continuous in probability on $[0,1]$. Then the process X is said to be *C-regular*, with the sequence $\overline{m} \equiv (m_n)_{n=0,1,\ldots}$ as a *modulus of C-regularity* if its family F of marginal distributions is C-regular with the sequence $\overline{m}$ as a modulus of C-regularity. □

We will prove that a process on $[0,1]$ is a.u. continuous iff it is C-regular.

Theorem 9.2.2. a.u. Continuity implies C-regularity. *Let (Ω, L, E) be an arbitrary sample space. Let $X : [0,1] \times \Omega \to S$ be an a.u. continuous process, with a modulus of a.u. continuity δ_{auc}. Then the process X is C-regular, with a modulus of C-regularity given by $\overline{m} \equiv (m_n)_{n=0,1,\ldots}$, where $m_0 \equiv 1$ and*

$$m_n \equiv [m_{n-1} \vee (-\log_2 \delta_{auc}(2^{-n}))]_1$$

for each $n \geq 1$.

Proof. 1. First note that X is continuous in probability. Let F denote its family of marginal distributions. Then $F|Q_\infty$ is the family of marginal distributions of the process $Z \equiv X|Q_\infty$. We will show that Z is C-regular.

2. To that end, let $n \geq 0$ be arbitrary. By Definition 6.1.2 of a.u. continuity, there exists a measurable set D_n with $P(D_n^c) < 2^{-n}$ such that for each $\omega \in D_n$ and for each $s,t \in [0,1]$ with $|t-s| < \delta_{auc}(2^{-n})$, we have

$$d(X(t,\omega), X(s,\omega)) \leq 2^{-n}. \tag{9.2.3}$$

3. Let $\beta_n > 2^{-n}$ be arbitrary and let

$$A_{t,s}^{(n)} \equiv (d(Z_t, Z_s) > \beta_n) \equiv (d(X_t, X_s) > \beta_n)$$

for each $s,t \in Q_\infty$. Define

$$C_n \equiv \bigcup_{t \in Q(m(n))} \bigcup_{s \in [t,t']Q(m(n+1))} A_{t,s}^{(n)} \cup A_{s,t'}^{(n)}, \tag{9.2.4}$$

where for each $t \in Q_{m(n)}$, we abuse notations and write $t' \equiv 1 \wedge (t + 2^{-m(n)}) \in Q_m$. Suppose, for the sake of a contradiction, that $P(D_n C_n) > 0$. Then there exists some $\omega \in D_n C_n$. Hence, by equality 9.2.4, there exists $t \in Q_{m(n)}$ and $s \in [t,t']Q_{m(n+1)}$ with

$$d(Z(t,\omega), Z(s,\omega)) \vee d(Z(s,\omega), Z(t',\omega)) > \beta_n. \tag{9.2.5}$$

It follows from $s \in [t,t']$ that

$$|s-t| \vee |s-t'| \leq 2^{-m(n)} < \delta_{auc}(2^{-n}), \tag{9.2.6}$$

whence

$$d(Z(t,\omega), Z(s,\omega)) \vee d(Z(s,\omega), Z(t',\omega)) \leq 2^{-n} < \beta_n,$$

contradicting inequality 9.2.5. We conclude that $P(D_n C_n) = 0$. Consequently,

$$P(C_n) = P(D_n \cup D_n^c)C_n = P(D_n^c C_n) \leq P(D_n^c) < 2^{-n}.$$

Thus the conditions in Definition 9.2.1 are satisfied for the process Z, the family $F|Q_\infty$, the family F, and the process X to be C-regular, all with modulus of C-regularity given by $\overline{m}$. □

The next theorem is the converse of Theorem 9.2.2.

Theorem 9.2.3. C-regularity implies a.u. continuity. *Let (Ω, L, E) be an arbitrary sample space. Let F be a C-regular family of consistent f.j.d.'s. Then there exists an a.u. continuous process $X : [0,1] \times \Omega \to S$ with marginal distributions given by F.*

Specifically, let $\overline{m} \equiv (m_n)_{n=0,1,...}$ be a modulus of C-regularity of F. Let $Z : Q_\infty \times \Omega \to S$ be an arbitrary process with marginal distributions given by $F|Q_\infty$. Let $\varepsilon > 0$ be arbitrary. Define $h \equiv [0 \vee (4 - \log_2 \varepsilon)]_1$ and $\delta_{auc}(\varepsilon, \overline{m}) \equiv 2^{-m(h)}$. Then $\delta_{auc}(\cdot, \overline{m})$ is a modulus of a.u. continuity of Z.

Moreover, the extension-by-limit $X \equiv \Phi_{Lim}(Z) : [0,1] \times \Omega \to S$ of the process Z to the full parameter set $[0,1]$ is a.u. continuous, with the same modulus of a.u. continuity $\delta_{auc}(\cdot, \overline{m})$, and with marginal distributions given by F.

Proof. 1. Note that $F|Q_\infty$ is C-regular, with $\overline{m}$ as a modulus of C-regularity. Let $Z : Q_\infty \times \Omega \to S$ be an arbitrary process with marginal distributions given by $F|Q_\infty$. We will verify that Z is a.u. continuous.

2. To that end, let $n \geq 0$ be arbitrary. Take any $\beta_n \in (2^{-n}, 2^{-n+1})$. Then, by Definition 9.2.1,

$$P(C_n) < 2^{-n}, \tag{9.2.7}$$

where

$$C_n \equiv \bigcup_{t \in Q(m(n))} \bigcup_{s \in [t,t']Q(m(n+1))} (d(Z_t, Z_s) > \beta_n) \cup (d(Z_s, Z_{t'}) > \beta_n). \tag{9.2.8}$$

As before, for each $t \in Q_{m(n)}$, we abuse notations and write $t' \equiv 1 \wedge (t + 2^{-m(n)}) \in Q_{m(n)}$.

2. Now define $D_n \equiv \left(\bigcup_{j=n}^{\infty} C_j\right)^c$. Then

$$P(D_n^c) \leq \sum_{j=n}^{\infty} P(C_j) < \sum_{j=n}^{\infty} 2^{-j} = 2^{-n+1}.$$

Let $\omega \in D_n$ be arbitrary.

3. Consider each $t \in Q_{m(n)}$. For each $s \in [t,t']Q_{m(n+1)}$, since $\omega \in C_n^c$, we have

$$d(Z(t,\omega), Z(s,\omega)) \leq \beta_n < 2^{-n+1}.$$

In short,

$$[t,t']Q_{m(n+1)} \subset (d(Z(\cdot,\omega), Z(t,\omega)) < 2^{-n+1}). \tag{9.2.9}$$

4. Repeating the argument in Step 3, with n replaced by $n+1$ and with t replaced by each $s \in [t,t']Q_{m(n+1)}$, we obtain

$$[s,s']Q_{m(n+2)} \subset (d(Z(\cdot,\omega), Z(s,\omega)) < 2^{-n}), \tag{9.2.10}$$

where $s' \equiv 1 \wedge (s + 2^{-m(n+1)}) \in Q_{m(n+1)}$. Since

$$[t,t']Q_{m(n+2)} = [t,t']Q_{m(n+1)} \cap \bigcup_{s \in [t,t']Q(m(n+1))} [s,s']Q_{m(n+2)},$$

relations 9.2.9 and 9.2.10 together yield

$$[t,t']Q_{m(n+2)} \subset (d(Z(\cdot,\omega), Z(t,\omega)) < 2^{-n+1} + 2^{-n}).$$

5. Inductively with $k = n+1, n+2, \ldots$, we obtain

$$[t,t']Q_{m(k)} \subset (d(Z(\cdot,\omega), Z(t,\omega)) < 2^{-n+1} + 2^{-n} + \cdots + 2^{-k+2})$$
$$\subset (d(Z(\cdot,\omega), Z(t,\omega)) < 2^{-n+2})$$

for each $k \geq n+1$. Therefore

$$[t,t']Q_\infty = [t,t'] \bigcup_{k=n+1}^{\infty} Q_{m(k)} \subset (d(Z(\cdot,\omega), Z(t,\omega)) < 2^{-n+2}).$$

In particular, $d(Z(t',\omega), Z(t,\omega)) < 2^{-n+2}$, so the last displayed condition implies

$$[t,t']Q_\infty \subset (d(Z(\cdot,\omega), Z(t,\omega)) \vee d(Z(\cdot,\omega), Z(t',\omega)) < 2^{-n+3}), \tag{9.2.11}$$

where $n \geq 1$, $\omega \in D_n$, and $t \in Q_{m(n)}$ are arbitrary.

6. Continuing with arbitrary $n \geq 1$, suppose $r, s \in Q_\infty$ are arbitrary such that

$$0 < s - r < 2^{-m(n)}. \tag{9.2.12}$$

Then there exist $t, u \in Q_{m(n)}$ with $t \leq u$ such that $r \in [t,t']$ and $s \in [u,u']$, where $u' \equiv 1 \wedge (u + 2^{-m(n)}) \in Q_{m(n)}$. If $t' < u$, then $s \geq u \geq t' + 2^{-m(n)} \geq r + 2^{-m(n)}$, which contradicts inequality 9.2.12. Hence $u \leq t'$. On the other hand, $t \leq u$ by the choice of t, u. Consequently, $u = t$ or $u = t'$.

At the same time, according to relation 9.2.11, we have

$$d(Z(r,\omega), Z(t,\omega)) \vee d(Z(r,\omega), Z(t',\omega)) < 2^{-n+3}.$$

Similarly,

$$d(Z(s,\omega), Z(u,\omega)) \vee d(Z(s,\omega), Z(u',\omega)) < 2^{-n+3}.$$

If $u = t$, then it follows that

$$d(Z(r,\omega), Z(s,\omega)) \leq d(Z(r,\omega), Z(t,\omega)) + d(Z(u,\omega), Z(s,\omega)) < 2^{-n+4}.$$

If $u = t'$, then similarly

$$d(Z(r,\omega), Z(s,\omega)) \leq d(Z(r,\omega), Z(t',\omega)) + d(Z(u,\omega), Z(s,\omega)) < 2^{-n+4}.$$

Summing up, for each $\omega \in D_n$, for each $n \geq 1$, and for each $r,s \in Q_\infty$ with $0 < s - r < 2^{-m(n)}$, we have

$$d(Z(r,\omega), Z(s,\omega)) < 2^{-n+4}. \tag{9.2.13}$$

By symmetry, the last inequality holds for each $\omega \in D_n$, for each $n \geq 1$, and for each $r,s \in Q_\infty$ with $|s - r| < 2^{-m(n)}$.

7. Now let $\varepsilon > 0$ be arbitrary. Let $n \equiv [0 \vee (4 - \log_2 \varepsilon)]_1$ and $\delta_{auc}(\varepsilon, \overline{m}) \equiv 2^{-m(n)}$, as in the hypothesis. In Step 6, we saw that the measurable set $D_n \equiv \left(\bigcup_{j=n}^{\infty} C_j\right)^c$ is such that $P(D_n^c) \leq 2^{-n+1} < \varepsilon$ and such that $d(Z(r,\omega), Z(s,\omega)) < 2^{-n+4} < \varepsilon$ for each $r,s \in Q_\infty$ with $|s - r| < \delta_{auc}(\varepsilon, \overline{m}) \equiv 2^{-m(n)}$. Thus the process Z is a.u. continuous, with $\delta_{auc}(\cdot, \overline{m})$ as a modulus of a.u. continuity of Z.

8. By Theorem 9.1.4, the extension-by-limit $X \equiv \Phi_{Lim}(Z)$ of the process Z to the full parameter set $[0,1]$ is a.u. continuous with the same modulus of a.u. continuity $\delta_{auc}(\cdot, \overline{m})$.

9. It remains to verify that the process X has marginal distributions given by the family F. To that end, let $r_1, \ldots, r_k$ be an arbitrary sequence in $[0,1]$, and let $s_1, \ldots, s_k$ be an arbitrary sequence in Q_∞. Let $f \in C_{ub}(S^k, d^k)$ be arbitrary. Then

$$F_{s(1),\ldots,s(k)} f = E(Z_{s(1)}, \ldots, Z_{s(k)}) = E(X_{s(1)}, \ldots, X_{s(k)}). \tag{9.2.14}$$

Now let $s_i \to r_i$ for each $i = 1, \ldots, k$. Then $F_{s(1),\ldots,s(k)} f \to F_{r(1),\ldots,r(k)} f$ because the family F is continuous in probability, and $E(X_{s(1)}, \ldots, X_{s(k)}) \to E(X_{r(1)}, \ldots, X_{r(k)})$ because the process X is a.u. continuous. Hence equality 9.2.14 yields

$$F_{r(1),\ldots,r(k)} f = E(X_{r(1)}, \ldots, X_{r(k)}), \tag{9.2.15}$$

where $r_1, \ldots, r_k$ is an arbitrary sequence in $[0,1]$, and where $f \in C_{ub}(S^k, d^k)$ is arbitrary. In short, the process X has marginal distributions given by the family F. The theorem is proved. □

Theorem 9.2.4. Continuity of extension-by-limit of C-regular processes. *Recall from Definition 6.4.2 the metric space $(\widehat{R}(Q_\infty \times \Omega, S), \rho_{Prob,Q(\infty)})$ of processes $Z : Q_\infty \times \Omega \to S$ with parameter set $Q_\infty \equiv \{t_0, t_1, \ldots\}$, sample space (Ω, L, E), and state space (S,d). Let $\widehat{R}_0$ be a subset of $\widehat{R}(Q_\infty \times \Omega, S)$ whose members are C-regular with a common modulus of C-regularity $\overline{m} \equiv (m_n)_{n=0,1,\ldots}$. Let $(\widehat{C}[0,1], \rho_{\widehat{C}[0,1]})$ be the metric space of a.u. continuous processes on $[0,1]$, as in Definition 9.1.1.*

Then the extension-by-limit

$$\Phi_{Lim} : (\widehat{R}_0, \rho_{Prob,Q(\infty)}) \to (\widehat{C}[0,1], \rho_{\widehat{C}[0,1]}),$$

as in Definition 9.1.3, is uniformly continuous with a modulus of continuity $\delta_{Creg2auc}(\cdot, \overline{m})$.

Proof. 1. Let $\varepsilon > 0$ be arbitrary. Define

$$j \equiv [0 \vee (6 - \log_2 \varepsilon)]_1,$$
$$h_j \equiv 2^{m(j)},$$

and

$$\delta_{Creg2auc}(\varepsilon, \overline{m}) \equiv 2^{-h(j)-2j-1}.$$

We will prove that $\delta_{Creg2auc}(\cdot, \overline{m})$ is a modulus of continuity of Φ_{Lim} on $\widehat{R}_0$.

2. Let $Z, Z' \in \widehat{R}_0$ be arbitrary and let $X \equiv \Phi_{Lim}(Z)$, $X' \equiv \Phi_{Lim}(Z')$. Suppose

$$\rho_{prob, Q(\infty)}(Z, Z') \equiv E \sum_{n=0}^{\infty} 2^{-n-1} \widehat{d}(Z_{t(n)}, Z'_{t(n)}) < \delta_{Creg2auc}(\varepsilon, \overline{m}). \quad (9.2.16)$$

We need to prove that $\rho_{\widehat{C}[0,1]}(X, X') < \varepsilon$.

3. By Step 6 of the proof of Theorem 9.2.3, there exist measurable sets D_j, D'_j with $P(D_j^c) \vee P(D_j'^c) < 2^{-j+1}$ such that

$$d(X_r, X_s) \vee d(X'_r, X'_s) = d(Z_r, Z_s) \vee d(Z'_r, Z'_s) \leq 2^{-j+4} \quad (9.2.17)$$

on $D_j D'_j$, for each $r, s \in Q_\infty$ with $|r - s| < 2^{-m(j)}$. Consider each $\omega \in D_j D'_j$ and $t \in [0, 1]$. Then there exists $s \in Q_{m(j)}$ such that $|t - s| < 2^{-m(j)}$. Letting $r \to t$ with $r \in Q_\infty$ and $|r - s| < 2^{-m(j)}$, inequality 9.2.17 yields

$$d(X_t(\omega), X_s(\omega)) \vee d(X'_t(\omega), X'_s(\omega)) \leq 2^{-j+4}.$$

Consequently,

$$|d(X_t(\omega), X'_t(\omega)) - d(X_s(\omega), X'_s(\omega))| < 2^{-j+5},$$

where $\omega \in D_j D'_j$ and $t \in [0, 1]$ are arbitrary. Therefore

$$\left| \sup_{t \in [0,1]} d(X_t, X'_t) - \bigvee_{s \in Q(m(j))} d(X_s, X'_s) \right| \leq 2^{-j+5} \quad (9.2.18)$$

on $D_j D'_j$. Note here that Lemma 9.1.2 earlier proved that the supremum is an r.r.v.

4. Separately, take any $\alpha \in (2^{-j}, 2^{-j+1})$ and define

$$A_j \equiv \left(\bigvee_{s \in Q(m(j))} d(X_s, X'_s) \leq \alpha \right). \quad (9.2.19)$$

Then inequality 9.2.18 and equality 9.2.19 together yield

$$G_j \equiv D_j D'_j A_j \subset \left(\sup_{t\in[0,1]} d(X_t, X'_t) \le 2^{-j+5} + 2^{-j+1} \right). \qquad (9.2.20)$$

5. By inequality 9.2.16, we have

$$\rho_{prob,Q(\infty)}(Z, Z') < \delta_{Creg2auc}(\varepsilon, \overline{m}) \equiv 2^{-h(j)-2j-1}, \qquad (9.2.21)$$

where $h_j \equiv 2^{m(j)}$. Hence

$$\begin{aligned} E \bigvee_{s\in Q(m(j))} \widehat{d}(X_s, X'_s) &= E \bigvee_{k=0}^{h(j)} \widehat{d}(X_{t(k)}, X'_{t(k)}) \\ &\le 2^{h(j)+1} E \sum_{k=0}^{h(j)} 2^{-k-1} \widehat{d}(X_{t(k)}, X'_{t(k)}) \\ &\le 2^{h(j)+1} E \sum_{k=0}^{h(j)} 2^{-k-1} \widehat{d}(Z_{t(k)}, Z'_{t(k)}) \\ &\le 2^{h(j)+1} \rho_{prob,Q(\infty)}(Z, Z') < 2^{-2j} < \alpha^2. \end{aligned} \qquad (9.2.22)$$

Chebychev's inequality therefore implies that

$$P(A_j^c) < \alpha < 2^{-j+1}.$$

Hence, in view of relation 9.2.20, we obtain

$$\begin{aligned} \rho_{\widehat{C}[0,1]}(X, X') &\equiv E \sup_{t\in[0,1]} \widehat{d}(X_t, X'_t) \le E 1_{G(j)} \sup_{t\in[0,1]} \widehat{d}(X_t, X'_t) + P(G_j^c) \\ &\le (2^{-j+5} + 2^{-j+1}) + P(A_j^c) + P(D_j^c) + P(D_j'^c) \\ &< (2^{-j+5} + 2^{-j+1}) + 2^{-j+1} + 2^{-j+1} + 2^{-j+1} < 2^{-j+6} < \varepsilon. \end{aligned}$$

Since $\varepsilon > 0$ is arbitrary, we see that $\delta_{Creg2auc}(\cdot, \overline{m})$ is a modulus of continuity of Φ_{Lim}. □

Theorems 9.2.3 and 9.2.4 can now be restated in terms of C-regular consistent families of f.j.d.'s.

Corollary 9.2.5. Construction of a.u. continuous process from C-regular family of f.j.d.'s. *Let*

$$(\Theta_0, L_0, I_0) \equiv \left([0,1], L_0, \int \cdot dx \right)$$

denote the Lebesgue integration space based on the interval $\Theta_0 \equiv [0,1]$. Let ξ be a fixed binary approximation of (S, d) relative to a reference point $x_\circ \in S$. Recall from Definition 6.2.12 the metric space $(\widehat{F}_{Cp}([0,1], S), \widehat{\rho}_{Cp,\xi,[0,1],Q(\infty)})$ of *consistent families of f.j.d.'s that are continuous in probability, with parameter set* $[0,1]$ *and state space* (S,d).

Let $\widehat{F}_0$ *be a subset of* $\widehat{F}_{Cp}([0,1],S)$ *whose members are C-regular and share a common modulus of C-regularity* $\overline{m} \equiv (m_n)_{n=0,1,2,...}$. *Define the restriction function* $\Phi_{[0,1],Q(\infty)} : \widehat{F}_0 \to \widehat{F}_0|Q_\infty$ *by* $\Phi_{[0,1],Q(\infty)}(F) \equiv F|Q_\infty$ *for each* $F \in \widehat{F}_0$. *Then the following conditions hold:*

1. The function

$$\Phi_{fjd,auc,\xi} \equiv \Phi_{Lim} \circ \Phi_{DKS,\xi} \circ \Phi_{[0,1],Q(\infty)} : (\widehat{F}_0, \widehat{\rho}_{Cp,\xi,[0,1],Q(\infty)}) \to (\widehat{C}[0,1], \rho_{\widehat{C}[0,1]}) \tag{9.2.23}$$

is well defined, where $\Phi_{DKS,\xi}$ *is the Daniell–Kolmogorov–Skorokhod extension constructed in Theorem 6.4.3, and where* Φ_{Lim} *is the extension-by-limit constructed in Theorem 9.2.3.*

2. For each consistent family $F \in \widehat{F}_0$, *the a.u. continuous process* $X \equiv \Phi_{fjd,auc,\xi}(F)$ *has marginal distributions given by* F.

3. Let $\widehat{F}_{0,\gamma}$ *be a subset of* $\widehat{F}_0$ *such that* $\{F_0; F \in \widehat{F}_{0,\gamma}\}$ *is tight with a certain modulus of tightness* γ. *Then the construction* $\Phi_{fjd,auc,\xi}$ *is uniformly continuous on the subset* $\widehat{F}_{0,\gamma}$.

Proof. 1. Let $F \in \widehat{F}_0$ be arbitrary. By hypothesis, F is C-regular, with $\overline{m}$ as a modulus of C-regularity. Since the process $Z \equiv \Phi_{DKS,\xi}(F|Q_\infty) : Q_\infty \times \Theta_0 \to S$ extends $F|Q_\infty$, Z is C-regular, with $\overline{m}$ as a modulus of C-regularity. In other words, $Z \in \widehat{R}_0$, where $\widehat{R}_0$ is the set of C-regular processes on Q_∞, with sample space (Θ_0, L_0, I_0) and whose members have $\overline{m}$ as a modulus of C-regularity. Hence the a.u. continuous process $X \equiv \Phi_{Lim}(Z)$ is well defined by Theorem 9.2.3, with $X|Q_\infty = Z$. Thus the composite mapping in equality 9.2.23 is well defined. Assertion 1 is verified.

2. Being C-regular, the family $F \in \widehat{F}_0$ is continuous in probability. Hence, for each $r_1, \ldots, r_n \in [0,1]$ and $f \in C_{ub}(S^n)$, we have

$$\begin{aligned} F_{r(1),\ldots,r(n)}f &= \lim_{s(i)\to r(i);\, s(i)\in Q(\infty);\, i=1,\ldots,n} F_{s(1),\ldots,s(n)}f \\ &= \lim_{s(i)\to r(i);\, s(i)\in Q(\infty);\, i=1,\ldots,n} I_0 f(Z_{s(1)}, \ldots, Z_{s(n)}) \\ &= \lim_{s(i)\to r(i);\, s(i)\in Q(\infty);\, i=1,\ldots,n} I_0 f(X_{s(1)}, \ldots, X_{s(n)}) \\ &= I_0 f(X_{r(1)}, \ldots, X_{r(n)}), \end{aligned}$$

where the last equality follows from the a.u. continuity of X. We conclude that F is the family of marginal distributions of X, proving Assertion 2.

3. Let $\widehat{F}_{0,\gamma}$ be an arbitrary subset of $\widehat{F}_0$ such that $\{F_0; F \in \widehat{F}_{0,\gamma}\}$ is tight, with a certain modulus of tightness γ. Consider each $F \in \widehat{F}_{0,\gamma}$. Then $X \equiv \Phi_{fjd,auc,\xi}(F)$ is a.u. continuous with a modulus of a.u. continuity $\delta_{auc}(\cdot, \overline{m})$.

Let $\varepsilon_0 > 0$ be arbitrary. Write $\varepsilon \equiv 2^{-1}\varepsilon_0$. Then there exists a measurable set A with $P(A^c) < \varepsilon$ such that for each $\omega \in A$, we have $d(X_t(\omega), X_s(\omega)) \le \varepsilon$ for each

$t, s \in [0,1]$ with $|t-s| < \delta_{auc}(\varepsilon, \overline{m})$. Take $n \geq 1$ so large that $n^{-1} < \delta_{auc}(\varepsilon, \overline{m})$. Separately, there exists $c > \gamma(\varepsilon)$ such that $P(C^c) < \varepsilon$, where $C \equiv (d(X_0, x_\circ) \leq c)$. Let $t \in [0,1]$ be arbitrary. Define

$$\beta(\varepsilon_0, t) \equiv \beta(\varepsilon_0, t, \gamma, \overline{m}) \equiv c + n\varepsilon.$$

Consider each $\omega \in AC$. Then

$$d(x_\circ, X_t(\omega)) \leq d(x_\circ, X_0(\omega)) + \sum_{i=1}^{n} d(X_{(i-1)t/n}(\omega), X_{it/n}(\omega))$$
$$\leq c + n\varepsilon \equiv \beta(\varepsilon_0, t).$$

In other words, $(d(x_\circ, X_t) > \beta(\varepsilon_0, t)) \subset A^c \cup C^c$. For each $k \geq 1$, define the function $h_k \equiv 1 \wedge (1 + k - d(\cdot, x_\circ))_+ \in C(S, d)$. Then, for each $k \geq \beta(\varepsilon_0, t)$, we have

$$(1 - h_k(X_t) > 0) \subset (d(X_t, x_\circ) > k) \subset (d(X_t, x_\circ) > \beta(\varepsilon_0, t)) \subset A^c \cup C^c.$$

Hence

$$F_t(1 - h_k) = E(1 - h_k(X_t)) \leq P(A^c \cup C^c) < 2\varepsilon \equiv \varepsilon_0.$$

Thus $\beta(\cdot, \cdot, \gamma, \overline{m})$ is a modulus of pointwise tightness of the family F, according to Definition 6.3.5. Summing up, $\widehat{F}_{0,\gamma} \subset \widehat{F}_\beta \equiv \widehat{F}_\beta([0,1], S)$.

4. Recall the metric space $(\widehat{R}(Q_\infty \times \Theta_0, S), \rho_{prob, Q(\infty)})$ of processes $Z : Q_\infty \times \Theta_0 \to S$. Then the function

$$\Phi_{[0,1], Q(\infty)} : (\widehat{F}_{0,\gamma}, \widehat{\rho}_{Cp,\xi,[0,1],Q(\infty)}) \to (\widehat{F}_{0,\gamma}|Q_\infty, \widehat{\rho}_{Marg,\xi,Q(\infty)}) \tag{9.2.24}$$

is an isometry, according to Definition 6.2.12. Separately, the Daniell–Kolmogorov–Skorokhod Extension

$$\Phi_{DKS,\xi} : (\widehat{F}_\beta|Q_\infty, \widehat{\rho}_{Marg,\xi,Q(\infty)}) \to (\widehat{R}(Q_\infty \times \Theta_0, S), \rho_{prob,Q(\infty)})$$

is uniformly continuous, according to Theorem 6.4.5. Hence, since $\widehat{F}_{0,\gamma} \subset \widehat{F}_\beta$, the function

$$\Phi_{DKS,\xi} : (\widehat{F}_{0,\gamma}|Q_\infty, \widehat{\rho}_{Marg,\xi,Q(\infty)}) \to (\widehat{R}(Q_\infty \times \Theta_0, S), \rho_{prob,Q(\infty)}) \tag{9.2.25}$$

is uniformly continuous. Moreover, by Step 1, we see that $\Phi_{DKS,\xi}(\widehat{F}_0|Q_\infty) \subset \widehat{R}_0$,where the $\widehat{R}_0$ is a set of C-regular processes on Q_∞ whose members share the common modulus of C-regularity $\overline{m}$. Therefore we have the uniformly continuous function

$$\Phi_{DKS,\xi} : (\widehat{F}_{0,\gamma}|Q_\infty, \widehat{\rho}_{Marg,\xi,Q(\infty)}) \to (\widehat{R}_0, \rho_{prob,Q(\infty)}). \tag{9.2.26}$$

5. Finally, Theorem 9.2.4 says that

$$\Phi_{Lim} : (\widehat{R}_0, \rho_{prob,Q(\infty)}) \to (\widehat{C}[0,1], \rho_{\widehat{C}[0,1]}) \tag{9.2.27}$$

is uniformly continuous. In view of expressions 9.2.24, 9.2.26, and 9.2.27, we see that the composite function

$$\Phi_{fjd,auc,\xi} \equiv \Phi_{Lim} \circ \Phi_{DKS,\xi} \circ \Phi_{[0,1],Q(\infty)} : (\widehat{F}_{0,\gamma}, \widehat{\rho}_{Cp,\xi,[0,1],Q(\infty)}) \\ \to (\widehat{C}[0,1], \rho_{\widehat{C}[0,1]}) \tag{9.2.28}$$

is uniformly continuous. Assertion 3 and the corollary are proved. □

9.3 a.u. Hoelder Process

In this section, let (S,d) be a locally compact metric space. We will prove a theorem, due to Kolmogorov and Chentsov, that gives a sufficient condition on pairwise joint distributions for the construction of an a.u. Hoelder continuous process $X : [0,1] \times \Omega \to S$, in a sense to be made precise presently.

Refer to Sections 9.1 and 9.2 and to Definitions 9.0.2 and 9.0.3 for notations and conventions, especially for the sets Q_k and Q_∞ of dyadic rationals in $[0,1]$, for each $k \geq 0$. Refer to Definition 9.1.3 for the operation Φ_{Lim} that extends, if possible, a process by limit.

Lemma 9.3.1. A sufficient condition on pair distributions for a.u. continuity. *Let $\kappa \geq 1$ be arbitrary. Let $\overline{\gamma} \equiv (\gamma_k)_{k=\kappa,\kappa+1,\ldots}$ and $\overline{\varepsilon} \equiv (\varepsilon_k)_{k=\kappa,\kappa+1,\ldots}$ be two arbitrary sequences of positive real numbers with $\sum_{k=\kappa}^{\infty} \gamma_k < \infty$ and $\sum_{k=\kappa}^{\infty} \varepsilon_k < \infty$.*

Let $Z : Q_\infty \times (\Omega, L, E) \to (S,d)$ be an arbitrary process such that for each $k \geq \kappa$ and for each $\alpha_k \geq \gamma_k$, we have

$$\sum_{t\in[0,1)Q(k)} P(d(Z_t, Z_{t+\Delta(k+1)}) > \alpha_k) \\ + \sum_{t\in[0,1)Q(k)} P(d(Z_{t+\Delta(k+1)}, Z_{t+\Delta(k)}) > \alpha_k) \leq 2\varepsilon_k. \tag{9.3.1}$$

Then the following conditions hold:

1. The extension-by-limit $X \equiv \Phi_{Lim}(Z) : [0,1] \times \Omega \to (S,d)$ is an a.u. continuous process, with a modulus of a.u. continuity $\delta_{auc}(\cdot, \overline{\gamma}, \overline{\varepsilon}, \kappa)$ defined as follows. Let $\varepsilon > 0$ be arbitrary. Take $n \geq \kappa$ so large that $12\sum_{k=n}^{\infty} \gamma_k \vee 2\sum_{k=n}^{\infty} \varepsilon_k < \varepsilon$. Define $\delta_{auc}(\varepsilon, \overline{\gamma}, \overline{\varepsilon}, \kappa) \equiv 2^{-n-1}$.

2. There exists a sequence $(D_n)_{n=\kappa,\kappa+1,\ldots}$ of measurable sets such that (i) *$D_\kappa \subset D_{\kappa+1} \subset \ldots$;* (ii) *for each $n \geq \kappa$, we have*

$$P(D_n^c) \leq 2\sum_{k=n}^{\infty} \varepsilon_k; \tag{9.3.2}$$

and (iii) *for each $n \geq \kappa$ and each $\omega \in D_n$, we have*

$$d(X_r(\omega), X_s(\omega)) < 12\sum_{k=n}^{\infty} \gamma_k \tag{9.3.3}$$

for each $r, s \in [0,1]$ with $|r - s| < 2^{-n-1}$.

Proof. 1. Let $r \in Q_\infty$ and $k \geq 1$ be arbitrary. There exists a unique $\overline{u}_k(r) \equiv \overline{u}(r,k) \in Q_k$ such that $r \in [\overline{u}_k(r), \overline{u}_k(r) + 2^{-k})$. In words, $\overline{u}_k : Q_\infty \to Q_k$ is the function that assigns to each $r \in Q_\infty$ the largest member in Q_k not to exceed r. Then either $r \in [\overline{u}_k(r), \overline{u}_k(r) + 2^{-k-1})$ or $r \in [\overline{u}_k(r) + 2^{-k-1}, \overline{u}_k(r) + 2^{-k})$. Note that $\overline{u}_k(r), \overline{u}_k(r) + 2^{-k-1} \in Q_{k+1}$. Hence we have either $\overline{u}_{k+1}(r) = \overline{u}_k(r)$ or $\overline{u}_{k+1}(r) = \overline{u}_k(r) + 2^{-k-1}$. In either case,

$$|\overline{u}_k(r) - \overline{u}_{k+1}(r)| \leq 2^{-k-1} \equiv \Delta_{k+1}. \tag{9.3.4}$$

Separately, if $r \in Q_k$, then $\overline{u}_k(r) = r$.

2. Let $Z : Q_\infty \times \Omega \to S$ be as given. Let $k \geq \kappa$ be arbitrary, and fix any $\alpha_k \in [\gamma_k, 2\gamma_k)$. Define

$$C_k \equiv \bigcup_{t \in [0,1)Q(k)} (d(Z_t, Z_{t+\Delta(k+1)}) > \alpha_k) \cup (d(Z_{t+\Delta(k+1)}, Z_{t+\Delta(k)}) > \alpha_k). \tag{9.3.5}$$

Then $P(C_k) \leq 2\varepsilon_k$, thanks to inequality 9.3.1 in the hypothesis. Let $\omega \in C_k^c$ be arbitrary. Consider each $u \in Q_k$ and $v \in Q_{k+1}$ with $|u - v| \leq \Delta_{k+1}$. There are three possibilities: (i′) $v = u$, (ii′) $v = u + \Delta_{k+1}$, or (iii′) $v = r + \Delta_{k+1}$ and $u = r + \Delta_k$ for some $r \in [0,1)Q_k$. In view of equality 9.3.5, each of Conditions (i′–iii′) yields

$$d(Z_u(\omega), Z_v(\omega)) \leq \alpha_k, \tag{9.3.6}$$

where $u \in Q_k$ and $v \in Q_{k+1}$ are arbitrary with $|u - v| \leq \Delta_{k+1}$, and where $\omega \in C_k^c$ and $k \geq \kappa$ are arbitrary.

3. Let $n \geq \kappa$ be arbitrary, but fixed until further notice. Define the measurable set

$$D_n \equiv \left(\bigcup_{k=n}^{\infty} C_k \right)^c.$$

Then $D_n \subset D_{n+1}$ and

$$PD_n^c \equiv P\left(\bigcup_{k=n}^{\infty} C_k \right) \leq \sum_{k=n}^{\infty} 2\varepsilon_k. \tag{9.3.7}$$

4. Consider each $\omega \in D_n$, $t \in [0,1)Q_n$, and $k \geq n$. Then $\omega \in D_n \subset C_k^c$. Let $r \in Q_\infty[t, t + 2^{-n})$ be arbitrary. According to inequalities 9.3.4 and 9.3.6, we have

$$d(Z_{\overline{u}(r,k)}(\omega), Z_{\overline{u}(r,k+1)}(\omega)) \leq \alpha_k.$$

The triangle inequality therefore yields, for each $i \geq n$,

$$d(Z_{\overline{u}(r,n)}(\omega), Z_{\overline{u}(r,i)}(\omega)) \leq \sum_{k=n}^{i-1} d(Z_{\overline{u}(r,k)}(\omega), Z_{\overline{u}(r,k+1)}(\omega)) \leq \sum_{k=n}^{\infty} \alpha_k. \tag{9.3.8}$$

At the same time, because $r \in [t, t+2^{-n})$ and $r \in [\overline{u}_n(r), \overline{u}_n(r)+2^{-n})$, the uniqueness of $\overline{u}_n(r)$ in Q_n implies that $t = \overline{u}_n(r)$. Moreover, because $r \in Q_\infty$, there exists $i \geq n$ such that $r \in Q_i$, whence $\overline{u}_i(r) = r$. Inequality 9.3.8 therefore leads to

$$d(Z_t(\omega), Z_r(\omega)) = d(Z_{\overline{u}(r,n)}(\omega), Z_{\overline{u}(r,i)}(\omega)) \leq \sum_{k=n}^{\infty} \alpha_k, \tag{9.3.9}$$

where $r \in Q_\infty[t, t+2^{-n})$, $t \in [0,1)Q_n$, and $\omega \in D_n$ are arbitrary.

5. Next, consider the endpoint $r = t + 2^{-n}$ of $[t, t+2^{-n})$. Then, because $\omega \in D_n \subset C_n^c$, the defining equality 9.3.5, with k replaced by n, implies that

$$d(Z_t(\omega), Z_{t+\Delta(n+1)}(\omega)) \vee d(Z_{t+\Delta(n+1)}(\omega), Z_{t+\Delta(n)}(\omega)) \leq \alpha_n.$$

Hence

$$\begin{aligned} d(Z_t(\omega), Z_r(\omega)) &= d(Z_t(\omega), Z_{t+\Delta(n)}(\omega)) \\ &\leq d(Z_t(\omega), Z_{t+\Delta(n+1)}(\omega)) + d(Z_{t+\Delta(n+1)}(\omega), Z_{t+\Delta(n)}(\omega)) \\ &\leq 2\alpha_n \leq 2\sum_{k=n}^{\infty} \alpha_k, \end{aligned}$$

where $r = t + 2^{-n}$. Combining with inequality 9.3.9, this yields

$$d(Z_t(\omega), Z_r(\omega)) \leq 2\sum_{k=n}^{\infty} \alpha_k, \tag{9.3.10}$$

where $r \in Q_\infty[t, t+2^{-n}]$, $t \in [0,1)Q_n$, and $\omega \in D_n$ are arbitrary. The same inequality 9.3.10 also holds when the condition $t \in [0,1)Q_n$ is relaxed to $t \in [0,1]Q_n = Q_n$, because if $t = 1$, then $r = 1$, and inequality 9.3.10 holds trivially.

6. Again, consider each $\omega \in D_n$. Write $m \equiv n+1$. Then $\omega \in D_m$. Let $r, s \in Q_\infty$ be arbitrary such that $|s - r| < 2^{-m} \equiv 2^{-n-1}$. Assume, without loss of generality, that $s \leq r$. Let $t \equiv \overline{u}_m(r) \in Q_m$. Then $r \in Q_\infty[t, t+2^{-m}]$. Hence, by inequality 9.3.10, we have

$$d(Z_t(\omega), Z_r(\omega)) \leq 2\sum_{k=m}^{\infty} \alpha_k. \tag{9.3.11}$$

Moreover, $s \in [t - 2^{-m}, t]$ or, $s \in [t, t+2^{-m}]$. Therefore, again by inequality 9.3.10, we have

$$d(Z_t(\omega), Z_s(\omega)) \leq 4\sum_{k=m}^{\infty} \alpha_k. \tag{9.3.12}$$

Combining inequalities 9.3.11, and 9.3.12, we obtain

$$d(Z_r(\omega), Z_s(\omega)) \leq 6\sum_{k=n+1}^{\infty} \alpha_k < 12\sum_{k=n}^{\infty} \gamma_k \tag{9.3.13}$$

for arbitrary $r, s \in Q_\infty$ with $|s - r| < 2^{-n-1}$, where $\omega \in D_n$ is arbitrary.

7. Since $P(D_n^c) \le \sum_{k=n}^{\infty} 2\varepsilon_k \to 0$ and $12\sum_{k=n}^{\infty} \gamma_k \to 0$ as $n \to \infty$, it follows that the process Z is a.u. continuous, with a modulus of a.u. continuity δ_{auc} defined as follows. Let $\varepsilon > 0$ be arbitrary. Let $n \equiv n(\varepsilon, \overline{\gamma}, \overline{\varepsilon}, \kappa) \ge \kappa$ be so large that

$$12\sum_{k=n}^{\infty} \gamma_k \vee 2\sum_{k=n}^{\infty} \varepsilon_k < \varepsilon.$$

Define

$$\delta_{auc}(\varepsilon) \equiv \delta_{auc}(\varepsilon, \overline{\gamma}, \overline{\varepsilon}, \kappa) \equiv 2^{-n-1}.$$

8. Theorem 9.1.4 then says that the extension-by-limit $X \equiv \Phi_{Lim}(Z) : [0,1] \times \Omega \to (S,d)$ is an a.u. continuous process, with the same modulus of a.u. continuity $\delta_{auc}(\cdot, \overline{\gamma}, \overline{\varepsilon}, \kappa)$. Moreover, inequality 9.3.13 implies that

$$d(X_r(\omega), X_s(\omega)) \le 12\sum_{k=n}^{\infty} \gamma_k, \tag{9.3.14}$$

where $r,s \in Q_\infty$ are arbitrary with $|r-s| < 2^{-n-1}$, and where $\omega \in D_n$ is arbitrary. Since X is a.u. continuous on $[0,1]$ and since Q_∞ is dense in $[0,1]$, it follows that inequality 9.3.14 holds for arbitrary $r,s \in [0,1]$ with $|r-s| < 2^{-n-1}$ and for arbitrary $\omega \in D_n$.

The lemma is proved. □

Definition 9.3.2. a.u. Hoelder continuous process. Let $a > 0$ be arbitrary. A process $X : [0,a] \times \Omega \to (S,d)$ is said to be *a.u. globally Hoelder*, or simply *a.u. Hoelder*, if there exists a constant $\theta > 0$ and for each $\varepsilon > 0$, there exists a measurable set D with $P(D^c) < \varepsilon$ and a real number $c_H(\varepsilon)$ such that for each $\omega \in D$, we have

$$d(X_r(\omega), X_s(\omega)) < c_H(\varepsilon)|r-s|^\theta \tag{9.3.15}$$

for each $r,s \in [0,a]$. The constant θ is called a *Hoelder exponent* of the process X, and the operation c_H is called an *a.u. Hoelder coefficient* of the process X.

We emphasize that θ and c_H are independent of r,s, which explains our use of the adverb *globally*.

Theorem 9.3.3. A sufficient condition on pair distributions for a.u. Hoelder continuity. *Let (S,d) be a locally compact metric space. Let $c_0, u, w > 0$ be arbitrary. Let $\theta > 0$ be arbitrary such that $\theta < u^{-1}w$. Then the following conditions hold:*

1. Suppose $Z : Q_\infty \times (\Omega, L, E) \to (S,d)$ is an arbitrary process such that

$$P(d(Z_r, Z_s) > b) \le c_0 b^{-u}|r-s|^{1+w} \tag{9.3.16}$$

for each $b > 0$, for each $r,s \in Q_\infty$. Then the extension-by-limit $X \equiv \Phi_{Lim}(Z) : [0,1] \times \Omega \to (S,d)$ is a.u. Hoelder with exponent θ, and with some a.u. Hoelder coefficient $c_H \equiv c_H(\cdot, c_0, u, w, \theta)$.

2. *Inequality 9.3.16 is satisfied if*

$$Ed(Z_r, Z_s)^u \leq c_0|r - s|^{1+w} \tag{9.3.17}$$

for each $r, s \in Q_\infty$.

3. *Suppose* F *is a consistent family of f.j.d.'s such that*

$$F_{r,s}1_{(d>b)} \leq c_0 b^{-u}|r - s|^{1+w} \tag{9.3.18}$$

for each $b > 0$, *for each* $r, s \in [0,1]$. *Then* F *is the family of marginal distributions of some process* $X : [0,1] \times \Omega \to (S,d)$ *that is a.u. Hoelder with exponent* θ, *and with some a.u. Hoelder coefficient* $c_H \equiv c_H(\cdot, c_0, u, w, \theta)$. *Inequality 9.3.18 is satisfied if*

$$F_{r,s}d^u \leq c_0|r - s|^{1+w} \tag{9.3.19}$$

for each $r, s \in [0,1]$.

Proof. Let $Z : Q_\infty \times (\Omega, L, E) \to (S,d)$ be an arbitrary process such that inequality 9.3.16 holds.

1. As an abbreviation, define the positive constants $a \equiv (w - \theta u)$, $c_1 \equiv 12(1 - 2^{-\theta})^{-1}$, and $c \equiv c_1 2^{2\theta}$. Fix $\kappa \equiv \kappa(c_0, a, w)$ so large that (i) $\kappa \geq 2 \vee 2c_0 2^{-w-1}$ and (ii) $2^{-ka}k^2 \leq 1$ for each $k \geq \kappa$. Thus K is determined by c_0, u, w, and θ. We will verify that the process X is a.u. Hoelder with Hoelder exponent θ.

2. To that end, let $k \geq \kappa$ be arbitrary. Define

$$\varepsilon_k \equiv c_0 2^{-w-1}k^{-2} \tag{9.3.20}$$

and

$$\gamma_k \equiv 2^{-kw/u}k^{2/u}. \tag{9.3.21}$$

Take any $\alpha_k \geq \gamma_k$ such that the set $(d(Z_t, Z_s) > \alpha_k)$ is measurable for each $t, s \in Q_\infty$. Let $t \in [0,1)Q_k$ be arbitrary. We estimate

$$\begin{aligned}
P(d(Z_t, Z_{t+\Delta(k+1)}) > \alpha_k) &\leq c_0\alpha_k^{-u}\Delta_{k+1}^{1+w} \\
&\leq c_0\gamma_k^{-u}2^{-(k+1)w}2^{-(k+1)} \\
&= c_0 2^{kw}k^{-2}2^{-(k+1)w}2^{-(k+1)} \\
&= c_0 2^{-w-1}k^{-2}2^{-k},
\end{aligned}$$

where the first inequality is thanks to inequality 9.3.16 in the hypothesis. Similarly,

$$P(d(Z_{t+\Delta(k+1)}, Z_{t+\Delta(k)}) > \alpha_k) \leq c_0 2^{-w-1}k^{-2}2^{-k}.$$

Combining, we obtain

$$\sum_{t\in[0,1)Q(k)} P(d(Z_t, Z_{t+\Delta(k+1)}) > \alpha_k) + \sum_{t\in[0,1)Q(k)} P(d(Z_{t+\Delta(k+1)}, Z_{t+\Delta(k)}) > \alpha_k)$$

$$\leq 2 \cdot 2^k(c_0 2^{-w-1}k^{-2}2^{-k}) = 2c_0 2^{-w-1}k^{-2} \equiv 2\varepsilon_k,$$

where $k \geq \kappa$ is arbitrary.

3. Since $\sum_{k=\kappa}^{\infty} \gamma_k < \infty$ and $\sum_{k=\kappa}^{\infty} \varepsilon_k < \infty$, the conditions in the hypothesis of Theorem 9.3.1 are satisfied by the objects $Z, (\gamma_k)_{k=\kappa,\kappa+1\ldots}, (\varepsilon_k)_{k=\kappa,\kappa+1\ldots}$. Accordingly, the extension-by-limit $X \equiv \Phi_{Lim}(Z) : [0,1] \times \Omega \to (S,d)$ is a.u. continuous.

4. Moreover, according to Assertion 2 of Lemma 9.3.1, there exists a sequence $(D_n)_{n=\kappa,\kappa+1,\ldots}$ of measurable sets such that (i) $D_\kappa \subset D_{\kappa+1} \subset \cdots$, (ii) for each $n \geq \kappa$, we have

$$P(D_n^c) \leq 2\sum_{k=n}^{\infty} \varepsilon_k; \tag{9.3.22}$$

and (iii) for each $n \geq \kappa$ and each $\omega \in D_n$, we have

$$d(X_r(\omega), X_s(\omega)) < 12\sum_{k=n}^{\infty} \gamma_k \tag{9.3.23}$$

for each $r,s \in [0,1]$ with $|r-s| \leq 2^{-n-1}$. Because the process X is a.u. continuous, we may assume that $X(\cdot,\omega)$ is continuous on $[0,1]$ for each $\omega \in D_n$, for each $n \geq \kappa$.

5. We will now estimate bounds for the partial sum on the right-hand side of each of the inequalities 9.3.22 and 9.3.23. To that end, consider each $n \geq \kappa$. Then

$$2\sum_{k=n}^{\infty} \varepsilon_k \equiv 2\sum_{k=n}^{\infty} c_0 2^{-w-1} k^{-2} \leq 2c_0 2^{-w-1} \int_{y=n-1}^{\infty} y^{-2} dy = 2c_0 2^{-w-1}(n-1)^{-1}. \tag{9.3.24}$$

At the same time,

$$\begin{aligned} 12\sum_{k=n}^{\infty} \gamma_k &\equiv 12\sum_{k=n}^{\infty} 2^{-kw/u} k^{2/u} \equiv 12\sum_{k=n}^{\infty} 2^{-k\theta u/u} 2^{-ka/u} k^{2/u} \\ &= 12\sum_{k=n}^{\infty} 2^{-k\theta} 2^{-ka/u} k^{2/u} \leq 12\sum_{k=n}^{\infty} 2^{-k\theta} \\ &= 12 \cdot 2^{-n\theta}(1-2^{-\theta})^{-1} \equiv c_1 2^{-n\theta}, \end{aligned} \tag{9.3.25}$$

where the second equality is because $a \equiv (w - \theta u)$, and where the inequality is because $2^{-ka} k^2 \leq 1$ for each $k \geq n \geq \kappa$.

6. Recall that $\kappa \geq 2 \vee 2c_0 2^{-w-1}$. Now let $\varepsilon > 0$ be arbitrary. Take $m \geq \kappa$ so large that

$$2c_0 2^{-w-1}(m-1)^{-1} < \varepsilon. \tag{9.3.26}$$

Thus m is determined by ε, κ, c_0, and w. In Step 1, we saw that κ is determined by c_0, θ, u, and w. Hence m is determined by $\varepsilon, \theta, u, c_0$, and w. It follows that $c_H \equiv c2^{(m+1)}$ is determined by $\varepsilon, \theta, u, c_0$, and w. Inequalities 9.3.22, 9.3.24, and 9.3.26 together imply that $P(D_m^c) \leq \varepsilon$. Consider each $\omega \in D_m \subset D_{m+1} \subset \ldots$.

Then, for each $n \geq m$, we have $\omega \in D_n$, whence inequalities 9.3.23 and 9.3.25 together imply that

$$d(X_r(\omega), X_s(\omega)) < c_1 2^{-n\theta} \tag{9.3.27}$$

for each

$$(r,s) \in G_n \equiv \{(r,s) \in Q_\infty^2 : 2^{-n-2} \leq |r-s| \leq 2^{-n-1}\}.$$

7. Hence

$$d(X_r(\omega), X_s(\omega)) < c_1 2^{2\theta} 2^{-(n+2)\theta} \leq c_1 2^{2\theta} |r-s|^\theta \equiv c|r-s|^\theta \tag{9.3.28}$$

for each $(r,s) \in G_n$, for each $n \geq m$.

8. Therefore

$$d(X_r(\omega), X_s(\omega)) < c|r-s|^\theta \tag{9.3.29}$$

for each

$$(r,s) \in \overline{G}_m \equiv \{(r,s) \in Q_\infty^2 : |r-s| \leq 2^{-m-1}\} = \bigcup_{n=m}^{\infty} G_n.$$

9. Let $r,s \in Q_\infty$ be arbitrary. Now write $n \equiv 2^{m+1} \geq m$. For each $i = 0,\ldots,n$, define $r_i \equiv (1 - in^{-1})r + in^{-1}s \in Q_\infty$. Then $r_0 = r$, $r_n = s$, and

$$|r_i - r_{i-1}| = n^{-1}|r-s| \equiv 2^{-m-1}|r-s| \leq 2^{-m-1}.$$

Hence $(r_{i-1}, r_i) \in \overline{G}_m$ for each $i = 1,\ldots,n$. Therefore, in view of inequality 9.3.29, we obtain

$$\begin{aligned} d(X_r(\omega), X_s(\omega)) &\leq \sum_{i=1}^{n} d(X_{r(i-1)}(\omega), X_{r(i)}(\omega)) < nc|r-s|^\theta \\ &\equiv 2^{m+1} c|r-s|^\theta \equiv c_H |r-s|^\theta, \end{aligned} \tag{9.3.30}$$

where $r,s \in Q_\infty$ and $\omega \in D_m$ are arbitrary. Because $X(\cdot,\omega)$ is continuous on $[0,1]$, inequality 9.3.30 implies that

$$d(X_r(\omega), X_s(\omega)) \leq c_H |r-s|^\theta, \tag{9.3.31}$$

for each $r,s \in [0,1]$, where $\omega \in D_m$ is arbitrary.

10. Thus the process X is a.u. Hoelder, with exponent θ and with coefficient c_H, as alleged. Assertion 1 is proved.

11. Assertion 2 follows from Chebychev's inequality.

12. Suppose F is a consistent family of f.j.d.'s such that

$$F_{r,s} 1_{(d>b)} \leq c_0 b^{-u} |r-s|^{1+w} \tag{9.3.32}$$

for each $b > 0$, for each $r,s \in [0,1]$. Then F is a consistent family of f.j.d.'s with state space S and parameter space $[0,1]$ that is continuous in probability.

Let $Z : Q_\infty \times (\Omega, L, E) \to (S,d)$ be an arbitrary process whose marginal distributions are given by $F|Q_\infty$. For example, we can take Z to be the Daniell–Kolmogorov–Skorokhod Extension of the family $F|Q_\infty$. Then inequality 9.3.32 shows that the a.u. continuous process Z satisfies inequality 9.3.16. Hence, by Assertion 1, the extension-by-limit $X \equiv \Phi_{Lim}(Z) : [0,1] \times \Omega \to (S,d)$ is a.u. Hoelder with exponent θ, and with a.u. Hoelder coefficient c_H. Let F' denote the family of marginal distributions of the process X. Then, since the process X is a.u. continuous, the family F' is continuous in probability. At the same time, for each sequence $s_1, \ldots, s_m \in Q_\infty$, the distribution $F_{s(1),\ldots,s(m)}$ is the joint distribution of the r.v.'s $Z_{s(1)}, \ldots, Z_{s(m)}$ by the construction of the process Z, while $F'_{s(1),\ldots,s(m)}$ is the joint distribution of the r.v.'s $X_{s(1)}, \ldots, X_{s(m)}$. Now $(X_{s(1)}, \ldots, X_{s(m)}) = (Z_{s(1)}, \ldots, Z_{s(m)})$ because $X \equiv \Phi_{Lim}(Z)$. We conclude that $F_{s(1),\ldots,s(m)} = F'_{s(1),\ldots,s(m)}$, where $m \geq 1$ and $s_1, \ldots, s_m \in Q_\infty$ are arbitrary. Therefore, by Assertion 2 of Lemma 6.2.11, we have $F = F'$. In other words, the marginal distributions of the a.u. Hoelder process X are given by the family F. Assertion 3 of the present theorem is proved. □

9.4 Brownian Motion

One application of Theorem 9.3.3 is the construction of the Brownian motion. In the this section, let the dimension $m \geq 1$ be arbitrary, but fixed.

Definition 9.4.1. Brownian motion in R^m. An a.u. continuous process $B : [0,\infty) \times (\Omega, L, E) \to R^m$ is called a *Brownian motion in* R^m if (i) $B_0 = 0$; (ii) for each sequence $0 \equiv t_0 \leq t_1 \leq \cdots \leq t_{n-1} \leq t_n$ in $[0,\infty)$, the r.v.'s $B_{t(1)} - B_{t(0)}, \ldots, B_{t(n)} - B_{t(n-1)}$ are independent; and (iii) for each $s,t \in [0,\infty)$, the r.v. $B_t - B_s$ is normal with mean 0 and covariance matrix $|t-s|I$. □

We first construct a Brownian motion in R^1. In the following, recall that $\overline{Q}_\infty$ stands for the set of dyadic rationals in $[0,\infty)$.

Theorem 9.4.2. Construction of Brownian motion in R. *Brownian motion in R exists. Specifically, the following conditions hold:*

1. Let $Z : \overline{Q}_\infty \times (\Omega, L, E) \to R$ be an arbitrary process such that (i) $Z_0 = 0$; (ii) *for each sequence $0 \equiv t_0 \leq t_1 \leq \cdots \leq t_{n-1} \leq t_n$ in $\overline{Q}_\infty$, the r.r.v.'s $Z_{t(1)} - Z_{t(0)}, \ldots, Z_{t(n)} - Z_{t(n-1)}$ are independent; and* (iii) *for each $s,t \in \overline{Q}_\infty$, the r.r.v. $Z_t - Z_s$ is normal with mean* 0 *and variance $|t-s|$. Then the extension-by-limit*

$$B \equiv \Phi_{Lim}(Z) : [0,\infty) \times \Omega \to R$$

is a Brownian motion.

2. For each $n \geq 1$ and for each $t_1, \ldots, t_n \in \overline{Q}_\infty$, define the f.j.d.

$$F_{t(1),\ldots,t(m)} \equiv \Phi_{0,\overline{\sigma}},$$

where

$$\overline{\sigma} \equiv [\sigma(t_k, t_j)]_{k=1,\ldots,n;\, j=1,\ldots,n} \equiv [t_k \wedge t_j]_{k=1,\ldots,n;\, j=1,\ldots,n}.$$

Then the family $F \equiv \{F_{t(1),\ldots,t(m)} : m \geq 1; t_1, \ldots, t_m \in [0, \infty)\}$ *of f.j.d.'s is consistent and is continuous in probability.*

3. Let $Z : \overline{Q}_\infty \times (\Omega, L, E) \to R$ *be an arbitrary process with marginal distributions given by the family* $F|\overline{Q}_\infty$*, where* F *is defined in Assertion 2. Then the extension-by-limit*

$$B \equiv \Phi_{Lim}(Z) : [0, \infty) \times \Omega \to R$$

is a Brownian motion.

Proof. For convenience, let $U, U_1, U_2, \ldots$ be an independent sequence of standard normal r.r.v.'s. on some probability space $(\widetilde{\Omega}, \widetilde{L}, \widetilde{E})$. Such a sequence can be seen to exist by taking $(\widetilde{\Omega}, \widetilde{L}, \widetilde{E})$ to be the infinite product of the probability space $(R, \widetilde{L}_0, \int \cdot d\Phi_{0,1})$, and taking $U, U_1, U_2, \ldots$ to be the successive coordinate functions on $\widetilde{\Omega}$.

1. Let $Z : \overline{Q}_\infty \times (\Omega, L, E) \to R$ be an arbitrary process such that Conditions (i–iii) hold. Let $b > 0$ and $s_1, s_2 \in \overline{Q}_\infty$ be arbitrary. Then, by Condition (iii), the r.r.v. $Z_{s(1)} - Z_{s(2)}$ is normal with mean 0 and variance $|s_1 - s_2|$. Consequently, by the formulas in Proposition 5.7.5 for moments of standard normal r.r.v.'s, we obtain $E(Z_{s(1)} - Z_{s(2)})^4 = 3|s_1 - s_2|^2$. Chebychev's inequality then implies that, for each $b > 0$, we have

$$\begin{aligned} P(|Z_{s(1)} - Z_{s(2)}| > b) &= P((Z_{s(1)} - Z_{s(2)})^4 > b^4) \\ &\leq b^{-4} E(Z_{s(1)} - Z_{s(2)})^4 = 3b^{-4}|s_1 - s_2|^2, \end{aligned} \tag{9.4.1}$$

where $s_1, s_2 \in \overline{Q}_\infty$ are arbitrary.

2. Let $N \geq 0$ be arbitrary and consider the shifted process $Y \equiv Z^N : Q_\infty \times (\Omega, L, E) \to R$ defined by $Z_s^N \equiv Z_{N+s}$ for each $s \in Q_\infty$. Note that $Z_0^N = Z_1^{N-1}$ if $N \geq 1$. Then, for each $b > 0$ and $s_1, s_2 \in Q_\infty$, we have

$$\begin{aligned} P(|Y_{s(1)} - Y_{s(2)}| > b) &\equiv P(|Z_{N+s(1)} - Z_{N+s(2)}| > b) \\ &\leq 3b^{-4}(|(N + s_1) - (N + s_2)|^2 = 3b^{-4}|s_1 - s_2|^2, \end{aligned}$$

where the inequality follows from inequality 9.4.1. Thus the process Y satisfies the hypothesis of Theorem 9.3.3, with $c_0 = 3$, $u = 4$, and $w = 1$. Accordingly, the extension-by-limit $W \equiv \Phi_{Lim}(Y) : [0, 1] \times \Omega \to R$ is a.u. Hoelder, and hence a.u. continuous. In particular, for each $t \in [N, N + 1]$, the limit

$$B_t \equiv \lim_{r \to t-N;\, r \in Q(\infty)} Z_r^N \equiv \lim_{r \to t-N;\, r \in Q(\infty)} Y_r \equiv W_{t-N}$$

exists as an r.r.v. In other words, $B|[N, N + 1] : [N, N + 1] \times \Omega \to R$ is a well-defined process. Moreover, since the process W is a.u. Hoelder, we see that $B|[N, N + 1]$ is a.u. Hoelder, where $N \geq 0$ is arbitrary. Combining, the process

$B : [0,\infty) \times \Omega \to R$ is a.u. continuous, in the sense of Definition 6.1.3. Note that $B_0 = Z_0 = 0$ by Condition (i).

3. Let the sequence $0 \equiv t_0 \leq t_1 \leq \cdots \leq t_{n-1} \leq t_n$ in $[0,\infty)$ and the sequence $0 \equiv s_0 \leq s_1 \leq \cdots \leq s_{n-1} \leq s_n$ in $\overline{Q}_\infty$ be arbitrary. Let $f_i \in C_{ub}(R)$ be arbitrary for each $i = 1,\ldots,n$. Then $s_i \in [N_i, N_i + 1]$ for some $N_i \geq 0$, for each $i = 0,\ldots,n$. We may assume that $0 = N_0 \leq N_1 \leq \cdots \leq N_n$. Therefore

$$\begin{aligned}
B_{s(i)} - B_{s(i-1)} &= \lim_{r\to s(i)-N(i); r\in Q(\infty)} Z^{N(i)}_{s(i)-N(i)} \\
&\quad - \lim_{r\to s(i-1)-N(i-1); r\in Q(\infty)} Z^{N(i-1)}_{s(i-1)-N(i-1)} \\
&= Z^{N(i)}_{s(i)-N(i)} - Z^{N(i-1)}_{s(i-1)-N(i-1)} \\
&= \left(Z^{N(i)}_{s(i)-N(i)} - Z^{N(i)}_0\right) + \left(Z^{N(i)-1}_1 - Z^{N(i)-1}_0\right) \\
&\quad + \cdots + \left(Z^{N(i-1)-1}_1 - Z^{N(i-1)-1}_0\right) \\
&\quad + \left(Z^{N(i-1)}_1 - Z^{N(i-1)}_{s(i-1)-N(i-1)}\right),
\end{aligned} \tag{9.4.2}$$

where, according to Conditions (ii) and (iii) in the hypothesis, the terms in parentheses are independent, normal, with mean 0 and variances $s_i - N_i, 1, \ldots, 1 - (s_{i-1} - N_{i-1})$, respectively. Hence $B_{s(i)} - B_{s(i-1)}$ is normal, with mean 0 and variance $s_i - s_{i-1}$. Moreover, in the special case where $s_1 = s \in \overline{Q}_\infty$ and $s_0 = 0$, equality 9.4.2 shows that $B_s = Z_s$, where $s \in \overline{Q}_\infty$ is arbitrary.

4. Therefore

$$\begin{aligned}
E\prod_{i=1}^n f_i(B_{s(i)} - B_{s(i-1)}) &= E\prod_{i=1}^n f_i(Z_{s(i)} - Z_{s(i-1)}) \\
&= \prod_{i=1}^n Ef_i(Z_{s(i)} - Z_{s(i-1)}) \\
&= \prod_{i=1}^n \int_R \Phi_{0,s(i)-s(i-1)}(du) f_i(u),
\end{aligned} \tag{9.4.3}$$

where the second and third equalities are due to Conditions (ii) and (iii), respectively. Now let $s_i \to t_i$ for each $i = 1,\ldots,n$. Since the process B is a.u. continuous, the left-hand side of equality 9.4.3 converges to $E\prod_{i=1}^n f_i(B_{t(i)} - B_{t(i-1)})$. At the same time, since, for each $i = 1,\ldots,n$, the integral

$$\int_R \Phi_{0,r}(du) f_i(u) = \widetilde{E} f(\sqrt{r}U)$$

is a continuous function of $r \in [0,\infty)$, the right-hand side of equality 9.4.3 converges to

$$\prod_{i=1}^n \int_R \Phi_{0,t(i)-t(i-1)}(du) f_i(u).$$

Combining, equality 9.4.3 leads to

$$E\prod_{i=1}^{n} f_i(B_{t(i)} - B_{t(i-1)}) = \prod_{i=1}^{n}\int_R \Phi_{0,t(i)-t(i-1)}(du) f_i(u).$$

Consequently, the r.r.v.'s $B_{t(1)} - B_{t(0)}, \ldots, B_{t(n)} - B_{t(n-1)}$ are independent, with normal distributions, with mean 0 and variances given by $t_1 - t_0, \ldots, t_n - t_{n-1}$, respectively.

All the conditions in Definition 9.4.1 have been verified for the process B to be a Brownian motion. Assertion 1 is proved.

5. To prove Assertion 2, define the function $\sigma : [0,\infty)^2 \to [0,\infty)$ by $\sigma(s,t) \equiv s \wedge t$ for each $(s,t) \in [0,\infty)^2$. Then the function σ is symmetric and continuous. We will verify that it is nonnegative definite in the sense of Definition 7.2.2. To that end, let $n \geq 1$ and $t_1, \ldots, t_n \in [0,\infty)$ be arbitrary. We need only show that the square matrix

$$\overline{\sigma} \equiv [\sigma(t_k, t_j)]_{k=1,\ldots,n;\, j=1,\ldots,n} \equiv [t_k \wedge t_j]_{k=1,\ldots,n;\, j=1,\ldots,n}$$

is nonnegative definite. Let $(\lambda_k, \ldots, \lambda_k) \in R^n$ be arbitrary. We wish to prove that

$$\sum_{k=1}^{n}\sum_{j=1}^{n} \lambda_k (t_k \wedge t_j)\lambda_j \geq 0. \tag{9.4.4}$$

First assume that $|t_k - t_j| > 0$ if $k \neq j$. Then there exists a permutation π of the indices $1, \ldots, n$ such that $t_{\pi(k)} \leq t_{\pi(j)}$ iff $k \leq j$. It follows that

$$\sum_{k=1}^{n}\sum_{j=1}^{n} \lambda_k (t_k \wedge t_j)\lambda_j = \sum_{k=1}^{n}\sum_{j=1}^{n} \lambda_{\pi(k)}(t_{\pi(k)} \wedge t_{\pi(j)})\lambda_{\pi(j)} \equiv \sum_{k=1}^{n}\sum_{j=1}^{n} \theta_k (s_k \wedge s_j)\theta_j, \tag{9.4.5}$$

where we write $s_k \equiv t_{\pi(k)}$ and $\theta_k \equiv \lambda_{\pi(k)}$ for each $k = 1, \ldots, n$. Recall the independent standard normal r.r.v.'s.$U_1, \ldots, U_n$ on the probability space $(\widetilde{\Omega}, \widetilde{L}, \widetilde{E})$. Thus $\widetilde{E}U_k U_j = 1$ or 0 according as $k = j$ or $k \neq j$. Define $V_k \equiv \sum_{i=1}^{k} \sqrt{s_i - s_{i-1}}U_i$ for each $k = 1, \ldots, n$, where $s_0 \equiv 0$. Then $\widetilde{E}V_k = 0$ and

$$\widetilde{E}V_k V_j = \sum_{i=1}^{k\wedge j}(s_i - s_{i-1})\widetilde{E}U_i^2 = \sum_{i=1}^{k\wedge j}(s_i - s_{i-1}) = s_{k\wedge j} = s_k \wedge s_j \tag{9.4.6}$$

for each $k, j = 1, \ldots, n$. Consequently, the right-hand side of equality 9.4.5 becomes

$$\sum_{k=1}^{n}\sum_{j=1}^{n} \theta_k (s_k \wedge s_j)\theta_j = \widetilde{E}\sum_{k=1}^{n}\sum_{j=1}^{n} \theta_k V_k V_j \theta_j = \widetilde{E}\left(\sum_{k=1}^{n} \theta_k V_k\right)^2 \geq 0.$$

Hence the sum on the left-hand side of equality 9.4.5 is nonnegative. In other words, inequality 9.4.4 is valid if the point $(t_1, \ldots, t_n) \in [0,\infty)^n$ is such that $|t_k - t_j| > 0$ if $k \neq j$. Since the set of such points is dense in $[0,\infty)^n$, inequality 9.4.4

holds, by continuity, for each $(t_1, \ldots, t_n) \in [0, \infty)^n$. In other words, the function $\sigma : [0, \infty)^2 \to [0, \infty)$ is nonnegative definite according to Definition 7.2.2.

6. Consider each $m \geq 1$ and each sequence $t_1, \ldots, t_m \in [0, \infty)$. Write the nonnegative definite matrix

$$\overline{\sigma} \equiv [\sigma(t_k, t_h)]_{k=1,\ldots,m;h=1,\ldots,m}, \tag{9.4.7}$$

and define

$$F_{t(1),\ldots,t(m)} \equiv \Phi_{0,\overline{\sigma}}, \tag{9.4.8}$$

where $\Phi_{0,\overline{\sigma}}$ is the normal distribution with mean 0 and covariance matrix $\overline{\sigma}$. Take any $M \geq 1$ so large that $t_1, \ldots, t_m \in [0, M]$. Proposition 7.2.3 says that the family

$$F^{(M)} \equiv \{F_{r(1),\ldots,r(m)} : m \geq 1; r_1, \ldots, r_m \in [0, M]\}$$

is consistent and is continuous in probability. Hence, for each $f \in C(R^n)$ and for each sequence mapping $i : \{1, \ldots, n\} \to \{1, \ldots, m\}$, we have

$$F_{t(1),\ldots,t(m)}(f \circ i^*) = F_{t(i(1)),\ldots,t(i(n))} f, \tag{9.4.9}$$

where the dual function $i^* : R^m \to R^n$ is defined by

$$i^*(x_1, \ldots, x_m) \equiv (x_{i(1)}, \ldots, x_{i(n)}). \tag{9.4.10}$$

7. To prove the remaining Assertion 3, let $Z : \overline{Q}_\infty \times (\Omega, L, E) \to R$ be an arbitrary process with marginal distributions given by the family $F|\overline{Q}_\infty$, where F is defined in Assertion 2. Such a process Z exists by the Daniell–Kolmogorov Theorem or the Daniell–Kolmogorov–Skorokhod Extension Theorem.

8. Let $t_1, t_2 \in \overline{Q}_\infty$ be arbitrary. Then the r.r.v.'s $Z_{t(1)}, Z_{t(2)}$ have a jointly normal distribution given by $F_{t(1),t(2)} \equiv \Phi_{0,\overline{\sigma}}$, where $\overline{\sigma} \equiv [t_r \wedge t_h]_{k=1,2;h=1,2}$. Hence

$$EZ_{t(1)}Z_{t(2)} = t_1 \wedge t_2.$$

It follows that $Z_{t(1)} - Z_{t(2)}$ is a normal r.r.v. with mean 0, and with variance given by

$$\begin{aligned} E(Z_{t(1)} - Z_{t(2)})^2 &= EZ_{t(1)}^2 + ZB_{t(2)}^2 - 2EZ_{t(1)}Z_{t(2)} \\ &= t_1 + t_2 - 2t_1 \wedge t_2 = |t_1 - t_2|. \end{aligned}$$

9. Now let $0 \equiv t_0 \leq t_1 \leq \cdots \leq t_{n-1} \leq t_n$ be arbitrary in $\overline{Q}_\infty$. Then the r.r.v.'s $Z_{t(1)}, \cdots, Z_{t(n)}$ have joint distribution $F_{t(1),\ldots,t(n)}$ according to Step 7. Hence $Z_{t(1)}, \cdots, Z_{t(n)}$ are jointly normal. Therefore the r.r.v.'s $Z_{t(1)} - Z_{t(0)}, \ldots, Z_{t(n)} - Z_{t(n-1)}$ are jointly normal. Moreover, for each $i, k = 1, \ldots, n$ with $i < k$, we have

$$\begin{aligned} &E(Z_{t(i)} - Z_{t(i-1)})(Z_{t(k)} - Z_{t(k-1)}) \\ &= EZ_{t(i)}Z_{t(k)} - EZ_{t(i)}Z_{t(k-1)} - EZ_{t(i-1)}Z_{t(k)} + EZ_{t(i-1)}Z_{t(k-1)} \\ &= t_t - t_i - t_{i-1} + t_{i-1} = 0. \end{aligned} \tag{9.4.11}$$

Thus the jointly normal r.r.v.'s $Z_{t(1)} - Z_{t(0)}, \ldots, Z_{t(n)} - Z_{t(n-1)}$ are pairwise uncorrelated. Hence, by Assertion 3 of Proposition 5.7.6, they are mutually independent. Summing up Steps 8 and 9, all of Conditions (i–iii) of Assertion 1 have been verified for the process Z. Accordingly, the extension-by-limit

$$B \equiv \Phi_{Lim}(Z) : [0,\infty) \times \Omega \to R$$

is a Brownian motion. Assertion 3 and the Theorem are proved. □

The following corollary is Levy's well-known result on the a.u. Hoelder continuity of a Brownian motion on each finite interval. A stronger theorem by Levy gives the best modulus of a.u. continuity of a Brownian motion, but shows that an arbitrary $\theta \in \left(0, \frac{1}{2}\right)$ is the best global Hoelder exponent that can be hoped for. Namely, a.u. Hoelder continuity for Brownian motion with global Hoelder exponent $\theta = \frac{1}{2}$ fails.

Corollary 9.4.3. Levy's Theorem: Brownian motion on a finite interval is a.u. Hoelder with any global Hoelder exponent less than $\frac{1}{2}$. *Let $B : [0,\infty) \times (\Omega, L, E) \to R$ be an arbitrary Brownian motion. Let $\theta \in \left(0, \frac{1}{2}\right)$ and $a > 0$ be arbitrary but fixed. Then $B|[0,a]$ is a.u. Hoelder, with Hoelder exponent θ and with some a.u. Hoelder coefficient $\widetilde{c}_H(\cdot, a, \theta)$. We emphasize that θ and c_H are independent of the time parameters $r,s \in [0,a]$. For that reason, θ and c_H may be called global coeffiecients.*

Proof. Since $\theta < \frac{1}{2}$, there exists $m \geq 0$ so large that $\theta < (2+2m)^{-1}m$. Consider the process $X : [0,1] \times (\Omega, L, E) \to R$ defined by $X_t \equiv B_{at}$ for each $t \in [0,1]$. Consider each $b > 0$ and each $r,s \in [0,1]$. Then the r.r.v. $X_s - X_r \equiv B_{as} - B_{ar}$ is normally distributed with mean 0 and variance $a|r-s|$. Therefore

$$E|X_s - X_r|^{2+2m} = E|U\sqrt{a|r-s|}|^{2+2m} = \overline{c}_0 a^{1+m}|r-s|^{1+m},$$

where U is a standard normal r.r.v. and $\overline{c}_0 \equiv EU^{2+2m}$. Thus the process $X|Q_\infty$ satisfies inequality 9.3.17 in the hypothesis of Theorem 9.3.3 with $u \equiv 2+2m$, $c_0 \equiv \overline{c}_0 a^{1+m}$, and $w \equiv m$. Note that $\theta < u^{-1}w$ by the choice of m. Hence, accordingly to Theorem 9.3.3, the process X is a.u. Hoelder, with exponent θ and with some a.u. Hoelder coefficient $\overline{c}_H \equiv \overline{c}_H(\cdot, a, \theta) \equiv \overline{c}_H(\cdot, c_0, u, w, \theta)$.

Let $\varepsilon > 0$ be arbitrary. Then, according to Definition 9.3.2, there exists a measurable set D with $P(D^c) < \varepsilon$ such that for each $\omega \in D$, we have

$$|X_r(\omega) - X_s(\omega)| < \overline{c}_H(\varepsilon)|r-s|^\theta \tag{9.4.12}$$

for each $r,s \in [0,1]$. Now consider each $\omega \in D$ and each $t,u \in [0,a]$ with $|t-u| < a$. Then inequality 9.4.12 yields

$$|B_t(\omega) - B_u(\omega)| \equiv |X_{t/a}(\omega) - X_{u/a}(\omega)| < \overline{c}_H(\varepsilon)|a^{-1}t - a^{-s}s|^\theta$$
$$= \overline{c}_H(\varepsilon)a^{-1}|t-s|^\theta. \tag{9.4.13}$$

Thus we see that the process $B|[0,a]$ is a.u. Hoelder, with Hoelder exponent θ and a.u. Hoelder coefficient $\widetilde{c}_H \equiv a^{-1}\overline{c}_H$, according to Definition 9.3.2, as alleged. □

Theorem 9.4.4. Construction of Brownian motion in R^m. *Brownian motion in R^m exists.*

Proof. Let $U_1,\ldots,U_m$ be an independent sequence of standard normal r.r.v.'s. on some probability space $(\widetilde{\Omega},\widetilde{L},\widetilde{E})$. Such a sequence exists by taking $(\widetilde{\Omega},\widetilde{L},\widetilde{E})$ to be the mth power of the probability space $(R,\widetilde{L}_0,\int\cdot d\Phi_{0,1})$, and taking $U_1,\ldots,U_m$ to be the coordinate mappings.

1. According to Theorem 9.4.2, there exists a Brownian motion

$$B:[0,\infty)\times(\Omega,L,E)\to R$$

with some sample space (Ω,L,E). Now define the mth direct product $(\overline{\Omega},\overline{L},\overline{E})\equiv(\Omega^m,L^{\otimes m},E^{\otimes m})$. Let $t,s\in[0,\infty)$, define

$$\overline{B}_t(\omega)\equiv(B_t(\omega_1),\ldots,B_t(\omega_m)) \tag{9.4.14}$$

for each $\omega\equiv(\omega_1,\ldots,\omega_m)\in\overline{\Omega}\equiv\Omega^m$. Then $\overline{B}_t$ is an r.v. on $(\overline{\Omega},\overline{L},\overline{E})$ with values in R^m, with $\overline{E}\overline{B}_t=0$. Thus $\overline{B}:[0,\infty)\times(\Omega,L,E)\to R$ is a process with values in R^m. Equality 9.4.14 says that the jth coordinate of $\overline{B}_t(\omega)$ is

$$(\overline{B}_t(\omega))_j=B_t(\omega_j)$$

for each $j=1,\ldots,m$.

2. Let $a>0$ be arbitrary. Then $B|[0,a]$ is a.u. Hoelder according to Corollary 9.4.3, and hence a.u. continuous. Let $\varepsilon>0$ be arbitrary. Then, by Definition 6.1.2, there exist $\delta_{auc}(\varepsilon)>0$ and a measurable set $D\subset\Omega$ with $P(D^c)<\varepsilon$ such that for each $\omega\in D$, we have (i) $domain(B(\cdot,\omega))=[0,a]$ and (ii) $|B(t,\omega)-B(s,\omega)|\le\varepsilon$ for each $t,s\in[0,a]$ with $|t-s|<\delta_{auc}(\varepsilon)$. Define

$$\overline{D}\equiv\bigcap_{i=1}^{m}\{(\omega_1,\ldots,\omega_m)\in\Omega^m:\omega_i\in D\}.$$

Then

$$\overline{P}(\overline{D}^c)\le\sum_{i=1}^{m}\overline{P}\{(\omega_1,\ldots,\omega_m)\in\Omega^m:\omega_i\in D^c\}=\sum_{i=1}^{m}P(D^c)<m\varepsilon,$$

where the equality is by Fubini's Theorem.

3. Consider each $\omega\in\overline{D}$. Then (i′) $domain(\overline{B}(\cdot,\omega))=\bigcap_{i=1}^{m}domain(B(\cdot,\omega_i))=[0,a]$ and (ii′)

$$\begin{aligned}|\overline{B}(t,\omega)-\overline{B}(s,\omega)|&\equiv|(B_t(\omega_1),\ldots,B_t(\omega_m))-(B_s(\omega_1),\ldots,B_s(\omega_m))|\\&\le\sum_{i=1}^{m}|B(t,\omega_i)-B(s,\omega_i)|\le m\varepsilon\end{aligned}$$

for each $t, s \in [0, a]$ with $|t - s| < \delta_{auc}(\varepsilon)$. Thus the conditions in Definition 6.1.2 are satisfied for the process $\overline{B}$ to be a.u. continuous on $[0, a]$, where $a > 0$ is arbitrary. Hence, according to Definition 6.1.3, the process $\overline{B}$ is a.u. continuous on $[0, \infty)$.

4. By the defining equality 9.4.14, we have $\overline{B}_0 = 0 \equiv (0, \ldots, 0) \in R^m$.

5. Let $0 \equiv t_0 \leq t_1 \leq \cdots \leq t_{n-1} \leq t_n$ in $[0, \infty)$ be arbitrary. Consider each $i = 1, \ldots, n$ and $j = 1, \ldots, m$. Let $f_{i,j} \in C_{ub}(R)$ be arbitrary. Define

$$V_{i,j}(\omega) \equiv B_{t(i)}(\omega_j) - B_{t(i-1)}(\omega_j)$$

for each $\omega \equiv (\omega_1, \ldots, \omega_m) \in \overline{\Omega} \equiv \Omega^m$. Then

$$\begin{aligned}(V_{i,1}, \ldots, V_{i,m})(\omega) &\equiv (B_{t(i)}(\omega_1) - B_{t(i-1)}(\omega_1), \ldots, B_{t(i)}(\omega_m) - B_{t(i-1)}(\omega_m)) \\ &= (\overline{B}(t_i) - \overline{B}(t_{i-1}))(\omega)\end{aligned}$$

for each $\omega \equiv (\omega_1, \ldots, \omega_m) \in \overline{\Omega} \equiv \Omega^m$. Hence

$$(V_{i,1}, \ldots, V_{i,m}) = \overline{B}(t_i) - \overline{B}(t_{i-1}). \tag{9.4.15}$$

Moreover,

$$\begin{aligned}&\overline{E}\prod_{j=1}^{m}\prod_{i=1}^{n} f_{i,j}(V_{i,j}) \\ &\equiv \int \overline{E}(d(\omega_1, \ldots, \omega_m)) \prod_{j=1}^{m}\prod_{i=1}^{n} f_{i,j}(V_{i,j}(\omega_1, \ldots, \omega_m)) \\ &\equiv \int \overline{E}(d(\omega_1, \ldots, \omega_m)) \prod_{j=1}^{m}\prod_{i=1}^{n} f_{i,j}(B_{t(i)}(\omega_j) - B_{t(i-1)}(\omega_j)) \\ &= \int \cdots \int E(d\omega_1) \cdots E(d\omega_m) \prod_{j=1}^{m}\prod_{i=1}^{n} f_{i,j}(B_{t(i)}(\omega_j) - B_{t(i-1)}(\omega_j)) \\ &= \prod_{j=1}^{m} \int E(d\omega_j) \prod_{i=1}^{n} f_{i,j}(B_{t(i)}(\omega_j) - B_{t(i-1)}(\omega_j)) \\ &\equiv \prod_{j=1}^{m} E \prod_{i=1}^{n} f_{i,j}(B_{t(i)} - B_{t(i-1)}) = \prod_{j=1}^{m}\prod_{i=1}^{n} E f_{i,j}(B_{t(i)} - B_{t(i-1)}),\end{aligned} \tag{9.4.16}$$

where we used Fubini's Theorem.

6. In the special case where $f_{i',j'} = 0$ for each $i' = 1, \ldots, n$, for each $j' = 1, \ldots, m$ such that $i' \neq i$ and $j' \neq j$, equality 9.4.16 reduces to

$$\overline{E} f_{i,j}(V_{i,j}) = E f_{i,j}(B_{t(i)} - B_{t(i-1)}). \tag{9.4.17}$$

Substituting this back into equality 9.4.16, we obtain

$$\overline{E}\prod_{j=1}^{m}\prod_{i=1}^{n} f_{i,j}(V_{i,j}) = \prod_{j=1}^{m}\prod_{i=1}^{n} \overline{E} f_{i,j}(V_{i,j}),$$

where $f_{i,j} \in C_{ub}(R)$ is arbitrary for each $i = 1,\ldots,n$, for each $j = 1,\ldots,m$. This implies that the r.r.v.'s $(V_{i,j})_{j=1,\ldots,m;i=1,\ldots,n}$ are mutually independent.

7. Now let $g_i \in C_{ub}(R^m)$ be arbitrary for each $i = 1,\ldots,n$. Then

$$\overline{E}\prod_{i=1}^{n} g_i(V_{i,1},\ldots,V_{i,m}) = \prod_{i=1}^{n} \overline{E} g_i(V_{i,1},\ldots,V_{i,m}).$$

It follows that the r.v.'s $(V_{1,1},\ldots,V_{1,m}),\ldots,(V_{n,1},\ldots,V_{n,m})$ are independent. Equivalently, in view of equality 9.4.15, the r.v.'s $\overline{B}_{t(1)} - \overline{B}_{t(0)},\ldots,\overline{B}_{t(n)} - \overline{B}_{t(n-1)}$ are independent, where $0 \equiv t_0 \le t_1 \le \cdots \le t_{n-1} \le t_n$ is an arbitrary sequence in $[0,\infty)$.

8. Let $s,t \in [0,\infty)$ be arbitrary. Let $n = 2$. Let $(t_0,t_1,t_2) \equiv (0,s,t)$. Let $j = 1,\ldots,m$ be arbitrary. Then equality 9.4.17 shows that the r.r.v. $V_{2,j}$ has the same distribution as the r.r.v. $B_{t(2)} - B_{t(1)} = B_t - B_s$. Hence the independent r.r.v.'s $V_{2,1},\ldots,V_{2,m}$ are normally distributed with mean 0 and variance $|t-s|$. Consequently, the r.v. $(V_{2,1},\ldots,V_{2,m})$ is normally distributed with mean $0 \in R^m$ and covariance matrix $|t-s|I$, where I is the $m \times m$ identity matrix. Therefore the r.v.

$$\overline{B}_t - \overline{B}_s \equiv \overline{B}_{t(2)} - \overline{B}_{t(1)} \equiv (V_{2,1},\ldots,V_{2,m})$$

is normal with mean $0 \in R^m$ and covariance matrix $|t-s|I$.

9. All the conditions in Definition 9.4.1 have been verified for the process $\overline{B}$ to be a Brownian motion. □ □

Corollary 9.4.5. Basic properties of Brownian motion in R^m. Let $B : [0,\infty) \times (\Omega, L, E) \to R^m$ *be an arbitrary Brownian motion in* R^m. *Then the following conditions hold:*

1. Let A be an arbitrary orthogonal $k \times m$ *matrix. Thus* $AA^T = I$ *is the* $k \times k$ *identity matrix. Then the process* $AB : [0,\infty) \times (\Omega, L, E) \to R^k$ *is a Brownian motion in* R^k.

2. Let b be an arbitrary unit vector. Then the process $b^T B : [0,\infty) \times (\Omega, L, E) \to R$ *is a Brownian motion in* R^1.

3. Suppose the process B is adapted to some filtration $\mathcal{L} \equiv \{L^{(t)} : t \in [0,\infty)\}$. *Let* $\gamma > 0$ *be arbitrary. Define the process* $\widetilde{B} : [0,\infty) \times (\Omega, L, E) \to R^m$ *by* $\widetilde{B}_t \equiv \gamma^{-1/2} B_{\gamma t}$ *for each* $t \in [0,\infty)$. *Then* $\widetilde{B}$ *is a Brownian motion in* R^m *adapted to the filtration* $\mathcal{L}_\gamma \equiv \{L^{(\gamma t)} : t \in [0,\infty)\}$.

Proof. 1. Let A be an orthogonal $k \times m$ matrix. Then trivially $AB_0 = A0 = 0 \in R^k$. Thus Condition (i) of Definition 9.4.1 holds for the process AB. Next, let the sequence $0 \equiv t_0 \le t_1 \le \cdots \le t_{n-1} \le t_n$ in $[0,\infty)$ be arbitrary. Then the r.v.'s $B_{t(1)} - B_{t(0)}, \ldots, B_{t(n)} - B_{t(n-1)}$ are independent. Hence the r.v.'s $A(B_{t(1)} - B_{t(0)}), \ldots, A(B_{t(n)} - B_{t(n-1)})$ are independent, establishing Condition (ii) of Definition 9.4.1 for the process AB. Now let $s,t \in [0,\infty)$ be arbitrary. Then the r.v. $B_t - B_s$ is normal with mean 0 and covariance matrix $|t-s|I$, where I stands for the $k \times k$ identity matrix. Hence $A(B_t - B_s)$ is normal with mean 0, with covariance matrix

$$E(A(B_t - B_s)(B_t - B_s)^T A^T) = A(E(B_t - B_s)(B_t - B_s)^T)A^T$$
$$= A(|t-s|IA^T) = |t-s|AA^T = |t-s|I.$$

This proves Condition (iii) of Definition 9.4.1 for the process AB. Assertion 1 is proved.

2. Let b be an arbitrary unit vector. Then b^T is a $1 \times m$ orthogonal matrix. Hence, according to Assertion 1, the process $b^T B : [0,\infty) \times (\Omega, L, E) \to R$ is a Brownian motion in R^1. Assertion 2 is proved.

3. Define the process $\widetilde{B} : [0,\infty) \times (\Omega, L, E) \to R^m$ by $\widetilde{B}_t \equiv \gamma^{-1/2} B_{\gamma t}$ for each $t \in [0,\infty)$. Trivially, Conditions (i) and (ii) of Definition 9.4.1 hold for the process $\widetilde{B}$. Let $s,t \in [0,\infty)$ be arbitrary. Then the r.v. $B_{\gamma t} - B_{\gamma s}$ is normal with mean 0 and covariance matrix $|\gamma t - \gamma s|I$. Hence the r.v. $\widetilde{B}_t - \widetilde{B}_s \equiv \gamma^{-1/2} B_{\gamma t} - \gamma^{-1/2} B_{\gamma s}$ has mean 0 and covariance matrix

$$(\gamma^{-1/2})^2 |\gamma t - \gamma s| I = |t-s|I.$$

Thus Condition (iii) of Definition 9.4.1 is also verified for the process $\widetilde{B}$ to be a Brownian motion. Moreover, for each $t \in [0,\infty)$, we have $\widetilde{B}_t \equiv \gamma^{-1/2} B_{\gamma t} \in L^{(\gamma t)}$. Hence the process $\widetilde{B}$ is adapted to the filtration $\mathcal{L}_\gamma \equiv \{L^{(\gamma t)} : t \in [0,\infty)\}$. Assertion 3 and the corollary are proved. □

9.5 a.u. Continuous Gaussian Process

In this section, we will restrict our attention to real-valued Gaussian processes with parameter set $[0,1]$.

Definition 9.5.1. Specification of a covariance function and a Gaussian process. In this section, let $\sigma : [0,1] \times [0,1] \to R$ be an arbitrary continuous symmetric nonnegative definite function. For each $r,s \in [0,1]$, define

$$\Delta_{r,s} \equiv \Delta\sigma(r,s) \equiv \sigma(r,r) + \sigma(s,s) - 2\sigma(r,s). \tag{9.5.1}$$

Let $X : [0,1] \times \Omega \to R$ be a measurable Gaussian process such that (i) $EX_r = 0$ and $EX_r X_s = \sigma(r,s)$ for each $r,s \in [0,1]$ and (ii) $X = \Phi_{Lim}(Z)$, where $Z \equiv X|Q_\infty : Q_\infty \times \Omega \to R$ is the restriction of X to the countable parameter set Q_∞. By Theorem 7.2.5 and its proof, such a process X exists.

Then $X_r - X_s$ is a normal r.r.v. with mean 0 and variance

$$E|X_r - X_s|^2 = \Delta_{r,s}. \tag{9.5.2}$$

Separately, for each $u > 0$, define the constant α_u to be the absolute uth moment of the standard normal distribution on R. □

The next theorem seems to be hitherto unknown in the probability literature.

Theorem 9.5.2. Sufficient condition for a Gaussian process to be a.u. Hoelder. *Suppose there exist constants $c_0, u, w > 0$ such that*

$$\alpha_u \Delta_{r,s}^u \leq c_0 |r - s|^{1+w} \tag{9.5.3}$$

for each $r, s \in [0, 1]$. Then the process X is a.u. globally Hoelder continuous with exponent θ, where $\theta < u^{-1}w$ is arbitrary.

Proof. We will give the proof only for the case where σ is positive definite. Let $r, s \in [0, 1]$ be arbitrary. Then, from the defining equalities 9.5.1 and 9.5.2, we see that $\Delta_{r,s}^{-1}(X_r - X_s)$ is a standard normal r.r.v. Hence its absolute uth moment is equal to the absolute constant α_u. Therefore

$$E|X_r - X_s|^u = \Delta_{r,s}^u E|\Delta_{r,s}^{-1}(X_r - X_s)|^u = \Delta_{r,s}^u \alpha_u \leq c_0 |r - s|^{1+w},$$

where the last inequality is by hypothesis. Hence

$$E|Z_r - Z_s|^u \leq c_0 |r - s|^{1+w}$$

for each $r, s \in Q_\infty$. Theorem 9.3.3 therefore implies that the process $\Phi_{Lim}(Z)$ is a.u. globally Hoelder with exponent θ, where $\theta < u^{-1}w$ is arbitrary. Since $X = \Phi_{Lim}(Z)$ by Condition (ii) in Definition 9.5.1, the present theorem is proved. □

Thus we have a condition in terms of a bound on $\Delta_{r,s}$ to guarantee a.u. Hoelder continuity. The celebrated paper by [Garsia, Rodemich, and Rumsey 1970] also gives a condition in terms of a bound on $\Delta_{r,s}$, under which the Gaussian process has an explicit modulus of a.u. continuity. The Garsia–Rodemich–Rumsey (GRR) proof shows that the partial sums of the Karhunen–Loeve expansion relative to σ are, under said condition, a.u. convergent to an a.u. continuous process.

We will quote the key real-variable lemma in [Garsia, Rodemich, and Rumsey 1970]. We will then present a proof, which is constructive in every detail, of their main theorem; our proof is otherwise essentially the proof in the cited paper. We dispense with the authors' use of a submartingale derived from the Karhunen–Loeve expansion, and dispense with their subsequent appeal to a version of the submartingale convergence theorem that asserts the a.u. convergence of each submartingale with bounded expectations. This version of the submartingale convergence implies the principle of infinite search and is not constructive.

In the place of the Karhunen–Loeve expansion and the use of submartingales, we will derive Borel–Cantelli estimates on conditional expectations, thus sticking to elementary time-domain analysis and obviating the need, for the present purpose, of more groundwork on spectral analysis of the covariance function. Note that the direct use of conditional expectations in relation to the Karhunen–Loeve expansion is mentioned in [Garsia, Rodemich, and Rumsey 1970] as part of a related result.

In the following discussion, recall that $\int_0^1 \cdot dp$ denotes the Riemann–Stieljes integration relative to an arbitrary distribution function p on $[0,1]$. Also, note that $Y \equiv X|Q_\infty : Q_\infty \times \Omega \to R$ is a centered Gaussian process with the continuous nonnegative definite covariance function σ.

Definition 9.5.3. Two auxiliary functions. Introduce the auxiliary function

$$\Psi(v) \equiv \exp\left(\frac{1}{4}v^2\right) \tag{9.5.4}$$

for each $v \in [0,\infty)$, with its inverse

$$\Psi^{-1}(u) \equiv 2\sqrt{\log u} \tag{9.5.5}$$

for each $u \in [1,\infty)$. □

Next we cite, without the proof from [Garsia, Rodemich, and Rumsey 1970], a remarkable real variable lemma. It is key to the GRR theorem.

Lemma 9.5.4. Garsia–Rodemich–Rumsey real-variable lemma. *Let the function Ψ and its inverse Ψ^{-1} be as in Definition 9.5.3. Let $\overline{p} : [0,1] \to [0,\infty)$ be an arbitrary continuous nondecreasing function with $\overline{p}(0) = 0$ that is increasing in some neighborhood of 0. Let f be an arbitrary continuous function on $[0,1]$ such that the function*

$$\Psi(\frac{|f(t)-f(s)|}{\overline{p}(|t-s|)})$$

of $(t,s) \in [0,1]^2$ is integrable, with

$$\int_0^1 \int_0^1 \Psi\left(\frac{|f(t)-f(s)|}{\overline{p}(|t-s|)}\right) dtds \leq B \tag{9.5.6}$$

for some $B > 0$. Then

$$|f(t)-f(s)| \leq 8\int_0^{|t-s|} \Psi^{-1}\left(\frac{4B}{u^2}\right) d\overline{p}(u) \tag{9.5.7}$$

for each $(t,s) \in [0,1]^2$.

Proof. See [Garsia, Rodemich, and Rumsey 1970]. The verification that their proof is constructive is left to the reader. □

Recall from Definition 9.0.2 some notations for dyadic rationals in $[0,1]$. For each $N \geq 0$, we have $p_N \equiv 2^N$, $\Delta_N \equiv 2^{-N}$, and the sets of dyadic rationals

$$Q_N \equiv \{t_0, t_1, \ldots, t_{p(N)}\} = \{q_{N,0}, \ldots, q_{N,p(N)}\} \equiv \{0, \Delta_N, 2\Delta_N, \ldots, 1\}$$

and

$$Q_\infty \equiv \bigcup_{N=0}^{\infty} Q_N = \{t_0, t_1, \ldots\}.$$

Recall that $[\cdot]_1$ is the operation that assigns to each $a \in R$ an integer $[a]_1 \in (a, a+2)$. Recall also the matrix notations in Definition 5.7.1 and the basic properties of conditional distributions established in Propositions 5.6.6 and 5.8.17. As usual, to lessen the burden on subscripts, we write the symbols x_y and $x(y)$ interchangeably for arbitrary expressions x and y.

Lemma 9.5.5. Interpolation of a Gaussian process by conditional expectations. *Let $Y : Q_\infty \times \Omega \to R$ be an arbitrary centered Gaussian process with a continuous positive definite covariance function σ. Thus $EY_tY_s = \sigma(t,s)$ and $E(Y_t - Y_s)^2 = \Delta\sigma(t,s)$ for each $t,s \in Q_\infty$. Let $\overline{p} : [0,1] \to [0,\infty)$ be an arbitrary continuous nondecreasing function such that*

$$\bigvee_{0 \le s,t \le 1;\, |s-t| \le u} (\Delta\sigma(s,t))^{1/2} \le \overline{p}(u)$$

for each $u \in [0,1]$. Then the following conditions hold:

1. Let $n \ge 0$ and $t \in [0,1]$ be arbitrary. Define the r.r.v.

$$Y_t^{(n)} \equiv E(Y_t | Y_{t(0)}, \ldots, Y_{t(n)}).$$

Then, for each fixed $n \ge 0$, the process $Y^{(n)} : [0,1] \times \Omega \to R$ is an a.u. continuous centered Gaussian process. Moreover, $Y_r^{(n)} = Y_r$ for each $r \in \{t_0, \ldots, t_n\}$. We will call the process $Y^{(n)}$ the interpolated approximation of Y by conditional expectations *on $\{t_0, \ldots, t_n\}$.*

2. For each fixed $t \in [0,1]$, the process $Y_t^{(\cdot)} : \{0,1,\ldots\} \times \Omega \to R$ is a martingale relative to the filtration $\mathcal{L} \equiv \{L(Y_{t(0)}, \ldots, Y_{t(n)}) : n = 0,1,\ldots\}$.

3. Let $m > n \ge 1$ be arbitrary. Define

$$Z_t^{(m,n)} \equiv Y_t^{(m)} - Y_t^{(n)} \in L(Y_{t(0)}, \ldots, Y_{t(m)})$$

for each $t \in [0,1]$. Let $\overline{\Delta} \in (0,1)$ be arbitrary. Suppose n is so large that the subset $\{t_0, \ldots, t_n\}$ is a $\overline{\Delta}$-approximation of $[0,1]$. Define the continuous nondecreasing function $\overline{p}_{\overline{\Delta}} : [0,1] \to [0,\infty)$ by

$$\overline{p}_{\overline{\Delta}}(u) \equiv 2\overline{p}(u) \wedge 2\overline{p}(\overline{\Delta})$$

for each $u \in [0,1]$. Then

$$E(Z_t^{(m,n)} - Z_s^{(m,n)})^2 \le \overline{p}_{\overline{\Delta}}^2(|t-s|)$$

for each $t,s \in [0,1]$.

Proof. Since $Y : Q_\infty \times \Omega \to R$ is centered Gaussian with covariance function σ, we have

$$E(Y_t - Y_s)^2 = EY_t^2 - 2EY_tY_s + EY_s^2 = \sigma(t,s) - 2\sigma(t,s) + \sigma(s,s) \equiv \Delta\sigma(t,s)$$

for each $t,s \in Q_\infty$.

1. Let $n \ge 0$ be arbitrary. Then the r.v. $U_n \equiv (Y_{t(0)}, \ldots, Y_{t(n)})$ with values in R^{n+1} is normal, with mean $0 \in R^{n+1}$ and the positive definite covariance matrix

$$\overline{\sigma}_n \equiv EU_nU_n^T = [\sigma(t_h, t_j)]_{h=0,\ldots;n;\, j=0,\ldots,n}.$$

For each $t \in [0,1]$, define

$$c_{n,t} \equiv (\sigma(t,t_0), \ldots, \sigma(t,t_n)) \in R^{n+1}.$$

Define the Gaussian process $\overline{Y}^{(n)} : [0,1] \times \Omega \to R$ by

$$\overline{Y}_t^{(n)} \equiv c_{n,t}^T \overline{\sigma}_n^{-1} U_n$$

for each $t \in [0,1]$, where the inverse matrix $\overline{\sigma}_n^{-1}$ is defined because the function σ is positive definite. Then, since $c_{n,t}$ is continuous in t, the process $\overline{Y}^{(n)}$ is a.u. continuous.

Moreover, for each $t \in Q_\infty$, the conditional expectation of Y_t given U_n is, according to Proposition 5.8.17, given by

$$E(Y_t | Y_{t(0)}, \ldots, Y_{t(n)}) = E(Y_t | U_n) = c_{n,t}^T \overline{\sigma}_n^{-1} U_n \equiv \overline{Y}_t^{(n)}, \tag{9.5.8}$$

whence $Y_t^{(n)} = \overline{Y}_t^{(n)}$. Since $\overline{Y}^{(n)}$ is an a.u. continuous and centered Gaussian process, so is $Y^{(n)}$. Assertion 1 is proved. Note that for each $r \in \{t_0, \ldots, t_n\}$ and for each $m \geq n$, we have $r \in \{t_0, \ldots, t_m\}$, whence $Y_r \in L(U_m)$ and

$$Y_r^{(m)} = E(Y_r | U_m) = Y_r, \tag{9.5.9}$$

where the second equality is by the properties of the conditional expectation.

2. Let $m > n \geq 1$ be arbitrary. Then, for each $t \in Q_\infty$, we have

$$E(Y_t^{(m)} | U_n) = E(E(Y_t | U_m) | U_n) = E(Y_t | U_n) = Y_t^{(n)},$$

where the first and third equalities are by equality 9.5.8, and where the second equality is because $L(U_n) \subset L(U_m)$. Hence, for each $V \in L(U_n)$, we have

$$EY_t^{(m)} V = EY_t^{(n)} V$$

for each $t \in Q_\infty$ and, by continuity, also for each $t \in [0,1]$. Thus $E(Y_t^{(m)} | U_n) = Y_t^{(n)}$ for each $t \in [0,1]$. We conclude that for each fixed $t \in [0,1]$, the process $Y_t^{(\cdot)} : \{0,1,\ldots\} \times \Omega \to R$ is a martingale relative to the filtration $\{L(U_n) : n = 0,1,\ldots\}$. Assertion 2 is proved.

3. Let $m > n \geq 1$ and $t,s \in [0,1]$ be arbitrary. Then

$$\begin{aligned} Z_t^{(m,n)} - Z_s^{(m,n)} &\equiv Y_t^{(m)} - Y_t^{(n)} - Y_s^{(m)} + Y_s^{(n)} \\ &= Y_t^{(m)} - Y_s^{(m)} - E(Y_t^{(m)} - Y_s^{(m)} | U_n). \end{aligned}$$

Hence, by Assertion 8 of Proposition 5.6.6, we have

$$E(Z_t^{(m,n)} - Z_s^{(m,n)})^2 \leq E(Y_t^{(m)} - Y_s^{(m)})^2.$$

Suppose $t,s \in Q_\infty$. Then equality 9.5.8 implies that

$$Y_t^{(m)} - Y_s^{(m)} = E(Y_t - Y_s | U_m). \tag{9.5.10}$$

Hence,

$$E(Z_t^{(m,n)} - Z_s^{(m,n)})^2 \leq E(Y_t^{(m)} - Y_s^{(m)})^2 \leq E(Y_t - Y_s)^2 = \Delta\sigma(t,s),$$

where $t,s \in Q_\infty$. are arbitrary, and where the second inequality is thanks to equality 9.5.10 and to Proposition 5.6.6. By continuity, we therefore have

$$E(Z_t^{(n,m)} - Z_s^{(n,m)})^2 \leq \Delta\sigma(t,s) \leq \overline{p}^2(|t-s|), \qquad (9.5.11)$$

where $t,s \in [0,1]$ are arbitrary.

Now let $\overline{\Delta} > 0$ be arbitrary. Suppose $n \geq 1$ is so large that the subset $\{t_0, \ldots, t_n\}$ is a $\overline{\Delta}$-approximation of $[0,1]$. Then there exist $t', s' \in \{t_0, \ldots, t_n\}$ such that $|t - t'| \vee |s - s'| < \overline{\Delta}$. Then equality 9.5.9 implies that

$$Z_{t'}^{(m,n)} \equiv Y_{t'}^{(m)} - Y_{t'}^{(n)} = Y_{t'} - Y_{t'} = 0, \qquad (9.5.12)$$

with a similar inequality for s'. Applying inequality 9.5.11 to t,t' in place of t,s, we obtain

$$E(Z_t^{(m,n)} - Z_{t'}^{(m,n)})^2 \leq \overline{p}^2(|t-t'|) \leq \overline{p}^2(\overline{\Delta}),$$

and a similar inequality for the pair s,s' in place of t,t'. In addition, equality 9.5.12 implies

$$Z_t^{(m,n)} - Z_s^{(m,n)} = (Z_t^{(m,n)} - Z_{t'}^{(m,n)}) - (Z_s^{(m,n)} - Z_{s'}^{(m,n)}).$$

Hence Minkowski's inequality yields

$$\begin{aligned}\sqrt{E(Z_t^{(m,n)} - Z_s^{(m,n)})^2} &\leq \sqrt{E(Z_t^{(m,n)} - Z_{t'}^{(m,n)})^2} + \sqrt{E(Z_s^{(m,n)} - Z_{s'}^{(m,n)})^2} \\ &\leq 2\overline{p}(\overline{\Delta}). \qquad (9.5.13)\end{aligned}$$

Combining inequalities 9.5.11 and 9.5.13, we obtain

$$E(Z_t^{(m,n)} - Z_s^{(m,n)})^2 \leq (2\overline{p}(|t-s|) \wedge 2\overline{p}(\overline{\Delta}))^2 \equiv \overline{p}_{\overline{\Delta}}^2(|t-s|).$$

Assertion 3 and the lemma are proved. □

The next lemma prepares for the proof of the GRR theorem.

Lemma 9.5.6. Modulus of a.u. continuity of an a.u. continuous Gaussian process. *Let $V : [0,1] \times \Omega \to R$ be an arbitrary a.u. continuous and centered Gaussian process, with a continuous covariance function σ. Suppose $\overline{p} : [0,1] \to [0,\infty)$ is a continuous function that is increasing in some neighborhood of 0, with $\overline{p}(0) = 0$, such that $\sqrt{-\log u}$ is integrable relative to the distribution function $\overline{p}$ on $[0,1]$. Thus*

$$\int_0^1 \sqrt{-\log u}\, d\overline{p}(u) < \infty. \qquad (9.5.14)$$

Suppose

$$\bigvee_{0 \le s,t \le 1;\, |s-t| \le u} \Delta\sigma(t,s) \le \overline{p}(u)^2 \tag{9.5.15}$$

for each $u \in [0,1]$. *Then there exists an integrable r.r.v.* B *with* $EB \le \sqrt{2}$ *such that*

$$|V(t,\omega) - V(s,\omega)| \le 16 \int_0^{|t-s|} \sqrt{\log\left(\frac{4B(\omega)}{u^2}\right)} d\overline{p}(u) \tag{9.5.16}$$

for each $t,s \in [0,1]$, *for each* ω *in the full set* $domain(B)$.

Proof. 1. By hypothesis, $\overline{p}$ is a nondecreasing function on $[0,1]$ that is increasing in some neighborhood of 0, with $\overline{p}(0) = 0$. It follows that $\overline{p}(u) > 0$ for each $u \in (0,1]$.

2. Define the full subset

$$D \equiv \{(t,s) \in [0,1]^2 : |t-s| > 0\}$$

of $[0,1]^2$ relative to the product Lebesgue integration. Because the process V is a.u. continuous, there exists a full set $A \subset \Omega$ such that $V(\cdot,\omega)$ is continuous on $[0,1]$, for each $\omega \in A$. Moreover, V is a measurable function on $[0,1] \times \Omega$. Define the function $U : [0,1]^2 \times \Omega \to R$ by

$$domain(U) \equiv D \times A$$

and

$$U(t,s,\omega) \equiv \Psi\left(\frac{|V(t,\omega) - V(s,\omega)|}{\overline{p}(|t-s|)}\right)$$

$$\equiv \exp\left(\frac{1}{4}\frac{(V(t,\omega) - V(s,\omega))^2}{\overline{p}(|t-s|)^2}\right) \tag{9.5.17}$$

for each $(t,s,\omega) \in domain(U)$. Then U is measurable on $[0,1]^2 \times \Omega$.

3. Using the Monotone Convergence Theorem, the reader can prove that the right-hand side of equality 9.5.17 is an integrable function on $[0,1]^2 \times \Omega$ relative to the product integration, with integral bounded by

$$\int_0^1 \int_0^1 \int_{-\infty}^{\infty} \frac{1}{\sqrt{2\pi}} \exp\left(-\frac{1}{4}u^2\right) du dt ds = \int_0^1 \int_0^1 \sqrt{2} dt ds = \sqrt{2}.$$

Therefore, by Fubini's Theorem, the function

$$B \equiv \int_0^1 \int_0^1 U dt ds \equiv \int_0^1 \int_0^1 \Psi\left(\frac{|V_t - V_s|}{\overline{p}(|t-s|)}\right) dt ds$$

on (Ω, L, E) is an integrable r.r.v, with expectation given by

$$E(B) \equiv E\int_0^1\int_0^1 U dtds \leq \sqrt{2}.$$

Consider each $\omega \in domain(B)$. Then

$$B(\omega) \equiv \int_0^1\int_0^1 \Psi\left(\frac{|V(t,\omega) - V(s,\omega)|}{\overline{p}(|t-s|)}\right) dtds. \tag{9.5.18}$$

In view of equality 9.5.18, Lemma 9.5.4 implies that

$$|V(t,\omega) - V(s,\omega)| \leq 8\int_0^{|t-s|} \Psi^{-1}\left(\frac{4B(\omega)}{u^2}\right) d\overline{p}(u)$$

$$\equiv 16\int_0^{|t-s|} \sqrt{\log\left(\frac{4B(\omega)}{u^2}\right)} d\overline{p}(u), \tag{9.5.19}$$

where the equality is by equality 9.5.5. The lemma is proved. □

Theorem 9.5.7. Garsia–Rodemich–Rumsey Theorem. *Let* $\overline{p} : [0,1] \to [0,\infty)$ *be a continuous increasing function with* $\overline{p}(0) = 0$, *such that* $\sqrt{-\log u}$ *is integrable relative to the distribution function* $\overline{p}$ *on* $[0,1]$. *Thus*

$$\int_0^1 \sqrt{-\log u}\, d\overline{p}(u) < \infty. \tag{9.5.20}$$

Let $\sigma : [0,1] \times [0,1] \to R$ *be an arbitrary symmetric positive definite function such that*

$$\bigvee_{0 \leq s,t \leq 1;\, |s-t| \leq u} (\Delta\sigma(s,t))^{1/2} \leq \overline{p}(u). \tag{9.5.21}$$

Then there exists an a.u. continuous centered Gaussian process $\overline{X} : [0,1]\times\Omega \to R$ *with* σ *as covariance function and an integrable r.r.v.* B *with* $EB \leq \sqrt{2}$ *such that*

$$|\overline{X}(t,\omega) - \overline{X}(s,\omega)| \leq 16\int_0^{|t-s|} \sqrt{\log\left(\frac{4B(\omega)}{u^2}\right)} d\overline{p}(u)$$

for each $t,s \in [0,1]$, *for each* $\omega \in domain(B)$.

Proof. 1. By hypothesis, $\overline{p}$ is an increasing function with $\overline{p}(0) = 0$. It follows that $\overline{p}(u) > 0$ for each $u \in (0,1]$.

2. Let

$$F^{\sigma} \equiv \Phi_{covar,fjd}(\sigma) \tag{9.5.22}$$

be the consistent family of normal f.j.d.'s on the parameter set $[0,1]$ associated with mean function 0 and the given covariance function σ, as defined in equalities 7.2.4 and 7.2.3 of and constructed in Theorem 7.2.5. Let $Y : Q_\infty \times \Omega \to R$ be an arbitrary process with marginal distributions given by $F^\sigma|Q_\infty$, the restriction of the family F^σ of the normal f.j.d.'s to the countable parameter subset Q_∞.

3. By hypothesis, the function $\overline{p}$ is continuous at 0, with $\overline{p}(0) = 0$. Hence there is a modulus of continuity $\delta_{\overline{p},0} : (0,\infty) \to (0,\infty)$ such that $\overline{p}(u) < \varepsilon$ for each u with $0 \leq u < \delta_{\overline{p},0}(\varepsilon)$, for each $\varepsilon > 0$.

4. Also by hypothesis, the function $\sqrt{-\log u}$ is integrable relative to $\int_0^1 \cdot d\overline{p}$. Hence there exists a modulus of integrability $\delta_{\overline{p},1} : (0,\infty) \to (0,\infty)$ such that

$$\int_{(0,c]} \sqrt{-\log u}\, d\overline{p}(u) < \varepsilon$$

for each $c \in [0,1]$ with $\int_{(0,c]} d\overline{p} = \overline{p}(c) < \delta_{\overline{p},1}(\varepsilon)$.

5. Let $k \geq 1$ arbitrary. Then, for each $u \in (0,1]$, we have $\log k^2 \geq 0$ and $-\log u \geq 0$, whence

$$\begin{aligned} \Psi^{-1}\left(\frac{k^2}{u^2}\right) &\equiv 2\sqrt{2\log k - 2\log u} \\ &= 2\sqrt{2}\sqrt{\log k - \log u} \leq 2\sqrt{2}(\sqrt{\log k} + \sqrt{-\log u}). \end{aligned} \tag{9.5.23}$$

The functions of u on both ends have domain $(0,1]$ and are continuous on $(0,1]$. Hence these functions are measurable relative to $\int_0^1 \cdot d\overline{p}$. Since the right-hand side of inequality 9.5.23 is an integrable function of u relative to the integration $\int_0^1 \cdot d\overline{p}$, so is the function on the left-hand side.

6. Define $m_0 \equiv 1$. Let $k \geq 1$ be arbitrary. In view of the conclusion of Step 5, there exists $m_k \equiv m_k(\delta_{\overline{p},0}, \delta_{\overline{p},1}) \geq m_{k-1}$ so large that

$$16 \int_0^{\Delta(m(k))} \Psi^{-1}\left(\frac{k^2}{u^2}\right) d\overline{p}(u) < k^{-2}, \tag{9.5.24}$$

where $\Delta_{m(k)} \equiv 2^{-m(k)}$.

7. Let $n \geq 0$ be arbitrary. Define, as in Lemma 9.5.5, the interpolated process $Y^{m(n)} : [0,1] \times \Omega \to R$ of $Y : Q_\infty \times \Omega \to R$ by conditional expectations on $\{t_0, \ldots, t_{m(n)}\}$. Lemma 9.5.5 implies that (i) $Y^{m(n)}$ is a centered Gaussian process, (ii) $Y^{m(n)}$ is a.u. continuous, and (iii) $Y_r^{m(n)} = Y_r$ for each $r \in \{t_0, \ldots, t_{m(n)}\}$. Consequently, the difference process

$$Z^{(n)} \equiv Z^{(m(n),m(n-1))} \equiv Y^{(m(n))} - Y^{(m(n-1))}$$

is a.u. continuous if $n \geq 1$. From Condition (iii), we have

$$Z_0^{(n)} \equiv Y_0^{(m(n))} - Y_0^{(m(n-1))} = Y_0 - Y_0 = 0$$

if $n \geq 1$.

8. Let $t,s \in [0,1]$ be arbitrary with $|t-s|>0$. Then, since $\{t_0,\ldots,t_{m(k-1)}\}$ is a $\Delta_{m(k-1)}$-approximation of $[0,1]$, we have, by Assertion 3 of Lemma 9.5.5,

$$E(Z_t^{(k)} - Z_s^{(k)})^2 \equiv E(Z_t^{(m(k),m(k-1))} - Z_s^{(m(k),m(k-1))})^2 \le \overline{p}_k^2(|t-s|), \tag{9.5.25}$$

where we define

$$\overline{p}_k(u) \equiv \overline{p}_{\Delta(m(k-1))}(u) \equiv 2\overline{p}(u) \wedge 2\overline{p}(\Delta_{m(k-1)})$$

for each $u \ge 0$. Note that $\overline{p}_k(u) = 2\overline{p}(u)$ for each $u \le \Delta_{m(k-1)}$, and $\overline{p}_k(u)$ is constant for each $u > \Delta_{m(k-1)}$. The definition of Riemann–Stieljes integrals then implies that for each nonnegative function f on $[0,1]$ that is integrable relative to the distribution function $\overline{p}$, we have

$$\begin{aligned}\int_0^1 f(u)d\overline{p}_k(u) &= \int_0^{\Delta(m(k-1))} f(u)d\overline{p}_k(u) \\ &= 2\int_0^{\Delta(m(k-1))} f(u)d\overline{p}(u) < \infty.\end{aligned} \tag{9.5.26}$$

In particular,

$$\int_0^1 \sqrt{-\log u}\,d\overline{p}_k(u) = 2\int_0^{\Delta(m(k-1))} \sqrt{-\log u}\,d\overline{p}(u) < \infty. \tag{9.5.27}$$

9. Inequalities 9.5.25 and 9.5.27 together imply that the a.u. continuous process $Z^{(k)}$ and the function $\overline{p}_k$ satisfy the conditions in the hypothesis of Lemma 9.5.6. Accordingly, there exists an integrable r.r.v. B_k with $EB_k \le \sqrt{2}$ such that

$$\begin{aligned}|Z^{(k)}(t,\omega) - Z^{(k)}(s,\omega)| &\le 16\int_0^{|t-s|} \sqrt{\log\left(\frac{4B_k(\omega)}{u^2}\right)}d\overline{p}_k(u) \\ &= 16\int_0^{|t-s|\wedge\Delta(m(k-1))} \sqrt{\log\left(\frac{4B_k(\omega)}{u^2}\right)}d\overline{p}(u)\end{aligned} \tag{9.5.28}$$

for each $t,s \in [0,1]$, for each $\omega \in domain(B_k)$.

10. Suppose $k \ge 2$. Let $\alpha_k \in (2^{-3}(k-1)^2, 2^{-2}(k-1)^2)$ be arbitrary, and define $A_k \equiv (B_k \le \alpha_k)$. Chebychev's inequality then implies that

$$P(A_k^c) \equiv P(B_k > \alpha_k) \le \alpha_k^{-1}\sqrt{2} < 2^3\sqrt{2}(k-1)^{-2}.$$

Consider each $\omega \in \bigcup_{n=k}^{n=\infty} A_n$. Let $n \ge k$ be arbitrary. Then inequality 9.5.28 implies that for each $t,s \in [0,1]$, we have

$$
\begin{aligned}
|Z_t^{(n)}(\omega) - Z_s^{(n)}(\omega)| &\leq 16 \int_0^{\Delta(m(n-1))} \sqrt{\log\left(\frac{4\alpha_n}{u^2}\right)} d\overline{p}(u) \\
&\leq 16 \int_0^{\Delta(m(n-1))} \sqrt{\log\left(\frac{(n-1)^2}{u^2}\right)} d\overline{p}(u) \\
&\equiv 16 \int_0^{\Delta(m(n-1))} 2^{-1}\Psi^{-1}\left(\frac{(n-1)^2}{u^2}\right) d\overline{p}(u) \\
&< 16 \int_0^{\Delta(m(n-1))} \Psi^{-1}\left(\frac{(n-1)^2}{u^2}\right) d\overline{p}(u) < (n-1)^{-2},
\end{aligned}
\tag{9.5.29}
$$

where the last inequality is by inequality 9.5.24. In particular, if we set $s = 0$ and recall that $Z_0^{(n)} = 0$, we obtain

$$
|Y_t^{(m(n))}(\omega) - Y_t^{(m(n-1))}(\omega)| \equiv |Z_t^{(n)}(\omega)| < (n-1)^{-2},
$$

where $\omega \in \bigcup_{n=k}^{n=\infty} A_n$ is arbitrary. Since $P(A_n^c) < 2^3\sqrt{2}(n-1)^{-2}$, we conclude that $Y_t^{(m(k))}$ converges a.u. to the limit r.r.v. $\overline{X}_t \equiv \lim_{k\to\infty} Y_t^{(m(k))}$. Thus we obtain the limiting process $\overline{X} : [0,1] \times \Omega \to R$.

11. We will next prove that the process $\overline{X}$ is a.u. continuous. To that end, note that since $Y^{(m(n))}$ is an a.u. continuous process for each $n \geq -1$ there exist a measurable set D_k with $P(D_k^c) < k^{-1}$ and a $\delta_k > 0$, such that for each $\omega \in D_k$, we have

$$
|\Sigma_{n=0}^{k-1}(Y_t^{(m(n))}(\omega) - Y_s^{(m(n))}(\omega))| < k^{-1}
$$

for each $t, s \in [0,1]$ with $|s - t| < \delta_k$. A similar inequality holds with n replaced by $n-1$. Separately, define the measurable set $C_k \equiv \bigcap_{n=k}^{\infty} A_n$. Then $P(C_k^c) \leq \sum_{n=k}^{\infty} 2^3\sqrt{2}(n-1)^{-2} \leq 2^3\sqrt{2}(k-1)^{-1}$.

12. Now consider each $\omega \in D_k C_k$, and each $t, s \in [0,1]$ with $|s - t| < \delta_k$. Then

$$
\overline{X}_t(\omega) = \left(\Sigma_{n=0}^{k+1} + \sum_{n=k}^{\infty}\right)(Y_t^{(m(n))}(\omega) - Y_t^{(m(n-1))}(\omega)),
$$

with a similar equality when t is replaced by s. Hence

$$
|\overline{X}_t(\omega) - \overline{X}_s(\omega)| \leq 2k^{-1} + 2\sum_{n=k}^{\infty}(n-1)^{-2} < 9k^{-1},
$$

where $\omega \in D_k C_k$ and $t, s \in [0,1]$ with $|s - t| < \delta_k$ are arbitrary. Since $P(D_k C_k)^c < 2k^{-1} + 2^3\sqrt{2}2k^{-1}$ and $9k^{-1}$ are arbitrarily small if $k \geq 2$ is sufficiently large, we see that $\overline{X} : [0,1] \times \Omega \to R$ is an a.u. continuous process. Consequently, the process $\overline{X}$ is continuous in probability.

13. Now we will verify that the process $\overline{X}$ is Gaussian, centered, with covariance function σ. Note that $\overline{X}|Q_\infty = Y$. Hence $\overline{X}|Q_\infty$ has marginal distributions given by the family $F^\sigma|Q_\infty$ of f.j.d.'s. Since the process $\overline{X}$ and the family F^σ are continuous in probability, and since the subset Q_∞ is dense in the parameter set $[0,1]$, it follows that $\overline{X}$ has marginal distributions given by the family F^σ. Thus $\overline{X}$ is Gaussian, centered, with covariance function σ.

14. In view of inequalities 9.5.20 and 9.5.21 in the hypothesis, the conditions in Lemma 9.5.6 are satisfied by the process $\overline{X}$ and the function $\overline{p}$. Hence Lemma 9.5.6 implies the existence of an integrable r.r.v. B with $EB \leq \sqrt{2}$ such that

$$|\overline{X}(t,\omega) - \overline{X}(s,\omega)| \leq 16 \int_0^{|t-s|} \sqrt{\log\left(\frac{4B(\omega)}{u^2}\right)} d\overline{p}(u) \tag{9.5.30}$$

for each $t,s \in [0,1]$, for each $\omega \in domain(B)$, as alleged. □

10

a.u. Càdlàg Process

In this chapter, let (S,d) be a locally compact metric space, with a binary approximation ξ relative to some fixed reference point $x_\circ$. As usual, write $\widehat{d} \equiv 1 \wedge d$. We will study processes $X : [0,\infty) \times \Omega \to S$ whose sample paths are right continuous with left limits, or càdlàg (a French acronym for "continue à droite, limite à gauche").

Classically, the proof of existence of such processes relies on Prokhorov's Relative Compactness Theorem. As discussed in the beginning of Chapter 9, this theorem implies the principle of infinite search. We will therefore bypass Prokhorov's theorem, in favor of direct proofs using Borel–Cantelli estimates.

Section 10.1 presents a version of Skorokhod's definition of càdlàg functions from $[0,\infty)$ to S. Each càdlàg function will come with a modulus of càdlàg, much as a continuous function comes with a modulus of continuity. In Section 10.2, we study a Skorokhod metric d_D on the space D of càdlàg functions.

In Section 10.3, we define an a.u. càdlàg process $X : [0,1]\times\Omega \to S$ as a process that is continuous in probability and that has, almost uniformly, càdlàg sample functions. In Section 10.4, we introduce a D-regular process $Z : Q_\infty \times \Omega \to S$, in terms of the marginal distributions of Z, where Q_∞ is the set of dyadic rationals in $[0,1]$. We then prove, in Sections 10.4 and 10.5, that a process $X : [0,1]\times\Omega \to S$ is a.u. càdlàg iff its restriction $X|Q_\infty$ is D-regular or, equivalently, iff X is the extension, by right limit, of a D-regular process Z. Thus we obtain a characterization of an a.u. càdlàg processes in terms of conditions on its marginal distributions. Equivalently, we have a procedure to construct an a.u. càdlàg process X from a consistent family F of f.j.d.'s that is D-regular. We will derive the modulus of a.u. càdlàg of X from the given modulus of D-regularity of F.

In Section 10.6, we will prove that this construction is metrically continuous, in epsilon–delta terms. Such continuity of construction seems to be hitherto unknown. In Section 10.7, we apply the construction to obtain a.u. càdlàg processes with strongly right continuous marginal distributions; in Section 10.8, to a.u. càdlàg martingales; and in Section 10.9, to processes that are right-Hoelder in a sense to be made precise there. In Section 10.10, we state the generalization of definitions and results in Sections 10.1–10.9, to the parameter interval $[0,\infty)$.

Finally, in Section 10.11, we will prove an abundance of first exit times for each a.u càdlàg process.

Before proceeding, we remark that our constructive method for a.u. càdlàg processes is by using certain accordion functions, defined in Definition 10.5.3, as time-varying boundaries for hitting times. This point will be clarified as we go along. This method was first used in [Chan 1974] to construct an a.u. càdlàg Markov process from a given strongly continuous semigroup.

Definition 10.0.1. Notations for dyadic rationals. For ease of reference, we restate the following notations in Definition 9.0.2 related to dyadic rationals. For each $m \geq 0$, define $p_m \equiv 2^m$, $\Delta_m \equiv 2^{-m}$; recall the enumerated sets of dyadic rationals

$$Q_m \equiv \{t_0, t_1, \ldots, t_{p(m)}\} = \{q_{m,0}, \ldots, q_{m,p(m)}\} \equiv \{0, \Delta_m, 2\Delta_m, \ldots, 1\} \subset [0,1],$$

and

$$Q_\infty \equiv \bigcup_{m=0}^{\infty} Q_m \equiv \{t_0, t_1, \ldots\}.$$

Thus

$$Q_m \equiv \{0, 2^{-m}, 2 \cdot 2^{-m}, \ldots, 1\}$$

is a 2^{-m}-approximation of $[0,1]$, with $Q_m \subset Q_{m+1}$, for each $m \geq 0$.

Moreover, for each $m \geq 0$, recall the enumerated sets of dyadic rationals

$$\overline{Q}_m \equiv \{u_0, u_1, \ldots, u_{p(2m)}\} \equiv \{0, 2^{-m}, 2 \cdot 2^{-m}, \ldots, 2^m\} \subset [0, 2^m]$$

and

$$\overline{Q}_\infty \equiv \bigcup_{m=0}^{\infty} \overline{Q}_m = \{u_0, u_1, \ldots\}. \square$$

10.1 Càdlàg Function

Recall some notations and conventions. To minimize clutter, a subscripted expression a_b will be written interchangeably with $a(b)$. For an arbitrary function x, we write $x(t)$ only with the explicit or implicit condition that $t \in domain(x)$. If $X : A \times \Omega \to S$ is an r.f., and if B is a subset of A, then $X|B \equiv X|(B \times \Omega)$ denotes the r.f. obtained by restricting the parameter set to B.

Definition 10.1.1. Pointwise left and right limits. Let Q be an arbitrary subset of $[0,\infty)$. Let the function $x : Q \to S$ be arbitrary.

The function x is said to be *right continuous* at a point $t \in domain(x)$ if $\lim_{r\to t; r\geq t} x(r) = x(t)$. The function x is said to have a *left limit* at a point $t \in Q$ if $\lim_{r\to t; r<t} x(r)$ exists.

Suppose, for each $t \in Q$ such that $\lim_{r\to t; r\geq t} x(r)$ exists, we have $t \in domain(x)$. Then we say that the function x is *right complete*. $\square$

Recall that the function x is said to be *continuous at t* if $t \in domain(x)$ and if $\lim_{r \to t} x(r) = x(t)$. Trivially, if x is continuous at t, then it is right continuous and has a left limit at t.

The next definition is essentially Skorokhod's characterization of càdlàg functions.

Definition 10.1.2. Càdlàg function on $[0,1]$**.** Let (S,d) be a locally compact metric space. Let $x : [0,1] \to S$ be a function such that $domain(x)$ contains all but countably many points in $[0,1]$. Suppose the following conditions are satisfied.

1. (Right continuity.) The function x is right continuous at each $t \in domain(x)$ and is continuous at $t = 1$.

2. (Right completeness.) Let $t \in [0,1]$ be arbitrary. If $\lim_{r \to t; r \geq t} x(r)$ exists, then $t \in domain(x)$.

3. (Approximation by step functions.) For each $\varepsilon > 0$, there exist $\delta_{cdlg}(\varepsilon) \in (0,1)$, $p \geq 1$, and a sequence

$$0 = \tau_0 < \tau_1 < \cdots < \tau_{p-1} < \tau_p = 1 \tag{10.1.1}$$

in $domain(x)$ such that (i) for each $i = 1, \ldots, p$, we have

$$\tau_i - \tau_{i-1} \geq \delta_{cdlg}(\varepsilon)$$

and (ii) for each $i = 0, \ldots, p-1$, we have

$$d(x, x(\tau_i)) \leq \varepsilon \tag{10.1.2}$$

on the interval $\theta_i \equiv [\tau_i, \tau_{i+1})$ or $\theta_i \equiv [\tau_i, \tau_{i+1}]$ depending on whether $i \leq p-2$ or $i = p-1$. We will call $(\tau_i)_{i=0,\ldots,p}$ a sequence of *ε-division points* of x with separation at least $\delta_{cdlg}(\varepsilon)$.

Then x said to be a *càdlàg* function on $[0,1]$ with values in S, with the operation δ_{cdlg} as a *modulus of càdlàg*.

We will let $D[0,1]$ denote the set of càdlàg functions. Two members of $D[0,1]$ are considered equal if they are equal as functions – i.e., if they have the same domain and have equal values at each point in the common domain. □

Note that Condition 3 implies that the endpoints $0, 1$ are in $domain(x)$. Condition 3 implies also that $p \leq \delta_{cdlg}(\varepsilon)^{-1}$. Let $x, y \in D[0,1]$ be arbitrary, with moduli of càdlàg $\delta_{cdlg}, \delta'_{cdlg}$, respectively. Then the operation $\delta_{cdlg} \wedge \delta'_{cdlg}$ is obviously a common modulus of càdlàg of x, y. The next lemma is a simple consequence of right continuity and generalizes its counterpart for $C[0,1]$.

Lemma 10.1.3. A càdlàg function is uniquely determined by its values on a dense subset of its domain. *Let $x, y \in D[0,1]$ be arbitrary. Let A be an arbitrary dense subset of $[0,1]$ that is contained in $B \equiv domain(x) \cap domain(y)$. Then the following conditions hold:*

1. Let $t \in B$ and $\alpha > 0$ be arbitrary. Then there exists $r \in [t, t+\alpha) \cap A$ such that

$$d(x(t),x(r)) \vee d(y(t),y(r)) \leq \alpha. \tag{10.1.3}$$

2. Let $f : S^2 \to R$ be a uniformly continuous function. Let $c \in R$ be arbitrary such that $f(x(r),y(r)) \leq c$ for each $r \in A$. Then $f(x,y) \leq c$. In other words, $f(x(t),y(t)) \leq c$ for each $t \in domain(x) \cap domain(y)$. The same assertion holds when "$\leq$" is replaced by "$\geq$" or by "$=$". In particular, if $d(x(r),y(r)) \leq \varepsilon$ for each $r \in A$, for some $\varepsilon > 0$, then $f(x,y) \leq \varepsilon$ on $domain(x) \cap domain(y)$.

3. Suppose $x(r) = y(r)$ for each $r \in A$. Then $x = y$. In other words, $domain(x) = domain(y)$ and $x(t) = y(t)$ for each $t \in domain(x)$.

4. Let $\lambda : [0,1] \to [0,1]$ be an arbitrary continuous and increasing function with $\lambda(0) = 0$ and $\lambda(1) = 1$. Then $x \circ \lambda \in D[0,1]$.

Proof. Let δ_{cdlg} be a common modulus of càdlàg of x and y.

1. Let $t \in B$ and $\alpha > 0$ be arbitrary. Let $(\tau_i)_{i=0,\ldots,p}$ and $(\tau'_i)_{i=0,\ldots,p'}$ be sequences of $\frac{\alpha}{2}$-division points of x and y, respectively, with separation of at least $\delta_{cdlg}(\frac{\alpha}{2})$. Then $\tau_{p-1} \vee \tau'_{p'-1} < 1$. Hence either (i) $t < 1$ or (ii) $\tau_{p-1} \vee \tau'_{p'-1} < t$. Consider case (i). Since x,y are right continuous at t according to Definition 10.1.2, and since A is dense in $[0,1]$, there exists r in $A \cap [t, 1 \wedge (t+\alpha))$ such that

$$d(x(t),x(r)) \vee d(y(t),y(r)) \leq \alpha,$$

as desired. Consider case (ii). Take $r \in (\tau_{p-1} \vee \tau'_{p'-1}, t) \cap A$. Then $t,r \in [\tau_{p-1},1] \cap [\tau'_{p'-1},1]$. Hence Condition 3 in Definition 10.1.2 implies that

$$d(x(t),x(r)) \leq d(x(\tau_{p-1}),x(t)) + d(x(\tau_{p-1}),x(r)) \leq \frac{\alpha}{2} + \frac{\alpha}{2} = \alpha.$$

Similarly, $d(y(t),y(r)) \leq \alpha$. Assertion 1 is proved.

2. Let $t \in B$ be arbitrary. By Assertion 1, for each $k \geq 1$, there exists $r_k \in [t, t+k^{-1})A$ such that

$$d(x(t),x(r_k)) \vee d(y(t),y(r_k)) \leq k^{-1}.$$

By hypothesis, $f(x(r_k),y(r_k)) \leq c$ because $r_k \in A$, for each $k \geq 1$. Consequently, by right continuity of x,y and continuity of f, we have

$$f(x(t),y(t)) = \lim_{k\to\infty} f(x(r_k),y(r_k)) \leq c.$$

3. By hypothesis, $d(x(r),y(r)) = 0$ for each $r \in A$. Consider each $t \in domain(x)$. We will first verify that $t \in domain(y)$. To that end, let $\varepsilon > 0$ be arbitrary. Since x is right continuous at t, there exists $c > 0$ such that

$$d(x(r),x(t)) < \varepsilon \tag{10.1.4}$$

for each $r \in [t,t+c) \cap domain(x)$. Consider each $s \in [t,t+c) \cap domain(y)$. Let $\alpha = (t+c-s) \wedge \varepsilon$. By Assertion 1 applied to the pair y,y in $D[0,1]$, there exists $r \in [s,s+\alpha) \cap A$ such that $d(y(s),y(r)) \leq \alpha \leq \varepsilon$. Then $r \in [t,t+c) \cap A$, whence inequality 10.1.4 holds. Combining,

$$d(y(s),x(t)) \leq d(y(s),y(r)) + d(y(r),x(r)) + d(x(r),x(t)) < \varepsilon + 0 + \varepsilon = 2\varepsilon.$$

Since $\varepsilon > 0$ is arbitrary, we see that $\lim_{s\to t;s>t} y(s)$ exists and is equal to $x(t)$. Hence the right completeness Condition 2 in Definition 10.1.2 implies that $t \in domain(y)$, as alleged. Condition 1 in Definition 10.1.2 then implies that $y(t) = \lim_{s\to t;s>t} y(s) = x(t)$. Since $t \in domain(x)$ is arbitrary, we conclude that $domain(x) \subset domain(y)$, and that $x = y$ on $domain(x)$. By symmetry, $domain(x) = domain(y)$.

4. Since λ is continuous and increasing, it has an inverse λ^{-1} that is also continuous and increasing, with some modulus of continuity $\overline{\delta}$. Let δ_{cdlg} be a modulus of càdlàg of x. We will prove that $x \circ \lambda$ is càdlàg, with $\delta_1 \equiv \overline{\delta} \circ \delta_{cdlg}$ as a modulus of càdlàg. To that end, let $\varepsilon > 0$ be arbitrary. Let

$$0 \equiv \tau_0 < \tau_1 < \cdots < \tau_{p-1} < \tau_p = 1 \tag{10.1.5}$$

be a sequence of ε-division points of x with separation at least $\delta_{cdlg}(\varepsilon)$. Thus, for each $i = 1,\ldots,p$, we have $\tau_i - \tau_{i-1} \geq \delta_{cdlg}(\varepsilon)$. Suppose $\lambda\tau_i - \lambda\tau_{i-1} < \overline{\delta}(\delta_{cdlg}(\varepsilon))$ for some $i = 1,\ldots,p$. Then, since $\overline{\delta}$ is a modulus of continuity of the inverse function λ^{-1}, it follows that

$$\tau_i - \tau_{i-1} = \lambda^{-1}\lambda\tau_i - \lambda^{-1}\lambda\tau_{i-1} < \delta_{cdlg}(\varepsilon),$$

which is a contradiction. Hence

$$\lambda\tau_i - \lambda\tau_{i-1} \geq \overline{\delta}(\delta_{cdlg}(\varepsilon)) \equiv \delta_1(\varepsilon)$$

for each $i = 1,\ldots,p$. Moreover, for each $i = 0,\ldots,p-1$, we have

$$d(x,x(\tau_i)) \leq \varepsilon \tag{10.1.6}$$

on the interval $\theta_i \equiv [\tau_i,\tau_{i+1})$ or $\theta_i \equiv [\tau_i,\tau_{i+1}]$ depending on whether $i \leq p-2$ or $i = p-1$. Since the function λ is increasing, it follows that

$$d(x \circ \lambda, x \circ \lambda(\tau_i)) \leq \varepsilon \tag{10.1.7}$$

on the interval $\theta_i' \equiv [\lambda^{-1}\tau_i,\lambda^{-1}\tau_{i+1})$ or $\theta_i' \equiv [\lambda^{-1}\tau_i,\lambda^{-1}\tau_{i+1}]$ depending on whether $i \leq p-2$ or $i = p-1$. Thus the sequence

$$0 = \lambda^{-1}\tau_0 < \lambda^{-1}\tau_1 < \cdots < \lambda^{-1}\tau_{p-1} < \lambda^{-1}\tau_p = 1$$

is a sequence of ε-division points of $x \circ \lambda$ with separation at least $\delta_1(\varepsilon)$. Condition 3 in Definition 10.1.2 has been proved for the function $x \circ \lambda$. In view of the monotonicity and continuity of the function λ, the other conditions can also be easily verified. Accordingly, the function $x \circ \lambda$ is càdlàg, with a modulus of càdlàg δ_1. Assertion 4 and the lemma are proved. □

Proposition 10.1.4. Each càdlàg function is continuous at all but countably many time points. *Let $x \in D[0,1]$ be arbitrary with a modulus of càdlàg δ_{cdlg}. Then the function x on $[0,1]$ is continuous at the endpoints 0 and 1.*

For each $k \geq 1$, let $(\tau_{k,i})_{i=0,\ldots,p(k)}$ be a sequence of k^{-1}-division points of x with separation of at least $\delta_{cdlg}(k^{-1})$. Then the following conditions hold:

1. Define the set $A \equiv \bigcap_{k=1}^{\infty} \bigcup_{i=0}^{p(k)-1} \theta_{k,i}$, where $\theta_{k,i} \equiv [\tau_{k,i}, \tau_{k,i+1})$ or $\theta_{k,i} \equiv [\tau_{k,i}, \tau_{k,i+1}]$ depending on whether $i = 0, \ldots, p_k - 2$ or $i = p_k - 1$. Then the set A contains all but countably many points in $[0,1]$ and is a subset of $domain(x)$.

2. Define the set $A' \equiv \bigcap_{k=1}^{\infty} \bigcup_{i=0}^{p(k)-1} \theta'_{k,i}$, where $\theta'_{k,i} \equiv [0, \tau_{k,1}]$ or $\theta'_{k,i} \equiv (\tau_{k,i}, \tau_{k,i+1}]$ depending on whether $i = 0$ or $i = 1, \ldots, p_k - 1$. Then the set A' contains all but countably many points in $[0,1]$ and the function x has a left limit at each $t \in A'$.

3. Define the set $A'' \equiv \bigcap_{k=1}^{\infty} \bigcup_{i=0}^{p(k)-1} (\tau_{k,i}, \tau_{k,i+1})$. Then the set A'' contains all but countably many points in $[0,1]$ and the function x is continuous at each $t \in A''$.

4. The function x is bounded on $domain(x)$. Specifically,

$$d(x_\circ, x(t)) \leq b \equiv \bigvee_{i=0}^{p(1)-1} d(x_\circ, x(\tau_{1,i})) + 1$$

for each $t \in domain(x)$, where $x_\circ$ is an arbitrary but fixed reference point in S.

Proof. By Definition 10.1.2, we have $1 \in domain(x)$. Condition 3 in Definition 10.1.2 implies that $0 = \tau_{1,0} \in domain(x)$ and that x is continuous at 0 and 1.

1. Let $t \in A$ be arbitrary. Let $k \geq 1$ be arbitrary. Then $t \in \theta_{k,i}$ for some $i = 0, \ldots, p_k - 1$. Let $\delta_0 \equiv \tau_{k,i+1} - t$ or $\delta_0 \equiv 2$ depending on whether $i = 0, \ldots, p_k - 2$ or $i = p_k - 1$. Then $domain(x) \cap [t, t + \delta_0)$ is a nonempty subset of $\theta_{k,i}$. Moreover, by Condition 3 of Definition 10.1.2, we have $d(x(r), x(\tau_{k,i})) \leq k^{-1}$ for each $r \in domain(x) \cap [t, t + \delta_0)$. Hence $d(x(r), x(s)) \leq 2k^{-1}$ for each $r, s \in domain(x) \cap [t, t + \delta_0)$. Since $2k^{-1}$ is arbitrarily small, and since the metric space (S, d) is complete, we see that $\lim_{r \to t; r \geq t} x(r)$ exists. The right completeness Condition 2 of Definition 10.1.2 therefore implies that $t \in domain(x)$. We conclude that $A \subset domain(x)$. Assertion 1 follows.

2. Let $t \in A'$ and $k \geq 1$ be arbitrary. Then $t \in \theta'_{k,i}$ for some $i = 0, \ldots, p_k - 1$. Let $\delta_0 \equiv 2$ or $\delta_0 \equiv t - \tau_{k,i}$ depending on whether $i = 0$ or $i = 1, \ldots, p_k - 1$. Then $domain(x) \cap (t - \delta_0, t)$ is a nonempty subset of $\theta_{k,i}$. Moreover, by Condition 3 of Definition 10.1.2, we have $d(x, x(\tau_{k,i})) \leq k^{-1}$ for each $r \in domain(x) \cap (t - \delta_0, t)$. An argument similar to that made in the previous paragraph then shows that $\lim_{r \to t; r < t} x(r)$ exists. Assertion 2 is verified.

3. Since $A'' \subset A$, we have $A'' \subset domain(x)$, thanks to Assertion 1. Consider each $t \in A''$. Let $k \geq 1$ be arbitrary. Then $t \in (\tau_{k,i}, \tau_{k,i+1})$ for some $i = 0, \ldots, p_k - 1$. Hence, by Condition 3 of Definition 10.1.2, we have $d(x(r), x(t)) \leq 2k^{-1}$ for each $r \in domain(x) \cap (\tau_{k,i}, \tau_{k,i+1})$. We conclude that the function x is continuous at t.

4. Finally, observe that each of the sets A, A', and A'' contains the metric complement of the countable subset $\{\tau_{k,i}\}$. Thus each contains all but countably

many points in $[0,1]$, and is dense in $[0,1]$. Now let $t \in A \subset domain(x)$ be arbitrary. Then $t \in \theta_{1,i}$ for some $i = 0, \ldots, p_1 - 1$. Hence

$$d(x_\circ, x(t)) \leq d(x_\circ, x(\tau_{1,i})) + d(x(\tau_{1,i}), x(t))) \leq b \equiv \bigvee_{j=0}^{p(1)-1} d(x_\circ, x(\tau_{1,j})) + 1.$$

Since A is dense in $[0,1]$ and since the function $d : S^2 \to R$ is uniformly continuous, Assertion 2 of Lemma 10.1.3 implies that $d(x_\circ, x(r)) \leq b$ for each $r \in domain(x)$. Assertion 4 and the lemma are proved. □

Proposition 10.1.5. a.u. Right continuity of càdlàg function. *Let $x \in D[0,1]$ be arbitrary, with a modulus of càdlàg δ_{cdlg}. For each $\alpha > 0$, let $h \equiv [2 + 0 \vee -\log_2 \alpha]_1$, and define*

$$\delta_{rc}(\alpha, \delta_{cdlg}) \equiv 2^{-h} \delta_{cdlg}(2^{-h}) > 0.$$

Let $\varepsilon > 0$ be arbitrary. Then there exists a Lebesgue measurable subset A of $domain(x)$ with Lebesgue measure $\mu(A) < \varepsilon$ such that for each $\alpha \in (0, \varepsilon)$, we have

$$d(x(t), x(s)) < \alpha$$

for each $t \in A^c \cap domain(x)$ and $s \in [t, t + \delta_{rc}(\alpha, \delta_{cdlg})) \cap domain(x)$. Note that the operation $\delta_{rc}(\cdot, \delta_{cdlg})$ is determined by δ_{cdlg}.

Proof. 1. Let $h \geq 0$ be arbitrary. Write $\alpha_h \equiv 2^{-h}$. Then, according to Condition (iii) of Definition 10.1.2, there exist an integer $p_h \geq 1$ and a sequence

$$0 = \tau_{h,0} < \tau_{h,1} < \cdots < \tau_{h,p-1} < \tau_{h,p(h)} = 1 \qquad (10.1.8)$$

in $domain(x)$, such that (i) for each $i = 1, \ldots, p_h$, we have

$$\tau_{h,i} - \tau_{h,i-1} \geq \delta_{cdlg}(\alpha_h) \qquad (10.1.9)$$

and (ii) for each $i = 0, \ldots, p_h - 1$, we have

$$d(x, x(\tau_{h,i})) \leq \alpha_h \qquad (10.1.10)$$

on the interval $\theta_{h,i} \equiv [\tau_{h,i}, \tau_{h,i+1})$ or $\theta_{h,i} \equiv [\tau_{h,i}, \tau_{h,i+1}]$ depending on whether $i \leq p_h - 2$ or $i = p_h - 1$.

2. Let $i = 0, \ldots, p_h - 1$ be arbitrary. Define

$$\overline{\theta}_{h,i} \equiv [\tau_{h,i}, \tau_{h,i+1} - \alpha_h(\tau_{h,i+1} - \tau_{h,i})) \subset \theta_{h,i}. \qquad (10.1.11)$$

Define $\overline{\theta}_h \equiv \bigcup_{i=0}^{p(h)-1} \overline{\theta}_{h,i}$. Then

$$\mu(\overline{\theta}_h) = \sum_{i=0}^{p(h)-1} \mu(\overline{\theta}_{h,i}) = \sum_{i=0}^{p(h)-1} (\tau_{h,i+1} - \tau_{h,i})(1 - \alpha_h) = 1 - \alpha_h,$$

whence $\mu \overline{\theta}_h{}^c = \alpha_h \equiv 2^{-h}$, where $h \geq 0$ is arbitrary.

3. Now let $\varepsilon > 0$ be arbitrary, and let $k \equiv [1 + 0 \vee -\log_2 \varepsilon]_1$. Define $A \equiv \bigcup_{h=k+1}^{\infty} \overline{\theta}_h{}^c$. Then

$$\mu(A) \leq \sum_{h=k+1}^{\infty} 2^{-h} = 2^{-k} < \varepsilon.$$

Consider each $t \in A^c \cap domain(x)$. Let $\alpha \in (0, \varepsilon)$ be arbitrary, and let $h \equiv [3 + 0 \vee -\log_2 \alpha]_1$. Then

$$h > 3 + 0 \vee -\log_2 \alpha > 3 + 0 \vee -\log_2 \varepsilon > k.$$

Hence $h \geq k+1$, and so $A^c \subset \overline{\theta}_h$. Therefore $t \in \overline{\theta}_{h,i} \equiv [\tau_{h,i}, \tau_{h,i+1} - \alpha_h(\tau_{h,i+1} - \tau_{h,i}))$ for some $i = 0, \ldots, p_h - 1$.

4. Now consider each $s \in [t, t + \delta_{rc}(\alpha, \delta_{cdlg})) \cap domain(x)$. Then

$$\begin{aligned}
\tau_{h,i} &\leq s \leq t + \delta_{rc}(\alpha, \delta_{cdlg}) \\
&< \tau_{h,i+1} - \alpha_h \cdot (\tau_{h,i+1} - \tau_{h,i}) + \delta_{rc}(\alpha, \delta_{cdlg}) \\
&\leq \tau_{h,i+1} - 2^{-h}\delta_{cdlg}(2^{-h}) + 2^{-h}\delta_{cdlg}(2^{-h}) = \tau_{h,i+1}.
\end{aligned}$$

Hence $s, t \in [\tau_{h,i}, \tau_{h,i+1})$. It follows that

$$d(x(s), x(\tau_{h,i})) \vee d(x(t), x(\tau_{h,i})) \leq \alpha_h$$

and therefore that

$$d(x(s), x(t)) \leq 2\alpha_h = 2^{-h+1} < \alpha.$$

□

The next proposition shows that if a function satisfies all the conditions in Definition 10.1.2 except perhaps the right completeness Condition 2, then it can be right completed and extended to a càdlàg function. This is analogous to the extension of a uniformly continuous function on a dense subset of [0, 1].

Proposition 10.1.6. Right-limit extension and càdlàg completion. *Let (S, d) be a locally compact metric space. Suppose $Q = [0, 1]$ or $Q = [0, \infty)$. Let $x : Q \to S$ be a function whose domain contains a dense subset A of Q, and which is right continuous at each $t \in domain(x)$. Define its right-limit extension $\overline{x} : Q \to S$ by*

$$domain(\overline{x}) \equiv \left\{ t \in Q;\ \lim_{r \to t; r \geq t} x(r) \text{ exists} \right\} \tag{10.1.12}$$

and by

$$\overline{x}(t) \equiv \lim_{r \to t; r \geq t} x(r) \tag{10.1.13}$$

for each $t \in domain(\overline{x})$. Then the following conditions hold:

1. The function $\overline{x}$ is right continuous at each $t \in domain(\overline{x})$.

2. Suppose $t \in Q$ is such that $\lim_{r \to t; r \geq t} \overline{x}(r)$ exists. Then $t \in domain(\overline{x})$.

3. Suppose $Q = [0, 1]$. Suppose, in addition, that $\delta_{cdlg} : (0, \infty) \to (0, \infty)$ is an operation such that x and δ_{cdlg} satisfy Conditions 1 and 3 in Definition 10.1.2.

Then $\overline{x} \in D[0,1]$. *Moreover,* $\overline{x}$ *has* δ_{cdlg} *as a modulus of càdlàg. Furthermore,* $x = \overline{x}|domain(x)$. *We will then call* $\overline{x}$ *the* càdlàg completion *of* x.

Proof. 1. Since, by hypothesis, x is right continuous at each $t \in domain(x)$, it follows from the definition of $\overline{x}$ that $domain(x) \subset domain(\overline{x})$ and that $\overline{x} = x$ on $domain(x)$. In other words, $x = \overline{x}|domain(x)$. Since $domain(x)$ contains the dense subset A of Q, so does $domain(\overline{x})$. Now let $t \in domain(\overline{x})$ and $\varepsilon > 0$ be arbitrary. Then, by the defining equality 10.1.13,

$$\overline{x}(t) \equiv \lim_{r \to t; r \geq t} x(r).$$

Hence there exists $\delta_0 > 0$ such that

$$d(\overline{x}(t), x(r)) \leq \varepsilon \tag{10.1.14}$$

for each $r \in domain(x) \cap [t, t + \delta_0)$. Let $s \in domain(\overline{x}) \cap [t, t + \delta_0)$ be arbitrary. Then, again by the defining equalities 10.1.12 and 10.1.13, there exists a sequence $(r_j)_{j=1,2,...}$ in $domain(x) \cap [s, t + \delta_0)$ such that $r_j \to s$ and $\overline{x}(s) = \lim_{j \to \infty} x(r_j)$. For each $j \geq 1$, we then have $r_j \in domain(x) \cap [t, t+\delta_0)$. Hence inequality 10.1.14 holds for $r = r_j$, for each $j \geq 1$. Letting $j \to \infty$, we therefore obtain $d(\overline{x}(t), \overline{x}(s)) \leq \varepsilon$. Since $\varepsilon > 0$ is arbitrary, we conclude that $\overline{x}$ is right continuous at each $t \in domain(x)$. Assertion 1 has been verified.

2. Next suppose $\lim_{r \to t; r \geq t} \overline{x}(r)$ exists. Then, since $x = \overline{x}|domain(x)$, the right limit

$$\lim_{r \to t; r \geq t} x(r) = \lim_{r \to t; r \geq t} \overline{x}(r)$$

exists. Hence $t \in domain(\overline{x})$ by the defining equality 10.1.12. Condition 2 of Definition 10.1.2 has been proved for $\overline{x}$.

3. Now let $\varepsilon > 0$ be arbitrary. Because $x = \overline{x}|domain(x)$, each sequence $(\tau_i)_{i=0,...,p}$ of ε-division points of x, with separation of at least $\delta_{cdlg}(\varepsilon)$, is also a sequence of ε-division points of $\overline{x}$ with separation of at least $\delta_{cdlg}(\varepsilon)$. Therefore Condition 3 in Definition 10.1.2 holds for $\overline{x}$ and the operation δ_{cdlg}. Summing up, the function $\overline{x}$ is càdlàg, with δ_{cdlg} as a modulus of càdlàg. □

The next definition introduces simple càdlàg functions as càdlàg completion of step functions.

Definition 10.1.7. Simple càdlàg function. Let $0 = \tau_0 < \cdots < \tau_{p-1} < \tau_p = 1$ be an arbitrary sequence of dyadic rationals in $[0,1]$ such that

$$\bigwedge_{i=1}^{p} (\tau_i - \tau_{i-1}) \geq \delta_0$$

for some $\delta_0 > 0$. Let $x_0, \ldots, x_{p-1}$ be an arbitrary sequence in S.

Define a function $z : [0,1] \to S$ by

$$domain(z) \equiv \bigcup_{i=0}^{p-1} \theta_i, \tag{10.1.15}$$

where $\theta_i \equiv [\tau_i, \tau_{i+1})$ or $\theta_i \equiv [\tau_i, \tau_{i+1}]$ depending on whether $i = 0, \ldots, p-2$ or $i = p-1$, and by

$$z(r) \equiv x_i$$

for each $r \in \theta_i$, for each $i = 0, \ldots, p-1$. Let $x \equiv \overline{z} \in D[0,1]$ be the càdlàg completion of z. Then x is called the *simple càdlàg function* determined by the pair of sequences $((\tau_i)_{i=0,\ldots,p-1}, (x_i)_{i=0,\ldots,p-1})$. In symbols, we then write

$$x \equiv \Phi_{smpl}((\tau_i)_{i=0,\ldots,p-1}, (x_i)_{i=0,\ldots,p-1})$$

or simply $x \equiv \Phi_{smpl}((\tau_i),(x_i))$ when the range of subscripts is understood. The sequence $(\tau_i)_{i=0,\ldots,p}$ is called the sequence of *division points* of the simple càdlàg function x. The next lemma verifies that x is a well-defined càdlàg function, with the constant operation $\delta_{cdlg}(\cdot) \equiv \delta_0$ as a modulus of càdlàg. □

Lemma 10.1.8. Simple càdlàg functions are well defined. *Use the notations and assumptions in Definition 10.1.7. Then z and δ_{cdlg} satisfy the conditions in Proposition 10.1.6. Accordingly, the càdlàg completion $\overline{z} \in D[0,1]$ of z is well defined.*

Proof. First note that $domain(z)$ contains the dyadic rationals in $[0,1]$. Let $t \in domain(z)$ be arbitrary. Then $t \in \theta_i$ for some $i = 0, \ldots, p-1$. Hence, for each $r \in \theta_i$, we have $z(r) \equiv x_i \equiv z(t)$. Therefore z is right continuous at t. Condition 1 in Definition 10.1.2 has been verified for z. The proof of Condition 3 in Definition 10.1.2 for z and δ_{cdlg} being trivial, the conditions in Proposition 10.1.6 are satisfied. □

Lemma 10.1.9. Insertion of division points leaves a simple càdlàg function unchanged. *Let $p \geq 1$ be arbitrary. Let $0 \equiv q_0 < q_1 < \cdots < q_p \equiv 1$ be an arbitrary sequence of dyadic rationals in $[0,1]$, with an arbitrary subsequenc $0 \equiv q_{i(0)} < q_{i(1)} < \cdots < q_{i(\kappa)} \equiv 1$. Let $(w_0, \ldots, w_{\kappa-1})$ be an arbitrary sequence in S. Let*

$$x \equiv \Phi_{smpl}((q_{i(k)})_{k=0,\ldots,\kappa-1}, (w_k)_{k=0,\ldots,\kappa-1}).$$

Let

$$y = \Phi_{smpl}((q_j)_{j=0,\ldots,p-1}, (x(q_j))_{j=0,\ldots,p-1}). \tag{10.1.16}$$

Then $x = y$.

Proof. Let $j = 0, \ldots, p-1$ and $t \in [q_j, q_{j+1})$ be arbitrary. Then $t \in [q_j, q_{j+1}) \subset [q_{i(k)}, q_{i(k-1)})$ for some unique $k = 0, \ldots, \kappa - 1$. Hence $y(t) = x(q_j) = w_k = x(t)$. Thus $y = x$ on the dense subset $\bigcup_{j=0}^{p-1}[q_j, q_{j+1})$ of $domain(y \cap domain(x)$. Hence, by Assertion 3 of Lemma 10.1.3, we have $y = x$. □

10.2 Skorokhod Space $D[0,1]$ of Càdlàg Functions

Following Skorokhod, via [Billingsley 1968], we proceed to define a metric on the space $D[0,1]$ of càdlàg functions. This metric is similar to the supremum metric in $C[0,1]$, except that it allows and measures a small error in time parameter, in terms of a continuous and increasing function $\lambda : [0,1] \to [0,1]$ with $\lambda(0) = 0$ and $\lambda(1) = 1$.

Let λ, λ' be any such continuous and increasing functions. We will write, as an abbreviation, λt for $\lambda(t)$, for each $t \in [0,1]$. We will write λ^{-1} for the inverse of λ, and $\lambda'\lambda \equiv \lambda' \circ \lambda$ for the composite function.

Definition 10.2.1. Skorokhod metric. Let Λ denote the set of continuous and increasing functions $\lambda : [0,1] \to [0,1]$ with $\lambda 0 = 0$ and $\lambda 1 = 1$, such that there exists $c > 0$ with

$$\left|\log \frac{\lambda t - \lambda s}{t - s}\right| \leq c \tag{10.2.1}$$

or, equivalently,

$$e^{-c}(t-s) \leq \lambda t - \lambda s \leq e^{c}(t-s),$$

for each $0 \leq s < t \leq 1$. We will call Λ the set of *admissible functions* on $[0,1]$.

Let $x, y \in D[0,1]$ be arbitrary. Let $A_{x,y}$ denote the set consisting of all pairs $(c,\lambda) \in [0,\infty) \times \Lambda$ such that (i) inequality 10.2.1 holds for each $0 \leq s < t \leq 1$, and (ii)

$$d(x(t), y(\lambda t)) \leq c \tag{10.2.2}$$

for each $t \in domain(x) \cap \lambda^{-1} domain(y)$.

Let

$$B_{x,y} \equiv \{c \in [0,\infty) : (c,\lambda) \in A_{x,y} \text{ for some } \lambda \in \Lambda\}.$$

Define the metric $d_{D[0,1]}$ on $D[0,1]$ by

$$d_{D[0,1]}(x,y) \equiv \inf B_{x,y}. \tag{10.2.3}$$

We will presently prove that $d_{D[0,1]}$ is well defined and is indeed a metric, called the *Skorokhod metric* on $D[0,1]$. When the interval $[0,1]$ is understood, we write d_D for $d_{D[0,1]}$. □

Intuitively, the number c bounds both (i) the error in the time measurement, represented by the distortion λ, and (ii) the supremum distance between the functions x and y when allowance is made for said error. Existence of the infimum in equality 10.2.3 would follow easily from the principle of infinite search. We will supply a constructive proof in Lemmas 10.2.4 through 10.2.8. Proposition 10.2.9 will then complete the proof that $d_{D[0,1]}$ is a metric. Finally, we will prove that the Skorokhod metric space $(D[0,1], d_D)$ is complete.

First two elementary lemmas.

Lemma 10.2.2. A sufficient condition for existence of infimum or supremum. *Let B be an arbitrary nonempty subset of R.*

1. Suppose for each $k \geq 0$, there exists $\alpha_k \in R$ such that (i) $\alpha_k \leq c + 2^{-k}$ *for each $c \in B$ and* (ii) $c \leq \alpha_k + 2^{-k}$ *for some $c \in B$. Then* $\inf B$ *exists and* $\inf B = \lim_{k\to\infty} \alpha_k$. *Moreover, $\alpha_k - 2^{-k} \leq \inf B \leq \alpha_k + 2^{-k}$ for each $k \geq 0$.*

2. Suppose for each $k \geq 0$, there exists $\alpha_k \in R$ such that (iii) $\alpha_k \geq c - 2^{-k}$ *for each $c \in B$ and* (iv) $c \geq \alpha_k - 2^{-k}$ *for some $c \in B$. Then* $\sup B$ *exists and* $\sup B = \lim_{k\to\infty} \alpha_k$. *Moreover, $\alpha_k - 2^{-k} \leq \sup B \leq \alpha_k + 2^{-k}$ for each $k \geq 0$.*

Proof. Let $h,k \geq 0$ be arbitrary. Then, by Condition (ii), there exists $c \in B$ such that $c \leq \alpha_k + 2^{-k}$. At the same time, by Condition (i), we have $\alpha_h \leq c + 2^{-h} \leq \alpha_k + 2^{-k} + 2^{-h}$. Similarly, $\alpha_k \leq \alpha_h + 2^{-h} + 2^{-k}$. Consequently, $|\alpha_h - \alpha_k| \leq 2^{-h} + 2^{-k}$. We conclude that the limit $\alpha \equiv \lim_{k\to\infty} \alpha_k$ exists. Let $c \in B$ be arbitrary. Letting $k \to \infty$ in Condition (i), we see that $\alpha \leq c$. Thus α is a lower bound for the set B. Suppose β is a second lower bound for B. By Condition (ii) there exists $c \in B$ such that $c \leq \alpha_k + 2^{-k}$. Then $\beta \leq c \leq \alpha_k + 2^{-k}$. Letting $k \to \infty$, we obtain $\beta \leq \alpha$. Thus α is the greatest lower bound of the set B. In other words, $\inf B$ exists and is equal to α, as alleged. Assertion 1 is proved. The proof of Assertion 2 is similar. □

Lemma 10.2.3. Logarithm of certain difference quotients. *Let*

$$0 = \tau_0 < \cdots < \tau_{p-1} < \tau_p \equiv 1$$

be an arbitrary sequence in $[0,1]$. *Suppose the function $\lambda \in \Lambda$ is linear on $[\tau_i, \tau_{i+1}]$, for each $i = 0, \ldots, p-1$. Then*

$$\sup_{0 \leq s < t \leq 1} \left| \log \frac{\lambda t - \lambda s}{t - s} \right| = \alpha \equiv \bigvee_{i=0}^{p-1} \left| \log \frac{\lambda \tau_{i+1} - \lambda \tau_i}{\tau_{i+1} - \tau_i} \right|. \tag{10.2.4}$$

Proof. Let $s,t \in A \equiv \bigcup_{i=0}^{p-1} (\tau_i, \tau_{i+1})$ be arbitrary, with $s < t$. Then $\tau_i < s < \tau_{i+1}$ and $\tau_j < t < \tau_{j+1}$ for some $i,j = 0, \ldots, p-1$ with $i \leq j$. Hence, in view of the linearity of λ on each of the intervals $[\tau_i, \tau_{i+1})$ and $[\tau_j, \tau_{j+1})$, we obtain

$$\begin{aligned}
\lambda t - \lambda s &= (\lambda t - \lambda \tau_j) + (\lambda \tau_j - \lambda \tau_{j-1}) + \cdots + (\lambda \tau_{i+2} - \lambda \tau_{i+1}) + (\lambda \tau_{i+1} - \lambda s) \\
&\leq e^{\alpha}(t - \tau_j) + e^{\alpha}(\tau_j - \tau_{j-1}) + \cdots + e^{\alpha}(\tau_j - \tau_{j-1}) + e^{\alpha}(\tau_{i+1} - s) \\
&= e^{\alpha}(t - s).
\end{aligned}$$

Similarly, $\lambda t - \lambda s \geq e^{-\alpha}(t-s)$. Hence

$$e^{-\alpha}(t-s) \leq \lambda t - \lambda s \leq e^{\alpha}(t-s),$$

where $s,t \in A$ with $s < t$ are arbitrary. Since A is dense in $[0,1]$, the last displayed inequality holds, by continuity, for each $s,t \in [0,1]$ with $s < t$. Equivalently,

$$\left|\log \frac{\lambda t - \lambda s}{t-s}\right| \leq \alpha \tag{10.2.5}$$

for each $s,t \in [0,1]$ with $s < t$. At the same time, for each $\varepsilon > 0$, there exists $j = 0,\ldots,m$ such that

$$\left|\log \frac{\lambda\tau_{j+1} - \lambda\tau_j}{\tau_{j+1} - \tau_j}\right| > \alpha - \varepsilon.$$

Thus

$$\alpha < \left|\log \frac{\lambda u - \lambda v}{u-v}\right| + \varepsilon, \tag{10.2.6}$$

where $u \equiv \tau_{j+1}$ and $v \equiv \lambda\tau_j$. Since $\varepsilon > 0$ is arbitrary, inequalities 10.2.5 and 10.2.6 together imply the desired equality 10.2.4, according to Lemma 10.2.2. □

Next, we prove some metric-like properties of the sets $B_{x,y}$ introduced in Definition 10.2.1.

Lemma 10.2.4. Metric-like properties of the sets $B_{x,y}$**.** *Let* $x,y,z \in D[0,1]$ *be arbitrary. Then* $B_{x,y}$ *is nonempty. Moreover, the following conditions hold:*

1. $0 \in B_{x,x}$. *More generally, if* $d(x,y) \leq b$ *for some* $b \geq 0$*, then* $b \in B_{x,y}$.

2. $B_{x,y} = B_{x,y}$.

3. Let $c \in B_{x,y}$ *and* $c' \in B_{y,z}$ *be arbitrary. Then* $c + c' \in B_{x,z}$. *Specifically, suppose* $(c,\lambda) \in A_{x,y}$ *and* $(c',\lambda') \in A_{y,z}$ *for some* $\lambda,\lambda' \in \Lambda$. *Then* $(c+c',\lambda\lambda') \in A_{x,z}$.

Proof. 1. Let $\lambda_0 : [0,1] \to [0,1]$ be the identity function. Then, trivially, λ_0 is admissible and $d(x, x \circ \lambda_0) = 0$. Hence $(0,\lambda_0) \in A_{x,x}$. Consequently, $0 \in B_{x,x}$. More generally, if $d(x,y) \leq b$, then for some $b \geq 0$, we have $d(x, y \circ \lambda_0) = d(x,y) \leq b$, whence $b \in B_{x,y}$.

2. Next consider each $c \in B_{x,y}$. Then there exists $(c,\lambda) \in A_{x,y}$, satisfying inequalities 10.2.1 and 10.2.2. For each $0 \leq s < t \leq 1$, if we write $u \equiv \lambda^{-1}t$ and $v \equiv \lambda^{-1}s$, then

$$\left|\log \frac{\lambda^{-1}t - \lambda^{-1}s}{t-s}\right| = \left|\log \frac{u-v}{\lambda u - \lambda v}\right| = \left|\log \frac{\lambda u - \lambda v}{u-v}\right| \leq c. \tag{10.2.7}$$

Separately, consider each $t \in domain(y) \cap (\lambda^{-1})^{-1}domain(x)$. Then $u \equiv \lambda^{-1}t \in domain(x) \cap \lambda^{-1}domain(y)$. Hence

$$d(y(t), x(\lambda^{-1}t)) = d(y(\lambda u), x(u)) \leq c. \tag{10.2.8}$$

Thus $(c, \lambda^{-1}) \in A_{y,x}$. Consequently $c \in B_{y,x}$. Since $c \in B_{x,y}$ is arbitrary, we conclude that $B_{x,y} \subset B_{y,x}$ and, by symmetry, that $B_{x,y} = B_{y,x}$.

3. Consider arbitrary $c \in B_{x,y}$ and $c' \in B_{y,z}$. Then $(c, \lambda) \in A_{x,y}$ and $(c', \lambda') \in A_{y,z}$ for some $\lambda, \lambda' \in \Lambda$. The composite function $\lambda'\lambda$ on $[0,1]$ then satisfies

$$\left|\log \frac{\lambda'\lambda t - \lambda'\lambda s}{t - s}\right| = \left|\log \frac{\lambda'\lambda t - \lambda'\lambda s}{\lambda t - \lambda s} + \log \frac{\lambda t - \lambda s}{t - s}\right| \leq c + c'.$$

Let

$$r \in A \equiv domain(x) \cap \lambda^{-1} domain(y) \cap (\lambda'\lambda)^{-1} domain(z)$$

be arbitrary. Then

$$d(x(r), z(\lambda'\lambda r)) \leq d(x(r), y(\lambda r)) + d(y(\lambda r), z(\lambda'\lambda r)) \leq c + c'.$$

By Proposition 10.1.4, the set A contains all but countably many points in $[0,1]$, Hence it is dense in $[0,1]$. It therefore follows from Assertion 2 of Lemma 10.1.3 that

$$d(x, z \circ (\lambda'\lambda)) \leq c + c'.$$

Combining, we see that $(c + c', \lambda'\lambda) \in B_{x,z}$. □

Definition 10.2.5. Simple càdlàg function and related notations.

1. Recall, from Definition 10.2.1, the set Λ of admissible functions on $[0,1]$. Let $\lambda_0 \in \Lambda$ be the identity function: $\lambda_0 t = t$ for each $t \in [0,1]$. Let $m' \geq m$ be arbitrary. Let $\Lambda_{m,m'}$ denote the finite subset of Λ consisting of functions λ such that (i) $\lambda Q_m \subset Q_{m'}$ and (ii) λ is linear on $[q_{m,i}, q_{m,i+1}]$ for each $i = 0, \ldots, p_m - 1$.

2. Let B be an arbitrary compact subset of (S,d), and let $\delta : (0,\infty) \to (0,\infty)$ be an arbitrary operation. Then $D_{B,\delta}[0,1]$ will denote the subset of $D[0,1]$ consisting of càdlàg functions x with values in the compact set B and with $\delta_{cdlg} \equiv \delta$ as a modulus of càdlàg.

3. Let $U \equiv \{u_1, \ldots, u_M\}$ be an arbitrary finite subset of (S,d). Then $D_{simple,m,U}[0,1]$ will denote the finite subset of $D[0,1]$ consisting of simple càdlàg functions with values in U and with $q_m \equiv (q_{m,0}, \ldots, q_{m,p(m)})$ as a sequence of division points. In symbols,

$$D_{simple,m,U}[0,1] \equiv \{\Phi_{smpl}((q_{m,i})_{i=0,\ldots,p(m)-1}, (x_i)_{i=0,\ldots,p(m)-1}) : x_i \in U \text{ for each } i = 0, \ldots, p_m - 1\}.$$

4. Let $\delta_{\log @1}$ be a modulus of continuity at 1 of the natural logarithm function log. Specifically, let $\delta_{\log @1}(\varepsilon) \equiv 1 - e^{-\varepsilon}$ for each $\varepsilon > 0$. Note that $0 < \delta_{\log @1} < 1$. □

Lemma 10.2.6. **$d_{D[0,1]}$ is well defined on the subset $D_{simple,m,U}[0,1]$.** *Let $M,m \geq 1$ be arbitrary. Let $U \equiv \{u_1,\ldots,u_M\}$ be an arbitrary finite subset of (S,d). Let $x,y \in D_{simple,m,U}[0,1]$ be arbitrary. Then $d_D(x,y) \equiv \inf B_{x,y}$ exists. Moreover, the following conditions hold:*

1. Take any $b > \bigvee_{i,j=0}^{M} d(u_i,u_j)$. Note that

$$x \equiv \Phi_{smpl}((q_{m,i})_{i=0,\ldots,p(m)-1},(x_i)_{i=0,\ldots,p(m)-1})$$

for some sequence $(x_i)_{i=0,\ldots,p(m)-1}$ in U. Similarly,

$$y \equiv \Phi_{smpl}((q_{m,i})_{i=0,\ldots,p(m)-1},(y_i)_{i=0,\ldots,p(m)-1})$$

for some sequence $(y_i)_{i=0,\ldots,p(m)-1}$ in U. Let $k \geq 0$ be arbitrary. Take $m_k \geq m$ so large that

$$2^{-m(k)} \leq 2^{-m-2}e^{-b}\delta_{\log @1}(2^{-k}) < 2^{-m-2}e^{-b}. \tag{10.2.9}$$

For each $\psi \in \Lambda_{m,m(k)}$, define

$$\beta_\psi \equiv \bigvee_{i=0}^{p(m)-1}\left(\left|\log\frac{\psi q_{m,i+1}-\psi q_{m,i}}{q_{m,i+1}-q_{m,i}}\right| \vee d(y(\psi q_{m,i}),x(q_{m,i})) \vee d(y(q_{m,i}),x(\psi^{-1}q_{m,i}))\right). \tag{10.2.10}$$

Then there exists $\psi_k \in \Lambda_{m,m(k)}$ with $(\beta_{\psi(k)},\psi_k) \in A_{x,y}$ such that $|d_D(x,y) - \beta_{\psi(k)}| \leq 2^{-k+1}$.

2. For each $k \geq 0$, we have $(d_D(x,y)+2^{-k+1},\psi_k) \in A_{x,y}$.

Proof. 1. Let M,m,U,x,y,b and k be as given. As an abbreviation, write $m' \equiv m_k$. Inequality 10.2.9 then implies that

$$(m'-m-2)-b > 0.$$

2. As an abbreviation, write $\varepsilon \equiv 2^{-k}$, $p \equiv p_m \equiv 2^m$, and

$$\tau_i \equiv q_{m,i} \equiv i2^{-m} \in Q_m \tag{10.2.11}$$

for each $i = 0,\ldots,p$. Similarly, write $n \equiv p_{m'} \equiv 2^{m'}$, $\Delta \equiv 2^{-m'}$ and

$$\eta_j \equiv q_{m',j} \equiv j2^{-m'} \in Q_{m'} \tag{10.2.12}$$

for each $j = 0,\ldots,n$. Then, by hypothesis, $x \equiv \Phi_{smpl}((\tau_i)_{i=0,\ldots,p-1},(x_i)_{i=0,\ldots,p-1})$ and

$$y \equiv \Phi_{smpl}((\tau_i)_{i=0,\ldots,p-1},(y_i)_{i=0,\ldots,p-1}).$$

By the definition of simple càdlàg functions, we have $x = x_i$ and $y = y_i$ on $[\tau_i,\tau_{i+1})$, for each $i = 0,\ldots,p-1$.

3. By hypothesis, $b > \bigvee_{i,j=0}^{M} d(u_i, u_j)$. Hence there exists $b' > 0$ such that

$$b > b' > \bigvee_{i,j=0}^{M} d(u_i, u_j) \geq \bigvee_{i,j=0}^{p-1} d(x_i, y_j) \geq d(x(t), y(t))$$

for each $t \in domain(x) \cap domain(y)$. Hence $b, b' \in B_{x,y}$ by Assertion 1 of Lemma 10.2.4.

4. Define

$$\alpha_k \equiv \bigwedge_{\psi \in \Lambda(m,m')} \beta_\psi,$$

where β_ψ has been defined for each $\psi \in \Lambda_{m,m'}$ in equality 10.2.10. We will prove that (i) $\alpha_k \leq c + 2^{-k}$ for each $c \in B_{x,y}$ and (ii) $c \leq \alpha_k + 2^{-k}$ for some $c \in B_{x,y}$. It will then follow from Lemma 10.2.2 that $\inf B_{x,y}$ exists and that $\inf B_{x,y} = \lim_{k \to \infty} \alpha_k$.

5. To prove Condition (i), we will first show that it suffices to prove that (iii) $\alpha_k \leq c + 2^{-k}$ for each $c \in B_{x,y}$ with $c < b$. Suppose Condition (iii) is proved. Then $\alpha_k \leq b' + 2^{-k}$. Now consider an arbitrary $c \in B_{x,y}$. Then either $c < b$ or $b' < c$. In the former case, we have $\alpha_k \leq c + 2^{-k}$ on account of Condition (iii). In the latter case, we have $\alpha_k \leq b' + 2^{-k} < c + 2^{-k}$. Combining, we see that Condition (i) follows from Condition (iii), in any case.

6. Proceed to prove Condition (iii), with $c \in B_{x,y}$ arbitrary and with $c < b$. Since $c \in B_{x,y}$, there exists $\lambda \in \Lambda$ such that $(c, \lambda) \in A_{x,y}$. In other words,

$$d(x, y \circ \lambda) \leq c \tag{10.2.13}$$

and, for each $0 \leq s < t \leq 1$,

$$\left| \log \frac{\lambda t - \lambda s}{t - s} \right| \leq c, \tag{10.2.14}$$

where the last inequality is equivalent to

$$e^{-c}(t - s) \leq \lambda t - \lambda s \leq e^{c}(t - s). \tag{10.2.15}$$

7. Now consider each $i = 1, \ldots, p - 1$. Then there exists $j_i = 1, \ldots, n - 1$ such that

$$\eta_{j(i)-1} < \lambda \tau_i < \eta_{j(i)+1}. \tag{10.2.16}$$

Either (i$'$)

$$d(y(\eta_{j(i)-1}), x(\tau_i)) < d(y(\eta_{j(i)}), x(\tau_i)) + \varepsilon \tag{10.2.17}$$

or (ii$'$)

$$d(y(\eta_{j(i)}), x(\tau_i)) < d(y(\eta_{j(i)-1}), x(\tau_i)) + \varepsilon. \tag{10.2.18}$$

In case (i′), define $\zeta_i \equiv \eta_{j(i)-1}$. In case (ii′), define $\zeta_i \equiv \eta_{j(i)}$. Then, in both cases (i′) and (ii′), inequality 10.2.16 implies that

$$\zeta_i - \Delta < \lambda\tau_i < \zeta_i + 2\Delta. \tag{10.2.19}$$

8. Moreover, inequalities 10.2.17 and 10.2.18 together imply that

$$d(y(\zeta_i), x(\tau_i)) \leq d(y(\eta_{j(i)-1}), x(\tau_i)) \wedge d(y(\eta_{j(i)}), x(\tau_i)) + \varepsilon. \tag{10.2.20}$$

At the same time, in view of inequality 10.2.16, there exists a point

$$s \in (\lambda\tau_i, \lambda\tau_{i+1}) \cap ((\eta_{j(i)-1}, \eta_{j(i)}) \cup (\eta_{j(i)}, \eta_{j(i)+1})).$$

Then $t \equiv \lambda^{-1}s \in (\tau_i, \tau_{i+1})$, whence $x(\tau_i) = x(t)$. Moreover, either $s \in (\eta_{j(i)-1}, \eta_{j(i)})$, in which case $y(\eta_{j(i)-1}) = y(s)$, or $s \in (\eta_{j(i)}, \eta_{j(i)+1})$, in which case $y(\eta_{j(i)}) = y(s)$. Here we used the fact that the simple càdlàg function y is constant over the interval (η_{h-1}, η_h), for each $h = 1, \ldots, n$. In both cases, inequality 10.2.20 yields

$$d(y(\zeta_i), x(\tau_i)) \leq d(y(s), x(\tau_i)) + \varepsilon = d(y(\lambda t), x(t)) + \varepsilon \leq c + \varepsilon, \tag{10.2.21}$$

where the last inequality is from inequality 10.2.13.

9. Now let $\zeta_0 \equiv 0$ and $\zeta_p \equiv 1$. Then, for each $i = 0, \ldots, p-1$, inequality 10.2.19 implies that

$$\begin{aligned}\zeta_{i+1} - \zeta_i &> (\lambda\tau_{i+1} - 2\Delta) - (\lambda\tau_i + \Delta) > \lambda\tau_{i+1} - \lambda\tau_i - 4\Delta \\ &\geq e^{-c}(\tau_{i+1} - \tau_i) - 4\Delta > e^{-b}2^{-m} - 2^{-m'+2} \geq 0,\end{aligned}$$

where the third inequality follows from inequality 10.2.15, where the fourth inequality is from the assumption $c < b$ in Condition (iii), and where the last inequality is from inequality 10.2.9. Thus we see that $(\zeta_i)_{i=0,\ldots,p}$ is an increasing sequence in $Q_{m'}$. As such, it determines a unique function $\psi \in \Lambda_{m,m'}$ that is linear on $[\tau_i, \tau_{i+1}]$ for each $i = 0, \ldots, p-1$, such that $\psi\tau_i \equiv \zeta_i$ for each $i = 0, \ldots, p$.

10. By Lemma 10.2.3, we then have

$$\begin{aligned}\sup_{0 \leq s < t \leq 1} \left|\log \frac{\psi t - \psi s}{t - s}\right| &= \bigvee_{i=0}^{p-1} \left|\log \frac{\psi\tau_{i+1} - \psi\tau_i}{\tau_{i+1} - \tau_i}\right| \\ &= \bigvee_{i=0}^{p-1} \left|\log\left(\frac{\lambda\tau_{i+1} - \lambda\tau_i}{\tau_{i+1} - \tau_i} \cdot \frac{\psi\tau_{i+1} - \psi\tau_i}{\lambda\tau_{i+1} - \lambda\tau_i}\right)\right| \\ &\leq \bigvee_{i=0}^{p-1} \left|\log \frac{\lambda\tau_{i+1} - \lambda\tau_i}{\tau_{i+1} - \tau_i}\right| + \bigvee_{i=0}^{p-1} \left|\log\left(\frac{\zeta_{i+1} - \zeta_i}{\lambda\tau_{i+1} - \lambda\tau_i}\right)\right| \\ &\leq c + \bigvee_{i=0}^{p-1} \left|\log\left(1 + \frac{(\zeta_{i+1} - \lambda\tau_{i+1}) - (\zeta_i - \lambda\tau_i)}{\lambda\tau_{i+1} - \lambda\tau_i}\right)\right| \\ &\equiv c + \bigvee_{i=0}^{p-1} |\log(1 + a_i)|,\end{aligned} \tag{10.2.22}$$

where, for each $i = 0, \ldots, p-1$, we define

$$a_i \equiv \frac{(\zeta_{i+1} - \lambda\tau_{i+1}) - (\zeta_i - \lambda\tau_i)}{\lambda\tau_{i+1} - \lambda\tau_i}$$

with

$$|a_i| \le \frac{2^{-m'+1} + 2^{-m'+1}}{\exp(-c)(\tau_{i+1} - \tau_i)} = e^c 2^{-m'+2+m} < e^b 2^{-m'+2+m} \le \delta_{\log @ 1}(\varepsilon).$$

Here, the first inequality is due to inequalities 10.2.19 and 10.2.15, and the third inequality is due to the first half of inequality 10.2.9. Hence, by the definition of the modulus of continuity $\delta_{\log @ 1}$, we obtain

$$\bigvee_{i=0}^{p-1} |\log(1 + a_i)| < \varepsilon.$$

Inequality 10.2.22 therefore yields

$$\sup_{0 \le s < t \le 1} \left| \log \frac{\psi t - \psi s}{t - s} \right| < c + \varepsilon. \tag{10.2.23}$$

11. Separately, by the definition of ψ, inequality 10.2.21 implies

$$\bigvee_{i=0}^{p-1} d(y(\psi\tau_i), x(\tau_i)) \equiv \bigvee_{i=0}^{p-1} d(y(\zeta_i), x(\tau_i)) \le c + \varepsilon. \tag{10.2.24}$$

12. We will also verify that

$$\bigvee_{j=0}^{p-1} d(y(\tau_j), x(\psi^{-1}\tau_j)) \le c + \varepsilon. \tag{10.2.25}$$

To that end, consider each $j = 1, \ldots, p-1$. Note that for each $i = 0, \ldots, p-1$, we have $\tau_j, \zeta_i \in Q_{m'}$. Hence there exists some $i = 0, \ldots, p-1$ such that

$$\tau_j \in [\zeta_i, \zeta_{i+1}) \equiv [\psi\tau_i, \psi\tau_{i+1}). \tag{10.2.26}$$

Either (i″) $\tau_j = \zeta_i$ or (ii″) $\tau_j \ge \zeta_i + \Delta$. First consider case (i″). Then

$$d(y(\tau_j), x(\psi^{-1}\tau_j)) = d(y(\zeta_i), x(\psi^{-1}\zeta_i)) = d(y(\zeta_i), x(\tau_i)) \le c + \varepsilon$$

by inequality 10.2.21. Now consider case (ii″). Then, in view of inequality 10.2.19, we obtain

$$\lambda\tau_i < \zeta_i + 2\Delta \le \tau_j + \Delta.$$

At the same time, relation 10.2.26 implies that $\tau_j < \zeta_{i+1}$. Hence

$$\tau_j \le \zeta_{i+1} - \Delta < \lambda\tau_{i+1},$$

where the inequality on the right-hand side is from inequality 10.2.19 applied to $l+1$. Combining the last two displayed inequalities, we see that there exists a point

$$s \in [\lambda\tau_i \vee \tau_j, \lambda\tau_{i+1} \wedge (\tau_j + \Delta)).$$

It follows that $s \in [\tau_j, \tau_j + \Delta) \subset [\zeta_i, \zeta_{i+1})$ and $r \equiv \lambda^{-1}s \in [\tau_i, \tau_{i+1})$. Therefore $y(s) = y(\tau_j)$ and $x(r) = x(\tau_i)$ by the definition of the simple càdlàg functions x and y. Moreover, inequality 10.2.26 yields

$$\psi^{-1}\tau_j \in [\tau_i, \tau_{i+1}),$$

whence $x(\psi^{-1}\tau_j) = x(\tau_i) = x(r)$. Combining, we obtain

$$d(y(\tau_j), x(\psi^{-1}\tau_j)) = d(y(s), x(r)) \equiv d(y(\lambda r), x(r)) \leq c + \varepsilon,$$

where the last inequality follows from inequality 10.2.13, and where $j = 1, \ldots, p-1$ is arbitrary. Thus we have verified inequality 10.2.25.

13. Inequalities 10.2.23, 10.2.24, and 10.2.25 together lead to

$$\beta_\psi \equiv \bigvee_{i=0}^{p-1} \left(\left| \log \frac{\psi\tau_{i+1} - \psi\tau_i}{\tau_{i+1} - \tau_i} \right| \vee d(y(\psi\tau_i), x(\tau_i)) \vee d(y(\tau_i), x(\psi^{-1}\tau_i)) \right) \leq c + \varepsilon.$$

Hence

$$\alpha_k \equiv \bigwedge_{\psi \in \Lambda(m,m')} \beta_\psi \leq c + \varepsilon,$$

where $c \in B_{x,y}$ is arbitrary, with $c < b$. Since $\Lambda_{m,m'}$ is a finite set, there exists $\psi_k \in \Lambda_{m,m'}$ such that

$$\alpha_k \leq \beta_{\psi(k)} < \alpha_k + \varepsilon \leq c + \varepsilon. \tag{10.2.27}$$

The desired Condition (iii) is established. According to Step 5, Condition (i) therefore follows.

14. It remains to prove Condition (ii). First note that, from inequality 10.2.27, we have

$$\bigvee_{i=0}^{p-1} \left(\left| \log \frac{\psi_k\tau_{i+1} - \psi_k\tau_i}{\tau_{i+1} - \tau_i} \right| \vee d(y(\psi_k\tau_i), x(\tau_i)) \vee d(y(\tau_i), x(\psi_k^{-1}\tau_i)) \right)$$
$$\equiv \beta_{\psi(k)} < \alpha_k + \varepsilon.$$

By Lemma 10.2.3, we have

$$\sup_{0 \leq s < t \leq 1} \left| \log \frac{\psi_k t - \psi_k s}{t - s} \right| = \bigvee_{i=0}^{p-1} \left| \log \frac{\psi_k\tau_{i+1} - \psi_k\tau_i}{\tau_{i+1} - \tau_i} \right| \leq \beta_{\psi(k)}. \tag{10.2.28}$$

15. We will show that $d(y \circ \psi_k, x) \leq \beta_{\psi(k)}$. To that end, consider each point t in the dense subset $A \equiv Q_{m'}^c \cup \psi_k^{-1} Q_{m'}^c$ of $[0,1]$. Then

$$t \in [\tau_i, \tau_{i+1}) \cap \psi_k^{-1}[\tau_j, \tau_{j+1}) \tag{10.2.29}$$

for some $i, j = 0, \ldots, p-1$. Either (i''') $\psi_k^{-1}\tau_j \leq \tau_i$ or (ii''') $\tau_i < \psi_k^{-1}\tau_j$. In case (i'''), we have $\tau_j \leq \psi_k\tau_i \leq \psi_k t < \tau_{j+1}$, whence

$$d(y(\psi_k t), x(t)) = d(y(\psi_k\tau_i), x(\tau_i)) \leq \beta_{\psi(k)},$$

where we used the fact that the simple càdlàg function y is constant over the interval $[\tau_j, \tau_{j+1})$. In case (ii'''), relation 10.2.29 implies that $\tau_i < \psi_k^{-1}\tau_j \leq t < \tau_{i+1}$, whence

$$d(y(\psi_k t), x(t)) = d(y(\tau_j), x(\psi_k^{-1}\tau_j)) \leq \beta_{\psi(k)},$$

where we used, once more, the properties of simple càdlàg functions. Since t is an arbitrary member of the dense subset A of $[0,1]$, we conclude that

$$d(y \circ \psi_k, x) \leq \beta_{\psi(k)}. \tag{10.2.30}$$

Combining with inequality 10.2.28, we see that $(\beta_{\psi(k)}, \psi_k) \in A_{x,y}$, whence $\beta_{\psi(k)} \in B_{x,y}$. Since $\beta_{\psi(k)} < \alpha_k + 2^{-k}$, this proves Condition (ii) in Step 4.

16. Since $k \geq 0$ is arbitrary, Conditions (i) and (ii) in Step 4 together then imply, by Lemma 10.2.2, that both $\lim_{k\to\infty} \alpha_k$ and $d_D(x,y) \equiv \inf B_{x,y}$ exist, and are equal to each other.

17. Moreover, $\alpha_k - 2^{-k} \leq \inf B_{x,y} \leq \alpha_k + 2^{-k}$ for each $k \geq 0$. Combined with inequality 10.2.27, this yields

$$\beta_{\psi(k)} - 2^{-k+1} \leq \inf B_{x,y} \leq \beta_{\psi(k)} + 2^{-k+1}. \tag{10.2.31}$$

Assertion 1 of the present lemma is proved.

18. Finally, since $(\beta_{\psi(k)}, \psi_k) \in A_{x,y}$, as observed in Step 15, and since $\beta_{\psi(k)} \leq \inf B_{x,y} + 2^{-k+1} \equiv d_D(x,y) + 2^{-k+1}$, we obtain $(d_D(x,y) + 2^{-k+1}, \psi_k) \in A_{x,y}$. Assertion 2 and the lemma are proved. □

The next lemma prepares us for a subsequent generalization of the Arzela–Ascoli Theorem to càdlàg functions.

Lemma 10.2.7. Preparatory lemma for the Arzela–Ascoli Theorem for $D[0,1]$. *Let B be an arbitrary compact subset of (S,d). Let $x \in D[0,1]$ be arbitrary with values in the compact set B and with a modulus of càdlàg δ_{cdlg}. Let $k \geq 0$ be arbitrary. Let $U \equiv U_k \equiv \{v_1, \ldots, v_M\}$ be a 2^{-k-1}-approximation of of B. Let*

$$m \equiv m(k, \delta_{cdlg}) \equiv [1 - \log_2(\delta_{\log @ 1}(2^{-k})\delta_{cdlg}(2^{-k-1}))]_1.$$

Then there exist an increasing sequence $(\eta_i)_{i=0,\ldots,n}$ in Q_m with $\eta_0 = 0$ and $\eta_n = 1$, and a sequence $(u_i)_{i=0,\ldots,n-1}$ in U_k, such that

$$2^{-k} \in B_{x,\overline{x}},$$

where

$$\overline{x} \equiv \Phi_{smpl}((\eta_i)_{i=0,\ldots,n-1}, (u_i)_{i=0,\ldots,n-1}).$$

Proof. 1. Let k, M, U, m be as given. Then

$$2^{-m} < 2^{-1}\delta_{\log @ 1}(2^{-k})\delta_{cdlg}(2^{-k-1}). \tag{10.2.32}$$

As an abbreviation, write $\varepsilon \equiv 2^{-k}$, $p \equiv p_m \equiv 2^m$, and $(q_0, \ldots, q_p) \equiv (0, 2^{-m}, 2^{-m+1}, \ldots, 1)$. By Definition 10.1.2, there exist $n \geq 1$ and a sequence of $2^{-1}\varepsilon$-division points $(\tau_i)_{i=0,\ldots,n}$ of the càdlàg function x, with separation at least $\delta_{cdlg}(2^{-1}\varepsilon)$. Thus

$$\bigwedge_{i=0}^{n-1}(\tau_{i+1} - \tau_i) \geq \delta_{cdlg}(2^{-1}\varepsilon). \tag{10.2.33}$$

Define $\eta_0 \equiv 0$ and $\eta_n \equiv 1$. Consider each $i = 1, \ldots, n-1$. Then there exists $j_i = 1, \ldots, p-1$ such that $\tau_i \in (q_{j(i)-1}, q_{j(i)+1})$. Define $\eta_i \equiv q_{j(i)} \in Q_m$. Then

$$\begin{aligned} |\eta_i - \tau_i| &< 2^{-m} < 2^{-1}\delta_{\log@1}(2^{-k})\delta_{cdlg}(2^{-k-1}) < 2^{-1}\delta_{cdlg}(2^{-k-1}) \\ &\equiv 2^{-1}\delta_{cdlg}(2^{-1}\varepsilon). \end{aligned} \tag{10.2.34}$$

2. In view of inequality 10.2.33, we have

$$\bigwedge_{i=0}^{n-1}(\eta_{i+1} - \eta_i) > \bigwedge_{i=0}^{n-1}(\tau_{i+1} - \tau_i) - \delta_{cdlg}(2^{-1}\varepsilon) \geq 0.$$

Thus $(\eta_i)_{i=0,\ldots,n}$ is an increasing sequence in $[0,1]$. Therefore we can define an increasing function $\psi \in \Lambda$ by (i) $\psi\tau_i = \eta_i$ for each $i = 0, \ldots, n$, and (ii) ψ is linear on $[\tau_i, \tau_{i+1}]$ for each $i = 1, \ldots, n-1$.

3. By Lemma 10.2.3,

$$\begin{aligned} \sup_{0 \leq s < t \leq 1} \left| \log \frac{\psi t - \psi s}{t - s} \right| &= \bigvee_{j=0}^{n-1} \left| \log \frac{\psi\tau_{i+1} - \psi\tau_i}{\tau_{i+1} - \tau_i} \right| \\ &= \bigvee_{i=0}^{n-1} \left| \log \frac{\eta_{i+1} - \eta_i}{\tau_{i+1} - \tau_i} \right| = \bigvee_{i=0}^{n-1} |\log(1 + a_i)|, \end{aligned} \tag{10.2.35}$$

where for each $i = 0, \ldots, n-1$, we define

$$a_i \equiv \frac{(\eta_{i+1} - \tau_{i+1}) - (\eta_i - \tau_i)}{\tau_{i+1} - \tau_i}$$

with

$$|a_i| \leq \frac{2^{-m+1}}{|\tau_{i+1} - \tau_i|} < 2^{-m+1}\delta_{cdlg}(2^{-1}\varepsilon)^{-1} < \delta_{\log@1}(\varepsilon).$$

Here, the first inequality is by the first part of inequality 10.2.34, the second inequality is by inequality 10.2.33, and the last inequality is from inequality 10.2.32. Hence, by the definition of the modulus of continuity $\delta_{\log@1}$ in Definition 10.2.5, we have $|\log(1 + a_i)| \leq \varepsilon$ for each $i = 0, \ldots, n-1$. Inequality 10.2.35 therefore reduces to

$$\sup_{0 \leq s < t \leq 1} \left| \log \frac{\psi t - \psi s}{t - s} \right| \leq \varepsilon. \tag{10.2.36}$$

4. Next, by hypothesis, the càdlàg function x has values in the compact set B and $U \equiv \{v_1,\ldots,v_M\}$ is an 2^{-k-1}-approximation of B. Hence, for each $i = 0,\ldots,n-1$, there exists $u_i \in U$ such that $d(u_i,x(\tau_i)) < 2^{-k-1} \equiv 2^{-1}\varepsilon$. Now define the simple càdlàg function

$$\overline{x} \equiv \Phi_{smpl}((\eta_i)_{i=0,\ldots,n-1},(u_i)_{i=0,\ldots,n-1}) \in D_{simple,m,U}[0,1].$$

We will prove that $(\varepsilon,\psi) \in A_{x,\overline{x}}$, and therefore that $\varepsilon \in B_{x,\overline{x}}$. To that end, consider each $i = 0,\ldots,n-1$ and each $t \in [\tau_i,\tau_{i+1})$. Then $\psi t \in [\eta_i,\eta_{i+1})$. Hence

$$\begin{aligned} d(x(t),\overline{x}(\psi t)) &\leq d(x(t),x(\tau_i)) + d(x(\tau_i),u_i) + d(u_i,\overline{x}(\eta_i)) \\ &\quad + d(\overline{x}(\eta_i),\overline{x}(\psi t)) < 2^{-1}\varepsilon + 2^{-1}\varepsilon + 0 + 0 = \varepsilon. \end{aligned}$$

Thus $d(x(t),\overline{x}(\psi t)) \leq \varepsilon$ for each t in the dense subset $A \equiv \bigcup_{i=0}^{m}[\tau_i,\tau_{i+1})$ of $[0,1]$. Therefore Assertion 2 of Lemma 10.1.3 says that

$$d(x,\overline{x} \circ \psi) \leq \varepsilon.$$

Combining with inequality 10.2.36, we see that $(\varepsilon,\psi) \in A_{x,\overline{x}}$. It follows that $\varepsilon \in B_{x,\overline{x}}$. The lemma is proved. □

The next lemma proves the existence of $\inf B_{x,y}$ for each $x,y \in D[0,1]$. The subsequent proposition then verifies that the Skorokhod metric is indeed a metric.

Lemma 10.2.8. The Skorokhod metric is well defined. *Let $x,y \in D[0,1]$ be arbitrary. Then the infimum $d_{D[0,1]}(x,y) \equiv \inf B_{x,y}$ in Definition 10.2.1 exists.*

Proof. 1. Let δ_{cdlg} be a common modulus of càdlàg of x and y. Let $k \geq 0$ be arbitrary and write $\varepsilon_k \equiv 2^{-k}$. Let

$$m \equiv m_k \equiv m(k,\delta_{cdlg}) \equiv [1 - \log_2(\delta_{\log@1}(2^{-k})\delta_{cdlg}(2^{-k-1}))]_1.$$

By Assertion 4 of Lemma 10.1.4, there exists $b \geq 0$ such that $d(x,x_\circ) \vee d(y,x_\circ) \leq b$. Since (S,d) is locally compact, the bounded set $(d(\cdot,x_\circ) \leq b)$ is contained in some compact subset B of S. Hence $x,y \in D_{B,\delta(cdlg)}[0,1]$.

2. Let $U_k \equiv \{v_{k,1},\ldots,v_{k,M(k)}\}$ be an 2^{-k-1}-approximation of the compact set B. Then, according to Lemma 10.2.7, there exist $\overline{x}_k,\overline{y}_k \in D_{simple,m(k),U(k)}[0,1]$ such that $\varepsilon_k \in B_{x,\overline{x}(k)}$ and $\varepsilon_k \in B_{y,\overline{y}(k)}$. At the same time, by Lemma 10.2.6, the infimum $\alpha_k \equiv \inf B_{\overline{x}(k),\overline{y}(k)}$ exists. Define $\overline{\alpha}_k \equiv \alpha_{k+2}$.

3. By the definition of the set $D_{simple,m(k),U(k)}[0,1]$, we have

$$\overline{x}_k \equiv \Phi_{smpl}((\eta_{k,i})_{i=0,\ldots,n(k)-1},(u_{k,i})_{i=0,\ldots,n(k)-1})$$

for some increasing sequence $(\eta_{k,i})_{i=0,\ldots,n(k)}$ in $Q_{m(k)}$, and some sequence $(u_{k,i})_{i=0,\ldots,n(k)-1}$ in U_k.

4. Now let $c \in B_{x,y}$ be arbitrary. Then, since $\varepsilon_k \in B_{x,\overline{x}(k)}$ and $\varepsilon_k \in B_{y,\overline{y}(k)}$, we have $c+\varepsilon_k+\varepsilon_k \in B_{\overline{x}(k),\overline{y}(k)}$ by Assertion 3 of Lemma 10.2.4. Hence $\alpha_k \leq c+2\varepsilon_k$, where $k \geq 0$ is arbitrary. Therefore (*) $\overline{\alpha}_k \equiv \alpha_{k+2} \leq c+2\varepsilon_{k+2} < c+2^{-k}$, where $c \in B_{x,y}$ is arbitrary.

5. Conversely, take any $\overline{c} \in B_{\overline{x}(k+2),\overline{y}(k+2)}$ such that $\overline{c} < \inf B_{\overline{x}(k+2),\overline{y}(k+2)} + \varepsilon_{k+2} \equiv \alpha_{k+2} + \varepsilon_{k+2}$. Then, since $\varepsilon_{k+2} \in B_{x,\overline{x}(k+2)}$ and $\varepsilon_{k+2} \in B_{y,\overline{y}(k+2)}$, we have $\overline{c} + \varepsilon_{k+2} + \varepsilon_{k+2} \in B_{x,y}$ by Assertion 3 of Lemma 10.2.4. Define

$$c \equiv \overline{c} + \varepsilon_{k+2} + \varepsilon_{k+2} < \alpha_{k+2} + 3 \cdot 2^{-k-2} \equiv \overline{\alpha}_k + 3 \cdot 2^{-k-2} < \overline{\alpha}_k + 2^{-k}.$$

Thus (**') $c \leq \overline{\alpha}_k + 2^{-k}$ for some $c \in B_{x,y}$

6. In view of Conditions (*) and (**) proved in Steps 4 and 5, respectively, we see that the conditions in Assertion 1 of Lemma 10.2.2 have been verified for the set $B \equiv B_{x,y}$ and the sequence $(\overline{\alpha}_k)_{k=0,1,\ldots}$. Accordingly, $\inf B_{x,y}$ exists and $\inf B_{x,y} = \lim_{k\to\infty} \overline{\alpha}_k$. Moreover, $\overline{\alpha}_k - 2^{-k} \leq \inf B_{x,y} \leq \overline{\alpha}_k + 2^{-k}$ for each $k \geq 0$. □

Proposition 10.2.9. The Skorokhod metric is indeed a metric. $(D[0,1], d_D)$ *is a metric space. Moreover, if* $x, y \in D[0,1]$ *are such that* $d(x,y) \leq c$ *on a dense subset of* $domain(x) \cap domain(y)$, *then* $d_D(x,y) \leq c$.

Proof. 1. Let $x, y, z \in D[0,1]$ be arbitrary. Let δ_{cdlg} be a common modulus of càdlàg of x, y and z. By Lemma 10.2.8, $d_D(x,y) \equiv \inf B_{x,y}$ exists. By Lemma 10.2.4, we have $0 \in B_{x,x}$ and $B_{x,y} = B_{x,y}$. It follows that $d_D(x,x) = 0$ and $d_D(x,y) = d_D(y,x)$. Now let $\varepsilon > 0$ be arbitrary. By the definition of the infimum, there exist $c \in B_{x,y}$ and $c' \in B_{y,z}$ such that $c < \inf B_{x,y} + \varepsilon \equiv d_D(x,y) + \varepsilon$ and $c' < \inf B_{y,z} + \varepsilon \equiv d_D(y,z) + \varepsilon$. Hence, again by Lemma 10.2.4, we have $c + c' \in B_{x,z}$, whence

$$d_D(x,z) \equiv \inf B_{x,z} \leq c + c' < d_D(x,y) + \varepsilon + d_D(y,z) + \varepsilon.$$

Since $\varepsilon > 0$ is arbitrary, it follows that $d_D(x,z) \leq d_D(x,y) + d_D(y,z)$.

2. It remains to prove that if $d_D(x,y) = 0$, then $x = y$. To that end, suppose $d_D(x,y) = 0$. Let $\varepsilon > 0$ be arbitrary. Let $(\tau_i)_{i=0,\ldots,p}$ and $(\eta_j)_{j=0,\ldots,n}$ be sequences of ε-division points of x, y, respectively. Note that $\tau_0 = \eta_0 = 0$ and $\tau_p = \eta_n = 1$. Let $m \geq 1 \vee \varepsilon^{-1/2}$ be arbitrary. Consider each $k \geq m \vee 2$. Then $\varepsilon_k \equiv k^{-2} \leq m^{-2} \leq \varepsilon$. Moreover, since $d_D(x,y) = 0 < \varepsilon_k$, we have $\varepsilon_k \in B_{x,y}$. Therefore there exists, by Definition 10.2.1, some $\lambda_k \in \Lambda$ such that

$$\left| \log \frac{\lambda_k r - \lambda_k s}{r - s} \right| \leq \varepsilon_k \tag{10.2.37}$$

for each $r, s \in [0,1]$ with $s \leq r$, and such that

$$d(x(t), y(\lambda_k t)) \leq \varepsilon_k \tag{10.2.38}$$

for each $t \in domain(x) \cap \lambda_k^{-1} domain(y)$. Then inequality 10.2.37 implies that

$$e^{-\varepsilon(k)} r \leq \lambda_k r \leq e^{\varepsilon(k)} r \tag{10.2.39}$$

for each $r \in [0,1]$. Define the subset

$$C_k \equiv \left(\bigcup_{j=0}^{n} [e^{-\varepsilon(k)}\eta_j, e^{\varepsilon(k)}\eta_j] \right)^c$$

$$= (e^{\varepsilon(k)}\eta_0, e^{-\varepsilon(k)}\eta_1) \cup (e^{\varepsilon(k)}\eta_1, e^{-\varepsilon(k)}\eta_2) \cup \cdots \cup (e^{\varepsilon(k)}\eta_{n-1}, e^{-\varepsilon(k)}\eta_n)$$

of $[0,1]$, where the superscript c signifies the measure-theoretic complement of a Lebesgue measurable set in [0,1]. Let μ denote the Lebesgue measure on $[0,1]$. Then

$$\mu(C_k) \geq 1 - \sum_{i=0}^{n} (e^{\varepsilon(k)}\eta_j - e^{-\varepsilon(k)}\eta_j)$$

$$\geq 1 - \sum_{j=0}^{n} (e^{\varepsilon(k)} - e^{-\varepsilon(k)}) \geq 1 - (n+1)(e^{2\varepsilon(k)} - 1) \geq 1 - 2(n+1)e\varepsilon_k$$

$$\equiv 1 - 2(n+1)ek^{-2},$$

where $k \geq m \vee 2$ is arbitrary, and where we used the elementary inequality $e^r - 1 \leq er$ for each $r \in (0,1)$. Now define

$$C \equiv \bigcup_{h=m}^{\infty} \bigcap_{k=h+1}^{\infty} C_k.$$

Then, for each $h \geq m$, we have

$$\mu(C^c) \leq \sum_{k=h+1}^{\infty} \mu(C_k^c) \leq \sum_{k=h+1}^{\infty} 2(n+1)ek^{-2} \leq 2(n+1)eh^{-1}.$$

Hence $\mu(C^c) = 0$ and C is a full subset of $[0,1]$. Consequently,

$$A \equiv C \cap domain(x) \cap domain(y) \cap \bigcap_{k=m}^{\infty} \lambda_k^{-1} domain(y)$$

is a full subset of $[0,1]$ and, as such, is dense in $[0,1]$. Now let $t \in A$ be arbitrary. Then $t \in C$. Hence there exists $h \geq m$ such that $t \in \bigcap_{k=h+1}^{\infty} C_k$. Consider each $k \geq h+1$. Then $t \in C_k$. Hence there exists $j = 0, \ldots, n-1$ such that $t \in (e^{\varepsilon(k)}\eta_j, e^{-\varepsilon(k)}\eta_{j+1})$. It then follows from inequality 10.2.39 that

$$\eta_j < e^{-\varepsilon(k)}t \leq \lambda_k t \leq e^{\varepsilon(k)}t < \eta_{j+1},$$

whence

$$\lambda_k t, t \in (\eta_j, \eta_{j+1}).$$

Since $(\eta_j)_{j=0,\ldots,n}$ is a sequence of ε-division points of the càdlàg function y, it follows that

$$d(y(\lambda_k t), y(t)) \leq d(y(\lambda_k t), y(\eta_j)) + d(y(\eta_j), y(t)) \leq 2\varepsilon.$$

Consequently,

$$d(x(t), y(t)) = d(x(t), y(\lambda_k t)) + d(y(\lambda_k t), y(t)) \leq \varepsilon_k + 2\varepsilon \leq 3\varepsilon,$$

where the first inequality is by inequality 10.2.38. Let $k \to \infty$ and then let $\varepsilon \to 0$. Then we obtain $d(x(t), y(t)) = 0$, where $t \in A$ is arbitrary. Summing up, $x = y$ on the dense subset A. Therefore Lemma 10.1.3 implies that $x = y$. We conclude that d_D is a metric.

3. Finally, suppose $x, y \in D[0,1]$ are such that $d(x(t), y(t)) \leq c$ for each t in a dense subset of $domain(x) \cap domain(y)$. Then $c \in B_{x,y}$ by Lemma 10.2.4. Hence $d_D(x,y) \equiv \inf B_{x,y} \leq c$. The proposition is proved. □

The next theorem is now a trivial consequence of Lemma 10.2.7.

Theorem 10.2.10. Arzela–Ascoli Theorem for $(D[0,1], d_D)$. *Let B be an arbitrary compact subset of (S,d). Let $k \geq 0$ be arbitrary. Let $U \equiv \{v_1, \ldots, v_M\}$ be a 2^{-k-1}-approximation of of B. Let $x \in D[0,1]$ be arbitrary with values in the compact set B and with a modulus of càdlàg δ_{cdlg}. Let $m \geq 1$ be so large that*

$$m \geq m(k, \delta_{cdlg}) \equiv [1 - \log_2(\delta_{\log @ 1}(2^{-k})\delta_{cdlg}(2^{-k-1}))]_1.$$

Then there exist an increasing sequence $(\eta_i)_{i=0,\ldots,n}$ in Q_m with $\eta_0 = 0$ and $\eta_n = 1$, and a sequence $(u_i)_{i=0,\ldots,n-1}$ in U such that

$$d_{D[0,1]}(x, \overline{x}) < 2^{-k},$$

where

$$\overline{x} \equiv \Phi_{smpl}((\eta_i)_{i=0,\ldots,n-1}, (u_i)_{i=0,\ldots,n-1}) \in D_{simple,m,U}[0,1].$$

Proof. Write $\overline{m} \equiv m(k, \delta_{cdlg})$. By Lemma 10.2.7, there exist an increasing sequence $(\eta_i)_{i=0,\ldots,n}$ in $Q_{\overline{m}} \subset Q_m$ with $\eta_0 = 0$ and $\eta_n = 1$, and a sequence $(u_i)_{i=0,\ldots,n-1}$ in U such that $2^{-k} \in B_{x,\overline{x}}$. At the same time, $d_{D[0,1]}(x, \overline{x}) \equiv \inf B_{x,\overline{x}}$, according to Definition 10.2.1. Hence $d_{D[0,1]}(x, \overline{x}) < 2^{-k}$, as desired. □

Theorem 10.2.11. The Skorokhod space is complete. *The Skorokhod space $(D[0,1], d_D)$ is complete.*

Proof. 1. Let $(y_k)_{k=1,2,\ldots}$ be an arbitrary Cauchy sequence in $(D[0,1], d_D)$. We need to prove that $d_D(y_k, y) \to 0$ for some $y \in D[0,1]$. Since $(y_k)_{k=1,2,\ldots}$ is Cauchy, it suffices to show that some subsequence converges. Hence, by passing to a subsequence if necessary, there is no loss in generality in assuming that

$$d_D(y_k, y_{k+1}) < 2^{-k} \tag{10.2.40}$$

for each $k \geq 1$. Let $\delta_k \equiv \delta_{cdlg,k}$ be a modulus of càdlàg of y_k, for each $k \geq 1$. For convenience, let $y_0 \equiv x_\circ$ denote the constant càdlàg function on $[0,1]$.

2. Let $k \geq 1$ be arbitrary. Then

$$d_D(y_0, y_k) \leq d_D(y_0, y_1) + 2^{-1} + \cdots + 2^{-k} < b_0 \equiv d_D(y_0, y_1) + 1.$$

Hence $d(x_\circ, y_k(t)) \leq b_0$ for each $t \in domain(y_k)$. Therefore the values of y_k are in some compact subset B of (S,d) that contains $(d(x_\circ,\cdot) \leq b_0)$. Define $b \equiv 2b_0 > 0$. Then $d(y_k(t), y_h(s)) \leq 2b_0 \equiv b$, for each $t \in domain(y_k)$, $s \in domain(y_h)$, for each $h,k \geq 0$.

3. Next, refer to Definition 9.0.2 for notations related to dyadic rationals, and refer to Definitions 10.2.1 and 10.2.5 for the notations related to the Skorokhod metric. In particular, for each $x, y \in D[0,1]$, recall the sets Λ, $\Lambda_{m,m'}$, $A_{x,y}$ and $B_{x,y}$. Thus $d_D(x,y) \equiv \inf B_{x,y}$. Let $\lambda_0 \in \Lambda$ denote the identity function on $[0,1]$.

4. The next two steps will approximate the given Cauchy sequence $(y_k)_{k=1,2,...}$ with a sequence $(\overline{x}_k)_{k=1,2,...}$ of simple càdlàg functions whose division points are dyadic rationals. To that end, fix an arbitrary $m_0 \equiv 0$ and, inductively for each $k \geq 1$, define

$$m_k \equiv [2 + m_{k-1} - \log_2(e^{-b}\delta_{\log @ 1}(2^{-k})\delta_{cdlg,k}(2^{-k-1}))]_1. \tag{10.2.41}$$

Then

$$m_k \geq [1 - \log_2(\delta_{\log @ 1}(2^{-k})\delta_{cdlg,k}(2^{-k-1}))]_1. \tag{10.2.42}$$

5. Define the constant càdlàg functions $\overline{x}_0 \equiv y_0 \equiv x_\circ$. Let $k \geq 1$ be arbitrary, and write $n_k \equiv p_{m(k)}$. Let $U_k \equiv \{v_{k,1}, \ldots, v_{k,M(k)}\}$ be a 2^{-k-1}-approximation of of B, such that $U_k \subset U_{k+1}$ for each $k \geq 1$. Then, in view of inequality 10.2.42, Theorem 10.2.10 (the Arzela–Ascoli Theorem for the Skorokhod space) applies to the càdlàg function y_k. It yields a simple càdlàg function

$$\overline{x}_k \equiv \Phi_{smpl}((\eta_{k,i})_{i=0,\ldots,n(k)-1}, (u_{k,i})_{i=0,\ldots,n(k)-1}) \in D_{simple,m(k),U(k)}[0,1] \tag{10.2.43}$$

such that

$$d_{D[0,1]}(y_k, \overline{x}_k) < 2^{-k}, \tag{10.2.44}$$

where $(\eta_{k,i})_{i=0,\ldots,n(k)}$ is an increasing sequence in $Q_{m(k)}$ with $\eta_{k,0} = 0$ and $\eta_{k,n(k)} = 1$, and where $(u_{k,i})_{i=0,\ldots,n(k)-1}$ is a sequence in U_k. Thus the sequence $(\eta_{k,i})_{i=0,\ldots,n(k)}$ comprises the division points of the simple càdlàg function $\overline{x}_k$.

6. Recall that

$$Q_{m(k)} \equiv \{q_{m(k),0}, q_{m(k),1}, \ldots, q_{m(k),p(m(k))}\} \equiv \{0, 2^{-m(k)}, 2 \cdot 2^{-m(k)}, \ldots, 1\}.$$

Hence $(\eta_{k,i})_{i=0,\ldots,n(k)}$ is a subsequence of $q_{m(k)} \equiv (q_{m(k),0}, q_{m(k),1}, \ldots, q_{m(k),p(m(k))})$. By Lemma 10.1.9, we may insert all points in $Q_{m(k)}$ that are not already in the sequence $(\eta_{k,i})_{i=0,\ldots,n(k)}$ into the latter, without changing the simple càdlàg function $\overline{x}_k$. Hence we may assume that

$$\eta_k \equiv (\eta_{k,0}, \eta_{k,1}, \ldots, \eta_{k,n(k)}) \equiv (q_{m(k),0}, q_{m(k),1}, \ldots, q_{m(k),p(m(k))}). \tag{10.2.45}$$

7. It follows from inequalities 10.2.40 and 10.2.44 that

$$\begin{aligned} d_D(\overline{x}_k, \overline{x}_{k+1}) &\leq d_D(\overline{x}_k, y_k) + d_D(y_k, y_{k+1}) + d_D(y_{k+1}, \overline{x}_{k+1}) \\ &< 2^{-k} + 2^{-k} + 2^{-k-1} < 2^{-k+2}. \end{aligned} \tag{10.2.46}$$

Moreover, since $\overline{x}_0 \equiv x_\circ \equiv y_0$, we have

$$\begin{aligned} d_D(x_\circ, \overline{x}_k) &\le d_D(y_0, y_k) + 2^{-k} \\ &< d_D(y_0, y_1) + d_D(y_1, y_2) + \cdots + d_D(y_{k-1}, y_k) + 2^{-k} \\ &< d_D(y_0, y_1) + 2^{-1} + \cdots + 2^{-(k-1)} + 2^{-k} < d_D(y_0, y_1) + 1 \equiv b_0. \end{aligned}$$

Hence $d(x_\circ, \overline{x}_k(t)) \le b_0$ for each $t \in domain(\overline{x}_k)$, where $k \ge 1$ is arbitrary. Consequently,

$$d(\overline{x}_k(t), \overline{x}_h(s)) \le 2b_0 \equiv b \tag{10.2.47}$$

for each $t \in domain(\overline{x}_k)$, $s \in domain(\overline{x}_h)$, for each $h, k \ge 0$.

8. Now let $k \ge 0$ be arbitrary. We will construct $\lambda_{k+1} \in \Lambda_{m(k+1),m(k+2)}$ such that

$$(2^{-k+2}, \lambda_{k+1}) \in A_{\overline{x}(k),\overline{x}(k+1)}. \tag{10.2.48}$$

First note, from equalities 10.2.43 and 10.2.45, that

$$\overline{x}_k \equiv \Phi_{smpl}((q_{m(k),i})_{i=0,\dots,p(m(k))-1}, (u_{k,i})_{i=0,\dots,p(m(k))-1})$$

for some sequence $(u_{k,i})_{i=0,\dots,p(m(k))-1}$ in $U_k \subset U_{k+1}$. Similarly,

$$\overline{x}_{k+1} \equiv \Phi_{smpl}((q_{m(k+1),i})_{i=0,\dots,p(m(k+1))-1}, (u_{k+1,i})_{i=0,\dots,p(m(k+1))-1})$$

for some sequence $(u_{k+1,i})_{i=0,\dots,p(m(k+1))-1}$ in U_{k+1}.

9. Separately, equality 10.2.41, where k is replaced by $k+2$, implies that

$$m_{k+2} \equiv [2 + m_{k+1} - \log_2(e^{-b}\delta_{\log@1}(2^{-k-2})\delta_{cdlg,k}(2^{-k-3}))]_1, \tag{10.2.49}$$

whence

$$2^{-m(k+2)} \le 2^{-m(k+1)-2}e^{-b}\delta_{\log@1}(2^{-k-2}). \tag{10.2.50}$$

Therefore we can apply Lemma 10.2.6, with m, m', k, x, y, U replaced by m_{k+1}, $m_{k+2}, k+2, \overline{x}_k, \overline{x}_{k+1}, U_{k+1}$, respectively, to yield some $\lambda_{k+1} \in \Lambda_{m(k+1),m(k+2)}$ such that

$$(d_D(\overline{x}_k, \overline{x}_{k+1}) + 2^{-k-1}, \lambda_{k+1}) \in A_{\overline{x}(k),\overline{x}(k+1)}. \tag{10.2.51}$$

From inequality 10.2.46, we have

$$d_D(\overline{x}_k, \overline{x}_{k+1}) + 2^{-k-1} < (2^{-k} + 2^{-k} + 2^{-k-1}) + 2^{-k-1} < 2^{-k+2},$$

so relation 10.2.51 trivially implies the desired relation 10.2.48.

10. From relation 10.2.48, it follows that

$$\left|\log\frac{\lambda_{k+1}t - \lambda_{k+1}s}{t-s}\right| \le 2^{-k+2} \tag{10.2.52}$$

for each $s, t \in [0,1]$ with $s < t$, and that

$$d(\overline{x}_{k+1}(\lambda_{k+1}t), \overline{x}_k(t)) \leq 2^{-k+2} \tag{10.2.53}$$

for each $t \in domain(\overline{x}_k) \cap \lambda_{k+1}^{-1} domain(\overline{x}_{k+1})$.

11. For each $k \geq 0$, define the composite admissible function

$$\mu_k \equiv \lambda_k \lambda_{k-1} \cdots \lambda_0 \in \Lambda.$$

We will prove that $\mu_k \to \mu$ uniformly on $[0, 1]$ for some $\mu \in \Lambda$. To that end, let $h > k \geq 0$ be arbitrary. By Lemma 10.2.4, relation 10.2.48 implies that

$$(2^{-k+2} + 2^{-k+1} + \cdots + 2^{-h+3}, \lambda_h \lambda_{h-1} \cdots \lambda_{k+1}) \in A_{\overline{x}(k), \overline{x}(h)}.$$

Hence, since $2^{-k+2} + 2^{-k+1} + \cdots + 2^{-h+3} < 2^{-k+3}$, we also have

$$(2^{-k+3}, \lambda_h \lambda_{h-1} \cdots \lambda_{k+1}) \in A_{\overline{x}(k), \overline{x}(h)}. \tag{10.2.54}$$

12. Let $t, s \in [0, 1]$ be arbitrary with $s < t$. Write $t' \equiv \mu_k t$ and $s' \equiv \mu_k s$. Then

$$\left|\log \frac{\mu_h t - \mu_h s}{\mu_k t - \mu_k s}\right| = \left|\log \frac{\lambda_h \lambda_{h-1} \cdots \lambda_{k+1} t' - \lambda_h \lambda_{h-1} \cdots \lambda_{k+1} s'}{t' - s'}\right| < 2^{-k+3}, \tag{10.2.55}$$

where the inequality is thanks to relation 10.2.54. Equivalently,

$$(t' - s') \exp(-2^{-k+3}) < \lambda_h \lambda_{h-1} \cdots \lambda_{k+1} t' - \lambda_h \lambda_{h-1} \cdots \lambda_{k+1} s'$$
$$< (t' - s') \exp(2^{-k+3}) \tag{10.2.56}$$

for each $t', s' \in [0, 1]$ with $s' < t'$. In the special case where $s = 0$, inequality 10.2.55 reduces to

$$|\log \mu_h t - \log \mu_k t| = \left|\log \frac{\mu_h t}{\mu_k t}\right| < 2^{-k+3}. \tag{10.2.57}$$

Hence the limit

$$\mu t \equiv \lim_{h \to \infty} \mu_h t$$

exists, where $t \in (0, 1]$ is arbitrary. Moreover, letting $k = 0$ and $h \to \infty$ in inequality 10.2.55, we obtain

$$\left|\log \frac{\mu t - \mu s}{t - s}\right| \leq 2^3 = 8. \tag{10.2.58}$$

Therefore μ is an increasing function that is uniformly continuous on $(0, 1]$. Furthermore,

$$te^{-8} \leq \mu t \leq te^8,$$

where $t \in (0, 1]$ is arbitrary. Hence μ can be extended to a continuous increasing function on $[0, 1]$, with $\mu 0 = 0$. Since $\mu_k 1 = 1$ for each $k \geq 0$, we have $\mu 1 = 1$. In view of inequality 10.2.58, we conclude that $\mu \in \Lambda$.

13. By letting $h \to \infty$ in inequality 10.2.55, we obtain

$$\left|\log \frac{\mu t - \mu s}{\mu_k t - \mu_k s}\right| \leq 2^{-k+3}, \tag{10.2.59}$$

where $t, s \in [0,1]$ are arbitrary with $s < t$. Replacing t, s by $\mu^{-1}t, \mu^{-1}s$, respectively, we obtain

$$\left|\log \frac{\mu_k \mu^{-1} t - \mu_k \mu^{-1} s}{t - s}\right| = \left|\log \frac{t - s}{\mu_k \mu^{-1} t - \mu_k \mu^{-1} s}\right| \leq 2^{-k+3}, \tag{10.2.60}$$

where $t, s \in [0,1]$ are arbitrary with $s < t$, and where $k \geq 0$ is arbitrary.

14. Recall from Steps 5 and 6 that for each $k \geq 0$, the sequence

$$\eta_k \equiv (\eta_{k,0}, \eta_{k,1}, \ldots, \eta_{k,n(k)}) \equiv q_{m(k)} \equiv (q_{m(k),0}, q_{m(k),1}, \ldots, q_{m(k),p(m(k))}) \tag{10.2.61}$$

is a sequence of division points of the simple càdlàg function $\overline{x}_k$. Define the set

$$A \equiv \bigcap_{h=0}^{\infty} \mu\mu_h^{-1}([\eta_{h,0}, \eta_{h,1}) \cup [\eta_{h,1}, \eta_{h,2}) \cup \cdots \cup [\eta_{h,n(h)-1}, \eta_{h,n(h)}]). \tag{10.2.62}$$

Then A contains all but countably many points in $[0,1]$, and is therefore dense in $[0,1]$.

15. Let $k \geq 0$ be arbitrary. Define the function

$$z_k \equiv \overline{x}_k \circ \mu_k \mu^{-1}. \tag{10.2.63}$$

Then $z_k \in D[0,1]$ by Assertion 4 of Lemma 10.1.3. Moreover,

$$z_{k+1} \equiv \overline{x}_{k+1} \circ \mu_{k+1}\mu^{-1} = \overline{x}_{k+1} \circ \lambda_{k+1}\mu_k \mu^{-1}. \tag{10.2.64}$$

Now consider each $r \in A$. Let $h \geq k$ be arbitrary. Then, by the defining equality 10.2.62 of the set A, there exists $i = 0, \ldots, n_h - 1$ such that

$$\mu_h \mu^{-1} r \in [\eta_{h,i}, \eta_{h,i+1}) \subset domain(\overline{x}_h). \tag{10.2.65}$$

Hence,

$$\lambda_{h+1}\mu_h \mu^{-1} r \in \lambda_{h+1}[\eta_{h,i}, \eta_{h,i+1}).$$

Moreover, from the defining equality 10.2.62, we have

$$r \in A \subset \mu\mu_h^{-1}(domain(\overline{x}_h)) = domain(\overline{x}_h \circ \mu_h \mu^{-1}) \equiv domain(z_h). \tag{10.2.66}$$

From equalities 10.2.64 and 10.2.63 and inequality 10.2.53, we obtain

$$d(z_{k+1}(r), z_k(r)) \equiv d(\overline{x}_{k+1}(\lambda_{k+1}\mu_k \mu^{-1} r), \overline{x}_k(\mu_k \mu^{-1} r)) \leq 2^{-k+2}. \tag{10.2.67}$$

16. Hence, since $r \in A$ and $k \geq 0$ are arbitrary, we conclude that $\lim_{k\to\infty} z_k$ exists on A. Define the function $z : [0,1] \to S$ by $domain(z) \equiv A$ and by $z(t) \equiv \lim_{k\to\infty} z_k(t)$ for each $t \in domain(z)$. Inequality 10.2.67 then implies that

$$d(z(r), z_k(r)) \leq 2^{-k+3}, \tag{10.2.68}$$

where $r \in A$ and $k \geq 0$ are arbitrary. We proceed to verify the conditions in Proposition 10.1.6 for the function z to have a càdlàg completion.

17. For that purpose, let $r \in A$ and $h \geq k$ be arbitrary. Then, as observed in Step 15, we have $r \in domain(z_h)$. Hence, since z_h is càdlàg, it is right continuous at r. Therefore there exists $c_h > 0$ such that

$$d(z_h(t), z_h(r)) < 2^{-h+3} \tag{10.2.69}$$

for each $t \in [r, r + c_h) \cap A$. In view of inequality 10.2.68, it follows that

$$d(z(t), z(r)) \leq d(z(t), z_h(t)) + d(z_h(t), z_h(r)) + d(z_h(r), z(r)) < 3 \cdot 2^{-h+3}$$

for each $t \in [r, r + c_h) \cap A$. Thus z is right continuous at each point $r \in A \equiv domain(z)$. Condition 1 in Definition 10.1.2 has been verified for the function z.

18. We will next verify Condition 3 in Definition 10.1.2 for the function z. To that end, let $\varepsilon > 0$ be arbitrary. Take $k \geq 0$ so large that $2^{-k+4} < \varepsilon$, and define $\delta(\varepsilon) \equiv \exp(-2^{-k+3})2^{-m(k)}$. Let $j = 0, \ldots, n_k$ be arbitrary. For brevity, define $\eta'_j \equiv \mu\mu_k^{-1}\eta_{k,j}$. By inequality 10.2.52, we have

$$\lambda_{h+1}t - \lambda_{h+1}s \geq (t - s)\exp(-2^{-h+2})$$

for each $s, t \in [0, 1]$ with $s < t$, for each $h \geq 0$. Hence, for each $j = 0, \ldots, n_k - 1$, we have

$$\begin{aligned}
\eta'_{j+1} - \eta'_j &\equiv \mu\mu_k^{-1}\eta_{k,j+1} - \mu\mu_k^{-1}\eta_{k,j} \\
&= \lim_{h\to\infty} (\mu_h\mu_k^{-1}\eta_{k,j+1} - \mu_h\mu_k^{-1}\eta_{k,j}) \\
&= \lim_{h\to\infty} (\lambda_h \cdots \lambda_{k+1}\eta_{k,j+1} - \lambda_h \cdots \lambda_{k+1}\eta_{k,j}) \\
&\geq \exp(-2^{-k+3})(\eta_{k,j+1} - \eta_{k,j}) = \exp(-2^{-k+3})2^{-m(k)} \equiv \delta(\varepsilon),
\end{aligned} \tag{10.2.70}$$

where the inequality is by the first half of inequality 10.2.56.

19. Now consider each $j = 0, \ldots, n_k - 1$ and each

$$t' \in domain(z) \cap [\eta'_j, \eta'_{j+1}) \equiv A \cap [\mu\mu_k^{-1}\eta_{k,j}, \mu\mu_k^{-1}\eta_{k,j+1}).$$

We will show that

$$d(z(t'), z(\eta'_j))) \leq \varepsilon. \tag{10.2.71}$$

To that end, write $t \equiv \mu^{-1}t'$, and write $s \equiv \mu_k t \equiv \mu_k\mu^{-1}t' \in [\eta_{k,j}, \eta_{k,j+1})$. Then

$$\overline{x}_k(s) = \overline{x}_k(\eta_{k,j}) \tag{10.2.72}$$

since $\overline{x}_k$ is a simple càdlàg function with $(\eta_{k,0}, \eta_{k,1}, \ldots, \eta_{k,n(k)})$ as a sequence of division points. Let $h > k$ be arbitrary, and define

$$r \equiv \mu_h \mu^{-1} t' = \mu_h t = \lambda_h \lambda_{h-1} \cdots \lambda_{k+1} s.$$

Then, according to the defining equality 10.2.63, we have

$$z_h(t') \equiv \overline{x}_h(\mu_h \mu^{-1} t') \equiv \overline{x}_h(r) \equiv \overline{x}_h(\lambda_h \lambda_{h-1} \cdots \lambda_{k+1} s).$$

Combining with equality 10.2.72, we obtain

$$d(z_h(t'), \overline{x}_k(\eta_{k,j})) = d(\overline{x}_h(\lambda_h \lambda_{h-1} \cdots \lambda_{k+1} s), \overline{x}_k(s)) \leq 2^{-k+3},$$

where the last inequality is a consequence of relation 10.2.54. Then

$$d(z_h(t'), z_h(\eta'_j)) \leq d(z_h(t'), \overline{x}_k(\eta_{k,j})) + d(z_h(\eta'_j), \overline{x}_k(\eta_{k,j})) \leq 2^{-k+4} < \varepsilon, \tag{10.2.73}$$

where $t' \in domain(z) \cap [\eta'_j, \eta'_{j+1})$ is arbitrary. Letting $h \to \infty$, we obtain the desired inequality 10.2.71. Inequalities 10.2.71 and 10.2.70 together say that $(\eta'_j)_{j=0,\ldots,n(k)}$ is a sequence of ε-division points for the function z, with separation of at least $\delta(\varepsilon)$. Condition 3 in Definition 10.1.2 has been verified for the objects z, η', and δ.

20. Thus Conditions 1 and 3 in Definition 10.1.2 have been verified for the objects z, η', and δ. Proposition 10.1.6 therefore says that (i′) the completion $y \in D[0,1]$ of z is well defined, (ii′) $y|domain(z) = z$, and (iii′) δ is a modulus of càdlàg of y.

21. Finally, we will prove that $d_D(y_h, y) \to 0$ as $h \to \infty$. To that end, let $h \geq 0$ and

$$r \in A \subset domain(z) \subset domain(y)$$

be arbitrary. By the Condition (ii′), we have $y(r) = z(r)$. Hence, by inequality 10.2.68,

$$d(y(r), z_h(r)) = d(z(r), z_h(r)) \leq 2^{-h+3}.$$

Consequently, since A is a dense subset of $[0,1]$, Lemma 10.1.3 says that

$$d(y(r), z_h(r)) \leq 2^{-h+3} \tag{10.2.74}$$

for each $r \in domain(y) \cap domain(z_h)$. In other words,

$$d(y(r), \overline{x}_h \circ \mu_h \mu^{-1}(r)) \leq 2^{-h+3} \tag{10.2.75}$$

for each $r \in domain(y) \cap \mu \mu_h^{-1} domain(\overline{x}_h)$.

22. Inequalities 10.2.75 and 10.2.60 together imply that $(2^{-h+3}, \mu_h \mu^{-1}) \in A_{y, \overline{x}(h)}$, whence $2^{-h+3} \in B_{y, \overline{x}(h)}$. Accordingly,

$$d_D(y, \overline{x}_h) \equiv \inf B_{y, \overline{x}(h)} \leq 2^{-h+3}.$$

Together with inequality 10.2.44, this implies

$$d_D(y, y_h) \leq 2^{-h+3} + 2^{-h},$$

where $h \geq 0$ is arbitrary. Thus $d_D(y_h, y) \to 0$ as $h \to \infty$.

23. Summing up, for each Cauchy sequence $(y_h)_{h=1,2,\ldots}$, there exists $y \in D[0,1]$ such that $d_D(y_h, y) \to 0$ as $h \to \infty$. In other words, $(D[0,1], d_D)$ is complete, as alleged. □

10.3 a.u. Càdlàg Process

Let (Ω, L, E) be a probability space, and let (S,d) be a locally compact metric space. Let $(D[0,1], d_D)$ be the Skorokhod space of càdlàg functions on the unit interval $[0,1]$ with values in (S,d), as introduced in Section 10.2.

Recall Definition 9.0.2 for notations related to the enumerated set of dyadic rationals Q_∞ in $[0,1]$.

Definition 10.3.1. Random càdlàg function. An arbitrary r.v. $Y : (\Omega, L, E) \to (D[0,1], d_D)$ with values in the Skorokhod space is called a *random càdlàg function* if, for each $\varepsilon > 0$, there exists a measurable set A with $P(A) < \varepsilon$ such that members of the family $\{Y(\omega) : \omega \in A^c\}$ of càdlàg functions share a common modulus of càdlàg. □

Of special interest is the subclass of the random càdlàg functions corresponding to a.u. càdlàg processes on $[0,1]$, defined next.

Definition 10.3.2. a.u. Càdlàg process. Let $X : [0,1] \times \Omega \to (S,d)$ be a stochastic process that is continuous in probability on $[0,1]$, with a modulus of continuity in probability δ_{Cp}. Suppose there exists a full set $B \subset \bigcap_{t \in Q(\infty)} domain(X_t)$ with the following properties:

1. (Right continuity.) For each $\omega \in B$, the function $X(\cdot, \omega)$ is right continuous at each $t \in domain(X(\cdot, \omega))$.

2. (Right completeness.) Let $\omega \in B$ and $t \in [0,1]$ be arbitrary. If $\lim_{r \to t; r \geq t} X(r, \omega)$ exists, then $t \in domain(X(\cdot, \omega))$.

3. (Approximation by step functions.) Let $\varepsilon > 0$ be arbitrary. Then there exist (i) $\delta_{aucl}(\varepsilon) \in (0,1)$, (ii) a measurable set $A \subset B$ with $P(A^c) < \varepsilon$, (iii) an integer $h \geq 1$, and (iv) a sequence of r.r.v.'s

$$0 = \tau_0 < \tau_1 < \cdots < \tau_{h-1} < \tau_h = 1 \tag{10.3.1}$$

such that, for each $i = 0, \ldots, h-1$, the function $X_{\tau(i)}$ is an r.v., and such that (v) for each $\omega \in A$, we have

$$\bigwedge_{i=0}^{h-1} (\tau_{i+1}(\omega) - \tau_i(\omega)) \geq \delta_{aucl}(\varepsilon) \tag{10.3.2}$$

with

$$d(X(\tau_i(\omega), \omega), X(\cdot, \omega)) \leq \varepsilon \tag{10.3.3}$$

on the interval $\theta_i(\omega) \equiv [\tau_i(\omega), \tau_{i+1}(\omega))$ or $\theta_i(\omega) \equiv [\tau_i(\omega), \tau_{i+1}(\omega)]$ depending on whether $0 \leq i \leq h-2$ or $i = h-1$.

Then the process $X : [0,1] \times \Omega \rightarrow S$ is called an *a.u. càdlàg process*, with δ_{Cp} as a modulus of continuity in probability and with δ_{aucl} as a *modulus of a.u. càdlàg*. We will let $\widehat{D}_{\delta(aucl),\delta(cp)}[0,1]$ denote the set of all such processes.

We will let $\widehat{D}[0,1]$ denote the set of all a.u. càdlàg processes. Two members X, Y of $\widehat{D}[0,1]$ are considered equal if there exists a full set B' such that for each $\omega \in B'$, we have $X(\cdot,\omega) = Y(\cdot,\omega)$ as functions on $[0,1]$. □

Lemma 10.3.3. a.u. Continuity implies a.u. càdlàg. *Let $X : [0,1] \times (\Omega, L, E) \rightarrow (S,d)$ be an arbitrary a.u. continuous process. Then X is a.u. càdlàg.*

Proof. Easy and omitted. □

Definition 10.3.4. Random càdlàg function from an a.u. càdlàg process. Let $X \in \widehat{D}[0,1]$ be arbitrary. Define a function

$$X^* : \Omega \rightarrow D[0,1]$$

by

$$domain(X^*) \equiv \{\omega \in \Omega : X(\cdot,\omega) \in D[0,1]\}$$

and by

$$X^*(\omega) \equiv X(\cdot,\omega)$$

for each $\omega \in domain(X^*)$. We will call X^* the *random càdlàg function* from the a.u. càdlàg process X. Proposition 10.3.5 (next) proves that the function X^* is well defined and is indeed a random càdlàg function. □

Proposition 10.3.5. Each a.u. càdlàg process gives rise to a random càdlàg function. Let $X \in \widehat{D}[0,1]$ *be an arbitrary a.u. càdlàg process, with some modulus of a.u. càdlàg δ_{aucl}. Then the following conditions hold:*

1. Let $\varepsilon > 0$ be arbitrary. Then there exists a measurable set A with $P(A) < \varepsilon$, such that members of the set $\{X^*(\omega) : \omega \in A^c\}$ of functions on $[0,1]$ are càdlàg functions that share a common modulus of càdlàg.

2. X^* is an r.v. with values in the complete metric space $(D[0,1], d_D)$. Thus X^* is a random càdlàg function.

Proof. 1. Define the full subset $B \equiv \bigcap_{t \in Q(\infty)} domain(X_t)$ of Ω. By Conditions 1 and 2 in Definition 10.3.2, members of the set $\{X(\cdot,\omega) : \omega \in B\}$ of functions satisfy the corresponding Conditions 1 and 2 in Definition 10.1.2.

2. Let $n \geq 1$ be arbitrary. By Condition 3 in Definition 10.3.2, there exist (i) $\delta_{aucl}(2^{-n}) > 0$, (ii) a measurable set $A_n \subset B$ with $P(A_n^c) < 2^{-n}$, (iii) an integer $h_n \geq 0$, and (iv) a sequence of r.r.v.'s

$$0 = \tau_{n,0} < \tau_{n,1} < \cdots < \tau_{n,h(n)} = 1 \tag{10.3.4}$$

such that for each $i = 0, \ldots, h_n - 1$, the function $X_{\tau(n,i)}$ is an r.v., and such that for each $\omega \in A_n$, we have

$$\bigwedge_{i=0}^{h(n)-1} (\tau_{n,i+1}(\omega) - \tau_{n,i}(\omega)) \geq \delta_{aucl}(2^{-n}) \tag{10.3.5}$$

with

$$d(X(\tau_{n,i}(\omega),\omega), X(\cdot,\omega)) \leq 2^{-n} \tag{10.3.6}$$

on the interval $\theta_{n,i}(\omega) \equiv [\tau_{n,i}(\omega), \tau_{n,i+1}(\omega))$ or $\theta_{n,i}(\omega) \equiv [\tau_{n,i}(\omega), \tau_{n,i+1}(\omega)]$ depending on whether $0 \leq i \leq h_n - 2$ or $i = h_n - 1$.

3. Let $\varepsilon > 0$ be arbitrary. Take $j \geq 1$ so large that $2^{-j} < \varepsilon$. Define $A \equiv B^c \cup \bigcup_{n=j+1}^{\infty} A_n^c$. Then $P(A) < \sum_{n=j+1}^{\infty} 2^{-n} = 2^{-j} < \varepsilon$. Consider each $\omega \in A^c$. Let $\varepsilon' > 0$ be arbitrary. Consider each $n \geq j+1$ so large that $2^{-n} < \varepsilon'$. Define $\delta_{cdlg}(\varepsilon') \equiv \delta_{aucl}(2^{-n})$. Then $\omega \in A_n$. Hence inequalities 10.3.5 and 10.3.6 hold, and imply that the sequence

$$0 = \tau_{n,0}(\omega) < \tau_{n,1}(\omega) < \cdots < \tau_{n,h(n)}(\omega) = 1 \tag{10.3.7}$$

is a sequence of ε'-division points of the function $X(\cdot,\omega)$, with separation of at least $\delta_{cdlg}(\varepsilon')$. Summing up, all the conditions in Definition 10.1.2 have been verified for the function $X^*(\omega) \equiv X(\cdot,\omega)$ to be càdlàg, with the modulus of càdlàg δ_{cdlg}, where $\omega \in A^c$ is arbitrary. Assertion 1 is proved.

4. Now let $n \geq 1$ be arbitrary. We will construct a random càdlàg function V_n as follows. Write d_{ecld} for the Euclidean metric on $(0,1)$. Fix $\overline{a}_0 \equiv 0$ and $\overline{a}_{h(n)} \equiv 1$. Let $(a_1, \ldots, a_{h(n)-1}) \in (0,1)^{h(n-1)}$ be arbitrary. As an abbreviation, write $\delta \equiv h_n^{-1}\delta_{aucl}(2^{-n}) > 0$. Define, for each $i = 1, \ldots, h_n - 1$,

$$\overline{a}_i \equiv f_i(a_1, \ldots, a_{h(n)-1}) \equiv i\delta \vee (a_0 \vee \cdots \vee a_i) \wedge (i\delta + 1 - h_n\delta).$$

Then $\overline{a}_{i+1} > \overline{a}_i + \delta$ for each $i = 1, \ldots, h_n - 1$. Hence $0 \equiv \overline{a}_0 < \overline{a}_1 < \cdots < \overline{a}_{h(n)} \equiv 1$. Therefore the simple càdlàg function

$$\Phi_{smpl}((\overline{a}_0, \overline{a}_1, \ldots, \overline{a}_{h(n)-1}), (x_0, \ldots, x_{h(n)-1})),$$

first introduced in Definition 10.1.7, is well defined for each

$$((a_1, \ldots, a_{h(n)-1}), (x_0, \ldots, x_{h(n)-1})) \in (0,1)^{h(n-1)} \times S^{h(n)-1}.$$

5. Thus we have a function

$$\Psi_n : ((0,1)^{h(n-1)} \times S^{h(n)-1}, d_{ecld}^{h(n)-1} \otimes d^{h(n)-1}) \to (D[0,1], d_D)$$

defined by

$$\Psi_n((a_1, \ldots, a_{h(n)-1}), (x_0, \ldots, x_{h(n)-1}))$$
$$\equiv \Phi_{smpl}((\overline{a}_0, \overline{a}_1, \ldots, \overline{a}_{h(n)-1}), (x_0, \ldots, x_{h(n)-1})),$$

for each $((a_1, \ldots, a_{h(n)-1}), (x_0, \ldots, x_{h(n)-1})) \in (0,1)^{h(n-1)} \times S^{h(n)-1}$. It can easily be proved that this function Ψ_n is uniformly continuous.

6. Hence Proposition 4.8.10 applies. It implies that the function

$$V_n \equiv \Psi_n((\tau_{n,1}, \ldots, \tau_{n,h(n)-1}), (X_{\tau(n,0)}, \ldots, X_{\tau(n,h(n)-1)})) :$$
$$(\Omega, L, E) \to (D[0,1], d_D)$$

is an r.v.

7. Define, for each $i = 1, \ldots, h_n - 1$, the r.r.v.

$$\overline{\tau}_{n,i} \equiv f_i(\tau_{n,1}, \ldots, \tau_{n,h(n)-1}).$$

Consider each $\omega \in A_n$. Let $i = 1, \ldots, h_n - 1$ be arbitrary. Then inequality 10.3.5 holds and implies that

$$\tau_{n,i}(\omega) \geq \delta_{aucl}(2^{-n}) > i\delta$$

and that

$$\tau_{n,i}(\omega) \leq \tau_{n,i+1}(\omega) - \delta_{aucl}(2^{-n}) \leq \cdots \leq \tau_{n,h(n)}(\omega) - (h_n - 1)\delta_{aucl}(2^{-n})$$
$$\leq 1 - (h_n - 1)i\delta = i\delta + 1 - h_n i\delta \leq i\delta + 1 - h_n\delta.$$

Therefore, in view of inequality 10.3.4, we have,

$$\overline{\tau}_{n,i}(\omega) \equiv i\delta \vee (\tau_{n,0}(\omega) \vee \cdots \vee \tau_{n,i}(\omega)) \wedge (i\delta + 1 - h_n\delta)$$
$$= i\delta \vee \tau_{n,i}(\omega) \wedge (i\delta + 1 - h_n\delta) = \tau_{n,i}(\omega).$$

8. Hence

$$V_n(\omega) \equiv \Psi_n((\tau_{n,1}(\omega), \ldots, \tau_{n,h(n)-1}(\omega)), (X_{\tau(n,0)}(\omega), \ldots, X_{\tau(n,h(n)-1)}(\omega)))$$
$$\equiv \Phi_{smpl}((\overline{\tau}_{n,0}(\omega), \ldots, \overline{\tau}_{n,h(n)-1}(\omega)), (X_{\tau(n,0)}(\omega), \ldots, X_{\tau(n,h(n)-1)}(\omega)))$$
$$\equiv \Phi_{smpl}((\tau_{n,0}(\omega), \ldots, \tau_{n,h(n)-1}(\omega)), (X_{\tau(n,0)}(\omega), \ldots, X_{\tau(n,h(n)-1)}(\omega))).$$

9. Now let $t \in [\tau_{n,i}(\omega), \tau_{n,i+1}(\omega))$ be arbitrary, for some $i = 0, \ldots, h_n - 1$. Then, by Definition 10.1.7 for the function Φ_{smpl}, we have

$$V_n(\omega)(t) = X_{\tau(n,i)}(\omega) \equiv X(\tau_{n,i}(\omega), \omega).$$

At the same time, inequality 10.3.6 implies that

$$d(X(\tau_{n,i}(\omega), \omega), X^*(\omega)(t)) \equiv d(X(\tau_{n,i}(\omega), \omega), X(t, \omega)) \leq 2^{-n}.$$

Combining, we obtain

$$d(V_n(\omega)(t), X^*(\omega)(t)) \leq 2^{-n} \tag{10.3.8}$$

for each t in $\bigcup_{i=0}^{h(n)-1} [\tau_{n,i}(\omega), \tau_{n,i+1}(\omega))$. Since this union is a dense subset of $[0, 1]$, inequality 10.3.8 holds for each $t \in domain(V_n(\omega)) \cap domain(X^*(\omega))$. It follows that

$$d_D(V_n(\omega), X^*(\omega)) \leq 2^{-n},$$

where $\omega \in A_n$ is arbitrary. Since $P(A_n^c) < 2^{-n}$, it follows that $V_n \to X^*$in probability, as functions with values in the complete metric space $(D[0,1], d_D)$. At the same time V_n is an r.v. for each $n \geq 1$. Therefore Assertion 2 of Proposition 4.9.3 says that X^* is an r.v. with values in $(D[0,1], d_D)$. Assertion 2 and the present proposition are proved. □

Definition 10.3.6. Metric for the space of a.u. càdlàg processes. Define the metric $\rho_{\widehat{D}[0,1]}$ on $\widehat{D}[0,1]$ by

$$\rho_{\widehat{D}[0,1]}(X,Y) \equiv \int E(d\omega)\widehat{d}_D(X(\cdot,\omega), Y(\cdot,\omega)) \equiv E\widehat{d}_D(X^*, Y^*) \qquad (10.3.9)$$

for each $X, Y \in \widehat{D}[0,1]$, where $\widehat{d}_D \equiv 1 \wedge d_D$. Lemma 10.3.7 (next) justifies the definition. □

Lemma 10.3.7. $\left(\widehat{D}[0,1], \rho_{\widehat{D}[0,1]}\right)$ **is a metric space.** *The function $\rho_{\widehat{D}[0,1]}$ is well defined and is a metric.*

Proof. Let $X, Y \in \widehat{D}[0,1]$ be arbitrary. Then, according to Proposition 10.3.5, the random càdlàg functions X^*, Y^* associated with X, Y, respectively, are r.v.'s with values in $(D[0,1], d_D)$. Therefore the function $\widehat{d}_D(X^*, Y^*)$ is an integrable r.r.v., and the defining equality 10.3.9 makes sense.

Symmetry and the triangle inequality can be trivially verified for $\rho_{\widehat{D}[0,1]}$. Now suppose $X = Y$ in $\widehat{D}[0,1]$. Then, by the definition of the set $\widehat{D}[0,1]$, we have $X(\cdot,\omega) = Y(\cdot,\omega)$ in $D[0,1]$, for a.e. $\omega \in \Omega$. Hence

$$\widehat{d}_D(X^*(\omega), Y^*(\omega)) \equiv \widehat{d}_D((X(\cdot,\omega), Y(\cdot,\omega)) = 0.$$

Thus $\widehat{d}_D(X^*, Y^*) = 0$ a.s. Consequently, $\rho_{\widehat{D}[0,1]}(X,Y) \equiv E\widehat{d}_D(X^*, Y^*) = 0$ according to the defining equality 10.3.9. The converse is proved similarly. Combining, we conclude that $\rho_{\widehat{D}[0,1]}$ is a metric. □

10.4 D-Regular Family of f.j.d.'s and D-Regular Process

In this and the following two sections, we construct an a.u. càdlàg process from a consistent family F of f.j.d.'s with the locally compact state space (S,d) and parameter set $[0,1]$ that satisfies a certain D-regularity condition, to be defined presently. The construction is by (i) taking any process $Z : Q_\infty \times \Omega \to S$ with marginal distributions given by $F|Q_\infty$ and (ii) extending the process Z to an a.u. càdlàg process $X : [0,1] \times \Omega \to S$ by taking right limits of sample paths. Step (i) can be done by the Daniell–Kolmogorov Extension or Daniell–Kolmogorov–Skorokhod Extension, for example. As a matter of fact, we can define D-regularity for F as D-regularity of any process Z with marginal distributions given by $F|Q_\infty$. The key Step (ii) then involves proving that a process $X : [0,1] \times \Omega \to S$ is a.u. càdlàg iff it is the right-limit extension of some D-regular process $Z : Q_\infty \times \Omega \to S$. In this section, we prove the "only if" part. In Section 10.5, we will prove the "if" part.

Definition 10.4.1. D-regular processes and D-regular families of f.j.d.'s with parameter set Q_∞. Let $Z : Q_\infty \times \Omega \to S$ be a stochastic process, with marginal distributions given by the family F of f.j.d.'s. Let $\overline{m} \equiv (m_n)_{n=0,1,...}$ be an increasing sequence of nonnegative integers. Suppose the following conditions are satisfied:

1. Let $n \geq 0$ be arbitrary. Let $\beta > 2^{-n}$ be arbitrary such that the set

$$A_{t,s}^{\beta} \equiv (d(Z_t, Z_s) > \beta) \tag{10.4.1}$$

is measurable for each $s,t \in Q_\infty$. Then

$$P(D_n) < 2^{-n}, \tag{10.4.2}$$

where we define the exceptional set

$$D_n \equiv \bigcup_{t \in Q(m(n))} \bigcup_{r,s \in (t,t')Q(m(n+1)); r \leq s} (A_{t,r}^{\beta} \cup A_{t,s}^{\beta})(A_{t,r}^{\beta} \cup A_{t',s}^{\beta})(A_{t',r}^{\beta} \cup A_{t',s}^{\beta}), \tag{10.4.3}$$

where for each $t \in Q_{m(n)}$ we abuse notations and write $t' \equiv 1 \wedge (t + 2^{-m(n)})$.

2. The process Z is continuous in probability on Q_∞, with a modulus of continuity in probability δ_{Cp}.

Then the process $Z : Q_\infty \times \Omega \to S$ and the family F of f.j.d.'s are said to be D-*regular*, with the sequence $\overline{m}$ as a *modulus of D-regularity* and with the operation δ_{Cp} as a modulus of continuity in probability. Let

$$\widehat{R}_{Dreg,\overline{m},\delta(Cp)}(Q_\infty \times \Omega, S)$$

denote the set of all such processes. Let $\widehat{R}_{Dreg}(Q_\infty \times \Omega, S)$ denote the set of all D-regular processes. Thus $\widehat{R}_{Dreg,\overline{m},\delta(Cp)}(Q_\infty \times \Omega, S)$ and $\widehat{R}_{Dreg}(Q_\infty \times \Omega, S)$ are subsets of the metric space $(\widehat{R}(Q_\infty \times \Omega, S), \widehat{\rho}_{Prob,Q(\infty)})$ introduced in Definition 6.4.2 and, as such, inherit the metric $\widehat{\rho}_{Prob,Q(\infty)}$. Thus we have the metric space

$$(\widehat{R}_{Dreg}(Q_\infty \times \Omega, S), \widehat{\rho}_{Prob,Q(\infty)}).$$

In addition, let

$$\widehat{F}_{Dreg,\overline{m},\delta(Cp)}(Q_\infty, S)$$

denote the set of all such families F of f.j.d.'s. Let $\widehat{F}_{Dreg}(Q_\infty, S)$ denote the set of all D-regular families of f.j.d.'s. Then $\widehat{F}_{Dreg,\overline{m},\delta(Cp)}(Q_\infty, S)$ and $\widehat{F}_{Dreg}(Q_\infty, S)$ are subsets of the metric space $(\widehat{F}(Q_\infty, S), \widehat{\rho}_{Marg,\xi,Q(\infty)})$ of consistent families of f.j.d.'s introduced in Definition 6.2.8, where the metric $\widehat{\rho}_{Marg,\xi,Q(\infty)}$ is defined relative to an arbitrarily given, but fixed, binary approximation $\xi \equiv (A_q)_{q=1,2,...}$ of (S,d). □

Condition 1 in Definition 10.4.1 is, in essence, equivalent to condition (13.10) in the key theorem 13.3 of [Billingsley 1999]. The crucial difference between our construction and Billingsley's theorem is that the latter relies on Prokhorov's

Theorem (theorem 5.1 in [Billingsley 1999]). As we observed earlier, Prokhorov's Theorem implies the principle of infinite search. This is in contrast to our simple and direct construction developed in this and the next section.

First we extend the definition of D-regularity to families of f.j.d.'s with a parameter interval $[0,1]$ that are continuous in probability on the interval.

Definition 10.4.2. D-regular families of f.j.d.'s with parameter interval $[0,1]$. Recall from Definition 6.2.12 the metric space $(\widehat{F}_{Cp}([0,1],S), \widehat{\rho}_{Cp,\xi,[0,1],Q(\infty)})$ of families of f.j.d.'s that are continuous in probability on $[0,1]$, where the metric is defined relative to the enumerated, countable, dense subset Q_∞ of $[0,1]$. Define two subsets of $\widehat{F}_{Cp}([0,1],S)$ by

$$\widehat{F}_{Dreg}([0,1],S) \equiv \{F \in \widehat{F}_{Cp}([0,1],S) : \ F|Q_\infty \in \widehat{F}_{Dreg}(Q_\infty,S)\}$$

and

$$\widehat{F}_{Dreg,\overline{m},\delta(Cp)}([0,1],S) \equiv \{F \in \widehat{F}_{Cp}([0,1],S) : F|Q_\infty \in \widehat{F}_{Dreg,\overline{m},\delta(Cp)}(Q_\infty,S)\}.$$

These subsets inherit the metric $\widehat{\rho}_{Cp,\xi,[0,1],Q(\infty)}$. □

We will prove that a process $X : [0,1] \times \Omega \to S$ is a.u càdlàg iff it is the extension by right limit of a D-regular process $Z : Q_\infty \times \Omega \to S$.

Theorem 10.4.3. Restriction of each a.u. càdlàg process to Q_∞ is D-regular. *Let $X : [0,1] \times \Omega \to S$ be an a.u càdlàg process with a modulus of a.u. càdlàg δ_{aucl} and a modulus of continuity in probability δ_{Cp}. Let $\overline{m} \equiv (m_n)_{n=0,1,2,\ldots}$ be an arbitrary increasing sequence of integers such that*

$$2^{-m(n)} < \delta_{aucl}(2^{-n-1}) \tag{10.4.4}$$

for each $n \geq 0$.

Then the process $Z \equiv X|Q_\infty$ is D-regular with a modulus of D-regularity $\overline{m}$ and with the same modulus of continuity in probability δ_{Cp}.

Proof. 1. By Definition 10.3.2, the a.u càdlàg process X is continuous in probability on $[0,1]$, with some modulus of continuity in probability δ_{Cp}. Hence so is Z on Q_∞, with the same modulus of continuity in probability δ_{Cp}. Consequently, Condition 2 in Definition 10.4.1 is satisfied.

2. Now let $n \geq 0$ be arbitrary. Write $\varepsilon_n \equiv 2^{-n}$. By Condition 3 in Definition 10.3.2, there exist (i) $\delta_{aucl}(2^{-1}\varepsilon_n) > 0$, (ii) a measurable set $A_n \subset B \equiv \bigcap_{t \in Q(\infty)} domain(X_t)$ with $P(A_n^c) < 2^{-1}\varepsilon_n$, (iii) an integer $h_n \geq 0$, and (iv) a sequence of r.r.v.'s

$$0 = \tau_{n,0} < \tau_{n,1} < \cdots < \tau_{n,h(n)-1} < \tau_{n,h(n)} = 1 \tag{10.4.5}$$

such that for each $i = 0, \ldots, h_n - 1$, the function $X_{\tau(n,i)}$ is an r.v., and such that for each $\omega \in A_n$, we have

$$\bigwedge_{i=0}^{h(n)-1} (\tau_{n,i+1}(\omega) - \tau_{n,i}(\omega)) \geq \delta_{aucl}(2^{-1}\varepsilon_n) \tag{10.4.6}$$

with

$$d(X(\tau_{n,i}(\omega),\omega), X(\cdot,\omega)) \leq 2^{-1}\varepsilon_n \tag{10.4.7}$$

on the interval $\theta_{n,i} \equiv [\tau_{n,i}(\omega), \tau_{n,i+1}(\omega))$ or $\theta_{n,i} \equiv [\tau_{n,i}(\omega), 1]$ depending on whether $i \leq h_n - 2$ or $i = h_n - 1$.

3. Take any $\beta > 2^{-n}$ such that the set

$$A^{\beta}_{t,s} \equiv (d(X_t, X_s) > \beta) \tag{10.4.8}$$

is measurable for each $s,t \in Q_\infty$. Let D_n be the exceptional set defined in equality 10.4.3.

Suppose, for the sake of a contradiction, that $P(D_n) > \varepsilon_n \equiv 2^{-n}$. Then $P(D_n) > \varepsilon_n > P(A^c_n)$ by Condition (ii). Hence $P(D_n A_n) > 0$. Consequently, there exists $\omega \in D_n A_n$. Since $\omega \in A_n$, inequalities 10.4.6 and 10.4.7 hold at ω. At the same time, since $\omega \in D_n$, there exist $t \in Q_{m(n)}$ and $r,s \in Q_{m(n+1)}$, with $t < r \leq s < t' \equiv t + 2^{-m(n)}$, such that

$$\omega \in (A^{\beta}_{t,r} \cup A^{\beta}_{t,s})(A^{\beta}_{t,r} \cup A^{\beta}_{t',s})(A^{\beta}_{t',r} \cup A^{\beta}_{t',s}). \tag{10.4.9}$$

4. Note that $\tau_{n,i+1}(\omega) - \tau_{n,i}(\omega) > t' - t$, for each $i = 0,\ldots,h_n - 1$. Hence there exists $i = 1,\ldots,h_n - 1$ such that $\tau_{n,i-1}(\omega) \leq t < t' \leq \tau_{n,i+1}(\omega)$. There are three possibilities regarding the order of r,s in relation to the points $\tau_{n,i-1}(\omega), \tau_{n,i}(\omega)$, and $\tau_{n,i+1}(\omega)$ in the interval $[0,1]$: (i′)

$$\tau_{n,i-1}(\omega) \leq t < r \leq s < \tau_{n,i}(\omega),$$

(i″)

$$\tau_{n,i-1}(\omega) \leq t < r < \tau_{n,i}(\omega) < s < t' \leq \tau_{n,i+1}(\omega),$$

and (iii′)

$$\tau_{n,i}(\omega) < r \leq s < t' \leq \tau_{n,i+1}(\omega).$$

In case (i′), we have, in view of inequality 10.4.7,

$$d(X_t(\omega), X_r(\omega)) \leq (d(X_{\tau(n,i-1)}(\omega), X_t(\omega)) + d(X_{\tau(n,i-1)}(\omega), X_r(\omega))) \leq \varepsilon_n < \beta$$

and similarly $d(X_t(\omega), X_s(\omega)) < \beta$. Hence $\omega \in (A^{\beta}_{t,r} \cup A^{\beta}_{t,s})^c$. Similarly, in case (ii′), we have

$$d(X_t(\omega), X_r(\omega)) \vee d(X_s(\omega), X_{t'}(\omega)) < \beta,$$

whence $\omega \in (A^{\beta}_{t,r} \cup A^{\beta}_{t',s})^c$. Likewise, in case (iii′), we have

$$d(X_{t'}(\omega), X_r(\omega)) \vee d(X_{t'}(\omega), X_s(\omega)) < \beta,$$

whence $\omega \in (A^{\beta}_{t',r} \cup A^{\beta}_{t',s})^c$. Combining, in each case, we have

$$\omega \in (A^{\beta}_{t,r} \cup A^{\beta}_{t,s})^c \cup (A^{\beta}_{t,r} \cup A^{\beta}_{t',s})^c \cup (A^{\beta}_{t',r} \cup A^{\beta}_{t',s})^c,$$

which contradicts relation 10.4.9. Thus the assumption that $P(D_n) > 2^{-n}$ leads to a contradiction. We conclude that $P(D_n) \leq 2^{-n}$, where $n \geq 0$ and $\beta > 2^{-n}$ are arbitrary. Hence the process $Z \equiv X|Q_\infty$ satisfies the conditions in Definition 10.4.1 to be D-regular, with the sequence $(m_n)_{n=0,1,2,\ldots}$ as a modulus of D-regularity. □

The converse to Theorem 10.4.3 will be proved in Section 10.5. From a D-regular family F of f.j.d.'s with parameter set $[0,1]$ we will construct an a.u. càdlàg process with marginal distributions given by the family F.

10.5 Right-Limit Extension of D-Regular Process Is a.u. Càdlàg

Refer to Definition 9.0.2 for the notations related to the enumerated set Q_∞ of dyadic rationals in the interval $[0,1]$. We proved in Theorem 10.4.3 that the restriction to Q_∞ of each a.u. càdlàg process on $[0,1]$ is D-regular. In this section, we will prove the converse, which is the main theorem in this chapter: the extension by right limit of each D-regular process on Q_∞ is a.u. càdlàg. Then we will prove the easy corollary that, given an D-regular family F of f.j.d.'s with parameter set $[0,1]$, we can construct an a.u. càdlàg process X with marginal distributions given by F and with a modulus of a.u. càdlàg in terms of the modulus of D-regularity of F.

We will use the following assumption and notations.

Definition 10.5.1. Assumption of a D-regular process. Recall that (S,d) is a locally compact metric space. Let $Z \in \widehat{R}_{Dreg,\overline{m},\delta(Cp)}(Q_\infty \times \Omega, S)$ be arbitrary but fixed for the remainder of this section. In other words, $Z : Q_\infty \times \Omega \to S$ is a fixed D-regular process with a fixed modulus of D-regularity $\overline{m} \equiv (m_k)_{k=0,1,\ldots}$ and with a fixed modulus of continuity in probability δ_{Cp}. □

Definition 10.5.2. Notation for the range of a sample function. Let $Y : Q \times \Omega \to S$ be an arbitrary process, and let $A \subset Q$ and $\omega \in \Omega$ be arbitrary. Then we write

$$Y(A,\omega) \equiv \{x \in S : x = Y(t,\omega) \text{ for some } t \in A\}. \quad \square$$

Definition 10.5.3. Accordion function. In the following discussion, unless otherwise specified, $(\beta_h)_{h=0,1,\ldots}$ will denote an arbitrary but fixed sequence of real numbers such that for each $k,h \geq 0$ with $k \leq h$, and for each $r,s \in Q_\infty$, we have (i)

$$\beta_h \in (2^{-h+1}, 2^{-h+2}), \tag{10.5.1}$$

(ii) the set

$$(d(Z_r, Z_s) > \beta_k + \cdots + \beta_h) \tag{10.5.2}$$

is measurable, and, in particular, (iii) the set

$$A_{r,s}^{\beta(h)} \equiv (d(Z_r, Z_s) > \beta_h) \tag{10.5.3}$$

is measurable.

Let $h, n \geq 0$ be arbitrary. Define

$$\beta_{n,h} \equiv \sum_{i=n}^{h} \beta_i, \tag{10.5.4}$$

where, by convention, $\sum_{i=n}^{h} \beta_i \equiv 0$ if $h < n$. Define

$$\beta_{n,\infty} \equiv \sum_{i=n}^{\infty} \beta_i. \tag{10.5.5}$$

Note that $\beta_{n,\infty} < \sum_{i=n}^{\infty} 2^{-i+2} = 2^{-n+3} \to 0$ as $n \to \infty$. For each subset A of Q_∞, write

$$A^- \equiv \{t \in Q_\infty : t \neq s \text{ for each } s \in A\},$$

the metric complement of A in Q_∞.

Let $s \in Q_\infty = \bigcup_{h=0}^{\infty} Q_{m(h)}$ be arbitrary. Define $\widehat{h}(s) \equiv h \geq 0$ to be the smallest integer such that $s \in Q_{m(h)}$. Let $n \geq 0$ be arbitrary. Define

$$\widehat{\beta}_n(s) \equiv \beta_{n,\widehat{h}(s)} \equiv \sum_{i=n}^{\widehat{h}(s)} \beta_i < \beta_{n,\infty}. \tag{10.5.6}$$

Note that, for each $u \in Q_{m(k)}$, we have $\widehat{h}(u) \leq k$, whence

$$\widehat{\beta}_n(u) \equiv \sum_{i=n}^{\widehat{h}(u)} \beta_i \leq \sum_{i=n}^{k} \beta_i \leq \beta_{n,k} < \beta_{n,\infty}. \tag{10.5.7}$$

Thus we have the functions

$$\widehat{h} : Q_\infty \to \{0, 1, 2, \ldots\}$$

and

$$\widehat{\beta}_n : Q_\infty \to (0, \beta_{n,\infty})$$

for each $n \geq 0$. These functions are defined relative to the sequences $(\beta_n)_{n=0,1,\ldots}$ and $(m_n)_{n=0,1,\ldots}$.

For want of a better name, we might call the function $\widehat{\beta}_n$ an *accordion function*, because its graph resembles a fractal-like accordion. It will furnish a time-varying boundary for some simple first exit times in the proof of the main theorem. Note that for arbitrary $s \in Q_{m(k)}$ for some $k \geq 0$, we have $\widehat{h}(s) \leq k$ and so $\widehat{\beta}_n(s) \leq \beta_{n,k}$. □

Definition 10.5.4. Some small exceptional sets. Let $k \geq 0$ be arbitrary. Then $\beta_{k+1} > 2^{-k}$. Hence, by the conditions in Definition 10.4.1 for $\overline{m} \equiv (m_h)_{h=0,1,\ldots}$ to be a modulus of D-regularity of the process Z, we have

$$P(D_k) \leq 2^{-k}, \tag{10.5.8}$$

where

$$D_k \equiv \bigcup_{u \in Q(m(k))} \bigcup_{r,s \in (u,u')Q(m(k+1));r \leq s} \left\{ (A_{u,r}^{\beta(k+1)} \cup A_{u,s}^{\beta(k+1)})(A_{u,r}^{\beta(k+1)} \cup A_{u',s}^{\beta(k+1)})(A_{u',r}^{\beta(k+1)} \cup A_{u',s}^{\beta(k+1)}) \right\}, \tag{10.5.9}$$

where, for each $u \in Q_{m(k)}$, we abuse notations and write $u' \equiv u + \Delta_{m(k)}$.

For each $h \geq 0$, define the small exceptional set

$$D_{h+} \equiv \bigcup_{k=h}^{\infty} D_k \tag{10.5.10}$$

with

$$P(D_{h+}) \leq \sum_{k=h}^{\infty} 2^{-k} = 2^{-h+1}. \tag{10.5.11}$$

□

Lemma 10.5.5. Existence of certain supremums as r.r.v.'s. *Let $Z : Q_\infty \times \Omega \to S$ be a D-regular process, with a modulus of D-regularity $\overline{m} \equiv (m_k)_{k=0,1,\ldots}$. Let $h \geq 0$ and $v, \overline{v}, v' \in Q_{m(h)}$ be arbitrary with $v \leq \overline{v} \leq v'$. Then the following conditions hold:*

1. For each $r \in [v, v']Q_\infty$, we have

$$d(Z_{\overline{v}}, Z_r) \leq \bigvee_{u \in [v,v']Q(m(h))} d(Z_{\overline{v}}, Z_u) + \widehat{\beta}_{h+1}(r) \tag{10.5.12}$$

on D_{h+}^c.

2. The supremum

$$\sup_{u \in [v,v']Q(\infty)} d(Z_v, Z_u)$$

exists as an r.r.v. Moreover,

$$0 \leq \sup_{u \in [v,v']Q(\infty)} d(Z_v, Z_u) - \bigvee_{u \in [v,v']Q(m(h))} d(Z_v, Z_u) \leq \beta_{h+1,\infty} \leq 2^{-h+4}$$

on D_{h+}^c, where $P(D_{h+}) \leq 2^{-h+1}$.

3. Write $\widehat{d} \equiv 1 \wedge d$. Then

$$0 \leq E \sup_{u \in [v,v']Q(\infty)} \widehat{d}(Z_v, Z_u) - E \bigvee_{u \in [v,v']Q(m(h))} \widehat{d}(Z_v, Z_u) \leq 2^{-h+5},$$

where $h \geq 0$ and $v, v' \in Q_{m(h)}$ are arbitrary with $v \leq v'$.

Proof. 1. First let $k \geq 0$ and $v, \overline{v}, v' \in Q_{m(k)}$ be arbitrary with $v \leq \overline{v} \leq v'$. Consider each $\omega \in D_k^c$. We will prove that

$$0 \leq \bigvee_{u \in [v,v']Q(m(k+1))} d(Z_{\overline{v}}(\omega), Z_u(\omega)) - \bigvee_{u \in [v,v']Q(m(k))} d(Z_{\overline{v}}(\omega), Z_u(\omega)) \leq \beta_{k+1}. \tag{10.5.13}$$

To that end, consider each $r \in [v, v']Q_{m(k+1)}$. Then $r \in [s, s + \Delta_{m(k)}]Q_{m(k+1)}$ for some $s \in Q_{m(k)}$ such that $[s, s + \Delta_{m(k)}] \subset [v, v']$. Write $s' \equiv s + \Delta_{m(k)}$. We need to prove that

$$d(Z_{\overline{v}}(\omega), Z_r(\omega)) \leq \bigvee_{u \in [v,v']Q(m(k))} d(Z_{\overline{v}}(\omega), Z_u(\omega)) + \beta_{k+1}. \tag{10.5.14}$$

If $r = s$ or $r = s'$, then $r \in [v, v']Q_{m(k)}$ and inequality 10.5.14 holds trivially. Hence we may assume that $r \in (s, s')Q_{m(k+1)}$. Since $\omega \in D_k^c$ by assumption, the defining equality 10.5.9 implies that

$$\omega \in (A_{s,r}^{\beta(k+1)})^c (A_{s,r}^{\beta(k+1)})^c \cup (A_{s,r}^{\beta(k+1)})^c (A_{s',r}^{\beta(k+1)})^c \cup (A_{s',r}^{\beta(k+1)})^c (A_{s',r}^{\beta(k+1)})^c.$$

Consequently, by the defining equality 10.5.3 for the sets in the last displayed expression, we have

$$d(Z_s(\omega), Z_r(\omega)) \wedge d(Z_{s'}(\omega), Z_r(\omega)) \leq \beta_{k+1}.$$

Hence the triangle inequality implies that

$$\begin{aligned}
d(Z_{\overline{v}}(\omega), Z_r(\omega)) &\leq (d(Z_s(\omega), Z_r(\omega)) + d(Z_{\overline{v}}(\omega), Z_s(\omega))) \bigwedge (d(Z_{s'}(\omega), Z_r(\omega)) \\
&\quad + d(Z_{\overline{v}}(\omega), Z_{s'}(\omega))) \\
&\leq (d(Z_s(\omega), Z_r(\omega)) + \bigvee_{u \in [v,v']Q(m(k))} d(Z_{\overline{v}}(\omega), Z_u(\omega))) \\
&\quad \bigwedge (d(Z_{s'}(\omega), Z_r(\omega)) + \bigvee_{u \in [v,v']Q(m(k))} d(Z_{\overline{v}}(\omega), Z_u(\omega))) \\
&\leq (\beta_{k+1} + \bigvee_{u \in [v,v']Q(m(k))} d(Z_{\overline{v}}(\omega), Z_u(\omega))) \\
&\quad \bigwedge (\beta_{k+1} + \bigvee_{u \in [v,v']Q(m(k))} d(Z_{\overline{v}}(\omega), Z_u(\omega))) \\
&= \beta_{k+1} + \bigvee_{u \in [v,v']Q(m(k))} d(Z_{\overline{v}}(\omega), Z_u(\omega)),
\end{aligned}$$

establishing inequality 10.5.14 for arbitrary $r \in [v,v']Q_{m(k+1)}$. The desired inequality 10.5.13 follows.

2. Now let $h \geq 0$ and $v, \overline{v}, v' \in Q_{m(h)}$ be arbitrary as given, with $v \leq \overline{v} \leq v'$. Consider each $\omega \in D_{h+}^c$. Then $\omega \in D_k^c$ for each $k \geq h$. Hence, inequality 10.5.13 can be applied repeatedly to $h, h+1, \ldots, k+1$, to yield

$$0 \leq \bigvee_{u \in [v,v']Q(m(k+1))} d(Z_{\overline{v}}(\omega), Z_u(\omega)) - \bigvee_{u \in [v,v']Q(m(h))} d(Z_{\overline{v}}(\omega), Z_u(\omega))$$

$$\leq \beta_{k+1} + \cdots + \beta_{h+1} = \beta_{h+1,k+1} < \beta_{h+1,\infty} < 2^{-h+2}. \tag{10.5.15}$$

3. Consider each $r \in [v,v']Q_\infty$. We will prove that

$$d(Z_{\overline{v}}(\omega), Z_r(\omega)) \leq \bigvee_{u \in [v,v']Q(m(h))} d(Z_{\overline{v}}(\omega), Z_u(\omega)) + \widehat{\beta}_{h+1}(r). \tag{10.5.16}$$

This is trivial if $r \in Q_{m(h)}$. Hence we may assume that $r \in Q_{m(k+1)}Q_{m(k)}^-$ for some $k \geq h$. Then $\widehat{h}(r) \equiv k+1$ and $\widehat{\beta}_h(r) = \beta_{h+1,k+1}$. Therefore the first half of inequality 10.5.15 implies that

$$d(Z_{\overline{v}}(\omega), Z_r(\omega)) \leq \bigvee_{u \in [v,v']Q(m(h))} d(Z_{\overline{v}}(\omega), Z_u(\omega)) + \beta_{h+1,k+1}$$

$$= \bigvee_{u \in [v,v']Q(m(h))} d(Z_{\overline{v}}(\omega), Z_u(\omega)) + \widehat{\beta}_{h+1}(r).$$

Inequality 10.5.16 is proved, where $\omega \in D_{h+}^c$ is arbitrary. Inequality 10.5.12 follows. Assertion 1 is proved.

4. Next consider the special case where $\overline{v} = v$. Since $P(D_{h+}) \leq 2^{-h+1}$, it follows from inequality 10.5.15 that the a.u. limit

$$Y_{v,v'} \equiv \lim_{k \to \infty} \bigvee_{u \in [v,v']Q(m(k+1))} d(Z_v, Z_u)$$

exists and is an r.r.v. Moreover, for each $\omega \in domain(Y_{v,v'})$, it is easy to verify that $Y_{v,v'}(\omega)$ gives the supremum $\sup_{u \in [v,v']Q(\infty)} d(Z_v(\omega), Z_u(\omega))$. Thus this supremum is defined and equal to the r.r.v. $Y_{v,v'}$ on a full set, and is therefore itself an r.r.v. Letting $k \to \infty$ in inequality 10.5.15, we obtain

$$0 \leq Y_{v,v'} - \bigvee_{u \in [v,v']Q(m(h))} d(Z_v, Z_u) \leq \beta_{h+1,\infty} \leq 2^{-h+4} \tag{10.5.17}$$

on $D_{h+}^c \cap domain(Y_{v,v'})$. This proves Assertion 2.

5. Write $\widehat{d} \equiv 1 \wedge d$. Then

$$0 \leq E \sup_{u \in [v,v']Q(\infty)} \widehat{d}(Z_v, Z_u) - E \bigvee_{u \in [v,v']Q(m(h))} \widehat{d}(Z_v, Z_u)$$

$$= E(1 \wedge Y_{v,v'} - 1 \wedge \bigvee_{u \in [v,v']Q(m(h))} d(Z_v, Z_u))$$

$$\leq E1_{D(h+)^c}(1 \wedge Y_{v,v'} - 1 \wedge \bigvee_{u \in [v,v']Q(m(h))} d(Z_v, Z_u)) + E1_{D(h+)}$$

$$\leq 2^{-h+4} + P(D_{h+}) \leq 2^{-h+4} + 2^{-h+1} < 2^{-h+5}.$$

Assertion 3 and the lemma are proved. □

Definition 10.5.6. Right-limit extension of a process with dyadic rational parameters. Recall the convention that if f is an arbitrary function, we write $f(x)$ only with the implicit or explicit condition that $x \in domain(f)$.

1. Let Q_∞ stand for the set of dyadic rationals in $[0,1]$. Let $Y : Q_\infty \times \Omega \to S$ be an arbitrary process. Define a function $X : [0,1] \times \Omega \to S$ by

$$domain(X) \equiv \{(r,\omega) \in [0,1] \times \Omega : \lim_{u \to r; u \geq r} Y(u,\omega) \text{ exists}\}$$

and by

$$X(r,\omega) \equiv \lim_{u \to r; u \geq r} Y(u,\omega) \tag{10.5.18}$$

for each $(r,\omega) \in domain(X)$. We will call

$$\Phi_{rLim}(Y) \equiv X$$

the right-limit extension of the process Y to the parameter set $[0,1]$.

2. Let $\overline{Q}_\infty$ stand for the set of dyadic rationals in $[0,\infty)$. Let $Y : \overline{Q}_\infty \times \Omega \to S$ be an arbitrary process. Define a function $X : [0,\infty) \times \Omega \to S$ by

$$domain(X) \equiv \left\{(r,\omega) \in [0,\infty) \times \Omega : \lim_{u \to r; u \geq r} Y(u,\omega) \text{ exists}\right\}$$

and by

$$X(r,\omega) \equiv \lim_{u \to r; u \geq r} Y(u,\omega) \tag{10.5.19}$$

for each $(r,\omega) \in domain(X)$. We will call

$$\overline{\Phi}_{rLim}(Y) \equiv X$$

the right-limit extension of the process Y to the parameter set $[0,\infty)$.

In general, the right-limit extension X need not be a well-defined process. Indeed, it need not be a well-defined function at all. □

In the following proposition, recall, as remarked after Definition 6.1.2, that continuity a.u. is a weaker condition than a.u. continuity.

Proposition 10.5.7. The right-limit extension of a D-regular process is a well-defined stochastic process and is continuous a.u. *Let Z be the arbitrary D-regular process as specified in Definition 10.5.1, along with the related objects specified in Definitions 10.5.3 and 10.5.4. Let $X \equiv \Phi_{rLim}(Z) : [0,1] \times \Omega \to S$ denote the right-limit extension of the process Z. Then the following conditions hold:*

1. Let $\varepsilon > 0$ be arbitrary. Fix any $n \geq 0$ so large that $2^{-n+5} \leq \varepsilon$. Take an arbitrary $N \geq n$ so large that

$$\Delta_{m(N)} \equiv 2^{-m(N)} < 2^{-2}\delta_{Cp}(2^{-2n+2}). \tag{10.5.20}$$

Define

$$\delta_{cau}(\varepsilon) \equiv \delta_{cau}(\varepsilon, \overline{m}, \delta_{Cp}) \equiv \Delta_{m(N)} > 0. \tag{10.5.21}$$

Then, for each $t \in [0,1]$, there exists an exceptional set $G_{t,\varepsilon}$ with $P(G_{t,\varepsilon}) < \varepsilon$ such that

$$d(X(t,\omega), X(t',\omega)) < \varepsilon$$

for each $t' \in [t - \delta_{cau}(\varepsilon), t + \delta_{cau}(\varepsilon)] \cap domain(X(\cdot,\omega))$, for each $\omega \in G_{t,\varepsilon}^c$.

2. For each $t \in [0,1]$, there exists a null set H_t such that for each $\omega \in H_t^c$, we have $t \in domain(X(\cdot,\omega))$ and $X(t,\omega) = \lim_{u \to t} Z(u,\omega)$.

3. There exists a null set H such that for each $\omega \in H^c$, we have $Q_\infty \subset domain(X(\cdot,\omega))$ and $X(\cdot,\omega)|Q_\infty = Z(\cdot,\omega)$.

4. The function X is a stochastic process that is continuous a.u., with δ_{cau} as modulus of continuity a.u.

5. Furthermore, the process X has the same modulus of continuity in probability δ_{Cp} as the process Z.

6. (Right continuity.) For each $\omega \in H^c$, the function $X(\cdot,\omega)$ is right continuous at each $t \in domain(X(\cdot,\omega))$.

7. (Right completeness.) For each $\omega \in H^c$ and for each $t \in [0,1]$ such that $\lim_{r \to t; r \geq t} X(r,\omega)$ exists, we have $t \in domain(X(\cdot,\omega))$.

Proof. 1. Let $\varepsilon > 0$ be arbitrary. Fix $n \geq 0$ and $N \geq n$ as in the hypothesis of Assertion 1. Write $\Delta \equiv \Delta_{m(N)} \equiv \delta_{cau}(\varepsilon)$. Recall that δ_{Cp} is the given modulus of continuity in probability of the process Z, and recall that $\beta_n \in (2^{-n+1}, 2^{-n+2})$ as in Definition 10.5.3. When there is little risk of confusion, suppress the subscript m_N, write $p \equiv p_{m(N)} \equiv 2^{m(N)}$, and write $q_i \equiv q_{m(N),i} \equiv i\Delta \equiv i2^{-m(N)}$ for each $i = 0, 1, \ldots, p$.

2. Consider each $t \in [0,1]$. Then there exists $i = 0, \ldots, p-2$ such that $t \in [q_i, q_{i+2}]$. The neighborhood $\theta_{t,\varepsilon} \equiv [t - \Delta, t + \Delta] \cap [0,1]$ of t in $[0,1]$ is a subset of $[q_{(i-1)\vee 0}, q_{(i+3)\wedge p}]$. Write $v \equiv v_t \equiv q_{(i-1)\vee 0}$ and $v' \equiv q_{(i+3)\wedge p}$. Then (i) $v, v' \in Q_{m(N)}$, (ii) $v < v'$, and (iii) the set

$$[v,v']Q_{m(N)} = \{q_{(i-1)\vee 0}, q_i, q_{i+1}, q_{i+2}, q_{(i+3)\wedge p}\}$$

contains four or five distinct and consecutive elements of $Q_{m(N)}$. Therefore, for each $u \in [v,v']Q_{m(N)}$, we have $|v - u| \leq 4\Delta < \delta_{Cp}(2^{-2n+2})$, whence

$$E1 \wedge d(Z_v, Z_u)) \leq 2^{-2n+2} < \beta_n^2$$

and, by Chebychev's inequality,

$$P(d(Z_v, Z_u) > \beta_n) \leq \beta_n.$$

Hence the measurable set

$$A_n \equiv A_{n,t} \equiv \bigcup_{u\in[v,v']Q(m(N))} (d(Z_v,Z_u) > \beta_n)$$

has probability bounded by $P(A_n) \leq 4\beta_n < 2^{-n+4.}$. Define $G_{t,\varepsilon} \equiv D_{n+} \cup A_{n,t}$. It follows that

$$P(G_{t,\varepsilon}) \leq P(D_{n+}) + P(A_n) < 2^{-n+1} + 2^{-n+4} < 2^{-n+5} \leq \varepsilon. \qquad (10.5.22)$$

3. Consider each $\omega \in G_{t,\varepsilon}^c = D_{n+}^c A_n^c$. Then, by the definition of the set A_n, we have

$$\bigvee_{u\in[v,v']Q(m(N))} d(Z(v,\omega),Z(u,\omega)) \leq \beta_n. \qquad (10.5.23)$$

At the same time, since $\omega \in D_{n+}^c \subset D_{N+}^c$, inequality 10.5.12 of Lemma 10.5.5 – where $h,\overline{v}$ are replaced by N,v, respectively – implies that, for each $r \in [v,v']Q_\infty$, we have

$$d(Z_r(\omega),Z_v(\omega)) \leq \widehat{\beta}_{N+1}(r) + \bigvee_{u\in[v,v']Q(m(N))} d(Z_v(\omega),Z_u(\omega))$$

$$= \widehat{\beta}_{N+1}(r) + \beta_n \leq \beta_{N+1,\infty} + \beta_n \leq \beta_{n+1,\infty} + \beta_n = \beta_{n,\infty},$$

where the second inequality is by inequality 10.5.23. Then the triangle inequality yields

$$d(Z(r,\omega),Z(r',\omega)) \leq 2\beta_{n,\infty} < 2^{-n+4} < \varepsilon \qquad (10.5.24)$$

for each $r,r' \in \theta_{t,\varepsilon}Q_\infty \subset [v,v']Q_\infty$, where $\omega \in G_{t,\varepsilon}^c$ is arbitrary.

4. Now write $\varepsilon_k \equiv 2^{-k}$ for each $k \geq 1$. Then the measurable set $G_\kappa^{(t)} \equiv \bigcup_{k=\kappa}^\infty G_{t,\varepsilon(k)}$ has probability bounded by $P(G_\kappa^{(t)}) \leq \sum_{k=\kappa}^\infty \varepsilon_k = 2^{-\kappa+1}$. Hence $H_t \equiv \bigcap_{\kappa=1}^\infty G_\kappa^{(t)}$ is a null set. Consider each $\omega \in H_t^c$. Then $\omega \in (G_\kappa^{(t)})^c$ for some $\kappa \geq 1$. Hence $\omega \in G_{t,\varepsilon(k)}^c$ for each $k \geq \kappa$. Inequality10.5.24 therefore implies that $\lim_{u\to t} Z(u,\omega)$ exists, whence the right limit $X(t,\omega)$ is well defined, with

$$X(t,\omega) \equiv \lim_{u\to t;u\geq t} Z(u,\omega) = \lim_{u\to t} Z(u,\omega). \qquad (10.5.25)$$

In short, $t \in domain(X(\cdot,\omega))$. This verifies Assertion 2. Assertion 1 then follows from inequality 10.5.24 by the right continuity of X.

5. Define the null set $H \equiv \bigcup_{u\in Q(\infty)} H_u$. Consider each $\omega \in H^c$ and each $u \in Q_\infty$. Then $\omega \in H_u^c$. Hence Assertion 2 implies that $u \in domain(X(\cdot,\omega))$, and that $X(u,\omega) = \lim_{v\to u} Z(v,\omega) = Z(u,\omega)$. In short, $X(\cdot,\omega)|Q_\infty = Z(\cdot,\omega)$ and $X(\cdot,\omega)|Q_\infty = Z(\cdot,\omega)$. Assertion 3 is proved.

6. For each $k \geq 1$, fix an arbitrary $r_k \in \theta_{t,\varepsilon(k)}Q_\infty$. Consider each $\kappa \geq 1$ and each $\omega \in G_\kappa^c$. Then for each $k \geq \kappa$, we have $r_k, r_\kappa \in \theta_{t,\varepsilon(\kappa)}Q_\infty$. So by inequality 10.5.24, where ε is replaced by ε_κ, we obtain

$$d(Z(r_k,\omega),Z(r_\kappa,\omega)) \leq \varepsilon_\kappa. \qquad (10.5.26)$$

Letting $k \to \infty$, this yields

$$d(X(t,\omega), Z(r_\kappa,\omega)) \leq \varepsilon_\kappa \equiv 2^{-k}, \tag{10.5.27}$$

where $\omega \in G_\kappa^c$ is arbitrary. Since $P(G_\kappa) \leq \varepsilon_\kappa \equiv 2^{-k}$, we conclude that $Z_{r(\kappa)} \to X_t$ a.u. Consequently, the function X_t is an r.v., where $t \in [0,1]$ is arbitrary. Thus the function $X \equiv \Phi_{rLim}(Z) : [0,1] \times \Omega \to S$ is a stochastic process.

7. Let $t' \in \theta_{t,\varepsilon} \equiv [t - \delta_{cau}(\varepsilon), t + \delta_{cau}(\varepsilon)] \cap domain(X(\cdot,\omega))$ be arbitrary. Letting $r \downarrow t$ and $r' \downarrow t'$ in inequality 10.5.24 while $r, r' \in \theta_{t,\varepsilon} Q_\infty$, we obtain $d(X(t,\omega), X(t',\omega)) < \varepsilon$, where $\omega \in G_{t,\varepsilon}^c$ is arbitrary. Since $t \in [0,1]$ is arbitrary, and since $P(G_{t,\varepsilon}) < \varepsilon$ is arbitrarily small, we see that the process X is continuous a.u. according to Definition 6.1.2, with δ_{cau} as a modulus of continuity a.u. Assertion 4 has been proved.

8. Next, we will verify that the process X has the same modulus of continuity in probability δ_{Cp} as the process Z. To that end, let $\varepsilon > 0$ be arbitrary, and let $t, s \in [0,1]$ be such that $|t - s| < \delta_{Cp}(\varepsilon)$. In Step 6 we saw that there exist sequences $(r_k)_{k=1,2,\ldots}$ and $(v_k)_{k=1,2,\ldots}$ in Q_∞ such that $r_k \to t$, $v_k \to s$, $Z_{r(k)} \to X_t$ a.u. and $Z_{v(k)} \to X_s$ a.u. Then, for sufficiently large $k \geq 0$ we have $|r_k - v_k| < \delta_{Cp}(\varepsilon)$, whence $E1 \wedge d(Z_{r(k)}, Z_{v(k)}) \leq \varepsilon$. The last cited a.u. convergence therefore implies that $E1 \wedge d(X_t, X_s) \leq \varepsilon$. Summing up, δ_{Cp} is a modulus of continuity of probability of the process X. Assertion 5 is proved.

9. Let $\omega \in H^c$ be arbitrary. Then $domain(Z(\cdot,\omega)) = Q_\infty$. Consider each $u \in Q_\infty$. Step 5 says that $u \in domain(X(\cdot,\omega))$ and that the function $Z(\cdot,\omega)$ is right continuous at u. Hence Proposition 10.1.6, where the functions $x, \overline{x}$ are replaced by $Z(\cdot,\omega)$ and $X(\cdot,\omega)$, respectively, implies that the function $X(\cdot,\omega)$ is right continuous at each $t \in domain(X(\cdot,\omega))$, and that for each $t \in [0,1]$ such that $\lim_{r \to t; r \geq t} X(r,\omega)$ exists, we have $t \in domain(X(\cdot,\omega))$. Assertions 6 and 7 have been proved. □

We now prove the main theorem of this section.

Theorem 10.5.8. The right-limit extension of a D-regular process is a.u. càdlàg. *Let Z be the arbitrary D-regular process as specified in Definition 10.5.1, along with the related objects in Definitions 10.5.3 and 10.5.4. Then the right-limit extension*

$$X \equiv \Phi_{rLim}(Z) : [0,1] \times \Omega \to S$$

is a.u. càdlàg.

Specifically, (i) *the process X has the same modulus of continuity in probability δ_{Cp} as the given D-regular process Z and* (ii) *it has a modulus of a.u. càdlàg $\delta_{aucl}(\cdot, \overline{m}, \delta_{Cp})$ defined as follows. Let $\varepsilon > 0$ be arbitrary. Let $n \geq 0$ be so large that $2^{-n+9} < \varepsilon$. Let $N \geq m_n + n + 6$ be so large that*

$$\Delta_{m(N)} \equiv 2^{-m(N)} < 2^{-2}\delta_{Cp}(2^{-2m(n)-2n-10}). \tag{10.5.28}$$

Define $\delta_{aucl}(\varepsilon, \overline{m}, \delta_{Cp}) \equiv \Delta_{m(N)}$. We emphasize that the operation $\delta_{aucl}(\cdot, \overline{m}, \delta_{Cp})$ depends only on $\overline{m}$ and δ_{Cp}.

Proof. We will refer to Definitions 10.5.3 and 10.5.4 for the properties of the objects $(\beta_h)_{h=0,1,...}, A_{r,s}^{\beta(h)}, \widehat{h}, \widehat{\beta}_n$ and D_k, D_{k+} relative to the D-regular process Z. Assertion 3 of Proposition 10.5.7 says that there exists a null set H such that $X(\cdot,\omega)|Q_\infty = Z(\cdot,\omega)$ for each $\omega \in H^c$. Assertion 6 of Proposition 10.5.7 says that for each $\omega \in H^c$, the function $X(\cdot,\omega)$ is right continuous at each $t \in domain$ $(X(\cdot,\omega))$. Moreover, Assertions 4 and 5 of Proposition 10.5.7 say that X is continuous a.u., with some modulus of continuity a.u.. δ_{cau}, and is continuous in probability with the same modulus of continuity in probability δ_{Cp} as the process Z. Refer to Proposition 8.1.13 for basic properties of the simple first exit times.

1. To start, let $\varepsilon > 0$ be arbitrary but fixed. Then let $n, N,$and $\delta_{aucl}(\varepsilon,\overline{m},\delta_{Cp})$ be fixed as described in the hypothesis. Let $i = 1,\ldots,p_{m(n)}$ be arbitrary but fixed until further notice. When there is little risk of confusion, we suppress references to n and i, and write simply

$$p \equiv p_{m(n)} \equiv 2^{m(n)},$$

$$\Delta \equiv \Delta_{m(n)} \equiv 2^{-m(n)},$$

$$t \equiv q_{i-1} \equiv q_{m(n),i-1} \equiv (i-1)2^{-m(n)},$$

and

$$t' \equiv q_i \equiv q_{m(n),i} \equiv i2^{-m(n)}.$$

Thus $t,t' \in Q_{m(n)}$ and $0 \le t < t' = t + \Delta \le 1$.

2. With n fixed, write $\varepsilon_n = 2^{-m(n)-n-1}$ and $\nu \equiv m_n + n + 6$. Then $N \ge \nu$, $2^{-\nu+5} = \varepsilon_n$ and

$$\Delta_{m(N)} \equiv 2^{-m(N)} < 2^{-2}\delta_{Cp}(2^{-2\nu+2}). \tag{10.5.29}$$

Hence

$$\delta_{cau}(\varepsilon_n,\overline{m},\delta_{Cp}) = \Delta_{m(N)},$$

where $\delta_{cau}(\cdot,\overline{m},\delta_{Cp})$ is the modulus of continuity a.u. of the process X defined in Proposition 10.5.7. Note that $\varepsilon_n \le \beta_n \in (2^{-n+1},2^{-n+2})$, where we recall the sequence $(\beta_h)_{h=0,1,...}$ specified relative to the process Z in Definition 10.5.3.

In the next several steps, we will prove that, except on a set of small probability, the set $Z([t,t']Q_\infty,\omega)$ is the union of two subsets $Z([t,\tau(\omega))Q_\infty,\omega)$ and $Z([\tau(\omega),t']Q_\infty,\omega)$, each of which is contained in a ball in (S,d) with small radius, where τ is some r.r.v. Recall here the notations from Definition 10.5.2 for the range of the sample function $Z(\cdot,\omega)$.

3. First introduce some simple first exit times of the process Z. Let $k \ge n$ be arbitrary. As in Definition 8.1.12, define the simple first exit time

$$\eta_k \equiv \eta_{k,i} \equiv \eta_{t,\widehat{\beta}(n),[t,t']Q(m(k))} \tag{10.5.30}$$

for the process $Z|[t,t']Q_{m(k)}$ to exit the time-varying $\widehat{\beta}_n$-neighborhood of Z_t. Note that the r.r.v. η_k has values in $[t+\Delta_{m(k)},t']Q_{m(k)}$. Thus

$$t+\Delta_{m(k)} \leq \eta_k \leq t'. \qquad (10.5.31)$$

In the case where $k=n$, this yields

$$\eta_n \equiv \eta_{n,i} = t' = q_{m(n),i}. \qquad (10.5.32)$$

Since $Q_{m(k)} \subset Q_{m(k+1)}$, the more frequently sampled simple first exit time η_{k+1} comes no later than η_k, according to Assertion 5 of Proposition 8.1.13. Hence

$$\eta_{k+1} \leq \eta_k. \qquad (10.5.33)$$

4. Let $\kappa \geq k \geq n$ be arbitrary. Consider each $\omega \in D^c_{k+} \subset D^c_{\kappa+}$. We will prove that

$$t \leq \eta_k(\omega) - 2^{-m(k)} \leq \eta_\kappa(\omega) - 2^{-m(\kappa)} \leq \eta_\kappa(\omega) \leq \eta_k(\omega). \qquad (10.5.34)$$

The first of these inequalities is from the first part of inequality 10.5.31. The third is trivial, and the last is by repeated applications of inequality 10.5.33. It remains only to prove the second inequality. To that end, write, as an abbreviation, $\alpha \equiv t$, $s \equiv \eta_k(\omega)$ and $\alpha' \equiv s - \Delta_{m(k)}$. Then $\alpha,t,s,\alpha' \in [t,t']Q_{m(k)}$.

Since $\alpha' < s \equiv \eta_k(\omega)$, the sample path $Z(\cdot,\omega)|[t,t']Q_{m(k)}$ has not exited the time-varying $\widehat{\beta}_n$-neighborhood of $Z_t(\omega)$ at time α'. In other words,

$$d(Z_t(\omega),Z_u(\omega)) \leq \widehat{\beta}_n(u) \leq \beta_{n,k} \qquad (10.5.35)$$

for each $u \in [t,\alpha']Q_{m(k)}$, where the last inequality is from equality 10.5.7 in Definition 10.5.3.

Next consider each $u \in [t,\alpha']Q_{m(\kappa)}Q^-_{m(k)}$, where $Q^-_{m(k)}$ is the metric complement of $Q_{m(k)}$. Then inequality 10.5.12 of Lemma 10.5.5, where $h,v,\overline{v},v',u,r$ are replaced by k,t,t,α',w,u, respectively, implies that

$$\begin{aligned} d(Z_t(\omega),Z_u(\omega)) &\leq \bigvee_{w\in[t,\alpha']Q(m(k))} d(Z_t(\omega),Z_w(\omega)) + \widehat{\beta}_{k+1}(u) \\ &\leq \beta_{n,k} + \widehat{\beta}_{k+1}(u) = \widehat{\beta}_n(u), \end{aligned}$$

Here the second inequality is from inequality 10.5.35 and the equality is thanks to the first half of inequality 10.5.6, while $u \in [t,\alpha']Q_{m(\kappa)}Q^-_{m(k)}$ is arbitrary. We can combine the last displayed inequality with inequality 10.5.35 to obtain

$$d(Z_t(\omega),Z_u(\omega)) \leq \widehat{\beta}_n(u) \qquad (10.5.36)$$

for each $u \in [t,\alpha']Q_{m(\kappa)}$. Thus the sample path $Z(\cdot,\omega)|[t,t']Q_{m(\kappa)}$ has not exited the time-varying $\widehat{\beta}_n$-neighborhood of $Z(t,\omega)$ at time $\alpha' \equiv \eta_k(\omega) - \Delta_{m(k)}$. In other words,

$$\eta_k(\omega) - \Delta_{m(k)} < \eta_\kappa(\omega).$$

Since both sides of this strict inequality are members of $Q_{m(\kappa)}$, it follows that

$$\eta_k(\omega) - \Delta_{m(k)} \leq \eta_\kappa(\omega) - \Delta_{m(\kappa)}.$$

Thus inequality 10.5.34 has been verified.

5. Inequality 10.5.34 immediately implies that $0 \leq \eta_k(\omega) - \eta_\kappa(\omega) \leq 2^{-m(k)}$, where $\kappa \geq k \geq n$ are arbitrary, and where $\omega \in D_{k+}^c$ is arbitrary. Hence the limit

$$\tau \equiv \tau_i \equiv \lim_{\kappa \to \infty} \eta_\kappa$$

exists uniformly on D_{k+}^c. Moreover, with $\kappa \to \infty$, inequality 10.5.34 implies that

$$t \leq \eta_k - 2^{-m(k)} \leq \tau \leq \eta_k \leq t' \tag{10.5.37}$$

on D_{k+}^c. Since $P(D_{k+}) \leq 2^{-k+3}$ is arbitrarily small, we conclude that

$$\eta_\kappa \downarrow \tau \quad \text{a.u.}$$

Therefore $\tau \equiv \tau_i$ is an r.r.v., with $t < \tau \leq t'$. Recall that $t = q_{i-1}$ and $t' = q_i$. This last equality can be rewritten as

$$q_{i-1} < \tau_i \leq q_i. \tag{10.5.38}$$

6. Now let $h \geq n$ be arbitrary, and let $\omega \in D_{h+}^c$ be arbitrary. Consider each $u \in [t, \tau(\omega))Q_\infty$. Then $u \in [t, \eta_k(\omega))Q_{m(k)}$ for some $k \geq h$. Hence, by the basic properties of the simple first exit time η_k, we have

$$d(Z(t,\omega), Z(u,\omega)) \leq \widehat{\beta}_n(u) < \beta_{n,\infty}. \tag{10.5.39}$$

Since $u \in [t, \tau(\omega))Q_\infty$ is arbitrary, inequality 10.5.39 implies

$$\begin{aligned} Z([q_{i-1}, \tau_i(\omega))Q_\infty, \omega) &\subset \{x \in S : d(x, Z(q_{i-1},\omega)) < \beta_{n,\infty}\} \\ &\subset \{x \in S : d(x, Z(q_{i-1},\omega)) < 2^{-n+3}\}. \end{aligned} \tag{10.5.40}$$

7. To obtain a similar bounding relation for the set $Z([\tau_i(\omega), q_i], \omega)$, we will first prove that

$$d(Z(w,\omega), Z(\eta_h(\omega),\omega)) \leq \beta_{h+1,\infty} \tag{10.5.41}$$

for each $w \in [\eta_{h+1}(\omega), \eta_h(\omega)]Q_\infty$. To that end, write, as an abbreviation, $u \equiv \eta_h(\omega) - \Delta_{m(h)}$, $r \equiv \eta_{h+1}(\omega)$, and $u' \equiv \eta_h(\omega)$. From inequality 10.5.34, where k, κ are replaced by $h, h+1$, respectively, we obtain $u < r \leq u'$. The desired inequality 10.5.41 holds trivially if $r = u'$. Hence we may assume that $u < r < u'$. Consequently, since u, u' are consecutive points in the set $Q_{m(h)}$ of dyadic rationals, we have $r \in Q_{m(h+1)}Q_{m(h)}^-$, whence $\widehat{\beta}_n(r) = \beta_{n,h+1}$ according to Definition 10.5.3. Moreover, since $u' \leq t'$ according to inequality 10.5.31, we have

$$\eta_{h+1}(\omega) \equiv r < u' \leq t'.$$

In words, the sample path $Z(\cdot,\omega)|[t,t']Q_{m(h+1)}$ successfully exits the time-varying $\widehat{\beta}_n$-neighborhood of $Z(t,\omega)$, for the first time at r. Therefore

$$d(Z(t,\omega),Z(r,\omega)) > \widehat{\beta}_n(r) = \beta_{n,h+1}. \tag{10.5.42}$$

However, since $u \in Q_{m(h)} \subset Q_{m(h+1)}$ and $u < r$, exit has not occurred at time u. Hence

$$d(Z(t,\omega),Z(u,\omega)) \le \widehat{\beta}_n(u) \le \beta_{n,h}. \tag{10.5.43}$$

Inequalities 10.5.42 and 10.5.43 together yield, by the triangle inequality,

$$d(Z(u,\omega),Z(r,\omega)) > \beta_{n,h+1} - \beta_{n,h} = \beta_{h+1}.$$

In other words,

$$\omega \in A_{u,r}^{\beta(h+1)} \equiv (d(Z_u,Z_r) > \beta_{h+1}), \tag{10.5.44}$$

where the equality is the defining equality 10.5.3 in Definition 10.5.3.

Now consider an arbitrary $s \in [r,u')Q_{m(h+1)}$. Then relation 10.5.44 implies, trivially, that

$$\omega \in A_{u,r}^{\beta(h+1)} \subset (A_{u,r}^{\beta(h+1)} \cup A_{u,s}^{\beta(h+1)})(A_{u,r}^{\beta(h+1)} \cup A_{u',s}^{\beta(h+1)}). \tag{10.5.45}$$

At the same time, $r,s \in (u,u')Q_{m(h+1)}$ with $r \le s$. Since $\omega \in D_{h+}^c \subset D_h^c$ and since $u,u' \in Q_{m(h)}$ with $u' \equiv u+\Delta_{m(h)}$, we can apply the defining formula 10.5.9 of the exceptional set D_h to obtain

$$\omega \in D_h^c \subset (A_{u,r}^{\beta(h+1)} \cup A_{u,s}^{\beta(h+1)})^c \cup (A_{u,r}^{\beta(h+1)} \cup A_{u',s}^{\beta(h+1)})^c \cup (A_{u',r}^{\beta(h+1)} \cup A_{u',s}^{\beta(h+1)})^c. \tag{10.5.46}$$

Relations 10.5.45 and 10.5.46 together then imply that

$$\omega \in (A_{u',r}^{\beta(h+1)} \cup A_{u',s}^{\beta(h+1)})^c \subset (A_{u',s}^{\beta(h+1)})^c.$$

Consequently, by the definition of the set $A_{u',s}^{\beta(h+1)}$, we have

$$d(Z(s,\omega),Z(u',\omega)) \le \beta_{h+1}, \tag{10.5.47}$$

where $s \in [r,u')Q_{m(h+1)}$ is arbitrary. Inequality 10.5.47 trivially holds for $s = u'$. Summing up,

$$\bigvee_{s\in[r,u']Q(m(h+1))} d(Z_{u'}(\omega),Z_s(\omega)) \le \beta_{h+1}. \tag{10.5.48}$$

Then, for each $w \in [r,u']Q_\infty \equiv [\eta_{h+1}(\omega),\eta_h(\omega)]Q_\infty$, we can apply inequality 10.5.12 of Lemma 10.5.5, where $h,v,v',\overline{v},r,u$ are replaced by $h+1,r,u',u',w,s$, respectively, to obtain

$$\begin{aligned} d(Z_{u'}(\omega),Z_w(\omega)) &\le \widehat{\beta}_{h+2}(w) + \bigvee_{s\in[r,u']Q(m(h+1))} d(Z_{u'}(\omega),Z_s(\omega)) \\ &\le \widehat{\beta}_{h+2}(w) + \beta_{h+1} \le \beta_{h+2,\infty} + \beta_{h+1} = \beta_{h+1,\infty}. \end{aligned} \tag{10.5.49}$$

Here, the second inequality is by inequality 10.5.48 and the third inequality is due to inequality 10.5.6 in Definition 10.5.3. In other words, inequality 10.5.41 is verified.

8. Proceed to prove that X_τ is a well-defined r.v. To that end, let $r' \in (\tau(\omega), \eta_h(\omega)]Q_\infty$ be arbitrary. Then $r' \in [\eta_{k+1}(\omega), \eta_k(\omega)]Q_\infty$ for some $k \geq h$. Hence

$$\begin{aligned} d(Z(r',\omega), Z(\eta_h(\omega),\omega)) &\leq d(Z(\eta_{k+1}(\omega),\omega), Z(r',\omega)) \\ &\quad + d(Z(\eta_{k+1}(\omega),\omega), Z(\eta_k(\omega),\omega)) \\ &\quad + \cdots + d(Z(\eta_{h+1}(\omega),\omega), Z(\eta_h(\omega),\omega) \\ &\leq 2\beta_{k+1,\infty} + \cdots + 2\beta_{h+1,\infty} < 2^{-h+5}, \end{aligned} \tag{10.5.50}$$

where the second inequality is by repeated applications of inequality 10.5.41, where $r' \in (\tau(\omega), \eta_h(\omega)]Q_\infty$, $\omega \in D_{h+}^c$, and $h \geq n$ are arbitrary.

9. Continuing, let $k \geq h$ be arbitrary. Define $\varepsilon'_k \equiv 2^{-k+6-m(k)}$. Then, by Assertion 1 of Proposition 10.5.7, there exists

$$\delta_k \equiv \delta_{cau}(\varepsilon'_k, \overline{m}, \delta_{Cp}) > 0 \tag{10.5.51}$$

such that for each $u \in [0,1]$, there exists an exceptional set $G_{u,\varepsilon'(k)}$ with $P(G_{u,\varepsilon'(k)}) < \varepsilon'_k$ such that

$$d(X(u,\omega), X(u',\omega)) < \varepsilon'_k \tag{10.5.52}$$

for each $u' \in [u - \delta_k, u + \delta_k] \cap domain(X(\cdot,\omega))$, for each $\omega \in G^c_{u,\varepsilon'(k)}$. Define the exceptional sets

$$B_k \equiv \bigcup_{u \in Q(m(k))} G_{u,\varepsilon'(k)} \tag{10.5.53}$$

and

$$B_{k+} \equiv \bigcup_{j=k}^{\infty} B_j. \tag{10.5.54}$$

Then

$$P(B_k) \leq \sum_{u \in Q(m(k))} \varepsilon'_k \equiv \sum_{u \in Q(m(k))} 2^{-k+6-m(k)} = (2^{m(k)} + 1)2^{-k+6-m(k)} \leq 2^{-k+7} \tag{10.5.55}$$

and

$$P(B_{k+}) \leq 2^{-k+8}. \tag{10.5.56}$$

Separately, define the full set

$$C \equiv \bigcap_{j=n}^{\infty} \bigcup_{u \in Q(m(j))} (\eta_j = u).$$

Then $P(C^c \cup B_{k+} \cup D_{k+}) < 2^{-k+8} + 2^{-k+4} < 2^{-k+9}$.

10. Next, with $k \geq h \geq n$ arbitrary, consider each $\omega \in CB_{h+}^c D_{h+}^c$. Then $\omega \in CB_{k+}^c D_{k+}^c \subset C$. Hence there exists $u \in Q_{m(k)}$ such that $\eta_k(\omega) = u$. Let $r' \in (\tau(\omega), \eta_k(\omega) + \delta_k)Q_\infty$ be arbitrary. Then either (i′) $r' \in (\tau(\omega), \eta_k(\omega)]Q_\infty$ or (ii′) $r' \in [\eta_k(\omega), \eta_k(\omega) + \delta_k)Q_\infty$.

In case (i′), we have

$$d(Z(r',\omega), Z(\eta_k(\omega),\omega)) < 2^{-k+5}, \tag{10.5.57}$$

according to inequality 10.5.50, where h is replaced by k. Consider case (ii′). Then $r' \in [u, u + \delta_k)Q_\infty$. At the same time, $\omega \in B_k^c \subset G_{u,\varepsilon'(k)}^c$. Hence

$$\begin{aligned} d(Z(\eta_k(\omega),\omega), Z(r',\omega)) &= d(Z(u,\omega), Z(r',\omega)) \\ &= d(X(u,\omega), X(r',\omega)) < \varepsilon'_k \equiv 2^{-k+6-m(k)} \leq 2^{-k+5}, \end{aligned}$$

where the first inequality is according to inequality 10.5.52. Summing up, in each of cases (i′) and (ii′), inequality 10.5.57 holds, where $r' \in (\tau(\omega), \eta_k(\omega) + \delta_k)Q_\infty$ is arbitrary. Therefore the triangle inequality yields

$$d(Z(r',\omega), Z(r'',\omega)) < 2^{-k+6} \tag{10.5.58}$$

for each $r', r'' \in (\tau(\omega), \eta_k(\omega) + \delta_k)Q_\infty$.

11. Now let $s', s'' \in [\tau(\omega), \eta_k(\omega) + \delta_k)Q_\infty$ be arbitrary. Let $(r'_j)_{j=1,2,\ldots}$ and $(r''_j)_{j=1,2,\ldots}$ be two sequences in $(\tau(\omega), \eta_k(\omega) + \delta_k)Q_\infty$ such that $r'_j \downarrow s'$ and $r''_j \downarrow s''$. Then

$$d(X(r'_j,\omega), X(r''_j,\omega)) = d(Z(r'_j,\omega), Z(r''_j,\omega)) < 2^{-k+6} \tag{10.5.59}$$

for each $j \geq 1$. At the same time, the process X is right continuous, according to Proposition 10.5.7. It follows that

$$d(X(s',\omega), X(s'',\omega)) \leq 2^{-k+6}, \tag{10.5.60}$$

where $s', s'' \in [\tau(\omega), \eta_k(\omega) + \delta_k)Q_\infty$ are arbitrary. Consequently,

$$d(Z(s',\omega), Z(s'',\omega)) = d(X(s',\omega), X(s'',\omega)) \leq 2^{-k+6}, \tag{10.5.61}$$

where $s', s'' \in [\tau(\omega), \tau(\omega) + \delta_k)Q_\infty$ are arbitrary, where $\omega \in C_h B_{h+}^c D_{h+}^c$ and $k \geq h$ are arbitrary. Thus $\lim_{s' \to \tau(\omega); s' \geq \tau(\omega)} Z(s',\omega)$ exists. In other words, $\tau(\omega) \in domain(X(\cdot,\omega))$, with $X(\tau(\omega),\omega) \equiv \lim_{s' \to \tau(\omega); s' \geq \tau(\omega)} Z(s',\omega)$. Equivalently, $\omega \in domain(X_\tau)$, with $X_\tau(\omega) \equiv \lim_{s' \to \tau(\omega); s' \geq \tau(\omega)} Z_{s'}(\omega)$. Moreover, setting $s' \equiv \eta_k(\omega)$ and letting $s'' \downarrow \tau(\omega)$ in inequality 10.5.60, we obtain

$$d(Z_{\eta(k)}(\omega), X_\tau(\omega)) \leq 2^{-k+6}. \tag{10.5.62}$$

Similarly, letting $s'' \downarrow \tau(\omega)$ in relation 10.5.60, we obtain

$$d(X_{s'}(\omega), X_\tau(\omega)) \leq 2^{-k+6}, \tag{10.5.63}$$

for each $s' \in [\tau(\omega), \tau(\omega) + \delta_k)Q_\infty$, where $\omega \in CB_{h+}^c D_{h+}^c$ is arbitrary.

12. Since $P(C^c \cup B_{h+} \cup D_{h+}) < 2^{-h+9}$ for each $h \geq n$, it follows that $Z_{\eta(k)} \to X_\tau$ a.u. Hence X_τ is a well-defined r.v., where $\tau \equiv \tau_i$, and where $i = 1, \ldots, p \equiv 2^{m(n)}$ is arbitrary.

13. Continue with $\varepsilon > 0$ arbitrary, and n, i, p, Δ, t, t' fixed accordingly. Consider the special case where $h = k = n$. Let $\omega \in CB_{n+}^c D_{n+}^c$ be arbitrary. By the defining equality 10.5.30 of $\eta_n \equiv \eta_{n,i}$, we see that the r.r.v. η_n has values in $(t,t']Q_{m(n)} = \{t'\}$, where the last equality is because $t \equiv q_{i-1}$ and $t' \equiv q_i$ are consecutive members of $Q_{m(n)}$. Therefore

$$[\tau(\omega), q_i]Q_\infty = [\tau(\omega), \eta_n(\omega)]Q_\infty \subset [\tau(\omega), \tau(\omega) + \delta_n)Q_\infty \cup (\tau(\omega), \eta_n(\omega)]Q_\infty.$$

According to inequality 10.5.62, we have

$$d(Z_{\eta(n)}(\omega), X_\tau(\omega)) \leq 2^{-n+6},$$

while, according to 10.5.63, we have

$$d(X_{s'}(\omega), X_\tau(\omega)) \leq 2^{-n+6}$$

for each $s' \in [\tau(\omega), \tau(\omega) + \delta_n)Q_\infty$. Combining, we obtain

$$d(X_{s'}(\omega), Z_{\eta(n)}(\omega)) \leq 2^{-n+7} \tag{10.5.64}$$

for each $s' \in [\tau(\omega), \tau(\omega) + \delta_n)Q_\infty$. Hence

$$\begin{aligned} Z([\tau_i(\omega), q_i]Q_\infty, \omega) &\subset Z([\tau(\omega), \tau(\omega) + \delta_n)Q_\infty, \omega) \cup Z((\tau(\omega), \eta_n(\omega)]Q_\infty, \omega) \\ &\subset \{x \in S : d(x, Z_{\eta(n)}(\omega)) \leq 2^{-n+8}\} \\ &= \{x \in S : d(x, Z(q_i, \omega)) \leq 2^{-n+8}\}, \end{aligned} \tag{10.5.65}$$

where $\omega \in CB_{n+}^c D_{n+}^c$ is arbitrary, and where the second containment relation is thanks to inequalities 10.5.64 and 10.5.50.

14. For convenience, define the constant r.r.v.'s $\tau_0 \equiv 0$ and $\tau_{p+1} \equiv 1$. Then relations 10.5.65 and 10.5.40 can be combined to yield

$$\begin{aligned} Z([\tau_{i-1}(\omega), \tau_i(\omega))Q_\infty, \omega) &= Z([\tau_{i-1}(\omega), q_{i-1}]Q_\infty, \omega) \cup Z([q_{i-1}, \tau_i(\omega))Q_\infty, \omega) \\ &\subset \{x \in S : d(x, Z(q_{i-1}, \omega)) \leq 2^{-n+8}\}. \end{aligned}$$

In addition, relation 10.5.65 gives

$$\begin{aligned} &Z([\tau_p(\omega), \tau_{p+1}(\omega)]Q_\infty, \omega) \\ &\quad = Z([\tau_p(\omega), q_p]Q_\infty, \omega) \subset \{x \in S : d(x, Z(q_p, \omega)) < 2^{-n+8}\}. \end{aligned}$$

Define the random interval $\theta_{i-1} \equiv [\tau_{i-1}, \tau_i)$ and $\theta_p \equiv [\tau_p, \tau_{p+1}]$. Then the last two displayed relations can be restated as

$$Z(\theta_j(\omega)Q_\infty, \omega) \subset \{x \in S : d(x, Z(q_j, \omega)) \leq 2^{-n+8}\} \tag{10.5.66}$$

for each $j = 0, \ldots, p$, where $\omega \in CB_{n+}^c D_{n+}^c$ is arbitrary.

15. We will next estimate a lower bound for lengths of the random intervals $\theta_0, \ldots, \theta_{p-1}, \theta_p$. With ε, n, N fixed as in the hypothesis, recall from Step 2 that $\varepsilon_n \equiv 2^{-m(n)-n-1} \leq 2^{-n-1} < \beta_n$ and $N \geq \nu = m_n + n + 6$. Then $2^{-\nu+5} = \varepsilon_n$ and

$$\overline{\Delta} \equiv \Delta_{m(N)} \equiv 2^{-m(N)} < 2^{-2}\delta_{Cp}(2^{-2\nu+2}). \tag{10.5.67}$$

Now write $\overline{q} \equiv 1 - \overline{\Delta}$, and note that

$$\delta_{cau}(\varepsilon_n) = \overline{\Delta} < 2^{-2}\delta_{Cp}(2^{-2\nu+2}).$$

Thus the conditions in Assertion 1 of Proposition 10.5.7 are satisfied with ε, n, t, t', replaced by $\varepsilon_n, \nu, r, r'$ respectively. Accordingly, there exists, for each $r \in [0,1]$, an exceptional set $\overline{G}_{r,\varepsilon(n)}$ with $P(\overline{G}_{r,\varepsilon(n)}) < \varepsilon_n$ such that for each $\omega \in \overline{G}^c_{r,\varepsilon(n)}$, we have

$$d(Z(r,\omega), Z(r',\omega)) < \varepsilon_n < \beta_n \tag{10.5.68}$$

for each $r' \in [r - \overline{\Delta}, r + \overline{\Delta}]Q_\infty$. Define the exceptional sets

$$\overline{C}_n \equiv \bigcup_{r \in Q(m(n))} \overline{G}_{r,\varepsilon(n)}, \tag{10.5.69}$$

$$\overline{C}_{n+} \equiv \bigcup_{k=n}^{\infty} \overline{C}_k, \tag{10.5.70}$$

and

$$\overline{A}_n \equiv CH^c B^c_{n+} D^c_{n+} \overline{C}^c_{n+},$$

where H is the null set introduced at the beginning of this proof. Then

$$P(\overline{C}_n) \leq \sum_{r \in Q(m(n))} \varepsilon_n < (2^{m(n)} + 1)\varepsilon_n \equiv (2^{m(n)} + 1)2^{-m(n)-n-1} \leq 2^{-n},$$

$$P(\overline{C}_{n+}) \leq 2^{-n+1},$$

and

$$P(\overline{A}^c_n) \leq 0 + 0 + 2^{-n+8} + 2^{-n+3} + 2^{-n+1} < 2^{-n+9} < \varepsilon.$$

16. Now let $\omega \in \overline{A}_n \equiv CH^c B^c_{n+} D^c_{n+} \overline{C}^c_{n+}$ be arbitrary, and let $s \in [t, t+\overline{\Delta})Q_\infty$ be arbitrary. Let $k \geq n$ be arbitrary. Then $\omega \in \overline{C}^c_n \subset \overline{G}^c_{t,\varepsilon(n)}$ because $t \in Q_{m(n)}$. Therefore inequality 10.5.68, with t in the place of r, holds:

$$d(Z(t,\omega), Z(r',\omega)) < \beta_n \tag{10.5.71}$$

for each $r' \in [t,s]Q_{m(k)} \subset (t, t + \overline{\Delta})Q_\infty$. Consider each $r' \in [t,s]Q_{m(k)}$. Then we have $r' \in Q^-_{m(n)}$ because $t \in Q_{m(n)}$ and $r' - t \leq s - t < \overline{\Delta} \leq 2^{m(n)}$. Hence equality 10.5.6 in Definition 10.5.3 applies with $\widehat{h}(r') \geq n+1$ and yields $\widehat{\beta}_n(r') \geq \beta_n + \beta_{n+1}$. Therefore inequality 10.5.71 implies

$$d(Z(t,\omega), Z(r',\omega)) < \widehat{\beta}_n(r') \tag{10.5.72}$$

for each $r' \in [t,s]Q_{m(k)}$. Inequality 10.5.72 says that for each $k \geq n$, the sample path $Z(\cdot,\omega)|[t,t']Q_{m(k)}$ stays within the time-varying $\widehat{\beta}_n$-neighborhood of $Z(t,\omega)$ up to and including time s. Hence, according to Assertion 4 of Proposition 8.1.13, the simple first exit time $\eta_k(\omega) \equiv \eta_{t,\widehat{\beta}(n),[t,t']Q(m(k))}(\omega)$ can come only after time s. In other words, $s < \eta_k(\omega)$. Letting $k \to \infty$, we therefore obtain $s \leq \tau(\omega)$. Since $s \in [t, t+\overline{\Delta})Q_\infty$ is arbitrary, it follows that $t + \overline{\Delta} \leq \tau(\omega)$. Therefore,

$$|\theta_{i-1}(\omega)| = \tau_i(\omega) - \tau_{i-1}(\omega) \geq \tau_i(\omega) - q_{i-1} \equiv \tau(\omega) - t \geq \overline{\Delta}, \qquad (10.5.73)$$

where $i = 1, \ldots, p$ is arbitrary.

17. The interval $\theta_p(\omega) \equiv [\tau_p(\omega), 1]$ remains, with a length possibly less than $\overline{\Delta}$. To deal with this nuisance, we will replace the r.r.v. τ_p with the r.r.v. $\overline{\tau}_p \equiv \tau_p \wedge (1 - \overline{\Delta})$, while keeping $\overline{\tau}_i \equiv \tau_i$ if $i \leq p-1$. For convenience, define the constant r.r.v.'s $\overline{\tau}_0 \equiv 0$ and $\overline{\tau}_{p+1} \equiv 1$. Define the random interval $\overline{\theta}_{i-1} \equiv [\overline{\tau}_{i-1}, \overline{\tau}_i)$ and $\overline{\theta}_p \equiv [\overline{\tau}_p, \overline{\tau}_{p+1}]$. First note that

$$\overline{\tau}_{j+1}(\omega) - \overline{\tau}_j(\omega) \equiv \tau_{j+1}(\omega) - \tau_j(\omega) \geq \overline{\Delta} \qquad (10.5.74)$$

for each $j = 0, \ldots, p-2$. Moreover, for $j = p-1$, we have

$$\begin{aligned}
\overline{\tau}_p(\omega) - \overline{\tau}_{p-1}(\omega) &\equiv \overline{\tau}_p(\omega) - \tau_{p-1}(\omega) \geq \overline{\tau}_p(\omega) - q_{p-1} \\
&\equiv \tau_p(\omega) \wedge (1 - \overline{\Delta}) - q_{p-1} \\
&= (\tau_p(\omega) - q_{p-1}) \wedge (1 - \overline{\Delta} - q_{p-1}) \\
&= (\tau_p(\omega) - q_{p-1}) \wedge (\Delta - \overline{\Delta}) \geq \overline{\Delta} \wedge \overline{\Delta} = \overline{\Delta}, \qquad (10.5.75)
\end{aligned}$$

where the last inequality follows from the second half of inequality 10.5.73 and from the inequality

$$\Delta - \overline{\Delta} \equiv 2^{-m(n)} - 2^{-m(N)} \geq 2^{-m(N)+1} - 2^{-m(N)} = 2^{-m(N)} \equiv \overline{\Delta}.$$

Furthermore, for $j = p$, we have

$$\overline{\tau}_{p+1}(\omega) - \overline{\tau}_p(\omega) \equiv 1 - \tau_p(\omega) \wedge (1 - \overline{\Delta}) \geq \overline{\Delta}. \qquad (10.5.76)$$

Combining inequalities 10.5.74, 10.5.75, and 10.5.76, we see that

$$\bigwedge_{j=0}^{p} (\overline{\tau}_{j+1}(\omega) - \overline{\tau}_j(\omega)) \geq \overline{\Delta} \equiv \delta_{aucl}(\varepsilon, \overline{m}, \delta_{Cp}), \qquad (10.5.77)$$

where $\omega \in \overline{A}_n$ is arbitrary.

18. We will now verify that relation 10.5.66 still holds when θ_j is replaced by $\overline{\theta}_j$, for each $j = 0, \ldots, p$. First note that for $j = 0, \ldots, p-1$, we have $\overline{\theta}_j \equiv [\overline{\tau}_j, \overline{\tau}_{j+1}) \subset [\tau_j, \tau_{j+1}) \equiv \theta_j$, whence, according to relation 10.5.66, we have

$$Z(\overline{\theta}_j(\omega)Q_\infty, \omega) = Z(\theta_j(\omega)Q_\infty, \omega) \subset \{x \in S : d(x, Z(q_j, \omega)) \leq 2^{-n+8}\}. \qquad (10.5.78)$$

19. We still need to verify a similar range-bound relation for $j = p$. To that end, let $r' \in \overline{\theta}_p(\omega)Q_\infty \equiv [\tau_p(\omega)\wedge(1-\overline{\Delta}),1]Q_\infty$ be arbitrary. Either (i″) $r' < 1-\overline{\Delta}$ or (ii″) $r' \geq 1-\overline{\Delta}$. Consider case (i″). Then the assumption $r' < \tau_p(\omega)$ would imply $r' < \tau_p(\omega) \wedge (1-\overline{\Delta})$, a contradiction. Therefore $r' \in [\tau_p(\omega),1]Q_\infty$, whence

$$d(Z(r',\omega),Z(1,\omega)) \leq 2^{-n+8}$$

by relation 10.5.66 with $j = p$. In case (ii″), because $1 \in Q_{m(n)}, r' \in [1-\overline{\Delta},1]Q_\infty$, and $\omega \in \overline{G}^c_{1,\varepsilon(n)}$, inequality 10.5.68 applies to $r \equiv 1$, yielding

$$d(Z(1,\omega),Z(r',\omega)) < \varepsilon_n < \beta_n < 2^{-n+8}.$$

Thus, in both cases (i″) and (ii″), we have $d(Z(r',\omega),Z(1,\omega)) \leq 2^{-n+8}$, where $r' \in \overline{\theta}_p(\omega)Q_\infty$ is arbitrary. We conclude that

$$Z(\overline{\theta}_p(\omega)Q_\infty,\omega) \subset \{x \in S : d(x,Z(q_p,\omega)) \leq 2^{-n+8}\}. \tag{10.5.79}$$

20. Inequalities 10.5.78 and 10.5.79 can be summarized as

$$Z(\overline{\theta}_j(\omega)Q_\infty,\omega) \subset \{x \in S : d(x,Z(q_j,\omega)) \leq 2^{-n+8}\} \tag{10.5.80}$$

for each $j = 1,\ldots,p$, where $\omega \in \overline{A}_n$ is arbitrary.

21. Separately, as observed at the beginning of this proof, the process X is continuous a.u., with modulus of continuity a.u. δ_{cau}. Hence, for each $h \geq 0$, there exists a measurable set $A_{\overline{q},h} \subset domain(X_{\overline{q}})$ with $P\left(A^c_{\overline{q},h}\right) < 2^{-h}$ such that for each $\omega' \in A_{\overline{q},h}$ and for each $u \in (\overline{q}-\delta_{cau}(2^{-h}),\overline{q}+\delta_{cau}(2^{-h})) \cap domain(X(\cdot,\omega'))$, we have

$$d(X(u,\omega'),X(\overline{q},\omega')) \leq 2^{-h}. \tag{10.5.81}$$

Take regular points b_h, c_h of the r.r.v. τ_p such that

$$\overline{q} - \delta_{cau}(2^{-h}) < b_h < \overline{q} < c_h < \overline{q} + \delta_{cau}(2^{-h})) \cap domain(X(\cdot,\omega)).$$

22. Now let $h \geq n$ be arbitrary. Define the measurable sets $A'_h \equiv (\tau_p \leq b_h)A_{\overline{q},h}$, $A''_h \equiv (b_h < \tau_p \leq c_h)A_{\overline{q},h}$, and $A'''_h \equiv (c_h < \tau_p)A_{\overline{q},h}$. Define the r.r.v. Y_h by $domain(Y_h) \equiv A'_h \cup A''_h \cup A'''_h \cup A^c_{\overline{q},h}$ and by

$$Y_h \equiv X_{\tau(p)}, X_{\overline{q}}, X_{\overline{q}}, X_{\overline{q}} \qquad \text{on } A'_h, A''_h, A'''_h, A^c_{\overline{q},h},$$

respectively. Then

$$\begin{aligned} P(d(X_{\overline{\tau}(p)},Y_h) > 2^{-h}) \leq{} & P(d(X_{\overline{\tau}(p)},Y_h) > 2^{-h})A'_h \\ & + P(d(X_{\overline{\tau}(p)},Y_h) > 2^{-h})A''_h \\ & + P(d(X_{\overline{\tau}(p)},Y_h) > 2^{-h})A'''_h + P\left(A^c_{\overline{q},h}\right). \end{aligned}$$

Consider the for summands on the right-hand side of this inequality. For the first summand, we have

$$(d(X_{\overline{\tau}(p)}, Y_h) > 2^{-h})A'_h \subset (d(X_{\overline{\tau}(p)}, X_{\tau(p)}) > 2^{-h})(\tau_p \leq b_h < \overline{q})$$
$$\subset (d(X_{\tau(p)}, X_{\tau(p)}) > 2^{-h})(\tau_p < \overline{q}) = \phi.$$

For the second summand, note that

$$(b_h < \overline{\tau}_p \leq c_h)A_{\overline{q},h} = (b_h < \overline{\tau}_p = \tau_p \wedge \overline{q} \leq c_h)A_{\overline{q},h}$$
$$\equiv (\overline{q} - \delta_{cau}(2^{-h}) < b_h < \overline{\tau}_p \leq c_h < \overline{q}$$
$$+ \delta_{cau}(2^{-h}))A_{\overline{q},h} \subset (d(X_{\overline{\tau}(p)}, X_{\overline{q}}) \leq 2^{-h}),$$

where the last inclusion relation is thanks to inequality 10.5.81, whence

$$(d(X_{\overline{\tau}(p)}, Y_h) > 2^{-h})A''_h \equiv (d(X_{\overline{\tau}(p)}, X_{\overline{q}}) > 2^{-h})(b_h < \overline{\tau}_p \leq c_h)A_{\overline{q},h}$$
$$\subset (d(X_{\overline{\tau}(p)}, X_{\overline{q}}) > 2^{-h})(d(X_{\overline{\tau}(p)}, X_{\overline{q}}) \leq 2^{-h}) = \phi.$$

For the third summand, we have

$$(d(X_{\overline{\tau}(p)}, Y_h) > 2^{-h})A'''_h = (d(X_{\overline{\tau}(p)}, X_{\overline{q}}) > 2^{-h})(\overline{q} < c_h < \tau_p)A_{\overline{q},h}$$
$$\subset (d(X_{\overline{\tau}(p)}, X_{\overline{q}}) > 2^{-h})(\overline{\tau}_p = \overline{q}) = \phi.$$

Combining, we see that

$$P(d(X_{\overline{\tau}(p)}, Y_h) > 2^{-h}) = 0 + 0 + 0 + P(A^c_{\overline{q},h}) \leq 2^{-h}.$$

Thus $Y_h \to X_{\overline{\tau}(p)}$ a.u. Consequently, $X_{\overline{\tau}(p)}$ is an r.v.

23. For each $j = 1, \ldots, p-1$, we have $\overline{\tau}_j = \tau_j$, whence $X_{\overline{\tau}(j)} = X_{\tau(j)}$, and so $X_{\overline{\tau}(j)}$ is an r.v. In view of the concluding remark of Step 22, we see that $X_{\overline{\tau}(j)}$ is an r.v. for each $j = 1, \ldots, p$.

24. Continuing, with $\omega \in \overline{A}_n \equiv CH^c B^c_{n+} D^c_{n+} \overline{C}^c_{n+}$ arbitrary, let $j = 1, \ldots, p$ be arbitrary. Inequality 10.5.80 and right continuity of $X(\cdot, \omega)$ imply that

$$d(X(r,\omega), X(q_j,\omega)) \leq 2^{-n+8} \tag{10.5.82}$$

for each $r \in \overline{\theta}_j(\omega) domain(X(\cdot,\omega))$. In particular,

$$d(X(\tau_j(\omega),\omega), X(q_j,\omega)) \leq 2^{-n+8}. \tag{10.5.83}$$

Hence the triangle inequality yields

$$d(X(\tau_j(\omega),\omega), X(r,\omega)) \leq 2^{-n+9} < \varepsilon, \tag{10.5.84}$$

for each $r \in \overline{\theta}_j(\omega) domain(X(\cdot,\omega))$, where $\omega \in \overline{A}_n$ is arbitrary.

25. Summing up, the process X is continuous in probability on $[0,1]$, with a modulus of continuity in probability δ_{Cp}. The set $\widetilde{B} \equiv H^c \cap \bigcap_{u \in Q(\infty)} domain(X_u)$ is a full set, where H is the null set defined at the beginning of this proof. Define $\widetilde{A}_n \equiv \overline{A}_n \widetilde{B}$. Assertions 6 and 7 of Proposition 10.5.7 say that for

each $\omega \in \widetilde{B} \subset H^c$, the function $X(\cdot,\omega)$ is right continuous at each $t \in domain(X(\cdot,\omega))$, and that for each $t \in [0,1]$ such that $\lim_{r\to t; r\geq t} X(r,\omega)$ exists, we have $t \in domain(X(\cdot,\omega))$. In other words, the right-continuity condition and the right-completeness condition in Definition 10.3.2 have been proved for the process X. Moreover, with $\varepsilon > 0$, we have constructed (i$'''$) $\delta_{aucl}(\varepsilon) \in (0,1)$, (ii$'''$) a measurable set $\widetilde{A}_n \subset \widetilde{B}$ with $P(\widetilde{A}_n^c) = P(\overline{A}_n^c) < \varepsilon$, (iii$'''$) an integer $p+1 \geq 1$, and (iv$'''$) a sequence of r.r.v.'s

$$0 \equiv \overline{\tau}_0 < \overline{\tau}_1 < \cdots < \overline{\tau}_p < \overline{\tau}_{p+1} \equiv 1$$

such that for each $i = 0,\ldots,p$, the function $X_{\overline{\tau}(i)}$ is an r.v., and such that (v$'''$) for each $\omega \in \widetilde{A}_n \subset \overline{A}_n$, we have, in view of inequalities 10.5.77 and 10.5.84,

$$\bigwedge_{i=0}^{p}(\overline{\tau}_{i+1}(\omega) - \overline{\tau}_i(\omega)) \geq \delta_{aucl}(\varepsilon) \tag{10.5.85}$$

with

$$d(X(\overline{\tau}_i(\omega),\omega), X(\cdot,\omega)) \leq \varepsilon \tag{10.5.86}$$

on the interval $\overline{\theta}_i(\omega) \equiv [\overline{\tau}_i(\omega),\overline{\tau}_{i+1}(\omega))$ or $\overline{\theta}_i(\omega) \equiv [\overline{\tau}_i(\omega),\overline{\tau}_{i+1}(\omega)]$ depending on whether $0 \leq i \leq p-1$ or $i = p$.

Thus all the conditions in Definition 10.3.2 have been verified for the process X. Accordingly, X is an a.u. càdlàg process that specifically satisfies Conditions (i) and (ii) in this theorem.

26. Note, for later reference, that inequality 10.5.38 implies

$$q_0 \equiv 0 < \tau_1 \leq q_1 < \tau_2 \leq q_2 \leq \cdots < \tau_{p-1} \leq q_{p-1} < \tau_p \leq q_p = 1. \tag{10.5.87}$$

By definition, $\overline{\tau}_i \equiv \tau_i$ if $i \leq p-1$, while $\overline{\tau}_p \equiv \tau_p \wedge (1-\overline{\Delta}) > q_{p-1}$ because $\tau_p > q_{p-1}$ and $(1-\overline{\Delta}) > (1-2^{-m(n)}) = q_{p-1}$. Therefore 10.5.87 yields

$$\overline{\tau}_0 \equiv q_0 \equiv 0 < \overline{\tau}_1 \leq q_1 < \overline{\tau}_2 \leq q_2 \leq \cdots < \overline{\tau}_{p-1} \leq q_{p-1} < \overline{\tau}_p \leq q_p = 1 \equiv \overline{\tau}_{p+1}.$$

It follows that $q_j \equiv j2^{-m(n)} \in \overline{\theta}_j$, for each $j = 0,\ldots,p+1$. □

10.6 Continuity of the Right-Limit Extension

We will prove that the right-limit extension of D-regular processes, defined in Section 10.5, is a continuous function.

Let (S,d) be a locally compact metric space. Refer to Definition 9.0.2 for notations related to the enumerated set $Q_\infty \equiv \{t_0,t_1,\ldots\}$ of dyadic rationals in the interval $[0,1]$ and its subset $Q_m \equiv \{q_{m,0},q_{m,1},\ldots,q_{m,p(m)}\} = \{t_0,\ldots,t_{p(m)}\}$ for each $m \geq 0$.

Recall from Definition 6.4.2 that $\widehat{R}(Q_\infty \times \Omega, S)$ denotes the space of stochastic processes with parameter set Q_∞, sample space (Ω,L,E), and state space (S,d), and that it is equipped with the metric $\widehat{\rho}_{Prob,Q(\infty)}$ defined by

$$\widehat{\rho}_{Prob,Q(\infty)}(Z,Z') \equiv \sum_{j=0}^{\infty} 2^{-j-1} E1 \wedge d(Z_{t(j)}, Z'_{t(j)}) \tag{10.6.1}$$

for each $Z, Z' \in \widehat{R}(Q_\infty \times \Omega, S)$.

Recall from Definitions 10.3.2 and 10.3.6 the metric space $(\widehat{D}[0,1], \rho_{\widehat{D}[0,1]})$ of a.u. càdlàg processes on the interval $[0,1]$, with

$$\rho_{\widehat{D}[0,1]}(X,X') \equiv \int E(d\omega)\widehat{d}_D(X(\cdot,\omega), X'(\cdot,\omega))$$

for each $X, X' \in \widehat{D}[0,1]$, where $\widehat{d}_D \equiv 1 \wedge d_D$, where d_D is the Skorokhod metric on the space $D[0,1]$ of càdlàg functions. Recall that $[\cdot]_1$ is an operation that assigns to each $a \in R$ an integer $[a]_1 \in (a, a+2)$.

Theorem 10.6.1. Continuity of the construction of a.u. càdlàg process by right-limit extension of D-regular process. *Let $\widehat{R}_{Dreg,\overline{m},\delta(Cp)}(Q_\infty \times \Omega, S)$ denote the subspace of $(\widehat{R}(Q_\infty \times \Omega, S), \widehat{\rho}_{Prob,Q(\infty)})$ whose members share some common modulus of continuity in probability δ_{Cp} and some common modulus of D-regularity $\overline{m} \equiv (m_n)_{n=0,1,\ldots}$. Let*

$$\Phi_{rLim} : (\widehat{R}_{Dreg,\overline{m},\delta(Cp)}(Q_\infty \times \Omega, S), \widehat{\rho}_{Prob,Q(\infty)}) \to (\widehat{D}_{\delta(aucl),\delta(Cp)}[0,1], \rho_{\widehat{D}[0,1]}) \tag{10.6.2}$$

be the right-limit extension as constructed in Theorem 10.5.8, where

$$(\widehat{D}_{\delta(aucl),\delta(Cp)}[0,1], \rho_{\widehat{D}[0,1]})$$

is defined as the metric subspace of a.u. càdlàg processes that share the common modulus of continuity in probability δ_{Cp}, and that share the common modulus of a.u. càdlàg $\delta_{aucl} \equiv \delta_{aucl}(\cdot, \overline{m}, \delta_{Cp})$ as defined in Theorem 10.5.8.

Then the function Φ_{rLim} is uniformly continuous, with a modulus of continuity $\delta_{rLim}(\cdot, \overline{m}, \delta_{Cp})$ that depends only on $\overline{m}$ and on δ_{Cp}.

Proof. 1. Let $\varepsilon_0 > 0$ be arbitrary. Write $\varepsilon \equiv 2^{-4}\varepsilon_0$. According to Theorem 10.5.8, the real number $\delta_{aucl}(\varepsilon, \overline{m}, \delta_{Cp}) > 0$ is defined as follows. Take any $n \geq 0$ so large that $2^{-n+9} < \varepsilon$. Take $N \geq m_n + n + 6$ so large that

$$\overline{\Delta} \equiv \Delta_{m(N)} \equiv 2^{-m(N)} < 2^{-2}\delta_{Cp}(2^{-2m(n)-2n-10}). \tag{10.6.3}$$

Then, we have

$$\delta_{aucl}(\varepsilon, \overline{m}, \delta_{Cp}) \equiv \Delta_{m(N)} \equiv 2^{-m(N)}.$$

2. Now take $k > N$ so large that

$$2^{-m(k)+2} < (1 - e^{-\varepsilon})\overline{\Delta}. \tag{10.6.4}$$

Write $p \equiv p_{m(n)} \equiv 2^{m(n)}$, $\widetilde{p} \equiv p_{m(k)} \equiv 2^{m(k)}$, and $\widetilde{\Delta} \equiv 2^{-m(k)}$. Define

$$\delta_{rLim}(\varepsilon_0, \overline{m}, \delta_{Cp}) \equiv \delta \equiv 2^{-1}\widetilde{p}^{-1}\varepsilon^2.$$

We will prove that the operation $\delta_{rLim}(\cdot,\overline{m},\delta_{Cp})$ is a modulus of continuity for the function Φ_{rLim} in expression 10.6.2.

3. To that end, let $Z,Z' \in \widehat{R}_{Dreg,\overline{m},\delta(Cp)}(Q_\infty \times \Omega, S)$ be arbitrary such that

$$\widehat{\rho}_{Prob,Q(\infty)}(Z,Z') \equiv E\sum_{j=0}^{\infty}2^{-j-1}\widehat{d}(Z_{t(j)},Z'_{t(j)}) < \delta \equiv \delta_{rLim}(\varepsilon_0,\overline{m},\delta_{Cp}). \tag{10.6.5}$$

Let $X \equiv \Phi_{rLim}(Z)$ and $X' \equiv \Phi_{rLim}(Z')$. By Theorem 10.5.8, X, X' are a.u. càdlàg processes. We need only verify that

$$\rho_{\widehat{D}[0,1]}(X,X') < \varepsilon_0. \tag{10.6.6}$$

4. Then, in view of the bound 10.6.5, Chebychev's inequality implies that there exists a measurable set G with $P(G) < \varepsilon$ such that for each $\omega \in G^c$, we have

$$\sum_{j=0}^{\infty}2^{-j-1}\widehat{d}(Z_{t(j)}(\omega),Z'_{t(j)}(\omega)) < \delta\varepsilon^{-1},$$

whence

$$\begin{aligned}
&\bigvee_{r\in Q(m(k))} d(Z(r,\omega),Z'(r,\omega))\\
&= \bigvee_{j=0}^{p(m(k))} d(Z_{t(j)}(\omega),Z'_{t(j)}(\omega))\\
&\leq \sum_{j=0}^{p(m(k))} d(Z_{t(j)}(\omega),Z'_{t(j)}(\omega)) \leq 2^{m(k))+1}\sum_{j=0}^{p(m(k))}2^{-j-1}\widehat{d}(Z_{t(j)}(\omega),Z'_{t(j)}(\omega))\\
&\leq 2^{m(k))+1}\sum_{j=0}^{\infty}2^{-j-1}\widehat{d}(Z_{t(j)}(\omega),Z'_{t(j)}(\omega)) \leq 2\widetilde{p}\delta\varepsilon^{-1}\\
&\equiv 2\widetilde{p}2^{-1}\widetilde{p}^{-1}\varepsilon^2\varepsilon^{-1} = \varepsilon.
\end{aligned} \tag{10.6.7}$$

5. According to Steps 25 and 26 in the proof of Theorem 10.5.8, we have constructed a measurable set $\widetilde{A}_n$ with $P(\widetilde{A}_n^c) = P(\overline{A}_n^c) < \varepsilon$ and a sequence of r.r.v.'s

$$\begin{aligned}
\overline{\tau}_0 \equiv q_0 \equiv 0 < \overline{\tau}_1 \leq q_1 < \overline{\tau}_2 \leq q_2 \leq \cdots < \overline{\tau}_{p-1}\\
\leq q_{p-1} < \overline{\tau}_p \leq q_p = 1 \equiv \overline{\tau}_{p+1},
\end{aligned} \tag{10.6.8}$$

where $q_i \equiv i2^{-m(n)} \in \overline{\theta}_i$ and $X_{\overline{\tau}(i)}$ is an r.v., for each $i = 0,\ldots,p$. Moreover, for each $\omega \in \widetilde{A}_n$, we have

$$\bigwedge_{i=0}^{p}(\overline{\tau}_{i+1}(\omega) - \overline{\tau}_i(\omega)) \geq \delta_{aucl}(\varepsilon) \equiv 2^{-m(N)} > 2^{-m(k)} \tag{10.6.9}$$

with

$$d(X(\overline{\tau}_i(\omega),\omega), X(\cdot,\omega)) \leq \varepsilon \tag{10.6.10}$$

on the interval $\overline{\theta}_i(\omega) \equiv [\overline{\tau}_i(\omega), \overline{\tau}_{i+1}(\omega))$ or $\overline{\theta}_i(\omega) \equiv [\overline{\tau}_i(\omega), \overline{\tau}_{i+1}(\omega)]$ depending on whether $0 \leq i \leq p-1$ or $i = p$.

6. In particular,

$$d(X(\overline{\tau}_i(\omega),\omega), Z(q_i,\omega)) \leq \varepsilon \tag{10.6.11}$$

for each $i = 0, \ldots, p$. Combined with inequality 10.6.10, this yields

$$X(\overline{\theta}_i(\omega),\omega) \subset (d(\cdot, Z(q_i,\omega)) \leq 2\varepsilon) \tag{10.6.12}$$

for each $i = 0, \ldots, p$, where $\omega \in \widetilde{A}_n$ is arbitrary.

7. Similarly, we can construct, relative to the processes Z' and X, a measurable set $\widetilde{A}'_n$ with $P(\widetilde{A}'_n)^c < \varepsilon$ and a sequence of r.r.v.'s

$$\begin{aligned} \overline{\tau}'_0 \equiv q_0 \equiv 0 < \overline{\tau}'_1 \leq q_1 < \overline{\tau}'_2 \leq q_2 \\ \leq \cdots < \overline{\tau}'_{p-1} \leq q_{p-1} < \overline{\tau}'_p \leq q_p = 1 \equiv \overline{\tau}'_{p+1}, \end{aligned} \tag{10.6.13}$$

where $q_i \equiv i2^{-m(n)} \in \overline{\theta}_i$ and $X_{\overline{\tau}(i)}$ is an r.v., for each $i = 0, \ldots, p$. Moreover, for each $\omega \in \widetilde{A}'_n$, we have

$$\bigwedge_{i=0}^{p} (\overline{\tau}'_{i+1}(\omega) - \overline{\tau}'_i(\omega)) > 2^{-m(k)} \tag{10.6.14}$$

with

$$d(X'(\overline{\tau}'_i(\omega),\omega), X'(\cdot,\omega)) \leq \varepsilon \tag{10.6.15}$$

with $\overline{\theta}'_i(\omega) \equiv [\overline{\tau}'_i(\omega), \overline{\tau}'_{i+1}(\omega))$ or $\overline{\theta}'_i(\omega) \equiv [\overline{\tau}'_i(\omega), \overline{\tau}'_{i+1}(\omega)]$ depending on whether $0 \leq i \leq p-1$ or $i = p$.

8. In particular,

$$(X'(\overline{\tau}'_i(\omega),\omega), X'(\cdot,\omega)) \leq \varepsilon \tag{10.6.16}$$

for each $i = 0, \ldots, p$. Combined with inequality 10.6.15, this yields

$$X(\overline{\theta}_i(\omega),\omega) \subset (d(\cdot, Z(q_i,\omega)) \leq 2\varepsilon) \tag{10.6.17}$$

for each $i = 0, \ldots, p$, where $\omega \in \widetilde{A}'_n$ is arbitrary.

9. Because X, X' are a.u. càdlàg processes, Proposition 10.3.5 says that there exists a full set $\overline{G}$ such that for each $\omega \in \overline{G}$, the functions $X(\cdot,\omega), X'(\cdot,\omega)$ are a.u. càdlàg functions on $[0,1]$.

9. Now consider each $\omega \in \overline{G}G^c\widetilde{A}_n\widetilde{A}'_n$. Let $i = 0, \ldots, p$ be arbitrary. Then $q_i \in Q_{m(n)} \subset Q_{m(k)}$. Hence inequality 10.6.7 implies that

$$d(Z(q_i,\omega), Z'(q_i,\omega)) \leq \bigvee_{r \in Q(m(k))} d(Z(r,\omega), Z'(r,\omega)) \leq \varepsilon. \tag{10.6.18}$$

To simplify notations, we will now suppress both the reference to ω and the overline, writing $\tau_i, \tau_i', x, x', z, z', \theta_i, \theta_i'$ for $\overline{\tau}_i(\omega), \overline{\tau}_i'(\omega), X(\cdot,\omega), X'(\cdot,\omega), Z(\cdot,\omega), Z'(\cdot,\omega), \overline{\theta}_i(\omega), \overline{\theta}_i'(\omega)$, respectively. Then inequality 10.6.18 can be rewritten as

$$d(z(q_i), z'(q_i)) \le \varepsilon \tag{10.6.19}$$

and inequality 10.6.12 can be rewritten as

$$x(\theta_i) \subset (d(\cdot, z(q_i)) \le 2\varepsilon). \tag{10.6.20}$$

Similarly,

$$x'(\theta_i') \subset (d(\cdot, z'(q_i)) \le 2\varepsilon). \tag{10.6.21}$$

10. Next, partition the set $\{0, \ldots, p+1\}$ into the union of two disjoint subsets A and B such that (i) $\{0, \ldots, p+1\} = A \cup B$, (ii) $|\tau_i - \tau_i'| < 2\widetilde{\Delta}$ for each $i \in A$, and (iii) $|\tau_i - \tau_i'| > \widetilde{\Delta}$ for each $i \in B$. Consider each $i \in B$. We will verify that $1 \le i \le p$ and that

$$d(z(q_i), z(q_{i-1})) \vee d(z'(q_i), z'(q_{i-1})) \le 6\varepsilon. \tag{10.6.22}$$

In view of Condition (iii), we may assume, without loss of generality, that $\tau_i' - \tau_i > \widetilde{\Delta} \equiv 2^{-m(k)}$. Then there exists $u \in [\tau_i, \tau_i')Q_{m(k)}$. Since $[\tau_0, \tau_0') = [0,0) = \phi$ and $[\tau_{p+1}, \tau_{p+1}') = [1,1) = \phi$, it follows that $1 \le i \le p$. Consequently, inequalities 10.6.8 and 10.6.13 together imply that

$$u \in [\tau_i, \tau_i') \subset [q_{i-1}, q_i) \subset [\tau_{i-1}', \tau_{i+1}).$$

Hence $u \in [\tau_i, \tau_{i+1}) \cap [\tau_{i-1}', \tau_i') = \theta_i \theta_{i-1}'$, and so $u \in \theta_i \theta_{i-1}' Q_{m(k)}$. Therefore, using inequalities 10.6.7, 10.6.19, 10.6.20, and 10.6.8, we obtain, respectively,

$$d(z(u), z'(u)) \le \varepsilon,$$
$$d(z(q_{i-1}), z'(q_{i-1})) \le \varepsilon,$$
$$z(u) \in x(\theta_i) \subset (d(\cdot, z(q_i)) \le 2\varepsilon),$$

and

$$z'(u) \in x'(\theta_{i-1}') \subset (d(\cdot, z'(q_{i-1})) \le 2\varepsilon). \tag{10.6.23}$$

The triangle inequality then yields

$$\begin{aligned} d(z(q_i), z(q_{i-1})) &\le d(z(q_i), z(u)) + d(z(u), z'(u)) \\ &\quad + d(z'(u), z'(q_{i-1})) + d(z'(q_{i-1}), z(q_{i-1})) \\ &\le 2\varepsilon + \varepsilon + 2\varepsilon + \varepsilon = 6\varepsilon. \end{aligned}$$

Similarly,

$$d(z'(q_i), z'(q_{i-1})) \le 6\varepsilon.$$

Thus inequality 10.6.22 is verified for each $i \in B$.

11. Now define an increasing function $\lambda : [0,1] \to [0,1]$ by $\lambda\tau_i \equiv \tau_i'$ or $\lambda\tau_i \equiv \tau_i$ depending on whether $i \in A$ or $i \in B$, for each $i = 0, \ldots, p+1$, and by linearity on $[\tau_i, \tau_{i+1}]$ for each $i = 0, \ldots, p$. Here we write $\lambda t \equiv \lambda(t)$ for each $t \in [0,1]$ for brevity. Then, in view of the definition of the index sets A and B, we have $|\tau_i - \lambda\tau_i| < 2\widetilde{\Delta}$ for each $i = 0, \ldots, p+1$. Now consider each $i = 0, \ldots, p$, and write

$$u_i \equiv \frac{\lambda\tau_{i+1} - \tau_{i+1} + \tau_i - \lambda\tau_i}{\tau_{i+1} - \tau_i}.$$

Then, since $\tau_{i+1} - \tau_i \geq 2^{-m(N)} \equiv \overline{\Delta}$ according to inequality 10.6.9, we have

$$\begin{aligned} |u_i| &\leq |\lambda\tau_{i+1} - \tau_{i+1}|\overline{\Delta}^{-1} + |\lambda\tau_i - \tau_i|\overline{\Delta}^{-1} \\ &\leq 2\widetilde{\Delta}\overline{\Delta}^{-1} + 2\widetilde{\Delta}\overline{\Delta}^{-1} = 2^{-m(k)+2}\overline{\Delta}^{-1} < (1 - e^{-\varepsilon}), \end{aligned}$$

where the last inequality is from inequality 10.6.4. Note that the function $\log(1+u)$ of $u \in [-1 + e^{-\varepsilon}, 1 - e^{-\varepsilon}]$ vanishes at $u = 0$; it has a positive first derivative and a negative second derivative on the interval. Hence the maximum of its absolute value is attained at the right endpoint of the interval and

$$|\log(1 + u_i)| \leq |\log(1 - 1 + e^{-\varepsilon})| = \varepsilon. \tag{10.6.24}$$

Lemma 10.2.3 therefore implies the bound

$$\begin{aligned} \sup_{0 \leq s < t \leq 1} \left| \log \frac{\lambda t - \lambda s}{t - s} \right| &= \bigvee_{i=0}^{p} \left| \log \frac{\lambda\tau_{i+1} - \lambda\tau_i}{\tau_{i+1} - \tau_i} \right| \\ &= \bigvee_{i=0}^{p} |\log(1 + u_i)| \leq \varepsilon < 11\varepsilon. \end{aligned} \tag{10.6.25}$$

12. We will next prove that

$$d(x, x' \circ \lambda) \leq 11\varepsilon \tag{10.6.26}$$

on $domain(x) \cap domain(x' \circ \lambda)$. To that end, let

$$v \in \bigcup_{i=0}^{p} [\tau_i, \tau_{i+1}) \cap \bigcup_{i=0}^{p} (\lambda^{-1}\tau_i', \lambda^{-1}\tau_{i+1}') \cap domain(x) \cap domain(x' \circ \lambda) \tag{10.6.27}$$

be arbitrary. We need to prove that $d(xv, x'\lambda v) \leq 9\varepsilon$. Note that $v \in [\tau_i, \tau_{i+1}) \equiv \theta_i$ for some $i = 0, \ldots, p$. There are four possible cases: (i′) $i, i+1 \in A$, (ii′) $i, i+1 \in B$, (iii′) $i \in A$ and $i+1 \in B$, and (iv′) $i \in B$ and $i+1 \in A$.

13. Consider case (i′), where $i, i+1 \in A$. Then $\lambda\tau_i \equiv \tau_i'$ and $\lambda\tau_{i+1} \equiv \tau_{i+1}'$. Hence

$$\lambda v \in [\lambda\tau_i, \lambda\tau_{i+1}) \subset [\tau_i', \tau_{i+1}') \equiv \theta_i'.$$

Therefore

$$d(x(v),x'(\lambda v)) \leq d(x(v),z(q_i)) + d(z(q_i),z'(q_i)) + d(z'(q_i),x'(\lambda v))$$
$$< 2\varepsilon + \varepsilon + 2\varepsilon = 5\varepsilon,$$

where the second inequality is due to inequalities 10.6.20, 10.6.19, and 10.6.21.

14. Next consider case (ii′), where $i, i+1 \in B$. Then, according to Step 10, we have $1 \leq i < i+1 \leq p$. Moreover, $\lambda\tau_i \equiv \tau_i$ and $\lambda\tau_{i+1} \equiv \tau_{i+1}$. Hence

$$\lambda v \in [\tau_i, \tau_{i+1}) \subset [q_{i-1}, q_{i+1}) \subset [\tau'_{i-1}, \tau'_{i+2}).$$

Since $\lambda v \neq \tau'_i$ and $\lambda v \neq \tau'_{i+1}$ by relation 10.6.27, we have subcases (ii′a) $\lambda v \in [\tau'_{i-1}, \tau'_i)$, (ii′b) $\lambda v \in [\tau'_i, \tau'_{i+1})$, and (ii′c) $\lambda v \in [\tau'_i, \tau'_{i+2})$ such that

$$\lambda v \in [\tau'_{i-1}, \tau'_i) \cup [\tau'_i, \tau'_{i+1}) \cup [\tau'_i, \tau'_{i+2}) \equiv \theta'_{i-1} \cup \theta'_i \cup \theta'_{i+1}.$$

In subcase (ii′a), we have

$$d(x(v),x'(\lambda v)) \leq d(x(v),z(q_i)) + d(z(q_i),z(q_{i-1})) + d(z(q_{i-1}),z'(q_{i-1}))$$
$$+ d(z'(q_{i-1}),x'(\lambda v))$$
$$\leq 2\varepsilon + 6\varepsilon + \varepsilon + 2\varepsilon = 11\varepsilon.$$

In subcase (ii′b), we have

$$d(x(v),x'(\lambda v)) \leq d(x(v),z(q_i)) + d(z(q_i),z'(q_i)) + d(z'(q_i),x'(\lambda v))$$
$$\leq 2\varepsilon + \varepsilon + 2\varepsilon = 5\varepsilon.$$

In subcase (ii′c), we have

$$d(x(v),x'(\lambda v)) \leq d(x(v),z(q_i)) + d(z(q_i),z(q_{i+1})) + d(z(q_{i+1}),z'(q_{i+1}))$$
$$+ d(z'(q_{i+1}),x'(\lambda v))$$
$$\leq 2\varepsilon + 6\varepsilon + \varepsilon + 2\varepsilon = 11\varepsilon.$$

Here we used inequalities 10.6.20, 10.6.22, 10.6.19, and 10.6.21.

15. Now consider case (iii′), where $i \in A$ and $i+1 \in B$. Then, according to Step 10, we have $i+1 \leq p$. Moreover, $\lambda\tau_i \equiv \tau'_i$ and $\lambda\tau_{i+1} \equiv \tau_{i+1}$. Hence

$$\lambda v \in [\tau'_i, \tau_{i+1}) \subset [\tau'_i, q_{i+1}) \subset [\tau'_i, \tau'_{i+2}).$$

Since $\lambda v \neq \tau'_{i+1}$ by relation 10.6.27, we have subcases (iii′a) $\lambda v \in [\tau'_i, \tau'_{i+1})$ and (iii′b) $\lambda v \in [\tau'_{i+1}, \tau'_{i+2})$.

In subcase (iii′a), we have

$$d(x(v),x'(\lambda v)) \leq d(x(v),z(q_i)) + d(z(q_i),z'(q_i)) + d(z'(q_i),x'(\lambda v))$$
$$\leq 2\varepsilon + \varepsilon + 2\varepsilon = 5\varepsilon.$$

In subcase (iii′b), we have

$$\begin{aligned} d(x(v),x'(\lambda v)) &\leq d(x(v),z(q_i)) + d(z(q_i),z(q_{i+1})) + d(z(q_{i+1}),z'(q_{i+1})) \\ &\quad + d(z'(q_{i+1}),x'(\lambda v)) \\ &\leq (2\varepsilon + 6\varepsilon + \varepsilon + 2\varepsilon) = 11\varepsilon, \end{aligned}$$

where we used relations 10.6.18 and 10.6.4, and inequalities 10.6.8 and 10.6.22.

16. Finally, consider case (iv′), where $i \in B$ and $i+1 \in A$. Then, according to Step 10, we have $1 \leq i \leq p$. Moreover, $\lambda\tau_i \equiv \tau_i$ and $\lambda\tau_{i+1} \equiv \tau'_{i+1}$. Hence

$$\lambda v \in [\tau_i, \tau'_{i+1}) \subset [q_{i-1}, \tau'_{i+1}) \subset [\tau'_{i-1}, \tau'_{i+1}).$$

Since $\lambda v \neq \tau'_i$ by relation 10.6.27, we have subcases (iv′a) $\lambda v \in [\tau'_{i-1}, \tau'_i)$ and (iv′b) $\lambda v \in [\tau'_i, \tau'_{i+1})$.

In subcase (iv′a), we have

$$\begin{aligned} d(x(v),x'(\lambda v)) &\leq d(x(v),z(q_{i-1})) + d(z(q_{i-1}),z'(q_i) + d(z'(q_i),x'(\lambda v)) \\ &\leq 2\varepsilon + \varepsilon + 2\varepsilon = 5\varepsilon. \end{aligned}$$

In case (iv′b), we have

$$\begin{aligned} d(x(v),x'(\lambda v)) &\leq d(x(v),z(q_i)) + d(z(q_i),z(q_{i+1})) \\ &\quad + d(z(q_{i+1}),z'(q_{i+1})) + d(z'(q_{i+1}),x'(\lambda v)) \\ &\leq (2\varepsilon + 6\varepsilon + \varepsilon + 2\varepsilon) = 11\varepsilon, \end{aligned}$$

where we used relations 10.6.18 and 10.6.4, and inequalities 10.6.8 and 10.6.22.

17. Summing up, we see that in each of the possible subcases (i′), (ii′a–c). (iii′a–b), and (iv′a–b), we have

$$d(x(v), x' \circ \lambda(v)) \leq 11\varepsilon, \tag{10.6.28}$$

where

$$v \in \bigcup_{i=0}^{p} [\tau_i, \tau_{i+1}) \cap \bigcup_{i=0}^{p} (\lambda^{-1}\tau'_i, \lambda^{-1}\tau'_{i+1}) \cap domain(x) \cap domain(x' \circ \lambda) \tag{10.6.29}$$

is arbitrary. Note that the set $\bigcup_{i=0}^{p}[\tau_i, \tau_{i+1}) \cap \bigcup_{i=0}^{p}(\lambda^{-1}\tau'_i, \lambda^{-1}\tau'_{i+1})$ contains all but finitely many points in the interval $[0,1]$, while the set $domain(x) \cap domain(x' \circ \lambda)$ contains all but countably many points in $[0,1]$, according to Proposition 10.1.4. Hence the set on the right-hand side of expression 10.6.29 contains all but countably many points in $[0,1]$, and is therefore dense in $domain(x) \cap domain(x' \circ \lambda)$. Consequently, by Lemma 10.1.3, inequality 10.6.28 extends to the desired inequality 10.6.26.

18. Inequalities 10.6.25 and 10.6.26 together show that the pair $(11\varepsilon, \lambda)$ satisfies the conditions in Definition 10.2.1 to be a member of the set $B_{x,x'}$ associated with

the càdlàg functions x, x', as introduced in Definition 10.2.1 for the Skorokhod metric d_D. Accordingly $d_D(x,x') \equiv \inf B_{x,x'} \leq 11\varepsilon$. Hence

$$\widehat{d}_D(X(\cdot,\omega),X'(\cdot,\omega)) \equiv 1 \wedge d_D(X(\cdot,\omega),X'(\cdot,\omega)) \equiv 1 \wedge d_D(x,x') \leq 11\varepsilon,$$

where $\omega \in \overline{G}G^c\widetilde{A}_n\widetilde{A}'_n$ is arbitrary. It follows that

$$\begin{aligned}
\rho_{\widehat{D}[0,1]}(X,X') &\equiv \int E(d\omega)\widehat{d}_D(X(\cdot,\omega),X'(\cdot,\omega)) \\
&\leq P(\overline{G}^c \cup G \cup \widetilde{A}_n^c \cup (\widetilde{A}'_n)^c) \\
&\quad + \int E(d\omega)\widehat{d}_D(X(\cdot,\omega),X'(\cdot,\omega))1_{\overline{G}G^c\widetilde{A}(n)\widetilde{A}'(n)} \\
&\leq P(G) + P(\widetilde{A}_n^c) + P((\widetilde{A}'_n)^c) + 11\varepsilon \leq 3\varepsilon + 11\varepsilon \\
&= 14\varepsilon < 2^4\varepsilon \equiv \varepsilon_0,
\end{aligned}$$

which proves inequality 10.6.6 and the theorem. □

The next corollary constructs an a.u. càdlàg process from a given D-regular family of f.j.d.'s. It shows that the construction is uniformly continuous on an arbitrary set of such D-regular families that share a common modulus of D-regularity and a common modulus of continuity in probability. For that purpose, let $\xi \equiv (A_q)_{q=1,2,\ldots}$ be an arbitrary but fixed binary approximation of the locally compact metric space (S,d) relative to some fixed reference point $x_\circ \in S$.

Recall from Definition 6.2.8 the metric space $(\widehat{F}(Q_\infty,S),\widehat{\rho}_{Marg,\xi,Q(\infty)})$ of consistent families of f.j.d.'s with parameter set Q_∞ and state space (S,d), where the marginal metric $\widehat{\rho}_{Marg,\xi,Q(\infty)}$ is defined relative to the binary approximation ξ of (S,d). Recall from Definition 10.4.1 its subset $\widehat{F}_{Dreg}(Q_\infty,S)$ consisting of D-regular families, and the subset $\widehat{F}_{Dreg,\overline{m},\delta(Cp)}(Q_\infty,S)$ consisting of D-regular families with some given modulus of D-regularity $\overline{m}$ and some given modulus of continuity in probability δ_{Cp}. Recall from Definition 10.4.2 the metric space $(\widehat{F}_{Dreg}([0,1],S),\widehat{\rho}_{Cp,\xi,[0,1],Q(\infty)})$ and its subset $\widehat{F}_{Dreg,\overline{m},\delta(Cp)}([0,1],S)$.

Corollary 10.6.2. Construction of a.u. càdlàg process on $[0,1]$ from a D-regular family of f.j.d.'s on $[0,1]$, and continuity of construction. *Let*

$$(\Theta_0,L_0,I_0) \equiv \left([0,1],L_0,\int \cdot dx\right)$$

denote the Lebesgue integration space based on the interval $\Theta_0 \equiv [0,1]$. Then the following conditions hold:

1. There exists a function

$$\begin{aligned}
\Psi_{aucl,\xi} &: (\widehat{F}_{Dreg,\overline{m},\delta(Cp)}([0,1],S),\widehat{\rho}_{Cp,\xi,[0,1],Q(\infty)}) \\
&\to (\widehat{D}_{\delta(aucl),\delta(Cp)}[0,1],\rho_{\widehat{D}[0,1]})
\end{aligned} \tag{10.6.30}$$

such that for each $F \in \widehat{F}_{Dreg,\overline{m},\delta(Cp)}([0,1],S)$*, the a.u. càdlàg process* $X \equiv \Phi_{aucl,\xi}(F)$ *has marginal distributions given by* F *and has the modulus of a.u. càdlàg* $\delta_{aucl}(\cdot,\overline{m},\delta_{Cp})$ *defined in Theorem 10.5.8.*

2. Let $\widehat{F}_{Dreg,\overline{m},\delta(Cp),\beta}$ *be an arbitrary subset of* $\widehat{F}_{Dreg,\overline{m},\delta(Cp)}([0,1],S)$ *that is pointwise tight with a certain modulus of pointwise tightness* β*. Then the function* $\Phi_{aucl,\xi}$ *is uniformly continuous on the subset* $\widehat{F}_{Dreg,\overline{m},\delta(Cp),\beta}$*, with a modulus of continuity*

$$\overline{\delta}_{aucl,\xi} \equiv \overline{\delta}_{aucl,\xi}(\cdot,\overline{m},\delta_{Cp},\ \|\xi\|\,,\beta)$$

that depends only on $\overline{m},\delta_{Cp}$, $\|\xi\|$, *and* β.

Proof. 1. Recall from Definition 6.2.12 the isometry

$$\Phi_{[0,1],Q(\infty)} : (\widehat{F}_{Cp}([0,1],S),\widehat{\rho}_{Cp,\xi,[0,1],Q(\infty)}) \to (\widehat{F}_{Cp}(Q_\infty,S),\widehat{\rho}_{Marg,\xi,Q(\infty)})$$

defined by $\Phi_{[0,1],Q(\infty)}(F) \equiv F|Q_\infty$ for each $F \in \widehat{F}_{Cp}([0,1]$. Since $\widehat{F}_{Dreg,\overline{m},\delta(Cp)}([0,1],S) \subset \widehat{F}_{Cp}([0,1],S)$, we have an isometry

$$\Phi_{[0,1],Q(\infty)} : (\widehat{F}_{Dreg,\overline{m},\delta(Cp)}([0,1],S),\widehat{\rho}_{Cp,\xi,[0,1],Q(\infty)}) \to (\widehat{F}_{Cp}(Q_\infty,S),\widehat{\rho}_{Marg,\xi,Q(\infty)}). \tag{10.6.31}$$

Moreover, for each $F \in \widehat{F}_{Dreg,\overline{m},\delta(Cp)}([0,1],S)$, we have $F|Q_\infty \in \widehat{F}_{Dreg,\overline{m},\delta(Cp)}(Q_\infty,S)$. Thus we obtain the isometry

$$\Phi_{[0,1],Q(\infty)} : (\widehat{F}_{Dreg,\overline{m},\delta(Cp)}([0,1],S),\widehat{\rho}_{Cp,\xi,[0,1],Q(\infty)}) \to (\widehat{F}_{Dreg,\overline{m},\delta(Cp)}(Q_\infty,S),\widehat{\rho}_{Marg,\xi,Q(\infty)}). \tag{10.6.32}$$

2. Let

$$\Phi_{DKS,\xi} : (\widehat{F}_{Dreg,\overline{m},\delta(Cp)}(Q_\infty,S),\widehat{\rho}_{Marg,\xi,Q(\infty)}) \to (\widehat{R}(Q_\infty \times \Theta_0,S),\widehat{\rho}_{Prob,Q(\infty)})$$

be the Daniell–Kolmogorov–Skorokhod Extension relative to the binary approximation ξ of (S,d). Let $F' \in \widehat{F}_{Dreg,\overline{m},\delta(Cp)}(Q_\infty,S)$ be arbitrary. Then the process $Z \equiv \Phi_{DKS,\xi}(F')$ has marginal distributions given by F'. Hence, by Definition 10.4.1, the process Z, like its family F' of marginal distributions, is D-regular, with a modulus of D-regularity $\overline{m}$ and with a modulus of continuity in probability δ_{Cp}. In other words, $\Phi_{DKS,\xi}(F') \in \widehat{R}_{Dreg,\overline{m},\delta(Cp)}(Q_\infty \times \Theta_0,S)$. Thus we obtain the function

$$\Phi_{DKS,\xi} : (\widehat{F}_{Dreg,\overline{m},\delta(Cp)}(Q_\infty,S),\widehat{\rho}_{Marg,\xi,Q(\infty)}) \to (\widehat{R}_{Dreg,\overline{m},\delta(Cp)}(Q_\infty \times \Theta_0,S),\widehat{\rho}_{Prob,Q(\infty)}). \tag{10.6.33}$$

3. By Theorem 10.6.1, the function

$$\Phi_{rLim} : (\widehat{R}_{Dreg,\overline{m},\delta(Cp)}(Q_\infty \times \Theta_0,S),\widehat{\rho}_{Prob,Q(\infty)}) \to (\widehat{D}_{\delta(aucl),\delta(Cp)}[0,1],\rho_{\widehat{D}[0,1]}) \tag{10.6.34}$$

is uniformly continuous, with a modulus of continuity $\delta_{rLim}(\cdot,\overline{m},\delta_{Cp})$ that depends only on $\overline{m}$ and δ_{Cp}.

4. The composite $\Phi_{aucl,\xi} \equiv \Phi_{rLim} \circ \Phi_{DKS,\xi} \circ \Phi_{[0,1],Q(\infty)}$ of the three functions 10.6.32, 10.6.33, and 10.6.34 satisfies the requirements in Assertion 1. Thus Assertion 1 is proved.

5. Proceed to prove continuity. Let $\widehat{F}_{Dreg,\overline{m},\delta(Cp),\beta}$ be an arbitrary subset of $\widehat{F}_{Dreg,\overline{m},\delta(Cp)}([0,1],S)$ that is pointwise tight with a certain modulus of pointwise tightness β. Then the function 10.6.32 yields the isometry

$$\Phi_{[0,1],Q(\infty)} : (\widehat{F}_{Dreg,\overline{m},\delta(Cp),\beta}, \widehat{\rho}_{Cp,\xi,[0,1],Q(\infty)}) \\ \to (\widehat{F}_{Dreg,\overline{m},\delta(Cp),\beta}|Q_\infty, \widehat{\rho}_{Marg,\xi,Q(\infty)}). \tag{10.6.35}$$

6. Because $\widehat{F}_{Dreg,\overline{m},\delta(Cp),\beta}|Q_\infty$ inherits the modulus of pointwise tightness β from $\widehat{F}_{Dreg,\overline{m},\delta(Cp),\beta}$, Theorem 6.4.5 says that the function

$$\Phi_{DKS,\xi} : (\widehat{F}_{Dreg,\overline{m},\delta(Cp),\beta}|Q_\infty, \widehat{\rho}_{Marg,\xi,Q(\infty)}) \\ \to (\widehat{R}_{Dreg,\overline{m},\delta(Cp)}(Q_\infty \times \Theta_0, S), \widehat{\rho}_{Prob,Q(\infty)}) \tag{10.6.36}$$

is uniformly continuous, with a modulus of continuity $\delta_{DKS}(\cdot, \|\xi\|, \beta)$.

7. Hence the composite

$$\Phi_{aucl,\xi} : (\widehat{F}_{Dreg,\overline{m},\delta(Cp)}([0,1],S), \widehat{\rho}_{Cp,\xi,[0,1],Q(\infty)}) \\ \to (\widehat{D}_{\delta(aucl),\delta(Cp)}[0,1], \rho_{\widehat{D}[0,1]})$$

of the uniformly continuous functions 10.6.35, 10.6.36, and 10.6.34 is uniformly continuous, with a composite modulus of continuity given by

$$\overline{\delta}_{aucl,\xi} \equiv \overline{\delta}_{aucl,\xi}(\cdot,\overline{m},\delta_{Cp}, \|\xi\|, \beta) \equiv \delta_{DKS}(\delta_{rLim}(\cdot,\overline{m},\delta_{Cp}), \|\xi\|, \beta).$$

The corollary is proved. □

10.7 Strong Right Continuity in Probability

Recall that (S,d) is a locally compact metric space.

In the previous sections, we proved that D-regularity is a necessary and sufficient condition for a family of f.j.d.'s to be extendable to an a.u. càdlàg process. We remarked that this method is a generalization of the treatment of a Markov process with a Markov semigroup in [Chan 1974]. In Chapter 11, we will make precise the notion of a Markov process with a Markov semigroup and will show, by way of D-regularity, that they are a.u. càdlàg. In preparation, we will introduce a sufficient condition for D-regularity, which is later verified for such Markov processes. In addition, this sufficient condition will reduce, in Section 10.8, to a simple condition for a martingale with parameter set $[0,1]$ to be a.u. càdlàg.

This sufficient condition will consist of two subconditions, to be defined presently. The first subcondition will be called, for lack of a better name, *strong right continuity in probability*. The second subcondition will be called *a.u. boundedness*. We emphasize that both subconditions are on f.j.d.'s, or equivalently, on finite samples of the process. We will note that the a.u. boundedness condition will always be satisfied if the locally compact metric state space (S,d) is bounded.

Recall Definition 9.0.2 for the enumerated set Q_∞ of dyadic rationals in $[0,1]$, and the enumerated subset

$$Q_h \equiv \{0, \Delta_h, 2\Delta_h, \ldots, 1\} = \{t_0, t_1, t_2, \ldots, t_{p(h)}\},$$

where $p_h \equiv 2^h$, $\Delta_h \equiv 2^{-h}$, and where $q_{h,i} \equiv i\Delta_h$ for each $i = 0, \ldots, p_h$, for each $h \geq 0$. Recall also the miscellaneous notations and conventions in Definition 9.0.3. In particular, $[\cdot]_1$ is an operation that assigns to each $a \in R$ an integer $[a]_1 \in (a, a+2)$. In addition, we will use the following notations.

Definition 10.7.1. Natural filtration and certain first exit times for a process sampled at regular intervals. In the remainder of this section, let $Z : Q_\infty \times (\Omega, L, E) \to (S,d)$ be an arbitrary process with marginal distributions given by the family F of f.j.d.'s. Let $h \geq 0$ be arbitrary. Considered the process $Z|Q_h$, which is the process Z sampled at the regular interval of Δ_h. Let

$$\mathcal{L}^{(h)} \equiv \{L^{(t,h)} : t \in Q_h\}$$

denote the natural filtration of the process $Z|Q_h$. In other words,

$$L^{(t,h)} \equiv L(Z_r : r \in [0,t]Q_h)$$

for each $t \in Q_h$. Let τ be an arbitrary simple stopping time with values in Q_h relative to the filtration $\mathcal{L}^{(h)}$. Define the probability space

$$L^{(\tau,h)} \equiv \{Y \in L : Y1_{(\tau \leq s)} \in L^{(s)} \text{ for each } s \in Q_h\}.$$

For each $t \in Q_h$ and $\alpha > 0$, recall, from Part 2 of Definition 8.1.12, the simple first exit time

$$\eta_{t,\alpha,[t,1]Q(h)} \equiv \sum_{r \in [t,1]Q(h)} r1_{(d(Z(t),Z(r))>\alpha)} \prod_{s \in [t,r)Q(h)} 1_{(d(Z(t),Z(s))\leq\alpha)}$$
$$+ \prod_{s \in [t,1]Q(h)} 1_{(d(Z(t),Z(s))\leq\alpha)} \tag{10.7.1}$$

for the process $Z|[t,1]Q_h$ to exit the closed α-neighborhood of Z_t. As usual, an empty product is equal to 1, by convention. Similarly, for each $\gamma > 0$, define the r.r.v.

$$\zeta_{h,\gamma} \equiv \sum_{r \in Q(h)} r1_{(d(x(\circ),Z(r))>\gamma)} \prod_{s \in [0,r)Q(h)} 1_{d(x(\circ),Z(s))\leq\gamma}$$
$$+ \prod_{s \in Q(h)} 1_{d(x(\circ),Z(s))\leq\gamma}, \tag{10.7.2}$$

where $x_\circ$ is an arbitrary but fixed reference point in the state space (S,d). It can easily be verified that $\zeta_{h,\gamma}$ is a simple stopping time relative to the filtration $\mathcal{L}^{(h)}$.

Intuitively, $\zeta_{h,\gamma}$ is the first time $r \in Q_h$ when the process $Z|Q_h$ is outside the bounded set $(d(x_\circ,\cdot) \leq \gamma)$, with $\zeta_{h,\gamma}$ set to 1 if no such $s \in Q_h$ exists.

Refer to Propositions 8.1.11 and 8.1.13 for basic properties of simple stopping times and simple first exit times. □

Definition 10.7.2. a.u. Boundedness on Q_∞. Let (S,d) be a locally compact metric space. Suppose the process $Z : Q_\infty \times (\Omega, L, E) \to (S,d)$ is such that for each $\varepsilon > 0$, there exists $\beta_{auB}(\varepsilon) > 0$ so large that

$$P\left(\bigvee_{r\in Q(h)} d(x_\circ, Z_r) > \gamma\right) < \varepsilon \tag{10.7.3}$$

for each $h \geq 0$, for each $\gamma > \beta_{auB}(\varepsilon)$. Then we will say that the process Z and the family F of its marginal distributions are *a.u. bounded*, with the operation β_{auB} as a *modulus of a.u. boundedness*, relative to the reference point $x_\circ \in S$. Note that this condition is trivially satisfied if $d \leq 1$, in which case we can take $\beta_{auB}(\varepsilon) \equiv 1$ for each $\varepsilon > 1$. □

Definition 10.7.3. Strong right continuity in probability on Q_∞. Let (S,d) be a locally compact metric space. Let $Z : Q_\infty \times (\Omega, L, E) \to (S,d)$ be an arbitrary process. Suppose that for each $\varepsilon, \gamma > 0$, there exists $\delta_{SRCp}(\varepsilon,\gamma) > 0$ such that for arbitrary $h \geq 0$ and $s, r \in Q_h$ with $s \leq r < s + \delta_{SRCp}(\varepsilon,\gamma)$, we have

$$\begin{aligned} P_A(d(Z_s, Z_r) > \alpha) &\leq \varepsilon, \\ A \in L^{(s,h)} &\equiv L(Z_r : r \in [0,s]Q_h) \end{aligned} \tag{10.7.4}$$

for each $\alpha > \varepsilon$ and for each $A \in L^{(s,h)}$ with $A \subset (d(x_\circ, Z_s) \leq \gamma)$ and $P(A) > 0$. Then we will say that the process Z is *strongly right continuous in probability*, with the operation δ_{SRCp} as a *modulus of strong right continuity in probability*.

An arbitrary consistent family F of f.j.d.'s, with state space (S,d) and with parameter set Q_∞ or $[0,1]$, is said to be *strongly right continuous in probability*, with the operation δ_{SRCp} as a *modulus of strong right continuity in probability*, if (i) it is continuous in probability and (ii) $F|Q_\infty$ is the family of marginal distributions of some process $Z : Q_\infty \times (\Omega, L, E) \to (S,d)$ that is strongly right continuous in probability, with δ_{SRCp} as a modulus of strong right continuity in probability.

Note that the operation δ_{SRCp} has two variables, but is independent of the sampling frequency h. □

This definition will next be restated without the assumption of $P(A) > 0$ or the reference to the probability P_A.

Lemma 10.7.4. Equivalent definition of strong right continuity in probability. *Let (S,d) be a locally compact metric space. Then a process $Z : Q_\infty \times (\Omega, L, E) \to (S,d)$ is strongly right continuous in probability, with a modulus of strong right continuity in probability δ_{SRCp}, iff for each $\varepsilon, \gamma > 0$, there exists*

$\delta_{SRCp}(\varepsilon,\gamma) > 0$ such that for arbitrary $h \geq 0$ and $s,r \in Q_h$ with $s \leq r < s + \delta_{SRCp}(\varepsilon,\gamma)$, we have

$$P(d(Z_s, Z_r) > \alpha; A) \leq \varepsilon P(A), \tag{10.7.5}$$

for each $\alpha > \varepsilon$ and for each $A \in L^{(s,h)}$ with $A \subset (d(x_\circ, Z_s) \leq \gamma)$.

Proof. Suppose Z is strongly right continuous in probability, with a modulus of strong right continuity in probability δ_{SRCp}. Let $\varepsilon,\gamma > 0$, $h \geq 0$, and $s,r \in Q_h$ be arbitrary with $s \leq r < s + \delta_{SRCp}(\varepsilon,\gamma)$. Consider each $\alpha > 0$ and $A \in L^{(s,h)}$ with $A \subset (d(x_\circ, Z_s) \leq \gamma)$. Suppose, for the sake of contradiction, that

$$P(d(Z_s, Z_r) > \alpha; A) > \varepsilon P(A).$$

Then $P(A) \geq P(d(Z_s, Z_r) > \alpha; A) > 0$. We can divide both sides of the last displayed inequality by $P(A)$ to obtain

$$P_A(d(Z_s, Z_r) > \alpha) \equiv P(A)^{-1} P(d(Z_s, Z_r) > \alpha; A) > \varepsilon,$$

which contradicts inequality 10.7.4 in Definition 10.7.3. Hence inequality 10.7.5 holds. Thus the "only if" part of the lemma is proved. The "if" part is equally straightforward, so its proof is omitted. □

Three more lemmas are presented next to prepare for the main theorem of this section. The first two are elementary.

Lemma 10.7.5. Minimum of a real number and a sum of two real numbers. *For each $a,b,c \in R$, we have $a \wedge (b + c) = b + c \wedge (a - b)$ or, equivalently, $a \wedge (b + c) - b = c \wedge (a - b)$.*

Proof. Write $c' \equiv a - b$. Then $a \wedge (b + c) = (b + c') \wedge (b + c) = b + c' \wedge c = b + (a - b) \wedge c$. □

Lemma 10.7.6. Function on two contiguous intervals. Let $Z : Q_\infty \times \Omega \to S$ *be an arbitrary process. Let $\alpha > 0$ and $\beta > 2\alpha$ be arbitrary, such that the set*

$$A^{\beta}_{r,s} \equiv (d(Z_r, Z_s) > \beta) \tag{10.7.6}$$

is measurable for each $r,s \in Q_\infty$. Let $\omega \in \bigcap_{r,s \in Q(\infty)} (A^{\beta}_{r,s} \cup (A^{\beta}_{r,s})^c)$ and let $h \geq 0$ be arbitrary. Let A_ω, B_ω be arbitrary intervals, with endpoints in Q_h, such that the right endpoint of A_ω is equal to the left endpoint of B_ω. Let $t,t' \in (A_\omega \cup B_\omega)Q_h$ be arbitrary such that $t < t'$. Suppose there exist $x_\omega, y_\omega \in S$ with

$$\bigvee_{r \in A(\omega)Q(h)} d(x_\omega, Z(r,\omega)) \vee \bigvee_{s \in B(\omega)Q(h)} d(y_\omega, Z(s,\omega)) \leq \alpha. \tag{10.7.7}$$

Then

$$\omega \in \bigcap_{r,s \in (t,t')Q(h);\, r \leq s} ((A^{\beta}_{t,r} \cup A^{\beta}_{t,s})(A^{\beta}_{t,r} \cup A^{\beta}_{t',s})(A^{\beta}_{t',r} \cup A^{\beta}_{t',s}))^c. \tag{10.7.8}$$

Proof. With ω fixed, write $z_r \equiv Z(r,\omega) \in S$ for each $r \in Q_\infty$, and write $A \equiv A_\omega$, $B \equiv B_\omega$, $x \equiv x_\omega$, and $y \equiv y_\omega$. Then inequality 10.7.7 can be restated as

$$\bigvee_{r \in AQ(h)} d(x,z_r) \vee \bigvee_{s \in BQ(h)} d(y,z_s) \leq \alpha. \tag{10.7.9}$$

Let $r,s \in (t,t')Q_h \subset AQ_h \cup BQ_h$ be arbitrary with $r \leq s$. By assumption, the endpoints of the intervals A and B are members of Q_h. Then, since the right endpoint of A is equal to the left endpoint of B, there are only three possibilities: (i) $t,r,s \in AQ_h$, (ii) $t,r \in AQ_h$ and $s,t' \in BQ_h$, or (iii) $r,s,t' \in BQ_h$. In case (i), inequality 10.7.9 implies that

$$\begin{aligned} d(z_t,z_r) \vee d(z_t,z_s) &\leq (d(x,z_t)+d(x,z_r)) \vee (d(x,z_t)+d(x,z_s)) \\ &\leq 2\alpha \vee 2\alpha = 2\alpha. \end{aligned}$$

Similarly, in case (ii), we have

$$\begin{aligned} d(z_t,z_r) \vee d(z_{t'},z_s) &\leq (d(x,z_t)+d(x,z_r)) \vee (d(y,z_{t'})+d(y,z_s)) \\ &\leq 2\alpha \vee 2\alpha = 2\alpha. \end{aligned}$$

Similarly, in case (iii), we have

$$\begin{aligned} d(z_{t'},z_r) \vee d(z_{t'},z_s) &\leq (d(y,z_{t'})+d(y,z_r)) \vee (d(y,z_{t'})+d(y,z_s)) \\ &\leq 2\alpha \vee 2\alpha = 2\alpha. \end{aligned}$$

Thus, in each case, we have

$$(d(z_t,z_r) \vee d(z_t,z_s)) \wedge (d(z_t,z_r) \vee d(z_{t'},z_s)) \wedge (d(z_{t'},z_r) \vee d(z_{t'},z_s)) \leq 2\alpha \leq \beta,$$

where $r,s \in (t,t')Q_h$ are arbitrary with $r \leq s$. Equivalently, the desired relation 10.7.8 holds. □

Next is the key lemma.

Lemma 10.7.7. Lower bound for mean waiting time before exit after a simple stopping time. *Let (S,d) be a locally compact metric space. Suppose the process $Z : Q_\infty \times \Omega \to S$ is strongly right continuous in probability, with a modulus of strong right continuity in probability δ_{SRCp}. Let $\varepsilon, \gamma > 0$ be arbitrary. Take any $m \geq 0$ so large that*

$$\Delta_m \equiv 2^{-m} < \delta_{SRCp}(\varepsilon,\gamma).$$

Let $h \geq m$ be arbitrary. Define the simple stopping time $\zeta_{h,\gamma}$ to be the first time when the process $Z|Q_h$ is outside the bounded set $(d(x_\circ,\cdot) \leq \gamma)$, as in Definition 10.7.1. Then the following conditions hold:

1. Let the point $t \in Q_h$ and $\alpha > \varepsilon$ be arbitrary. Let $\overline{\eta}$ be an arbitrary simple stopping time with values in $[t,t+\Delta_m]Q_h$ relative to the natural filtration $\mathcal{L}^{(h)}$ of the process $Z|Q_h$. Let $A \in L^{(\overline{\eta},h)}$ be an arbitrary measurable set such that $A \subset (d(x_\circ,Z_{\overline{\eta}}) \leq \gamma)$. Then

$$P(d(Z_{\overline{\eta}}, Z_{1\wedge(t+\Delta(m))}) > \alpha; A) \leq \varepsilon P(A) \tag{10.7.10}$$

for each $\alpha > \varepsilon$.

2. Suppose $\varepsilon \leq 2^{-2}$. *Let* $\alpha > 2\varepsilon$ *be arbitrary. As an abbreviation, write* $\eta_t \equiv \eta_{t,\alpha,[t,1]Q(h)}$ *for each* $t \in Q_h$. *Let* τ *be an arbitrary simple stopping time with values in* Q_h, *relative to the filtration* $\mathcal{L}^{(h)}$. *Then the r.r.v.*

$$\overline{\eta}_\tau \equiv \sum_{t\in Q(h)} (\eta_t 1_{\eta(t)<\zeta(h,\gamma)} + 1_{\zeta(h,\gamma)\leq\eta(t)})1_{(\tau=t)} \tag{10.7.11}$$

is a simple stopping time with values in Q_h *relative to the filtration* $\mathcal{L}^{(h)}$. *Moreover, we have the upper bound*

$$P(\overline{\eta}_\tau < 1 \wedge (\tau + \Delta_m)) \leq 2\varepsilon \tag{10.7.12}$$

for the probability of a quick first exit after τ, *and we have the lower bound*

$$E(\overline{\eta}_\tau - \tau) \geq 2^{-1}E((1-\tau) \wedge \Delta_m) \tag{10.7.13}$$

for the mean waiting time before exit after τ. *We emphasize that the upper bound 10.7.12 and the lower bound 10.7.13 are independent of the sampling frequency* h.

Proof. 1. Let $\alpha > \varepsilon$ be arbitrary. Let $t \in Q_h$, the simple stopping time $\overline{\eta}$ with values in $[t, t+\Delta_m]Q_h$, and the set $A \in L^{(\overline{\eta},h)}$ with $A \subset (d(x_\circ, Z_{\overline{\eta}}) \leq \gamma)$ be as given. Write $r \equiv 1 \wedge (t + \Delta_m)$. Then $\overline{\eta}$ has values in $[t,r]Q_h$. Let $s \in [t,r]Q_h$ be arbitrary. Then

$$s \leq r \leq t + \Delta_m < s + \delta_{SRCp}(\varepsilon, \gamma)$$

and $(\overline{\eta} = s; A) \in L^{(s,h)}$. Therefore, we can apply inequality 10.7.5 in Lemma 10.7.4 to the modulus of strong right continuity δ_{SRCp}, the points $s, r \in Q_h$, and the measurable set $(\overline{\eta} = s; A) \in L^{(s,h)}$ to obtain

$$P(d(Z_s, Z_r) > \alpha; \overline{\eta} = s; A) \leq \varepsilon P(\overline{\eta} = s; A),$$

where $s \in [t,r]Q_h$ is arbitrary. Consequently,

$$P(d(Z_{\overline{\eta}}, Z_r) > \alpha; A) = \sum_{s\in[t,r]Q(h)} P(d(Z_s, Z_r) > \alpha; \overline{\eta} = s; A)$$

$$\leq \sum_{s\in[t,r]Q(h)} \varepsilon P(\overline{\eta} = s; A) = \varepsilon P(A).$$

Assertion 1 is proved.

2. To prove Assertion 2, suppose $\varepsilon \leq 2^{-2}$, and let $\alpha > 2\varepsilon$ be arbitrary. First consider each $t \in Q_h$. Then, as a special case of the defining equality 10.7.11 in the hypothesis, we have the simple stopping time

$$\overline{\eta}_t \equiv \sum_{t\in Q(h)} (\eta_t 1_{\eta(t)<\zeta(h,\gamma)} + 1_{\zeta(h,\gamma)\leq\eta(t)})$$

$$= \sum_{u\in[t,1]Q(h)} u 1_{\eta(t)=u;u<\zeta(h,\gamma)} + \sum_{u\in[t,1]Q(h)} 1_{\eta(t)=u;u\geq\zeta(h,\gamma)}, \tag{10.7.14}$$

which has values in $[t,1]Q_h$. As consequences of this equality, note that (i′) if $s \in [0,t)Q_h$, then, trivially,

$$1_{(\overline{\eta}(t)=s)} = 0 \in L^{(s,h)};$$

(ii′) if $s \in [t,1)Q_h$, then

$$1_{(\overline{\eta}(t)=s)} = 1_{\eta(t)=s;s<\zeta(h,\gamma)} \in L^{(s,h)}$$

because η_t and $\zeta_{h,\gamma}$ are simple stopping times with values in Q_h relative to the filtration $\mathcal{L}^{(h)}$; and (iii′) if $s = 1$, then

$$1_{(\overline{\eta}(t)=s)} = 1_{\eta(t)=s;s\geq\zeta(h,\gamma)} = 1_{\eta(t)=s} \in L^{(s,h)}.$$

Combining cases (i′–iii′), we see that $1_{(\overline{\eta}(t)=s)} \in L^{(s,h)}$ for each $s \in Q_h$. Thus $\overline{\eta}_t$ is a simple stopping time with values in $[t,1]Q_h$ relative to the filtration $\mathcal{L}^{(h)}$. Intuitively, $\overline{\eta}_t$ is the first time $s \in [t,1]Q_h$ when the process $Z|Q_h$ exits the α-neighborhood of Z_t while staying in the γ-neighborhood of $x_\circ$ over the entire time interval $[t,s]Q_h$; $\overline{\eta}_t$ is set to 1 if no such time s exists.

Continuing, observe that

$$(\overline{\eta}_t < 1) \subset (\overline{\eta}_t = \eta_t < \zeta_{h,\gamma} \leq 1) \tag{10.7.15}$$

by the defining equality 10.7.14.

Define $\overline{\eta} \equiv \overline{\eta}_t \wedge r$, where $r \equiv 1\wedge(t+\Delta_m)$. Then $r-t = (1-t)\wedge\Delta_m$. Moreover, the simple stopping time $\overline{\eta}$ has values in $[t,r]Q_h$ relative to the filtration $\mathcal{L}^{(h)}$. Let $A \in L^{(t,h)}$ be arbitrary. Then $A \in L^{(t,h)} \subset L^{(\overline{\eta},h)}$. We estimate an upper bound for the probability

$$\begin{aligned}
P(\overline{\eta}_t < r; A) &\leq P(\overline{\eta} = \overline{\eta}_t < 1; A) \\
&\leq P(\overline{\eta}_t < 1; \overline{\eta} = \overline{\eta}_t = \eta_t < \zeta_{h,\gamma} \leq 1; A) \\
&= P(\eta_t < 1; \overline{\eta} = \overline{\eta}_t = \eta_t < \zeta_{h,\gamma} \leq 1; A) \\
&\leq P(d(Z_t, Z_{\eta(t)}) > \alpha; \overline{\eta} = \overline{\eta}_t = \eta_t < \zeta_{h,\gamma} \leq 1; A) \\
&= P(d(Z_t, Z_{\overline{\eta}}) > \alpha; \overline{\eta} = \overline{\eta}_t = \eta_t < \zeta_{h,\gamma} \leq 1; A) \\
&\leq P(d(Z_t, Z_{\overline{\eta}}) > \alpha; t < \zeta_{h,\gamma}; A) \\
&\leq P(d(Z_t, Z_{\overline{\eta}}) > \alpha; d(x_\circ, Z_t) \leq \gamma; A) \\
&\leq P(d(Z_t, Z_{\overline{\eta}}) > \alpha; d(Z_t, Z_r) \leq 2^{-1}\alpha; d(x_\circ, Z_t) \leq \gamma; A) \\
&\quad + P(d(Z_t, Z_r) > 2^{-1}\alpha; d(x_\circ, Z_t) \leq \gamma; A) \\
&\leq P(d(Z_{\overline{\eta}}, Z_r) > 2^{-1}\alpha; d(x_\circ, Z_t) \leq \gamma; A) \\
&\quad + P(d(Z_t, Z_r) > 2^{-1}\alpha; d(x_\circ, Z_t) \leq \gamma; A).
\end{aligned} \tag{10.7.16}$$

Here, the second inequality is thanks to relation 10.7.15, the third inequality is by the definition of the simple stopping time time η_t, and the fifth inequality is by the definition of the simple stopping time $\zeta_{h,\gamma}$. Since $2^{-1}\alpha > \varepsilon$, we have

$$P(d(Z_{\overline{\eta}}, Z_r) > 2^{-1}\alpha; d(x_\circ, Z_t) \leq \gamma; A) \leq \varepsilon P(d(x_\circ, Z_t) \leq \gamma; A)$$

by applying inequality 10.7.10 where α, A are replaced by $2^{-1}\alpha$, $(d(x_\circ, Z_t) \leq \gamma; A)$, respectively. Similarly, we have

$$P(d(Z_t, Z_r) > 2^{-1}\alpha; d(x_\circ, Z_t) \leq \gamma; A) \leq \varepsilon P(d(x_\circ, Z_t) \leq \gamma; A)$$

by applying inequality 10.7.10 where $\overline{\eta}, \alpha, A$ are replaced by $t, 2^{-1}\alpha$, $(d(x_\circ, Z_t) \leq \gamma; A)$, respectively. Combining, inequality 10.7.16 can be continued to yield

$$P(\overline{\eta}_t < r; A) \leq 2\varepsilon P(d(x_\circ, Z_t) \leq \gamma; A) \leq 2\varepsilon P(A). \tag{10.7.17}$$

Consequently,

$$\begin{aligned} E(\overline{\eta}_t - t; A) \geq E(r - t; \overline{\eta}_t \geq r; A) &= (r - t)P(\overline{\eta}_t \geq r; A) \\ &= ((1 - t) \wedge \Delta_m)(P(A) - P(\overline{\eta}_t < r; A)) \\ &\geq ((1 - t) \wedge \Delta_m)(P(A) - 2\varepsilon P(A)) \\ &\geq ((1 - t) \wedge \Delta_m)2^{-1}P(A), \end{aligned} \tag{10.7.18}$$

where the second inequality is by inequality 10.7.17, and where the last inequality is because $1 - 2\varepsilon \geq 2^{-1}$ by the assumption that $\varepsilon \leq 2^{-2}$.

3. To complete the proof of Assertion 2, let the simple stopping time τ be arbitrary, with values in Q_h relative to the filtration $\mathcal{L}^{(h)}$. Then

$$\begin{aligned} P(\overline{\eta}_\tau < 1 \wedge (\tau + \Delta_m)) &= \sum_{t \in Q(h)} P(\overline{\eta}_t < 1 \wedge (t + \Delta_m); \tau = t) \\ &\leq \sum_{t \in Q(h)} 2\varepsilon P(\tau = t) = 2\varepsilon, \end{aligned} \tag{10.7.19}$$

where the inequality is by applying inequality 10.7.17 to the measurable set $A \equiv (\tau = t) \in L^{(t,h)}$, for each $t \in Q_h$. Similarly,

$$\begin{aligned} E(\overline{\eta}_\tau - \tau) &= \sum_{t \in Q(h)} E(\overline{\eta}_t - t; \tau = t) \\ &\geq \sum_{t \in Q(h)} ((1 - t) \wedge \Delta_m)2^{-1}P(\tau = t) \\ &= 2^{-1}E((1 - \tau) \wedge \Delta_m), \end{aligned} \tag{10.7.20}$$

where the first inequality is by inequality 10.7.18. Summing up, inequalities 10.7.19 and 10.7.20 yield, respectively, the desired inequalities 10.7.12 and 10.7.13. The lemma is proved. □

Theorem 10.7.8. Strong right continuity in probability and a.u. boundedness together imply D-regularity and extendability by the right limit to an a.u. càdlàg process. *Let (S, d) be a locally compact metric space. Suppose an arbitrary process $Z : Q_\infty \times (\Omega, L, E) \to S$ is* (i) *a.u. bounded, with a modulus of a.u.*

boundedness β_{auB}, *and* (ii) *strongly right continuous in probability, with a modulus of strong right continuity in probability* δ_{SRCp}. *Then the following conditions hold:*

1. The process Z is D-regular, with a modulus of D-regularity $\overline{m} \equiv \overline{m}(\beta_{auB}, \delta_{SRCp})$ *and a modulus of continuity in probability* $\delta_{Cp}(\cdot, \beta_{auB}, \delta_{SRCp})$.

2. The right-limit extension $X \equiv \Phi_{rLim}(Z) : [0,1] \times \Omega \to S$ *is an a.u. càdlàg process, with the same modulus of continuity in probability* $\delta_{Cp}(\cdot, \beta_{auB}, \delta_{SRCp})$ *and a modulus of a.u. càdlàg* $\delta_{aucl}(\cdot, \beta_{auB}, \delta_{SRCp})$. *Recall here the right-limit extension* Φ_{rLim} *from Definition 10.5.6.*

Proof. 1. Condition (i), the a.u. boundedness condition in the hypothesis, says that for each $\varepsilon > 0$, we have

$$P\left(\bigvee_{u \in Q(h)} d(x_\circ, Z_u) > \gamma\right) < \varepsilon \tag{10.7.21}$$

for each $h \geq 0$, for each $\gamma > \beta_{auB}(\varepsilon)$.

2. Let $\varepsilon > 0$ and $\alpha \in (2^{-2}\varepsilon, 2^{-1}\varepsilon)$ be arbitrary. Take an arbitrary

$$\gamma \in (\beta_{auB}(2^{-2}\varepsilon), \beta_{auB}(2^{-2}\varepsilon) + 1).$$

Define

$$m \equiv [-\log_2(1 \wedge \delta_{SRCp}(2^{-2}\varepsilon, \gamma))]_1$$

and

$$\delta_{Cp}(\varepsilon) \equiv \delta_{Cp}(\varepsilon, \beta_{auB}, \delta_{SRCp}) \equiv \Delta_m \equiv 2^{-m} < \delta_{SRCp}(2^{-2}\varepsilon, \gamma).$$

We will verify that $\delta_{Cp}(\cdot, \beta_{auB}, \delta_{SRCp})$ is a modulus of continuity in probability of the process Z. For each $h \geq 0$, define the measurable set

$$\overline{G}_h \equiv \left(\bigvee_{u \in Q(h)} d(x_\circ, Z_u) > \gamma\right).$$

Then, since $\gamma > \beta_{auB}(2^{-2}\varepsilon)$, inequality 10.7.21, with ε replaced by $2^{-2}\varepsilon$, implies that

$$P(\overline{G}_h) < 2^{-2}\varepsilon, \tag{10.7.22}$$

where $h \geq 0$ is arbitrary.

3. Consider each $s, r \in Q_\infty$ with $|s - r| < \delta_{Cp}(\varepsilon)$. Define the measurable set

$$D_{s,r} \equiv (d(Z_s, Z_r) \leq \alpha) \subset (d(Z_s, Z_r) \leq 2^{-1}\varepsilon). \tag{10.7.23}$$

First assume that $s \leq r$. Then

$$s \leq r < s + \delta_{Cp}(\varepsilon) < s + \delta_{SRCp}(2^{-2}\varepsilon, \gamma). \tag{10.7.24}$$

Take $h \geq m$ so large that $s,r \in Q_h$. Since $\alpha > 2^{-2}\varepsilon$, we can apply inequality 10.7.5 in Lemma 10.7.4, where ε and A are replaced by $2^{-2}\varepsilon$ and $(d(x_\circ, Z_s) \leq \gamma)$, respectively, to obtain

$$P(D^c_{s,r}\overline{G}^c_h) \leq P(d(Z_s,Z_r) > \alpha; d(x_\circ, Z_s) \leq \gamma)$$
$$\leq 2^{-2}\varepsilon P(d(x_\circ, Z_s) \leq \gamma) \leq 2^{-2}\varepsilon.$$

Combining with inequality 10.7.22, we obtain

$$P(D^c_{s,r}) \leq P(D^c_{s,r}\overline{G}^c_h) + P(\overline{G}_h) \leq 2^{-2}\varepsilon + 2^{-2}\varepsilon = 2^{-1}\varepsilon. \tag{10.7.25}$$

Consequently, in view of relation 10.7.23, we see that

$$E1 \wedge d(Z_s,Z_r) \leq P(D^c_{s,r}) + E(1 \wedge d(Z_s,Z_r); D_{s,r}) \leq P(D^c_{s,r}) + 2^{-1}\varepsilon \leq \varepsilon.$$

By symmetry, the same inequality holds for each $s,r \in Q_\infty$ with $|s-r| < \delta_{Cp}(\varepsilon)$, where $\varepsilon > 0$ is arbitrary. Hence the process Z is continuous in probability, with a modulus of continuity in probability $\delta_{Cp} \equiv \delta_{Cp}(\cdot, \beta_{auB}, \delta_{SRCp})$. Thus the process Z satisfies Condition 2 of Definition 10.4.1.

4. It remains to prove Condition 1 in Definition 10.4.1 for Z to be D-regular. To that end, define $m_{-1} \equiv 0$. Let $n \geq 0$ be arbitrary, but fixed until further notice. Write $\varepsilon_n \equiv 2^{-n}$. Take any

$$\gamma_n \in (\beta_{auB}(2^{-3}\varepsilon_n), \beta_{auB}(2^{-3}\varepsilon_n) + 1).$$

Define the integer

$$m_n \equiv [m_{n-1} \vee -\log_2(1 \wedge \delta_{SRCp}(2^{-m(n)-n-6}\varepsilon_n, \gamma_n))]_1. \tag{10.7.26}$$

We will prove that the increasing sequence $\overline{m} \equiv (m_k)_{k=0,1,\ldots}$ is a modulus of D-regularity of the process Z. As an abbreviation, define $K_n \equiv 2^{m(n)+n+3}$, $h_n \equiv m_{n+1}$. Define the measurable set

$$G_n \equiv \left(\bigvee_{u \in Q(h(n))} d(x_\circ, Z_u) > \gamma_n \right). \tag{10.7.27}$$

Then, since $\gamma_n > \beta_{auB}(2^{-3}\varepsilon_n)$, inequality 10.7.21 implies that $P(G_n) < 2^{-3}\varepsilon_n$. Moreover,

$$h_n \equiv m_{n+1} > m_n \geq n \geq 0.$$

Equality 10.7.26 can be rewritten as

$$m_n \equiv [m_{n-1} \vee -\log_2(1 \wedge \delta_{SRCp}(K_n^{-1}2^{-3}\varepsilon_n, \gamma_n))]_1. \tag{10.7.28}$$

Then

$$\Delta_{m(n)} \equiv 2^{-m(n)} < \delta_{SRCp}(K_n^{-1}2^{-3}\varepsilon_n, \gamma_n). \tag{10.7.29}$$

5. Now let $\beta > \varepsilon_n$ be arbitrary such that the set

$$A^{\beta}_{r,s} \equiv (d(Z_r, Z_s) > \beta) \tag{10.7.30}$$

is measurable for each $r, s \in Q_\infty$. Define the exceptional set

$$D_n \equiv \bigcup_{t \in Q(m(n))} \bigcup_{r,s \in (t,t')Q(m(n+1)); r \le s} (A^{\beta}_{t,r} \cup A^{\beta}_{t,s})(A^{\beta}_{t,r} \cup A^{\beta}_{t',s})(A^{\beta}_{t',r} \cup A^{\beta}_{t',s}), \tag{10.7.31}$$

where for each $t \in [0,1)Q_{m(n)}$, we write $t' \equiv t + 2^{-m(n)}$. It remains only to prove that $P(D_n) < 2^{-n}$, as required in Condition 1 of Definition 10.4.1.

6. For that purpose, let the simple first stopping time $\zeta \equiv \zeta_{h(n),\gamma(n)}$ be as in Definition 10.7.1. Thus ζ is the first time $s \in Q_{h(n)}$ when the process $Z|Q_{h(n)}$ is outside the γ_n-neighborhood of the reference point $x_\circ$, with ζ set to 1 if no such $s \in Q_h$ exists. Take any

$$\alpha_n \in (2^{-1}\varepsilon_n, 2^{-1}\beta).$$

This is possible because $\beta > \varepsilon_n$ by assumption.

7. Now let $t \in Q_{h(n)}$ be arbitrary, but fixed until further notice. Define the simple first exit time

$$\eta_t \equiv \eta_{t,\alpha(n),[t,1]Q(h(n))}$$

and define the simple stopping time

$$\overline{\eta}_t \equiv \eta_t 1_{\eta(t)<\zeta} + 1_{\zeta \le \eta(t)} = \sum_{s\in[t,1]Q(h(n))} s 1_{\eta(t)=s; s<\zeta} + \sum_{s\in[t,1]Q(h(n))} s 1_{\eta(t)=s; \zeta\le s} \tag{10.7.32}$$

as a special case of formula 10.7.11 in Lemma 10.7.7. Define the trivial simple stopping time $\tau_0 \equiv 0$. Let $k = 1, \ldots, K_n$ be arbitrary. Define the simple stopping time

$$\tau_k \equiv \overline{\eta}_{\tau(k-1)} \equiv \sum_{u\in Q(h(n))} \overline{\eta}_u 1_{(\tau(k-1)=u)}$$

relative to the natural filtration $\mathcal{L}^{(h)}$ of the process $Z|Q_h$. Then $\tau_k \ge \tau_{k-1}$. Intuitively, $\tau_1, \tau_2, \ldots, \tau_{K(n)}$ are the successive stopping times when the process $Z|Q_h$ moves away from the previous stopping state $Z_{\tau(k-1)}$ by a distance of more than α_n while staying within the bounded set $(d(x_\circ, \cdot) < \gamma_n)$. In view of the trivial inequality $2^{-3}\varepsilon_n \le 2^{-2}$ and the bound

$$\alpha_n > 2^{-1}\varepsilon_n > 2(2^{-3}\varepsilon_n),$$

we can apply Part 2 of Lemma 10.7.7, where $\varepsilon, \gamma, m, h, \tau, \Delta$ are replaced by

$$2^{-3}\varepsilon_n, \gamma_n, m_n, h_n, \tau_{k-1}, \Delta_{m(n)},$$

respectively, to obtain the bounds

$$P(\overline{\eta}_{\tau(k-1)} < 1 \wedge (\tau_{k-1} + \Delta_{m(n)})) \leq 2(2^{-3}\varepsilon_n) \tag{10.7.33}$$

and

$$E(\overline{\eta}_{\tau(k-1)} - \tau_{k-1}) \geq 2^{-1}E((1 - \tau_{k-1}) \wedge \Delta_{m(n)}). \tag{10.7.34}$$

Hence

$$\begin{aligned} P(\tau_k - \tau_{k-1} &< (1 - \tau_{k-1}) \wedge \Delta_{m(n)}) \\ &\equiv P(\overline{\eta}_{\tau(k-1)} < \tau_{k-1} + (1 - \tau_{k-1}) \wedge \Delta_{m(n)})) \\ &= P(\overline{\eta}_{\tau(k-1)} < 1 \wedge (\tau_{k-1} + \Delta_{m(n)})) \leq 2(2^{-3}\varepsilon_n), \end{aligned} \tag{10.7.35}$$

where the inequality is due to inequality 10.7.33. Similarly,

$$E(\tau_k - \tau_{k-1}) \equiv E(\overline{\eta}_{\tau(k-1)} - \tau_{k-1}) \geq 2^{-1}E((1 - \tau_{k-1}) \wedge \Delta_{m(n)}), \tag{10.7.36}$$

where the inequality is due to inequality 10.7.34. Recall here that $k = 1, \ldots, K_n$ is arbitrary. Consequently,

$$\begin{aligned} 1 \geq E(\tau_{K(n)}) &= \sum_{k=1}^{K(n)} E(\tau_k - \tau_{k-1}) \geq 2^{-1} \sum_{k=1}^{K(n)} E((1 - \tau_{k-1}) \wedge \Delta_{m(n)}) \\ &\geq 2^{-1} \sum_{k=1}^{K(n)} E((1 - \tau_{K(n)-1}) \wedge \Delta_{m(n)}) \\ &= 2^{-1}K_n E((1 - \tau_{K(n)-1}) \wedge \Delta_{m(n)}) \\ &\geq 2^{-1}K_n E((1 - \tau_{K(n)-1}) \wedge \Delta_{m(n)}; 1 - \tau_{K(n)-1} > \Delta_{m(n)}) \\ &= 2^{-1}K_n E(\Delta_{m(n)}; 1 - \tau_{K(n)-1} > \Delta_{m(n)}) \\ &= 2^{-1}K_n \Delta_{m(n)} P(1 - \tau_{K(n)-1} > \Delta_{m(n)}) \\ &= 2^{-1}K_n \Delta_{m(n)} P(\tau_{K(n)-1} < 1 - \Delta_{m(n)}), \end{aligned}$$

where the second inequality is from inequality 10.7.36. Dividing by $2^{-1}K_n\Delta_{m(n)}$, we obtain

$$P(\tau_{K(n)-1} < 1 - \Delta_{m(n)}) < 2K_n^{-1}\Delta_{m(n)}^{-1} \equiv 2 \cdot 2^{-m(n)-n-3}2^{m(n)} = 2^{-2}\varepsilon_n. \tag{10.7.37}$$

At the same time,

$$\begin{aligned} P(\tau_{K(n)} < 1; \tau_{K(n)-1} \geq 1 - \Delta_{m(n)}) &\equiv P(\overline{\eta}_{\tau(K(n)-1)} < 1; \tau_{K(n)-1} \geq 1 - \Delta_{m(n)}) \\ &\leq P(\overline{\eta}_{\tau(K(n))-1)} < 1 \wedge (\tau_{K(n)-1} + \Delta_{m(n)})) \\ &\leq 2(2^{-3}\varepsilon_n) = 2^{-2}\varepsilon_n, \end{aligned} \tag{10.7.38}$$

where the last inequality is by applying inequality 10.7.35 to $k = K_n$.

8. Next define the exceptional set

$$H_n \equiv (\tau_{K(n)} < 1).$$

Then, combining inequalities 10.7.38 and 10.7.37, we obtain

$$\begin{aligned} P(H_n) &\equiv P(\tau_{K(n)} < 1) \\ &\leq P(\tau_{K(n)} < 1; \tau_{K(n)-1} \geq 1 - \Delta_{m(n)}) + P(\tau_{K(n)-1} < 1 - \Delta_{m(n)}) \\ &\leq 2^{-2}\varepsilon_n + 2^{-2}\varepsilon_n = 2^{-1}\varepsilon_n. \end{aligned}$$

Summing up, except for the small exceptional set $G_n \cup H_n$, there are at most the finite number K_n of simple stopping times $0 < \tau_1 < \cdots < \tau_{K(n)} = 1$, each of which is the first time in $Q_{h(n)}$ when the process Z strays from the previous stopped state by a distance greater than α_n, while still staying in the bounded set $(d(x_\circ, \cdot) < \gamma_n)$. At the same time, inequality 10.7.35 says that each waiting time $\tau_k - \tau_{k-1}$ exceeds a certain lower bound with some probability close to 1. We wish, however, to be able to say that with some probability close to 1, all these K_n waiting times simultaneously exceed a certain lower bound. For that purpose, we will relax the lower bound and specify two more small exceptional sets, as follows.

9. Define two more exceptional sets,

$$B_n \equiv \bigcup_{k=1}^{K(n)} (\tau_k - \tau_{k-1} < (1 - \tau_{k-1}) \wedge \Delta_{m(n)}) \tag{10.7.39}$$

and

$$C_n \equiv \bigcup_{k=1}^{K(n)} (d(Z_{\tau(k-1)\vee(1-\Delta(m(n))}, Z_1) > \alpha_n), \tag{10.7.40}$$

and proceed to estimate $P(B_n)$ and $P(C_n)$.

10. To estimate $P(C_n)$, let $k = 1, \ldots, K_n$ be arbitrary. Trivially,

$$\alpha_n > 2^{-1}\varepsilon_n > 2K_n^{-1}2^{-3}\varepsilon_n.$$

Recall from inequality 10.7.29 that

$$\Delta_{m(n)} \equiv 2^{-m(n)} < \delta_{SRCp}(K_n^{-1}2^{-3}\varepsilon_n, \gamma_n).$$

As an abbreviation, write $t \equiv 1 - \Delta_{m(n)}$. Define the simple stopping time

$$\widetilde{\eta} \equiv \tau_{k-1} \vee t \equiv \tau_{k-1} \vee (1 - \Delta_{m(n)})$$

with values in $[t,1]Q_{h(n)}$ relative to the filtration $\mathcal{L}^{(h(n))}$. Then $(d(x_\circ, Z_{\widetilde{\eta}}) \leq \gamma_n) \in L^{(\widetilde{\eta}, h(n))}$. Hence we can apply relation 10.7.10 in Lemma 10.7.7, where $\varepsilon, \gamma, \alpha, m, h, t, r, \overline{\eta}, \tau, A$ are replaced by $K_n^{-1}2^{-3}\varepsilon_n, \gamma_n, \alpha_n, m_n, h_n, t, 1, \widetilde{\eta}, \tau_{k-1}$, $(d(x_\circ, Z_{\widetilde{\eta}}) \leq \gamma_n)$, respectively, to obtain

$$P(d(Z_{\widetilde{\eta}}, Z_1) > \alpha_n; d(x_\circ, Z_{\widetilde{\eta}}) \leq \gamma_n) \leq K_n^{-1}2^{-3}\varepsilon_n P(d(x_\circ, Z_{\widetilde{\eta}}) \leq \gamma_n) \leq K_n^{-1}2^{-3}\varepsilon_n. \tag{10.7.41}$$

Recalling the defining equalities 10.7.40 and 10.7.27 for the measurable sets C_n and G_n, respectively, we can now estimate

$$P(C_n G_n^c) \equiv P\left(\bigcup_{k=1}^{K(n)} (d(Z_{\widetilde{\eta}}, Z_1) > \alpha_n; \bigvee_{u \in Q(h(n))} d(x_\circ, Z_u) \leq \gamma_n\right)$$

$$\leq P\left(\bigcup_{k=1}^{K(n)} (d(Z_{\widetilde{\eta}}, Z_1) > \alpha_n; d(x_\circ, Z_{\widetilde{\eta}}) \leq \gamma_n)\right)$$

$$\leq \sum_{k=1}^{K(n)} K_n^{-1} 2^{-3} \varepsilon_n = 2^{-3} \varepsilon_n,$$

where the last inequality is by inequality 10.7.41

11. To estimate $P(B_n)$, apply relation 10.7.12 in Lemma 10.7.7, where $\varepsilon, \gamma, \alpha, m, h, \tau$ are replaced by $K_n^{-1} 2^{-3} \varepsilon_n, \gamma_n, \alpha_n, m_n, h_n, \tau_{k-1}$, respectively, to obtain

$$P(\overline{\eta}_{\tau(k-1)} < 1 \wedge (\tau_{k-1} + \Delta_{m(n)})) \leq 2K_n^{-1} 2^{-3} \varepsilon_n. \tag{10.7.42}$$

Since $\overline{\eta}_{\tau(k-1)} \equiv \tau_k$ and

$$1 \wedge (\tau_{k-1} + \Delta_{m(n)}) = \tau_{k-1} + \Delta_{m(n)} \wedge (1 - \tau_{k-1})$$

according to Lemma 10.7.5, inequality 10.7.42 is equivalent to

$$P(\tau_k - \tau_{k-1} < \Delta_{m(n)} \wedge (1 - \tau_{k-1})) \leq 2K_n^{-1} 2^{-3} \varepsilon_n. \tag{10.7.43}$$

Recalling the defining equality 10.7.39 for the measurable sets B_n, we obtain

$$P(B_n) \equiv P\left(\bigcup_{k=1}^{K(n)} (\tau_k - \tau_{k-1} < \Delta_n \wedge (1 - \tau_{k-1}))\right) \leq \sum_{k=1}^{K(n)} 2K_n^{-1} 2^{-3} \varepsilon_n = 2^{-2} \varepsilon_n.$$

Combining, we see that

$$P(G_n \cup H_n \cup B_n \cup C_n) = P(G_n \cup H_n \cup B_n \cup C_n G_n^c)$$
$$\leq 2^{-3} \varepsilon_n + 2^{-1} \varepsilon_n + 2^{-2} \varepsilon_n + 2^{-3} \varepsilon_n = \varepsilon_n.$$

12. Finally, we will prove that $D_n \subset G_n \cup H_n \cup B_n \cup C_n$. To that end, consider each $\omega \in G_n^c H_n^c B_n^c C_n^c$. Then, since $\omega \in G_n^c$, we have $\bigvee_{t \in Q(h(n))} d(x_\circ, Z_t(\omega)) \leq \gamma_n$. Consequently, $\zeta_{h(n), \gamma(n)}(\omega) = 1$ according to the defining equality 10.7.2. Hence, by the defining equality 10.7.32, we have

$$\overline{\eta}_t(\omega) \equiv \eta_t(\omega) 1_{\eta(t,\omega) < 1} + 1_{1 \leq \eta(t,\omega)} = \eta_t(\omega) \tag{10.7.44}$$

for each $t \in Q_{h(n)}$. Separately, since $\omega \in H_n^c$, we have $\tau_{K(n)}(\omega) = 1$. Let $k = 1, \ldots, K_n$ be arbitrary. Write $t \equiv \tau_{k-1}(\omega)$. Then

$$\tau_k(\omega) \equiv \overline{\eta}_{\tau(k-1)}(\omega) = \overline{\eta}_t(\omega) = \eta_t(\omega) \equiv \eta_{t, \alpha(n), [t,1] Q(h(n))}(\omega),$$

where the third equality is by equality 10.7.44. Hence the basic properties of the simple first exit time $\eta_{t,\alpha(n),[t,1]Q(h(n))}$ imply that

$$d(Z(t,\omega),Z(u,\omega)) \le \alpha_n \tag{10.7.45}$$

for each $u \in [t,\tau_k(\omega))Q_h$. In other words,

$$d(Z(\tau_{k-1}(\omega),\omega),Z(u,\omega)) \le \alpha_n \tag{10.7.46}$$

for each $u \in [\tau_{k-1}(\omega),\tau_k(\omega))Q_h$, where $k = 1,\ldots,K_n$ is arbitrary.

Next, let $t \in [0,1)Q_{m(n)}$ be arbitrary, and write $t' \equiv t + \Delta_{m(n)}$. Consider the following two possible cases (i′) and (ii′) regarding the number of members in the sequence $\tau_1(\omega),\ldots,\tau_{K(n)-1}(\omega)$ that are in the interval $(t,t']$.

(i′). Suppose the interval $(t,t']$ contains two or more members in the sequence $\tau_1(\omega),\ldots,\tau_{K-1}(\omega)$. Then there exists $k = 1,\ldots,K_n - 1$ such that

$$\tau_{k-1}(\omega) \le t < \tau_k(\omega) \le \tau_{k+1}(\omega) \le t'. \tag{10.7.47}$$

It follows that

$$\Delta_{m(n)} \equiv t' - t > \tau_{k+1}(\omega) - \tau_k(\omega) \ge \Delta_{m(n)} \wedge (1-\tau_k(\omega)), \tag{10.7.48}$$

where the last inequality is because $\omega \in B_n^c$. Consequently, $\Delta_{m(n)} > 1 - \tau_k(\omega)$. Hence $\Delta_{m(n)} \wedge (1-\tau_k(\omega)) = 1-\tau_k(\omega)$. Therefore the second half of inequality 10.7.48 yields

$$\tau_{k+1}(\omega) - \tau_k(\omega) \ge 1 - \tau_k(\omega),$$

which is equivalent to $1 = \tau_{k+1}(\omega)$. Hence inequality 10.7.47 implies that $t' = 1$ and $\tau_k(\omega) > t = 1 - \Delta_{m(n)}$. It follows that

$$d(Z_{\tau(k)}(\omega),Z_1(\omega)) = d(Z_{\tau(k)\vee(1-\Delta(m(n))}(\omega),Z_1(\omega)) \le \alpha_n, \tag{10.7.49}$$

where the inequality is because $\omega \in C_n^c$. At the same time, because $1 = \tau_{k+1}(\omega)$ and $\omega \in G_n^c H_n^c B_n^c C_n^c$, we have

$$\begin{aligned}[]
[\tau_k(\omega),1)Q_{h(n)} &= [\tau_k(\omega),\tau_{k+1}(\omega))Q_{h(n)} \\
&\equiv [\tau_k(\omega),\overline{\eta}_{\tau(k)}(\omega))Q_{h(n)} = [\tau_k(\omega),\eta_{\tau(k)}(\omega))Q_{h(n)},
\end{aligned}$$

where the last equality follows from equality 10.7.44, where t is replaced by $\tau_k(\omega)$. The basic properties of the simple first exit time $\eta_{\tau(k)}$ imply that

$$d(Z(\tau_k(\omega),\omega),Z(u,\omega)) \le \alpha_n \tag{10.7.50}$$

for each $u \in [\tau_k(\omega),1)Q_{h(n)}$. Combining with inequality 10.7.49 for the endpoint $u = 1$, we see that

$$d(Z(\tau_k(\omega),\omega),Z(u,\omega)) \le \alpha_n \tag{10.7.51}$$

for each $u \in [\tau_k(\omega),1]Q_{h(n)}$. Note that $\beta > 2\alpha_n$ and $t,t' \in [\tau_{k-1}(\omega),\tau_k(\omega)) \cup [\tau_k(\omega),t']$, with $t' = 1$. In view of inequalities 10.7.46 and 10.7.51, we can apply Lemma 10.7.6, where $\alpha, h, A_\omega, B_\omega$ t, t' x_ω, y_ω are replaced by α_n, h_n, $[\tau_{k-1}(\omega), \tau_k(\omega)), [\tau_k(\omega), t']$, t, t', $Z(\tau_{k-1}(\omega), \omega)$, $Z(\tau_k(\omega), \omega)$, respectively, to obtain

$$\omega \in \bigcap_{r,s \in (t,t')Q(h(n)); r \leq s} ((A^\beta_{t,r} \cup A^\beta_{t,s})(A^\beta_{t,r} \cup A^\beta_{t',s})(A^\beta_{t',r} \cup A^\beta_{t',s}))^c$$

$$\equiv \bigcap_{r,s \in (t,t')Q(m(n+1)); r \leq s} ((A^\beta_{t,r} \cup A^\beta_{t,s})(A^\beta_{t,r} \cup A^\beta_{t',s})(A^\beta_{t',r} \cup A^\beta_{t',s}))^c \subset D^c_n. \tag{10.7.52}$$

(ii′) Now suppose the interval $(t, t']$ contains zero or one member in the sequence $\tau_1(\omega), \ldots, \tau_{K(n)-1}(\omega)$. Either way,

$$t, t' \in [\tau_{k-1}(\omega), \tau_k(\omega)) \cup [\tau_k(\omega), \tau_{k+1}(\omega)].$$

Then inequality 10.7.46 holds for both k and $k+1$. Hence we can apply Lemma 10.7.6, where $\alpha, h, A_\omega, B_\omega$ t, t' x_ω, y_ω are replaced by α_n, h_n, $[\tau_k(\omega), \tau_{k+1}(\omega)), [\tau_{k-1}(\omega), \tau_k(\omega))$, t, t', $Z(\tau_{k-1}(\omega), \omega)$, $Z(\tau_k(\omega), \omega)$, respectively, to again obtain relation 10.7.52.

13. Summing up, we have proved that $\omega \in D^c_n$ for each $\omega \in G^c_n H^c_n B^c_n C^c_n$. Consequently, $D_n \subset G_n \cup H_n \cup B_n \cup C_n$, whence

$$P(D_n) \leq P(G_n \cup H_n \cup B_n \cup C_n) < \varepsilon_n \equiv 2^{-n},$$

where $n \geq 0$ is arbitrary. Condition 1 of Definition 10.4.1 has also been verified. Accordingly, the process Z is D-regular, with the sequence $\overline{m} \equiv (m_n)_{n=0,1,\ldots}$ as a modulus of D-regularity, and with a modulus of continuity in probability $\delta_{Cp} \equiv \delta_{Cp}(\cdot, \beta_{auB}, \delta_{SRCp})$. Assertion 1 is proved.

14. Therefore, by Theorem 10.5.8, the right-limit extension $X \equiv \Phi_{rLim}(Z) : [0,1] \times \Omega \to S$ is a.u. càdlàg, with the same modulus of continuity in probability $\delta_{Cp} \equiv \delta_{Cp}(\cdot, \beta_{auB}, \delta_{SRcp})$, and with a modulus of a.u. càdlàg

$$\delta_{aucl}(\cdot, \overline{m}, \delta_{Cp}) \equiv \delta_{aucl}(\cdot, \beta_{auB}, \delta_{SRCp}).$$

Assertion 2 is proved. □

10.8 Sufficient Condition for an a.u. Càdlàg Martingale

Using Theorem 10.7.8 from Section 10.7, we will prove a sufficient condition for a martingale $X : [0,1] \times \Omega \to R$ to be equivalent to one that is a.u. càdlàg.

To that end, recall from Definition 8.3.3 the special convex function $\overline{\lambda} : R \to R$, defined by

$$\overline{\lambda}(x) \equiv 2x + (e^{-|x|} - 1 + |x|) \tag{10.8.1}$$

for each $x \in R$. Theorem 8.3.4 says that the function $\overline{\lambda}$ is increasing and strictly convex, with

$$|x| \leq |\overline{\lambda}(x)| \leq 3|x| \tag{10.8.2}$$

for each $x \in R$.

Also recall Definition 9.0.2 for the enumerated set $Q_\infty \equiv \{t_0, t_1, \ldots\}$ of dyadic rationals in $[0,1]$, and the enumerated subset

$$Q_h \equiv \{0, \Delta_h, 2\Delta_h, \ldots, 1\} = \{t_0, t_1, t_2, \ldots, t_{p(h)}\},$$

where $p_h \equiv 2^h$, $\Delta_h \equiv 2^{-h}$, for each $h \geq 0$.

Lemma 10.8.1. Each martingale on Q_∞ is a.u. bounded. *Let $Z : Q_\infty \times \Omega \to R$ be a martingale relative to some filtration $\mathcal{L} \equiv \{L^{(t)} : t \in Q_\infty\}$. Suppose $b > 0$ is an upper bound of $E|Z_0| \vee E|Z_1| \leq b$. Then the process Z is a.u. bounded in the sense of Definition 10.7.2, with a modulus of a.u. boundedness $\beta_{auB} \equiv \beta_{auB}(\cdot, b)$.*

Proof. Let $\varepsilon > 0$ be arbitrary. Take an arbitrary $\alpha \in (2^{-2}\varepsilon, 2^{-1}\varepsilon)$. Take an arbitrary real number $K > 0$ so large that

$$6b < \frac{1}{6}\alpha^3 K \exp(-3K^{-1}b\alpha^{-1}). \tag{10.8.3}$$

Such a real number K exists because the right-hand side of the inequality 10.8.3 is arbitrarily large for sufficiently large K. Define

$$\beta_{auB}(\varepsilon) \equiv \beta_{auB}(\varepsilon, b) \equiv b\alpha^{-1} + K\alpha.$$

Then, by inequality 10.8.2, we have

$$\begin{aligned} E\overline{\lambda}(K^{-1}Z_1) - E\overline{\lambda}(K^{-1}Z_0) &\leq E|\overline{\lambda}(K^{-1}Z_1)| + E|\overline{\lambda}(K^{-1}Z_0)| \\ &\leq 3E|K^{-1}Z_1| + 3E|K^{-1}Z_0| \\ &\leq 3K^{-1}b + 3K^{-1}b = 6K^{-1}b \\ &< \frac{1}{6}\alpha^3 \exp(-3(E|K^{-1}Z_0| \vee E|K^{-1}Z_1|)\alpha^{-1}), \end{aligned} \tag{10.8.4}$$

where the third inequality is by hypothesis, and the fourth inequality is a consequence of inequality 10.8.3.

Let $h \geq 0$ be arbitrary. Then we can apply Theorem 8.3.4 to the martingale $K^{-1}Z|Q_h : Q_h \times \Omega \to R$, where ε is replaced by α, to obtain

$$P\left(\bigvee_{r \in Q(h)} |Z_r - Z_0| > K\alpha\right) = P\left(\bigvee_{r \in Q(h)} |K^{-1}Z_r - K^{-1}Z_0| > \alpha\right) < \alpha. \tag{10.8.5}$$

Separately, Chebychev's inequality implies that

$$P(|Z_0| > b\alpha^{-1}) \leq b^{-1}\alpha E|Z_0| \leq \alpha.$$

Combining with inequality 10.8.5, we obtain

$$P\left(\bigvee_{r\in Q(h)}|Z_r| > K\alpha + b\alpha^{-1}\right) \le P\left(\bigvee_{r\in Q(h)}|Z_r - Z_0| > K\alpha\right)$$
$$+ P(|Z_0| > b\alpha^{-1}) < 2\alpha < \varepsilon.$$

Consequently, $P(\bigvee_{r\in Q(h)}|Z_r| > \gamma) < \varepsilon$ for each $\gamma > K\alpha + b\alpha^{-1} \equiv \beta_{auB}(\varepsilon)$, where $h \ge 0$ is arbitrary. In other words, the process Z is a.u. bounded in the sense of Definition 10.7.2, with β_{auB} as a modulus of a.u. boundedness relative to the reference point 0 in R. □

Lemma 10.8.2. Martingale after an event observed at time t. *Let $Z : Q_\infty \times (\Omega, L, E) \to R$ be a martingale relative to some filtration $\mathcal{L} \equiv \{L^{(t)} : t \in Q_\infty\}$. Let $t \in Q_\infty$ and $A \in L^{(t)}$ be arbitrary with $P(A) > 0$. Recall from Definition 5.6.4 the conditional probability space (Ω, L, E_A). Then the process*

$$Z|[t,1]Q_\infty : [t,1]Q_\infty \times (\Omega, L, E_A) \to R$$

is a martingale relative to the filtration $\mathcal{L}$.

Proof. Consider each $s, r \in [t,1]Q_\infty$ with $s \le r$. Let $U \in L^{(s)}$ be arbitrary, with U bounded. Then $U1_A \in L^{(s)}$. Hence, since $Z : Q_\infty \times (\Omega, L, E) \to R$ is a martingale relative to the filtration $\mathcal{L}$, we have

$$E_A(Z_rU) \equiv P(A)^{-1}E(Z_rU1_A) = P(A)^{-1}E(Z_sU1_A) \equiv E_A(Z_sU),$$

where $Z_s \in L^{(s)}$. Hence $E_A(Z_r|L^{(s)}) = Z_s$, where $s, r \in [t,1]Q_\infty$ are arbitrary with $s \le r$. Thus the process

$$Z|[t,1]Q_\infty : [t,1]Q_\infty \times (\Omega, L, E_A) \to R$$

is a martingale relative to the filtration $\mathcal{L}$. □

Theorem 10.8.3. Sufficient condition for a martingale on Q_∞ to have an a.u. càdlàg martingale extension to $[0,1]$. *Let $Z : Q_\infty \times \Omega \to R$ be an arbitrary martingale relative to some filtration $\mathcal{L} \equiv \{L^{(t)} : t \in Q_\infty\}$. Suppose the following conditions are satisfied:*

(i). There exists $b > 0$ such that $E|Z_1| \le b$.

(ii). For each $\alpha, \gamma > 0$, there exists $\overline{\delta}_{Rcp}(\alpha,\gamma) > 0$ such that for each $h \ge 1$ and $t, s \in Q_h$ with $t \le s < t + \overline{\delta}_{Rcp}(\alpha,\gamma)$, and for each $A \in L^{(t,h)}$ with $A \subset (|Z_t| \le \gamma)$ and $P(A) > 0$, we have

$$E_A|Z_s| - E_A|Z_t| \le \alpha \tag{10.8.6}$$

and

$$|E_Ae^{-|Z(s)|} - E_Ae^{-|Z(t)|}| \le \alpha. \tag{10.8.7}$$

Then the following conditions hold:

1. The martingale Z is D-regular, with a modulus of continuity in probability $\delta_{Cp} \equiv \delta_{Cp}(\cdot, b, \overline{\delta}_{Rcp})$ and a modulus of D-regularity $\overline{m} \equiv \overline{m}(b, \overline{\delta}_{Rcp})$.

2. Let $X \equiv \Phi_{rLim}(Z) : [0,1] \times \Omega \to R$ be the right-limit extension of Z. Then X is an a.u. càdlàg martingale relative to the right-limit extension $\mathcal{L}^+ \equiv \{L^{(t+)} : t \in [0,1]\}$ of the given filtration $\mathcal{L}$, with $\delta_{Cp}(\cdot, b, \overline{\delta}_{Rcp})$ as a modulus of continuity in probability, and with a modulus of a.u. càdlàg $\delta_{aucl}(\cdot, b, \overline{\delta}_{Rcp})$.

3. Recall from Definition 6.4.2 the metric space $(\widehat{R}(Q_\infty \times \Omega, R), \widehat{\rho}_{Prob,Q(\infty)})$ of stochastic processes with parameter set Q_∞, where sequential convergence relative to the metric $\widehat{\rho}_{Prob,Q(\infty)}$ is equivalent to convergence in probability when stochastic processes are viewed as r.v.'s. Let $\widehat{R}_{Mtgl,b,\overline{\delta}(Rcp)}(Q_\infty \times \Omega, R)$ denote the subspace of $(\widehat{R}(Q_\infty \times \Omega, R), \widehat{\rho}_{Prob,Q(\infty)}))$ consisting of all martingales $Z : Q_\infty \times \Omega \to R$ satisfying Conditions (i–ii) *with a common bound $b > 0$ and a common operation $\overline{\delta}_{Rcp}$. Then the right-limit extension*

$$\Phi_{rLim} : (\widehat{R}_{Mtgl,b,\overline{\delta}(Rcp)}(Q_\infty \times \Omega, R), \widehat{\rho}_{Prob,Q(\infty)})) \\ \to (\widehat{D}_{\delta(aucl),\delta(cp)}[0,1], \rho_{\widehat{D}[0,1]}) \tag{10.8.8}$$

is a well-defined uniformly continuous function, where

$$(\widehat{D}_{\delta(aucl),\delta(cp)}[0,1], \rho_{\widehat{D}[0,1]})$$

is the metric space of a.u. càdlàg processes that share the common modulus of continuity in probability $\delta_{Cp} \equiv \delta_{Cp}(\cdot, b, \overline{\delta}_{Rcp})$, and that share the common modulus of a.u. càdlàg $\delta_{aucl} \equiv \delta_{aucl}(\cdot, \overline{m}, \delta_{Cp})$. Moreover, the mapping Φ_{rLim} has a modulus of continuity $\delta_{rLim}(\cdot, b, \overline{\delta}_{Rcp})$ depending only on b and $\overline{\delta}_{Rcp}$.

Proof. 1. Because Z is a martingale, we have $E|Z_0| \le E|Z_1|$ by Assertion 4 of Proposition 8.2.2. Hence $E|Z_0| \vee E|Z_1| = E|Z_1| \le b$. Therefore, according to Lemma 10.8.1, the martingale Z is a.u. bounded, with the modulus of a.u. boundedness $\beta_{auB} \equiv \beta_{auB}(\cdot, b)$, in the sense of Definition 10.7.2.

2. Let $\varepsilon, \gamma > 0$ be arbitrary. Define

$$\alpha \equiv \alpha_{\varepsilon,\gamma} \equiv 1 \wedge \frac{1}{12}\varepsilon^3 \exp(-3(\gamma \vee b + 1)\varepsilon^{-1}).$$

Take $\kappa \ge 0$ so large that $2^{-\kappa} < \overline{\delta}_{Rcp}(\alpha, \gamma)$, where $\overline{\delta}_{Rcp}$ is the operation given in the hypothesis. Define

$$\delta_{SRCp}(\varepsilon, \gamma) \equiv \delta_{SRCp}(\varepsilon, \gamma, \overline{\delta}_{Rcp}) \equiv 2^{-\kappa} < \overline{\delta}_{Rcp}(\alpha, \gamma).$$

We will show that the operation δ_{SRCp} is a modulus of strong right continuity in probability of the process Z in the sense of Definition 10.7.3.

3. To that end, let $h \ge 0$ and $t, s \in Q_h$ be arbitrary with $t < s < t + \delta_{SRCp}(\varepsilon, \gamma)$. Then

$$t < s < t' \equiv t + \delta_{SRCp}(\varepsilon, \gamma) \equiv t + 2^{-\kappa} < t + \overline{\delta}_{Rcp}(\alpha, \gamma).$$

Hence inequalities 10.8.6 and 10.8.7 hold. Moreover, $2^{-h} \leq s - t < 2^{-\kappa}$, whence $h > \kappa$. Now let $A \in L^{(t,h)}$ be arbitrary with $A \subset (|Z_t| \leq \gamma)$ and $P(A) > 0$. By Lemma 10.8.2, the process

$$Z|[t,1]Q_\infty : [t,1]Q_\infty \times (\Omega, L, E_A) \rightarrow R \tag{10.8.9}$$

is a martingale relative to the filtration $\mathcal{L}$. Hence the process

$$Z|[t,t']Q_h : [t,t']Q_\infty \times (\Omega, L, E_A) \rightarrow R \tag{10.8.10}$$

is also a martingale relative to the filtration $\mathcal{L}$. Consequently, $E_A Z_{t'} = E_A Z_t$ and $E_A|Z_{t'}| \geq E_A|Z_t|$. Moreover,

$$E_A|Z_t| \equiv E(|Z_t|; A)/P(A) \leq E(\gamma; A)/P(A) = \gamma.$$

Taken together, equality 10.8.1 and inequalities 10.8.6 and 10.8.7 imply that

$$\begin{aligned} E_A\overline{\lambda}(Z_{t'}) - E_A\overline{\lambda}(Z_t) &\leq |E_A e^{-|Z(t')|} - E_A e^{-|Z(t)|}| + |E_A|Z_{t'}| - E_A|Z_t|| \\ &= |E_A e^{-|Z(t')|} - E_A e^{-|Z(t)|}| + E_A|Z_{t'}| - E_A|Z_t| \\ &\leq \alpha + \alpha = 2\alpha \leq \frac{1}{6}\varepsilon^3 \exp(-3(b \vee \gamma + 1)\varepsilon^{-1}) \\ &< \frac{1}{6}\varepsilon^3 \exp(-3b\varepsilon^{-1}) \\ &\leq \frac{1}{6}\varepsilon^3 \exp(-3(E_A|Z_t| \vee E_A|Z_{t'}|)\varepsilon^{-1}). \end{aligned}$$

With this inequality, we can apply Assertion 2 of Theorem 8.3.4, with Q, X replaced by $[t,t']Q_h$, $Z|[t,t']Q_h$, respectively, to obtain the bound

$$P_A(|Z_t - Z_u| > \varepsilon) < \varepsilon \tag{10.8.11}$$

for each $u \in [t, t + \delta_{SRCp}(\varepsilon, \gamma))Q_h \subset [t,t']Q_h$, where $\varepsilon > 0$, $\gamma > 0$, $h \geq 0$, and

$$A \in L^{(t,h)} \equiv L(Z_r : r \in [0,t]Q_h)$$

are arbitrary with $A \subset (|X_t| \leq \gamma)$ and $P(A) > 0$. Thus the process Z is strongly right continuous in probability in the sense of Definition 10.7.3, with the operation δ_{SRCp} as a modulus of strong right continuity in probability. In Step 1, we observed that the process Z is a.u. bounded, with the modulus of a.u. boundedness $\beta_{auB} \equiv \beta_{auB}(\cdot, b)$, in the sense of Definition 10.7.2.

4. Thus the process Z satisfies the conditions of Theorem 10.7.8. Accordingly, the process Z is D-regular, with a modulus of D-regularity $\overline{m} \equiv \overline{m}(b, \overline{\delta}_{Rcp}) \equiv \overline{m}(\beta_{auB}, \delta_{SRCp})$, and with a modulus of continuity in probability $\delta_{Cp}(\cdot, b, \overline{\delta}_{Rcp}) \equiv \delta_{Cp}(\cdot, b, \delta_{SRCp})$. In addition, Theorem 10.7.8 says that its right-limit extension $X \equiv \Phi_{rLim}(Z)$ is an a.u. càdlàg process, with the same modulus of continuity in probability $\delta_{Cp}(\cdot, b, \overline{\delta}_{Rcp})$, and with the modulus of a.u. càdlàg $\delta_{aucl}(\cdot, b, \overline{\delta}_{Rcp}) \equiv \delta_{aucl}(\cdot, \beta_{auB}, \delta_{SRCp})$. Assertions 1 is proved. Assertion 2 is proved except that we still need to verify that X is a martingale relative to the filtration $\mathcal{L}^+$.

5. To that end, let $t, r \in [0,1]$ be arbitrary with $t < r$ or $r = 1$. Take any sequence $(u_k)_{k=1,2,...}$ in $[t,r)Q_\infty$ such that $u_k \to t$. Since the process Z is continuous in probability, the sequence $(Z_{u(k)})_{k=1,2,...}$ of r.r.v.'s is Cauchy in probability. Hence, according to Proposition 4.9.4, there exists a subsequence $(Z_{u(k(n))})_{n=1,2,...}$ such that $Y \equiv \lim_{n\to\infty} Z_{u(k(n))}$ is an r.r.v., with $Z_{u(k(n))} \to Y$ a.u. Now consider each $\omega \in A \equiv domain(X_t) \cap domain(Y)$. Then by the definition of the right-limit extension, $\lim_{v\in[t,\infty)Q(\infty)} Z_v(\omega)$ exists and equals $X_t(\omega)$. It follows that $\lim_{n\to\infty} Z_{u(k(n))}(\omega) = X_t(\omega)$. At the same time, $Y(\omega) \equiv \lim_{n\to\infty} Z_{u(k(n))}(\omega)$. Hence $X_t = Y$ on the full set A. Separately, Assertion 5 of Proposition 8.2.2 implies that the family

$$H \equiv \{Z_u : u \in Q_\infty\} = \{Z_t : t \in [0,1]Q_\infty\}$$

is uniformly integrable. It follows from Proposition 5.1.13 that the r.r.v. Y is integrable, with $E|Z_{u(k(n))} - Y| \to 0$. Hence the r.r.v. X_t is integrable, with $E|Z_{u(k(n))} - X_t| \to 0$. Consequently, since $Z_{u(k(n))} \in L^{(u(k(n))} \subset L^{(r)}$, we have $X_t \in L^{(r)}$, where $r \in [0,1]$ is arbitrary with $t < r$ or $r = 1$. It follows that $X_t \in L^{(t+)}$, where $t \in [0,1]$ is arbitrary. In short, the process X is adapted to the filtration $\mathcal{L}^+$.

6. We will next show that the process X is a martingale relative to the filtration $\mathcal{L}^+$. To that end, let $t, s \in [0,1]$ be arbitrary with $t < s$. Now let $r, u \in Q_\infty$ be arbitrary such that $t < r \leq u$ and $s \leq u$. Let the indicator $Y \in L^{(t+)}$ be arbitrary. Then $Y, X_t \in L^{(t+)} \subset L^{(r)}$. Hence, since Z is a martingale relative to the filtration $\mathcal{L}$, we have

$$EYZ_r = EYZ_u.$$

Let $r \downarrow t$ and $u \downarrow s$. Then $E|Z_r - X_t| = E|X_r - X_t| \to 0$ and $E|Z_u - X_s| = E|X_u - X_s| \to 0$. It then follows that

$$EYX_t = EYX_s, \tag{10.8.12}$$

where $t, s \in [0,1]$ are arbitrary with $t < s$. Consider each $t, s \in [0,1]$ with $t \leq s$. Suppose $EYX_t \neq EYX_s$. If $t < s$, then equality 10.8.12 would hold, which is a contradiction. Hence $t = s$. Then trivially $EYX_t = EYX_s$, again a contradiction. We conclude that $EYX_t = EYX_s$ for each $t, s \in [0,1]$ with $t \leq s$, and for each indicator $Y \in L^{(t+)}$. Thus X is a martingale relative to the filtration $\mathcal{L}^+$. Assertion 2 is proved.

7. Assertion 3 remains. Since $Z \in \widehat{R}_{Mtgl,b,\overline{\delta}(Rcp)}(Q_\infty \times \Omega, R)$ is arbitrary, we proved in Step 5 that

$$\widehat{R}_{Mtgl,b,\overline{\delta}(Rcp)}(Q_\infty \times \Omega, R) \subset \widehat{R}_{Dreg,\overline{m},\delta(Cp)}(Q_\infty \times \Omega, S).$$

At the same time, Theorem 10.6.1 says that the right-limit extension function

$$\begin{aligned}\Phi_{rLim} &: (\widehat{R}_{Dreg,\overline{m},\delta(Cp)}(Q_\infty \times \Omega, R), \widehat{\rho}_{Prob,Q(\infty)}) \\ &\to (\widehat{D}_{\delta(aucl),\delta(Cp)}[0,1], \rho_{\widehat{D}[0,1]})\end{aligned} \tag{10.8.13}$$

is uniformly continuous, with a modulus of continuity $\delta_{rLim}(\cdot,\overline{m},\delta_{Cp}) \equiv \delta_{rLim}(\cdot,b,\overline{\delta}_{Rcp})$. Assertion 3 and the theorem are proved. $\square$

Theorem 10.8.3 can be restated in terms of the continuous construction of a.u. càdlàg martingales $X:[0,1]\times\Omega\rightarrow R$ from their marginal distributions. More precisely, let ξ be an arbitrary but fixed binary approximation of R relative to $0\in R$. Recall from Definition 6.2.12 the metric space $(\widehat{F}_{cp}([0,1],R),\widehat{\rho}_{Cp,\xi,[0,1],Q(\infty)})$ of consistent families of f.j.d.'s on $[0,1]$ that are continuous in probability. Let $\widehat{F}_1 \equiv \widehat{F}_{Mtgl,b,\overline{\delta}(Rcp)}([0,1],R)$ be the subspace of consistent families F such that $F|Q_\infty$ gives the marginal distributions of some martingale $Z \in \widehat{R}_{Mtgl,b,\overline{\delta}(Rcp)}(Q_\infty\times\Omega,R)$. Let $F\in\widehat{F}_1$ and $t\in Q_\infty$ be arbitrary. Then $E|Z_t|\leq E|Z_1|\leq b$. Hence, by Lemma 5.3.7, the r.r.v. Z_t has modulus of tightness $\beta(\cdot,t)$ relative to $0\in R$ defined by $\beta(\varepsilon,t)\equiv b\varepsilon^{-1}$ for each $\varepsilon>0$. In other words, the distribution F_t has modulus of tightness β. Thus $\widehat{F}_1|Q_\infty$ is pointwise tight with a modulus of pointwise tightness β. Now let

$$(\Theta_0,L_0,I_0)\equiv\left([0,1],L_0,\int\cdot dx\right)$$

denote the Lebesgue integration space based on the interval $[0,1]$. Recall from Theorem 6.4.5 that the Daniell–Kolmogorov–Skorokhod Extension

$$\Phi_{DKS,\xi}:(\widehat{F}_1|Q_\infty,\widehat{\rho}_{Marg,\xi,Q(\infty)})\rightarrow(\widehat{R}(Q_\infty\times\Theta_0,R),\rho_{Q(\infty)\times\Theta(0),R})$$

is uniformly continuous, with a modulus of continuity $\delta_{DKS}(\cdot,\|\xi\|,\beta)$ dependent only on $\|\xi\|,b$. Hence the composite mapping

$$\Phi_{rLim}\circ\Phi_{DKS,\xi}:(\widehat{F}_{Mtgl,b,\overline{\delta}(Rcp)}([0,1],R)|Q_\infty,\widehat{\rho}_{Marg,\xi,Q(\infty)})$$
$$\rightarrow(\widehat{D}_{\delta(aucl),\delta(cp)}[0,1],\rho_{\widehat{D}[0,1]})$$

is uniformly continuous, with a modulus of continuity depending only on $b,\overline{\delta}_{Rcp},\|\xi\|$.

10.9 Sufficient Condition for Right-Hoelder Process

In this section, $\widehat{F}$ will denote a set of consistent families F of f.j.d.'s with parameter set $[0,1]$, and with a locally compact state space (S,d). We will give a sufficient condition, in terms of triple joint distributions, for a member $F\in\widehat{F}$ to be D-regular. Theorem 10.5.8 will then be applied to construct a corresponding a.u. càdlàg process.

As an application, we will prove that under an additional continuity condition on the double joint distributions, the a.u. càdlàg process so constructed is right-Hoelder, in a sense made precise in the following discussion. This result seems to be new.

In this section, for each family $F \in \widehat{F}$, and for each f.j.d. $F_{r(1),\dots,r(n)}$ in the family F, we will use the same symbol $F_{r(1),\dots,r(n)}$ for the associated probability function. For example, we write $F_{r,s,t}(A) \equiv F_{r,s,t}(1_A)$ for each subset A of S^3 that is measurable relative to a triple joint distribution $F_{r,s,t}$.

Recall Definition 9.0.2 for the notations associated with the enumerated sets $(Q_k)_{k=0,1,\dots}$ and Q_∞ of dyadic rationals in $[0,1]$. In particular, $p_k \equiv 2^k$ and $\Delta_k \equiv 2^{-k}$ for each $k \geq 0$. Separately, $(\Theta_0, L_0, E_0) \equiv ([0,1], L_0, \int \cdot dx)$ denotes the Lebesgue integration space based on the interval $\Theta_0 \equiv [0,1]$. This will serve as a sample space. Let $\xi \equiv (A_q)_{q=1,2,\dots}$ be an arbitrary but fixed binary approximation of the locally compact metric space (S,d) relative to some fixed reference point $x_\circ \in S$. Recall the operation $[\cdot]_1$ that assigns to each $a \in R$ an integer $[a]_1 \in (a, a+2)$.

The next theorem is in essence due to Kolmogorov.

Theorem 10.9.1. Sufficient condition for D-regularity in terms of triple joint distributions. *Let F be an arbitrary consistent family of f.j.d.'s with parameter set $[0,1]$ and state space (S,d). Suppose F is continuous in probability. Let $Z : Q_\infty \times (\Omega, L, E) \to S$ be an arbitrary process with marginal distributions given by $F|Q_\infty$. Suppose there exist two sequences $(\gamma_k)_{k=0,1\dots}$ and $(\alpha_k)_{k=0,1,\dots}$ of positive real numbers such that* (i) $\sum_{k=0}^{\infty} 2^k \alpha_k < \infty$ *and* $\sum_{k=0}^{\infty} \gamma_k < \infty$; (ii) *the set*

$$A_{r,s}^{\prime(k)} \equiv (d(Z_r, Z_s) > \gamma_{k+1}) \tag{10.9.1}$$

is measurable for each $s,t \in Q_\infty$, for each $k \geq 0$; and (iii)

$$P(A_{v,v''}^{\prime(k)} A_{v'',v'}^{\prime(k)}) < \alpha_k,$$

where $v' \equiv v + \Delta_k$ and $v'' \equiv v + \Delta_{k+1} = v' - \Delta_{k+1}$, for each $v \in [0,1)Q_k$, for each $k \geq 0$.

Then the family $F|Q_\infty$ of f.j.d.'s is D-regular. Specifically, let $m_0 \equiv 0$. For each $n \geq 1$, let $m_n \geq m_{n-1} + 1$ be so large that $\sum_{k=m(n)}^{\infty} 2^k \alpha_k < 2^{-n}$ and $\sum_{k=m(n)+1}^{\infty} \gamma_k < 2^{-n-1}$. Then the sequence $(m_n)_{n=0,1,\dots}$ is a modulus of D-regularity of the family $F|Q_\infty$.

Proof. 1. Let $k \geq 0$ be arbitrary. Define

$$D_k' \equiv \bigcup_{v \in [0,1)Q(k)} A_{v,v''}^{\prime(k)} A_{v'',v'}^{\prime(k)}. \tag{10.9.2}$$

Then $P(D_k') \leq 2^k \alpha_k$ according to Condition (iii) in the hypothesis.

2. Inductively, for each $n \geq 0$, take any

$$\beta_n \in (2^{-n}, 2^{-n+1}) \tag{10.9.3}$$

such that, for each $s,t \in Q_\infty$, and for each $k = 0, \dots, n$, the sets

$$(d(Z_t, Z_s) > \beta_k + \cdots + \beta_n) \tag{10.9.4}$$

and

$$A_{t,s}^{(n)} \equiv (d(Z_t, Z_s) > \beta_{n+1}) \tag{10.9.5}$$

are measurable. Note that $\beta_{n,\infty} \equiv \sum_{k=n}^{\infty} \beta_k \le \sum_{k=n}^{\infty} 2^{-k+1} = 2^{-n+2}$ for each $n \ge 0$.

3. Let $n \ge 0$ be arbitrary, but fixed until further notice. For ease of notations, suppress some symbols signifying dependence on n and write $q_i \equiv q_{m(n),i} \equiv 2^{-p(m(n))}$ for each $i = 0, \ldots, p_{m(n)}$. Let $\beta > 2^{-n} > \beta_{n+1}$ be arbitrary such that the set

$$A_{t,s}^{\beta} \equiv (d(Z_t, Z_s) > \beta) \tag{10.9.6}$$

is measurable for each $s,t \in Q_\infty$. Define the exceptional set

$$D_n \equiv \bigcup_{t\in[0,1)Q(m(n))} \bigcup_{r,s\in(t,t')Q(m(n+1));r\le s} (A_{t,r}^{\beta} \cup A_{t,s}^{\beta})(A_{t,r}^{\beta} \cup A_{t',s}^{\beta})(A_{t',r}^{\beta} \cup A_{t',s}^{\beta})$$

$$\subset \bigcup_{t\in[0,1)Q(m(n))} \bigcup_{r,s\in(t,t')Q(m(n+1));r\le s} (A_{t,r}^{(n)} \cup A_{t,s}^{(n)})(A_{t,r}^{(n)} \cup A_{t',s}^{(n)})(A_{t',r}^{(n)} \cup A_{t',s}^{(n)}), \tag{10.9.7}$$

where for each $t \in [0,1)Q_{m(n)}$, we write $t' \equiv t + \Delta_{m(n)} \in Q_{m(n)}$. To verify Condition 1 in Definition 10.4.1 for the sequence $(m)_{n=0,1,\ldots}$ and the process Z, we need only show that

$$P(D_n) \le 2^{-n}. \tag{10.9.8}$$

4. To that end, consider each $\omega \in \left(\bigcup_{k=m(n)}^{m(n+1)} D'_k\right)^c$. Let $t \in [0,1)Q_{m(n)}$ be arbitrary, and write $t' \equiv t + \Delta_{m(n)}$. We will show inductively that for each $k = m_n, \ldots, m_{n+1}$, there exists $r_k \in (t,t']Q_k$ such that

$$\bigvee_{u\in[t,r(k)-\Delta(k)]Q(k)} d(Z_t(\omega), Z_u(\omega)) \le \sum_{j=m(n)+1}^{k} \gamma_j \tag{10.9.9}$$

and such that

$$\bigvee_{v\in[r(k),t']Q(k)} d(Z_v(\omega), Z_{t'}(\omega)) \le \sum_{j=m(n)+1}^{k} \gamma_j, \tag{10.9.10}$$

where an empty sum, by convention, is equal to 0. Start with $k = m_n$. Define $r_k \equiv t'$, whence $r_k - \Delta_k = t$. Then inequalities 10.9.9 and 10.9.10 are trivially satisfied.

Suppose, for some $k = m_n, \ldots, m_{n+1} - 1$, we have constructed $r_k \in (t,t']Q_k$ that satisfies inequalities 10.9.9 and 10.9.10. According to the defining equality 10.9.2, we have

$$\omega \in D_k'^{c} \subset (A'^{(k)}_{r(k)-\Delta(k),r(k)-\Delta(k+1)})^c \cup (A'^{(k)}_{r(k)-\Delta(k+1),r(k)})^c.$$

Hence, by the defining equality 10.9.1, we have (i′)

$$d(Z_{r(k)-\Delta(k)}(\omega), Z_{r(k)-\Delta(k+1)}(\omega)) \leq \gamma_{k+1} \tag{10.9.11}$$

or (ii′)

$$d(Z_{r(k)-\Delta(k+1)}(\omega), Z_{r(k)}(\omega)) \leq \gamma_{k+1}. \tag{10.9.12}$$

In case (i′), define $r_{k+1} \equiv r_k$. In case (ii′), define $r_{k+1} \equiv r_k - \Delta_{k+1}$.

We wish to prove inequalities 10.9.9 and 10.9.10 for $k+1$. To prove inequality 10.9.9 for $k+1$, consider each

$$\begin{aligned} u &\in [t, r_{k+1} - \Delta_{k+1}]Q_{k+1} \\ &= [t, r_k - \Delta_k]Q_k \cup [t, r_k - \Delta_k]Q_{k+1}Q_k^c \cup (r_k - \Delta_k, r_{k+1} - \Delta_{k+1}]Q_{k+1}. \end{aligned} \tag{10.9.13}$$

Suppose $u \in [t, r_k - \Delta_k]Q_k$. Then inequality 10.9.9 in the induction hypothesis trivially implies

$$d(Z_t(\omega), Z_u(\omega)) \leq \sum_{j=m(n)+1}^{k+1} \gamma_j. \tag{10.9.14}$$

Suppose next that $u \in [t, r_k - \Delta_k]Q_{k+1}Q_k^c$. Then $u \leq r_k - \Delta_k - \Delta_{k+1}$. Let $v \equiv u - \Delta_{k+1}$ and $v' \equiv v + \Delta_k = u + \Delta_{k+1}$. Then $v \in [0,1)Q_k$, so the defining inequality 10.9.2 implies that

$$\omega \in D_k'^{\,c} \subset (A_{v,u}'^{(k)})^c \cup (A_{u,v'}'^{(k)})^c,$$

and therefore that

$$d(Z_u(\omega), Z_v(\omega)) \wedge d(Z_u(\omega), Z_{v'}(\omega)) \leq \gamma_{k+1}.$$

It follows that

$$\begin{aligned} d(Z_u(\omega), Z_t(\omega)) &\leq (d(Z_u(\omega), Z_v(\omega)) + d(Z_v(\omega), Z_t(\omega))) \wedge (d(Z_u(\omega), Z_{v'}(\omega)) \\ &\quad + d(Z_{v'}(\omega), Z_t(\omega))) \\ &\leq \gamma_{k+1} + d(Z_v(\omega), Z_t(\omega)) \vee d(Z_{v'}(\omega), Z_t(\omega)) \\ &\leq \gamma_{k+1} + \sum_{j=m(n)+1}^{k} \gamma_j = \sum_{j=m(n)+1}^{k+1} \gamma_j, \end{aligned}$$

where the last inequality is due to inequality 10.9.9 in the induction hypothesis. Thus we have also verified inequality 10.9.14 for each $u \in [t, r_k - \Delta_k]Q_{k+1}Q_k^c$. Now suppose $u \in (r_k - \Delta_k, r_{k+1} - \Delta_{k+1}]Q_{k+1}$. Then

$$r_{k+1} > r_k - \Delta_k + \Delta_{k+1} = r_k - \Delta_{k+1},$$

which, by the definition of r_{k+1}, rules out case (ii′). Hence case (i′) must hold, where $r_{k+1} \equiv r_k$. Consequently,

$$u \in (r_k - \Delta_k, r_k - \Delta_{k+1}]Q_{k+1} = \{r_k - \Delta_{k+1}\},$$

so $u = r_k - \Delta_{k+1}$. Let $v \equiv r_k - \Delta_k$. Then inequality 10.9.11 implies that

$$d(Z_v(\omega), Z_u(\omega)) \equiv d(Z_{r(k)-\Delta(k)}(\omega), Z_{r(k)-\Delta(k+1)}(\omega)) \leq \gamma_{k+1}. \tag{10.9.15}$$

Hence

$$\begin{aligned} d(Z_u(\omega), Z_t(\omega)) &\leq d(Z_u(\omega), Z_v(\omega)) + d(Z_v(\omega), Z_t(\omega)) \\ &\leq \gamma_{k+1} + d(Z_v(\omega), Z_t(\omega)) \leq \gamma_{k+1} + \sum_{j=m(n)+1}^{k} \gamma_j = \sum_{j=m(n)+1}^{k+1} \gamma_j, \end{aligned}$$

where the last inequality is due to inequality 10.9.9 in the induction hypothesis. Combining, we see that inequality 10.9.14 holds for each $u \in [t, r_{k+1} - \Delta_{k+1}] Q_{k+1}$.

Similarly, we can verify that

$$d(Z_u(\omega), Z_{t'}(\omega)) \leq \sum_{j=m(n)+1}^{k+1} \gamma_j \tag{10.9.16}$$

for each $u \in [r_{k+1}, t']Q_{k+1}$. Summing up, inequalities 10.9.9 and 10.9.10 hold for $k+1$. Induction is completed. Thus inequalities 10.9.9 and 10.9.10 hold for each $k = m_n, \ldots, m_{n+1}$.

5. Continuing, let $r, s \in (t, t')Q_{m(n+1)}$ be arbitrary with $r \leq s$. Write $k \equiv m_{n+1}$. Then $r, s \in (t, t')Q_k$, and, by the construction in Step 4, we have $r_k \in (t, t']Q_k$. Hence there are three possibilities: (i″) $r, s \in [t, r_k - \Delta_k]Q_k$, (ii″) $r \in [t, r_k - \Delta_k]Q_k$ and $s \in [r_k, t']Q_k$, or (iii″) $r, s \in [r_k, t']Q_k$. In case (i″), inequality 10.9.9 applies to r, s, yielding

$$d(Z_t(\omega), Z_r(\omega)) \vee d(Z_t(\omega), Z_s(\omega)) \leq \sum_{j=m(n)+1}^{k} \gamma_j < 2^{-n-1} < \beta_{n+1}.$$

Hence $\omega \in (A_{t,r}^{(n)} \cup A_{t,s}^{(n)})^c \subset D_n^c$. In case (ii″), inequalities 10.9.9 and 10.9.10, apply to r and s, respectively, yielding

$$d(Z_t(\omega), Z_r(\omega)) \vee d(Z_s(\omega), Z_{t'}(\omega)) \leq \sum_{j=m(n)+1}^{k} \gamma_j < 2^{-n-1} < \beta_{n+1}.$$

In other words, $\omega \in (A_{t,r}^{(n)} \cup A_{s,t'}^{(n)})^c \subset D_n^c$. Similarly, in case (iii″), we can prove that $\omega \in (A_{r,t'}^{(n)} \cup A_{s,t'}^{(n)})^c \subset D_n^c$.

6. Summing up, we have shown that $\omega \in D_n^c$ where $\omega \in \left(\bigcup_{k=m(n)}^{m(n+1)} D_k'\right)^c$ is arbitrary. Consequently, $D_n \subset \bigcup_{k=m(n)}^{m(n+1)} D_k'$. Hence

$$P(D_n) \leq \sum_{k=m(n)}^{m(n+1)} P(D'_k) \leq \sum_{k=m(n)}^{\infty} 2^k \alpha_k < 2^{-n},$$

where the second inequality follows from the observation in Step 1, and where $n \geq 0$ is arbitrary. This proves inequality 10.9.8 and verifies Condition 1 in Definition 10.4.1 for the sequence $(m)_{n=0,1,...}$ and the process Z. At the same time, since the family F is continuous in probability by hypothesis, Condition 2 of Definition 10.4.1 is also satisfied for $F|Q_\infty$ and for Z. We conclude that the family $F|Q_\infty$ of f.j.d.'s is D-regular, with the sequence $(m)_{n=0,1,...}$ as modulus of D-regularity. □

Definition 10.9.2. Right-Hoelder process. Let $C_0 \geq 0$ and $\lambda > 0$ be arbitrary. Let $X : [0,1] \times (\Omega, L, E) \to S$ be an arbitrary a.u. càdlàg process. Suppose, for each $\varepsilon > 0$, there exist (i) $\widetilde{\delta} > 0$ and (ii) a measurable subset $B \subset \Omega$ with $P(B^c) < \varepsilon$, and (iii) for each $\omega \in B$, there exists a Lebesgue measurable subset $\widetilde{\theta}(\omega)$ of $[0,1]$ with Lebesgue measure $\mu\widetilde{\theta}_k(\omega)^c < \varepsilon$ such that for each $t \in \widetilde{\theta}(\omega) \cap domain(X(\cdot,\omega))$ and for each $s \in [t, t+\widetilde{\delta}) \cap domain(X(\cdot,\omega))$, we have

$$d(X(t,\omega), X(s,\omega)) \leq C_0(s-t)^\lambda.$$

Then the a.u. càdlàg process X is said to be *right-Hoelder*, with *right-Hoelder exponent* λ, and with *right-Hoelder coefficient* C_0.□

Assertion 3 of the next theorem, concerning right-Hoelder processes, seems hitherto unknown. Assertion 2 is in essence due to [Chentsov 1956].

Theorem 10.9.3. Sufficient condition for a right-Hoelder process. *Let $u \geq 0$ and $w, K > 0$ be arbitrary. Let F be an arbitrary consistent family of f.j.d.'s with parameter set* $[0,1]$ *and state space* (S,d). *Suppose F is continuous in probability, with a modulus of continuity in probability δ_{Cp}, and suppose*

$$F_{t,r,s}\{(x,y,z) \in S^3 : d(x,y) \wedge d(y,z) > b\} \leq b^{-u}(Ks - Kt)^{1+w} \qquad (10.9.17)$$

for each $b > 0$ and for each $t \leq r \leq s$ in $[0,1]$. *Then the following conditions hold:*

1. The family $F|Q_\infty$ is D-regular.

2. There exists an a.u. càdlàg process $X : [0,1] \times \Omega \to S$ with marginal distributions given by the family F, and with a modulus of a.u. càdlàg dependent only on u, w and the modulus of continuity in probability δ_{Cp}.

3. Suppose, in addition, that there exists $\overline{\alpha} > 0$ such that

$$F_{t,s}(\widehat{d}) \leq |Ks - Kt|^{\overline{\alpha}} \qquad (10.9.18)$$

for each $t,s \in [0,1]$. Then there exist constants $\lambda(K,u,w,\overline{\alpha}) > 0$ and $C_0(K,u,$ $w,\alpha) > 0$ such that each a.u. càdlàg process

$$X : [0,1] \times \Omega \to S \qquad (10.9.19)$$

with marginal distribution given by the family F is right-Hoelder, with right-Hoelder exponent λ and right-Hoelder constant C_0.

Proof. Let $u \geq 0$ and $w, K > 0$, and the family F be as given.

1. Let $Z : Q_\infty \times \Omega \to S$ be an arbitrary process with marginal distributions given by $F|Q_\infty$. Define $u_0 \equiv u + 2^{-1}$ and $u_1 \equiv u + 1$. Then $\gamma_0 \equiv 2^{-w/u(0)} < \gamma_1 \equiv 2^{-w/u(1)}$. Take an arbitrary $\gamma \in (\gamma_0, \gamma_1)$ such that the subset

$$A'^{(k)}_{r,s} \equiv (d(Z_r, Z_s) > \gamma^{k+1})$$
$$= ((d(Z_r, Z_s))^{1/(k+1)} > \gamma) \tag{10.9.20}$$

of Ω is measurable for each $r, s \in Q_\infty$ and for each $k \geq 0$. Then, since $2^{-w/v}$ is a strictly increasing continuous function of $v \in (u_0, u_1)$ with range (γ_0, γ_1), there exists a unique

$$v \in (u_0, u_1) \equiv (u + 2^{-1}, u + 1)$$

such that

$$\gamma = 2^{-w/v}. \tag{10.9.21}$$

Note that $0 < \gamma < 1$ and that

$$0 < \gamma^{-u} 2^{-w} = 2^{uw/v - w} < 2^{w-w} = 1. \tag{10.9.22}$$

2. Let $k \geq 0$ be arbitrary. Define the positive real numbers

$$\gamma_k \equiv \gamma^k$$

and

$$\alpha_k \equiv K^{1+w} 2^{-k} \gamma^{-(k+1)u} 2^{-kw}.$$

Then

$$\sum_{j=0}^{\infty} \gamma_j = \sum_{j=0}^{\infty} \gamma^j < \infty$$

and, in view of inequality 10.9.22,

$$\sum_{j=0}^{\infty} 2^j \alpha_j = K^{1+w} \gamma^{-u} \sum_{j=0}^{\infty} (\gamma^{-u} 2^{-w})^j < \infty.$$

Let $t \in [0,1)Q_k$ be arbitrary. Write $t' \equiv t + \Delta_k$ and $t'' \equiv t + \Delta_{k+1} = t' - \Delta_{k+1}$. Then

$$P(A'^{(k)}_{t,t''} A'^{(k)}_{t'',t'}) = P(d(Z_t, Z_{t''}) \wedge d(Z_{t''}, Z_{t'}) > \gamma^{k+1})$$
$$\leq \gamma^{-(k+1)u} (Kt' - Kt)^{1+w} = \gamma^{-(k+1)u} (K\Delta_k)^{1+w}$$

$$= \gamma^{-(k+1)u}(K2^{-k})^{1+w}$$

$$= K^{1+w}2^{-k}\gamma^{-(k+1)u}2^{-kw} \equiv \alpha_k, \tag{10.9.23}$$

where the inequality follows from inequality 10.9.17 in the hypothesis.

3. Thus we have verified the conditions in Theorem 10.9.1 for the consistent family $F|Q_\infty$ of f.j.d.'s to be D-regular, with a modulus of D-regularity $(m_n)_{n=0,1,\dots}$ dependent only on the sequences $(\alpha_k)_{k=0,1,\dots}$. and $(\gamma_k)_{k=0,1,\dots}$, which, in turn, depend only on the constants K,u,w. Assertion 1 is proved.

4. Corollary 10.6.2 can now be applied to construct an a.u. càdlàg process $X : [0,1]\times\Omega \to S$ with marginal distributions given by the family F, and with a modulus of a.u. càdlàg depending only on the modulus of D-regularity $(m_n)_{n=0,1,\dots}$ and the modulus of continuity in probability δ_{Cp}. Thus the process X has a modulus of a.u. càdlàg depending only on the constants K,u,w and the modulus of continuity in probability δ_{Cp}. Assertion 2 is proved.

5. Proceed to prove Assertion 3. Suppose, in addition, the positive constant $\overline{\alpha}$ is given and satisfies inequality 10.9.18. Let $X : [0,1] \times \Omega \to S$ be an arbitrary a.u. càdlàg process with marginal distributions given by the family F. Such a process exists by Assertion 2. Consider the D-regular processes $Z \equiv X|Q_\infty$. Then X is equal to the right-limit extension of Z. Use the notations in Steps 1–2. We need to show that X is right-Hoelder. As an abbreviation, define the constants

$$c_0 \equiv w + (1+w)\log_2 K - \log_2(1-\gamma^{-u}2^{-w}),$$

$$c \equiv w(1-uv^{-1}) = w - uwv^{-1} > 0,$$

$$c_1 \equiv -\log_2(1-\gamma) > 0,$$

$$c_2 \equiv -\log_2\gamma = wv^{-1} > 0,$$

$$\kappa_0 \equiv \kappa_0(K,u,w) \equiv [(c^{-1}c_0 - 1) \vee c_2^{-1}(c_1+1)]_1 \geq 1, \tag{10.9.24}$$

and

$$\kappa \equiv \kappa(K,u,w) \equiv [c^{-1} \vee c_2^{-1}]_1 \geq 1. \tag{10.9.25}$$

6. Define the constant integer valued coefficients

$$b_0 \equiv \kappa_0 + (\kappa+1)8 + 6 + [2 + \overline{\alpha}\log_2 K + (2\kappa_0 + 2(\kappa+1)8 + 10)]_1 \tag{10.9.26}$$

and

$$b_1 \equiv \kappa_0 + (\kappa+1) + 6 + [\overline{\alpha}^{-1}2(\kappa+1)]_1. \tag{10.9.27}$$

Define

$$\lambda \equiv (\kappa b_1 + 2^{-1})^{-1} \tag{10.9.28}$$

and write, as an abbreviation,

$$\eta \equiv 2^{-1} = \lambda^{-1} - \kappa b_1.$$

7. Define $m_0 \equiv 0$. Let $n \geq 1$ be arbitrary. Define

$$m_n \equiv \kappa_0 + \kappa n. \tag{10.9.29}$$

Then $m_n \geq m_{n-1} + 1$. Moreover,

$$\begin{aligned}
\log_2 \sum_{h=m(n)}^{\infty} 2^h \alpha_h &= \log_2 \sum_{h=m(n)}^{\infty} 2^h K^{1+w} 2^{-h} \gamma^{-(h+1)u} 2^{-hw} \\
&= \log_2 \gamma^{-(m(n)+1)u} 2^{-m(n)w} K^{1+w} (1 - \gamma^{-u} 2^{-w})^{-1} \\
&= \log_2 2^{(w/v)(m(n)+1)u} 2^{-m(n)w} K^{1+w} (1 - \gamma^{-u} 2^{-w})^{-1} \\
&= -(m_n + 1)(w - uwv^{-1}) + w + (1+w)\log_2 K \\
&\quad - \log_2 (1 - \gamma^{-u} 2^{-w}) \\
&\equiv -(\kappa_0 + \kappa n + 1)c + c_0 < -((c^{-1}c_0 - 1) + \kappa n + 1)c + c_0 \\
&= -\kappa n c < -c^{-1} n c = -n.
\end{aligned}$$

where the last two inequalities are thanks to the defining equalities 10.9.24 and 10.9.25. Hence

$$\sum_{k=m(n)}^{\infty} 2^k \alpha_k < 2^{-n}. \tag{10.9.30}$$

Similarly,

$$\begin{aligned}
\log_2 \sum_{k=m(n)+1}^{\infty} \gamma_k < \log_2 \sum_{k=m(n)}^{\infty} \gamma^k &= \log_2 \gamma^{m(n)} (1-\gamma)^{-1} \\
&= m_n \log_2 \gamma - \log_2 (1-\gamma) \\
&\equiv -(\kappa_0 + \kappa n) c_2 + c_1 \\
&= -\kappa_0 c_2 - \kappa c_2 n + c_1 < -c_2^{-1}(c_1 + 1) c_2 \\
&\quad - \kappa c_2 n + c_1 = -1 - \kappa c_2 n < -1 - n,
\end{aligned}$$

where the first inequality is thanks to equality 10.9.24. Hence

$$\sum_{k=m(n)+1}^{\infty} \gamma_k < 2^{-n-1}. \tag{10.9.31}$$

In view of inequalities 10.9.23, 10.9.30, and 10.9.31, Theorem 10.9.1 applies, and says that the sequence $\overline{m} \equiv (m_n)_{n=0,1,\ldots}$ is a modulus of D-regularity of the family $F|Q_\infty$ and of the process Z.

8. By the hypothesis of Assertion 3, the operation $\overline{\delta}_{Cp}$ defined by

$$\overline{\delta}_{Cp}(\varepsilon) \equiv K^{-\overline{\alpha}} \varepsilon \tag{10.9.32}$$

for each $\varepsilon > 0$ is a modulus of continuity in probability of the family F of f.j.d.'s and of the D-regular process

$$Z: Q_\infty \times \Omega \to S. \tag{10.9.33}$$

9. Theorem 10.5.8 says that the right-limit extension X of Z has a modulus of a.u. càdlàg δ_{aucl} defined as follows. Let $\varepsilon > 0$ be arbitrary. Let $n \geq 0$ be so large that $2^{-n+9} < \varepsilon$. Let $N > m_n + n + 6$ be so large that

$$\Delta_{m(N)} \equiv 2^{-m(N)} < 2^{-2}\overline{\delta}_{Cp}(2^{-2m(n)-2n-10}). \tag{10.9.34}$$

Define

$$\delta_{aucl}(\varepsilon, \overline{m}, \overline{\delta}_{Cp}) \equiv \Delta_{m(N)}. \tag{10.9.35}$$

10. Now fix

$$\overline{k} \equiv [2 - 2\log_2(1 - 2^{-\eta})]_1.$$

Let $k \geq \overline{k}$ be arbitrary, but fixed until further notice. Then

$$k - 2 > 2\log_2(1 - 2^{-\eta}).$$

Write $\varepsilon_k \equiv 2^{-k+2}$, $\delta_k \equiv 2^{-(k-2)\eta}$, $n_k \equiv k + 8$, and

$$\beta \equiv (2\kappa_0 + 2(\kappa + 1))\overline{\alpha}^{-1}.$$

Then $2^{-n(k)+9} < \varepsilon_k$. Define

$$N_k \equiv b_0 + b_1 k. \tag{10.9.36}$$

Then

$$N_k > b_0 > \kappa_0 + (\kappa + 1)(k + 8) + 6 = \kappa_0 + (\kappa + 1)n_k + 6$$
$$= \kappa_0 + \kappa n_k + n_k + 6 = m_{n(k)} + n_k + 6$$

At the same time,

$$\begin{aligned} N_k \equiv b_0 + b_1 k &> 2 + \overline{\alpha}\log_2 K + (2\kappa_0 + 2(\kappa + 1)8 + 10) \\ &\quad + 2(\kappa + 1)k \\ &= 2 + \overline{\alpha}\log_2 K + (2\kappa_0 + 2(\kappa + 1)(k + 8) + 10) \\ &= 2 + \overline{\alpha}\log_2 K + (2\kappa_0 + 2\kappa n_k + 2n_k + 10) \\ &= 2 + \overline{\alpha}\log_2 K + (2m_{n(k)} + 2n_k + 10). \end{aligned}$$

Hence

$$-m(N_k) \leq -N_k < -2 - \overline{\alpha}\log_2 K - (2m_{n(k)} + 2n_k + 10).$$

Consequently,

$$\begin{aligned} \Delta_{m(N(k))} &\equiv 2^{-m(N(k)_k)} < 2^{-2}K^{-\overline{\alpha}}2^{-2m(n(k))-2n(k)-10} \\ &= 2^{-2}\overline{\delta}_{Cp}(2^{-2m(n(k))-2n(k)-10}), \end{aligned} \tag{10.9.37}$$

where the last equality is from the defining equality 10.9.32. Therefore

$$\delta_{aucl}(\varepsilon_k, \overline{m}, \overline{\delta}_{Cp}) \equiv \Delta_{m(N(k))}$$

according to the defining equality 10.9.35.

11. Hence, by Condition 3 in Definition 10.3.2 of a modulus of a.u. càdlàg, there exist an integer $\widetilde{h}_k \geq 1$, a measurable set A_k with

$$P(A_k^c) < \varepsilon_k \equiv 2^{-k+2},$$

and a sequence of r.r.v.'s

$$0 = \tau_{k,0} < \tau_{k,1} < \cdots < \tau_{k,\widetilde{h}(k)-1} < \tau_{k,\widetilde{h}(k)} = 1, \tag{10.9.38}$$

such that for each $i = 0, \ldots, \widetilde{h}_k - 1$, the function $X_{\tau(k,i)}$ is an r.v., and such that for each $\omega \in A_k$, we have

$$\bigwedge_{i=0}^{\widetilde{h}(k)-1} (\tau_{k,i+1}(\omega) - \tau_{k,i}(\omega)) \geq \delta_{aucl}(\varepsilon_k, \overline{m}, \overline{\delta}_{Cp}) \tag{10.9.39}$$

with

$$d(X(\tau_{k,i}(\omega), \omega), X(\cdot, \omega)) \leq \varepsilon_k \tag{10.9.40}$$

on the interval $\theta_{k,i}(\omega) \equiv [\tau_{k,i}(\omega), \tau_{k,i+1}(\omega))$ or $\theta_{k,i}(\omega) \equiv [\tau_{k,i}(\omega), \tau_{k,i+1}(\omega)]$ depending on whether $i \leq \widetilde{h}_k - 2$ or $i = \widetilde{h}_k - 1$.

12. Consider each $\omega \in A_k$ For each $i = 0, \ldots, \widetilde{h}_k - 1$, define the subinterval

$$\overline{\theta}_{k,i}(\omega) \equiv [\tau_{k,i}(\omega), \tau_{k,i+1}(\omega) - \delta_k \cdot (\tau_{k,i+1}(\omega) - \tau_{k,i}(\omega))]$$

of $\theta_{k,i}(\omega)$, with Lebesgue measure

$$\mu\overline{\theta}_{k,i}(\omega) = (1 - \delta_k)(\tau_{k,i+1}(\omega) - \tau_{k,i}(\omega)).$$

Note that the intervals $\theta_{k,0}(\omega), \ldots, \theta_{k,\widetilde{h}(k)-1}(\omega)$ are mutually exclusive. Therefore

$$\mu\left(\bigcup_{i=0}^{\widetilde{h}(k)-1} \overline{\theta}_{k,i}(\omega)\right)^c = \sum_{i=0}^{\widetilde{h}(k)-1} \delta_k \cdot (\tau_{k,i+1}(\omega) - \tau_{k,i}(\omega)) = \delta_k. \tag{10.9.41}$$

13. Define the measurable subset

$$B_k \equiv \bigcap_{h=k}^{\infty} A_h. \tag{10.9.42}$$

Then

$$P(B_k^c) \leq \sum_{h=k}^{\infty} P(A_h^c) < \sum_{h=k}^{\infty} 2^{-h+2} = 2^{-k+3}. \tag{10.9.43}$$

Consider each $\omega \in B_k$. Then $\omega \in A_h$ for each $h \geq k$. Define the Lebesgue measurable subset

$$\tilde{\theta}_k(\omega) \equiv \bigcap_{h=k}^{\infty} \bigcup_{i=0}^{\tilde{h}(h)-1} \overline{\theta}_{h,i}(\omega) \tag{10.9.44}$$

of $[0,1]$. The Lebesgue measure $\mu(\tilde{\theta}_k(\omega)^c)$ is bounded by

$$\mu(\tilde{\theta}_k(\omega)^c) \leq \sum_{h=k}^{\infty} \mu\left(\bigcup_{i=0}^{\tilde{h}(h)-1} \overline{\theta}_{h,i}(\omega)\right)^c$$

$$= \sum_{h=k}^{\infty} \delta_h \equiv \sum_{h=k}^{\infty} 2^{-(h-2)\eta} = 2^{-(k-2)\eta}(1-2^{-\eta})^{-1}, \qquad (10.9.45)$$

where the first equality is from equality 10.9.41. Consider each $t \in \tilde{\theta}_k(\omega) \cap domain(X(\cdot,\omega))$.

14. Let

$$s \in \bigcup_{h=k}^{\infty} (t + \delta_{h+1}\Delta_{m(N(h+1))}, t + \delta_h \Delta_{m(N(h))}) \cap domain(X(\cdot,\omega)) \qquad (10.9.46)$$

be arbitrary. Then

$$s \in (t + \delta_{h+1}\Delta_{m(N(h+1))}, t + \delta_h \Delta_{m(N(h))}) \qquad (10.9.47)$$

for some $h \geq k \geq k_0$. Since $t \in \tilde{\theta}_k(\omega) \subset \bigcup_{i=0}^{\tilde{h}(h)-1} \overline{\theta}_{h,i}(\omega)$, there exists $i = 0, \cdots, \tilde{h}_h - 1$ such that

$$t \in \overline{\theta}_{h,i}(\omega) \equiv [\tau_{h,i}(\omega), \tau_{h,i+1}(\omega) - \delta_h \cdot (\tau_{h,t+1}(\omega) - \tau_{h,i}(\omega))].$$

It follows that

$$\begin{aligned}
s \in (t, t + \delta_h \Delta_{m(N(h))}) &\subset [\tau_{h,i}(\omega), \tau_{h,t+1}(\omega) \\
&\quad - \delta_h \cdot (\tau_{h,t+1}(\omega) - \tau_{h,i}(\omega)) + \delta_h \Delta_{m(N(h))}) \\
&= [\tau_{h,i}(\omega), \tau_{h,i+1}(\omega) - \delta_h \cdot (\tau_{h,i+1}(\omega) - \tau_{h,i}(\omega)) + \delta_h \delta_{aucl}(\varepsilon_h, \overline{m}, \overline{\delta}_{Cp})) \\
&\subset [\tau_{h,t}(\omega), \tau_{h,i+1}(\omega) - \delta_h \cdot (\tau_{h,i+1}(\omega) - \tau_{h,i}(\omega)) \\
&\quad + \delta_h \cdot (\tau_{h,i+1}(\omega) - \tau_{h,i}(\omega))) \\
&= [\tau_{h,i}(\omega), \tau_{h,i+1}(\omega)) \subset \theta_{h,i}(\omega),
\end{aligned}$$

where the second set-inclusion relation is due to inequality 10.9.39. Hence inequality 10.9.40 implies that

$$d(X(\tau_{h,i}(\omega),\omega), X(t,\omega)) \vee d(X(\tau_{h,i}(\omega),\omega), X(s,\omega)) \leq \varepsilon_h, \qquad (10.9.48)$$

and therefore that

$$d(X(t,\omega), X(s,\omega)) \leq 2\varepsilon_h = 2^{-h+3}. \qquad (10.9.49)$$

At the same time, relation 10.9.47 implies

$$s - t \geq \delta_{h+1}\Delta_{m(N(h+1))} \equiv 2^{-\eta(h-1)}2^{-m(N(h+1))}. \qquad (10.9.50)$$

Inequalities 10.9.49 and 10.9.50 then lead to

$$d(X(t,\omega), X(s,\omega))(s-t)^{-\lambda} \leq 2^{-h+3}(2^{\eta(h-1)}2^{m(N(h+1))})^{\lambda}. \qquad (10.9.51)$$

We will show that the right-hand side of inequality 10.9.51 is bounded by a constant. To that end, note that by the definition of λ in Step 6, we have

$$\lambda(\eta + \kappa b_1) = 1.$$

Consequently, the exponent on the right-hand side of inequality 10.9.51 is equal to

$$\begin{aligned}
&-h + 3 + \lambda(\eta(h-1) + m(N_{h+1})) \\
&= -h + 3 + \lambda(\eta(h-1) + \kappa_0 + \kappa N_{h+1}) \\
&= -h + 3 + \lambda(\eta(h-1) + \kappa_0 + \kappa b_0 + \kappa b_1(h+1)) \\
&= -h + 3 + \lambda(-\eta + \kappa_0 + \kappa b_0 + \kappa b_1) + \lambda(\eta + \kappa b_1)h \\
&= -h + 3 + \lambda(-\eta + \kappa_0 + \kappa b_0 + \kappa b_1) + h \\
&= 3 + \lambda(-\eta + \kappa_0 + \kappa b_0 + \kappa b_1) \equiv \tilde{c}.
\end{aligned}$$

Hence inequality 10.9.51 yields

$$d(X(t,\omega), X(s,\omega))(s-t)^{-\lambda} \leq 2^{-h+3}(2^{\eta(h-1)}2^{m(N(h+1))})^{\lambda} < C_0 \equiv 2^{\tilde{c}}.$$

Multiplying this inequality by $(s-t)^{\lambda}$ we see that

$$d(X(t,\omega), X(s,\omega)) \leq C_0(s-t)^{\lambda}, \tag{10.9.52}$$

where $\omega \in B_k$, $t \in \widetilde{\theta}_k(\omega) \cap domain(X(\cdot,\omega))$ and

$$s \in G \equiv \bigcup_{h=k}^{\infty}(t + \delta_{h+1}\Delta_{m(N(h+1))}, t + \delta_h \Delta_{m(N(h))}) \cap domain(X(\cdot,\omega))$$

are arbitrary. Since the set G is a dense subset of $[t, t + \delta_k \Delta_{m(N(k))}) \cap domain(X(\cdot,\omega))$, the right continuity of the function $X(\cdot,\omega)$ implies that inequality 10.9.52 holds for each $\omega \in B_k$, $t \in \widetilde{\theta}_k(\omega) \cap domain(X(\cdot,\omega))$ and $s \in [t, t + \delta_k \Delta_{m(N(k))}) \cap domain(X(\cdot,\omega))$. Since $P(B_k^c)$ and $\mu(\widetilde{\theta}_k(\omega)^c)$ are arbitrarily small for sufficiently large k, the process X is right-Hoelder by Definition 10.9.2. □

Corollary 10.9.4 (next) generalizes the Theorem 10.9.3. The proof of its Assertion 1 by means of a deterministic time scaling is a new and simpler proof of Theorem 13.6 of [Billingsley 1999]. Its Assertion 2, regarding the right-Hoelder property, is hitherto unknown.

Corollary 10.9.4. Sufficient condition for a time-scaled right-Hoelder process. *Let $u \geq 0$ and $w > 0$ be arbitrary. Let $G : [0,1] \to [0,1]$ be a nondecreasing continuous function. Let F be an arbitrary consistent family of f.j.d.'s with parameter set $[0,1]$ and state space (S,d). Suppose F is continuous in probability, and suppose*

$$F_{t,r,s}\{(x,y,z) \in S^3 : d(x,y) \wedge d(y,z) > b\} \leq b^{-u}(G(s) - G(t))^{1+w} \tag{10.9.53}$$

for each $b > 0$ and for each $t \leq r \leq s$ in $[0,1]$. Then the following conditions hold:

1. There exists an a.u. càdlàg process $Y : [0,1] \times \Omega \to S$ with marginal distributions given by the family F.

2. Suppose, in addition, that there exists $\overline{\alpha} > 0$ such that

$$F_{t,s}(\widehat{d}) \leq |G(s) - G(t)|^{\overline{\alpha}} \tag{10.9.54}$$

for each $t,s \in [0,1]$. Then $Y(r,\omega) = X(h(r),\omega)$ for each $(r,\omega) \in domain(Y)$, for some right-Hoelder process X and for some continuous increasing function $h : [0,1] \to [0,1]$.

Proof. Write $a_0 \equiv G(0)$ and $a_1 \equiv G(1)$. Write $K \equiv a_1 - a_0 + 1 > 0$. Define the continuous increasing function $h : [0,1] \to [0,1]$ by $h(r) \equiv K^{-1}(G(r) - a_0 + r)$ for each $r \in [0,1]$. Then its inverse $g \equiv h^{-1}$ is also continuous and increasing. Moreover, for each $s,t \in [0,1]$ with $t \leq s$, we have

$$\begin{aligned} G(g(s)) - G(g(t)) &= (Kh(g(s)) + a_0 - g(s)) - (Kh(g(t)) + a_0 - g(t)) \\ &= (Ks - Kt) - (g(s) - g(t)) \leq Ks - Kt. \end{aligned} \tag{10.9.55}$$

1. Since the family F of f.j.d.'s is, by hypothesis, continuous in probability, the reader can verify that the singleton set $\{F\}$ satisfies the conditions in the hypothesis of Theorem 7.1.7. Accordingly, there exists a process $V : [0,1] \times \Omega \to S$ with marginal distributions given by the family F. Define the function $U : [0,1] \times \Omega \to S$ by

$$domain(U) \equiv \{(t,\omega) : (g(t),\omega) \in domain(V)\}$$

and by

$$U(t,\omega) \equiv V(g(t),\omega) \tag{10.9.56}$$

for each $(t,\omega) \in domain(U)$. Then $U(t,\cdot) \equiv V(g(t),\cdot)$ is an r.v. for each $t \in [0,1]$. Thus U is a stochastic process. Let F' denote the family of its marginal distributions. Then, for each $b > 0$ and for each $t \leq r \leq s$ in $[0,1]$, we have

$$\begin{aligned} &F'_{t,r,s}\{(x,y,z) \in S^3 : d(x,y) \wedge d(y,z) > b) \\ &= P(d(U_t,U_r) \wedge d(U_r,U_s) > b) \\ &\equiv P(d(V_{g(t)},V_{g(r)}) \wedge d(V_{g(r)},V_{g(s)}) > b) \\ &= F_{g(t),g(r),g(s)}\{(x,y,z) \in S^3 : d(x,y) \wedge d(y,z) > b) \\ &\leq b^{-u}(G(g(s)) - G(g(t)))^{1+w} \leq b^{-u}(Ks - Kt)^{1+w}, \end{aligned}$$

where the next-to-last inequality follows from inequality 10.9.53 in the hypothesis, and the last inequality is from inequality 10.9.55. Thus the family F' and the constants K,u,w satisfy the hypothesis of Theorem 10.9.3. Accordingly, there exists an a.u. càdlàg process $X : [0,1] \times \Omega \rightarrow S$ with marginal distributions given by the family F'. Now define a process $Y : [0,1] \times \Omega \rightarrow S$ by

$$domain(Y) \equiv \{(r,\omega) : (h(r),\omega) \in domain(X)\}$$

and by

$$Y(r,\omega) \equiv X(h(r),\omega)$$

for each $(r\omega) \in domain(Y)$. Because the function h is continuous, it can be easily verified that the process Y is a.u. càdlàg. Moreover, in view of the defining equality 10.9.56, we have

$$V(r,\omega) = U(h(r),\omega)$$

for each $(r,\omega) \in domain(V)$. Since the processes X and U share the same marginal distributions given by the family F', the last two displayed equalities imply that the processes Y and V share the same marginal distributions. Since the process V has marginal distributions given by the family F, so does the a.u. càdlàg process Y. Assertion 1 of the corollary is proved.

2. Suppose, in addition, that there exists $\overline{\alpha} > 0$ such that inequality 10.9.54 holds for each $t,s \in [0,1]$. Then

$$\begin{aligned} F'_{t,s}(\widehat{d}) &= E(\widehat{d}(V_{g(t)}, V_{g(s)})) = F_{g(t),g(s)}(\widehat{d}) \\ &\leq |G(g(s)) - G(g(t))|^{\overline{\alpha}} \leq |Ks - Kt|^{\overline{\alpha}}, \end{aligned}$$

where the first inequality is from inequality 10.9.54 in the hypothesis, and where the last inequality is from inequality 10.9.55. Thus the family F' of marginal distributions of the process X and the constants $K,\overline{\alpha}$, satisfy the hypothesis of Assertion 3 of Theorem 10.9.3. Accordingly, the a.u. càdlàg process $X : [0,1] \times \Omega \rightarrow S$ is right-Hoelder. The corollary is proved. □

10.10 a.u. Càdlàg Process on $[0,\infty)$

In the preceding sections, a.u. càdlàg processes are constructed with the unit interval $[0,1]$ as the parameter set. We now generalize to the parameter interval $[0,\infty)$ by constructing the process piecewise on unit subintervals $[M, M+1]$, for $M = 0,1,\ldots$, and then stitching the results back together.

For later reference, we will state several related definitions and will extend an arbitrary D-regular process on $\overline{Q}_\infty$ to an a.u. càdlàg process on $[0,\infty)$. Recall here that Q_∞ and $\overline{Q}_\infty \equiv \{u_0, u_1, \ldots\}$ are the enumerated sets of dyadic rationals in $[0,1]$ and $[0,\infty)$, respectively.

Let (Ω, L, E) be a sample space. Let (S,d) be a locally compact metric space, which will serve as the state space. Let $\xi \equiv (A_q)_{q=1,2,\ldots}$ be a binary approximation of (S,d). Recall that $D[0,1]$ stands for the space of càdlàg functions on $[0,1]$, and that $\widehat{D}[0,1]$ stands for the space of a.u. càdlàg processes on $[0,1]$.

Definition 10.10.1. Skorokhod metric space of càdlàg functions on $[0,\infty)$. Let $x:[0,\infty)\to S$ be an arbitrary function whose domain contains the enumerated set $\overline{Q}_\infty$ of dyadic rationals. Let $M\geq 0$ be arbitrary. Then the function x is said to be *càdlàg* on the interval $[M,M+1]$ if the shifted function $x^M:[0,1]\to S$, defined by $x^M(t)\equiv x(M+t)$ for each t with $M+t\in domain(x)$, is a member of $D[0,1]$. The function $x:[0,\infty)\to S$ is said to be *càdlàg* if it is càdlàg on the interval $[M,M+1]$ for each $M\geq 0$. We will write $D[0,\infty)$ for the set of càdlàg functions on $[0,\infty)$.

Recall, from Definition 10.2.1, the Skorokhod metric $d_{D[0,1]}$ on $D[0,1]$. Define the *Skorokhod metric* on $D[0,\infty)$ by

$$d_{D[0,\infty)}(x,y)\equiv\sum_{M=0}^{\infty}2^{-M-1}(1\wedge d_{D[0,1]}(x^M,y^M))$$

for each $x,y\in D[0,\infty)$. We will call $(D[0,\infty),d_{D[0,\infty)})$ the *Skorokhod space* on $[0,\infty)$. The reader can verify that this metric space is complete. □

Definition 10.10.2. D**-regular processes with parameter set** $\overline{Q}_\infty$. Let $Z:\overline{Q}_\infty\times\Omega\to S$ be a stochastic process. Recall from Definition 10.4.1 the metric space $(\widehat{R}_{Dreg}(Q_\infty\times\Omega,S),\widehat{\rho}_{Prob,Q(\infty)})$ of D-regular processes with parameter set Q_∞.

Suppose, for each $M\geq 0$, (i) the process $Z|\overline{Q}_\infty[0,M+1]$ is continuous in probability, with a modulus of continuity in probability $\delta_{Cp,M}$, and (ii) the shifted process $Z^M:Q_\infty\times\Omega\to S$, defined by $Z^M(t)\equiv Z(M+t)$ for each $t\in Q_\infty$, is a member of the space $\widehat{R}_{Dreg}(Q_\infty\times\Omega,S)$ with a modulus of D-regularity $\overline{m}_M$.

Then the process $Z:\overline{Q}_\infty\times\Omega\to S$ is said to be D-regular, with a modulus of continuity in probability $\widetilde{\delta}_{Cp}\equiv(\delta_{Cp,M})_{M=0,1,\ldots}$ and a modulus of D-regularity $\widetilde{m}\equiv(\overline{m}_M)_{M=0,1,\ldots}$. Let $\widehat{R}_{Dreg}(\overline{Q}_\infty\times\Omega,S)$ denote the set of all D-regular processes with parameter set $\overline{Q}_\infty$. Let $\widehat{R}_{Dreg,\widetilde{\delta}(Cp),\widetilde{m}.}(\overline{Q}_\infty\times\Omega,S)$ denote the subset whose members share the common modulus of continuity in probability $\widetilde{\delta}_{Cp}$ and the common modulus of D-regularity $\widetilde{m}\equiv(\overline{m}_M)_{M=0,1,\ldots}$. If, in addition, $\delta_{Cp,M}=\delta_{Cp,0}$ and $\overline{m}_M=\overline{m}_0$ for each $M\geq 0$, then we say that the process Z *time-uniformly D-regular* on $\overline{Q}_\infty$. □

Definition 10.10.3. Metric space of a.u. càdlàg processes on $[0,\infty)$. Let

$$X:[0,\infty)\times(\Omega,L,E)\to(S,d)$$

be an arbitrary process. Suppose, for each $M\geq 0$, (i) the process $X|[0,M+1]$ is continuous in probability, with a modulus of continuity in probability $\delta_{Cp,M}$, and (ii) the shifted process $X^M:[0,1]\times\Omega\to S$, defined by $X(t)\equiv X(M+t)$ for each $t\in[0,\infty)$, is a member of the space $\widehat{D}[0,1]$ with some modulus of a.u. càdlàg δ^M_{aucl}.

Then the process X is said to be *a.u. càdlàg* on the interval $[0,\infty)$, with a modulus of a.u. càdlàg $\widetilde{\delta}_{aucl}\equiv(\delta^M_{aucl})_{M=0,1,\ldots}$ and a modulus of continuity in probability $\widetilde{\delta}_{Cp} \equiv (\delta_{Cp}^M)_{M=0,1,...}$. If, in addition, $\delta_{aucl}^M = \delta_{aucl}^0$ and $\delta_{Cp}^M = \delta_{Cp}^0$ for each $M \geq 0$, then we say that the process is *time-uniformly a.u. càdlàg* on the interval $[0,\infty)$.

We will write $\widehat{D}[0,\infty)$ for the set of a.u. càdlàg processes on $[0,\infty)$, and equip it with the metric $\widetilde{\rho}_{\widehat{D}[0,\infty)}$ defined by

$$\widetilde{\rho}_{\widehat{D}[0,\infty)}(X,X') \equiv \rho_{Prob,\overline{Q}(\infty)}(X|\overline{Q}_\infty, X'|\overline{Q}_\infty)$$

for each $X, X' \in \widehat{D}[0,\infty)$, where, according to Definition 6.4.2, we have

$$\rho_{Prob,\overline{Q}(\infty)}(X|\overline{Q}_\infty, X'|\overline{Q}_\infty) \equiv E\sum_{n=0}^{\infty} 2^{-n-1}(1 \wedge d(X_{u(n)}, X'_{u(n)})). \quad (10.10.1)$$

Thus

$$\widetilde{\rho}_{\widehat{D}[0,\infty)}(X,X') \equiv E\sum_{n=0}^{\infty} 2^{-n-1}(1 \wedge d(X_{u(n)}, X'_{u(n)})$$

for each $X, X' \in \widehat{D}[0,\infty)$.

Let $\widehat{D}_{\widetilde{\delta}(aucl),\widetilde{\delta}(Cp)}[0,\infty)$ denote the subspace of the metric space $(\widehat{D}[0,\infty), \widetilde{\rho}_{\widehat{D}[0,\infty)})$ whose members share a common modulus of continuity in probability $\widetilde{\delta}_{Cp} \equiv (\delta_{Cp,M})_{M=0,1,...}$ and share a common modulus of a.u. càdlàg $\widetilde{\delta}_{aucl} \equiv (\delta_{aucl}(\cdot, \overline{m}_M, \delta_{Cp,M}))_{M=0,1,...}$. □

Definition 10.10.4. Random càdlàg function from an a.u. càdlàg process on $[0,\infty)$**.** Let $X \in \widehat{D}[0,\infty)$ be arbitrary. Define a function

$$X^* : \Omega \to D[0,\infty)$$

by

$$domain(X^*) \equiv \{\omega \in \Omega : X(\cdot,\omega) \in D[0,\infty)\}$$

and by

$$X^*(\omega) \equiv X(\cdot,\omega)$$

for each $\omega \in domain(X^*)$. We will call X^* the *random càdlàg function* from the a.u. càdlàg process X. The reader can prove that the function X^* is a well-defined r.v. with values in the Skorokhod metric space $(D[0,\infty), d_{D[0,\infty)})$ of càdlàg functions. □

Lemma 10.10.5. a.u. Continuity implies a.u. càdlàg, on $[0,\infty)$**.** *Let* $X : [0,\infty) \times (\Omega, L, E) \to (S,d)$ *be an arbitrary a.u. continuous process. Then* X *is a.u. càdlàg. If, in addition,* X *is time-uniformly a.u. continuous, then* X *is time-uniformly a.u. càdlàg.*

Proof. Straightforward and omitted. □

Corollary 10.10.6. Brownian motion in R^m is time-uniformly a.u. continuous on $[0,\infty)$. *Let $\overline{B} : [0,\infty) \times (\Omega, L, E) \to R^m$ be an arbitrary Brownian motion. Then $\overline{B}$ is time-uniformly a.u. continuous, and hence time-uniformly a.u. càdlàg.*

Proof. Straightforward and omitted. □

In the following, recall the right-limit extension mappings Φ_{rLim} and $\overline{\Phi}_{rLim}$ from Definition 10.5.6.

Lemma 10.10.7. The right-limit extension of a D-regular process on $\overline{Q}_\infty$ is continuous in probability on $[0,\infty)$. *Suppose*

$$Z : \overline{Q}_\infty \times (\Omega, L, E) \to (S,d)$$

is an arbitrary D-regular process on $\overline{Q}_\infty$, with a modulus of continuity in probability $\widetilde{\delta}_{Cp} \equiv (\delta_{Cp,M})_{M=0,1,\dots}$. Then the right-limit extension

$$X \equiv \overline{\Phi}_{rLim}(Z) : [0,\infty) \times (\Omega, L, E) \to (S,d)$$

of Z is a well-defined process that is continuous in probability, with the same modulus of continuity in probability $\widetilde{\delta}_{Cp} \equiv (\delta_{Cp,M})_{M=0,1,\dots}$ as Z.

Proof. Let $N \geq 0$ be arbitrary. Consider each $t,t' \in [0,N+1]$. Let $(r_k)_{k=1,2,\dots}$ be a sequence in $[0,N+1]\overline{Q}_\infty$ such that $r_k \downarrow t$. Since the process Z is continuous in probability on $[0,N+1]\overline{Q}_\infty$, with a modulus of continuity in probability $\delta_{Cp,N}$, it follows that

$$E(1 \wedge d(Z_{rk)}, Z_{r(h)})) \to 0$$

as $k,h \to 0$. By passing to a subsequence if necessary, we can assume

$$E(1 \wedge d(Z_{r(k)}, Z_{r(k+1)})) \leq 2^{-k}$$

for each $k \geq 1$. Then

$$E(1 \wedge d(Z_{r(k)}, Z_{r(h)})) \leq 2^{-k+1} \tag{10.10.2}$$

for each $h \geq k \geq 1$. It follows that $Z_{r(k)} \to U_t$ a.u. for some r.v. U_t. Consequently, $Z_{rk)}(\omega) \to U_t(\omega)$ for each ω in some full set B. Consider each $\omega \in B$. Since $X \equiv \overline{\Phi}_{rLim}(Z)$ and since $r_k \downarrow t$ with $Z_{r(k)}(\omega) \to U_t(\omega)$, we see that $\omega \in domain(X_t)$ and $X_t(\omega) = U_t(\omega)$. In short, $X_t = U_t$ a.s. Consequently, X_t is an r.v., where $t \in [0,N+1]$ and $N \geq 0$ are arbitrary. Since $[0,\infty) = \bigcup_{M=0}^{\infty}[0,M+1]$, it follows that X_u is an r.v. for each $u \in [0,\infty)$. Thus $X : [0,\infty) \times (\Omega, L, E) \to (S,d)$ is a well-defined process.

Letting $h \to \infty$ in equality 10.10.2, we obtain

$$E(1 \wedge d(Z_{r(k)}, X_t)) = E(1 \wedge d(Z_{r(k)}, U_t)) \leq 2^{-k+1}$$

for each $k \geq 1$. Similarly, we can construct a sequence $(r'_k)_{k=1,2,\dots}$ in $[0,N+1]\overline{Q}_\infty$ such that $r'_k \downarrow t'$ and such that

$$E(1 \wedge d(Z_{r'(k)}, X_{t'})) \leq 2^{-k+1}.$$

Now let $\varepsilon > 0$ be arbitrary, and suppose $t, t' \in [0, N+1]$ are such that $|t - t'| < \delta_{Cp,N}(\varepsilon)$. Then $|r_k - r'_k| < \delta_{Cp,N}(\varepsilon)$, whence

$$E(1 \wedge d(Z_{r(k)}, Z_{r'(k)})) \leq \varepsilon$$

for sufficiently large $k \geq 1$. Combining, we obtain

$$E(1 \wedge d(X_t, X_{t'})) \leq 2^{-k+1} + \varepsilon + 2^{-k+1}$$

for sufficiently large $k \geq 1$. Hence

$$E(1 \wedge d(X_t, X_{t'})) \leq \varepsilon,$$

where $t, t' \in [0, N+1]$ are arbitrary with $|t - t'| < \delta_{Cp,N}(\varepsilon)$. Thus we have verified that the process X is continuous in probability, with a modulus of continuity in probability $\widetilde{\delta}_{Cp} \equiv (\delta_{Cp.N})_{N=0,1,\ldots}$, which is the same as the modulus of continuity in probability of the given D-regular process. □

Theorem 10.10.8. The right-limit extension of a D-regular process on $\overline{Q}_\infty$ is an a.u. càdlàg process on $[0, \infty)$. *Suppose*

$$Z : \overline{Q}_\infty \times (\Omega, L, E) \to (S, d)$$

is an arbitrary D-regular process on $\overline{Q}_\infty$, with a modulus of continuity in probability $\widetilde{\delta}_{Cp} \equiv (\delta_{Cp,M})_{M=0,1,\ldots}$, and with a modulus of D-regularity $\widetilde{m} \equiv (\overline{m}_M)_{M=0,1,\ldots}$. In symbols, suppose

$$Z \in \widehat{R}_{Dreg,\widetilde{\delta}(Cp),\widetilde{m}.}(\overline{Q}_\infty \times \Omega, S).$$

Then the right-limit extension

$$X \equiv \overline{\Phi}_{rLim}(Z) : [0, \infty) \times (\Omega, L, E) \to (S, d)$$

of Z is an a.u. càdlàg process, with the same modulus of continuity in probability $\widetilde{\delta}_{Cp} \equiv (\delta_{Cp,M})_{M=0,1,\ldots}$ as Z, and with a modulus of a.u. càdlàg $\widetilde{\delta}_{aucl} \equiv \widetilde{\delta}_{aucl}(\widetilde{m}, \widetilde{\delta}_{Cp})$. In other words,

$$X \equiv \overline{\Phi}_{rLim}(Z) \in \widehat{D}_{\widetilde{\delta}(aucl,\widetilde{m},\widetilde{\delta}(Cp)),\widetilde{\delta}(Cp)}[0, \infty) \subset \widehat{D}[0, \infty).$$

Proof. 1. Lemma 10.10.7 says that the process X is continuous in probability, with the same modulus of continuity in probability $\widetilde{\delta}_{Cp} \equiv (\delta_{Cp,M})_{M=0,1,\ldots}$ as Z. Let $N \geq 0$ be arbitrary. Then $Z^N : Q_\infty \times (\Omega, L, E) \to (S, d)$ is a D-regular process with a modulus of continuity in probability $\delta_{Cp,N}$ and a modulus of D-regularity $\overline{m}_N$. Theorem 10.5.8 therefore implies that the right-limit extension process

$$Y_N \equiv \Phi_{rLim}(Z^N) : [0, 1] \times (\Omega, L, E) \to (S, d)$$

is a.u. càdlàg, with the same modulus of continuity in probability $\delta_{Cp,N}$ and a modulus of a.u. càdlàg $\delta_{aucl}(\cdot, \overline{m}_N, \delta_{Cp,N})$. Separately, Proposition 10.5.7 implies that the process Y_N is continuous a.u., with a modulus of continuity a.u. $\delta_{cau}(\cdot,\overline{m}_N,\delta_{Cp,N})$. Here the reader is reminded that continuity a.u. is not to be confused with the stronger condition of a.u. continuity. Note that $Y_N = Z^N$ on Q_∞ and $X = Z$ on $\overline{Q}_\infty$.

2. Let $k \geq 1$ be arbitrary. Define

$$\delta_k \equiv \delta_{cau}(2^{-k},\overline{m}_N,\delta_{Cp,N}) \wedge \delta_{cau}(2^{-k},\overline{m}_{N+1},\delta_{Cp,N+1}) \wedge 2^{-k}.$$

Then, since $\delta_{cau}(\cdot,\overline{m}_N,\delta_{Cp,N})$ is a modulus of continuity a.u. of the process Y_N, there exists, according to Definition 6.1.2, a measurable set $D_{1,k} \subset domain(Y_{N,1})$ with $P(D_{1,k}^c) < 2^{-k}$ such that for each $\omega \in D_{1,k}$ and for each $r \in domain(Y_N(\cdot,\omega))$ with $|r-1| < \delta_k$, we have

$$d(Y_N(r,\omega),Z(N+1,\omega)) = d(Y_N(r,\omega),Y_N(1,\omega)) \leq 2^{-k}. \qquad (10.10.3)$$

Likewise, since $\delta_{cau}(\cdot,\overline{m}_{N+1},\delta_{Cp,N+1})$ is a modulus of continuity a.u. of the process Y_{N+1}, there exists, according to Definition 6.1.2, a measurable set $D_{0,k} \subset domain(Y_{N+1,0})$ with $P(D_{0,k}^c) < 2^{-k}$ such that for each $\omega \in D_{0,k}$ and for each $r \in domain(Y_{N+1}(0,\omega))$ with $|r-0| < \delta_k$, we have

$$d(Y_{N+1}(r,\omega),Z(N+1,\omega)) = d(Y_{N+1}(r,\omega),Y_{N+1}(0,\omega)) \leq 2^{-k}. \qquad (10.10.4)$$

Define $D_{k+} \equiv \bigcap_{h=k}^{\infty} D_{1,h}D_{0,h}$ and $B \equiv \bigcup_{\kappa=1}^{\infty} D_{\kappa+}$. Then $P(D_{k+}^c) < 2^{-k+2}$. Hence $P(B) = 1$. In words, B is a full set.

3. Consider each $t \in [N,N+1)$. Since $Y_N \equiv \Phi_{rLim}(Z^N)$ and $X \equiv \overline{\Phi}_{rLim}(Z)$, we have

$$\begin{aligned}
domain(Y_{N,t-N}) &\equiv \left\{\omega \in \Omega : \lim_{s\to t-N;s\in[t-N,\infty)Q(\infty)} Z_s^N(\omega) \text{ exists}\right\} \\
&= \left\{\omega \in \Omega : \lim_{s\to t-N;s\in[t-N,1]Q(\infty)} Z_s^N(\omega) \text{ exists}\right\} \\
&= \left\{\omega \in \Omega : \lim_{s\to t-N;s\in[t-N,1]Q(\infty)} Z(N+s,\omega) \text{ exists}\right\} \\
&= \left\{\omega \in \Omega : \lim_{r\to t;r\in[t,N+1]\overline{Q}(\infty)} Z(r,\omega) \text{ exists}\right\} \\
&= \left\{\omega \in \Omega : \lim_{r\to t;r\in[t,\infty)\overline{Q}(\infty)} Z(r,\omega) \text{ exists}\right\} \\
&\equiv domain(X_t),
\end{aligned}$$

because each limit that appears in the previous equality exists iff all others exist, in which case they are equal. Hence

$$X_t(\omega) = \lim_{r\to t;r\in[t,\infty)\overline{Q}(\infty)} Z(r,\omega) = \lim_{s\to t-N;s\in[t-N,\infty)Q(\infty)} Z_s^N(\omega) = Y_{N,t-N}(\omega)$$

for each $\omega \in domain(Y_{N,t-N})$. Thus the two functions X_t and $Y_{N,t-N}$ have the same domain and have equal values on the common domain. In short,

$$X_t = Y_{N,t-N}, \qquad (10.10.5)$$

where $t \in [N, N+1)$ is arbitrary. As for the endpoint $t = N+1$, we have, trivially,

$$X_{N+1} = Z_{N+1} = Z_1^N = Y_{N,1} = Y_{N,(N+1)-1}.$$

Hence

$$X_t = Y_{N,t-N} \tag{10.10.6}$$

for each $t \in [N, N+1) \cup \{N+1\}$.

4. We wish to extend equality 10.10.6 to each $t \in [N, N+1]$. To that end, consider each

$$t \in [N, N+1].$$

We will prove that

$$X_t = Y_{N,t-N} \tag{10.10.7}$$

on the full set B.

5. Let $\omega \in B$ be arbitrary. Suppose $\omega \in domain(X_t)$. Then, since $X \equiv \overline{\Phi}_{rLim}(Z)$, the limit $\lim_{r \to t; r \in [t,\infty)\overline{Q}(\infty)} Z(r,\omega)$ exists and is equal to $X_t(\omega)$. Consequently, each of the following limits

$$\lim_{r \to t; r \in [t, N+1]\overline{Q}(\infty)} Z(r,\omega) = \lim_{N+s \to t; s \in [t-N, 1]\overline{Q}(\infty)} Z(N+s,\omega)$$

$$= \lim_{s \to t-N; s \in [t-N, 1]\overline{Q}(\infty)} Z(N+s,\omega)$$

$$= \lim_{s \to t-N; s \in [t-N, 1]Q(\infty)} Z_s^N(\omega)$$

exists and is equal to $X_t(\omega)$. Since $Y_N \equiv \Phi_{rLim}(Z^N)$, the existence of the last limit implies that $\omega \in domain(Y_{N,t-N})$ and $Y_{N,t-N}(\omega) = X_t(\omega)$. Thus

$$domain(X_t) \subset domain(Y_{N,t-N}) \tag{10.10.8}$$

and

$$X_t = Y_{N,t-N} \tag{10.10.9}$$

on $domain(X_t)$. We have proved half of the desired equality 10.10.7.

6. Conversely, suppose $\omega \in domain(Y_{N,t-N})$. Then $y \equiv Y_{N,t-N}(\omega) \in S$ is defined. Hence, since $Y_N \equiv \Phi_{rLim}(Z^N)$, the limit

$$\lim_{s \to t-N; s \in [t-N,\infty)Q(\infty)} Z^N(s,\omega)$$

exists and is equal to y. Let $\varepsilon > 0$ be arbitrary. Then there exists $\delta' > 0$ such that

$$d(Z^N(s,\omega), y) < \varepsilon$$

for each

$$s \in [t-N,\infty)Q_\infty = [t-N,\infty)[0,1]\overline{Q}_\infty$$

such that $s - (t - N) < \delta'$. In other words,

$$d(Z(u,\omega), y) < \varepsilon \tag{10.10.10}$$

for each $u \in [t, t + \delta')[N, N+1]\overline{Q}_\infty$.

7. Recall the assumption that $\omega \in B \equiv \bigcup_{\kappa=1}^{\infty} D_{\kappa+}$. Therefore there exists some $\kappa \geq 1$ such that $\omega \in D_{\kappa+} \equiv \bigcap_{k=\kappa}^{\infty} D_{1,k} D_{0,k}$. Take $k \geq \kappa$ so large that $2^{-k} < \varepsilon$. Then $\omega \in D_{1,k} D_{0,k}$. Therefore, for each $r \in domain(Y_{N+1}(0,\omega))$ with $|r - 0| < \delta_k$, we have, according to inequality 10.10.4,

$$d(Y_{N+1}(r,\omega), Z(N+1,\omega)) \leq 2^{-k} < \varepsilon. \tag{10.10.11}$$

Similarly, for each $r \in domain(Y_N(\cdot,\omega))$ with $|r - 1| < \delta_k$, we have, according to inequality 10.10.3,

$$d(Y_N(r,\omega), Z(N+1,\omega)) \leq 2^{-k} < \varepsilon. \tag{10.10.12}$$

8. Now let $u, v \in [t, t + \delta_k \wedge \delta')\overline{Q}_\infty$ be arbitrary with $u < v$. Then $u, v \in [t, t + \delta_k \wedge \delta')[N, \infty)\overline{Q}_\infty$. Since u, v are dyadic rationals, there are three possibilities: (i) $u < v \leq N+1$, (ii) $u \leq N+1 < v$, or (iii) $N+1 < u < v$. Consider case (i). Then $u, v \in [t, t+\delta')[N, N+1]\overline{Q}_\infty$. Hence inequality 10.10.10 applies to u and v and yields

$$d(Z(u,\omega), y) \vee d(Z(v,\omega), y) < \varepsilon,$$

whence

$$d(Z(u,\omega), Z(v,\omega)) < 2\varepsilon.$$

Next consider case (ii). Then $|(u - N) - 1| < v - t < \delta_k$. Hence, by inequality 10.10.12, we obtain

$$\begin{aligned} d(Z(u,\omega), Z(N+1,\omega)) &\equiv d(Z^N(u - N,\omega), Z(N+1,\omega)) \\ &= d(Y_N(u - N,\omega), Z(N+1,\omega)) < \varepsilon. \end{aligned} \tag{10.10.13}$$

Similarly, $|(v - (N+1)) - 0| < v - t < \delta_k$. Hence, by inequality 10.10.11, we obtain

$$\begin{aligned} d(Z(v,\omega), Z(N+1,\omega)) &\equiv d(Z^{N+1}(v - (N+1),\omega), Z(N+1,\omega)) \\ &= d(Y_{N+1}(v - (N+1),\omega), Z(N+1,\omega)) < \varepsilon. \end{aligned} \tag{10.10.14}$$

Combining 10.10.13 and 10.10.14, we obtain

$$d(Z(u,\omega), Z(v,\omega)) < 2\varepsilon$$

in case (ii) as well.

Now consider case (iii). Then $|(u - (N+1)) - 0| < v - t < \delta_k$ and $|(v - (N+1)) - 0| < v - t < \delta_k$. Hence inequality 10.10.11 implies

$$d(Z(u,\omega), Z(N+1,\omega)) = d(Z^{N+1}(u-(N+1),\omega), Z(N+1,\omega))$$
$$= d(Y_{N+1}(u-(N+1),\omega), Z(N+1,\omega)) < \varepsilon \tag{10.10.15}$$

and, similarly,

$$d(Z(v,\omega), Z(N+1,\omega)) < \varepsilon. \tag{10.10.16}$$

Hence

$$d(Z(u,\omega), Z(v,\omega)) < 2\varepsilon$$

in case (iii) as well.

9. Summing up, we see that $d(Z(u,\omega), Z(v,\omega)) < 2\varepsilon$ for each $u,v \in [t, t+\delta_k \wedge \delta')\overline{Q}_\infty$ with $u < v$. Since $\varepsilon > 0$ is arbitrary, we conclude that $\lim_{u \to t; u \in [t,\infty)\overline{Q}(\infty)} Z(u,\omega)$ exists. Thus

$$(t,\omega) \in domain(\overline{\Phi}_{rLim}(Z)) \equiv domain(X).$$

In other words, we have $\omega \in domain(X_t)$, where $\omega \in B \cap domain(Y_{N,t-N})$ is arbitrary. Hence

$$B \cap domain(Y_{N,t-N}) \subset domain(X_t) \subset domain(Y_{N,t-N}),$$

where the second inclusion is by relation 10.10.8 in Step 5. Consequently,

$$B \cap domain(Y_{N,t-N}) = B \cap domain(X_t),$$

while, according to equality 10.10.9,

$$X^N_{t-N} = X_t = Y_{N,t-N} \tag{10.10.17}$$

on $B \cap domain(X_t)$, In other words, on the full subset B, we have $X^N_{t-N} = Y_{N,t-N}$ for each $t \in [N, N+1]$. Equivalently, $X^N = Y_N$ on the full subset B. Since the process

$$Y_N \equiv \Phi_{rLim}(Z^N) : [0,1] \times (\Omega, L, E) \to (S,d)$$

is a.u. càdlàg with a modulus of a.u. càdlàg $\delta_{aucl}(\cdot, \overline{m}_N, \delta_{Cp,N})$, the same is true for the process X^N. Thus, by Denition 10.10.3, the process X is a.u. càdlàg, with a modulus of continuity in probability $\widetilde{\delta}_{Cp} \equiv (\delta_{Cp,M})_{M=0,1,\ldots}$, and a modulus of a.u. càdlàg $\widetilde{\delta}_{aucl} \equiv (\delta_{aucl}(\cdot, \overline{m}_M, \delta_{Cp,M}))_{M=0,1,\ldots}$. In other words,

$$X \in \widehat{D}_{\widetilde{\delta}(aucl),\widetilde{\delta}(Cp)}[0,\infty).$$

The theorem is proved. □

The next theorem is straightforward, and is proved here for future reference.

Theorem 10.10.9. $\overline{\Phi}_{rLim}$ is an isometry on a properly restricted domain. *Recall from Definition 10.10.2 the metric space $(\widehat{R}_{Dreg,\widetilde{\delta}(Cp),\widetilde{m}.}(\overline{Q}_\infty \times \Omega, S), \widehat{\rho}_{Prob,\overline{Q}(\infty)})$ of D-regular processes whose members Z share a given modulus of continuity in probability $\widetilde{\delta}_{Cp} \equiv (\delta_{Cp,M})_{M=0,1,...}$ as well as a given modulus of D-regularity $\widetilde{m} \equiv (\overline{m}_M)_{M=0,1,...}$.*

Recall from Definition 10.10.3 the metric space $(\widehat{D}[0,\infty), \widetilde{\rho}_{\widehat{D}[0,\infty)})$ of a.u. càdlàg processes on $[0,\infty)$, where

$$\widetilde{\rho}_{\widehat{D}[0,\infty)}(X,X') \equiv \rho_{Prob,\overline{Q}(\infty)}(X|\overline{Q}_\infty, X'|\overline{Q}_\infty)$$

for each $X, X' \in \widehat{D}[0,\infty)$.

Then the mapping

$$\overline{\Phi}_{rLim} : (\widehat{R}_{Dreg,\widetilde{\delta}(Cp),\widetilde{m}.}(\overline{Q}_\infty \times \Omega, S), \widehat{\rho}_{Prob,\overline{Q}(\infty)})$$
$$\to \widehat{D}_{\widetilde{\delta}(aucl,\widetilde{m},\widetilde{\delta}(Cp)),\widetilde{\delta}(Cp)}[0,\infty) \subset (\widehat{D}[0,\infty), \widetilde{\rho}_{\widehat{D}[0,\infty)})$$

is a well-defined isometry on its domain, where the modulus of a.u. càdlàg $\widetilde{\delta}_{aucl} \equiv \widetilde{\delta}_{aucl}(\widetilde{m}, \widetilde{\delta}_{Cp})$ is defined in the following proof.

Proof. 1. Let $Z \in \widehat{R}_{Dreg,\widetilde{\delta}(Cp),\widetilde{m}.}(\overline{Q}_\infty \times \Omega, S)$ be arbitrary. In other words,

$$Z, : \overline{Q}_\infty \times (\Omega, L, E) \to (S,d)$$

is a D-regular process, with a modulus of continuity in probability $\widetilde{\delta}_{Cp} \equiv (\delta_{Cp,M})_{M=0,1,...}$ and a modulus of D-regularity $\widetilde{m} \equiv (\overline{m}_M)_{M=0,1,...}$. Consider each $N \geq 0$. Then the shifted processes $Z^N : Q_\infty \times (\Omega, L, E) \to (S,d)$ is D-regular, with a modulus of continuity in probability $\delta_{Cp,N}$ and a modulus of D-regularity $\overline{m}_N$. In other words,

$$Z^N \in (\widehat{R}_{Dreg,\overline{m},\delta(Cp)}(Q_\infty \times \Omega, S), \widehat{\rho}_{Prob,Q(\infty)}).$$

Hence, by Theorem 10.6.1, the process $Y^N \equiv \Phi_{rLim}(Z^N)$ is a.u. càdlàg, with a modulus of continuity in probability $\delta_{Cp,N}$ and a modulus of a.u. càdlàg $\delta_{aucl}(\cdot, \overline{m}_N, \delta_{Cp,N})$. It is therefore easily verified that $X \equiv \overline{\Phi}_{rLim}(Z)$ is a well-defined process on $[0,\infty)$, with

$$X^N \equiv \overline{\Phi}_{rLim}(Z)^N = \Phi_{rLim}(Z^N)$$

for each $N \geq 0$. In other words, $\overline{\Phi}_{rLim}(Z) \equiv X \in \widehat{D}[0,\infty)$. Hence the function $\overline{\Phi}_{rLim}$ is well defined on $\widehat{R}_{Dreg,\widetilde{\delta}(Cp),\widetilde{m}.}(\overline{Q}_\infty \times \Omega, S)$. In other words,

$$X \equiv \overline{\Phi}_{rLim}(Z) \in \widehat{D}_{\widetilde{\delta}(aucl,),\widetilde{\delta}(Cp)}[0,\infty),$$

where $\widetilde{\delta}_{Cp} \equiv (\delta_{Cp,M})_{M=0,1,...}$ and $\widetilde{\delta}_{aucl} \equiv (\delta_{aucl}(\cdot, \overline{m}_M, \delta_{Cp,M}))_{M=0,1,...}$.

2. It remains to prove that the function $\overline{\Phi}_{rLim}$ is uniformly continuous on its domain. To that end, let $Z, Z' \in \widehat{R}_{Dreg,\widetilde{\delta}(Cp),\widetilde{m}.}(\overline{Q}_\infty \times \Omega, S)$ be arbitrary. Define $X \equiv \overline{\Phi}_{rLim}(Z)$ and $X' \equiv \overline{\Phi}_{rLim}(Z')$ as in Step 1. Then

$$\widetilde{\rho}_{\widehat{D}[0,\infty)}(X,X') \equiv E\sum_{n=0}^{\infty}2^{-n-1}(1\wedge d(X_{u(n)},X'_{u(n)}))$$

$$= E\sum_{n=0}^{\infty}2^{-n-1}(1\wedge d(Z_{u(n)},Z'_{u(n)})) \equiv \widehat{\rho}_{Prob,\overline{Q}(\infty)}(Z,Z').$$

Hence the function $\overline{\Phi}_{rLim}$ is an isometry on its domain. □

10.11 First Exit Time for a.u. Càdlàg Process

In this section, let $X : [0,\infty) \times (\Omega, L, E) \to (S,d)$ be an arbitrary a.u. càdlàg process that is adapted to some right continuous filtration $\mathcal{L}$. Let f be an arbitrary bounded and uniformly continuous function on (S,d). In symbols, $f \in C_{ub}(S,d)$.

Definition 10.11.1. First exit time. Let $N \geq 1$ be arbitrary. Suppose τ is a stopping time relative to $\mathcal{L}$, with values in $(0,N]$, such that the function X_τ is a well-defined r.v. relative to $L^{(\tau)}$, where $L^{(\tau)}$ is the probability subspace first introduced in Definition 8.1.9. Suppose also that for each ω in some full set, we have

(i) $f(X(\cdot,\omega)) < a$ on the interval $[0,\tau(\omega))$ and
(ii) $f(X(\tau(\omega),\omega)) \geq a$ if $\tau(\omega) < N$.

Then we say that τ is the *first exit time* in $[0,M]$ of the open subset $(f < a)$ by the process X, and define $\overline{\tau}_{f,a,N} \equiv \overline{\tau}_{f,a,N}(X) \equiv \tau$. □

Note that there is no requirement that the process X ever actually exits $(f < a)$. Observation stops at time N if exit does not occur by then. The next lemma makes precise some intuition.

Lemma 10.11.2. Basics of first exit times. *Let $a \in (a_0,\infty)$ be such that the first exit times $\overline{\tau}_{f,a,N} \equiv \overline{\tau}_{f,a,N}(X)$ and $\overline{\tau}_{f,a,M} \equiv \overline{\tau}_{f,a,M}(X)$ exist for some $N \geq M$. Then the following conditions hold:*

1. $\overline{\tau}_{f,a,M} \leq \overline{\tau}_{f,a,N}$.
2. $(\overline{\tau}_{f,a,M} < M) \subset (\overline{\tau}_{f,a,N} = \overline{\tau}_{f,a,M})$.
3. $(\overline{\tau}_{f,a,N} \leq r) = (\overline{\tau}_{f,a,M} \leq r)$ for each $r \in (0,M)$.

Proof. 1. Let $\omega \in domain(\overline{\tau}_{f,a,M}) \cap domain(\overline{\tau}_{f,a,N})$ be arbitrary. For the sake of a contradiction, suppose $t \equiv \overline{\tau}_{f,a,M}(\omega) > s \equiv \overline{\tau}_{f,a,N}(\omega)$. Then $s < \overline{\tau}_{f,a,M}(\omega)$. Hence we can apply Condition (i) of Definition 10.11.1 to the first exit time $\overline{\tau}_{f,a,M}$, obtaining $f(X_s(\omega)) < a$. At the same time, $\overline{\tau}_{f,a,N}(\omega) < t \leq N$. Hence we can apply Condition (ii) to the first exit time $\overline{\tau}_{f,a,N}$, obtaining $f(X_{\overline{\tau}(f,a,N)}(\omega)) \geq a$. In other words, $f(X_s(\omega)) \geq a$, which is a contradiction. We conclude that $\overline{\tau}_{f,a,N}(\omega) \geq \overline{\tau}_{f,a,M}(\omega)$, where $\omega \in domain(\overline{\tau}_{f,a,M}) \cap domain(\overline{\tau}_{f,a,M})$ is arbitrary. Assertion 1 is proved.

2. Next, suppose $t \equiv \overline{\tau}_{f,a,M}(\omega) < M$. Then Condition (ii) of Definition 10.11.1 implies that $f(X_t(\omega)) \geq a$. For the sake of a contradiction, suppose $t < \overline{\tau}_{f,a,N}(\omega)$. Then Condition (i) of Definition 10.11.1 implies that $f(X_t(\omega)) < a$, which is a contradiction. We conclude that $\overline{\tau}_{f,a,M}(\omega) \equiv t \geq \overline{\tau}_{f,a,N}(\omega)$. Combining with Assertion 1, we see that $\overline{\tau}_{f,a,M}(\omega) = \overline{\tau}_{f,a,N}(\omega)$. Assertion 2 is proved.

3. Note that

$$\begin{aligned}(\overline{\tau}_{f,a,N} \leq r) \subset (\overline{\tau}_{f,a,M} \leq r) &= (\overline{\tau}_{f,a,M} \leq r)(\overline{\tau}_{f,a,M} < M)\\ &\subset (\overline{\tau}_{f,a,M} \leq r)(\overline{\tau}_{f,a,M} = \overline{\tau}_{f,a,N})\\ &\subset (\overline{\tau}_{f,a,M} \leq r)(\overline{\tau}_{f,a,N} \leq r) = (\overline{\tau}_{f,a,N} \leq r),\end{aligned} \tag{10.11.1}$$

where we used the just established Assertions 1 and 2 repeatedly. Since the leftmost set and the rightmost set in relation 10.11.1 are the same, all inclusions therein can be replaced by equality. Assertion 3 and the lemma are proved. □

Proposition 10.11.3. Basic properties of stopping times relative to a right continuous filtration. *All stopping times in the following discussion will be relative to some given filtration $\mathcal{L} \equiv \{L^{(t)} : t \in [0,\infty)\}$ and will have values in $[0,\infty)$.*

Let τ', τ'' be arbitrary stopping times. Then the following conditions hold:

1. (Approximating stopping time by stopping times that have regularly spaced dyadic rational values.) Let τ be an arbitrary stopping time relative to the filtration $\mathcal{L}$. Then for each regular point t of the r.r.v. τ, we have $(\tau < t), (\tau = t) \in L^{(t)}$. Moreover, there exists a sequence $(\eta_h)_{h=0,1,\ldots}$ of stopping times relative to the filtration $\mathcal{L}$, such that for each $h \geq 0$, (i) *the stopping time η_h has positive values in the enumerated set*

$$\widetilde{Q}_h \equiv \{s_0, s_1, s_2, \ldots\} \equiv \{0, \Delta_h, 2\Delta_h, \ldots\};$$

(ii)

$$\tau + 2^{-h-1} < \eta_h < \tau + 2^{-h+2}; \tag{10.11.2}$$

and (iii) *for each $j \geq 1$, there exists a regular point r_j of τ with $r_j < s_j$ such that $(\eta_h \leq s_j) \in L^{(r(j))}$.*

2. (Construction of stopping time as right limit of sequence of stopping times.) Suppose the filtration $\mathcal{L}$ is right continuous. Suppose $(\eta_h)_{h=0,1,\ldots}$ is a sequence of stopping times relative to $\mathcal{L}$, such that for some r.r.v. τ, we have (i′) *$\tau \leq \eta_h$ for each $h \geq 0$ and* (ii)′ *$\eta_h \to \tau$ in probability. Then τ is a stopping time.*

3. Suppose the filtration $\mathcal{L}$ is right continuous. Then $\tau' \wedge \tau''$, $\tau' \vee \tau''$, and $\tau' + \tau''$ are stopping times.

4. If $\tau' \leq \tau''$ then $L^{(\tau')} \subset L^{(\tau''')}$.

5. Suppose the filtration $\mathcal{L}$ is right continuous. Suppose τ is a stopping time. Define the probability subspace $L^{(\tau+)} \equiv \bigcap_{s>0} L^{(\tau+s)}$. Then $L^{(\tau+)} = L^{(\tau)}$, where $L^{(\tau)}$ is the probability space first introduced in Definition 8.1.9.

Proof. 1. Suppose τ is a stopping time. Let t be an arbitrary regular point of the r.r.v. τ. Then, according to Definition 5.1.2 there exists an increasing sequence $(r_k)_{k=1,2,\ldots}$ of regular points of τ such that $r_k \uparrow t$ and $P(\tau \leq r_k) \uparrow P(\tau < t)$. Consequently, $E|1_{\tau \leq r(k)} - 1_{\tau < t}| \to 0$. Since $1_{\tau \leq r(k)} \in L^{(r(k))} \subset L^{(t)}$ for each $k \geq 1$, we conclude that $1_{\tau < t} \in L^{(t)}$. In other words, $(\tau < t) \in L^{(t)}$. Therefore $(\tau = t) = (\tau \leq t)(\tau < t)^c \in L^{(t)}$.

2. Separately, let $h \geq 0$ be arbitrary. For convenience, define $r_0 \equiv -\Delta_h$. For each $j \geq 1$, take a regular point

$$r_j \in ((j-1)\Delta_h, (j-2^{-1})\Delta_h) = (s_{j-1}, s_j - 2^{-1}\Delta_h) \subset (s_{j-1}, s_j)$$

of the r.r.v. τ. Then $s_{j-1} < r_j < r_{j+1} < s_{j+1} - 2^{-1}\Delta_h$, whence $r_j - r_{j-1} < 3 \cdot 2^{-1}\Delta_h$, for each $j \geq 1$. Moreover, the r.r.v.

$$\eta_h \equiv \sum_{j=1}^{\infty} s_j 1_{(r(j-1)<\tau \leq r(j))}$$

has positive values in $\widetilde{Q}_h$. Furthermore, for each $j \geq 1$, we have, on the measurable set $(r_{j-1} < \tau \leq r_j)$, the inequality

$$\begin{aligned} \tau + 2^{-1}\Delta_h \leq r_j + 2^{-1}\Delta_h < s_j = \eta_h < r_{j+1} &< r_{j+1} + (\tau - r_{j-1}) \\ &= \tau + (r_{j+1} - r_{j-1}) < \tau + 3\Delta_h < \tau + 2^{-h+2}. \end{aligned} \tag{10.11.3}$$

3. The set $\bigcup_{j=1}^{\infty}(r_{j-1} < \tau \leq r_j)$ is a full set, because τ is a nonnegative r.r.v. and because $r_j \uparrow \infty$ as $j \to \infty$. Inequality 10.11.3 therefore implies that

$$\tau + 2^{-h-1} < \eta_h < \tau + 2^{-h+2}. \tag{10.11.4}$$

Condition (ii) of Assertion 1 follows.

4. Consider each possible value s_j of the r.r.v. η_h, for some $j \geq 1$. Then

$$(\eta_h = s_j) = (r_{j-1} < \tau \leq r_j) = (\tau \leq r_j)(\tau \leq r_{j-1})^c \in L^{(r(j))} \subset L^{(s(j))},$$

because τ is a stopping time relative to $\mathcal{L}$ and because r_{j-1} and r_j are regular points of τ. Thus η_h is a stopping time, with positive values in $\widetilde{Q}_h$. Therefore the sequence $(\eta_h)_{h=0,1,\ldots}$ satisfies Conditions (i–iii) in Assertion 1, as desired.

5. To prove Assertion 2, suppose the filtration $\mathcal{L}$ is right continuous and $(\eta_h)_{h=0,1,\ldots}$ is a sequence of stopping times such that (i′) $\tau \leq \eta_h$ for each $h \geq 0$ and (ii′) $\eta_h \to \tau$ in probability. Let $t \in R$ be a regular point of the r.r.v. τ. Consider each $r > t$. Let $\varepsilon > 0$ be arbitrary. Let $s \in (t, r)$ be an arbitrary regular point of all the r.r.v.'s in the countable set $\{\tau, \eta_0, \eta_1, \ldots\}$. Then, by Conditions (i′) and (ii′), there exists $k \geq 0$ such that

$$P(s < \eta_h)(\tau \leq t) \leq P(s - t < \eta_h - \tau) \leq \varepsilon$$

for each $h \geq k$. Hence, for each $h \geq k$, we have

$$E|1_{\tau \leq t} - 1_{\eta(h) \leq s}| \leq E1_{\tau \leq t < s < \eta(h)} \leq \varepsilon.$$

Therefore $E|1_{\tau \le t} - 1_{\eta(h) \le s}| \to 0$ as $h \to \infty$. Because η_h is a stopping time, we have $1_{\eta(h) \le s} \in L^{(s)} \subset L^{(r)}$. Hence $1_{\tau \le t} \in L^{(r)}$, where $r > t$ is arbitrary. Thus

$$1_{\tau \le t} \in \bigcap_{r > t} L^{(r)} \equiv L^{(t+)} = L^{(t)},$$

where the last equality is thanks to the assumption that the filtration $\mathcal{L}$ is right continuous. We have verified that τ is a stopping time relative to the filtration $\mathcal{L}$. Assertion 2 is proved.

6. To prove Assertion 3, again suppose the filtration $\mathcal{L}$ is right continuous. By hypothesis, τ', τ' are stopping times. Consider each regular point $t \in R$ of the r.r.v. $\tau' \wedge \tau''$. Let $r > t$ be arbitrary. Take any common regular point $s > t$ of the three r.r.v.'s $\tau', \tau', \tau' \wedge \tau''$. Then $(\tau' \le s) \cup (\tau' > s)$ and $(\tau'' \le s) \cup (\tau'' > s)$ are full sets. Hence

$$(\tau' \wedge \tau'' \le s) = (\tau' \le s) \cup (\tau'' \le s) \in L^{(s)} \subset L^{(r)}. \tag{10.11.5}$$

Now let $s \downarrow t$. Then, since t is a regular point of the r.r.v. $\tau' \wedge \tau''$, we have $E 1_{(\tau' \wedge \tau'' \le s)} \downarrow E 1_{(\tau' \wedge \tau'' \le t)}$. Then $E|1_{(\tau' \wedge \tau'' \le s)} - 1_{(\tau' \wedge \tau'' \le t)}| \to 0$. Consequently, since $1_{(\tau' \wedge \tau'' \le s)} \in L^{(r)}$ according to relation 10.11.5, it follows that $1_{(\tau' \wedge \tau'' \le t)} \in L^{(r)}$, where $r > t$ is arbitrary. Hence

$$1_{(\tau' \wedge \tau'' \le t)} \in \bigcap_{r \in (t, \infty)} L^{(r)} \equiv L^{(t+)} = L^{(t)},$$

where the last equality is again thanks to the right continuity of the filtration $\mathcal{L}$. Thus $\tau' \wedge \tau''$ is a stopping time relative to $\mathcal{L}$. Similarly, we can prove that $\tau' \vee \tau''$ and $\tau' + \tau''$ are stopping times relative to $\mathcal{L}$. Assertion 3 is verified.

7. Suppose the stopping times τ', τ'' are such that $\tau' \le \tau''$. Let $Y \in L^{(\tau')}$ be arbitrary. Consider each regular point t'' of the stopping time τ''. Take an arbitrary regular point t' of τ' such that $t' \ne t''$. Then

$$\begin{aligned} Y 1_{(\tau'' \le t'')} &= Y 1_{(t' < \tau' \le \tau'' \le t'')} + Y 1_{(\tau' \le t' < \tau'' \le t'')} \\ &\quad + Y 1_{(\tau' \le \tau'' \le t' < t'')} + Y 1_{(\tau' \le \tau'' \le t'' < t')}. \end{aligned}$$

Consider the first summand in the last sum. Since $Y \in L^{(\tau')}$ by assumption, we have $Y 1_{(t' < \tau')} \in L^{(t')} \subset L^{(t'')}$. At the same time, $1_{(\tau'' \le t'')} \in L^{(t'')}$. Hence $Y 1_{(t' < \tau')} 1_{(\tau'' \le t'')} \in L^{(t'')}$. Similarly, all the other summands in the last sum are members of $L^{(t'')}$. Consequently, the sum $Y 1_{(\tau'' \le t'')}$ is a member of $L^{(t'')}$, where t'' is an arbitrary regular point of the stopping time τ''. In other words, $Y \in L^{(\tau'')}$. Since $Y \in L^{(\tau')}$ is arbitrary, we conclude that $L^{(\tau')} \subset L^{(\tau'')}$, as alleged in Assertion 4.

8. Proceed to prove Assertion 5. Suppose the filtration $\mathcal{L}$ is right continuous. Suppose τ is a stopping time relative to the right continuous filtration $\mathcal{L}$. Let $Y \in L^{(\tau+)} \equiv \bigcap_{s>0} L^{(\tau+s)}$ be arbitrary. Let t be an arbitrary regular point of the stopping time τ. Consider each $s > 0$. Then $Y \in L^{(\tau+s)}$. At the same time $1_{(\tau \le t)} \in L^{(t)} \subset L^{(t+s)}$. Hence $Y 1_{(\tau \le t)} \in L^{(t+s)}$. Consequently,

$$Y1_{(\tau \leq t)} \in \bigcap_{s>0} L^{(t+s)} \equiv L^{(t+)} = L^{(t)},$$

where the last equality is because the filtration $\mathcal{L}$ is right continuous. Thus $Y \in L^{(\tau)}$ for each $Y \in L^{(\tau+)}$. Therefore $L^{(\tau+)} \subset L^{(\tau)}$. In the other direction, we have, trivially, $L^{(\tau)} \subset L^{(\tau+)}$. Summing up, $L^{(\tau)} = L^{(\tau+)}$ as alleged in Assertion 5. The proposition is proved. □

11

Markov Process

In this chapter, we will construct an a.u. càdlàg Markov process from a given Markov semigroup of transition distributions, and show that the construction is a continuous mapping.

Definition 11.0.1. Specification of state space. In this chapter, let (S,d) be a locally compact metric space, with an arbitrary, but fixed reference point. Let $\xi \equiv (A_k)_{k=1,2,...}$ be a given binary approximation of (S,d) relative to $x_\circ$. Let

$$\pi \equiv (\{g_{k,x} : x \in A_k\})_{k=1,2,...}$$

be the partition of unity of (S,d) determined by ξ, as in Definition 3.3.4. □

Definition 11.0.2. Notations for dyadic rationals. Recall from Definition 9.0.2 the notations related to the enumerated set $Q_\infty \equiv \{t_0, t_1, \ldots\}$ of dyadic rationals in the interval $[0,1]$, and the enumerated set $\overline{Q}_\infty \equiv \{u_0, u_1, \ldots\}$ of dyadic rationals in the interval $[0,\infty)$.

In particular, for each $m \geq 0$, recall the notations $p_m \equiv 2^m$, $\Delta_m \equiv 2^{-m}$, and the sets

$$Q_m \equiv \{t_0, t_1, \ldots, t_{p(m)}\} = \{q_{m,0}, \ldots, q_{m,p(m)}\} = \{0, \Delta_m, 2\Delta_m, 3\Delta_m, \ldots, 1\},$$
$$\overline{Q}_m \equiv \{u_0, u_1, \ldots, u_{p(2m)}\} \equiv \{0, 2^{-m}, 2 \cdot 2^{-m}, \ldots, 2^m\} \subset [0, 2^m],$$
$$\widetilde{Q}_m \equiv \{0, \Delta_m, 2\Delta_m, \ldots\} \equiv \{0, 2^{-m}, 2 \cdot 2^{-m}, \ldots\} \subset [0, \infty),$$

and

$$\overline{Q}_\infty \equiv \bigcup_{m=0}^{\infty} \overline{Q}_m \equiv \{u_0, u_1, \ldots\}.$$

Unless otherwise specified, we will let Q denote one of the three parameter sets $\{0, 1, \ldots\}$, $\overline{Q}_\infty$, or $[0,\infty)$.

We will also use the miscellaneous notations and conventions in Definition 9.0.3. In addition, to ease the burden on notations, for real-valued expressions a, b, c, we will write the expressions $a = b \pm c$ interchangeably with $|a - b| \leq c$.

11.1 Markov Process and Strong Markov Process

In this section, we define a Markov process and a strong Markov process. From Sections 8.1, recall the definitions and basic properties of filtration, stopping time, and related objects.

Definition 11.1.1. Markov process and strong Markov process. Let (S,d) be an arbitrary locally compact metric space. Let Q denote one of the three parameter sets $\{0,1,\ldots\}$, $\overline{Q}_\infty$, or $[0,\infty)$. Let $\mathcal{L} \equiv \{L^{(t)} : t \in Q\}$ denote an arbitrary filtration of the sample space (Ω, L, E). Let

$$X : Q \times (\Omega, L, E) \to (S,d)$$

be an arbitrary process that is adapted to the filtration $\mathcal{L}$.

1. (Markov process.) Suppose, for each $t \in Q$, for each nondecreasing sequence $t_0 \equiv 0 \le t_1 \le \cdots \le t_m$ in Q, and for each function $f \in C(S^{m+1}, d^{m+1})$, we have (i) the conditional expectation $E(f(X_{t+t(0)}, X_{t+t(1)}, \ldots, X_{t+t(m)})|L^{(t)})$ exists, (ii) the conditional expectation $E(f(X_{t+t(0)}, X_{t+t(1)}, \ldots, X_{t+t(m)})|X_t)$ exists, and (iii) the two are equal:

$$\begin{aligned} &E(f(X_{t+t(0)}, X_{t+t(1)}, \ldots, X_{t+t(m)})|L^{(t)}) \\ &\quad = E(f(X_{t+t(0)}, X_{t+t(1)}, \ldots, X_{t+t(m)})|X_t). \end{aligned} \tag{11.1.1}$$

Then the process X is called a *Markov process* relative to the filtration $\mathcal{L}$. We will refer to Conditions (i–iii) as the *Markov property*. In the special case where $\mathcal{L}$ is the natural filtration of the process X, we will omit the reference to $\mathcal{L}$ and simply say that X is a *Markov process.*

2. (Strong Markov process.) Suppose, for each stopping time τ with values in Q relative to the filtration $\mathcal{L}$, the following two conditions hold:

2-1. The function X_τ is a well-defined r.v. relative to $L^{(\tau)}$ with values in (S,d), where the probability subspace $L^{(\tau)}$ and the function X_τ were introduced in Definition 8.1.9.

2-2. For each nondecreasing sequence $t_0 \equiv 0 \le t_1 \le \cdots \le t_m$ in Q, and for each function $f \in C(S^{m+1}, d^{m+1})$, we have (i) the conditional expectation

$$E(f(X_{\tau+t(0)}, X_{\tau+t(1)}, \ldots, X_{\tau+t(m)})|L^{(\tau)})$$

exists, (ii) the conditional expectation $E(f(X_{\tau+t(0)}, X_{\tau+t(1)}, \ldots, X_{\tau+t(m)})|X_\tau)$ exists, and (iii) the two r.r.v.'s are equal:

$$\begin{aligned} &E(f(X_{\tau+t(0)}, X_{\tau+t(1)}, \ldots, X_{\tau+t(m)})|L^{(\tau)}) \\ &\quad = (f(X_{\tau+t(0)}, X_{\tau+t(1)}, \ldots, X_{\tau+t(m)})|X_\tau). \end{aligned} \tag{11.1.2}$$

Then the process X is called a *strong Markov process* relative to the filtration $\mathcal{L}$.

Since each constant time $t \in Q$ is a stopping time, each strong Markov process relative to $\mathcal{L}$ is a Markov process relative to $\mathcal{L}$. □

11.2 Transition Distribution

In the next several sections, we will define and construct a Markov process from a semigroup of transition functions, also defined presently. For ease of presentation, we will assume a compact state space. This is no loss of generality, because a process with a locally compact metric state space can then be constructed and studied as a process with sample paths embedded in the one-point compactification of the locally compact space. As an example, Feller processes with a general locally compact metric state space will be constructed in a later section. In the meantime, our assumption of compactness simplifies proofs.

Definition 11.2.1. Transition distribution. Let (S_0,d_0) and (S_1,d_1) be compact metric spaces, with $d_0 \le 1$ and $d_1 \le 1$, and with fixed reference points $x_{0,\circ}$ and $x_{1,\circ}$, respectively. Let

$$T : C(S_1,d_1) \to C(S_0,d_0)$$

be an arbitrary nonnegative linear function. Write

$$T^x \equiv T(\cdot)(x) : C(S_1,d_1) \to R$$

for each $x \in S_0$. Suppose (i) for each $x \in S_0$, the function T^x is a distribution on (S_1,d_1), in the sense of Definition 5.2.1, and (ii) for each $f \in C(S_1,d_1)$ with a modulus of continuity δ_f, the function $Tf \in C(S_0,d_0)$ has a modulus of continuity $\alpha(\delta_f) : (0,\infty) \to (0,\infty)$ that depends only on δ_f, and otherwise not on the function f.

Then the function T is called a *transition distribution* from (S_0,d_0) to (S_1,d_1). The operation α is called a *modulus of smoothness of the transition distribution.* □

Lemma 11.2.2. Composite transition distribution. *For each $j = 0,1,2$, let (S_j,d_j) be a compact metric space with $d_i \le 1$. For each $j = 0,1$, let $T_{j,j+1}$ be a transition distribution from (S_j,d_j) to (S_{j+1},d_{j+1}), with modulus of smoothness $\alpha_{j,j+1}$. Then the composite function*

$$T_{0,2} \equiv T_{0,1}T_{1,2} \equiv T_{0,1} \circ T_{1,2} : C(S_2,d_2) \to C(S_0,d_0)$$

is a transition distribution from (S_0,d_0) to (S_2,d_2), with a modulus of smoothness $\alpha_{0,2}$ defined by

$$\alpha_{0,2}(\delta_f) \equiv \alpha_{0,1}(\alpha_{1,2}(\delta_f)) : (0,\infty) \to (0,\infty)$$

for each modulus of continuity δ_f.

We will call $T_{0,2} \equiv T_{0,1}T_{1,2}$ the composite transition distribution *of $T_{0,1}$ and $T_{1,2}$, and call $\alpha_{0,2} \equiv \alpha_{0,1} \circ \alpha_{1,2}$ the* composite modulus of smoothness *of $T_{0,1}T_{1,2}$.*

Proof. 1. Being the composite of two linear functions, the function $T_{0,2}$ is linear. Let $f \in C(S_2,d_2)$ be arbitrary, with a modulus of continuity δ_f. Then $T_{1,2}f \in C(S_1,d_1)$ has a modulus of continuity $\alpha_{1,2}(\delta_f)$. Hence the function $T_{0,1}(T_{1,2}f)$ has a modulus of continuity $\alpha_{0,1}(\alpha_{1,2}((\delta_f))$.

2. Consider each $x \in S_0$. Suppose $f \in C(S_2,d_2)$ is such that $T_{0,2}^x f > 0$. Then $T_{0,1}^x T_{1,2} f = T_{0,2}^x f > 0$. Therefore, since $T_{0,1}^x$ is a distribution, there exists $y \in S_1$ such that $T_{1,2}^y f \equiv T_{1,2} f(y) > 0$. Since $T_{1,2}^y$ is a distribution, there exists, in turn, some $z \in S_2$ such that $f(z) > 0$. Thus $T_{0,2}^x$ is an integration on (S_2,d_2) in the sense of Definition 4.2.1. Since $1 \in C(S_2,d_2)$ and $d_2 \leq 1$, it follows that $T_{0,2}^x$ is a distribution on (S_2,d_2) in the sense of Definition 5.2.1, where $x \in S_0$ is arbitrary. The conditions in Definition 11.2.1 have been verified for $T_{0,2}$ to be a transition distribution. □

By Definition 11.2.1, the domain and range of a transition distribution are spaces of continuous functions. We next extend both to spaces of integrable functions. Recall that for each $x \in S$, we let δ_x denote the distribution concentrated at x.

Proposition 11.2.3. Complete extension of a transition distribution relative to an initial distribution. *Let (S_0,d_0) and (S_1,d_1) be compact metric spaces, with $d_0 \leq 1$ and $d_1 \leq 1$. Let T be a transition distribution from (S_0,d_0) to (S_1,d_1). Let E_0 be a distribution on (S_0,d_0). Define the composite function*

$$E_1 \equiv E_0 T : C(S_1) \to R \tag{11.2.1}$$

Then the following conditions hold:

1. $E_1 \equiv E_0 T$ is a distribution on (S_1,d_1).

2. For each $i = 0,1$, let $(S_i, L_{E(i)}, E_i)$ be the complete extension of $(S_i, C(S_i), E_i)$. Let $f \in L_{E(1)}$ be arbitrary. Define the function Tf on S_0 by

$$domain(Tf) \equiv \{x \in S_0 : f \in L_{\delta(x)T}\} \tag{11.2.2}$$

and by

$$(Tf)(x) \equiv (\delta_x T) f \tag{11.2.3}$$

for each $x \in domain(Tf)$. Then (i) *$Tf \in L_{E(0)}$ and* (ii) *$E_0(Tf) = E_1 f \equiv (E_0 T) f$.*

3. Moreover, the extended function

$$T : L_{E(1)} \to L_{E(0)},$$

thus defined, is a contraction mapping and hence continuous relative to the norm $E_1|\cdot|$ on $L_{E(1)}$ and the norm $E_0|\cdot|$ on $L_{E(0)}$.

Proof. 1. By the defining equality 11.2.2, the function E_1 is clearly linear and nonnegative. Suppose $E_1 f \equiv E_0 T f > 0$ for some $C(S_1,d_1)$. Then, since E_0 is a distribution, there exists $x \in S_0$ such that $T^x f > 0$. In turn, since T^x is a distribution, there exists $y \in S_1$ such that $f(y) > 0$. Thus $E_1$is an integration. Since $1 \in C(S_1,d_1)$ and $d_1 \leq 1$, it follows that E_1 is a distribution.

2. For each $i = 0,1$, let $(S_i, L_{E(i)}, E_i)$ be the complete extension of $(S_i, C(S_i), E_i)$. Let $f \in L_{E(1)}$ be arbitrary. Define the function Tf on S_0 by equalities 11.2.2 and 11.2.3. Then, by Definition 4.4.1 of complete extensions,

there exists a sequence $(f_n)_{n=1,2,...}$ in $C(S_1)$ such that (i′) $\sum_{n=1}^{\infty} E_0T|f_n| = \sum_{n=1}^{\infty} E_1|f_n| < \infty$, (ii′)

$$\left\{x \in S_1 : \sum_{n=1}^{\infty} |f_n(x)| < \infty\right\} \subset domain(f),$$

and (iii′) $f(x) = \sum_{n=1}^{\infty} f_n(x)$ for each $x \in S_1$ with $\sum_{n=1}^{\infty} |f_n(x)| < \infty$. Condition (i′) implies that the subset

$$D_0 \equiv \left\{x \in S_0 : \sum_{n=1}^{\infty} T|f_n|(x) < \infty\right\} \equiv \left\{x \in S_0 : \sum_{n=1}^{\infty} T^x|f_n| < \infty\right\}$$

of the probability space $(S_0, L_{E(0)}, E_0)$ is a full subset. It implies also that the function $g \equiv \sum_{n=1}^{\infty} Tf_n$, with $domain(g) \equiv D_0$, is a member of $L_{E(0)}$. Now consider an arbitrary $x \in D_0$. Then

$$\sum_{n=1}^{\infty} |T^x f_n| \leq \sum_{n=1}^{\infty} T^x|f_n| < \infty. \tag{11.2.4}$$

Together with Condition (iii′), this implies that $f \in L_{\delta(x)T}$, with $(\delta_x T)f = \sum_{n=1}^{\infty}(\delta_x T)f_n$. Hence, according to the defining equalities 11.2.2 and 11.2.3, we have $x \in domain(Tf)$, with

$$(Tf)(x) \equiv (\delta_x T)f = \left(\sum_{n=1}^{\infty} Tf_n\right)(x) = g(x).$$

Thus $Tf = g$ on the full subset D_0 of $(S_0, L_{E(0)}, E_0)$. Since $g \in L_{E(0)}$, it follows that $Tf \in L_{E(0)}$. The desired Condition (i) is verified. Moreover,

$$E_0(Tf) = E_0 g \equiv E_0 \sum_{n=1}^{\infty} Tf_n = \sum_{n=1}^{\infty} E_0 Tf_n = E_0 T \sum_{n=1}^{\infty} f_n = E_1 f,$$

where the third and fourth equalities are both justified by Condition (i′). The desired Condition (ii) is also verified. Thus Assertion 2 is proved.

3. Let $f \in L_{E(1)}$ be arbitrary. Then, in the notations of Step 2, we have

$$E_0|Tf| = E_0|g| = \lim_{N\to\infty} E_0 \left|\sum_{n=1}^{N} Tf_n\right| = \lim_{N\to\infty} E_0 \left|T \sum_{n=1}^{N} f_n\right|$$

$$\leq \lim_{N\to\infty} E_0 T \left|\sum_{n=1}^{N} f_n\right| = \lim_{N\to\infty} E_1 \left|\sum_{n=1}^{N} f_n\right| = E_1|f|.$$

In short, $E_0|Tf| \leq E_1|f|$. Thus the mapping $T : L_{E(1)} \to L_{E(0)}$ is a contraction, as alleged in Assertion 3. □

Definition 11.2.4. Convention regarding automatic completion of a transition distribution. We hereby make the convention that, given each transition distribution T from a compact metric space (S_0, d_0) to a compact metric space (S_1, d_1)

with $d_0 \leq 1$ and $d_1 \leq 1$, and given each initial distribution E_0 on (S_0, d_0), the transition distribution

$$T : C(S_1, d_1) \to C(S_0, d_0)$$

is automatically completely extended to the nonnegative linear function

$$T : L_{E(1)} \to L_{E(0)}$$

in the manner of Proposition 11.2.3, where $E_1 \equiv E_0 T$ and $(S_i, L_{E(i)}, E_i)$ is the complete extension of $(S_i, C(S_i), E_i)$, for each $i = 0, 1$,

Thus Tf is integrable relative to E_0 for each integrable function f relative to $E_0 T$, with $(E_0 T)f = E_0(Tf)$. In the special case where E_0 is the point mass distribution δ_x concentrated at some $x \in S_0$, we have $x \in domain(Tf)$ and $T^x f = (Tf)^x$, for each integrable function f relative to T^x. □

Lemma 11.2.5. One-step transition distribution at step *m*. *Let (S, d) be a compact metric space with $d \leq 1$. Let T be a transition distribution from (S, d) to (S, d), with a modulus of smoothness α_T. Define ${}^1T \equiv T$. Let $m \geq 2$ and $f \in C(S^m, d^m)$ be arbitrary. Define a function $({}^{m-1}T)f$ on S^{m-1} by*

$$(({}^{m-1}T)f)(x_1, \ldots, x_{m-1}) \equiv \int T^{x(m-1)}(dx_m) f(x_1, \ldots, x_{m-1}, x_m) \quad (11.2.5)$$

for each $x \equiv (x_1, \ldots, x_{m-1}) \in S^{m-1}$. Then the following conditions hold:

1. If $f \in C(S^m, d^m)$ has values in $[0,1]$ and has a modulus of continuity δ_f, then the function $({}^{m-1}T)f$ is a member of $C(S^{m-1}, d^{m-1})$, with values in $[0,1]$, and has the modulus of continuity

$$\widetilde{\alpha}_{\alpha(T)}(\delta_f) : (0, \infty) \to (0, \infty)$$

defined by

$$\widetilde{\alpha}_{\alpha(T)}(\delta_f)(\varepsilon) \equiv \alpha_T(\delta_f)(2^{-1}\varepsilon) \wedge \delta_f(2^{-1}\varepsilon) \quad (11.2.6)$$

for each $\varepsilon > 0$.

2. For each $x \equiv (x_1, \ldots, x_{m-1}) \in S^{m-1}$, the function

$${}^{m-1}T : C(S^m, d^m) \to C(S^{m-1}, d^{m-1})$$

is a transition distribution from (S^{m-1}, d^{m-1}) to (S^m, d^m), with modulus of smoothness $\widetilde{\alpha}_{\alpha(T)}$.

We will call ${}^{m-1}T$ the one-step transition distribution at step m according to T.

Proof. 1. Let $f \in C(S^m, d^m)$ be arbitrary, with values in $[0,1]$ and with a modulus of continuity δ_f. Since T is a transition distribution, $T^{x(m-1)}$ is a distribution on (S, d) for each $x_{m-1} \in S$. Hence the integration on the right-hand side of equality 11.2.5 makes sense and has values in $[0,1]$. Therefore the left-hand side is well defined and has values in $[0,1]$. We need to prove that the function $({}^{m-1}T)f$ is a continuous function.

2. To that end, let $\varepsilon > 0$ be arbitrary. Let $x \equiv (x_1, \ldots, x_{m-1}), x' \equiv (x'_1, \ldots, x'_{m-1}) \in S^{m-1}$ be arbitrary such that

$$d^{m-1}(x, x') < \widetilde{\alpha}_{\alpha(T)}(\delta_f)(\varepsilon).$$

As an abbreviation, write $y \equiv x_{m-1}$ and $y' \equiv x'_{m-1}$.

With x, y fixed, the function $f(x, \cdot)$ on S also has a modulus of continuity δ_f. Hence the function $Tf(x, \cdot)$ has a modulus of continuity $\alpha_T(\delta_f)$, by Definition 11.2.1. Therefore, since

$$d(y, y') \leq d^{m-1}(x, x') < \widetilde{\alpha}_{\alpha(T)}(\delta_f)(\varepsilon) \leq \alpha_T(\delta_f)(2^{-1}\varepsilon),$$

where the last inequality is by the defining equality 11.2.6, it follows that

$$|(Tf(x, \cdot))(y) - (Tf(x, \cdot))(y')| < 2^{-1}\varepsilon.$$

In other words,

$$\int T^y(dz) f(x, z) = \int T^{y'}(dz) f(x, z) \pm 2^{-1}\varepsilon. \tag{11.2.7}$$

At the same time, for each $z \in S$, since

$$d^m((x, z), (x', z)) = d^{m-1}(x, x') < \delta_f(2^{-1}\varepsilon),$$

we have $|f(x, z) - f(x', z)| < 2^{-1}\varepsilon$. Hence

$$\int T^{y'}(dz) f(x, z) = \int T^{y'}(dz)(f(x', z) \pm 2^{-1}\varepsilon) = \int T^{y'}(dz) f(x', z) \pm 2^{-1}\varepsilon.$$

Combining with equality 11.2.7, we obtain

$$\int T^y(dz) f(x, z) = \int T^{y'}(dz)(f(x', z) \pm \varepsilon.$$

In view of the defining equality 11.2.5, this is equivalent to

$$((^{m-1}T) f)(x) = ((^{m-1}T) f)(x') \pm \varepsilon.$$

Thus $(^{m-1}T) f$ is continuous, with a modulus of continuity $\widetilde{\alpha}_{\alpha(T)}(\delta_f)$. Assertion 1 is proved.

3. By linearity, we see that $(^{m-1}T) f \in C(S^{m-1}, d^{m-1})$ for each $f \in C(S^m, d^m)$. Therefore the function

$$^{m-1}T : C(S^m, d^m) \to C(S^{m-1}, d^{m-1})$$

is well defined. It is clearly linear and nonnegative from the defining formula 11.2.5. Consider each $x \equiv (x_1, \ldots, x_{m-1}) \in S^{m-1}$. Suppose $(^{m-1}T)^x f \equiv \int T^{x(m-1)}(dy) f(x, y) > 0$. Then, since $T^{x(m-1)}$ is a distribution, there exists $y \in S$ such that $f(x, y) > 0$. Hence $(^{m-1}T)^x$ is an integration on (S^m, d^m) in the sense of Definition 4.2.1. Since $d^m \leq 1$ and $1 \in C(S^m, d^m)$, the function $(^{m-1}T)^x$ is a distribution on (S^m, d^m) in the sense of Definition 5.2.1. We have verified all the conditions in Definition 11.2.1 for ^{m-1}T to be a transition distribution. Assertion 2 is proved. □

11.3 Markov Semigroup

Recall that Q denotes one of the three parameter sets $\{0,1,\ldots\}$, $\overline{Q}_\infty$, or $[0,\infty)$. Let (S,d) be a compact metric space with $d \leq 1$. As discussed at the beginning of Section 11.2, the assumption of compactness is no loss of generality.

Definition 11.3.1. Markov semigroup. Let (S,d) be a compact metric space with $d \leq 1$. Unless otherwise specified, the symbol $\|\cdot\|$ will stand for the supremum norm for the space $C(S,d)$. Let $\mathbf{T} \equiv \{T_t : t \in Q\}$ be a family of transition distributions from (S,d) to (S,d), such that T_0 is the identity mapping. Suppose the following three conditions are satisfied:

1. (Smoothness.) For each $N \geq 1$, for each $t \in [0,N]Q$, the transition distribution T_t has some modulus of smoothness $\alpha_{\mathbf{T},N}$, in the sense of Definition 11.2.1. Note that the modulus of smoothness $\alpha_{\mathbf{T},N}$ is dependent on the finite interval $[0,N]$, but is otherwise independent of t.

2. (Semigroup property.) For each $s,t \in Q$, we have $T_{t+s} = T_t T_s$.

3. (Strong continuity.) For each $f \in C(S,d)$ with a modulus of continuity δ_f and with $\|f\| \leq 1$, and for each $\varepsilon > 0$, there exists $\delta_{\mathbf{T}}(\varepsilon,\delta_f) > 0$ so small that for each $t \in [0,\delta_{\mathbf{T}}(\varepsilon,\delta_f))Q$, we have

$$\|f - T_t f\| \leq \varepsilon. \tag{11.3.1}$$

Note that this strong continuity condition is trivially satisfied if $Q = \{0,1,\ldots\}$.

Then we call the family $\mathbf{T}$ a *Markov semigroup* of transition distributions with state space (S,d) and parameter space Q. For short, we will simply call $\mathbf{T}$ a *semigroup*. The operation $\delta_{\mathbf{T}}$ is called a *modulus of strong continuity* of $\mathbf{T}$. The sequence $\alpha_{\mathbf{T}} \equiv (\alpha_{\mathbf{T,N}})_{N=1,2,\ldots}$ is called the *modulus of smoothness of the semigroup* $\mathbf{T}$.

We will let $\mathscr{T}$ denote the set of semigroups $\mathbf{T}$ with state space (S,d) and with parameter set Q. □

The next lemma strengthens the continuity of T_t at $t = 0$ to uniform continuity over $t \in Q$.

Lemma 11.3.2. Uniform strong continuity on the parameter set. *Let (S,d) be a compact metric space with $d \leq 1$. Suppose Q is one of the three parameter sets $\{0,1,\ldots\}$, $\overline{Q}_\infty$, or $[0,\infty)$. Let $\mathbf{T}$, $\overline{\mathbf{T}}$ be arbitrary semigroups with state space (S,d) and parameter space Q, and with a common modulus of strong continuity $\delta_{\mathbf{T}}$. Then the following conditions hold:*

1. Let $f \in C(S,d)$ be arbitrary, with a modulus of continuity δ_f and with $|f| \leq 1$. Let $\varepsilon > 0$ and $r,s \in Q$ be arbitrary, with $|r-s| < \delta_{\mathbf{T}}(\varepsilon,\delta_f)$. Then $\|T_r f - T_s f\| \leq \varepsilon$.

2. Let $\iota : (0,\infty) \to (0,\infty)$ denote the identity operation, defined by $\iota(\varepsilon) \equiv \varepsilon$ for each $\varepsilon > 0$. Let $\varepsilon > 0$ and $t \in [0,\delta_{\mathbf{T}}(\varepsilon,\iota))Q$ be arbitrary. Then $\|T_t(d(x,\cdot)) - d(x,\cdot)\| \leq \varepsilon$ for each $x \in S$.

3. *Suppose $T_r = \overline{T}_r$ for each $r \in Q'$, for some dense subset of Q' of Q. Then $T_t = \overline{T}_t$ for each $t \in Q$. In short, $\mathbf{T} = \overline{\mathbf{T}}$.*

Proof. 1. The proof for the case where $Q = \{0, 1, \ldots\}$ is trivial and is omitted.

2. Suppose $Q = \overline{Q}_\infty$ or $Q = [0, \infty)$. Let $\varepsilon > 0$ and $r, s \in Q$ be arbitrary with $0 \leq s - r < \delta_{\mathbf{T}}(\varepsilon, \delta_f)$. Then, for each $x \in S$, we have

$$|T_s^x f - T_r^x f| = |T_r^x(T_{s-r} f - f)| \leq \|T_{s-r} f - f\| \leq \varepsilon,$$

where the equality is by the semigroup property, where the first inequality is because T_r^x is a distribution on (S, d), and where the last inequality is by the definition of $\delta_{\mathbf{T}}$ as a modulus of strong continuity. Thus

$$\|T_r f - T_s f\| \leq \varepsilon. \tag{11.3.2}$$

3. Let $\varepsilon > 0$ and $r, s \in \overline{Q}_\infty$ be arbitrary with $|s - r| < \delta_{\mathbf{T}}(\varepsilon, \delta_f)$. Either $0 \leq s - r < \delta_{\mathbf{T}}(\varepsilon, \delta_f)$, in which case inequality 11.3.2 holds according to Step 2, or $0 \leq r - s < \delta_{\mathbf{T}}(\varepsilon, \delta_f)$, in which case inequality 11.3.2 holds similarly. Thus Assertion 1 is proved if $Q = \overline{Q}_\infty$.

4. Now suppose $Q = [0, \infty)$. Let $\varepsilon > 0$ and $r, s \in [0, \infty)$ be arbitrary with $|r - s| < \delta_{\mathbf{T}}(\varepsilon, \delta_f)$. Let $\varepsilon' > 0$ be arbitrary. Let $t, v \in \overline{Q}_\infty$ be arbitrary such that (i) $r \leq t < r + \delta_{\mathbf{T}}(\varepsilon', \delta_f)$, (ii) $s \leq v < s + \delta_{\mathbf{T}}(\varepsilon', \delta_f)$, and (iii) $|t - v| < \delta_{\mathbf{T}}(\varepsilon, \delta_f)$. Then, according to inequality 11.3.2 in Step 2, we have $\|T_t f - T_r f\| \leq \varepsilon'$ and $\|T_v f - T_s f\| \leq \varepsilon'$. According to Step 3, we have $\|T_t f - T_v f\| \leq \varepsilon$. Combining, we obtain

$$\|T_r f - T_s f\| \leq \|T_t f - T_r f\| + \|T_s f - T_v f\| + \|T_t f - T_v f\| < \varepsilon' + \varepsilon' + \varepsilon.$$

Letting $\varepsilon' \to 0$, we obtain $\|T_r f - T_s f\| \leq \varepsilon$. Thus Assertion 1 is also proved for the case where $Q = [0, \infty)$.

5. To prove Assertion 2, consider each $x \in S$. Then the function $f_x \equiv (1 - d(\cdot, x)) \in C(S, d)$ is nonnegative and has a modulus of continuity $\delta_{f(x)}$ defined by $\delta_{f(x)}(\varepsilon) \equiv \iota(\varepsilon) \equiv \varepsilon$ for each $\varepsilon > 0$. Now let $\varepsilon > 0$ be arbitrary. Let $t \in [0, \delta_{\mathbf{T}}(\varepsilon, \iota))Q$ be arbitrary. Then $t < \delta_{\mathbf{T}}(\varepsilon, \delta_{f(x)})$. Hence, by Condition 3 of Definition 11.3.1, we have

$$\|T_t(d(x, \cdot)) - d(x, \cdot)\| = \|T_t f_x - f_x\| \leq \varepsilon.$$

Assertion 2 is proved.

6. Let $\overline{\mathbf{T}}$ and Q' be as given in Assertion 3. Consider each $t \in Q$. Let $(r_k)_{k=1,2,\ldots}$ be a sequence in Q' such that $r_k \to t$. Let $f \in C(S, d)$ be arbitrary. Then $T_{r(k)} f = \overline{T}_{r(k)} f$ for each $k \geq 1$, by hypothesis. At the same time, $\left\|T_{r(k)} f - T_t f\right\| \to 0$ by Assertion 1. Similarly, $\left\|\overline{T}_{r(k)} f - \overline{T}_t f\right\| \to 0$. Combining, $T_t f = \overline{T}_t f$, where $f \in C(S, d)$ is arbitrary. Thus $T_t = \overline{T}_t$ as transition distributions. Assertion 3 and the lemma are proved. □

11.4 Markov Transition f.j.d.'s

In this section, we will define a consistent family of f.j.d.'s generated by an initial distribution and a semigroup. The parameter set Q is assumed to be one of the three sets $\{0, 1, \ldots\}$, $\overline{Q}_\infty$, or $[0, \infty)$. We will refer loosely to the first two as the metrically discrete parameter sets. For ease of presentation, we assume that the state space (S, d) is compact with $d \leq 1$.

Let $\xi \equiv (A_k)_{k=1,2,\ldots}$ be a binary approximation of (S, d) relative to $x_\circ$. Let

$$\pi \equiv (\{g_{k,x} : x \in A_k\})_{k=1,2,\ldots}$$

be the partition of unity of (S, d) determined by ξ, as in Definition 3.3.4.

Definition 11.4.1. Family of transition f.j.d.'s generated by an initial distribution and a Markov semigroup. Let Q be one of the three sets $\{0, 1, \ldots\}$, $\overline{Q}_\infty$, or $[0, \infty)$. Let $\mathbf{T}$ be an arbitrary Markov semigroup, with the compact state space (S, d) where $d \leq 1$, and with parameter set Q. Let E_0 be an arbitrary distribution on (S, d). For arbitrary $m \geq 1$, $f \in C(S^m, d^m)$, and nondecreasing sequence $r_1 \leq \cdots \leq r_m$ in Q, define

$$F^{E(0),\mathbf{T}}_{r(1),\ldots,r(m)} f \equiv \int E_0(dx_0) \int T^{x(0)}_{r(1)}(dx_1) \int T^{x(1)}_{r(2)-r(1)}(dx_2) \cdots \times \int T^{x(m-1)}_{r(m)-r(m-1)}(dx_m) f(x_1, \ldots, x_m). \tag{11.4.1}$$

In the special case where $E_0 \equiv \delta_x$ is the distribution that assigns probability 1 to some point $x \in S$, we will simply write

$$F^{*,\mathbf{T}}_{r(1),\ldots,r(m)}(x) \equiv F^{x,\mathbf{T}}_{r(1),\ldots,r(m)} \equiv F^{\delta(x),\mathbf{T}}_{r(1),\ldots,r(m)}. \tag{11.4.2}$$

The next theorem will prove that $F^{*,\mathbf{T}}_{r(1),\ldots,r(m)} : C(S^m, d^m) \to C(S, d)$ is then a well-defined transition distribution.

1. An arbitrary consistent family

$$\{F_{r(1),\ldots,r(m)} f : m \geq 0; r_1, \ldots, r_m \text{ in } Q\}$$

of f.j.d.'s satisfying Condition 11.4.1 is said to be *generated by the initial distribution E_0 and the semigroup* $\mathbf{T}$.

2. An arbitrary process

$$X : Q \times (\Omega, L, E) \to (S, d),$$

whose marginal distributions are given by a consistent family generated by the initial distribution E_0 and the semigroup $\mathbf{T}$, will itself be called *a process generated by the initial distribution E_0 and semigroup* $\mathbf{T}$. We will see later that such processes are Markov processes.

3. In the special case where $E_0 \equiv \delta_x$, the qualifier "generated by the initial distribution E_0" will be replaced simply by "generated by the initial state x." □

Theorem 11.4.2. Construction of a family of transition f.j.d.'s from an initial distribution and semigroup. *Let (S, d) be a compact metric space with $d \leq 1$.*

Let Q be one of the three sets $\{0,1,\ldots\}$, $\overline{Q}_\infty$, or $[0,\infty)$. Let $\mathbf{T}$ be an arbitrary semigroup with state space (S,d), with parameter set Q, with a modulus of strong continuity $\delta_\mathbf{T}$, and with a modulus of smoothness $\alpha_\mathbf{T} \equiv (\alpha_{\mathbf{T},\mathbf{N}})_{N=1,2,\ldots}$, in the sense of Definition 11.3.1. Then the following conditions hold:

1. Let the sequence $0 \equiv r_0 \leq r_1 \leq \cdots \leq r_m$ in Q be arbitrary. Then the function

$$F^{*,\mathbf{T}}_{r(1),\ldots,r(m)} : C(S^m,d^m) \to C(S,d),$$

as defined by equality 11.4.2 in Definition 11.4.1, is a well-defined transition distribution. Specifically, it is equal to the composite transition distribution

$$F^{*,\mathbf{T}}_{r(1),\ldots,r(m)} = ({}^1T_{r(1)-r(0)})({}^2T_{r(2)-r(1)})\cdots({}^mT_{r(m)-r(m-1)}), \tag{11.4.3}$$

where the factors on the right hand side are one-step transition distributions defined in Lemma 11.2.5.

2. Let the sequence $0 \equiv r_0 \leq r_1 \leq \cdots \leq r_m$ in Q be arbitrary. Let $N \geq 1$ be so large that $r_m \leq N$. Let $\widetilde{\alpha}_{\alpha(T,N)}$ be the modulus of smoothness of the one-step transition distribution

$${}^iT_{r(i)-r(i-1)} : C(S^i,d^i) \to C(S^{i-1},d^{i-1})$$

for each $i = 1,\ldots,m$, as constructed in Lemma 11.2.5. Then the transition distribution $F^{*,\mathbf{T}}_{r(1),\ldots,r(m)}$ has a modulus of smoothness

$$\widetilde{\alpha}^{(m)}_{\alpha(\mathbf{T},N)} \equiv \widetilde{\alpha}_{\alpha(\mathbf{T},N)} \circ \cdots \circ \widetilde{\alpha}_{\alpha(\mathbf{T},N)}.$$

3. For each $m \geq 1$ and $\varepsilon > 0$, there exists $\delta_m(\varepsilon,\delta_f,\delta_\mathbf{T},\alpha_\mathbf{T}) > 0$ such that for each $f \in C(S^m,d^m)$ with values in $[0,1]$ and a modulus of continuity δ_f, and for arbitrary nondecreasing sequences $r_1 \leq \cdots \leq r_m$ and $s_1 \leq \cdots \leq s_m$ in Q with

$$\bigvee_{i=1}^m |r_i - s_i| < \delta_m(\varepsilon,\delta_f,\delta_\mathbf{T},\alpha_\mathbf{T}),$$

we have

$$\left\| F^{*,\mathbf{T}}_{r(1),\ldots,r(m)} f - F^{*,\mathbf{T}}_{s(1),\ldots s(m)} f \right\| \leq \varepsilon.$$

4. Suppose $Q = \{0,1,\ldots\}$ or $Q = \overline{Q}_\infty$. Let E_0 be an arbitrary distribution on (S,d). Then the family

$$F^{E(0),\mathbf{T}} \equiv \{F^{E(0),\mathbf{T}}_{r(1),\ldots,r(m)} f : m \geq 1; r_1 \leq \cdots \leq r_m \text{ in } Q\}$$

can be uniquely extended to a consistent family

$$\Phi_{Sg,fjd}(E_0,\mathbf{T}) \equiv F^{E(0),\mathbf{T}} \equiv \{F^{E(0),\mathbf{T}}_{r(1),\ldots,r(m)} f : m \geq 1; r_1,\ldots,r_m \text{ in } Q\}$$

of f.j.d.'s with parameter set Q and state space (S,d), with said family $F^{E(0),\mathbf{T}}$ being continuous in probability, with a modulus of continuity in probability $\delta_{Cp}(\cdot,\delta_\mathbf{T},\alpha_\mathbf{T})$. Moreover, the consistent family $\Phi_{Sg,fjd}(E_0,\mathbf{T})$ is generated by the *initial distribution* E_0 *and the semigroup* $\mathbf{T}$, *in the sense of Definition* 11.4.1. *Thus, in the special case where* $Q = \{0, 1, \ldots\}$ *or* $Q = \overline{Q}_\infty$, *we have a mapping*

$$\Phi_{Sg, fjd} : \widehat{J}(S,d) \times \mathscr{T} \to \widehat{F}_{Cp}(Q,S),$$

where $\widehat{J}(S,d)$ *is the space of distributions* E_0 *on* (S,d), *where* $\mathscr{T}$ *is the space of semigroups with state space* (S,d) *and parameter set* Q, *and where* $\widehat{F}_{Cp}(Q,S)$ *is the set of consistent families of f.j.d.'s with parameter set* Q *and state space* S, *whose members are continuous in probability.*

5. *Suppose* $Q = \{0, 1, \ldots\}$ *or* $Q = \overline{Q}_\infty$, *and consider the special case where* $E_0 \equiv \delta_x$ *for some* $x \in S$. *Write*

$$\Phi_{Sg, fjd}(x, \mathbf{T}) \equiv \Phi_{Sg, fjd}(\delta_x, \mathbf{T}) \equiv F^{x,\mathbf{T}}.$$

Then we have a function

$$\Phi_{Sg, fjd} : S \times \mathscr{T} \to \widehat{F}_{Cp}(Q,S).$$

Proof. 1. Let E_0 be an arbitrary distribution on (S,d). Let the sequence $0 \equiv r_0 \leq r_1 \leq \cdots \leq r_m$ in Q be arbitrary. Consider each $f \in C(S^m, d^m)$. By the defining equality 11.4.1, we have

$$F^{E(0),\mathbf{T}}_{r(1),\ldots,r(m)} f = \int E_0(dx_0) \int T^{x(0)}_{r(1)}(dx_1) \int T^{x(1)}_{r(2)-r(1)}(dx_2) \cdots \times \int T^{x(m-1)}_{r(m)-r(m-1)}(dx_m) f(x_1, \ldots, x_m). \tag{11.4.4}$$

By the defining equality 11.2.5 of Lemma 11.2.5, the rightmost integral is equal to

$$\int T^{x(m-1)}_{r(n)-r(n-1)}(dx_m) f(x_1, \ldots, x_{m-1}, x_m) \equiv ((^{m-1}T_{r(m)-r(m-1)}) f)(x_1, \ldots, x_{m-1}). \tag{11.4.5}$$

Recursively backward, Equality 11.4.4 becomes

$$F^{E(0),\mathbf{T}}_{r(1),\ldots,r(m)} f = \int E_0(dx_0)(^0T_{r(1)-r(0)}) \cdots (^{m-1}T_{r(m)-r(m-1)}) f.$$

In particular,

$$F^{x,\mathbf{T}}_{r(1),\ldots,r(m)} f = (^0T^x_{r(1)-r(0)}) \cdots (^{m-1}T_{r(m)-r(m-1)}) f.$$

for each $x \in S$. In other words,

$$F^{*,\mathbf{T}}_{r(1),\ldots,r(m)} = (^1T_{r(1)-r(0)})(^2T_{r(2)-r(1)}) \cdots (^mT_{r(m)-r(m-1)}). \tag{11.4.6}$$

2. Now Lemma 11.2.5 says the factors $(^1T_{r(1)-r(0)}), (^2T_{r(2)-r(1)}), \ldots, (^mT_{r(m)-r(m-1)})$ on the right-hand side are a transition distribution with the common modulus of smoothness $\widetilde{\alpha}_{\alpha(T,N)}$ defined therein. Hence, by repeated applications of Lemma 11.2.2, the composite $F^{*,\mathbf{T}}_{r(1),\ldots,r(m)}$ is a transition distribution, with a modulus of smoothness that is the m-fold composite operation

$$\widetilde{\alpha}_{\alpha(T,N)}^{(m)} \equiv \widetilde{\alpha}_{\alpha(\mathbf{T},N)} \circ \cdots \circ \widetilde{\alpha}_{\alpha(\mathbf{T},N)}.$$

Assertions 1 and 2 of the present theorem are proved.

3. Proceed to prove Assertion 3 by induction on $m \geq 1$. Consider each $f \in C(S^m, d^m)$ with values in $[0,1]$ and a modulus of continuity δ_f. Let $\varepsilon > 0$ be arbitrary. In the case where $m = 1$, define $\delta_1 \equiv \delta_1(\varepsilon, \delta_f, \delta_{\mathbf{T}}, \alpha_{\mathbf{T}}) \equiv \delta_{\mathbf{T}}(\varepsilon, \delta_f)$. Suppose r_1, s_1 in Q are such that

$$|r_1 - s_1| < \delta_1(\varepsilon, \delta_f, \delta_{\mathbf{T}}, \alpha_{\mathbf{T}}) \equiv \delta_{\mathbf{T}}(\varepsilon, \delta_f).$$

Then

$$\left\| F_{r(1)}^{*,\mathbf{T}} f - F_{s(1)}^{*,\mathbf{T}} f \right\| = \left\| {}^1T_{r(1)-r(0)} f - {}^1T_{s(1)-s(0)} f \right\| = \left\| T_{r(1)} f - T_{s(1)} f \right\| \leq \varepsilon,$$

where the inequality is by Lemma 11.3.2. Assertion 3 is thus proved for the starting case $m = 1$.

4. Suppose, inductively, for some $m \geq 2$, that the operation

$$\delta_{m-1}(\cdot, \delta_f, \delta_{\mathbf{T}}, \alpha_{\mathbf{T}})$$

has been constructed with the desired properties. Define

$$\delta_m(\varepsilon, \delta_f, \delta_{\mathbf{T}}, \alpha_{\mathbf{T}}) \equiv 2^{-1}\delta_{m-1}(2^{-1}\varepsilon, \delta_f, \delta_{\mathbf{T}}, \alpha_{\mathbf{T}}) \wedge \delta_{m-1}(\varepsilon, \delta_f, \delta_{\mathbf{T}}, \alpha_{\mathbf{T}}). \tag{11.4.7}$$

Suppose

$$\bigvee_{i=1}^{m} |r_i - s_i| < \delta_m(\varepsilon, \delta_f, \delta_{\mathbf{T}}, \alpha_{\mathbf{T}}). \tag{11.4.8}$$

Define the function

$$h \equiv ({}^2T_{r(2)-r(1)}) \cdots ({}^mT_{r(m)-r(m-1)}) f \in C(S, d). \tag{11.4.9}$$

Then, by the induction hypothesis for an $(m-1)$-fold composite, the function h has modulus of continuity $\delta_1(\cdot, \delta_f, \delta_{\mathbf{T}}, \alpha_{\mathbf{T}})$. We emphasize here that as $m \geq 2$, the modulus of smoothness of the one-step transition distribution ${}^2T_{r(2)-r(1)}$ on the right-hand side of equality 11.4.9 actually depends on the modulus $\alpha_{\mathbf{T}}$, according to Lemma 11.2.5. Hence the modulus of continuity of function h indeed depends on $\alpha_{\mathbf{T}}$, which justifies the notation.

At the same time, inequality 11.4.8 and the defining equality 11.4.7 together imply that

$$|r_1 - s_1| < \delta_m(\varepsilon, \delta_f, \delta_{\mathbf{T}}, \alpha_{\mathbf{T}}) \leq \cdots \leq \delta_1(\varepsilon, \delta_f, \delta_{\mathbf{T}}, \alpha_{\mathbf{T}}).$$

Hence

$$\left\| ({}^1T_{r(1)-r(0)})h - ({}^1T_{s(1)-s(0)})h \right\| = \left\| F_{r(1)}^{*,\mathbf{T}} h - F_{s(1)}^{*,\mathbf{T}} h \right\| \leq 2^{-1}\varepsilon, \tag{11.4.10}$$

where the inequality is by the induction hypothesis for the starting case where $m = 1$.

5. Similarly, inequality 11.4.8 and the defining equality 11.4.7 together imply that

$$\bigvee_{i=2}^{m} |(r_i - r_{i-1}) - (s_i - s_{i-1})| \leq 2\bigvee_{i=1}^{m} (r_i - s_i| < \delta_{m-1}(2^{-1}\varepsilon, \delta_f, \delta_{\mathbf{T}}, \alpha_{\mathbf{T}}).$$

Hence

$$\begin{aligned}
&\|(^2T_{r(2)-r(1)}) \cdots (^mT_{r(m)-r(m-1)})f - (^2T_{s(2)-s(1)}) \cdots (^mT_{s(m)-s(m-1)})f\| \\
&\equiv \left\| F^{*,\mathbf{T}}_{r(2),\ldots,r(m)}f - F^{*,\mathbf{T}}_{s(2),\ldots,s(m)}f \right\| < 2^{-1}\varepsilon, \qquad (11.4.11)
\end{aligned}$$

where the inequality is by the induction hypothesis for the case where $m-1$.

6. Combining, we estimate, for each $x \in S$, the bound

$$\begin{aligned}
&|F^{x,\mathbf{T}}_{r(1),\ldots,r(m)}f - F^{x,\mathbf{T}}_{s(1),\ldots,s(m)}f| \\
&= |(^1T^x_{r(1)-r(0)})(^2T_{r(2)-r(1)}) \cdots (^mT_{r(m)-r(m-1)})f \\
&\quad - (^1T^x_{s(1)-s(0)})(^2T_{s(2)-s(1)}) \cdots (^mT_{s(m)-s(m-1)})f| \\
&\equiv |(^1T^x_{r(1)-r(0)})h - (^1T^x_{s(1)-s(0)})(^2T_{s(2)-s(1)}) \cdots (^mT_{s(m)-s(m-1)})f| \\
&\leq |(^1T^x_{r(1)-r(0)})h - (^1T^x_{s(1)-s(0)})h| \\
&\quad + |(^1T^x_{s(1)-s(0)})h - (^1T^x_{s(1)-s(0)})(^2T_{s(2)-s(1)}) \cdots (^mT_{s(m)-s(m-1)})f| \\
&\leq 2^{-1}\varepsilon + |(^1T^x_{s(1)-s(0)})(^2T_{r(2)-r(1)}) \cdots (^mT_{r(m)-r(m-1)})f \\
&\quad - (^1T^x_{s(1)-s(0)})(^2T_{s(2)-s(1)}) \cdots (^mT_{s(m)-s(m-1)})f| \\
&\leq 2^{-1}\varepsilon + (^1T^x_{s(1)-s(0)})|(^2T_{r(2)-r(1)}) \cdots (^mT_{r(m)-r(m-1)})f \\
&\quad - (^2T_{s(2)-s(1)}) \cdots (^mT_{s(m)-s(m-1)})f| \\
&< 2^{-1}\varepsilon + 2^{-1}\varepsilon = \varepsilon,
\end{aligned}$$

where the second inequality is by inequality 11.4.10, and where the last inequality is by inequality 11.4.11. Since $x \in S$ is arbitrary, it follows that

$$\left\| F^{*,\mathbf{T}}_{r(1),\ldots,r(m)}f - F^{*,\mathbf{T}}_{s(1),\ldots s(m)}f \right\| \leq \varepsilon.$$

Induction is completed, and Assertion 3 is proved.

7. To prove Assertion 4, assume that $Q = \{0, 1, \ldots\}$ or $Q = \overline{Q}_\infty$. We need to prove that the family

$$\left\{ F^{E(0),\mathbf{T}}_{r(1),\ldots,r(m)} : m \geq 1; r_1, \ldots, r_m \in Q; r_1 \leq \cdots \leq r_m \right\}$$

can be uniquely extended to a consistent family

$$\left\{ F^{E(0),\mathbf{T}}_{s(1),\ldots,s(m)} : m \geq 1; s_1, \ldots, s_m \in Q \right\}$$

of f.j.d.'s with parameter set Q. We will give the proof only for the case where $Q = \overline{Q}_\infty$, with the case of $\{0, 1, \ldots\}$ being similar. Assume in the following that $Q = \overline{Q}_\infty$.

8. Because $F^{*,\mathbf{T}}_{r(1),\ldots,r(m)}$ is a transition distribution, Proposition 11.2.3 says that the composite function

$$F^{E(0),\mathbf{T}}_{r(1),\ldots,r(m)} = E_0 F^{*,\mathbf{T}}_{r(1),\ldots,r(m)} = E_0({}^1T_{r(1)-r(0)})({}^2T_{r(2)-r(1)}) \cdots ({}^mT_{r(m)-r(m-1)}) \tag{11.4.12}$$

is a distribution on (S^m, d^m), for each $m \geq 1$, and for each sequence $0 \equiv r_0 \leq r_1 \leq \cdots \leq r_m$ in $\overline{Q}_\infty$.

9. To proceed, let $m \geq 2$ and $r_1, \ldots, r_m \in \overline{Q}_\infty$ be arbitrary, with $r_1 \leq \cdots \leq r_m$. Let $n = 1, \ldots, m$ be arbitrary. Define the sequence

$$\kappa \equiv \kappa_{n,m} \equiv (\kappa_1, \ldots, \kappa_{m-1}) \equiv (1, \ldots, \widehat{n}, \ldots, m),$$

where the caret on the top of an element in a sequence signifies the omission of that element in the sequence. Let $\kappa^* \equiv \kappa^*_{n,m} : S^m \to S^{m-1}$ denote the dual function of the sequence κ, defined by

$$\kappa^*(x_1, \ldots, x_m) \equiv \kappa^*(x) \equiv x \circ \kappa = (x_{\kappa(1)}, \ldots, x_{\kappa(m-1)}) = (x_1, \ldots, \widehat{x_n}, \ldots, x_m)$$

for each $x \equiv (x_1, \ldots, x_m) \in S^m$. Let $f \in C(S^{m-1})$ be arbitrary. We will prove that

$$F^{E(0),\mathbf{T}}_{r(1),\ldots,\widehat{r(n)},\cdots,r(m)} f = F^{E(0),\mathbf{T}}_{r(1),\ldots,r(m)} f \circ \kappa^*_{n,m}. \tag{11.4.13}$$

To that end, note that equality 11.4.4 yields

$$\begin{aligned} F^{E(0),\mathbf{T}}_{r(1),\ldots,\widehat{r(n)}\cdots,r(m)} f \equiv & \int E_0(dx_0) \int T^{x(0)}_{r(1)}(dx_1) \cdots \int T^{x(n-1)}_{r(n+1)-r(n-1)}(dy_n) \\ & \times \left\{ \int T^{y(n)}_{r(n+2)-r(n+1)}(dy_{n+1}) \cdots \right. \\ & \left. \times \int T^{y(m-2)}_{r(m)-r(m-1)}(dy_{m-1}) f(x_1, \ldots, x_{n-1}, y_n, \ldots, y_{m-1}) \right\} \end{aligned}$$

For each fixed $(x_1, \ldots, x_{n-1})$, the expression in braces is a continuous function of the one variable y_n. Call this function $g_{x(1),\ldots,x(n-1)} \in C(S,d)$. Then the last displayed equality can be continued as

$$\begin{aligned} &\equiv \int E_0(dx_0) \int T^{x(0)}_{r(1)}(dx_1) \cdots \left(\int T^{x(n-1)}_{r(n+1)-r(n-1)}(dy_n) g_{x(1),\ldots,x(n-1)}(y_n) \right) \\ &= \int E_0(dx_0) \int T^{x(0)}_{r(1)}(dx_1) \cdots \\ &\quad \times \left(\int T^{x(n-1)}_{r(n)-r(n-1)}(dx_n) \int T^{x(n)}_{r(n+1)-r(n)}(dy_n) g_{x(1),\ldots,x(n-1)}(y_n) \right), \end{aligned}$$

where the last equality is thanks to the semigroup property of $\mathbf{T}$. Combining, we obtain

$$F^{E(0),\mathbf{T}}_{r(1),\ldots,\widehat{r(n)},\ldots,r(m)}f = \int E_0(dx_0)\int T^{x(0)}_{r(1)}(dx_1)\cdots\int T^{x(n-1)}_{r(n)-r(n-1)}(dx_n)$$

$$\times \int T^{x(n)}_{r(n+1)-r(n)}(dy_n)g_{x(1),\ldots,x(n-1)}(y_n)$$

$$\equiv \int E_0(dx_0)\int T^{x(0)}_{r(1)}(dx_1)\cdots\int T^{x(n-1)}_{r(n)-r(n-1)}(dx_n)$$

$$\times \int T^{x(n)}_{r(n+1)-r(n)}(dy_n)\left\{\int T^{y(n)}_{r(n+2)-r(n+1)}(dy_{n+1})\cdots\right.$$

$$\left.\times \int T^{y(m-2)}_{r(m)-r(m-1)}(dy_{m-1})f(x_1,\ldots,x_{n-1},y_n,\ldots,y_{m-1})\right\}$$

$$= \int E_0(dx_0)\int T^{x(0)}_{r(1)}(dx_1)\cdots\int T^{x(n-1)}_{r(n)-r(n-1)}(dx_n)$$

$$\times \int T^{x(n)}_{r(n+1)-r(n)}(dx_{n+1})\left\{\int T^{x(n+1)}_{r(n+2)-r(n+1)}(dx_{n+2})\cdots\right.$$

$$\left.\times \int T^{x(m-1)}_{r(m)-r(m-1)}(dx_m)f(x_1,\ldots,x_{n-1},x_{n+1},\ldots,x_m)\right\}$$

$$= F^{E(0),\mathbf{T}}_{r(1),\ldots,r(m)}(f\circ\kappa^*_{n,m}),$$

where the third equality is by a trivial change of the dummy integration variables $y_n,\ldots,y_{m-1}$ to $x_{n+1},\ldots,x_m$, respectively. Thus equality 11.4.13 has been proved for the family

$$\left\{F^{E(0),\mathbf{T}}_{r(1),\ldots,r(m)} : m\geq 1; r_1,\ldots,r_m\in Q; r_1\leq\cdots\leq r_m\right\}$$

of f.j.d.'s. Consequently, the conditions in Lemma 6.2.3 are satisfied, to yield a unique extension of this family to a consistent family

$$\left\{F^{E(0),\mathbf{T}}_{s(1),\ldots,s(m)} : m\geq 0; s_0,\ldots,s_m\in Q\right\}$$

of f.j.d.'s with parameter set Q. Finally, equality 11.4.4 says that $F^{E(0),\mathbf{T}}$ is generated by the initial distribution E_0 and the semigroup $\mathbf{T}$.

10. It remains to verify that the family $F^{E(0),\mathbf{T}}$ is continuous in probability. To that end, let $\varepsilon>0$ be arbitrary. Consider the function $\widehat{d}\equiv 1\wedge d\in C(S^2,d^2)$, with a modulus of continuity given by the identity operation ι. Consider each $r,s\in Q$ with $|s-r|<\delta_{Cp}(\varepsilon,\delta_{\mathbf{T}},\alpha_{\mathbf{T}})\equiv\delta_2(\varepsilon,\iota,\delta_{\mathbf{T}},\alpha_{\mathbf{T}})$. First assume $r\leq s$. From Assertion 3, we then have

$$\left|F^{*,\mathbf{T}}_{r,s}\widehat{d}-F^{*,\mathbf{T}}_{r,r}\widehat{d}\right|\leq\varepsilon.$$

At the same time, by Definition 11.4.1, we have

$$F_{r,r}^{x,\mathbf{T}}\widehat{d} = \int T_r^x(dx_1)\int T_0^{x(1)}(dx_2)\widehat{d}(x_1,x_2) = \int T_r^x(dx_1)\widehat{d}(x_1,x_1) = 0. \tag{11.4.14}$$

Hence

$$F_{r,s}^{E(0),\mathbf{T}}\widehat{d} = E_0 F_{r,s}^{*,\mathbf{T}}\widehat{d} = E_0\left|F_{r,s}^{*,\mathbf{T}}\widehat{d} - F_{r,r}^{*,\mathbf{T}}\widehat{d}\right| \le \varepsilon.$$

Similarly, we can prove the same inequality in the case where $r \ge s$. Since $\varepsilon > 0$ is arbitrary, we conclude that the family $F^{E(0),\mathbf{T}}$ is continuous in probability, with a modulus of continuity in probability $\delta_{Cp}(\varepsilon,\delta_{\mathbf{T}},\alpha_{\mathbf{T}})$. This proves Assertions 4. Assertion 5 is a special case of Assertion 4. □

Corollary 11.4.3. Continuity of Markov f.j.d.'s. *Let (S,d) be a compact metric space with $d \le 1$. Let $\mathbf{T}$ be an arbitrary semigroup with state space (S,d), with parameter set $[0,\infty)$, with a modulus of strong continuity $\delta_{\mathbf{T}}$, and with a modulus of smoothness $\alpha_{\mathbf{T}} \equiv (\alpha_{\mathbf{T,N}})_{N=1,2,\ldots}$, in the sense of Definition 11.3.1. Let $m \ge 1$ and $\varepsilon > 0$ be arbitrary. Then there exists $\delta > 0$ such that for each $f \in C(S^m,d^m)$ with values in $[0,1]$ and a modulus of continuity δ_f, and for arbitrary nondecreasing sequences $r_1 \le \cdots \le r_m$ and $s_1 \le \cdots \le s_m$ in $[0,\infty)$, we have*

$$\left|F_{r(1),\ldots,r(m)}^{x,\mathbf{T}}f - F_{s(1),\ldots,s(m)}^{y,\mathbf{T}}f\right| < \varepsilon$$

provided that $d(x,y) \vee \bigvee_{i=1}^m |r_i - s_i| < \delta$. In short, the function $F_{r(1),\ldots,r(m)}^{x,\mathbf{T}}f$ of

$$(x,r_1,\ldots,r_m) \in S \times \{(r_1,\ldots,r_m) \in [0,\infty)^m : r_1 \le \cdots \le r_m\}$$

is uniformly continuous relative to the metric $d \otimes d_{ecld}^m$, where d_{ecld} is the Euclidean metric.

Proof. Let $m \ge 1$ and $\varepsilon > 0$. Write $\varepsilon_0 \equiv 2^{-1}\varepsilon$. There is no loss of generality to assume that the function $f \in C(S^m,d^m)$ has values in $[0,1]$ and a modulus of continuity δ_f. Then Assertion 3 of Theorem 11.4.2 yields some $\delta_m(\varepsilon_0,\delta_f,\delta_{\mathbf{T}},\alpha_{\mathbf{T}}) > 0$ such that for arbitrary nondecreasing sequences $r_1 \le \cdots \le r_m$ and $s_1 \le \cdots \le s_m$ in $[0,\infty)$ with

$$\bigvee_{i=1}^{m} |r_i - s_i| < \delta_1 \equiv \delta_m(\varepsilon_0,\delta_f,\delta_{\mathbf{T}},\alpha_{\mathbf{T}}),$$

we have

$$\left\|F_{r(1),\ldots,r(m)}^{*,\mathbf{T}}f - F_{s(1),\ldots s(m)}^{*,\mathbf{T}}f\right\| \le \varepsilon_0.$$

Hence, for each $x,y \in S$, we have

$$|F_{r(1),\ldots,r(m)}^{x,\mathbf{T}}f - F_{s(1),\ldots,s(m)}^{x,\mathbf{T}}f| \le \varepsilon_0, \tag{11.4.15}$$

with a similar inequality for y in the place of x.

At the same time, Assertion 1 of Theorem 11.4.2 says that $F^{*,\mathbf{T}}_{r(1),\ldots,r(m)}$ is a transition function. Hence the function $F^{*,\mathbf{T}}_{r(1),\ldots,r(m)}f$ is a member of $C(S,d)$. Therefore, there exists $\delta_2 > 0$ such that

$$\left|F^{x,\mathbf{T}}_{r(1),\ldots,r(m)}f - F^{y,\mathbf{T}}_{r(1),\ldots,r(m)}f\right| < \varepsilon_0,$$

provided that $d(x,y) < \delta_2$. Combining with inequality 11.4.15, we obtain

$$\left|F^{x,\mathbf{T}}_{r(1),\ldots,r(m)}f - F^{y,\mathbf{T}}_{s(1),\ldots,s(m)}f\right|$$
$$\leq \left|F^{x,\mathbf{T}}_{r(1),\ldots,r(m)}f - F^{y,\mathbf{T}}_{r(1),\ldots,r(m)}f\right| + \left|F^{x,\mathbf{T}}_{r(1),\ldots,r(m)}f - F^{y,\mathbf{T}}_{s(1),\ldots,s(m)}f\right| < 2\varepsilon_0 = \varepsilon$$

provided that $d(x,y) \vee \bigvee_{i=1}^{m} |r_i - s_i| < \delta \equiv \delta_1 \wedge \delta_2$. The corollary is proved. □

11.5 Construction of a Markov Process from a Semigroup

In this section, we construct a Markov process from a Markov semigroup and an initial state. The next theorem gives the construction for the discrete parameter set $\{0,1,\ldots\}$ or $\overline{Q}_\infty$. Subsequently, Theorem 11.5.6 will do the same for the parameter set $[0,\infty)$, with the resulting Markov process having the additional property being a.u. càdlàg.

First some notations.

Definition 11.5.1. Notations for two natural filtrations. Let $X : [0,\infty) \times (\Omega, L, E) \rightarrow (S,d)$ be an arbitrary process that is continuous in probability, whose state space (S,d) is a locally compact metric space. Let $Z \equiv X|\overline{Q}_\infty : \overline{Q}_\infty \times (\Omega, L, E) \rightarrow (S,d)$. In this section, we will then use the following notations:

1. $\mathcal{L}_X \equiv \{L^{(X,t)} : t \in [0,\infty)\}$ denotes the natural filtration of the process X, where for each $t \in [0,\infty)$, we have the probability subspace $L^{(X,t)} \equiv L(X_r : r \in [0,t])$.

2. $\mathcal{L}_Z \equiv \{L^{(Z,t)} : t \in \overline{Q}_\infty\}$ denotes the natural filtration of the process Z, where for each $t \in \overline{Q}_\infty$, we have the probability subspace $L^{(Z,t)} \equiv L(Z_r : r \in [0,t]\overline{Q}_\infty)$. □

Lemma 11.5.2. Two natural filtrations. *Let $X : [0,\infty) \times (\Omega, L, E) \rightarrow (S,d)$ be an arbitrary process that is continuous in probability, whose state space (S,d) is a locally compact metric space. Let $Z \equiv X|\overline{Q}_\infty : \overline{Q}_\infty \times (\Omega, L, E) \rightarrow (S,d)$. Let the filtrations $\mathcal{L}_X$ and $\mathcal{L}_Z$ be as in Definition 11.5.1. Then for each $t \in \overline{Q}_\infty$, we have $L^{(Z,t)} = L^{(X,t)}$.*

Proof. Let $t \in \overline{Q}_\infty$ be arbitrary. Then

$$L^{(Z,t)} \equiv L(Z_r : r \in [0,t]\overline{Q}_\infty) = L(X_r : r \in [0,t]\overline{Q}_\infty)$$
$$\subset L(X_r : r \in [0,t]) \equiv L^{(X,t)}. \qquad (11.5.1)$$

Conversely, let $Y \in L^{(X,t)}$ be arbitrary. Then, for each $\varepsilon > 0$, there exists $r_1, \ldots, r_n \in [0,t]$ and $g \in C(S^n, d^n)$ such that

$$E|Y - g(X_{r(1)}, \ldots, X_{r(n)})| < \varepsilon.$$

By continuity in probability, there exists $s_1, \ldots, s_n \in [0,t]\overline{Q}_\infty$ such that

$$E|g(X_{r(1)}, \ldots, X_{r(n)}) - g(Z_{s(1)}, \ldots, Z_{s(n)})|$$
$$= E|g(X_{r(1)}, \ldots, X_{r(n)}) - g(X_{s(1)}, \ldots, X_{s(n)})| < \varepsilon.$$

Combining, we have $E|Y - Y'| < 2\varepsilon$, where $Y' \equiv g(Z_{s(1)}, \ldots, Z_{s(n)}) \in L^{(Z,t)}$, and where $\varepsilon > 0$ is arbitrary. Since $L^{(Z,t)}$ is complete, it follows that $Y \in L^{(Z,t)}$, where $Y \in L^{(X,t)}$ is arbitrary. Thus $L^{(X,t)} \subset L^{(Z,t)}$. It follows, in view of relation 11.5.1 in the opposite direction, that $L^{(X,t)} = L^{(Z,t)}$. □

Lemma 11.5.3. Approximation of certain stopping times. *Let $X : [0,\infty) \times (\Omega, L, E) \to (S,d)$ be an arbitrary process that is continuous in probability, whose state space (S,d) is a locally compact metric space. Let $\overline{\mathcal{L}}_X \equiv \{L^{(X,t+)} : t \in [0,\infty)\}$ denote the right-limit extension of the natural filtration $\mathcal{L}_X \equiv \{L^{(X,t)} : t \in [0,\infty)\}$. Define $Z \equiv X|\overline{Q}_\infty : \overline{Q}_\infty \times (\Omega, L, E) \to (S,d)$ and let $\mathcal{L}_Z \equiv \{L^{(Z,t)} : t \in \overline{Q}_\infty\}$ denote the natural filtration of the process Z.*

Suppose τ is a stopping time relative to $\overline{\mathcal{L}}_X$. Then there exists a sequence $(\eta_h)_{h=0,1,\ldots}$ of stopping times relative to $\mathcal{L}_Z$, such that for each $h \geq 0$, (i) *the stopping time η_h has positive values in the enumerated set*

$$\widetilde{Q}_h \equiv \{s_0, s_1, s_2, \ldots\} \equiv \{0, \Delta_h, 2\Delta_h, \ldots\}$$

and (ii)

$$\tau + 2^{-h-1} < \eta_h < \tau + 2^{-h+2}. \tag{11.5.2}$$

Proof. 1. Since τ is a stopping time relative to $\overline{\mathcal{L}}_X$, there exists, by Assertion 1 of Proposition 10.11.3, a sequence $(\eta_h)_{h=0,1,\ldots}$ of stopping times relative to $\mathcal{L}_Z$, such that for each $h \geq 0$, (i) the stopping time η_h has positive values in the enumerated set

$$\widetilde{Q}_h \equiv \{s_0, s_1, s_2, \ldots\} \equiv \{0, \Delta_h, 2\Delta_h, \ldots\},$$

(ii)

$$\tau + 2^{-h-1} < \eta_h < \tau + 2^{-h+2}, \tag{11.5.3}$$

and (iii) for each $j \geq 1$, there exists a regular point r_j of τ, with $r_j < s_j$ such that $(\eta_h \leq s_j) \in L^{(X,r(j)+)}$.

2. Consider the stopping time η_h, for each $h \geq 0$. By Condition (iii), for each possible value s_j of the r.r.v. η_h, we have

$$(\eta_h \leq s_j) \in L^{(X,r(j)+)} \equiv \bigcap_{u > r(j)} L^{(X,u)} \subset L^{(X,s(j))} = L^{(Z,s(j))},$$

where the last equality is due to Lemma 11.5.2. Thus η_h is actually a stopping time relative to the filtration $\mathcal{L}_Z$. The present lemma is proved. □

Theorem 11.5.4. Existence of a Markov process with a given semigroup, and with discrete parameters. *Let (S,d) be a compact metric space with $d \leq 1$. Suppose Q is one of the two parameter sets $\{0,1,\ldots\}$ or $\overline{Q}_\infty$. Let*

$$(\Omega, L, E) \equiv (\Theta_0, L_0, I_0) \equiv \left([0,1], L_0, \int \cdot dx\right)$$

denote the Lebesgue integration space based on the unit interval $[0,1]$.

Let $\mathbf{T} \equiv \{T_t : t \in Q\}$ be an arbitrary Markov semigroup with state space (S,d), and with a modulus of strong continuity $\delta_{\mathbf{T}}$. Let $x \in S$ be arbitrary. Let

$$F^{x,\mathbf{T}} \equiv \Phi_{Sg,fjd}(x,\mathbf{T})$$

be the corresponding consistent family of f.j.d.'s constructed in Theorem 11.4.2. Let

$$Z^{x,\mathbf{T}} \equiv \overline{\Phi}_{DKS,\xi}(F^{x,\mathbf{T}}) : Q \times (\Omega, L, E) \to S$$

be the Compact Daniell–Kolmogorov–Skorokhod Extension of the consistent family $F^{x,\mathbf{T}}$, as constructed in Theorem 6.4.3.

Then the following conditions hold:

1. The process $Z \equiv Z^{x,\mathbf{T}}$ is generated by the initial state x and semigroup $\mathbf{T}$, in the sense of Definition 11.4.1. Moreover, the process $Z^{x,\mathbf{T}}$ is continuous in probability.

2. The process $Z \equiv Z^{x,\mathbf{T}}$ is a Markov process relative to its natural filtration $\mathcal{L}_Z \equiv \{L^{(Z,t)} : t \in Q\}$, where

$$L^{(Z,t)} \equiv L(Z_s : s \in [0,t]Q\} \subset L$$

for each $t \in Q$. Specifically, let the nondecreasing sequence $0 \equiv s_0 \leq s_1 \leq \cdots \leq s_m$ in Q, the function $f \in C(S^{m+1}, d^{m+1})$, and $t \in Q$ be arbitrary. Then

$$\begin{aligned} &E(f(Z_{t+s(0)}, Z_{t+s(1)}, \ldots, Z_{t+s(m)})|L^{(Z,t)}) \\ &\quad = E(f(Z_{t+s(0)}, Z_{t+s(1)}, \ldots, Z_{t+s(m)})|Z_t) = F^{Z(t),\mathbf{T}}_{s(0),\ldots,s(m)}(f) \qquad (11.5.4) \end{aligned}$$

as r.r.v.'s, where $F^{*,\mathbf{T}}_{s(0),\ldots,s(m)}$ is the transition distribution in Definition 11.4.1.

3. The process $Z^{x,\mathbf{T}}$ has a modulus of continuity in probability $\delta_{Cp,\delta(\mathbf{T})}$ that is completely determined by $\delta_{\mathbf{T}}$, and is independent of x. Hence so does the family $F^{x,\mathbf{T}}$ of the marginal distributions of $Z^{x,\mathbf{T}}$.

Proof. 1. By Theorem 6.4.3, the process $Z \equiv Z^{x,\mathbf{T}} \equiv \overline{\Phi}_{DKS,\xi}(F^{x,\mathbf{T}})$ has marginal distributions given by the family $F^{x,\mathbf{T}}$. Moreover, Theorem 11.4.2 says that the consistent family $F^{x,\mathbf{T}}$ is generated by the initial state x and the semigroup $\mathbf{T}$, in the sense of Definition 11.4.1, and that the family $F^{x,\mathbf{T}}$ is continuous in probability. Hence the process $Z^{x,\mathbf{T}|Q}$ is continuous in probability. Assertion 1 is proved.

2. Let $t \in Q$ be arbitrary. Let $0 \equiv r_0 \leq r_1 \leq \cdots \leq r_n \equiv t$ and $0 \equiv s_0 \leq s_1 \leq \cdots \leq s_m$ be arbitrary sequences in Q. Write $r_{n+j} \equiv t + s_j$ for each $j = 0, \ldots, m$. Thus $s_j = r_{n+j} - r_n$ for each $j = 0, \ldots, m$. Consider each $f \in C(S^{m+1}, d^{m+1})$. Let $h \in C(S^{n+1}, S^{n+1})$ be arbitrary. Then

$$
\begin{aligned}
&Eh(Z_{r(0)}, \ldots, Z_{r(n)}) f(Z_{r(n)}, \ldots, Z_{r(n+m)}) \\
&= \int F^{x,\mathbf{T}}_{r(0),\ldots,r(n+m)} d(x_0, \ldots, x_{n+m}) h(x_0, \ldots, x_n) f(x_n, \ldots, x_{n+m}) \\
&\equiv \int T^{x}_{r(1)}(dx_1) \int T^{x(1)}_{r(2)-r(1)}(dx_2) \cdots \\
&\quad \times \int T^{x(n+m-1)}(dx_{n+m}) h(x_1, \ldots, x_n) f(x_n, \ldots, x_{n+m}) \\
&= \int T^{x}_{r(1)}(dx_1) \int T^{x(1)}_{r(2)-r(1)}(dx_2) \cdots \\
&\quad \times \int T^{x(n-1)}_{r(n)-r(n-1)}(dx_n) h(x_1, \ldots, x_n) \\
&\quad \times \left\{ \int T^{x(n)}_{r(n+1)-r(n)}(dx_{n+1}) \cdots \right. \\
&\quad \left. \times \int T^{x(n+m-1)}_{r(n+m)-r(n+m-1)}(dx_{n+m}) f(x_n, x_{n+1} \ldots, x_{n+m}) \right\}.
\end{aligned}
\tag{11.5.5}
$$

The term inside the braces in the last expression is, by changing the names of the dummy integration variables, equal to

$$
\begin{aligned}
&\int T^{x(n)}_{r(n+1)-r(n)}(dy_1) \int T^{y(1)}_{r(n+2)-r(n+1)}(dy_2) \cdots \\
&\int T^{y(m-1)}_{r(n+m)-r(n+m-1)}(dy_m) f(x_n, y_1, \ldots, y_m) \\
&= \int T^{x(n)}_{s(1)}(dy_0) \int T^{y(1)}_{s(2)}(dy_2) \cdots \int T^{y(m-1)}_{s(m)}(dy_m) f(x_n, y_1, \ldots, y_m) \\
&= F^{x(n),\mathbf{T}}_{s(1),\ldots,s(m)} f.
\end{aligned}
$$

Substituting back into equality 11.5.5, we obtain

$$
\begin{aligned}
&Eh(Z_{r(0)}, \ldots, Z_{r(n)}) f(Z_{r(n)}, \ldots, Z_{r(n+m)}) \\
&= \int T^{x}_{r(1)}(dx_1) \int T^{x(1)}_{r(2)-r(1)}(dx_2) \cdots \int T^{x(n-1)}_{r(n)-r(n-1)}(dx_n) F^{x(n),\mathbf{T}}_{s(1),\ldots,s(m)} f \\
&= Eh(Z_{r(0)}, \ldots, Z_{r(n)}) F^{Z(r(n)),\mathbf{T}}_{s(1),\ldots,s(m)} f,
\end{aligned}
\tag{11.5.6}
$$

where $F^{*,\mathbf{T}}_{s(1),\ldots,s(m)} f \in C(S, d)$ because $F^{*,\mathbf{T}}_{s(1),\ldots,s(m)}$ is a transition distribution according to Assertion 1 of Theorem 11.4.2.

3. Next note that the set of r.r.v.'s $h(Z_{r(0)},\ldots,Z_{r(n)})$, with arbitrary $0 \equiv r_0 \leq r_1 \leq \cdots \leq r_n \equiv t$ and arbitrary $h \in C(S^{n+1},d^{n+1})$, is dense in $L^{(Z,t)}$relative to the norm $E|\cdot|$. Hence equality 11.5.6 implies, by continuity relative to the norm $E|\cdot|$, that

$$EYf(Z_{r(n)},\ldots,Z_{r(n+m)}) = EYF^{Z(r(n)),\mathbf{T}}_{s(0),\ldots,s(m)}(f) \tag{11.5.7}$$

for each $Y \in L^{(x,t)}$. Since $F^{Z(r(n)),\mathbf{T}}_{s(0),\ldots,s(m)}(f) = F^{Z(t),\mathbf{T}}_{s(0),\ldots,s(m)}(f) \in L^{(Z,t)}$, it follows that

$$E(f(Z_{r(n)},\ldots,Z_{r(n+m)})|L^{(Z,t)}) = F^{Z(t),\mathbf{T}}_{s(0),\ldots,s(m)}(f) \tag{11.5.8}$$

or, equivalently,

$$E(f(Z_t,Z_{t+s(1)},\ldots,Z_{t+s(m)})|L^{(Z,t)}) = F^{Z(t),\mathbf{T}}_{s(0),\ldots,s(m)}(f). \tag{11.5.9}$$

In the special case where Y is arbitrary in $L(Z_t) \subset L^{(Z,t)}$, inequality 11.5.7 holds, whence

$$E(f(Z_t,Z_{t+s(1)},\ldots,Z_{t+s(m)})|Z_t) = F^{Z(t),\mathbf{T}}_{s(0),\ldots,s(m)}(f). \tag{11.5.10}$$

Equalities 11.5.9 and 11.5.10 together prove the desired equality 11.5.4 in Assertion 2.

4. Assertion 3 remains. We will give the proof only in the case where $Q = \overline{Q}_\infty$, with the case where $Q = \{0,1,\ldots\}$ being trivial. To that end, let $\varepsilon > 0$ be arbitrary. Let $t \in \overline{Q}_\infty$ be arbitrary with

$$t < \delta_{Cp,\delta(\mathbf{T})}(\varepsilon) \equiv \delta_{\mathbf{T}}(\varepsilon,\iota).$$

Then, by Assertion 2 of Lemma 11.3.2, we have

$$\|T_t(\widetilde{d}(x,\cdot)) - \widehat{d}(x,\cdot)\| = \|T_t(d(x,\cdot)) - d(x,\cdot)\| \leq \varepsilon, \tag{11.5.11}$$

for each $x \in S$. Now let $r_1,r_2 \in \overline{Q}_\infty$ be arbitrary with $|r_2 - r_1| < \delta_{Cp,\delta(\mathbf{T})}(\varepsilon)$. Then

$$\begin{aligned}
E\widehat{d}(Z_{r(1)},Z_{r(2)}) &= E\widehat{d}(Z_{r(1)\wedge r(2)},Z_{r(1)\vee r(2)}) = F^{x,\mathbf{T}}_{r(1)\wedge r(2),r(1)\vee r(2)}\widehat{d}\\
&= \int T^x_{r(1)\wedge r(2)}(dx_1)\int T^{x(1)}_{r(1)\vee r(2)-r(1)\wedge r(2)}(dx_2)\widehat{d}(x_1,x_2)\\
&= \int T^x_{r(1)\wedge r(2)}(dx_1)\int T^{x(1)}_{|r(2)-r(1)|}(dx_2)\widehat{d}(x_1,x_2)\\
&= \int T^x_{r(1)\wedge r(2)}(dx_1)T^{x(1)}_{|r(2)-r(1)|}\widehat{d}(x_1,\cdot)\\
&\leq \int T^x_{r(1)\wedge r(2)}(dx_1)(\widehat{d}(x_1,x_1)+\varepsilon)\\
&= \int T^x_{r(1)\wedge r(2)}(dx_1)\varepsilon = \varepsilon,
\end{aligned}$$

where the inequality is by applying inequality 11.5.11 to $t \equiv |r_2 - r_1|$ and $x = x_1$. Thus we have shown that $\delta_{Cp,\delta(\mathbf{T})} \equiv \delta_{\mathbf{T}}(\cdot,\iota)$ is a modulus of continuity in probability for the process Z, according to Definition 6.1.3. Note here that the operation $\delta_{Cp,\delta(\mathbf{T})}$ depends only on $\delta_{\mathbf{T}}$; it is independent of x. Assertion 3 is proved. □

The next proposition implies that, given a Markov semigroup $\mathbf{T} \equiv \{T_t : t \in [0,\infty)\}$, the Markov process $Z \equiv Z^{x,\mathbf{T}|\overline{Q}(\infty)} : \overline{Q}_\infty \times (\Omega, L, E) \to (S,d)$ can be extended by right limit to an a.u. càdlàg process $X^{x,\mathbf{T}} : [0,\infty) \times (\Omega, L, E) \to (S,d)$. A subsequent theorem then says that the resulting process $X^{x,\mathbf{T}}$ is Markov.

Proposition 11.5.5. Markov process with a semigroup on dyadic rationals is D-regular, and is extendable to a time-uniformly a.u. càdlàg process. *Let (S,d) be a compact metric space with $d \leq 1$. Let $\mathbf{T} \equiv \{T_t : t \in [0,\infty)\}$ be an arbitrary Markov semigroup with state space (S,d) and with a modulus of strong continuity $\delta_{\mathbf{T}}$. Let*

$$(\Omega, L, E) \equiv (\Theta_0, L_0, I_0) \equiv \left([0,1], L_0, \int \cdot dx\right)$$

denote the Lebesgue integration space based on the unit interval $[0,1]$. Let $x \in S$ be arbitrary. Let

$$Z \equiv Z^{x,\mathbf{T}|\overline{Q}(\infty)} : \overline{Q}_\infty \times (\Omega, L, E) \to (S,d)$$

denote the process generated by the initial state x and semigroup $\mathbf{T}|\overline{Q}_\infty$, in the sense of Definition 11.4.1, as constructed in Theorem 11.5.4.

Let $N \geq 0$ be arbitrary. Define the shifted process $Z^N : Q_\infty \times \Omega \to S$ by $Z^N(t,\cdot) \equiv Z(N+t,\cdot)$ for each $t \in Q_\infty$. Then the following conditions hold:

1. The process Z^N is continuous in probability, with a modulus of continuity in probability $\delta_{Cp,\delta(\mathbf{T})} \equiv \delta_{Cp}(\cdot,\delta_{\mathbf{T}})$ that is completely determined by $\delta_{\mathbf{T}}$ and is independent of x and N.

2. The process Z^N is strongly right continuous in probability, in the sense of Definition 10.7.3, with a modulus of strong right continuity in probability given by the operation $\delta_{SRCp,\delta(\mathbf{T})}$ defined by

$$\delta_{SRCp,\delta(\mathbf{T})}(\varepsilon,\gamma) \equiv \delta_{Cp}(\varepsilon^2,\delta_{\mathbf{T}})$$

for each $\varepsilon,\gamma > 0$. Note that $\delta_{SRCp,\delta(\mathbf{T})}(\varepsilon,\gamma)$ is completely determined by $\delta_{\mathbf{T}}$ and is independent of x, N and γ.

3. The process $Z^N : Q_\infty \times \Omega \to S$ is D-regular, with a modulus of D-regularity $\overline{m}_{\delta(\mathbf{T})}$ that is completely determined by $\delta_{\mathbf{T}}$.

4. The process $Z : \overline{Q}_\infty \times (\Omega, L, E) \to (S,d)$ is time-uniformly D-regular in the sense of Definition 10.10.2, with a modulus of continuity in probability $\widetilde{\delta}_{Cp,\delta(\mathbf{T})} \equiv (\delta_{Cp,\delta(\mathbf{T})}, \delta_{Cp,\delta(\mathbf{T})}, \ldots)$ and a modulus of D-regularity $\widetilde{m}_{\delta(\mathbf{T})} \equiv (\overline{m}_{\delta(\mathbf{T})}, \overline{m}_{\delta(\mathbf{T})}, \ldots)$.

5. The right-limit extension

$$X \equiv \overline{\Phi}_{rLim}(Z) : [0,\infty) \times (\Omega, L, E) \to (S,d)$$

is a time-uniformly a.u. càdlàg process in the sense of Definition 10.10.3, with a modulus of a.u. càdlàg $\widetilde{\delta}_{aucl,\delta(\mathbf{T})} \equiv (\delta_{aucl,\delta(\mathbf{T})}, \delta_{aucl,\delta(\mathbf{T})}, \ldots)$ *and a modulus of continuity in probability* $\widetilde{\delta}_{Cp,\delta(\mathbf{T})}$. *Moreover, both of these two moduli are completely determined by* $\delta_{\mathbf{T}}$. *Here we recall the mapping* $\overline{\Phi}_{rLim}$ *from Definition 10.5.6. In the notations of Definition 10.10.3, we have*

$$X \equiv \overline{\Phi}_{rLim}(Z) \in \widehat{D}_{\widetilde{\delta}(aucl,\delta(\mathbf{T})),\widetilde{\delta}(Cp,\delta(\mathbf{T}))}[0,\infty).$$

Proof. 1. By hypothesis, the process Z has initial state x and Markov semigroup $\mathbf{T}|\overline{Q}_\infty$. In other words, the process Z has marginal distributions given by the consistent family $F^{x,\mathbf{T}|\overline{Q}(\infty)}$ of f.j.d.'s. as constructed in Theorem 11.5.4. Therefore, by Assertion 3 of Theorem 11.5.4, the process Z is continuous in probability, with a modulus of continuity in probability $\delta_{Cp,\delta(\mathbf{T})} \equiv \delta_{Cp}(\cdot,\delta_{\mathbf{T}})$ that is is completely determined by $\delta_{\mathbf{T}}$.

2. Consequently, the shifted process

$$Y \equiv Z^N : Q_\infty \times \Omega \to S$$

is continuous in probability, with a modulus of continuity in probability $\delta_{Cp,\delta(\mathbf{T})} \equiv \delta_{Cp}(\cdot,\delta_{\mathbf{T}})$ that is completely determined by $\delta_{\mathbf{T}}$. Assertion 1 is proved.

3. To prove Assertion 2, let $\varepsilon,\gamma > 0$ be arbitrary. Define

$$\delta_{SRCp,\delta(\mathbf{T})}(\varepsilon,\gamma) \equiv \delta_{\mathbf{T}}(\varepsilon^2,\iota), \tag{11.5.12}$$

where the operation ι is defined by $\iota(\varepsilon') \equiv \varepsilon'$ for each $\varepsilon' > 0$. We will show that the operation $\delta_{SRCp,\delta(\mathbf{T})}$ is a modulus of strong right continuity of the process Y. To that end, let $h \geq 0$ and $s,r \in Q_h$ be arbitrary, with $s \leq r < s + \delta_{SRCp,\delta(\mathbf{T})}(\varepsilon,\gamma)$. Recall the enumerated set

$$Q_h = \{0,\Delta_h,2\Delta_h,\ldots,1\} \equiv \{q_{h,0},\ldots,q_{h,p(h)}\},$$

where $\Delta_h \equiv 2^{-h}$, $p_h \equiv 2^h$, and $q_{h,i} \equiv i\Delta_h$, for each $i = 0,\ldots,p_h$. Then $s = q_{h,i}$ and $r = q_{h,j}$ for some $i,j = 0,\ldots,p_h$ with $i \leq j$. Now let $g \in C(S^{i+1},d^{i+1})$ be arbitrary. Then

$$\begin{aligned}
&Eg(Y_{q(h,0)},\ldots,Y_{q(h,i)})d(Y_s,Y_r)\\
&= Eg(Y_{q(h,0)},\ldots,Y_{q(h,i)})d(Y_{q(h,i)},Y_{q(h,j)})\\
&\equiv Eg(Z_{N+q(h,0)},\ldots,Z_{N+q(h,i)})d(Z_{N+q(h,i)s},Z_{N+q(h,i)})\\
&= Eg(Z_{N+0\Delta},\ldots,Z_{N+i\Delta})d(Z_{N+i\Delta},Z_{N+j\Delta})\\
&= E(g(Z_{N+0\Delta},\ldots,Z_{N+i\Delta})E(d(Z_{N+i\Delta},Z_{N+j\Delta})|L^{(Z,N+i\Delta)}))\\
&= E\Big(g(Z_{N+0\Delta},\ldots,Z_{N+i\Delta})F^{Z(N+i\Delta),\mathbf{T}}_{0,(j-i)\Delta}d\Big),
\end{aligned} \tag{11.5.13}$$

where the fourth equality is because $g(Z_{N+0\Delta},\ldots,Z_{N+i\Delta}) \in L^{Z,(N+i\Delta)}$, and where the fifth equality is by equality 11.5.4 in Theorem 11.5.4. Continuing, let $z \in S$ be arbitrary. Since

$$(j-i)\Delta = r - s \in [0, \delta_{SRCp,\delta(\mathbf{T})}(\varepsilon,\gamma)) \equiv [0, \delta_{\mathbf{T}}(\varepsilon^2,\iota)),$$

we have, according to Assertion 2 of Lemma 11.3.2, the bound

$$\left\| T_{(j-i)\Delta}(d(z,\cdot)) - d(z,\cdot) \right\| \le \varepsilon^2. \tag{11.5.14}$$

In particular,

$$F^{z,\mathbf{T}}_{0,(j-i)\Delta} d = T^{z}_{(j-i)\Delta} d(z,\cdot) = |T^{z}_{(j-i)\Delta} d(z,\cdot) - d(z,z)| \le \varepsilon^2, \tag{11.5.15}$$

where $z \in S$ is arbitrary. Hence equality 11.5.13 can be continued to yield

$$Eg(Y_{q(h,0)},\ldots,Y_{q(h,i)})d(Y_s,Y_r) \le \varepsilon^2 Eg(Z_{N+0\Delta},\ldots,Z_{N+i\Delta}),$$

where $g \in C(S^{i+1},d^{i+1})$ is arbitrary. It follows that

$$EUd(Y_s,Y_r) \le \varepsilon^2 EU \tag{11.5.16}$$

for each $U \in L(Y_{q(h,0)},\ldots,Y_{q(h,i)})$. Now let $\gamma > 0$ be arbitrary, and take an arbitrary measurable set

$$A \in L(Y_r : r \in [0,s]Q_h) = L(Y_{q(h,0)},\ldots,Y_{q(h,i)}) \tag{11.5.17}$$

with $A \subset (d(x_\circ,Y_s) \le \gamma)$ and $P(A) > 0$. Let $U \equiv 1_A$ denote the indicator of A. Then the membership relation 11.5.17 is equivalent to

$$U \equiv 1_A \in L(Y_{q(h,0)},Y_{q(h,1)},\ldots,Y_{q(h,i)}).$$

Hence equality 11.5.16 applies, to yield

$$E1_A d(Y_s,Y_r) \le \varepsilon^2 E1_A. \tag{11.5.18}$$

Dividing by $P(A)$, we obtain

$$E_A d(Y_s,Y_r) \le \varepsilon^2,$$

where E_A is the conditional expectation given the event A. Chebychev's inequality therefore implies

$$P_A(d(Y_s,Y_r) > \alpha) \le \varepsilon$$

for each $\alpha > \varepsilon$. Here $h \ge 0, \varepsilon, \gamma > 0$, and $s,r \in Q_h$ are arbitrary with $s \le r < s + \delta_{SRCp,\delta(\mathbf{T})}(\varepsilon,\gamma)$. Summing up, the process $Y \equiv Z^N$ is strongly right continuous in the sense of Definition 10.7.3, with a modulus of strong right continuity $\delta_{SRCp} \equiv \delta_{SRCp,\delta(\mathbf{T})}$. Assertion 2 is proved.

4. Proceed to prove Assertion 3. To that end, recall that, by hypothesis, $d \le 1$. Hence the process $Y \equiv Z^N : Q_\infty \times \Omega \to (S,d)$ is trivially a.u. bounded, with a modulus of a.u. boundedness $\beta_{auB} \equiv 1$. Combining with Assertion 2, we see that the conditions for Theorem 10.7.8 are satisfied for the process

$$Y \equiv Z^N : Q_\infty \times \Omega \to (S,d)$$

to be D-regular, with a modulus of D-regularity $\overline{m}_{\delta(\mathbf{T})} \equiv \overline{m}(1, \delta_{SRCp,\delta(\mathbf{T})})$ and a modulus of continuity in probability $\delta_{Cp,\delta(\mathbf{T})} \equiv \delta_{Cp}(\cdot, 1, \delta_{SRCp,\delta(\mathbf{T})})$. Note that both moduli are completely determined by $\delta_{\mathbf{T}}$. Assertion 3 of the present proposition is proved.

5. Since the moduli $\delta_{Cp,\delta(\mathbf{T})}$ and $\overline{m}_{\delta(\mathbf{T})}$ are independent of the integer $N \geq 0$, we see that the process

$$Z : \overline{Q}_\infty \times \Omega \to (S,d)$$

is time-uniformly D-regular in the sense of Definition 10.10.2, with a modulus of continuity in probability $\widetilde{\delta}_{Cp,\delta(\mathbf{T})} \equiv (\delta_{Cp,\delta(\mathbf{T})}, \delta_{Cp,\delta(\mathbf{T})}, \ldots)$ and a modulus of D-regularity $\widetilde{m}_{\delta(\mathbf{T})} \equiv (\overline{m}_{\delta(\mathbf{T})}, \overline{m}_{\delta(\mathbf{T})}, \ldots)$. Assertion 4 of the present proposition is verified.

6. In view of Assertion 3 of the present proposition, Assertion 2 of Theorem 10.7.8 says that the right-limit extension $\Phi_{rLim}(Z^N)$ is an a.u. càdlàg process, with the same modulus of continuity in probability $\delta_{Cp}(\cdot, 1, \delta_{SRCp})$ as Z^N, and with a modulus of a.u. càdlàg

$$\delta_{aucl,\delta(\mathbf{T})} \equiv \delta_{aucl}(\cdot, 1, \delta_{SRCp,\delta(\mathbf{T})}) \equiv \delta_{aucl}(\cdot, \beta_{auB}, \delta_{SRCp,\delta(\mathbf{T})}).$$

Consider the right-limit extension process

$$X \equiv \overline{\Phi}_{rLim}(Z) : [0,\infty) \times (\Omega, L, E) \to (S,d).$$

Consider each $N \geq 0$. Then $X^N = \Phi_{rLim}(Z^N)$ on the interval $[0,1)$. Near the endpoint 1, things are a bit more complicated. Recall that since $\Phi_{rLim}(Z^N)$ is a.u. càdlàg, it is continuous a.u. on $[0,1]$. Hence, for a.e. $\omega \in \Omega$, the function $Z(\cdot,\omega)$ is continuous at $1 \in [0,1]$. Therefore $X^N = \Phi_{rLim}(Z^N)$ on the interval $[0,1]$. We saw earlier that the process $\Phi_{rLim}(Z^N)$ is a.u. càdlàg, with a modulus of a.u. càdlàg $\delta_{aucl,\delta(\mathbf{T})}$. Hence $X^N$is a.u. càdlàg, with the same modulus of a.u. càdlàg $\delta_{aucl,\delta(\mathbf{T})}$.

7. As an immediate consequence of Assertion 4 of the present proposition, the process $Z|]0,N+1]\overline{Q}_\infty$ is continuous in probability, with a modulus of continuity in probability $\delta_{Cp,\delta(\mathbf{T})}$. It follows that the process $X|[0,N+1]$ is continuous in probability, with the same modulus of continuity in probability $\delta_{Cp,\delta(\mathbf{T})}$.

8. Summing up the results in Steps 6 and 7, we see that the process X is time-uniformly a.u. càdlàg in the sense of Definition 10.10.3, with a modulus of continuity in probability $\widetilde{\delta}_{Cp,\delta(\mathbf{T})} \equiv (\delta_{Cp,\delta(\mathbf{T})}, \delta_{Cp,\delta(\mathbf{T})}, \ldots)$ and a modulus of a.u. càdlàg $\widetilde{\delta}_{aucl,\delta(\mathbf{T})} \equiv (\delta_{aucl,\delta(\mathbf{T})}, \delta_{aucl,\delta(\mathbf{T})}, \ldots)$. Assertion 5 and the proposition are proved. □

In the following, to minimize notational clutter, we will write $a = b \pm c$ to mean $|a - b| \leq c$, for arbitrary real-valued expressions a, b, c.

Theorem 11.5.6. Existence; construction of an a.u. càdlàg Markov process from an initial state and a semigroup, with parameter set $[0,\infty)$. *Let* (S,d) *be a compact metric space with* $d \leq 1$. *Let* $\mathbf{T} \equiv \{T_t : t \in [0,\infty)\}$ *be an arbitrary Markov semigroup with state space* (S,d) *and a modulus of strong continuity* $\delta_{\mathbf{T}}$. *Let*

$$(\Omega, L, E) \equiv (\Theta_0, L_0, I_0) \equiv ([0,1], L_0, \int \cdot dx)$$

denote the Lebesgue integration space based on the unit interval Θ_0. *Let* $x \in S$ *be arbitrary. Let*

$$Z \equiv Z^{x,\mathbf{T}|\overline{Q}(\infty)} : \overline{Q}_\infty \times (\Omega, L, E) \to (S,d)$$

denote the Markov process generated by the initial state x *and semigroup* $\mathbf{T}|\overline{Q}_\infty$, *in the sense of Definition 11.4.1, as constructed in Theorem 11.5.4. Let* $\mathcal{L}_Z \equiv \{L^{(Z,t)} : t \in \overline{Q}_\infty\}$ *be the natural filtration of the process* $Z \equiv Z^{x,\mathbf{T}|\overline{Q}(\infty)}$. *Let*

$$X \equiv X^{x,\mathbf{T}} \equiv \overline{\Phi}_{rLim}(Z) : [0,\infty) \times \Omega \to (S,d)$$

be the right-limit extension of the process Z. *Let* $\mathcal{L}_X \equiv \{L^{(X,t)} : t \in [0,\infty)\}$ *denote the natural filtration of the process* $X \equiv X^{x,\mathbf{T}}$.

Then the following conditions hold:

1. The function $X \equiv X^{x,\mathbf{T}}$ *is a time-uniformly a.u. càdlàg process, with a modulus of continuity in probability* $\widetilde{\delta}_{Cp,\delta(\mathbf{T})} \equiv (\delta_{Cp,\delta(\mathbf{T})}, \delta_{Cp,\delta(\mathbf{T})}, \ldots)$ *and a modulus of a.u. càdlàg* $\widetilde{\delta}_{aucl,\delta(\mathbf{T})} \equiv (\delta_{aucl,\delta(\mathbf{T})}, \delta_{aucl,\delta(\mathbf{T})}, \ldots)$. *Note that both moduli are completely determined by* $\delta_{\mathbf{T}}$.

2. The marginal distributions of the process $X^{x,\mathbf{T}}$ *is given by the family* $F^{x,\mathbf{T}}$ *of f.j.d.'s generated by the initial state* x *and the semigroup* $\mathbf{T}$, *in the sense of Definition 11.4.1.*

3. The process $X \equiv X^{x,\mathbf{T}}$ *is Markov relative to its natural filtration* $\mathcal{L}_X \equiv \{L^{(X,t)} : t \in [0,\infty)\}$. *Specifically, let* $v \geq 0$ *and let* $t_0 \equiv 0 \leq t_1 \leq \cdots \leq t_m$ *be an arbitrary sequence in* $[0,\infty)$, *with* $m \geq 1$. *Let* $f \in C_{ub}(S^{m+1}, d^{m+1})$ *be arbitrary. Then*

$$\begin{aligned} &E(f(X_{v+t(0)}, \ldots, X_{v+t(m)})|L^{(X,v)}) \\ &= E(f(X_{v+t(0)}, \ldots, X_{v+t(m)})|X_v) = F^{X(v),\mathbf{T}}_{0,t(1)\cdots,t(m)} f, \end{aligned} \tag{11.5.19}$$

where $F^{*,\mathbf{T}}_{0,t(1)\cdots,t(m)}$ *is the transition distribution in Definition 11.4.1.*

Proof. 1. Let $x \in S$ be arbitrary. Note that $\mathbf{T}|\overline{Q}_\infty$ is a Markov semigroup with a parameter set $\overline{Q}_\infty$ and a modulus of strong continuity $\delta_{\mathbf{T}}$. Assertion 5 of Proposition 11.5.5 therefore applies; it says that the right-limit extension $X^{x,\mathbf{T}} \equiv \overline{\Phi}_{rLim}(Z^{x,\mathbf{T}|\overline{Q}(\infty)})$ is a time-uniformly a.u. càdlàg process in the sense of Definition 10.10.3, with a modulus of a.u. càdlàg $\widetilde{\delta}_{aucl,\delta(\mathbf{T})} \equiv (\delta_{aucl,\delta(\mathbf{T})}, \delta_{aucl,\delta(\mathbf{T})}, \ldots)$ that is completely determined by $\delta_{\mathbf{T}}$, and with a modulus of continuity in probability $\widetilde{\delta}_{Cp,\delta(\mathbf{T})} \equiv (\delta_{Cp,\delta(\mathbf{T})}, \delta_{Cp,\delta(\mathbf{T})}, \ldots)$. Assertion 1 of the present theorem is proved.

2. To prove Assertion 2, note that, by Assertion 4 of Theorem 11.4.2, the consistent family $F^{x,\mathbf{T}|\overline{Q}(\infty)}$ is generated by the initial state x and the semigroup $\mathbf{T}|\overline{Q}_{\infty}$. Hence, for each sequence $0 \equiv r_0 \leq r_1 \leq \cdots \leq r_m$ in $\overline{Q}_{\infty}$ and for each $f \in C(S^m, d^m)$, we have

$$
\begin{aligned}
&F^{x,\mathbf{T}}_{r(1),\ldots,r(m)} f \\
&\quad \equiv Ef(X_{r(1)}, \ldots, X_{r(m)}) = Ef(Z_{r(1)}, \ldots, Z_{r(m)}) = F^{x,\mathbf{T}|\overline{Q}(\infty)}_{r(1),\ldots,r(m)} f \\
&\quad = \int T^{x}_{r(1)}(dx_1) \int T^{x(1)}_{r(2)-r(1)}(dx_2) \cdots \int T^{x(m-1)}_{r(m)-r(m-1)}(dx_m) f(x_1, \ldots, x_m).
\end{aligned}
\tag{11.5.20}
$$

Because the process X is continuous in probability, and because the semigroup $\mathbf{T}$ is strongly continuous, this equality extends to

$$
\begin{aligned}
&F^{x,\mathbf{T}}_{r(1),\ldots,r(m)} f \\
&\quad = Ef(X_{r(1)}, \ldots, X_{r(m)}) \\
&\quad = \int T^{x}_{r(1)}(dx_1) \int T^{x(1)}_{r(2)-r(1)}(dx_2) \cdots \int T^{x(m-1)}_{r(m)-r(m-1)}(dx_m) f(x_1, \ldots, x_m)
\end{aligned}
\tag{11.5.21}
$$

for each sequence $0 \equiv r_0 \leq r_1 \leq \cdots \leq r_m$ in $[0, \infty)$. Thus the marginal distributions of the process X is given by the family $F^{x,\mathbf{T}}$, and the latter is generated by the initial state x and the semigroup $\mathbf{T}$, in the sense of Definition 11.4.1. Assertion 2 is proved.

3. To prove Assertion 3, first let $\mathcal{L}_Z \equiv \{L^{(Z,t)} : t \in \overline{Q}_{\infty}\}$ be the natural filtration of the process $Z \equiv Z^{x,\mathbf{T}|\overline{Q}(\infty)}$. Assertion 2 of Theorem 11.5.4 says that the process Z is a Markov process relative to the filtration $\mathcal{L}_Z$. Specifically, let the nondecreasing sequence $0 \equiv s_0 \leq s_1 \leq \cdots \leq s_m$ in $\overline{Q}_{\infty}$, the function $f \in C(S^{m+1}, d^{m+1})$, and the point $t \in \overline{Q}_{\infty}$ be arbitrary. Then Assertion 2 of Theorem 11.5.4 says that

$$
\begin{aligned}
&E(f(Z_{t+s(0)}, Z_{t+s(1)}, \ldots, Z_{t+s(m)})|L^{(Z,t)}) \\
&\quad = E(f(Z_{t+s(0)}, Z_{t+s(1)}, \ldots, Z_{t+s(m)})|Z_t)) = F^{Z(t),\mathbf{T}|\overline{Q}(\infty)}_{s(0),\ldots,s(m)}(f)
\end{aligned}
\tag{11.5.22}
$$

as r.r.v.'s, where $F^{*,\mathbf{T}|\overline{Q}(\infty)}_{s(0),\ldots,s(m)}$ is the transition distribution as in Definition 11.4.1.

4. Next, let $\mathcal{L}_X \equiv \{L^{(X,t)} : t \in [0, \infty)\}$ be the natural filtration of the process X. Let $t \in \overline{Q}_{\infty}$ be arbitrary. Take any sequence $0 \equiv r_0 \leq r_1 \leq \cdots \leq r_n \leq \cdots \leq r_{n+m}$ in $\overline{Q}_{\infty}$ such that $r_n = t$. Let the function $g \in C(S^{n+1}, d^{n+1})$ be arbitrary,

with values in $[0,1]$ and a modulus of continuity δ_g. Let $f \in C(S^{m+1}, d^{m+1})$ be arbitrary, with a modulus of continuity δ_f. Write

$$(s_0, s_1, \ldots, s_m) \equiv (0, r_{n+1} - r_n, \ldots, r_m - r_n).$$

Then

$$(r_n, \ldots, r_m) = (t + s_0, \ldots, t + s_m).$$

Moreover, $g(Z_{r(0)}, \ldots, Z_{r(n)}) \in L^{(Z,r(n))}$. Hence equality 11.5.22 yields

$$\begin{aligned} &Eg(Z_{r(0)}, \ldots, Z_{r(n)}) f(Z_{r(n)}, \ldots, Z_{r(n+m)}) \\ &\quad = E(g(Z_{r(0)}, \ldots, Z_{r(n)}) E(f(Z_{t+s(0)}, \ldots, Z_{t+s(m)}) | L^{(Z,r(n))})) \\ &\quad = Eg(Z_{r(0)}, \ldots, Z_{r(n)}) F^{Z(r(n)), \mathbf{T}|\overline{Q}(\infty)}_{s(0) \cdots, s(m)}(f). \\ &\quad = Eg(Z_{r(0)}, \ldots, Z_{r(n)}) F^{Z(r(n)), \mathbf{T}|\overline{Q}(\infty)}_{0, r(n+1)-r(n), \ldots, r(n+m)-r(n)}(f). \end{aligned}$$

Since $X \equiv \overline{\Phi}_{rLim}(Z)$, we have $X_r = Z_r$ for each $r \in \overline{Q}_\infty$. Hence the last displayed equality is equivalent to

$$\begin{aligned} &Eg(X_{r(0)}, \ldots, X_{r(n)}) f(X_{r(n)}, \ldots, X_{r(n+m)}) \\ &\quad = Eg(X_{r(0)}, \ldots, X_{r(n)}) F^{X(r(n)), \mathbf{T}}_{0, r(n+1)-r(n), \ldots, r(n+m)-r(n)}(f), \end{aligned} \tag{11.5.23}$$

where $F^{*, \mathbf{T}}_{0, r(n+1)-r(n), \ldots, r(n+m)-r(n)}$ is the transition distribution as in Definition 11.4.1.

5. Now let $v \geq 0$ and the sequence $t_0 \equiv 0 \leq t_1 \leq \cdots \leq t_m$ in $[0, \infty)$ be arbitrary. Let $n \geq 1$ and $v_0 \equiv 0 \leq v_1 \leq \cdots \leq v_{n-1}$ in $[0, v]$ be arbitrary. Define $v_{n+i} \equiv v + t_i$ for each $i = 0, \ldots, m$. Thus $v_n \equiv v$ and

$$v_0 \equiv 0 \leq v_1 \leq \cdots \leq v_{n+m}.$$

Fix any integer $N \geq 0$ so large that $v_{n+m} \in [0, N-1]$. Take any sequence

$$0 \equiv r_0 \leq r_1 \leq \cdots \leq r_n \leq \cdots \leq r_{n+m} \quad \text{in} \quad \overline{Q}_\infty[0, N],$$

and let $r_i \downarrow v_i$ for each $i = 0, \ldots, n + m$. Then the left-hand side of equality 11.5.23 converges to the limit

$$Eg(X_{v(0)}, \ldots, X_{v(n)}) f(X_{v(n)}, \ldots, X_{v(n+m)}),$$

thanks to the continuity in probability of the process $X|[0, N+1]$.

6. Consider the right-hand side of equality 11.5.23. Let $\varepsilon > 0$ be arbitrary. As observed in Step 1, the process $X|[0, N+1]$ has a modulus of continuity in probability $\delta_{Cp, \delta(\mathbf{T})}$. Consequently, there exists $\delta_0 > 0$ so small that

$$E|g(X_{r(0)}, \ldots, X_{r(n)}) - g(X_{v(0)}, \ldots, X_{v(n)})| < \varepsilon \tag{11.5.24}$$

provided that $\bigvee_{i=0}^{n}(r_i - v_i) < \delta_0$.

7. Separately, Assertion 3 of Theorem 11.4.2 implies that there exists $\delta_{m+1}(\varepsilon,\delta_f,\delta_{\mathbf{T}},\alpha_{\mathbf{T}})>0$ such that for arbitrary nondecreasing sequences $0\equiv\overline{r}_0\leq\overline{r}_1\leq\cdots\leq\overline{r}_m$ and $0\equiv\overline{s}_0\leq\overline{s}_1\leq\cdots\leq\overline{s}_m$ in $[0,\infty)$ with

$$\bigvee_{i=0}^{m}|\overline{r}_i-\overline{s}_i|<\delta_{m+1}(\varepsilon,\delta_f,\delta_{\mathbf{T}},\alpha_{\mathbf{T}}),$$

we have

$$\left\|F^{*,\mathbf{T}}_{\overline{r}(0),\overline{r}(1),\ldots,\overline{r}(m)}f-F^{*,\mathbf{T}}_{\overline{s}(0),\overline{s}(1),\ldots\overline{s}(m)}f\right\|<\varepsilon. \tag{11.5.25}$$

8. Now suppose

$$\bigvee_{i=0}^{n+m}(r_i-v_i)<2^{-1}\delta_{m+1}(\varepsilon,\delta_f,\delta_{\mathbf{T}},\alpha_{\mathbf{T}})\wedge\delta_{Cp,\delta(\mathbf{T})}(\varepsilon\delta_f(\varepsilon))\wedge\delta_0. \tag{11.5.26}$$

Write $\overline{r}_i\equiv r_{n+i}-r_n$ and $\overline{s}_i\equiv t_i=v_{n+i}-v_n$, for each $i=0,\ldots,m$. Then $\overline{r}_i,\overline{s}_i\in[0,N]$ for each $i=0,\ldots,m$, with

$$\bigvee_{i=0}^{m}|\overline{r}_i-\overline{s}_i|\leq 2\bigvee_{i=0}^{n+m}(r_i-v_i)<\delta_{m+1}(\varepsilon,\delta_f,\delta_{\mathbf{T}},\alpha_{\mathbf{T}}),$$

where the last equality is by inequality 11.5.26. Therefore inequality 11.5.25 holds. As an abbreviation, define $h\equiv F^{*,\mathbf{T}}_{\overline{r}(0),\overline{r}(1),\ldots,\overline{r}(m)}f$ and $\overline{h}\equiv F^{*,\mathbf{T}}_{\overline{s}(0),\overline{s}(1),\ldots\overline{s}(m)}f$. Then inequality 11.5.25 can be rewritten as

$$\left\|h-\overline{h}\right\|<\varepsilon. \tag{11.5.27}$$

9. Since $0\equiv\overline{r}_0\leq\overline{r}_1\leq\cdots\leq\overline{r}_m$ is a sequence in $[0,N]$, Theorem 11.4.2 implies that the transition distribution $F^{*,\mathbf{T}}_{r(1),\ldots,r(m)}$ has a modulus of smoothness given by the m-fold composite product operation

$$\widetilde{\alpha}^{(m)}_{\alpha(\mathbf{T},N),\xi}\equiv\widetilde{\alpha}_{\alpha(\mathbf{T},N),\xi}\circ\cdots\circ\widetilde{\alpha}_{\alpha(\mathbf{T},N),\xi},$$

where each factor on the right-hand side is as defined in Lemma 11.2.5. Hence the function $h\equiv F^{*,\mathbf{T}}_{\overline{r}(0),\overline{r}(1),\ldots,\overline{r}(m)}f$ is a member of $C(S,d)$, with values in $[0,1]$ and a modulus of continuity $\delta_h\equiv\widetilde{\alpha}^{(m)}_{\alpha(\mathbf{T},N),\xi}(\delta_f)$.

10. Now the bound 11.5.26 implies

$$\bigvee_{i=0}^{n+m}(r_i-v_i)<\delta_{Cp,\delta(\mathbf{T})}(\varepsilon\delta_h(\varepsilon))\equiv\delta_{Cp,\delta(\mathbf{T})}(\varepsilon\widetilde{\alpha}^{(m)}_{\alpha(\mathbf{T},N),\xi}(\delta_f)(\varepsilon)). \tag{11.5.28}$$

In particular, we have $r_n-v_n<\delta_{Cp,\delta(\mathbf{T})}(\varepsilon\delta_h(\varepsilon))$. Hence $Ed(X_{r(n)},X_{v(n)})<\varepsilon\delta_h(\varepsilon)$ by the definition of $\delta_{Cp,\delta(\mathbf{T})}$ as a modulus of continuity in probability of the process $X|[0,N+1]$. Therefore Chebychev's inequality yields a measurable set $A\subset\Omega$ with $EA^c<\varepsilon$ such that

$$A\subset(d(X_{r(n)},X_{v(n)})<\delta_h(\varepsilon)). \tag{11.5.29}$$

11. Since the function h has modulus of continuity δ_h, relation 11.5.29 immediately implies that

$$A \subset (d(X_{r(n)}, X_{v(n)}) < \delta_h(\varepsilon)) \subset (|h(X_{r(n)}) - h(X_{v(n)})| < \varepsilon). \qquad (11.5.30)$$

As a result, we obtain the estimate

$$\begin{aligned}
&Eg(X_{r(0)}, \ldots, X_{r(n)}) F^{X(r(n)), \mathbf{T}}_{0, r(n+1)-r(n), \ldots, r(n+m)-r(n)} f \\
&\quad \equiv (Eg(X_{r(0)}, \ldots, X_{r(n)}) h(X_{r(n)}) 1_A) + (Eg(X_{r(0)}, \ldots, X_{r(n)}) h(X_{r(n)}) 1_{A^c}) \\
&\quad = (Eg(X_{r(0)}, \ldots, X_{r(n)}) h(X_{r(n)}) 1_A) \pm \varepsilon \\
&\quad = (Eg(X_{r(0)}, \ldots, X_{r(n)}) h(X_{v(n)}) 1_A) \pm \varepsilon \pm \varepsilon, \qquad (11.5.31)
\end{aligned}$$

where the second equality is because $EA^c < \varepsilon$, and where the third inequality is thanks to relation 11.5.30. At this point, note that the bound 11.5.26 implies also that $\bigvee_{i=0}^{n}(r_i - v_i) < \delta_0$. Hence inequality 11.5.24 holds, and leads to

$$E(g(X_{r(0)}, \ldots, X_{r(n)}) h(X_{v(n)}) 1_A) = (Eg(X_{v(0)}, \ldots, X_{v(n)}) h(X_{v(n)}) 1_A) \pm \varepsilon.$$

Therefore equality 11.5.31 can be continued, to yield

$$\begin{aligned}
&Eg(X_{r(0)}, \ldots, X_{r(n)}) F^{X(r(n)), \mathbf{T}}_{0, r(n+1)-r(n), \ldots, r(n+m)-r(n)} f \\
&\quad = (Eg(X_{v(0)}, \ldots, X_{v(n)}) h(X_{v(n)}) 1_A) \pm 3\varepsilon \\
&\quad = (Eg(X_{v(0)}, \ldots, X_{v(n)}) h(X_{v(n)}) 1_A) \\
&\qquad + (Eg(X_{v(0)}, \ldots, X_{v(n)}) h(X_{v(n)}) 1_{A^c}) \pm 4\varepsilon \\
&\quad = Eg(X_{v(0)}, \ldots, X_{v(n)}) h(X_{v(n)}) \pm 4\varepsilon \\
&\quad = Eg(X_{v(0)}, \ldots, X_{v(n)}) \overline{h}(X_{v(n)}) \pm \varepsilon \pm 4\varepsilon \\
&\quad = Eg(X_{v(0)}, \ldots, X_{v(n)}) \overline{h}(X_{v(n)}) \pm 5\varepsilon,
\end{aligned}$$

where we used the condition that the functions f, g, h have values in $[0, 1]$, and where the fourth equality is thanks to equality 11.5.27. Summing up,

$$\begin{aligned}
&\Big| Eg(X_{r(0)}, \ldots, X_{r(n)}) F^{X(r(n)), \mathbf{T}}_{0, r(n+1)-r(n), \ldots, r(n+m)-r(n)} f \\
&\qquad - Eg(X_{v(0)}, \ldots, X_{v(n)}) \overline{h}(X_{v(n)}) \Big| \le 5\varepsilon
\end{aligned}$$

provided that the bound 11.5.26 is satisfied. Since $\varepsilon > 0$ is arbitrarily small, we have proved the convergence of the right-hand side of equality 11.5.23 with, specifically,

$$\begin{aligned}
&Eg(X_{r(0)}, \ldots, X_{r(n)}) F^{X(r(n)), \mathbf{T}}_{0, r(n+1)-r(n), \ldots, r(n+m)-r(n)} f \\
&\qquad \to Eg(X_{v(0)}, \ldots, X_{v(n)}) \overline{h}(X_{v(n)}),
\end{aligned}$$

as $r_i \downarrow v_i$ for each $i = 0, \ldots, n+m$. In view of the convergence of the left-hand side of the same equality 11.5.23, as observed in Step 5, the two limits are equal. Namely,

$$Eg(X_{v(0)},\ldots,X_{v(n)})f(X_{v(n)},\ldots,X_{v(n+m)}) = Eg(X_{v(0)},\ldots,X_{v(n)})\overline{h}(X_{v(n)}), \tag{11.5.32}$$

where the function $g \in C(S^{n+1}, d^{n+1})$ is arbitrary with values in $[0,1]$, and where the integer $n \geq 1$ and the sequence $0 \equiv v_0 \leq v_1 \leq \cdots \leq v_{n-1} \leq v_n = v$ are arbitrary. In other words,

$$EYf(X_{v(n)},\ldots,X_{v(n+m)}) = EY\overline{h}(X_v)$$

for each r.r.v. Y in the linear subspace

$$G_v \equiv \{g(X_{v(0)},\ldots,X_{v(n-1)},X_{v(n)}) : g \in C(S^{n+1},d^{n+1}); \\ v_0 \leq v_1 \leq \cdots \leq v_{n-1} \leq v_n = v\}.$$

Since the linear subspace G_v is dense in the space $L^{(X,v)}$ relative to the norm $E|\cdot|$, it follows that

$$EYf(X_{v(n)},\ldots,X_{v(n+m)}) = EY\overline{h}(X_v) \tag{11.5.33}$$

for each $Y \in L^{(X,v)}$. Since $\overline{h}(X_v) \in G_v \subset L^{(X,v)}$, we obtain the conditional expectation

$$E(f(X_{v(n)},\ldots,X_{v(n+m)})|L^{(X,v)}) = \overline{h}(X_v) \equiv F^{X(v),\mathbf{T}}_{\overline{s}(0),\overline{s}(1),\ldots\overline{s}(m))}f. \tag{11.5.34}$$

In the special case of an arbitrary $Y \in L(X_v) \subset L^{(X,v)}$, equality 11.5.33 implies that

$$E(f(X_{v(n)},\ldots,X_{v(n+m)})|X_v) = \overline{h}(X_v) \equiv F^{X(v),\mathbf{T}}_{\overline{s}(0),\overline{s}(1),\ldots\overline{s}(m))}f. \tag{11.5.35}$$

Recall that $v_{n+i} \equiv v + t_i$ and $\overline{s}_i \equiv t_i$, for each $i = 0,\ldots,m$, and recall that $t_0 \equiv 0$. Equalities 11.5.34 and 11.5.35 can then be rewritten as

$$E(f(X_{v+t(0)},\ldots,X_{v+t(m)})|L^{(X,v)}) = F^{X(v),\mathbf{T}}_{0,t(1),\ldots,t(m)}f \tag{11.5.36}$$

and

$$E(f(X_{v+t(0)},\ldots,X_{v+t(m)})|X_v) = F^{X(v),\mathbf{T}}_{0,t(1),\ldots,t(m)}, \tag{11.5.37}$$

respectively. The last two equalities together yield the desired equality 11.5.19. Assertion 3 has been verified. Assertion 3 and the theorem are proved. □

11.6 Continuity of Construction

In this section, we will prove that the construction in Theorem 11.5.6 of a Markov process with parameter set $[0,\infty)$, from an initial state x and a semigroup $\mathbf{T}$, is uniformly metrically continuous over each subspace of semigroups $\mathbf{T}$ whose members share a common modulus of strong continuity and share a common modulus of smoothness.

First we specify a compact state space, and define a metric on the space of Markov semigroups.

Definition 11.6.1. Specification of state space, its binary approximation, and partition of unity. In this section, unless otherwise specified, (S,d) will denote a given compact metric space, with $d \leq 1$ and a fixed reference point $x_\circ \in S$. Recall that $\xi \equiv (A_k)_{k=1,2,...}$ is a binary approximation of (S,d) relative to $x_\circ$, and that

$$\pi \equiv (\{g_{k,x} : x \in A_k\})_{k=1,2,...}$$

is the partition of unity of (S,d) determined by ξ, as in Definition 3.3.4. Recall that $|A_k|$ denotes the number of elements in the metrically discrete finite subset $A_k \subset S$, for each $k \geq 1$.

For each $n \geq 1$, let ξ^n denote the nth power of ξ, and let $\pi^{(n)}$denote the corresponding partition of unity for (S^n,d^n). Thus, for each $n \geq 1$, the sequence $\xi^n \equiv (A_k^{(n)})_{k=1,2,...}$ is the product binary approximation for (S^n,d^n) relative to the reference point $x_\circ^{(n)} \equiv (x_\circ, \ldots x_\circ) \in S^n$, and

$$\pi^{(n)} \equiv \left(\{g_{k,x}^{(n)} : x \in A_k^{(n)}\}\right)_{k=1,2,...}$$

is the partition of unity of (S^n,d^n) determined by the binary approximation $\xi \equiv (A_k)_{k=1,2,...} \equiv (A_k^{(1)})_{k=1,2,...}$ of (S,d). For each $k \geq 1$, the set $A_k^{(n)}$ is a 2^k-approximation of the bounded subset

$$(d^n(\cdot,(x_\circ,\ldots,x_o) \leq 2^k) \subset S^n.$$

To lessen the burden of subscripts, we write $A_k^{(n)}$ and $A_{n,k}$ interchangeably, for each $n \geq 1$.

Recall from Definition 11.0.2 the enumerated set $\overline{Q}_\infty \equiv \{u_0,u_1,\ldots\}$ of dyadic rationals. □

Definition 11.6.2. Metric on the space of Markov semigroups. Let (S,d) be the specified compact metric space, with $d \leq 1$. Suppose $Q = \overline{Q}_\infty$or $Q = [0,\infty)$. Let $\mathscr{T}$ be set of Markov semigroups on the parameter set Q and with the compact metric state space (S,d). For each $n \leq 0$, write $\Delta_n \equiv 2^{-n}$. Define the metric $\rho_{\mathscr{T},\xi}$ on the set $\mathscr{T}$ by

$$\rho_{\mathscr{T},\xi}(\mathbf{T},\overline{\mathbf{T}}) \equiv \sum_{n=0}^{\infty} 2^{-n-1} \sum_{k=1}^{\infty} 2^{-k}|A_k|^{-1} \sum_{z\in A(k)} \left\| T_{\Delta(n)}g_{k,z} - \overline{T}_{\Delta(n)}g_{k,z}\right\| \tag{11.6.1}$$

for arbitrary members $\mathbf{T} \equiv \{T_t : t \in Q\}$ and $\overline{\mathbf{T}} \equiv \{\overline{T}_t : t \in Q\}$ of the set $\mathscr{T}$. Here $\|\cdot\|$ stands for the supremum norm on $C(S,d)$. The next lemma proves that $\rho_{\mathscr{T}}$ is indeed a metric. Note that $\rho_{\mathscr{T}} \leq 1$.

Let $(S \times \mathscr{T},d \otimes \rho_{\mathscr{T}})$ denote the product metric metric space of (S,d) and $(\mathscr{T},\rho_{\mathscr{T}})$. For each $\mathbf{T} \equiv \{T_t : t \in Q\} \in \mathscr{T}$, define the semigroup $\mathbf{T}|\overline{Q}_\infty \equiv \{T_t : t \in \overline{Q}_\infty\}$. Then

$$\rho_{\mathscr{T},\xi}(\mathbf{T}|\overline{Q}_\infty,\overline{\mathbf{T}}|\overline{Q}_\infty) = \rho_{\mathscr{T},\xi}(\mathbf{T},\overline{\mathbf{T}}).$$

In other words, the mapping

$$\Psi : (\mathscr{T}, \rho_{\mathscr{T},\xi}) \to (\mathscr{T}|\overline{Q}_\infty, \rho_{\mathscr{T},\xi}),$$

defined by $\Psi(\mathbf{T}) \equiv \mathbf{T}|\overline{Q}_\infty$ for each $\mathbf{T} \in \mathscr{T}$, is an isometry. Note here that we abuse notations and omit the reference to the parameter set Q or $\overline{Q}_\infty$ in the symbol $\rho_{\mathscr{T},\xi}$. □

Lemma 11.6.3. $\rho_{\mathscr{T},\xi}$ **is a metric.** *The function* $\rho_{\mathscr{T}} \equiv \rho_{\mathscr{T},\xi}$ *is indeed a metric.*

Proof. Symmetry and triangle inequality for the function $\rho_{\mathscr{T}}$ are obvious from the defining equality 11.6.1. Suppose $\mathbf{T}, \overline{\mathbf{T}} \in \mathscr{T}$ are such that $\rho_{\mathscr{T}}(\mathbf{T}, \overline{\mathbf{T}}) = 0$. Consider each $r \in \overline{Q}_\infty$. Then $r = \kappa\Delta_n$ for some $n, \kappa \geq 0$, while equality 11.6.1 implies that

$$\sum_{k=1}^{\infty} 2^{-k}|A_k|^{-1} \sum_{z \in A(k)} \left\| T_{\Delta(n)} g_{k,z} - \overline{T}_{\Delta(n)} g_{k,z} \right\| = 0. \tag{11.6.2}$$

Now let $x \in S$ be arbitrary. Then

$$\sum_{k=1}^{\infty} 2^{-k}|A_k|^{-1}|T^x_{\Delta(n)} g_{k,z} - \overline{T}^x{}_{\Delta(n)} g_{k,z}| = 0. \tag{11.6.3}$$

In other words, $\rho_{Dist,\xi}(T^x_{\Delta(n)}, \overline{T}^x{}_{\Delta(n)}) = 0$, where $\rho_{Dist,\xi}$ is the distribution metric introduced in Definition 5.3.4. It follows that $T^x_{\Delta(n)} = \overline{T}^x{}_{\Delta(n)}$ as distributions. Since $x \in S$ is arbitrary, we see that $T_{\Delta(n)} = \overline{T}_{\Delta(n)}$ as transition distributions. Consequently, by the semigroup property, we obtain

$$T_r = T_{\kappa\Delta(n)} = T_{\Delta(n)} \circ \cdots \circ T_{\Delta(n)} = \overline{T}_{\Delta(n)} \circ \cdots \circ \overline{T}_{\Delta(n)} = \overline{T}_{\kappa\Delta(n)} = \overline{T}_r.$$

where $r \in \overline{Q}_\infty$. is arbitrary. Since $\overline{Q}_\infty$ is dense in $[0,\infty)$, Assertion 3 of Lemma 11.3.2 implies that $\mathbf{T} = \overline{\mathbf{T}}$. Summing up, $\rho_{\mathscr{T}}$ is a metric. □

Lemma 11.6.4. Continuity of construction of a family of transition f.j.d.'s from an initial state and semigroup. *Let* (S,d) *be the specified compact metric space, with* $d \leq 1$. *Let* $\mathscr{T}(\overline{\delta},\overline{\alpha})$ *be an arbitrary family of Markov semigroups with parameter set* $\overline{Q}_\infty$ *and state space* (S,d), *such that all its members* $\mathbf{T} \in \mathscr{T}(\overline{\delta},\overline{\alpha})$ *share a common modulus of strong continuity* $\delta_{\mathbf{T}} = \overline{\delta}$ *and a common modulus of smoothness* $\alpha_{\mathbf{T}} = \overline{\alpha}$, *in the sense of Definition 11.3.1. Thus* $\mathscr{T}(\overline{\delta},\overline{\alpha})$ *is a subset of of the metric space* $(\mathscr{T}, \rho_{\mathscr{T},\xi})$ *introduced in Definition 11.6.2 and, as such, inherits the metric* $\rho_{\mathscr{T},\xi}$. *Recall from Definition 6.2.8 the metric space* $(\widehat{F}(\overline{Q}_\infty, S), \widehat{\rho}_{Marg,\xi,\overline{Q}(\infty)})$ *of consistent families of f.j.d.'s with parameter set* $\overline{Q}_\infty$ *and state space* (S,d). *Recall the subspace* $\widehat{F}_{Cp}(\overline{Q}_\infty, S)$ *consisting of members that are continuous in probability, as in Definition 6.2.10.*

Then the mapping

$$\Phi_{Sg,fjd} : (S \times \mathscr{T}(\overline{\delta},\overline{\alpha}), d \otimes \rho_{\mathscr{T},\xi}) \to (\widehat{F}_{Cp}(\overline{Q}_\infty, S), \widehat{\rho}_{Marg,\xi,\overline{Q}(\infty)}),$$

constructed in Assertion 5 of Theorem 11.4.2, is uniformly continuous, with a modulus of continuity $\delta_{Sg,fjd}(\cdot,\overline{\delta},\overline{\alpha}, \|\xi\|)$ determined by the moduli $\overline{\delta},\overline{\alpha}$ and by the modulus of local compactness $\|\xi\| \equiv (|A_n|)_{n=1,2,...}$ of (S,d).

Proof. 1. Let $(x,\mathbf{T}), (\overline{x},\overline{\mathbf{T}}) \in S \times \mathscr{T}(\overline{\delta},\overline{\alpha})$ be arbitrary, but fixed. As an abbreviation, write $F \equiv F^{x,\mathbf{T}} \equiv \Phi_{Sg,fjd}(x,\mathbf{T})$ and $\overline{F} \equiv F^{\overline{x},\overline{\mathbf{T}}} \equiv \Phi_{Sg,fjd}(\overline{x},\overline{\mathbf{T}})$. Define the distance

$$\overline{\rho}_0 \equiv (d \otimes \rho_{\mathscr{T}})((x,\mathbf{T}),(\overline{x},\overline{\mathbf{T}})) \equiv d(x,\overline{x}) \vee \rho_{\mathscr{T}}(\mathbf{T},\overline{\mathbf{T}}). \tag{11.6.4}$$

2. Let $\varepsilon_0 > 0$ be arbitrary, but fixed until further notice. Take $M \geq 0$ so large that $2^{-p(2M)-1} < 3^{-1}\varepsilon_0$, where $p_{2M} \equiv N \equiv 2^{2M}$. As an abbreviation, also write $\Delta \equiv \Delta_M \equiv 2^{-M}$ and $K \equiv N+1$. Then $2^{-K} = 2^{-N-1} < 3^{-1}\varepsilon_0$. Moreover, in the notations of Definitions 11.0.2, we have

$$\overline{Q}_M \equiv \{u_0, u_1, \ldots, u_{p(2M)}\} \equiv \{u_0, u_1, \ldots, u_N\} = \{0, \Delta, 2\Delta, \ldots, N\Delta\} \tag{11.6.5}$$

as enumerated sets. Thus $u_n = n\Delta$ for each $n = 0, \ldots, N$.

3. By Definition 6.2.8, we have

$$\begin{aligned}\widehat{\rho}_{Marg,\xi,\overline{Q}(\infty)}(F,\overline{F}) &\equiv \sum_{n=0}^{\infty} 2^{-n-1}\rho_{Dist,\xi^{n+1}}(F_{u(0),\ldots,u(n)}, \overline{F}_{u(0),\ldots,u(n)}) \\ &\leq \sum_{n=0}^{N} 2^{-n-1}\rho_{Dist,\xi^{n+1}}(F_{u(0),\ldots,u(n)}, \overline{F}_{u(0),\ldots,u(n)}) + 2^{-N-1} \\ &\leq \sum_{n=0}^{N} 2^{-n-1}\rho_{Dist,\xi^{n+1}}(F_{0,\Delta,\ldots,n\Delta}, \overline{F}_{0,\Delta,\ldots,n\Delta}) + 3^{-1}\varepsilon_0.\end{aligned} \tag{11.6.6}$$

For each $n \geq 0$, the metric $\rho_{Dist,\xi^{n+1}}$ was introduced in Definition 5.3.4 for the space of distributions on (S^{n+1}, d^{n+1}), where it is observed that $\rho_{Dist,\xi^{n+1}} \leq 1$ and that sequential convergence relative to $\rho_{Dist,\xi^{n+1}}$ is equivalent to weak convergence.

4. We will prove that the summand indexed by n in the last sum in equality 11.6.6 is bounded by $2^{-n-1}2 \cdot 3^{-1}\varepsilon_0$, provided that the distance $\overline{\rho}_0$ is sufficiently small. To that end, let $n = 0, \ldots, N$ be arbitrary, but fixed until further notice. Then the summand is bounded by

$$\rho_{Dist,\xi^{n+1}}(F_{0,\Delta,\ldots,n\Delta}, \overline{F}_{0,\Delta,\ldots,n\Delta})$$

$$\equiv \sum_{k=1}^{\infty} 2^{-k}|A_{n+1,k}|^{-1} \sum_{y \in A(n+1,k)} |F_{0,\Delta,\ldots,n\Delta}g_{k,y}^{(n+1)} - \overline{F}_{0,\Delta,\ldots,n\Delta}g_{k,y}^{(n+1)}|$$

$$\leq \sum_{k=1}^{K} 2^{-k}|A_{n+1,k}|^{-1} \sum_{y\in A(n+1,k)} |F_{0,\Delta,\ldots,n\Delta} g_{k,y}^{(n+1)} - \overline{F}_{0,\Delta,\ldots,n\Delta}) g_{k,y}^{(n+1)}| + 2^{-K}$$

$$< \sum_{k=1}^{K} 2^{-k}|A_{n+1,k}|^{-1} \sum_{y\in A(n+1,k)} |F_{0,\Delta,\ldots,n\Delta} g_{k,y}^{(n+1)} - \overline{F}_{0,\Delta,\ldots,n\Delta}) g_{k,y}^{(n+1)}| + 3^{-1}\varepsilon_0, \tag{11.6.7}$$

where the first equality is by Definition 5.3.4 of the distribution metric $\rho_{Dist,\xi^{n+1}}$, and where for each $k \geq 1$ and for each $y \in A_{n+1,k}$, the basis function $g_{k,y}^{(n+1)} \in C(S^{n+1}, d^{n+1})$ is from the partition of unity

$$\pi^{(n+1)} \equiv (\{g_{k,y}^{(n+1)} : y \in A_{n+1,k}\})_{k=1,2,\ldots}$$

of (S^{n+1}, d^{n+1}) relative to the 2^{-k}-approximation $A_{n+1,k}$ of the metric space (S^{n+1}, d^{n+1}), as specified in Definition 11.6.1.

5. Consider each $k = 0, \ldots, K$ and each $y \in A_{n+1,k}$ on the left-hand side of inequality 11.6.7. Write $g = g_{k,y}^{(n+1)}$. Then by equality 11.4.1 of Definition 11.4.1, we obtain

$$F_{0,\Delta,\ldots,n\Delta} g_{k,y}^{(n+1)} = F_{0,\Delta,\ldots,n\Delta}^{x,\mathbf{T}} g$$

$$\equiv \int \delta_x(dx_0) \int T_0^{x(0)}(dx_1) \int T_\Delta^{x(1)}(dx_2) \int T_\Delta^{x(2)}(dx_3) \cdots$$

$$\times \int T_\Delta^{x(n)}(dx_{n+1}) g(x_1, x_2, \ldots, x_{n+1})$$

$$= \int \delta_x(dx_0) \int T_\Delta^{x(0)}(dx_2) \int T_\Delta^{x(2)}(dx_3) \cdots \int T_\Delta^{x(n)}(dx_{n+1}) g(x_0, x_2, \ldots, x_{n+1})$$

$$= \int \delta_x(dz_0) \int T_\Delta^{z(0)}(dz_1) \int T_\Delta^{z(1)}(dz_2) \cdots \int T_\Delta^{z(n-1)}(dz_n) g(z_0, z_1, \ldots, z_n)$$

$$\equiv \delta_x({}^1T_\Delta)({}^2T_\Delta)\cdots({}^nT_\Delta) g$$

$$\equiv \delta_x({}^1T_\Delta)({}^2T_\Delta)\cdots({}^nT_\Delta) g_{k,y}^{(n+1)},$$

where the factors ${}^iT_\Delta$ on the right-hand side are one-step transition distributions according to T_Δ, as defined in Lemma 11.2.5. A similar equality holds for $\overline{F}_{0,\Delta,\ldots,n\Delta}$ and $\overline{\mathbf{T}}$. We will prove that

$$\left|\delta_x({}^1T_\Delta)({}^2T_\Delta)\cdots({}^nT_\Delta) g_{k,y}^{(n+1)} - \delta_{\overline{x}}({}^1\overline{T}_\Delta)({}^2\overline{T}_\Delta))\cdots({}^n\overline{T}_\Delta) g_{k,y}^{(n+1)}\right| \leq 3^{-1}\varepsilon_0 \tag{11.6.8}$$

for each $k = 0, \ldots, K$ and for each $y \in A_{n+1,k}$.

6. To that end, define the finite family

$$G_n = \left\{g_{k,y}^{(n+1)} : k = 0, \ldots, K \text{ and } y \in A_{n+1,k}\right\} \subset C(S^{n+1}, d^{n+1}).$$

For each $p = 0, \ldots, n-1$, define the finite family

$$G_p \equiv \{({}^{p+1}\overline{T}_\Delta) \cdots ({}^{n}\overline{T}_\Delta) g_{k,y}^{(n+1)} : k = 0, \ldots, K \text{ and } y \in A_{n+1,k}\} \\ \subset C(S^{p+1}, d^{p+1}).$$

7. We will prove by backward induction that for each $p = n, n-1, \ldots, 0$, there exists an operation $\rho_p : (0,\infty) \to (0,\infty)$ and an operation $\widetilde{\delta}_p : (0,\infty) \to (0,\infty)$ such that (i) $\widetilde{\delta}_p$ is a modulus of continuity of each of the members of G_p and (ii) if $p < n$, then for each $\varepsilon' > 0$, we have

$$\bigvee_{g \in G(p)} \left\|({}^{p}T_\Delta)g - ({}^{p}\overline{T}_\Delta)g\right\| < \varepsilon'$$

provided that the distance $\overline{\rho}_0$ is bounded by

$$\overline{\rho}_0 < \rho_p(\varepsilon').$$

8. Start with $p = n$. Arbitrarily define $\rho_n \equiv 1$. Then the operation ρ_p trivially satisfies Condition (ii) in Step 7. Consider each $g \in G_p$. Then $g = g_{k,y}^{(n+1)}$ for some $k = 0, \ldots, K$ and $y \in A_{n+1,k}$. According to Proposition 3.3.3, the basis function $g_{k,y}^{(n+1)} \in C(S^{n+1}, d^{n+1})$ has a Lipschitz constant $2 \cdot 2^k \leq 2^{K+1} \equiv 2^{N+2}$ and has values in $[0,1]$. Thus the function g has a modulus of continuity $\widetilde{\delta}_n$ defined by $\widetilde{\delta}_n(\varepsilon') \equiv 2^{-N-2}\varepsilon'$ for each $\varepsilon' > 0$. Since $g \in G_n$ is arbitrary, the modulus $\widetilde{\delta}_n$ satisfies Condition (i) in Step 7. The pair $\widetilde{\delta}_n, \rho_n$ has been constructed to satisfy Conditions (i) and (ii), in the case where $p = n$.

9. Suppose, for some $p = n, n-1, \ldots, 1$, the pair of operations $\widetilde{\delta}_p, \rho_p$ has been constructed to satisfy Conditions (i) and (ii) in Step 7. Proceed to construct the pair $\widetilde{\delta}_{p-1}, \rho_{p-1}$. Consider each

$$g \in G_{p-1} \equiv \{({}^{p}\overline{T}_\Delta)) \cdots ({}^{n}\overline{T}_\Delta) g_{k,y}^{(n+1)} : k = 0, \ldots, K \text{ and } y \in A_{n+1,k}\} \\ = ({}^{p}\overline{T}_\Delta) G_p \subset C(S^p, d^p).$$

Then $g = ({}^{p}\overline{T}_\Delta)\overline{g}$ for some $\overline{g} \in G_p$. By Condition (ii) in the backward induction hypothesis, the function $\overline{g}$ has a modulus of continuity $\widetilde{\delta}_p$. Lemma 11.2.5 therefore says that the function $g = ({}^{p}\overline{T}_\Delta)\overline{g}$ has a modulus of continuity given by

$$\widetilde{\delta}_{p-1} \equiv \widetilde{\alpha}_{\alpha(\overline{T}(\Delta)),\xi}(\widetilde{\delta}_p),$$

where the modulus of smoothness $\widetilde{\alpha}_{\alpha(\overline{T}(\Delta)),\xi}$ is as defined in Lemma 11.2.5. Since $g \in G_{p-1}$ is arbitrary, we see that $\widetilde{\delta}_{p-1}$ satisfies Condition (i) in Step 7. It remains to construct $\rho_{p-1}$to satisfy Condition (ii) in Step 7.

10. To that end, let $\varepsilon' > 0$ be arbitrary. Take $h \geq 1$ so large that

$$2^{-h} < \frac{1}{2}\widetilde{\delta}_{p-1}(3^{-2}\varepsilon'). \tag{11.6.9}$$

Define

$$\rho_{p-1}(\varepsilon') \equiv \rho_p(3^{-1}\varepsilon') \wedge 2^{-M-1}2^{-h}|A_h|^{-1}(3^{-1}\varepsilon'). \tag{11.6.10}$$

Consider each $g \in G_{p-1}$. By Step 9, the function g has a modulus of continuity $\widetilde{\delta}_{p-1}$. Let $(w_1,\ldots,w_{p-1}) \in S^{p-1}$ be arbitrary, and consider the function $\overline{f} \equiv g(w_1,\ldots,w_{p-1},\cdot) \in C(S,d)$. Because $d \leq 1$ by assumption, we have

$$S \subset (d(\cdot,x_\circ) \leq 1) \subset (d(\cdot,x_\circ) \leq 2^h),$$

whence, trivially, the function $\overline{f} \in C(S,d)$ has the set $(d(\cdot,x_\circ) \leq 2^h)$ as support. Moreover, the function $\overline{f}$ has the same modulus of continuity $\delta_{\overline{f}} \equiv \widetilde{\delta}_{p-1}$ as the function g. Thus the conditions in Assertion 4 of Proposition 3.3.6, where n,k,f,x,g_x,ε are replaced by $1,h,\overline{f},z,g_{h,z},3^{-1}\varepsilon'$, respectively, are satisfied. Accordingly,

$$\left\|\overline{f} - \sum_{z\in A(h)} \overline{f}(z)g_{h,z}\right\| \leq 3^{-1}\varepsilon' \tag{11.6.11}$$

on S. Consequently, since T_Δ is a contraction mapping, we have

$$\left\|T_\Delta\overline{f} - \sum_{z\in A(h)} \overline{f}(z)T_\Delta g_{h,z}\right\| \leq 3^{-1}\varepsilon'. \tag{11.6.12}$$

Similarly,

$$\left\|\overline{T}_\Delta\overline{f} - \sum_{z\in A(h)} \overline{f}(z)\overline{T}_\Delta g_{h,z}\right\| \leq 3^{-1}\varepsilon'. \tag{11.6.13}$$

11. Now suppose $\overline{\rho}_0 < \rho_{p-1}(\varepsilon')$. Then

$$\sum_{j=0}^{\infty} 2^{-j-1} \sum_{h'=1}^{\infty} 2^{-h'}|A_{h'}|^{-1} \sum_{z\in A(h')} \left\|T_{\Delta(j)}g_{h',z} - \overline{T}_{\Delta(j)}g_{h',z}\right\|$$
$$\equiv \rho_{\mathscr{T},\xi}(\mathbf{T},\overline{\mathbf{T}}) \leq \overline{\rho}_0 < \rho_{p-1}(\varepsilon'),$$

where the first inequality is by equality 11.6.4. Consequently,

$$2^{-M-1}2^{-h}|A_h|^{-1} \sum_{z\in A(h)} \left\|T_{\Delta(M)}g_{h,z} - \overline{T}_{\Delta(M)}g_{h,z}\right\| < \rho_{p-1}(\varepsilon').$$

Recall that $\Delta \equiv \Delta_M \equiv 2^{-M}$. The last displayed inequality can be rewritten as

$$2^{-M-1}2^{-h}|A_h|^{-1} \sum_{z\in A(h)} \left\|T_\Delta g_{h,z} - \overline{T}_\Delta g_{h,z}\right\| < \rho_{p-1}(\varepsilon'). \tag{11.6.14}$$

Therefore, since the function $\overline{f}$ has values in $[0, 1]$, we have

$$\left\| \sum_{z \in A(h)} \overline{f}(z) T_\Delta g_{h,z} - \sum_{z \in A(h)} \overline{f}(z) \overline{T}_\Delta g_{h,z} \right\| \leq \sum_{z \in A(h)} \left\| \overline{f}(z) T_\Delta g_{h,z} - \overline{f}(z) \overline{T}_\Delta g_{h,z} \right\|$$

$$\leq \sum_{z \in A(h)} \left\| T_\Delta g_{h,z} - \overline{T}_\Delta g_{h,z} \right\|$$

$$< 2^{M+1} 2^h |A_h| \rho_{p-1}(\varepsilon') \leq 3^{-1} \varepsilon', \tag{11.6.15}$$

where the third inequality is by inequality 11.6.14, and where the last inequality follows from the defining formula 11.6.10.

12. Combining inequalities 11.6.15, 11.6.13, and 11.6.12, we obtain

$$\left\| T_\Delta \overline{f} - \overline{T}_\Delta \overline{f} \right\| \leq \varepsilon'.$$

In other words,

$$\left\| T_\Delta g(w_1, \ldots, w_{p-1}, \cdot) - \overline{T}_\Delta g(w_1, \ldots, w_{p-1}, \cdot) \right\| \leq \varepsilon',$$

where $(w_1, \ldots, w_{p-1}) \in S^p$is arbitrary. Consequently,

$$\left\| {}^{p-1}T_\Delta g - {}^{p-1}\overline{T}_\Delta g \right\| \leq \varepsilon'.$$

where $\varepsilon' > 0$ and $g \in G_{p-1}$ are arbitrary, provided that $\overline{\rho}_0 < \rho_{p-1}(\varepsilon')$.

13. In short, the operation ρ_{p-1} has been constructed to satisfy Condition (ii) in Step 7. The backward induction is completed, and we have obtained the pair $(\widetilde{\delta}_p, \rho_p)$ for each $p = n, n-1, \ldots, 0$ to satisfy Conditions (i) and (ii) in Step 7. In particular, we obtained the pair $(\widetilde{\delta}_0, \rho_0)$. of operations.

14. Now let $\varepsilon' \equiv 3^{-1}(n+2)^{-1}\varepsilon_0$. Suppose

$$\overline{\rho}_0 < \delta_{Sg,fjd}(\varepsilon_0, \overline{\delta}, \overline{\alpha}, \|\xi\|) \equiv \rho_0(\varepsilon') \wedge \widetilde{\delta}_0(\varepsilon').$$

Let $p = 1, \ldots n$ and $g \in G_{p,}$ be arbitrary. Then

$$({}^p\overline{T}_\Delta) g \in G_{p-1}, \tag{11.6.16}$$

Moreover, as a result of the monotonicity of the operations ρ_p in the defining formula 11.6.10, we have $\overline{\rho}_0 < \rho_p(\varepsilon')$. Hence, by Condition (ii) in Step 7, we have

$$({}^pT_\Delta) g = ({}^p\overline{T}_\Delta) g \pm \varepsilon'.$$

Therefore, since the transition distributions $({}^1T_\Delta), \ldots, ({}^{p-1}T_\Delta)$ are contraction mappings, it follows that

$$\begin{aligned} &({}^1T_\Delta)({}^2T_\Delta) \cdots ({}^{p-2}T_\Delta)({}^{p-1}T_\Delta)({}^pT_\Delta) g \\ &= ({}^1T_\Delta)({}^2T_\Delta) \cdots ({}^{p-2}T_\Delta)({}^{p-1}T_\Delta)({}^p\overline{T}_\Delta) g \pm \varepsilon'. \end{aligned} \tag{11.6.17}$$

15. Finally, let $k = 0, \ldots, K$ and $y \in A_{n+1,k}$. be arbitrary. Let $g \equiv g_{k,y}^{(n+1)} \in G_n$. Hence, using equality 11.6.17 repeatedly, we obtain

$$
\begin{aligned}
F_{0,\Delta,\ldots,n\Delta}^{x,\mathbf{T}} g &= \delta_x({}^1T_\Delta)({}^2T_\Delta)\cdots({}^{n-1}T_\Delta)({}^nT_\Delta)g \\
&= \delta_x({}^1T_\Delta)({}^2T_\Delta)\cdots({}^{n-1}T_\Delta)({}^nT_\Delta)({}^{n+1}\overline{T}_\Delta)g \pm \varepsilon' \\
&= \delta_x({}^1T_\Delta)({}^2T_\Delta)\cdots({}^{n-1}T_\Delta)({}^n\overline{T}_\Delta)({}^{n+1}\overline{T}_\Delta)g \pm 2\varepsilon' \\
&= \cdots \\
&= \delta_x({}^1\overline{T}_\Delta)({}^2\overline{T}_\Delta)\cdots({}^{n-1}T_\Delta)({}^n\overline{T}_\Delta)({}^{n+1}\overline{T}_\Delta)g \pm (n+1)\varepsilon' \\
&= F_{0,\Delta,\ldots,n\Delta}^{x,\overline{\mathbf{T}}} g \pm (n+1)\varepsilon'. \qquad (11.6.18)
\end{aligned}
$$

16. Moreover, by Condition (i) in Step 7, the function

$$F_{0,\Delta,\ldots,n\Delta}^{*,\mathbf{T}} g = ({}^1T_\Delta)({}^2T_\Delta)\cdots({}^{n-1}T_\Delta)({}^nT_\Delta)g \in G_0$$

has a modulus of continuity $\widetilde{\delta}_0$. Therefore

$$F_{0,\Delta,\ldots,n\Delta}^{x,\mathbf{T}} g = F_{0,\Delta,\ldots,n\Delta}^{\overline{x},\mathbf{T}} g \pm \varepsilon'$$

because

$$d(x,\overline{x}) \leq \overline{\rho}_0 < \delta_{Sg,fjd}(\varepsilon_0,\overline{\delta},\overline{\alpha},\ \|\xi\|) \equiv \rho_0(\varepsilon') \wedge \widetilde{\delta}_0(\varepsilon') \leq \widetilde{\delta}_0(\varepsilon').$$

By symmetry,

$$F_{0,\Delta,\ldots,n\Delta}^{x,\overline{\mathbf{T}}} g = F_{0,\Delta,\ldots,n\Delta}^{\overline{x},\overline{\mathbf{T}}} g \pm \varepsilon'. \qquad (11.6.19)$$

Equalities 11.6.18 and 11.6.19 together imply

$$
\begin{aligned}
F_{0,\Delta,\ldots,n\Delta}^{x,\mathbf{T}} g &= F_{0,\Delta,\ldots,n\Delta}^{x,\overline{\mathbf{T}}} g \pm (n+1)\varepsilon' = F_{0,\Delta,\ldots,n\Delta}^{\overline{x},\overline{\mathbf{T}}} g \pm (n+1)\varepsilon' \pm \varepsilon' \\
&= F_{0,\Delta,\ldots,n\Delta}^{\overline{x},\overline{\mathbf{T}}} g \pm (n+2)\varepsilon' \equiv F_{0,\Delta,\ldots,n\Delta}^{\overline{x},\overline{\mathbf{T}}} g \pm 3^{-1}\varepsilon_0.
\end{aligned}
$$

Equivalently,

$$|\delta_x({}^1T_\Delta)({}^2T_\Delta)\cdots({}^nT_\Delta)g_{k,y}^{(n+1)} - \delta_{\overline{x}}({}^1\overline{T}_\Delta)({}^2\overline{T}_\Delta))\cdots({}^n\overline{T}_\Delta)g_{k,y}^{(n+1)}| \leq 3^{-1}\varepsilon_0, \qquad (11.6.20)$$

where $k = 0, \ldots, K$ and $y \in A_{n+1,k}$. are arbitrary. The desired equality 11.6.8 follows for each $n = 0, \ldots, N$. Inequalities 11.6.7 and 11.6.6 then imply that

$$
\begin{aligned}
&\widehat{\rho}_{Marg,\xi,\overline{Q}(\infty)}(\Phi_{Sg,fjd}(x,\mathbf{T}), \Phi_{Sg,fjd}(\overline{x}, \widehat{\ \mathbf{T}\ })) \\
&= \widehat{\rho}_{Marg,\xi,\overline{Q}(\infty)}(F,\overline{F}) \leq 3^{-1}\varepsilon_0 + 3^{-1}\varepsilon_0 + 3^{-1}\varepsilon_0 = \varepsilon_0,
\end{aligned}
$$

provided that the distance $\overline{\rho}_0 \equiv (d \otimes \rho_{\mathscr{T}})((x,\mathbf{T}),(\overline{x},\overline{\mathbf{T}}))$ is bounded by

$$\overline{\rho}_0 < \delta_{Sg,fjd}(\varepsilon_0,\overline{\delta},\overline{\alpha},\ \|\xi\|).$$

Summing up, $\delta_{Sg,fjd}(\cdot,\overline{\alpha},\ \|\xi\|)$ is a modulus of continuity of the mapping $\Phi_{Sg,fjd}$. The theorem is proved. □

Following is the main theorem of this section.

Theorem 11.6.5. Construction of a time-uniformly a.u. càdlàg Markov process from an initial state and Markov semigroup on $[0,\infty)$, and continuity of said construction. *Let (S,d) be the specified compact metric space, with $d \leq 1$. Let $\mathscr{T}(\overline{\delta},\overline{\alpha})$ be an arbitrary family of Markov semigroups with parameter set $[0,\infty)$ and state space (S,d), such that all its members $\mathbf{T} \in \mathscr{T}(\overline{\delta},\overline{\alpha})$ share a common modulus of strong continuity $\delta_{\mathbf{T}} \equiv \overline{\delta}$ and a common modulus of smoothness $\alpha_{\mathbf{T}} \equiv \overline{\alpha}$, in the sense of Definition 11.3.1. Thus $\mathscr{T}(\overline{\delta},\overline{\alpha})$ is a subset of the metric space $(\mathscr{T},\rho_{\mathscr{T},\xi})$ introduced in Definition 11.6.2 and, as such, inherits the metric $\rho_{\mathscr{T},\xi}$. Separately, recall the space $\widehat{F}(\overline{Q}_\infty,S)$ of consistent families of f.j.d.'s, equipped with the marginal metric $\widehat{\rho}_{Marg,\xi,\overline{Q}(\infty)}$ as in Definition 6.2.8. Similarly, recall from Definition 6.2.10 the space $\widehat{F}_{Cp}([0,\infty),S)$ of consistent families of f.j.d.'s that are continuous in probability, equipped with the metric $\widehat{\rho}_{Cp,\xi,[0,\infty),\overline{Q}(\infty)}$ introduced in Definition 6.2.12. Then the following conditions hold:*

1. There exists a uniformly continuous mapping

$$\Phi_{Sg,fjd} : (S \times \mathscr{T}(\overline{\delta},\overline{\alpha})|\overline{Q}_\infty, d \otimes \rho_{\mathscr{T},\xi}) \rightarrow (\widehat{F}_{Cp}(\overline{Q}_\infty,S),\widehat{\rho}_{Marg,\xi,\overline{Q}(\infty)})$$

such that for each $(x,\mathbf{T}|\overline{Q}_\infty) \in S \times \mathscr{T}(\overline{\delta},\overline{\alpha})|\overline{Q}_\infty$, the family $F^{x,\mathbf{T}|\overline{Q}(\infty)} \equiv \overline{\Phi}_{Sg,fjd}(x,\mathbf{T}|\overline{Q}_\infty)$ of f.j.d.'s is generated by the initial state x and the semigroup $\mathbf{T}|\overline{Q}_\infty$, in the sense of Definition 11.4.1. *In particular, for each fixed $\mathbf{T} \in \mathscr{T}(\overline{\delta},\overline{\alpha})$, the function*

$$F^{*,\mathbf{T}|\overline{Q}(\infty)} \equiv \overline{\Phi}_{Sg,fjd}(\cdot,\mathbf{T}|\overline{Q}_\infty) : (S,d) \rightarrow (\widehat{F}_{Cp}(\overline{Q}_\infty,S),\widehat{\rho}_{Marg,\xi,\overline{Q}(\infty)})$$

is uniformly continuous. Consequently, for each $m \geq 1$, $f \in C(S^m,d^m)$, and for each nondecreasing sequence $r_1 \leq \cdots \leq r_m$ in $\overline{Q}_\infty$, the function

$$F^{*,\mathbf{T}|\overline{Q}(\infty)}_{r(1),\ldots,r(m)} f : (S,d) \rightarrow R$$

is uniformly continuous on (S,d).

2. There exists a uniformly continuous mapping

$$\Phi_{Sg,CdlgMrkv} : (S \times \mathscr{T}(\overline{\delta},\overline{\alpha}), d \otimes \rho_{\mathscr{T},\xi}) \rightarrow (\widehat{D}[0,\infty),\widetilde{\rho}_{\widehat{D}[0,\infty)}),$$

where $(\widehat{D}[0,\infty),\widetilde{\rho}_{\widehat{D}[0,\infty)})$ is the metric space of an a.u. càdlàg process with parameter set $[0,\infty)$, as defined in Definition 10.10.3, such that for each $(x,\mathbf{T}) \in S \times \mathscr{T}(\overline{\delta},\overline{\alpha})$, the a.u. càdlàg process $X^{x,\mathbf{T}} \equiv \Phi_{Sg,CdlgMrkv}(x,\mathbf{T})$ is generated by the initial state x and the semigroup $\mathbf{T}$, in the sense of Definition 11.4.1. *For each $x \in S$, the family $F^{x,\mathbf{T}}$ of marginal distributions of $X^{x,\mathbf{T}}$ is therefore generated by the initial state x and the semigroup $\mathbf{T}$. Specifically,*

$$\Phi_{Sg,CdlgMrkv} \equiv \overline{\Phi}_{rLim} \circ \overline{\Phi}_{DKS,\xi} \circ \Phi_{Sg,fjd} \circ \overline{\Psi},$$

where the component mappings on the right-hand side will be defined precisely in the proof.

Moreover, $X^{x,\mathbf{T}}$ *has a modulus of a.u. càdlàg* $\widetilde{\delta}_{aucl,\overline{\delta}} \equiv (\delta_{aucl,\overline{\delta}}, \delta_{aucl,\overline{\delta}}, \ldots)$ *and a modulus of continuity in probability* $\widetilde{\delta}_{Cp,\overline{\delta}} \equiv (\delta_{Cp,\overline{\delta}}, \delta_{Cp,\overline{\delta}}, \ldots)$*. In other words,* $X^{x,\mathbf{T}} \in \widehat{D}_{\widetilde{\delta}(aucl,\overline{\delta}),\widetilde{\delta}(Cp,\overline{\delta})}[0,\infty)$ *is time-uniformly a.u. càdlàg. Hence the function* $\Phi_{Sg,CdlgMrkv}$ *has a range in* $\widehat{D}_{\widetilde{\delta}(aucl,\overline{\delta}),\widetilde{\delta}(Cp,\overline{\delta})}[0,\infty)$ *and can be regarded as a uniformly continuous mapping*

$$\Phi_{Sg,CdlgMrkv} : (S \times \mathscr{T}(\overline{\delta},\overline{\alpha}), d \otimes \rho_{\mathscr{T},\xi}) \to \widehat{D}_{\widetilde{\delta}(aucl,\overline{\delta}),\widetilde{\delta}(Cp,\overline{\delta})}[0,\infty).$$

Note that the moduli $\widetilde{\delta}_{Cp,\overline{\delta}}$ *and* $\widetilde{\delta}_{aucl,\overline{\delta}}$ *are completely determined by* $\overline{\delta} \equiv \delta_{\mathbf{T}}$.

3. Let $(x,\mathbf{T}) \in S \times \mathscr{T}(\overline{\delta},\overline{\alpha})$ *be arbitrary. Then the a.u. càdlàg process*

$$X \equiv X^{x,\mathbf{T}} \equiv \Phi_{Sg,CdlgMrkv}(x,\mathbf{T}) : [0,\infty) \times (\Omega, L, E) \to (S,d)$$

is Markov relative to the right-limit extension $\overline{\mathcal{L}}_X \equiv \{\overline{L}^{(X,t)} : t \in [0,\infty)\}$ *of its natural filtration* $\mathcal{L}_X \equiv \{L^{(X,t)} : t \in [0,\infty)\}$*. More precisely, let the nondecreasing sequence* $0 \equiv s_0 \leq s_1 \leq \cdots \leq s_m$ *in* $[0,\infty)$*, the function* $f \in C(S^{m+1}, d^{m+1})$*, and* $t \geq 0$ *be arbitrary. Then*

$$\begin{aligned} &E(f(X_{t+s(0)}, X_{t+s(1)}, \ldots, X_{t+s(m)})|\overline{L}^{(X,t)}) \\ &\quad = E(f(X_{t+s(0)}, X_{t+s(1)}, \ldots, X_{t+s(m)})|X_t) = F^{X(t),\mathbf{T}}_{s(0),\ldots,s(m)}(f) \qquad (11.6.21) \end{aligned}$$

as r.r.v.'s.

Proof. Let $(\Omega, L, E) \equiv (\Theta_0, L_0, I_0) \equiv ([0,1], L_0, \int \cdot dx)$ denote the Lebesgue integration space based on the unit interval Θ_0.

1. Let $\mathbf{T} \in \mathscr{T}(\overline{\delta},\overline{\alpha})$ be arbitrary. Then, since $\mathbf{T}$ is a Markov semigroup with parameter set $[0,\infty)$, its restriction $\mathbf{T}|\overline{Q}_\infty$ satisfies the conditions in Definition 11.3.1 to be a Markov semigroup with parameter set $\overline{Q}_\infty$, with the same modulus of strong continuity $\delta_{\mathbf{T}} = \overline{\delta}$ and the same modulus of smoothness $\alpha_{\mathbf{T}} = \overline{\alpha}$.

2. Define the set $\mathscr{T}(\overline{\delta},\overline{\alpha})|\overline{Q}_\infty \equiv \{\mathbf{T}|\overline{Q}_\infty : \mathbf{T} \in \mathscr{T}(\overline{\delta},\overline{\alpha})\}$. Then the sets $\mathscr{T}(\overline{\delta},\overline{\alpha})$ and $\mathscr{T}(\overline{\delta},\overline{\alpha})|\overline{Q}_\infty$ both inherit the metric $\rho_{\mathscr{T},\xi}$ introduced in Definition 11.6.2. As observed in the previous paragraph, members of the family $\mathscr{T}(\overline{\delta},\overline{\alpha})|\overline{Q}_\infty$ share the common moduli $\overline{\delta}$ and $\overline{\alpha}$. As observed in Definition 11.6.2, the mapping

$$\Psi : (\mathscr{T}(\overline{\delta},\overline{\alpha}), \rho_{\mathscr{T},\xi}) \to (\mathscr{T}(\overline{\delta},\overline{\alpha})|\overline{Q}_\infty, \rho_{\mathscr{T},\xi}),$$

defined by $\Psi(\mathbf{T}) \equiv \mathbf{T}|\overline{Q}_\infty$ for each $\mathbf{T} \in \mathscr{T}(\overline{\delta},\overline{\alpha})$, is an isometry. Hence the mapping

$$\overline{\Psi} : (S \times \mathscr{T}(\overline{\delta},\overline{\alpha}), d \otimes \rho_{\mathscr{T},\xi}) \to (S \times \mathscr{T}(\overline{\delta},\overline{\alpha})|\overline{Q}_\infty, d \otimes \rho_{\mathscr{T},\xi}),$$

defined by $\overline{\Psi}(x,\mathbf{T}) \equiv (x, \Psi(\mathbf{T}))$ for each $(x,\mathbf{T}) \in S \times \mathscr{T}(\overline{\delta},\overline{\alpha})$, is uniformly continuous.

3. Separately, Lemma 11.6.4 says that the mapping

$$\Phi_{Sg,fjd} : (S \times \mathscr{T}(\overline{\delta},\overline{\alpha})|\overline{Q}_\infty, d \otimes \rho_{\mathscr{T},\xi}) \to (\widehat{F}_{Cp}(\overline{Q}_\infty, S), \widehat{\rho}_{Marg,\xi,\overline{Q}(\infty)}),$$

constructed in Theorem 11.4.2, is uniformly continuous, with a modulus of continuity $\delta_{Sg,fjd}(\cdot,\overline{\delta},\overline{\alpha},\ \|\xi\|)$ completely determined by the moduli $\overline{\delta},\overline{\alpha}$ and the modulus of local compactness $\|\xi\| \equiv (|A_n|)_{n=1,2,\ldots}$ of (S,d). Assertion 1 is proved.

4. Moreover, Theorem 6.4.4 says that the Compact Daniell–Kolmogorov–Skorokhod Extension

$$\overline{\Phi}_{DKS,\xi} : (\widehat{F}(\overline{Q}_\infty, S), \widehat{\rho}_{Marg,\xi,\overline{Q}(\infty)}) \to (\widehat{R}(\overline{Q}_\infty \times \Theta_0, S), \widehat{\rho}_{Prob,\overline{Q}(\infty)})$$

is uniformly continuous with a modulus of continuity $\overline{\delta}_{DKS}(\cdot,\ \|\xi\|)$ dependent only on $\|\xi\|$.

5. Combining, we see that the composite mapping $\overline{\Phi}_{DKS,\xi} \circ \Phi_{Sg,fjd} \circ \overline{\Psi}$ is uniformly continuous. Now consider the range of this composite mapping. Specifically, take an arbitrary $(x,\mathbf{T}) \in S \times \mathscr{T}(\overline{\delta},\overline{\alpha})$ and consider the image process

$$Z \equiv \overline{\Phi}_{DKS,\xi}(\Phi_{Sg,fjd}(\overline{\Psi}(x,\mathbf{T}))).$$

Write

$$F \equiv F^{x,\mathbf{T}|\overline{Q}(\infty)} \equiv \Phi_{Sg,fjd}(x,\mathbf{T}|\overline{Q}_\infty)$$

for the consistent family constructed in Theorem 11.4.2, generated by the initial state x and the semigroup $\mathbf{T}|\overline{Q}_\infty$, in the sense of Definition 11.4.1. Thus $Z \equiv \overline{\Phi}_{DKS,\xi}(F^{x,\mathbf{T}|\overline{Q}(\infty)})$ and, by the definition of the mapping $\overline{\Phi}_{DKS,\xi}$, the process Z has marginal distributions given by the family $F^{x,\mathbf{T}|\overline{Q}(\infty)}$. In particular $Z_0 = x$.

6. The semigroup $\mathbf{T}|\overline{Q}_\infty$ has a modulus of strong continuity $\overline{\delta}$. At the same time, Assertion 1 of Theorem 11.5.4 implies that the process $Z \equiv \overline{\Phi}_{DKS,\xi}(F^{x,\mathbf{T}|\overline{Q}(\infty)})$ is generated by the initial state x and semigroup $\mathbf{T}|\overline{Q}_\infty$, in the sense of Definition 11.4.1. Hence, by Assertion 4 of Proposition 11.5.5, the process Z is time-uniformly D-regular in the sense of Definition 10.10.2, with some modulus of continuity in probability $\widetilde{\delta}_{Cp,\overline{\delta}} \equiv (\delta_{Cp,\overline{\delta}}, \delta_{Cp,\overline{\delta}}, \ldots)$ and some modulus of D-regularity $\widetilde{m}_{\overline{\delta}} \equiv (\overline{m}_{\overline{\delta}}, \overline{m}_{\overline{\delta}}, \ldots)$. In the notations of Definition 10.10.2, we thus have $Z \in \widehat{R}_{Dreg,\widetilde{\delta}(Cp,\overline{\delta}),\widetilde{m}(\overline{\delta})}.(\overline{Q}_\infty \times \Omega, S)$. Summing up, we see that the range of the composite mapping $\overline{\Phi}_{DKS,\xi} \circ \Phi_{Sg,fjd} \circ \overline{\Psi}$ is contained in the subset $\widehat{R}_{Dreg,\widetilde{\delta}(Cp,\overline{\delta}),\widetilde{m}(\overline{\delta})}.(\overline{Q}_\infty \times \Omega, S)$ of $(\widehat{R}(\overline{Q}_\infty \times \Theta_0, S), \widehat{\rho}_{Prob,\overline{Q}(\infty)})$. Hence we have the uniformly continuous mapping

$$\begin{aligned}\overline{\Phi}_{DKS,\xi} \circ \Phi_{Sg,fjd} \circ \overline{\Psi} &: (S \times \mathscr{T}(\overline{\delta},\overline{\alpha}), d \otimes \rho_{\mathscr{T},\xi}) \\ &\to (\widehat{R}_{Dreg,\widetilde{\delta}(Cp,\overline{\delta}),\widetilde{m}(\overline{\delta})}.(\overline{Q}_\infty \times \Omega, S), \widehat{\rho}_{Prob,\overline{Q}(\infty)}).\end{aligned}$$

7. By Assertion 5 in Proposition 11.5.5, the right-limit extension

$$X \equiv \overline{\Phi}_{rLim}(Z) : [0,\infty) \times (\Omega, L, E) \to (S,d)$$

is a time-uniformly a.u. càdlàg process on $[0,\infty)$, in the sense of Definition 10.10.3, with some modulus of a.u. càdlàg $\widetilde{\delta}_{aucl,\overline{\delta}} \equiv (\delta_{aucl,\overline{\delta}}, \delta_{aucl,\overline{\delta}}, \ldots)$, and with the same modulus of continuity in probability $\widetilde{\delta}_{Cp,\delta}$ as Z. In short,

$$X \equiv \overline{\Phi}_{rLim}(Z) \in \widehat{D}_{\widetilde{\delta}(aucl,\delta(\mathbf{T})),\widetilde{\delta}(Cp,\delta(\mathbf{T}))}[0,\infty).$$

8. Since $(x,\mathbf{T}) \in S \times \mathscr{T}(\overline{\delta},\overline{\alpha})$ is arbitrary, the composite mapping $\overline{\Phi}_{rLim} \circ \overline{\Phi}_{DKS,\xi} \circ \Phi_{Sg,fjd} \circ \overline{\Psi}$ is well defined. We have already seen that the mapping $\overline{\Phi}_{DKS,\xi} \circ \Phi_{Sg,fjd} \circ \overline{\Psi}$ is uniformly continuous. In addition, Theorem 10.10.9 says that

$$\overline{\Phi}_{rLim} : (\widehat{R}_{Dreg,\widetilde{\delta}(Cp,\overline{\delta}),\widetilde{m}(\overline{\delta}).}(\overline{Q}_\infty \times \Omega, S), \widehat{\rho}_{Prob,\overline{Q}(\infty)}) \to (\widehat{D}[0,\infty), \widetilde{\rho}_{\widehat{D}[0,\infty)})$$

is an isometry. Combining, we see that the composite construction mapping

$$\Phi_{Sg,CdlgMrkv} \equiv \overline{\Phi}_{rLim} \circ \overline{\Phi}_{DKS,\xi} \circ \Phi_{Sg,fjd} \circ \overline{\Psi} : (S \times \mathscr{T}(\overline{\delta},\overline{\alpha}), d \otimes \rho_{\mathscr{T},\xi})$$
$$\to (\widehat{D}[0,\infty), \widetilde{\rho}_{\widehat{D}[0,\infty)})$$

is well defined and is uniformly continuous. Moreover, $X|\overline{Q}_\infty = Z$ has marginal distributions given by the family $F^{x,\mathbf{T}|\overline{Q}(\infty)}$.

9. Let $m \geq 1$, $f \in C(S^m,d^m)$, $s_1 \leq \cdots \leq s_m$ in $[0,\infty)$, and $r_1 \leq \cdots \leq r_m$ in $\overline{Q}_\infty$. Then, because $Z \equiv \overline{\Phi}_{DKS,\xi}(F^{x,\mathbf{T}|\overline{Q}(\infty)})$ is generated by the initial state x and semigroup $\mathbf{T}|\overline{Q}_\infty$, we have

$$Ef(X_{r(1)},\ldots,X_{r(m)})$$
$$= \int T^x_{r(1)}(dx_1) \int T^{x(1)}_{r(2)-r(1)}(dx_2) \cdots \int T^{x(m-1)}_{r(m)-r(m-1)}(dx_m) f(x_1,\ldots,x_m). \tag{11.6.22}$$

Now let $r_i \downarrow s_i$ for each $i = 1,\ldots,m$. Then the left-hand side of equality 11.6.22 converges to $Ef(X_{s(1)},\ldots,X_{s(m)})$ because the process X is continuous in probability. At the same time, by Assertion 1 of Lemma 11.3.2, the right-hand side of equality 11.6.22 converges to

$$\int T^x_{s(1)}(dx_1) \int T^{x(1)}_{s(2)-s(1)}(dx_2) \cdots \int T^{x(m-1)}_{s(m)-s(m-1)}(dx_m) f(x_1,\ldots,x_m).$$

Consequently,

$$Ef(X_{s(1)},\ldots,X_{s(m)})$$
$$= T^x_{s(1)}(dx_1) \int T^{x(1)}_{s(2)-s(1)}(dx_2) \cdots \int T^{x(m-1)}_{s(m)-s(m-1)}(dx_m) f(x_1,\ldots,x_m).$$

where $m \geq 1$, $f \in C(S^m,d^m)$, and $s_1 \leq \cdots \leq s_m$ in $[0,\infty)$ are arbitrary. Thus the process X is generated by the initial state x and semigroup $\mathbf{T}$, according to Definition 11.4.1. Assertion 2 of the present theorem is proved.

10. Finally, note that Assertion 3 of the present theorem is merely a restatement of Assertions 3 and 4 of Theorem 11.5.6. The present theorem is proved. □

11.7 Feller Semigroup and Feller Process

In the previous sections, we constructed and studied Markov processes with a compact metric state space. In the present section, we will construct and study Feller processes, which are Markov processes with a locally compact metric state space.

Definition 11.7.1. Specification of locally compact state space and related objects. In this section, let (S,d) be a locally compact metric space, as specified in Definition 11.0.1, along with related objects, including a reference point $x_\circ \in S$ and a binary approximation ξ. In addition, for each $n \geq 0$ and for each $y \in S$, define the function

$$h_{y,n} \equiv (1 \wedge (1 + n - d(\cdot, y))_+ \in C(S,d).$$

Then, for each fixed $y \in S$, we have $h_{y,n} \uparrow 1$ as $n \to \infty$, uniformly on compact subsets of (S,d). Define

$$\overline{h}_{y,n} \equiv 1 - h_{y,n} \in C_{ub}(S,d).$$

The continuous functions $h_{y,n}$ and $\overline{h}_{y,n}$ will be surrogates for the indicators $1_{(d(y,\cdot)\leq n)}$ and $1_{(d(y,\cdot)>n)}$, respectively.

Let $(\overline{S},\overline{d})$ be a one-point compactification of the metric space (S,d), where $\overline{d} \leq 1$ and where Δ is called the point at infinity. For ease of reference, we list here almost verbatim the conditions from Definition 3.4.1 for the one-point compactification $(\overline{S},\overline{d})$.

1. $S \cup \{\Delta\}$ is a dense subset of $(\overline{S},\overline{d})$. Moreover, $\overline{d} \leq 1$.

2. For each compact subset K of (S,d), there exists $c > 0$ such that $\overline{d}(x,\Delta) \geq c$ for each $x \in K$.

3. Let K be an arbitrary compact subset of (S,d). Let $\varepsilon > 0$ be arbitrary. Then there exists $\delta_K(\varepsilon) > 0$ such that for each $y \in K$ and $z \in S$ with $\overline{d}(y,z) < \delta_K(\varepsilon)$, we have $d(y,z) < \varepsilon$. In particular, the identity mapping $\bar{\iota} : (S,\overline{d}) \to (S,d)$ is uniformly continuous on each compact subset of S.

4. The identity mapping $\iota : (S,d) \to (S,\overline{d})$, defined by $\iota(x) \equiv x$ for each $x \in S$, is uniformly continuous on (S,d). In other words, for each $\varepsilon > 0$, there exists $\delta_{\overline{d}}(\varepsilon) > 0$ such that $\overline{d}(x,y) < \varepsilon$ for each $x,y \in S$ with $d(x,y) < \delta_{\overline{d}}(\varepsilon)$.

5. For each $n \geq 1$, we have

$$(d(\cdot,x_\circ) > 2^{n+1}) \subset (\overline{d}(\cdot,\Delta) \leq 2^{-n}). \quad \square$$

Separately, refer to Definition 11.0.2 for notations related to the enumerated sets $\widetilde{Q}_0,\widetilde{Q}_1 \cdots, \overline{Q}_\infty$ of dyadic rationals in $[0,\infty)$, and to the enumerated sets $Q_0,Q_1 \cdots, Q_\infty$ of dyadic rationals in $[0,1]$.

Definition 11.7.2. Feller semigroup. Let $\mathbf{V} \equiv \{V_t : t \in [0,\infty)\}$ be an arbitrary family of nonnegative linear mappings V_t from $C_{ub}(S,d)$ to $C_{ub}(S,d)$ such that V_0 is the identity mapping. Suppose, for each $t \in [0,\infty)$ and for each $y \in S$, the function

$$V_t^y \equiv V_t(\cdot)(y) : C_{ub}(S,d) \to R$$

is a distribution on the locally compact space (S,d). Suppose, in addition, the following four conditions are satisfied:

1. (Smoothness.) For each $N \geq 1$, for each $t \in [0,N]$, and for each $f \in C_{ub}(S,d)$ with a modulus of continuity δ_f and with $|f| \leq 1$, the function

$V_t f \in C_{ub}(S,d)$ has a modulus of smoothness $\alpha_{\mathbf{V},N}(\delta_f)$ that depends on N and on δ_f, and that is otherwise independent of the function f.

2. (Semigroup property.) For each $s,t \in [0,\infty)$, we have $V_{t+s} = V_t V_s$.

3. (Strong continuity.) For each $f \in C_{ub}(S,d)$ with a modulus of continuity δ_f and with $|f| \leq 1$, and for each $\varepsilon > 0$, there exists $\delta_{\mathbf{V}}(\varepsilon,\delta_f) > 0$ so small that for each $t \in [0,\delta_{\mathbf{V}}(\varepsilon,\delta_f))$, we have

$$|f - V_t f| \leq \varepsilon \tag{11.7.1}$$

as functions on (S,d).

4. (Non-explosion.) For each $N \geq 1$, for each $t \in [0,N]$, and for each $\varepsilon > 0$, there exists an integer $\kappa_{\mathbf{V},N}(\varepsilon) > 0$ so large that if $n \geq \kappa_{\mathbf{V},N}(\varepsilon)$, then

$$V_t^y \overline{h}_{y,n} \leq \varepsilon$$

for each $y \in S$.

Then we call the family $\mathbf{V}$ a *Feller semigroup*. The operation $\delta_{\mathbf{V}}$ is called a *modulus of strong continuity* of $\mathbf{V}$. The sequence $\alpha_{\mathbf{V}} \equiv (\alpha_{\mathbf{V},N})_{N=1,2,\ldots}$ of operations is called a *modulus of smoothness of* $\mathbf{V}$. The sequence $\kappa_{\mathbf{V}} \equiv (\kappa_{\mathbf{V},N})_{N=1,2,\ldots}$ of operations is called *a modulus of non-explosion* of $\mathbf{V}$. □

Lemma 11.7.3. Each Markov semigroup, with a compact state space, is a Feller semigroup. *Each Markov semigroup* $\mathbf{T}$, *with a compact state space, in the sense of Definition 11.3.1, is a Feller semigroup.*

Proof. Straightforward and omitted. □

Consequently, all the following results for Feller semigroups and Feller processes will be applicable to Markov processes with Markov semigroups.

Definition 11.7.4. Family of transition distributions generated by Feller semigroup. Suppose, for each $x \in S$, we are given a consistent family $F^{x,\mathbf{V}}$ of f.j.d.'s with state space (S,d) and parameter set $[0,\infty)$.

1. Suppose, for arbitrary $m \geq 1$, $f \in C_{ub}(S^m,d^m)$, and nondecreasing sequence $r_1 \leq \cdots \leq r_m$ in $[0,\infty)$, we have, for each $x \in S$,

$$F^{x,\mathbf{V}}_{r(1),\ldots,r(m)} f$$
$$= \int V^x_{r(1)}(dx_1) \int V^{x(1)}_{r(2)-r(1)}(dx_2) \cdots \int V^{x(m-1)}_{r(m)-r(m-1)}(dx_m) f(x_1,\ldots,x_m). \tag{11.7.2}$$

Then, for each $x \in S$, the family $F^{x,\mathbf{V}}$ is called the *consistent family of f.j.d.'s generated by the initial state x and Feller semigroup* $\mathbf{V}$.

2. Suppose, in addition, for arbitrary $m \geq 1$, $f \in C_{ub}(S^m,d^m)$, and sequence $r_1 \leq \cdots \leq r_m$ in $[0,\infty)$, the function $F^{*,\mathbf{V}}_{r(1),\ldots,r(m)} f : (S,d) \to R$, defined by $(F^{*,\mathbf{V}}_{r(1),\ldots,r(m)} f)(x) \equiv F^{x,\mathbf{V}}_{r(1),\ldots,r(m)} f$ for each $x \in S$, is uniformly continuous and

bounded or, in symbols, $F^{*,\mathbf{V}}_{r(1),\ldots,r(m)}f \in C_{ub}(S,d)$. Then $\{F^{*,\mathbf{V}}_{s(1),\ldots,s(m)} : m \geq 1;$ $s_1,\ldots,s_m \in [0,\infty)\}$ is called the *family of transition distributions generated by the Feller semigroup* $\mathbf{V}$. □

To use the results developed in previous sections for Markov semigroups and Markov processes, where the state space is assumed to be compact, we embed each given Feller semigroup on the locally compact state space (S,d) into a Markov semigroup on the one-point compactification $(\overline{S},\overline{d})$ state space, as follows.

Lemma 11.7.5. Compactification of a Feller semigroup into a Markov semigroup with a compact state space. *Let* $\mathbf{V} \equiv \{V_t : t \in [0,\infty)\}$ *be an arbitrary Feller semigroup on the locally compact metric space* (S,d)*, with moduli* $\delta_{\mathbf{V}}$, $\alpha_{\mathbf{V}}$, $\kappa_{\mathbf{V}}$ *as in Definition 11.7.2. Then there exists a Markov semigroup* $\mathbf{T} \equiv \{T_t : t \in [0,\infty)\}$ *with state space* $(\overline{S},\overline{d})$*, such that*

$$(T_t g)(\Delta) \equiv T_t^{\Delta} g \equiv g(\Delta) \tag{11.7.3}$$

and

$$(T_t g)(y) \equiv T_t^{y} g \equiv \int_{z\in\overline{S}} T_t^{y}(dz)g(z) \equiv \int_{z\in S} V_t^{y}(dz)g(z) \tag{11.7.4}$$

for each $y \in S$*, for each* $g \in C(\overline{S},\overline{d})$*, for each* $t \in [0,\infty)$*. Equality 11.7.4 is equivalent to*

$$T_t^{y} g \equiv V_t^{y}(g|S) \equiv V_t(g|S)(y)$$

for each $y \in S$*, for each* $g \in C(\overline{S},\overline{d})$*, for each* $t \in [0,\infty)$*. Moreover,* S *is a full subset, and* $\{\Delta\}$ *is a null subset, of* $\overline{S}$ *relative to the distribution* T_t^{y}*, for each* $t \in [0,\infty)$.

For want of a better name, such a Markov semigroup $\mathbf{T}$ *will be called a* compactification of the Feller semigroup $\mathbf{V}$.

Proof. Let $t \in [0,\infty)$ be arbitrary. Let $N \geq 1$ be such that $t \in [0,N]$. Let $g \in C(\overline{S},\overline{d})$ be arbitrary, with a modulus of continuity $\overline{\delta}_g$. There is no loss of generality in assuming that g has values in $[0,1]$. As an abbreviation, write $f \equiv g|S \in C_{ub}(S,d)$.

Let $\varepsilon > 0$ be arbitrary.

1. Let $\delta_{\overline{d}}$ be the operation listed in Condition 4 of Definition 11.7.1. Consider arbitrary points $y,z \in S$ with $d(y,z) < \delta_{\overline{d}}(\overline{\delta}_g(\varepsilon))$. Then, according to Condition 4 of Definition 11.7.1, we have $\overline{d}(y,z) < \overline{\delta}_g(\varepsilon)$. Hence

$$|f(y) - f(z)| = |g(y) - g(z)| < \varepsilon.$$

Thus the function $f \equiv g|S$ has a modulus of continuity $\delta_{\overline{d}} \circ \overline{\delta}_g$. Therefore, according to the definition of the modulus of smoothness $\alpha_{\mathbf{V}}$ in Definition 11.7.2, the function $V_t f \in C_{ub}(S,d)$ has a modulus of continuity $\alpha_{\mathbf{V},N}(\delta_{\overline{d}} \circ \overline{\delta}_g)$.

2. Let $k \geq 0$ be so large that

$$2^{-k+1} < \overline{\delta}_g(\varepsilon). \tag{11.7.5}$$

Define

$$n \equiv 2^{k+1} \vee \kappa_{\mathbf{V},N}(\varepsilon).$$

Then, by the definition of $\kappa_{\mathbf{V},N}$, we have

$$0 \leq V_t^y \overline{h}_{y,n} \leq \varepsilon \tag{11.7.6}$$

for each $y \in S$.

3. By Condition 5 of Definition 11.7.1, we have, for each $u \in S$, if $d(u,x_\circ) > n \geq 2^{k+1}$, then $\overline{d}(u,\Delta) \leq 2^{-k} < \overline{\delta}_g(\varepsilon)$, whence $|f(u) - g(\Delta)| = |g(u) - g(\Delta)| \leq \varepsilon$.

4. Take an arbitrary $a \in (2n+1, 2n+2)$ such that the set $K \equiv \{u \in S : d(x_\circ,u) \leq a\}$ is a compact subset of (S,d). Define

$$K' \equiv \{u \in S : d(x_\circ,u) > 2n+1\}. \tag{11.7.7}$$

Then $S = K \cup K'$. Recall here that $S \cup \{\Delta\}$ is a dense subset of $(\overline{S},\overline{d})$ by Condition 1 of Definition 11.7.1. At the same time, by Condition 2 of Definition 11.7.1, there exists $c_K > 0$ such that $\overline{d}(x,\Delta) \geq c_K$ for each $x \in K$.

5. Define $\varepsilon' \equiv \alpha_{\mathbf{V},N}(\delta_{\overline{d}} \circ \overline{\delta}_g)(\varepsilon)$. Then, by Condition 3 of Definition 11.7.1, there exists $\delta_K(\varepsilon') > 0$ such that for each $y \in K$ and $z \in S$ with

$$\overline{d}(y,z) < \delta_K(\varepsilon'),$$

we have $d(y,z) < \varepsilon' \equiv \alpha_{\mathbf{V},N}(\delta_{\overline{d}} \circ \overline{\delta}_g)(\varepsilon)$, whence, in view of the last statement in Step 1, we have

$$|V_t^y f - V_t^z f| < \varepsilon. \tag{11.7.8}$$

6. Now define

$$\overline{\delta}(\varepsilon) \equiv \delta_K(\varepsilon') \wedge c_K$$

and consider each $y,z \in S \cup \{\Delta\}$ with $\overline{d}(y,z) < \overline{\delta}(\varepsilon)$. We will verify that

$$|(T_t g)(y) - (T_t g)(z)| \leq 4\varepsilon. \tag{11.7.9}$$

To that end, note that there are five possibilities: (i) $y \in K$ and $z \in S$, (ii) $y \in K$ and $z = \Delta$, (iii) $y,z \in K'$, (iv) $y \in K'$ and $z = \Delta$, and (v) $y = z = \Delta$. Consider case (i). Then inequality 11.7.8 holds. Moreover, the left-hand side of inequality 11.7.9 is equal to the left-hand side of inequality 11.7.8. Therefore the desired inequality 11.7.9 holds. Next consider case (ii). Then $y \in K$ and $z = \Delta$. Hence $\overline{d}(y,z) \geq c_K \geq \overline{\delta}(\varepsilon)$, which is a contradiction. Therefore case (ii) can be ruled out.

7. Suppose $y \in K'$. Then $d(x_\circ,y) > 2n+1$. Therefore, for each point $u \in S$ with $h_{y,n}(u) > 0$, we have $d(y,u) < n+1$, and so

$$d(u,x_\circ) \geq d(x_\circ,y) - d(y,u) > (2n+1) - (n+1) = n.$$

In view of Step 3, it follows that for each point $u \in S$ with $h_{y,n}(u) > 0$, we have $|f(u) - g(\Delta)| \leq \varepsilon$. Therefore $|fh_{y,n} - g(\Delta)h_{y,n}| \leq \varepsilon$ on S. Consequently,

$$\begin{aligned} 0 \leq (T_t g)(y) \equiv V_t^y f &= V_t^y h_{y,n} f + V_t^y \overline{h}_{y,n} f \leq V_t^y h_{y,n} f + V_t^y \overline{h}_{y,n} \\ &\leq V_t^y h_{y,n} f + \varepsilon \leq V_t^y (g(\Delta) h_{y,n} + \varepsilon) + \varepsilon \\ &= g(\Delta) V_t^y h_{y,n} + 2\varepsilon \\ &\leq g(\Delta) + 2\varepsilon, \end{aligned} \tag{11.7.10}$$

where we have used equality 11.7.6.

8, To continue, consider case (iii). Then, with z in the role of y in the previous paragraph, we can prove, similarly, that $0 \leq (T_t g)(z) \leq g(\Delta) + 2\varepsilon$, Combining with equality 11.7.10, we again obtain inequality 11.7.9. Now consider case (iv). Then equality 11.7.10 holds, while $(T_t g)(z) \equiv g(\Delta)$. Combining, we obtain, once more, inequality 11.7.9. Finally, consider case (v). Then $(T_t g)(y) \equiv g(\Delta) \equiv (T_t g)(z)$. Hence inequality 11.7.9 trivially holds.

9. Summing up, inequality 11.7.9 holds for each $y, z \in S \cup \{\Delta\}$ with $\overline{d}(y,z) < \overline{\delta}(\varepsilon)$, where $\varepsilon > 0$ is arbitrary. Thus the function $T_t g$ is continuous on the dense subset S of $(\overline{S}, \overline{d})$, with a modulus of continuity $\alpha_{\mathbf{T,N}}(\overline{\delta}_g)$ defined by

$$\alpha_{\mathbf{T},N}(\overline{\delta}_g) \equiv \widehat{\alpha}(\overline{\delta}_g, \alpha_{\mathbf{V},N}) \equiv \overline{\delta} \equiv c_K \wedge \delta_K \circ \alpha_{\mathbf{V},N} \circ \delta_{\overline{d}}(\overline{\delta}_g), \tag{11.7.11}$$

where $\overline{\delta}_g$ is the modulus of continuity of the arbitrary function $g \in C(\overline{S}, \overline{d})$.

10. Hence $T_t g$ can be extended to a continuous function $T_t g \in C(\overline{S}, \overline{d})$, with the modulus of continuity $\alpha_{\mathbf{T},N}(\overline{\delta}_g)$. Thus $T_t : C(\overline{S}, \overline{d}) \to C(\overline{S}, \overline{d})$ is a well-defined function. By the defining equality 11.7.4, it is a nonnegative linear function, with $T_t 1 = 1$. Hence, for each $y \in \overline{S}$, the linear and nonnegative function T_t^y is an integration with $T_t^y 1 = 1$. Moreover, for each $t \in [0, \infty)$, for each $N \geq 1$ such that $t \in [0, N]$, and for each $g \in C(\overline{S}, \overline{d})$ with modulus of continuity $\overline{\delta}_g$, the function $T_t g$ has a modulus of continuity $\alpha_{\mathbf{T},N}(\overline{\delta}_g)$. We conclude that T_t is a transition distribution from $(\overline{S}, \overline{d})$ to $(\overline{S}, \overline{d})$, where $N \geq 1$ and $t \in [0, N]$ are arbitrary. It is also clear from the defining equality 11.7.4 that T_0 is the identity mapping.

11. It remains to verify the conditions in Definition 11.3.1 for the family $\mathbf{T} \equiv \{T_t : t \in [0, \infty)\}$ to be a Markov semigroup. The smoothness condition follows immediately from the first sentence in Step 10, which says that the operation $\alpha_{\mathbf{T},N}$ is a modulus of smoothness for the transition distribution T_t, for each $N \geq 1$, for each $t \in [0, N]$.

12. For the semigroup property, consider each $s, t \in [0, \infty)$. Let $y \in S$ be arbitrary. Then, by inequality 11.7.6, we have

$$T_s^y h_{y,k} = V_s^y h_{y,k} \uparrow 1 \tag{11.7.12}$$

and $h_{y,k} \uparrow 1_S$, as $k \to \infty$. Consequently, S is a full subset and $\{\Delta\}$ is a null subset of $\overline{S}$ relative to the distribution T_s^y. Hence $g|S$ is equal to the r.r.v. g on a full set, and is itself an r.r.v. relative to T_s^y, with

$$T_t^y g = T_t^y(g|S) \equiv V_t^y(g|S), \tag{11.7.13}$$

where $y \in S$ and $g \in C(\overline{S},\overline{d})$ are arbitrary. Therefore

$$(T_t g)|S = V_t(g|S), \tag{11.7.14}$$

where $g \in C(\overline{S},\overline{d})$ is arbitrary. Equality 11.7.13, with $T_t g, s$ in the roles of g, t, respectively, implies that

$$T_s^y(T_t g) = V_s^y((T_t g)|S) = V_s^y(V_t(g|S)) = V_{s+t}^y(g|S), \tag{11.7.15}$$

where the second equality is from equality 11.7.14, and where the last equality is by the semigroup property of the Feller semigroup V. Applying equality 11.7.15, with t,s replaced by $0, t+s$, respectively, we obtain $T_{s+t}^y(g) = V_{s+t}^y(g|S)$. Substituting back into equality 11.7.15, we have

$$T_s^y(T_t g) = T_{s+t}^y(g),$$

where $y \in S$ is arbitrary. At the same time, the defining equality 11.7.3 implies that

$$T_s^\Delta(T_t g) \equiv (T_t g)(\Delta) \equiv g(\Delta) = T_{s+t}^\Delta(g).$$

Thus we have proved that $T_s(T_t g) = T_{s+t}(g)$ on the dense subset $S \cup \{\Delta\}$ of $(\overline{S},\overline{d})$. Hence, by continuity, $T_s(T_t g) = T_{s+t}(g)$, where $g \in C(\overline{S},\overline{d})$ is arbitrary. The semigroup property is proved for the family $\mathbf{T}$.

13. It remains to verify strong continuity of the family $\mathbf{T}$. To that end, let $\varepsilon > 0$ be arbitrary, and let $g \in C(\overline{S},\overline{d})$ be arbitrary, with a modulus of continuity $\overline{\delta}_g$ and $\|g\| \leq 1$. Define

$$\delta_{\mathbf{T}}(\varepsilon, \overline{\delta}_g) \equiv \widehat{\delta}(\varepsilon, \overline{\delta}_g, \delta_{\mathbf{V}}) \equiv \delta_{\mathbf{V}}(\varepsilon, \delta_{\overline{d}} \circ \overline{\delta}_g). \tag{11.7.16}$$

We will prove that

$$\|g - T_t g\| \leq \varepsilon, \tag{11.7.17}$$

provided that $t \in [0, \delta_{\mathbf{T}}(\varepsilon, \overline{\delta}_g))$. First note that, by the defining equality 11.7.3, we have

$$g(\Delta) - (T_t g)(\Delta) = 0. \tag{11.7.18}$$

Next, recall from Step 1 that the function $g|S$ has a modulus of continuity $\delta_{\overline{d}} \circ \overline{\delta}_g$. Hence, by the strong continuity of the Feller semigroup, there exists $\delta_{\mathbf{V}}(\varepsilon, \delta_{\overline{d}} \circ \overline{\delta}_g) > 0$ so small that for each $t \in [0, \delta_{\mathbf{V}}(\varepsilon, \delta_{\overline{d}} \circ \overline{\delta}_g))$, we have

$$|(g|S) - V_t(g|S)| \leq \varepsilon \tag{11.7.19}$$

as functions on S. Then, for each $y \in S$, we have

$$|T_t^y g - g(y)| \equiv |V_t^y(g|S) - g(y)| \leq \varepsilon,$$

where the inequality is from inequality 11.7.19. Combining with equality 11.7.18, we obtain $|T_t g - g| \leq \varepsilon$ on the dense subset $S \cup \{\Delta\}$ of $(\overline{S}, \overline{d})$. Hence, by continuity, we have

$$\|T_t g - g\| \leq \varepsilon,$$

where $t \in [0, \delta_{\mathbf{T}}(\varepsilon, \overline{\delta}_g))$ and $g \in C(\overline{S}, \overline{d})$ are arbitrary, with a modulus of continuity $\overline{\delta}_g$ and $\|g\| \leq 1$. Thus we have also verified the strong continuity condition in Definition 11.3.1 for the family $\mathbf{T} \equiv \{T_t : t \in [0,\infty)\}$ to be a Markov semigroup. □

Definition 11.7.6. Feller process. Let $\mathbf{V} \equiv \{V_t : t \in [0,\infty)\}$ be an arbitrary Feller semigroup on the locally compact metric space (S, d). Let (Ω, L, E) be an arbitrary probability space. For each $x \in S$, let

$$U^{x,\mathbf{V}} : [0,\infty) \times (\Omega, L, E) \to (S, d)$$

be a process such that

1. $U^{x,\mathbf{V}}$ is a.u. càdlàg and
2. $U^{x,\mathbf{V}}$ has marginal distributions given by the family $F^{x,\mathbf{V}}$ of f.j.d.'s generated by the initial state x and Feller semigroup $\mathbf{V}$, in the sense of Definition 11.7.4.

Then the triple

$$((S,d), (\Omega, L, E), \{U^{x,\mathbf{V}} : x \in S\})$$

is called a *Feller process* with the Feller semigroup $\mathbf{V}$. □

From a given Feller semigroup $\mathbf{V}$ with locally compact state space (S, d), the preceding theorem constructed a compactification Markov semigroup $\mathbf{T}$ with compact state space $(\overline{S}, \overline{d})$. For each $x \in S$, we can construct an a.u. càdlàg strong Markov process $X^{x,\mathbf{T}}$ with compact state space $(\overline{S}, \overline{d})$, from which we can extract a Feller process with the given locally compact state space along with all the nice sample properties. Thus the next theorem proves the existence of a Feller process with an arbitrarily given Feller semigroup.

Theorem 11.7.7. Construction of a Feller process and Feller transition f.j.d.'s from a Feller semigroup. *Let* $\mathbf{V} \equiv \{V_t : t \in [0,\infty)\}$ *be an arbitrary Feller semigroup on the locally compact metric space* (S, d), *with moduli* $\delta_{\mathbf{V}}$, $\alpha_{\mathbf{V}}$, $\kappa_{\mathbf{V}}$ *as in Definition 11.7.2. Let* $\mathbf{T} \equiv \{T_t : t \in [0,\infty)\}$ *be a compactification of* $\mathbf{V}$, *in the sense of Lemma 11.7.5. Let* $x \in S$ *be arbitrary. Let*

$$X \equiv X^{x,\mathbf{T}} : [0,\infty) \times (\Omega, L, E) \to (S, \overline{d})$$

be the a.u. càdlàg Markov process generated by the initial state x and semigroup $\mathbf{T}$, *as constructed in Theorem 11.6.5. Then the following conditions hold:*

1. For each $t \in [0,\infty)$, *we have* $P(X_t \in S) = 1$.

2. Let $M \geq 1$ *be arbitrary. For each* $\varepsilon_0 > 0$, *there exists* $\beta \equiv \beta_M(\varepsilon_0) > 0$ *such that for each* $h \geq 0$ *we have*

$$P\left(\bigcup_{v\in[0,M]\widetilde{Q}(h)}(d(x,X_v)>\beta)\right)\leq\varepsilon_0. \tag{11.7.20}$$

We emphasize that the bound β_M exists regardless of how large h is.

3. Let $t\in[0,\infty)$ be arbitrary. Define the function $U_t^{x,\mathbf{V}}:(\Omega,L,E)\to(S,d)$ by $domain(U_t^{x,\mathbf{V}})\equiv D_t^{x,\mathbf{T}}\equiv(X_t^{x,\mathbf{T}}\in S)$ and by $U_t^{x,\mathbf{V}}(\omega)\equiv X_t^{x,\mathbf{T}}(\omega)$ for each $\omega\in domain(U_t^{x,\mathbf{V}})$. More succinctly, we define

$$U_t^{x,\mathbf{V}}\equiv\iota(X_t^{x,\mathbf{T}}|D_t^{x,\mathbf{T}}), \tag{11.7.21}$$

where the identity mapping $\bar{\iota}:(S,\bar{d})\to(S,d)$, defined by $\bar{\iota}(y)\equiv y$ for each $y\in S$, is uniformly continuous on each compact subset of (S,d), according to Condition 3 of Definition 11.7.1. Then $U_t^{x,\mathbf{V}}=X_t^{x,\mathbf{T}}a.s.$ Moreover, the function

$$U\equiv U^{x,\mathbf{V}}:[0,\infty)\times(\Omega,L,E)\to(S,d)$$

is a well-defined process.

4. The process $U^{x,\mathbf{V}}$ has marginal distributions given by the consistent family $F^{x,\mathbf{V}}$ of f.j.d.'s generated by the initial state x and the Feller semigroup $\mathbf{V}$, *in the sense of Definition 11.7.4. In particular, for arbitrary $m\geq 1$, $f\in C_{ub}(S^m,d^m)$, and nondecreasing sequence $r_1\leq\cdots\leq r_m$ in $[0,\infty)$, we have*

$$Ef(U_{r(1)},\ldots,U_{r(m)})=F_{r(1),\ldots,r(m)}^{x,\mathbf{T}}f=F_{r(1),\ldots,r(m)}^{x,\mathbf{V}}f.$$

5. The process $U^{x,\mathbf{V}}$ is time-uniformly a.u. càdlàg in the sense of Definition 10.10.1, with a modulus of continuity in probability $(\delta_{Cp,\delta(\mathbf{V})},\delta_{Cp,\delta(\mathbf{V})},\ldots)$ and a modulus of a.u. càdlàg $(\delta_{aucl,\delta(\mathbf{V})},\delta_{aucl,\delta(\mathbf{V})},\ldots)$ that are completely determined by $\delta_{\mathbf{V}}$ and independent of x.

6. The triple

$$((S,d),(\Omega,L,E),\{U^{x,\mathbf{V}}:x\in S\})$$

is a Feller process, with the Feller semigroup $\mathbf{V}$.

Proof. 1. Lemma 11.7.5 says that S is a full subset, and $\{\Delta\}$ is a null subset, of $\overline{S}$ relative to the distribution T_t^x, whence $P(X_t\in S)=T_s^x 1_S=1$, where $x\in S$ and $t\in[0,\infty)$ are arbitrary. Assertion 1 follows.

2. Let $t\geq 0$ and $M\geq 1$ be arbitrary such that $t\in[0,M]$. Let $\varepsilon_0>0$ be arbitrary. Write $\varepsilon\equiv 2^{-1}\varepsilon_0$. Take any $n\geq\kappa_{\mathbf{V},M}(\varepsilon)$ and any $b\geq n+1$. Then, by the non-explosion condition in Definition 11.7.2, we have

$$V_t^y\overline{h}_{y,n}\leq\varepsilon$$

for each $y\in S$. Define the function $g_n\in C_{ub}(S^2,d^2)$ by

$$g_n(y,z)\equiv\overline{h}_{y,n}(z)$$

for each $(y,z) \in S^2$. Then, using equality 11.4.1 in Definition 11.4.1, we obtain

$$F_{0,t}^{y,\mathbf{T}} g_n = \int T_0^y(dx_1) \int T_t^{x(1)}(dx_2) g_n(x_1,x_2) = \int T_0^y(dx_1) T_t^{x(1)} g_n(x_1,\cdot)$$
$$= T_t^y g_n(y,\cdot) = V_t^y g_n(y,\cdot) \equiv V_t^y \overline{h}_{y,n} \leq \varepsilon, \tag{11.7.22}$$

where we used the fact that T_0 is the identity mapping, and where $y \in S$ and $t \in [0,M]$ are arbitrary. Moreover, $T_M^x \overline{h}_{x,n} = V_M^x \overline{h}_{x,n} \leq \varepsilon$. Consequently,

$$P(d(x,X_M) > b) \leq E\overline{h}_{x,n}(X_M) = T_M^x \overline{h}_{x,n} \leq \varepsilon. \tag{11.7.23}$$

3. Next, take any $\beta > 2b$ such that $(d(x,\cdot) \leq \beta)$ is a compact subset of (S,d). Consider each $h \geq 0$. Let $\eta \equiv \eta_{0,\beta,[0,M]\widetilde{Q}(h)}(X|[0,M]\widetilde{Q}_h)$ be the simple first exit time for the process $X|[0,M]\widetilde{Q}_h$ to exit the β-neighborhood of X_0, in the sense of Definition 8.1.12. Then

$$\begin{aligned}
P&\left(\bigcup_{v\in[0,M]\widetilde{Q}(h)} (d(x,X_v) > \beta)\right) \\
&\leq P(d(x,X_\eta) > \beta) \\
&\leq P(d(x,X_\eta) > \beta; d(x,X_M) \leq b) + P(d(x,X_M) > b) \\
&\leq P(d(x,X_\eta) > \beta; d(x,X_M) \leq b) + \varepsilon \\
&= \sum_{v\in[0,M]\widetilde{Q}(h)} P(d(x,X_v) > \beta, \eta = v; d(x,X_M) \leq b) + \varepsilon \\
&\leq \sum_{v\in[0,M]\widetilde{Q}(h)} P(d(X_v,X_M) > \beta - b, \eta = v) + \varepsilon \\
&\leq \sum_{v\in[0,M]\widetilde{Q}(h)} P(d(X_v,X_M) > b, \eta = v) + \varepsilon \\
&\leq \sum_{v\in[0,M]\widetilde{Q}(h)} E(\overline{h}_{X(v),n}(X_M), \eta = v) + \varepsilon \\
&\equiv \sum_{v\in[0,M]\widetilde{Q}(h)} E(g_n(X_v,X_M), \eta = v) + \varepsilon \\
&= \sum_{v\in[0,M]\widetilde{Q}(h)} E\big(F_{0,M-v}^{X(v),\mathbf{T}} g_n, \eta = v\big) + \varepsilon, \\
&\leq \sum_{v\in[0,M]\widetilde{Q}(h)} E(\varepsilon; \eta = v) + \varepsilon = 2\varepsilon \equiv \varepsilon_0,
\end{aligned} \tag{11.7.24}$$

where the last inequality is thanks to inequality 11.7.22; where the third inequality is by inequality 11.7.23; where the sixth inequality is because $b \geq n+1$, whence $\overline{h}_{X(v),n}(X_M) > 1_{b<d(X(M),X(v))}$; and where the third-to-last equality is by equality 11.4.1 for the Markov property of the process $X^{x,\mathbf{T}}$ in Definition 11.4.1. Assertion 2 is proved.

4. Next, let $t \in [0,\infty)$ be arbitrary. According to the defining equality 11.7.21 in the hypothesis, the function $U_t^{x,\mathbf{V}}$ is a continuous function ι of the r.v. $X_t^{x,\mathbf{T}}|D_t^{x,\mathbf{T}}$, where ι is uniformly continuous on compact subsets. Therefore $U_t^{x,\mathbf{V}}$ is an r.v. Assertion 3 is proved.

5. Now let $m \geq 1$, $f \in C_{ub}(S^m, d^m)$, and nondecreasing sequence $r_1 \leq \cdots \leq r_m$ in $[0,\infty)$, be arbitrary. Then

$$
\begin{aligned}
Ef(U_{r(1)},\ldots,U_{r(m)}) &= Ef(X_{r(1)},\ldots,X_{r(m)}) = F^{x,\mathbf{T}}_{r(1),\ldots,r(m)}f \\
&= \int_{x(1)\in\overline{S}} T^x_{r(1)}(dx_1) \int_{x(2)\in\overline{S}} T^{x(1)}_{r(2)-r(1)}(dx_2)\cdots \\
&\quad \int_{x(m)\in\overline{S}} T^{x(m-1)}_{r(m)-r(m-1)}(dx_m) f(x_1,\ldots,x_m) \\
&= \int_{x(1)\in S} V^x_{r(1)}(dx_1) \int_{x(2)\in\overline{S}} T^{x(1)}_{r(2)-r(1)}(dx_2)\cdots \\
&\quad \int_{x(m)\in\overline{S}} T^{x(m-1)}_{r(m)-r(m-1)}(dx_m) f(x_1,\ldots,x_m) \\
&= \int_{x(1)\in S} V^x_{r(1)}(dx_1) \int_{x(2)\in S} V^{x(1)}_{r(2)-r(1)}(dx_2)\cdots \\
&\quad \int_{x(m)\in\overline{S}} T^{x(m-1)}_{r(m)-r(m-1)}(dx_m) f(x_1,\ldots,x_m) \\
&= \cdots \\
&= \int_{x(1)\in S} V^x_{r(1)}(dx_1) \int_{x(2)\in S} V^{x(1)}_{r(2)-r(1)}(dx_2)\cdots \\
&\quad \int_{x(m)\in\overline{S}} V^{x(m-1)}_{r(m)-r(m-1)}(dx_m) f(x_1,\ldots,x_m) \\
&= F^{x,\mathbf{V}}_{r(1),\ldots,r(m)}f.
\end{aligned}
\tag{11.7.25}
$$

Thus

$$Ef(U_{r(1)},\ldots,U_{r(m)}) = F^{x,\mathbf{T}}_{r(1),\ldots,r(m)}f = F^{x,\mathbf{V}}_{r(1),\ldots,r(m)}f,$$

as alleged in Assertion 4.

6. By Theorem 11.6.5, the process

$$X^{x,\mathbf{T}} : [0,\infty) \times (\Omega, L, E) \to (S, \overline{d})$$

is time-uniformly a.u. càdlàg, with a modulus of a.u. càdlàg $\widetilde{\delta}_{aucl,\delta(\mathbf{T})} \equiv (\delta_{aucl,\delta(\mathbf{T})}, \delta_{aucl,\delta(\mathbf{T})}, \ldots)$ and a modulus of continuity in probability $\widetilde{\delta}_{Cp,\delta(\mathbf{T})} \equiv (\delta_{Cp,\delta(\mathbf{T})}, \delta_{Cp,\delta(\mathbf{T})}, \ldots)$ that are completely determined by the modulus of strong continuity $\delta_{\mathbf{T}}$, and that are independent of x.

7. Now consider the process

$$U^{x,\mathbf{V}} : [0,\infty) \times (\Omega, L, E) \to (S, d).$$

Then, since $X^{x,\mathbf{T}}$ is a.u. càdlàg, there exists a full set B such that $X^{x,\mathbf{T}}(\cdot,\omega)$ is right continuous on $domain(X^{x,\mathbf{T}}(\cdot,\omega))$ for each ω in B. At the same time, equality 11.7.21 implies that for each ω in the full set $D \equiv \bigcap_{t\in\overline{Q}(\infty)} D_t^{x,\mathbf{T}}$, we have $U^{x,\mathbf{V}}(t,\omega) \equiv \bar{\iota}(X^{x,\mathbf{T}}(t,\omega))$ for each $t \in \overline{Q}_\infty \cap domain(X^{x,\mathbf{T}}(\cdot,\omega))$. Hence, by the continuity of the mapping $\bar{\iota}$ and by the right continuity of $X^{x,\mathbf{T}}(\cdot,\omega)$, we have $U^{x,\mathbf{V}}(t,\omega) \equiv \bar{\iota}(X^{x,\mathbf{T}}(t,\omega))$ for each $t \in domain(X^{x,\mathbf{T}}(\cdot,\omega))$. In short, we have a.s. $U^{x,\mathbf{V}} \equiv \bar{\iota}(X^{x,\mathbf{T}})$. Hence, by the continuity of the mapping $\bar{\iota}$, the process $U^{x,\mathbf{V}}$ is time-uniformly a.u. càdlàg, with a modulus of a.u. càdlàg and a modulus of continuity in probability that are completely determined by the modulus of strong continuity $\delta_{\mathbf{T}}$, and that are independent of x.

8. At the same time, by equality 11.7.16 of Lemma 11.7.5, the modulus $\delta_{\mathbf{T}}$ is, in turn, completely determined by the modulus of strong continuity $\delta_{\mathbf{V}}$. Combining, we see that the process $U^{x,\mathbf{V}}$ is time-uniformly a.u. càdlàg, with a modulus of a.u. càdlàg $\widetilde{\delta}_{aucl,\delta(\mathbf{V})} \equiv (\delta_{aucl,\delta(\mathbf{V})}, \delta_{aucl,\delta(\mathbf{V})}, \ldots)$ and a modulus of continuity in probability $\widetilde{\delta}_{Cp,\delta(\mathbf{V})} \equiv (\delta_{Cp,\delta(\mathbf{V})}, \delta_{Cp,\delta(\mathbf{V})}, \ldots)$ that are completely determined by the modulus of strong continuity $\delta_{\mathbf{V}}$, and that are independent of x. Assertion 5 has been verified.

9. Assertions 4 and 5 show that the conditions in Definition 11.7.6 are satisfied for the triple $((S,d),(\Omega,L,E),\{U^{x,\mathbf{V}} : x \in S\})$ to be a Feller process with Feller semigroup $\mathbf{V}$, as alleged in Assertion 6. □

Corollary 11.7.8. Continuity of Feller f.j.d.'s. *Let $m \geq 1$ and $f \in C(S^m,d^m)$ be arbitrary. Then the function $F_{r(1),\ldots,r(m)}^{x,\mathbf{V}} f$ of*

$$(x,r_1,\ldots,r_m) \in (S,d) \times \{(r_1,\ldots,r_m) \in [0,\infty)^m : r_1 \leq \cdots \leq r_m\}$$

is uniformly continuous relative to the metric $d \otimes d_{ecld}^m$, where d_{ecld} is the Euclidean metric on $[0,\infty)$.

Proof. 1. By Corollary 11.4.3, the function $F_{r(1),\ldots,r(m)}^{x,\mathbf{T}} f$ of

$$(x,r_1,\ldots,r_m) \in (\overline{S},\overline{d}) \times \{(r_1,\ldots,r_m) \in [0,\infty)^m : r_1 \leq \cdots \leq r_m\}$$

is uniformly continuous relative to the metric $\overline{d} \otimes d_{ecld}^m$. Recall from Definition 11.7.1 that the identity mapping $\iota : (S,d) \to (S,\overline{d})$, defined by $\iota(x) \equiv x$ for each $x \in S$, is uniformly continuous on (S,d). Hence the function $F_{r(1),\ldots,r(m)}^{\iota(x),\mathbf{T}} f$ of

$$(x,r_1,\ldots,r_m) \in (S,d) \times \{(r_1,\ldots,r_m) \in [0,\infty)^m : r_1 \leq \cdots \leq r_m\}$$

is uniformly continuous relative to the metric $d \otimes d_{ecld}^m$.

2. By Assertion 4 of Theorem 11.7.7, for each $x \in S$, we have

$$F_{r(1),\ldots,r(m)}^{\iota(x),\mathbf{T}} f = F_{r(1),\ldots,r(m)}^{x,\mathbf{T}} f = F_{r(1),\ldots,r(m)}^{x,\mathbf{V}} f.$$

Combining, we see that the function $F_{r(1),\ldots,r(m)}^{x,\mathbf{V}} f$ of

$$(x,r_1,\ldots,r_m) \in (S,d) \times \{(r_1,\ldots,r_m) \in [0,\infty)^m : r_1 \leq \cdots \leq r_m\}$$

is uniformly continuous. □

11.8 Feller Process Is Strongly Markov

Definition 11.8.1. Specification of parameter sets. Refer to Definition 11.0.2 for notations related to the sets $\overline{Q}_0,\overline{Q}_1,\ldots,\overline{Q}_\infty$ of dyadic rationals in $[0,\infty)$, the sets $\widetilde{Q}_0,\widetilde{Q}_1,\ldots$ of dyadic rationals in $[0,\infty)$, and the sets $Q_0,Q_1,\ldots,Q_\infty$ of dyadic rationals in $[0,1]$. In particular, for each $h \geq 0$, recall the enumerated set $\widetilde{Q}_h \equiv \{0,\Delta_h,2\Delta_h,\ldots\}$, where $\Delta_h \equiv 2^{-h}$. Recall the right-limit extension functions Φ_{rLim} and $\overline{\Phi}_{rLim}$ from Definition 10.5.6. □

Definition 11.8.2. Specification of locally compact state space, Feller semigroup, and related processes and filtrations. In this section, let (S,d) be a given locally compact metric space. Let $(\overline{S},\overline{d})$ denote a one-point compactification of (S,d). Let $\mathbf{V}$ be an arbitrary Feller semigroup with parameter set $[0,\infty)$ and state space (S,d), with a modulus of strong continuity $\delta_{\mathbf{V}}$, a modulus of smoothness $\alpha_{\mathbf{V}}$, and a modulus of non-explosion $\kappa_{\mathbf{V}}$, in the sense of Definition 11.7.2. Let $x \in S$ be arbitrary.

(i) Let $((S,d),(\Omega,L,E),\{U^{y,\mathbf{V}} : y \in S\})$ be an arbitrary Feller process with the Feller semigroup $\mathbf{V}$, in the sense of Definition 11.7.6.

(ii) Let $F^{x,\mathbf{V}}$ denote the family of transition f.j.d.'s, with parameter set $[0,\infty)$, generated by the initial state x and the Feller semigroup $\mathbf{V}$, as introduced in Definition 11.7.4, and whose existence was proved in Theorem 11.7.7.

(iii) Let $Z \equiv Z^{x,\mathbf{V}} \equiv U^{x,\mathbf{V}}|\overline{Q}_\infty : \overline{Q}_\infty \times (\Omega,L,E) \to (S,d)$ be the restriction of the process $U \equiv U^{x,\mathbf{V}} : [0,\infty)\times(\Omega,L,E) \to (S,d)$ to the countable parameter set $\overline{Q}_\infty$.

(iv) Let $\mathcal{L}_Z \equiv \{L^{(Z,t)} : t \in \overline{Q}_\infty\}$ denote the natural filtration of the process $Z \equiv Z^{x,\mathbf{V}}$. Let $\mathcal{L}_U \equiv \{L^{(U,t)} : t \in [0,\infty)\}$ denote the natural filtration of the process $U \equiv U^{x,\mathbf{V}}$. Let $\overline{\mathcal{L}}_U \equiv \{\overline{L}^{(U,t)} : t \in [0,\infty)\}$ denote the right-limit extension of $\mathcal{L}_U$, where $\overline{L}^{(U,t)} \equiv \bigcap_{s\in(t,\infty)} L^{(U,s)}$ for each $t \in [0,\infty)$. Note that, to lessen the load on notations, we have suppressed the reference to the initial state x for these filtrations. □

Theorem 11.8.3. Feller processes are time-uniformly and state-uniformly a.u. càdlàg. *Let $x \in S$ be arbitrary. Then the process $U \equiv U^{x,\mathbf{V}} : [0,\infty) \times (\Omega,L,E) \to (S,d)$ is time-uniformly a.u. càdlàg, with a modulus of continuity of probability and a modulus of a.u. càdlàg that are completely determined by $\delta_{\mathbf{V}}$ and independent of x. On account of this independence of the initial state, we say that the Feller process is* state-uniformly a.u. càdlàg.

Proof. 1. Let

$$((S,d),(\widetilde{\Omega},\widetilde{L},\widetilde{E}),\{\widetilde{U}^{x,\mathbf{V}} : x \in S\})$$

be the Feller process constructed in Theorem 11.7.7, relative to the Feller semigroup $\mathbf{V}$. We note the small risk of confusion here because $\widetilde{\Omega},\widetilde{L},\widetilde{E},\widetilde{U}$ were designated by Ω,L,E,U, respectively, in Theorem 11.7.7. Then Assertion 5 of

Theorem 11.7.7 says that the process $\widetilde{U} \equiv \widetilde{U}^{x,\mathbf{V}} : [0,\infty) \times (\widetilde{\Omega},\widetilde{L},\widetilde{E}) \to (S,d)$ is time-uniformly a.u. càdlàg, with a modulus of continuity of probability $(\widetilde{\delta}_{Cp,\delta(\mathbf{V})},\widetilde{\delta}_{Cp,\delta(\mathbf{V})},\ldots)$ and a modulus of a.u. càdlàg $(\widetilde{\delta}_{aucl,\delta(\mathbf{V})},\widetilde{\delta}_{aucl,\delta(\mathbf{V})},\ldots)$ that are completely determined by $\delta_{\mathbf{V}}$ and independent of x. Hence, for each $N \geq 0$, the shifted process $\widetilde{U}^N : [0,1] \times (\widetilde{\Omega},\widetilde{L},\widetilde{E}) \to (S,d)$ defined by $\widetilde{U}_t^N \equiv \widetilde{U}_{N+t}$ for each $t \in [0,1]$, is a.u. càdlàg with a modulus of a.u. càdlàg $\widetilde{\delta}_{aucl,\delta(\mathbf{V})}$ independent of x and N. Hence, by Theorem 10.4.3, the restricted process $\widetilde{U}^N|Q_\infty : Q_\infty \times (\widetilde{\Omega},\widetilde{L},\widetilde{E}) \to (S,d)$ is D-regular, with a modulus of continuity in probability $\delta_{Cp,\delta(\mathbf{V})} \equiv \widetilde{\delta}_{Cp,\delta(\mathbf{V})}$ and a modulus of D-regularity $\overline{m}_{\delta(\mathbf{V})} \equiv (m_{\delta(\mathbf{V}),n})_{n=0,1,2,\ldots}$, where

$$2^{-m(\delta(\mathbf{V}),n)} < \delta_{aucl,\delta(\mathbf{V})}(2^{-n-1}), \tag{11.8.1}$$

for each $n \geq 0$. As an abbreviation, write $\overline{m} \equiv (m_n)_{n=0,1,2,\ldots} \equiv \overline{m}_{\delta(\mathbf{V})} \equiv (m_{\delta(\mathbf{V}),n})_{n=0,1,2,\ldots}$. There is no loss in generality to assume that $\overline{m}$ is an increasing sequence of positive integers.

2. Since $U^N|Q_\infty$ and $\widetilde{U}^N|Q_\infty$ are equivalent (i.e., they share the same marginal distributions), we infer that $U^N|Q_\infty : Q_\infty \times (\Omega,L,E) \to (S,d)$ also is D-regular, with a modulus of continuity in probability $\delta_{Cp,\delta(\mathbf{V})}$ and a modulus of D-regularity $\overline{m}_{\delta(\mathbf{V})} \equiv (m_{\delta(\mathbf{V}),n})_{n=0,1,2,\ldots}$. At the same time, $U^N = \Phi_{rLim}(U^N|Q_\infty)$. Hence, by Theorem 10.5.8, the process $U^N : [0,1] \times (\Omega,L,E) \to (S,d)$ is a.u. càdlàg, with a modulus of a.u. càdlàg $\delta_{aucl,\delta(\mathbf{V})} \equiv \delta_{aucl}(\cdot,\overline{m}_{\delta(\mathbf{V})},\delta_{Cp,\delta(\mathbf{V})})$ that is completely determined by $\delta_{\mathbf{V}}$. Therefore the process $U : [0,\infty) \times (\widetilde{\Omega},\widetilde{L},\widetilde{E}) \to (S,d)$ is time-uniformly a.u. càdlàg, with a modulus of a.u. càdlàg $(\delta_{aucl,\delta(\mathbf{V})},\delta_{aucl,\delta(\mathbf{V})},\ldots)$ that is completely determined by $\delta_{\mathbf{V}}$. The theorem is proved. □

Lemma 11.8.4. Markov property when restricted to $\overline{Q}_\infty$. *The process*

$$Z \equiv Z^{x,\mathbf{V}} \equiv U^{x,\mathbf{V}}|\overline{Q}_\infty : \overline{Q}_\infty \times (\Omega,L,E) \to (S,d)$$

is Markov relative to its natural filtration $\mathcal{L}_Z \equiv \{L^{(Z,t)} : t \in \overline{Q}_\infty\}$.

Specifically, let $t \in \overline{Q}_\infty$ be arbitrary. Then

$$\begin{aligned}&E(f(U_{t+r(0)},\ldots,U_{t+r(m)})|L^{(Z,t)})\\&\quad = E(f(U_{t+r(0)},\ldots,U_{t+r(m)})|U_t) = F^{U(t),\mathbf{V}}_{r(0),\ldots,r(m)}(f),\end{aligned} \tag{11.8.2}$$

for each nondecreasing sequence $0 \equiv r_0 \leq r_1 \leq \cdots \leq r_m$ in $\overline{Q}_\infty$, for each $f \in C_{ub}(S^{m+1},d^{m+1})$.

Proof. Let $t \in \overline{Q}_\infty$, the sequence $0 \equiv r_0 \leq r_1 \leq \cdots \leq r_m$ in $\overline{Q}_\infty$, and the function $f \in C_{ub}(S^{m+1},d^{m+1})$ be arbitrary. Consider each

$$\begin{aligned}Y \in G \equiv \{&g(U_{v(1)},\ldots,U_{v(n)}) : n \geq 1; g \in C_{ub}(S^n,d^n);\\&v_1 \leq \cdots \leq v_n \equiv t \text{ in } \overline{Q}_\infty\}.\end{aligned}$$

Then $Y = g(U_{v(1)},\ldots,U_{v(n)})$ for some $n \geq 1, g \in C_{ub}(S^n,d^n)$, and some sequence $v_1 \leq \cdots \leq v_n \equiv t$ in $\overline{Q}_\infty$. Hence

$$\begin{aligned}
&EYf(U_{t+r(0)},\ldots,U_{t+r(m)}) \\
&\quad = Eg(U_{v(1)},\ldots,U_{v(n)})f(U_{t+r(0)},\ldots,U_{t+r(m)}) \\
&\quad = \int V^x_{v(1)}(dx_1)\int V^{x(1)}_{v(2)-r(1)}(dx_2)\cdots\int V^{x(n-1)}_{v(n)-r(n-1)}(dx_n) \\
&\qquad \int V^{x(n)}_{t+r(0)-v(n)}(dx_{n+1})\cdots\int V^{x(m)}_{t+r(m)-(t+r(m-1))}(dx_{m+1})g(x_1,\ldots x_n) \\
&\qquad \times f(x_{n+1},\ldots,x_{n+m+1}) \\
&\quad = \int V^x_{v(1)}(dx_1)\int V^{x(1)}_{v(2)-r(1)}(dx_2)\cdots\int V^{x(n-1)}_{v(n)-r(n-1)}(dx_n)g(x_1,\ldots x_n) \\
&\qquad \left\{\int V^{x(n)}_{t+r(0)-v(n)}(dx_{n+1})\cdots\int V^{x(m)}_{t+r(m)-v(0)}(dx_{m+1})f(x_{n+1},\ldots,x_{n+m+1})\right\} \\
&\quad = \int V^x_{v(1)}(dx_1)\int V^{x(1)}_{v(2)-r(1)}(dx_2)\cdots\int V^{x(n-1)}_{v(n)-r(n-1)}(dx_n)g(x_1,\ldots x_n) \\
&\qquad \times \left\{F^{x(n),\mathbf{V}}_{r(0),\ldots,r(m)}f\right\} \\
&\quad = Eg(U_{v(1)},\ldots,U_{v(n)})\left\{F^{U(v(n)),\mathbf{V}}_{r(0),\ldots,r(m)}f\right\} \\
&\quad = Eg(U_{v(1)},\ldots,U_{v(n)})\left\{F^{U(t),\mathbf{V}}_{r(0),\ldots,r(m)}f\right\} \\
&\quad = EY\left\{F^{U(t),\mathbf{V}}_{r(0),\ldots,r(m)}f\right\}.
\end{aligned}$$

Summing up,

$$EYf(U_{t+r(0)},\ldots,U_{t+r(m)}) = EY\{F^{U(t),\mathbf{V}}_{r(0),\ldots,r(m)}f\}, \tag{11.8.3}$$

for each $Y \in G$. Since the set

$$G \equiv \{g(U_{v(1)}),\ldots,U_{v(n)}) : n \geq 1; g \in C_{ub}(S^n,d^n); v_1 \leq \cdots \leq v_n \equiv t \text{ in } \overline{Q}_\infty\}$$

is dense in

$$L^{(Z,t)} \equiv L(Z : u \in [0,t]\overline{Q}_\infty)$$

relative to L_1-convergence with respect to the expectation E, equality 11.8.3 holds also for each $Y \in L^{(Z,t)}$. In particular, it holds for the special case where $Y \in L(Z_t) = L(U_i)$. We conclude that

$$\begin{aligned}
&L(f(U_{t+r(0)},U_{t+r(1)},\ldots,U_{t+r(m)})|L^{(Z,t)}) \\
&\qquad = E(f(U_{t+r(0)},U_{t+r(1)},\ldots,U_{t+r(m)})|U_t) = F^{U(t),\mathbf{V}}_{r(0),\ldots,r(m)}(f).
\end{aligned} \tag{11.8.4}$$

The lemma is proved. □

Lemma 11.8.5. a.u. Right continuity at each stopping time with regularly spaced dyadic values. *Recall the abbreviations* $U \equiv U^{x,\mathbf{V}}$ *and* $\mathcal{L} \equiv \mathcal{L}_U \equiv \{L^{(U,t)} : t \in [0,\infty)\}$, *where* $\mathcal{L}_U$ *is the natural filtration of the process* U. *Then there exists an increasing sequence* $(m_k)_{k=1,2,\ldots}$ *of positive integers such that for each stopping time* η *relative to the filtration* $\mathcal{L}$, *with values in* $\widetilde{Q}_h$ *for some* $h \geq 0$, *the following conditions hold:*

1. For each $k \geq 1$, *the function*

$$V_{\eta,k} \equiv \sup_{u \in [0,\Delta(m(k))]} \widehat{d}(U_\eta, U_{\eta+u})$$

is a well-defined r.r.v. Recall here that $\Delta_{m(k)} \equiv 2^{-m(k)}$ *and that* $\widehat{d} \equiv 1 \wedge d$.

2. For each $k \geq 1$, *we have*

$$EV_{\eta,k} \leq 2^{-k+1}.$$

Proof. 1. Let $x \in S$ and $N \geq 0$ be arbitrary. By Theorem 11.8.3, the shifted process $U^N : [0,1] \times (\Omega, L, E) \to (S,d)$ is a.u. càdlàg, with a modulus of a.u. càdlàg $\delta_{aucl,\delta(\mathbf{V})}$ independent of N and of the initial state x. By Theorem 10.4.3, the restricted process $U^N|Q_\infty$ is D-regular, with a modulus of D-regularity $\overline{m}' \equiv (m'_k)_{k=1,2,\ldots} \equiv (m'_{\delta(\mathbf{V}),k})_{k=1,2,\ldots}$ that is completely determined by $\delta_{aucl,\delta(\mathbf{V})}$. Hence, by Definition 10.3.2 for a.u. càdlàg processes, there exists a full set $B \subset \bigcap_{t \in Q(\infty)} domain(U_t)$ such that for each $\omega \in B$, the function $U(\cdot,\omega)$ is right continuous at each $t \in domain(U(\cdot,\omega))$. Fix an arbitrary increasing sequence of positive integers $\overline{m} \equiv (m_k)_{k=1,2,\ldots} \equiv (m_{\delta(\mathbf{V}),k})_{k=1,2,\ldots}$ such that

$$2^{-m(k)} < \delta_{aucl,\delta(\mathbf{V})}(2^{-k})$$

for each $k \geq 1$. By replacing m_k with $m_k \vee m'_k$ for each $k \geq 1$, we may assume that the sequence $\overline{m}$ is also a modulus of D-regularity of the restricted process $U^N|Q_\infty$.

2. Let $k \geq 1$ be arbitrary, but fixed until further notice. By Condition 3 in Definition 10.3.2, there exist a measurable set A with $P(A^c) < 2^{-k}$ and an r.r.v. τ_1 with values in $[0,1]$ such that for each $\omega \in A$, we have

$$\tau_1(\omega) \geq \delta_{aucl,\delta(\mathbf{V})}(2^{-k}) > 2^{-m(k)} \equiv \Delta_{m(k)} \tag{11.8.5}$$

and

$$d(U(0,\omega), U(\cdot,\omega)) \leq 2^{-k} \tag{11.8.6}$$

on the interval $\theta_0(\omega) \equiv [0,\tau_1(\omega))$. Inequalities 11.8.6 and 11.8.5 together imply that for each $\omega \in A$, we have

$$d(U_0(\omega), U_u(\omega)) \leq 2^{-k} \tag{11.8.7}$$

for each $u \in [0,\Delta_{m(k)}] \cap domain(U(\cdot,\omega))$.

3. Consider each $\omega \in AB$ and each $\kappa \geq k$. Write $J_\kappa = J_{\kappa,k} = 2^{m(\kappa)-m(k)}$. Then $J_\kappa \Delta_{m(\kappa)} = \Delta_{m(k)}$. Since $\omega \in B$, we have $j\Delta_{m(\kappa)} \in [0,\Delta_{m(k)}] \cap domain(U(\cdot,\omega))$ for each $j = 0,\ldots,J_\kappa$. Therefore, according to inequality 11.8.7, we have

$$\bigvee_{j=0}^{J(\kappa)} \widehat{d}(U_0, U_{j\Delta(m(\kappa))})1_A \leq 2^{-k}$$

on AB, where we recall that $\widehat{d} \equiv 1 \wedge d$. Define the function $f_\kappa \in C_{ub}(S^{J(\kappa)+1}, d^{J(\kappa)+1})$ by

$$f_\kappa(x_0, x_1, \ldots, x_{J(\kappa)}) \equiv \bigvee_{j=0}^{J(\kappa)} \widehat{d}(x_0, x_j)$$

for each $(x_0, x_1, \ldots, x_{J(\kappa)}) \in S^{J(\kappa)+1}$. Then

$$\begin{aligned} & Ef_\kappa(U_0, U_{\Delta(m(\kappa))}, \ldots, U_{J(\kappa)\Delta(m(\kappa))}) \\ & = E \bigvee_{j=0}^{J(\kappa)} \widehat{d}(U_0, U_{j\Delta(m(\kappa))}) \\ & \leq E \bigvee_{j=0}^{J(\kappa)} \widehat{d}(U_0, U_{j\Delta(m(\kappa))})1_{AB} + P((AB)^c) \leq 2^{-k} + 2^{-k} = 2^{-k+1}, \end{aligned} \tag{11.8.8}$$

In terms of marginal distributions of the process $U \equiv U^{x,\mathbf{V}}$, inequality 11.8.8 can be rewritten as

$$F^{x,\mathbf{V}}_{0,\Delta(m(\kappa)),\ldots,J(\kappa)\Delta(m(\kappa))} f_\kappa \leq 2^{-k+1}, \tag{11.8.9}$$

where $x \in S$ is arbitrary.

4. Separately, because $U|Q_\infty$ is D-regular, with the sequence $\overline{m}$ as a modulus of D-regularity, Assertion 2 of Lemma 10.5.5, where Z, v, v' are replaced by $U|Q_\infty, 0, \Delta_{m(k)}$, respectively, implies that the supremum

$$V_{0,k,\infty} \equiv \sup_{u \in [0,\Delta(m(k))]Q(\infty)} \widehat{d}(U_0, U_u) = 1 \wedge \sup_{u \in [0,\Delta(m(k))]Q(\infty)} d(U_0, U_u)$$

is a well-defined r.r.v.

5. Recall that $\kappa \geq k$ is arbitrary. Assertion 3 of Lemma 10.5.5, where Z, v, v', h are replaced by $U|Q_\infty, 0, \Delta_{m(k)}, \kappa$, respectively, says that

$$0 \leq E \sup_{u \in [0,\Delta(m(k))]Q(\infty)} \widehat{d}(U_0, U_u) - E \bigvee_{u \in [0,\Delta(m(k))]Q(m(\kappa))} \widehat{d}(U_0, U_u) \leq 2^{-\kappa+5}. \tag{11.8.10}$$

As an abbreviation, define the r.r.v.

$$\begin{aligned} V_{0,k,\kappa} & \equiv \bigvee_{u \in [0,\Delta(m(k))]Q(m(\kappa))} \widehat{d}(U_0, U_u) \\ & = \bigvee_{j=0}^{J(\kappa)} \widehat{d}(U_0, U_{j\Delta(m(\kappa))}) \equiv f_\kappa(U_0, U_{\Delta(m(\kappa))}, \ldots, U_{J(\kappa)\Delta(m(\kappa))}). \end{aligned} \tag{11.8.11}$$

Then inequality 11.8.10 can be rewritten compactly as

$$0 \le EV_{0,k,\infty} - EV_{0,k,\kappa} \le 2^{-\kappa+5}, \tag{11.8.12}$$

where $\kappa \ge k$ is arbitrary. Consequently, for each $\kappa' \ge \kappa \ge k$, we have

$$0 \le EV_{0,k,\kappa'} - EV_{0,k,\kappa} \le 2^{-\kappa+5}. \tag{11.8.13}$$

Equivalently, for each $\kappa' \ge \kappa \ge k$, we have

$$0 \le F^{x,\mathbf{V}}_{0,\Delta(m(\kappa')),\dots,J(\kappa')\Delta(m(\kappa'))} f_{\kappa'} - F^{x,\mathbf{V}}_{0,\Delta(m(\kappa)),\dots,J(\kappa)\Delta(m(\kappa))} f_\kappa \le 2^{-\kappa+5}, \tag{11.8.14}$$

where $x \in S$ is arbitrary.

6. Now let η be an arbitrary stopping time with values in $\widetilde{Q}_h$ for some $h \ge 0$, relative to the filtration $\mathcal{L} \equiv \mathcal{L}_Z \equiv \{L^{(Z,t)} : t \in \overline{Q}_\infty\}$. Generalize the defining equality 11.8.11 for $V_{0,k,\kappa}$ to

$$\begin{aligned} V_{\eta,k,\kappa} &\equiv \bigvee_{u\in[0,\Delta(m(k))]Q(m(\kappa))} \widehat{d}(U_\eta, U_{\eta+u}) \\ &\equiv \bigvee_{j=0}^{J(\kappa)} \widehat{d}(U_\eta, U_{\eta+j\Delta(m(\kappa))}) \equiv f_\kappa(U_\eta, U_{\eta+\Delta(m(\kappa))}, \dots, U_{\eta+J(\kappa)\Delta(m(\kappa))}). \end{aligned} \tag{11.8.15}$$

Then, $1_{(\eta=t)} \in L^{(Z,t)}$ for each $t \in \widetilde{Q}_h$, whence

$$\begin{aligned} EV_{\eta,k,\kappa} &= \sum_{t\in\widetilde{Q}(h)} Ef_\kappa(U_\eta, U_{\eta+\Delta(m(\kappa))}, \dots, U_{\eta+J(\kappa)\Delta(m(\kappa))})1_{(\eta=t)} \\ &= \sum_{t\in\widetilde{Q}(h)} Ef_\kappa(U_t, U_{t+\Delta(m(\kappa))}, \dots, U_{t+J(\kappa)\Delta(m(\kappa))})1_{(\eta=t)} \\ &= \sum_{t\in\widetilde{Q}(h)} EF^{U(t),\mathbf{V}}_{0,\Delta(m(\kappa)),\dots,J(\kappa)\Delta(m(\kappa))}(f_\kappa)1_{(\eta=t)} \\ &= \sum_{t\in\widetilde{Q}(h)} EF^{U(\eta),\mathbf{V}}_{0,\Delta(m(\kappa)),\dots,J(\kappa)\Delta(m(\kappa))}(f_\kappa)1_{(\eta=t)} \\ &= EF^{U(\eta),\mathbf{V}}_{0,\Delta(m(\kappa)),\dots,J(\kappa)\Delta(m(\kappa))}(f_\kappa), \end{aligned} \tag{11.8.16}$$

where the third equality is by Lemma 11.8.4. Hence

$$\begin{aligned} 0 &\le EV_{\eta,k,\kappa'} - EV_{\eta,k,\kappa} \\ &= E\Big(F^{U(t),\mathbf{V}}_{0,\Delta(m(\kappa)),\dots,J(\kappa')\Delta(m(\kappa'))}(f_{\kappa'}) - F^{U(t),\mathbf{V}}_{0,\Delta(m(\kappa)),\dots,J(\kappa)\Delta(m(\kappa))}(f_\kappa)\Big) \\ &\le 2^{-\kappa+5}, \end{aligned}$$

where the equality is by equality 11.8.16, where the inequality is by inequality 11.8.14, and where κ', κ are arbitrary with $\kappa' \ge \kappa \ge k$.

7. Consequently, $EV_{\eta,k,\kappa}$ converges. By the Monotone Convergence Theorem, we have

$$E|V_{\eta,k,\kappa} - W| \to 0$$

for some r.r.v. $W \in L$, as $\kappa \to \infty$. Moreover, $V_{\eta,k,\kappa} \uparrow W$ a.u. as $\kappa \to \infty$. In particular, $V_{\eta,k,\kappa}(\omega) \uparrow W(\omega)$ as $\kappa \to \infty$, for each ω in some full set B'. By replacing B' with BB' if necessary, we may assume that $B' \subset B$. Then, for each $\omega \in B'$, the supremum

$$\sup_{u\in[0,\Delta(m(k))]Q(\infty)} \widehat{d}(U(\eta(\omega),\omega), U(\eta(\omega)+u,\omega)) = W(\omega)$$

exists. Since $\omega \in B$, the function $U(\cdot,\omega)$ is right continuous on its domain. Hence the last displayed equality implies

$$\sup_{v\in[0,\Delta(m(k))]} \widehat{d}(U(\eta(\omega),\omega), U(\eta(\omega)+v,\omega)) = W(\omega).$$

Therefore, in terms of the function $V_{\eta,k}$ defined in the hypothesis, we have

$$V_{\eta,k}(\omega) \equiv \sup_{v\in[0,\Delta(m(k))]} \widehat{d}(U_\eta(\omega), U_{\eta+v}(\omega)) = W(\omega),$$

where $\omega \in B' \cap domain(W)$ is arbitrary. We conclude that $V_{\eta,k} = W$ a.s., and therefore that $V_{\eta,k} \in L$ is a well-defined integrable r.r.v. Assertion 1 is proved. Moreover,

$$EV_{\eta,k} = EW = \lim_{\kappa\to\infty} EV_{\eta,k,\kappa} = \lim_{\kappa\to\infty} EF^{U(\eta),\mathbf{T}}_{0,\Delta(m(\kappa)),\ldots,J(\kappa)\Delta(m(\kappa))} f_\kappa \le 2^{-k+1},$$

where the third equality is by equality 11.8.16, and where the inequality follows from inequality 11.8.9. Assertion 2 and the lemma are proved. □

Lemma 11.8.6. Observability of Feller process at stopping times. *Recall the abbreviations $U \equiv U^{x,\mathbf{V}}$ and $Z \equiv Z^{x,\mathbf{V}} \equiv U^{x,\mathbf{V}}|\overline{Q}_\infty$. Recall the natural filtration $\mathcal{L} \equiv \mathcal{L}_U \equiv \{L^{(U,t)} : t \in [0,\infty)\}$ of the process U, and the natural filtration $\mathcal{L}_Z \equiv \{L^{(Z,t)} : t \in \overline{Q}_\infty\}$ of the process Z. In addition, let $\overline{\mathcal{L}} \equiv \overline{\mathcal{L}}_U \equiv \{\overline{L}^{(U,t)} : t \in [0,\infty)\}$ be the right-limit extension of $\mathcal{L}_U$. Let τ be an arbitrary stopping time relative to the right continuous filtration $\overline{\mathcal{L}} \equiv \overline{\mathcal{L}}_U$. Then the following conditions hold.*

1. The function U_τ is a well-defined r.v. which is measurable relative to the probability subspace

$$\overline{L}^{(U,\tau)} \equiv \{Y \in L : Y1_{(\tau\le t)} \in \overline{L}^{(U,t)} \text{ for each regular point } t \in [0,\infty) \text{ of } \tau\}, \tag{11.8.17}$$

introduced in Definition 8.1.9.

2. There exists a nonincreasing sequence $(\eta_h)_{h=0,1,\ldots}$ of stopping times, relative to the filtration $\mathcal{L}_Z \equiv \{L^{(Z,t)} : t \in \overline{Q}_\infty\}$, such that for each $h \ge 0$, the r.r.v. η_h has values in $\widetilde{Q}_h$ and such that

$$\tau + 2\Delta_h < \eta_h < \tau + 2\Delta_{h-1}. \qquad (11.8.18)$$

Then $U_{\eta(h)}$ is an r.v. for each $h \geq 0$, and $U_{\eta(h)} \rightarrow U_\tau$ a.u. as $h \rightarrow \infty$. Moreover, for each ω in some full set A, the function $U(\cdot,\omega)$ is right continuous at $\tau(\omega)$.

3. More generally, let $m \geq 1$ and $r_0 \leq \cdots \leq r_m$ be arbitrary. Then, for each $i = 0, \cdots, m, U_{\tau+r(i)}$ is a well-defined r.v. that is measurable relative to the probability subspace $\overline{L}^{(U,\tau+r(i))}$. Moreover, for each $i = 0, \cdots, m$ and for each $h \geq 0$, (i) there exists $u_{h,i} \in (r_i, r_i + (m+2)2^{-h}) \cap \widetilde{Q}_h$, (ii) $u_{h,0} < \cdots < u_{h,m}$, and (iii) $U_{\eta(h)+u(h,i)} \rightarrow U_{\tau+r(i)}$ a.u. as $h \rightarrow \infty$.

Proof. 1. For each $h \geq 0$, recall the set

$$\widetilde{Q}_h \equiv \{0, \Delta_h, 2\Delta_h, \cdots\} \equiv \{0, 2^{-h}, 2 \cdot 2^{-h}, \cdots\} \subset [0,\infty).$$

Let the increasing sequence $(m_k)_{k=1,2,\cdots} \equiv (m_{\delta(\mathbf{V}),k})_{k=1,2,\cdots}$ of positive integers be as constructed in Lemma 11.8.5 relative to Feller semigroup $\mathbf{V}$.

2. Let $h \geq 0$ be arbitrary. Because the set $\widetilde{Q}_h$ is countable, there exists $\alpha_h \in (\Delta_h, 2\Delta_h)$ such that the set $(\tau + \alpha_h \leq s)$ is measurable for each $s \in \widetilde{Q}_h$. Define the topping time

$$\eta_h \equiv \sum_{s \in \widetilde{Q}(h)}^{\infty} (s + \Delta_h) 1_{s-\Delta(h)<\tau+\alpha(h)\leq s}. \qquad (11.8.19)$$

Then

$$\tau + \alpha_h + \Delta_h \leq \eta_h < \tau + \alpha_h + 2\Delta_h \qquad (11.8.20)$$

and

$$\tau + 2\Delta_h < \eta_h < \tau + 2\Delta_{h-1}. \qquad (11.8.21)$$

Hence $\eta_h \rightarrow \tau$ uniformly. Moreover, it follows immediately from equality 11.8.21 that $\eta_{h+1} < \eta_h$.

3. Let $k \geq 1$ be arbitrary. Define

$$h_k \equiv m_{k+2} + 2$$

and

$$\delta_k \equiv \Delta(m_{k+2}) \equiv 2^{-m(k+2)}.$$

Define the stopping time

$$\zeta_k \equiv \eta_{h(k)}.$$

with values in $\widetilde{Q}_{n(h(k))}$, relative to the filtration $\mathcal{L}_Z$. Hence Lemma 11.8.5, where η, k are replaced by $\zeta_{k+1}, k+2$, respectively, implies that the function

$$W_k \equiv V_{\zeta(k+1),k+2} \equiv \sup_{u \in [0,\Delta(m(k+2))]} \widehat{d}(U_{\zeta(k+1)}, U_{\zeta(k+1)+u})$$

is a well-defined r.r.v., with

$$EW_k \equiv EV_{\zeta(k+1),k+2} \leq 2^{-k-1}, \tag{11.8.22}$$

where $k \geq 1$ is arbitrary.

4. Then inequality 11.8.21 implies that

$$\tau < \eta_{h(k)} \equiv \zeta_k < \tau + 2^{-h(k)+2} = \tau + 2^{-m(k+2)} = \tau + \delta_k, \tag{11.8.23}$$

whence $\zeta_k \downarrow \tau$ uniformly as $k \to \infty$. Moreover,

$$\zeta_k < \tau + 2^{-h(k)+2} < \zeta_{k+1} + 2^{-h(k)+2} \equiv \zeta_{k+1} + 2^{-m(k+2)} \equiv \zeta_{k+1} + \delta_k, \tag{11.8.24}$$

where $k \geq 1$ is arbitrary. Hence. recursively, we obtain

$$\begin{aligned}(\tau, \tau + \delta_k]\overline{Q}_\infty &\subset (\tau, \zeta_{k+1} + \delta_k]\overline{Q}_\infty \\ &= \cdots \cup [\zeta_{k+3}, \zeta_{k+2}]\overline{Q}_\infty \cup [\zeta_{k+2}, \zeta_{k+1}]\overline{Q}_\infty \cup [\zeta_{k+1}, \zeta_{k+1}+\delta_k]\overline{Q}_\infty \\ &\subset \cdots \cup [\zeta_{k+3}, \zeta_{k+3} + \delta_{k+2}]\overline{Q}_\infty \cup [\zeta_{k+2}, \zeta_{k+2} + \delta_{k+1}]\overline{Q}_\infty \\ &\quad \cup [\zeta_{k+1}, \zeta_{k+1} + \delta_k]\overline{Q}_\infty \\ &\equiv \bigcup_{j=k}^{\infty} [\zeta_{j+1}, \zeta_{j+1} + \delta_j]\overline{Q}_\infty, \end{aligned} \tag{11.8.25}$$

where the first equality is because $\zeta_k \downarrow \tau$ uniformly.

5. Separately, for each $j \geq 0$, take any $\varepsilon_j \in (2^{-j/2-1}, 2^{-j/2}]$ such that the set $A_j \equiv (W_j > \varepsilon_j)$ is measurable. Then

$$P(A_j) \leq \varepsilon_j^{-1} EW_j \leq \varepsilon_j^{-1} 2^{-j-1} < 2^{j/2+1} 2^{-j-1} = 2^{-j/2},$$

where the first inequality is by Chebychev's inequality, and where the second inequality is by inequality 11.8.22. Therefore, we can define the measurable set $A_{k+} \equiv \bigcup_{j=k}^{\infty} A_j$, with

$$P(A_{k+}) < \sum_{j=k}^{\infty} P(A_j) < \sum_{j=k}^{\infty} 2^{-j/2} = 2^{-k/2}(1 - 2^{-1/2})^{-1} < 2^{-k/2+4}.$$

Hence we have the full set

$$A \equiv \bigcup_{\kappa=0}^{\infty} A_{\kappa+}^c.$$

6. Consider each $\omega \in A_{k+}^c$. Consider each

$$t \in (\tau(\omega), \tau(\omega) + \delta_k]\overline{Q}_\infty.$$

Then, in view of relation 11.8.25, there exists $j \geq k$ such that

$$t \in [\zeta_{j+1}(\omega), \zeta_{j+1}(\omega) + \delta_j]\overline{Q}_\infty \subset [\zeta_{j+1}(\omega), \zeta_{j+1}(\omega) + \Delta_{m(j+2)}]\overline{Q}_\infty.$$

Hence

$$\widehat{d}(U_{\zeta(j+1)}(\omega), U_t(\omega)) \leq \sup_{u \in [0, \Delta(m(j+2))]} \widehat{d}(U_{\zeta(j+1)}(\omega), U_{\zeta(j+1)+u}(\omega))$$
$$\equiv W_j(\omega) \leq \varepsilon_j. \tag{11.8.26}$$

Consequently,

$$\widehat{d}(U_t(\omega), U_{\zeta(k)}(\omega)) \leq \widehat{d}(U_t(\omega), U_{\zeta(j+1)}(\omega)) + \widehat{d}(U_{\zeta(j+1)}(\omega), U_{\zeta(j)}(\omega))$$
$$+ \cdots + \widehat{d}(U_{\zeta(k+1)}(\omega), U_{\zeta(k)}(\omega))$$
$$\leq \varepsilon_j + \varepsilon_j + \varepsilon_{j-1} + \cdots + \varepsilon_k < 2\sum_{j=k}^{\infty} \varepsilon_j,$$

where $t \in (\tau(\omega), \tau(\omega) + \delta_k]\overline{Q}_\infty$ is arbitrary. Now the set $(\tau(\omega), \tau(\omega) + \delta_k]\overline{Q}_\infty$ is a dense subset of $[\tau(\omega), \tau(\omega) + \delta_k] \cap domain(U(\cdot, \omega))$. Moreover, the function $U(\cdot, \omega)$ is right continuous on *domain* $(U(\cdot, \omega))$ by Assertion 1 of Proposition 10.1.2, where $x, \overline{x}$ are replaced by $U(\cdot, \omega)|\overline{Q}_\infty, U(\cdot, \omega)$, respectively. It follows that

$$\widehat{d}(U_r(\omega), U_{\zeta(k+1)}(\omega)) \leq 2\sum_{j=k}^{\infty} \varepsilon_j \tag{11.8.27}$$

for each $r \in [\tau(\omega), \tau(\omega) + \delta_k]$. Thus $\lim_{r \to \tau(\omega); r \geq \tau(\omega)} U(\tau(\omega), \omega)$ exists and equals some $\widehat{U}(\omega) \in S$. Therefore, by Assertion 2 of Proposition 10.1.2, we have $\tau(\omega) \in domain(U(\cdot, \omega))$. Since, as observed earlier, $U(\cdot, \omega)$ is right continuous on $domain(U(\cdot, \omega))$, it follows that

$$U_\tau(\omega) \equiv U(\tau(\omega), \omega) = \lim_{r \to \tau(\omega); r \geq \tau(\omega)} U(r, \omega) = \widehat{U}(\omega) \tag{11.8.28}$$

Hence we can let $r \downarrow \tau(\omega)$ in inequality 11.8.27 to obtain

$$\widehat{d}(U_\tau(\omega), U_{\zeta(k+1)}(\omega)) \leq 2\sum_{j=k}^{\infty} \varepsilon_j, \tag{11.8.29}$$

where $\omega \in A_{k+}^c$ is arbitrary, and where $P(A_{k+}^c) < 2^{-k/2+4}$. We conclude that $U_{\zeta(k)} \to U_\tau$ a.u. as $k \to \infty$. It follows that U_τ is an r.v., and that

$$U_\tau \equiv \widehat{U} \tag{11.8.30}$$

on the full set A.

7. We will next verify that U_τ is measurable relative to the probability subspace $\overline{L}^{(U,\tau)}$. To that end, let $s > 0$ be arbitrary. Take $k \geq 0$ so large that $\delta_k < s$. Then, by inequality 11.8.24, we have

$$\zeta_k < \tau + \delta_k < \tau + s,$$

whence the r.v. $U_{\zeta(k)}$ is measurable relative to the probability subspace $L^{(U,\tau+s)}$. Consequently, as $k \to \infty$, the limiting r.v. U_τ is measurable relative to $L^{(\tau+s)}$.

Let $f \in C(S,d)$ be arbitrary. It follows that $f(U_\tau) \in L^{(U,\tau+s)}$. Consider each regular point $t \in [0,\infty)$ of the stopping time τ. Then $t+s \in [0,\infty)$ is a regular point of the stopping time $\tau + s$. Hence $f(U_\tau)1_{(\tau+s\leq t+s)} \in L^{(U,t+s)}$. In other words, $f(U_\tau)1_{(\tau\leq t)} \in L^{(U,t+s)}$, where $s>0$ is arbitrary. Consequently,

$$f(U_\tau)1_{(\tau\leq t)} \in \bigcap_{s>0} L^{(U,t+s)} \equiv L^{(U,t+)} \equiv \overline{L}^{(U,t)}. \tag{11.8.31}$$

In view of relation 11.8.31, we can apply the defining equality 11.8.17 to the stopping time τ to obtain $f(U_\tau) \in \overline{L}^{(U_\tau)}$. Since $f \in C(S,d)$ is arbitrary, we conclude that the r.v. U_τ is measurable relative to $\overline{L}^{(U_\tau)}$. Assertion 1 is proved.

8. Let $k \geq 0$ be arbitrary. Then $\tau \leq \eta_K < \tau + \delta_k$ for sufficiently large $K \geq 0$. Hence, by inequality 11.8.27, we have

$$\widehat{d}(U_{\eta(\kappa)}, U_{\zeta(k+1)}) \leq 2\sum_{j=k}^{\infty} \varepsilon_j \tag{11.8.32}$$

on A_{k+}^c. Combining with inequality 11.8.29, we obtain

$$\widehat{d}(U_{\eta(\kappa)}, U_\tau) \leq 2\sum_{j=k}^{\infty} \varepsilon_j \tag{11.8.33}$$

on A_{k+}^c, for sufficiently large $\kappa \geq 0$. Since $P(A_{k+})$ is arbitrarily small, we conclude that $U_{\eta(\kappa)} \to U_\tau$ *a.u.* as $\kappa \to \infty$. Furthermore, equality 11.8.28 shows that $U(\cdot,\omega)$ is right continuous at $\tau(\omega)$. Assertion 2 is verified.

9. To prove Assertion 3, let $(r_i)_{0,\cdots,m}$ be an arbitrary nondecreasing sequence in $[0,\infty)$. Let $i = 0,\cdots,m$ be arbitrary. Consider the stopping time $\tau + r_i$. By Assertion 2, there exists a full set A_i such that for each $\omega \in A_i$, the function $U(\cdot,\omega)$ is right continuous at $\tau(\omega)+r_i$. Note that the set $B_{h,i} \equiv (r_i, r_i + (m+2)2^{-h})\widetilde{Q}_h$ contains at least $m+1$ members. Let $u_{h,i}$ be the i-th member of $B_{h,i}$. Suppose $i \geq 1$. Then $u_{h,i}$ is at least equal to the i-th member of $B_{h,i-1}$. Hence it is greater than the $(i-1)$-st member of $B_{h,i-1}$. In other words, $u_{h,i} > u_{h,i-1}$. Conditions (i) and (ii) of Assertion 3 are proved. Since $\eta_h + u_{h,i} \downarrow \tau + r_i$, the right continuity of U at $\tau(\omega)+r_i$ implies that $U_{\eta(h)+u(h,i)} \to U_{\tau+r(i)}$ a.u. as $h \to \infty$. Condition (iii) of Assertion 3 is verified. The lemma is proved.

Theorem 11.8.7. Feller processes are strongly Markov. *Let $x \in S$ be arbitrary. Recall the abbreviations $U \equiv U^{x,\mathbf{V}}$ and $Z \equiv Z^{x,\mathbf{V}} \equiv U^{x,\mathbf{V}}|\overline{Q}_\infty$. Recall the natural filtration $\mathcal{L} \equiv \mathcal{L}_U \equiv \{L^{(U,t)} : t \in [0,\infty)\}$ of the process U, and the natural filtration $\mathcal{L}_Z \equiv \{L^{(Z,t)} : t \in \overline{Q}_\infty\}$ of the process Z. In addition, let $\overline{\mathcal{L}} \equiv \overline{\mathcal{L}}_U \equiv \{\overline{L}^{(U,t)} : t \in [0,\infty)\}$ be the right-limit extension of $\mathcal{L}_U$.*

Then the process $U \equiv U^{x,\mathbf{V}}$ is strongly Markov relative to the right-continuous filtration $\overline{\mathcal{L}}$. More precisely, let τ be an arbitrary stopping time relative to the right-continuous filtration $\overline{\mathcal{L}} \equiv \overline{\mathcal{L}}_U$. Then

$$E(f(U_{\tau+r(0)},\cdots,U_{\tau+r(m)})|\overline{L}^{(U,\tau)})$$
$$= E(f(U_{\tau+r(0)},\cdots,U_{\tau+r(m)})|U_\tau) = F^{U(\tau),\mathbf{V}}_{r(0),\ldots,r(m)}(f), \qquad (11.8.34)$$

for each nondecreasing sequence $0 \equiv r_0 \le r_1 \le \cdots \le r_m$, *for each* $f \in C_{ub}(S^{m+1}, d^{m+1})$.

Proof. 1. Let the nondecreasing sequence $0 \equiv r_0 \le r_1 \le \cdots \le r_m$, be arbitrary. Let $f \in C_{ub}(S^{m+1}, d^{m+1})$ be arbitrary. By Lemma 11.8.6, there exists a nonincreasing sequence $(\eta_h)_{h=0,1,\ldots}$ of stopping times, relative to the filtration $\mathcal{L}_Z$, such that for each $h \ge 0$, the r.r.v. η_h has values in $\widetilde{Q}_h$, and such that

$$\tau + 2\Delta_h < \eta_h < \tau + 2\Delta_{h-1}. \qquad (11.8.35)$$

2. Consider each $i = 0,\cdots,m$. According to Assertion 3 of Lemma 11.8.6, there exists a sequence $(u_{h,i})_{h=0,1,\ldots}$ such that for each $h \ge 0$, we have (i) $u_{h,i} \in (r_i, r_i+(m+2)2^{-h}) \cap \widetilde{Q}_h$, (ii) $u_{h,0} < \cdots < u_{h,m}$, and (iii) $U_{\eta(h)+u(h,i)} \to U_{\tau+r(i)}$ *a.u.* as $h \to \infty$.

3. Consider each $Y \in \overline{L}^{(U,\tau)}$. Recall, from Definition 8.1.9, that

$$\overline{L}^{(U,\tau)} \equiv \left\{Y \in L : Y1_{(\tau\le t)} \in \overline{L}^{(U,t)} \text{ for each regular point } t \in [0,\infty) \text{ of } \tau\right\}.$$

Consider each $h \ge 0$. Take any regular point t of the r.r.v. τ. Then

$$Y1_{(\tau\le t)} \in \overline{L}^{(U,t)} = \bigcap_{u\in(t,\infty)} L^{(U,u)}$$
$$\subset \bigcap_{u\in(t,\infty)\overline{Q}(\infty)} L^{(U,u)} = \bigcap_{u\in(t,\infty)\overline{Q}(\infty)} L^{(Z,u)}. \qquad (11.8.36)$$

where the last equality is due to Lemma 11.5.2.

4. Now let $u \in \widetilde{Q}_h$ be arbitrary such that $P(\eta_{h+2} = u) > 0$. Then, on the set $(\eta_{h+2} = u)$, we have, by inequality 11.8.35,

$$\tau + 2\Delta_{h+2} < \eta_{h+2} = u.$$

Consequently, $u - 2\Delta_{h+2} > 0$. Take a regular point $t \in (u - 2\Delta_{h+2}, u)$ of the r.r.v. τ. Then, on $(\eta_{h+2} = u)$, we have

$$\tau < u - 2\Delta_{h+2} < t.$$

It follows that

$$1_{\eta(h+2)=u} = 1_{\tau\le t}1_{\eta(h+2)=u}.$$

Because $t < u$ by the choice of t, we have $Y1_{(\tau\le t)} \in L^{(Z,u)}$ according to relation 11.8.36. Hence

$$Y1_{\eta(h+2)=u} = Y1_{\tau\le t}1_{\eta(h+2)=u} \in L^{(Z,u)},$$

where $u \in \widetilde{Q}_{n(h+2)}$ is arbitrary such that $P(\eta_{h+2} = u) > 0$.

5. Next, write $k \equiv h+2$. Then

$$\begin{aligned}
&Ef(U_{\eta(k)+u(k,0)},\cdots,U_{\eta(k)+u(k,m)})Y \\
&= \sum_{u\in\widetilde{Q}(k)} Ef(U_{\eta(k)+u(k,0)},\cdots,U_{\eta(k)+u(k,m)})Y1_{(\eta(k)=u)} \\
&= \sum_{u\in\widetilde{Q}(k)} Ef(U_{u+u(k,0)},\cdots,U_{u+u(k,m)})Y1_{(\eta(k)=u)} \\
&= \sum_{u\in\widetilde{Q}(k)} EF^{U(u),\mathbf{V}}_{u(k,0),\cdots,u(k,m)}(f)Y1_{(\eta(k)=u)} \\
&= \sum_{u\in\widetilde{Q}(k)} EF^{U(\eta(k)),\mathbf{V}}_{u(k,0),\cdots,u(k,m))}(f)Y1_{(\eta(k)=u)} \\
&= EF^{U(\eta(k)),\mathbf{V}}_{u(k,0),\cdots,u(k,m))}(f)Y\sum_{u\in\widetilde{Q}(k)} 1_{(\eta(k)=u)} \\
&= EF^{U(\eta(k)),\mathbf{V}}_{u(k,0),\cdots,u(k,m))}(f)Y, \qquad (11.8.37)
\end{aligned}$$

where the third equality is by applying relation 11.8.4 to the r.v. $Y1(\eta(k)=u) \in L^{(Z,u)}$.

6. Now let $h\to\infty$. Then $U_{\eta(k)+u(h,j)} \to U_{\tau+r(j)}$ a.u., for each $j=0,\cdots,m$. Hence, in view of the boundedness and uniform continuity of the function f, the left-hand side of equality 11.8.38 converges to $Ef(U_{\tau+r(0)},\cdots U_{\tau+r(m)})Y$.

7. Consider the right-hand side, Consider first the case where $f \in C(S^{m+1}, d^{m+1})$. Then Corollary 11.7.8 says that the function $F^{x,\mathbf{V}}_{r(0),\cdots,r(m)}f$ of

$$(x,r_0,\cdots,r_m) \in S\times\{(r_0,\cdots,r_m)\in[0,\infty)^{m+1} : r_0\le\cdots\le r_m\}$$

is continuous relative to the metric $d\otimes d^{m+1}_{ecld}$, where d_{ecld} is the Euclidean metric. Hence

$$F^{U(\eta(k)),\mathbf{V}}_{u(k,0),\cdots,u(k,m))}(f)Y \to F^{U(\tau),\mathbf{V}}_{r(0),\cdots,r(m)}f \quad \text{a.u.} \qquad (11.8.38)$$

as $k\to\infty$. Therefore

$$E(F^{U(\eta(k)),\mathbf{V}}_{u(k,0),\cdots,u(k,m)}f)Y \to E(F^{U(\tau),\mathbf{V}}_{r(0),\cdots,r(m)}f)Y. \qquad (11.8.39)$$

Combining with Step 7, we see that as $k\to\infty$, equality 11.8.38 yields

$$Ef(U_{\tau+r(0)},\cdots,U_{\tau+r(m)})Y = E(F^{U(\tau),\mathbf{V}}_{r(0),\cdots,r(m)}f)Y, \qquad (11.8.40)$$

where $f \in C(S^{m+1},d^{m+1})$. is arbitrary. Since $C(S^{m+1},d^{m+1})$ is dense in C_{ub} (S^{m+1},d^{m+1}) relative to the L_1-norm with respect to each distribution, we see that equality 11.8.40 holds for each $C_{ub}(S^{m+1},d^{m+1})$, where $Y\in\overline{L}^{(U,\tau)}$ is arbitrary. Thus

$$E(f(U_{\tau+r(0)},\cdots,U_{\tau+r(m)})|\overline{L}^{(U,\tau)}) = F^{U(\tau),\mathbf{V}}_{r(0),\cdots,r(m)}f. \qquad (11.8.41)$$

In particular, equality 11.8.40 holds for each $Y \in \overline{L}(U_\tau) \subset \overline{L}^{(U,\tau)}$. Hence

$$E(f(U_{\tau+r(0)}, \cdots, U_{\tau+r(m)})|U_\tau) = F^{U(\tau),\mathbf{V}}_{r(0),\cdots,r(m)} f. \quad (11.8.42)$$

The desired equality 11.8.34 has been verified. The theorem is proved. □

11.9 Abundance of First Exit Times

In this section, let the arbitrary Feller process $((S,d),(\Omega,L,E),\{U^{y,\mathbf{V}} : y \in S\})$ be as specified, along with related objects, in Definitions 11.8.2 and 11.8.1. In particular, we have $x \in S$, $U \equiv U^{x,\mathbf{V}}$, and $\overline{\mathcal{L}} \equiv \overline{L}_U \equiv \{\overline{L}^{(U,t)} : t \in [0,\infty)\}$ is the right-limit extension of the natural filtration $\mathcal{L}_U$ of U. In addition, let $f \in C_{ub}(S,d)$ and $a_0 \in R$ be arbitrary but fixed, such that $a_0 \geq f(x)$. Let δ_f be a modulus of continuity of the function f. Let $M \geq 1$ be an arbitrary integer.

Recall Definition 10.11.1 of first exit times $\overline{\tau}_{f,a,N}(U)$ related to the right continuous filtration $\overline{\mathcal{L}} \equiv \overline{L}_U$, and recall Lemma 10.11.2 for their basic properties.

Theorem 11.9.1. Abundance of first exit times for Feller process. *There exists a countable subset G of R such that for each $a \in (a_0,\infty)G_c$, the first exit time $\overline{\tau}_{f,a,M}(U)$ exists relative to the filtration $\overline{\mathcal{L}}$. Here G_c denotes the metric complement of G in R.*

Proof. 1. Consider the process

$$U \equiv U^{x,\mathbf{V}} : [0,\infty) \times (\Omega,L,E) \to (S,d).$$

Define the process $Z \equiv U|\overline{Q}_\infty$.

2. Let $N = 0,\ldots,M-1$ be arbitrary. Consider the shifted processes U^N and Z^N. By assumption, the process U is a.u. càdlàg. Hence the process $U^N : [0,1] \times (\Omega,L,E) \to (S,d)$ is a.u. càdlàg, with some modulus of continuity of probability $\delta_{Cp} \equiv \delta_{Cp,N}$, and modulus of a.u. càdlàg $\delta_{aucl} \equiv \delta_{aucl,N}$. Therefore, by Theorem 10.4.3, the restricted process $Z^N = U^N|Q_\infty : Q_\infty \times (\Omega,L,E) \to (S,d)$ is D-regular with some modulus of D-regularity $\overline{m} \equiv (m_k)_{k=0,1,\ldots}$.

3. Let $w \in [0,M]\overline{Q}_\infty$ be arbitrary. Define the function

$$V_w \equiv \sup_{u\in[0,w]} f(U_u) : (\Omega,L,E) \to R.$$

In the following steps, we will prove that (i) V_w is a well-defined r.r.v., whence (ii) the function $V : [0,M]\overline{Q}_\infty \times (\Omega,L,E) \to R$ is a well-defined process. The motivation for this alleged process V is that the process U exits the open subset $(f < a)$ approximately when V exceeds the level a, and the fact that as a nondecreasing real-valued process with a countable parameter set, V is simpler to handle.

4. First note that Definition 10.3.2 says that there exists a full set $B \subset \bigcap_{t\in Q(\infty)} domain(U_t^N)$ with the following properties. For each $\omega \in B$, the function $U^N(\cdot,\omega)$ satisfies the right-continuity condition and the right-completeness

condition in Definition 10.3.2. Moreover, for each $k \geq 0$ and $\varepsilon_k > 0$, there exist (i′) $\delta_k \equiv \delta_{aucl,N}(\varepsilon_k) > 0$, (ii′) a measurable set $A_k \subset B$ with $P(A_k^c) < \varepsilon_k$, (iii′) an integer $h_k \geq 1$, and (iv′) a sequence of r.r.v.'s

$$0 = \tau_{k,0} < \tau_{k,1} < \cdots < \tau_{k,h(k)-1} < \tau_{k,h(k)} = 1, \tag{11.9.1}$$

such that for each $i = 0, \ldots, h_k - 1$, the function $U^N_{\tau(k,i)}$ is an r.v., and such that (v′) for each $\omega \in A_k$, we have

$$\bigwedge_{i=0}^{h(k)-1} (\tau_{k,i+1}(\omega) - \tau_{k,i}(\omega)) \geq \delta_k \tag{11.9.2}$$

with

$$d(U^N(\tau_{k,i}(\omega),\omega), U^N(\cdot,\omega)) \leq \varepsilon_k \tag{11.9.3}$$

on the interval $\theta_{k,i}(\omega) \equiv [\tau_{k,i}(\omega), \tau_{k,i+1}(\omega))$ or $\theta_{k,i}(\omega) \equiv [\tau_{k,i}(\omega), \tau_{k,i+1}(\omega)]$ depending on whether $0 \leq i \leq h_k - 2$ or $i = h_k - 1$.

5. For each $k \geq 0$, let

$$\varepsilon_k \equiv 2^{-k} \wedge 2^{-2}\delta_f(2^{-k}).$$

Then Conditions (i′–v′) in Step 4 hold. Now define $n_{-1} \equiv 0$. Inductively, for each $k \geq 0$, fix an integer $n_k \geq n_{k-1} + 1$ so large that

$$2^{-n(k)} < \delta_k \equiv \delta_{aucl,N}(\varepsilon_k).$$

6. Now let $s \in Q_\infty$ be arbitrary. We will show that $\sup_{u\in[0,s]} f(U_u^N) \in L^{(N+s)}$. If $s = 0$, then

$$\sup_{u\in[0,s]} f(U_u^N) = f(U_0^N) = Y_{N,0} \equiv f(U_N) \in L^{(N+s)}.$$

Hence we may proceed with the assumption that $s > 0$. To that end, take $k \geq 0$ so large that $s \in (0,1]Q_{n(k)}$. Consider each $\omega \in A_k$. Let

$$r \in G_{k,\omega} \equiv [0,s] \cap \bigcup_{i=0}^{h(k)-1} \theta_{k,i}(\omega) \cap Q_\infty$$

be arbitrary. Then there exists $i = 0, \ldots, h_k - 1$ such that $r \in [0,s] \cap \theta_{k,i}(\omega)$. According to 11.9.2, we have

$$|\theta_{k,i}(\omega)| \equiv \tau_{k,i+1}(\omega) - \tau_{k,i}(\omega) \geq \delta_k > 2^{-n(k)},$$

so there exists some $t \in \theta_{k,i}(\omega)Q_{n(k)}$. Since the set $Q_{n(k)}$ is discrete, we have either $t \leq s$ or $s < t$.

7. Consider first the case where $t \leq s$. Then $t \in [0,s]Q_{n(k)}$. From Step 6, note that $t, r \in \theta_{k,i}(\omega)$. Hence

$$d(U^N(\tau_{k,i}(\omega),\omega), U^N(t,\omega)) \leq \varepsilon_k \leq 2^{-2}\delta_f(2^{-k})$$

and

$$d(U^N(\tau_{k,i}(\omega),\omega),U^N(r,\omega)) \leq \varepsilon_k \leq 2^{-2}\delta_f(2^{-k}),$$

according to inequality 11.9.3. Consequently, the triangle inequality implies that

$$d(U_t^N(\omega),U_r^N(\omega)) \equiv d(U^N(t,\omega),U^N(r,\omega)) \leq 2^{-1}\delta_f(2^{-k}) < \delta_f(2^{-k}),$$

where we recall that δ_f is a modulus of continuity of the function f. Therefore

$$f(U_r^N(\omega)) < f(U_t^N(\omega)) + 2^{-k} \leq \bigvee_{u\in[0,s]Q(n(k))} f(U_u^N(\omega)) + 2^{-k}, \quad (11.9.4)$$

where the last inequality is because $t \in [0,s]Q_{n(k)}$.

8. Now consider the other case, where $t > s$. Then $s \in [r,t)Q_{n(k)} \subset \theta_{k,i}(\omega)$, because $r,t \in \theta_{k,i}(\omega)$. Thus $r,s \in \theta_{k,i}(\omega)$. Hence

$$d(U^N(\tau_{k,i}(\omega),\omega),U^N(s,\omega)) \leq \varepsilon_k \leq 2^{-2}\delta_f(2^{-k})$$

and

$$d(U^N(\tau_{k,i}(\omega),\omega),U^N(r,\omega)) \leq \varepsilon_k \leq 2^{-2}\delta_f(2^{-k}),$$

according to inequality 11.9.3. Consequently,

$$d(U_s^N(\omega),U_r^N(\omega)) \equiv d(U^N(s,\omega),U^N(r,\omega)) \leq 2^{-1}\delta_f(2^{-k}) < \delta_f(2^{-k}).$$

Therefore

$$f(U_r^N(\omega)) < f(U_s^N(\omega)) + 2^{-k} \leq \bigvee_{u\in[0,s]Q(n(k))} f(U_u^N(\omega)) + 2^{-k}, \quad (11.9.5)$$

where the last inequality is because $s \in [0,s]Q_{n(k)}$. Combining inequalities 11.9.4 and 11.9.5, we see that

$$f(U_r^N(\omega)) \leq \bigvee_{u\in[0,s]Q(n(k))} f(U_u^N(\omega)) + 2^{-k}, \quad (11.9.6)$$

where $r \in G_{k,\omega}$ is arbitrary. Since the set $G_{k,\omega}$ is dense in

$$G_\omega \equiv [0,s] \cap domain(U^N(\cdot,\omega)),$$

inequality 11.9.6 holds for each $r \in G_\omega$, thanks to the right continuity of the function $U^N(\cdot,\omega)$. In particular, it holds for each $r \in [0,s]Q_{n(k+1)} \subset G_\omega$, where $\omega \in A_k$ is arbitrary. Thus

$$\bigvee_{r\in[0,s]Q(n(k+1))} f(U_r^N) \leq \bigvee_{u\in[0,s]Q(n(k))} f(U_u^N) + 2^{-k}$$

on A_k. Consequently,

$$0 \leq \bigvee_{r\in[0,s]Q(n(k+1))} f(U_r^N) - \bigvee_{u\in[0,s]Q(n(k))} f(U_u^N) \leq 2^{-k}$$

on A_k, where $P(A_k^c) < \varepsilon_k \leq 2^{-k}$ and $k \geq 0$ is arbitrary. Since $\sum_{\kappa=0}^{\infty} 2^{-\kappa} < \infty$, it follows that the a.u.- and L_1-limit

$$Y_{N,s} \equiv \lim_{\kappa \to \infty} \bigvee_{u \in [0,s]Q(n(\kappa))} f(U_u^N) \tag{11.9.7}$$

exists as an r.r.v., where $s \in Q_\infty$ is arbitrary.

9. The equality 11.9.7 implies that for each ω in the full set $domain(Y_{N,s})$, the supremum

$$\sup_{u \in [0,s]Q(\infty)} f(U_u^N(\omega))$$

exists and is given by $Y_{N,s}(\omega)$. Now the function $U^N(\cdot,\omega)$ is right continuous for each ω in the full set B. Hence, by right continuity, we have

$$\sup_{u \in [0,s]} f(U_u^N) = \sup_{u \in [0,s]Q(\infty)} f(U_u^N) = Y_{N,s} \tag{11.9.8}$$

on the full set $B \cap domain(Y_{N,s})$. Therefore $\sup_{u \in [0,s]} f(U_u^N)$ is a well-defined r.r.v., where $N = 0, \ldots, M-1$ and $s \in Q_\infty$are arbitrary.

10. Moreover, from equality 11.9.7, we see that $Y_{N,s}$ is the L_1-limit of a sequence in $L^{(N+s)}$. Hence,

$$\sup_{u \in [0,s]} f(U_u^N) \in L^{(N+s)} \tag{11.9.9}$$

for each $s \in Q_\infty$. Equivalently,

$$\sup_{u \in [N,w]} f(U_u) \in L^{(w)} \tag{11.9.10}$$

for each $w \in [N, N+1]\overline{Q}_\infty$, where $N = 0, \ldots, M-1$ is arbitrary.

11. Now let $w \in [0, M]\overline{Q}_\infty$ be arbitrary. Then $w \in [N, N+1]\overline{Q}_\infty$ for some $N = 0, \ldots, M-1$. Write $s \equiv w - N \in Q_\infty$. There are two possibilities: (i$''$) $N = 0$, in which case $s \equiv w$ and

$$\sup_{u \in [0,w]} f(U_u) = \sup_{u \in [0,w]} f(U_u^0) = \sup_{u \in [0,s]} f(U_u^0) \in L^{(s)} = L^{(w)}, \tag{11.9.11}$$

where the set membership is by relation 11.9.9, or (ii$''$) $N \geq 1$, in which case the function

$$\sup_{u \in [0,w]} f(U_u)$$

$$= \sup_{u \in [0,1]} f(U_u) \vee \sup_{u \in [1,2]} f(U_u) \vee \cdots \vee \sup_{u \in [N-1,N]} f(U_u) \vee \sup_{u \in [N,w]} f(U_u)$$

is a member of $L^{(w)}$, thanks to relation 11.9.9 and to the filtration relation

$$L^{(1)} \subset L^{(2)} \subset \cdots \subset L^{(N)} \subset L^{(w)}.$$

Summing up, we conclude that the function

$$V_w \equiv \sup_{u \in [0,w]} f(U_u) \tag{11.9.12}$$

is an r.r.v. in $L^{(w)}$, for each $w \in [0,M]\overline{Q}_\infty$. This verifies Conditions (i–ii) in Step 3, and proves that the function $V : [0,M]\overline{Q}_\infty \times (\Omega, L, E) \to R$ is a nondecreasing real-valued process adapted to the filtration $\mathcal{L}$.

12. Since the set $\{V_w : w \in [0,M]\overline{Q}_\infty\}$ of r.r.v.'s is countable, there exists a countable subset G of R such that each point $a \in G_c$ is a continuity point of the r.r.v. V_w for each $w \in [0,M]\overline{Q}_\infty$. Here

$$G_c \equiv \{a \in R : |a-b| > 0 \quad \text{for each } b \in G\}$$

denotes the metric complement of G.

13. Consider each $a \in (a_0,\infty)G_c$. Thus a is a continuity point of the r.r.v. V_w for each $w \in [0,M]\overline{Q}_\infty$. Hence the set $(V_w < a)$ is measurable, for each $w \in [0,M]\overline{Q}_\infty$. Now let $k \geq 0$ be arbitrary. Recall that $\Delta_{n(k)} \equiv 2^{-n(k)}$ and that $\widetilde{Q}_{n(k)} \equiv \{0, \Delta_{n(k)}, 2\Delta_{n(k)}, \ldots\}$. Define the r.r.v.

$$\eta_k \equiv \sum_{u \in (0,M)\widetilde{Q}(n(k))} u 1_{(V(u) \geq a)} \prod_{w \in (0,u)\widetilde{Q}(n(k))} 1_{(V(w)<a)}$$
$$+ M \prod_{w \in (0,M]\widetilde{Q}(n(k))} 1_{(V(w)<a)}. \tag{11.9.13}$$

In words, η_k is the first time in $(0,M]\widetilde{Q}_{n(k)}$ for the real-valued nondecreasing process V to exit the interval $(-\infty,a)$, with η_k set to M if no such time exists. Note that η_k is an r.r.v. with values in the finite set $(0,M]\widetilde{Q}_{n(k)}$. Moreover, from the defining equality 11.9.13, we see that $(\eta_k = u) \in L^{(u)}$ for each $u \in (0,M]\widetilde{Q}_{n(k)}$. Thus η_k is a simple stopping time relative to the filtration $\mathcal{L}$.

14. Using the monotonicity of the process V, we will next verify that $\eta_k - \Delta_{n(k)} \leq \eta_{k+1} - \Delta_{n(k+1)}$. To that end, let

$$\omega \in \bigcap_{j=0}^{\infty} domain(\eta_j)$$

be arbitrary, and suppose, for the sake of a contradiction, that $w \equiv \eta_k(\omega) - \Delta_{n(k)} > \eta_{k+1}(\omega) - \Delta_{n(k+1)}$. Then $w \in (0,M)\widetilde{Q}_{n(k)}$ and $w < \eta_k(\omega)$. The assumption that $V_w(\omega) \geq a$ then implies, by the defining equality 11.9.13, that $\eta_k(\omega) \leq w$, a first contradiction. Hence $V_w(\omega) < a$. Moreover, $w \in (0,M)\widetilde{Q}_{n(k+1)}$ and $w \geq \eta_{k+1}(\omega)$. Hence $M > \eta_{k+1}(\omega)$ and so $V_{\eta(k+1)}(\omega) \geq a$, by the defining equality 11.9.13 applied to $k+1$. Therefore, due to the monotonicity of the process V, we obtain $V_w(\omega) \geq V_{\eta(k+1)}(\omega) \geq a$, again a contradiction. We conclude that $\eta_k - \Delta_{n(k)} \leq \eta_{k+1} - \Delta_{n(k+1)}$.

15. In the opposite direction, suppose $\eta_{k+1}(\omega) > \eta_k(\omega)$. Since $\eta_k(\omega) \in \widetilde{Q}_{n(k+1)}$, it follows from the defining equality 11.9.13, applied to $k+1$, that

$V_{\eta(k)}(\omega) \leq a$. At the same time, $M > \eta_k(\omega)$, whence $V_{\eta(k)}(\omega) > a$ by equality 11.9.13. This, again, is a contradiction. Thus we conclude that $\eta_{k+1}(\omega) \leq \eta_k(\omega)$.

16. Combining, we see that on the full set $\bigcap_{j=0}^{\infty} domain(\eta_j)$, we have

$$\eta_k - \Delta_{n(k)} \leq \eta_{k+1} - \Delta_{n(k+1)} < \eta_{k+1} \leq \eta_k \tag{11.9.14}$$

and, iterating inequality 11.9.14,

$$\eta_k - \Delta_{n(k)} \leq \eta_\kappa - \Delta_{n(\kappa)} \leq \eta_\kappa \leq \eta_k \tag{11.9.15}$$

for each $\kappa \geq k+1$. Since $\Delta_{n(k)} \equiv 2^{-n(k)} \to 0$, it follows that $\eta_\kappa \downarrow \tau$ uniformly on the full set $\bigcap_{j=0}^{\infty} domain(\eta_j)$, for some r.r.v. τ with

$$\eta_k - \Delta_{n(k)} \leq \tau \leq \eta_k, \tag{11.9.16}$$

where $k \geq 0$ is arbitrary. Consequently, Assertion 2 of Proposition 10.11.3 implies that τ is a stopping time relative to the right continuous filtration $\overline{\mathcal{L}}$.

17. It remains to verify that the stopping time τ is a first exit time in the sense of Definition 10.11.1. In view of the right continuity of the filtration $\overline{\mathcal{L}}$, Assertion 1 of Lemma 11.8.6 says that U_τ is an r.v. Assertion 2 of Lemma 11.8.6 then says that for each ω in some full set A, the function $U(\cdot,\omega)$ is right continuous at $\tau(\omega)$. Hence, for each

$$\omega \in \widehat{D} \equiv domain(U_\tau) \cap \bigcap_{j=0}^{\infty} domain(\eta_j) \cap A,$$

we have $U(\eta_k(\omega),\omega) \to U(\tau(\omega),\omega)$.

18. Consider each $\omega \in \widehat{D}$. Let $t \in domain(U(\cdot,\omega)) \cap [0,\tau(\omega))$ be arbitrary. Then $t < \tau(\omega) - 2^{-n(k)}$ for some sufficiently large $k \geq 0$. In view of inequality 11.9.16, it follows that

$$t < \tau(\omega) - 2^{-n(k)} \leq \eta_k(\omega) - \Delta_{n(k)}.$$

Hence $w \equiv \eta_k(\omega) - \Delta_{n(k)} \in (t, \eta_k(\omega))\widetilde{Q}_{n(k)}$. Consequently, the defining equality 11.9.13 implies that $V_w(\omega) < a$. Therefore

$$f(U(t,\omega)) \leq \sup_{u\in[0,w]} f(U_u(\omega)) \equiv V_w(\omega) < a,$$

where $t \in domain(U(\cdot,\omega)) \cap [0,\tau(\omega))$ is arbitrary, and where ω is arbitrary in the full set $\widehat{D}$. Condition (i) of Definition 10.11.1 is proved for the stopping time τ to be the first exit time in $[0,M]$ of the open subset $(f < a)$.

19. Proceed to verify Condition (ii) of Definition 10.11.1. To that end, let $k \geq 0$ be arbitrary. The defining equality 11.9.13 says that the stopping time η_k has values in the finite set $(0,M]\widetilde{Q}_{n(k)}$. Let μ_k denote the number of elements in this finite set $(0,M]\widetilde{Q}_{n(k)}$. Consider each $t \in (0,M]\widetilde{Q}_{n(k)}$. By assumption, we have $a \in G_c$. Hence a is a continuity point of V_t. Therefore there exists $\varepsilon'_k < \mu_k^{-1}\alpha_k$ so small that $P(a \leq V_t < a + \varepsilon'_k) < \mu_k^{-1}2^{-k}$. Define

$$D_k \equiv \bigcap_{t \in (0,M]\widetilde{Q}(n(k))} (a \le V_t < a + \varepsilon'_k)^c.$$

Then

$$P(D_k^c) < \mu_k \mu_k^{-1} 2^{-k} = 2^{-k}.$$

Hence

$$D \equiv \bigcap_{j=0}^{\infty} \bigcup_{k=j}^{\infty} D_k$$

is a full set.

20. Next, take a sequence $(\alpha_k)_{k=0,1,\ldots}$ such that $\alpha_k \downarrow 0$. Let $k \ge 0$ be arbitrary. Define the set

$$A_k \equiv (\tau < M - \alpha_k) \cap (f(U_\tau) < a - \alpha_k).$$

Then $A_0 \subset A_1 \subset \cdots$. Consider each

$$\omega \in \overline{D} \equiv domain(U_\tau) \cap \bigcap_{j=0}^{\infty} domain(\eta_j) \cap A \cap D \cap \bigcup_{j=0}^{\infty} A_j.$$

Then $\omega \in A_j$ for some $j \ge 0$. Hence $\tau(\omega) < M - \alpha_j$ and $f(U_\tau(\omega)) < a - \alpha_j$. Since $\omega \in A$, the function $U(\cdot,\omega)$ is right continuous at $\tau(\omega)$. Hence there exists $k \ge j$ so large that for each $\kappa \ge k$, we have $\tau(\omega) \le \eta_\kappa(\omega) < M - \alpha_j$ and $f(U_{\eta(\kappa)}(\omega)) < a - \alpha_j$.

21. Consider each $\kappa \ge k$. Then, since $\eta_\kappa(\omega) < M$, we have $V_{\eta(\kappa)}(\omega) \ge a$ according to the defining equality 11.9.13. At the same time, because $\omega \in D \subset \bigcup_{k'=\kappa}^{\infty} D_{k'}$, there exists $k' \ge \kappa$ such that

$$\omega \in D_{k'} \equiv \bigcap_{t \in (0,M]\widetilde{Q}(n(k'))} (a \le V_t < a + \varepsilon'_{k'})^c$$

$$\subset \bigcap_{t \in (0,M]\widetilde{Q}(n(\kappa))} (a \le V_t < a + \varepsilon'_{k'})^c. \tag{11.9.17}$$

Because $\eta_\kappa(\omega) < M$, we have $V_{\eta(\kappa)}(\omega) \ge a$ according to the defining equality 11.9.13. On the other hand, $t \equiv \eta_\kappa(\omega) \in (0,M]\widetilde{Q}_{n(\kappa)}$. Hence $V_t(\omega) < a$ or $a + \varepsilon'_{k'} \le V_t(\omega)$ according to relation 11.9.17. Combining, we infer that $a + \varepsilon'_{k'} \le V_t(\omega)$. In other words,

$$\sup_{u \in [0,t]\overline{Q}(\infty)} f(U_u(\omega)) \ge a + \varepsilon'_{k'}.$$

Hence there exists $u_\kappa \in [0,t]\overline{Q}_\infty \equiv [0,\eta_\kappa(\omega)]\overline{Q}_\infty$ such that

$$f(U_{u(\kappa)}(\omega)) \ge a, \tag{11.9.18}$$

where $\kappa \ge k$ is arbitrary.

22. Suppose, for the sake of a contradiction, that $u_\kappa \in [0, \tau(\omega))$. Then $f(U_{u(\kappa)}(\omega)) < a$, according to Condition (i) established previously, which is a contradiction to inequality 11.9.18. Therefore $u_\kappa \geq \tau(\omega)$, whence $u_\kappa \in [\tau(\omega), \eta_\kappa(\omega)]\overline{Q}_\infty$. Consequently, $u_\kappa \to \tau(\omega)$ with $u_\kappa \geq \tau(\omega)$ and $f(U_{u(\kappa)}(\omega)) \geq a$, as $\kappa \to \infty$. Since $\omega \in A$, the function $U(\cdot, \omega)$ is right continuous at $\tau(\omega)$. It follows that $f(U_\tau(\omega)) \geq a$, where ω is arbitrary in the full set $\overline{D}$. Condition (ii) of Definition 10.11.1 is also verified. Accordingly, $\overline{\tau}_{f,a,M}(U) \equiv \tau$ is the first exit time in $[0, M]$, by the process U, of the open subset $(f < a)$. The theorem is proved. □

11.10 First Exit Time for Brownian Motion

Definition 11.10.1. Specification of a Brownian motion. In this section, let $m \geq 1$ be arbitrary. Let $B : [0, \infty) \times (\Omega, L, E) \to R^m$ be an arbitrary but fixed Brownian motion in the sense of Definition 9.4.1, where $B_0 = 0$. For each $x \in R^m$, define the process $B^x \equiv x + B : [0, \infty) \times (\Omega, L, E) \to R^m$ with initial state x. □

In the following, let d denote the Euclidean metric on R. We will prove that the triple

$$((R^m, d^m), (\Omega, L, E), \{B^x : x \in R^m\})$$

is a Feller process relative to a certain Feller semigroup $\mathbf{V}$. Then the Brownian motion can inherit Theorem 11.9.1, with an abundance of first exit times. These exit times are useful in the probability analysis of classical potential theory.

Definition 11.10.2. Specification of some independent standard normal r.v.'s. Let $U, U_1, U_2, \ldots$ be an arbitrary independent sequence of R^m-valued standard normal r.v.'s on some probability space $(\widetilde{\Omega}, \widetilde{L}, \widetilde{E})$. Thus each of these r.v.'s has distribution $\Phi_{0,I}$, where I is the $m \times m$ identical matrix. The only purpose for these r.v.'s is to provide compact notations for the normal distributions. Such a sequence can be constructed by taking a product probability space and using Fubini's Theorem. □

Proceed to construct the desired Feller semigroup $\mathbf{V}$.

Lemma 11.10.3. Brownian semigroup. *Let $t \geq 0$ be arbitrary. Define the function*

$$V_t : C_{ub}(R^m, d^m) \to C_{ub}(R^m, d^m)$$

by

$$V_t(f)(x) \equiv V_t^x(f) \equiv \widetilde{E} f(x + \sqrt{t}U) = \int_{R^m} \Phi_{0,I}(du) f(x + \sqrt{t}u) \qquad (11.10.1)$$

for each $x \in R^m$, for each $f \in C_{ub}(R^m, d^m)$. Then the family $\mathbf{V} \equiv \{V_t : t \in [0, \infty)\}$ is a Feller semigroup in the sense of Definition 11.7.2. For lack of a better name, we will call $\mathbf{V}$ the Brownian semigroup.

Proof. We need to verify the conditions in Definition 11.7.2 for the family $\mathbf{V}$ to be a Feller semigroup. First, note from equality 11.10.1 that for each $t \in [0,\infty)$ and for each $x \in R^m$, the function V_t^x is the distribution on (R^m, d^m) induced by the r.v. $x + \sqrt{t}U$. In particular, V_t^x is a distribution on (R^m, d^m).

1. Let $N \geq 1$, $t \in [0,N]$, and $f \in C_{ub}(R^m, d^m)$ be arbitrary with a modulus of continuity δ_f and with $|f| \leq 1$. Consider the function $V_t f$ defined in equality 11.10.1. Let $\varepsilon > 0$ be arbitrary. Let $x, y \in R^m$ be arbitrary with $|x - y| < \delta_f(\varepsilon)$. Then

$$|(x + \sqrt{t}U) - (y + \sqrt{t}U)| \leq |x - y| < \delta_f(\varepsilon).$$

Hence

$$\begin{aligned}|V_t(f)(x) - V_t(f)(y)| &\equiv |\widetilde{E} f(x + \sqrt{t}U) - \widetilde{E} f(y + \sqrt{t}U))| \\ &\leq \widetilde{E}|f(x + \sqrt{t}U) - f(y + \sqrt{t}U))| \leq \widetilde{E}\varepsilon = \varepsilon.\end{aligned}$$

Thus $V_t(f)$ has the same modulus of continuity δ_f as the function f. In other words, the family $\mathbf{V}$ has a modulus of smoothness $\alpha_{\mathbf{V}} \equiv (\iota, \iota, \ldots)$, where ι is the identity operation ι. The smoothness condition, Condition 1 of Definition 11.7.2, has been verified for the family $\mathbf{V}$. In particular, we have a function $V_t : C_{ub}(R^m, d^m) \to C_{ub}(R^m, d^m)$.

2. We will next prove that $V_t^x(f)$ is continuous in t. To that end, let $\varepsilon > 0$, $x \in R^m$ be arbitrary. Note that the standard normal r.v. U is an r.v. Hence there exists $a \equiv a(\varepsilon) > 0$ so large that

$$\widetilde{E} 1_{(|U| > a(\varepsilon))} < 2^{-1}\varepsilon. \tag{11.10.2}$$

Note also that $\sqrt{t}$ is a uniformly continuous function of $t \in [0,\infty)$, with some modulus of continuity δ_{sqrt}. Let $t, s \geq 0$ be such that

$$|t - s| < \delta_{\mathbf{V}}(\varepsilon, \delta_f) \equiv \delta_{sqrt}(a(\varepsilon)^{-1}\delta_f(2^{-1}\varepsilon)). \tag{11.10.3}$$

Then $|\sqrt{t} - \sqrt{s}| < a(\varepsilon)^{-1}\delta_f(2^{-1}\varepsilon)$. Hence, on the measurable subset $(|U| \leq a(\varepsilon))$ of $(\widetilde{\Omega}, \widetilde{L}, \widetilde{E})$, we have

$$|(x + \sqrt{t}U) - (x + \sqrt{s}U)| = |\sqrt{t} - \sqrt{s}| \cdot |U| < \delta_f(2^{-1}\varepsilon),$$

whence $|f(x + \sqrt{t}U) - f(x + \sqrt{s}U)| < 2^{-1}\varepsilon$. Therefore

$$\begin{aligned}|V_t^x(f) - V_s^x(f)| &\leq \widetilde{E}|f(x + \sqrt{t}U) - f(x + \sqrt{s}U)| \\ &\leq \widetilde{E}|f(x + \sqrt{t}U) - f(x + \sqrt{s}U)| 1_{(|U| \leq a(\varepsilon))} + \widetilde{E} 1_{(|U| > a(\varepsilon))} \\ &\leq 2^{-1}\varepsilon \widetilde{E} 1_{(|U| \leq a(\varepsilon))} + \widetilde{E} 1_{(|U| > a(\varepsilon))} \\ &\leq 2^{-1}\varepsilon + 2^{-1}\varepsilon = \varepsilon.\end{aligned} \tag{11.10.4}$$

Thus the function $V_{t\cdot}^x f$ is uniformly continuous in $t \in [0,\infty)$, with a modulus of continuity $\delta_{\mathbf{V}}(\cdot, \delta_f)$ independent of x. In the special case where $s = 0$, we have $V_s^x(f) = \widetilde{E} f(x + \sqrt{0}U) = f(x)$, whence inequality 11.10.4 yields

$$|V_t(f) - f| \leq \varepsilon$$

for each $t \in [0, \delta_{\mathbf{V}}(\varepsilon, \delta_f))$. The strong-continuity condition, Condition 3 in Definition 11.7.2, is thus verified for the family $\mathbf{V}$, with $\delta_{\mathbf{V}}$ being the modulus of strong continuity.

3. Proceeding to Condition 2, the semigroup property, let $t, s \geq 0$ be arbitrary. First assume that $s > 0$. Then $(t+s)^{-\frac{1}{2}}(\sqrt{t}W_1 + \sqrt{s}W_2)$ is a standard normal r.v. with values in R^m. Hence, for each $x \in R^m$, we have

$$\begin{aligned}
V_{t+s}^x(f) &= \widetilde{E}f(x + \sqrt{t+s}U) = \widetilde{E}f(x + \sqrt{t}W_1 + \sqrt{s}W_2) \\
&= \int_{w(1)\in R^m} \int_{w(2)\in R^m} \varphi_{0,t}(w_1)\varphi_{0,s}(w_2) f(x + w_1 + w_2) dw_1 dw_2 \\
&= \int_{w(2)\in R^m} \left\{ \int_{w(1)\in R^m} \varphi_{0,t}(w_1) f(x + w_1 + w_2) dw_1 \right\} \varphi_{0,s}(w_2) dw_2 \\
&= \int_{w(2)\in R^m} \{(V_t f)(x + w_2)\} \varphi_{0,s}(w_2) dw_2 \\
&= V_s^x(V_t f),
\end{aligned} \tag{11.10.5}$$

provided that $s \in (0, \infty)$. At the same time, inequality 11.10.4 shows that both ends of equality 11.10.5 are continuous functions of $s \in [0, \infty)$. Hence equality 11.10.5 can be extended, by continuity, to

$$V_{t+s}^x(f) = V_s^x(V_t f)$$

for each $s \in [0, \infty)$. It follows that $V_{s+t} = V_s V_t$ for each $t, s \in [0, \infty)$. Thus the semigroup property, Condition 2 in Definition 11.7.2, is proved for the family $\mathbf{V}$.

4. It remains to verify the nonexplosion condition, Condition 4 in Definition 11.7.2. To that end, let $N \geq 1$, $t \in [0, N]$, and $\varepsilon > 0$ be arbitrary. Let $n \geq \kappa_{\mathbf{V},N}(\varepsilon) \equiv N^{-1/2}a(\varepsilon)$ be arbitrary. Define the functions

$$h_{x,n} \equiv (1 \wedge (1 + n - d^m(\cdot, x))_+ \in C(R^m, d^m)$$

and

$$\overline{h}_{x,n} \equiv 1 - h_{x,n} \in C_{ub}(R^m, d^m).$$

Then

$$\begin{aligned}
V_t^x \overline{h}_{x,n} &\equiv \widetilde{E}\overline{h}_{x,n}(x + \sqrt{t}U)) \leq \widetilde{E}1_{(d^m(x+\sqrt{t}U, x)\geq n)} \\
&= \widetilde{E}1_{(|\sqrt{t}U|\geq n)} \leq \widetilde{E}1_{(|\sqrt{N}U|\geq n))} \leq \widetilde{E}1_{(|U|\geq a(\varepsilon))} < 2^{-1}\varepsilon < \varepsilon,
\end{aligned}$$

where the next-to-last inequality is inequality 11.10.2. Condition 4 in Definition 11.7.2 is verified for the family $\mathbf{V}$.

5. Summing up, all the conditions in Definition 11.7.2 hold for the family $\mathbf{V}$ to be a Feller semigroup with state space (R^m, d^m). □

Theorem 11.10.4. Brownian motion in R^m as a Feller process. *Let* $\mathbf{V}$ *be the* m*-dimensional Brownian semigroup constructed in Lemma 11.10.3. Then the triple* $((R^m, d^m), (\Omega, L, E), \{B^x : x \in R^m\})$, *where* $B^x \equiv x + B$ *for each* $x \in R^m$, *is a* Feller process *with the Feller semigroup* $\mathbf{V}$, *in the sense of Definition 11.7.6.*

Proof. 1. Let $x \in R^m$ be arbitrary. Then, according to Corollary 10.10.6, the process $B : [0,\infty) \times (\Omega, L, E) \to (R^m, d^m)$ is time-uniformly a.u. continuous. It follows that the process $B^x \equiv x + B$ is time-uniformly a.u. càdlàg. Let $F^{x,\mathbf{V}}$ be the family of f.j.d.'s generated by the initial state x and Feller semigroup $\mathbf{V}$, as defined in 11.7.4 and constructed in Assertion 4 of Theorem 11.7.7. Let

$$((R^m, d^m), (\widehat{\Omega}, \widehat{L}, \widehat{E}), \{U^{y,\mathbf{V}} : y \in R^m\})$$

be an arbitrary Feller process, with the Feller semigroup $\mathbf{V}$. Then the family $F^{x,\mathbf{V}}$ gives the marginal distributions of the process

$$U^{x,\mathbf{V}} : [0,\infty) \times (\widehat{\Omega}, \widehat{L}, \widehat{E}) \to (R^m, d^m).$$

2. Now let $n \geq 1$, $f \in C_{ub}((R^m)^n, (d^m)^n)$, and nondecreasing sequence $r_1 \leq \cdots \leq r_n$ in $\overline{Q}_\infty$ be arbitrary. We proceed to prove, by induction on $n \geq 1$, that

$$F^{x,\mathbf{V}}_{r(1),\ldots,r(n)} f = Ef(B^x_{r(1)}, \ldots, B^x_{r(n)}). \tag{11.10.6}$$

To start, suppose $n = 1$. Then, by the defining equality 11.10.1 for $\mathbf{V}$, we have

$$\begin{aligned} F^{x,\mathbf{V}}_{r(1)} f &= \int V^x_{r(1)}(dx_1) f(x_1) = V_{r(1)}(f)(x) = \widetilde{E} f(x + \sqrt{r_1} U) \\ &= Ef(x + B_{r(1)}) \equiv Ef(B^x_{r(1)}), \end{aligned}$$

which proves equality 11.10.6 for $n = 1$.

3. Inductively, suppose $n \geq 2$ and equality 11.10.6 has been proved for $n - 1$. Define, for each $(x_1, \ldots, x_{n-1}) \in R^n$,

$$\begin{aligned} g(x_1, \ldots, x_{n-1}) &\equiv \widetilde{E} f(x_1, \ldots, x_{n-1}, x_{n-1} + \sqrt{r_n - r_{n-1}} U) \\ &= V^{x(n-1)}_{r(n)-r(n-1)} f(x_1, \ldots, x_{n-1}, \cdot) \\ &\equiv \int V^{x(n-1)}_{r(n)-r(n-1)}(dx_n) f(x_1, \ldots, x_{n-1}, x_n), \end{aligned} \tag{11.10.7}$$

where the second equality is by the defining equality 11.10.1 for $\mathbf{V}$, and where the third equality is by equality 11.7.2 of Definition 11.7.4. Then

$$\begin{aligned} &Ef(B^x_{r(1)}, \ldots, B^x_{r(n-1)}, B^x_{r(n)}) \\ &= Ef(B^x_{r(1)}, \ldots, B^x_{r(n-1)}, B^x_{r(n-1)} + (B^x_{r(n)} - B^x_{r(n-1)})) \end{aligned}$$

$$= E\widetilde{E}f(B^x_{r(1)},\ldots,B^x_{r(n-1)},B^x_{r(n-1)}+\sqrt{r_n-r_{n-2}}U)$$
$$= Eg(B^x_{r(1)},\ldots,B^x_{r(n-1)})$$
$$= F^{x,\mathbf{V}}_{r(1),\ldots,r(n-1)}g$$
$$= \int V^x_{r(1)}(dx_1)\cdots\int V^{x(n-2)}_{r(n-1)-r(n-2)}(dx_{n-1})g(x_1,\ldots,x_{n-1})$$
$$= \int V^x_{r(1)}(dx_1)\cdots\int V^{x(n-2)}_{r(n-1)-r(n-2)}(dx_{n-1})$$
$$\int V^{x(n-1)}_{r(n)-r(n-1)}(dx_n)f(x_1,\ldots,x_{n-1},x_n)$$
$$= F^{x,\mathbf{V}}_{r(1),\ldots,r(n)}f,$$

where the third equality is by the first part of equality 11.10.7, where the fourth equality is by equality 11.10.6 in the induction hypothesis, where the fifth and the last equalities are by equality 11.7.2 in Definition 11.7.4, and where the sixth equality is again thanks to equality 11.10.7.

4. Induction is completed. Equality 11.10.6 has been proved for an arbitrary nondecreasing sequence $r_1 \leq \cdots \leq r_n$ in $\overline{Q}_\infty$. By consistency, it follows that

$$F^{x,\mathbf{V}}_{r(1),\ldots,r(n)}f = Ef(B^x_{r(1)},\ldots,B^x_{r(n)}) \tag{11.10.8}$$

for each sequence $r_1,\ldots,r_n$ in $\overline{Q}_\infty$. Since $\overline{Q}_\infty$ is dense in $[0,\infty)$, equality 11.10.8 extends, by continuity, to each sequence $r_1,\ldots,r_n$ in $[0,\infty)$. Thus $F^{x,\mathbf{V}}$ is the family of marginal distributions of the process B^x, where $x \in R^m$ is arbitrary.

5. In addition, $F^{x,\mathbf{V}}$ gives the marginal distributions of the process $U^{x,\mathbf{V}}$, for each $x \in R^m$. At the same time, Condition 2 of Definition 11.7.6, applied to the Feller process $((R^m,d^m),(\widehat{\Omega},\widehat{L},\widehat{E}),\{U^{x,\mathbf{V}} : x \in R^m\})$, yields the continuity of the function $F^{*,\mathbf{V}}_{s(1),\ldots,s(n)}f \in C_{ub}(R^m,d^m)$. Said Condition 2 is thus satisfied also by the triple $((R^m,d^m),(\Omega,L,E),\{B^x : x \in R^m\})$ relative to the semigroup $\mathbf{V}$.

6, Now let $x \in R^m$ be arbitrary. Then $B^x \equiv x+B$ is a.u. continuous, and hence a.u. càdlàg. Thus the triple $((R^m,d^m),(\Omega,L,E),\{B^x : x \in R^m\})$ satisfies all the conditions of Definition 11.7.6. Accordingly, it is a Feller process with the Feller semigroup $\mathbf{V}$. □

Definition 11.10.5. Specification of some filtrations. Let $\mathcal{L} \equiv \mathcal{L}_B \equiv \{L^{(B,t)} : t \in [0,\infty)\}$ be the natural filtration of the process B. Let $\overline{\mathcal{L}} \equiv \overline{\mathcal{L}}_B \equiv \{\overline{L}^{(B,t)} : t \in [0,\infty)\}$ be the right-limit extension of $\mathcal{L}_B$. □

Theorem 11.10.6. Brownian motion is strongly Markov. *For each $x \in R^m$, the process B^x is strongly Markov relative to the right continuous filtration $\overline{\mathcal{L}}$. More precisely, let τ be an arbitrary stopping time relative to $\overline{\mathcal{L}} \equiv \overline{\mathcal{L}}_B$. Then*

$$E\Big(f(B^x_{\tau+r(0)},\ldots,B^x_{\tau+r(m)})|\overline{L}^{(B,\tau)}\Big)$$
$$= E(f(B^x_{\tau+r(0)},\ldots,B^x_{\tau+r(m)})|B^x_\tau) = F^{B^x(\tau),\mathbf{V}}_{r(0),\ldots,r(m)}(f), \tag{11.10.9}$$

for each nondecreasing sequence $0 \equiv r_0 \leq r_1 \leq \cdots \leq r_m$*, for each* $f \in C_{ub}(S^{m+1}, d^{m+1})$.

Proof. The present theorem is an immediate corollary of Theorem 11.8.7 $\square$

Theorem 11.10.7. Abundance of first exit times for Brownian motion. *Let* $x \in R^m$ *be arbitrary. Let* $f \in C_{ub}(R^m, d^m)$ *and* $a_0 \in R$ *be arbitrary, such that* $a_0 \geq f(x)$*. Then there exists a countable subset* H *of* R *such that for each* $a \in (a_0, \infty)H_c$ *and for each* $M \geq 1$*, the first exit time* $\overline{\tau}_{f,a,M}(U)$ *exists relative to the filtration* $\overline{\mathcal{L}}$*. Here* H_c *denotes the metric complement of* H *in* R*.*

Proof. Let $M \geq 1$ be arbitrary. Theorem 11.9.1 says that there exists a countable subset G_M of R such that for each $a \in (a_0, \infty)G_{M,c}$, the first exit time $\overline{\tau}_{f,a,M}(B^x)$ exists relative to the filtration $\overline{\mathcal{L}}$. Then the countable set $H \equiv \bigcup_{M=1}^{\infty} G_M$ has the desired properties. $\square$

An application of Theorem 11.10.7 is where $f \equiv \overline{a} \wedge |\cdot| \in C_{ub}(R^m, d^m)$ for some $\overline{a} > 0$, and where $x \in R^m$ and $a_0 \in R$ are such that $|x| < a_0 < \overline{a}$. Then, for each $a \in (a_0, \overline{a})H_c$, the stopping time $\overline{\tau}_{f,a,M}(B^x)$ is the first exit time, on or before M, for the process B^x to exit the sphere in R^m of center 0 and radius a.

We end this section, and this book, with the remark, without giving any proof, that in the case of a Brownian motion, (i) the exceptional set H can be taken to be empty and (ii) $\overline{\tau}_{f,a}(B^x) \equiv \lim_{M \to \infty} \overline{\tau}_{f,a,M}(B^x)$ exists a.u. Hence, if the starting state x is in the sphere with center $0 \in R^m$ and arbitrary radius a, then, a.s., the process B^x exits said sphere at some finite time $\overline{\tau}_{f,a}(B^x)$.

Appendices

Several times in this book, we apply the method of change of integration variables for the calculation of Lebesgue integrals in R^n. In these two appendices, we prove those theorems, and only those theorems, that justify the several applications. In the proofs of the following theorems, we assume no prior knowledge of Euclidean geometry. We will use some matrix algebra and calculus, and will use theorems on metric spaces and integration spaces that are treated in the first four chapters of this book. Note that the following theorems are used only in Chapter 5 and later.

Appendix A

Change of Integration Variables

In this appendix, let $n \geq 1$ be an arbitrary, but fixed, integer. As usual, we write a_b and $a(b)$ interchangeably for arbitrary expressions a, b. We write AB and $A \cap B$ interchangeably for arbitrary sets A, B. Recall that for each locally compact metric space (S, d), we write $C(S, d)$ or $C(S)$ for the space of continuous functions on S with compact support. At some small risk of confusion, for real-valued expressions a, b, and c, we will write $a = b \pm c$ to mean $|a - b| \leq c$.

Definition A.0.1. Some notations for Lebesgue integrations and for matrices. Let μ_1 and μ denote the measures with respect to the Lebesgue integrations $\int \cdot dx$ and $\int \cdots \int \cdot dx_1 \dots dx_n \equiv \left(\int \cdot dx\right)^{\otimes n}$ in R^1 and R^n, respectively. Unless otherwise specified, all measure-theoretic terms will be relative to these Lebesgue integrations. Moreover, when the risk of confusion is low, we will write $\int_{x \in A} f(x)dx$ for $\int \cdots \int 1_A(x_1, \dots, x_n) f(x_1, \dots, x_n) dx_1 \dots dx_n$, for each integrable function f and measurable integrator 1_A on R^n.

Let $m \geq 1$ be arbitrary. We will write $M_{n \times m}$ for the set of all $n \times m$ matrices. For each $\overline{\alpha} \in M_{n \times m}$, define $|\overline{\alpha}| \equiv \bigvee_{i=1}^{n} \bigvee_{j=1}^{m} |\overline{\alpha}_{i,j}|$ and $\|\overline{\alpha}\| \equiv \sqrt{\sum_{i=1}^{n} \sum_{j=1}^{m} |\overline{\alpha}_{i,j}|^2}$. Note that $|\cdot|$ and $\|\cdot\|$ are equivalent norms for $M_{n \times m}$ because $|\overline{\alpha}| \leq \|\overline{\alpha}\| \leq \sqrt{nm}|\overline{\alpha}|$. Unless otherwise indicated, the space $M_{n \times m}$ is equipped with the norm $|\cdot|$, and with the metric associated with this norm. Continuity on $M_{n \times m}$ means continuity relative to the latter norm and metric.

Each point $x \equiv (x_1, \dots, x_n) \in R^n$, will be identified with the $n \times 1$ matrix $\begin{bmatrix} x_1 \\ \vdots \\ x_n \end{bmatrix}$, also called a column vector. Thus $|x| \equiv \bigvee_{i=1}^{n} |x_i|$ and $\|x\| \equiv \sqrt{\sum_{i=1}^{n} |x_i|^2}$. Moreover, $|\cdot|$ and $\|\cdot\|$ are equivalent norms on R^n.

The *determinant* of $\overline{\alpha}$ is defined as

$$\det \overline{\alpha} \equiv \sum_{\sigma \in \Pi} \operatorname{sign}(\sigma) \overline{\alpha}_{1,\sigma(1)} \overline{\alpha}_{2,\sigma(2)} \dots \overline{\alpha}_{n,\sigma(n)}, \tag{A.0.1}$$

where Π is the set of all permutations on $\{1,\ldots,n\}$, and where $\text{sign}(\sigma)$ is $+1$ or -1 depending on whether σ is an even or odd permutation, for each $\sigma \in \Pi$.

Suppose $n \geq 2$. For each $i, j = 1, \ldots, n$, let $\overline{\alpha}'_{i,j}$ denote the $(n-1) \times (n-1)$ matrix obtained by deleting the i-th row and j-th column from $\overline{\alpha}$. Then the number $\det \overline{\alpha}'_{i,j}$ is called the (i, j)-*minor* of $\overline{\alpha}$. The number $(-1)^{i+j} \det \overline{\alpha}'_{i,j}$ is called the (i, j)-*cofactor* of $\overline{\alpha}$.

Finally, if $\varphi(x)$ is a given expression of x for each member x of a set G, then we write $x \to \varphi(x)$ for the function $\overline{\varphi}$ defined on G by $\overline{\varphi}(x) \equiv \varphi(x)$ for each $x \in G$. For example, the expression $x \to x + \sin x$ will stand for the function $\overline{\varphi}$ defined on R by $\overline{\varphi}(x) \equiv x + \sin x$ for each $x \in R$. Likewise, $\overline{\alpha} \to \det \overline{\alpha}$ will stand for the function $\overline{\varphi}$ defined on $M_{n\times n}$ by $\overline{\varphi}(\overline{\alpha}) \equiv \det \overline{\alpha}$, for each $\overline{\alpha} \in M_{n\times n}$. We will write $\overline{\alpha}\cdot$ for the function $x \to \overline{\alpha}x$ on R^n, for each $\overline{\alpha} \in M_{n\times n}$. Thus $\overline{\alpha}\cdot$ is the function of left multiplication with the matrix $\overline{\alpha}$.

Lemma A.0.2. Matrix basics. *Let $\overline{\alpha}, \overline{\beta} \in M_{n\times n}$ be arbitrary. Then the following conditions hold:*

1. $(\det \overline{\alpha}\overline{\beta}) = (\det \overline{\alpha}) \cdot (\det \overline{\beta})$.

2. The function $\overline{\alpha} \to \det \overline{\alpha}$ is uniformly continuous on compact subsets of $M_{n\times n}$. If $|\overline{\alpha}| \leq b$ for some $b \geq 0$, then $|\det \overline{\alpha}| \leq n!\, b^n$.

3. **Cramer's rule.** *Suppose $n \geq 2$. Suppose $|\det \overline{\alpha}| > 0$. Then the inverse matrix $\overline{\alpha}^{-1}$ is well defined and is given by*

$$(\overline{\alpha}^{-1})_{i,j} \equiv (\det \overline{\alpha})^{-1}(-1)^{i+j} \det \overline{\alpha}'_{j,i}, \tag{A.0.2}$$

for each $i, j = 1, \ldots, n$.

4. Let $c > 0$ be arbitrary. Define

$$M_{n\times n,c} \equiv \{\overline{\alpha} \in M_{n\times n} : |\det \overline{\alpha}| \geq c\}.$$

Then the function $\overline{\alpha} \to \overline{\alpha}^{-1}$ is uniformly continuous on compact subsets of $M_{n\times n,c}$.

Proof. 1. For Assertions 1 and 3, consult any textbook on matrix algebra.

2. Assertion 2 follows from the defining equality A.0.1, which also says that $\det \overline{\alpha}$ is a polynomial in the entries of $\overline{\alpha}$, whence $\overline{\alpha} \to \det \overline{\alpha}$ is a uniformly continuous function of these entries when these entries are restricted to a compact subset of R. In short, the function $\overline{\alpha} \to \det \overline{\alpha}$ is uniformly continuous on compact subsets of $M_{n\times n}$ relative to the norm $|\cdot|$ on $M_{n\times n}$.

3. Let $i, j = 1, \ldots, n$ be arbitrary. Then each entry of the $(n-1) \times (n-1)$ matrix $\overline{\alpha}'_{j,i}$ is equal to some entry of $\overline{\alpha}$. Hence each entry of the matrix $\overline{\alpha}'_{j,i}$ is a uniformly continuous function on $M_{n\times n}$. Therefore the function $\overline{\alpha} \to \overline{\alpha}'_{j,i}$ is a uniformly continuous function from $M_{n\times n}$ to $M_{(n-1)\times(n-1)}$. Since $\overline{\gamma} \to \det \overline{\gamma}$ is, in turn, a uniformly continuous function on compact subsets of $M_{(n-1)\times(n-1)}$, we see that $\overline{\alpha} \to \det \overline{\alpha}'_{j,i}$ is a uniformly continuous function on compact subsets of $M_{n\times n}$, for each $i, j = 1, \ldots, n$. At the same time, Assertion 2 says that $\overline{\alpha} \to \det \overline{\alpha}$ is a uniformly continuous function on compact subsets of $M_{n\times n}$. Hence the function $\overline{\alpha} \to (\det \overline{\alpha})^{-1}$ is uniformly continuous on compact subsets

of $M_{n\times n,c}$. Combining, we see that the right-hand side of equality A.0.2 is a uniformly continuous function on compact subsets of $M_{n\times n,c}$. Hence so is the left-hand side of equality A.0.2. Thus the function $\overline{\alpha} \to (\overline{\alpha}^{-1})_{i,j}$ is uniformly continuous on compact subsets of $M_{n\times n,c}$. Therefore the function $\overline{\alpha} \to \overline{\alpha}^{-1}$ is uniformly continuous on compact subsets of $M_{n\times n,c}$. □

Definition A.0.3. (n-interval). In the following discussion, any open subinterval of R, whether proper or not, will be called an open 1-*interval*. Let $a, b \in R$ be arbitrary with $a < b$.Then the intervals $[a,b]$ and $[a,b)$ will be called, respectively, closed and half-open 1-*intervals*. Each open, closed, and half-open 1-interval will also simply be called a 1-*interval*, and denoted by Δ. If Δ is a proper subinterval, with endpoints $a \leq b$ in R, then $|\Delta| \equiv b - a > 0$ will be called the *length* of the 1-*interval* Δ. Note that, as defined here, each closed or half-open interval is proper and has a well-defined length.

More generally, a product subset $\Delta \equiv \prod_{i=1}^{n} \Delta_i$ of R^n, where Δ_i is a 1-interval for each $i = 1, \ldots, n$, will be called an n-*interval*. The intervals $\Delta_1, \ldots, \Delta_n$ in R are then called the *factors* of the n-interval Δ.

Suppose all the factors are proper. Then the n-interval Δ is said to be proper. The real number $|\Delta| \equiv \bigvee_{i=1}^{n} |\Delta_i| > 0$ will be called the *length* of the n-*interval* Δ. The real number $\|\Delta\| \equiv \sqrt{\sum_{i=1}^{n} |\Delta_i|^2}$ will then be called the diameter of Δ. The *center* x of the n-*interval* Δ is then defined to be $x = (x_1, \ldots, x_n)$, where x_i is the midpoint of Δ_i for each $i = 1, \ldots, n$.

If all the factors of the n-interval Δ are closed/half-open/open 1-intervals, then Δ is called a closed/half-open/open n-*interval*. □

Lemma A.0.4. Proper n-intervals are Lebesgue integrable. *Each proper n-interval $\Delta \equiv \prod_{i=1}^{n} \Delta_i$ is Lebesgue integrable, with $\mu(\Delta) = \prod_{i=1}^{n} |\Delta_i|$.*

Proof. 1. Consider the case where $n = 1$, and where $\Delta = [a,b]$. For each $k \geq 1$, define the continuous function h_k on R by (i) $h_k(a) \equiv h_k(b) \equiv 1$; (ii) $h_k(a - k^{-1}) \equiv h_k(b + k^{-1}) \equiv 0$; (iii) h_k is linear on each of the intervals $[a - k^{-1}, a]$, $[a,b]$, and $[b, b + k^{-1}]$; and (iv) h_k vanishes off the interval $[a - k^{-1}, b + k^{-1}]$. Then the Riemann integral $\int h_k(x)dx$ is nonincreasing with k, and converges to $b - a$ as $k \to \infty$. Hence, by the Monotone Convergence Theorem, the indicator $1_{[a,b]}$ is Lebesgue integrable, with integral $b - a$. In other words, the interval $[a,b]$ is Lebesgue integrable, with Lebesgue measure $|\Delta|$.

2. Next, with $n = 1$, consider the half-open interval $[a,b)$. Then, for each $k \geq 1$, the interval $[a, a \vee (b - k^{-1})]$ is integrable, with measure $(b - a - k^{-1})_+$ which converges to $b - a$ as $k \to \infty$. Hence, again by the Monotone Convergence Theorem, the interval $[a,b)$ is integrable, with Lebesgue integral $b - a$. Similarly, the interval (a,b) is integrable, with Lebesgue integral $|\Delta| = b - a$.

3. Consider now the general case where $n \geq 2$. Then $\Delta \equiv \prod_{i=1}^{n} \Delta_i$ of R^n, where Δ_i is a proper 1-interval for each $i = 1, \ldots, n$. If $n = 2$, then $1_\Delta \equiv \bigotimes_{i=1}^{n} 1_{\Delta(i)}$ is a simple function relative to $\left(\int \cdot dx\right)^{\otimes n} \equiv \int \cdots \int \cdot dx_1 \ldots dx_n$ in the sense

of Definition 4.10.2, whence $\prod_{i=1}^{2} \Delta_i$ is an integrable set in R^2. Repeating this argument, we can prove, inductively, that $\prod_{i=1}^{n} \Delta_i$ is an integrable subset of R^n, where $n \geq 2$ is arbitrary, and that, by Fubini's Theorem,

$$\int \cdots \int 1_\Delta dx_1 \ldots dx_n = \prod_{i=1}^{n} \int 1_{\Delta(i)} dx = \prod_{i=1}^{n} |\Delta_i|.$$

Equivalently, $\mu(\Delta) = \prod_{i=1}^{n} |\Delta_i|$. The lemma is proved. □

We need some more elementary (no pun intended) results in matrix algebra.

Definition A.0.5. Elementary matrix. An $n \times n$ matrix $T \equiv [T_{i,j}]_{i,j=1,\ldots,n}$ is called an *elementary matrix* if one of the following three conditions holds:

1. **Row swapping.** Suppose there exist $\bar{i}, \bar{j} \in \{1, \ldots, n\}$ with $\bar{i} < \bar{j}$ such that, for each $i, j = 1, \ldots, n$, we have (i) $T_{i,j} = 1$, if $i = j \neq \bar{i}$ and $i = j \neq \bar{j}$; (ii) $T_{i,j} = 1$, if $(i, j) = (\bar{i}, \bar{j})$ or $(j, i) = (\bar{i}, \bar{j})$; and (iii) $T_{i,j} = 0$, if $i \neq j$, $(i, j) \neq (\bar{i}, \bar{j})$, and $(i, j) \neq (\bar{j}, \bar{i})$. In other words, T is obtained by swapping the $\bar{i}$-th row and the $\bar{j}$-th row of the identity matrix I. We will then write $T \equiv T_{Swp,\bar{i},\bar{j}}$.

The left multiplication by $T_{Swp,\bar{i},\bar{j}}$ to any $n \times k$ matrix A then produces a matrix B that is obtainable by swapping the $\bar{i}$-th row and the $\bar{j}$-th row in the matrix A. Moreover, $\det T_{Swp,\bar{i},\bar{j}} = -1$, whence $\det(T_{Swp,\bar{i},\bar{j}} A) = -\det A$ for each $n \times n$ matrix A.

2. **Multiplication of one row by a nonzero constant**. Suppose there exist $\bar{i} \in \{1, \ldots, n\}$ and $c \in R$ with $|c| > 0$ such that, for each $i, j = 1, \ldots, n$, we have (i) $T_{i,j} = 1$, if $i = j \neq \bar{i}$; (ii) $T_{i,j} = c$, if $i = j = \bar{i}$; and (iii) $T_{i,j} = 0$ if $i \neq j$. In other words, T is obtained by multiplying the $\bar{i}$-th row of the identity matrix I with the constant c. We will then write $T \equiv T_{Mult,\bar{i},c}$.

The left multiplication by $T_{Mult,\bar{i},c}$ to any $n \times k$ matrix A then produces a matrix B that is obtainable by multiplying the $\bar{i}$-th row of the matrix A by the constant c. It follows that $\det T_{Mult,\bar{i},c} = c$, whence $\det(T_{Mult,\bar{i},c} A) = c \det A$ for each $n \times n$ matrix A.

3. **Adding a constant multiple of one row to another.** Suppose there exist $\bar{i}, \bar{j} \in \{1, \ldots, n\}$ with $\bar{i} \neq \bar{j}$, and $c \in R$ such that, for each $i, j = 1, \ldots, n$, we have (i) $T_{i,j} = 1$ if $i = j$; (ii) $T_{i,j} = c$, if $i \neq j$ and $(i, j) = (\bar{i}, \bar{j})$; and (iii) $T_{i,j} = 0$ if $i \neq j$ and $(i, j) \neq (\bar{i}, \bar{j})$. In other words, T is obtained by adding c times the $\bar{i}$-th row of the identity matrix I to the $\bar{j}$-th row. We will then write $T \equiv T_{Add,c,\bar{i},\bar{j}}$.

The left multiplication by $T_{Add,c,\bar{i},\bar{j}}$ to any $n \times k$ matrix A then produces a matrix B that is obtainable by adding c times the $\bar{i}$-th row of the matrix A to the $\bar{j}$-th row. It follows that $\det T_{Add,c,\bar{i},\bar{j}} = 1$, whence $\det(T_{Add,c,\bar{i},\bar{j}} A) = \det A$ for each $n \times n$ matrix A.

An elementary matrix whose entries are rational numbers is called a *rational elementary matrix*. For example, $T_{Add,c,\bar{i},\bar{j}}$ is rational iff c is rational. □

Proposition A.0.6. Gauss–Jordan method of finding the inverse of an $n \times n$ matrix. *Let $\overline{\alpha}$ be an $n \times n$ matrix with $|\det \overline{\alpha}| > 0$. Then there exists a finite* *sequence* $(T^{(k)})_{k=0,\ldots,\kappa}$ *of elementary matrices, where* $\kappa \geq 0$ *and where* $T^{(0)} = I$*, such that* $\overline{\alpha}^{-1} = T^{(\kappa)} \ldots T^{(0)}$*. If, in addition, all the entries of the matrix* $\overline{\alpha}$ *are rational, then there exists a finite sequence* $(T^{(k)})_{k=0,\ldots,\kappa}$ *of rational elementary matrices, where* $\kappa \geq 0$ *and where* $T^{(0)} = I$*, such that* $\overline{\alpha}^{-1} = T^{(\kappa)} \ldots T^{(0)}$*. Consequently, the inverse matrix* $\overline{\alpha}^{-1}$ *has rational entries iff the matrix* $\overline{\alpha}$ *does.*

Proof. See any matrix algebra text on Gauss–Jordan elimination. □

Lemma A.0.7. Change of integration variables in R^1, by a multiplication and a shift. *Let* $f \in C(R)$ *be arbitrary. Let* $v, c \in R$ *be arbitrary with* $|c| > 0$*. Then* (i)

$$\int f(x)dx = \int f(cu)|c|du$$

and (ii)

$$\int f(x)dx = \int f(v+u)du.$$

Proof. 1. First assume that $c > 0$. By Definitions 4.9.9 and 4.1.2, the Lebesgue integral of f is the limit of Riemann sums. Specifically,

$$\begin{aligned}
\int f(x)dx &\equiv \lim_{\alpha(0)\to-\infty;\alpha(n)\to+\infty;\ \bigvee_{i=1}^{n}(\alpha(i)-\alpha(i-1))\to 0} \\
&\quad \times \left\{ \sum_{i=1}^{n} f(\alpha_i)(\alpha_i - \alpha_{i-1}) : \alpha_0 < \cdots < \alpha_n \right\} \\
&= \lim_{\alpha(0)/c\to-\infty;\alpha(n)/c\to+\infty;\ \bigvee_{i=1}^{n}(\alpha(i)/c-\alpha(i-1)/c)\to 0} \\
&\quad \times \left\{ \sum_{i=1}^{n} f(cc^{-1}\alpha_i)(cc^{-1}\alpha_i - cc^{-1}\alpha_{i-1}) : c^{-1}\alpha_0 < \cdots < c^{-1}\alpha_n \right\} \\
&= \lim_{\beta(0)\to-\infty;\beta(n)\to+\infty;\ \bigvee_{i=1}^{n}(\beta(i)-\beta(i-1))\to 0} \\
&\quad \times \left\{ \sum_{i=1}^{n} f(c\beta_i)c(\beta_i - \beta_{i-1}) : \beta_0 < \cdots < \beta_n \right\} \\
&\equiv \int f(cu)cdu,
\end{aligned}$$

as alleged in Assertion (i). In the case where $c < 0$, Assertion (i) can be proved similarly.

2. Assertion (ii) can be proved similarly. □

A special case of the next theorem proves, in terms of Lebesgue measures, the intuitive notion of areas being invariant under Euclidean transformations. Note that we assume no prior knowledge of Euclidean geometry.

Theorem A.0.8. Change of integration variables in R^n by a matrix multiplication and a shift. *Let g be an arbitrary function* on R^n. *Let $\overline{v} \in R^n$ and the $n \times n$-matrix $\overline{\alpha}$ be arbitrary, with $|\det \overline{\alpha}| > 0$. Then* the following conditions hold:

1. The function g is integrable on R^n iff the function $x \to g(\overline{\alpha} \cdot x)$ is integrable on R^n, in which case

$$|\det \overline{\alpha}| \int \cdots \int g(\overline{\alpha} \cdot x) dx_1 \ldots dx_n = \int \cdots \int g(u) du_1 \ldots du_n.$$

2. The function g is integrable on R^n iff the function $x \to g(\overline{v} + x)$ is integrable on R^n, in which case

$$\int \cdots \int g(\overline{v} + x) dx_1 \ldots dx_n = \int \cdots \int g(u) du_1 \ldots du_n.$$

Here we write $x \equiv (x_1, \ldots, x_n)$ and $u \equiv (u_1, \ldots, u_n)$.

Proof. We will prove only Assertion 1. The proof of Assertion 2 is similar. First assume that $\overline{\alpha}$ has rational entries.

1. Then, by Proposition A.0.6, there exists a finite sequence $(T^{(k)})_{k=0,\ldots,\kappa}$ of elementary matrices with rational entries, where $\kappa \geq 0$ and where $T^{(0)} = I$, such that

$$\overline{\alpha}^{-1} = T^{(\kappa)} \ldots T^{(0)}. \tag{A.0.3}$$

2. Suppose, for some $k = 1, \ldots, \kappa$, we have $T^{(k)} = T_{Add,c,\overline{i},\overline{j}}$ for some c. Then c is a rational number. Either $c = 0$ or $c \neq 0$. If $c = 0$, then $T_{Add,c,\overline{i},\overline{j}} = I$, whence the factor $T^{(k)}$can be removed from the right-hand side of equality A.0.3. If $c \neq 0$, then

$$T_{Add,c,\overline{i},\overline{j}} = T_{Mult,\overline{i},1/c} T_{Add,1,\overline{i},\overline{j}} T_{Mult,\overline{i},c}, \tag{A.0.4}$$

whence the factor $T^{(k)}$ in the right-hand side of equality A.0.3 can be removed and replaced by the right-hand side of equality A.0.4. Summing up, we may assume that, if $T^{(k)} = T_{Add,c,\overline{i},\overline{j}}$ for some $k = 1, \ldots, \kappa$, then $T^{(k)} = T_{Add,1,\overline{i},\overline{j}}$. We proceed to prove Assertion 1, by induction on κ.

3. For that purpose, first assume that $g \in C(R^n)$. Then there exists $M \geq 0$ such that $[-M, M]^n$ is a support of the function g. Define the continuous function $f \equiv g \circ \overline{\alpha} \cdot$ on R^n. Suppose $|f(x)| > 0$ for some $x \in R^n$. Then $|g(\overline{\alpha} \cdot x)| > 0$. Hence $y \equiv \overline{\alpha} \cdot x \in [-M, M]^n$. Consequently, $|y| \leq M$. Therefore

$$|x| = |\overline{\alpha}^{-1} \cdot y| \leq \bigvee_{i=1}^{n} \left(\sum_{j=1}^{n} |\overline{\alpha}_{i,j}^{-1}| \cdot |y_j| \right) \leq \bigvee_{i=1}^{n} \left(\sum_{j=1}^{n} |\overline{\alpha}_{i,j}^{-1}| \right) \cdot |y|$$

$$\leq \bigvee_{i=1}^{n} (n|\overline{\alpha}^{-1}|) \cdot |y| \leq \overline{M} \equiv n|\overline{\alpha}^{-1}| \cdot M.$$

In short, the continuous function f is supported by the compact subset $[-\overline{M},\overline{M}]^n$. Thus $f \in C(R^n)$. Consequently, f is integrable. In other words, the function $x \to g(\overline{\alpha} \cdot x)$ is integrable on R^n.

To start the induction, consider the case where $\kappa = 1$. According to Definition A.0.5 and the previous paragraph, the rational elementary matrix $T^{(\kappa)}$ can be one of three kinds: (i) $T^{(\kappa)} \equiv T_{Swp,\overline{i},\overline{j}}$ for some $\overline{i},\overline{j} \in \{1,\ldots,n\}$ with $\overline{i} < \overline{j}$; (ii) $T^{(\kappa)} \equiv T_{Mult,\overline{i},c}$ for some $\overline{i} \in \{1,\ldots,n\}$, for some rational number $c \neq 0$; and (iii) $T^{(k)} = T_{Add,1,\overline{i},\overline{j}}$.

4. Consider Case (i), where $T^{(\kappa)} \equiv T_{Swp,\overline{i},\overline{j}}$ for some $\overline{i},\overline{j} \in \{1,\ldots,n\}$ with $\overline{i} < \overline{j}$. For ease of notations, there is no loss of generality to assume that $\overline{i} = 1$ and $\overline{j} = 2$. Then

$$\int_{x(1)\in R}\int_{x(2)\in R}\int_{x(3)\in R}\cdots\int_{x(n)\in R} g(\overline{\alpha}\cdot x)dx_1dx_2dx_3\ldots dx_n$$

$$\equiv \int_{x(1)\in R}\int_{x(2)\in R}\int_{x(3)\in R}\cdots\int_{x(n)\in R} f(x_1,x_2,x_3,\ldots,x_n)dx_1dx_2dx_3\ldots dx_n$$

$$= \int_{x(2)\in R}\int_{x(1)\in R}\int_{x(3)\in R}\cdots\int_{x(n)\in R} f(x_1,x_2,x_3,\ldots,x_n)dx_2dx_1dx_3\ldots dx_n$$

$$= \int_{u(1)\in R}\int_{u(2)\in R}\int_{u(3)\in R}\cdots\int_{u(n)\in R} f(u_2,u_1,u_3,\ldots,u_n)du_1du_2du_3\ldots du_n$$

$$\equiv \int_{u(1)\in R}\int_{u(2)\in R}\int_{u(3)\in R}\cdots\int_{u(n)\in R} f(T_{Swp,1,2}(u_1,u_2,u_3,\ldots,u_n)) \times du_1du_2du_3\ldots du_n$$

$$\equiv \int_{u(1)\in R}\int_{u(2)\in R}\int_{u(3)\in R}\cdots\int_{u(n)\in R} f(T_{Swp,1,2}(u_1,u_2,u_3,\ldots,u_n))\cdot|\det T_{Swp,1,2}|\cdot du_1du_2du_3\ldots du_n$$

$$\equiv \int_{u(1)\in R}\int_{u(2)\in R}\int_{u(3)\in R}\cdots\int_{u(n)\in R} \times f(\overline{\alpha}^{-1}(u_1,u_2,u_3,\ldots,u_n))\cdot|\det\overline{\alpha}^{-1}|\cdot du_1du_2du_3\ldots du_n$$

$$\equiv \int_{u(1)\in R}\int_{u(2)\in R}\int_{u(3)\in R}\cdots\int_{u(n)\in R} g(u_1,u_2,u_3,\ldots,u_n)\cdot|\det\overline{\alpha}^{-1}| \times du_1du_2du_3\ldots du_n,$$

where the second equality is by Fubini's Theorem, where the third equality is by a change of the names in the dummy variables, and where the third-to-last equality is because $\det T_{Swp,\overline{i},\overline{j}} = -1$. This proves Assertion 1 in Case (i) if $\kappa = 1$, assuming $g \in C(R^n)$, and assuming $\overline{\alpha}$ has rational entries.

6. With $\kappa = 1$, next consider Case (ii), where $T^{(\kappa)} \equiv T_{Mult,\bar{i},c}$ for some $\bar{i} \in \{1,\ldots,n\}$ and for some rational number $c \neq 0$. For ease of notations, there is no loss of generality to assume that $\bar{i} = 1$. Then

$$\int_{x(1)\in R}\int_{x(2)\in R}\cdots\int_{x(n)\in R} g(\overline{\alpha}\cdot x)dx_1dx_2\ldots dx_n$$

$$\equiv \int_{x(1)\in R}\int_{x(2)\in R}\cdots\int_{x(n)\in R} f(x_1,x_2,\ldots,x_n)dx_1dx_2\ldots dx_n$$

$$= \int_{x(2)\in R}\cdots\int_{x(n)\in R}\left\{\int_{x(1)\in R} f(x_1,x_2,\ldots,x_n)dx_1\right\}dx_2\ldots dx_n$$

$$= \int_{x(2)\in R}\cdots\int_{x(n)\in R}\left\{\int_{u(1)\in R} f(cu_1,x_2,\ldots,x_n)|c|du_1\right\}dx_2\ldots dx_n$$

$$= \int_{u(1)\in R}\int_{x(2)\in R}\cdots\int_{x(n)\in R} f(cu_1,x_2,\ldots,x_n)|c|du_1dx_2\ldots dx_n$$

$$= \int_{u(1)\in R}\int_{u(2)\in R}\cdots\int_{u(n)\in R} f(T_{Mult,1,c}(u_1,u_2,\ldots,u_n))$$
$$\times |\det T_{Mult,1,c}|du_1du_2\ldots du_n$$

$$= |\det\overline{\alpha}^{-1}|\int_{u(1)\in R}\int_{u(2)\in R}\cdots\int_{u(n)\in R} f(\overline{\alpha}^{-1}\cdot(u_1,u_2,\ldots,u_n))du_1du_2\ldots du_n$$

$$= |\det\overline{\alpha}^{-1}|\int_{u(1)\in R}\int_{u(2)\in R}\cdots\int_{u(n)\in R} g(u_1,u_2,\ldots,u_n)du_1du_2\ldots du_n,$$

where the third equality is by Lemma A.0.7, and where the third-to-last equality is because $\det T_{Mult,\bar{i},c} = c$. This proves Assertion 1 in Case (ii) if $\kappa = 1$, assuming $f \in C(R^n)$, and assuming $\overline{\alpha}$ has rational entries.

7. Continuing with $\kappa = 1$, consider Case (iii), where $T^{(\kappa)} = T_{Add,1,\bar{i},\bar{j}}$ for some $\bar{i},\bar{j} \in \{1,\ldots,n\}$. For ease of notations, there is no loss of generality to assume that $\bar{i} = 1$ and $\bar{j} = 2$. Then

$$\int_{x(1)\in R}\int_{x(2)\in R}\int_{x(3)\in R}\cdots\int_{x(n)\in R} g(\overline{\alpha}\cdot x)dx_2dx_3\ldots dx_n$$

$$\equiv \int_{x(1)\in R}\int_{x(2)\in R}\int_{x(3)\in R}\cdots\int_{x(n)\in R} f(x_1,x_2,x_3,\ldots,x_n)dx_1dx_2dx_3\ldots dx_n$$

$$= \int_{x(1)\in R}\int_{x(3)\in R}\cdots\int_{x(n)\in R}\left\{\int_{x(2)\in R} f(x_1,x_2,x_3,\ldots,x_n)dx_2\right\}dx_1dx_3\ldots dx_n$$

$$= \int_{x(1)\in R}\int_{x(3)\in R}\cdots\int_{x(n)\in R}\left\{\int_{u(2)\in R} f(x_1,x_1+u_2,x_3,\ldots,x_n)du_2\right\}$$
$$\times dx_1dx_3\ldots dx_n$$

$$= \int_{u(1)\in R}\int_{u(3)\in R}\cdots\int_{u(n)\in R}\left\{\int_{u(2)\in R} f(u_1,u_1+u_2,u_3,\ldots,u_n)du_2\right\}$$
$$\times du_1du_3\ldots du_n$$

$$= \int_{u(1)\in R} \int_{u(2)\in R} \int_{u(3)\in R} \cdots \int_{u(n)\in R} f(T_{Add,1,1,2}(u_1, u_2, u_3, \ldots, u_n))$$

$$\times du_1 du_2 du_3 \ldots du_n$$

$$= |\det T_{Add,1,1,2}| \int_{u(1)\in R} \cdots \int_{u(n)\in R} f(T_{Add,1,1,2}(u_1, \ldots, u_n)) du_1 \ldots du_n$$

$$= |\det \overline{\alpha}^{-1}| \int_{u(1)\in R} \cdots \int_{u(n)\in R} f(\overline{\alpha}^{-1} \cdot (u_1, \ldots, u_n)) du_1 \ldots du_n$$

$$= |\det \overline{\alpha}^{-1}| \int_{u(1)\in R} \cdots \int_{u(n)\in R} g(u_1, \ldots, u_n) du_1 \ldots du_n,$$

where the third equality is by Lemma A.0.7, and where the third-to-last equality is because $\det T_{Add,1,1,2} = 1$. Summing up, Assertion 1 is proved if $\kappa = 1$, assuming $g \in C(R^n)$ and assuming $\overline{\alpha}$ has rational entries.

8. Note that, assuming $\kappa = 1$, Assertion 1 says that the Lebesgue integration $\int \cdot d\mu$ on R^n agrees on $C(R^n)$ with the completion of the integration $\int \cdot d\overline{\mu}$ defined by $\int g d\overline{\mu} \equiv |\det \overline{\alpha}| \int g(\overline{\alpha} \cdot x) dx$ for each $g \in C(R^n)$. Hence the completions are equal. In particular, a function g on R^n is integrable relative to $\int \cdot d\mu$ iff it is integrable relative to $\int \cdot d\overline{\mu}$, in which case

$$\int g d\mu = \int g d\overline{\mu} \equiv |\det \overline{\alpha}| \int g(\overline{\alpha} \cdot x) dx.$$

Assertion 1 is thus proved in the case where $\kappa = 1$, assuming $\overline{\alpha}$ has rational entries.

9. Still assuming $\overline{\alpha}$ has rational entries, suppose, inductively, that Assertion 1 has been proved for $\kappa = 1, \ldots, \overline{\kappa} - 1$ for some $\overline{\kappa} \geq 2$. Now suppose

$$\overline{\alpha}^{-1} = T^{(\overline{\kappa})} \ldots T^{(0)}. \tag{A.0.5}$$

Define the $n \times n$ matrix $\widetilde{\alpha}$ by its inverse

$$\widetilde{\alpha}^{-1} = T^{(\overline{\kappa}-1)} \ldots T^{(0)}. \tag{A.0.6}$$

Then

$$\overline{\alpha}^{-1} = T^{(\overline{\kappa})} \widetilde{\alpha}^{-1}.$$

Let g be an arbitrary integrable function on R^n. Then, in view of equality A.0.6, the induction hypothesis applies to $\kappa = \overline{\kappa} - 1$, and implies that the function $\widetilde{g} \equiv g \circ \widetilde{\alpha}$ is integrable on R^n, with

$$\int \cdots \int \widetilde{g}(u) du_1 \ldots du_n = |\det \widetilde{\alpha}|^{-1} \int \cdots \int \widetilde{g}(\widetilde{\alpha}^{-1} \cdot v) dv_1 \ldots dv_n. \tag{A.0.7}$$

In turn, the induction hypothesis applies to $\kappa = 1$, with $(T^{(\overline{\kappa})})^{-1}$ in place of $\overline{\alpha}$, and implies that

$$|\det T^{(\overline{\kappa})}|^{-1} \int \cdots \int \widetilde{g}((T^{(\overline{\kappa})})^{-1} \cdot x) dx_1 \ldots dx_n = \int \cdots \int \widetilde{g}(u) du_1 \ldots du_n. \tag{A.0.8}$$

Combining equalities A.0.8 and A.0.7, we obtain

$$|\det T^{(\overline{\kappa})}|^{-1}\int\cdots\int g\circ\widetilde{\alpha}((T^{(\overline{\kappa})})^{-1}\cdot x)dx_1\ldots dx_n$$
$$=|\det\widetilde{\alpha}|^{-1}\int\cdots\int g\circ\widetilde{\alpha}(\widetilde{\alpha}^{-1}\cdot v)dv_1\ldots dv_n.$$

Since $|\det T^{(\overline{\kappa})}|\cdot|\det\widetilde{\alpha}|^{-1}=|\det\overline{\alpha}|^{-1}$, it follows that

$$\int\cdots\int g(\widetilde{\alpha}\cdot(T^{(\overline{\kappa})})^{-1}\cdot x)dx_1\ldots dx_n=|\det\overline{\alpha}|^{-1}\int\cdots\int g(v)dv_1\ldots dv_n.$$

Equivalently,

$$\int\cdots\int g(\overline{\alpha}\cdot x)dx_1\ldots dx_n=|\det\overline{\alpha}|^{-1}\int\cdots\int g(v)dv_1\ldots dv_n.$$

Induction is completed, and Assertion 1 is proved with the additional assumption that the matrix $\overline{\alpha}$ has rational entries.

10. Now let $\widehat{\alpha}$ be an arbitrary $n\times n$ matrix, with entries not necessarily rational. We proceed to prove that Assertion 1 remains valid.

Let $g\in C(R^n)$ be arbitrary, with compact support K. Then the real-valued function $v\to g(\widehat{\alpha}\cdot v)$ is also a continuous function, with compact support $\widehat{K}\equiv\widehat{\alpha}^{-1}\cdot K$. Take any $b,c\in R$ such that $b>\max|g|\vee|\det\widehat{\alpha}|$ and $|\det\widehat{\alpha}|>c>0$. Let $\varepsilon>0$ be arbitrary. By Assertion 4 of Lemma A.0.2, the function $\overline{\alpha}\to\overline{\alpha}^{-1}$ is uniformly continuous on

$$M_{n\times n,c,b}\equiv\{\overline{\alpha}\in M_{n\times n,c}:|\det\widehat{\alpha}|\leq b\}.$$

Since $b>|\det\widehat{\alpha}|>c$, there exists an $n\times n$ matrix $\overline{\alpha}\in M_{n\times n,c,b}$ with rational entries and with $|\overline{\alpha}-\widehat{\alpha}|$ so small that (i) $|\det\overline{\alpha}^{-1}-\det\widehat{\alpha}^{-1}|<\varepsilon$; (ii) $|g(\overline{\alpha}\cdot v)-g(\widehat{\alpha}\cdot v)|<\varepsilon$ for each $v\in R^n$; and (iii) there exists an integrable compact subset $\widetilde{K}$ of R^n that is a support of both the function $g\circ\widehat{\alpha}\cdot$ and the function $g(\overline{\alpha}\cdot x)$. By the last statement in Step 9, we have

$$\int\cdots\int g(\overline{\alpha}\cdot x)dx_1\ldots dx_n=|\det\overline{\alpha}|^{-1}\int\cdots\int g(v)dv_1\ldots dv_n.$$

Conditions (i) and (iii) therefore lead to

$$\int\cdots\int_{x\in\widetilde{K}} g(\overline{\alpha}\cdot x)dx_1\ldots dx_n=(|\det\widehat{\alpha}^{-1}|\pm\varepsilon)\int\cdots\int g(v)dv_1\ldots dv_n.$$

Condition (ii) then implies

$$\int\cdots\int_{x\in\widetilde{K}}(g(\widehat{\alpha}\cdot x)\pm\varepsilon)dx_1\ldots dx_n=(|\det\widehat{\alpha}^{-1}|\pm\varepsilon)\int\cdots\int g(v)dv_1\ldots dv_n$$

Letting $\varepsilon\to 0$, we obtain

$$\int\cdots\int g(\widehat{\alpha}\cdot x)dx_1\ldots dx_n=|\det\widehat{\alpha}^{-1}|\cdot\int\cdots\int g(v)dv_1\ldots dv_n,$$

where $g \in C(R^n)$ is arbitrary. Since $C(R^n)$ is dense in the space of integrable functions, the last displayed equality holds for each integrable function g. Since $\widehat{\alpha}$ is an arbitrary $n \times n$ matrix with $|\det\widehat{\alpha}| > 0$, Assertion 1 is proved. The proof of Assertion 2 is similar. □

Corollary A.0.9. Integrability of convolution of two integrable functions. *Let L stand for the Lebesgue integrable functions on R^n. Let $f, g \in L$ be arbitrary. The* convolution *$f \star g : R^n \to R$ is the function defined by*

$$domain(f \star g) \equiv \{x \in R^n : f(x - \cdot)g \in L\},$$

and by $(f \star g)(x) \equiv \int_{u \in R^n} f(x - y)g(y)dy$ for each $x \in domain(f \star g)$.

Then $f \star g \in L$.

Proof. 1. First assume that $f, g \in C(R^n)$. Define $\widetilde{f}(x,y) \equiv f(x - y)$ and $\widetilde{g}(x,y) \equiv g(y)$. Then $\widetilde{f}\widetilde{g} \in C(R^{2n})$. Consequently, $\widetilde{f}\widetilde{g}$ is Lebesgue integrable on R^{2n}. Therefore, by Fubini's Theorem, the function $\widetilde{f}(x,\cdot)\widetilde{g}(x,\cdot)$ is a member of L, for each x in some full subset D of R^n, with

$$\begin{aligned}\int\int \widetilde{f}(x,y)\widetilde{g}(x,y)dxdy &= \int\left(\int \widetilde{f}(x,y)\widetilde{g}(x,y)dx\right)dy\\ &\equiv \int\left(\int f(x-y)dx\right)g(y)dy\\ &= \int\left(\int f(x)dx\right)g(y)dy = \int f(x)dx \cdot \int g(y)dy,\end{aligned} \tag{A.0.9}$$

where the third equality is thanks to Theorem A.0.8. Note that the set $\{\widetilde{f}(x,\cdot)\widetilde{g}(x,\cdot) : x \in D\}$ is dense in the metric subspace

$$A \equiv \{\widetilde{f}(x,\cdot)\widetilde{g}(x,\cdot) : x \in R^n\} \equiv \{f(x - \cdot)g : x \in R^n\}$$

of $C(R^n)$, relative to the supremum norm. Hence equality A.0.9 implies that each member of A is integrable and also satisfies

$$\int\int \widetilde{f}(x,y)\widetilde{g}(x,y)dxdy = \int f(x)dx \cdot \int g(y)dy.$$

In other words, for each $x \in R^n$, we have

$$x \in domain(f \star g) \equiv \{x \in R^n : f(x - \cdot)g \in L\},$$

and, in addition,

$$(f \star g)(x) \equiv \int_{u \in R^n} f(x - u)g(u)du = \int f(x)dx \cdot \int g(y)dy, \tag{A.0.10}$$

where $f, g \in C(R^n)$ are arbitrary.

2. Next let $g \in C(R^n)$ be arbitrary with $\int g(y)dy > 0$. Define

$$L_g \equiv \{f \in L : f \star g \in L\}.$$

Then $C(R^n) \subset L_g \subset L$ according to Step 1. Define the functions $I_1, I_2 : C(R^n) \to R$ by

$$I_1(f) \equiv \int \left(\int f(x-y)dx \right) g(y)dy$$

and

$$I_2(f) \equiv \int f(x)dx \cdot \int g(y)dy,$$

for each $f \in C(R^n)$. Equality A.0.10 shows that $(R^n, C(R^n), I_1)$ and $(R^n, C(R^n), I_2)$ are equal as integration spaces. Hence a function f is integrable relative to the integration I_1 iff it is integrable relative to I_2. Since I_2 is a positive constant multiple of the Lebesgue integration, the completion L of $C(R^n)$ relative to the Lebesgue integration is equal to the completion of $C(R^n)$ relative to I_2. At the same time, since $C(R^n) \subset L_g \subset L$, we have $L_g = L$.

Thus we see that each $f \in L$ is a member of L_g, and that $I_1(f) = I_2(f)$ for each $f \in L$. Summing up, for each $f \in L$ and $g \in C(R^n)$ with $\int g(y)dy > 0$, we have $f \star g \in L$ and

$$\int (f \star g)(y)dy = \int f(x)dx \cdot \int g(y)dy. \tag{A.0.11}$$

3. Now let $f \in L$ be arbitrary with $\int f(x)dx > 0$, but fixed. Then equality A.0.11 holds for each $g \in C(R^n)$ with $\int g(y)dy > 0$, Hence, by linearity and continuity, it holds for each $g \in C(R^n)$. Define the functions $I_3, I_4 : C(R^n) \to R$ by

$$I_3(g) \equiv \int \left(\int f(x-y)dx \right) g(y)dy$$

and

$$I_4(g) \equiv \int f(x)dx \cdot \int g(y)dy, \tag{A.0.12}$$

for each $g \in C(R^n)$. Then equality A.0.11 shows that $(R^n, C(R^n), I_3)$ and $(R^n, C(R^n), I_4)$are equal as integration spaces. Since $\int f(x)dx > 0$, we see from relation A.0.12 that

$$(R^n, C(R^n), I_4) = \left(R^n, C(R^n), \int f(x)dx \cdot \int \cdot dy \right).$$

Hence a function g is integrable relative to I_3 iff it is integrable relative to $\int f(x)dx \cdot \int \cdot dy$, in which case

$$\begin{aligned} \left(\int f(x-y)dx \right) g(y)dy &= I_3(g) \\ &= I_4(g) = \int f(x)dx \cdot \int g(y)dy. \end{aligned} \tag{A.0.13}$$

Now let $g \in L$ be arbitrary. Then g is integrable relative to $\int f(x)dx \cdot \int \cdot dy$. Therefore g is integrable relative to I_3. In other words, $\left(\int f(x - \cdot)dx \right) g \in L$, or, equivalently, $f \star g \in L$, where $g, f \in L$ are arbitrary, with $\int f(x)dx > 0$. By linearity and continuity, we conclude that $f \star g \in L$ for arbitrary $g, f \in L$. The corollary is proved. □

Theorem A.0.10. Change of integration variables in R^1. *Let B be an open interval in R. Suppose:*

1. $\beta : B \to R$ is a diferentiable function whose derivative $\frac{d\beta}{du} : B \to R$ is uniformly continuous on each compact subset $K \subset B$.

2. For each compact subset $K \subset B$, there exists some $c > 0$ such that $\left|\frac{d\beta}{du}(v)\right| \geq c$ for each $v \in K$.

Then, for an arbitrary function f on R, the function $1_{\beta(B)} \cdot f$ is integrable relative to the Lebesgue integration iff *the function $1_B \cdot (f \circ \beta) \cdot \left|\frac{d\beta}{du}\right|$ is integrable relative to the Lebesgue integration, in which case*

$$\int_{x\in\beta(B)} f(x)dx = \int_{u\in B} f(\beta(u)) \cdot \left|\frac{d\beta}{du}(u)\right| du. \tag{A.0.14}$$

Proof. 1. Let $K \equiv [a,b]$ be an arbitrary closed interval with $K \subset B$. Then $|\frac{d\beta}{du}(v)| \geq c$ for each $v \in K$, for some $c > 0$. Hence, there is no loss of generality to assume that $\frac{d\beta}{du} \geq c$ on K. Thus β is an increasing function on the interval K, whence $\beta(K) = [\beta(a), \beta(b)]$ is a closed interval. Then

$$\begin{aligned}
\int_{\beta(u)\in\beta(K)} \left|\frac{d\beta}{du}(u)\right| \cdot du &= \int_{u\in K} \frac{d\beta}{du}(u)du \\
&= \int_{u=a}^{b} \frac{d\beta}{du}(u)du = \beta(b) - \beta(a) \\
&= \int_{x=\beta(a)}^{\beta(b)} dx = \int 1_{\beta(K)}(x)dx,
\end{aligned}$$

where the third equality is by the Second Fundamental Theorem of Calculus, which can be found in textbooks on advanced calculus or in [Bishop 1967].

2. Next, let $K \equiv [a,b]$ be an arbitrary closed interval. Let $(K_i)_{i=1,2,\ldots}$ be an arbitrary sequence of closed intervals with $K_i \subset K_{i+1} \subset B$ for each $i = 1, 2, \ldots$, such that (i) $\bigcup_{i=1}^{\infty} K_i = B$ and (ii) $\int_{u\in K(i)} G(u)du \uparrow \int_{u\in B} G(u)du$ as $i \to \infty$, for each nonnegative integrable function G on R. Consider each $i \geq 1$. Then $KK_i \subset B$, Hence, by Step 1 of this proof, we have

$$\int_{\beta(u)\in\beta(KK(i))} \left|\frac{d\beta}{du}(u)\right| du = \int 1_{\beta(KK(i))}(x)dx.$$

Consequently,

$$\begin{aligned}
\int_{u\in K(i)} 1_K(u) \cdot \left|\frac{d\beta}{du}(u)\right| du &= \int_{u\in KK(i)} \left|\frac{d\beta}{du}(u)\right| du = \int_{\beta(u)\in\beta(KK(i))} \left|\frac{d\beta}{du}(u)\right| du \\
&= \int 1_{\beta(KK(i))}(x)dx = \int 1_{\beta(K(i))}(x) 1_{\beta(K)}(x)dx.
\end{aligned} \tag{A.0.15}$$

Let $i \to \infty$. Then, since $1_K |\frac{d\beta}{du}|$ is integrable on R, Condition (ii) implies that the left-hand side of equality A.0.15 converges monotonically to $\int_{u\in B} 1_K(u) \cdot |\frac{d\beta}{du}(u)|du$. Hence so does the right-hand side of equality A.0.15. Therefore, by the Monotone Convergence Theorem, the function $1_{\beta(B)}1_{\beta(K)}$ is integrable on R and

$$\int_{u\in B} 1_K(u) \cdot \left|\frac{d\beta}{du}(u)\right| du = \int 1_{\beta(B)}(x)1_{\beta(K)}(x)dx = \int_{x\in\beta(B)} 1_{\beta(K)}(x)dx,$$

where $K \equiv [a,b]$ is an arbitrary closed interval.

3. Next, let $K \equiv [a,b)$ be an arbitrary half-open interval where $a < b < \infty$. Let $(b_j)_{j=1,2,\ldots}$ be an increasing sequence in $[a,b)$ such that $b_j \uparrow b$, and let $K'_j \equiv [a,b_j] \subset [a,b)$ for each $j \geq 1$. Then $K'_j \subset K'_{j+1}$ for each $j \geq 1$. Moreover, $\bigcup_{j=1}^{\infty} K'_j = K$. Furthermore, by Step 2 of this proof, we have

$$\int_{u\in B} 1_{K'(j)}(u) \cdot \left|\frac{d\beta}{du}(u)\right| du = \int_{x\in\beta(B)} 1_{\beta(K'(j))}(x)dx \tag{A.0.16}$$

for each $j \geq 1$. The Dominated Convergence Theorem implies that the left-hand side of equality A.0.16 converges to $\int_{u\in B} 1_K(u) \cdot |\frac{d\beta}{du}(u)|du$ monotonically. Therefore, so does the right-hand side. The Monotone Convergence Theorem then, in turn, implies that the function

$$1_{\beta(B)}1_{\beta(K)} = 1_{\beta(B)}1_{\beta\left(\bigcup_{j=1}^{\infty} K'(j)\right)} = 1_{\beta(B)}1_{\bigcup_{j=1}^{\infty} \beta(K'(j))}$$

is integrable, with integral

$$\int_{x\in\beta(B)} 1_{\beta(K)}(x)dx = \int_{u\in B} 1_K(u) \cdot \left|\frac{d\beta}{du}(u)\right| du,$$

where $K \equiv [a,b)$ is an arbitrary half-open interval.

4. Now let $f \in C(R)$ be arbitrary. Then $f \circ \beta \in C(R)$ has support in some finite half-open interval $[a,b)$. Let $\kappa \geq 1$ be arbitrary. Then we can partition $[a,b)$ into a finite number $m_\kappa \geq 1$ of disjoint half-open intervals $\widetilde{K}_{\kappa,1}, \ldots, \widetilde{K}_{\kappa,m(\kappa)}$, each so small that $|f - c_{\kappa,j}| < 2^{-\kappa}1_{\beta([a,b])}$ on $\beta(\widetilde{K}_{\kappa,j})$, for each $j = 1, \ldots, m_\kappa$. Define the function

$$f_\kappa \equiv \sum_{j=1}^{m(\kappa)} c_{\kappa,j}1_{\beta(\widetilde{K}(\kappa,j))}.$$

Then $|f - f_\kappa| < 2^{-\kappa}1_{\beta([a,b])}$. Moreover,

$$f_\kappa \circ \beta \equiv \sum_{j=1}^{m(\kappa)} c_{\kappa,j}1_{\beta(\widetilde{K}(\kappa,j))} \circ \beta = \sum_{j=1}^{m(\kappa)} c_{\kappa,j}1_{\widetilde{K}(\kappa,j)}$$

and

$$|f \circ \beta - f_\kappa \circ \beta| < 2^{-\kappa}1_{[a,b]}.$$

Since $\widetilde{K}_{\kappa,1}, \ldots, \widetilde{K}_{\kappa,m(\kappa,)}$ are half-open intervals, we have, by Step 3,

$$\int_{x\in\beta(B)} 1_{\beta(\widetilde{K}(\kappa,j))}(x)dx = \int_{u\in B} 1_{\widetilde{K}(\kappa,j)}(u)\cdot\left|\frac{d\beta}{du}(u)\right| du$$

for each $1,\ldots,m_\kappa$. Hence

$$\int_{x\in\beta(B)} \sum_{j=1}^{m(\kappa)} c_{\kappa,j} 1_{\beta(\widetilde{K}(\kappa,j))}(x)dx = \int_{u\in B} \sum_{j=1}^{m(\kappa)} c_{\kappa,j} 1_{\widetilde{K}(\kappa,j)}(u)\left|\frac{d\beta}{du}(u)\right| du.$$

Equivalently,

$$\int_{x\in\beta(B)} f_\kappa(x)dx = \int_{u\in B} f_\kappa(\beta(u))\cdot\left|\frac{d\beta}{du}(u)\right| du.$$

Letting $\kappa\to\infty$, we obtain

$$\int_{x\in\beta(B)} f(x)dx = \int_{u\in B} f(\beta(u))\cdot\left|\frac{d\beta}{du}(u)\right| du,$$

where $f\in C(R)$ is arbitrary.

5. Now define the functions $I_1, I_2 : C(R)\to R$ by

$$I_1(f)\equiv\int_{x\in\beta(B)} f(x)dx$$

and

$$I_2(f)\equiv\int_{u\in B} f(\beta(u))\cdot\left|\frac{d\beta}{du}(u)\right| du,$$

for each $f\in C(R)$. Then both I_1 and I_2 are integrations on R, in the sense of Definition 4.2.1. Proposition 4.3.3 says that $(R,C(R),I_1)$ and $(R,C(R),I_2)$ are integration spaces. Moreover, Step 4 of this proof shows that $I_1 = I_2$. Hence the complete extensions of $(R,C(R),I_1)$ and $(R,C(R),I_2)$ are equal. In other words, a function f on R is integrable relative to I_1 iff f is integrable relative to I_2. Hence a function f is such that $1_{\beta(B)}f$ is integrable relative to the Lebesgue integral iff the function $1_B\cdot(f\circ\beta)\cdot\left|\frac{d\beta}{du}\right|$ is integrable relative to the Lebesgue integral, in which case

$$\int_{x\in\beta(B)} f(x)dx = \int_{u\in B} f(\beta(u))\cdot\left|\frac{d\beta}{du}(u)\right| du.$$

The theorem is proved. □

The next lemma will be key to our subsequent proof of Theorem A.0.12 for the change of integration variables from rectangular to polar coordinates in the half plane. The proof of the lemma is longer than a few lines because, alas, Theorem A.0.12 is yet to be established.

Lemma A.0.11. Lebesgue measure of certain fan-shaped subsets of a half disk.

1. For each $r > 0$, the half disk

$$D_r \equiv \{(x,y) \in (0,\infty) \times R : \sqrt{x^2+y^2} < r\}$$

is integrable, with measure $2^{-1}\pi r^2$.

2. For each $r, s > 0$ with $s < r$, the subset

$$D_{s,r} \equiv \{(x,y) \in (0,\infty) \times R : s \leq \sqrt{x^2+y^2} < r\}$$

is integrable, with measure $2^{-1}\pi(r^2 - s^2)$.

3. For each $r, s > 0$ with $s < r$, and for each $u, v \in \left(-\frac{\pi}{2}, \frac{\pi}{2}\right)$ with $u < v$, the subset

$$A_{s,r,u,v} \equiv \left\{(x,y) \in D_{s,r} : u < \arctan\frac{y}{x} < v\right\} \tag{A.0.17}$$

is integrable, with measure

$$\mu(A_{s,r,u,v}) = 2^{-1}(v-u)(r^2-s^2). \tag{A.0.18}$$

Proof. 1. First note that the set $B_0 \equiv \{(x,y) : x > 0\} = (0,\infty) \times R$ is a measurable set in R^2 because its indicator $1_{B(0)} = 1_{(0,\infty)} \otimes 1_R$ is the product of two measurable functions on R^2, according to Assertion 1 of Theorem 4.10.10. Separately, the real-valued function $(x,y) \to \sqrt{x^2+y^2}$ is continuous, and hence measurable on R^2. Thus, according to Proposition 4.8.14, there exists a countable subset $\overline{A} \subset (0,\infty)$ such that the set

$$\overline{B}_t \equiv \left\{(x,y) : \sqrt{x^2+y^2} < t\right\} \subset [-M,M] \times [-M,M]$$

is integrable for each $t \in [-M,M] \cap \overline{A}_c$, for each $M \geq 1$, where $\overline{A}_c$ is the metric complement in $(0,\infty)$ of the set $\overline{A}$. Let $t \in \overline{A}_c$ be arbitrary. Then $D_t = B_0\overline{B}_t$. Hence D_t is an integrable subset of $(0,\infty) \times R$. Using Fubini's Theorem, we compute

$$\mu(D_t) = \int_{x\in(0,t)} \left(\int_{y\in\left(-\sqrt{t^2-x^2},\sqrt{t^2-x^2}\right)} dy\right) dx = 2\int_{x\in(0,t)} \left(\sqrt{t^2-x^2}\right) dx$$

$$= 2\int_{t\sin u\in(0,t)} \left(\sqrt{t^2-t^2\sin^2 u}\right)(t\cos u)du = 2t^2 \int_{u\in(0,\pi/2)} (\cos^2 u)du$$

$$= 2t^2 \int_{u=0}^{\pi/2} \frac{1+\cos 2u}{2} du = t^2\left(u + \frac{1}{2}\sin 2u\right)\Big|_{u-0}^{\pi/2} = t^2\frac{\pi}{2},$$

where the third equality is by a change of integration variables $x = t\sin u$ in R^1, justified by Theorem A.0.10, and where the sixth equality is by a second change of integration variables in R^1, also justified by Theorem A.0.10.

2. Now consider each $r > 0$. Let $(t_k)_{k=1,2,...}$ be an increasing sequence in $\overline{A}_c$ such that $t_k \uparrow r$. Then, according to Step 1, $D_{t(k)}$ is an integrable set, with $D_{t(k)} \subset D_{t(k+1)}$for each $k \geq 1$, and with $\mu(D_{t(k)}) = t_k^2\frac{\pi}{2} \uparrow r^2\frac{\pi}{2}$. At the same time,

$D_r = \bigcup_{k=1}^{\infty} D_{t(k)}$. Hence, by the Monotone Convergence Theorem, the set D_r is an integrable set, with $\mu(D_r) = r^2 \frac{\pi}{2}$. Assertion 1 is proved.

3. Assertion 2 follows from $D_{s,r} = D_r D_s^c$ and from

$$\mu(D_{s,r}) = \mu(D_r D_s^c) = \mu(D_r) - \mu(D_s) = 2^{-1}\pi r^2 - 2^{-1}\pi s^2.$$

4. It remains to prove Assertion 3. To that end, let $r, s > 0$ be arbitrary with $s < r$. Define the function $\gamma : R^2 \to \left(-\frac{\pi}{2}, \frac{\pi}{2}\right)$ by (i) $\gamma(x,y) \equiv \arctan \frac{y}{x}$ for each (x,y) with $x > 0$ and (ii) $\gamma(x,y) \equiv 0$ for each (x,y) with $x \leq 0$. Thus the function γ is defined a.e. on R^2. Separately, let $k \geq 0$ be arbitrary. Define the function $\gamma_k : R^2 \to \left(-\frac{\pi}{2}, \frac{\pi}{2}\right)$ by (i′) $\gamma_k(x,y) \equiv \gamma(x,y)$ for each (x,y) with $x \geq 2^{-k}$; (ii′) $\gamma_k(x,y) \equiv 0$ for each (x,y) with $x \leq 2^{-k-1}$; and (iii′)

$$\gamma_k(x,y) \equiv \frac{(x - 2^{-k-1})}{(2^{-k} - 2^{-k-1})}\gamma(2^{-k}, y)$$

for each (x,y) with $2^{-k-1} < x < 2^{-k}$. Then γ_k is uniformly continuous on R^2, and hence measurable. At the same time, by Assertion 2, the set $D_{s,r}$ is an integrable subset of R^2. Hence the product $\gamma_k 1_{D(s,r)}$ is an integrable function. Moreover,

$$(|\gamma_k 1_{D(s,r)} - \gamma 1_{D(s,r)}| > 0) \subset [0, 2^{-k}] \times [-r, r],$$

where $\mu([0,2^{-k}] \times [-r,r]) = 2^{-k+1}r \to 0$ as $k \to \infty$. Thus $\gamma_k 1_{D(s,r)} \to \gamma 1_{D(s,r)}$ in measure as $k \to \infty$. Since $|\gamma 1_{D(s,r)}| \leq \pi 1_{D(s,r)}$, and $|\gamma_k 1_{D(s,r)}| \leq \pi 1_{D(s,r)}$ for each $k \geq 0$, the Dominated Convergence Theorem applies; it implies that the function $\gamma 1_{D(s,r)}$ is integrable.

5. Hence, according to Assertion 1 of Proposition 4.8.14, all but countably many $t \in R$ are continuity points of the function $\gamma 1_{D(s,r)}$ relative to $D_{s,r}$. In other words, there exists a countable subset A of R such that each point t in the metric complement A_c is a a continuity point of the function $\gamma 1_{D(s,r)}$ relative to the integrable set $D_{s,r}$. Therefore Assertion 3 of Proposition 4.8.14 implies that, for each $\overline{u}, \overline{v} \in A_c$, the sets $D_{s,r} \cap (\gamma 1_{D(s,r)} \leq \overline{u})$ and $D_{s,r} \cap (\gamma 1_{D(s,r)} < \overline{v})$ are integrable. Consequently, the set

$$\begin{aligned}&\{(x,y) \in D_{s,r} : \overline{u} < \gamma(x,y) 1_{D(s,r)}(x,y) < \overline{v})\\ &= \left\{(x,y) \in D_{s,r} : \overline{u} < \arctan \frac{y}{x} < \overline{v}\right) \equiv A_{s,r,\overline{u},\overline{v}}\end{aligned}$$

is integrable, where $\overline{u}, \overline{v} \in A_c$ are arbitrary with $\overline{u} < \overline{v}$.

6. Next, let $u, v \in \left(-\frac{\pi}{2}, \frac{\pi}{2}\right)$ be arbitrary with $u < v$, such that $A_{s,r,u,v}$ is an integrable set. Take an arbitrary $\lambda \in \left(-\frac{\pi}{2} - u, \frac{\pi}{2} - v\right)$. Then $v + \lambda < \frac{\pi}{2}$ and $u + \lambda > -\frac{\pi}{2}$. Consider the matrix $\overline{\alpha} \equiv \begin{bmatrix} \cos\lambda, & -\sin\lambda \\ \sin\lambda, & \cos\lambda \end{bmatrix}$. Then $\det \overline{\alpha} = 1$ and $\overline{\alpha}^{-1} = \begin{bmatrix} \cos\lambda, & \sin\lambda \\ -\sin\lambda, & \cos\lambda \end{bmatrix}$. We will show that

$$\overline{\alpha} \cdot A_{s,r,u,v} = A_{s,r,u+\lambda,v+\lambda}. \tag{A.0.19}$$

This follows from

$$
\begin{aligned}
&\overline{\alpha} \cdot A_{s,r,u,v} \\
&\quad \equiv \{\overline{\alpha} \cdot (x,y) : (x,y) \in A_{s,r,u,v}\} \\
&\quad \equiv \bigcup_{(x,y)\in A(s,r,u,v)} \{(\overline{x},\overline{y}) \in R^2 : (x,y) = \overline{\alpha}^{-1} \cdot (\overline{x},\overline{y}); \\
&\qquad\qquad s \leq \sqrt{x^2+y^2} < r; x \sin u < y \cos u; y \cos v < x \sin v\} \\
&\quad = \bigcup_{(x,y)\in A(s,r,u,v)} \{(\overline{x},\overline{y}) \in R^2 : (x,y) = (\overline{x}\cos\lambda + \overline{y}\sin\lambda, \; -\overline{x}\sin\lambda + \overline{y}\cos\lambda); \\
&\qquad\qquad s \leq \sqrt{\overline{x}^2+\overline{y}^2} < r; (\overline{x}\cos\lambda + \overline{y}\sin\lambda)\sin u \\
&\qquad\qquad\quad < (-\overline{x}\sin\lambda + \overline{y}\cos\lambda)\cos u; \\
&\qquad\qquad (-\overline{x}\sin\lambda + \overline{y}\cos\lambda)\cos v < (\overline{x}\cos\lambda + \overline{y}\sin\lambda)\sin v\} \\
&\quad = \bigcup_{(x,y)\in A(s,r,u,v)} \{(\overline{x},\overline{y}) \in D_{s,r} : (x,y) \\
&\qquad\qquad = (\overline{x}\cos\lambda + \overline{y}\sin\lambda, \; -\overline{x}\sin\lambda + \overline{y}\cos\lambda); \\
&\qquad\qquad \overline{x}\cos\lambda\sin u + \overline{x}\sin\lambda\cos u < \overline{y}\cos\lambda\cos u - \overline{y}\sin\lambda\sin u; \\
&\qquad\qquad \overline{y}\cos\lambda\cos v - \overline{y}\sin\lambda\sin v < \overline{x}\cos\lambda\sin v + \overline{x}\sin\lambda\cos v\} \\
&\quad = \bigcup_{(x,y)\in A(s,r,u,v)} \{(\overline{x},\overline{y}) \in D_{s,r} : (x,y) \\
&\qquad\qquad = (\overline{x}\cos\lambda + \overline{y}\sin\lambda, \; -\overline{x}\sin\lambda + \overline{y}\cos\lambda); \\
&\qquad\qquad \overline{x}\sin(\lambda+u) < \overline{y}\cos(\lambda+u); \overline{y}\cos(\lambda+v) < \overline{x}\sin(\lambda+v)\} \\
&\quad = \bigcup_{(x,y)\in A(s,r,u,v)} \Big\{(\overline{x},\overline{y}) \in D_{s,r} : (x,y) = (\overline{x}\cos\lambda + \overline{y}\sin\lambda, \\
&\qquad\qquad -\overline{x}\sin\lambda + \overline{y}\cos\lambda); \tan(\lambda+u) < \frac{\overline{y}}{\overline{x}} < \tan(\lambda+v)\Big\} \\
&\quad = \bigcup_{(x,y)\in A(s,r,u,v)} \{(\overline{x},\overline{y}) \in A_{s,r,\lambda+u,\lambda+v} : (x,y) \\
&\qquad\qquad = (\overline{x}\cos\lambda + \overline{y}\sin\lambda, \; -\overline{x}\sin\lambda + \overline{y}\cos\lambda)\} \\
&\quad = A_{s,r,\lambda+u,\lambda+v}.
\end{aligned}
\tag{A.0.20}
$$

Equality A.0.19 has been verified, where $\lambda \in \left(-\frac{\pi}{2} - u, \frac{\pi}{2} - v\right)$ is arbitrary, assuming that $A_{s,r,u,v}$ is an integrable set.

At the same time, since, by assumption, $A_{s,r,u,v}$ is an integrable set, its indicator $g \equiv 1_{A(s,r,u,v)}$ is an integrable function on R^2. Hence, according to Assertion 1 of Theorem A.0.8, the function $g \circ \overline{\alpha}^{-1} \cdot$ is integrable on R^2, with

$$|\det \overline{\alpha}^{-1}| \cdot \int\int g(\overline{\alpha}^{-1} \cdot (x, y))dxdy = \int\int g(\widetilde{u}, \widetilde{v})d\widetilde{u}d\widetilde{v}. \qquad \text{(A.0.21)}$$

Consequently, the indicator

$$1_{\overline{\alpha} \cdot A(s,r,u,v)} = 1_{A(s,r,u,v)} \circ \overline{\alpha}^{-1} \equiv g \circ \overline{\alpha}^{-1}$$

is integrable. In other words, the set $\overline{\alpha} \cdot A_{s,r,u,v}$ is an integrable set. Since $|\det \overline{\alpha}^{-1}| = 1$, equality A.0.21 simplifies to

$$\int\int 1_{A(s,r,u,v)}(\overline{\alpha}^{-1} \cdot (x, y))dxdy = \int\int 1_{A(s,r,u,v)}(\widetilde{u}, \widetilde{v})d\widetilde{u}d\widetilde{v}.$$

Equivalently,

$$\int\int 1_{\overline{\alpha} \cdot A(s,r,u,v)}(x, y)dxdy = \int\int 1_{A(s,r,u,v)}(\widetilde{u}, \widetilde{v})d\widetilde{u}d\widetilde{v}$$

or

$$\mu(\overline{\alpha} \cdot A_{s,r,u,v}) = \mu(A_{s,r,u,v}). \qquad \text{(A.0.22)}$$

Moreover, equality A.0.19 says that

$$\overline{\alpha} \cdot A_{s,r,u,v} = A_{s,r,u+\lambda,v+\lambda}. \qquad \text{(A.0.23)}$$

Hence $A_{s,r,u+\lambda,v+\lambda}$ is an integrable set. Equalities A.0.22 and A.0.23 together yield

$$\mu(A_{s,r,u+\lambda,v+\lambda}) = \mu(A_{s,r,u,v}), \qquad \text{(A.0.24)}$$

where $\lambda \in \left(-\frac{\pi}{2} - u, \frac{\pi}{2} - v\right)$ is arbitrary, assuming that $u, v \in \left(-\frac{\pi}{2}, \frac{\pi}{2}\right)$ are such that $u < v$ and $A_{s,r,u,v}$ is an integrable set.

7. Continuing with the same assumptions as in Step 6, proceed to estimate a bound for $\mu(A_{s,r,u,v})$. For that purpose, define the midpoint $\lambda \equiv -2^{-1}(u + v)$ of the interval $\left(-\frac{\pi}{2} - u, \frac{\pi}{2} - v\right)$. Then $\lambda \in \left(-\frac{\pi}{2} - u, \frac{\pi}{2} - v\right)$. Write $\varepsilon \equiv \tan(2^{-1}(v - u))$. Then

$$\begin{aligned}
&A_{s,r,u+\lambda,v+\lambda} \\
&\equiv \left\{(x, y) \in D_{s,r} : \tan(u + \lambda) < \frac{y}{x} < \tan(v + \lambda)\right\} \\
&\equiv \left\{(x, y) \in D_{s,r} : \tan(u - 2^{-1}(u + v)) < \frac{y}{x} < \tan(v - 2^{-1}(u + v))\right\} \\
&= \left\{(x, y) \in D_{s,r} : -\varepsilon < \frac{y}{x} < \varepsilon\right\} \\
&= \{(x, y) \in D_{s,r} : -\varepsilon x < y < \varepsilon x\} \subset [0, r) \times (-\varepsilon r, \varepsilon r).
\end{aligned}$$

Hence, using equality A.0.24, we obtain the bound

$$\mu(A_{s,r,u,v}) = \mu(A_{s,r,u+\lambda,v+\lambda}) \le 2r^2\varepsilon$$
$$\equiv 2r^2 \tan(2^{-1}(v-u)), \tag{A.0.25}$$

assuming that $u, v \in \left(-\frac{\pi}{2}, \frac{\pi}{2}\right)$ are such that $u < v$ and $A_{s,r,u,v}$ is an integrable set.

8. Next, let $u, v \in \left(-\frac{\pi}{2}, \frac{\pi}{2}\right)$ be arbitrary with $u < v$. We will show that $A_{s,r,u,v}$ is an integrable set. Recall the countable exceptional set A defined in Step 5. Take a decreasing sequence $(\overline{u}_h)_{h=1,2,\ldots}$ in A_c such that $\overline{u}_h > u$ and $\tan(2^{-1}(\overline{u}_h - u)) \to 0$ as $h \to \infty$. Take an increasing sequence $(\overline{v}_i)_{i=1,2,\ldots}$ in A_c such that $\overline{v}_i < v$ and $\tan(2^{-1}(v - \overline{v}_i)) \to 0$ as $i \to \infty$. Since $u < v$, we may assume, without loss of generality, that $u_h < v_i$ for each $h, i \ge 1$. According to the conclusion of Step 5, $A_{s,r,\overline{u}(h),\overline{v}(i)}$ is then an integrable set, for each $h, i \ge 1$. Let $h \ge 1$ be arbitrary. Note that for each $i' \ge i+2$, we have $A_{s,r,\overline{u}(h),\overline{v}(i)} \subset A_{s,r,\overline{u}(h),\overline{v}(i')}$ and because $(\overline{u}_h, \overline{v}_{i'}) \subset (\overline{u}_h, \overline{v}_i) \cup (\overline{v}_{i-1}, \overline{v}_{i'})$, we have

$$A_{s,r,\overline{u}(h),\overline{v}(i')} \subset A_{s,r,\overline{u}(h),\overline{v}(i)} \cup A_{s,r,\overline{v}(i-1),\overline{v}(i')}.$$

Hence

$$\mu(A_{s,r,\overline{u}(h),\overline{v}(i')}) - \mu(A_{s,r,\overline{u}(h),\overline{v}(i)}) \le \mu(A_{s,r,\overline{v}(i-1),\overline{v}(i')})$$
$$\le 2r^2 \tan(2^{-1}(\overline{v}_{i'} - \overline{v}_{i-1}))$$
$$\le 2r^2 \tan(2^{-1}(v - \overline{v}_{i-1})) \to 0,$$

as $i' \ge i+2 \to \infty$, where the second inequality is by inequality A.0.25. Consequently, by the Monotone Convergence Theorem, the union

$$A_{s,r,\overline{u}(h),v} = \bigcup_{i=1}^{\infty} A_{s,r,\overline{u}(h),\overline{v}(i)}$$

is an integrable set, where $h \ge 1$ is arbitrary.

Similarly, note that $A_{s,r,\overline{u}(h),v} \subset A_{s,r,\overline{u}(h'),v}$ and that

$$A_{s,r,\overline{u}(h'),v} \subset A_{s,r,\overline{u}(h'),\overline{u}(h-1)} \cup A_{s,r,\overline{u}(h),v}$$

because $(\overline{u}_{h'}, v) \subset (\overline{u}_{h'}, \overline{u}_{h-1}) \cup (\overline{u}_h, v)$, for each $h \ge 1$ and $h' \ge h+2$. Hence

$$\mu(A_{s,r,\overline{u}(h'),v}) - \mu(A_{s,r,\overline{u}(h),v}) \le \mu(A_{s,r,\overline{u}(h'),\overline{u}(h-1)})$$
$$\le 2r^2 \tan(2^{-1}(\overline{u}_{h-1} - \overline{u}_{h'}))$$
$$\le 2r^2 \tan(2^{-1}(\overline{u}_{h-1} - u)) \to 0,$$

as $h' \ge h+2 \to \infty$, where the second inequality is by inequality A.0.25. Consequently, by the Monotone Convergence Theorem, the union

$$A_{s,r,u,v} = \bigcup_{h=1}^{\infty} A_{s,r,\overline{u}(h),v}$$

is an integrable set, where $r, s > 0$ are arbitrary with $s < r$, and where $u, v \in \left(-\frac{\pi}{2}, \frac{\pi}{2}\right)$ are arbitrary with $u < v$.

9. Proceed to prove equality A.0.18, first with the additional assumption that $v, u \in \left(-\frac{\pi}{2}, \frac{\pi}{2}\right)$ are rational multiples of $\frac{\pi}{2}$. Take an arbitrarily large $q \geq 1$, such that $u = -\frac{\pi}{2} + jq^{-1}\frac{\pi}{2}$ and $v = -\frac{\pi}{2} + kq^{-1}\frac{\pi}{2}$ for some $j, k \geq 1$ with $1 \leq j < k \leq 2q-1$. For each $i = 1, \ldots, 2q-1$, define $\eta_i \equiv -\frac{\pi}{2} + iq^{-1}\frac{\pi}{2}$. Thus $u = \eta_j$ and $v = \eta_k$. As an abbreviation, write

$$\overline{u} \equiv \eta_1 \equiv -\frac{\pi}{2} + q^{-1}\frac{\pi}{2}$$

and

$$\overline{v} \equiv \eta_{2q-1} \equiv \frac{\pi}{2} - q^{-1}\frac{\pi}{2}.$$

Then

$$\cos \overline{u} \equiv \cos\left(-\frac{\pi}{2} + q^{-1}\frac{\pi}{2}\right) = \sin\left(q^{-1}\frac{\pi}{2}\right)$$

and

$$\cos \overline{v} = \cos \overline{u} = \sin\left(q^{-1}\frac{\pi}{2}\right).$$

Define $\lambda \equiv q^{-1}\frac{\pi}{2}$, and define the matrix $\overline{\alpha} \equiv \left[\begin{smallmatrix} \cos\lambda, & -\sin\lambda \\ \sin\lambda & \cos\lambda \end{smallmatrix}\right]$. Then $\eta_{i+1} - \eta_i = \lambda$ for each $i = 1, \ldots, 2q-2$, Moreover, $\det \overline{\alpha} = 1$. Consider each $i = 1, \ldots, 2q-3$. Then $\lambda + \eta_i = \eta_{i+1}$ and $\lambda + \eta_{i+1} = \eta_{i+2}$. Hence

$$\begin{aligned}
\mu(A_{s,r,\eta(i),\eta(i+1)}) &= \int\int 1_{A(s,r,\eta(i),\eta(i+1))}(x)dxdy \\
&= |\det \overline{\alpha}|^{-1} \int\int 1_{A(s,r,\eta(i),\eta(i+1)}(\overline{\alpha}^{-1} \cdot (\widetilde{u}, \widetilde{v}))d\widetilde{u}d\widetilde{v} \\
&= \int\int 1_{\overline{\alpha}\cdot(A(s,r,\eta(i),\eta(i+1))}(\widetilde{u}, \widetilde{v})d\widetilde{u}d\widetilde{v} \\
&= \int\int 1_{A(s,r,\lambda+\eta(i),\lambda+\eta(i+1)}(\widetilde{u}, \widetilde{v})d\widetilde{u}d\widetilde{v} \\
&= \int\int 1_{A(s,r,\eta(i+1),\eta(i+2))}(\widetilde{u}, \widetilde{v})d\widetilde{u}d\widetilde{v} \\
&= \mu(A_{s,r,\eta(i+1),\eta(i+2)}),
\end{aligned} \tag{A.0.26}$$

where the second equality is by Assertion 1 of Theorem A.0.8, and where the fourth equality is by the recently proved equality A.0.19.

10. Based on the defining equality A.0.17, the sets $A_{s,r,\eta(1),\eta(2)}, \ldots, A_{s,r,\eta(2q-2),\eta(2q-1)}$ are mutually exclusive. Hence

$$\begin{aligned}
\sum_{i=1}^{2q-2} \mu(A_{s,r,\eta(i),\eta(i+1)}) &= \mu\left(\bigcup_{i=1}^{2q-2} A_{s,r,\eta(i),\eta(i+1)}\right) \\
&\leq \mu(D_{s,r}) = 2^{-1}\pi(r^2 - s^2).
\end{aligned} \tag{A.0.27}$$

Consider each $i = 1, \ldots, 2q-2$. Let $u'_i, u''_i \in \left(-\frac{\pi}{2}, \frac{\pi}{2}\right)$ with $u'_i < \eta_i < u''_i$ be arbitrary such that $\tan(2^{-1}(u''_i - u'_i)) < q^{-2}$. Let $v'_i, v''_i \in \left(-\frac{\pi}{2}, \frac{\pi}{2}\right)$ with $v'_i < \eta_{i+1} < v''_i$ be arbitrary such that $\tan(2^{-1}(v''_i - v'_i)) < q^{-2}$. Then

$$A_{s,r,u'(i),v''(i)} \subset A_{s,r,\eta(i),\eta(i+1)} \cup A_{s,r,u'(i),u''(i)} \cup A_{s,r,v'(i),v''(i)},$$

whence

$$\begin{aligned}
\mu(A_{s,r,u'(i),v''(i)}) &\le \mu(A_{s,r,\eta(i),\eta(i+1)}) + \mu(A_{s,r,u'(i),u''(i)}) + \mu(A_{s,r,v'(i),v''(i)}) \\
&\le \mu(A_{s,r,\eta(i),\eta(i+1)}) + 2r^2 \tan(2^{-1}(u''_i - u'_i)) \\
&\quad + 2r^2 \tan(2^{-1}(v''_i - v'_i)) \\
&\le \mu(A_{s,r,\eta(i),\eta(i+1)}) + 4r^2 q^{-2}, \qquad (A.0.28)
\end{aligned}$$

where the second inequality is by inequality A.0.25. At the same time, $(\eta_j, \eta_k) \subset \bigcup_{i=j}^{k-1} (u'_i, v''_i)$. Therefore

$$A_{s,r,\eta(j),\eta(k)} \subset \bigcup_{i=j}^{k-1} A_{s,r,u'(i),v''(i)},$$

whence

$$\mu(A_{s,r,\eta(j),\eta(k)}) \le \sum_{i=j}^{k-1} \mu(A_{s,r,u'(i),v''(i)}. \qquad (A.0.29)$$

Combining,

$$\begin{aligned}
\sum_{i=j}^{k-1} \mu(A_{s,r,\eta(i),\eta(i+1)}) &= \mu\left(\bigcup_{i=j}^{k-1} A_{s,r,\eta(i),\eta(i+1)}\right) \le \mu(A_{s,r,\eta(j),\eta(k)}) \\
&\le \sum_{i=j}^{k-1} \mu(A_{s,r,u'(i),v''(i)}) \le \sum_{i=j}^{k-1} \mu(A_{s,r,\eta(i),\eta(i+1)}) \\
&\quad + (k-j)4r^2 q^{-2},
\end{aligned}$$

where the equality is because the members of the union are mutually exclusive, where the second inequality is inequality A.0.29, and where the last inequality is thanks to inequality A.0.28. Thus

$$\mu(A_{s,r,\eta(j),\eta(k)}) = \sum_{i=j}^{k-1} \mu(A_{s,r,\eta(i),\eta(i+1)}) \pm (k-j)4r^2 q^{-2}. \qquad (A.0.30)$$

11. In the special case where $j = 1$ and $k = 2q-1$, inequality A.0.30 yields

$$\begin{aligned}
\mu(A_{s,r,\overline{u},\overline{v}}) &\equiv \mu(A_{s,r,\eta(1),\eta(2q-1)}) \\
&= \sum_{i=1}^{2q-2} \mu(A_{s,r,\eta(i),\eta(i+1)}) \pm (2q-2)4r^2 q^{-2}. \qquad (A.0.31)
\end{aligned}$$

At the same time, $A_{s,r,\overline{u},\overline{v}} \subset D_{s,r}$. Moreover,

$$D_{s,r} \subset A_{s,r,\overline{u},\overline{v}} \cup ([0, r\cos\overline{v}] \times [-r,r]) \quad \text{a.e.}$$

Hence

$$\mu(D_{s,r}) = \mu(A_{s,r,\overline{u},\overline{v}}) \pm 2r^2 \cos\overline{v}$$
$$= \sum_{i=1}^{2q-2} \mu(A_{s,r,\eta(i),\eta(i+1)}) \pm (2q-2)4r^2 q^{-2} \pm 2r^2 \sin\left(q^{-1}\frac{\pi}{2}\right),$$

where the last equality is due to equality A.0.31. Equivalently,

$$\sum_{i=1}^{2q-2} \mu(A_{s,r,\eta(i),\eta(i+1)}) = \mu(D_{s,r}) \pm (2q-2)4r^2 q^{-2} \pm 2r^2 \sin\left(q^{-1}\frac{\pi}{2}\right)$$
$$= 2^{-1}\pi(r^2 - s^2) \pm (2q-2)4r^2 q^{-2} \pm 2r^2 \sin\left(q^{-1}\frac{\pi}{2}\right).$$

Since all the summands on the left-hand side are equal, according to equality A.0.26, it follows that

$$\mu(A_{s,r,\eta(i),\eta(i+1)}) = (2q-2)^{-1}2^{-1}\pi(r^2 - s^2)$$
$$\pm 4r^2 q^{-2} \pm (2q-2)^{-1}2r^2 \sin\left(q^{-1}\frac{\pi}{2}\right), \qquad \text{(A.0.32)}$$

for each $i = 1, \ldots, 2q-2$. Consequently,

$$\mu(A_{s,r,u,v}) = \mu(A_{s,r,\eta(j),\eta(k)}) = \sum_{i=j}^{k-1} \mu(A_{s,r,\eta(i),\eta(i+1)}) \pm (k-j)4r^2 q^{-2}$$
$$= (k-j) \cdot \left\{(2q-2)^{-1}2^{-1}\pi(r^2 - s^2) \pm 4r^2 q^{-2}\right.$$
$$\left.\pm (2q-2)^{-1}2r^2 \sin\left(q^{-1}\frac{\pi}{2}\right)\right\} \pm (k-j)4r^2 q^{-2}$$
$$= (k-j)q^{-1}\frac{\pi}{2} \cdot \left\{q(2q-2)^{-1}(r^2 - s^2) \pm 4r^2 q^{-1} 2\pi^{-1}\right.$$
$$\left.\pm q2\pi^{-1}(2q-2)^{-1}2r^2 \sin\left(q^{-1}\frac{\pi}{2}\right)\right\}$$
$$\pm (k-j)q^{-1}\frac{\pi}{2} \cdot q2\pi^{-1}4r^2 q^{-2}$$
$$= (v-u) \cdot \left\{q(2q-2)^{-1}(r^2 - s^2) \pm 4r^2 q^{-1} 2\pi^{-1}\right.$$
$$\left.\pm q2\pi^{-1}(2q-2)^{-1}2r^2 \sin\left(q^{-1}\frac{\pi}{2}\right)\right\}$$
$$\pm (v-u) \cdot q2\pi^{-1}4r^2 q^{-2}, \qquad \text{(A.0.33)}$$

where the second equality is equality A.0.30, and where the third equality is from equality A.0.32. Since $q \geq 1$ is arbitrary, we can let $q \to \infty$. After all the vanishing terms on the right-hand side of equality A.0.33 drop out, we obtain

$$\mu(A_{s,r,u.v}) = \frac{1}{2}(v-u)(r^2-s^2),$$

provided that v, u are rational multiples of $\frac{\pi}{2}$.

12. Finally, let $u, v \in \left(-\frac{\pi}{2}, \frac{\pi}{2}\right)$ be arbitrary, not necessarily multiples of $\frac{\pi}{2}$, with $v > u$. Let $(\overline{v}_i)_{i=1,2,...}$ be an increasing sequence in $\left(-\frac{\pi}{2}, \frac{\pi}{2}\right)$ such that $\overline{v}_i$ is a rational multiple of $\frac{\pi}{2}$ for each $i \geq 1$, and such that $\overline{v}_i \uparrow v$. Similarly, let $(\overline{u}_h)_{h=1,2,...}$ be a decreasing sequence in $\left(-\frac{\pi}{2}, \frac{\pi}{2}\right)$ such that $\overline{u}_h$ is a rational multiple of $\frac{\pi}{2}$ for each $h \geq 1$, and such that $\overline{u}_h \downarrow u$. Let $h \geq 1$ be arbitrary. Then, by the preceding paragraph, we have

$$\mu(A_{s,r,\overline{u}(h),\overline{v}(i)}) = \frac{1}{2}(\overline{v}_i - \overline{u}_h)(r^2-s^2)$$

for each $i \geq 1$. Let $i \to \infty$. Then the Monotone Convergence Theorem implies that the set $A_{s,r,\overline{u}(h),v}$ is integrable, with measure

$$\mu(A_{s,r,\overline{u}(h),v}) = \frac{1}{2}(v - \overline{u}_h)(r^2-s^2).$$

Now let $h \to \infty$. Then the Monotone Convergence Theorem implies that the set $A_{s,r,u.v}$ is integrable, with measure

$$\mu(A_{s,r,u.v}) = \frac{1}{2}(v-u)(r^2-s^2),$$

which is the desired equality A.0.18. Assertion 3 and the lemma are proved. □

Theorem A.0.12. Change of integration variables from rectangular to polar coordinates in the half plane. *Let $H \equiv (0,\infty) \times R$ denote the open half plane, equipped with the Euclidean metric. Let g be an arbitrary function on H. Then g is integrable on H relative to the Lebesgue integration iff the function $(r,u) \to rg(r\cos u, r\sin u)$ is integrable on $(0,\infty) \times \left(-\frac{\pi}{2}, \frac{\pi}{2}\right)$, relative to the Lebesgue integration, in which case*

$$\int_{x=0}^{\infty}\int_{y=-\infty}^{\infty} g(x,y)dxdy = \int_{r=0}^{\infty}\int_{u=-\frac{\pi}{2}}^{\frac{\pi}{2}} rg(r\cos u, r\sin u)drdu.$$

Proof. 1. Define the function $\beta : [0,\infty) \times R \to R^2$ by $\beta(r,u) \equiv (r\cos u, r\sin u)$ for each $(r,u) \in [0,\infty) \times R$. Then the function β is uniformly continuous, with some modulus of continuity δ_β. Moreover,

$$\beta\left(\sqrt{x^2+y^2}, \arctan\frac{y}{x}\right) = (x,y) \tag{A.0.34}$$

for each $(x,y) \in (0,\infty) \times R$.

2. First, let $g \in C(H)$ be arbitrary, with some modulus of continuity δ_g. Take $M > \overline{x} > 0$ such that g has $[\overline{x}, M] \times [-M, \overline{M}]$ as support. Write $\overline{r} \equiv \sqrt{2}M$ and $\overline{s} \equiv \arccos\frac{\overline{r}}{\overline{x}}$. Then $[\overline{x}, M] \times [-M, \overline{M}] \subset D_{\overline{s},\overline{r}}$. Hence there exists $\overline{u}, \overline{v} \in \left(-\frac{\pi}{2}, \frac{\pi}{2}\right)$ such that

$$[\overline{x}, M] \times [-M, \overline{M}] \subset A_{\overline{s},\overline{r},\overline{u},\overline{v}} \subset D_{\overline{s},\overline{r}},$$

whence

$$\int_{x=0}^{\infty}\int_{y=-\infty}^{\infty} g(x,y)dxdy = \int\int_{(x,y)\in A(\overline{s},\overline{r},\overline{u},\overline{v})} g(x,y)dxdy. \tag{A.0.35}$$

3. Let $\varepsilon > 0$ be arbitrary. Take $q \geq 1$ so large that

$$\delta_r \equiv (\overline{r}-\overline{s})q^{-1} < \frac{\sqrt{2}}{2}\delta_\beta(\delta_g(\varepsilon))$$

and

$$\delta_u \equiv (\overline{v}-\overline{u})q^{-1} < \frac{\sqrt{2}}{2}\delta_\beta(\delta_g(\varepsilon)).$$

For each $h = 0,\ldots,q$, define $r_h \equiv \overline{s} + h\delta_r$. For each $i = 0,\ldots,q$, define $u_i \equiv \overline{u} + i\delta_u$. Then

$$\bigcup_{h=0}^{q-1}\bigcup_{i=0}^{q-1} A_{r(h),r(h+1),u(i),u(i+1)} \subset A_{\overline{s},\overline{r},\overline{u},\overline{v}}, \tag{A.0.36}$$

while according to Assertion 3 of Lemma A.0.11, we have

$$\begin{aligned}\mu\left(\bigcup_{h=0}^{q-1}\bigcup_{i=0}^{q-1} A_{r(h),r(h+1),u(i),u(i+1)}\right) &= \sum_{h=0}^{q-1}\sum_{i=0}^{q-1}\mu(A_{r(h),r(h+1),u(i),u(i+1)})\\ &= \sum_{h=0}^{q-1}\sum_{i=0}^{q-1} 2^{-1}(u_{i+1}-u_i)(r_{h+1}^2-r_h^2)\\ &= 2^{-1}(\overline{v}-\overline{u})(\overline{r}^2-\overline{s}^2)\\ &= \mu(A_{\overline{s},\overline{r},\overline{u},\overline{v}}).\end{aligned}$$

Thus the union on the left-hand side of relation A.0.36 is actually a full subset of the right-hand side. Consequently,

$$\begin{aligned}&\int\int_{(x,y)\in A(\overline{s},\overline{r},\overline{u},\overline{v})} g(x,y)dxdy\\ &= \int\int_{(x,y)\in\bigcup\bigcup A(r(h),r(h+1),u(i),u(i+1))} g(x,y)dxdy\\ &= \sum_{h=0}^{q-1}\sum_{i=0}^{q-1}\int\int_{(x,y)\in A(r(h),r(h+1),u(i),u(i+1))} g(x,y)dxdy,\end{aligned}$$

Combining with equality A.0.35, we obtain

$$\int_{x=0}^{\infty}\int_{y=-\infty}^{\infty} g(x,y)dxdy = \sum_{h=0}^{q-1}\sum_{i=0}^{q-1}\iint_{(x,y)\in A(r(h),r(h+1),u(i),u(i+1))} g(x,y)dxdy. \tag{A.0.37}$$

4. Let $h, i = 1, \ldots, q-1$ be arbitrary. Define $x_{h,i} \equiv r_h \cos u_i$ and $y_{h,i} \equiv r_h \sin u_i$. Consider each $(x, y) \in A_{r(h),r(h+1),u(i),u(i+1)}$. Then, by the defining equality A.0.17, we have

$$\sqrt{x^2+y^2}, \sqrt{x_{h,i}^2+y_{h,i}^2} \in [r_h, r_{h+1})$$

and

$$\arctan\frac{y}{x}, \arctan\frac{y_{h,i}}{x_{h,i}} \in [u_i u_{i+1}).$$

Hence

$$\left\|\left(\sqrt{x^2+y^2}, \arctan\frac{y}{x}\right) - \left(\sqrt{x_{h,i}^2+y_{h,i}^2}. \arctan\frac{y_{h,i}}{x_{h,i}}\right)\right\|$$
$$< \sqrt{(r_{h+1}-r_h)^2+(u_{i+1}-u_i)^2} = \sqrt{\delta_r^2+\delta_u^2} < \delta_\beta(\delta_g(\varepsilon)).$$

Therefore

$$\left\|(x,y)-(x_{h,i},y_{h,i})\right\|$$
$$= \left\|\beta\left(\sqrt{x^2+y^2}, \arctan\frac{y}{x}\right) - \beta\left(\sqrt{x_{h,i}^2+y_{h,i}^2}. \arctan\frac{y_{h,i}}{x_{h,i}}\right)\right\| < \delta_g(\varepsilon),$$

where the equality is thanks to equality A.0.34. Consequently,

$$|g(x,y)-g(x_{h,i},y_{h,i})| < \varepsilon.$$

5. Equality A.0.37 therefore yields

$$\int_{x=0}^{\infty}\int_{y=-\infty}^{\infty} g(x,y)dxdy$$
$$= \sum_{h=0}^{q-1}\sum_{i=0}^{q-1}\int\int_{(x,y)\in A(r(h),r(h+1),u(i),u(i+1))} (g(x_{h,i},y_{h,i}) \pm \varepsilon)dxdy$$
$$= \sum_{h=0}^{q-1}\sum_{i=0}^{q-1}\int\int_{(x,y)\in A(r(h),r(h+1),u(i),u(i+1))} (g(r_h\cos u_i, r_h \sin u_i) \pm \varepsilon)dxdy$$
$$= \sum_{h=0}^{q-1}\sum_{i=0}^{q-1}\mu(A_{r(h),r(h+1),u(i),u(i+1)})(g(r_h\cos u_i, r_h \sin u_i) \pm \varepsilon)dxdy$$
$$= \sum_{h=0}^{q-1}\sum_{i=0}^{q-1}2^{-1}(u_{i+1}-u_i)(r_{h+1}^2-r_h^2)(g(r_h\cos u_i, r_h \sin u_i) \pm \varepsilon)$$
$$= \sum_{h=0}^{q-1}\sum_{i=0}^{q-1}(u_{i+1}-u_i)(r_{h+1}-r_h)\left(\frac{r_{h+1}+r_h}{2}\right)(g(r_h\cos u_i, r_h \sin u_i) \pm \varepsilon)$$
$$= \sum_{h=0}^{q-1}\sum_{i=0}^{q-1}\int_{r(h)}^{r(h+1)}\int_{u(i)}^{u(i+1)}(r_h \pm \varepsilon)(g(r_h\cos u_i, r_h \sin u_i) \pm \varepsilon)drdu$$

$$\equiv \sum_{h=0}^{q-1}\sum_{i=0}^{q-1}\int_{r(h)}^{r(h+1)}\int_{u(i)}^{u(i+1)}(r\pm 2\varepsilon)(g(r\cos u, r\sin u)\pm 2\varepsilon)drdu$$

$$\equiv \int_{r=0}^{\infty}\int_{u=-\frac{\pi}{2}}^{\frac{\pi}{2}}(r\pm 2\varepsilon)(g(r\cos u, r\sin u)\pm 2\varepsilon)drdu,$$

where $\varepsilon > 0$ is arbitrarily small. Letting $\varepsilon \to 0$, we obtain

$$\int_{x=0}^{\infty}\int_{y=-\infty}^{\infty} g(x,y)dxdy = \int_{r=0}^{\infty}\int_{u=-\frac{\pi}{2}}^{\frac{\pi}{2}} rg(r\cos u, r\sin u)drdu, \qquad \text{(A.0.38)}$$

where $g \in C(H)$ is arbitrary.

3. Define the functions $I_1, I_2 : C(H) \to R$ by

$$I_1(g) \equiv \int_{x=0}^{\infty}\int_{y=-\infty}^{\infty} g(x,y)dxdy$$

and

$$I_2(g) \equiv \int_{r=0}^{\infty}\int_{u=-\frac{\pi}{2}}^{\frac{\pi}{2}} rg(r\cos u, r\sin u)drdu,$$

for each $g \in C(H)$. Then each of I_1 and I_2 is an integration on H, in the sense of Definition 4.2.1. Proposition 4.3.3 says that $(H, C(H), I_1)$ and $(H, C(H), I_2)$ are integration spaces. Moreover, equality A.0.38 implies that $I_1 = I_2$. Hence the complete extensions of $(H, C(H), I_1)$ and $(H, C(H), I_2)$ are equal. In other words, a function g on H is integrable relative to I_1 iff g is integrable relative to I_2. Thus a function g on H is integrable relative to the Lebesgue integral iff the function $(r,\theta) \to rg(r\cos u, r\sin u)$ is integrable relative to the Lebesgue integral, in which case

$$\int_{x=0}^{\infty}\int_{y=-\infty}^{\infty} g(x,y)dxdy = \int_{r=0}^{\infty}\int_{u=-\frac{\pi}{2}}^{\frac{\pi}{2}} rg(r\cos u, r\sin u)drdu.$$

The theorem is proved. □

Corollary A.0.13. Integral of a function related to the normal p.d.f. The function $(x,y) \to e^{-(x^2+y^2)/2}$ on R^2 is Lebesgue integrable, with

$$\int_{x=-\infty}^{\infty}\int_{y=-\infty}^{\infty} e^{-(x^2+y^2)/2}dxdy = 2\pi.$$

Proof. 1. The function $f : R \to R$ defined by $f(x) \equiv e^{-x^2/2}$ for each $x \in R$ is integrable relative to the Lebesgue integration. Hence, by Fubini's Theorem, the function $f \otimes f$ is integrable on R^2. In other words, the function $(x,y) \to e^{-(x^2+y^2)/2}$ is integrable relative to Lebesgue integration on R^2. Moreover,

$$
\begin{aligned}
&\int_{x=-\infty}^{\infty}\int_{y=-\infty}^{\infty} e^{-(x^2+y^2)/2}dxdy \\
&= \int_{y=-\infty}^{\infty}\left(\int_{x=-\infty}^{\infty} e^{-(x^2+y^2)/2}dx\right)dy \\
&= \int_{y=-\infty}^{\infty}\left(\int_{x=-\infty}^{0} e^{-(x^2+y^2)/2}dx + \int_{x=0}^{\infty} e^{-(x^2+y^2)/2}dx\right)dy \\
&= \int_{y=-\infty}^{\infty}\left(\int_{z=0}^{\infty} e^{-(z^2+y^2)/2}dz + \int_{x=0}^{\infty} e^{-(x^2+y^2)/2}dx\right)dy \\
&= 2\int_{y=-\infty}^{\infty}\int_{x=0}^{\infty} e^{-(x^2+y^2)/2}dxdy \\
&= 2\int_{u=-\frac{\pi}{2}}^{\frac{\pi}{2}}\int_{r=0}^{\infty} e^{-(r^2\cos^2 u+r^2\sin^2 u)/2}rdrdu \\
&= 2\int_{u=-\frac{\pi}{2}}^{\frac{\pi}{2}}\int_{r=0}^{\infty} e^{-r^2/2}rdrdu = 2\int_{u=-\frac{\pi}{2}}^{\frac{\pi}{2}}\int_{s=0}^{\infty} e^{-s}dsdu \\
&= 2\int_{u=-\frac{\pi}{2}}^{\frac{\pi}{2}} 1du = 2\int_{u=-\frac{\pi}{2}}^{\frac{\pi}{2}} 1du = 2\pi,
\end{aligned}
$$

where the third equality is by the change of integration variables $z = -x$ in R^1, as justified by Theorem A.0.10; where the seventh equality is by the change of integration variables $s = r^2/2$ in R^1, again as justified by Theorem A.0.10; and where the fifth equality is thanks to Theorem A.0.12. The corollary is proved. □

Appendix B

Taylor's Theorem

For ease of reference, we cite here Taylor's Theorem from [Bishop and Bridges 1985].

Theorem B.0.1. Taylor's Theorem. *Let D be a nonempty open interval in R. Let f be a complex-valued function on D. Let $n \geq 0$ be arbitrary. Suppose f has continuous derivatives up to order n on D. For $k = 1, \ldots, n$ write $f^{(k)}$ for the k-th derivative of f. Let $t_0 \in D$ be arbitrary, and define*

$$r_n(t) \equiv f(t) - \sum_{k=0}^{n} f^{(k)}(t_0)(t - t_0)^k / k!$$

for each $t \in D$. Then the following conditions hold:

1. If $|f^{(n)}(t) - f^{(n)}(t_0)| \leq M$ on D for some $M > 0$, then $|r_n(t)| \leq M|t - t_0|^n / n!$

2. $r_n(t) = o(|t - t_0|^n)$ as $t \to t_0$. More precisely, suppose $\delta_{f,n}$ is a modulus of continuity of $f^{(n)}$ at the point t_0. Let $\varepsilon > 0$ be arbitrary. Then $|r_n(t)| < \varepsilon |t - t_0|^n$ for each $t \in R$ with $|t - t_0| < \delta_{f,n}(n!\,\varepsilon)$.

3. If $f^{(n+1)}$ exists on D and $|f^{(n+1)}| \leq M$ for some $M > 0$, then $|r_n(t)| \leq M|t - t_0|^{n+1} / (n + 1)!$

Proof. See [Bishop and Bridges 1985]. □

References

[Aldous 1978] Aldous, D.: Stopping Times and Tightness, *Annals of Probability*, Vol. 6, no. 2, 335–340, 1978

[Billingsley 1968] Billingsley, P.: *Convergence of Probability Measures.* New York: John Wiley & Sons, 1968

[Billingsley 1974] Billingsley, P.: Conditional Distributions and Tightness, *Annals of Probability*, Vol. 2, no. 3, 480–485, 1974

[Billingsley 1999] Billingsley, P.: *Convergence of Probability Measures* (2nd ed.). New York: John Wiley & Sons, 1999

[Bishop 1967] Bishop, E.: *Foundations of Constructive Analysis.* New York, San Francisco, St. Louis, Toronto, London, and Sydney: McGraw-Hill, 1967

[Bishop and Bridges 1985] Bishop, E., and Bridges, D.: *Constructive Analysis.* Berlin, Heidelberg, New York, and Tokyo: Springer, 1985

[Bishop and Cheng 1972] Bishop, E., and Cheng, H.: Constructive Measure Theory, AMS Memoir no. 116, 1972

[Blumenthal and Getoor 1968] Blumenthal, R. M., and R. K. Getoor: *Markov Processes and Potential Theory*, New York and London: Academic Press, 1968

[Chan 1974] Chan, Y. K.: Notes on Constructive Probability Theory, *Annals of Probability*, Vol. 2, no. 1, 51–75, 1974

[Chan 1975] Chan, Y. K.: A Short Proof of an Existence Theorem in Constructive Measure Theory, *Proceedings of the American Mathematical Society*, Vol. 48, no. 2, 435–437, 1975

[Chentsov 1956] Chentsov, N.: Weak Convergence of Stochastic Processes with Trajectories have No Discontinuities of the Second Kind, *Theory of Probability & Its Applications*, Vol. 1, no. 1, 140–144, 1956.

[Chung 1968] Chung, K. L.: *A Course in Probability Theory*. New York, Chicago, San Francisco, and Atlanta: Harcourt, Brace & World, 1968

[Doob 1953] Doob, J. L.: *Stochastic Processes*. New York, London, and Sydney: John Wiley & Sons, 1953

[Durret 1984] Durrett, R.: *Brownian Motion and Martingales in Analysis*. Belmont: Wadsworth, 1984

[Feller I 1971] Feller, W.: *An Introduction to Probability and Its Applications*, Vol. 1 (3rd ed.). New York: John Wiley & Sons, 1971

[Feller II 1971] Feller, W.: *An Introduction to Probability and Its Applications*, Vol. 2 (2nd ed.). New York: John Wiley & Sons, 1971

[Garsia, Rodemich, and Rumsey 1970] Garsia, A. M., Rodemich, E., and Rumsey, H. Jr.: A Real Variable Lemma and the Continuity of Paths of Some Gaussian Processes, *Indiana University Mathematics Journal*, Vol. 20, no. 6, 565–578, 1970.

[Grigelionis 1973] Grigelionis, B.: On the Relative Compactness of Sets of Probability Measures in $D_{[0,\infty)}(\mathfrak{X})$, *Mathematical Transactions of the Academy of Sciences of the Lithuanian SSR*, Vol. 13, no. 4, 576–586, 1973

[Kolmogorov 1956] Kolmogorov, A. N.: Asymptotic Characteristics of Some Completely Bounded Metric Spaces, *Proceedings of the USSR Academy of Sciences*, Vol. 108, no. 3, 585–589, 1956

[Loeve 1960] Loeve, M.: *Probability Theory* (2nd ed.). Princeton, Toronto, New York, and London: Van Nostrand, 1960

[Lorentz 1966] Lorentz, G.G.: Metric Entropy and Approximation. *Bulletin of the American Mathematical Society*, Vol. 72, no. 6, 903–937, 1966

[Mines, Richman, and Ruitenburg 1988] Mines, R., Richman, F., and Ruitenburg, W.: *A Course in Constructive Algebra*. New York: Springer, 1988

[Neveu 1965] Neveu, J.: *Mathematical Foundations of the Calculus of Probability* (translated by A. Feinstein). San Francisco, London, and Amsterdam: Holden-Day, 1965

[Pollard 1984] Pollard, D.: *Convergence of Stochastic Processes*. New York, Berlin, Heidelberg, and Tokyo: Springer, 1984

[Potthoff 2009] Potthoff, J.: Sample Properties of Random Fields, I, Separability and Measurability, *Communications on Stochastic Analysis*, Vol. 3, no. 1, 143–153, 2009

[Potthoff 2009-2] Potthoff, J.: Sample Properties of Random Fields, II, Continuity, *Communications on Stochastic Analysis*, Vol. 3, no. 3, 331–348, 2009

[Potthoff 2009-3] Potthoff, J.: Sample Properties of Random Fields, III, Differentiability, *Communications on Stochastic Analysis*, Vol. 4, no. 3, 335–353, 2010

[Richman 1982] Richman, F.: Meaning and Information in Constructive Mathematics. *American Mathematical Monthly*, Vol. 89, 385–388 (1982)

[Ross 2003] Ross, S.: *Introduction to Probability Models.* San Diego, London, and Burlington: Academic Press, 2003

[Skorokhod 1956] Skorokhod, A. V.: Limit Theorems for Stochastic Processes. *Theory of Probability and Its Applications*, Vol. I, no. 3, 1956

[Stolzenberg 1970] Stolzenberg, G.: Review of “Foundation of Constructive Analysis.” *Bulletin of the American Mathematical Society*, Vol. 76, 301–323, 1970

Index

Printed in the United States
by Baker & Taylor Publisher Services